# HANDBOOK OF THE NEUROSCIENCE OF AGING

# HANDBOOK OF THE NEUROSCIENCE OF AGING

EDITORS

PATRICK R. HOF
Mount Sinai School of Medicine
Department of Neuroscience
New York, NY
USA

CHARLES V. MOBBS
Mount Sinai School of Medicine
Department of Neuroscience
New York, NY
USA

ELSEVIER

AMSTERDAM • BOSTON • HEIDELBERG • LONDON • NEW YORK • OXFORD
PARIS • SAN DIEGO • SAN FRANCISCO • SINGAPORE • SYDNEY • TOKYO
Academic Press is an imprint of Elsevier

ACADEMIC
PRESS

Academic Press is an imprint of Elsevier
32 Jamestown Road, London NW1 7BY, UK
30 Corporate Drive, Suite 400, Burlington, MA 01803, USA
525 B Street, Suite 1900, San Diego, CA 92101-4495, USA



British Library Cataloguing in Publication Data
A catalogue record for this book is available from the British Library

Library of Congress Catalog Number: 2009923451

ISBN: 978-0-12-374898-0

For information on all Elsevier publications
visit our website at www.elsevierdirect.com

PRINTED AND BOUND IN ITALY
09 10 11 12 13 10 9 8 7 6 5 4 3 2 1

**Cover photo:** Lateral views of postmortem human brains showing the left hemisphere. The top left specimen was obtained from a cognitively normal 82 years old individual, the middle one from a neurologically intact centenarian, and the last one from an 80 years old patient with severe Alzheimer's disease. The cortical atrophy in the Alzheimer's disease case is generalized. The preservation of the normally aging brains is in contrast remarkable. Courtesy of Dr. Constantin Bouras, Department of Psychiatry, University of Geneva School of Medicine, Geneva, Switzerland.

# TABLE OF CONTENTS

## SECTION X: MODELS



## SECTION XI: INTERVENTIONS

# CONTRIBUTORS

**J Alberch**
University of Barcelona, Barcelona, Spain

**M Albert**
The Johns Hopkins University, Baltimore, MD, USA

**L J P Altmann**
University of Florida, Gainesville, FL, USA

**S Ancoli-Israel**
University of California at San Diego, San Diego, CA, USA

**M Artal-Sanz**
Foundation for Research and Technology, Heraklion, Crete, Greece

**K H Ashe**
University of Minnesota, Minneapolis, MN, USA

**G Bartzokis**
University of California School of Medicine, Los Angeles, CA, USA

**M Baudry**
University of Southern California, Los Angeles, CA, USA

**G Baum**
Rush University Medical Center, Chicago, IL, USA

**M G Baxter**
Oxford University, Oxford, UK

**E Bellavista**
University of Bologna, Bologna, Italy

**G M Bernal**
Chicago Medical School at Rosalind Franklin University of Medicine and Science, North Chicago, IL, USA

**A Björklund**
Wallenberg Neuroscience Center, Lund, Sweden

**J C Blanks**
Florida Atlantic University, Boca Raton, FL, USA

**I Blasko**
Innsbruck Medical University, Innsbruck, Austria

**P Bomont**
University of California at San Diego, La Jolla, CA, USA

**A M Brickman**
Columbia University, New York, NY, USA

**P Brundin**
Wallenberg Neuroscience Center, Lund, Sweden

**A M Buchan**
University of Oxford, Oxford, UK

**N J Cairns**
Washington University School of Medicine, St. Louis, MO, USA

**J M Canals**
University of Barcelona, Barcelona, Spain

**S L Carroll**
The University of Alabama, Birmingham, AL, USA

**J-H J Cha**
MassGeneral Institute for Neurodegenerative Disease, Charlestown, MA, USA

**E-Y Chen**
Rush University Medical Center, Chicago, IL, USA

**M F Chesselet**
The David Geffen School of Medicine at UCLA, Los Angeles, CA, USA

**P Clavero**
University Clinic and Medical School, University of Navarra, Pamplona, Spain

**D W Cleveland**
University of California at San Diego, La Jolla, CA, USA

**A M Comi**
Kennedy Krieger Institute, Johns Hopkins University, Baltimore, MD, USA

**V Conte**
University of Milan, Milan, Italy

**G Cortopassi**
University of California at Davis, Davis, CA, USA

**C W Cotman**
University of California, Irvine, CA, USA

**T Cowen**
University College London, London, UK

**T M Dawson,**
The Johns Hopkins University School of Medicine, Baltimore, MD, USA

**V L Dawson**
The Johns Hopkins University School of Medicine, Baltimore, MD, USA

**M T R De Cristofaro**
University of Florence, Florence, Italy

**E R de Kloet**
LACDR, Leiden University Medical Center, Leiden, The Netherlands

**C K Dorey**
Florida Atlantic University, Boca Raton, FL, USA

**R L Doty**
University of Pennsylvania, Philadelphia, PA, USA

**C Ferrarese**
University of Milan-Bicocca, Monza, Italy

**S P Finklestein**
Biotrofix, Inc., Needham, MA, USA

**D Fox**
Rush University Medical Center, Chicago, IL, USA

**C Franceschi**
University of Bologna, Bologna, Italy

**I G Gazaryan**
Weill Medical College of Cornell University, New York, NY, USA

**M Goedert**
Medical Research Council Laboratory of Molecular Biology, Cambridge, UK

**L S B Goldstein**
University of California at San Diego, La Jolla, CA, USA

**E Golob**
Tulane University, New Orleans, LA, USA

**R G Goya**
National University of La Plata, La Plata, Argentina

**C L Grady**
University of Toronto, Toronto, ON, Canada

**M B Graeber**

**L T Grinberg**
Faculdade de Medicina da Universidade de São Paulo and Instituto Israelita de Ensino e Pesquisa Albert Einstein, Sao Paulo, Brazil

**B Grubeck-Loebenstein**
Institute for Biomedical Aging Research of the Austrian Academy of Sciences, Innsbruck, Austria

**J Guridi**
University Clinic and Medical School, University of Navarra, Pamplona, Spain

**C Hamani**
University of Toronto, Toronto, ON, Canada

**H Hampel**
Ludwig-Maximilian University, Munich, Germany; and Trinity College Dublin, The Adelaide and Meath Hospital, Incorporating the National Children's Hospital, Dublin, Republic of Ireland

**H Heinsen**
University Würzburg, Würzburg, Germany

**C Henchcliffe**
Weill Medical College of Cornell University, New York, NY, USA

**C B Herenu**
National University of La Plata, La Plata, Argentina

**R O Heuckeroth**
Washington University School of Medicine, St. Louis, MO, USA

**M A Hickey**
The David Geffen School of Medicine at UCLA, Los Angeles, CA, USA

**P R Hof**
Mount Sinai School of Medicine, New York, NY, USA

**D M Holtzman**
Washington University, St. Louis, MO, USA

**L C Hoyte**
University of Oxford, Oxford, UK

**S Humbert**
Institut Curie and CNRS UMR 146, Orsay, France

**C Humpel**
Innsbruck Medical University, Innsbruck, Austria

**J S Janowsky**
Oregon Health and Science University, Portland, OR, USA

**M V Johnston**
Kennedy Krieger Institute, Johns Hopkins University, Baltimore, MD, USA

**D M Keenan**
University of Virginia, Charlottesville, VA, USA

**K B Kegel**
MassGeneral Institute for Neurodegenerative Disease, Charlestown, MA, USA

**S Kemper**
University of Kansas, Lawrence, KS, USA

**G Kempermann**
Max Delbrück Center for Molecular Medicine (MDC) Berlin-Buch, Berlin, Germany

**E A Kensinger**
Boston College, Chestnut Hill, MA, USA

**S G Kernie**
The University of Texas Southwestern Medical Center, Dallas, TX, USA

**M A Kliem**
Emory University, Atlanta, GA, USA

**J Koenigsknecht-Talboo**
Washington University, St. Louis, MO, USA

**J H Kordower**
Rush University Medical Center, Chicago, IL, USA

**L Korhonen**
Minerva Foundation Institute for Medical Research, Helsinki, Finland

**D K Lahiri**
Indiana University School of Medicine, Indianapolis, IN, USA

**A W Laxton**
University of Toronto, Toronto, ON, Canada

**V M-Y Lee**
University of Pennsylvania School of Medicine, Philadelphia, PA, USA

**Y-I Lee**
The Johns Hopkins University School of Medicine, Baltimore, MD, USA

**M S Levine**
Semel Neuroscience Institute, The David Geffen School of Medicine at UCLA, Los Angeles, CA, USA

**C Li**
Weill Medical College of Cornell University, New York, NY, USA

**D Lindholm**
Minerva Foundation Institute for Medical Research, Helsinki, Finland

**P Y Liu**
Mayo Clinic, Rochester, MN, USA

**E H Lo**
Harvard Medical School, Boston, MA, USA

**L Longhi**
University of Milan, Milan, Italy

**A M Lozano**
University of Toronto, Toronto, ON, Canada

**P H Lu**
University of California School of Medicine, Los Angeles, CA, USA

**E MacDonald**
University of Toronto, Toronto, ON, Canada

**B Maloney**
Indiana University School of Medicine, Indianapolis, IN, USA

**C B Mantilla**
Mayo Clinic College of Medicine, Rochester, MN, USA

**A Martin**
National Institute of Mental Health, Bethesda, MD, USA

**J M Matsubara**
University Clinic and Medical School, University of Navarra, Pamplona, Spain

**R B McDonald**
University of California, Davis, CA, USA

**T K McIntosh**
Media NeuroConsultants, Inc., Media, PA, USA

**C Milligan**
Wake Forest University School of Medicine, Winston-Salem, NC, USA

**C V Mobbs**
Mount Sinai School of Medicine, New York, NY, USA

**P I Moreira**
University of Coimbra, Coimbra, Portugal

**T E Morgan**
University of Southern California, Los Angeles, CA, USA

**J E Morley**
St. Louis University and Veterans Affairs Medical Center, St. Louis, MO, USA

**E Moro**
University of Toronto, Toronto, ON, Canada

**L Mosconi**
New York University School of Medicine, New York, NY, USA

**P J Muchowski**
Gladstone Institute of Neurological Disease, University of California, San Francisco, CA, USA

**B Nacmias**
University of Florence, Florence, Italy

**A Nehra**
Mayo Clinic, Rochester, MN, USA

**J A Obeso**
University Clinic and Medical School, University of Navarra, Pamplona, Spain

**R Oppenheim**
Wake Forest University School of Medicine, Winston-Salem, NC, USA

**T Pasternak**
University of Rochester, Rochester, NY, USA

**E Pérez-Navarro**
University of Barcelona, Barcelona, Spain

**G Perry**
University of Texas at San Antonio, San Antonio, TX, USA

**D A Peterson**
Chicago Medical School at Rosalind Franklin University of Medicine and Science, North Chicago, IL, USA

**M K Pichora-Fuller**
University of Toronto, Mississauga, ON, Canada

**S M Pincus**
Guilford, CT, USA

**E J Ploran**
University of Pittsburgh, Pittsburgh, PA, USA

**B R Postle**
University of Wisconsin – Madison, Madison, WI, USA

**V Prakash Reddy**
University of Missouri-Rolla, Rolla, MO, USA

**H Pratt**
Technion – Israel Institute of Technology, Technion City, Israel

**D L Price**
The Johns Hopkins University, Baltimore, MD, USA

**A Prigione**
University of California at Davis, Davis, CA, USA

**A Pupi**
University of Florence, Florence, Italy

**N Putkonen**
Minerva Foundation Institute for Medical Research, Helsinki, Finland

**N E Rance**
University of Arizona College of Medicine, Tucson, AZ, USA

**P R Rapp**
Mount Sinai School of Medicine, New York, NY, USA

**R R Ratan**
Weill Medical College of Cornell University, New York, NY, USA

**J M Ren**
Biotrofix, Inc., Needham, MA, USA

**C A Reynolds**
University of California at Riverside, Riverside, CA, USA

**M Rissling**
University of California at San Diego, San Diego, CA, USA

**D R Roalf**
Oregon Health and Science University, Portland, OR, USA

**M C Rodriguez-Oroz**
University Clinic and Medical School, University of Navarra, Pamplona, Spain

**K A Roth**
University of Alabama at Birmingham, Birmingham, AL, USA

**S Roy**
University of Pennsylvania School of Medicine, Philadelphia, PA, USA

**A Russo-Neustadt**
California State University, Los Angeles, CA, USA

**F Saudou**
Institut Curie and CNRS UMR 146, Orsay, France

**A V Savonenko**
The Johns Hopkins University, Baltimore, MD, USA

**T Schallert**
University of Texas at Austin, Austin, TX, USA

**A B Scheibel**
Brain Research Institute UCLA, Los Angeles, CA, USA

**H M Schipper**
McGill University, Montreal, QC, Canada

**C Schwarz**
New York Presbyterian Hospital – Weill Cornell Medical Center, New York, NY, USA

**J J Shacka**
University of Alabama at Birmingham, Birmingham, AL, USA

**M C Shake**
University of Illinois at Urbana–Champaign, Champaign, IL, USA

**R Sherrington**
Unité de Recherche en Neuroscience, Ste-Foy, QC, Canada

**G C Sieck**
Mayo Clinic College of Medicine, Rochester, MN, USA

**A B Singhal**
Harvard Medical School, Boston, MA, USA

**C T Siwak-Tapp**
University of California at Irvine, Irvine, CA, USA

**A D Smith**
University of Pittsburgh, Pittsburgh, PA, USA

**M A Smith**
Case Western Reserve University, Cleveland, OH, USA

**K E Soderstrom**
Rush University Medical Center, Chicago, IL, USA

**W E Sonntag**
University of Oklahoma Health Science Center, Oklahoma City, OK, USA

**S Sorbi**
University of Florence, Florence, Italy

**P H St. George-Hyslop**
University of Toronto, Toronto, ON, Canada

**A Starr**
University of California at Irvine, Irvine, CA, USA

**Y Stern**
Columbia University, New York, NY, USA

**E A L Stine-Morrow**
University of Illinois at Urbana–Champaign, Champaign, IL, USA

**N Stocchetti**
University of Milan, Milan, Italy

**G Y Sun**
University of Missouri, Columbia, MO, USA

**W A Suzuki**
New York University, New York, NY, USA

**E Syková**
Institute of Experimental Medicine, Academy of Sciences of the Czech Republic, and Charles University, Prague, Czech Republic

**P Takahashi**
Mayo Clinic, Rochester, MN, USA

**M G Tansey**
The University of Texas Southwestern Medical Center, Dallas, TX, USA

**P D Tapp**
University of California at Irvine, Irvine, CA, USA

**S H Tariq**
St. Louis University and Veterans Affairs Medical Center, St. Louis, MO, USA

**N Tavernarakis**
Foundation for Research and Technology, Heraklion, Crete, Greece

**S J Teipel**
University of Rostock, Rostock, Germany

**B Teter**
University of California, Los Angeles, CA, USA

**P M Thompson**
UCLA School of Medicine, Los Angeles, CA, USA

**A W Toga**
UCLA School of Medicine, Los Angeles, CA, USA

**J Q Trojanowski**
University of Pennsylvania School of Medicine, Philadelphia, PA, USA

**J C Troncoso**
The Johns Hopkins University, Baltimore, MD, USA

**K Troulinaki**
Foundation for Research and Technology, Heraklion, Crete, Greece

**J D Veldhuis**
Mayo Clinic, Rochester, MN, USA

**M V Vitiello**
University of Washington, Seattle, WA, USA

**M E Wheeler**
University of Pittsburgh, Pittsburgh, PA, USA

**T Wichmann**
Emory University, Atlanta, GA, USA

**P C Wong**
The Johns Hopkins University, Baltimore, MD, USA

**W G Wood**
University of Minnesota School of Medicine, Minneapolis, MN, USA

**E R Zanier**
University of Milan, Milan, Italy

**C Zhu**
The David Geffen School of Medicine at UCLA, Los Angeles, CA, USA

**X Zhu**
Case Western Reserve University, Cleveland, OH, USA

**M J Zigmond**
University of Pittsburgh, Pittsburgh, PA, USA

# PREFACE

The present volume is the result of two parallel processes that have ultimately converged. First, we edited a similar volume, *Functional Neurobiology of Aging* (Hof and Mobbs, 2001) in response to a growing need, not the least of which was our own, for an authoritative comprehensive text on this area for use in advanced seminar classes, and by new investigators, in this rapidly growing field. In teaching our own course on the neurobiology of aging, at the Mount Sinai School of Medicine, we found that simply reviewing the primary literature was often unsatisfactory because suitable background information was difficult to obtain, often leaving students struggling with the primary material. Students studying neuropathology, neurology, or gerontology often encountered similar difficulties. Even those of us who specialize in one area or another in the neurobiology of aging often need an introduction to related areas outside our expertise, but obtaining such overviews often entails a time-consuming internet search, with no assurance of quality control. *Functional Neurobiology of Aging* was developed to meet such needs, and that the book was well received indicated there was indeed a demand for such a volume. As noted by Dr. Caleb Finch in the Foreword to that book, the book's most immediate predecessor was James Birren's *The Psychology of Aging* (Prentice Hall, 1964), so it might seem as if a book published in 2001 would maintain important relevance for at least a decade or so. Unfortunately, however, it has become increasingly clear that in this rapidly advancing area, revisions would be required to maintain relevance. Nevertheless, the magnitude of such a task would be daunting, and we hesitated to commit to such until absolutely necessary.

In the meantime, a truly heroic project was being undertaken at Elsevier: the publication of a comprehensive and authoritative *Encyclopedia of Neuroscience* (Elsevier, 2009). Indeed, this work is so comprehensive that it occupies a full ten volumes and is available primarily in electronic-only form. While the availability of such a resource is indeed to be celebrated, there are certain practical disadvantages. First, the very comprehensiveness of the *Encyclopedia of Neuroscience* poses a problem for efficiently accessing more specialized information. For example, if a student or researcher new to a specialized area in the area of the neurobiology of aging were to attempt to obtain an overview of Alzheimer's disease in relation to normal cognitive aging, several different chapters scattered throughout the work would have to be accessed and these chapters would not necessary be organized logically in the specific context of aging. Furthermore, for many purposes an actual book (as old-fashioned as that is becoming) is a more efficient means of accessing information, especially introductory information, a reason that books are still with us.

These two processes eventually converged into the idea that the editors of the original *Functional Neurobiology of Aging* would work to reorganize relevant material from the *Encyclopedia of Neuroscience*, as well as relevant reviews from other sources, in particular from *Functional Neurobiology of Aging*, to produce a new volume that would be at least as useful as the original book was for students and professionals. The task has become more urgent as the average age of populations throughout the world, increases rapidly: the oldest segments of society are also the most rapidly growing, and the prevalence of Alzheimer's disease in that elderly segment of the population may be as high as 40-50 % (Wang and Ding, 2008).

An obvious effect of the provenance of this volume is that we now refer to the *Neuroscience of Aging*, rather than the *Neurobiology of Aging*. This apparently semantic shift was motivated by the shift in the term "Neuroscience" to indicate the systematic analysis of neuron-based processes in the context of well-developed concepts such as synaptic plasticity, structure-function relationships, and other key concepts that form the conceptual basis of this discipline.

The present volume generally follows the organizational philosophy of *Functional Neurobiology of Aging*, although most of the material is new, thus necessitating a somewhat different structure. Two key concepts guiding the organization of the material are, first, that there are key mechanistic distinctions between pathological and non-pathological impairments associated with aging, and, consequently, that these impairments must be kept conceptually distinct, often in separate chapters; and second, that age-related impairments can be most easily understood according to function. The distinction between pathological processes and non-pathological impairments associating with normal aging was at one time controversial, but the utility of this distinction has largely been established among neuroscientists studying the aging brain. The importance of distinguishing between these two aspects of the aging brain may be vividly illustrated by the example of age-related changes in hippocampal and cortical areas associated with cognitive impairments, one of the major problems that attract research into the aging brain. A long-standing assumption, typified by comments in a recent book discussing normal brain aging, is that "Some estimates have the loss [of neurons] as one hundred thousand neurons per day after age thirty" (Kausler et al., 2007, p. 65). While not unusual, such an estimate would suggest that between ages 30 and 80, 1.5 billion neurons would be lost, or about 10% of all neurons. Because it is to be expected that loss of neurons would be non-uniform across different brain areas, as with neurodegenerative neuronal loss, some brain areas would be expected, by this estimate, to sustain even much greater loss of neurons. However, over the last decade it has become increasingly clear that such massive neuron loss, or, in most brain areas, any significant neuron loss, is not observed during normal aging (Morrison and Hof, 1997; Hof and Morrison, 2004), even when cognitive impairments are clearly demonstrable (Rapp and Gallagher, 1996). In contrast, major loss of neurons in well-defined areas in the hippocampus and neocortex are a defining feature of Alzheimer's disease, well correlated with loss of cognitive function, even though Alzheimer's disease can be defined according to characteristic clinical presentation and, more definitely, according to the burden of plaques and tangles in those brain areas (see article on **Erectile Dysfunction**, this volume). The importance of distinguishing between normal and pathological aging are far more than merely semantic. For example, the estimate that on the average some hundred thousand neurons per day are lost with age arose, in large part, by combining changes from individuals during normal aging (which are minimal) and changes in individuals suffering from neurodegenerative diseases (which can be massive). One implication of such a conflation is that, if one lives long enough, the development of Alzheimer's disease would be almost inevitable, viz., Alzheimer's disease is simply an accelerated form of normal aging. However, it is now clear that the pathological processes that drive neurodegeneration in Alzheimer's disease are largely distinct from those that occur during normal aging (see articles on pages 381, 389, 395 and 403), and in fact in the oldest-old population (individuals over 95 years of age) cognitive impairments, while common, are relatively less likely to be due to Alzheimer's disease (Giannakopoulos et al., 1996; von Gunten et al., 2005).

The present book is divided into five large-scale parts. The first part comprises general processes that plausibly apply to many or all aspects of the aging brain, primarily during normal aging. Thus, we begin with an Introduction with a very general overview of the normal aging brain. The following three sections describe structural (Section II), mechanistic (Section III), and molecular (Section IV) processes that contribute to functional impairments occurring during normal aging. The second large part comprises descriptions of functional impairments during normal aging in four main functional domains of the brain, sensory (Section V), motor (Section VI), cognition (Section VII), and neuroendocrine (Section VIII). Section IX addresses neurodegenerative diseases whose incidence increases with age, Section X describes animal models for both normal and pathological aging of the brain, and Section XI ends the book with the hopeful topic of potential interventions to ameliorate brain aging and prevent age-related brain ilnesses. It is hoped that this organizational structure will facilitate not only understanding the separate topics in depth, but also clarify the more general aspects of the aging brain.

*See also:* Alzheimer's Disease: An Overview; Alzheimer's Disease: Neurodegeneration; Axonal Transport and Alzheimer's Disease; Erectile Dysfunction.

## References

Birren JE (1964) *The Psychology of Aging*. New York: Prentice Hall.

Giannakopoulos P, Hof PR, Kövari E, et al. (1996) Distinct patterns of neuronal loss and Alzheimer's disease lesion distribution in elderly individuals older than 90 years. *Journal of Neuropathology and Experimental Neurology* 55: 1210–1220.

Hof PR and Mobbs CV (eds.) (2001) *Functional Neurobiology of Aging*. San Diego: Academic Press.
Hof PR and Morrison JH (2004) The aging brain: morphomolecular senescence of cortical circuits. *Trends in Neuroscience* 27: 607–613.
Kausler DH, Kausler BC, and Krupsaw JA (2007) *The Essential Guide to Aging in the Twenty-First Century: Mind, Body, and Behavior.* St. Louis, University of Missouri Press.
Morrison JH and Hof PR (1997) Life and death of neurons in the aging brain. *Science* 278: 412–419.
Rapp PR and Gallagher M (1996) Preserved neuron number in the hippocampus of aged rats with spatial learning deficits. *Proceedings of the National Academy of Sciences of the USA* 93: 9926–9930.
von Gunten A, Kövari E, Rivara CB, et al. (2005) Stereologic analysis of hippocampal Alzheimer disease pathology in the oldest-old: evidence for sparing of the entorhinal cortex and CA1 field. *Experimental Neurology* 193: 198–206.
Wang XP and Ding HL (2008) Alzheimer's disease: epidemiology, genetics, and beyond. *Neuroscience Bulletin* 24: 105–109.

Patrick R. Hof and Charles V. Mobbs
Department of Neuroscience,
Mount Sinai School of Medicine,
New York,
NY 10029

# Environment versus Heredity in Normal and Pathological Aging of Neurological Functions

**C V Mobbs and P R Hof**, Mount Sinai School of Medicine, New York, NY, USA



## Normal versus Pathological Aging

In addressing age-related impairments in brain function, a central issue is whether extreme impairments in function, for example, the debilitating loss of memory function in Alzheimer's disease (AD) observed in only relatively few patients, represent only an 'exaggerated' form of impairments observed in most or all elderly individuals. Similar hypotheses could be suggested for the extreme loss of motor function observed in Parkinson's disease, or the extreme loss of visual function observed in glaucoma. Such hypotheses suggest that patients with AD are only at the extreme end of a normal distribution of symptoms and that as patients age, they universally drift toward the extreme set of symptoms. On the other hand, careful analysis indicates that in fact AD is characterized by a set of pathologies, especially neuronal loss, that are quite distinct from those observed in age-matched, nondiseased brains, whereas nonpathological age-related impairments in memory function are generally not due to neuronal loss. Similarly, nonpathological age-related impairments in memory are easily distinguished from impairments associated with AD, using functional memory tests.

Distinguishing age-related diseases from more universal age-related impairments, as indicated with AD, requires the same analytic tools that allow the diagnosis of any disease, in particular the existence of separate populations characterized by a bimodal distribution of some set of traits. In studying age-related impairments, an additional tool useful in distinguishing between normal and pathological impairments is the analysis of rate-specific incidence rates. In both humans and animals, mortality rates increase exponentially throughout most of the lifespan, but late in life the acceleration of mortality rate begins to decrease, to the extent that mortality rate becomes constant in very old populations. Similarly, the incidence of neurological diseases increases exponentially with age, but incidence peaks at some late age, the peak age depending on the disease. For example, the incidence of Parkinson's disease peaks at ~75 years of age, the incidence of Huntington's disease peaks at ~40 years of age, and the incidence of AD appears to peak at ~90 years of age.

The distinction between normal and pathological aging has important mechanistic implications, as an analysis of incidence rate demonstrates. For example, the incidence rate of familial forms of age-related diseases generally peaks at earlier ages than the non-familial or sporadic forms. Thus, mutations in the parkin gene lead to a form of Parkinson's disease whose incidence peaks at ~20 years of age and mutations in α-synuclein lead to a form of Parkinson's disease whose incidence peaks at ~50 years of age. On the other hand, the incidence rate of Parkinson's disease in the population as a whole peaks at ~70 years. Similarly, the peak incidence of forms of AD caused by mutations occurs before the incidence of AD as a whole, and centenarians exhibit a lower incidence of AD than individuals between 70 and 80.

Further support for the conclusion that age-related diseases are mechanistically distinct from normal age-related diseases is that although most animal models, in particular rodents, exhibit age-related impairments similar to those observed in humans (e.g., in memory and motor impairments), animal models do not normally exhibit age-related diseases that are observed in humans unless genetically modified to express alleles that cause human diseases (and even then the phenotypes observed in rodents rarely reproduce the exact pathologies observed in human disease). Therefore, neurological diseases are not merely accentuated forms of 'normal' age-related changes but are due to distinct pathological processes; conversely, the more universal (but milder) impairments observed during aging are not likely to be due to the same mechanisms as those which cause diseases.

## Heredity versus Environment

As the cumulative effect of environment increases with age, whereas genotype stays constant, it would be tempting to assume that the relative importance of genotype decreases with age. However, this is only the case after about age 70 for most diseases. For age-related phenotypes that are inherited, and even for age-related phenotypes in which the environment plays a major role, the importance of genotype actually increases with age until about age 70 after which the importance of genotype begins to decline.

For example, Huntington's disease is an autosomal dominant neurodegenerative disease whose age-specific incidence peaks at ~40 years of age. Since the genetic mutation causing Huntington's disease has been cloned, it is now possible to demonstrate that monozygotic twins are essentially 100% concordant in the development of Huntington's disease if they live past 40, demonstrating the primary contribution of genotype to the risk of developing Huntington's disease. Nevertheless, under 20 years of age, there is essentially no phenotype of the Huntington allele. Therefore, Huntington's disease represents an extreme form of the coupling between genotype and age-related phenotype, in which the coupling increases from very low below the age of 20 years to essentially 100% concordance by age 70 years. Similarly, twin and kindred studies have demonstrated a significant genetic contribution to the risk of developing AD, and as with Huntington's disease, the contribution of genotype (e.g., the effect of ApoE genotype) to AD phenotype increases with age up to the age of 70, after which it begins to decrease.

Since the effect of genotype on performance in standardized intelligence tests has been examined in great detail in young populations, several studies have used similar methodologies to assess the effects of aging on the contribution of genotype to performance on these tests. For example, several studies suggest that the average heritability of general cognitive function peaks at ~80% at 65 years of age and then begins to decline. Interestingly, the heritability of general cognitive function during aging is greater than the heritability of any of the functions reflected by subscales, which has been interpreted to indicate that the 'nature of the genetic influence in the cognitive domain appears to be more general than specific.' Thus, in contrast to the effect of age on the heritability of general cognitive function, heritability on memory function alone is reported to be stable with age.

Considering that cognitive function is defined by experience, it may seem surprising that the heritability of cognitive function can increase with age at all (although decreasing after the age of 70). One resolution of this apparent paradox is that the influence of experience may greatly depend on genotype; that is, the effect of experience may be enhanced by genotype. For example, in a twin study examining the effect of genotype on the acquisition of a motor skill, while genotype influenced the initial performance of the skill, the effect of genotype on the enhancement of the skill by practice was even greater. These investigators concluded that 'the effect of practice is to decrease the effect of environmental variation (previous learning) and increase the relative strength of genetic influences on motor performance'.

Similarly, the heritability of body mass index is ~46% in males aged 46–59 years old, and 61% in males 60–76 years old. Work in rodents suggests a possible mechanism. In mice, most strains exhibit about the same body weight when given a low-fat diet, but on a high-fat diet, some strains become massively obese, while others do not. Thus in the presence of one environment (characterized by a low-fat diet), there is little effect of genotype on body weight, whereas in a different environment (characterized by a high-fat diet), the effect of genotype (which initially is very small) increases substantially.

Age-related impairments in sensory functions may also involve the genetic exacerbation of environmental insults. For example, some mouse strains become progressively deaf with age, whereas others do not. However, the allele that causes progressive deafness also greatly potentiates noise-induced hearing loss. Thus, while the effect of genotype on hearing function clearly increases with age in mice, part of the mechanism by which this occurs may involve exacerbation of effects.

## Central versus Peripheral Impairments

Four major neurological functions are particularly susceptible to age-related impairments: cognitive, motor, sensory, and neuroendocrine. However, not all impairments in these functions are due to normal age-related impairments in brain function, and indeed brain function is surprisingly robust in association with normal processes of aging, as indicated earlier, although this is not the case in many age-related diseases.

For example, as indicated earlier, central substrates subserving memory functions (e.g., hippocampal and neocortical neurons) are relatively robust during aging. For example, as discussed earlier, in the absence of pathology, there is substantial heterogeneity in decline in memory functions in both humans and animals such that for memory functions independent of speed, many individuals of even advanced age show no impairments. Even in those individuals demonstrating memory impairments, the impairments are relatively modest and unlikely to substantially impair independent living, and are not associated with neuron loss. In contrast, in humans with age-related pathology, for example AD, major neuron loss occurs in association with a much more robust loss of function, generally leading to a loss of independence. Nevertheless, almost by definition, age-related impairments in cognitive function are entirely due to central impairments, whether those impairments are purely functional, as in normal aging, or due to cell loss, as with pathologies such as AD.

With motor, sensory, and endocrine functions, in which functionality necessarily requires intimate interactions between the brain and peripheral systems, it is perhaps the peripheral systems that are most impaired during normal aging, although central impairments may play a more important role in pathology. For example, one of the most robust normal age-related impairments is sarcopenia, or loss of muscle mass. During normal aging, muscle strength decreases progressively until in men and women both muscle strength and muscle mass are reduced by ~30%, a far greater loss of function and cellularity than occurs in the brain in normal aging. This normal loss of muscle mass can be compensated for by exercise but cannot be prevented: even weight lifters exhibit the same progressive loss of muscle mass, although starting at a higher initial value. Although for most people this loss of strength does not cause a major loss of independence, the loss of strength can require changing to jobs that demand less strength. In older patients, the loss of strength can lead to falls, in turn leading to bone fractures that ultimately hinder mobility, but in general, sarcopenia is a relatively minor cause of morbidity. In contrast age-related diseases impacting motor systems, especially Parkinson's disease and Huntington's disease, are entirely central and produce much more devastating impairment in motor function (in the case of Huntington's disease, eventually death).

Of the senses, visual and auditory functions are the most adversely impacted during normal aging. A universal age-related change is presbyopia, a gradual loss of the ability to focus on near objects. This universal condition appears to be caused largely by a progressive loss of flexibility in the crystalline lens and a weakening of the ciliary muscles that focus the lens, and thus appears to be entirely driven by peripheral impairments. Presbyopia, while a nuisance for reading and other close work, does not lead to blindness. Some age-related pathology, such as glaucoma and cataracts, can cause blindness, but again these impairments are due to peripheral processes. Similarly, the loss of auditory acuity, or presbycusis, that occurs progressively in aging, is also caused by a variety of age-related impairments in the auditory canal, but not in the brain itself, and therefore, is also driven primarily by peripheral mechanisms. There may be some modest impairment in central processing of auditory inputs, but even when present, they would not be the main cause of deafness. Age-related impairments in other sensory modalities, smell, taste, and touch, can be demonstrated under laboratory conditions, but very rarely do these impairments produce substantial impact on normal functioning. Thus, with respect to age-related sensory function, central impairments are only very rarely involved, in stark contrast to cognitive impairments.

As with sensory functions, most endocrine functions are generally adequate during aging to maintain appropriate homeostasis. Thus, absent trauma or tumor, thyroid, adrenal, and testicular functions are largely as functional in elderly individuals as in younger ones, although as with sensory functions impairments in these endocrine functions can be demonstrated under laboratory conditions. The major exception, of course, is female reproduction, which becomes completely impaired after menopause around the age of 50 years. In women, menopause is almost entirely caused by a peripheral mechanism, ovarian depletion due to ongoing atresia beginning at birth, leading to a complete loss of estrogen secretion after menopause. Although impairments in the neuroendocrine control of ovarian function (mediated by hypothalamic release of luteinizing hormone releasing hormone and pituitary release of luteinizing hormone) can be sometimes observed under laboratory conditions, there is no evidence that these subtle central impairments play an important role in the loss of reproduction in women. In contrast, such central impairments may play a more important role in the loss of reproductive function in female rodents, which, in contrast to women, continue to secrete substantial levels of estradiol after menopause. Although the loss of estradiol after menopause clearly predisposes to osteoporosis, and may predispose to other age-related impairments as well, estrogen replacement may entail risks of its own; the value of estrogen replacement therapy is therefore currently unclear. Aging also reduces the secretion of growth hormone and dihydroepiandosterone (DHEA), although not as dramatically as estradiol after menopause. The causes of the reduction in secretion of these hormones are not understood and could involve either central or peripheral mechanisms. As with estradiol, current clinical evidence does not support replacing growth hormone or DHEA in elderly patients.

A final endocrine system that exhibits important age-related impairments is the insulin system regulating glucose and lipid homeostasis. The insulin system has not classically been considered significantly under neuroendocrine control, but substantial recent evidence suggests that although glucose robustly stimulates insulin secretion in isolated pancreatic beta cells, in intact individuals neuronal systems (primarily parasympathetic) play a major modulatory role in regulating insulin secretion. In any case, it is clear that pathology in the insulin system, expressed most simply as an increase in blood glucose (expressed clinically as the incidence of Type II diabetes, clinically defined as fasting plasma glucose greater than 150 mg $dl^{-1}$), increases substantially with age, due to a combination of impaired insulin secretion and impaired insulin sensitivity. The mechanisms mediating these related impairments are not known, but genetic causes of Type II diabetes are always due to neuroendocrine impairments (e.g., to loss of leptin or melanocortin, or their receptors). Furthermore, the increase in Type II diabetes with age is overwhelmingly associated with an increase in obesity with age, which again is generally thought to be due to neuroendocrine impairments. While as with other functions the insulin system incurs pathological impairments with age, more common and possibly universal age-related impairments in the insulin system, manifest as an increase in insulin resistance (the metabolic syndrome), are not always easily distinguished from Type II diabetes.

## Conclusions

This brief review demonstrates the utility of distinguishing between pathological processes that occur during aging and more general, even universal processes that occur with age. While this distinction was once controversial, it is now clear that such a distinction is essential to understand the mechanisms mediating specific age-related impairments as are demonstrated by chapters throughout this book.

*See also:* Aging and Memory in Humans; Autonomic Neuroplasticity and Aging; Brain Composition: Age-Related Changes; Brain Volume: Age-Related Changes; Gene Expression in Normal Aging Brain; Hormones and Memory; Neuroendocrine Aging: Hypothalamic-Pituitary-Gonadal Axis in Women; Neuroendocrine Aging: Pituitary-Adrenal Axis; Neuroendocrine Aging: Pituitary-Gonadal Axis in Males; Neuroimmune System: Aging.

## Further Reading

Curtsinger JW, Fukui HH, Townsend DR, and Vaupel JW (1992) Demography of genotypes: Failure of the limited life-span paradigm in *Drosophila melanogaster. Science* 258: 461–463.

Erway LC, Shiau YW, Davis RR, and Krieg EF (1996) Genetics of age-related hearing loss in mice. III. Susceptibility of inbred and F1 hybrid strains to noise-induced hearing loss. *Hearing Research* 93: 181–187.

Finkel D and McGue M (1998) Age differences in the nature and origin of individual differences in memory: A behavior genetic analysis. *International Journal of Aging and Human Development* 47: 217–239.

Gallagher M, Landfield PW, McEwen B, et al. (1996) Hippocampal neurodegeneration in aging. *Science* 274: 484–485.

Greenamyre JT and Shoulson I (1994) *Huntington's Disease.* In: Calne DB (ed.) *Neurodegenerative Diseases*, pp. 685–704. Philadelphia: W.B. Saunders.

Herskind AM, McGue M, Sorensen TI, and Harvald B (1996) Sex and age specific assessment of genetic and environmental influences on body mass index in twins. *International Journal of Obesity Related Metabolic Disorders* 20: 106–113.

Langston JW (1998) Epidemiology versus genetics in Parkinson's disease: Progress in resolving an age-old debate. *Annals of Neurology* 44: S45–S52.

Lautenschlager NT, Cupples LA, Rao VS, et al. (1996) Risk of dementia among relatives of Alzheimer's disease patients in the MIRAGE study: What is in store for the oldest old? *Neurology* 46: 641–650.

Martilla RJ (1987) *Epidemiology.* In: Koller W (ed.) *Handbook of Parkinson's Disease*, pp. 50–55. New York: Marcel Dekker.

Mohs RC, Knopman D, Petersen RC, et al. (1997) Development of cognitive instruments for use in clinical trials of antidementia drugs: Additions to the Alzheimer's Disease Assessment Scale that broaden its scope. The Alzheimer's Disease Cooperative Study. *Alzheimer Disease and Associated Disorders* 11: S13–S21.

Morrison JH and Hof PR (1997) Life and death of neurons in the aging brain. *Science* 278: 412–419.

Plomin R, Pedersen NL, Lichtenstein P, and McClearn GE (1994) Variability and stability in cognitive abilities are largely genetic later in life. *Behavioral Genetics* 24: 207–215.

Rapp PR and Gallagher M (1996) Preserved neuron number in the hippocampus of aged rats with spatial learning deficits. *Proceedings of the National Academy of Sciences USA* 93: 9926–9930.

Rapp PR, Stack EC, and Gallagher M (1999) Morphometric studies of the aged hippocampus: I. Volumetric analysis in behaviorally characterized rats. *Journal of Comparative Neurology* 403: 459–470.

Riggs JE (1993) The Gompertz function: Distinguishing mathematical from biological limitations. *Mechanisms of Ageing and Development* 69: 33–36.

Riggs JE (2001) Age-specific rates of neurological disease. In: Hof PR and Mobbs CV (eds.) *Functional Neurobiology of Aging*, pp. 3–12. San Diego: Academic Press.

Vaupel JW, Carey JR, Christensen K, et al. (1998) Biodemographic trajectories of longevity. *Science* 280: 855–860.

# Aging of the Brain

**A B Scheibel**, Brain Research Institute UCLA,
Los Angeles, CA, USA



## Possible Causal Factors

Experience indicates that all living things have finite lives, whether they be the ephemeral summer of the butterfly, a thousand days for the rat, the reported hundred-year-long lives of large tortoises, or the multi-thousand-year histories of giant redwoods and bristle-cone pines. The concept of genetic programming of the aging process seemed to receive support from Hayflick's studies of the late 1950s and early 1960s, in which young, actively dividing connective tissue cells raised *in vitro* appeared limited to a certain maximum number (50) of mitotic divisions, after which degenerative changes invariably set in. In retrospect, these classic experiments now seem less convincing, because of methodologic errors involved, and, although the concept of genetic control remains intuitively appealing, the case is considered far from proven. Nonetheless, the capacity for base excision DNA repair has been shown to be reduced in the brains of aging primates. Furthermore, the activity of polymerase beta in brain cells appears compromised with age. These findings may be bellwethers for future more widely ranging age-related changes in genomic capabilities.

A presently favored alternative theory operating either independently or, more likely, in tandem with the proposed genetic constraints envisages the possibility of progressive damage to the cell from external or internal factors, resulting in a cumulative pattern of dysfunctions. For example, toxic factors within the environment (i.e., chemical carcinogens and background radiation of terrestrial or galactic origin) may slowly affect the cell DNA content and protein-synthesizing machinery, leading to errors in synthesis with resultant progressive changes in cell structure and function. This summation of many small developing errors in the synthetic and enzymatic machinery of the cell (such as the progressive degradation of base excision DNA repair mentioned above) is conceived as mounting to a point beyond which conditions for cell life become impossible (error-catastrophe theory).

Obviously, this putative process might complement and potentiate a set of existing genetic instructions also directed toward the cell's eventual demise. However, normal oxidative metabolic activity of the cell may itself result in cumulative damage. Of particular interest is the development of a family of free radicals of oxygen (e.g., singlet oxygen, hydrogen peroxide, superoxide and hydroxyl radicals) that, through crosslinkage or cleavage reactions, may permanently alter the structure of the cell. In support of this are observations that mitochondrial free radical generation is lower in long-lived species than in short-lived ones. The varied but often overlapping phenomena of repeated stress/glucocorticoid-induced cell damage, especially in the hippocampus, excitotoxic activity of neurotransmitters like glutamate, and programmed cell death (apoptosis) may all contribute to the process. These ideas are still in early stages of development and exemplify the questions that surround the processes of aging in general, and of the nervous system in particular.

## Structural Changes

A number of changes, both gross and microscopic, have been reported in the brains of the aged, although variation is the rule rather than the exception. The brain itself is often somewhat reduced in size and weight, especially beyond the eighth decade of life. Decrements of 10–15% are frequently quoted, although interindividual differences are large. Some aged brains show mild to moderate gyral atrophy and sulcal widening, but this is not the rule. The surrounding meninges are frequently more opaque and milky in appearance than those of the young brain, and may be adherent to underlying cortex. Isolated deposits of calcium are found in and around the pacchionian granulations near the vertex of the hemisphere. The ventricular system may show alterations in silhouette, and computed axial tomography and magnetic resonance imaging (MRI) scans often show enlarged ventricular shadows. A large and growing series of studies using T1- and T2-weighted MRIs reveal considerable variation in the aged brain. Although some apparently normal individuals in their seventh and eighth decades show signs of cortical atrophy, ventricular enlargement, and even textural changes in the cerebral white matter near the ventricles (leukoairosis), others show little or no discernible alteration. Recent MRI studies indicate that a slowly developing loss of gray matter, starting as early as the third decade, is accompanied by increases in white matter volume into the mid- or late 40s, followed then by degeneration. Similar studies at the regional level suggest decrement in size of the hippocampus bilaterally in men but not in women. Localized periventricular white matter changes, in particular, are increasingly thought to result from underlying processes such as hypertension or diabetes, and to have little or no relevance to normal brain aging.

Microscopically, a group of well-known stigmata is seen in routine Nissl or reduced silver-stained preparations. Their incidence varies rather widely among individuals, although there is a general tendency toward increased frequency with age. Neuronal cell bodies gradually accumulate masses of refractile granules with high lipid content, known as lipofuscins. The significance of these so-called aging pigments is not clear, and they have been found as early as the tenth year of life in cells of the inferior olive. The present consensus is that they represent the remains of lysosomal and mitochondrial membranes that have accumulated, due perhaps to gradual failure of mechanisms for turnover and reutilization. As such, they do not necessarily constitute a direct threat to the neuron except to the extent that they preempt increasing amounts of cytoplasmic space previously used for synthesis of glycoproteins, lipoproteins, and neurotransmitters. One interesting and opposing view maintains that the appreciable content of myoglobin and respiratory enzymes allows lipofuscin granules to serve a positive role in providing energy to neurons under conditions of low oxygen tension. Coarser vacuolization of the cytoplasm, usually concentrated in the area of development of the apical dendrite shaft in cortical pyramidal cells, is known as 'granulovacuolar degeneration of Simchowitz' and is of equally unknown etiology.

The neurofibrillary tangle is a structural alteration of neuronal cytoplasm that may involve soma, dendrites, and axon. Initially recognized by Alzheimer and Simchowitz as one of two defining microcriteria of individuals dying with dementia, it is also found in very small numbers in the normal aged. Electron microscope study reveals that the tangle consists of paired helical filaments, about 22 nm in width, which braid loosely around each other with a characteristic series of partial twists, each 80 nm from the next. Tangles are now believed to result from abnormal phosphorylation of tau protein, which is necessary to normal polymerization–depolymerization mechanisms of microtubule-associated proteins. Microtubules provide the infrastructure that facilitates virtually all movement of neurotransmitters, structural proteins, enzymes and endocellular organelles within the neuron and its extensive processes (axonal and dendritic transport). It is therefore easy to imagine how the buildup of neurofibrillary tangles can progressively starve and throttle the neuron. The appearance of tangles (tau aggregates) in the frontal lobe appears to correlate positively with cognitive loss. Experimental neurofibrillary tangle development is being extensively studied in primates, in small laboratory animals, and even in *Drosophila*.

Accompanying these characteristically intracellular alterations is the development of small foci of destruction in the surrounding neuropil, the senile plaque of Alzheimer. These are classically described as containing central cores of Congo red-positive, amyloid-like material, surrounded by radial auras of degenerating dendritic and axonal tissue and a halo of microglia. They constitute the second component of the two major histopathologic criteria of dementia and, like neurofibrillary tangles, are present in more restricted number in the brains of most aged individuals. The density of senile plaques has been said to correlate positively with the degree of cognitive impairment in dementia, although these data have recently been subject to question.

Intensive biochemical and immunocytologic analyses of these structures are presently under way. It is now believed that beta amyloid (a 40- to 42-amino-acid peptide) may represent an unusual by-product of a larger, normally present molecule, called 'amyloid precursor protein', and that production and deposition may result from aberrant enzymatic (secretase) activity. Some data suggest that accumulations of beta amyloid may drive molecular cascades that potentiate neurodegenerative changes, thereby representing a direct threat to the integrity of surrounding neuronal structures. Several experimental interventions are being explored in an effort to prevent its development or mitigate its effects.

All the histopathologic changes so far described appear most intensively in the limbic system, especially the entorhinal cortex and hippocampus. They are also found widely throughout the rest of central nervous system, however, including cerebral neocortex, diencephalon, brain stem, and spinal cord. Since interindividual variation is the rule in the aged brain, these descriptions represent a distillation and summary rather than the expected picture of the brain of any one aged individual.

The problem of neuronal loss has been hotly debated and is still not settled. The few quantitative studies dating from the late 1950s and 1960s suggested neuron loss of up to 30% in the normal aging cerebral neocortex. More detailed recent investigations, based on larger numbers of cases, indicate that such levels of neuronal loss, in normal aging at least, may have been excessive, based as they were on the study of a few areas in a limited number of brains. More broadly sampled data suggest a pattern of modest loss of neurons, primarily large cortical cells, with compensatory dendritic growth in adjacent neurons. The cholinergic cell masses of the basal forebrain and the noradrenergic cells of the locus coeruleus undergo undoubted change during the aging process, but even here, declining function may be more an expression of waning metabolic vigor and decreased axodendritic dimensions than of massive neuronal loss. This is in sharp contrast to dementing syndromes such as Alzheimer's disease, in which more than half of the complement of locus coeruleus cells may disappear.

Carefully controlled studies indicate that in the rat the only significant cerebral cell losses occur during the first 100 days of life in what might be considered a period of adaptation and fine-tuning to the environment. After this, there is little further discernable neuronal loss, even at 900 days of age, when rats are beginning to die of natural causes. Although the weight of evidence from both animal and human studies now increasingly points to negligible neuronal loss during the normal, uncomplicated aging process, a large number of cells appear to undergo changes in the dendritic (and axonal) extensions of the soma. Many neurons show progressive restriction and atrophy of their more peripheral dendrite branches and, especially in cortical pyramidal cells, among the basilar shafts. Accompanying this is irregular loss of dendrite spines and frequent beaded swellings along the remaining dendritic branches. These changes can be related in general terms to progressive loss of protein-synthesizing capabilities due, perhaps in part, to increasing incursions upon cytoplasmic space by lipofuscin deposits and neurofibrillary tangles. However, it has also been shown that the potential for neuronal growth is not lost during aging. Accompanying the progressive destruction of some dendritic systems, it appears that other neurons grow further dendritic extensions, thereby increasing their available synaptic areas (**Figure 1**). The concept of two types of neuronal response to aging, one involving dendritic retraction and one reactive dendritic expansion, brings with it a number of exciting implications for providing more effective and fulfilling lives for the elderly.

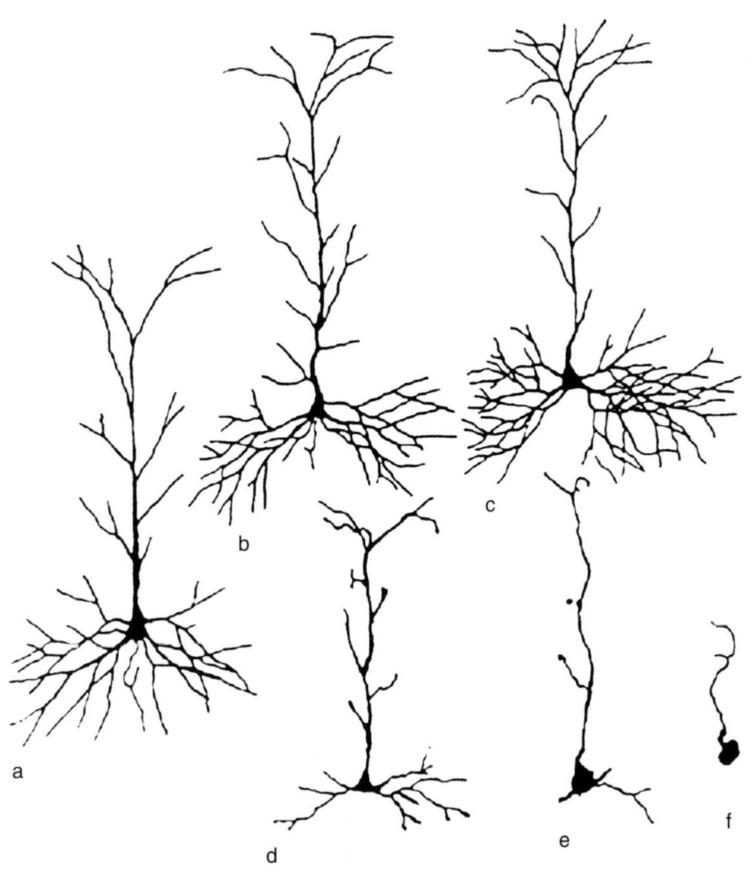


**Figure 1** Two possible patterns of age-related alterations in cortical pyramidal cells. The normal mature neuron (a) may show regressive dendritic changes characterized by loss of basilar dendritic branches and eventual loss of the entire dendritic tree (d, e, f). Other neurons (b, c) may show progressive increase in dendritic branching. Drawing based on Golgi impregnations.

During the first half of the twentieth century, brain aging and dementia were usually associated with visible changes in the major arteries of the central nervous system. Cerebral arteriosclerosis was considered a necessary accompaniment to and, in fact, virtually synoymous with aging and senility. Careful studies during the 1950s showed the inaccuracy of such concepts, and today, large-vessel disease is seldom considered to contribute significantly to the picture of general brain aging. Recent investigations with positron emission tomography scanning methods and xenon clearance techniques have again called attention to the relative adequacy of blood flow and oxygen and glucose metabolism in the healthy aging brain. The scanning electron microscope and immunohistochemical studies have, coincidentally, focused interest on the microcirculation (capillary bed) of the brain and on the plexus of neural fibers that normally innervate their walls. Subtle changes in this delicate but all-pervasive system may prove to have significant impact on the maintained vigor or decline of brain structure and function.

## Biochemical and Metabolic Changes

It is becoming clear that the aging process entails a broad range of biochemical and metabolic changes. Among these, alterations in neurotransmitter content and activity figure prominently. As many as 90% of neurons are involved, directly or indirectly, in cholinergic mechanisms, synthesizing, transporting, and releasing – or else being synaptically dependent upon – acetylcholine. Cholinergic systems are highly energy dependent, and age-related decreases in several important glycolytic enzymes have been reported. A link might thus conceivably be postulated between altered glycolytic energy mechanisms and cognitive function.

The most important concentration of acetylcholine-rich neurons is found in the basal forebrain area, in and around the nucleus basalis of Meynert and the ventral pallidum. Most studies indicate mild to moderate age-related cell loss in these areas. Associated with (although not entirely dependent upon) this loss are decreases in acetylcholine content in various portions of the brain, lower levels of the synthesizing enzyme, choline acetyltransferase, and decreased numbers of acetylcholine receptor binding sites. The relevance of these alterations to cognitive changes in the elderly, such as impairment of recent memory function, is not entirely clear. Of uncertain import, also, is the apparent loss of fluidity of the cytoplasmic membrane of the cell as the lipoprotein structure changes, a possible result of progressive diminution of choline content. Age-related increases in membrane microviscosity may bring with them a significant train of sequelae including, initially, a higher capacity for receptor binding followed by enhanced rates of receptor loss with eventual overall reduction in receptor binding affinity. Mechanisms responsible for these changes remain uncertain, but it has been postulated that increased microviscosity of the cell membrane leads first to increased exposure of those receptors out of the plane of the membrane, after which the uncovered receptors are progressively sloughed off into the surrounding medium. This suggests that membrane-bound proteins such as receptors and transporters are maintained in dynamic equilibrium by the extent of fluidity of the surrounding membrane lipids. As one example, immunohistochemical studies of a glucose transporter (GLUT 3) revealed significant loss (46%) in the hippocampal-dentate gyrus of the aged rat. Attempts to forestall or modify age-related losses in membrane fluidity through administration of active lipid fractions have been under investigation.

Catecholaminergic systems also show moderate to marked alteration in the brain. Levels of both dopamine and norepinephrine synthesis decrease, as do the numbers of adrenergic and dopaminergic receptors. Marked attenuation of norepinephrine synthesis in the hypothalamus may, in turn, be responsible, at least in part, for decrease in the synthesis of certain hypothalamic hormones.

Significant cell loss has been reported in the locus coeruleus, the single most important source of brain parenchymal norepinephrine. The substantia nigra and adjacent ventral tegmental area (VTA), major sources of dopamine, also undergo histologic alterations that include microvascular changes, iron deposition, and neuronal loss. The serotonin-rich systems of the raphe seem, on the other hand, to be somewhat less vulnerable. Resultant maintenance of minimally disturbed titers of serotonin in a setting of falling catecholaminergic levels may conceivably be related to a high incidence of disturbed sleep patterns and depression in the aged.

A broad range of age-related neuroendocrine changes is known to exist. Although such alterations are documented by innumerable anecdotal observations, regional study of the mechanisms involved is in its early phases. For instance, the importance of the ovary as a pacing factor in the aging female rat is receiving documentation in many laboratories. At an equally obvious level, the high likelihood of occurrence of benign prostatic hypertrophy and prostatic cancer in the aged male is being related in meaningful fashion to age-related alterations in testosterone metabolism that may lead to unbalanced growth stimulation. In a more global sense, the entire aging process can apparently be slowed by early restriction of nutritional input, which seems to delay pubescence and maturation. Finally, increasing imbalance among the various neurotransmitters, and in particular among those that are aminergic, appears to exert progressive impact upon the hypothalamus, pituitary, and pineal gland, leading to cascades of dysfunction that manifest themselves at every level of organization.

Although the picture presented appears to emphasize regressive elements, a much more positive picture of the normal aging brain can now be painted. Accumulated experience can help make up for age-related loss in the speed of cerebral processing or of memory recall.

Experimental evidence conclusively demonstrates the continued plasticity of even the very old brain. Continual environmental enrichment and challenge appear to enhance cognitive power in the aged. In fact, it has been shown that a subset of those genes whose expression is adversely affected by aging is oppositely affected by exposure to enriched environments. Under conditions of optimally maintained health, increasing numbers of individuals in their 70s and 80s and beyond continue to hold responsible positions, or otherwise distinguish themselves in commerce, the arts, and certain aspects of the sciences. In every case, the touchstone appears to be activity, involvement, and purpose.

*See also:* Aging of the Brain and Alzheimer's Disease; Alzheimer's Disease: An Overview; Cognition in Aging and Age-Related Disease; Gene Expression in Normal Aging Brain; Metal Accumulation During Aging.

## Further Reading

Barja G and Herrero A (2000) Oxidative damage to mitochondrial DNA is inversely related to maximum life span in the heart and brain of mammals. *FASEB Journal* 14(2): 312–318.

Buell S and Coleman P (1979) Dendritic growth in the aged human brain and failure of growth in senile dementia. *Science* 206: 854–856.

Coffey CE, Wilkinson WE, Parashos IA, et al. (1992) Quantitative cerebral anatomy of the aging human brain: A cross-sectional study using magnetic resonance imaging. *Neurology* 42(3 pt. 1): 527–536.

Flood D and Coleman C (1988) Neuron numbers and sizes in aging brains: Comparisons of human, monkey and rodent data. *Neurobiology of Aging* 9: 453–463.

Lee J, Duan W, Long JM, Ingram DK, and Mattson MP (2000) Dietary restriction increases the number of newly generated neural cells and induces BDNF expression in the dentate gyrus of rats. *Journal of Molecular Neuroscience* 15(2): 99–108.

Rowe J and Kahn R (1987) Human aging: Usual and successful. *Science* 237: 143–149.

Salmon E, Maquet P, Sadzot B, Degueldre C, Lemaire C, and Franck G (1991) Decrease of frontal metabolism demonstrated by positron emission tomography in a population of healthy elderly volunteers. *Acta Neurologica Belgica* 5: 288–295.

Scheibel A (1996) Structural and functional changes in the aging brain. In: Birren J and Schaie K (eds.) *Handbook of the Psychology of Aging,* 4th edn., pp. 105–128. San Diego: Academic Press.

Ylikowski R, Ylikowski A, Erkinjuntti T, Sulkava R, Raininko R, and Tilvis R (1993) White matter changes in healthy elderly persons correlate with attention and speed of mental processing. *Archives of Neurology* 50: 818–824.

# Functional Neuroimaging Studies of Aging

**C L Grady**, University of Toronto, Toronto, ON, Canada



## Introduction

Age differences in cognitive function have been studied for many years, and of all the cognitive domains, memory has probably been studied the most extensively. Compared to young adults, older adults have particular difficulty with episodic memory, defined as the conscious recollection of events that have occurred in a person's experience. In the laboratory, these age differences in episodic memory are seen in a reduced ability to learn and retrieve lists of stimuli, such as words or pictures of common objects. Older adults have particular difficulty in retrieving accompanying information about stimuli learned during the course of an experiment, so-called source memory. For example, when learning a list of words presented verbally by either a male or a female speaker, young adults are readily able to recall both the words and whether they were spoken by a man or woman, whereas older adults will have more difficulty remembering the speaker's gender than the words. Older adults also recall fewer details about real-life, autobiographical memories, and substantial age-related declines are seen on working memory tasks (i.e., tasks that require holding information on line for brief periods of time). In addition to these differences in memory, older adults also show reductions in inhibitory function, reflected in an increased vulnerability to distracting information, as well as lower performance on tests tapping other types of executive function. In contrast, semantic memory, or the accumulation of knowledge about the world, is maintained or even increased in older adults, compared to younger adults. Some aspects of emotional function also are preserved with age, although identifying and remembering negative emotional material are sometimes reduced in older adults. Thus, not all aspects of cognitive function are affected adversely by older age.

In recent years, functional neuroimaging has been used to explore brain activity accompanying cognitive tasks in young and old adults, and to understand the neural mechanisms underlying behavioral differences. Memory has received the most attention in this work, similar to the predominance of memory studies in the behavioral literature. Nevertheless, other cognitive domains have been studied, and some of the fundamental processes contributing to cognitive aging are becoming clear.

## Methodological Issues in Studying Aging

Prior to reviewing the results of neuroimaging studies of cognitive aging, it is useful to consider some of the issues that one may encounter when using this methodology to study older adults. The typical techniques used are either positron emission tomography (PET) or, more commonly today, functional magnetic resonance imaging (fMRI). Both of these techniques are based on task-dependent modulation of hemodynamic function, such as blood flow or oxygenation. Therefore, age-related changes in the vasculature might influence the coupling between neural activity and hemodynamic functional measures obtained in the elderly. Although the shape of the hemodynamic response function, as measured with fMRI, is similar in young and old adults, there is reduced signal-to-noise in the elderly. In addition, older adults undergo brain shrinkage, or atrophy, and some loss of white-matter integrity, both of which could impact functional measures independently of any changes in neural activity. Despite these difficulties, several robust and replicable age-related differences in brain activity have been noted, suggesting that these potential artifacts may not pose too severe a problem.

When brain activity in young and older adults is compared on a task, there are at least three possible outcomes in any given brain area: (1) young and old groups could have equivalent brain activity, (2) older adults could show less activity, or (3) old adults could show greater activity. Reduced activity in the elderly can reasonably be assumed to reflect a reduced level of functioning, particularly when accompanied by poorer performance on the task. Equivalent activity is generally considered evidence for spared function in the elderly. The major challenge has been to interpret increased recruitment of task-related brain areas, or activation of unique brain regions in the elderly.

Several analytic approaches have been applied to the study of brain mechanisms underlying cognitive aging. The most common approach has been a straightforward contrast of brain activity between young and old groups of participants and interpretation of the observed differences in activity in the context of behavioral performance on the tasks. Much of this work has focused on the frontal lobes, which are thought to be particularly vulnerable to the effects of aging. However, this type of univariate approach is often plagued by issues related to multiple comparisons and statistical power, which are compounded when more than one group of individuals is studied. Other approaches that utilize multivariate analyses to identify whole-brain patterns of functional changes, and hence avoid problems with multiple comparisons, have been used with some success in the aging field. Functional connectivity is another type of multivariate approach that focuses on cognitive networks (i.e., groups of regions working in an integrated fashion during cognitive tasks). Connectivity involves assessing how activity in a given region correlates with activity in other areas of the brain during a task. This type of correlation can be done on a pair-wise region basis or across the whole brain. Thus, connectivity approaches emphasize the functional interactions among brain areas and the ways in which these interactions mediate cognitive processing, rather than focusing on the activity in any individual brain region. This 'whole-brain' type of approach has revealed that a number of brain areas, including the frontal lobes, appear to be critical sites of age-related change in function.

## Reduced Brain Activity in Older Adults

Reduced brain activity in older adults typically has been noted during encoding, or learning new material. Younger adults have increased activity in left prefrontal cortex when learning new stimuli, and a common finding is that older adults show less activity in this area during encoding compared to younger adults, across a variety of stimulus types (e.g., words, faces, objects; see **Figure 1**). The age-related reduction of activity in left prefrontal cortex can be ameliorated to some extent by using encoding tasks that promote deep, or semantic, processing of the to-be-remembered information. Both young and old adults show more left prefrontal activity during a semantic encoding task, compared to a task that does not require a focus on the semantic aspects of the stimulus, and this increase in older adults can be

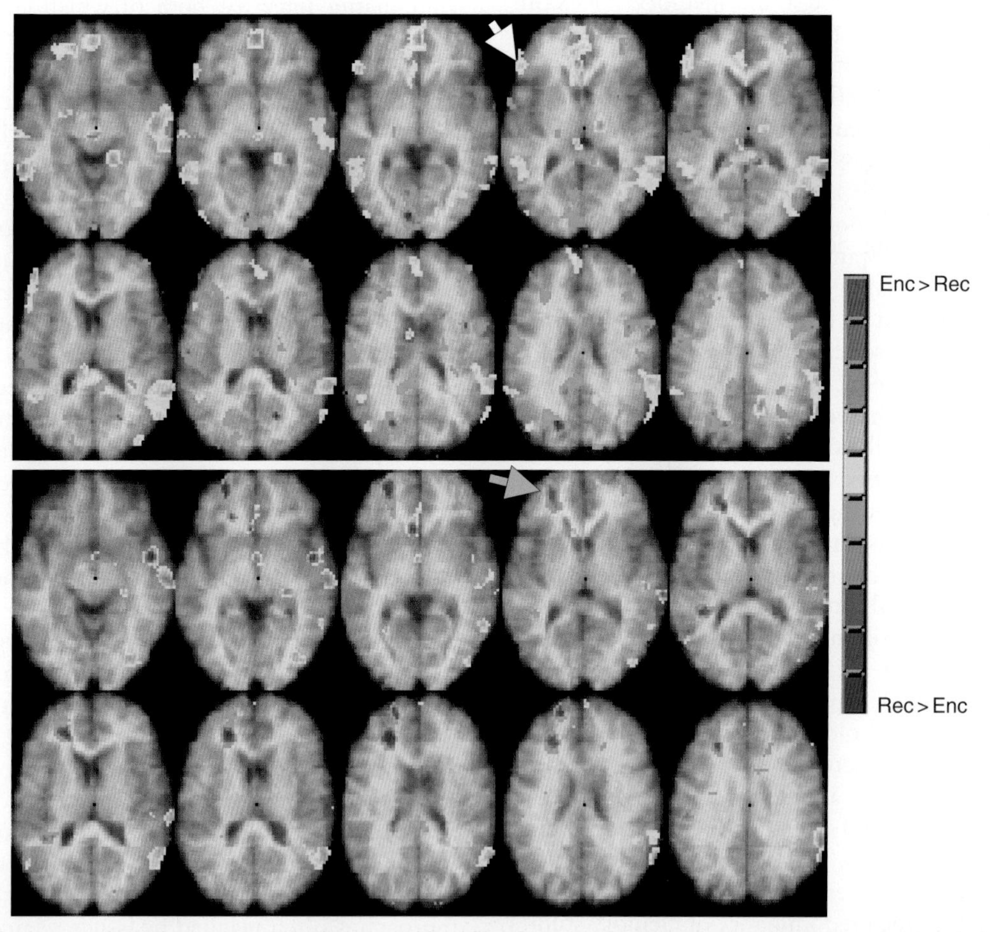


**Figure 1** The images depict brain areas with changes in blood flow (measured with positron emission tomography) during encoding (Enc) and recognition (Rec) of faces in a group of young adults (top panel, $n = 12$; mean age 23 years) and a group of older adults (bottom panel, $n = 11$; mean age 70 years). Results are shown on images from a standard structural magnetic resonance image. Yellow/red areas are those with more activity during encoding, compared to recognition, and blue areas indicate more activity during recognition. Young adults show increased activity during encoding in left inferior prefrontal cortex (white arrow), but older adults do not. In contrast, old adults have more activity during recognition, compared to young adults, in widespread areas of left prefrontal cortex (cyan arrow).

large enough to remove any age difference. This pattern of results is consistent with behavioral data suggesting that older adults may not spontaneously engage effective encoding strategies – that is, those involving semantic processing by left inferior frontal cortex – but can use such strategies when required to do so.

In contrast to the left frontal activity seen during encoding, during memory retrieval, when people are attempting to recall or recognize previously learned information, frontal activity often is found predominately in the right hemisphere. Although this right frontal activity is often of the same magnitude regardless of age, older adults do show reduced activity under some conditions. When the retrieval task places a heavy demand on memory search, without providing cues for retrieval, older adults show a reduction in right prefrontal activation. Retrieval of source memory, such as whether a word was spoken by a man or a woman, also places significant demand on memory search and typically activates prefrontal regions in young adults. Consistent with their reduced source memory, older adults have less activation of prefrontal cortex during source retrieval compared to their younger counterparts.

Another area of memory that has been examined fairly extensively in the aging field is that of working memory. A number of working memory tasks have been used, including delayed match-to-sample and *n*-back tasks. Regardless of task, these studies have generally found less prefrontal activity in older adults. When the different phases of the working memory task have been examined separately (memory set presentation, delay, and probe), age differences have been seen primarily during the presentation of the probe – that is, when the memory decision is being made. This suggests that some aspects of working memory, such as rehearsal, might not be affected by age. Some recent work has shown that older adults are able to recruit prefrontal cortex at lower levels of working memory task load, but that this declines at higher task loads, whereas younger adults continue to increase prefrontal recruitment with increasing load. This finding is similar to that noted earlier for episodic retrieval, as both would suggest that tasks with high processing demands are likely to be associated with reduced brain activity in older adults, and potentially poorer performance as well.

In addition to prefrontal regions, less activity during encoding and/or recognition memory in older adults also has been found in other areas important for memory, including the hippocampus and the parahippocampal gyrus, and in visual areas of cortex (**Figure 2**). Interestingly, when brain activity has been compared across different kinds of tasks, such as working memory and episodic retrieval, similar age reductions in brain activity have been found in visual cortex and hippocampus. Such a result is consistent with the idea that there is a 'common cause' of cognitive aging, and indicates that reductions in both

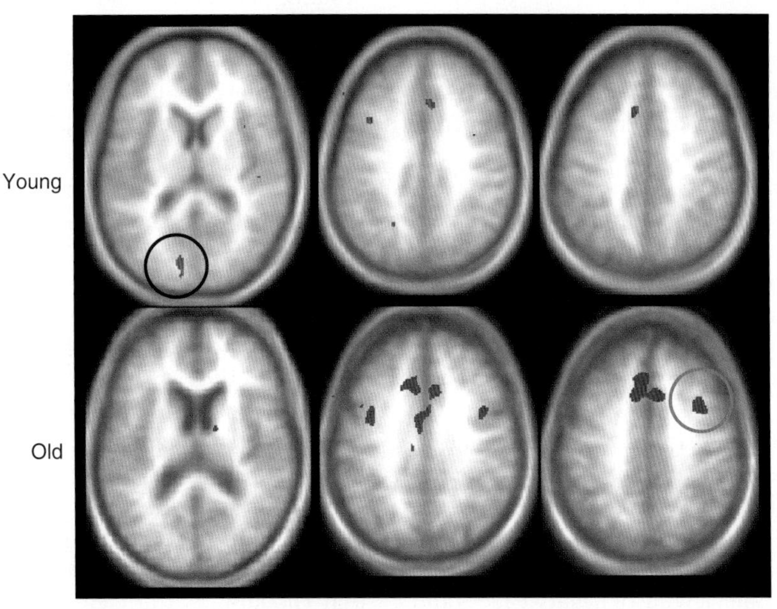


**Figure 2** The images depict brain areas with changes in blood oxygenation (measured with functional magnetic resonance imaging) during recognition of auditorily presented words in a group of young adults ($n = 12$; mean age 26 years) and a group of older adults ($n = 12$; mean age 71 years). Results are shown in red on mean structural magnetic resonance images for the young adults (top) and the older adults (bottom). Young adults had greater activity in a region of left extrastriate cortex (black circle), whereas older adults had more activity in prefrontal regions (one of which is indicated by the blue circle).

sensory processing and memory encoding could be involved in such a common cause. One thing to keep in mind, however, particularly when considering age differences in sensory cortex, is that less activity during a particular cognitive task may mean that older adults have more activity in these areas during the baseline condition. Thus, when activity during the baseline is subtracted from task activity, older adults will have a smaller task-related increase, but this does not mean that the region cannot be activated. Rather, the older adults are starting out from a higher level of activity. Such a phenomenon has been observed in visual cortex.

Functional neuroimaging has also been used to study executive functions, thought to be mediated primarily by frontal cortex, in older adults. One such function, inhibition, has been studied using tasks that require inhibiting a prepotent response or suppressing information irrelevant to the task. Consistent with their poorer performance, older adults have less activity in frontal regions thought to mediate inhibitory processes on such tasks. Activity in frontal regions also is reduced with age during performance of the Wisconsin Card Sorting Task, a standard instrument measuring concept formation and cognitive flexibility. Other brain regions also show age differences during tasks requiring inhibitory processing. Young adults either enhance or suppress activity in regions of occipital cortex that respond selectively to different types of objects depending on whether the objects are attended or ignored. In contrast, older adults show enhanced activity in these areas, but fail to reduce activity in object-specific regions when viewing objects that should be ignored because they are irrelevant to the task at hand. Thus, age reductions in the ability to ignore irrelevant and distracting information may be the result of both top-down (i.e., frontal) and bottom-up (occipital) mechanisms.

Emotional processing and emotional enhancement of memory have been studied for a number of years using animal models, but only recently has investigation of this area of cognition been explored with functional neuroimaging. Consistent with animal work showing critical involvement of the amygdala in emotion, young adults show increased activity in the amygdala when emotional stimuli, particularly negative items, are viewed. This activity also is related to later memory for this material, even when memory is tested months after initial exposure to the stimuli. Older adults show less amygdala activity compared to younger adults during viewing of negative faces and scenes, consistent with their reduced ability to both identify and remember negatively valenced stimuli.

Finally, there has been recent interest in a group of brain areas that have more activity when people are not carrying out a specific task, but are merely monitoring their internal and external milieus, the so-called 'default mode' regions. Young adults show marked reductions of activity in these default areas when carrying out a variety of tasks, whereas older adults have much less change in these areas during tasks. In fact, there appears to be linear reduction in this default region 'suppression' over the adult life span, suggesting a gradual, age-related reduction in the ability to suspend non-task-related, or default, activity and to engage areas for carrying out memory tasks. Thus, it may be just as important to look at age-related differences in activity reductions as it is to explore differences in activity increases.

## Increased Brain Activity in Older Adults

It is perhaps surprising that, in addition to reduced brain activity in some areas, older adults often show increased brain activity, compared to young adults. Additional recruitment of brain areas by older adults typically is found in the frontal lobes, either in areas of frontal cortex not active in young adults, or in homologous regions in the opposite hemisphere from regions active in the young. During episodic memory retrieval older adults frequently show more extensive activation of prefrontal regions compared to younger adults, sometimes in those left hemisphere frontal areas that are less active during encoding in older adults. Greater prefrontal activation during memory retrieval has been found in older adults for a variety of stimuli, both verbal and nonverbal (**Figures 1** and **2**). Another way of looking at the brain activity that supports retrieval is to see how activity during encoding of items that are later correctly remembered differs from that for encoding of items that are forgotten, known as the 'subsequent memory effect.' That is, what is the brain activity, during learning, that predicts whether a stimulus will be successfully remembered later? Using the subsequent memory effect, a number of investigators have found that activity in left inferior prefrontal cortex and the left medial temporal region in young adults is greater during encoding of subsequently remembered stimuli. A similar pattern has been found in old adults; however, older adults also have an association between more activity in right prefrontal cortex and subsequent recognition. Thus, older adults can show a more extensive pattern of frontal activity during encoding, as they do during retrieval, but only when that encoding is sufficient to support later retrieval. A similar pattern has been found during some working memory tasks, where frontal activation often is unilateral in young adults, but in both hemispheres in older adults.

There is some evidence for greater brain activity in older adults during memory retrieval tasks in areas outside of the frontal lobes, such as medial temporal cortex. In particular, studies of recognition memory, such as recognition for words presented earlier in the experiment, have shown more activation of parahippocampal regions in older adults, compared to younger adults. Indeed, parahippocampal cortex in young adults often shows less activity when previously seen stimuli are presented during a recognition task, and this reduction in activity is thought to represent a signal of familiarity with the stimulus. Increased activity in older adults in the parahippocampal gyrus may still indicate familiarity, but the nature of the signal may undergo a change with age. Interestingly, although hippocampal activity during memory retrieval is often reduced with aging, bilateral activation of the hippocampus has been reported in older adults during an autobiographical memory retrieval task, whereas young adults had only left hippocampal activation. This bilateral activation is similar to the bilateral prefrontal activity seen in older adults during other memory tasks, and may support those aspects of autobiographical retrieval which are maintained in the elderly.

Increased prefrontal activation in older adults also has been found in other cognitive domains, although less information is available than for memory. For example, there are reports of more prefrontal activity in older adults when carrying out what one might consider relatively simple motor tasks. This may indicate that even sensorimotor tasks require more cognitive control on the part of older adults.

So, what could be the impact of this increased brain activity on behavioral performance in older adults? There have been suggestions that additional recruitment of task-related brain areas, or engagement of unique areas not active in young adults, compensates for age-related changes in brain structure and function. Compensation has been invoked to explain greater prefrontal activity in older adults when they are able to perform a particular task, as well as younger adults. Additional support for this idea has been found in studies showing that older adults who perform better on memory tasks recruit prefrontal cortex to a greater degree than do those performing less well. Older adults also may have more prefrontal activity than do young adults when brain activity is examined just for those task trials where participants respond correctly. Such findings suggest that correct performance may be supported by this overrecruitment of frontal regions. Another approach has been to determine how brain activity is related to task performance by correlating these two measures in groups of young and old adults. Increased prefrontal activity is often correlated with better performance on memory tasks in older adults, but rarely in young adults. In fact, in young adults prefrontal activity is usually correlated with poorer task performance, indicating that with age, the impact of brain activity on memory performance is altered.

On the other hand, some of this increased prefrontal activity in the elderly may reflect greater need or use of frontally mediated executive functions at lower levels of task demand than would be necessary for activation of this area in young adults. In this case the increased activity in prefrontal cortex might not be related to performance on the particular task, but would reflect a type of nonselective recruitment in older adults, perhaps related to task difficulty. Examples of this kind of scenario have been observed, and include increased activity in widely distributed areas of both lateral and medial portions of prefrontal cortex in older adults during tasks on which they perform worse than young adults (**Figure 2**). These findings suggest that while additional recruitment of prefrontal cortex in older adults may often be compensatory, it is not necessarily so.

## Different Cognitive Networks in Older Adults

A relatively new approach to studying brain function is the focus on functional connectivity, or how activity in a given brain area is correlated with activity elsewhere in the brain, to identify networks of interacting regions. Studies using this approach have provided intriguing evidence that young and older adults recruit different networks, even when the measured behavior is the same. For example, when pictures are used as stimuli in memory tasks, they are recognized well by both young and old adults. Encoding of these pictures also is associated with increased activity in medial temporal regions, including the hippocampus, in both young and old groups. However, the areas that are functionally connected to the hippocampus during encoding vary with age. In young adults, during picture encoding, hippocampal activity is correlated with activity in ventral prefrontal and visual regions, whereas older adults show correlations between hippocampal activity and dorsolateral prefrontal and parietal regions (**Figure 3**). Of particular interest is that these age-specific networks are associated with better memory performance. That is, those individuals in each age group who engage the specific network to a greater degree are the ones who remember the greatest number of pictures.

Different functional connectivity of medial temporal regions in older adults, compared to young adults, also has been found during recognition tasks.

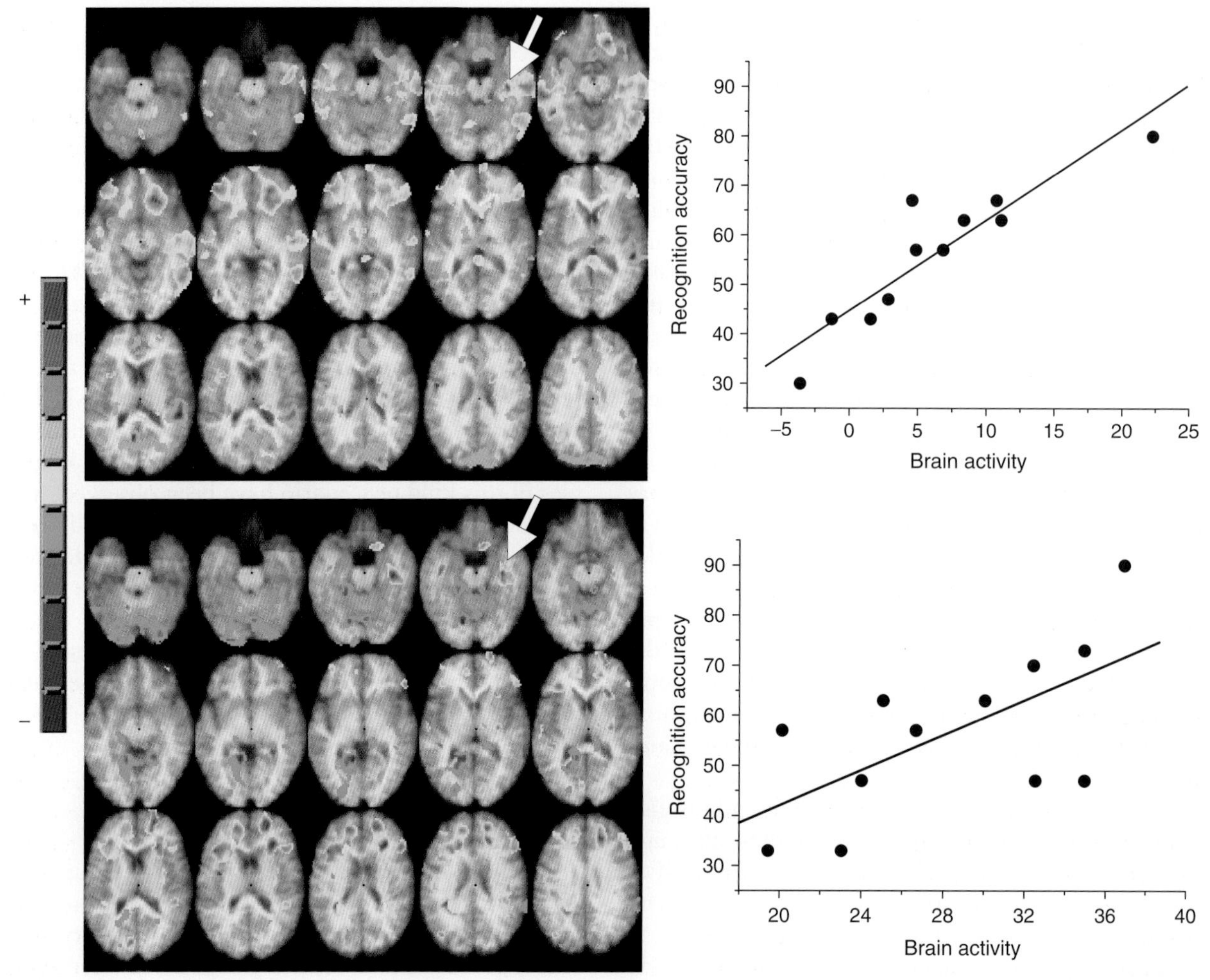


**Figure 3** The images depict brain areas where blood flow (measured with positron emission tomography) during picture encoding was significantly correlated with flow in right hippocampus (white arrows). Results are from a group of young adults (top, $n = 12$; mean age 23 years) and a group of older adults (bottom, $n = 12$; mean age 66 years) and are shown on a standard structural magnetic resonance image. Yellow/red areas are those with positive correlations and blue areas are those with negative correlations. In young adults hippocampal activity was positively correlated with activity in ventral prefrontal regions, whereas in old adults hippocampal activity was positively correlated with activity in dorsolateral frontal cortex. The graphs show the scatterplots of activity in these age-specific networks, and recognition accuracy. As activity increased during encoding in the hippocampus, and the areas positively correlated with it, subsequent recognition accuracy increased in both groups.

A recent study found that during recognition, the hippocampus showed stronger functional connections with cortical regions in young adults, whereas the entorhinal cortex showed stronger connections in older adults. This finding was consistent with evidence that these two medial temporal areas mediate recollection and familiarity, respectively, two aspects of memory that tap different qualitative experiences, and with the many studies showing age differences in recollection but not familiarity. Thus, age differences in recollective processes mediated by the hippocampus may be offset by more reliance on familiarity mediated by medial temporal cortex adjacent to the hippocampus. Similar evidence for age differences in cognitive networks involving the hippocampus and its interactions with the cortex has been found in experiments tapping working memory as well. All of these studies that focus on network approaches to image analysis have provided evidence that aging results in the modification of large-scale network operations, in addition to affecting activity in specific brain regions, which in turn has an impact on memory ability.

## Genetic Factors in Brain Aging

Recently there has been interest in using functional neuroimaging to study the genetic basis for cognitive behaviors, including memory. Most of this work has been done in young adults, by dividing healthy samples of people according to different genetic alleles that bear some relation to risk for psychiatric diseases, such as schizophrenia. This work has shown that activity in prefrontal cortex and medial temporal areas differs across genotypes, and that these differences are associated with differences in personality traits, such as risk taking, and in memory function. Little work in this area has been applied to the study of cognitive aging, but a few studies have looked at the effect of *APOE* genotype on brain function and memory in older adults. *APOE* is one of the genes that confers risk for Alzheimer's disease when the *APOE* ε4 allele is present. Several studies have shown that middle-aged and older individuals who have the ε4 allele, but are nevertheless asymptomatic, have reduced resting metabolic activity in parieto-temporal cortex compared to those without the ε4 allele. Follow-up assessments showed that cognitive decline after 2 years was greatest in those *APOE* ε4 carriers with the lowest metabolism in parietal and temporal regions at baseline. In addition, older *APOE* ε4 carriers have faster cognitive decline than do noncarriers, and show a relation between subjective memory complaints and decline in temporal lobe metabolism over time. Interestingly, it was found with fMRI that activation of prefrontal and medial temporal areas was increased in *APOE* ε4 carriers, compared to noncarriers, but only during memory tasks. This finding suggests that the cognitive and neural effort necessary for memory function varies according to one's genetic makeup, and that this effort may also reflect the likelihood of cognitive decline in later years. All of these studies indicate that cognitive changes noted in older adults, even in those individuals who appear to be cognitively normal for their age, may be influenced by genetic factors. In addition, it is possible that genetic differences underlie some of the variability that has been found across neuroimaging studies of older adults.

## Conclusions

It is clear that functional neuroimaging has opened a window into cognitive aging, shedding light on the roles of brain areas, such as prefrontal cortex and the hippocampus, in age-related differences in cognitive performance. However, a number of questions remain. For example, the extent to which alterations of brain activity or brain networks seen in older adults are due to changes in the function of the brain regions, or to changes in the white-matter tracts that connect the regions, is unknown. Communication between cortical regions could be reduced or altered because the neurons in the regions have altered function or because the fibers that connect them are damaged or less efficient. Changes with age in white matter, and the impact of these on cognition, have been documented, but how these changes affect functional connectivity of gray-matter regions during cognitive tasks is not well understood. Recent evidence indicates that age differences in white matter are more prominent in frontal than in posterior brain areas, and that white-matter integrity of the frontal lobes is related to activation in prefrontal gray-matter regions in older adults. These structural changes are consistent with age changes in frontal activity that suggest nonselectivity, but not with evidence that frontal cortex is most often recruited for compensatory purposes. Clearly we have much to learn about the complex interplay between brain structure and function, and how this interaction is affected by age.

There are other factors that could influence brain activity in older and younger adults and these are just now beginning to receive attention. An example is education, which has been shown to have a protective effect against decline in cognitive function in older adults. There is some evidence that education, or other indices of intellectual achievement, are related differently to brain activity in young versus older adults, such that education is positively correlated with frontal activity only in older adults. This finding is consistent with the evidence of greater frontal recruitment by older adults and suggests that frontal cortex is engaged particularly by those older adults who are highly educated. Thus, it may be that compensatory brain activity in older adults is most likely to be seen in those with higher education, but more work needs to be done in this area. Other variables, such as personality traits or mood, have yet to be examined, but as age differences have been found on these measures, these factors also could be influencing brain activity in older adults.

Finally, it is important to note that it is possible to identify a difference in functional connectivity, or network interactions, involving a particular brain area even when no change in mean activation is seen in this area. This indicates that a failure to find differences in brain activity between younger and older adults in a given region on a particular task does not mean that the functional communication of that region with other area is the same as well. Thus, one cannot always assume that the function of a brain area is unchanged with age merely by assessing its mean activity level. A more detailed examination of the dynamic interactions among brain regions, in conjunction with underlying structural changes and assessment of the impact of network activity on behavioral performance, will likely be necessary to ultimately understand the impact of aging on cognitive function.

*See also:* Aging of the Brain; Aging and Memory in Humans; Cognition in Aging and Age-Related Disease; Declarative Memory System: Anatomy; Episodic Memory.

## Further Reading

Anderson ND and Grady CL (2004) Functional imaging in healthy aging and Alzheimer's disease. In: Cronin-Golomb A, and Hof P (eds.) *Vision in Alzheimer's Disease*, pp. 62–95. Basel: Karger.

Bookheimer SY, Strojwas MH, Cohen MS, et al. (2000) Patterns of brain activation in people at risk for Alzheimer's disease. *New England Journal of Medicine* 343: 450–456.

Buckner RL (2004) Memory and executive function in aging and AD: Multiple factors that cause decline and reserve factors that compensate. *Neuron* 44: 195–208.

Cabeza R, Nyberg L, and Park D (eds.) (2005) *Cognitive Neuroscience of Aging*. Oxford: Oxford University Press.

Ercoli L, Siddarth P, Huang SC, et al. (2006) Perceived loss of memory ability and cerebral metabolic decline in persons with the apolipoprotein E-IV genetic risk for Alzheimer disease. *Archives of General Psychiatry* 63: 442–448.

Gazzaley A, Cooney JW, Rissman J, et al. (2005) Top-down suppression deficit underlies working memory impairment in normal aging. *Nature Neuroscience* 8: 1298–1300.

Grady CL and Craik FIM (2000) Changes in memory processing with age. *Current Opinion in Neurobiology* 10: 224–231.

Grady CL, McIntosh AR, and Craik FI (2003) Age-related differences in the functional connectivity of the hippocampus during memory encoding. *Hippocampus* 13: 572–586.

Gutchess AH, Welsh RC, Hedden T, et al. (2005) Aging and the neural correlates of successful picture encoding: Frontal activations compensate for decreased medial temporal activity. *Journal of Cognitive Neuroscience* 17: 84–96.

Morcom AM, Good CD, Frackowiak RS, et al. (2003) Age effects on the neural correlates of successful memory encoding. *Brain* 126: 213–229.

Nordahl CW, Ranganath C, Yonelinas AP, et al. (2006) White matter changes compromise prefrontal cortex function in healthy elderly individuals. *Journal of Cognitive Neuroscience* 18: 418–429.

Rajah MN and D'Esposito M (2005) Region-specific changes in prefrontal function with age: A review of PET and fMRI studies on working and episodic memory. *Brain* 128: 1964–1983.

Small GW, Ercoli LM, Silverman DH, et al. (2000) Cerebral metabolic and cognitive decline in persons at genetic risk for Alzheimer's disease. *Proceedings of the National Academy of Sciences of the United States of America* 97: 6037–6042.

Stern Y, Habeck C, Moeller J, et al. (2005) Brain networks associated with cognitive reserve in healthy young and old adults. *Cerebral Cortex* 15: 394–402.

# Aging: Extracellular Space

**E Syková**, Institute of Experimental Medicine, Academy of Sciences of the Czech Republic, and Charles University, Prague, Czech Republic



## Gross Anatomic Aging Changes: Extracellular Space

Aging, Alzheimer's disease, and many degenerative diseases are accompanied by serious cognitive deficits, particularly impaired learning and memory loss. This decline in old age is a consequence of changes in brain anatomy, morphology, and volume, as well as functional deficits. Nervous tissue, particularly in the hippocampus and cortex, is subject to various degenerative processes, including a decreased number and efficacy of synapses, a decrease in transmitter release, neuronal loss, a decreased number of dendritic spines, astrogliosis, changes in astrocytic morphology, demyelination, deposits of $\beta$-amyloid, changes in extracellular matrix proteins, and, consequently, changes in the brain's diffusion parameters. These age-related changes in the morphology of neural elements have no universal pattern across the entire brain. Despite the fact that aging is associated with cognitive impairment, in most brain areas neuronal loss does not have a significant role in the observed deficits. It has been suggested that rather small, region-specific changes in dendritic branching and spine density are more characteristic changes in neuronal morphology.

## Diffusion and Extrasynaptic Transmission

The gross anatomic changes and other subtle morphological changes not only affect the efficacy of signal transmission at synapses, but also the function of glia and extrasynaptic ('volume') transmission mediated by the diffusion of transmitters and other substances through the volume of the extracellular space (ECS). This mode of communication by diffusion (i.e., without synapses) provides a mechanism of long-range information processing in functions such as vigilance, sleep, chronic pain, hunger, depression, long-term potentiation (LTP), long-term depression (LTD), memory formation, and other plastic changes in the central nervous system (CNS). Neurons interact both by synapses and by the diffusion of ions and neurotransmitters in the extracellular space. Since glial cells do not have synapses, their communication with neurons is only mediated by the diffusion of ions and neuroactive substances in the ECS. Neurons and glia release ions, transmitters, and various other neuroactive substances into the ECS. Substances released nonsynaptically diffuse through the ECS and bind to extrasynaptic, usually high-affinity, binding sites located on neurons, axons, and glial cells.

Diffusion in the ECS is critically dependent on the structure and physicochemical properties of the ECS – the nerve cell microenvironment. These properties vary, however, around each cell and in different brain regions. Certain synapses ('private synapses') or even whole neurons are clearly tightly ensheathed by glial processes and by the extracellular matrix, forming so-called perineuronal nets, while others are left more 'naked.' The 'open' synapses are more easily reached by molecules diffusing in the ECS (**Figure 1**). On the other hand, many mediators, including glutamate and $\gamma$-aminobutyric acid (GABA), bind to high-affinity binding sites located on nonsynaptic parts of the membranes of neurons and glia. Mediators that escape from the synaptic clefts at an activated synapse, particularly following repetitive stimulation, diffuse in the ECS and can cross-react with receptors in nearby synapses. This phenomenon, called 'cross-talk' between synapses by the 'spillover' of a transmitter (e.g., glutamate, GABA, glycine), has been proposed to account for LTP, LTD, and memory formation in the rat hippocampus. The cross-talk between synapses, and the efficacy and directionality of volume transmission, are critically dependent on the diffusion properties of the ECS.

There is increasing evidence that changes in neuron–glia interaction (e.g., glial coverage and/or the retraction of glial processes from synapses) occur in many brain regions during physiological and pathological functional changes. The glial environment of neurons is likely to be a key factor in the regulation of intersynaptic communication mediated by glutamate. For example, most synaptically released glutamate is taken up by high-affinity transporters, such as GLT-1 and GLAST, that are located on surrounding astrocytes. Moreover, glial cells represent a diffusion barrier in the ECS, hindering the movement of neuroactive substances within the tissue. Long-term changes in the physical and chemical parameters of the ECS accompany many physiological (e.g., development, aging, lactation) and pathological (e.g., ischemia, epilepsy, hydrocephalus, demyelination, CNS trauma) states.

## Measurements of the Diffusion Parameters of the ECS

The diffusion of substances in a free medium, such as water or diluted agar, is described by Fick's laws.

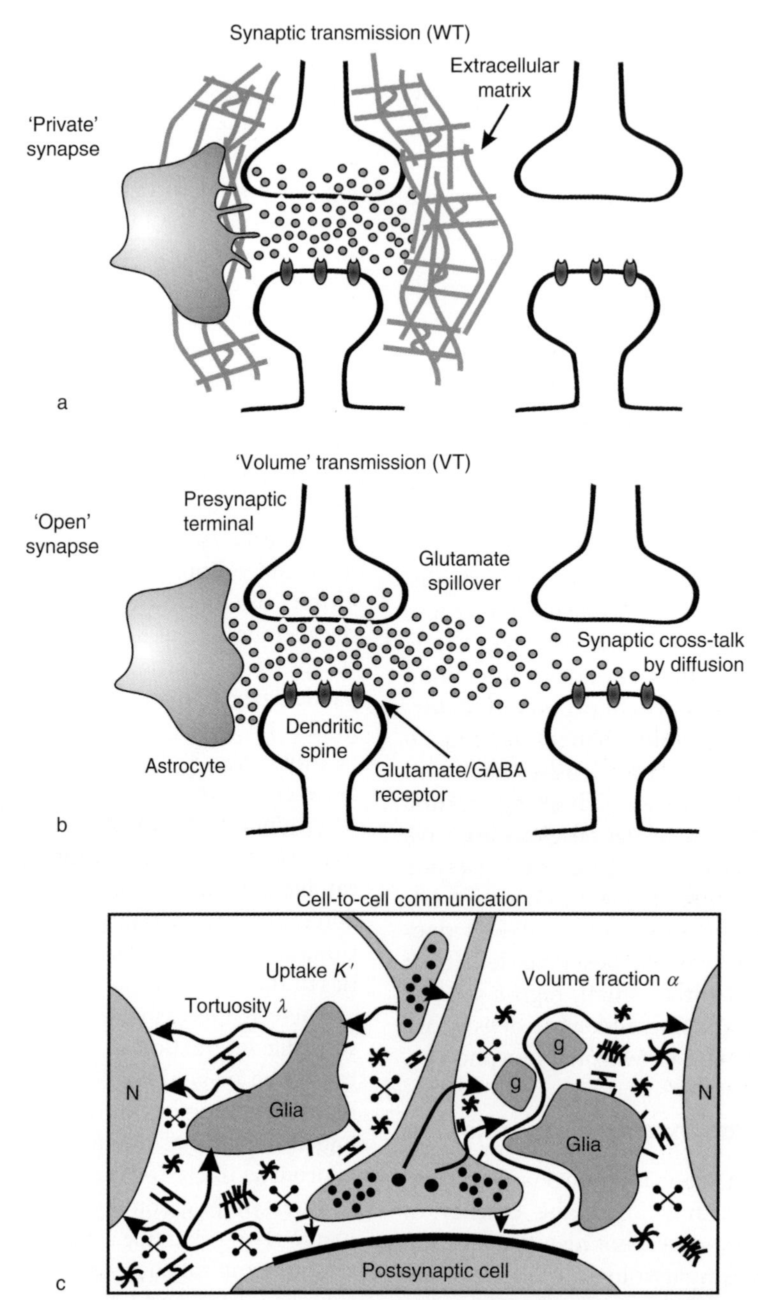


**Figure 1** Concept of synaptic 'wired' transmission (WT) and extrasynaptic 'volume' transmission (VT). (a) Closed synapses are typical of synaptic transmission. The synapse is tightly ensheathed by glial processes and the extracellular matrix, forming perineuronal or perisynaptic nets. (b) An open synapse is typical of volume transmission. It allows the escape of transmitter – for example, glutamate and $\gamma$-aminobutyric acid (GABA) – from the synaptic cleft (spillover), diffusion in the extracellular space, and binding to receptors on nearby synapses. This phenomenon is known as 'cross-talk' between synapses. Spillover may also lead to plastic changes, inducing the formation of new synapses or eliciting the rearrangement of astrocytic processes around the synapse and the formation of diffusion barriers. (c) Schematic of central nervous system architecture, showing how it is composed of neurons (N), axons, glial cells (g), cellular processes, molecules of the extracellular matrix, and intercellular channels between the cells. The architecture affects the movement (diffusion) of substances in the brain, which is critically dependent on channel size, extracellular space tortuosity, and cellular uptake. Adapted from Syková E (2001) Glial diffusion barriers during aging and pathological states. *Progress in Brain Research* 132: 339–363.

In contrast to a free medium, the diffusion of any substance in the ECS is hindered by the ECS size, the presence of various obstacles, and also by cellular uptake. To take these factors into account, it was necessary to modify Fick's original diffusion equations. First, diffusion in the CNS is constrained by the restricted volume of the tissue available for the diffusing particles – that is, by the extracellular space volume fraction ($\alpha$), which is a dimensionless quantity and is defined as the ratio between the volume of the ECS and the total volume of the tissue. It is now evident that the ECS in the adult brain amounts to about 20% of the total brain volume (i.e., $\alpha = 0.2$). Second, the free diffusion coefficient ($D$) in the brain is reduced by the tortuosity factor ($\lambda$). ECS tortuosity is defined as $\lambda = (D/\mathrm{ADC})^{0.5}$, where $D$ is the free diffusion coefficient and ADC is the apparent diffusion coefficient in the brain. As a result of tortuosity, $D$ is reduced to an apparent diffusion coefficient $\mathrm{ADC} = D/\lambda^2$. Thus, any substance diffusing in the ECS is hindered by membrane obstructions, glycoproteins, macromolecules of the extracellular matrix, charged molecules, and fine neuronal and glial cell processes. Third, substances released into the ECS are transported across membranes by nonspecific concentration-dependent uptake ($k'$). In many cases however, these substances are transported by energy-dependent uptake systems that obey nonlinear kinetics. When these three factors ($\alpha$, $\lambda$, and $k'$) are incorporated into Fick's law, diffusion in the CNS is described fairly satisfactorily.

At the present time, the real-time iontophoretic method is used to determine the absolute values of the ECS diffusion parameters and their dynamic changes in nervous tissue *in vitro* as well as *in vivo*. The second method that is also frequently used to study ECS volume and geometry is diffusion-weighted magnetic resonance imaging (DW-MRI). DW-MRI provides information only about the apparent diffusion coefficient of water. A relationship between an increase in the *ADC* of water and a decrease in ECS volume fraction has recently been found during pathological states as well as during aging.

The real-time iontophoretic method (**Figure 2(a)**) uses ion-selective microelectrodes to measure the diffusion of ions to which the cell membranes are relatively impermeable – for example, tetraethylammonium and tetramethylammonium ions ($TEA^+$, $TMA^+$) or choline. These substances are injected into the nervous tissue by pressure or by iontophoresis from an electrode aligned parallel to a double-barreled ion-selective microelectrode (ISM) at a fixed distance. Usually, such an electrode array is made by gluing together an iontophoretic pipette and a $TMA^+$-sensitive ISM with a tip separation of 130–200 μm. In the case of iontophoretic application, $TMA^+$ is released into the ECS by applying a current step of +100 nA with a duration of 40–80 s. The released $TMA^+$ is recorded with the $TMA^+$-sensitive ISM as a diffusion curve, which is then transferred to a computer to calculate $\alpha$, $\lambda$, and $k'$ from the modified diffusion equation. Values of the ECS volume, ADC, tortuosity, and nonspecific cellular uptake are extracted by a nonlinear curve-fitting simplex algorithm applied on the diffusion curves.

By introducing the tortuosity factor into diffusion measurements in nervous tissue, it soon became evident that diffusion is not uniform in all directions and is affected by the presence of diffusion barriers, including neuronal and glial processes, myelin sheaths, macromolecules, and molecules with fixed negative surface charges. This so-called anisotropic diffusion preferentially channels the movement of substances in the ECS in one direction, (e.g., along axons) and may, therefore, be responsible for a certain degree of specificity in volume transmission. Diffusion anisotropy was found in the CNS in the molecular and granular layers of the cerebellum, in the hippocampus, and in the auditory but not in the somatosensory cortex, and a number of studies have revealed that it is present in the myelinated white matter of the corpus callosum or spinal cord. It was shown that diffusion anisotropy in white matter increases during development. At first, diffusion in unmyelinated tissue is isotropic; it becomes more anisotropic as myelination progresses.

## ECS Diffusion Parameters during Aging

Aging is not only accompanied by a decrease in the efficacy of signal transmission at synapses, but also by changes in the ECS diffusion parameters and consequently by a deficit in extrasynaptic transmission. Using the $TMA^+$ method, the ECS diffusion parameters were measured in the cortex, corpus callosum, and hippocampus (CA1, CA3, and dentate gyrus) of aged rats. The measurements revealed a significant decrease in ECS volume fraction. One of the explanations as to why $\alpha$ in the cortex, corpus callosum, and hippocampus of senescent rats and mice is significantly lower than in young adults could be astrogliosis in the aged brain. Increased glial fibrillary acidic protein (GFAP) staining and an increase in the size and fibrous character of astrocytes have been found in the cortex, corpus callosum, and hippocampus of senescent rats. Other changes could account for the decreases in $\lambda$ and for the disruption of tissue anisotropy. In the hippocampus in the CA1 and CA3 regions, as well as in the dentate gyrus, changes in the arrangement of fine astrocytic processes were regularly found. These are normally organized in parallel in the $x$–$y$ plane, but this organization totally disappears during aging. Moreover, decreased staining for chondroitin sulfate proteoglycans and for fibronectin suggests a loss of extracellular matrix macromolecules (**Figure 3**).

If diffusion in a particular brain region is anisotropic, then the correct value of the ECS volume

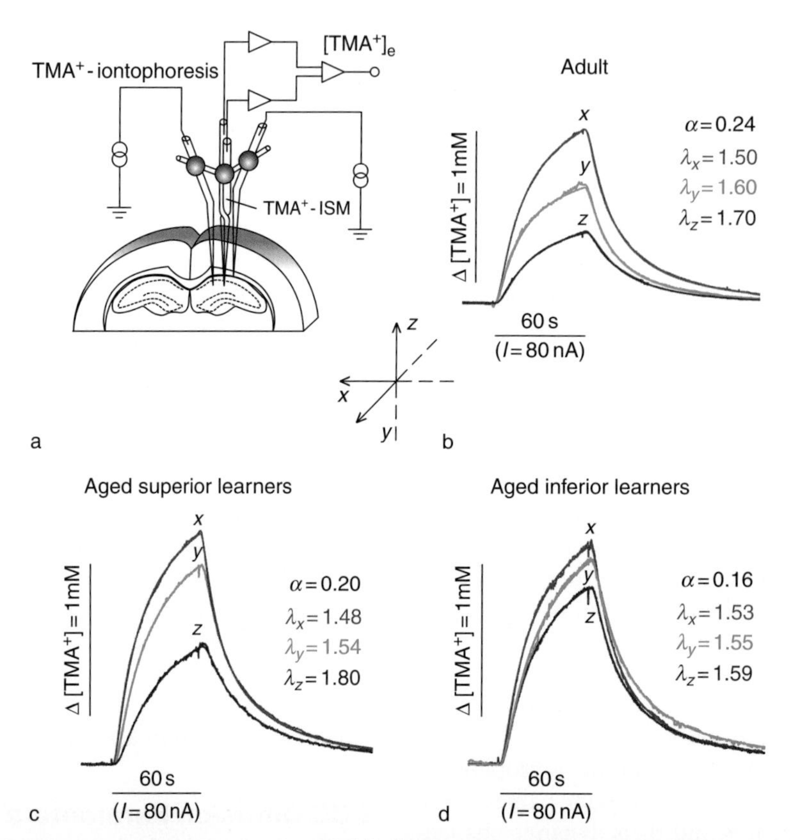


**Figure 2** Experimental setup: tetramethylammonium ion ($TMA^+$) diffusion curves and typical extracellular space diffusion parameters $\alpha$ (volume fraction) and $\lambda$ (tortuosity) in the hippocampus dentate gyrus of a young adult rat and an aged rat with memory impairment (rats were tested in a Morris water maze). (a) Schema of the experimental arrangement. A $TMA^+$-selective double-barreled ion-selective microelectrode (ISM) was glued to a bent iontophoresis microelectrode. The separation between electrode tips was 130–200 μm. (b) Anisotropic diffusion in the dentate gyrus of a young adult rat required measurements of $TMA^+$ diffusion curves (concentration–time profiles) along three orthogonal axes (*x*, mediolateral; *y*, rostrocaudal; *z*, dorsoventral). The slower rises in the *z* axis (compared to the *y* axis) and in the *y* axis (compared to the *x* axis) indicate a higher tortuosity and more restricted diffusion. The amplitudes of the curves show that the $TMA^+$ concentration, at approximately the same distance from the tip of the iontophoresis electrode, is much higher along the *x* axis than along the *y* axis and is even higher than along the *z* axis ($\lambda_x$, $\lambda_y$, $\lambda_z$). Note that the actual extracellular space volume fraction $\alpha$ is about 0.2 and can be calculated only when measurements are done along the *x*, *y*, and *z* axes. (c, d) Volume fraction and anisotropy decrease during aging. Note the greater decrease in volume fraction and that the anisotropy is almost lost in an aged rat with memory impairment (d). Note that the diffusion curves are higher, showing that $\alpha$ is smaller, and that their rise and decay times are longer when $\lambda$ (diffusion barriers) increases. Adapted from Syková E, Mazel T, Hasenöhrl RU, et al. (2002) Learning deficits in aged rats related to decrease in extracellular volume and loss of diffusion anisotropy in hippocampus. *Hippocampus* 12: 469–479.

fraction cannot be calculated from measurements done only in one direction. For anisotropic diffusion, the diagonal components of the tortuosity tensor are not equal, and generally its nondiagonal components need not be zero. Nevertheless, if a suitable referential frame is chosen (i.e., if we measure in three privileged orthogonal directions), neglecting the nondiagonal components becomes possible, and the correct value of the ECS size can thus be determined. Therefore, $TMA^+$ diffusion was measured in the ECS independently along three orthogonal axes (*x*, transversal; *y*, sagittal; *z*, vertical) in the cortex, corpus callosum, and hippocampus of aged rats and mice. These studies revealed that the mean ECS volume fraction $\alpha$ is significantly lower in aged rats (26–32 months old), ranging from 0.17 to 0.19, than in young adults (3–4 months old), in which $\alpha$ ranges from 0.21 to 0.22. Nonspecific uptake $k'$ is also significantly lower in aged rats. From **Figure 2** it is evident that the diffusion curves for the hippocampus are larger in the aged rat than in the young one – that is, the space available for $TMA^+$ diffusion is smaller. Although the mean $\alpha$ values along the *x* axis were not significantly different between young and aged rats, the values were significantly lower in aged rats along the *y* and *z* axes. This means that there is a loss of anisotropy in the aging hippocampus, particularly in the CA3 region and the dentate gyrus.

The three-dimensional pattern of diffusion away from a point source can be illustrated by

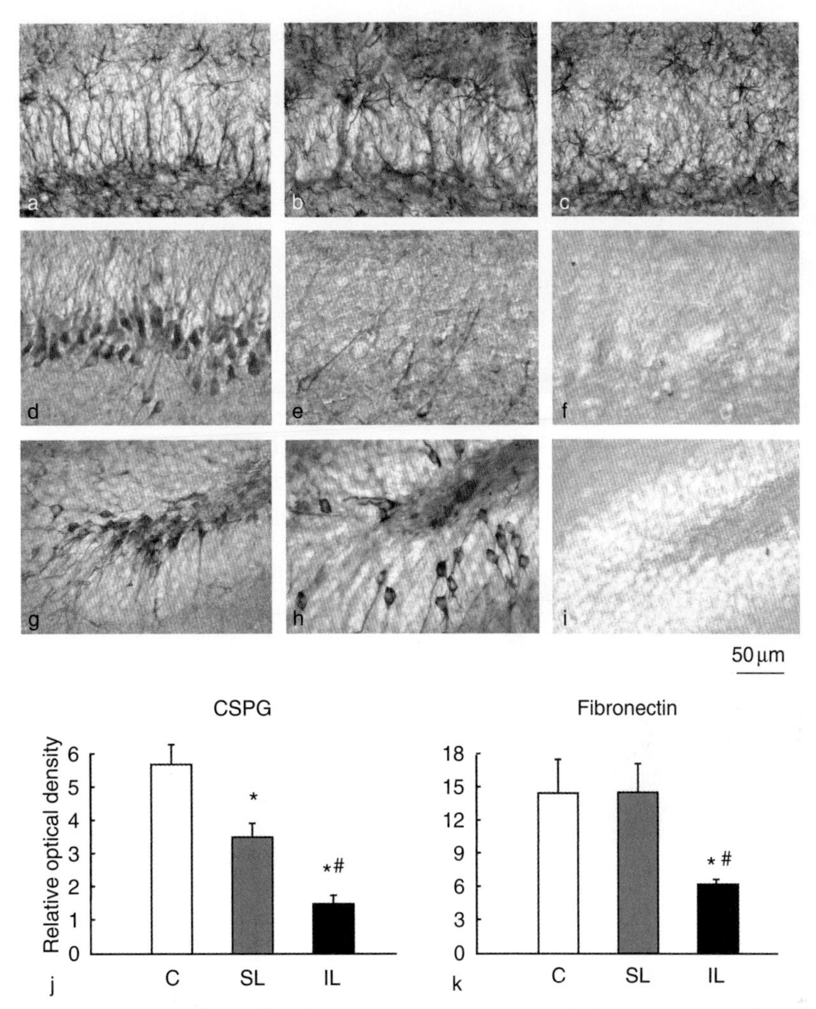


**Figure 3** Structural changes in aged superior and inferior learners. (a–c) Astrocytes in the dentate gyrus (inner blade) of an adult rat, an aged superior learner (SL), and an aged inferior learner (IL). Note the loss of radial organization of the astrocytic processes in the aged IL. Staining for chondroitin sulfate proteoglycans (CSPG) in the hippocampal CA3 region (d–f), and for fibronectin in the dentate gyrus (g–i), shows a decrease in perineuronal staining in the aged SL and a loss in the aged IL. The graphs show the relative optical densities of CSPG (j) and fibronectin cell attachment fragment (k) immunoreactivity. Note the significant decrease in CSPG immunoreactivity between each group (C, control) and the decrease in fibronectin immunoreactivity in the aged IL only. *significant difference ($p < 0.05$) compared to control (adults). #significant difference ($p < 0.05$) compared to aged superior learners. Adapted from Syková E, Mazel T, Hasenöhrl RU, et al. (2002) Learning deficits in aged rats related to decrease in extracellular volume and loss of diffusion anisotropy in hippocampus. *Hippocampus* 12: 469–479.

constructing isoconcentration spheres (isotropic diffusion) or ellipsoids (anisotropic diffusion) for extracellular $TMA^+$ concentration. The surfaces in **Figure 4** represent the locations where the $TMA^+$ concentration first reached 1 mM, 60 s after its application in the center. The ellipsoid in the hippocampus of the young adult rat reflects the different abilities of substances to diffuse along the $x$, $y$, and $z$ axes, while the sphere in the hippocampus of the aged rat shows isotropic diffusion. The smaller ECS volume fraction in aged rats is reflected in the sphere's being larger than the ellipsoid. Indeed, we found that there is a significant decrease in the ADC of many neuroactive substances, including the ADC of water (**Figure 5**), in the aging brain, which accompanies a decrease in ECS volume.

The hippocampus is well known for its role in memory formation, especially declarative memory. Indeed, it has been found that the degree of learning deficit during aging correlates with changes in $\alpha$, $\lambda$, and $k'$. It is therefore reasonable to assume that diffusion anisotropy leads to a certain degree of specificity in extrasynaptic communication. There is a significant difference between mildly and severely behaviorally impaired rats (rats were tested in a Morris water maze), which is particularly apparent in the hippocampus. The ECS in the dentate gyrus of severely impaired rats (bad learners) is significantly smaller

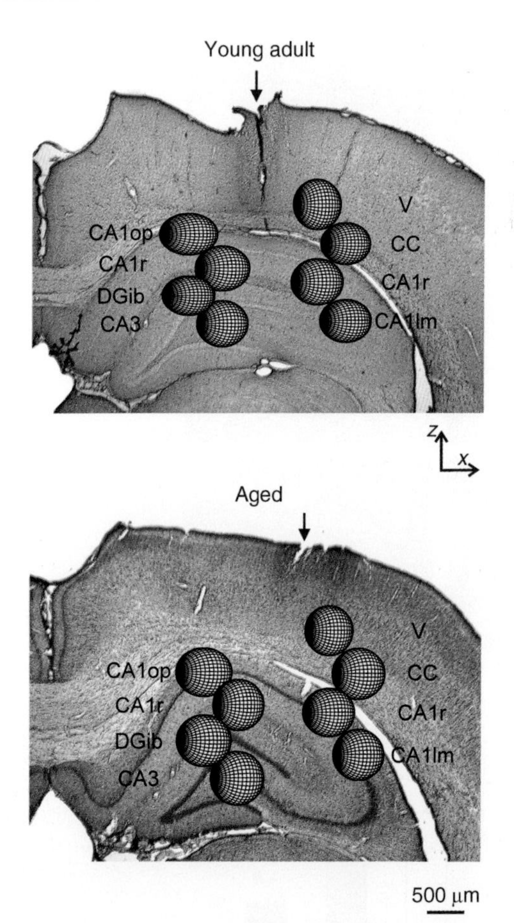


**Figure 4** Diffusion parameters in a young adult (3 months old) rat and in an aged (28 months old) rat with a learning deficit in the Morris water maze. Data were recorded in anesthetized animals; microelectrode tracks were verified after the experiments (see arrows); isoconcentration surfaces for a 1 mM tetramethylammonium ion concentration contour 60 s after the onset of an 80 nA iontophoretic pulse. The surfaces were generated using the actual values of volume fraction and tortuosity. The ellipsoids represent anisotropic diffusion in a young adult rat. The larger sphere in an aged rat corresponds to isotropic diffusion and to a small extracellular space volume fraction. It demonstrates that diffusion from any given source will lead to a higher concentration of substances in the surrounding tissue and a larger action radius in aged rats than in young adults. Anisotropy is almost lost in the aged rat. Adapted from Syková E, Mazel T, Vargová L, et al. (2000) Extracellular space diffusion and pathological states. *Progress in Brain Research* 125: 155–178.

than in mildly impaired rats (good learners). Also, anisotropy in the hippocampus of bad learners, particularly in the dentate gyrus, is much reduced, while a substantial degree of anisotropy is still present in aged rats with good learning performance (**Figures 2** and **3**). Anisotropy might be important for extrasynaptic transmission by channeling the flux of substances in a preferential direction, and its loss may severely disrupt extrasynaptic communication in the CNS, which has been suggested to play an important role in memory formation.

The ensuing decrease in the ECS size could be explained by the disappearance of a significant part of the extracellular matrix, partial neuronal degeneration and loss, a decrease in the number of dendritic processes, demyelination, and changes in glia. Indeed, a decrease in fibronectin and chondroitin sulfate proteoglycan staining was found in the hippocampus of mildly impaired aged rats and almost a complete loss of staining was seen in severely impaired aged rats (**Figure 3**). Chondroitin sulfate proteoglycans participate in multiple cellular processes, such as axonal outgrowth, axonal branching, and synaptogenesis, which are important for the formation of memory traces. The observed loss of anisotropy in senescent rats could therefore lead to impaired cortical and, particularly, hippocampal function. Moreover, the decrease in ECS volume could be responsible for the greater susceptibility of the aged brain to pathological events, particularly to ischemia, and for the poorer outcome of clinical therapy and the more limited recovery of affected tissue after anesthesia or any severe insult.

The decrease in ECS volume fraction during normal aging might be attributed to the loss of extracellular matrix molecules such as chondroitin sulfate proteoglycans, fibronectin, tenascin-R, or highly sialylated cell adhesion molecules, such as polysialylated neural cell adhesion molecules (PSA-NCAMs). The extracellular matrix molecules act like a sponge, binding a large number of water molecules and tending to occupy a lot of space, due to the mutual repulsion of their numerous negatively charged residues. Thus, their loss during aging might lead to a decrease in ECS volume. ECS shrinkage might also be related to the general decrease in hydration during aging. Life can be viewed as a process during which the highly hydrated state of embryos and neonates is transformed into a gradually more and more dehydrated one. Among other things, this dehydration is associated with an increase in colloid density, resulting in a decrease in catalytic enzyme activity and possibly contributing to the accumulation and deposition of certain substances, both intra- and extracellularly. A reduction in ECS volume would thus contribute, together with neuronal shrinkage, to the decrease in cortical volume observed during aging. These changes might be partially compensated for by concomitant reactive gliosis and glial cell hypertrophy.

Significant differences in ECS volume fraction were also found in aged female mice compared to age-matched males; an even more pronounced difference was found for $ADC_W$. This was probably the first observation of a gender difference in these two parameters. However, other differences have also been described: female mice have higher numbers of astrocytes and microglia in the hippocampus than

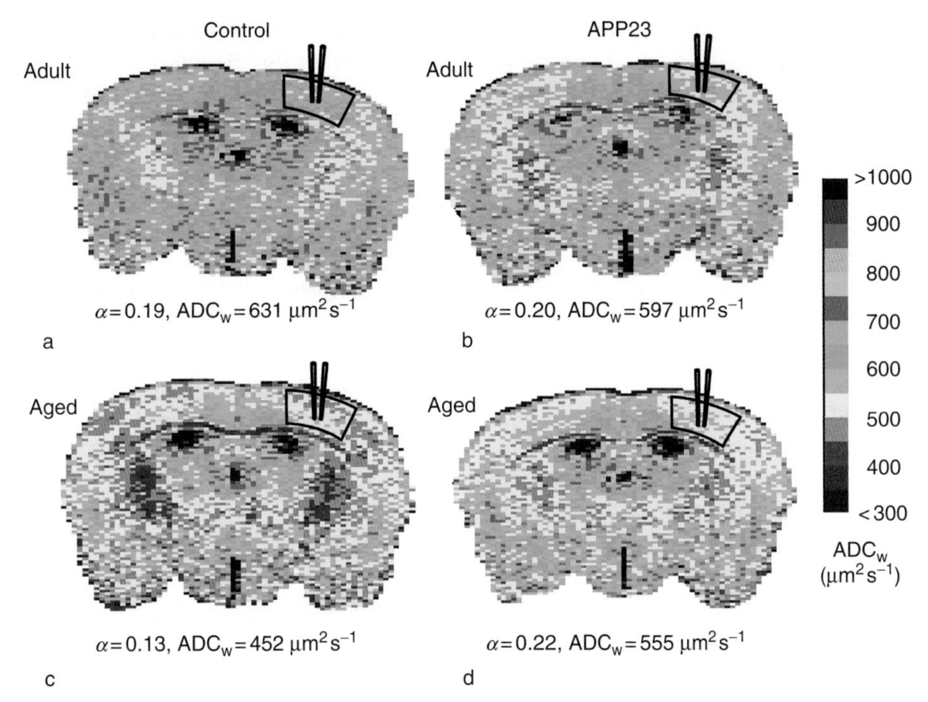


**Figure 5** Apparent (water) diffusion coefficient ($ADC_W$) maps acquired in the brain of control and APP23 female mice. The mean value of $ADC_W$ was calculated in the areas indicated. In the corresponding region, tetramethylammonium ion measurements were performed on the same animals a few days later. The mean values of the $ADC_W$ and the extracellular space volume fraction ($\alpha$) are given below each map. There were no differences between adult control (a) and adult transgenic (b) mice. (a, c) A decrease in $ADC_W$ and $\alpha$ was found during aging in control mice. In aged APP23 mice (d), there was an increase in both $ADC_W$ and $\alpha$ when compared with age-matched control mice (c). Adapted from Syková E, Voříšek I, Antonova T, et al. (2005) Changes in extracellular space size and geometry in APP23 transgenic mice – A model of Alzheimer's disease. *Proceedings of the National Academy of Sciences of the United States of America* 102: 479–484. 

males do, and this difference increases with age. These changes might be related to the decrease in 17$\beta$-estradiol after estropause, though there is no direct evidence for this. Changes in diffusion parameters may serve as an important indicator of pathological processes, and therefore diffusion changes have been studied under many experimental pathological states using the TMA method as well as diffusion-weighted MRI, which is important for diagnostic purposes.

### Changes in ECS Diffusion Parameters in an APP23 Mouse Model of Alzheimer's Disease

Neuronal loss and changes in synaptic transmission are generally considered the main reasons for memory impairment in Alzheimer's disease. However, changes in extrasynaptic transmission might also be important. Changes in ECS diffusion parameters were found in the cortex of transgenic APP23 mice, in which overproduction of mutated human amyloid precursor protein (APP) leads to amyloid plaque formation. In these animals, TMA measurements were compared with data obtained in the same animals using diffusion-weighted magnetic resonance imaging to measure the apparent diffusion coefficient of water ($ADC_W$) in the tissue (**Figure 5**). Measurements were performed *in vivo* in 6- to 8-month-old and 17- to 25-month-old hemizygous APP23 mice and in age-matched controls. In 6- to 8-month-old APP23 mice, the ECS volume fraction and $ADC_W$ were not significantly different from these factors in age-matched controls (mean $\alpha$ was 0.20, $ADC_W$ was 618 $\mu m^2 s^{-1}$). Aging in 17- to 25-month-old controls was accompanied by decreases in the ECS volume fraction (to 0.15) and $ADC_W$ (to 549 $\mu m^2 s^{-1}$) decreases significantly greater in females than in males, but there was no change in tortuosity. In aged, 17- to 25-month-old APP23 mice the ECS volume fraction increased to 0.22, and ECS tortuosity and $ADC_W$ increased compared to age-matched controls. These results confirm that in mice as well as in rats, aging leads to a decrease in ECS volume. In contrast, the deposition of $\beta$-amyloid is associated with an increase in ECS volume fraction, tortuosity, and $ADC_W$. The impaired navigation observed in transgenic APP23 mice in the Morris water maze correlated with their amyloid plaque load, which was twice as high in females (20%) as in males (10%). Obviously, the volume of the amyloid plaques, which are located extracellularly, is not included in the ECS volume fraction as measured by the TMA method. If the ECS volume fraction is about 20% and the amyloid plaque load in old female transgenic mice is also about 20%, the real ECS volume is the sum of both – about 40% of the total tissue volume.

Amyloid deposits, together with altered ECS diffusion properties, may therefore account for the impaired synaptic as well as extrasynaptic volume transmission and spatial cognition observed in old female transgenic mice. The enlargement of the ECS will contribute to a decrease in the extracellular concentrations of many important neuroactive substances, such as acetylcholine, dopamine, and serotonin. The functional radius of protein and peptide molecules, including nerve growth factors, will be compromised. It is therefore quite possible that there is a link between ECS volume, diffusion changes, and behavioral deficits in aged patients with Alzheimer's disease.

## Concluding Remarks

The results obtained by diffusion measurements indicate that the degree of learning impairment in aged rats and mice correlates with changes in the ECS diffusion parameters. The changes in diffusion parameters during aging can therefore have an important functional significance. Anisotropy, which, particularly in the hippocampus and corpus callosum, may help to facilitate the diffusion of neurotransmitters and neuromodulators to regions occupied by their high-affinity extrasynaptic receptors, might have crucial importance for the specificity of signal transmission. The importance of anisotropy for the 'spill-over' of glutamate, for 'cross-talk' between synapses, and for LTP and LTD has been proposed. The observed loss of anisotropy in senescent rats could therefore lead to impaired cortical and, particularly, hippocampal function. The decrease in ECS size could be responsible for the greater susceptibility of the aged brain to pathological events, the poorer outcome of clinical therapy, and the more limited recovery of affected tissue after various insults.

*See also:* Aging of the Brain and Alzheimer's Disease; Alzheimer's Disease: An Overview; Alzheimer's Disease: MRI Studies; Animal Models of Alzheimer's Disease; Brain Volume: Age-Related Changes; Functional Neuroimaging Studies of Aging; Synaptic Plasticity: Learning and Memory in Normal Aging.

## Further Reading

Fuxe K, and Agnati LF (eds.) (1991) *Volume Transmission in the Brain. Novel Mechanisms for Neural Transmission.* New York: Raven Press.

Nicholson C and Phillips JM (1981) Ion diffusion modified by tortuosity and volume fraction in the extracellular microenvironment of the rat cerebellum. *Journal of Physiology (London)* 321: 225–257.

Nicholson C and Syková E (1998) Extracellular space structure revealed by diffusion analysis. *Trends in Neuroscience* 21: 207–215.

Syková E (1992) Ionic and volume changes in the microenvironment of nerve and receptor cells. In: Ottoson D (ed.) *Progress in Sensory Physiology*, pp. 1–167. Heidelberg: Springer.

Syková E (1997) The extracellular space in the CNS: Its regulation, volume and geometry in normal and pathological neuronal function. *Neuroscientist* 3: 28–41.

Syková E (2001) Glia and extracellular space diffusion parameters in the injured and aging brain. In: de Vellis J (ed.) *Neuroglia in the Aging Brain*, pp. 77–98. Totowa: Humana Press.

Syková E (2001) Glial diffusion barriers during aging and pathological states. *Progress in Brain Research* 132: 339–363.

Syková E (2003) Diffusion parameters of the extracellular space. *Israel Journal of Chemistry* 43: 55–69.

Syková E (2004) Extrasynaptic volume transmission and diffusion parameters of the extracellular space. *Neuroscience* 129: 861–876.

Syková E, Mazel T, Hasenöhrl RU, et al. (2002) Learning deficits in aged rats related to decrease in extracellular volume and loss of diffusion anisotropy in hippocampus. *Hippocampus* 12: 469–479.

Syková E, Mazel T, Vargová L, et al. (2000) Extracellular space diffusion and pathological states. *Progress in Brain Research* 125: 155–178.

Syková E, Voříšek I, Antonova T, et al. (2005) Changes in extracellular space, size, and geometry in APP23 transgenic mice – A model of Alzheimer's disease. *Proceedings of the National Academy of Sciences of the United States of America* 102: 479–484.

Zoli M, Jansson A, Syková E, et al. (1999) Intercellular communication in the central nervous system. The emergence of the volume transmission concept and its relevance for neuropsychopharmacology. *Trends in Pharmacological Sciences* 20: 142–150.

# Brain Volume: Age-Related Changes

**G Bartzokis and P H Lu**, University of California School of Medicine, Los Angeles, CA, USA



## Introduction

The popular concept of postnatal human development separated into three stages (development/maturation, adulthood/stability, and aging/degeneration) may be appropriate to human physical characteristics such as height but has distorted the understanding of human brain biology. Over the life span, the human brain is in a continuous state of change roughly defined as periods of development/maturation followed by degeneration/repair. The popular concept of a stable, unchanging 'adult' life stage for the brain is misleading and biologically invalid (**Figures 1, 2,** and **3**). The concept of a stable adult stage (defined as ages 18–55) is valid only when whole brain volume is considered (**Figure 1(c)**) but fails to capture the underlying infrastructural changes to our brain's 'Internet' (see section titled 'The myelin model of the human brain' and **Figures 1(b), 2, 3,** and **4(b)**). Our very nature as a species depends on optimizing the performance of our brain's Internet, and this performance is dependent on the extensive and protracted process of myelination (see next section and **Figure 5**).

Magnetic resonance imaging (MRI) can provide images with appropriate resolution and tissue contrasts to quantify *in vivo* the brain's developmental and subsequent degenerative changes. MRI images can measure white matter volumes. When myelin-sensitive inversion-recovery and heavily T1-weighted MRI sequence images are used, they likely include some of the deeper layers of cortical gray matter as myelination spreads into the cortex (lower panels of **Figure 2**). Thus, these imaging methods measure white matter (WM) volumes that can be better conceptualized as 'myelinated WM volume,' which includes deeper layers of cortex. Not surprisingly, these imaging methods have replicated and extended postmortem observations using myelin stains (**Figures 2** and **3**). In other words, such sequences are the MRI equivalent of myelin stains, which is the likely reason MRI studies using these methods correspond so closely with postmortem myelin stain data (**Figures 2** and **3**). Using such images, MRI can detect the shift of the gray matter–myelinated WM boundary into cortical gray matter caused by the process of 'intracortical' myelination (lower panels of **Figure 2**). This intracortical myelination, together with continued myelination of unmyelinated subcortical fibers (**Figure 6(a)**) and addition of myelin lamellae in the subcortical myelinated fibers, form the myelinated WM volume. The term 'myelinated WM volume' may therefore be a more appropriate description of the volume measured with inversion-recovery and heavily T1-weighted sequences than simply 'WM volume.' Although this way of conceptualizing MRI data helps explain the similarities between MRI data and postmortem studies, the usual term of 'WM volume' will be used throughout this article except for specific discussion of the impact of intracortical myelination (e.g., **Figure 2**).

In humans, the disproportional contributor to the enlargement of the late-maturing structures such as the frontal and temporal lobes is their higher WM and myelin content compared to other species (**Figure 5**). Going against popular belief, recent studies show that compared to primates, adult humans have neither disproportionately enlarged frontal lobes for our overall brain size nor disproportionate increase in gray matter. Rather, human brain has disproportionately more (approximately 20%) WM. These findings are consistent with the overall evolutionary trend to have increasing cognitive abilities associated with disproportionate increases in glia rather than neuron numbers. During brain evolution, glial cell numbers have increased disproportionately compared to neurons; from 10–20% of all brain cells in the nematode to 25% in the fly, 65% in the rodent, and 90% in human. The higher number of oligodendrocytes results in the disproportionately higher percentage of brain dry weight accounted for by myelin in humans (30% higher), compared with rodents.

## The Myelin Model of the Human Brain

The human brain is unique in its high myelin content as well as its long developmental (myelination) phase. On average, the myelination phase continues until approximately age 50–60 (**Figures 1** and **3**) and thus spans our entire evolved life span (before modern medicine, few individuals lived beyond that age). The popular analogy of the brain to a computer is misleading as it ignores the WM (axons, or 'wires,' plus their myelin, or 'insulation') that composes approximately half the adult human brain. A more useful analogy for the human brain is thus the Internet, where speed, bandwidth, and an online connection are the paramount attributes of optimal function. All these attributes of the brain's Internet depend on myelin.

Axon myelination results in saltatory conduction of action potentials that markedly increases (>tenfold) signal transmission speed. Speed makes it

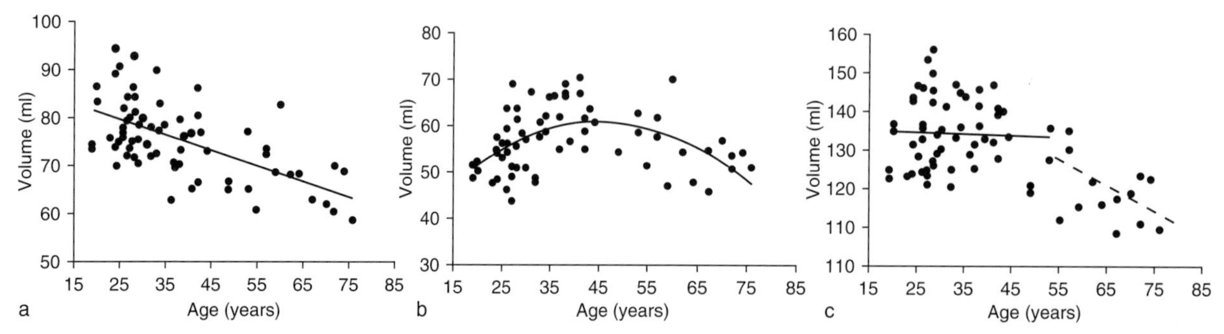


**Figure 1** MRI volumes in frontal plus temporal lobes of healthy individuals across the life span. (a) Gray matter. Linear: $r = -0.62$, $p < 0.0001$; Quadratic: not significant (NS), $p > 0.94$. (b) White matter. Linear: NS, $p > 0.84$; Quadratic: $t = 4.56$, $p < 0.0001$, Peak = 45 years. (c) Gray plus white matter. Linear correlations for adults (18–54) and older age (55–76), confirming stable adult brain volume (gray plus white matter) observed in postmortem studies. See Miller AK, Alston RL, and Corsellis JA (1980) Variation with age in the volumes of grey and white matter in the cerebral hemispheres of man: Measurements with an image analyser. *Neuropathology and Applied Neurobiology* 6: 119–132. For details on methods, see Bartzokis G, Beckson M, Lu PH, Nuechterlein KH, Edwards N, and Mintz J (2001) Age-related changes in frontal and temporal lobe volumes in men: A magnetic resonance imaging study. *Archives of General Psychiatry* 58: 461–465.

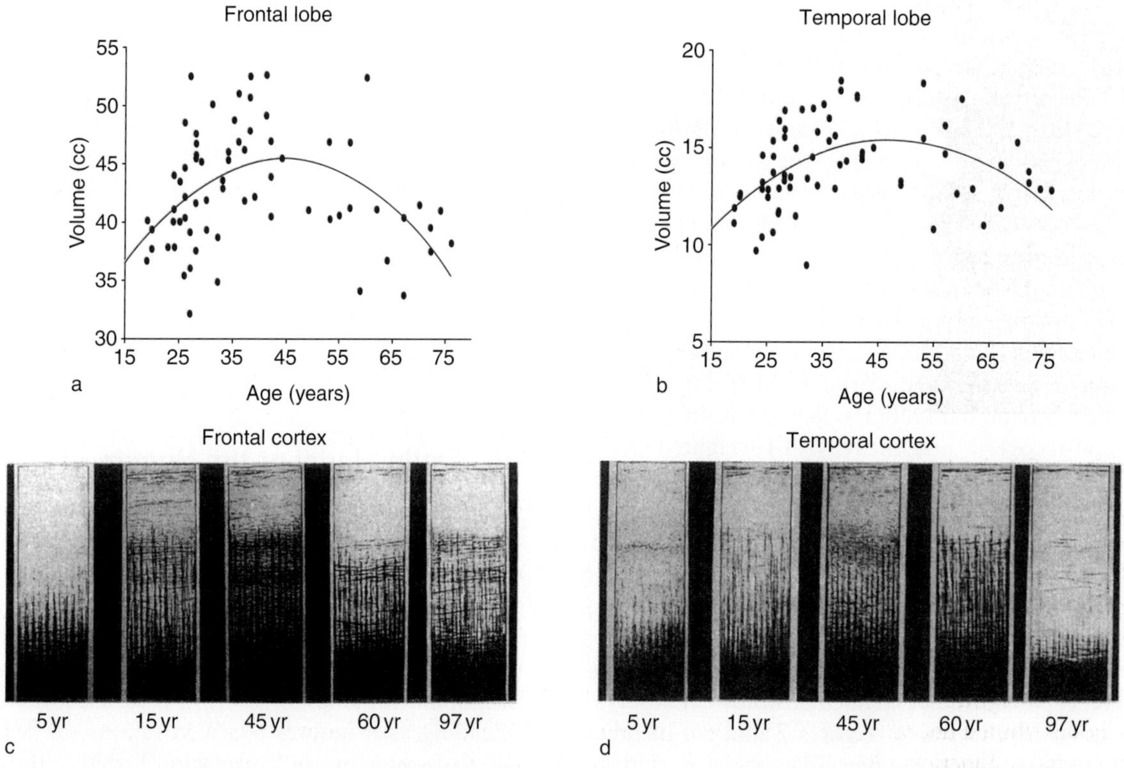


**Figure 2** Quadratic (inverted U) myelination trajectories of human brain over the life span. Myelination ($y$-axis) vs. age ($x$-axis) in (a) frontal and (b) temporal lobes of normal individuals. Figures (a) and (b) are *in vivo* MRI data reproduced from Bartzokis G, Beckson M, Lu PH, Nuechterlein KH, Edwards N, and Mintz J (2001) Age-related changes in frontal and temporal lobe volumes in men: A magnetic resonance imaging study. *Archives of General Psychiatry* 58: 461–465, copyright ©(2001) American Medical Association. Figures (c) and (d) are, respectively, frontal and temporal cortex postmortem intracortical myelin stain data from Kaes, 1907; adapted and reproduced in Kemper T (1994) Neuroanatomical and neuropathological changes during aging and dementia. In: Albert M and Knoefel J (eds.) *Clinical Neurology of Aging*, 2nd edn., pp. 3–67. New York: Oxford University Press, with permission from Oxford University Press. The data depicted in (a) and (b) were acquired some 100 years later than the data depicted in (c) and (d), yet the two samples of normal individuals show remarkably similar myelination trajectories in the two regions. Note that different brain regions have significantly different myelination trajectories even when the regions are similar, as is the case with these two association regions. Frontal lobe peak myelination is reached at age 45, but peak myelination is reached even later in the temporal lobe.

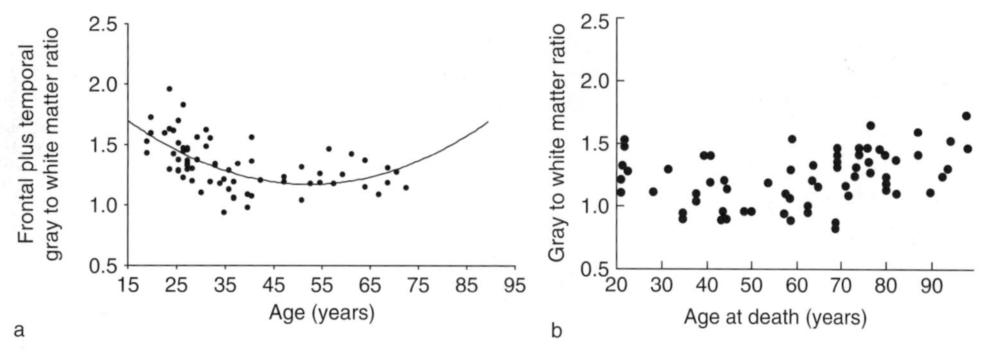


**Figure 3** (a) *In vivo* evaluation of frontal plus temporal gray to white matter ratio in healthy individuals. (b) Postmortem evaluation of whole brain gray to white matter ratio. Adapted from Miller AK, Alston RL, and Corsellis JA (1980) Variation with age in the volumes of grey and white matter in the cerebral hemispheres of man: Measurements with an image analyser. *Neuropathology and Applied Neurobiology* 6: 119–132, with permission from Blackwell Publishing.

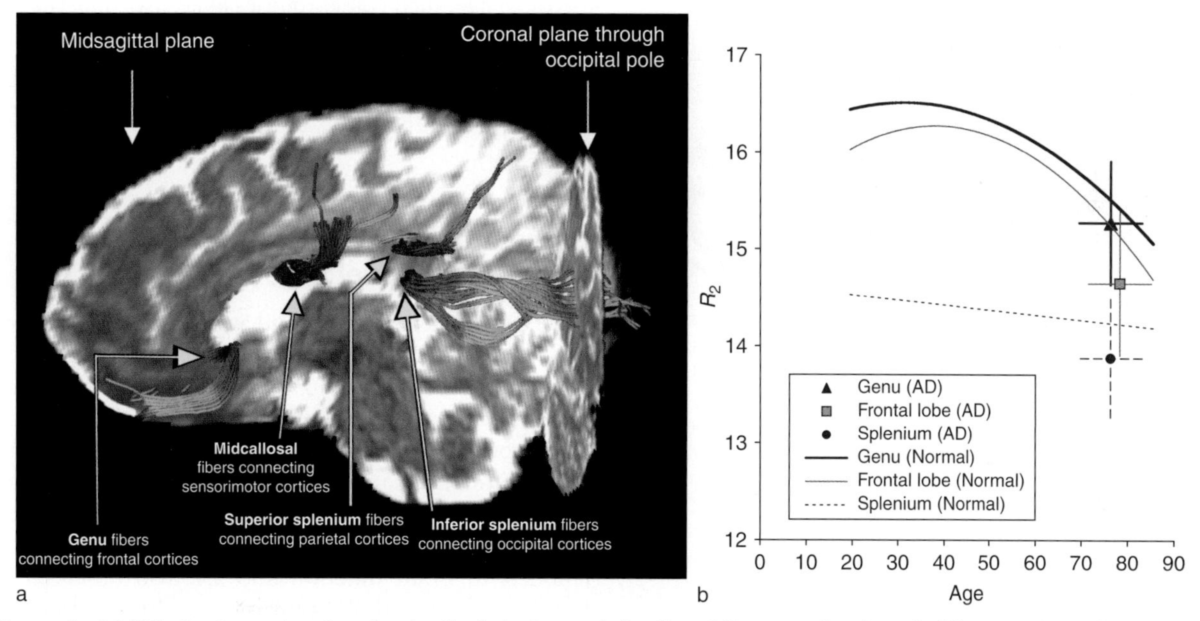


**Figure 4** (a) Diffusion tensor imaging showing the trajectory and direction of fibers passing through different regions of the corpus callosum. The directions are coded as colors, with brown representing fibers traveling directly out of (perpendicular to) the page (where the arrows are pointing), green representing the direction that is parallel and horizontal to the plane of the page, and purple representing the direction of perpendicular and vertical to the page. (b) Transverse relaxation rate ($R_2$) vs. age in healthy and Alzheimer's disease subjects. Normal, healthy controls ($n = 252$, regression lines depicted); AD, Alzheimer's disease patients ($n = 36$, means and standard deviations depicted).

possible to integrate information across the highly spatially distributed neural networks that underlie our higher cognitive functions. Myelination also decreases the refractory time (time needed for repolarization before a new action potential can be supported by the axon) by as much as 34-fold. Thus, if the brain were to be compared to the Internet, myelination not only increases its speed (e.g., transforming it from a telephone line system to a T1 line system) but also its 'bandwidth,' increasing the actual amount of information that can be transmitted per unit of time. This allows myelinated axons to support 'high-frequency bursts' of signals, on which processing speed and long-term potentiation depend, and further facilitates information processing by allowing for precise temporal coding of neuronal activity. However, until a network or circuit is fully myelinated (from neuron body to neuron body), it will not be 'online' to immediately affect ongoing cognition and behavior. **Figure 2** demonstrates that from a biological standpoint, becoming a healthy middle-aged adult involves intracortical myelination all the way to the neuron body that brings networks or circuits online. Thus, humans' extensive process of myelination markedly increases cognitive

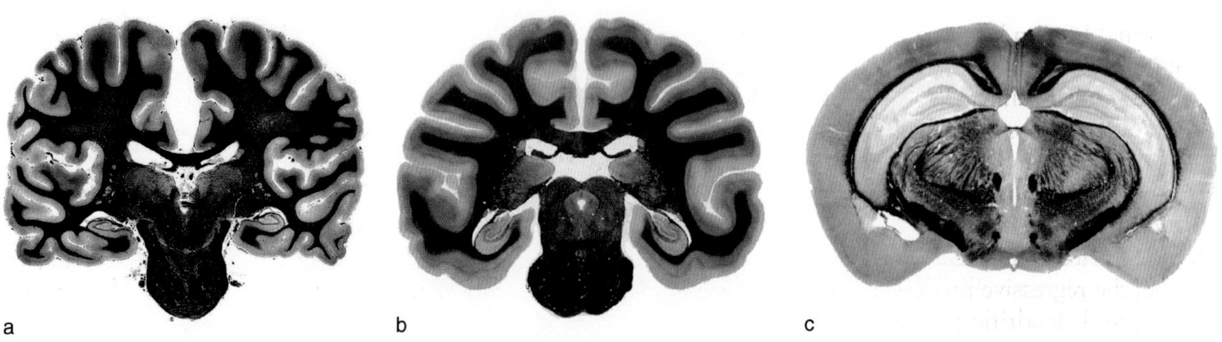


**Figure 5** Coronal myelin stains of (a) human, (b) chimpanzee, and (c) rodent brains. Brains are not to scale. Chimpanzee and rodent brain are enlarged to approximate human brain in order to more easily demonstrate the striking differences in the proportion of myelin (stained black) in each brain. Human brain has approximately 20% more white matter than the chimpanzee's.

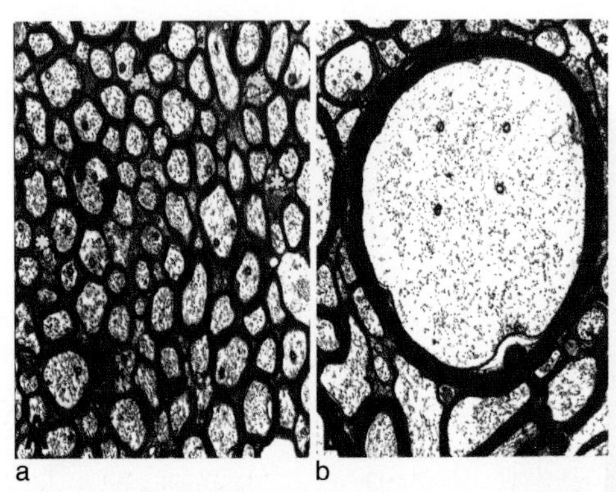


**Figure 6** Corpus callosum regional differences in size and myelination of axons. (a) Genu: many small, thinly myelinated axons and clusters of unmyelinated axons (20–30% of axons in this structure connecting late-myelinating frontal regions). (b) Larger, more heavily myelinated axons and far fewer unmyelinated axons (<7%; none in this image) observed in regions like ventral splenium which connect primary sensory (visual) early myelinating regions. Reproduced from Lamantia AS and Rakic P (1990) Cytological and quantitative characteristics of four cerebral commissures in the rhesus monkey. *Journal of Comparitive Neurology* 291: 520–537, with permission from John Wiley & Sons, Inc.

information-processing speed and underlies many of our brain's unique capabilities, such as language, memory, inhibitory control, and higher cognitive functions. This interpretation suggests that the brain could experience neurodevelopmental arrest, even in adulthood, if pathological states (e.g., brain trauma, schizophrenia, severe stress, substance abuse, etc.) alter the normal age-related pattern of structural changes.

The widely distributed neural networks on which the brain's organization is based make speed of processing paramount for higher cognitive functions. The impact of myelin on information processing speed makes the production and maintenance of myelin essential for normal brain function. The vulnerability of the oligodendrocytes and the myelin they produce to genetic and environmental insults likely also underlies our brain's unique susceptibility to highly prevalent and often unique human disorders of development, ranging from autism and learning disabilities to schizophrenia. In older age, such vulnerabilities likely also underlie the exponential increase in degenerative disorders such as Alzheimer's and Parkinson's diseases, whose prevalence is such that they may be considered 'normal' (i.e., affect the majority of the population above the age of 90).

## Functional Implications of the Myelin Model

The view of development and aging as a continuum of interactive structural and functional developmental trajectories fosters the examination of multiple factors that promote or inhibit myelination throughout the protracted process of human brain development. Quadratic-like patterns of change over the life span similar to the myelination trajectory have been demonstrated in neurophysiologic and cognitive models of brain function and are also observed in studies of clinical symptomatology in a variety of neuropsychiatric disorders.

Such quadratic-like life-span trajectories have been repeatedly documented in age-related changes of cognitive speed that increase throughout childhood and early adulthood and then decrease with an accelerating rate in old age. Salthouse delineated the hypothesis that a basic parameter such as speed is directly related to biological factors and is essential for higher-order cognitive processing. Similar quadratic trajectories are also apparent in paradigms that test inhibitory control. These quadratic trajectories also resemble the trajectory of some neurophysiologic (electroencephalogram) parameters. The myelin model described above would suggest that the principal underlying biological factor to consider as an explanation for all such life-span trajectories is the process of myelination (**Figures 2, 3,** and **4**).

In the context of the myelin model, the occurrence of all these changes in the structure and function of the normal human brain peaking at roughly the same midlife stage can be viewed as indirect evidence that after infancy, myelination is the central biologic process underlying human aging-related changes. It is important to point out that the age-related cognitive improvements in processing speed and inhibitory control peak in adulthood. Thus, these developmental trajectories do not temporally 'fit' (by more than a decade) the regressive processes of neuronal, synaptic, axonal, and dendritic pruning and elimination that occur primarily before mid-adolescence.

## Brain Development during Childhood to Adolescence

The focus on myelin and myelination is not intended to diminish the importance of other early brain developmental processes such as neurogenesis (primarily intrauterine but whose disruption can cause catastrophic abnormalities that are usually evident at birth or in early infancy). The focus on myelin is especially not intended to dismiss the drastic processes of neuronal, synaptic, axonal, and dendritic pruning and elimination that appear to reduce the 'connectivity' of the child's brain by as much as two-thirds as it develops into adulthood. These processes, which have been hypothesized to affect several childhood disorders, are in large part completed by puberty or mid-adolescence.

These regressive pruning processes occur in concert with a precisely regulated but highly vulnerable and sometimes ignored process of myelination. The regressive pruning processes occur in a heterochronous pattern in primary process areas (motor, sensory) before association areas (frontal, temporal, and parietal lobes; see example in **Figure 2**). This pattern of heterochronous, volume-reducing, regressive pruning matches the pattern of the heterochronous, volume-increasing, developmental process of myelination (**Figure 1**). The human skull becomes rigid in late childhood, eliminating the possibility of further expansion of the brain. Thus, it can be argued that the regressive pruning processes of childhood described above occur in large part to provide the 'room' (intracranial volume) and possibly other resources necessary to support the crucial process of myelination.

How can it be that a process that produces such tremendous losses of synapses and axons does not result in dementia in this young age range? To return to the Internet analogy, the pruning of the neurons, and synaptic loss is equivalent to the reduction in capability of the computers of the Internet (e.g., reduced random access memory). The dramatic gains in processing speed (transmission speed, bandwidth, and online connectivity) and other possible benefits, such as reduced energy consumption, provided by myelination likely more than compensate (functionally) for the regressive pruning processes. Instead of reducing 'connectivity' between different parts of the brain suggested by the regressive pruning processes, the myelination process markedly increases the functional potential (connectivity) of the remaining neural networks and circuits. It is thus possible to continue human cognitive, emotional, and behavioral development without the loss, and in fact with the addition, of functional capacity that in middle age culminates in healthy adults with better emotional regulation and inhibitory control underlying what is commonly referred to as wisdom.

Studies examining brain development in children and adolescents have demonstrated increasing WM volume and decreasing gray matter volume in later childhood and adolescence. These findings were recently confirmed in prospective MRI studies of brain development which demonstrated that WM volume increases linearly between ages 4 and 20, a period characterized by increased myelination. Unlike the WM changes, cortical gray matter changes were quadratic rather than linear and exhibited region-specific patterns of change (heterochronous) within this age span. Gray matter reached maximum volume at about age 12 in the frontal lobes but not until age 16 in the temporal lobes. The myelination process continues this regional developmental trajectory, resulting in the same regional pattern of maximal myelination reached 30 years later in the frontal lobe (in the mid-40s) and in the temporal lobe (in the late 40s; **Figure 2**).

## Brain Development Continues during Adulthood until Middle Age

The history of research on age-related changes in adult human brain structure has been immersed in confusion and controversy. Much of the confusion stemmed from the long-standing dogma that in normal adulthood (18–55 years of age), large reductions in cortical neuron numbers occur. This erroneous conclusion was based on data showing that neuronal density in young adults was higher than in middle-aged and older adults. The age-related reductions in cortical neuronal density were subsequently shown to be caused by an age-related tissue fixation shrinkage artifact that caused cortices of younger individuals to shrink more than those of older ones. Together with the assumption that neuronal density was equivalent to number, the dogma took hold that beginning in young adulthood, neuronal numbers decreased with age. Recent postmortem data based on unbiased stereologic assessments have demonstrated that there is minimal, if any, neuronal cell loss or cortical thickness decrement before the age of 55.

The confusion was also promoted by the underappreciation of the fact that in primate brain, postnatal development is primarily driven by the protracted process of myelination. Unlike other brain cell lines, oligodendrocyte numbers continue to increase into old age to support the process of subcortical and intracortical myelination. The excess cortical shrinkage of younger brains that occurs on fixation is likely due to the lower levels of intracortical myelin at younger ages (**Figure 2**). Intracortical myelination increases during adulthood to peak in the mid- to late fifth decade of life in association regions such as the frontal and temporal lobes (**Figure 2**), and the presence of lipid-rich myelin reduces the amount of dehydration-related shrinkage that occurs in older compared to younger brains. The lower the myelin content of any cortex, the more shrinkage would be expected on fixation and the higher the neuronal density (as opposed to numbers) would be.

With the advent of MRI, the dogma was reinforced that in healthy aging, intracortical neuronal numbers and presumably gray matter thickness decline beginning in the teens. This occurred because the vast majority of imaging studies observed an apparent reduction in gray matter volume in adulthood (ages 18–55). The two largely erroneous conclusions (the excess cortical shrinkage of younger brains that occurs with fixation and the imaging findings of an apparent cortical volume loss with age) are likely both due to the process of increasing intracortical myelination. Thus, as **Figure 1** demonstrates, the apparent decrease in gray matter volume from ages 18 to 55 (**Figure 1(a)**) is offset by the increase in myelinated WM volume (**Figure 1(b)**), to result in a stable brain volume in this age range (**Figure 1(c)**), as established by postmortem studies. The shift in the gray–white border created by intracortical myelination (**Figures 1** and **2**) can also help explain why some studies of structure–function relationships in healthy developing individuals report improving function with faster rates of gray matter volume decline. From the framework of the myelin model, such findings can best be interpreted as improved function with increasing intracortical myelination.

With few exceptions, however, most of the early MRI studies reported that WM volume remains constant into the seventh decade, as if it were unaffected by the aging process. This is in contrast to later imaging studies using improved methods to demonstrate continual WM expansion into middle age. The failure of early imaging studies to detect age-related WM volume increase in adulthood may be accounted for by many methodological issues. These include the use of axial images aimed at studying the brain in its entirety rather than focusing on regions such as the frontal and temporal lobes, which complete maturational changes later (in middle age) than the primary sensorimotor and occipital areas (**Figures 7(a)** and **7(b)**). The age-related trajectory of structural changes can also depend on the age range examined because different regions differ in their age-related trajectory of myelination (**Figures 2** and **7(a)**). Finally, the use of automated procedures for gray and WM segmentation and suboptimal contrast can contribute to misclassification of tissues, and slice thickness greater than 3 mm can increase partial volume effects.

The difference in gray–white contrast obtained with various imaging parameters deserves special focus as it can help clarify discrepant imaging results. This focus can also highlight how consistent imaging and postmortem results can be when appropriate imaging contrasts are employed. When myelin tissue stains or myelin-sensitive brain imaging sequences are used to assess whole brain changes with age, the process of myelination (including intracortical myelination) results in similar quadratic (inverted U) age-related changes in the gray to WM ratio in the adult age range (**Figure 3**). As outlined above, the process of intracortical myelination can alter both the amount of fixation shrinkage of the cortex and the location of the gray–white border detected on brain imaging studies (**Figure 2**). Since most imaging sequences are sensitive (in varying degrees) to myelin and water content, they will detect an apparent age-related decrease in cortical volume as the process of age-related intracortical myelination causes the border between gray and WM to 'track' the process of intracortical myelination and be placed higher into the cortex (**Figure 2**). The slope of the apparent gray matter volume decline will depend on the region being evaluated (e.g., the average of the heterogeneous cortical areas included in the measure – see **Figure 7**), and the age ranges examined as different regions differ in their age-related pattern of myelination. Thus, many factors can influence imaging findings, but in general, inversion recovery and heavily T1-weighted images are most strongly influenced by the myelin content of the tissues.

**Figure 3** demonstrates how myelin-tracking postmortem and *in vivo* imaging of two samples of normal individuals over the life span produce consistent results. Regression lines from the 1980 'gold standard' myelin stain study of AK Miller and colleagues evaluating whole brain gray/WM ratio (**Figure 3(b)**) estimate a value of approximately 1.3 for the gray/WM ratio of healthy individuals at the age of 20. In MRI measures that were limited to the evaluation of frontal plus temporal regions, a higher ratio would

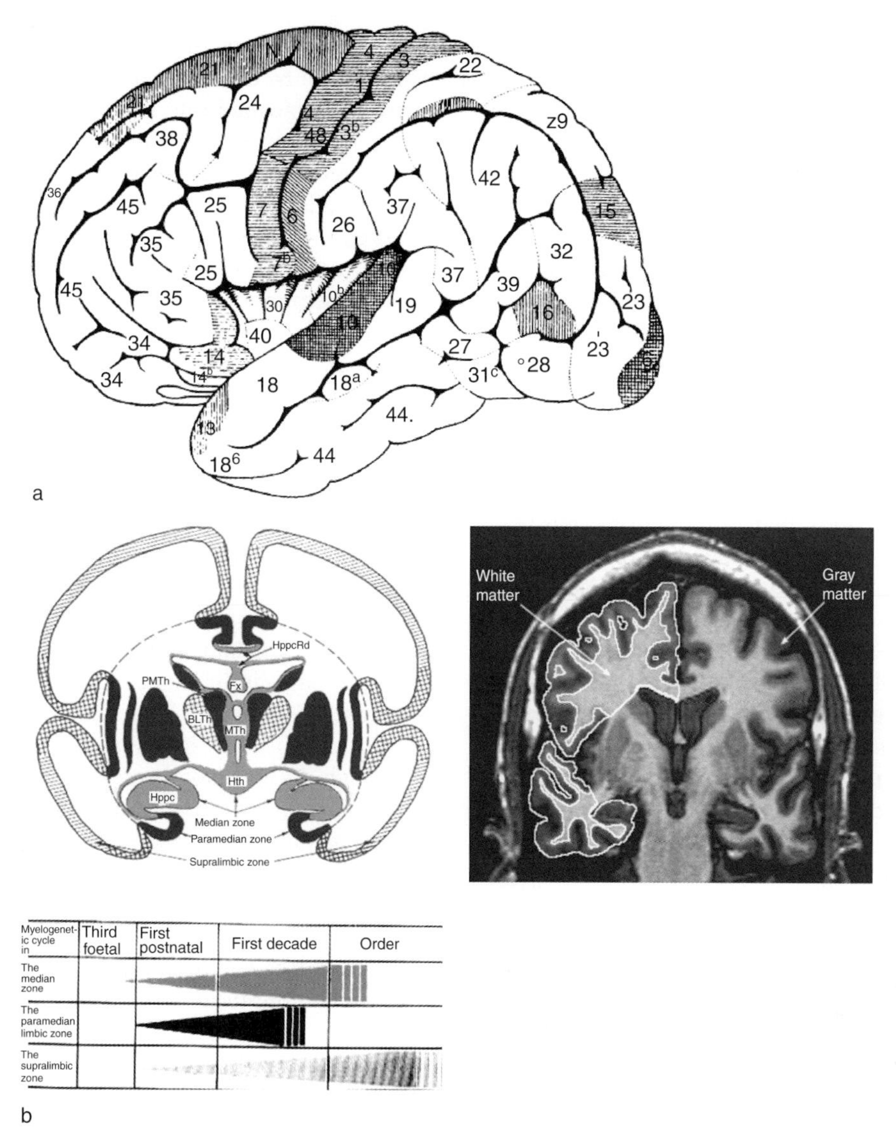


**Figure 7** (a) The myelogenetic fields numbered according to the time of myelination. From Flechsig, 1920; reproduced in Meyer A (1981) Paul Flechsig's system of myelogenetic cortical localization in the light of recent research in neuroanatomy and neurophysiology: Part II. *Canadian Journal of Neurological Sciences* 8: 95–104. This is one of the few works attempting to map the sequence of myelination. It illustrates the developmental heterogeneity (heterochronic timing) of brain myelination, with sensory and motor regions (shaded) myelinating first (in very early childhood) and temporal and frontal regions myelinating last – confirmed by Yakovlev PI and Lecours AR (1967) *Regional Development of the Brain in Early Life.* Boston, MA: Blackwell Scientific. (b, left) The image shows three zones of cerebral myelination which differ (are heterochronic) by age at myelination. Medial zone, light gray; paramedial zone, dark; dorsolateral zone, opercular and paralimbic areas – cross hatching; supralimbic association areas – diagonal hatching. (Left) Adapted from Yakovlev PI and Lecours AR (1967) *Regional Development of the Brain in Early Life.* Boston, MA: Blackwell Scientific, with permission from Blackwell Publishing. Showing the 'supralimbic association areas' (diagonal hatching of the frontal and temporal cortices) as these regions continue myelinating into 'older age.' This is graphically depicted in the lower left graph where the cortical supralimbic zones (lowest horizontal graph) continue myelinating well beyond the first decade. On the right, MRI image used to measure myelinated white matter volume in the same frontal and temporal 'supralimbic association areas', demonstrating myelination continues late into the fifth decade (also see **Figure 2**). In the MRI image the outer border of the brain and 'islands' of lighter material seen in the frontal gray matter are cerebrospinal fluid areas that are eliminated from computation using a coregistered T2-weighted image. Inversion recovery MRI reproduced from Bartzokis G, Beckson M, Lu PH, Nuechterlein KH, Edwards N, and Mintz J (2001) Age-related changes in frontal and temporal lobe volumes in men: A magnetic resonance imaging study. *Archives of General Psychiatry* 58: 461–465, copyright © (2001) American Medical Association.

be expected since, as described in myelin stain studies (**Figures 5** and **7**), myelination is continuing in these lobes beyond age 20 although it is complete in many other primary brain regions (e.g., motor, visual). From inversion recovery MRI images, a gray/WM ratio of 1.55 for the frontal plus temporal lobes at age 20 (estimated from regression lines – **Figure 3(a)**) is thus consistent with the gold standard myelin stain postmortem gray/WM ratio results (**Figure 3(b)**). In addition, the minimum (trough) gray/WM ratio is reached at age 50 in the postmortem whole brain data and at age 54.6 for *in vivo* data. The trough is again strikingly consistent given that the brain frontal and temporal lobe regions continue to develop in adulthood, which would tend to push the trough to older ages.

Thus, during healthy postadolescent development and maturation of the prefrontal and association areas, apparent gray matter volume reduction observed on imaging studies occurs in concert with an expansion in myelinated WM volume that continues into middle age (**Figure 2**), when peak myelin brain levels are reached. Myelin-sensitive imaging sequences produce data that are strikingly similar to postmortem myelin stain data (**Figures 2** and **3**). The myelinated WM expansion volumetrically cancels out the apparent gray matter reduction; thus, both imaging and postmortem studies show minimal, if any, changes in total brain (gray plus WM) volume in normal individuals during adulthood (18–55 years of age) (**Figure 1(c)**). In older age, the cortex undergoes additional changes consisting primarily of shrinkage of large neurons and increase in the proportion of small neurons. What effect such changes, combined with the increased number of oligodendrocytes and intracortical myelin growth and eventual loss, have on imaging results remains to be fully elucidated.

The biology of brain myelination shows that the life span can be appropriately conceptualized as dynamic, roughly quadratic (inverted U) trajectories that differ (are heterochronic – have different developmental/age trajectories) by brain region (**Figure 2**) and function. The trajectory of intracortical myelination with a shift in the gray/WM separation can help explain the quadratic trajectories in life span WM volumes (**Figure 2**). It can also help explain the complementary, U-shaped gray matter life span trajectories that emerge when people in their ninth and tenth decades are included in aging studies. At advanced ages, the loss of intracortical myelin with a shift of the gray–white border toward the WM could make the cortex appear to be getting 'thicker' or maintaining its volume with age while the overall WM volume continues to decline. This trend will also create a U-shaped curve in a gray–white ratio analysis (**Figure 3**), where the trough represents the point at which loss in myelinated WM volume surpasses the change in gray matter volume produced (at least in part) by the loss of intracortical myelin.

## Aging and Myelin Breakdown

In addition to gross volume changes in myelinated WM, the loss of myelin integrity in older age has profound effects on brain function and age-related cognitive decline in human and nonhuman primates. The breakdown in the structural integrity of myelin sheaths can be measured indirectly *in vivo* with MRI using multiple methods. One such method uses transverse relaxation rate ($R_2$) measures ($R_2$ is directly associated with transverse relaxation times [$T_2$] by the formula $R_2 = 1/T_2 \times 1000$). These measures are markedly sensitive to small changes in the amount of tissue water. Ultrastructural electron microscopy studies demonstrate that age-related myelin breakdown results in microvacuolations consisting of splits of myelin sheath layers that create microscopic fluid-filled spaces and increase MRI-'visible' water, thus decreasing $R_2$. These microvacuolations are ultrastructurally very similar to reversible myelinopathies produced by certain toxins. Animal studies have confirmed that this type of myelin breakdown can be detected with MRI in circumscribed susceptible regions and that the histopathologic changes produced by toxins as well as the recovery process can be tracked by changes in MRI relaxation rates. In asymptomatic individuals, focal WM regions of $R_2$ decrease ($T_2$ increase) have also been associated with myelin pallor on histological stains (**Figure 8**). Although $R_2$ has not been directly correlated with myelin breakdown due to normal aging (as opposed to the reversible toxin-induced myelin breakdown and focal regions described above), in humans and primates, healthy aging is not associated with neuronal loss. On the other hand, the process of myelin breakdown and loss has been repeatedly demonstrated. In this article, the term 'myelin breakdown' will therefore be used to refer to the $R_2$ findings.

In addition to being heterochronic, the brain's myelination process is highly heterogeneous (**Figures 2** and **7**). For example, the myelination process for primary sensory and motor regions such as the visual system occurs primarily in infancy and consists of large axons myelinated very heavily by a single oligodendrocyte per myelin segment (**Figures 6(b)** and **4(a)**). In the late-myelinating association regions such as frontal lobes and corpus callosum regions that connect them (such as the genu – **Figures 6(a)** and **4(a)**), the myelination process continues into early middle age (**Figure 4(b)**) and consists of myelination of much smaller fibers by thinner myelin sheaths (**Figure 6(a)**). In these late-myelinating regions, one oligodendrocyte myelinates multiple axons (as many as 50–60 in the latest myelinating regions in the frontal and temporal lobes).

Peters and others have shown in both primate animal models and human postmortem and *in vivo* data that myelination is followed by a degenerative process consisting of progressive myelin breakdown and loss (**Figures 2** and **4**). Published observations from a primate model of aging show that myelin breakdown begins to occur while active myelination and myelin volume expansion is still continuing. The same sequence has been shown to occur in human aging. Analysis of data for 68 normal men for whom both frontal lobe WM (FWM) volume (**Figure 2**) and relaxation rate (a measure sensitive to changes in tissue integrity which produce increased tissue water – **Figure 4(b)** were available showed that, as would be expected, the beginning of WM integrity decline in Fwm (at age 38) (**Figure 4(b)**) precedes the age at which Fwm volume decreases begin (at age 45; **Figure 2**).

These imaging data are also consistent with postmortem data from human brain that show a 45% loss of myelinated fiber length during aging, affecting primarily thinner and less heavily myelinated fibers such as the ones seen in the frontal lobe WM and genu of the corpus callosum (**Figure 6(a)**). These declines are also consistent with age-related ferritin iron declines in these WM regions. It is postulated that age-related myelin breakdown and oligodendrocyte loss (**Figure 8**) release the considerable iron stores of oligodendrocytes and their myelin, which are then redistributed and contribute to the age-related increase in brain iron levels observed in heavily-myelinated subcortical gray matter regions.

## Functional Implications of Age-Related Myelin Breakdown

Neuroimaging and postmortem data have demonstrated that the process of myelination peaks in middle age, followed by myelin breakdown and loss (**Figures 2, 3,** and **6(b)**). Age-related slowing in speed of information processing, particularly after middle age, has been repeatedly documented in the neuropsychological literature and is a well-accepted clinical phenomenon. Salthouse and others have demonstrated that the slowing in cognitive processing speed (CPS) with age likely underlies the age-related decline observed in most, if not, all human cognitive functions.

Investigations of CPS performance across the life span reveal a trajectory that parallels myelination and subsequent myelin breakdown. The similarity in these trajectories supports the hypothesis that in older age, myelin breakdown may be the cellular mechanism underlying the age-related decline in cognitive speed. Thus, the uniquely extensive myelination of the human brain makes myelin maintenance and repair especially critical for sustaining our high CPS. Myelin breakdown may result in a progressive 'disconnection' of widely distributed neural networks that eventually contributes to age-related cognitive decline and Alzheimer's disease (AD).

Apolipoprotein E (ApoE) 4 genotype, coded on chromosome 19, is second only to advancing age in its influence as a risk factor for developing AD. Epidemiological studies have demonstrated that ApoE4 genotype shifts the age of onset for AD by more than a decade, thus accounting for the vast majority of observed cases of AD. Conversely, the onset age is delayed in carriers of the ApoE2 allele.

The physiologic role of ApoE suggests that myelin may be a central biological link between the presence of this gene and AD. In nonhuman primates, myelin

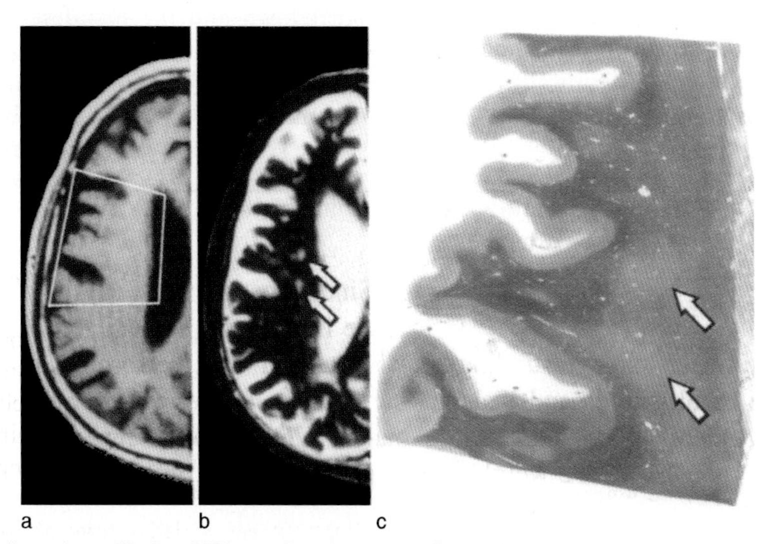


**Figure 8** White matter 'hyperintensities' on MRI are often the center of much larger areas of diffuse myelin breakdown. Small patchy hyperintense white matter lesions on T2-weighted MRI (arrows, (b)) are often the very center of much more widespread and diffuse regions of myelin breakdown, as shown in the postmortem myelin stain of (c) – see arrows. The panels are different MRI images through the same slice of tissue; (a) is a T1-weighted image, (b) is a T2-weighted image, and (c) is a histologic section (luxol fast blue staining for myelin) through the same slice of the same brain (showing the boxed region in left panel). Adapted from Takao M, Koto A, Tanahashi N, Fukuuchi Y, Takagi M, and Morinaga S (1999) Pathologic findings of silent hyperintense white matter lesions on MRI. *Journal of the Neurological Sciences* 167: 127–131, with permission from Elsevier.

repair and remyelination is an essential process that continues into old age. ApoE is the primary transporter of endogenously produced lipids such as cholesterol and sulfatide, entirely produced by oligodendrocytes and essential for myelin production and function. ApoE coordinates mobilization and transport of such lipids for uptake and use in repair, growth, and maintenance of myelin as well as other membranes, including synapses and dendrites essential for brain plasticity and learning. ApoE is essential to the extensive process of 'recycling' lipids such as cholesterol released when damaged myelin is degraded and supplying these lipids for rapid membrane repair and biogenesis. This process of 'remyelination' has been demonstrated in the primate model. The quantity of ApoE molecules necessary for this recycling process is lowest in ApoE4 carriers and highest in individuals with the ApoE2 allele. Thus, it has been hypothesized and imaging data have supported that the disruption of myelin repair secondary to insufficient ApoE molecules results in greater age-related myelin breakdown and contributes to the association between ApoE4 and AD as well as the associated influence of ApoE genotype on age-related cognitive decline.

Studies testing the myelin model have demonstrated that healthy older individuals possessing the ApoE4 genotype displayed more myelin breakdown with age. This relationship was absent for participants with the ApoE2 genotype, and ApoE3 individuals displayed an intermediate slope of decline. This genetic effect was specific to late myelinating regions and was not observed in early myelinating regions such as the splenium of the corpus callosum (**Figures 6** and **4**). Furthermore, CPS performance (as assessed by Trails A and Digit Symbol neurocognitive tasks) was significantly associated with $R_2$ in late myelinating regions (frontal lobe WM, $p < 0.0001$, and genu, $p < 0.004$) but not in the early myelinating splenium region ($p > 0.2$). These data suggest that myelin breakdown in healthy older individuals underlies the age-related cognitive decline that ultimately progresses to AD.

## Conclusions

Postmortem and *in vivo* imaging evidence can be integrated to suggest that an understanding of brain aging and the aging-related processes of myelin production and subsequent breakdown may be directly relevant to deepening our understanding of the biology and function of the human brain. The lifelong quadratic trajectories of myelination and eventual myelin breakdown and loss of different networks, circuits, and regions can help explain the lifelong trajectories of cognitive, behavioral, and emotional changes that define our everyday lives, from the impulsiveness of youth to the wisdom of older ages. More important, the focus on the myelination trajectories of the human brain may help explain many of the unique but highly prevalent and devastating diseases that plague our species over its entire life span.

This myelin model of the human brain posits that the uniquely pervasive and protracted myelination process creates a continuum of vulnerability among the myelin-producing oligodendrocytes and their myelin that culminates in the latest developing oligodendrocytes. Genetic and/or environmental effects that affect myelin development and breakdown will manifest as risk factors (or protective factors) for both developmental disorders such as autism and schizophrenia and aging-related disorders such as AD and Parkinson's disease.

The creation and maintenance of myelin is at the core of the earliest changes involved in both postnatal brain development and, in older age, the aging process itself and the subsequent emergence of degenerative disorders such as AD. Myelin deficits disrupt brain functions that depend on highly synchronized timing of high-frequency bursts of action potentials and eventually result in functional 'disconnections' of association cortical regions. The eventual progression to frank neuronal loss results in permanent deficits. This model suggests that many pathological states (e.g., genetic and hormonal states, head trauma, hypertension, hypercholesterolemia, and substance abuse) can affect the normal age-related pattern of myelin development, resulting in developmental disorders that at older ages can also affect the pattern of myelin breakdown and the manifestations of degenerative disorders. This life-span model of a seamless transition from brain development and its dynamic interaction with degenerative processes can be directly tested using prospective *in vivo* imaging studies.

The focus on myelination provides a conceptual framework for considering and developing novel myelin-centered treatments with wide spectra of efficacy that likely will extend to many disorders that share derangements in myelin health. The model suggests that the current focus on neuronal neurotransmitter 'imbalances' that much of neuropsychopharmacology is attempting to correct may be too narrow. It suggests that expanding the research focus to include interventions that affect vulnerable structural developmental processes and specifically the process of myelination may provide opportunities for novel and powerful interventions. Interventions that affect myelination may be effective in strengthening inhibitory controls of cognition in developmental disorders of younger ages as well as providing preventive opportunities for age-related degenerative disorders. As briefly outlined above, the tools (imaging, genetic, molecular, clinical, etc.) to perform such medication development work directly in humans are either already available or being developed at a rapid pace. The model suggests that interceding early in dysregulated developmental and accelerated degenerative trajectories may increase the effectiveness of treatments, decrease the need for more aggressive interventions later, and thus be accomplished with reduced risk to patients.

*See also:* Aging of the Brain; Aging: Extracellular Space; Alzheimer's Disease: MRI Studies; Cognition in Aging and Age-Related Disease; Functional Neuroimaging Studies of Aging; Lipids and Membranes in Brain Aging.

## Further Reading

Bartzokis G (2004) Age-related myelin breakdown: A developmental model of cognitive decline and Alzheimer's disease. *Neurobiology of Aging* 25: 5–18.

Bartzokis G (2004) Quadratic trajectories of brain myelin content: Unifying construct for neuropsychiatric disorders. *Neurobiology of Aging* 25: 49–62.

Bartzokis G, Beckson M, Lu PH, Nuechterlein KH, Edwards N, and Mintz J (2001) Age-related changes in frontal and temporal lobe volumes in men: A magnetic resonance imaging study. *Archives of General Psychiatry* 58: 461–465.

Bartzokis G, Minztz J, Marx P, et al. (1993) Reliability of *in vivo* volume measures of hippocampus and other brain structures using MRI. *Magnetic Resonance Imaging* 11: 993–1006.

Bartzokis G, Sultzer D, Lu PH, Nuechterlein KH, Mintz J, and Cummings J (2004) Heterogeneous age-related breakdown of white matter structural integrity: Implications for cortical 'Disconnection' in aging and Alzheimer's disease. *Neurobiology of Aging* 25: 843–851.

Connor JR and Menzies SL (1996) Relationship of iron to oligodendrocytes and myelination. *Glia* 17: 83–93.

Giedd JN, Blumenthal J, Jeffries NO, et al. (1999) Brain development during childhood and adolescence: A longitudinal MRI study. *Nature Neuroscience* 2: 861–863.

Jernigan TL and Gamst AC (2005) Changes in volume with age-consistency and interpretation of observed effects. *Neurobiology of Aging* 26: 1271–1274.

Kemper T (1994) Neuroanatomical and neuropathological changes during aging and dementia. In: Albert M and Knoefel J (eds.) *Clinical Neurology of Aging*, 2nd edn., pp. 3–67. New York: Oxford University Press.

Kretschmann H-J, Tafesse U, and Herrmann A (1982) Different volume changes of cerebral cortex and white matter during biological histological preparation. *Microscopica Acta* 86: 13–24.

Lintl P and Braak H (1983) Loss of intracortical myelinated fibers: A distinctive age-related alteration in the human striate area. *Acta Neuropathologica* 61: 178–182.

Marner L, Nyengaard JR, Tang Y, and Pakkenberg B (2003) Marked loss of myelinated nerve fibers in the human brain with age. *Journal of Comparative Neurology* 462: 144–152.

Miller AK, Alston RL, and Corsellis JA (1980) Variation with age in the volumes of grey and white matter in the cerebral hemispheres of man: Measurements with an image analyser. *Neuropathology and Applied Neurobiology* 6: 119–132.

Oldendorf WH and Oldendorf W Jr. (1988) *Basics of Magnetic Resonance Imaging*. Boston: Martinus Nijhof.

Peters A, Morrison JH, Rosene DL, and Hyman BT (1998) Are neurons lost from the primate cerebral cortex during normal aging? *Cerebral Cortex* 8: 295–300.

Peters A and Sethares C (2002) Aging and the myelinated fibers in prefrontal cortex and corpus callosum of the monkey. *Journal of Comparative Neurology* 442: 277–291.

Salthouse TA (2000) Aging and measures of processing speed. *Biological Psychology* 54: 35–54.

Schaie KW (2005) What can we learn from longitudinal studies of adult development? *Research in Human Development* 2: 133–158.

Takao M, Koto A, Tanahashi N, Fukuuchi Y, Takagi M, and Morinaga S (1999) Pathologic findings of silent hyperintense white matter lesions on MRI. *Journal of the Neurological Sciences* 167: 127–131.

Yakovlev PI and Lecours AR (1967) *Regional Development of the Brain in Early Life*. Boston, MA: Blackwell Scientific.

# Brain Composition: Age-Related Changes

**V Prakash Reddy**, University of Missouri-Rolla, Rolla, MO, USA
**G Perry**, University of Texas at San Antonio, San Antonio, TX, USA
**M A Smith**, Case Western Reserve University, Cleveland, OH, USA



At the cellular level, age-related neurological diseases including Alzheimer's disease (AD), Parkinson's disease, and Down syndrome share the characteristics of oxidative stress. Oxidation of lipids, DNA/RNA, and proteins produces a variety of toxic compounds, protein modifications, and DNA damage. AD is the leading cause of senile dementia, and there are more than four million AD patients in the United States alone. It is estimated that by 2050, fourteen million people will be inflicted by this disease. The familial form of AD worldwide accounts for 5–10% of the cases, whereas the sporadic form ranges from 90% to 95%. The brains of sporadic as well as familial cases of AD are characterized by extensive neuronal loss, deposition of extracellular senile plaques consisting of amyloid-$\beta$ (A$\beta$) peptide aggregates, and intracellular neurofibrillary tangles (NFTs) consisting of hyperphosphorylated tau proteins.

A$\beta$40–42 peptides (A$\beta$40–42) are formed *in vivo* through the consecutive action of $\beta$-secretase and $\gamma$-secretase on A$\beta$ protein precursor (A$\beta$PP), whose physiological function is not yet clear. Presenilin 1 and 2 are the intermembrane aspartyl proteases that process many diverse proteins in the lipid bilayers and form the active centers of the $\gamma$-secretase enzyme. Transition metal-mediated oxidative stress plays a key role in the formation of A$\beta$ aggregates and NFTs. Further, cell cycle re-entry followed by abnormal mitotic alterations in some neuronal cells (less than one in 10 000 at a given time) leads to their apoptosis.

Protein modifications as a result of glycation and lipid peroxidations also play a major role in the regulation of oxidative stress at sites of the lesions of AD. Antioxidants such as vitamin C and $\alpha$-tocopherol (vitamin E) as well as nonsteroidal anti-inflammatory drugs (NSAIDs) such as aspirin have limited protective effects against AD. In addition, metal ion chelators may also have beneficial effects for the control of the progression of neurodegenerative diseases. Among the most commonly used drugs for the treatment of the symptoms of AD are Donepezil and galantamine (also called as Razadyne; galantamine hydrobromide), which are acetylcholinesterase (AChE) inhibitors. Donepezil shows concentration-dependent inhibition of cellular proliferation in human neuroblastoma cell lines, and thus approaches in targeting cell cycle abnormalities for the treatment of AD are promising. Cholesterol and other lipids also aid in the secretion of A$\beta$ from A$\beta$PP. Thus, cholesterol/lipid-lowering drugs are also being considered for the treatment of AD. Statins inhibit enzymes involved in the biosynthesis of cholesterol, and in addition show anti-inflammatory effects and show other beneficial effects by modulating $\alpha$-secretase and $\beta$-secretase enzymes. Further, stereoisomerization of the naturally occurring L-amino acid residues of the long-lived proteins *in vivo* to D-amino acids results in the formation of proteolysis-resistant toxic peptides which may also contribute the etiology of the age-related diseases. To date, treatment effectiveness has been modest and variable among patients. While any improvement or delay of disease progression is viewed positively by families and caregivers, all drug treatments require careful monitoring and consideration of the risks and side effects.

## Lipids

Lipid oxidation (also called lipid peroxidation) has consequences in aging and age-related diseases including neurological diseases such as AD and Parkinson disease. Membrane- and lipoprotein-associated polyunsaturated fatty acids upon reactive oxygen species (ROS) or reactive nitrogen species (RNS)-mediated oxidations give rise to a variety of toxic molecules including malondialdehyde, glyoxal, acrolein, *trans*-4-hydroxy-2-nonenal (HNE), and *trans*-4-oxo-2-nonenal (ONE). HNE and ONE are particularly toxic compounds to the surrounding biomacromolecules, as they can react relatively readily with cysteine-derived thiol groups and $\varepsilon$-amino groups of lysine units, which may result in enzyme deactivation. They can also deplete levels of intracellular glutathione, a cellular antioxidant, by Michael reactions involving the thiol group of the glutathione. Similarly, HNE deactivates glutathione peroxidase enzyme, which is responsible for detoxifying hydrogen peroxide. HNE and ONE also react with amino groups of nucleic acid-bases in DNA forming a variety of adducts, resulting in transversion mutations. HNE-derived propano adducts have been shown to exist in cerebellum and cortex of cases of AD (**Figure 1**).

There is strong evidence for the modulation of the activity of the $\gamma$-secretase as a function of the concentration of cholesterol and other lipids: higher cholesterol levels lead to the enhanced formation of

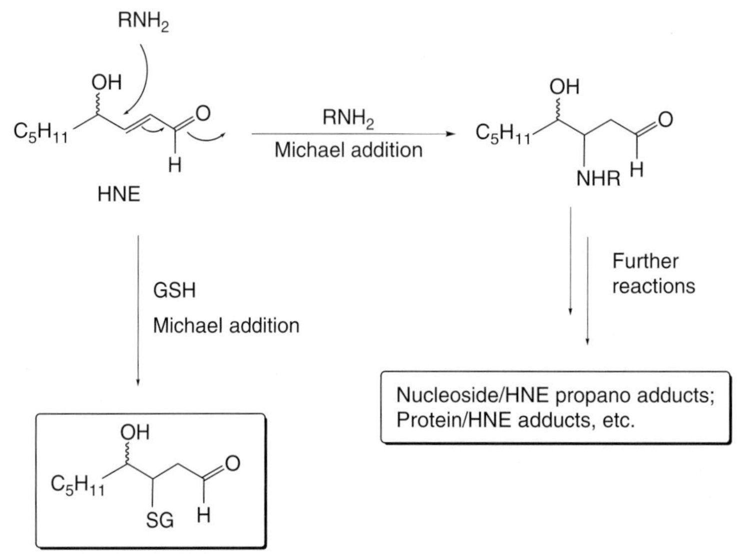

**Figure 1** HNE-derived propano adducts are found in the brain of patients with AD.

the A$\beta$ peptide and vice versa. Hypercholesterolemia is a risk factor for AD, as elevated concentrations of cholesterol are found in the brains of AD patients during the progression of AD, and thus cholesterol-lowering drugs are currently being developed for the treatment of AD. Statins (3-hydroxy-3-methylglutaryl-CoA reductase inhibitors) have well-established cholesterol-lowering properties and inhibit the A$\beta$-induced expression of interleukin-1$\beta$ and inducible nitric oxide synthase (iNOS). Statins also have anti-inflammatory properties that are independent of their cholesterol-lowering effects. Although at present there are no reliable clinical data to show the effectiveness of statins in AD therapy, they are expected to attenuate the symptoms of AD by cholesterol-lowering as well as anti-inflammatory properties. Apolipoprotein E (ApoE) deficiency in AD brains impairs the lipid transport system, thereby altering lipid homeostasis in cases of AD. The cholesterol-lowering drug Probucol increases the concentrations of cerebrospinal fluid (CSF) ApoE levels and decreases the CSF A$\beta$1–42 levels.

## Metals and Oxidations

Transition metal ions, for example, $Fe^{2+}$ and $Cu^{+}$, although essential in maintaining the central nervous system, induce neuronal damage when present in excess quantities. $Fe^{2+}$ and $Cu^{+}$ are sources of ROS (e.g., hydrogen peroxide, hydroxyl radical, and superoxide radical anion) and RNS (e.g., nitrous acid, nitric oxide, peroxynitrite) through Fenton and related reactions. Metal ion-catalyzed reduction of molecular oxygen gives the superoxide radical anions, which are further reduced to hydrogen peroxide ($H_2O_2$). Although $O_2^{\bullet -}$ and $H_2O_2$ by themselves are relatively nontoxic, $H_2O_2$ may lead to the production of hydroxyl radicals (HO•) through metal ion-catalyzed Fenton reactions. The hydroxyl radicals are highly toxic to neurons and the surrounding medium. A$\beta$ may be a source of the ROS, as it binds to $Cu^{2+}$ and $Fe^{3+}$. In the presence of biological reducing agents, $Cu^{2+}$ and $Fe^{3+}$ are reduced to $Cu^{+}$ and $Fe^{2+}$, which are mediators of the Fenton reaction forming hydroxyl radicals (**Figure 2**). In agreement with this, when A$\beta$ is pretreated with the iron chelator desferrioxamine, neuronal toxicity is significantly reduced. In various animal models, the addition of trace amounts of copper, but not zinc or aluminum,-resulted in increased A$\beta$ deposition. However, copper ions bound to A$\beta$ can cause conformational changes that confer to it a superoxide dismutase activity, maintaining oxidative balance. Thus, the role of A$\beta$ in the cause of oxidative stress in AD may be dependent on the metal ion homoeostasis in the surrounding medium. It is notable that although aggregated A$\beta$ peptide in combination with associated metal ions can induce oxidative stress, the formation of A$\beta$ itself is triggered by enhanced cellular oxidative stress, through activation of $\beta$-secretase and expression of A$\beta$PP by the activation of the AP-1 transcription factor.

There is abundant evidence showing that the oxidative stress mediated by ROS and RNS precedes the onset of the formation of A$\beta$ and NFT in cases of AD. Serving as markers of oxidative damage during aging and in age-related diseases such as AD, Parkinson's disease, and Down syndrome, 8-hydroxydeoxyguanosine (8-OHdG), 8-hydroxyguanosine (8-OHG), protein carbonyls, and nitrotyrosine are abundantly formed in neuronal cells, decades before the onset of the disease hallmarks, such as A$\beta$ plaques or NFTs. Abnormal mitochondria are the source of the majority of the ROS produced in cells, and metal ions are intrinsically involved in their formation. Iron and copper ions have been demonstrated histochemically to be localized in A$\beta$ plaques and NFTs, the main constituents of the brains of AD. Desferrioxamine ($Fe^{3+}$ chelator), tetrathiomolybdate ($Cu^{2+}$ chelator), and mercaptopropanol ($Pb^{2+}$ and $Hg^{2+}$ chelator) suppress the expression of A$\beta$PP, thus lowering the A$\beta$ peptide secretion. Desferrioixamine and clioquinol, a $Cu^{2+}$ chelator, have found limited therapeutic applications for AD (**Figure 3**). Oxidative stress may contribute to the irreversible protein aggregation associated with neuronal degeneration. For example, HNE binding to phosphorylated tau protein facilitates the *in vitro* formation of filaments. Also, oxidative stress and reactive oxygen species bring about the covalent cross-linking of tau filaments to form large aggregates that are resistant to proteolytic cleavage.

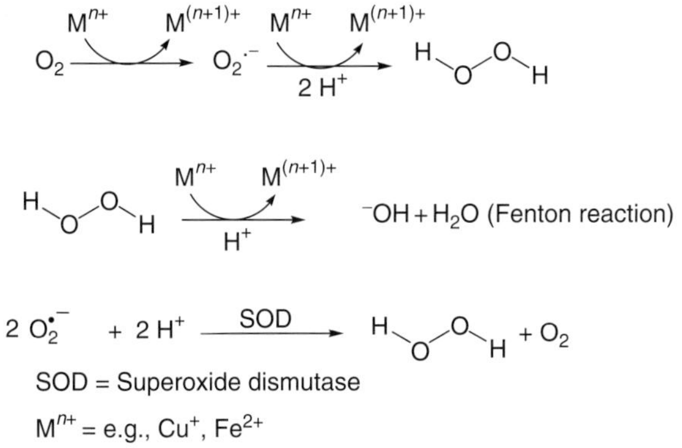

$$O_2 \xrightarrow{M^{n+} \rightarrow M^{(n+1)+}} O_2^{\cdot -} \xrightarrow[2\,H^+]{M^{n+} \rightarrow M^{(n+1)+}} H_2O_2$$

$$H_2O_2 \xrightarrow[H^+]{M^{n+} \rightarrow M^{(n+1)+}} {}^{\cdot}OH + H_2O \text{ (Fenton reaction)}$$

$$2\,O_2^{\cdot -} + 2\,H^+ \xrightarrow{SOD} H_2O_2 + O_2$$

SOD = Superoxide dismutase

$M^{n+}$ = e.g., $Cu^+$, $Fe^{2+}$

**Figure 2** Reactions involving the detoxification of reactive oxygen species (ROS).

## Racemization

The amino acid residues of long-lived proteins in biological tissues are progressively racemized with time, and the racemization process may be abnormally accelerated in age-related diseases. Long-lived proteins such as lens crystallin show increased amounts of D-amino acid residues, which is due to the slow racemization of the L-amino acids over time. Aspartic acid residues are especially prone to racemization. D-Asp concentrations increase with age in tissues such as dentine and may serve as estimates of the age of the individuals. In lens crystallin proteins, amino acid residues Asp151 and Asp58 are relatively more stereochemically labile. Racemization of the native L-amino acids in proteins results in the change of their conformations, resistance to proteolytic degradation, and loss of enzyme activity of the involved enzymes.

There are several reports of racemization of the Ser and Asp residues of the insoluble A$\beta$1–40 peptides *in vivo* during the aging process, over a 20–30 year duration. The secondary structures of the [D-Asp23] A$\beta$20–29 and [D-Ser26]A$\beta$20–29 have been shown to differ from the corresponding unracemized peptides by vibrational circulation dichroism. Purified amyloid was found to contain relatively large amounts of D-aspartate and D-serine, which may be derived from the long-lived A$\beta$PP. The A$\beta$ peptide with racemized Ser or Asp residues, but not the unracemized peptide, is hydrolyzed by chymotrypsin-like enzymes to form soluble peptide fragments, for example, [D-Ser26]A$\beta$25–35, which are neurotoxic and are involved in the degeneration of hippocampal neurons in AD brains. The latter peptides are also resistant to brain proteinases.

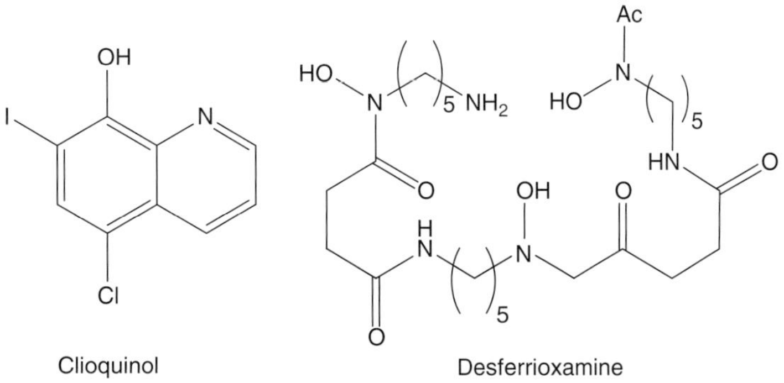

Clioquinol    Desferrioxamine

**Figure 3** The metal chelators, clioquinol, and deserioxamine, have both shown promise in clinical trials for the treatment of AD.

## Conclusions

Aging and age-related diseases, in particular AD, have oxidative stress as an important pathogenic factor. ROS, derived through abnormal mitochondrial metabolism involving transition metal ions, and lipoxidation-derived aldehyde products (also implicating transition metal ions in their generation) oxidatively modify cellular proteins and exert DNA/RNA damage. Oxidative stress precedes the formation of the A$\beta$ plaques and NFTs in AD, which in turn, through associated transition metal ions, further exacerbate the oxidative stress. Further, racemization of amino acids in the long-lived proteins, for example, A$\beta$PP in AD, may result in the formation of proteolytically stable, toxic peptides. Antioxidants and agents that would sequester the ROS and lipid peroxidation endproducts may be useful targets in the identification of drugs for age-related diseases.

*See also:* Aging of the Brain and Alzheimer's Disease; Alzheimer's Disease: An Overview; Lipids and Membranes in Brain Aging; Metal Accumulation During Aging.

## Further Reading

Bosch-Morell F, Flohe L, Marin N, and Romero FJ (1999) 4-hydroxynonenal inhibits glutathione peroxidase: Protection by glutathione. *Free Radical Biology and Medicine* 22: 1383–1387.

Friguet B, Stadtman ER, and Szweda LI (1994) Modification of glucose-6-phosphate dehydrogenase by 4-hydroxy-2-nonenal. Formation of cross-linked protein that inhibits the multicatalytic protease. *Journal of Biological Chemistry* 269: 21639–21643.

Grimm MOW, Grimm HS, Paetzold AJ, et al. (2005) Regulation of cholesterol and sphingomyelin metabolism by amyloid-$\beta$ and presenilin. *Nature Cell Biology* 7: 1118–1123.

Perry G, Castellani RJ, Hirai K, and Smith MA (1998) Reactive oxygen species mediate cellular damage in Alzheimer disease. *Journal of Alzheimers Disease* 1: 45–55.

Poirier J (2005) Apolipoprotein E, cholesterol transport and synthesis in sporadic Alzheimer's disease. *Neurobiology of Aging* 26: 355–361.

Ritz-Timme S and Collins MJ (2002) Racemization of aspartic acid in human proteins. *Ageing Research Reviews* 1: 43–59.

Selkoe DJ (2004) Cell biology of protein misfolding: The examples of Alzheimer's and Parkinson's diseases. *Nature Cell Biology* 6: 1054–1061.

Shapira R, Austin GE, and Mirra SS (1988) Neuritic plaque amyloid in Alzheimer's disease is highly racemized. *Journal of Neurochemistry* 50: 69–74.

Sjoegren M, Mielke M, Gustafson D, Zandi P, and Skoog I (2006) Cholesterol and Alzheimer's disease – Is there a relation? *Mechanisms of Ageing and Development* 127: 138–147.

Sortino MA, Frasca G, Chisari M, et al. (2004) Novel neuronal targets for the acetylcholinesterase inhibitor Donepezil. *Neuropharmacology* 47: 1198–1204.

Sparks DL, Friedland R, Petanceska S, et al. (2006) Trace copper levels in the drinking water, but not zinc or aluminum influence CNS Alzheimer-like pathology. *Journal of Nutrition, Health, and Aging* 10: 247–254.

Takeda A, Smith MA, Avila J, et al. (2000) In Alzheimer's disease, heme oxygenase is coincident with Alz50: An epitope of tau induced by 4-hydroxy-2-nonenal modification. *Journal of Neurochemistry* 75: 1234–1241.

Webber KM, Raina AK, Marlatt MW, et al. (2005) The cell cycle in Alzheimer disease: A unique target for neuropharmacology. *Mechanisms of Ageing and Development* 126: 1019–1025.

Zhu X, Raina AK, Perry G, and Smith MA (2004) Alzheimer's disease: The two-hit hypothesis. *Lancet Neurology* 3: 219–226.

# Glial Cells: Microglia during Normal Brain Aging

M B Graeber



This article is reproduced from the previous edition © 2004, Elsevier B.V.

## Background

Microglia were first described at the turn of the nineteenth century, independently by the neuropathologists Franz Nissl (in Munich) and William Ford Robertson (in Edinburgh). In Spain, in 1913, the Nobel prize-winning physician and histologist Ramón y Cajal described these cells as a *tercer elemento* ('third element') among central nervous system cells. However, it was Cajal's student, Pio del Rio-Hortega, who, around 1920, coined the term 'microglia' and conducted the first systematic studies on this cell type. Many of the observations of Rio-Hortega were of fundamental character and are still valid.

## Morphology

After staining with Rio-Hortega's silver carbonate method, resting microglia show characteristic elongated, almost bipolar cell bodies with spinelike processes arising at a right angle and carrying short branches (**Figure 1**). The nucleus is large in comparison with the small perikaryon and contains heterochromatin apposed to the nuclear envelope. In electron micrographs, microglia show a dense cytoplasm, vimentin-positive intermediate filaments, a few microtubules, and prominent cisterns of granular endoplasmic reticulum, especially when activated, which run in parallel to the main axis of the cell (**Figure 2**). Lysosomes and Golgi complexes are also present.

## Cellular Markers

After their discovery and description, there was decades-long controversy surrounding microglial identity and nature; the uncertainty about these cells, which lasted until the 1980s, can be explained by the remarkable phenotypic plasticity of microglia. Their appearance ranges from that of highly ramified cells in normal central nervous system (CNS) tissue (**Figure 1**) to rod cells in chronically inflamed cerebral cortex (**Figure 3**) to full-blown macrophages in tissue-destructive pathologies (**Figure 4**). The problem of defining the nature of microglia was solved when lectin and immunocytochemical markers became available, and when experimental models, including bone marrow chimeras, unequivocally demonstrated that microglia are closely related to cells of the peripheral monocyte–macrophage system. However, no marker possesses true, singular microglia specificity, and all known histological microglia markers are only microglia specific in the sense that they do not label other glia or neurons. Other macrophages that normally occur within the bony confinements of the CNS – for example, perivascular, plexus, and meningeal macrophages – typically cross-react with microglial detection methods, but many of the molecules expressed by the former are absent from microglia. Thus, the immunophenotype of microglial cells appears to be 'downregulated.' Before monoclonal antibodies became available, enzymes such as thiamine pyrophosphatase and nucleoside diphosphatase were used to identify microglial cells in tissue sections. Furthermore, the ectoenzyme 5′-nucleotidase is positive in the plasmalemma of microglia and strongly expressed in activated cells, but is not cell type specific. Nonspecific esterase, a monocytic marker, is negative.

## Distribution

Resting microglia account for up to 20% of all glial cells in the CNS. Their almost uniform distribution varies rather little between brain regions, although there appear to be local differences in the expression of some molecules, such as major histocompatibility complex (MHC) class II antigens, which are present at higher levels in white matter and in the aged brain.

## Origin

During vascularization of the CNS, cells related to the mononuclear phagocyte system invade the central nervous tissue and settle as resting microglia. They form the resident macrophage precursor cell pool of the brain, spinal cord, and retina. Renewal of juxtavascular (perivascular) and parenchymal microglia from the bone marrow occurs at an extremely low rate in the adult. Perivascular macrophages ('perivascular cells'), which are separated from the brain parenchyma by a basement membrane, are anatomically, immunophenotypically, and functionally distinct from microglia.

## Function

The main function of microglia in the normal brain and spinal cord appears to lie in the formation of a network of highly responsive surveillance and defense cells. Accordingly, microglia serve as 'sensors of threats' (a term introduced by GW Kreutzberg) to the CNS. Microglia are potentially cytotoxic, but healthy microglia do not attack their microenvironment except in the case of CNS infections. Thus, the metaphor of microglial 'guardians' that are armed to defend seems more appropriate than that of harmful 'killer cells.' In contrast to neuroectodermal glia, microglial cells are not electrotonically coupled via gap junctions. This may explain their very localized involvement in disease processes, which is of great diagnostic use. Antigen presentation is a likely function of microglia. Complement receptors are quickly upregulated in response to various stimuli. The mitotic activity of microglia is regulated by macrophage and granulocyte/macrophage colony-stimulating factors (M-CSFs and GM-CSFs) and the appropriate receptors. Other cytokines, such as interleukins (IL-1, IL-6, and IL-10) and transforming growth factor-$\beta$ (TGF-$\beta$1), function as important signals in the control of microglial activation. Brain macrophages of microglial origin have a strong respiratory burst activity, meaning that they can produce oxygen radicals.

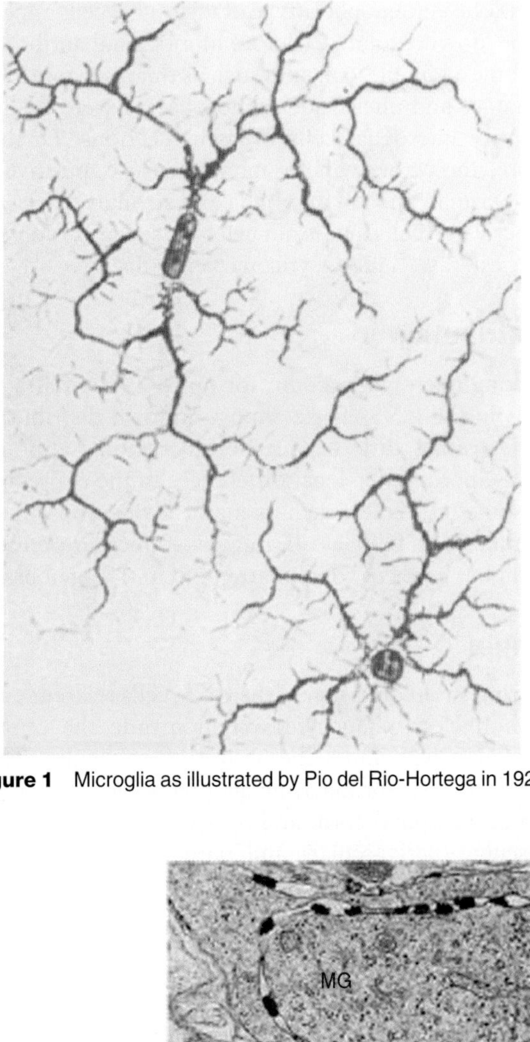

**Figure 1** Microglia as illustrated by Pio del Rio-Hortega in 1920.

## Pathology

More than a dozen names for microglia document their involvement in neuropathologies such as inflammation and vascular disease, and in removal of tissue debris after trauma, pointing to the phagocytic nature of the reactive microglial cells. Terms still in use are foam cells, gitter cells, rod cells, and fat granule cells. The term 'ameboid microglia' should be reserved to describe microglia as they appear in developing tissue. It is thought that microglial phagocytes become motile and migrate toward lesion sites, where they are loaded with tissue debris, which they then transport to the meninges or vascular system. Perivascular cuffs of fat-laden phagocytes can be seen many months and even years after acute tissue destruction. Axotomy of motor nerves leads not only to retrograde changes of the motor neurons but also to proliferation of perineuronal microglial cells. After a few days these microglia are found covering most of the surface of the regenerating motor neurons, thus displacing the synaptic terminals ('synaptic stripping'). In this situation no phagocytosis is observed. Based on observations under various pathological conditions, a concept of a graded response of microglial cells has been developed. The diagram in **Figure 5** describes the stepwise activation of microglia from a resting to a phagocytic state.

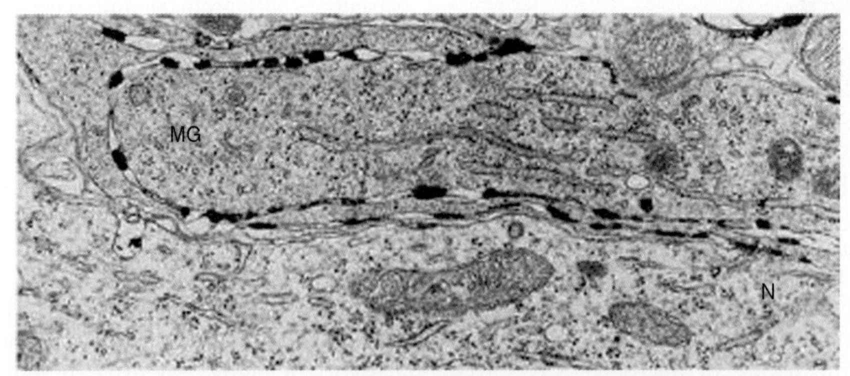


**Figure 2** Activated perineuronal microglia (MG) cell covering the surface of a chromatolytic facial motor neuron (N) in the rat. The MG has a dense cytoplasm and an active rough endoplasmic reticulum arranged in parallel broad cisterns. Electron-dense reaction product indicates the presence of ecto-5′-nucleotidase in the plasmalemma of the MG. Electron micrograph 1:40 000. From Kreutzberg GW (1999) Microglia. In: Adelman G and Smith BH (eds.) *Encyclopedia of Neuroscience*, 2nd edn. Amsterdam: Elsevier.

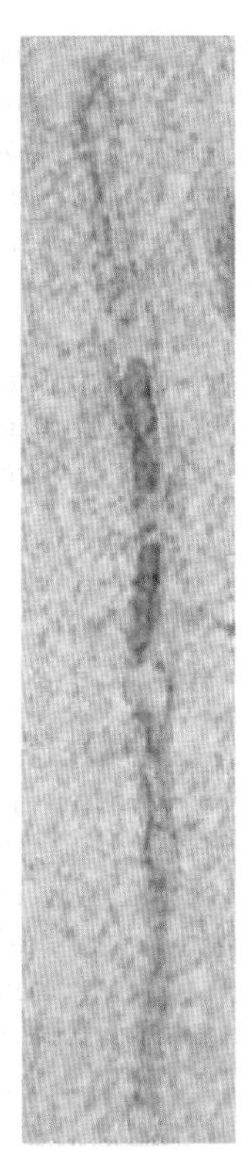

**Figure 3** Rod cells in the human cerebral cortex, stained for major histocompatibility complex class II molecules, in a case of subacute sclerosing panencephalitis, a condition in which rod cells are typically found.

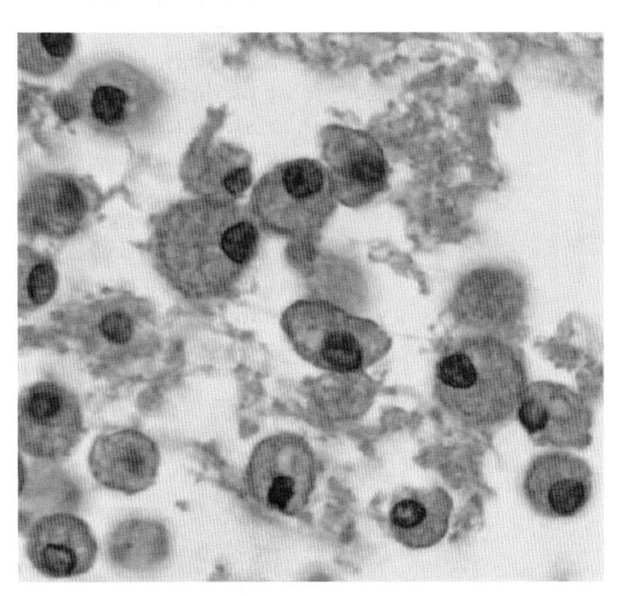

**Figure 4** Brain macrophages in a cerebral infarct. Human brain; hematoxylin–eosin staining.

**Figure 5** Microglial activation cascade. Activated microglia (middle) usually show a shortened, stout process but can also have the appearance of rod cells (**Figure 3**), depending on the anatomical region in which the cells reside. Activated microglia are not necessarily phagocytic (**Figure 2**).

*See also:* Glial Cells: Astrocytes and Oligodendrocytes During Normal Brain Aging.

## Further Reading


Angelov DN, Walther M, Streppel M, et al. (1998) The cerebral perivascular cells. *Advances in Anatomy, Embryology & Cell Biology* 147: 1–87.

Flügel A, Bradl M, Kreutzberg GW, et al. (2001) Transformation of donor-derived bone marrow precursors into host microglia during autoimmune CNS inflammation and during the retrograde response to axotomy. *Journal of Neuroscience Research* 66: 74–82.

Graeber MB, Kreutzberg GW, and Streit WJ (eds.) (1993) *Microglia: Special issue* 7: 1–119.

Kreutzberg GW (1996) Microglia: A sensor for pathological events in the CNS. *Trends in Neurosciences* 19: 312–318.

Perry VH (1994) *Macrophages and the Nervous System.* Austin: R. G. Landes Co.

Rio-Hortega Pdel (1932) Microglia. In: Penfield W (ed.) *Cytology and Cellular Pathology of the Nervous System*, vol. 2, pp. 483–534. New York: Hafner [1965 facsimile of 1932 edition].

# Glial Cells: Astrocytes and Oligodendrocytes during Normal Brain Aging

**I Blasko and C Humpel**, Innsbruck Medical University, Innsbruck, Austria
**B Grubeck-Loebenstein**, Institute for Biomedical Aging Research of the Austrian Academy of Sciences, Innsbruck, Austria



## Overview

Astrocytes are the most abundant cells in the brain, and together with oligodendrocytes, they are multifunctional housekeeping cells that allow neurons to become progressively specialized for the tasks of information processing. Astrocytes not only organize the structural architecture of the brain, but also its communication pathways and plasticity. Neurons co-cultured with astrocytes develop approximately sevenfold more synapses and display a sevenfold increase in synaptic efficacy, compared with neurons cultured in the absence of astrocytes. Astrocytes modulate synaptic strength and activity and regulate neurogenesis of the adult human brain. Oligodendrocytes are responsible for the production and maintenance of myelin, which is essential for normal brain function and signal transmission speed. A series of recent studies clearly demonstrate reciprocal paracrine interactions between astrocytes, endothelial cells, and ependymal cells and also between oligodendrocytes and neurons. In fact, astrocyte-derived signals induce neuronal differentiation of oligodendrocyte precursor cells. Coincident with these findings, it has become obvious that both cell types are substantially influenced during the aging process.

## Role of Astrocytes in the Aging Process

### Astrocytes in the Aging Brain

In the past decade considerable attention has been focused on understanding the role of astrocytes in aging and age-related neurodegenerative disorders. During normal aging, astrocytes display increased activation ('reactive gliosis'), with progressive increases of glial fibrillary acidic protein (GFAP) beginning before midlife (**Figure 1(a)**). Similar to the concept of age-related microglial dystrophy, analogous astrocytic dystrophy might be relevant in normal aging of astrocytes. Neurite outgrowth by embryonic neurons is markedly decreased when co-cultured with confluent astrocytes derived from old (24 months) versus young (3 months) animals. Overexpression of GFAP in young astrocytes models the effects of aging by reducing neurite outgrowth. These results implicate increased astrocytic GFAP expression that leads to functional defects, which may trigger neuronal dysfunction and support the development of neuronal atrophy during normal aging.

The ability of astrocytes to protect neurons or stimulate neurogenesis is compromised during aging. It is well known that activated astrocytes can produce and release cytokines, but they also produce neurotrophins such as nerve growth factor (NGF) and can enhance the survival of neurons in the brain (**Figure 1(c)**). There is evidence that multiple hippocampal stem cell/progenitor proliferation factors exhibit an early decline during aging, an event that seems to be linked to age-related alterations in hippocampal astrocytes. Neurogenesis in the dentate gyrus plays an important role for learning and memory and declines dramatically by middle age. Studies in rat astrocytes show that several growth factors – fibroblast growth factor-2 (FGF-2), insulin-like growth factor-1, and vascular endothelial growth factor – decline in concentration considerably by middle age but remain at steady levels between middle age and old age. Decreased concentrations of FGF-2 during aging are associated with decreased numbers of FGF-2-positive astrocytes. Quantification of GFAP and FGF-2-positive cells reveals that aging does not decrease the total number of astrocytes, but that some astrocytes express lower FGF-2 levels by middle age. Thus, dramatically decreased neurogenesis by middle age is likely linked to reduced concentrations of growth factors in the hippocampus.

Astrocytes regulate synthesis and extracellular levels of the excitatory amino acid glutamate via the glutamine cycle and strongly contribute to brain homeostasis and regeneration. Neurotrophic factors, transporter molecules, and enzymes involved in the metabolism of excitatory amino acids play a role in neuroprotection of neurons by regulating the production of potentially toxic substances (e.g., radicals). It has been found that mitochondrial membrane potentials in astrocytes isolated from old mice are more depolarized than are those from their younger counterparts and display a greater sensitivity to the oxidant hydrogen peroxide. ATP-induced $Ca^{2+}$ responses in astrocytes isolated from old mice are consistently larger in amplitude and more frequently oscillatory than is seen in younger astrocytes, which may be attributable to lower mitochondrial $Ca^{2+}$ sequestration. Finally, NGF-differentiated PC12

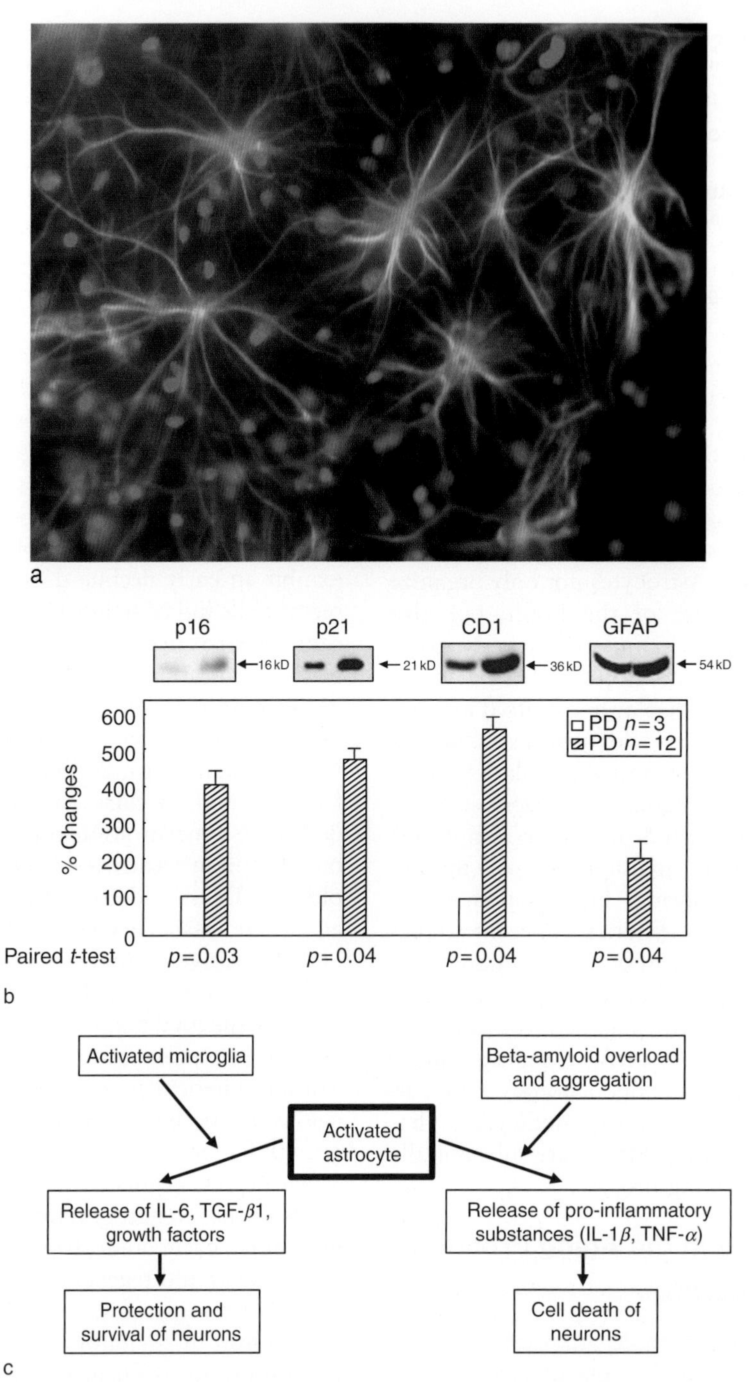


**Figure 1** (a) Fluorescence immunohistochemistry of astrocytes in the brain expressing high levels of glial fibrillary acidic protein (GFAP). The cells were counterstained with the nuclear dye 4′,6-diamidino-2-phenylindole (blue). (b) Expression pattern of the senescence markers p16, p21, cyclin D1 (CD1), and the differentiation marker GFAP. Each blot shown in the upper part represents one characteristic Western blot experiment. The bars in the lower part represent mean ± SEM ($n = 3$ in each group). The cells from early passage (population doubling (PD) number 3) and late passage (PD 12) are compared. Marker expression levels in PD 3 were considered as 100%. PD 12 results are presented as percent change compared to early passage cells. The levels of significance between both groups are indicated (Student's $t$-test). (c) Schematic role of activated astrocytes during aging and neurodegeneration. Activated microglia stimulate astrocytes to produce growth factors and cytokines, which support survival of neurons (IL-6, interleukin-6; TGF-$\beta$1, transforming growth factor-$\beta$1; TNF-$\alpha$, tumor necrosis factor-$\alpha$). $\beta$-Amyloid overload and aggregation induce astrocytes to release proinflammatory cytokines, which induce neuronal cell death. (b) Reproduced from Blasko I, Stampfer-Kountchev M, Robatscher P, et al. (2004) How chronic inflammation can affect the brain and support the development of Alzheimer's disease in old age: The role of microglia and astrocytes. *Aging Cell* 3(4): 169–176, with permission from Blackwell Publishing.

cells co-cultured with old astrocytes are significantly more sensitive to hydrogen peroxide treatment than are co-cultures of NGF-differentiated PC12 cells with young astrocytes.

### Senescence of Astrocytes

Cellular senescence is a stress response phenomenon resulting in a permanent withdrawal from the cell cycle and the appearance of distinct morphological and functional changes. As astrocytes can be easily passaged in cell culture, replicative senescence of primary human astrocytes has been well studied. In analogy to the established '*in vitro* senescence' models of fibroblasts and endothelial cells, astrocytes isolated from postmortem brains of Alzheimer' disease patients proliferate rigorously in culture and undergo six to eight population doublings (PDs) before astrocytes stop to divide. Growth arrest can be reached after 9–10 PDs, and astrocytes are therefore referred to as 'late-passage astrocytes' after 8 PDs. The quantification of cellular GFAP content reveals that late-passage astrocytes express more GFAP in comparison to early-passage astrocytes (**Figure 1(b)**). In agreement with this observation, a higher GFAP content is found *in vivo* in aged rats and mice in comparison to their young littermates. Additionally, late-passage astrocytes exhibit a significantly increased expression of the senescent markers p16, p21, and cyclin D1 (**Figure 1(b)**). The number of PDs of gray-matter astrocytes obtained from frontal cortex is higher, compared to the number from white-matter astrocytes. Whether this increased growth rate depends on *in vivo* degenerative or inflammatory processes is not yet precisely known. In contrast to fibroblasts, but in accordance with endothelial cells, astrocytes are prone to undergo apoptosis after having reached the end of their replicative life span. Interestingly, in late-passage astrocytes, tumor necrosis factor-$\alpha$ and interferon-$\gamma$ stimulate the production of $\beta$-amyloid, demonstrating that stimulation of $\beta$-amyloid production by inflammatory products does not decrease during aging.

### Astrocytes and Age-Related Neurodegenerative Diseases

Astrocytes greatly outnumber microglia in the brain and play a more critical role in the development of age-related diseases than previously thought. Many metabolic products, depending on their aggregation and deposition status, can activate astrocytes. It is well known that astrocytes are involved in the degradation of damaging products. A disturbed balance between the production and the degradation of toxic metabolites triggers chronic inflammatory processes during aging. Chronically activated microglia and astrocytes can damage neurons via release of highly toxic products such as reactive oxygen intermediates, nitric oxide, inflammatory cytokines, proteolytic enzymes, complement factors, or excitatory amino acids (**Figure (1c)**). In addition, astrocytes appear to be first-line phagocytic cells of degenerating myelin in cerebral cortex, and microglia cells are not activated until degeneration of entire nerve fibers proceeds (see later).

Inflammation in the aging brain is most likely the result of accumulation of toxic metabolites or their decreased degradation. Glial products such as interleukin-1$\beta$ or tumor necrosis factor-$\alpha$ induce the synthesis of $\beta$-amyloid binding proteins in astrocytes, microglia, or neurons. The chronic release of proinflammatory cytokines in the aging brain is likely to maintain a chronic acute-phase protein secretion favoring the formation of $\beta$-amyloid fibrils. The astrocytic synthesis of amyloid precursor protein (APP) can be stimulated by these proinflammatory cytokines through regulation of gene transcription at the promoter level. In combination, cytokines such as interleukin-1$\beta$ or tumor necrosis factor-$\alpha$ and interferon-$\gamma$ affect the metabolism of APP and stimulate the production of $\beta$-amyloid (1–40) and $\beta$-amyloid (1–42) peptides. This molecular mechanism is complex, but it seems likely that the maturation (i.e., proper glycosylation) of APP is disturbed. The chemokine monocyte chemoattractant protein-1 (MCP-1) is believed to play an important role in astrocytosis, as MCP-1 levels are increased in aging brain. It is also of interest that adult astrocytes do not respond to stimulation with $\beta$-amyloid by increasing their release of MCP-1 in the same way as young control cells do.

Astrocytosis is a typical morphological feature of the Alzheimer's disease brain and represents either proliferation of astrocytes, in an effort to replace dying neurons, or a reaction to degrade the increasing amounts of toxic $\beta$-amyloid peptides. There are clear indications that age-related changes of astrocytes may influence the development of age-related neurodegenerative disorders such as Alzheimer's disease, Huntington's disease, or Parkinson's disease. On the other hand, age-related degenerative changes can modulate astrocytes and their supportive function for neurons. Little is known about the molecular mechanisms that underlie these age-related changes in astrocytes. Alzheimer's disease may coincide with the decreased capacity of aged glial cells to degrade and/or store toxic products. It is now well understood that an early and invariant step in the pathogenesis of Alzheimer's disease is the overproduction and/or decreased degradation of $\beta$-amyloid, which leads to the oligomerization of $\beta$-amyloid, both of which are regarded as pathogenetic hallmarks of Alzheimer's disease. In contrast to neonatal astrocytes, adult astrocytes efficiently clear exogenously added and surface-bound $\beta$-amyloid and are capable of removing $\beta$-amyloid deposited in transgenic APP overexpressing mice. In addition, in young animals, astrocytes produce low amounts of the $\beta$-amyloid-degrading enzyme neprilysin, but older animals have increased levels of both $\beta$-amyloid peptides and neprilysin.

## Role of Oligodendrocytes in the Aging Process

Oligodendrocytes are highly specialized cells that support between 1 and 40 separate myelin sheaths, depending on the size of the axons with which they associate. The production and maintenance of myelin are essential for normal brain function, because myelination results in more than a tenfold increase of signal transmission speed. Faster conduction velocity facilitates information flow by allowing for precise temporal coding of high-frequency bursts of neuronal activity. The human brain, which is approximately 2% of the body weight, contains approximately 25% of the body's membrane cholesterol. Therefore, the high lipid and cholesterol content of myelin sheaths is extremely important. The total length of myelinated axons is reduced by 27–45% in old age in humans, mostly because of loss of fibers with a small diameter, which myelinate later in development.

Oligodendrocytes are the most vulnerable cells in the brain. Myelin production continues with age and this is evident when the average thickness of myelin sheaths in monkey visual cortex is compared in young and old animals. The mean number of lamellae in the sheaths in young and in old monkeys is 5.6 and 7.0, respectively. An increase in the number of myelin sheaths requires an active incorporation of newly synthesized myelin components. Incorporation and disappearance of the components are compartmentalized into two pools, rapidly and slowly exchanging, and there is evidence that through aging the metabolic turnover in rapidly exchanging myelin is slightly accelerated. Remyelination in the brain is very low due to the expression of several axonal growth-inhibiting molecules. However, astrocytes and macrophages influence remyelination and create together an environment that allows regenerative processes, by producing multiple signaling molecules. Measurement of white-matter volume demonstrates that frontal lobe white-matter volume of healthy men continues to increase into the fifth decade of life, reaching a maximum volume at the age of about 45 years, and then declining. The question of whether there is a turnover of oligodendrocytes has not yet been fully resolved, but, in contrast to neurons, there is some evidence that the number of oligodendrocytes increases.

Oligodendrocytes are markedly affected in the aging cortex, displaying enhanced dense cellular inclusions, swellings of processes, an enhanced turnover of myelin, splitting of myelin sheaths, and enhanced lysosomes and vacuoles. Furthermore, the formation of 'balloons' is found; these are round or spherical cavities and they bulge out and expand the myelin, representing degenerative alterations. The enhanced formation of sheaths with redundant myelin points to uncontrolled production of myelin during the aging process. Impairment of learning and memory is in proportion to the severity of myelin sheath degeneration. Interestingly, in experimental models of glucose applications, oligodendrocytes undergo a senescent program, displaying elongation and thickening of processes and shortening of telomeres. Aging affects several key genes that play a role in the myelin formation, most importantly the genes encoding myelin basic protein (MBP) and proteolipid protein (PLP). Constitutive levels of both of these proteins are decreased during aging. In addition, Gtx, a homeodomain transcription factor that regulates MBP and PLP, is reduced during the aging process. In this context, progesterone and progesterone derivatives counteract the age-associated abnormalities of myelin production and enhance myelin formation by regulating MBP gene expression. In fact, promyelination is enhanced twofold after injury, and interaction of macrophages and astrocytes, mediators of successful myelination, loses its precision as the brain ages. Studies have indicated that steroid hormones and their derivatives have beneficial effects on the age-related decline in maintenance of myelin integrity, and partially reverse the decline in the efficiency of myelin repair.

Oligodendrocytes are particularly susceptible to damage by toxic stimuli, including proinflammatory cytokines, nitric oxide, enhanced calcium, glutamate excitotoxicity, oxidative stress, or complement from macrophages. These degenerative events lead to changes in conduction velocity, formation of redundant myelin, and enhanced thickness of myelin sheaths. In addition, oligodendrocytes contain the highest iron content of all brain cell types, and as much as 70% of brain iron is associated with myelin. Cholesterol- and lipid-synthesizing enzymes in oligodendrocytes require iron for proper function, and age-related increases in iron levels in brain are established characteristics of aging. Since the levels of soluble $\beta$-amyloid peptides increase with age, and metals such as iron, copper, and zinc can promote $\beta$-amyloid oligomerization and toxicity, these parameters are likely of high importance in the development of Alzheimer's disease.

The susceptibility of myelination processes over a lifetime to heterogeneous risk factors such as brain trauma, vascular disease, hypertension, nutritional deficiencies, hypercholesterolemia, and reduced hormone levels could explain the pattern of myelin breakdown in aging and age-related diseases. Age-related myelin breakdown is regionally heterogeneous, and differences in myelin properties make later-myelinating regions more susceptible to normal aging processes. Little information is available on the role of oligodendrocytes as myelin-producing cells in neurodegenerative pathologies such as Parkinson's or Alzheimer's disease. The best studied disease in regard to oligodendrocytes is multiple sclerosis, due to the severe damage to oligodendrocytes, but the majority of studies during aging have focused on Alzheimer's disease. In Alzheimer's disease, myelin breakdown is globally exacerbated, consistent with an extracellular deleterious process such as $\beta$-amyloid peptide toxicity. In addition, myelin breakdown during aging is associated with apolipoprotein E, which points to an additional role in the initiation of Alzheimer's disease.

*See also:* Glial Cells: Microglia During Normal Brain Aging.

## Further Reading

Ando S, Tanaka Y, Toyoda Y, et al. (2003) Turnover of myelin lipids in aging brain. *Neurochemical Research* 28: 5–13.

Bartzokis G (2004) Age-related myelin breakdown: A developmental model of cognitive decline and Alzheimer's disease. *Neurobiology of Aging* 25: 5–18.

Blasko I, Stampfer-Kountchev M, Robatscher P, et al. (2004) How chronic inflammation can affect the brain and support the development of Alzheimer's disease in old age: The role of microglia and astrocytes. *Aging Cell* 3(4): 169–176.

Franklin RJM, Zhao C, and Sim FJ (2002) Ageing and CNS remyelination. *NeuroReport* 13: 923–928.

Peters A (2002) The effects of normal aging on myelin and nerve fibers: A review. *Journal of Neurocytology* 31: 581–593.

Goldman S (2003) Glia as neural progenitor cells. *Trends in Neurosciences* 26: 590–596.

Pfrieger FW and Barres BA (1997) Synaptic efficacy enhanced by glial cells *in vitro*. *Science* 277: 1684–1687.

Rozovsky I, Wei M, Morgan TE, et al. (2005) Reversible age impairments in neurite outgrowth by manipulations of astrocytic GFAP. *Neurobiology of Aging* 26: 705–715.

Song H, Stevens CF, and Gage FH (2002) Astroglia induce neurogenesis from adult neural stem cells. *Nature* 417: 39–44.

Wyss-Coray T, Loike JD, Brionne TC, et al. (2003) Adult mouse astrocytes degrade amyloid-beta *in vitro* and *in situ*. *Nature Medicine* 9: 453–457.

# Wallerian Degeneration

**S L Carroll**, The University of Alabama, Birmingham, AL, USA



## Introduction

The term Wallerian degeneration refers to a series of stereotyped, synchronized changes that occur in the distal segment of a peripheral nerve after this structure has been separated from its associated neuronal cell bodies by trauma or, less commonly, an ischemic event. Wallerian degeneration is named after Augustus V Waller, the scientist who first described this process in 1850. Over the decades since its initial description, it has been recognized that Wallerian degeneration shares many features with axonal degeneration, a common form of axonopathy that can be produced by numerous diseases, drugs, and toxins. Given these similarities and the fact that Wallerian degeneration is initiated by a single, precisely timed event (the severing of the nerve), Wallerian degeneration has proven to be an extremely useful experimental model that has provided valuable new insights into the mechanisms underlying axonal degeneration. Below, the morphological events that occur during Wallerian degeneration, the biochemical alterations that accompany these morphological changes, and how these changes prepare the nerve for subsequent regeneration, are considered. Then, the Wallerian degeneration in the peripheral nervous system (PNS) is compared to Wallerian degeneration in the injured central nervous system (CNS), highlighting the similarities and differences in these processes. Finally, the recent studies of the Wallerian degeneration slow (*Wld$^s$*) mouse, a spontaneously arising mutant that has fundamentally altered our understanding of axonal degeneration by demonstrating that this process is actively mediated by an axon death program distinct from apoptosis, are discussed.

## Morphologic Changes during Wallerian Degeneration of Peripheral Nerve

The majority of studies of Wallerian degeneration have been performed using rodents (primarily rats and mice) as experimental models. Consequently, the timing of events described below is that observed in these animal models unless otherwise indicated. Wallerian degeneration has been studied in man and the findings compared to observations made in rodent models. The primary difference noted between Wallerian degeneration in man and in rodents is that many of the morphologic alterations described below take significantly longer to develop in human nerves, a feature that may be related to their much larger size. Other than these differences in timing, Wallerian degeneration in humans, rats, and mice is highly similar.

### Axonal Changes Following Nerve Transection

For a short period after nerve transection, axon segments distal to the injury site remain intact. Although separated from the cell body of the neuron, these distal axon segments also remain capable of transmitting electrical impulses for 48–96 h (4–10 days in man, with the ability to conduct current typically lasting longer in lengthier distal nerve segments). Despite this continuing electrical responsiveness, morphologic changes begin to appear in the distal axon segment quite rapidly. Within hours, membranous organelles, glycogen, and atypical swollen mitochondria, which are normally transported along the axon, begin to accumulate in paranodal regions and adjacent to the lesion site on both the proximal and distal sides. Lysosomes also begin to accumulate in the paranodal and lesional regions by 2 h after injury. The accumulation of organelles in severed axon segments adjacent to the cut site is of a magnitude sufficient to enlarge the axon, thereby stretching and thinning the adjacent myelin sheath; however, this dilatation is typically of a lesser degree than is characteristic of the axonal spheroids that develop following interruption of a CNS axon. Within 12–24 h, axonal microtubules begin to become disorganized, heralding the dissolution of this portion of the axonal cytoskeleton. One to two days post injury, calcium ions enter the axon segment via calcium-specific channels and activate calcium-dependent calpain proteases which degrade the axonal neurofilaments. As these morphologic changes, which are known as granular disintegration, progress, the normal axonal cytoskeleton, axolemma, and axonal organelles are degraded, converting the axoplasm into a watery fluid filled with amorphous and granular debris. At the time granular disintegration occurs, the myelin sheath surrounding the axon, although stretched, is still intact.

Although the sequence of morphologic changes described above is widely accepted, the spatiotemporal pattern with which the final breakdown of the axon occurs has long been controversial. In large part, this reflects the technical difficulty of reliably distinguishing individual axons within a nerve during a prolonged period of observation and the fact that when axonal degradation occurs, it occurs quite rapidly. Early studies produced contradictory results, with some investigators reporting that distal axon segments degenerate in a proximal-to-distal direction and others reporting that degeneration occurred either distally to proximally or simultaneously throughout the entire length of the axon. It was also suggested that the type of experimental injury used affected the pattern, with crush injury reported to produce degeneration that proceeded proximally to distally and cut injury resulting in simultaneous degradation of the entire distal axon segment. These issues have recently been clarified by studies performed with transgenic mice that express either yellow fluorescent protein (YFP) or green fluorescent protein (GFP) in small subsets of axons. When nerves in these animals were transected and axonal degeneration followed by imaging of nerve segments, it was found that axons transition from a fully intact state to a fully degraded state quite rapidly, with axons identifiable in a partially degraded state for less than an hour. Sequential imaging of the partially degraded axons indicated that degeneration occurred in a distal to proximal direction after a cut injury, while axon degeneration after a crush injury proceeds proximally to distally.

### Schwann Cell Changes Following Nerve Transection

In noninjured peripheral nerve, myelinated and unmyelinated Schwann cells are mitotically quiescent differentiated elements whose primary function is to facilitate axonal conduction. Following injury, however, these glia dedifferentiate, proliferate, and activate a genetic program that enhances and guides the growth of regenerating axonal sprouts through the distal nerve segment and back to their original peripheral targets. It has long been believed that these changes in the post axotomy phenotype of Schwann cells are initiated by a positive signal sent from the injured axon. At this time, however, the nature of this axon-derived signal is unknown.

In the early stages of Wallerian degeneration, myelinated Schwann cells mediate the initial breakdown of their myelin sheath and associated distal axon segments. Minutes after nerve transection, widening of Schmidt–Lanterman incisures (channels in Schwann cells that run transversely across spiraled compact myelin layers and allow these layers to communicate with one another) can be observed in myelinated Schwann cells immediately distal to the injury site; by 24–36 h after injury, this same alteration is evident in Schwann cells throughout the more distal portions of the nerve segment. As the Schmidt–Lanterman incisures open up, the cytoplasm of myelinated Schwann cells begins to swell and compress associated axons. Twenty-four to forty-eight hours after nerve transection, myelin destruction begins. By light microscopy, the initial sign that myelin destruction is underway is the fragmentation of the myelin sheath into myelin ovoids that are contained in compartments in the Schwann cell cytoplasm known as digestion chambers (**Figure 1**). As Wallerian degeneration proceeds, digestion chambers are often found to also contain fragments of axoplasm. In subsequent days and weeks, myelin ovoids are further fragmented and cleared as macrophages begin to assist in the clearance of myelin and axonal debris (see below). Degeneration in unmyelinated Schwann cells occurs in a similar fashion except, of course, that myelin ovoids are not seen.

During development, Schwann cells that do not establish axonal contact undergo apoptosis. However, at least during the first few weeks post axotomy, apoptosis is not evident in the population of adult Schwann cells within denervated nerve segments.

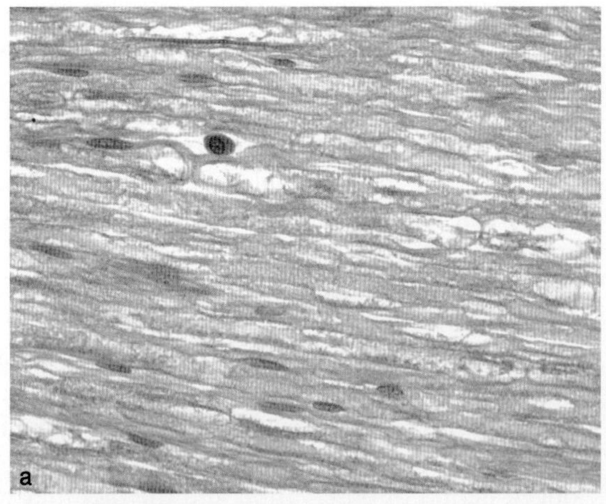


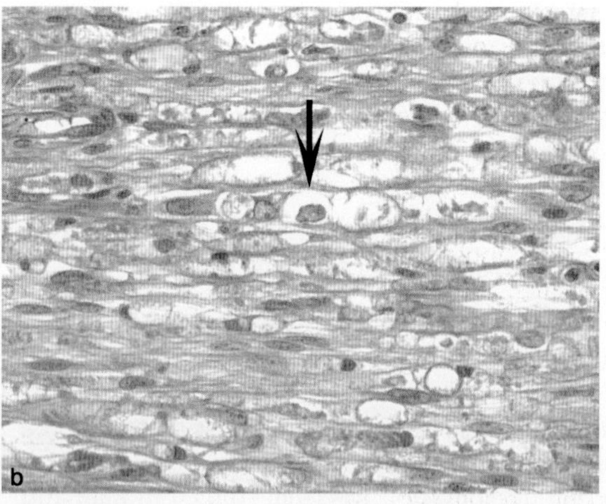


**Figure 1** Morphologic changes in peripheral nerve undergoing Wallerian degeneration. (a) Noninjured rat sciatic nerve. Note the relatively low cellularity and the orderly alignment of axons and their associated myelin sheaths. (b) Rat sciatic nerve distal to a site of crush injury performed 5 days previously. Cellularity is now increased relative to the noninjured nerve illustrated in (a) as the result of macrophage infiltration and Schwann cell proliferation. The arrow indicates one of the numerous Schwann cell digestion chambers evident in this field. A myelin ovoid is present within the indicated digestion chamber (the red roughly oval mass).

Indeed, adult Schwann cells, which heretofore have been mitotically quiescent, begin to proliferate rapidly within the confines of their original basal lamina. In rodents, Schwann cell mitogenesis peaks 3–4 days post axotomy and gradually declines thereafter, with only scattered proliferating Schwann cells evident in the distal nerve segment by 18 days after injury. As a result of this Schwann cell proliferation and accompanying macrophage infiltration (see below), the distal nerve segment becomes obviously hypercellular when examined by light microscopy. The proliferating Schwann cells layer upon one another within their basal lamina to form structures known as bands of Büngner or bands of von Büngner. The bands of Büngner are critically important for subsequent axonal regeneration as they form a substrate that will provide both structural and biochemical support for newly elaborated axonal sprouts as they elongate (see below).

### Macrophage Recruitment and Action Following Nerve Transection

Injuries occurring at most sites in the body elicit an acute inflammatory response involving polymorphonuclear leukocytes which is subsequently replaced by a chronic inflammatory infiltrate rich in lymphocytes. Distal nerve segments undergoing Wallerian degeneration, however, typically contain only occasional polymorphonuclear leukocytes and lymphocytes. Instead, macrophages are the primary inflammatory cell type found in degenerating peripheral nerve. Studies using chimeric rats in which the grafted marrow cells have been GFP-tagged have demonstrated that resident macrophages are normally present in the endoneurium and that these cells become activated during Wallerian degeneration. Indeed, resident tissue macrophages are more abundant in peripheral nerve than is commonly appreciated, representing an estimated 2–9% of the endoneurial cell population in noninjured nerve. The action of the activated resident macrophages is complemented by an influx of large numbers of hematogenous macrophages that are recruited into degenerating distal nerve segments by a variety of signals including degenerating myelin (likely the most important signal), complement components, adhesion molecules, and factors derived from injured axons. Macrophage infiltration is also facilitated by opening of the blood–nerve barrier; this is triggered by the release of histamine, serotonin, and other substances from resident mast cells, which proliferate *in situ* during Wallerian degeneration. Magnetic resonance imaging studies following the fate of circulating macrophages tagged with superparamagnetic iron oxide particles and analyses of chimeric mice with GFP-tagged bone marrow cells demonstrate that hematogenous macrophages begin to infiltrate distal nerve segments 1 day after injury, with infiltration peaking 4 days post axotomy and markedly declining by 8 days after injury. Macrophage infiltration begins at the lesion site and proceeds in a proximodistal direction, paralleling the breakdown of the myelin sheath.

Once in the distal nerve segment, macrophages replace Schwann cells as the primary phagocytes. Unlike Schwann cells, whose phagocytosis of myelin is lectin mediated and opsonin independent, macrophage phagocytosis of myelin debris is dependent upon their expression of complement receptor type 3 and the presence of complement component C3 on degenerating myelin sheaths. Macrophages can phagocytose myelin debris lying free in extracellular spaces or they can remove this debris directly from the Schwann cell's cytoplasm via a poorly understood mechanism. Macrophages also facilitate the clearance of myelin debris by synthesizing and secreting apolipoprotein E, which scavenges lipid products for future reutilization. After macrophages finish clearing myelin and axon debris from the distal nerve segment, they are eliminated from the nerve both by local apoptosis and by re-entering the circulation and then migrating to regional lymph nodes and the spleen.

### Neuronal Regenerative Responses Following Nerve Transection

In contrast to development, where loss of neuronal contact with its peripheral target frequently results in neuronal apoptosis, adult neurons that have undergone axonal transection most commonly survive. Shortly after nerve transection, the neuronal Nissl substance (rough endoplasmic reticulum) begins to disperse (a process known as chromatolysis), heralding a fundamental shift in their metabolism as they prepare for regeneration. Then, as the bands of Büngner and their associated supportive microenvironment form, neurons begin to put out regenerative sprouts from the stump of the axonal segment still associated with the neuronal cell body and from nearby nodes of Ranvier. Neurons typically put out many more sprouts than will be needed, with up to 25 sprouts arising from a single axon. These sprouts will, if possible, extend across the gap in the injured nerve to establish contact with the bands of Büngner in the distal nerve segment. Once this contact is established, the axonal sprouts begin to elongate along either the surface of Schwann cells within the band of Büngner or along the inner surface of the basal lamina that forms the boundary of these structures. It is clear that Schwann cells play an important supportive role in this process; irradiating distal nerve segments so as to kill Schwann cells within the bands of Büngner markedly impairs axonal regeneration. On the other hand, it is also evident that Schwann cells are not the sole element within the distal nerve segment that promotes axonal regeneration. Regenerating axons do continue to grow along the basal lamina in irradiated nerve segments, albeit with a reduced rate of extension. Furthermore, *in vivo* imaging of GFP- and YFP-tagged axons regenerating after a crush injury (a injury which minimally disrupts the basal lamina) has provided an elegant demonstration that physical cues, likely provided at least in part by the basal lamina, allow regenerating axons to precisely retrace their original pathway and reform appropriate synaptic contacts with skeletal muscle.

As axonal sprouts re-establish contact with Schwann cells, the Schwann cells respond to axonal cues by beginning anew to elaborate myelin sheaths. Since multiple axonal sprouts are initially present and elongating along the band of Büngner, this results in the formation of groups of small, thinly myelinated axons (regenerative clusters) within a single basal lamina. With time, one of these axons will become dominant and the other sprouts will regress. The mechanisms responsible for precisely matching axon-Schwann cell numbers during regeneration are not understood.

## Molecular Alterations Occurring during Wallerian Degeneration of Peripheral Nerve

Although it may seem paradoxical for a process that is described as a degeneration, Wallerian degeneration is accompanied by the increased expression of a number of molecules in the distal nerve segment. Over the last three decades, a variety of techniques have been used to identify transcripts whose expression is altered by nerve injury, including direct examination of individual candidate mRNAs, subtractive and differential hybridization and, more recently, microarray analyses. When examined at the tissue level, changes detected by these techniques reflect the sum of multiple alterations in gene expression occurring in multiple cell types endogenous to the nerve such as Schwann cells, fibroblasts, resident tissue macrophages, mast cells, and perineurial cells. Many of the apparent alterations in gene expression occurring during the first 7–8 days post axotomy also reflect the increase in the number of hematogenous macrophages migrating into the distal nerve segment. However, the majority of the axotomy-responsive genes identified to date are of Schwann cell origin. Axotomy-responsive genes expressed by Schwann cells encode multiple functional classes of proteins. These proteins include factors that promote demyelination, macrophage recruitment and macrophage activation, molecules that encourage axonal regeneration such as neurotrophic factors and cell adhesion molecules, and, as remyelination begins to occur, proteins necessary for myelin structure and synthesis.

By 2 days after nerve transection, myelinating Schwann cells have markedly decreased their expression of mRNAs encoding myelin proteins such as peripheral myelin protein-22 (PMP-22), myelin protein zero ($P_0$), myelin basic protein, myelin-associated glycoprotein (MAG), periaxin, and proteolipid protein (PLP). Following this decrease in the expression of myelin protein transcripts, Schwann cells in the distal nerve segment dedifferentiate to a state characterized by increased expression of the low affinity neurotrophin receptor ($p75^{LNTR}$), the intermediate filament glial fibrillary acidic protein, and the cell adhesion molecules L1 and neural cell adhesion molecule (NCAM) and decreased expression of Pax3, a transcription factor that represses myelin gene expression, and the transcription factor Krox-20. The phenotype of these dedifferentiated Schwann cells is similar to that of premyelinating and nonmyelinating Schwann cells. Because of this, the phenotype of dedifferentiated Schwann cells is often equated with that of Schwann cells during development. Such comparisons are potentially hazardous. While it is true that these cell types demonstrate many similarities, it is also clear that there are some important differences between dedifferentiated adult Schwann cells and developing Schwann cells (e.g., the ability of adult Schwann cells to survive when axon contact is lost).

As Schwann cells in the distal nerve segment dedifferentiate, they also begin to elaborate a variety of cytokines capable of recruiting macrophages such as interleukin-1$\beta$ (IL-1$\beta$), stromal cell-derived factor 1 (SDF-1), tumor necrosis factor-$\alpha$ (TNF-$\alpha$), macrophage inflammatory protein-1$\alpha$ (MIP-1$\alpha$), and monocyte chemoattractant protein-1 (MCP-1). As macrophages are recruited into the injured nerve segment, they too begin to elaborate TNF$\alpha$, interleukin-1$\alpha$ (IL-1$\alpha$), and IL-1$\beta$. Injecting neutralizing antibodies that inhibit the action of MCP-1 or MIP-1$\alpha$ both reduces macrophage recruitment into degenerating peripheral nerve and impairs phagocytosis of myelin debris. Although IL-1$\beta$ neutralizing antibodies do not effectively inhibit macrophage recruitment, they do similarly interfere with myelin phagocytosis. The role played by other cytokines expressed in degenerating nerve is still incompletely understood.

Dedifferentiated Schwann cells in degenerated peripheral nerve also begin to express a variety of neurotrophic factors such as nerve growth factor (NGF), brain-derived neurotrophic factor (BDNF), leukemia inhibitory factor, fibroblast growth factor-2 (FGF-2), insulin-like growth factor I (IGF-I) and its receptor, and glial cell-derived neurotrophic factor (GDNF) and its GFR$\alpha$ receptor subunit. These molecules are thought to provide trophic and tropic support for regenerating axons, enticing and guiding them along the bands of Büngner so that physiologic function can be restored. Dedifferentiated Schwann cells likewise begin to express increased levels of cell surface molecules such as NCAM, L1, ninjurin-1, ninjurin-2, tenascin, and N-cadherin. These cell adhesion molecules mediate intercellular interactions and thus likely play a role in guide elongating axons along the length of denervated nerve segments.

Some recent observations have provided insight into the identity of the molecules that trigger demyelination in axotomized peripheral nerve. Proteins in the neuregulin-1 (NRG-1) family of growth and differentiation factors potently regulate the proliferation, survival, migration, and differentiation of developing Schwann cells *in utero* and during early postnatal life. Schwann cell expression of class II NRG-1 isoforms (also known as glial growth factors; GGFs) and their erbB receptors (erbB2 and erbB3) is induced in injured sciatic nerve beginning at 3 days post axotomy, a time that coincides with the onset of Schwann cell proliferation induced by nerve injury, and continuing to at least 30 days post axotomy. Tyrosine phosphorylation (an indirect indication of activation) of erbB2 and erbB3 becomes evident in nerve with a similar time course. Based on these observations, it was initially suggested that increased class II NRG-1 expression and erbB receptor activation promoted Schwann cell mitogenesis in injured nerve and potentially had other effects such as triggering demyelination. Consistent with these observations, transgenic mice overexpressing the class II NRG-1 isoform GGF$\beta$3 in myelinating Schwann cells demonstrate Schwann cell hyperplasia and develop a demyelinating hypertrophic neuropathy. While GGF$\beta$3 clearly is promitogenic for Schwann cells in adult nerve, it may not be required for proliferation as transgenic mice with an inducible genetic ablation of erbB2 in adult Schwann cells do not demonstrate a decrease in post axotomy Schwann cell proliferation. However, it is quite clear that treating myelinated Schwann cell–neuron co-cultures with GGF induces demyelination in a concentration-dependent manner, consistent with the phenotype of GGF$\beta$3 overexpressing mice. Intriguingly, the effects that NRG-1 exerts on Schwann cells may be distinct for different NRG-1 subfamilies. In contrast to the demyelination observed in transgenic mice overexpressing class II NRG-1 in Schwann cells, transgenic mice overexpressing class III NRG-1 in neurons develop hypermyelination rather than demyelination. In further support of the hypothesis that NRG-1 effects on myelination are subfamily specific, transgenic mice overexpressing class I NRG-1 in neurons show no abnormalities of myelination. These latter studies have focused on NRG-1 effects on myelination during development; it remains to be determined whether class III NRG-1 isoforms are critical for the re-establishment of myelination in regenerating nerve.

## A Comparison of Wallerian Degeneration in the PNS and CNS

Transection of axons whose length lies entirely within the CNS also results in degeneration of the distal nerve segment. However, both the specifics of this CNS process and the ultimate outcome are clearly distinct from those seen with Wallerian degeneration in the PNS. Because of this, purists argue that it is not appropriate to refer to posttraumatic axonal degeneration in the CNS as Wallerian degeneration. In contrast to this perspective, a growing body of evidence indicates that the mechanisms underlying the active degeneration of transected axon segments (see below) are similar in the PNS and CNS. Consequently, it is increasingly clear that there are at least some important similarities between Wallerian degeneration in the PNS and in the CNS.

In both the PNS and the CNS, degeneration of the severed distal axon segment occurs before the destruction of myelin sheaths associated with the degenerating axon. However, thereafter these processes diverge. Wallerian degeneration in the CNS is slower than Wallerian degeneration in the PNS. Furthermore, oligodendrocytes, unlike Schwann cells, do not form structures equivalent to the bands of Büngner and thus fail to establish a microenvironment supportive of axonal regeneration. This latter difference clearly reflects the distinct nature of the relevant glia (and the presence of inhibitory molecules such as Nogo in the CNS) rather than differences in the PNS and CNS neuronal populations; in a classic set of experiments, Aguayo and colleagues showed that CNS neurons will extend processes readily into peripheral nerve grafts transplanted into a CNS injury site. Myelin debris is also removed more slowly during Wallerian degeneration occurring in the CNS. This may be because the macrophages responding to CNS Wallerian degeneration are not hematogenously derived and thus likely differ functionally from the macrophages responding to nerve injury. Given these differences, it seems most appropriate to equate the similarities in axonal degeneration seen in CNS and PNS Wallerian degeneration and to be extremely cautious when comparing other aspects of Wallerian degeneration in the PNS and CNS.

## Recent Findings Indicating that Axonal Degeneration Results from Active Self-Destruction Rather than Simple Atrophy of Severed Axons

For decades after the initial description of Wallerian degeneration, it was thought that degeneration of distal axon segments occurred because axonal transport had been interrupted, resulting in atrophy that was secondary to the disruption of metabolic communication between the axon and the neuron cell body. This notion was challenged by the unexpected discovery of a spontaneous mouse mutant that demonstrated abnormal Wallerian degeneration in both the PNS and the CNS. In these Wallerian degeneration slow mice (*Wld*$^s$ mice, previously known as C57BL/Ola mice), the distal segments of transected axons remain viable and capable of conducting action potentials for up to 3 weeks. Experiments with bone marrow chimeric and sciatic nerve chimeric mice have demonstrated that prolonged axon survival in *Wld*$^s$ mice results from an alteration intrinsic to the axon itself. The *Wld*$^s$ mutation is inherited as an autosomal dominant trait that maps to the distal arm of chromosome 4. An 85 kb tandem triplication present at this site encodes a chimeric nuclear protein in which the $NH_2$-terminal 70 amino acids of the 1173-residue-long E4b ubiquitination factor (UbE4b; also known as ubiquitin fusion degradation protein 2a (Ufd2a)) is fused in frame to the entire coding sequence of the NAD+ synthesizing enzyme nicotinamide adenylyltransferase 1 (NMNAT1). Expression of this chimeric gene under the control of a $\beta$-actin promoter confers the *Wld*$^s$ phenotype on wild-type mice. Considered together, these findings indicate that axonal degeneration results from the action of an active axon death program rather simply a passive atrophic process. The localization of the UbE4b/NMNAT1 fusion protein to neuronal nuclei implies that this mutant protein acts indirectly via other as yet unknown molecules that are localized to axons.

In addition to its intriguing effects on Wallerian degeneration, the phenotype of the *Wld*$^s$ mutant mouse also has important implications for the mechanisms underlying other nervous system disease states in which the axon of a diseased neuron progressively degenerates from the distal portions of the axon toward the cell body. These dying back axonal degenerations occur in a variety of medically important peripheral neuropathies (e.g., diabetic neuropathy) and have been suggested to contribute to neurodegeneration in a variety of CNS diseases such as Alzheimer's disease, Parkinson's disease, and motor neuron disease. Consistent with this latter hypothesis, dying back axonal degeneration is delayed in progressive motor neuronopathy (*pmn*) mice crossed to *Wld*$^s$ mice. Similar results have been obtained by crossing *Wld*$^s$ mice to gracile axonal dystrophy (*gad*) mice or mice with a null mutation of the $P_0$ gene (a model of hereditary motor and sensory neuropathy IB/Charcot–Marie–Tooth disease type 1B), two other models thought to involve a dying back axonopathy. The *Wld*$^s$ gene also confers resistance to peripheral neuropathy induced by vincristine toxicity. The ability of the *Wld*$^s$ mutation to inhibit the progress of these PNS and CNS diseases thus provides a clear mechanistic link between Wallerian degeneration and dying back axonopathies.

Several lines of evidence indicate that the axon death program is distinct from the apoptosis program typically associated with neuronal cell death. Culture systems in which the distal axons of wild-type sympathetic neurons are locally deprived of NGF results in degeneration of the distal axon while the remainder of the axon and the cell body survive. In contrast, in *Wld*$^s$ sympathetic neurons deprived of NGF, axons remain viable long after the cell body has undergone an apoptotic death. These observations are consistent with the finding that inhibitors of caspase action do not prevent Wallerian degeneration and that activated caspase-3 and cleavage products resulting from caspase action are not found in transected axons. In addition, mutations that prevent apoptosis such as overexpression of the anti-apoptotic molecule Bcl2 or genetic ablation of the pro-apoptotic molecules Bax and Bak do not prevent axonal degeneration. Finally, the UbE4b/NMNAT1 fusion protein encoded by the *Wld*$^s$ gene bears no structural similarity to any of the known anti-apoptotic proteins.

At present, little is known about the mechanism or the downstream molecules involved in the axon death program. Initial attempts to establish the function mediated by the UbE4b/NMNAT1 fusion protein essential for delaying Wallerian degeneration focused on the possibility that this protein modulates the action of the ubiquitin–proteasome system. Consistent with this hypothesis, inhibiting the action of the ubiquitin–proteasome system pharmacologically or by overexpression of the ubiquitin protease UBP2 has been found to delay Wallerian degeneration in cultured sympathetic neurons and in transected optic nerve *in vivo*. However, it was subsequently found that overexpressing the UbE4b fragment from the UbE4b/NMNAT1 fusion protein in dorsal root ganglion neurons was not sufficient to delay axonal degeneration following transection or vincristine treatment. In contrast, both overexpression of NMNAT1 and administration of NAD prior to injury delays axonal degeneration induced by both of these treatments. Both NMNAT1 overexpression and pretreatment with NAD seem to act through the NAD-dependent deacetylase SIRT1 as inhibiting the action of this deacetylase by RNA interference or with a small molecular inhibitor blocks the delay in Wallerian degeneration; enhancing SIRT1 action with a specific activator similarly delays Wallerian degeneration. Of note, SIRT1, like the UbE4b/NMNAT1 fusion protein, is expressed in the nucleus. Consequently, the axonal effectors cooperating with the UbE4b/NMNAT1 fusion protein to delay Wallerian degeneration are as yet completely unknown.

*See also:* Proteasome Role in Neurodegeneration.

## Further Reading

Araki T, Nagarajan R, and Milbrandt J (2001) Identification of genes induced in peripheral nerve after injury: Expression profiling and novel gene discovery. *Journal of Biological Chemistry* 276: 34131–34141.

Araki T, Saski Y, and Milbrandt J (2004) Increased nuclear NAD biosynthesis and SIRT1 activation prevent axonal degeneration. *Science* 305: 1010–1013.

Beirowski B, Adalbert R, Wagner D, et al. (2005) The progressive nature of Wallerian degeneration in wild-type and slow Wallerian degeneration (Wlds) nerves. *BMC Neuroscience* 6: 6.

Bendszus M and Stoll G (2003) Caught in the act: *In vivo* mapping of macrophage infiltration in nerve injury by magnetic resonance imaging. *Journal of Neuroscience* 23: 10892–10896.

Coleman M (2005) Axon degeneration mechanisms: Commonality amid diversity. *Nature Reviews Neuroscience* 6: 889–898.

Guertin AD, Zhang DP, Mak KS, Alberta JA, and Kim HA (2005) Microanatomy of axon/glial signaling during Wallerian degeneration. *Journal of Neuroscience* 25: 3478–3487.

Huijbregts RPH, Roth KA, Schmidt RE, and Carroll SL (2003) Hypertrophic neuropathies and malignant peripheral nerve sheath tumors in transgenic mice overexpressing glial growth factor-$\beta 3$ in myelinating Schwann cells. *Journal of Neuroscience* 23: 7269–7280.

Kerschensteiner M, Schwab ME, Lichtman JW, and Misgeld T (2005) *In vivo* imaging of axonal degeneration and regeneration in the injured spinal cord. *Nature Medicine* 11: 572–577.

Kuhlmann T, Bitsch A, Stadelmann C, Siebert H, and Brück W (2001) Macrophages are eliminated from the injured peripheral nerve via local apoptosis and circulation to regional lymph nodes and the spleen. *Journal of Neuroscience* 21: 3401–3408.

Mack TGA, Reiner M, Beirowski B, et al. (2001) Wallerian degeneration of injured axons and synapses is delayed by a *Ube4b/Nmnat* chimeric gene. *Nature Neuroscience* 4: 1199–1206.

Michailov GV, Sereda MW, Brinkmann BG, et al. (2004) Axonal neuregulin-1 regulates myelin sheath thickness. *Science* 304: 700–703.

Mueller M, Leonhard C, Wacker K, et al. (2003) Macrophage response to peripheral nerve injury: The quantitative contribution of resident and hematogenous macrophages. *Laboratory Investigation* 83: 175–185.

Nguyen QT, Sanes JR, and Lichtman JW (2002) Pre-existing pathways promote precise projection patterns. *Nature Neuroscience* 5: 861–867.

Zanazzi G, Einheber S, Westreich R, et al. (2001) Glial growth factor/neuregulin inhibits Schwann cell myelination and induces demyelination. *Journal of Cell Biology* 152: 1289–1299.

Zhai Q, Wang J, Kim A, et al. (2003) Involvement of the ubiquitin–proteasome system in the early stages of Wallerian degeneration. *Neuron* 39: 217–225.

# Synaptic Plasticity: Neuronogenesis and Stem Cells in Normal Brain Aging

**G M Bernal and D A Peterson**, Chicago Medical School at Rosalind Franklin University of Medicine and Science, North Chicago, IL, USA



## Introduction

Although the postnatal brain had long been thought to lack the capacity to generate new neurons, the past decade's research has radically changed this assumption with ample evidence that new neurons are generated in some regions of the adult brain. This process is referred to as neuronogenesis for the specific production of neurons and more broadly as neurogenesis when all neural cell types may be generated. In many cases the terms neuronogenesis and neurogenesis are used interchangeably, as they will in this article. Neural stem cells and neurogenesis are restricted to specific areas in the adult brain: the subventricular zone/rostral migratory stream/olfactory bulb system and the dentate gyrus of the hippocampus. This spatial restriction means that most of the brain does not support neurogenesis under normal conditions and is thus referred to as nonneurogenic. Understanding the capacity of different regions to support survival and integration of neural stem cells will be of importance for the potential use of stem cells for brain repair, as it is this wider, nonneurogenic portion of the brain that will most frequently need to be the target of repair strategies.

Within the young adult brain, neurogenesis routinely contributes large numbers of new neurons to the olfactory bulb and small numbers of new neurons to the dentate gyrus. Many studies have shown that the rate of neurogenesis in the young adult brain is not fixed, but subject to stimulation by experimental manipulation. Neuropathological studies of the aging central nervous system (CNS) frequently utilize young adult, rather than aged, animals in experimental models or transgenic mutations of senescence-associated diseases, such as Parkinson's, Alzheimer's, or Huntington's diseases. Less is known about the underlying changes leading to the decline in normal (nonpathological) brain aging, yet this characterization will be necessary to interpret the contribution of specific disease pathology to impaired brain function. Here we look at the effects of normal aging on neurogenesis in the adult mammalian brain, the process in which resident neural stem cells give rise to new functional neurons that are incorporated into the circuitry. We discuss how age-induced changes may modulate neural stem cell activity resulting in a significant decline in neurogenesis and the implications of this for the aging brain.

## Characterization of Neural Stem Cells

Many postnatal mammalian tissue systems contain cells that possess the features consistent with the definition of a stem cell, that is, the capacity for unlimited self-renewal and the ability of their progeny to differentiate into different cellular types. Stem cells present during embryonic development have considerable flexibility in sculpting tissues; however, this capacity may be restricted in postnatal tissue. In fact, it is a matter of debate as to whether stem cells found in certain adult tissues, such as the CNS, are truly stem cells or would be better termed progenitor cells, reflecting their more restricted proliferate capacity and differentiation potential to the same tissue in which they reside.

At present, there is no single definitive marker by which neural stem cells in tissue can be identified. Identification of stem/progenitor cells in adult tissue is usually accomplished through labeling the proliferating cells and assessing their subsequent lineage outcome. Most *in vivo* studies of neurogenesis have used administration of the thymidine analog bromodeoxyuridine (BrdU) to identify new cells. The duration of BrdU delivery and the interval between administration and examination are both of critical importance in interpreting the cell type under study. For example, sustained delivery of BrdU over several days labels a population of proliferating cells, but obscures the ability to distinguish whether those proliferating cells are slowly dividing or rapidly dividing, or whether the labeled cohort contains a heterogeneous population of both cell types. Therefore, many studies with longer BrdU administration do not provide information discriminating between slowly cycling stem cells and rapidly cycling progenitor cells. The ability to ascertain the frequency of reentry into cell cycle is of importance for understanding the contribution of this parameter to age-related changes in neurogenesis.

Another important parameter to establish when assessing neurogenesis is to extend beyond the quantification of changes in BrdU-positive cell number by also assessing their fate to determine the frequency at which they turn into neurons. As no single definitive marker exists for neural stem cells, identification of progression through lineage commitment is accomplished by combinatorial expression of various markers. There has been considerable progress in defining the sequential expression of lineage markers in the young adult CNS, but our knowledge of whether the same sequence of markers and duration of their expression are invariably expressed in the aging brain is less than complete. In addition, parameters of neurogenesis differ between mice and rats, suggesting that species differences should be kept in mind when extrapolating results from animal studies.

## Neurogenesis Is Decreased with Aging

Neurogenesis does persist in the aged brain, including that of elderly humans. Quantitative studies evaluating age-related changes have consistently reported lower levels of neurogenesis in both the subventricular zone and the dentate gyrus that become statistically significant by mid-age. In aged dentate gyrus of various rat or mouse strains, neurogenesis has been reported to be at the level of 10–20% of young adult animals. Olfactory bulb neurogenesis is significantly reduced to approximately 30% of neuronal production in the young brain and is accompanied by decreased proliferation in both the subventricular zone and the rostral migratory stream to approximately 20–40% of levels in the young brain. The decline of hippocampal neurogenesis is accompanied by a decrease in cell proliferation in addition to a decline in the number of cells expressing doublecortin, a marker associated with early neuronal lineage commitment. However, despite the age-related reduction in neurogenesis, the capacity of newly generated cells in the aged brain to migrate, survive, and differentiate into mature neurons is maintained. Furthermore, early work demonstrated that the age-related decline in dentate neurogenesis could be partially reversed in some conditions, such as environmental enrichment or alteration of glucocorticoid levels. Those reports suggest that neurogenesis in the aged brain is not absolutely diminished, but there remains a capacity to elevate neurogenesis, at least in the hippocampus.

Despite the reduced proliferation of new cells in the aged hippocampus, the proportion of those newly generated cells that differentiate into neurons is comparable to the young hippocampus. One interpretation of these data was that there are fewer neural stem cells in the aged brain generating the basal level of cell production. Studies using BrdU labeling of cell proliferation could not address this possibility, as they fail to discriminate between slowly dividing stem cells and rapidly dividing progenitor cells. However, recent work assessing expression of a stem cell marker, Sox2, in combination with other lineage markers indicates that the subpopulation of Sox2-positive cells that correspond most closely to the likely neural stem cell population is not reduced in the aging brain. These results suggest that the age-related reduction in neurogenesis is not due to there being fewer neural stem cells, but rather that these cells may be more quiescent in the aged CNS. In addition to quiescent stem cells producing fewer new cells in the aged brain, the maturation of newly generated neurons appears delayed and the extent of their dendritic arborization is less in the aged hippocampus.

## Modulation of Neurogenesis in the Aged Brain

The observed age-related decrease in neurogenesis may result from a combination of environmental changes and cell-intrinsic limitations with increasing age. The quiescence of neural stem cells in the aged hippocampus may be a cell-autonomous property, but proliferation in aged hippocampus can be stimulated under certain conditions. Thus, there is substantial evidence that environmental factors play an important role in the age-related reduction in neurogenesis.

Overexpression of trophic factors has been shown to enhance neurogenesis in the young brain. Protein levels for some of the key trophic factors (such as brain-derived neurotrophic factor (BDNF), fibroblast growth factor-2 (FGF-2), insulin-like growth factor 1 (IGF-1), and vascular endothelial growth factor (VEGF)) promoting neural stem cell proliferation, differentiation, and survival show a decline as early as mid-age, correlating with the observed decrease in the proliferation of stem cells and neurogenesis in the aged brain. These changes may be accounted for by alterations in gene expression and/or posttranslational modification. Enhanced availability of trophic factors has been explored as an attractive target for therapeutic manipulation to promote neurogenesis in the young brain. Increased neurogenesis in the young brain has been reported following delivery of BDNF, FGF-2, IGF-1, VEGF, and epidermal growth factor (EGF). The aged brain, likewise, can be stimulated to increase cell proliferation and possibly neurogenesis following delivery of FGF-2, BDNF, IGF-1, and heparin-binding EGF (HB-EGF). However, in neurogenic regions, any enhancement of neurogenesis appears to be modest and does not restore neurogenesis to the level seen in the young brain. These results suggest that the decline of trophic factor support alone may not be responsible for age-related changes.

Trophic support of cell proliferation and differentiation may be regulated by the availability of both ligand and the receptors to bind to those ligands. Changes in the expression of trophic factor receptors, either on the neural stem cells themselves or on other cells in the neurogenic niche, may contribute to reduced neurogenesis in the aged brain. Relatively little is known, at present, about possible changes in receptor expression on neural stem cells themselves, but there are indications that alteration of receptor expression occurs in other types of cells residing in the neurogenic niche, suggesting possible secondary effects on neural stem cell proliferation. Age-related changes in receptor expression would support possible cell-autonomous mechanisms for modulating neurogenesis through downstream alteration of the local environment. For example, there is reduced expression of FGFR-2 on astrocytes of the aged neurogenic niche, which may reflect an impairment of these cells to respond to FGF-2 signaling and/or in their autocrine regulation to produce FGF-2. This may, in turn, reduce the paracrine effect of FGF-2 on neural stem cells in the aged neurogenic niche.

Structural plasticity of dendrites and spines in the aging hippocampus can be significantly modified by alteration of hormone levels. Neurogenesis may also be modulated by other factors such as hormones and inflammatory molecules that show age-related changes. Despite the strong evidence for age-related hormone-induced plasticity in the hippocampus, relatively few studies have directly examined modulation of neurogenesis in this context (**Table 1**). Changes in glucocorticoid regulation do appear to influence neurogenesis in the aging brain. Stress has a potent effect on neurogenesis in both the young and aged brain, although the precise linkage with glucocorticoid as the proximal mechanism of that change is a matter of debate. Unfortunately, few studies on the important estrogen-related changes in the aging brain have investigated neurogenesis. There is evidence that neurosteroids, such as allopregnanolone, can promote neurogenesis in an Alzheimer's disease (AD) mouse model, although these mice were still relatively young. The role of hormonal modulation is an active area of investigation with important clinical implications, and assessment of neurogenesis is likely to be incorporated into future reports.

## Implications of Decreased Neurogenesis in the Aging Brain

While numerous studies have demonstrated that neurogenesis can be enhanced, the advancement of our understanding about the biology regulating neurogenesis has been hampered by an inability to experimentally block neurogenesis. In this regard, studying age-related neurogenesis provides a natural model of reduction that may reveal the key regulatory elements for sustaining the generation of new neurons.

**Table 1** Reported effects of hormonal alteration on neural stem cell proliferation and neurogenesis in the hippocampal dentate gyrus

| *Hormone* | *Natural age-related change* | *Experimental manipulation of hormonal levels* | *Outcomes* |
|---|---|---|---|
| Corticosterone | Increased | *Rodents* | |
| | | Intact young (exogenous corticosterone) | Decreased neurogenesis |
| | | Intact young (following various stressors) | Decreased cell proliferation and/or neurogenesis |
| | | Intact aged | Decreased neurogenesis |
| | | Intact aged (following psychosocial stress) | Additional decrease in neurogenesis |
| | | Adrenalectomy (decreased corticosterone) | |
| | | Early postnatal; tested young and aged | No effect on neurogenesis at either age |
| | | Young | Increased neurogenesis |
| | | Mid-aged- tested when aged | Increased neurogenesis |
| | | Aged | Increased neurogenesis (reversed by corticosterone) |
| | | *Primates* | |
| | | Young adult (following psychosocial stress) | Decreased cell proliferation |
| Gonadal hormones (estrogen) | Decreased in females | *Young female rodents* | |
| | | Ovariectomy (OVX; results in decreased estrogen): | |
| | | OVX | Decreased cell proliferation |
| | | OVX + single estradiol administration | At 1 week, increased cell proliferation<br>At 4 weeks, no effect on cell proliferation |
| | | OVX + single estradiol and progesterone administration | No effect on cell proliferation |
| | | OVX + chronic or cyclic estradiol administration | No effect on cell proliferation |
| | | *Aged female rodents* | |
| | | OVX ± chronic or cyclic estradiol administration | No effect on neurogenesis reported |
| | | *Aged female primates* | |
| | | OVX ± chronic or cyclic estradiol administration | No effect on neurogenesis reported |

While individual new neurons are generated in the adult brain that assume functional characteristics of mature neurons, the extent to which their integration in a larger neuronal network directly contributes to a functional outcome remains unclear. Numerous studies have sought to establish a link between the generation and addition of new neurons in the olfactory bulb and hippocampus and functional output from those systems. With its natural reduction in neurogenesis, the study of aging animals may provide one approach to addressing the question of function. Unfortunately, studies attempting to correlate the level of impairment in hippocampal-dependent learning tasks and the extent of age-related reduction in neurogenesis have yielded mixed results. Some results have suggested that the aged animals most impaired in behavioral performance had fewer new hippocampal neurons, and other studies have indicated that animals in this group had the highest levels of neurogenesis. However, stimulation of hippocampal neurogenesis in aged animals, such as enriched environment or running, has been reported to correlate with improved performance in hippocampal-dependent tasks. It remains to be demonstrated if this correlation of stimulated neurogenesis and improved performance is causal (i.e., due to increased neurogenesis) or if it is a secondary effect to other biological changes, such as improved blood flow, elevation of circulating cytokines, etc. Although models of hippocampal-dependent learning are relevant to assessing function of hippocampal neurogenesis, there is no evidence to date suggesting that neurogenesis plays a role in age-related mild cognitive impairment in humans.

The incidence of anosmia (decreased or absent sense of smell) is increased in elderly patients with attendant risks for malnutrition and food poisoning. Newly generated olfactory bulb neurons in young animals have been shown to integrate and possibly participate in processing of olfactory input, and it had been postulated that the age-related reduction of olfactory bulb neurogenesis may contribute to the development of anosmia. However, it has been reported recently that humans lack a defined rostral migratory stream, making the extent to which stem cells in the subventricular zone contribute to olfactory bulb neurogenesis in humans unclear. This report also creates uncertainty as to the extent to which age-reduced neurogenesis may contribute to anosmia in elderly humans.

The aging brain is vulnerable to a variety of insults and neurodegenerative disease. With our understanding that neurogenesis persists in the adult and aging brain, there has been enthusiasm for utilizing neural stem cells for structural brain repair, where dysfunctional or dead neurons are replaced by grafted neural stem cells. Much of the work in grafting of neural stem cells has been performed in young animal models. It will be important to assess the physiological changes in the aged brain that may constrain the ability of grafted neural stem cells to survive, differentiate, and integrate into the aged brain. In particular, more information is needed about the ability of nonneurogenic regions of the aged brain to support neural stem cell survival and differentiation, as it is primarily the nonneurogenic regions that will be the target of stem cell therapy. Establishment of these parameters will be critical for discriminating between nonpathological aging and the contribution of pathology in animal models of neurodegenerative disease. Further study of the regulation of neurogenesis in the aged brain should provide useful insights into the development of therapeutic strategies for repairing the aged brain.

*See also:* Neurogenesis in the Intact Adult Brain.

## Further Reading

Bernal GM and Peterson DA (2004) Neural stem cells as therapeutic agents for age-related brain repair. *Aging Cell* 3: 345–351.

Carson CT, Aigner S, and Gage FH (2006) Stem cells: The good, bad and barely in control. *Nature Medicine* 12: 1237–1238.

Eriksson PS, Perfilieva E, Bjork-Eriksson T, et al. (1998) Neurogenesis in the adult human hippocampus. *Nature Medicines* 4: 1313–1317.

Hattiangady B and Shetty AK (2006) Aging does not alter the number or phenotype of putative stem/progenitor cells in the neurogenic region of the hippocampus. *Neurobiology of Aging* Nov 6 [Epub ahead of print].

Jin K, Sun Y, Xie L, et al. (2003) Neurogenesis and aging: FGF-2 and HB-EGF restore neurogenesis in hippocampus and subventricular zone of aged mice. *Aging Cell* 2: 175–183.

Kempermann G, Kuhn HG, and Gage FH (1998) Experience-induced neurogenesis in the senescent dentate gyrus. *Journal of Neuroscience* 18: 3206–3212.

Rao MS, Hattiangady B, Abdel-Rahman A, Stanley DP, and Shetty AK (2005) Newly born cells in the ageing dentate gyrus display normal migration, survival and neuronal fate choice but endure retarded early maturation. *European Journal of Neuroscience* 21: 464–476.

Shetty AK, Hattiangady B, and Shetty GA (2005) Stem/progenitor cell proliferation factors FGF-2, IGF-1, and VEGF exhibit early decline during the course of aging in the hippocampus: Role of astrocytes. *Glia* 51: 173–186.

van Praag H, Shubert T, Zhao C, and Gage FH (2005) Exercise enhances learning and hippocampal neurogenesis in aged mice. *Journal of Neuroscience* 25: 8680–8685.

# Synaptic Plasticity: Learning and Memory in Normal Aging

**M Baudry**, University of Southern California, Los Angeles, CA, USA



## LTP/LTD and Learning and Memory

Since the first publication reporting the existence of long-term potentiation (LTP) by Bliss and Lomo, there has been a plethora of papers, books, symposia, and meetings devoted to the discussion of the relationship between this molecular/cellular form of synaptic plasticity and some behavioral forms of learning and memory. Although the issue may occasionally still be debated, there is little doubt that LTP indeed represents one of the cellular mechanisms that is used in a number of neuronal networks for the storage of new information. This has been particularly evidenced in the hippocampal formation, to which a wealth of pharmacological and genetic manipulations have been applied to attempt to demonstrate that LTP induction and maintenance are causally related to memory formation and duration. According to this body of work, LTP in hippocampus would be ideally suited for the formation of episodic or declarative memory. LTP has been observed in a large variety of mammalian species, including humans, clearly indicating the universality of this phenomenon.

On the other hand, the relationship between long-term depression (LTD) and learning and memory has not been as clearly established until now. The best case for such an association has been made in cerebellar learning, in which LTD of synaptic transmission at the parallel fibers to Purkinje cell synapses is likely to participate in certain forms of motor learning. In other brain structures, the potential link between LTD and learning of specific types of information is much more tenuous, and further work is needed to establish convincingly that LTD is necessary and sufficient for any particular form of learning. Note that another form of synaptic plasticity referred to as depotentiation has also been discussed as a potential mechanism for memory erasure. Whereas LTP is generally elicited by trains of high-frequency stimulation, depotentiation is generally elicited by trains of low-frequency stimulation delivered to pathways previously potentiated. Whether this form of plasticity is related to 'forgetting' is still a matter of speculation.

This article focuses on age-related alterations in LTP/LTD, on the one hand, and on learning and memory, on the other hand, and we argue that age-related cellular modifications leading to changes in LTP/LTD properties are also responsible for age-related changes in properties of some forms of memory. We first summarize the main findings regarding age-related changes in learning and memory. Changes in LTP/LTD properties are then reviewed. Finally, we discuss the potential cellular origins of age-related modifications in LTP/LTD and illustrate how these changes reflect basic age-related changes in cellular function.

## Learning and Memory and Aging

It has long been recognized that aging is associated with a loss of learning and memory; this has led to the definition of age-associated memory impairment and has been a strong stimulus for the development of new drugs that could remedy this impairment first in the extreme case of this condition – that is, Alzheimer's disease – but also in its milder version, known as mild cognitive impairment (MCI). However, it is only recently that this view has been modified with the appreciation that the decline in learning and memory is in fact taking place throughout the life span and is initiated very soon after the peak in learning and memory at the beginning of adulthood. It has been argued that mechanisms of learning and memory are critical during the developmental period when the organism needs to adapt to the environment, whereas the focus of adulthood is on the transmission of genes to the next generation. The extreme form of this argument is that evolution of memory was associated with that of the aging process. Such a view is similar to the idea that plasticity and age-related deterioration of neuronal function involve the opposite requirements for neurons to be stable to resist injuries and insults and to be plastic to continue the adaptation of the organism to a continuously changing environment. According to this view, the existence of mismatch between resources and requirements to satisfy these two sets of constraints results in the progressive impairment of synaptic plasticity during aging.

A wealth of experiments have documented age-related impairments in learning and memory in a wide variety of tasks in experimental animals as well as in humans. Barnes and colleagues in the late 1970s used learning to escape a circular platform to show age-related impairment in spatial learning in rats. Using the Morris water maze to study behavioral learning, numerous laboratories have since reported that cohorts of aged rats can be subdivided into learning-impaired and learning-nonimpaired animals. Reasons for this dichotomy are not clear, although it is intimately related to a similar dichotomy in plasticity properties at the cellular level. Age-related learning deficits have also been reported in mice both in spatial tasks and in fear conditioning. Trace eyeblink conditioning also requires hippocampal function, and impairment of this form of learning has been found in aged mice, rats, rabbits, and humans. Much less has been done regarding progressive memory decline in experimental animals during the life span, although results indicate that environmental events in early life can produce learning impairment in middle-aged rats.

Age-related impairment in delay eyeblink conditioning, a cerebellar-dependent form of learning, has also been found and, interestingly, is already present in middle-aged mice. It is also present in humans and, in fact, it might even be more dramatic than impairment in a verbal task, which would be considered more declarative and therefore more hippocampus dependent.

## Age-Related Changes in LTP/LTD

Although it was initially thought that cognitive decline with aging was due to massive neuronal loss and alterations in neuronal morphology, particularly dendritic branching, reexamination of this issue has led to the conclusion that aging is not associated with a general loss of neurons, dendritic alterations, or loss of synaptic contacts. Likewise, aging is not associated with gross alterations in biophysical properties of neurons in hippocampus. The most significant change in membrane biophysics associated with aging is an increased conductance for calcium, possibly due to an increased number of L-type voltage-dependent calcium channels. As discussed later, this change in calcium conductance may have a critical role in age-related changes in synaptic plasticity and learning and memory.

Although the intimate details of the mechanisms involved in LTP and LTD are not completely known, there is a growing consensus for the major events responsible for the induction and expression of the long-term increase (LTP) or decrease (LTD) in synaptic transmission elicited by the specific patterns of electrical stimulation required to produce them (**Figure 1**). In the case of LTP, brief bursts of electrical stimulation result in sufficient postsynaptic depolarization to relieve NMDA receptors from the voltage-dependent magnesium blockade of their associated channels. In turn, activation of NMDA receptors results in calcium influx in postsynaptic spines and the stimulation of a host of calcium-dependent processes, including calcium-dependent protein kinases (in particular, calcium/calmodulin protein kinase (CamKII)), calcium-dependent proteases, and calcium-dependent phospholipases. The combined action of these processes results in changes in spine cytoskeleton, the insertion of AMPA receptors in postsynaptic density, and the modification of cell adhesion proteins involved in pre/postsynaptic interactions. It has thus been proposed that dendritic spines modify their structures providing for increased insertion of AMPA receptors or, conversely, that a regulated process of receptor insertion imposes increased surface area for the postsynaptic density. Moreover, integrin activation has been proposed to represent a form of consolidation because interfering with this class of adhesion proteins within a short time after tetanus is capable of preventing LTP maintenance. It has also been proposed that protein synthesis as well as gene transcription participate in the late phase of activity-dependent changes in synaptic structure and function, although the exact nature of the genes activated and the proteins translated remains unknown. Numerous studies point to a critical role for brain-derived nerve growth factor (BDNF) in the translation of brief bursts of electrical activity into long-lasting changes in synaptic efficacy. BDNF would be released as a result of NMDA receptor activation and activate its membrane-associated receptor, trkB, in an autocrine-type manner. Stimulation of trkB would trigger a number of signaling pathways and, in particular, by regulating protein synthesis localized in dendrites, BDNF appears to be a regulator of synaptic consolidation. It is also noteworthy to mention the existence of non-NMDA receptor-dependent LTP that requires activation of the voltage-dependent calcium channels for its induction.

Although the detailed mechanisms underlying LTD are not as clearly established as those for LTP, several of the same processes participating in LTP have also been shown to be activated in LTD. Two major forms of LTD have been particularly well studied – an NMDA receptor-dependent form and a metabotropic glutamate receptor (mGluR)-dependent one. In both cases, internalization of AMPA receptors appears to be responsible for the maintenance of decreased synaptic efficacy.

Several studies have indicated alterations in both LTP and LTD properties in old animals, although there are differences in the details of these age-related modifications depending on the anatomical regions evaluated and the protocols used to elicit LTP or LTD. The first studies performed approximately 30 years ago indicated that following a strong stimulus to elicit LTP at the synapses of the perforant path to dentate gyrus or to CA3 pathways, LTP induction was similar

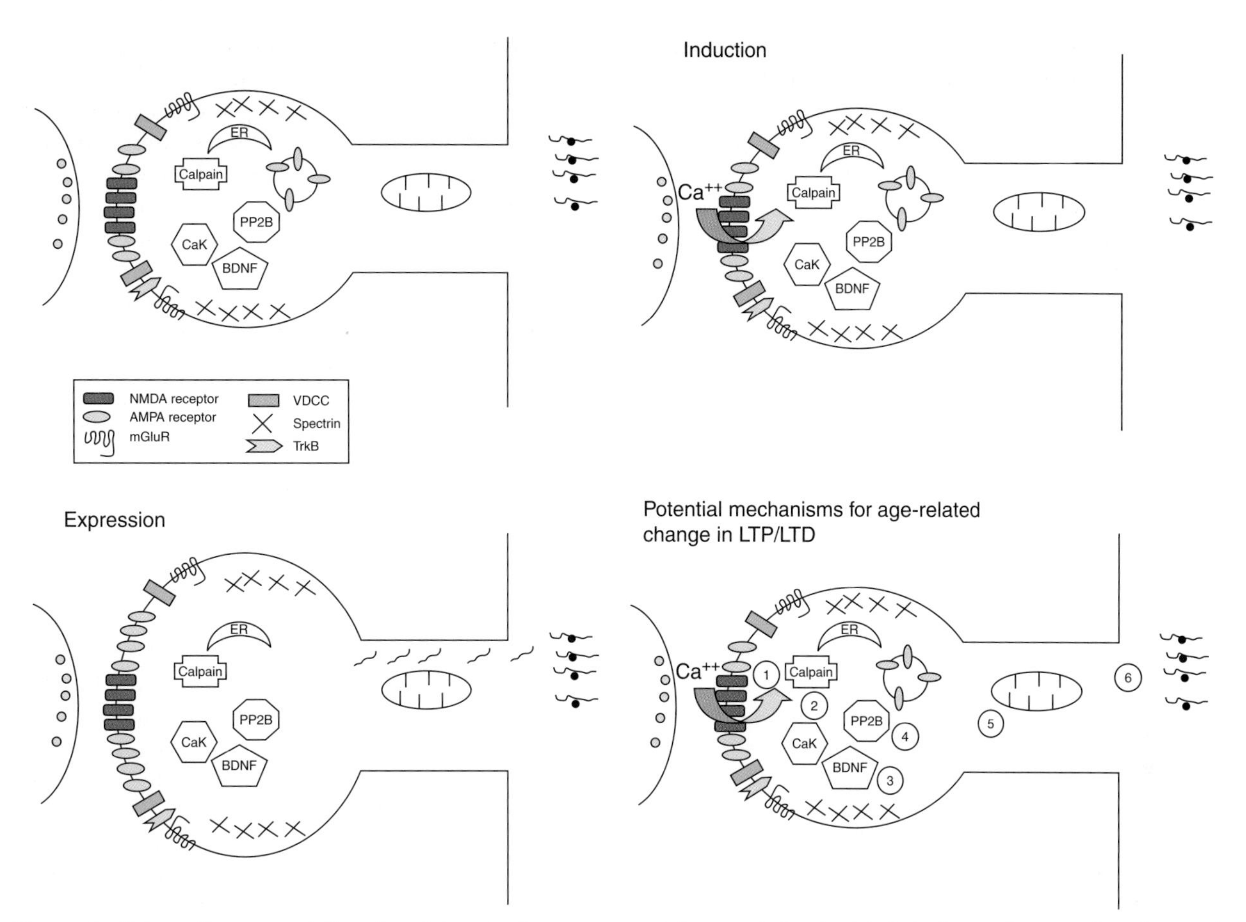


**Figure 1** Schematic representation of a glutamatergic synapse and the events involved in induction and expression of synaptic plasticity and potential mechanisms accounting for age-related impairment in synaptic plasticity. (Top left) Dendritic spine with AMPA and NMDA receptors as well as mGluRs and voltage-dependent calcium channels (VDCC). Also represented are calmodulin kinase (CaK), calcineurin (PP2B), and calpain. In addition, an intracellular pool of AMPA receptors is shown. BDNF and the trkB receptors are also present postsynaptically, as are mitochondria and endoplasmic reticulum (ER). Finally, ribosomes with mRNAs are represented in the base of the spine. (Top right) LTP/LTD induction. An influx of calcium is the major trigger for induction of synaptic plasticity. Depending on the amplitude and duration of the calcium signal, LTP or LTD will be elicited due to changes in the balance between protein kinase and protein phosphatase activity and calpain activation. Stimulation of BDNF release and activation of trkB also participate in the regulation of the changes in synaptic structure and function. Note that VDCC, ER, and mitochondria will participate in the regulation of the calcium signal generated by the stimulation protocol. (Bottom left) LTP expression. Increase in synaptic efficacy is due to an increase in the available number of AMPA receptors produced by the insertion of AMPA receptors in the postsynaptic membrane. This effect might be associated with a change in the morphology of the dendritic spine (note the enlarged spine head). Local translation of mRNAs might provide for newly synthesized proteins that could stabilize the modifications produced by proteolytic activation. (Bottom right) Potential mechanisms responsible for age-related impairment in synaptic plasticity. 1, Changes in NMDA receptor function or signaling; 2, changes in calcium regulation (VDCC, ER, and mitochondria); 3, changes in BDNF signaling; 4, changes in calcineurin activity; 5, changes in free radical production or elimination; 6, changes in protein synthesis.

in aged and young rats but decayed faster in the former than in the latter. On the other hand, altered LTP induction was observed at the Schaffer collateral to CA1 pathway. More recent studies have used induction protocols more closely related to natural stimulation frequencies, such as the so-called theta burst stimulation (TBS), in which brief bursts of high-frequency stimulation (generally 100 Hz) are delivered at the frequency of the theta rhythm (5–7 Hz). Under these conditions, LTP induction was impaired at the Schaffer collateral to CA1 synapses. Interestingly, using a 5-Hz stimulation protocol to induce LTP revealed the same dichotomy between subgroups of rats previously mentioned, with rats exhibiting learning impairment in the Morris water maze also exhibiting decreased amplitude of potentiation at these synapses. In addition, another form of NMDA receptor-dependent LTP elicited by prolonged paired-pulse stimulation delivered at 1 Hz appears in aged animals. Finally, one study indicated that, similar to the gradually occurring learning deficits over the life span, deficits in LTP can also be found in middle-aged rats; thus, although LTP induction was not modified in the basal dendrites of

CA1 pyramidal neurons of 7- to 10-month-old rats, it failed to stabilize and excitatory postsynaptic potentials returned to baseline levels within 30 min after TBS.

Age-related changes in LTD are almost the mirror images of those found for LTP. Thus, whereas LTD is difficult to elicit in hippocampal slices prepared from adult and middle-aged rats using prolonged trains of low-frequency stimulation (generally 1 Hz for 15 min), it is readily produced in hippocampal slices from aged rats. Again, an interesting dichotomy was reported regarding LTD properties in hippocampus of learning-impaired versus learning-unimpaired aged rats: learning-unimpaired rats 'switched' from NMDA receptor-dependent LTD to non-NMDA receptor-dependent LTD. No data are available regarding potential age-related changes in LTD in cerebellum.

## Potential Cellular Mechanisms Underlying Age-Related Changes in LTP/LTD

### Changes in NMDA Receptors

In view of the critical role of NMDA receptors in the induction of LTP and LTD, numerous studies have evaluated potential changes in NMDA receptor properties with age. Although several developmental changes in NMDA receptor properties have been described, it does not appear that changes in NMDA receptors account for age-related changes in LTP/LTD. However, because the NMDA receptor is part of a large complex of more than 200 proteins, it is plausible that changes in the organization/function of this complex take place during aging and modify some intracellular signaling cascades associated with activation of the NMDA receptors.

### Changes in Calcium Regulation

As mentioned previously, one of the most conspicuous age-related changes in neuron biophysical properties consists of a change in calcium regulation and, in particular, in an increased number of L-type voltage-dependent calcium channels. Although the increase in calcium channels does not affect basal intracellular calcium concentration, it produces a significantly higher calcium concentration as a result of synaptic activity in neurons from aged rats compared to that in young rats. In turn, this results in increased afterhyperpolarization (AHP) resulting from the activation of calcium-dependent potassium channels. Interestingly, the amplitude of the AHP is a good predictor of learning ability in rabbits, and learning results in decreased amplitude of the AHP. AHP decreases excitability and may thus be responsible for the age-related decrease in LTP because increased AHP would decrease the ability of trains of stimulation to depolarize the postsynaptic membrane sufficiently to activate the NMDA receptors. Results indicate that calcium-induced calcium release from intracellular stores is actually responsible for the increased AHP observed in aged animals, further incriminating calcium dysregulation as the primary cause for age-related impairment in synaptic plasticity. Interestingly, pharmacological or genetic manipulations directed at calcium-activated potassium channels eliminate age-related impairment in LTP and in cognitive function.

### Changes in BDNF Signaling

As discussed previously, BDNF has been implicated as a critical regulator of synaptic efficacy in hippocampus and some age-related alterations in synaptic plasticity could be due to age-related changes in BDNF levels or signaling. Conflicting reports regarding changes in expression and levels of BDNF have been published, but the overall view is that there are no significant changes in the levels of BDNF in hippocampus between 2 and 24 months of age. On the other hand, there appears to be a consensus regarding age-related decrease in trkB, the major receptor for BDNF, in hippocampus. Accordingly, BDNF-mediated effects should be significantly reduced in old animals. Understanding the precise role that this decreased signaling plays in age-related decrease in LTP/LTD requires further investigation. Interestingly, whereas increased physical activity does not seem to reverse age-related decrease in BDNF signaling, enriched environment does seem to do so.

### Changes in Calcineurin

Calcineurin is a calcium-dependent phosphatase that also plays a critical role in synaptic plasticity because the balance between protein kinase and protein phosphatase activity is a major determinant of the direction and amplitude of activity-dependent changes in synaptic efficacy. It is clear that calcineurin activity is increased with aging, and this has major consequences for the regulation of the state of phosphorylation of several targets of this phosphatase at synapses. Previous results have shown that calcineurin inhibition results in LTP that occludes tetanus-induced potentiation. Moreover, in aged rats, inhibition of calcium release from intracellular stores results in LTP, an effect that is blocked by calcineurin inhibition. It has thus been proposed that calcineurin indirectly regulates the phosphorylation of voltage-dependent calcium channels, thereby contributing to the regulation of cell excitability and LTP induction.

### Role of Oxygen Radicals

One of the key contributors to age-related changes in cellular function is oxidative stress, and it has been repeatedly shown that aging is associated with increased protein oxidation and lipid peroxidation as a result of increased formation of reactive oxygen species in the absence of increased levels of detoxifying enzymes (superoxide dismutase, catalase, and peroxidases) and scavengers of free radicals. In recent years, it has been proposed that free radicals exhibit opposite effects on synaptic plasticity, LTP in particular, in adult and aged rats. In the adult, low levels of reactive oxygen species, specifically $H_2O_2$, would be required for the normal induction of LTP, whereas high concentrations would be detrimental. Moreover, the effects of low concentrations of $H_2O_2$ on LTP would be mediated through the stimulation of calcineurin and the regulation of voltage-dependent calcium channels. On the other hand, reactive oxygen species would be mostly detrimental to LTP induction in the aged animals. In this case, high concentrations of $H_2O_2$ are presumed to stimulate calcium release from internal stores, resulting in activation of calcineurin and thus interfering with LTP induction. Furthermore, reactive oxygen species have been shown to interfere with proteolytic mechanisms such as the proteosome and the lysosomes. Studies have shown a critical role for the proteosome/ubiquitin proteolytic pathway in synaptic plasticity.

### Changes in Protein Synthesis

Whereas aging does not appear to be associated with major changes in LTP-induced gene expression, the transcription of c-fos appears to be decreased in CA1 pyramidal neurons from old rats. As discussed previously, BDNF enhances translation rate in dendrites, and the decrease in BDNF responsiveness with aging is likely to affect protein synthesis, thereby further contributing to age-related impairment in synaptic plasticity.

## Conclusions

It is clear that the data strengthen the argument that LTP and, to some extent, LTD represent cellular correlates of learning and memory. In fact, in several instances, the parallel is striking, as in the studies demonstrating the dichotomy in subgroups of rats between learning-impaired and learning-unimpaired with the similar dichotomy with impaired LTP or normal LTP. The data also provide better evidence for a role of LTD in learning and memory; in particular, in young rats, there is a significant correlation between the amplitude of NMDA receptor-dependent LTD and an index of performance in the learning task. Interestingly, in aged rats, successful learners switch from NMDA receptor-dependent LTD to non-NMDA receptor-dependent LTD. As more information has become available regarding the cellular and molecular mechanisms underlying LTP/LTD, it has become easier to assess which components of the complex machinery linking selective pattern of electrical activity to long-lasting changes in synaptic structure and function are exhibiting age-related alterations. Thus, it has become clear that since LTP/LTD use a number of basic mechanisms of cell biology involved in cell motility, cell signaling, and cell communication, age-related changes in basic cell function have significant impact on the features of LTP/LTD in aged organisms.

*See also:* Aging of the Brain; Aging: Brain Potential Measures and Reaction Time Studies; Cognition in Aging and Age-Related Disease.

## Further Reading

Barnes CA (1979) Memory deficits associated with senescence: A neurophysiological and behavioral study in the rat. *Journal of Comparative Physiology and Psychology* 93: 74–104.

Barnes CA (2001) Plasticity in the aging central nervous system. *International Review of Neurobiology* 45: 339–354.

Barnes CA (2003) Long-term potentiation and the ageing brain. *Philosophical Transactions of the Royal Society of London* 358: 765–772.

Baudry M and Thompson RF (1999) Synaptic plasticity: From molecules to behavior. In: Baudry M, Thompson RF, and Davis JL (eds.) *Advances in Synaptic Plasticity*, pp. 299–319. Cambridge, MA: MIT Press.

Bear MF and Malenka RC (1994) Synaptic plasticity: LTP and LTD. *Current Opinion in Neurobiology* 4: 389–399.

Bergado JA and Almaguer W (2002) Aging and synaptic plasticity: A review. *Neural Plasticity* 9: 217–232.

Blank T, Nijholt I, Kye MJ, Radulovic J, and Spiess J (2003) Small-conductance, $Ca^{2+}$-activated $K^+$ channel SK3 generates age-related memory and LTP deficits. *Nature Neuroscience* 6: 911–912.

Bramham CR and Messaoudi E (2005) BDNF function in adult synaptic plasticity: The synaptic consolidation hypothesis. *Progress in Neurobiology* 76: 99–125.

Burke SN and Barnes CA (2006) Neural plasticity in the ageing brain. *Nature Reviews Neuroscience* 7: 30–40.

Chapman PF (2005) Cognitive aging: Recapturing the excitation of youth? *Current Biology* 15: R31–R33.

Cooke SF and Bliss TVP (2006) Plasticity in the human central nervous system. *Brain* 129: 1659–1673.

Davis HP, Small SA, Stern Y, et al. (2003) Acquisition, recall, and forgetting of verbal information in long-term memory by young, middle-aged, and elderly individuals. *Cortex* 39: 1063–1091.

de Magalhaes JP and Sandberg A (2005) Cognitive aging as an extension of brain development: A model linking learning, brain plasticity, and neurodegeneration. *Mechanisms of Ageing and Development* 126: 1026–1033.

Driscoll I and Sutherland RJ (2005) The aging hippocampus: Navigating between rat and human experiments. *Review of Neuroscience* 16: 87–121.
Foster TC and Kumar A (2002) Calcium dysregulation in the aging brain. *Neuroscientist* 8: 297–301.
Kamsler A and Segal M (2003) Hydrogen peroxide modulation of synaptic plasticity. *Journal of Neuroscience* 23: 269–276.
Landfield PW (1994) Increased hippocampal $Ca^{2+}$ channel activity in brain aging and dementia. Hormonal and pharmacologic modulation. *Annals of the New York Academy of Sciences* 747: 351–364.
Lee HK, Min SS, Gallagher M, and Kirkwood A (2005) NMDA receptor-independent long-term depression correlates with successful aging in rats. *Nature Neuroscience* 8: 1657–1659.
Lu T, Pan Y, Kao SY, et al. (2004) Gene regulation and DNA damage in the ageing human brain. *Nature* 429: 883–891.
Malinow R and Malenka RC (2002) AMPA receptor trafficking and synaptic plasticity. *Annual Review of Neuroscience* 25: 103–126.
Mattson M (ed.) (2002) *Diet–Brain Connection: Impact on Memory, Mood, Aging and Disease*. Dordrecht: Kluwer.
Murphy GG, Fedorov NB, Giese KP, et al. (2004) Increased neuronal excitability, synaptic plasticity, and learning in aged Kvbeta1.1 knockout mice. *Current Biology* 14: 1907–1915.
Pfeiffer BE and Huber KM (2006) Current advances in local protein synthesis and synaptic plasticity. *Journal of Neuroscience* 26: 7147–7150.
Serrano F and Klann E (2004) Reactive oxygen species and synaptic plasticity in the aging hippocampus. *Ageing Research Reviews* 3: 431–443.
Soule J, Messaoudi E, and Bramham CR (2006) Brain-derived neurotrophic factor and control of synaptic consolidation in the adult brain. *Biochemical Society Transactions* 34: 600–604.
Thibault O, Hadley R, and Landfield PW (2001) Elevated postsynaptic $[Ca^{2+}]_i$ and L-type calcium channel activity in aged hippocampal neurons: Relationship to impaired synaptic plasticity. *Journal of Neuroscience* 21: 9744–9756.
Tombaugh GC, Rowe WB, Chow AR, Michael TH, and Rose GM (2002) Theta-frequency synaptic potentiation in CA1 *in vitro* distinguishes cognitively impaired from unimpaired aged Fischer 344 rats. *Journal of Neuroscience* 22: 9932–9940.
Woodruff-Pak DS and Finkbiner RG (1995) Larger nondeclarative than declarative deficits in learning and memory in human aging. *Psychology of Aging* 10: 416–426.

# Lipids and Membranes in Brain Aging

**W G Wood**, University of Minnesota School of Medicine, Minneapolis, MN, USA
**G Y Sun**, University of Missouri, Columbia, MO, USA



## Introduction

Biological membranes have multiple roles in cells. At a macro level, the membrane acts as a barrier between the cell interior and extracellular environment and between the cytoplasm and intracellular organelles. Viewed at the molecular level, membranes are composed of lipids and proteins and in some instances are in dynamic interaction with each other. How membranes are organized has evolved from the early notion in the 1850s describing the membrane as a relatively thin ion-impermeable layer covering cells and subsequent work showing that such a membrane contains lipids based on lipid solubility experiments. Later evidence in the 1920s suggested a bilayer of lipids with a hydrophobic interior and hydrophilic part of the lipid in contact with the extracellular and intracellular aqueous environment. The idea that proteins may be associated with the membrane was introduced in the 1930s, and later work introduced the concept that the membrane was a single lipid layer with proteins or what was referred to as the unit membrane model. In 1972, Singer and Nicolson proposed the fluid mosaic model of membranes, with lipids and proteins randomly distributed with lateral and rotational diffusion. Being a major component of the membrane, lipids were viewed as being important in maintaining the fluidity of the bilayer. An optimal fluidity was thought to be required for numerous membrane functions, such as transport, signal transduction, exocytosis, and endocytosis. Changes in lipid composition, such as an increase in cholesterol content, would reduce fluidity and alter activity of membrane function, which could be detrimental to the cell. That the state of the membrane was important to cell function has been proposed to be a factor in several different pathophysiological conditions ranging from alcoholism to Zellweger's syndrome. Changes in membrane fluidity and lipid composition have previously been proposed to be major contributors to cellular dysfunction observed with increasing age. This article critically examines aging and brain membranes from the perspectives of changes in fluidity, lipid composition, and membrane domains and identifies areas in which further study is needed.

## Brain Membrane Fluidity

Fluidity is a broad term used to describe the lateral motion of lipids in membranes. Several studies have reported a reduction in fluidity of brain membranes with increasing age. However, reduced membrane fluidity in aged membranes has not been consistently observed. The lack of consistency in membrane fluidity and aging studies could possibly be attributable to different techniques (electron spin resonance, fluorescence spectroscopy, and nuclear magnetic resonance), probes residing in different membrane locations (DPH, pyrene, 5-doxyl stearic acid), purity of the membrane preparation (whole brain homogenate, synaptosomes, synaptic plasma membranes in different brain regions), and different species used. Regardless of whether there are aged differences in membrane fluidity, one conclusion regarding the extant studies on fluidity of brain membranes and aging is that most have been limited in their technological sophistication and precision. For example, fluidity actually consists of two components which have been characterized as rate of motion, a dynamic component, and the extent of motion, a static component. Parameters such as limiting anisotropy and order parameter refer to static components, whereas rotational correlation time and rotational rate denote dynamic components. Studies examining brain membrane fluidity in aged organisms have not distinguished between dynamic and static components. Furthermore, the majority of studies on fluidity of brain membranes of aged organisms have used techniques which provide an estimate of the overall average fluidity of the membrane in contrast to more circumscribed areas of the membrane. There may be specific differences in fluidity of the aged brain membrane which would not be detected by conventional techniques, and this issue is discussed later. Another limitation of studies on membrane fluidity and aging is that changes in fluidity on membrane and cell functions in general are not well understood.

## Brain Membrane Cholesterol Levels

Lipids are a major contributor to the fluidity of membranes, and this has been demonstrated experimentally in both biological and model membranes. Cholesterol, which accounts for more than 25% of the total lipid of synaptic plasma membranes (SPMs), is a key factor in determining membrane fluidity. Above the phase transition of the lipid environment, reducing membrane cholesterol increases fluidity, whereas the opposite effect is observed when membrane cholesterol levels are increased. Age differences in brain membrane cholesterol levels have not been consistently observed. This observation applies to both human and animal studies. Even in those studies reporting an increase in cholesterol levels with increasing age, the mechanism for such an increase has not been forthcoming. The biosynthesis of cholesterol requires more than 30 enzymes and several cofactors. To date, there has been very little work on the brain cholesterol synthetic pathway in aging organisms.

Cholesterol is derived from mevalonate, which is also the precursor of various compounds including isoprenoids. There is great interest in the role of isoprenoids such as geranylgeranyl pyrophosphate (GGPP) and farnesyl pyrophosphate (FPP), which prenylate various small GTPases, such as Rho, Rac, and Ras. These proteins are involved in a variety of cell signaling functions. GGPP and FPP levels have not been reported in brain, but an isoprenoid whose abundance has been reported is dolichol. Dolichols are long-chain polyisoprenoids derived from FPP. These compounds are ubiquitous among body organs, with the highest levels observed in testis, kidney, brain, and various exocrine glands. Dolichol can be in a free form or phosphorylated. Dolichol phosphates are intermediates in the synthesis of oligosaccharide groups of glycoproteins. The cellular function of free dolichols is not known. Free dolichol increased fluidity of SPMs, and it has been detected in membranes in which it might function to regulate membrane fluidity. There is a marked increase of free dolichol in human brain tissue with aging and in SPMs of aged mice compared to younger mice. Increased free dolichol content has also been observed in brain tissue from individuals with Alzheimer's disease and individuals with neuronal lipofuscinosis. The precise function of free dolichol and the consequences of elevated dolichol levels observed with increasing age and certain neurodegenerative diseases have not been established.

## Brain Membrane Phospholipids and Fatty Acids

In order of abundance, the membrane phospholipids in brain are phosphatidylcholine (PC), phosphatidylethanolamine (PE), phosphatidylserine (PS), phosphatidylinositol (PI), and cardiolipin (CL). Neural membranes are also enriched in ether phospholipids (i.e., phospholipids with alkyl or alkenyl linkage in the *sn*-1 position). Whereas most diacyl-type phospholipids are synthesized in the endoplasmic reticulum, ether phospholipids are synthesized in peroxisomes and changes in levels of ether lipids are associated with peroxisomal disorders. Cardiolipin, an unusual phospholipid with two glycerol backbones and four fatty acids, is exclusively present in the mitochondrial membrane. Cardiolipin plays an important role in regulating respiratory chain activity in mitochondria. Phospholipids in brain membranes are enriched in plasmalogens of the ethanolamine type (alkenyl-acyl-gly cerophosphoethanolamine(PEpl)). PEpl is especially enriched in myelin, comprising more than 60% of total ethanolamine phospholipids. Plasmalogens may act as endogenous antioxidants. Although phosphatidylinositol (PI) comprises only 4% of total phospholipid, it is metabolically active and is converted to phosphatidylinositol-4-phosphate (PIP) and phosphatidylinositol-4,5-bisphosphates (PIP2) by kinases. PIP2 plays an important role in the stimulus–signal coupling reactions that lead to generation of second messengers. Stimulation of many G-protein-coupled receptors (GPCRs) leads to activation of phospholipase C (PLC), which hydrolyzes PIP2 to form diacylglycerol (DG), a second messenger for activation of protein kinase C (PKC) and inositol trisphosphate (IP3), a second messenger for mobilization of intracellular $Ca^{2+}$ stores. Another PI kinase, PI3K, adds a phosphate group to the third position of PIP2 to form PIP3. Activation of PI3K by tyrosine kinase and/or kinases associated with GPCRs has been implicated in signaling pathways associated with cell growth, proliferation, survival, and plasticity.

### Fatty Acids of Brain Phospholipids

Phospholipids in neural membranes are enriched in polyunsaturated fatty acids such as 20:4n-6 (AA) and 22:6n-3 (DHA). Each type of phospholipid is composed of unique patterns of acyl or alkenyl groups, and these patterns may vary depending on their subcellular localization. For example, myelin is known for its high content of PEpl, and this phospholipid is enriched with monounsaturated fatty acids such as 18:1 and 20:1. In fact, 20:1 is an unusual fatty acid almost exclusively found in the myelin membrane. On the other hand, fatty acids of PEpl in the nerve ending fraction (synaptosome) are enriched in 20:4n-6 and 22:6n-3. Earlier studies showed an age-dependent increase in monounsaturated fatty acids (e.g., 18:1 n-9 and 20:1 n-9 (~43%)) and a decrease in polyunsaturated fatty acids (e.g., 22:6 n-3 (~14%)) in mouse brain PE between 3 and 26 months of age. Similar changes are found in PE in the brain of rhesus monkey. Studies on the glutathione redox state in mouse brain showed an age-dependent decrease in the ratio of reduced to oxidized glutathione (GSH/GSSG) between 3 and 21 months of age. Whether the age-dependent decrease in polyunsaturated fatty acids is associated with the decrease in GSH remains to be investigated. Since levels of PEpl and 22:6n-3 decrease with normal aging and with memory impairment, considerable interest has focused on whether dietary supplementation of fish oil which is enriched in n-3 fatty acids may help to reverse the age-related deficits and improve memory and cognitive functions.

The fatty acid profile of PS is enriched in 22:6 n-3, which is predominantly coupled with 16:0 in the *sn*-1 position. On the other hand, fatty acids in PI, as well as poly-PI are enriched in 20:4 n-6, which is linked to 18:0 in the *sn*-1 position. Consequently, when PIP2 is hydrolyzed by PLC to form DG and IP3, the fatty acid of this DG pool is enriched in 18:0/20:4 and is different from the DG pools associated with PL biosynthesis. Unlike fatty acids of phospholipids in other organs, brain fatty acids have low levels of 18:2 n-6, although this fatty acid is enriched in CL. The unique presence of CL in mitochondria suggests its role in mitochondrial membrane function. There is evidence that in cerebral ischemia, CL in mitochondria decreases together with release of cytochrome c. Because these anionic phospholipids are present in minute amounts in the brain, little is known about age-dependent changes of these phospholipids.

### Enzymes Regulating Membrane Phospholipids

Fatty acids in the *sn*-2 position of phospholipids are maintained in a dynamic equilibrium, regulated by phospholipases $A_2$ ($PLA_2$) and lysophospholipid acyltransferases, constituting the deacylation–reacylation cycle. It is important to note that whereas hydrolysis of fatty acids by $PLA_2$ releases fatty acids from phospholipids, the reacylation process involves ATP-dependent conversion of a fatty acid to its acyl-CoA and transfer of acyl-CoA to lysophospholipids. Although the number of studies on the structure and function of different types of $PLA_2$ has increased considerably in the past decade, studies regarding lysophospholipid acyltransferases are still lacking. Besides PLC, phospholipids are also hydrolyzed by phospholipase D (PLD), resulting in formation of phosphatidic acid (PA). In the presence of ethanol, PLD may mediate transphosphatidylation of PC to form phosphatidylethanol, which is a marker of ethanol metabolism.

Turnover of fatty acids among different phospholipids is dependent not only on the type of phospholipids but also on the types of membranes in different cell types. Thus, it is conceivable that fatty acids in myelin phospholipids undergo slower turnover compared to synaptic membranes. Age-related decrease in turnover of fatty acids in SPM phospholipids may provide an explanation for the decrease in synaptic functions during aging.

#### Phospholipase(s) $A_2$

$PLA_2$s are enzymes for cleaving fatty acids in the *sn*-2 position of phospholipids. These enzymes are ubiquitous in mammalian cells. Besides their role in maintenance of cell membrane phospholipids, they also play an important role in providing lipid mediators such as arachidonic acid (AA), which serves as the precursor for synthesis of a number of eicosanoids. Hydrolysis of phospholipids by $PLA_2$s also produces lysophospholipids which possess detergent-like properties and can alter membrane micro domains. Among more than 20 different types of $PLA_2$s, studies have focused on three groups, namely the calcium- and kinase-dependent group IV cytosolic $PLA_2$ ($cPLA_2$), the $Ca^{2+}$-independent group VI $PLA_2$ ($iPLA_2$), and the $Ca^{2+}$-dependent secretory $PLA_2$ ($sPLA_2$).

$cPLA_2$ is composed of three isoforms: $\alpha$, $\beta$, and $\gamma$. In brain, most studies have focused on $cPLA_2$-$\alpha$, an 85-kDa protein comprising a $Ca^{2+}$ binding C2 domain and a catalytic domain with several putative phosphorylation sites for mitogen-activated protein kinase (MAPK), PKC, and $Ca^{2+}$/calmodulin-dependent protein kinase II. Upon binding with $Ca^{2+}$ and phosphorylation by protein kinases, $cPLA_2$-$\alpha$ is translocated to the membrane and targets phospholipids, especially those containing AA. Consequently, $cPLA_2$-$\alpha$ is activated by metabotropic and ionotropic receptors and releases AA for biosynthesis of eicosanoids. However, the mechanism for the coupling between $cPLA_2$ and cyclooxygenases (COX) remains a subject for further investigation. $cPLA_2$s have been shown to target a number of intracellular membranes, including the Golgi complex, plasma membranes, and nuclear membranes. The putative pleckstrin homology (PH) domain in $cPLA_2$ suggests possible interaction with acidic phospholipids such as PIP2 which are localized in plasma membranes. There is evidence that the translocation process may involve binding to cytoskeletal proteins and intermediate filaments such as vimentin and actin.

Secretory $PLA_2$s are low-molecular-weight proteins (~14 kDa). More than ten subtypes of $sPLA_2$s are present in the mammalian system, including the brain. In recent years, considerable attention has been given to the $sPLA_2$-IIA in the cardiovascular system because of its association with inflammatory responses in diseases such as arthritis, atherosclerosis, and sepsis. Indeed, $sPLA_2$ is an independent risk factor for cardiovascular diseases. Although studies have implicated the increase in $sPLA_2$-IIA in reactive astrocytes in response to cerebral ischemia, studies demonstrating $sPLA_2$ in human brain and in relation with other neurodegenerative diseases have not been forthcoming. In astrocytes, mRNA expression for $sPLA_2$-IIA along with COX-2 and inducible nitric oxide synthase are stimulated by proinflammatory cytokines (tumor necrosis factor-$\alpha$, interleukin-1$\beta$, and interferon-$\gamma$) and lipopolysaccharides, and the increase in inflammatory response is coupled with production of prostaglandin $E_2$ and nitric oxide.

Since $sPLA_2$-IIA is induced by proinflammatory cytokines, it needs to be determined whether its release from glial cells can cause neuronal damage. Studies using extracellular forms of $sPLA_2$ have clearly demonstrated the ability of $sPLA_2$-IIA to cause neuronal excitotoxicity and apoptosis by enhancing the ionotropic receptor and $Ca^{2+}$ channel activities. Other studies have shown that secreted $sPLA_2$-IIA can perturb cellular membranes, especially those undergoing apoptosis. In PC12 cells, lysophospholipids produced by $sPLA_2$-IIA were shown to enhance neurite outgrowth, further implicating an important role of this enzyme in neurodegenerative diseases.

$iPLA_2$ consists of two major subtypes, VIA and VIB. $iPLA_2$-VIA comprises 85–88 kDa proteins with three splice variants – VIA-1, VIA-2, and VIA-3 – all containing multiple ankyrin domains, suggesting its ability to interact with other proteins. $iPLA_2$-VIA shows no apparent specificity for the fatty acids in the *sn*-2 position or the type of phospholipid headgroups. This enzyme can efficiently hydrolyze oxidized phospholipids. In agreement with its major role in maintenance of membrane phospholipid homeostasis, $iPLA_2$ activity is frequently proportional with the level of lysoPC upon inhibition by BEL, a hydrophobic serine-reactive inhibitor specific for $iPLA_2$. Nevertheless, there is increasing evidence that this enzyme may also play a role in stimulus–response coupling activity and in the release of AA for prostaglandin biosynthesis.

### Oxidative Stress, Aging, and Membrane Phospholipids

Due to the high oxygen consumption, the enrichment of polyunsaturated fatty acids, and the relatively low activity of antioxidative defense enzymes, cells in brain are highly susceptible to oxidative insults, which are important underlying factors for the pathogenesis of a number of age-related neurodegenerative diseases. Several cellular enzymes, including enzymes in the respiratory chain in mitochondria, xanthine/xanthine oxidase, myloperoxidase, cytochrome P450 in cell cytoplasm, monoamine oxidases, cyclooxygenases, and NADPH oxidase, can catalyze molecular oxygen to produce reactive oxygen species (ROS) such as superoxide anion ($O_2^-$) and $H_2O_2$. In the presence of metal ions, $Fe^{2+}$, and $Cu^{2+}$, $H_2O_2$ can be further converted to hydroxyl radical (OH) through the Fenton reaction. Hydroxyl radicals can react with polyunsaturated fatty acids in membrane phospholipids forming peroxyl radical (ROO) and propagate the chain reaction of lipid peroxidation. External factors such as radiation, drugs, and disease states can upset the oxidative–antioxidative equilibrium, rendering cells to become pro-oxidative. More studies should be directed at examining the mechanism(s) leading to the increase in oxidized lipids in brain, especially under pathological conditions.

Among the phospholipids, plasmalogens appear to be most susceptible to oxidative stress. There is considerable evidence that production of ROS and lipid peroxidative products in brain increases with age and that the age-related increase in oxidative stress may enhance neuroinflammatory responses in the aged population. An obvious decrease in plasmalogens was observed in Alzheimer's brain and the decrease could be detected even at early stages of the disease. Lipid peroxidation products such as 4-hydroxynonenal (HNE) can form adducts with proteins and cause neurotoxicity. Thus, the increase in HNE has been regarded as an important contributor to the pathogenesis of many neurodegenerative diseases, including Alzheimer's disease and stroke.

## Brain Lipid Domains

The preceding discussion examined membranes from the perspective of broad changes in fluidity and lipid composition in contrast to specific areas or domains of membranes. Bretscher and others in the 1970s postulated and demonstrated that the two leaflets which form the bilayer are asymmetric in their lipid distribution. **Figure 1(a)** is a model of SPM and it can be seen that the outer or exofacial leaflet is enriched in phosphatidylcholine and sphingomyelin. In contrast, the inner or cytofacial leaflet is enriched in phosphatidylethanolamine, phosphatidylserine, phosphatidylinositol, and cholesterol. This lipid asymmetry confers differences in the charge properties and fluidity of the two leaflets. The exofacial leaflet is considered to be neutral or zwitterionic, whereas the cytofacial leaflet is more negatively charged due to the enrichment of the anionic phospholipids phosphatidylserine and phosphatidylinositol. Mechanisms involved in the maintenance of phospholipid asymmetry are not completely understood. There is evidence that at least three different proteins may be involved in promoting phospholipid asymmetry: an ATP-dependent transporter (flippases) directed at the cytofacial leaflet, an ATP-dependent transporter (floppase) directed at the exofacial leaflet, and an ATP-independent transporter (scramblase) interacting with both leaflets. Phospholipid asymmetry is thought to be important in such cell signaling functions as $Ca^{2+}$ homeostasis and receptor-mediated signal transduction. A loss of phospholipid asymmetry has been observed in sickle cell anemia,

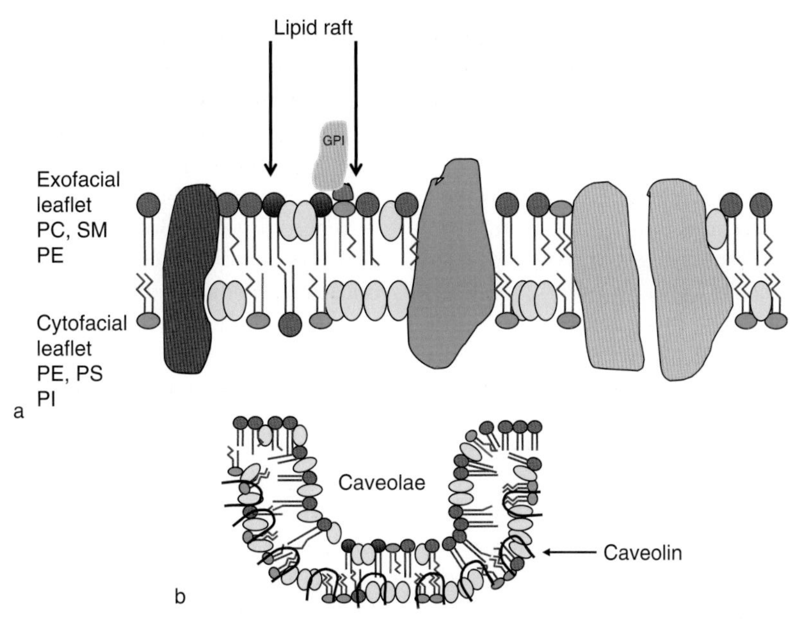


**Figure 1** Plasma membrane domains. (a) The two leaflets of the plasma membrane showing the asymmetric distribution of cholesterol and phospholipids. Large globular structures represent proteins. Cholesterol (denoted by yellow structure), phosphatidylinositol (PI), phosphatidylethanolamine (PE), and phosphatidylserine (PS) are enriched in the cytofacial leaflet. Phosphatidylcholine (PC) and sphingomyelin (SM) are in abundance in the exofacial leaflet, and there is some PC in the cytofacial leaflet. A lipid raft is denoted in the exofacial leaflet (a) containing a glycosylphosphatidyl inositol-anchored protein (GPI). (b) A model of caveolae enriched in cholesterol, sphingolipids, and caveolin. Reproduced from Wood WG, Igbavboa U, Eckert GP, and Müller WE (2007) Cholesterol – a Janus faced molecule in the central nervous system. In: Lajtha A (ed.) *Handbook of Neurochemistry and Molecular Neurobiology*, 3rd edn., pp. 151–170. Heidelberg, Germany: Springer. With kind permission of Springer Science and Business Media.

sperm cell maturation, and apoptosis. In human aging erythrocytes whose life span in the body is approximately 120 days, phospholipid asymmetry is reduced. There have been a few reports on phospholipid asymmetry in brain membranes whose distribution is similar to that of nonbrain membranes, but there have been no reported studies on phospholipid asymmetry in brain membranes of aging organisms.

### Exofacial and Cytofacial Leaflet Fluidity

There is a major difference in fluidity of the exofacial leaflet compared to the cytofacial leaflet. The exofacial leaflet is markedly more fluid, as determined by limiting anisotropy and rotational relaxation time of DPH in SPM, and is more easily perturbed than is the cytofacial leaflet. For example, 25 mM ethanol significantly increased fluidity of the exofacial leaflet but had no effect on the cytofacial leaflet even with an ethanol concentration as high as 400 mM. Chronic ethanol consumption reduced fluidity of the SPM exofacial leaflet producing resistance to ethanol perturbation, demonstrating cellular membrane tolerance. The asymmetry in fluidity of the two leaflets is diminished with increasing age in mouse SPM. In mice 3 or 4 months old, the exofacial leaflet was 21% more fluid than the cytofacial leaflet; in mice 14 or 15 months old, it was 9% more fluid; and in mice 24 or 25 months old, differences between the two leaflets in fluidity were indistinguishable.

### Transbilayer Cholesterol Distribution

The major difference in fluidity between SPM exofacial and cytofacial leaflets is associated with the transbilayer cholesterol distribution (TCD). **Figure 1(a)** shows that cholesterol is asymmetrically distributed between the two leaflets. In young mice, approximately 15% of SPM cholesterol resides in the exofacial leaflet and the remaining 85% in the cytofacial leaflet. TCD is not static but, rather, can be altered by different treatment conditions. Chronic ethanol consumption, apolipoprotein E (APOE) and its isoforms, and HMG-CoA reductase inhibitors (statins) all have been shown to alter TCD. Increasing age had a striking effect on SPM TCD. The exofacial leaflet of aged mice (24 or 25 months old) showed a twofold increase in cholesterol compared to that of younger mice. Mice 14 or 15 months of age were intermediate in their cholesterol distribution with respect to the young and aged mice. Interestingly, the increase in exofacial leaflet cholesterol of the aged mice was not the result of an increase in total SPM cholesterol. There were no significant differences in total SPM cholesterol among the three age groups. A valuable conclusion which can be drawn from this study is that striking changes in neuronal membrane cholesterol homeostasis can occur in the absence of changes in total membrane cholesterol levels.

The mechanism for maintaining TCD is not well understood. Several potential candidates have been proposed, including APOE, the low-density lipoprotein receptor (LDLR), sterol carrier protein-2 (SCP-2), polyunsaturated fatty acids, fatty acid-binding proteins (FABPs), and caveolin. However, it is worth mentioning that even when some of these potential mechanisms are experimentally altered, the cytofacial leaflet continues to maintain the greater amount of cholesterol compared with the exofacial leaflet. From these results, one conclusion is that there may be multiple mechanisms which are involved in the regulation of TCD. These mechanisms might involve multiple sites, such as the surface of the exofacial leaflet (APOE, LDLR, and caveolin), within the plasma membrane (fatty acid composition) and cytofacial leaflet interaction with intracellular proteins (SCP-2 and FABPs). A question which has not been addressed is the functionality of changes in TCD. In the case of chronic ethanol consumption, the redistribution of cholesterol may be adaptive and diminish partitioning of ethanol into the cell, in turn altering cell function. With respect to the redistribution of cholesterol in aged SPM, it is not known if such changes are adaptive or have pathophysiological consequences.

### Lipid Rafts and Caveolae

Two other lipid domains which have received substantial attention are lipid rafts and caveolae. In this discussion, lipid rafts refer to structures enriched in cholesterol, glycosphingolipids, and glycosylphosphatidylinositol-anchored proteins and not containing the caveolae protein, caveolin, as shown in **Figure 1(a)**. Other names for lipid rafts include detergent-insoluble glycolipid-enriched membranes and detergent-resistant membranes. Caveolae as seen in **Figure 1(b)** are defined as membrane invaginations containing the protein caveolin and are enriched in cholesterol and glycosphingolipids. There has been considerable interest in these lipid domains, and several reviews have been published. In contrast, there has been very little advancement in the aging field regarding brain lipid rafts and caveolae. Levels of the lipid raft markers flotillin-1 and flotillin-2 did not differ in brain lipid rafts prepared from a whole brain homogenate of young and aged rats. In contrast, a reduction of flotillin-1 levels as well as alkaline phosphatase activity was observed with increasing age in synaptosomal lipid rafts of aged mice expressing human apoE3 and apoE4 compared with young apoE3 mice. Lipid rafts of young apoE4 mice were more similar to those of older mice compared to young apoE3 mice for flotillin-1 levels and alkaline phosphatase activity. Cholesterol and sphingomyelin levels of lipid rafts did not differ between the young apoE3 and apoE 4 mice, but levels of those lipids did increase with age. Both increasing age and the apoE4 genotype are risk factors for Alzheimer's disease, and both have similar effects on brain lipid raft protein markers. There is evidence that lipid rafts may play a role in the processing of amyloid $\beta$-protein, which is thought to be a causal factor in the pathophysiology of Alzheimer's disease.

Superoxide dismutase-1, which is associated with membrane caveolae, was found to be reduced in aortic tissue of aged rats compared to younger animals. Caveolin-1 is a major protein of caveolae, and it has been demonstrated that the life span of caveolin-1 null mice is reduced by approximately 50% compared with wild-type mice and caveolin-1 heterozygous (+/−) mice. Brain caveolin-1 levels were found to be decreased in aged rats compared to younger rats.

The concept of lipid rafts, like that of the membrane, has been evolving. At a Keystone Symposium on Lipid Rafts and Cell Function in 2006, a consensus was reached that the term 'lipid rafts' should be changed to 'membrane rafts' recognizing the role of raft proteins, and that caveolae should be considered a membrane raft. Another issue which was discussed at that meeting and that has been proposed by others was that some membrane rafts may be transient with a lifetime of milliseconds or less, in contrast to a static structure. Such a short lifetime could pose problems for isolation of rafts as commonly done using density gradient centrifugation. An understanding of membrane rafts in aging organisms will require conceptual and technological approaches which are state of the art in the membrane raft field.

## Conclusions

A broad conclusion which can be reached is that studies on age differences in membrane organization have mostly been approached from the standpoint of the Singer–Nicolson fluid mosaic membrane model. However, modification of this model has been proposed by different groups suggesting a more mosaic-like structure which is heterogeneous, with lipid–lipid, protein–protein, and lipid–protein interactions, in contrast to a fluid membrane in which proteins are unrestricted in their lateral movement. The majority of studies on age differences in membranes, whether fluidity or lipid composition, have examined bulk changes in the membrane. Although this approach provides an overview of the membrane environment, it does not address the vast heterogeneity of the membrane. There is a need to study both macro and micro membrane domains in aging organisms and the functional consequences of membrane domains on cellular function.

*See also:* Aging of the Brain; Brain Composition: Age-Related Changes; Glial Cells: Microglia During Normal Brain Aging; Glial Cells: Astrocytes and Oligodendrocytes During Normal Brain Aging; Metal Accumulation During Aging; Synaptic Plasticity: Neuronogenesis and Stem Cells in Normal Brain Aging.

## Further Reading


Datta DP (1987) *A Comprehensive Introduction to Membrane Biochemistry.* Madison, WI: Floral.

Edidin M (2003) The state of lipid rafts: From model membranes to cells. *Annual Review of Biophysics and Biomolecular Structure* 32: 257–283.

Jacobson K, Sheets ED, and Simson R (1995) Revisiting the fluid mosaic model of membranes. *Science* 268: 1441–1442.

Lommerse PHM, Spaink HP, and Schmidt T (2004) *In vivo* plasma membrane organization: Results of biophysical approaches. *Biochimica et Biophysica Acta* 1664: 119–131.

Packer L, Deamer DW, and Heath RL (1967) Regulation and deterioration of structure in membranes. *Advances in Gerontology Research* 2: 77–120.

Pike LJ (2006) Rafts defined: A report on the Keystone Symposium on Lipid Rafts and Cell Function. *Journal of Lipid Research* 47: 1597–1598.

Schroeder F (1984) Role of membrane lipid asymmetry in aging. *Neurobiology of Aging* 5: 323–333.

Sun AY and Sun GY (1979) Neurochemical aspects of the membrane hypothesis of aging. *Interdisciplinary Topics in Gerontology* 15: 34–53.

Sun GY, Xu J, Jensen MD, and Simonyi A (2004) Phospholipase A2 in the central nervous system: Implications for neurodegenerative diseases. *Journal of Lipid Research* 45: 205–213.

Vereb G, Szöllösi J, Matkó J, et al. (2003) Dynamic, yet structured: The cell membrane three decades after the Singer–Nicolson model. *Proceedings of the National Academy of Sciences of the United States of America* 100: 8053–8058.

Verkleij AJ and Post JA (2000) Membrane phospholipid asymmetry and signal transduction. *Journal of Membrane Biology* 178: 1–10.

Wood WG, Eckert GP, Igbavboa U, and Müller WE (2003) Amyloid beta-peptide interactions with membranes and cholesterol: Causes or casualties of Alzheimer's disease. *Biochimica et Biophysica Acta* 1610: 281–290.

Wood WG, Igbavboa U, Eckert GP, and Müller WE (2007) Cholesterol – a Janus faced molecule in the central nervous system. In: Lajtha A (ed.) *Handbook of Neurochemistry and Molecular Neurobiology,* 3rd edn., pp. 151–170. Heidelberg, Germany: Springer.

Wood WG, Schroeder F, Igbavboa U, Avdulov NA, and Chochina SV (2002) Brain membrane cholesterol domains, aging, and amyloid beta-peptides. *Neurobiology of Aging* 23: 685–694.

Zs-Nagy I (1979) The role of membrane structure and function in cellular aging: A review. *Mechanisms of Ageing and Development* 9: 237–246.

# Metal Accumulation during Aging

**H M Schipper**, McGill University, Montreal, QC, Canada



## Introduction

Central nervous system (CNS) senescence has been implicated as the most robust risk factor for a host of human neurodegenerative disorders, including sporadic (nonfamilial) Alzheimer's disease (AD), Parkinson's disease (PD), and amyotrophic lateral sclerosis (ALS). Thus, a thorough understanding of intrinsic brain aging mechanisms and their relationship to intervening neuropathological processes is essential for the development of therapeutic modalities that successfully prevent, ameliorate, interrupt, or reverse neuronal attrition and clinical decline in these conditions. Over the past two decades, oxidative stress (OS) and free radical damage, altered patterns of transition metal (especially iron) deposition, and mitochondrial insufficiency (bioenergetic failure) have been abundantly documented in the aging and degenerating mammalian CNS. Although the tendency has been to investigate these changes as disparate histopathological features of neural senescence and disease, we hypothesize that OS, abnormal iron mobilization, and mitochondrial dysfunction combine inextricably to yield a single neuropathological 'lesion' that serves as a vital link between normal brain aging physiology and neurodegeneration. In this article, support for this unifying perspective is reviewed based on evidence adduced from investigations of rodent astroglial senescence, experimental models of aging-associated neurodegenerative disorders, and human neuropathological studies. Several scientific and clinical implications of this formulation are considered.

## Free Radicals, Transition Metals, and OS

Free radicals are atoms containing unpaired electrons in their outer orbitals. In the course of normal aging, and to a greater extent in disease states, free radicals and other reactive oxygen species (ROS), reactive nitrogen species (RNS), and reactive sulfur species (RSS) generated endogenously or derived from environmental sources may overwhelm tissue enzymatic and nonenzymatic antioxidant defenses (a condition broadly referred to as 'OS') and incur molecular damage to various cellular substrates, including proteins, lipids, carbohydrates, and nucleic acids. Among the transition metals (d-block elements listed between Group IIA and Group IIB of the periodic table), iron, copper, manganese, mercury, and cadmium are redox-active and are capable of mediating significant injury upon deposition within neural tissues. Ferrous iron ($Fe^{2+}$) and cuprous copper ($Cu^{+}$), the predominant transition metals in mammalian tissues, play pivotal roles in cellular redox chemistry by reducing $H_2O_2$ to the highly cytotoxic hydroxyl ($OH^\bullet$) radical (Fenton catalysis) or by behaving as nonenzymatic peroxidases that bioactivate benign compounds into toxic free radical intermediates. Within the CNS, ROS in turn may promote aberrant cell signaling and electrophysiological derangements; bioenergetic failure; proteosomal dysfunction, protein aggregation, and inclusion formation; protoxin bioactivation; and neuronal synaptolysis, apoptosis, and necrosis. Pathological transition metal deposition and attendant OS constitute an important pathway for cellular dysfunction and death (and hence a potential therapeutic target) in a host of human neurological (**Table 1**) and medical disorders, and may contribute to the senescence-dependant decline in cognition and other CNS activities. This article focuses primarily on mechanisms of brain iron sequestration and the relationship of the latter to the prevailing 'free radical–mitochondrial' theory of aging in the context of human neurodegenerative disorders.

## The Free Radical–Mitochondrial Theory of Aging

In 1956, Denham Harman first proposed a 'free radical' theory of aging which posits that "free radical reactions are involved in the aging changes associated with the environment, disease and intrinsic aging processes." The theory gained momentum upon recognition that mitochondria are both a leading source of ROS generation and a prime target of oxidative molecular damage in aging eukaryotic cells. The free radical–mitochondrial theory states that oxidative damage to mitochondria, instigated by intrinsic metabolic processes and/or environmental insults, leads to cascading infidelity of electron transport and a self-propagating spiral of increased ROS generation within the inner mitochondrial membrane. The latter, in turn, elicits bioenergetic deficits (ATP loss and failure of ATP-dependent processes) and progressive tissue aging. Evidence supporting this theory has been garnered from a broad range of experimental and observational studies involving intact animal and *in vitro* models that span the entire phylogenetic spectrum. The theory suggests that (1) any injury or disease that exacerbates mitochondrial ROS production may accelerate senescence of the

**Table 1** Redox-active transition metals implicated in the etiopathogenesis of human nervous system conditions

| Iron | Copper |
|---|---|
| CNS senescence | Wilson's disease |
| Cerebral hemorrhage | Menkes disease |
| Alzheimer's disease | Acquired copper deficiency |
| Parkinson's disease | Alzheimer's disease? |
| Amyotrophic lateral sclerosis | Creutzfeldt-Jakob disease? |
| Huntington's disease | **Manganese** |
| Progressive supranuclear palsy | Parkinsonism (manganism) |
| Multiple sclerosis | Hepatic encephalopathy? |
| Corticobasal degeneration | **Mercury** |
| Friedreich ataxia | Inorganic mercury poisoning |
| X-linked sideroblastic anemia with ataxia | Minamata disease |
| PANK-2 deficiency (formerly Hallervorden-Spatz)[a] | **Cadmium** |
| Aceruloplasminemia | Toxic encephalopathy |
| Neuroferritinopathy | Peripheral neuropathy |
| Superficial siderosis | |
| Restless legs syndrome | |
| HIV-1 encephalitis | |
| Hemochromatosis? | |

[a]PANK-2, Pantothenate kinase type 2.
Adapted from Schipper HM (2004) Redox neurology: Visions of an emerging subspecialty. *Annals of the New York Academy of Sciences* 1012: 342–355, with permission.

affected tissue or organ and (2) normal, aging tissues compromised by a progressively unfavorable mosaic of healthy and ROS-generating mitochondria (termed 'heteroplasmy') may become increasingly vulnerable to aging-associated conditions such as atherosclerosis, neurodegeneration, and cancer.

## Iron Deposition in the Aging CNS

Iron is indispensable for normal neural development and physiology and contributes to a diverse spectrum of cellular functions, including myelination, biogenic amine metabolism, electron transport, antioxidant enzyme activity, and cytokinesis. To facilitate these vital functions, while at the same time limiting the metal's propensity to mediate toxic Fenton reactions (*vide supra*), iron homeostasis in neural and peripheral tissues is tightly controlled by an orchestra of proteins governing its absorption, valence configuration, extracellular transport, cellular flux, chemical signaling activities, and intracellular storage (**Table 2**). Despite this elaborate regulation, iron progressively accumulates in the mammalian CNS as a function of advancing age, and a portion of the metal in the brain parenchyma and cerebrospinal fluid (CSF) is maintained in a redox-active state. In the normal human CNS, the sequestered metal exhibits an affinity for the basal ganglia, hippocampus, certain cerebellar nuclei, and other, largely subcortical, brain regions. In these aging brain loci, the iron deposits co-register with multiple markers of oxidative substrate modifications and genetic, biochemical, and morphological indices of enhanced mitochondrial damage. Moreover, there appears to be a fairly consistent exacerbation of these co-localizing pathological features in most, if not all, of the major aging-related human neurodegenerative disorders and in some inflammatory, ischemic, and infectious CNS conditions.

**Table 2** Some key proteins implicated in mammalian iron homeostasis

| *Protein name* | *Function* |
|---|---|
| ATP-binding cassette protein, ABC7 | Nonheme Fe export from mitochondria(?) |
| ABC-me (mitochondrial erythroid) | Mitochondrial transport related to heme synthesis |
| Ceruloplasmin | Ferroxidase activity/cellular Fe export |
| Divalent metal transporter (DMT1)/Nramp2 | Membrane transport for $Fe^{2+}$ |
| Duodenal cytochrome *b* | Ferric reductase (provides $Fe^{2+}$ for DMT1 in duodenum) |
| Ferritin (H and L) | Cellular Fe storage |
| Ferrochelatase | $Fe^{2+}$ insertion into protoporphyrin IX (heme synthesis) |
| Ferroportin 1/MTP1/Ireg1 | Cellular Fe export |
| Frascati | Mitochondrial Fe transport |
| Frataxin | [Fe–S] cluster synthesis |
| Heme oxygenase (1 and 2) | Recycling of heme Fe/cell survival |
| Hemojuvelin (repulsive guidance molecule C) | Unknown/cardiac and striated muscle Fe metabolism(?) |
| Hepcidin | Inhibitor of intestinal Fe absorption and macrophage heme Fe recycling |
| Hephaestin | Ferroxidase activity/enterocyte Fe export |
| Hereditary hemochromatosis protein | Modulates tissue Fe uptake via interaction with TfR(?) |
| Iron regulatory protein (1 and 2) | Cellular Fe sensors/regulators of Fe uptake and storage proteins |
| Lactoferrin | Fe-binding protein/antibacterial and antiviral activities |
| Melanotransferrin | Unknown/Fe binding protein |
| Mitochondrial ferritin | Mitochondrial Fe storage(?) |
| Sideroflexin 1 | Mitochondrial transport related to Fe metabolism |
| Steap3 | Endosomal ferrireductase |
| Transferrin | Plasma Fe(III) carrier |
| Transferrin receptor (TfR1 and 2) | Membrane receptor for diferric-Tf |

Adapted from Ponka P (2004) Hereditary causes of disturbed iron homeostasis in the central nervous system. *Annals of the New York Academy of Sciences* 1012: 267–281, with permission.

## Iron Deposition in Alzheimer's and Parkinson's Diseases

Aging remains the most robust risk factor thus far identified for the development of sporadic (nonfamilial) AD and idiopathic PD, two of the most common and devastating neurodegenerations in aging populations of the industrialized world. AD is a dementing condition characterized by progressive neuronal degeneration, gliosis, and the accumulation of extracellular amyloid (senile plaques) and intraneuronal inclusions (neurofibrillary tangles) in discrete regions of the basal forebrain, hippocampus, and association cortices. Idiopathic PD is an extrapyramidal (movement) disorder featuring gradual attrition of dopaminergic neurons in the substantia nigra pars compacta, formation of intraneuronal fibrillar inclusions (Lewy bodies) in this cell population, and variable depletion of noradrenaline and serotonin in other brain stem nuclei. Oxidative damage and mitochondrial insufficiency have been amply documented in the affected brain of AD and PD patients and in transgenic mouse models of these conditions. Neuropathological and magnetic resonance imaging studies have demonstrated augmented iron deposition in the cerebral cortex and basal ganglia of AD and PD patients, respectively. In both the AD and PD brain, increased expression of the intracellular iron storage protein, ferritin, parallels the distribution of the excess iron and largely implicates nonneuronal (glial, endothelial) cellular compartments. In AD, moreover, redox-active iron co-localizes to senile plaques and neurofibrillary tangles and may contribute to the biogenesis of these hallmark inclusions via free radical mechanisms. Interestingly, local concentrations of transferrin-binding sites in these neurodegenerative conditions remain unchanged or vary inversely with the elevated iron stores. Thus, although important for normal iron delivery to most systemic tissues, the transferrin pathway of iron mobilization appears to contribute little to pathological brain iron trapping in these aging-related neurodegenerations. As discussed later, experiments conducted in our laboratory have implicated the enzyme heme oxygenase-1 (HO-1) as a putative transducer of aberrant mitochondrial iron trapping under conditions of OS that likely prevail in the senescent CNS and, to a greater extent, in AD- and PD-affected neural tissues. Other iron-mobilizing proteins, such as lactoferrin, melanotransferrin (p97), ceruloplasmin, and the mutant hereditary hemochromatosis protein (encoded by the *HFE* gene), may also participate in the pathogenesis of these common neurodegenerative diseases.

## Iron Deposition in Other Human CNS Disorders

Although most vigorously investigated in AD and PD, OS, mitochondrial lesions, and disturbances in iron homeostasis have also been described in other aging-related human neurodegenerative conditions, including ALS, progressive supranuclear palsy, corticobasal degeneration, Huntington's disease, and aceruloplasminemia. This pathological triad may also compromise neural tissues affected by multiple sclerosis (MS), HIV-1 encephalitis, ischemic or hemorrhagic stroke, and cerebral trauma. The extent to which transferrin-dependent and -independent pathways contribute to pathological iron mobilization in these diverse conditions remains to be determined.

## Hypothesis

On the basis of their frequent co-occurrence within the aging and diseased neuraxis, we posit that OS, augmented iron deposition, and mitochondrial insufficiency represent a single cohesive neuropathological 'lesion,' with the advent of one constituent obligating the appearance of the others. If confirmed, this formulation may (1) establish a novel conceptual bridge between normal brain senescence and neurodegeneration in the context of the free radical–mitochondrial theory of aging, (2) impact our understanding of the pathogeneses and regional predilections of various senescence-dependant human neurological afflictions, and (3) possibly inform strategies for therapeutic intervention that target the molecular pathways serving to bind the components of this tripartite lesion. In the following section, a model of glial senescence is presented that lends support to our central hypothesis and provides a useful experimental paradigm for delineation of key events favoring pathological iron deposition and mitochondrial injury in the aging and degenerating CNS.

## Glial Iron Sequestration in Brain Aging and Disease

### Astrocytes

Astrocytes, the most abundant and diversified of the neuroglia, subserve a broad range of adaptive functions. Under certain circumstances, however, astrocytes may mediate dystrophic effects within the CNS and thereby contribute to neurological decline. Examples of the latter include the bioactivation of proneurotoxins; the release of excitotoxic amino acids following tissue hypoxia, metal exposure, and OS; formation of epileptogenic scar tissue in response to

CNS injury; and neoplastic transformation and malignant behavior. Astrocyte hypertrophy, accumulation of glial fibrillary acidic protein (GFAP)-positive intermediate filaments, and possibly hyperplasia (reactive gliosis) are characteristic pathological features of the major aging-related neurodegenerative disorders including AD, PD, Huntington's disease, and ALS. Gliosis also occurs, to a lesser extent, in the course of normal brain aging. Increases in monoamine oxidase B (MAO-B) activity accompany reactive gliosis in the aging mammalian brain, in AD, and in experimental models of PD. Under these circumstances, neuronal injury may be fostered by excessive $H_2O_2$ derived from the accelerated oxidative deamination of dopamine (DA) and other monoamines.

### Iron Deposition in Subcortical Astroglia

In the course of normal aging, a subpopulation of astrocytes residing within the mammalian hippocampus, basal ganglia and other subcortical brain regions progressively accumulates cytoplasmic granules replete with ferrous iron and occasionally other transition metals such as copper and chromium. For reasons discussed earlier, the accumulation of these iron-rich inclusions may render the senescent nervous system increasingly prone to free radical-mediated degeneration. We have determined that the gliosomes are derived from damaged, metal-laden mitochondria engaged in a complex macroautophagic process. In turn, the gliosomes may give rise to corpora amylacea (CA), glycoproteinaceous inclusions that accumulate in astrocytes and the extracellular space in subpial and periventricular regions of the aging and degenerating human brain. In rodents, the appearance of this senescent glial phenotype can be accelerated *in vivo* and/or *in vitro* by dietary iron and copper supplementation, cranial X-irradiation, mechanical trauma, cysteamine (CSH) administration, and, in certain hypothalamic nuclei, chronic estrogenization. We have also garnered evidence that intracellular OS is a 'final common pathway' leading to the transformation of normal astrocyte mitochondria to peroxidase-positive inclusions *in vitro* and in the intact aging brain. Thus, the pathological triad of OS, iron sequestration, and mitochondrial damage is fully manifest in the development of this senescent gliopathy. The topography and intensity of endogenous glial peroxidase activity (iron-laden mitochondria) may denote CNS regions that are particularly susceptible to OS during normal aging and under pathological conditions.

In rat astroglial cultures, the pro-oxidant effects of CSH, hydrogen peroxide, menadione, DA, $\beta$-amyloid$_{40/42}$, tumor necrosis factor-$\alpha$ (TNF-$\alpha$), and interleukin-1$\beta$ (IL-1$\beta$) promote the uptake of non-transferrin-derived $^{59/55}$Fe, but not diferric-transferrin-derived $^{59/55}$Fe, by the mitochondrial compartment. These findings are commensurate with the fact that pathological iron accumulation in the degenerating human CNS appears to be a transferrin-independent process. Moreover, the targeting of excess cellular iron to mitochondria may readily explain why bioenergetic failure is a ubiquitous feature of neural tissues exhibiting iron overload. As discussed in the following section, the heme-degrading enzyme, HO-1, may be a pivotal factor promoting the establishment of this aging-associated gliopathy.

## Heme Oxygenase-1: Transducer of Mitochondrial Iron Deposition in 'Stressed' Astroglia

### Heme Oxygenase

Heme oxygenases reside within the endoplasmic reticulum where they catalyze, in collaboration with reduced nicotinamide adenine dinucleotide phosphate (NADPH) cytochrome P450 reductase, the oxidative cleavage of heme to biliverdin, free ferrous iron, and carbon monoxide (CO). Biliverdin is metabolized further to the bile pigment, bilirubin, via biliverdin reductase (**Figure 1**). At least two isoforms of heme oxygenase, HO-1 (also called HSP32) and HO-2, are expressed in mammalian cells. Basal HO-1 expression in the normal rodent brain is restricted to small groups of neurons and neuroglia, whereas HO-2 protein is widely distributed throughout the neuraxis. Diverse response elements in the mammalian *ho-1* promoter and enhancer regions render the gene highly inducible by numerous pro-oxidant and inflammatory stimuli, including $H_2O_2$, transition metals, ultraviolet light, T helper cell (Th1) cytokines, prostaglandins, lipopolysaccharide, dopamine, $\beta$-amyloid, and the natural substrate, heme. By promoting the conformational modification and accelerated degradation of vulnerable tissue hemoproteins, SW Ryter and RM Tyrrel surmised that OS may transiently increase the intracellular 'free heme pool.' In these stressed cells, the upregulation of HO-1 may confer protection by accelerating the conversion of pro-oxidant heme to the radical-scavenging bile pigments, biliverdin and bilirubin. In many tissues overexpressing HO-1, potential toxicity resulting from the intracellular liberation of heme-derived ferrous iron is mitigated by co-induction of apoferritin synthesis, a major iron sequestration pathway. However, the Piantadosi lab and others have shown that heme-derived iron and CO may at times exacerbate intracellular OS and substrate damage by facilitating ROS

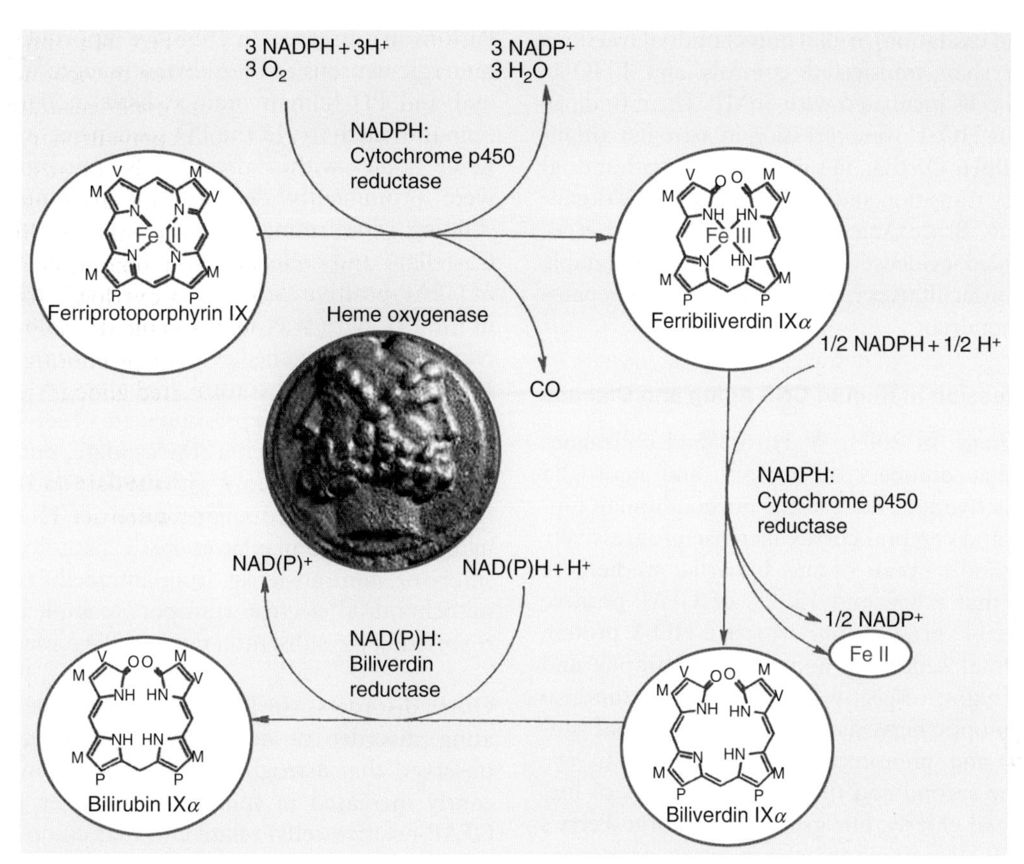


**Figure 1** The heme catabolic pathway. The heme products – ferrous iron (Fe II), carbon monoxide, and biliverdin/bilirubin – may behave as either pro-oxidants or antioxidants, accounting for the disparate influences of heme oxygenase expression on cellular survival (symbolized by Janus faces). In the structures depicted, M is methyl, V is vinyl, and P is propionate. Adapted from Ryter SW and Tyrrell RM (2000) The heme synthesis and degradation pathways: Role in oxidant sensitivity. Heme oxygenase has both pro- and antioxidant properties. *Free Radical Biology & Medicine* 28: 289–309, with permission.

production within the mitochondrial and other subcellular compartments.

### HO-1 and Mitochondrial Iron Sequestration

In rat astroglial cultures, we have observed that HO-1 mRNA, protein, and/or activity levels increase within hours of treatment with CSH, DA, $\beta$-amyloid$_{40/42}$, TNF-$\alpha$, or IL-1$\beta$. After 3–6 days of exposure to these stimuli, uptake of non-transferrin-derived $^{59/55}$Fe (but not diferric transferrin-derived iron) is significantly augmented by the mitochondrial compartment relative to untreated control cultures. Coadministration of antioxidants (ascorbate, melatonin, or *trans*-resveratrol) suppresses the HO-1 response to CSH, DA, and the proinflammatory cytokines, indicating that OS is a common mechanism mediating glial *ho-1* gene induction under these experimental conditions. HO-1 overexpression appears necessary and sufficient for the mitochondrial iron trapping in oxidatively challenged astroglia because the latter can be (1) attenuated by tin mesoporphyrin (SnMP), a competitive inhibitor of heme oxygenase activity, or dexamethasone (DEX), a transcriptional suppressor of the *ho-1* gene and (2) recapitulated by transient transfection of the human (h) *ho-1* gene.

### HO-1, Intracellular OS, and the Mitochondrial Permeability Transition

We have demonstrated that mitochondrial iron trapping by hHO-1-transfected glia and cells exposed to DA, TNF-$\alpha$, or IL-1$\beta$ is abrogated by co-treatment with cyclosporin A or trifluoperazine, inhibitors of the mitochondrial permeability transition pore. These findings suggest that, in 'stressed' astroglia, engagement of this membrane megachannel facilitates the import of cytosolic iron (and possibly other metal ions) to the mitochondrial matrix. We determined that (1) treatment with SnMP or antioxidants (ascorbate, melatonin or resveratrol) blocks the compensatory induction of the redox-sensitive *mnsod* gene in astrocytes challenged with DA or transiently transfected with hHO-1 cDNA and (2) hHO-1 transfection significantly augments levels of protein carbonyls (protein oxidation), 8-epiPGF2$\alpha$ (lipid peroxidation), and 8-OHdG (nucleic acid oxidation) in glial mitochondrial fractions relative to sham-transfected controls and hHO-1-transfected cells incubated with SnMP. These findings indicate that HO-1 overexpression in astroglia amplifies intracellular OS that, in turn, favors mitochondrial permeability transition and iron influx into this organelle. Of note, Bruce Ames' group and U Rauen et al. have provided evidence that increases in chelatable cytosolic iron facilitate permeability transition in hepatocyte mitochondria.

### HO-1 Expression in Human CNS Aging and Disease

**Normal aging** In 2003, W Hirose and colleagues reported that numbers of neurons and neuroglia immunoreactive for HO-1 in the normal human hippocampus and cerebral cortex increase progressively between 3 and 84 years of age. In earlier studies, we calculated that 6.8% and 12.8% of GFAP-positive astrocytes co-express immunoreactive HO-1 protein in the normal senescent human hippocampus and substantia nigra, respectively. HO-1 expression also increases in optic nerve astrocytes and in retinal ganglion cells and photoreceptors of the human eye between the second and the seventh decades of life. In the normal elderly, our group and George Perry's lab noted HO-1 staining to be particularly prominent in choroid plexus epithelial cells, ependymocytes, cerebrovascular endothelial cells, many glial and extracellular CA, and the olfactory neuroepithelium. On the basis of the previously discussed evidence and the tenets of the free radical–mitochondrial theory of aging, we propose that a self-reinforcing cascade of ambient OS, glial HO-1 expression, and mitochondrial iron deposition may contribute to the gradual depletion of high-energy substrates in aging human neural tissues.

**Alzheimer's disease** In the AD hippocampus, we have calculated that ~86% of GFAP-positive astrocytes exhibit HO-1 immunoreactivity, whereas only 6–7% of hippocampal astroglia in age-matched normal control tissue express this stress protein. In collaboration with David Bennett's group in Chicago, we observed significant induction of astroglial HO-1 expression in the brains of individuals with amnestic mild cognitive impairment (a harbinger of incipient AD), indicating that glial HO-1 upregulation is an early event in the pathogenesis of this dementing disorder. To the extent that our *in vitro* data can be extrapolated to the intact human brain, it is conceivable that the perturbed CNS iron homeostasis and mitochondriopathy characteristic of AD may be downstream effects of chronic HO-1 overproduction in the diseased neural tissues.

**Parkinson's disease** In 1998, we reported that dopaminergic neurons of the substantia nigra in both normal and PD human brain exhibit moderate HO-1 immunoreactivity. In the PD specimens, cytoplasmic Lewy bodies within affected dopaminergic neurons were prominently decorated with annular HO-1 staining, confirming an earlier observation by R Castellani and colleagues. In our study, the fraction of GFAP-positive astrocytes expressing HO-1 protein in the PD nigra was substantially increased (77.1%) relative to age-matched control preparations (12.8%). In both the PD and control specimens, relatively low levels of glial HO-1 expression were observed in other subcortical nuclei, such as the caudate, putamen, and globus pallidus. In accord with the data and our central hypothesis, sustained augmentation of HO-1 activity in nigral astroglia may promote the pathological deposition of nontransferrin iron, intracellular OS, and mitochondrial electron transport (complex I) deficits reported in the substantia nigra of PD patients.

**Other disorders** In MS, an autoimmune demyelinating disorder of central white matter, we have observed that astroglial HO-1 expression is significantly increased in spinal cord plaques (57.3% of GFAP-positive cells) relative to that computed in the spinal white matter of normal individuals (15.4%). As argued in the cases of AD and PD, enhanced intraglial heme degradation catalyzed by HO-1 may contribute to the aberrant iron deposits and mitochondrial lesions reported in human MS and in rodent models of this condition. Robust HO-1 induction has also been documented in the brains of patients with Pick disease, corticobasal degeneration, progressive supranuclear palsy, ischemic and hemorrhagic stroke, and *Plasmodium falciparum* malaria. The triad of oxidative substrate damage, deregulated iron homeostasis, and bioenergetic failure occurs in most, if not all, of these conditions and may be secondary to the antecedent overexpression of HO-1 in 'stressed' astroglia.

## Conclusions and Implications

The mechanisms responsible for the augmented deposition of iron and other redox-active transition metals in the aging and diseased CNS remain incompletely understood. On the basis of a considerable body of neuropathological and experimental data, we propose that oxidative molecular damage, the deposition of nontransferrin iron (and possibly other metals), and mitochondrial insufficiency in these conditions are causally interdependent phenomena and inextricable components of a single cytopathological 'lesion.' This central hypothesis predicts that iron homeostasis will invariably be perturbed in CNS tissues harboring mitochondrial lesions, that neural tissues manifesting iron overload will exhibit obligatory bioenergetic failure, and that at least some of the aberrant metal deposits will co-localize to the mitochondrial compartment. We have reviewed evidence that induction of the heme-degrading enzyme, HO-1, in astroglia stressed by aging or disease may represent a 'final common pathway' for the sequestration of nontransferrin iron by the mitochondrial compartment and transformation of these organelles into peroxidase-positive inclusions and CA, structural biomarkers of the senescent mammalian CNS. Given the exquisite sensitivity of the glial *ho-1* gene to ambient OS and other noxious stimuli, it is not surprising that pathological iron deposition and mitochondrial damage are a fairly consistent response of neural tissues to a broad range of genetic and acquired metabolic disturbances. Thus, OS accruing from an excessive amyloid burden or the elaboration of proinflammatory cytokines may be responsible for glial HO-1 induction, iron sequestration, and bioenergetic failure in the AD forebrain. Dopamine-derived ROS, 1-methyl-4-phenyl-1,2,3,6-tetrahydropyridine (MPTP)-like xenobiotics, endoneurotoxins, nitric oxide, and Th1 cytokines are plausible inducers

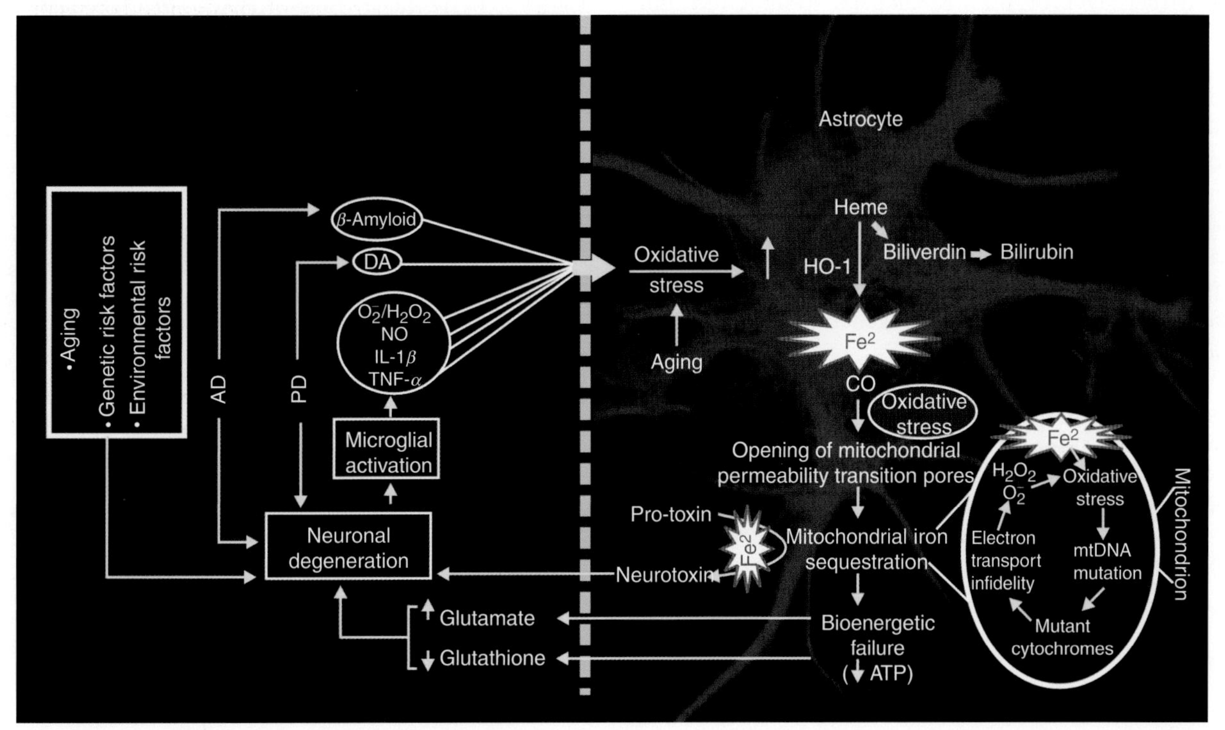


**Figure 2** A model for pathological iron deposition, OS, and mitochondrial insufficiency in brain aging and disease. Extracellular pro-oxidant stressors and intrinsic aging processes upregulate heme oxygenase-1 (HO-1) in astrocytes, liberating heme-derived free ferrous iron ($Fe^2$) and carbon monoxide (CO). The latter generate intraglial OS that promotes opening of mitochondrial permeability transition pores and influx of (non-transferrin-derived) iron into the mitochondrial matrix. The redox-active iron augments oxidative damage to the electron transport chain machinery (mitochondrial DNA (mtDNA), cytochromes), resulting in a vicious spiral of mitochondrial ROS generation and injury in accord with the prevailing free radical–mitochondrial theory of aging. This progressive state of astroglial mitochondrial heteroplasmy endangers nearby neuronal constituents by several mechanisms: (1) The mitochondrial ferrous iron catalyzes the bioactivation of protoxins (e.g., catechols, 1-methyl-4-phenyl-1,2,3,6-tetrahydropyridine (MPTP)) to potent neurotoxins (*o*-semiquinones, $MPP^+$). The latter may be extruded to the extracellular space and exert direct dystrophic effects on susceptible neuronal targets. Diminished ATP levels accruing from glial bioenergetic failure may further predispose to neuronal degeneration by (2) augmenting extracellular glutamate concentrations (excitotoxicity), (3) curtailing glutathione (GSH) delivery to the neuronal compartment, and (4) limiting the degradation of potentially toxic protein aggregates by the ubiquitin–proteosome system. Neuronal degeneration stimulates microglial activation, resulting in the release of ROS, nitric oxide (NO), and proinflammatory cytokines. The latter further induce the *ho-1* gene in indigent astroglia, completing a self-sustaining loop of pathological cellular interactions, perpetuating oxidative damage and mitochondrial insufficiency within senescent and degenerating neural tissues. Genetic and environmental risk factors may confer disease specificity by superimposing unique pathological signatures on this core lesion. In the case of AD and PD, for example, β-amyloid deposition in the hippocampus and dopamine (DA) release from dying nigrostriatal projections, respectively, feed into the pathological cascade by stimulating HO-1 activity in regional astroglia. Progressive derangement of astroglial iron mobilization and mitochondrial insufficiency would then drive the primary neurodegenerative process regardless of its etiology. $H_2O_2$, hydrogen peroxide; IL-1β, interleukin-1β; TNF-α, tumor necrosis factor-α. Adapted from Schipper HM (2004) Brain iron deposition and the free radical-mitochondrial theory of aging. *Aging Research Reviews* 3: 265–301.

of the astroglial heme oxygenase–mitochondrial iron axis in the PD nigra, and iron-mediated mitochondrial injury in MS may be secondary to the upregulation of glial HO-1 by IL-1$\beta$, TNF-$\alpha$, and/or myelin basic protein within the affected white matter.

Importantly, neuron-like PC12 cells co-cultured with rat astrocytes replete with mitochondrial iron (induced by CSH pretreatment) or overexpressing the human *ho-1* gene were found to be far more vulnerable to OS-related killing than were PC12 cells grown atop control ('iron-poor') astroglia. Thus, the accumulation of iron-laden, metabolically deficient astroglia may seriously compromise the integrity and viability of nearby neuronal constituents and thereby promote neurological dysfunction and disease. In our model, pathological, mutually reinforcing glial–neuronal interactions may perpetuate oxidative damage, pathological iron deposition, and mitochondrial insufficiency within senescent and degenerating neural tissues long after initiating insults may have dissipated (**Figure 2**). As a putative transducer of mitochondrial iron deposition in 'stressed' astroglia, the augmented glial HO-1 activity may provide a rational target for therapeutic intervention in a host of human neurodegenerative and neuroinflammatory conditions.

*See also:* Aging of the Brain; Aging of the Brain and Alzheimer's Disease; Alzheimer's Disease: Neurodegeneration; Brain Composition: Age-Related Changes; Oxidative Damage in Neurodegeneration and Injury; Parkinson's Disease: Alpha-Synuclein and Neurodegeneration.

## Further Reading

Connor JR (1997) *Metals and Oxidative Damage in Neurological Disorders.* New York: Plenum Press.

Dennery PA (2000) Regulation and role of heme oxygenase in oxidative injury. *Current Topics in Cellular Regulation* 36: 181–199.

Dwyer BE, Takeda A, Zhu X, et al. (2005) Ferric cycle activity and Alzheimer disease. *Current Neurovascular Research* 2: 261–267.

Halliwell B and Gutteridge JMC (2007) *Free Radicals in Biology and Medicine.* Oxford: Oxford University Press.

Heales SJ, Lam AA, Duncan AJ, et al. (2004) Neurodegeneration or neuroprotection: The pivotal role of astrocytes. *Neurochemical Research* 29: 513–519.

Jenner P (2003) Oxidative stress in Parkinson's disease. *Annals of Neurology* 53(supplement 3): S26–S36; discussion S36–S28.

Ponka P (2004) Hereditary causes of disturbed iron homeostasis in the central nervous system. *Annals of the New York Academy of Sciences* 1012: 267–281.

Reddy PH and Beal MF (2005) Are mitochondria critical in the pathogenesis of Alzheimer's disease? *Brain Research Reviews* 49: 618–632.

Ryter SW and Tyrrell RM (2000) The heme synthesis and degradation pathways: Role in oxidant sensitivity. Heme oxygenase has both pro- and antioxidant properties. *Free Radical Biology & Medicine* 28: 289–309.

Schipper HM (2004) Heme oxygenase-1: Transducer of pathological brain iron sequestration under oxidative stress. *Annals of the New York Academy of Sciences* 1012: 84–93.

Schipper HM (2004) Brain iron deposition and the free radical-mitochondrial theory of ageing. *Ageing Research Reviews* 3: 265–301.

Schipper HM (2004) Redox neurology: Visions of an emerging subspecialty. *Annals of the New York Academy of Sciences* 1012: 342–355.

Schipper HM (2004) Heme oxygenase expression in human central nervous system disorders. *Free Radical Biology & Medicine* 37: 1995–2011.

# Oxidative Damage in Neurodegeneration and Injury

**I G Gazaryan and R R Ratan**, Weill Medical College of Cornell University, New York, NY, USA



## Introduction

The etiology of dysfunction and cell death in neurodegenerative diseases remains enigmatic. Most neurodegenerative conditions are insidious in onset and run a gradually progressive and painfully inexorable course for those afflicted and their families. Prototypic neurodegenerative conditions include Alzheimer's disease (AD), Huntington's disease (HD), Parkinson's disease (PD), and amyotrophic lateral sclerosis (ALS). By contrast, acute neurodegenerations are monophasic and characterized by stroke and spinal cord injury. Many neurodegenerative conditions are linked to inherited mutations, resulting in accumulation of the damaged proteins or their wrongly processed variants. Sporadic neurodegenerative conditions are also associated with accumulations of misfolded proteins and their aggregates. Downstream of these events, metal-catalyzed oxidation leading to oxidative stress has been implicated as a common final pathway of injury (**Figure 1**).

## Aging and Neurodegeneration

### Mitochondria as a Major Source of Superoxide under Normal Conditions

Neurodegeneration is directly modified by cell aging. Some theories invoke the mitochondria as the major sites of generation of deleterious free radicals that promote aging. The chemistry of reactive species that could result in oxidative damage under normal physiological conditions is the same for any cell. The major species in cells is the superoxide radical, produced as a by-product of the mitochondrial respiration, in particular by complex I and complex III of the respiratory chain. Complex III breaks a two-electron reducing equivalent into two single-electron reducing equivalents inside the membrane (in what is known as the Q cycle), and the resulting quinone radical reacts with oxygen, giving rise to the superoxide radical. This side reaction is known as 'mitochondrial leakage' and comprises up to 3–4% of the total oxygen consumption. If in a lifetime we consume nearly 60 000 l of oxygen per kilogram of body weight, it means that we produce approximately 2000 l of superoxide per kilogram of body weight. What happens to the released superoxide and how does a cell handle the consequent 'oxidative' load?

### Antioxidant/Antiaging Mechanisms: Superoxide Scavenging

In both mitochondria and the cytosol, superoxide is scavenged and converted into hydrogen peroxide and oxygen by superoxide dismutase (SOD), although the nature of SOD in mitochondria and the cytosol is different (i.e., MnSOD and CuZnSOD, respectively). The formed hydrogen peroxide is further decomposed into water with the help of catalase or is reduced by the glutathione peroxidase/glutathione system. Very recent studies show that the thioredoxin reductase/thioredoxin system is also capable of reducing hydrogen peroxide to water.

There are a number of theoretical reasons why the nervous system in general and neurons in particular are under the biggest load of oxidative stress under basal conditions. Neurons have a very high metabolic rate in order to meet the demands of electrical signaling. Specifically, our brain uses 20% of the total oxygen consumed, although the brain is only 2% of the body weight. An expected corollary of high oxygen utilization is a tenfold higher level of radical production. Paradoxically, despite higher radical production, catalase is absent in brain mitochondria. This is not the case for heart mitochondria.

Given the higher level of radical production and the absence of catalase in brain mitochondria, what are the mechanisms at play inside neuronal mitochondria to neutralize the formed hydrogen peroxide and keep oxidative damage to tolerable levels? In addition to glutathione peroxidase, the only other known system that neutralizes hydrogen peroxide is that of thioredoxin and peroxiredoxin. Reduced thioredoxin provides peroxiredoxin with reducing equivalents to reduce hydrogen peroxide to water. Oxidized thioredoxin is reduced back by a mitochondrial thioredoxin reductase, selenocysteine flavodithiol oxidoreductase. Why mitochondrial catalase is absent in neurons is not well understood, but could be related to neuronal signaling functions of peroxide that are not necessary in other tissues. The absence of catalase from neuronal mitochondria places the burden of neutralizing hydrogen peroxide generated in mitochondria squarely on the shoulders of the thiol-based detoxification systems, particularly glutathione, thioredoxin, and peroxiredoxin. A causal role for mitochondrial peroxide in aging has been supported by the generation of transgenic mice that overexpress catalase in the nucleus, the peroxisome, and the mitochondria. Significant extension of life span was only seen in the animals in which catalase was overexpressed in the mitochondria. The extension of life span was associated with improvement

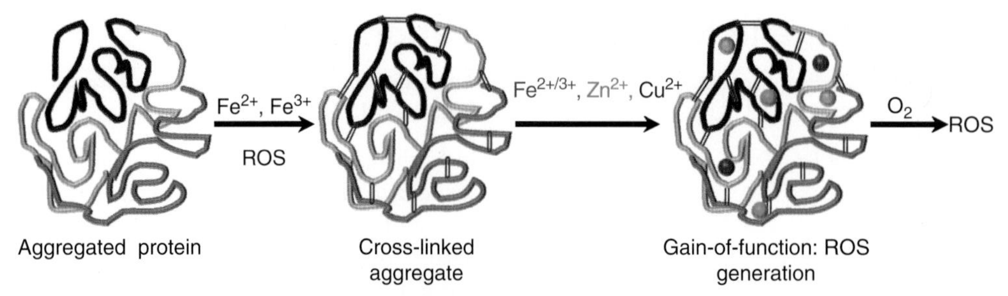


**Figure 1** Widely accepted hypothetic gain-of-function transformation of cross-linked aggregates of misfolded proteins. Confirmed for AD, PD, and ALS, and supposed for HD. ROS, Reactive oxygen species.

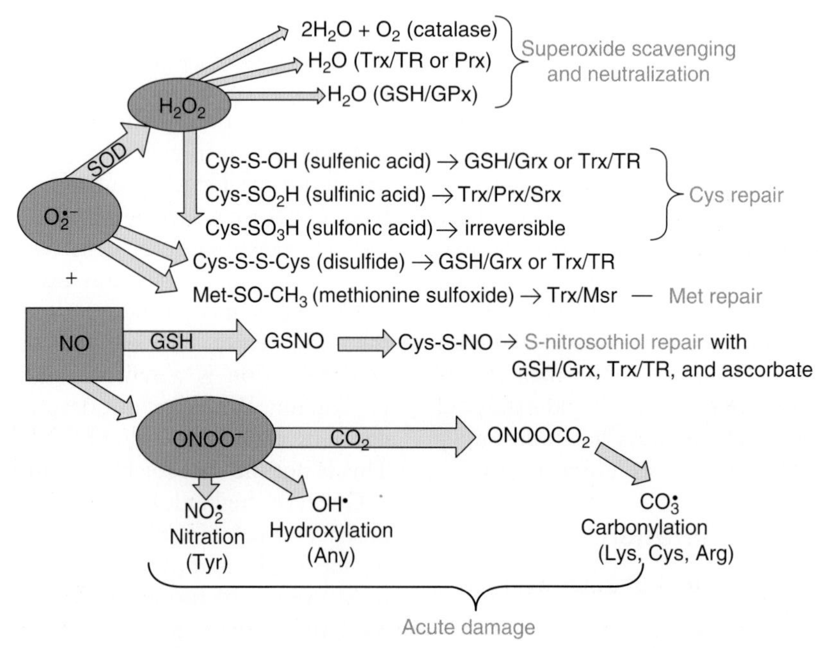


**Figure 2** Reactive oxygen/nitrogen species generation and neutralization reactions and systems: Upper level: Superoxide scavenging reactions catalyzed by superoxide dismutase (SOD), catalase, thioredoxin reductase (TR), peroxiredoxin (Prx), and glutathione peroxidase (GPx) and mediated by thioredoxin (Trx) and glutathione (GSH). Center: Oxidation states of cysteine and methionine and their repair; glutathione repairs most of the oxidative modifications by glutathionylation, followed by glutathione removal with the glutaredoxin (Grx) system; methionine is repaired with the methionine sulfoxide reductase (Msr)/Trx system; sulfonic acid modification is irreversible, while sulfinic acid can be repaired with the use of a recently characterized ATP- and $Mg^{2+}$-dependent enzyme, sulfiredoxin (Srx), which is capable of reducing the sulfinic form of peroxiredoxin. Lower level: Peroxynitrite-generated radicals cause largely nonrepairable damage to proteins, DNA, and lipids.

in cardiac function, decreased development of cataracts, and diminished mitochondrial DNA mutations. It is unclear to what extent the increased life span reflects improvements in brain function.

### Antioxidant/Antiaging Mechanisms: Repair of Oxidized Residues

Unscavenged radicals are thermodynamically driven to oxidize cellular constituents, including protein cysteines and methionines. These critical amino acid residues can be subsequently reduced only with the help of specialized enzyme systems (**Figure 2**). Cysteines are reduced with a glutaredoxin/glutathione system, and methinones are reduced with a methionine sulfoxide reductase (Msr) system. The Msr system includes two enzymes, only one of which belongs to the same enzyme class as glutathione reductase and thioredoxin reductase.

The glutaredoxin/glutathione enzyme system has beneficial effects on functional activity of a number of proteins, including the thiol-containing mitochondrial complex I, the inhibition of which is caused by administration of 1-methyl-4-phenyl-1,2,3,6-tetrahydropyridine (MPTP), a neurotoxin that produces PD-like symptoms in primates, including humans. Interestingly, glutathione depletion is an early feature of

**Table 1** Changes in key proteins regulating the cellular redox status in disease, injury, and aging

| *Protein/enzyme* | *AD (brain)* | *PD (substantia nigra)* | *ALS (spinal cord)* | *Injury, excitotoxicity (brain)* | *Aging (brain)* |
|---|---|---|---|---|---|
| Trx1 | Protein ↓ | n.a. | n.a. | n.a. | n.a. |
| TR | Activity ↑ | n.a. | n.a. | n.a. | n.a. |
| MsrA | Activity ↓ | n.a. | n.a. | n.a. | Activity ↓, protein ↓, mRNA ↓ |
| Prx1 | Protein unchanged | Protein ↓ | n.a. | n.a. | n.a. |
| Prx2 | Protein ↑ | Protein ↓ | Protein ↓ | Protein ↑ | Protein ↓ |
| Prx3 | Protein ↓ | Protein ↓ | | Protein ↓ | Protein ↓ |
| GPx1 | mRNA ↑ | mRNA ↓ | Protein ↓ | n.a. | Activity ↑ |
| GR | mRNA ↑ | n.a. | n.a. | n.a. | Activity ↓ |
| GCL | n.a. | Activity unchanged | n.a. | n.a. | Activity ↓, protein ↓ |
| Grx1 | Protein ↑ | n.a. | n.a. | Protein ↑ | n.a., ↓ in interfibrillary mitochondria |
| SOD1 | Protein ↑ | n.a. | Activity ↓ | Protein ↑ | n.a. |
| NADH:UQ reductase | Protein ↓ | Protein ↓ | n.a. | n.a. | n.a. |

Trx1, thioredoxin; TR, thioredoxin reductase; MsrA, methionine sulfoxide reductase A; Prx1–3, peroxiredoxins; GPx1, glutathione peroxidase 1; GR, glutathione reductase; GCL, glutamate cysteine ligase; Grx1, glutaredoxin 1; SOD1, superoxide dismutase; NADH: UQ reductase, NAD(P)H:ubiquinone oxidoreductase (antioxidant response element-dependent, two-electron reduction of quinones); n.a., not available.

PD and appears to be disease specific. The depletion of glutathione may explain defects in complex I function found in sporadic forms of PD.

Msr, another important enzyme in repair of proteins, is proposed to play a central role in neurodegeneration and aging. Genetically engineered organisms overexpressing the enzyme live longer. On the contrary, the reduced or suppressed activity of the enzyme results in shortened life span, hippocampal degeneration, and increased sensitivity to oxidative stress. The enzyme can repair oxidized methionines and restore the function of many important proteins – for example, calmodulin. Msrs act on oxidized calmodulin and repair all eight methionine sulfoxide residues initially present in the inactive protein.

As mentioned, neuronal mitochondria protect themselves against hydrogen peroxide formed inside the matrix solely with the thiol-dependent enzyme systems. Moreover, repair systems also largely depend on reduced thiols (see earlier). In many cases, the catalytically important enzyme thiols are shown to be protected against oxidative damage by glutathionylation. Glutathionylation is a posttranslational, reversible redox modification of proteins by thiol/disulfide exchange. Indeed, glutathione used to be a dominant theme in brain neurodegeneration, and only recently the significance of thioredoxin-based systems has been fully appreciated. Measurements of reduced nicotinamide adenine dinucleotide phosphate (NADPH)-dependent thiol reducing activity in brain mitochondria show that glutathione reductase and thioredoxin reductase are equally important. Thiol-dependent systems of hydrogen peroxide neutralization and subsequent repair are also targets for oxidative modification (**Table 1**), and this obviously creates a threshold of oxidative damage that can be handled by neuronal mitochondria. Once reached, it results in mitochondrial failure and subsequent cell death by either apoptotic or necrotic pathways.

The delicate balance existing under normal conditions between oxidant production and antioxidant defensive mechanisms (**Figure 3**) can either be slowly shifted with aging or disturbed at once upon acute injury to produce oxidative stress. The adaptive response to tissue damage includes the activation of transcription factors such as nuclear factor-kappa B (NF-κB), among others, to switch on the inflammation response to limit damage and promote repair.

## Injury and Neurodegeneration

The most common form of neurodegeneration occurs after an acute injury. The causes of neuronal injury are many, and include stroke, epilepsy, trauma, DNA damage from radiation, chemotherapeutic agents, and exposure to environmental neurotoxins. Acute neuronal damage involves a complex combination of processes, including excitotoxicity, inflammation, necrosis, and apoptosis.

### Excitotoxicity

Excitotoxicity is a particularly important event that initiates acute neurodegeneration. Accumulation of glutamate within the synaptic cleft leads to $Ca^{2+}$

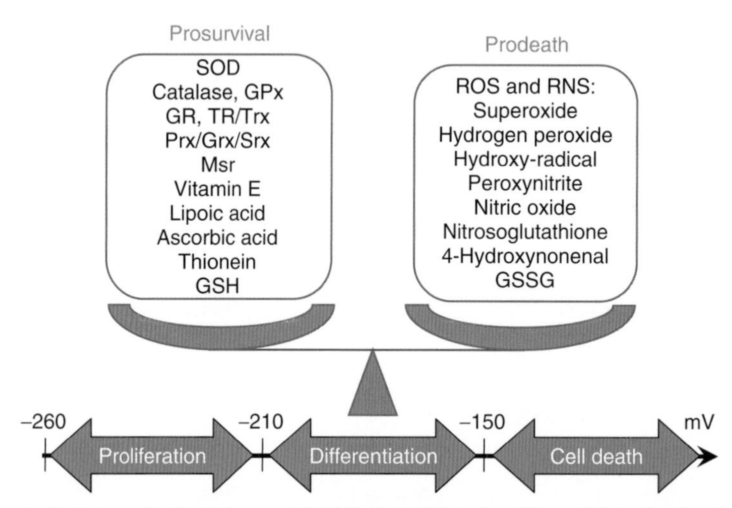


**Figure 3** Delicate balance between prodeath and prosurvival factors determines the cell fate *in vivo*. For cultured nerve cells, the depletion of intracellular redox potential, shown as that for the glutathione/oxidized glutathione (GSSG/GSH) couple, below a certain level determines whether the cells will differentiate or die. SOD, superoxide dismutase; ROS, reactive oxygen species; RNS, reactive nitrogen species; GPx, glutathione peroxidase; GR, glutathione reductase; TR/Trx, thioredoxin reductase/thioredoxin; Prx/Grx/Srx, peroxiredoxin/glutaredoxin/sulfiredoxin; Msr, methionine sulfoxide reductase.

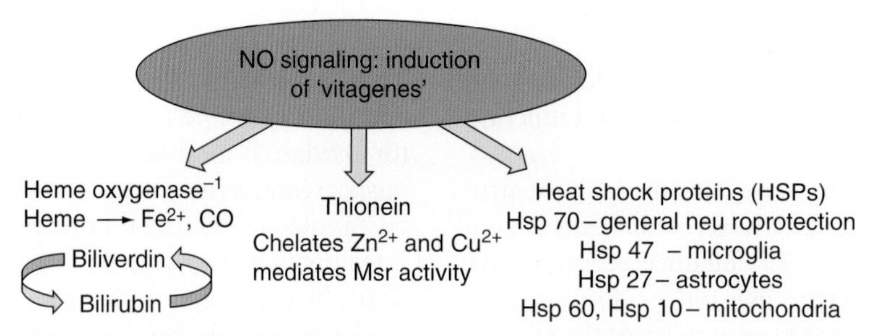


**Figure 4** Prosurvival role of nitric oxide (NO): induction of synthesis of heme oxygenase, which provides cells with an antioxidant bilirubin/biliverdin/biliverdin reductase system; synthesis of thioneins, which chelate metal ions and in addition can mediate/enhance methionine sulfoxide reduction by methionine sulfoxide reductase (Msr) and thioredoxin (Trx); and synthesis of heat shock proteins.

influx via hyperactivation of *N*-methyl-D-aspartate receptors, voltage-gated $Ca^{2+}$ channels, and nonspecific cation conductances. The latter nonspecific cation channels are represented by the recently discovered transient receptor potential (melastatin) (TRPM) ion channels, among which TRPM7 and TRPM2, shown to be permeable for $Ca^{2+}$ and inhibited by gadolinium, appear to play a critical role in anoxic cell death.

Excess $Ca^{2+}$ via ionotropic glutamate receptors results in the initiation of neurotoxic signaling cascade by activating calmodulin and neuronal nitric oxide synthase (nNOS). Current opinion holds that the intracellular redox state is the critical factor determining whether in brain cells NO is toxic or protective (**Figure 4**). NO is known to signal the induction of heme oxygenase-1, which is considered as a prosurvival enzyme. It degrades heme, yielding ferrous iron, carbon monoxide (CO), and biliverdin, which cycles between the oxidized and reduced form, bilirubin. The latter exhibits strong antioxidant properties. Biliverdin reductase is present in brain in large functional excess, suggesting that such redox cycling amplifies antioxidant effects of heme oxygenase expression. On the other hand, NO also reacts with glutathione, generating nitrosoglutathione, which is able to modify and thus inactivate protein thiols. However, the most dangerous species – peroxynitrite – is generated via the direct interaction of superoxide radical and NO, a comparatively stable and harmless molecule. Peroxynitrite is extremely reactive and capable of nitrating tyrosine residues in proteins (**Figure 2**). A downstream effect of peroxynitrite production is the activation of TRPM channels. Molecular deletion of TRPM2 or TRPM7 channels renders neurons resistant to hypoxia/aglycemia. Whether peroxynitrite induces activation of TRPM channels via nitration of its critical components or some upstream modulatory protein is unclear. Protein nitration has been shown to

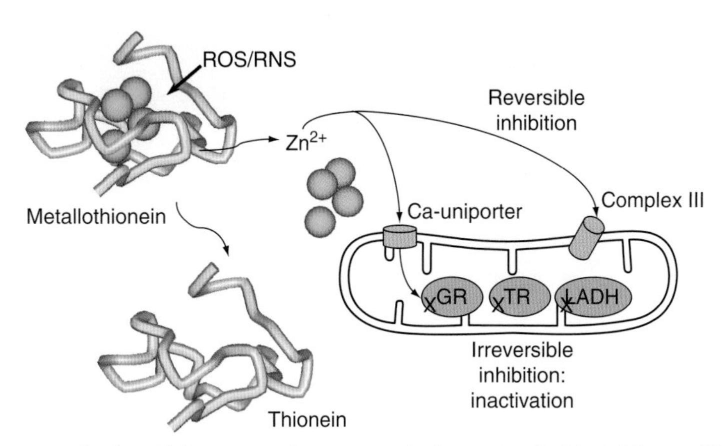


**Figure 5** Zn-induced damage to mitochondrial enzymes of energy production and antioxidant defense. GR, glutathione reductase; TR, thioredoxin; LADH, lipoamide dehydrogenase.

take place in brain injuries and in some but not all neurodegenerative diseases.

Superoxide- and especially peroxynitrite-induced modifications of the metallothionein thiols result in the release of zinc. Neurodegenerative diseases are characterized by a mobilization of intracellular zinc. The latter has been shown to mediate NO-induced neuronal death by directly affecting mitochondrial respiration through inhibiting complex III. In addition, it has recently been shown that zinc is capable of entering mitochondria and directly inactivating lipoamide dehydrogenase, the terminal enzyme of major multienzyme energy-producing complexes, as well as glutathione reductase and thioredoxin reductase. The inactivation is irreversible, and only the newly synthesized proteins transported into mitochondria may compensate for the damage (**Figure 5**). Thus, acute oxidative damage, resulting in massive release of intracellular zinc, will cause mitochondrial failure and cell death.

### Inflammation

Oxidative stress can directly or indirectly initiate an inflammatory cascade. Collective evidence from many recent studies suggests that increased phospholipase A2 (PLA2) activity plays a central role in acute inflammatory responses in the brain as well as in oxidative damage associated with AD, PD, and multiple sclerosis (MS). PLA2 contributes to pathogenesis of neuroinflammation by attacking neural membrane phospholipids, to yield arachidonic acid and lysophospholipids. These are subsequently metabolized to a variety of proinflammatory lipid mediators such as prostaglandins, leukotrienes, thromboxanes, and platelet-activating factor. Arachidonic acid metabolism is also one of the major sources of oxidative damage (**Figure 6**). Arachidonic acid undergoes catalytic oxidation by cyclooxygenases 1 and 2, and, in addition, can react with superoxide or peroxynitrite to generate one of the most potent modifying agents, 4-hydroxy-2-nonenal (HNE). The latter can modify lysine, cysteine, and histidine residues in proteins and can bind to free amino acids and deoxyguanosine. Immunostaining for HNE-modified proteins shows that such modification is characteristic of acute oxidative stress, occurring during brain and spinal cord injury.

Inflammation is characterized with the increased deposition of iron. Iron, especially in the ferrous or unbound form, is able to catalyze the formation of free radicals and could be a cause of neuronal injury. Depletion of antioxidants in the brain, rise in iron-dependent oxidative stress, and monoamine oxidase B (MAO B) activity are key characteristics of aging that contribute to the onset of neurodegenerative disorders. MAO B is a mitochondrial flavin-dependent enzyme that catalyzes oxidative deamination of neurotransmitters and exogenous arylalkylamines. MPTP, an impurity in synthetic heroin, being activated by MAO B, affords a widely used chemically induced model of PD. $MPP^{+}$, the toxic product of MPTP conversion, increases superoxide formation by suppressing activity of NADH dehyrogenase (complex I) and increasing leak of electrons to oxygen. Superoxide is then available to attack Fe/S cluster proteins such as aconitase. Destabilization of cytosolic as well as mitochondrial aconitase results in an increase in iron regulatory protein-1 (IRP-1), an RNA-binding protein that signals cellular iron deficiency. This leads to the paradoxical and maladaptive increase in iron in the cell. The mechanisms underlying iron cellular toxicity are only beginning to emerge. A representative trend in neuroprotective drug development is to combine an iron chelator and a monoamine oxidase inhibitor in the same compound (see later). Iron chelators have recently been shown to

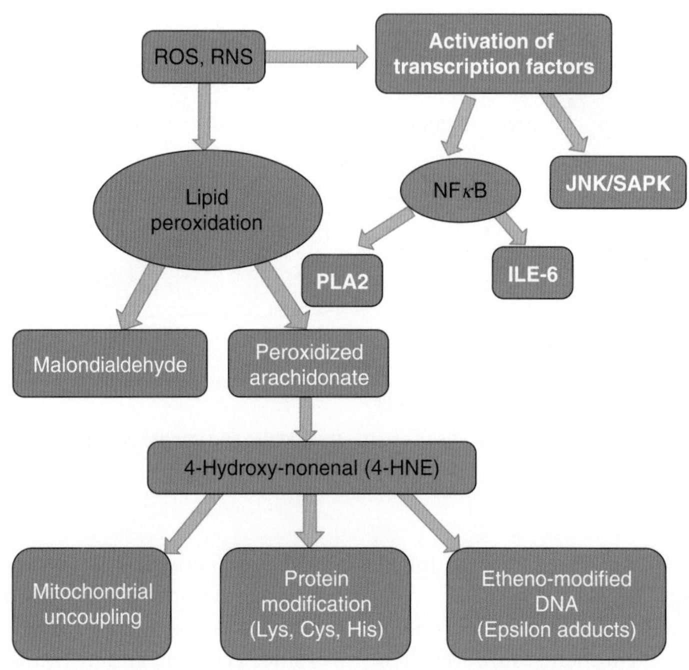


**Figure 6** 4-Hydroxynonenal, the most damaging product of lipid peroxidation. Phospholipase A2 (PLA2) inhibitors were shown to be neuroprotective. ROS, reactive oxygen species; RNS, reactive nitrogen species; NFκB, nuclear factor-kappa B; JNK/SAPK, c-Jun N-terminal kinase/stress-activated protein kinase; ILE-6, interleukin-6.

have the ability to induce adaptive gene expression via the stabilization of factors such as hypoxia-inducible factor-1 (HIF-1). HIF is a heterodimeric transcriptional complex that mediates the induction of more than 70 genes involved in hypoxic compensation, including those for vascular endothelial growth factor (VEGF) and erythropoietin (Epo).

Activated neutrophils and macrophages generate widespread secondary damage at the traumatic site by releasing cytokines and free radicals. They produce inducible NO synthase and NAD(P)H oxidase (Phox), generating nonmitochondrial superoxide; both enzymes cooperate to generate peroxynitrite and thus expose cells to further oxidative damage. In addition to peroxynitrite, leukocytes possess myeloperoxidase, which generates hypochlorite from hydrogen peroxide and chloride anion. Hypochlorite is a strong oxidizing agent capable of chlorinating protein residues and oxidizing membrane lipids. In all cases of rodent neurodegeneration, medications reducing inflammatory responses were shown to exhibit beneficial effects on the disease progression.

## Apoptosis

The discovery that oxidative stress can trigger a program of neuronal cell death with features of apoptosis was significant in several aspects. First, it showed that oxidative stress does not always result in random and disordered cell damage. Second, it demonstrated the possibility of a free radical triggering an endogenous program of cell suicide.

Oxidative stress has been shown to activate a host of downstream signaling pathways leading to apoptosis. In some schemes, the c-Jun N-terminal kinase (JNK) signaling pathway is activated, resulting in Bax translocation, cytochrome *c* release, and apoptosis. In other schemes, Erk activation or poly(ADP-ribose) polymerase (PARP) activation leads to translocation of an apoptosis inducing factor and a caspase-independent cell death. Oxidative stress has been implicated in activation of a cell cycle resulting in cell death, although two studies in which oxidative stress has been induced by downregulating antioxidant defenses failed to demonstrate a protective effect of cell cycle inhibitors.

The superoxide released by mitochondria is directed both 'in' to and 'out' of the mitochondrial inner membrane and, therefore, is capable of damaging mitochondrial DNA, matrix proteins, inner membrane proteins and lipids, and cytosolic proteins and nuclear DNA. With respect to DNA damage, more than 1000 DNA-damaging events occur in each mammalian cell every day, due to replication errors and cellular metabolism. To cope with the deleterious consequences of DNA lesions, cells are equipped with efficient defense mechanisms to remove DNA damage by DNA repair pathways, control cell cycle progression,and eliminate damaged cells via apoptosis. The complicated network of DNA repair mechanisms includes base excision repair, transcription-coupled repair, global genome repair, mismatch repair, homologous recombination, and nonhomologous end-joining damage. Evolution has overlaid the core cell cycle machinery with a series of surveillance pathways termed 'cell cycle checkpoints.' Checkpoints in proliferating cells tightly control progress through the cell cycle; cells may be arrested at any of the checkpoints, and either DNA will be repaired or cells will die by apoptosis. It appears that apoptosis induced by oxidative damage may be promoted by multiple pathways, both cell cycle dependent and independent. The determining factor for the pathways induced is an area of active exploration (**Figure** 7).

## Nature of Aggregated Deposits in Neurodegenerative Diseases

### Alzheimer's Disease

First described by Alois Alzheimer in 1906, AD has two characteristic 'deposits,' tangles and plaques. The latter are found in brain parenchyma and in blood vessels and are formed from amyloid mixed with inflammatory cells. Amyloid presents cross-linked A$\beta$-peptides formed via wrong processing of amyloid precursor protein (APP). Genetic AD is associated with mutations in APP or in presenilins (1 and 2), the proteins responsible for APP processing. Neurofibrillary tangles can be formed from the oxidatively cross-linked intraneuronal tau protein and occur mainly in large pyramidal neurons. The oxidative stress results in more and more dense plaques and tangles. Once formed, tangles and plaques directly and indirectly enhance the oxidative stress. The direct pathway represents the oxidative activity of metal-bound ($Zn^{2+}$, $Fe^{2+}$, or $Cu^{2+}$) plaques and tangles. The indirect pathway includes the activation of microglial cells, which in turn generate inflammatory messengers, putting the brain on permanent 'red alert.' This indirect pathway had led some to call AD 'arthritis of brain.' Thus oxidative stress may precipitate neuropathological hallmarks of AD or may be a consequence of these neuropathological changes. Disappointing therapeutic results with antioxidants in this disease could reflect an inability to adequately neutralize damaging radicals in all of the cell types and subcellular compartments afflicted in AD.

### Parkinson's Disease

PD is a common neurodegenerative disorder affecting 1% of the population over the age of 65 years. The disease is characterized by the loss of dopaminergic neurons in the substantia nigra pars compacta, as well as by the presence of Lewy body inclusions in these cells. Five genes have been identified that are associated with relatively rare familial PD, by either autosomal dominant or recessive transmission. These include genes encoding $\alpha$-synuclein and dardarin/leucine-rich kinase 2 (LRRK2), which are associated with dominantly inherited PD, and genes for parkin, DJ-1, and pten-induced kinase 1 (PINK1), which cause recessively inherited PD. $\alpha$-Synuclein readily aggregates and is a major fibrillar component of Lewy bodies.

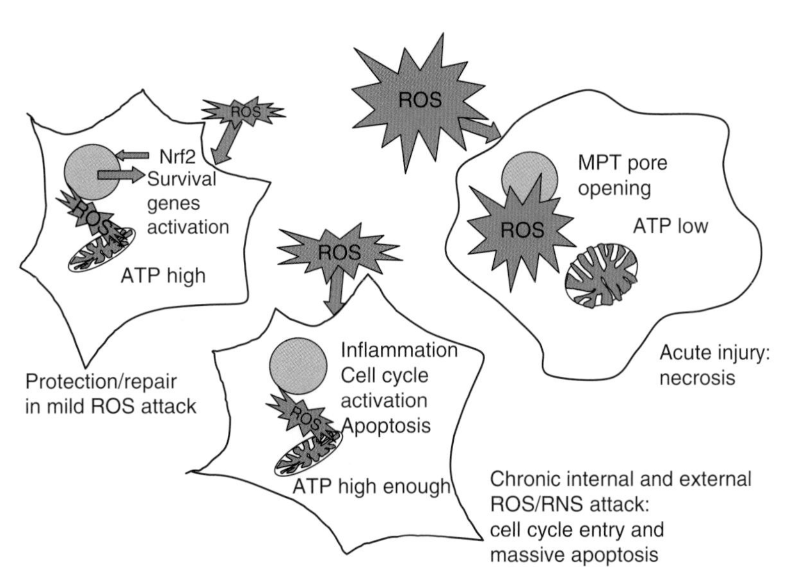


**Figure 7** Cell fate depends on the level of oxidative stress. ROS, reactive oxygen species; MPT, mitochondrial permeability transition; Nrf2, nuclear factor E2-related factor 2; RNS, reactive nitrogen species.

Aggregation of α-synuclein is enhanced with tissue transglutaminase: Lewy bodies in PD patients are positively immunostained with antibodies recognizing isodipeptide bonds, a marker of tissue transglutaminase cross-linking. Phosphorylation also promotes α-synuclein aggregation. Cultured cells overexpressing α-synuclein generate reactive oxygen species (ROS). One mechanism by which aggregated synuclein leads to ROS generation in PD is via trapping trace metals into the aggregates. These metals can then easily accept electrons from reducing substances such as superoxide or glutathione and transfer them to oxygen to form superoxide radical. DJ-1, encoded by a gene that when mutated is associated with autosomal recessive parkinsonism, has antioxidant properties which are lost if the mutation destroys its dimeric structure. The monomer redistributes from cytosol to the mitochondria and nucleus.

### Polyglutamine Diseases

Polyglutamine diseases (the CAG trinucleotide repeat/polyglutamine diseases) are characterized by the occurrence of protein aggregates within neurons. The most well characterized among ten known diseases of this origin are HD, dentatorubral and pallidoluysian atrophy (DRPLA), spinal and bulbar muscular atrophy, and multiple forms of spinocerebellar ataxia. Each disease is caused by a distinct gene product with expanded polyglutamine repeats.

### Huntington's Disease

The hallmarks of the genetic disorder known as HD are a progressive chorea combined with dementia. Historically, the lurching madness was mistaken for possession by witchcraft (some of the Salem witches burnt in 1693 may have actually had HD). HD is caused by a single, dominant gene (i.e., only one copy is sufficient to cause the disease), unlike most genetic disorders, which are recessive (i.e., two 'bad' copies are needed to cause the disease). The average onset is from 35 to 40 years of age. Huntingtin is a large (350 kDa) protein of unknown function; once the polyglutamine tail crosses the threshold of 38 residues, the mutant begins to form aggregates. The latter precipitate in cytosol and also form nuclear inclusions. The mechanisms by which huntingtin aggregates launch the disease are still disputable. It is supposed to be a gain-of-function event. It is documented that mutant huntingtin can interfere with gene expression that is associated with adaptation to oxidative stress or mitochondrial dysfunction. Thus mutant huntingtin may be directly toxic to mitochondria and this toxicity may be sustained by mutant huntingtin's suppression of compensatory gene expression. While defects in energy metabolism are widely documented in human HD and associated animal models, the only evidence for oxidative stress is oxidative DNA damage.

### Amyotrophic Lateral Sclerosis

ALS (Lou Gehrig's disease) is a chronic, progressive disease marked by gradual degeneration of motor neurons. Most incidences are sporadic, with ~10% of cases inherited in an autosomal dominant manner. The cause of ALS is unknown; however, in 2% of all cases (20% of familial disease) the primary cause is a mutation of the CuZnSOD (*SOD1*) gene. Approximately 117 mutations have been identified; they occur on the protein surface and all yield dominantly inherited disease. As in AD and PD, intracellular and extracellular aggregates are present in ALS, and the aggregates are intensely immunoreactive with antibodies to SOD1, supporting the theory that mutant SOD1 is toxic because of aggregation and coaggregation with other essential proteins. Peripherin, a neuronal intermediate filament protein, is associated with inclusion bodies in motor neurons of patients with ALS. Proteomics study has shown the upregulation of peripherin in ALS patients, and the downregulation of mitochondrial membrane transport and respiratory chain proteins. Study of transgenic mice that express the SOD1 mutation has shown that the mutant protein aggregates onto mitochondria only in cells within the spinal cord. This finding explains the selectivity of SOD1 mutations to affect motor neurons. Theories to explain ALS include oxidative and nitrosative stress, aggregation, mitochondrial damage, and glutamate and microglial cell-mediated damage.

## Neuroprotection Strategies: Problems and Perspectives

By the time a patient is diagnosed as having a neurological illness, extensive neuronal damage has usually already occurred. Consequently, there is a great need for the discovery of biomarkers that allow early diagnosis and intervention. Common features among neurodegenerative diseases – genetic mutations (SOD1, parkin, huntingtin), protein misfolding and aggregation (Lewy bodies, amyloid plaques, tangles), and mitochondrial dysfunction and apoptosis – have implications for disease prevention and for development of effective therapies. Despite common final pathways, it is unlikely that a single drug or targeting a single mechanism will be sufficient to halt neurodegenerative processes.

Chronic neurodegeneration is age related, and thus delay in biological aging will decrease the occurrence of age-related diseases, with resulting prolongation of a healthy life span. Available evidence suggests that to delay aging one has to maintain healthy mitochondria and reduce oxidative stress. One of the proposed approaches is to maintain or recover the activity of the so-called vitagenes. In this regard, the positive effect of heme oxygenase is linked to the production of bilirubin and biliverdin (see earlier), the administration of which, after the first few weeks of life, in doses slightly above normal levels, results in cytoprotective effects. Also, positive effects of CO administration in doses comparable to that of heavy smoker have been shown in suppression of ischemia–reperfusion injury.

Another approach to activate vitagenes is administration of nutritional antioxidants. Curcumin, the most prevalent nutritional and medicinal compound used by the Indian population, has the potential to inhibit lipid peroxidation and efficiently neutralize reactive oxygen and nitrogen species. It has been recently shown that curcumin inhibits NF-κB activation and induces heme oxygenase-1. Administration of curcumin for 6 months resulted in a suppression of indices of inflammation and oxidative damage in the brain of an AD transgenic mouse model (APPSw). Caffeic acid phenethyl ester, an active component of propolis, has been shown to induce heme oxygenase-1 in astroglial cells. Gene induction in both cases occurs through the antioxidant response element (ARE), and this led to the conclusion that the increased expression of genes regulated by the ARE may provide the CNS with protection against oxidative stress. Indeed, numerous studies have supported a role for ARE activators in the prophylaxis against acute and chronic neurodegenerative conditions.

Anti-inflammatory and anti-apoptotic treatments have also shown benefit for many forms of neurodegeneration, although it is also becoming clear that parts of the inflammatory response must be maintained to facilitate repair. A combinatorial approach that highlights the promise of this strategy is the synergistic neuroprotective effects of minocycline (MC) and glatiramer acetate (GA) in a MS model. MC showed remarkable benefit in animal models of MS, PD, HD, ALS, stroke, and spinal cord injury, and in preliminary clinical data for 36-month human trials of MS. MC has a role in interdicting a host of deleterious pathways, including inhibiting matrix metalloproteases (MMPs), chelating $Ca^{2+}$ divalent metals, and inhibiting p38 mitogen-activated protein kinase (MAPK), reducing microglial activation and inhibiting cytochrome *c* release from mitochondria. GA polarizes T cells toward T helper type 2 (Th2) status. GA-reactive Th2 cells cross the blood–brain barrier, and upon reactivation by myelin antigens, release Th2 cytokines. In the CNS, Th2 cells confer neuroprotection, as demonstrated in animal models of PD, ALS, optic nerve crush, ocular glutamate excitotoxicity, etc.

A recent development in chelators involves the design and synthesis of multifunctional drugs that have the ability to bind iron, inhibit a particular enzyme, and exhibit antioxidant properties (free radical scavengers). Metal (in particular, iron) chelation therapy has been proposed as a way of reducing the level of redox-active metals in neurodegenerative diseases. The green tea catechin epigallocatechin gallate (EGCG; known for its iron-chelating and antioxidant properties), the antibiotic iron chelator clioquinol, and intracerebroventricularly injected desferal (DFO) are potent neuroprotective agents. Obviously, iron chelation targets not only unbound iron and that in the aggregated deposits, but, more significantly, also iron dioxygenases such as HIF prolyl hydroxylase. The latter is an emerging target for neuroprotection, although HIF may not be the only substrate of this enzyme. More than 70 genes of putative nonheme iron oxygenases have been identified in the human genome, but only a number have the physiological functions ascribed. An assumption on the uniqueness of an inhibitory action of a particular drug selected among others may be wrong, if only one enzyme candidate has been tested. The best example is probably EGCG, which targets HIF prolyl hydroxylase, MAO B, and MICAL ('molecule interacting with Cas ligand'), a flavin monooxygenase implicated in axonal guidance and highly homologous to MAO B. The recently discovered M30 as an antioxidant/chelator/MAO inhibitor has also been shown to stimulate neurite outgrowth and thus may actually target MICAL as well.

## Conclusions

Therapies targeted at reducing oxidative damage in the nervous system must achieve several goals in order to be effective. First, they must interdict pathological oxidant interactions without affecting physiological signaling by radicals. Over the past two decades, peroxide and nitric oxide have been shown to play the role of messengers in the nervous system. Antioxidants that inadvertently abrogate these signaling functions would not be desirable. Second, antioxidants or, alternatively, effective repair strategies must be augmented in distinct cell types and subcellular compartments. Oxidative and nitrosative stress are not an undifferentiated whole and are mediated by distinct species produced in distinct cellular compartments and distinct cell types. The great challenge has been to divine a multimodal strategy that inhibits a cassette of targets without the expected toxicity that arises as the specificity of the therapy decreases. We propose that understanding endogenous homeostatic pathways for protecting against oxidative and nitrosative stress is the way forward. These homeostatic pathways involve the activation of preexisting proteins as well as *de novo* gene expression. Small molecules that activate homeostatic responses to oxidant stress are expected to reap large therapeutic benefits. Evidence that such an approach is effective and safe continues to emerge from preclinical studies. The ultimate proof will be the demonstration of neuroprotection in a human clinical trial. Such success will also provide long overdue evidence supporting a role for oxidative damage in human neurological disease.

*See also:* Alzheimer's Disease: Neurodegeneration; Apoptosis in Neurodegenerative Disease; Axonal Transport and Neurodegenerative Diseases; Huntington's Disease: Neurodegeneration; Inflammation in Neurodegenerative Disease and Injury; Parkinson's Disease: Alpha-Synuclein and Neurodegeneration.

## Further Reading

Aarts M, Iihara K, Wei WL, et al. (2003) A key role for TRPM7 channels in anoxic neuronal death. *Cell* 115: 863–877.

Becker EB and Bonni A (2004) Cell cycle regulation of neuronal apoptosis in development and disease. *Progress in Neurobiology* 72: 1–25.

Binda C, Hubalek F, Li M, et al. (2006) Structure of the human mitochondrial monoamine oxidase B: New chemical implications for neuroprotectant drug design. *Neurology* 67: S5–S7.

Calabrese V, Butterfield DA, Scapagnini G, et al. (2006) Redox regulation of heat shock protein expression by signaling involving nitric oxide and carbon monoxide: Relevance to brain aging, neurodegenerative disorders, and longevity. *Antioxidants & Redox Signaling* 8: 444–477.

Dhib-Jalbut S, Arnold DL, Cleveland DW, et al. (2006) Neurodegeneration and neuroprotection in multiple sclerosis and other neurodegenerative diseases. *Journal of Neuroimmunology* 176: 198–215.

Farooqui AA, Ong WY, and Horrocks LA (2006) Inhibitors of brain phospholipase A2 activity: Their neuropharmacological effects and therapeutic importance for the treatment of neurologic disorders. *Pharmacological Reviews* 58: 591–620.

Fratelli M, Demol H, Puype M, et al. (2002) Identification by redox proteomics of glutathionylated proteins in oxidatively stressed human T lymphocytes. *Proceedings of the National Academy of Sciences of the United States of America* 99: 3505–3510.

Ghezzi P and Bonetto V (2003) Redox proteomics: Identification of oxidatively modified proteins. *Proteomics* 3: 1145–1153.

Jenner P, Dexter DT, Sian J, et al. (1992) Oxidative stress as a cause of nigral cell death in Parkinson's disease and incidental Lewy body disease. *Annals of Neurology* 32: S82–S87.

Johnson MD, Yu LR, Conrads TP, et al. (2005) The proteomics of neurodegeneration. *American Journal of Pharmacogenomics* 5: 259–270.

Langley B and Ratan RR (2004) Oxidative stress-induced death in the nervous system: Cell cycle dependent or independent? *Journal of Neuroscience Research* 77: 621–629.

Lehtinen MK, Yuan Z, Boag PR, et al. (2006) A conserved MST-FOXO signaling pathway mediates oxidative-stress responses and extends life span. *Cell* 125: 987–1001.

Maher P (2006) Redox control of neural function: Background, mechanisms, and significance. *Antioxidants & Redox Signaling* 8: 1941–1970.

Ong WY and Farooqui AA (2005) Iron, neuroinflammation, and Alzheimer's disease. *Journal of Alzheimer's Disease* 8: 183–200; discussion 209–215.

Poole LB, Karplus PA, and Claiborne A (2004) Protein sulfenic acids in redox signaling. *Annual Review of Pharmacology and Toxicology* 44: 325–347.

Schwartz EI, Smilenov LB, Price MA, et al. (2007) Cell cycle activation in postmitotic neurons is essential for DNA repair. *Cell Cycle* 6: 318–329.

Siddiq A, Aminova LR, and Ratan RR (2007) Hypoxia inducible factor prolyl 4-hydroxylase enzymes: Center stage in the battle against hypoxia, metabolic compromise and oxidative stress. *Neurochemical Research* 32: 931–946.

Singh IN, Sullivan PG, Deng Y, et al. (2006) Time course of post-traumatic mitochondrial oxidative damage and dysfunction in a mouse model of focal traumatic brain injury: Implications for neuroprotective therapy. *Journal of Cerebral Blood Flow & Metabolism* 26: 1407–1418.

Wang JY, Wen LL, Huang YN, et al. (2006) Dual effects of antioxidants in neurodegeneration: Direct neuroprotection against oxidative stress and indirect protection via suppression of glia-mediated inflammation. *Current Pharmaceutical Deign* 12: 3521–3533.

Zheng H, Gal S, Weiner LM, et al. (2005) Novel multifunctional neuroprotective iron chelator-monoamine oxidase inhibitor drugs for neurodegenerative diseases: *In vitro* studies on antioxidant activity, prevention of lipid peroxide formation and monoamine oxidase inhibition. *Journal of Neurochemistry* 95: 68–78.

# Protein Folding and the Role of Chaperone Proteins in Neurodegenerative Disease

**P J Muchowski**, Gladstone Institute of Neurological Disease, University of California, San Francisco, CA, USA



## Introduction

Many neurodegenerative disorders are characterized by the accumulation of misfolded proteins in or adjacent to neurons and glia. Misfolded proteins can adopt distinct conformational states that exert abnormal and even pathological activities. Molecular chaperones are proteins that mediate the proper folding of other proteins and appear to have other important functions in the healthy and diseased brain. One essential function is to ensure that misfolded proteins do not accumulate in the brain. Increased expression of molecular chaperones is protective in numerous animal models of neurodegenerative disease. A small molecule that induces the expression of molecular chaperones is being evaluated in a clinical trial for amyotrophic lateral sclerosis (ALS). The molecular mechanisms that enable molecular chaperones to combat the toxicity of misfolded proteins *in vivo* are not well understood. However, their ability to modulate conformational states of disease-related proteins suggests that they prevent early aberrant protein interactions that trigger pathogenic cascades.

## Protein Misfolding, Aggregation, and Molecular Chaperones

The amino acid sequence of a protein contains all of the information necessary for it to fold into a functional, three-dimensional structure *in vitro*. Within cells, however, protein folding occurs in a complex environment. The cellular cytoplasm is chock full of proteins, nucleotides, sugars, lipids, and organelles. As they attempt to fold in this dense milieu, proteins can easily misfold and aggregate.

Because the proper folding (or conformation) of proteins is critical to life, a class of 'helper' proteins called molecular chaperones has evolved. These proteins prevent inappropriate interactions within and between nonnative polypeptides, enhance the efficiency of *de novo* protein folding, and promote the refolding of proteins that have become misfolded as a result of cellular stress. Molecular chaperones function at essentially any cellular location where a folding or unfolding protein chain may interact inappropriately (for example, as a protein chain is translocated into mitochondria). Under certain conditions, chaperones interact with components of the ubiquitin–proteasome pathway and lysosome-mediated autophagy to degrade proteins they cannot refold.

What exactly are molecular chaperones? Many are 'heat shock' proteins (HSPs). When cells are stressed, such as with increased temperature, the transcription of HSPs is quickly induced, while generalized protein synthesis is turned off. HSPs are classified into six main families based on their approximate molecular weight. The number of human diseases linked to mutations in molecular chaperones is large and growing.

## Protein Aggregation and Neurodegeneration

Molecular chaperones, together with the ubiquitin–proteasome pathway and lysosome-mediated autophagy, constitute a 'quality control' system that normally prevents the accumulation of misfolded proteins. However, since misfolded proteins are thought to accumulate as a result of normal aging, additional insults, such as mutations, can easily saturate the quality control system. Interestingly, neurons and other postmitotic cells appear to be especially vulnerable to the effects of misfolded proteins. It is now abundantly clear that saturation of the protein quality control machinery is a factor in many neurodegenerative diseases, including Alzheimer's disease (AD), Parkinson's disease (PD), Huntington's disease (HD) and related polyglutamine (polyQ) disorders, ALS, and the prion encephalopathies.

The major pathological feature of neurodegeneration associated with misfolded proteins is the accumulation of aggregated protein in intra- or extracellular deposits, generically referred to as amyloid and known as senile plaques and neurofibrillary tangles in AD, Lewy bodies and neurites in PD, and inclusion bodies in HD and other diseases in which the protein aggregates inside cells. The aggregated proteins are composed of fibrillar material, as shown by electron microscopy, and have similar chemical structures, consisting of $\beta$-sheets perpendicular to the fibril axis, that produce a characteristic cross-$\beta$ X-ray diffraction pattern. They also share biochemical properties, such as insolubility in detergents, resistance to proteases, and the ability to bind lipophilic dyes such as congo red.

Diverse neurodegenerative diseases are characterized by accumulation of amyloid-like fibrils and likely have common pathogenic mechanisms. But are the amyloid fibrils themselves toxic? Increasing evidence suggests that fibrillar aggregates are inert or perhaps even protective, rather than being directly pathogenic. For example, senile plaques are found in persons without clinical symptoms of AD, and in cases of AD, the severity of dementia does not correlate well with plaque density. In PD, neuropathological analyses show that neurons that contain Lewy bodies are healthier than cells that do not, indicating that inclusion bodies might be protective. In support of this hypothesis, the formation of inclusion bodies by proteins with expanded polyQ repeats is a regulated cellular process that requires an intact microtubule cytoskeleton. This process might have evolved as a protective mechanism to sequester toxic, misfolded protein entities that would otherwise disrupt cellular homeostasis.

If amyloid fibrils do not trigger pathogenic cascades in neurodegenerative diseases, then what is the toxic culprit? In the past decade, detailed structural and biochemical studies of purified proteins have shown that the assembly of aggregating proteins into amyloid fibrils is a complex process, involving a number of different intermediates. Intense research has focused on the pathogenic role of these so-called prefibrillar or protofibrillar amyloid intermediates, also called soluble oligomers or, in the case of AD, amyloid $\beta$-peptide (A$\beta$)-derived diffusible ligands (ADDLs). Atomic force microscopy has been used to characterize several metastable structures, such as spherical oligomers, protofibrils, and porelike annular structures, which might be on a pathway to fibril formation. These small, potentially diffusible assemblies, rather than mature amyloid fibrils, might trigger neuronal dysfunction by initiating a cascade that culminates in neuronal death. For example, injecting soluble oligomers of A$\beta$ into the brain decreases long-term potentiation, and it is thought that the oligomers specifically target proteins localized to synapses. Antibodies have been generated that specifically recognize aggregating proteins solely in their oligomeric forms. Incubation of oligomers with these antibodies blocks their subsequent toxicity to cultured cells, supporting the hypothesis that soluble oligomers are the toxic culprit. However, our understanding of soluble oligomers is largely based on *in vitro* and cell culture studies. Little is known about their *in vivo* role for the simple reason that we lack the tools to visualize their structure *in vivo*. The development of such tools will allow one to carefully characterize from a temporal and spatial perspective the role of oligomers in pathogenesis.

## Molecular Chaperones and Neurodegeneration

Since molecular chaperones normally function in protein folding, it is not surprising that they can be found, together with components of the ubiquitin–proteasome system, by immunohistological analysis of sections from diseased brains at sites of aggregate deposition, such as in senile plaques and inclusion bodies. In several studies, sophisticated fluorescence imaging techniques showed that molecular chaperones are not bound irreversibly to intracellular aggregates. Rather, they quickly cycle on and off the aggregates, apparently in an effort to refold the misfolded protein, to transfer it to the ubiquitin–proteasome system, or to help sequester oligomeric intermediates into fibrillar inclusion bodies. Future studies will be required to discriminate among these and other possibilities.

Compelling evidence to support a critical role for chaperones in neurodegenerative diseases comes from recent *in vivo* studies that show that overexpression of the chaperone Hsp70 or Hsp40 suppresses neurodegeneration in fruit fly and transgenic mouse models of HD and PD. While initial studies were based on a candidate gene approach, unbiased genetic screens in *Saccharomyces cerevisiae, Drosophila melanogaster,* and *Caenorhabditis elegans* models of polyQ aggregation have further substantiated the importance of molecular chaperones and protein folding in regulating polyQ aggregation and toxicity. Importantly, overexpression of chaperones also suppresses neurodegeneration *in vivo* in transgenic mouse models of PD and polyQ diseases.

## Mechanisms of Chaperone-Mediated Neuroprotection

Although chaperones are clearly protective in animal models of neurodegenerative disease, the mechanisms that underlie protection are unclear. They can directly influence the folding state or conformation of disease-associated proteins that aggregate. They might alter a particular conformational state that is highly toxic, exerting a protective effect simply by decreasing its abundance. Another possibility is that molecular chaperones may cooperate with the ubiquitin–proteasome and lysosome degradation systems to help degrade aggregation-prone proteins; however, experimental evidence in animal models that directly supports this mechanism of action is lacking. The chaperones likely facilitate neuroprotection through mechanisms that are both dependent on, and independent of, their protein folding activity (**Figure 1**). In addition to cooperating with the protein degradation machinery, they can also block apoptotic signaling pathways. Ultimately, the collective effects of multiple direct and indirect chaperone actions might be essential to impede disease pathogenesis.

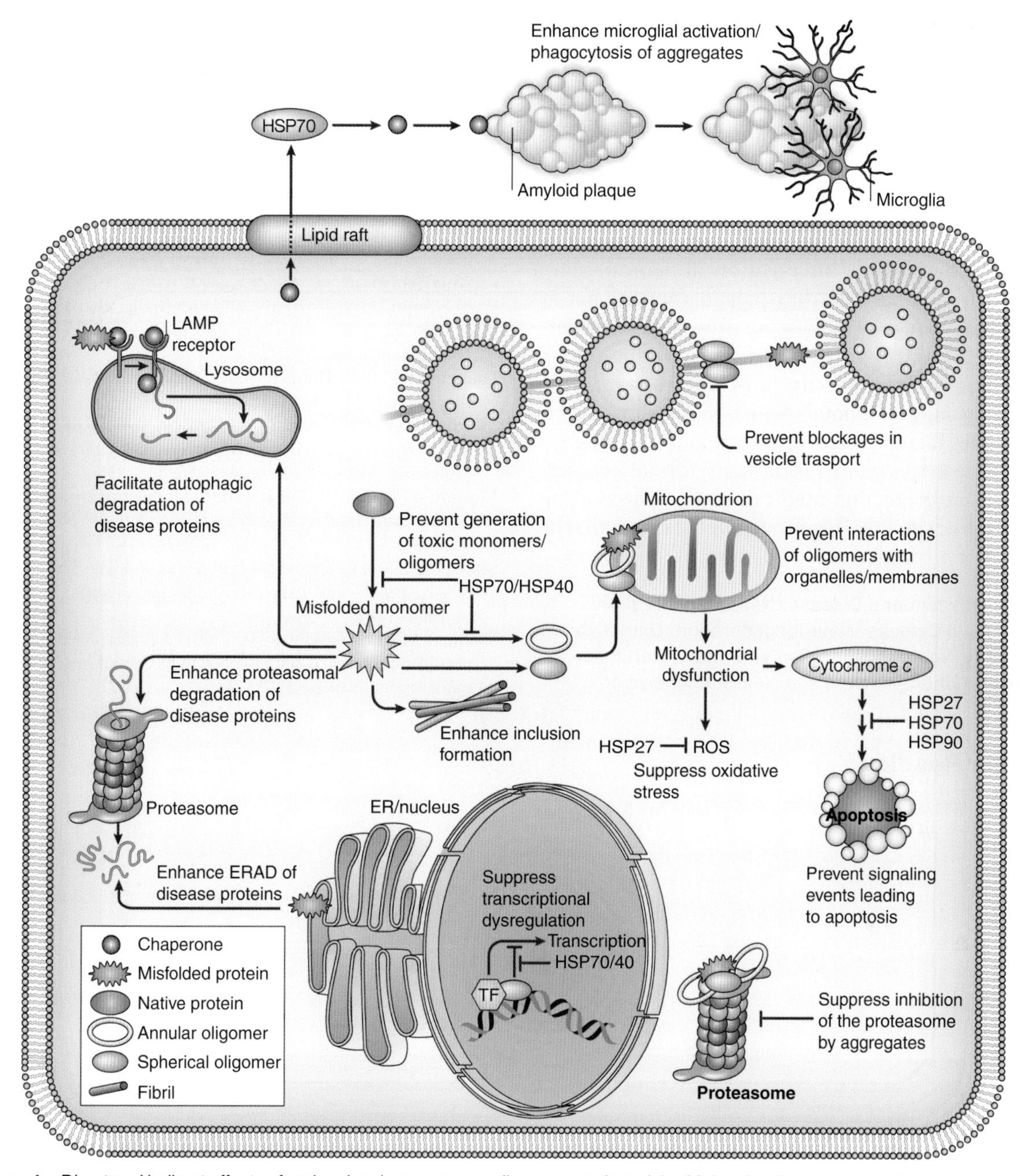


**Figure 1** Direct and indirect effects of molecular chaperones on disease protein toxicity. Molecular chaperones might prevent toxicity by preventing inappropriate protein interactions, by facilitating disease protein degradation/sequestration, and/or by blocking downstream signaling events that lead to neuronal dysfunction and apoptosis. HSP, heat shock protein; LAMP, lysosome-associated membrane protein; ROS, reactive oxygen species; ER, endoplasmic reticulum; ERAD, endoplasmic-reticulum-associated degradation. Reproduced from Muchowski PJ and Wacker JL (2005) Modulation of neurodegeneration by molecular chaperones. *Nature Reviews Neuroscience* 6: 11–22, with permission.

## Molecular Chaperones as Potential Drug Targets

It is clear from animal studies that increased expression of molecular chaperones might prove beneficial for the treatment of neurodegenerative diseases. However, finding drugs that increase the expression of proteins in specific cells and tissues will be a major challenge. Furthermore, continuous upregulation of molecular chaperones might cause undesirable side effects, including cancer. Nonetheless, a small molecule that induces expression of HSPs has shown promise in an animal model of ALS and is being evaluated in a phase II clinical trial.

## Conclusions

Protein folding in neurons mediated by molecular chaperones plays a critical role in normal brain function and in neurodegenerative diseases. The mechanism by which chaperone activity facilitates neuroprotection is being studied intensively in many laboratories, but deep insights are still lacking. In principle, molecular chaperones are an attractive target for therapeutic intervention in neurodegenerative diseases, because they presumably can act at the earliest change of the disease process, by preventing aggregation-prone proteins from adopting a toxic conformation. Interestingly, most studies of molecular chaperones in animal models have shown that chaperones facilitate neuroprotection in the absence of a visible effect on inclusion-body formation, suggesting that protection might arise from decreasing formation of small, diffusible aggregate assemblies or perhaps of misfolded conformations of monomers.

*See also:* Alzheimer's Disease: Neurodegeneration; Huntington's Disease: Neurodegeneration; Oxidative Damage in Neurodegeneration and Injury; Parkinson's Disease: Alpha-Synuclein and Neurodegeneration.

## Further Reading

Bukau B and Horwich AL (1998) The Hsp70 and Hsp60 chaperone machines. *Cell* 92: 351–366.

Caughey B and Lansbury PT (2003) Protofibrils, pores, fibrils, and neurodegeneration: Separating the responsible protein aggregates from the innocent bystanders. *Annual Review of Neuroscience* 26: 267–298.

Cuervo AM and Dice JF (1998) Lysosomes, a meeting point of proteins, chaperones, and proteases. *Journal of Molecular Medicine* 76: 6–12.

Dobson CM (2003) Protein folding and misfolding. *Nature* 426: 884–890.

Glabe CG (2006) Common mechanisms of amyloid oligomer pathogenesis in degenerative disease. *Neurobiology of Aging* 27: 570–575.

Hartl FU and Hayer-Hartl M (2002) Molecular chaperones in the cytosol: From nascent chain to folded protein. *Science* 295: 1852–1858.

Lindquist S (1986) The heat-shock response. *Annual Review of Biochemistry* 55: 1151–1191.

Morimoto RI and Santoro MG (1998) Stress-inducible responses and heat shock proteins: New pharmacologic targets for cytoprotection. *Nature Biotechnology* 16: 833–838.

Muchowski PJ and Wacker JL (2005) Modulation of neurodegeneration by molecular chaperones. *Nature Reviews Neuroscience* 6: 11–22.

Selkoe DJ (2004) Cell biology of protein misfolding: The examples of Alzheimer's and Parkinson's diseases. *Nature Cell Biology* 6: 1054–1061.

Sherman MY and Goldberg AL (2001) Cellular defenses against unfolded proteins: A cell biologist thinks about neurodegenerative diseases. *Neuron* 29: 15–32.

# Role of NO in Neurodegeneration

**Y-I Lee, T M Dawson, and V L Dawson**, The Johns Hopkins University School of Medicine, Baltimore, MD, USA



## Introduction

Nitric oxide (NO) was first identified as endothelium-derived relaxing factor. NO is produced by many different cells in multicellular organisms, where it acts as an unprecedented biological messenger molecule. NO is synthesized by the enzymes NO synthases (NOSs), which catalyze the essential amino acid L-arginine into NO and L-citrulline. In the central nervous system (CNS), NO is an unusual neuronal messenger molecule in that conventional neurotransmitters are released by exocytosis from the nerve terminal, but because of its small size and ability to move through lipids, NO reaches its targets through diffusion. NO forms covalent and noncovalent linkages with protein and nonprotein targets. Conventional neurotransmitters undergo reversible interactions with cell surface receptors. NO is inactivated by diffusion away from its targets, by forming covalent linkages to the superoxide anion, or by scavenger proteins. In contrast, conventional neurotransmitters are terminated by presynaptic reuptake or enzymatic degradation. Because NO does not use conventional means for control of its biological actions, NO depends upon its small size, reactivity, and diffusibility more than any other biological molecule to exert its biological effects. NO is involved in a wide range of physiological processes, and under conditions of excess formation in the presence of superoxide anion it can mediate a series of pathophysiological actions. The multifaceted actions of NO in brain physiology include mechanisms related to development, synaptic function and neural plasticity, and neuronal survival and neuronal cell death.

## NO in the Nervous System

NO is produced by a group of enzymes designated NOS. There are three members of the NOS family: neuronal NOS (nNOS), endothelial NOS (eNOS), and inducible NOS (iNOS). A fourth family member was proposed, mitochondrial NOS (mtNOS), but studies indicate that mtNOS is nNOS$\alpha$ – it is coded by the nNOS gene, and nNOS knockout mice do not have mtNOS. All the NOSs share between 50% and 60% sequence homology. In the nervous system, nNOS is expressed in neurons, eNOS is predominantly expressed in the endothelium of blood vessels, and iNOS is expressed in glial cells. nNOS and eNOS are $Ca^{2+}$-calmodulin-dependent enzymes that are constitutively expressed, although both isoforms can be induced following a variety of physiological and pathological stimuli. In contrast, iNOS does not require $Ca^{2+}$-calmodulin for activation and its regulation depends on *de novo* synthesis. Despite its highly diffusible nature, NO can exert very specific effects within the CNS and peripheral nervous system (PNS). Under physiological conditions, NO facilitates neurotransmitter release and uptake via neuron-glial communication, modulating the release of neurotransmitters, including glutamate, $\gamma$-aminobutyric acid (GABA), substance P, and acetylcholine. nNOS and eNOS are regulated via posttranslational modification, including phosphorylation, and nNOS levels are dynamically regulated through transcription. Cerebral blood flow is regulated through NO released from endothelial cells as well as autonomic nerves within the adventitia. NO may also regulate cerebral blood flow through activity-dependent activation of NOS in neurons that influence small cerebral arterioles. In the brain, nNOS is linked to the postsynaptic membrane near *N*-methyl-D-aspartate (NMDA) receptors via the postsynaptic density protein-95, and nNOS function is regulated through its interactions with the protein inhibitor of NOS (PIN) and carboxyl-terminal PDZ ligand of NOS (CAPON) (**Figure 1**). In the gut, NO functions as the nonadrenergic, noncholinergic (NANC) neurotransmitter where it mediates relaxation of smooth muscles associated with peristalsis. Penile erections are mediated, in part, through the neurotransmitter action of NO.

The diffusion properties and the short half-life of NO make it an ideal candidate for playing a role in nervous system morphogenesis and synaptic plasticity. In the nervous system, NO was first recognized as a messenger molecule that mediates increases in cyclic GMP (cGMP) levels that occur after activation of glutamate receptors, particularly those of the NMDA subtype. In this capacity, NO modulates long-term potentiation (LTP), which is essential for learning and memory. LTP is inhibited by administration of NOS blockers or genetic deletion of NOS isoforms, confirming the role of NO in synaptic plasticity. Nerve growth factor (NGF)-induced differentiation of pheochromacytoma cells (PC12) has been shown to be NO dependent with NO acting as a cytostasis factor. NO may play a role in the developmental processes of the nervous system, in that it is transiently expressed in the cerebral cortical plate during critical periods of neuronal development. NO is also critical for the neurite outgrowth and differentiation in murine cortical

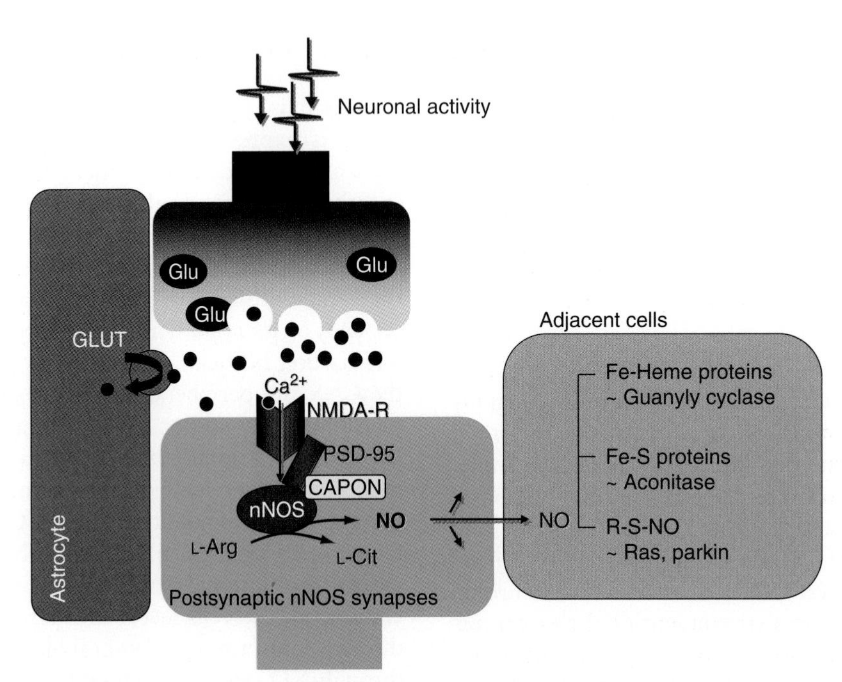


**Figure 1** NO signaling in the CNS. Glutamate activation of NMDA receptor results in stimulation of neuronal NO synthase (nNOS) which is functionally coupled to the NMDA receptor by PSD-95, a scaffolding protein. The tethering of nNOS near the NMDA receptor facilitates its activation by $Ca^{2+}$ resulting in the generation of NO. CAPON binds nNOS and regulates its interaction with PSD-95 and thus regulates the generation of NO. NO diffuses between cells and can activate a number of cell signaling pathways in adjacent cells. NO can react with the heme center of soluble guanylyl cyclase (GC) activating GC to produce cyclic GMP (cGMP). NO can react with iron–sulfur (Fe–S)-containing proteins such as aconitase to regulate their activity. NO may S-nitrosylate (R-S-NO) proteins such as Ras or parkin, thus regulating complex cell signal cascades.

neurons. NO may play a role in the molecular maturation of adult spinal cord motor neurons and also in the initial pattern of connections in ocular dominance columns in the developing visual system. It has also been suggested to play a prominent role in the activity-dependent establishment of connections in both the developing and regenerating olfactory neuronal epithelium. Thus, NO may play an important role in synaptic refinement in the developing nervous system.

In the immune system, NO generated by iNOS is released in bursts of high concentrations and is a key cytotoxic weapon for the destruction of bacteria. The CNS immune cell counterparts, microglia and astrocytes, also have iNOS, which generates a burst of NO in response to injury and can contribute to neural injury. The expression of iNOS occurs following trauma, following stroke, or during neurodegenerative diseases and is thought to contribute to neural injury.

Although NOS is the main NO source, in some special situations this molecule can be synthesized by other mechanisms. NO can be produced by the xanthine oxidase pathway or by $H_2O_2$ and L-arginine in a nonenzymatic reaction or by the reduction of nitrites in acid and reducing conditions, such as occurs in ischemic tissue.

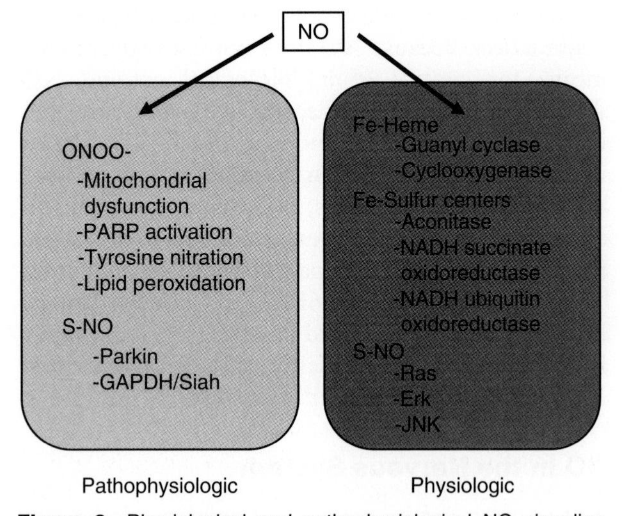


**Figure 2** Physiological and pathophysiological NO signaling events.

## NO Molecular Targets

There are several molecular mechanisms by which NO elicits cellular signaling (**Figure 2**). The unique chemical properties of NO facilitate its role as an important signaling molecule in the biological system by reacting directly with biological molecules or indirectly following reaction with other reactive oxygen species. NO reacts with the ferrous ion in the heme complex in enzymes such as guanylate cyclase (GC), P450, and hemoglobin to form an iron nitrosyl complex. The binding of NO induces a conformational change in the protein structure activating these proteins. The prototype reaction is that of NO on GC leading to the production of cyclic guanosine-3′,5′-monophosphate (cGMP). cGMP is a second messenger that activates protein kinases. In the nervous system, NO was first recognized as a messenger molecule that regulates cGMP accumulation through activation of glutamate receptors, particularly those of the NMDA subtype. cGMP, in turn, binds to and activates a cGMP-dependent protein kinase (G kinase), which phosphorylates specific proteins on serine or threonine residues and also activates phosphodiesterase and ion channels. NO increases prostaglandin production by activation of cyclo-oxygenase, another heme-containing enzyme. Numerous enzymes (including *cis*-aconitase, NADH succinate oxidoreductase, and NADH ubiquinone oxidoreductase) that contain nonheme iron in iron sulfur clusters are also regulated by NO. NO also stimulates the RNA-binding function of the iron-responsive element binding protein in brain slices while diminishing its aconitase activity.

In recent years, the importance of protein S-nitrosylation of cysteine thiols by NO in physiological and pathophysiological actions has been appreciated. S-nitrosylation is a posttranslational protein modification that increases or decreases the activity of the target protein. Some known biological targets for S-nitrosylation include glyceraldehyde-3-phosphate dehydrogenase (GAPDH), caspases, transglutaminase, aromatases, mitochondrial aldehyde dehydrogenase, cyclooxygenase-2, G-proteins, p21-Ras, RAC1, cdc42, Dexras1, ERK, JNK, p38, and PKBa/AKT1. S-nitrosylation of the monomeric GTPase Ras activates this enzyme, triggering the MAPK and PI3K pathways in an NO-dependent and cGMP-independent manner. A recent study reports that NO donors promote HIF-1$\alpha$ accumulation via an increase in transcription and translation. It is not yet known how this occurs, but it is possible that these effects are due to activation of Ras and PI3K through NO-mediated nitrosylation of Ras. NO has inhibitory effects on cell cycle progression and promotes cell differentiation, in part due to activation of the retinoblastoma gene product (pRb) by S-nitrosylation. Nuclear factor kappa beta (NF-$\kappa\beta$), a transcription factor that induces iNOS expression, is inhibited by S-nitrosylation. Thus, NO inhibition of NF-k$\beta$ could be an autoregulatory feedback mechanism to regulate NO production. Another autoregulatory mechanism is the prevention of eNOS dimerization by S-nitrosylation, resulting in decreased eNOS activity. S-nitrosylation is also involved in the pathological activation of matrix metalloproteases in stroke and neurodegenerative diseases. NO regulates NMDA receptor activity through nitrosylation, thus reducing calcium permeability. One of the important features of S-nitrosylation is that the modification is reversible, and a specific protein sequence of acidic–basic motifs seems to be necessary. As usual, the modification is also selective, since only a specific cysteine is generally nitrosylated in a protein.

Oxygen and other free radicals can also react with NO to form reactive nitrogen oxide species and exertits effects on different biological molecules. Superoxide anion can react with NO to form peroxynitrite, a potent oxidant with many cellular targets. Peroxynitrite damages proteins by reacting with the tyrosine residues to form 3-nitrotyrosine. It can also induce lipid peroxidation and cause DNA damage. It is thought that many of the acute toxic effects of NO are mediated by peroxynitrite.

## Neurotoxic Effects of NO

In the CNS, glutamate excitotoxicity due to activation of the NMDA receptor is mediated in large part by NO. Under physiological conditions, the binding of glutamate activates NMDA receptors in the postsynaptic neuron that result in $Ca^{2+}$ influx, which in turn activates nNOS. Then NO either via cGMP or by S-nitrosylation suppresses NMDA receptor activity, thereby limiting further $Ca^{2+}$ influx and thus protecting neurons. However, under pathophysiological conditions such as occur during stroke, trauma, or neurodegenerative stress, NO participates in neuronal damage through disruption of this system and generation of peroxynitrite that activates subsequent cell death cascades. NO regulation of the release of glutamate contributes to the initial toxic events. When NO concentrations are low, there is a decrease in glutamate release despite elevated cGMP levels. However, under increased NO production and excessive increases in cGMP levels, the inhibitory effect on glutamate release is reversed, and glutamate release is increased. Under the pathological conditions of excess glutamate release and NMDA receptor activation, activation of NOS leads to cell death. Excitotoxicity can be blocked by several classes of NOS inhibitors or by nNOS gene deletion. nNOS knockout mice are resistant to focal and transient global ischemia, suggesting a role for nNOS in CNS injury. The neuronal toxicity of NO is mediated by peroxynitrite, which inhibits mitochondrial respiration and manganese superoxide dismutage (MnSOD) activity, thus establishing a feed-forward loop for peroxynitrite generation. Peroxynitrite damages DNA, which activates the nuclear enzyme poly(ADP-ribose) polymerase (PARP) that leads to the release of the cell death effector apoptosis-inducing factor (AIF) from mitochondria. AIF can induce phosphatidyl serine exposure on the plasma membrane, and in the nucleus it facilitates DNA cleavage and nuclear condensation (**Figure 3**). In addition, glutamate receptor-mediated excitotoxicity activates parallel cellular responses, resulting in the activation of proteolytic enzymes, lipid peroxidation, ROS and RNS formation, and activation of programmed cell death cascades. Under some conditions, NO can activate the ceramide–Jun kinase pathway and the NF-κβ–Jun kinase pathway, which regulate the expression of a large number of genes involved in both neuroprotection and neurotoxicity. The biological significance of NO stimulation of these pathways depends on the cell type and the environment. NO can also play a protective role via the blocking of caspases (cysteine–aspartate proteases), which form a complex family of proteolytic enzymes. It is not yet clear whether NO acts as a master regulator, determining which cell death pathway will be activated in neuronal injury, but these data are suggestive that NO production could direct neurons away from a classic apoptotic type of cell death and toward an excitotoxic form of cell death acutely and perhaps activate other apoptotic cascades in the late remodeling phase of neuronal injury and recovery.

Oxidative and nitrosative stress plays a pivotal role in ischemic-reperfusion injury during stroke. The increase of NO production in the presence of increased superoxide anion leads to brain damage. Brain ischemia triggers a cascade of events, including membrane depolarization, increase in intracellular $Ca^{2+}$ concentrations via increase of extracellular glutamate concentration, and overstimulation of NMDA receptors, leading to activation of calcium-dependent nNOS. Released NO rapidly reacts with superoxide produced in excess during reperfusion resulting in peroxynitrite formation, and nitrosylation or nitration of proteins. Furthermore, ischemia or reperfusion eventually induces the expression of iNOS. This isoform is not normally present in the CNS, but it can be detected after inflammatory, infectious, or ischemic damage, as well as in the normal aging brain. Activation of nNOS or induction of iNOS mediates ischemic brain damage through mechanisms previously discussed. On the other hand, eNOS is thought to act as a neuroprotective agent by enhancing blood supply to the injured tissue. Therefore, NO generated by nNOS or iNOS contributes to injury following stroke, whereas NO generated by eNOS is protective by maintaining blood flow (**Figure 4**).

## NO and Neurodegenerative Diseases

Abnormal activation of glutamate neurotransmission may contribute to neurodegenerative processes and

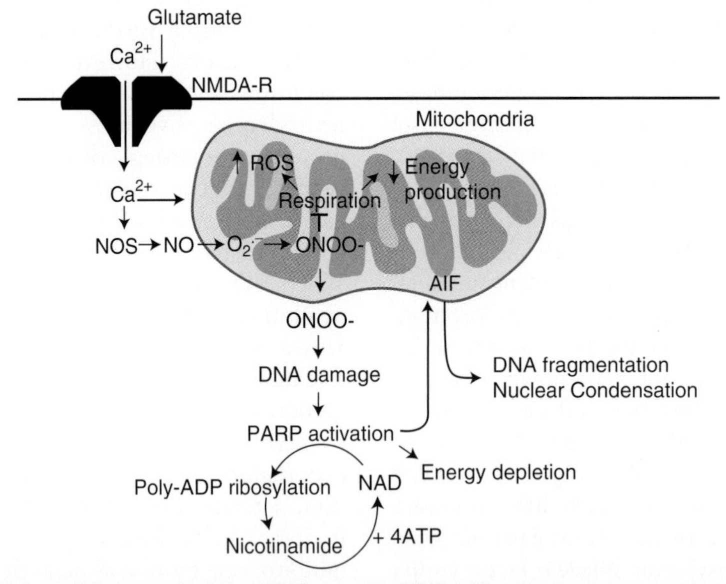


**Figure 3** Mechanism of NO-mediated neurotoxicity. NMDA receptor (NMDA-R) activation causes an increase in intracellular calcium levels, which activates NOS, as well as other potential free radical-generating enzymes. These free radicals then damage DNA and also inhibit mitochondrial function, leading to more free radical formation. The damaged DNA activates PARP-1, which transfers ADP-ribose groups to nuclear proteins consuming NAD. NAD is resynthesized from nicotinamide (NAm), a reaction that consumes four high-energy equivalents of ATP. PARP activation also elicits the release and translocation of AIF from the mitochondria into the nucleus, initiating a caspase-independent cascade of cell death.

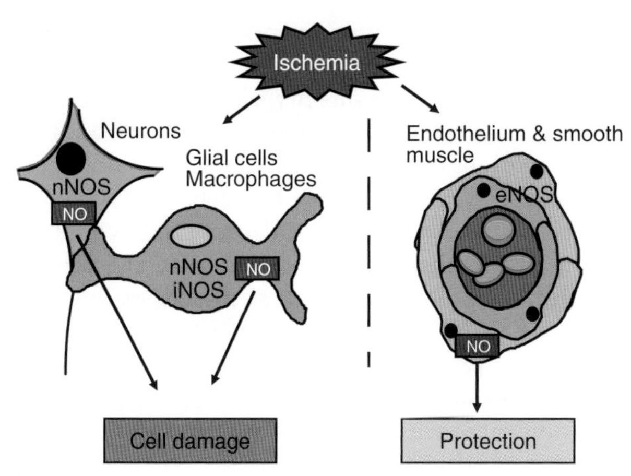


**Figure 4** Dual roles for NO in brain ischemia. NO is involved in the mechanisms of neurotoxicity after cerebral ischemia. Following a stroke, the generation of NO can be either beneficial or detrimental to the brain depending on the which NOS isoform is activated and in which cellular compartment. NO generated from eNOS in endothelial cells provides protection by inducing vasodilation. Inhibition of eNOS can drastically increase stroke injury. NO produced by nNOS in neurons or iNOS in glia results in the formation of toxic peroxynitrite and cellular damage. Inhibition of nNOS or iNOS greatly reduces brain injury following stroke.

diseases such as Parkinson's disease (PD), Alzheimer's disease (AD), amyotrophic lateral sclerosis (ALS), and multiple sclerosis (MS). Neurodegenerative disorders are usually marked by a progressive loss of a selective group of neurons and are associated with aging. One of the common features of neurodegeneration is the presence of protein aggregates and increased indices of oxidative stress.

PD is a common neurodegenerative disorder characterized by impairment in motor function. It is another common age-related neurodegenerative condition associated with selective loss of dopaminergic neurons in the substantia nigra of the midbrain region. In PD, postmortem pathological studies have revealed mitochondria complex I dysfunction, reduced glutathione and ferritin levels, increased lipid peroxidation, nitrotyrosine immunoreactivity, and increased levels of iron. Genetically engineered animals reveal that NO generated by both iNOS and nNOS contributes to pathology. In the experimental 1-methyl-4-phenyl-1,2,3,6-tetrahydropyridine (MPTP) intoxication model of PD, NOS inhibition in wild-type mice or gene deletion of the iNOS or nNOS gene protects dopaminergic neurons. In PD patients, polymorphonuclear cells have increased expression of nNOS and increased NO production and accumulation of nitrotyrosine-containing proteins, reflecting an increase in nitrosative stress in these patients. Glial-derived neuronal factor (GDNF) is protective in animal models of PD, although its effect on patients remains controversial. GDNF action leads to the inhibition of nNOS activity and neurotoxicity. *R*-(−)-deprenyl (deprenyl) slows the progression of disability in early PD. Deprenyl prevents S-nitrosylation of GAPDH, which blocks the binding of GAPDH to Siah, preventing the GADH/Siah death cascade. In MPTP-treated mice, low doses of deprenyl prevent binding of GAPDH and Siah1 in the dopamine-enriched corpus striatum and it is neuroprotective.

PD is mostly sporadic, but rare familial cases are also found. Mutations in $\alpha$-synuclein, parkin, DJ-1, PINK1, and LRRK2 have been linked to rare familial forms of PD. Point mutations or replication of $\alpha$-synuclein results in autosomal dominant PD, which prompted intense investigation into the actions and pathobiology of this protein. The hallmark lesions of PD, the Lewy body, contain $\alpha$-synuclein that is modified by nitration of tyrosine residues. It is possible that dityrosine cross-linking can generate stable oligomers that lead to aggregates. Both wild-type and disease-causing $\alpha$-synuclein mutants can be nitrated, and the nitration induces the formation of $\alpha$-synuclein inclusions with similar biochemical characteristics to protein extracted from human PD tissue. In addition to nitration, nitrosylation may also play a key role in both sporadic and familial PD, as there is a greater than twofold increase in S-nitrosylated proteins in PD. Parkin is an E3 ligase that adds ubiquitin on specific protein substrates by both K48 and K63 linkages. Disease-causing mutations and loss-of-function mutants lead to the accumulation of toxic parkin substrates in both mouse and humans. Parkin can be S-nitrosylated. This modification results in loss of function as an E3 ligase and inhibition of the protective function of parkin. In sporadic cases of PD, parkin is S-nitrosylated, suggesting that in the more common form of PD, parkin function may be compromised. While parkin and DJ-1 do not interact physiologically, under conditions of oxidative and nitrosative stress there is a gain of function and these two proteins interact. It is possible that this interaction results in the sequestering of DJ-1 into insoluble aggregates terminating its normal biological actions. The pathophysiological significance of this interaction between DJ-1 and parkin is not yet known but is an area of active scientific investigation. Actions of oxidative and nitrosative stress on PINK1 or LRRK2 await investigation.

AD is the most common chronic progressive neurodegenerative condition; it was originally described by Alzheimer about a century ago. Brains from AD patients are characterized by extracellular aggregates of amyloid $\beta$-peptide (A$\beta$) that form the neuritic plaques and intracellular neurofibrillary tangles due to the hyperphosphorylation of tau protein. NO can contribute to AD neurodegeneration in several ways. Aβ fibrils are toxic, in part, by inducing ROS formation, which can produce peroxynitrite by reacting with NO. Increased expression of iNOS and eNOS in astrocytes has been associated with the presence of neuritic plaques indicating increased nitrosative stress in AD. Additional studies indicate that Aβ is critically involved in the induction of an inflammatory response through activation of astrocytes and microglia, eliciting subsequent production of inflammatory cytokines and neurotoxic factors, including NO, superoxide, and prostaglandins (via COX-2). Aβ deposition in brain vessels correlates with a decrease in the expression of eNOS in endothelial cells that can lead to decreased regulation of cerebral blood flow. In cell culture and experimental animal models, Aβ elicits NO-mediated excitotoxicity, neuroinflammation, and oxidative stress. Similar stress events are observed in patient tissue, implicating a role for NO in the progression of AD.

ALS is a neurodegenerative disease characterized by selective death of motor neurons in the cerebral cortex, brain stem, and spinal cord. The role of NO in ALS is controversial. There is evidence that nitrotyrosination induces motor neuron death and that there is irreversible inhibition of the mitochondrial respiratory chain in these cells, suggesting a role for NO. However, mice transgenic for the human disease-causing mutation in SOD-1 when crossed with the nNOS knockout mice did not derive any benefit and progressed to disease and death. On the other hand, treatment with a nonselective NOS inhibitor slightly decreased motor neuron degeneration in an ALS mouse model. It is possible that NO from different isoforms could contribute to progressive neurodegeneration in ALS and exacerbate disease progression. In postmortem tissue, there are reactive astrocytes that contain protein inclusions, express iNOS and COX-2, display nitrotyrosine immunoreactivity, and have a loss of the glutamate transporter, EAAT2. The combination of excitotoxicity and nitrosative stress could contribute to the development and acceleration of the disease.

## Conclusion

NO is a multifunctional molecule that is needed for physiological functions, especially in the brain. NO has clearly revolutionized our thought about aspects of neuronal transmission. It is capable of regulating various body functions directly, via a cGMP-mediated mechanism or a cGMP-independent mechanism. NO is an important mediator of a variety of physiological effects in the nervous system; however, excessive NO production under pathological conditions could be destructive to the brain. Understanding the role of NO in these processes as well as the disposition and activation of NO will lead to a better understanding of the basic processes underlying normal and pathological neuronal functions, and hopefully to the development of selective protective therapeutic agents against neurodegenerative diseases.

*See also:* Alzheimer's Disease: Neurodegeneration; Parkinson's Disease: Alpha-Synuclein and Neurodegeneration.

## Further Reading

Ahern GP, Klyachko VA, and Jackson MB (2002) cGMP and S-nitrosylation: Two routes for modulation of neuronal excitability by NO. *Trends Neuroscience* 25: 510–517.

Chung KK, Dawson TM, and Dawson VL (2005) Nitric oxide, S-nitrosylation and neurodegeneration. *Cellular and Molecular Biology* 51: 247–254.

Contestabile A and Ciani E (2004) Role of nitric oxide in the regulation of neuronal proliferation, survival and differentiation. *Neurochemistry International* 45: 903–914.

Dawson VL and Dawson TM (2004) Deadly conversations: Nuclear-mitochondrial cross-talk. *Journal of Bioenergetics and Biomembranes* 36: 287–294.

Emerit J, Edeas M, and Bricaire F (2004) Neurodegenerative diseases and oxidative stress. *Biomedicine and Pharmacotherapy* 58: 39–46.

Esplugues JV (2002) NO as a signaling molecule in the nervous system. *British Journal of Pharmacology* 135: 1079–1095.

Guix FX, Uribesalgo I, Coma M, and Munoz FJ (2005) The physiology and pathophysiology of nitric oxide in the brain. *Progress in Neurobiology* 76: 126–152.

Moro MA, Cardenas A, Hurtado O, Leza JC, and Lizasoain I (2004) Role of nitric oxide after brain ischemia. *Cell Calcium* 36: 265–275.

Thippeswamy T, Mckay JS, Quinn JP, and Morris R (2006) Nitric oxide, biological double-faced janus – Is this good or bad? *Histology and Histopathology* 21: 445–458.

# Axonal Transport and Neurodegenerative Diseases

**S Roy, V M-Y Lee, and J Q Trojanowski**, University of Pennsylvania School of Medicine, Philadelphia, PA, USA



## Introduction

The primary functions of the neuron are to receive, process, and transmit information. A typical neuron has multiple dendrites and a single elongated process, the axon. The dendrites and the cell bodies play a role in the collection and processing of information, and the axon is responsible for the transmission of information to other neurons via synapses. Whereas dendrites are usually in close proximity to the neuronal cell bodies and have some capacity for protein synthesis, axons can extend up to tremendous lengths, sometimes several feet, and are generally devoid of the protein synthetic machinery. As a result, axonal proteins are synthesized in the cell bodies and subsequently transported into the axons and synapses. This process is called axonal transport. Axonal transport is essential for the survival of axons and synapses, and it occurs throughout the life of the neuron.

The unique architecture of the axon combined with its dependence on the remotely located cell body for growth and maintenance makes it especially vulnerable to a variety of insults. Not surprisingly, axonal and synaptic defects are characteristic features of most neurodegenerative diseases, and it was long suspected that defects in axonal transport could play a major role in these diseases. Although correlative evidence linking axonal transport defects and neurodegenerative diseases was well known, recent experimental evidence from several laboratories suggests that many neurodegenerative diseases may be a direct consequence of altered or defective axonal transport. In this article, we first highlight the basic principles of the mechanisms of axonal transport. Then, we review some of the evidence that links axonal transport to neurodegenerative diseases, including pathogenic mutations in human genes that encode proteins involved in axonal transport and studies of animal models of disease showing impairments of axonal transport.

## Basic Mechanisms of Axonal Transport

Although many specific details of axonal transport mechanisms are not fully understood, the basic principles are well established. In general, to move any cargo, two components are required – motors to move the cargo and rails to transport it. In addition, motor proteins often require adaptor/linker proteins to attach the motors to the cargo. For the rails, the neurons use the structural network of the cell, the cytoskeleton. Most long-range transport is dependent on the microtubules, but actin filaments also play a role in short-range movements. Microtubules in axons are polarized (unlike dendrites), allowing a directional bias in the movement of the motors and their cargo. For the motors, the neurons have an abundance of small molecular machines, known as kinesins and dyneins, moving the cargo mainly anterogradely (away from the cell body to the axon tip) or retrogradely (toward the cell body), respectively (**Figure 1**). The anterograde transport consists of components required for maintaining synapses, cytoskeletal proteins required for maintaining axonal structure and function, as well as a variety of other proteins required for neuronal function, including mitochondria, various chaperones, and glycolytic enzymes. The retrograde transport consists mainly of endosomal/lysosomal organelles that carry degraded proteins back to the cell bodies for degradation and also neurotrophic factors required for neuronal survival. In reality, however, most transport is bidirectional, with a bias either in the anterograde or in the retrograde direction.

Our understanding of the mechanisms of transport has increased dramatically during the past few decades due to advances in the development of model systems to dissect basic transport mechanisms and advances in biochemistry and genetics that have led to the identification and characterization of the molecular motors and their cellular functions. Model systems developed to visualize the transport process revealed that the bulk of proteins are transported anterogradely in two broad, discrete groups. Whereas some of the proteins are transported rapidly, at rates of 100–400 mm $day^{-1}$ (1–5 μm $s^{-1}$), quickly reaching up to the tip of the axon, others move at rates that are several orders of magnitude slower, at approximately 0.2–5 mm $day^{-1}$ (0.0002–0.05 μm $s^{-1}$). These two components are called fast axonal transport and slow axonal transport, respectively. Whereas fast transport mainly comprises vesicular cargo, including synaptic proteins, ion channels, and other components, slow axonal transport is primarily composed of transported cytoskeletal proteins, mainly microtubules, neurofilaments, and actin, along with many additional cytosolic proteins involved in neuronal homeostasis. Visualization of slow transport of the cytoskeleton revealed that despite moving with overall distinct dynamics, both slow and fast transport use similar

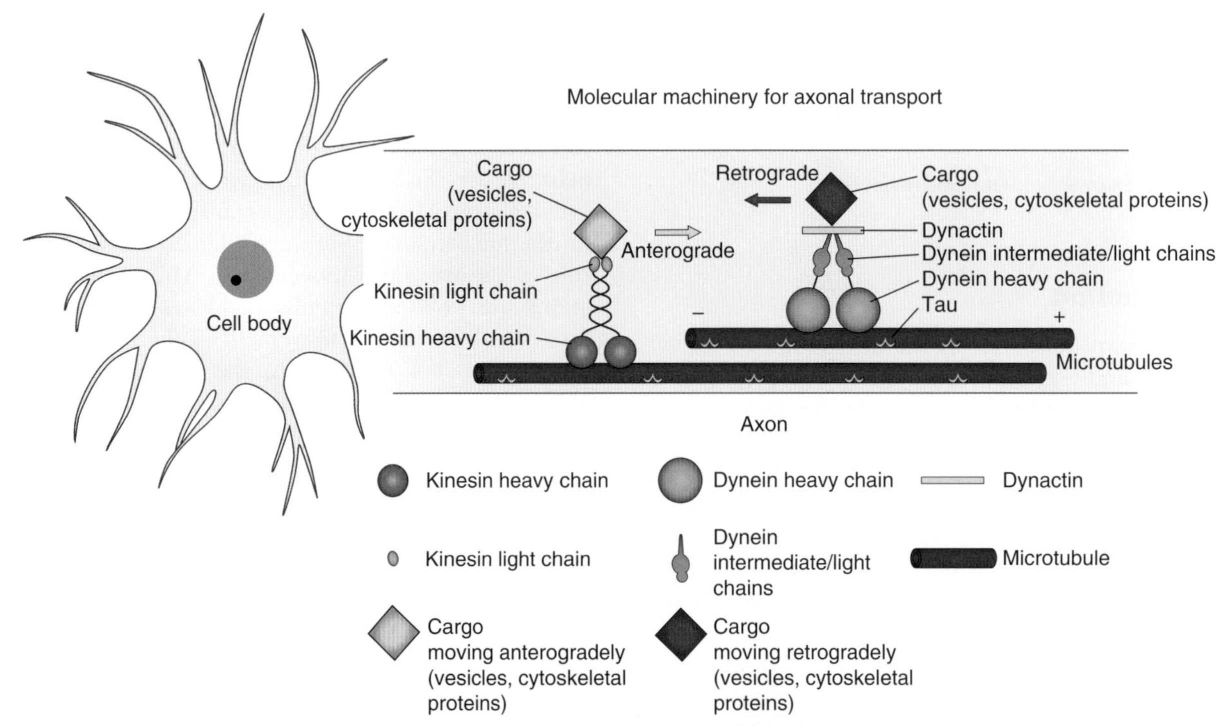


**Figure 1** Basic principles of axonal transport. Cargo is transported anterogradely and retrogradely by the molecular motors kinesin and dynein, respectively. A variety of additional adaptor/linker proteins may also help in binding kinesin to the cargo (not shown). Dynactin, a multiprotein complex, links dynein to its cargo and is thought to increase the efficiency of the motor. Long-range movement occurs along cytoskeletal 'tracks,' mainly microtubules.

underlying mechanisms. These studies showed that slow cargo moves rapidly like fast cargo, but unlike fast cargo, it pauses for prolonged times during transit. Thus, the overall movement is infrequent and intermittent, leading to a slow overall rate of the transported cytoskeletal population. It has also been shown that slow and fast transport share similar molecular motors.

Progress in biochemistry and genetics has led to the identification of approximately 45 members of the kinesin superfamily (Kifs), grouped into 14 subfamilies, and two members of the cytoplasmic dynein family. Besides this large array of motor proteins, there are also a number of linker/adaptor proteins responsible for binding cargo to the motor proteins that have been identified. This heterogeneity is thought to play a major role in the recognition of specific motors to their cargo. Many excellent reviews on the subject are available.

## Axonal Transport and Neurodegenerative Diseases

The following developments have dramatically highlighted the role of axonal transport disruptions in human neurodegenerative diseases: (1) the discovery of motor protein mutations in human neurodegenerative diseases, (2) axonal transport defects in animal and *in vitro* cellular models of neurodegenerative diseases, and (3) newly discovered roles in axonal transport regulation for known pathogenic proteins involved in neurodegenerative diseases. With a focus on specific diseases, we discuss some of the recent finings linking axonal transport to neurodegenerative diseases.

### Kinesin Mutation in Hereditary Spastic Paraplegia

Hereditary spastic paraplegias (HSPs), also known as familial spastic paraplegias, represent an autosomal dominant inherited group of neurodegenerative diseases. Patients present in their thirties or forties with symptoms in the lower limbs, with gradual proximal spread of symptoms. Neuropathologically, a distal axonopathy is seen, with severe degeneration and gliosis of the distal corticospinal tracts and relative sparing of the tracts in the brain stem and proximal cord. Due to the peculiar distal-to-proximal 'dying back' axonopathy observed in this disease, it had been hypothesized that dysfunctions in axonal transport leading to selective damage of the distant portions of the axons may be responsible for the pathogenesis of HSPs.

Indeed, a missense mutation in one of the genes encoding a major kinesin protein (the gene for kinesin heavy chain Kif 5a) was found in a family with HSP. The same mutation was found in all affected members of the family, as well as in some presymptomatic members. This mutation occurs within a functional motor domain of the kinesin protein and a homologous mutation in yeast has been found to decouple kinesin binding to microtubules, highlighting the functional role of the kinesin mutation in the pathology of HSPs.

### Dynactin Mutations in Distal Spinal and Bulbar Muscular Atrophy

Dynactin is a large protein complex linked to the retrograde motor dynein, and it is thought to link the motor to its cargo and/or increase the processivity of the motor. Animal models disrupting the dynein/dynactin complex develop late-onset motor neuron degeneration, and missense mutations in a dynein subunit cause progressive motor neuron degeneration in mice. Due to the central role of the dynein/dynactin complex in axonal transport, and based on data from the animal studies, it was hypothesized that mutations in the dynein/dynactin complex could play a role in neurodegeneration. Indeed, mutations in the gene encoding a subunit ($p150^{Glued}$) of the dynactin complex have been reported in a family with the neurodegenerative disease distal spinal and bulbar muscle atrophy (SBMA). In familial SBMA cases, the disease is transmitted in an autosomal dominant fashion and is manifested as is a primary lower motor-type neuropathy with patients presenting in their thirties, often with breathing difficulty due to vocal fold paralysis, which later leads to weakness in the face and distal extremities. Neuron loss as well as inclusions containing dynein and dynactin were also seen in autopsy studies from one patient. The mutation reduces the binding affinity of dynactin to microtubules and also causes subtle defects in dynein function.

### Kinesin Mutations in Charcot–Marie–Tooth Disease

Charcot–Marie–Tooth (CMT) disease comprises a heterogeneous group of inherited peripheral neuropathies characterized by motor and sensory deficits, often presenting in young adults as tingling, numbness, and loss of deep reflexes. The progression of the disease varies among individuals, with symptoms ranging from mild neuropathy to complete disability. Two basic forms can be recognized, with primary demyelinating (CMT1) or axonal (CMT2) types of degeneration predominating. Various genes have been implicated in CMT syndromes, including several genes known to play a role in myelination (PMP22 and MPZ) and genes for gap junction proteins (Connexin 32). However, in a remarkable series of studies, it was shown that mutations in a kinesin subunit protein (Kif 1B beta) can lead to an axonal type of CMT (CMT type 2A).

While studying animal models of kinesin knockout mice, it was found that heterozygous knockout mice for one of the kinesins, Kif 1B, developed progressive muscle weakness with normal motor nerve conduction velocities, symptoms resembling axonal type CMT, CMT type 2. Incidentally, the gene for CMT type 2A had been mapped to the same interval as the gene for Kif 1B, and genomic analysis of pedigrees with CMT type 2A revealed that these patients had a mutation in the Kif 1B gene. It was further shown that the mutant motor protein may not tightly bind to microtubules, thus suggesting a loss of function of the Kif 1B protein in patients with CMT type 2A.

### Axonal Transport Defects in Alzheimer's Disease and Other Tauopathies

The story of the role of axonal transport in Alzheimer's disease (AD) is a rapidly developing one. Neuropathologically, the two hallmarks of AD are deposits of fibrillar A$\beta$ into diffuse and neuritic plaques in the extracellular space, and filamentous accumulations of tau proteins as neurofibrillary tangles and neuropil threads within neurons and their processes. Amyloidogenic A$\beta$ peptides are generated by proteolytic processing of the A$\beta$ precursor proteins (APPs), conveyed by fast axonal transport. Several enzymes are involved in the proteolytic processing of APPs to A$\beta$, including the gamma- and beta-secretases (presenilins and BACE, respectively). Human mutations of both APPs and presenilins are seen in familial AD cases, highlighting the critical roles of these two proteins in AD. On the other hand, tau is a microtubule binding protein thought to stabilize microtubules *in vivo*.

The accumulation of tau in neuronal cell bodies and axons, as well as axonal swellings seen in AD, prompted the notion that axonal transport failure was important in the pathogenesis of this disease. Recent observations indicate that defects in axonal transport can occur long before the onset of severe symptoms in animal models of AD as well as in patients with AD. Interestingly, global reduction of axonal transport by reduction of kinesin levels can also exacerbate AD-type pathology in mouse models, further highlighting the role of diminished axonal transport in the pathogenesis of this disease.

The previous observations suggest that axonal transport defects play a role in the pathogenesis of AD. In addition, many key proteins directly involved in AD are also thought to play roles in the regulation of axonal transport. Studies of several different model systems have suggested that APP may act as a receptor for kinesin, thus proposing a direct link between a pathogenic protein and the axonal transport machinery. However, other studies have been unable to confirm these results. It has also been proposed that APPs may be transported in a vesicular complex containing presenilins and BACE, the gamma- and beta-secretase enzymes, and altered processing of APPs within this complex can lead to exacerbation of AD. These findings suggest that misregulation of APPs, either directly from known APP mutations (as in familial AD) or indirectly via proteins associating/interacting with APPs, can lead to disruption of fast axonal transport in general, thus leading to axonal depletion of critical components and neurodegeneration.

Another interesting line of evidence highlighting the role of presenilins in the regulation of fast axonal transport in AD derives from studies of presenilin mutant mice. As mentioned previously, presenilins are proteins responsible for regulated proteolysis of APPs, and mutations in presenilins are seen in most cases of early familial AD. Several studies indicate that presenilins interact with glycogen synthase kinase-3$\beta$ (GSK-3$\beta$). GSK-3$\beta$ is a kinase with many different roles, including the phosphorylation of kinesin light chains, and it has been shown that GSK-3$\beta$-mediated phosphorylation of kinesin light chains led to detachment of the kinesin motor from the cargo, thus preventing further transport of cargo. By using transgenic presenilin mutant mice, it was also shown that mutant or absent presenilin increased GSK-3$\beta$ levels, thereby phosphorylating kinesin light chains, detaching the kinesins from their cargo, and impairing axonal transport.

Roles in axonal transport regulation have also been assigned to tau. Because tau binds to microtubules and is thought to stabilize them, it was proposed that dysfunctions in tau can destabilize microtubules and lead to a failure of axonal transport. Indeed, human tau mutations impair the ability of tau to bind to and stabilize microtubules, and tau overexpression in cellular models can lead to defects in axonal transport. In addition, the mechanisms causing the accumulation of tau in AD are also beginning to be understood. Defective axonal transport of disease-associated mutant tau has also been demonstrated in mouse models, providing experimental evidence for a mechanism of tau accumulation.

A final line of evidence suggesting a role for axonal transport defects in AD comes from studies of ApoE4, a gene whose allelic state is associated with an increased risk for AD. Mice expressing human ApoE4 exhibit defects in axonal transport, and the receptor for ApoE4, ApoER2, binds to JIP1/2, a protein that appears to mediate the binding of APPs to kinesin I. Thus, it can be postulated that overexpression of ApoE4 protein can lead to misregulation of JIP1/2-mediated binding of kinesin to APPs, leading to defects in fast axonal transport.

Stabilization of microtubules is also being explored as a therapeutic option in AD. The proof of principle for this concept was demonstrated in a study that showed that a microtubule-stabilizing drug (paclitaxel) could ameliorate the neurodegenerative phenotype in transgenic tau mice by offsetting the loss of tau function by stabilizing the microtubules and correcting the fast axonal transport defects in these mice.

### Axonal Transport and Huntington's Disease

Neuropathologically, Huntington's disease is characterized by atrophy and degeneration of striatal neurons, with aggregates of pathological polyglutamine containing the protein huntingtin. Huntingtin is a predominantly cytoplasmic protein that associates with vesicles and moves in the fast axonal transport component. Although it has been known for several years that polyglutamine repeats within the huntingtin protein cause a gain of deleterious function leading to neurodegeneration, the exact pathogenic role of the repeats is unclear. By infusing huntingtin-harboring pathological polyglutamine repeats into a model for studies of fast axonal transport, it was shown that fast axonal transport was specifically inhibited by pathologically expanded polyglutamine repeats (but not normal proteins), along with inhibition of neurite extension in cultured cells. Furthermore, disruption of the *Drosophila* huntingtin gene also caused axonal transport defects. These findings lend support to a model in which aggregates of polyglutamine repeats disrupt fast axonal transport. Whether the disruption of axonal transport is a direct effect of the polyglutamine repeats or a secondary phenomenon remains to be established.

### Axonal Transport and Amyotrophic Lateral Sclerosis

The histopathologic observation of prominent neurofilament-rich inclusions in the axons of spinal motor neurons of patients with amyotrophic lateral sclerosis (ALS) led to the hypothesis that disrupted axonal transport of proteins may play a role in the pathogenesis of the disease. However, the first direct evidence that axonal transport is disrupted in ALS awaited the development of transgenic mouse models of familial ALS based on expression of mutant superoxide dismutase (SOD-1) protein mutant mice that replicate several key aspects of ALS. Studies of these SOD-1 transgenic mouse models of ALS showed that the transport of neurofilament proteins was retarded in these animals, even before the mice were symptomatic, thereby implicating impaired axonal transport as an early deficit in ALS. Mutations in the p150 subunit of dynactin have also been reported in ALS patients.

### Axonal Transport and Synucleinopathies

Synucleinopathies, also known as α-synucleinopathies, are a group of neurodegenerative disorders in which the primary pathology is the intracytoplasmic accumulation of α-synuclein primarily in neurons and, in some cases, glial cells. These disorders include Parkinson's disease, dementia with Lewy bodies, the Lewy body variant of AD, multiple system atrophy, and neurodegeneration with brain iron accumulation. In familial forms of Parkinson's disease, autosomal dominant missense mutations in genes encoding for α-synuclein are seen, suggesting a role of α-synuclein in the pathogenesis of these disorders. α-Synuclein is a highly conserved protein belonging to a multigene family that includes β-synuclein and γ-synuclein. α-Synuclein is strongly expressed in neurons, highly enriched in presynaptic terminals, and transported predominantly in the slow component. Axonal transport abnormalities of α-synuclein have been proposed in synucleinopathies, based on the observation that axonal α-synuclein pathology is pronounced in the disease and also on experimental evidence suggesting that α-synuclein may play a role in transport of presynaptic vesicles. Age-related retardation in the normal transport of α-synuclein was also seen in a study. Collectively, these studies propose a model in which age-related retardation of α-synuclein transport leads to accumulations of the protein over time, predisposing to the α-synuclein pathology in axons. Although these findings are interesting, many questions remain unanswered. The physiological role of synuclein is far from clear, and much work needs to be done to identify the role of synuclein in neurodegenerative disorders.

## Conclusions and Future Directions for Research

A growing body of evidence implicates axonal transport defects in the etiology and pathogenesis of neurodegenerative diseases. Although motor protein defects in neurodegenerative diseases are direct evidence for this, it is likely that many other disease proteins are directly or indirectly linked to the complicated machinery of axonal transport. Thus, studies on the molecular mechanisms of transport are necessary to facilitate our understanding of the role of axonal transport in these diseases. Since pathogenic proteins in AD have roles in the regulation of axonal transport, and these proteins have been extensively studied, the time is perhaps ripe to uncover the links between AD and impaired axonal transport. With greater understanding of the other neurodegenerative diseases, it is likely that many more links to axonal transport will be uncovered. Another exciting but largely neglected avenue of research is drug discovery efforts to counteract axonal transport impairments as therapeutic interventions for neurodegenerative diseases. Indeed, if axonal transport defects are shown to be part of a common mechanism of disease in many neurodegenerative disorders, the discovery of drugs that modulate axonal transport could result in the development of important therapeutic interventions for the treatment of these disorders.

*See also:* Axonal Transport and Alzheimer's Disease; Axonal Transport and Huntington's Disease.

## Further Reading

Brown A (2003) Axonal transport of membranous and nonmembranous cargoes: A unified perspective. *Journal of Cell Biology* 160: 817–821.

Chevalier-Larsen E and Holzbaur EL (2006) Axonal transport and neurodegenerative disease. *Biochimica et Biophysica Acta* 1762: 1094–1108.

Duncan JE and Goldstein LS (2006) The genetics of axonal transport and axonal transport disorders. *PLoS Genetics* 2: e124.

Hirokawa N and Takemura R (2005) Molecular motors and mechanisms of directional transport in neurons. *Nature Reviews Neuroscience* 6: 201–214.

Munch C, Rosenbohm A, Sperfeld AD, et al. (2005) Heterozygous R1101K mutation of the DCTN1 gene in a family with ALS and FTD. *Annals of Neurology* 58: 777–780.

Munch C, Sedlmeier R, Meyer T, et al. (2004) Point mutations of the p150 subunit of dynactin (DCTN1) gene in ALS. *Neurology* 63: 724–726.

Puls I, Jonnakuty C, LaMonte BH, et al. (2003) Mutant dynactin in motor neuron disease. *Nature Genetics* 33: 455–456.

Puls I, Oh SJ, Sumner CJ, et al. (2005) Distal spinal and bulbar muscular atrophy caused by dynactin mutation. *Annals of Neurology* 57: 687–694.

Reid E, Kloos M, Ashley-Koch A, et al. (2002) A kinesin heavy chain (KIF5A) mutation in hereditary spastic paraplegia (SPG10). *American Journal of Human Genetics* 71: 1189–1194.

Roy S, Zhang B, Lee VM, and Trojanowski JQ (2005) Axonal transport defects: A common theme in neurodegenerative diseases. *Acta Neuropathologica (Berlin)* 109: 5–13.

Trojanowski JQ, Smith AB, Huryn D, and Lee VM (2005) Microtubule-stabilising drugs for therapy of Alzheimer's disease and other neurodegenerative disorders with axonal transport impairments. *Expert Opinion on Pharmacotherapy* 6: 683–686.

Zhao C, Takita J, Tanaka Y, et al. (2001) Charcot–Marie–Tooth disease type 2A caused by mutation in a microtubule motor KIF1Bbeta. *Cell* 105: 587–597.

# Inflammation in Neurodegenerative Disease and Injury

**M G Tansey and S G Kernie**, The University of Texas Southwestern Medical Center, Dallas, TX, USA



## Overview

### Features of Neuroinflammation

Inflammation can be defined in clinical, pathological, and molecular terms. The four cardinal signs of inflammation described by Celsius in AD 25, dolor, tumor, calor, and rubor (i.e., pain, swelling, heat, and redness), are not present in patients with neurodegenerative disease. Neuropathological and neuroradiological studies indicate that inflammation in neurodegenerative diseases is characterized by a glial response that in certain diseases may be present prior to significant loss of neurons. It was once believed that the blood–brain barrier (BBB) prevented access of immune cells to the brain and, as a result, the immune and central nervous system were relatively independent of each other. However, it has become clear that the permeability of the BBB can be regulated under normal conditions and may increase or become dysregulated in disease states. The inflammatory reaction in the brain associated with most acute or chronic neurodegenerative diseases is often termed 'neuroinflammation.' Neuroinflammation consists mainly of an innate immune response involving activation of glial cells and central production of cytokines, chemokines, prostaglandins, complement cascade proteins, and reactive oxygen species and reactive nitrogen species (ROS/RNS) in response to a central or peripheral immune challenge. The features of neuroinflammation include the presence of reactive astrocytes and activated microglial cells in the central nervous system (CNS). In certain neurodegenerative diseases, infiltration of lymphocytes (B and T cells) and polymorphonuclear cells occurs when the re is extensive breakdown of the BBB.

**Glia: Microglia and astrocytes** Microglia are the monocyte-derived resident macrophages of the brain responsible for immune surveillance. An initial physical or pathogenic event in the CNS is expected to elicit activation of microglia and secretion of factors that promote neuronal survival, with the purpose of limiting injury, protecting vulnerable neuronal populations, and aiding in repair processes. When the permeability of the BBB increases, recruitment of peripheral macrophages to the CNS results in expansion of the brain microglia population. Activated microglia produce ROS and RNS such as nitric oxide (NO) and secrete prostaglandins, chemokines, and cytokines, as well as neurotrophic factors (including glial-derived neurotrophic factor family ligands, or GFLs) that promote neuronal survival. If glial activation persists for extended periods, a cycle of chronic neuroinflammation may result and lead to elevation of local levels of cytokines, chemokines, prostaglandins, and other inflammatory mediators that can contribute to tissue damage. The mechanisms that regulate microglial activities in the brain are poorly understood. In recent years, a more thorough understanding of the functional link between microglial activation markers and specific cellular functions has evolved. New information about the mechanisms that regulate the balance of neuroprotective microglial activities (e.g., neurotrophic factor production, phagocytosis of debris, amyloid or senile plaque removal) and neurotoxic microglial activities (i.e., ROS/RNS, prostaglandin, cytokine, and chemokine overproduction) will be required to understand how dysregulation of this balance hastens the death of vulnerable neuronal populations.

A second type of glial cell, the astrocyte, resides in the CNS and is important for regulation of the permeability of the blood–brain barrier. The cellular processes of astrocytes make intimate contact with essentially all areas of the brain and are functionally coupled to neurons, oligodendrocytes, and other astrocytes. This tight coupling enables astrocytes to regulate the homeostatic environment (including regulation of the excitatory neurotransmitter glutamate) and promote proper functioning of the neuronal network. Inflammation in the CNS may disrupt this process either transiently or permanently, as cytokines have important roles in reprogramming gene expression in glia (both in microglia and in astrocytes) during injury and recovery. Upregulation of glial fibrillary acidic protein (GFAP) in astrocytes denotes reactive astrogliosis in the CNS and may be part of a response to acute injury or neurodegeneration that can lead to permanent scarring and tissue damage. It is becoming increasingly evident that astrocytes also have important roles in neurological disorders. Mouse models of disease have revealed astrocyte-specific pathologies that contribute to neurodegeneration. Development of animal models in which astrocyte-specific proteins and pathways have been manipulated has contributed to our understanding of the normal biology of astrocytes. It is expected that development of targeted therapies for pathways in which astrocytes participate to the detriment of neurons will be possible in humans and should ultimately influence the clinical treatment of neurodegenerative disorders.

**Cytokines, chemokines, and lymphokines** Cytokines, chemokines, and lymphokines are multifunctional immunoregulatory proteins secreted by cells of the immune system. Within the CNS, cytokines, most notably tumor necrosis factor-$\alpha$ (TNF-$\alpha$), the interleukin-1 (IL-1) and interleukin-6 (IL-6) families, interferon-$\gamma$ (IFN-$\gamma$), and transforming growth factor-$\beta$ (TGF-$\beta$) act in context-dependent ways to modulate inflammatory processes that affect the permeability of the BBB. Cytokines can exert neuroprotective effects independent of their immunoregulating properties and some of them (e.g., TNF-$\alpha$) have been shown to modulate neurotransmission. However, under certain conditions, cytokines and chemokines promote apoptosis of neurons, oligodendrocytes, and astrocytes as well as damage to myelinated axons. The outcome of cytokine action in the brain depends on the dynamics, cellular source, degree of compartmentalization of cytokine release, pathophysiological context, and presence of co-expressed factors. The levels of several cytokines (including TNF-$\alpha$, IL-1, and IL-6) become elevated in a wide range of CNS disorders, including ischemia, trauma, multiple sclerosis, Alzheimer's disease (AD), and Parkinson's disease (PD). While there is no evidence to support a role for any of them in triggering any of these diseases, cytokine-driven neuroinflammation and neurotoxicity may modify disease progression in a number of neurodegenerative diseases (see later).

### What Triggers Neuroinflammation?

Neuroinflammation may be triggered by protein aggregation and formation of inclusions arising from mutations (i.e., $\alpha$-synuclein) or disruption of the ubiquitin–proteasome system (UPS), immunologic challenges (bacterial or viral infections), neuronal injury (brain trauma or stroke), or other epigenetic factors. Chronic inflammatory syndromes (rheumatoid arthritis, Crohn's disease, and multiple sclerosis) and environmental toxins (pesticides and particulate matter) are the most notable epigenetic factors likely to trigger neuroinflammation. These insults can increase the permeability of the BBB to allow infiltration of lymphocytes and macrophages into the brain parenchyma. The immune reactions initiated by viruses and bacteria may contribute to latent vulnerabilities which could become manifest with future immunologic challenges. Depending on context, duration, and type of inflammatory response, inflammation may be detrimental or beneficial to the individual. During trauma to the CNS, acute inflammatory responses are mounted to limit injury and to aid in neuronal repair. However, when inflammation becomes chronic, it may lead to tissue damage and is likely to be detrimental to neuronal survival. Acute injury in most tissues is followed by release of histamine, with resulting increases in vascular permeability (i.e., exudation of fluid into the injured tissue and neutrophil infiltration). In the brain, such a response could result in increased intracranial pressure and would be detrimental within the confines of the skull. Thus, it is possible that the downgraded response of resident microglia is an evolved response that permits the CNS to respond to injury in a way that will result in minimal brain damage.

### How Does Neuroinflammation Affect Neurons?

At the molecular level, acute inflammation in the CNS is believed to involve the neuroprotective actions of microglia, and to be beneficial because it acts to limit tissue damage (as in ischemic stroke). In contrast, chronic neuroinflammation is more likely to involve microglial activities that promote ROS/RNS generation, enhance the unfolded protein response (UPR), and contribute to neurotoxicity. Inflammatory challenges may trigger clinical symptoms of neurological or psychiatric conditions. High levels of certain microglial-derived inflammatory mediators (e.g., TNF-$\alpha$) often lead to enhanced oxidative stress in neurons because they contribute to glutathione depletion. In addition,, several of the downstream signaling pathways (including caspases) activated by inflammatory factors may converge to elicit mitochondrial dysfunction and death of vulnerable neuronal populations through apoptosis or other forms of programmed cell death.

## Evidence for Inflammation in Neurodegenerative Diseases

Neurodegenerative diseases are characterized by the loss of specific neuronal populations and often by intraneuronal as well as extracellular accumulation of fibrillary materials. Formation of intracellular inclusion bodies may result from abnormal protein–protein interactions, aberrant protein folding, and/or dysregulation of the UPS. These conditions, often referred to as 'proteinopathies,' are thought to have a central role in the neuronal dysfunction that characterizes several neurodegenerative diseases. One common way in which a number of divergent molecular or cellular events (e.g., mutations, oxidation, protein misfolding, truncation, or aggregation) may all contribute to death of neurons is via activation of resident microglial populations in specific brain regions. If the host is unable to eliminate the initiating trigger (i.e., proteinopathy caused by a mutation), chronic activation of microglia and chronic neuroinflammation-derived oxidative stress may contribute to progressive neuronal dysfunction. If neuroinflammation is in fact contributing to neurodegeneration of vulnerable populations, then timely use of certain anti-inflammatory drugs should offer neuroprotection and limit progression of disease. However, human clinical trials with anti-inflammatory drugs have had very limited success in the management of several neurodegenerative diseases to date. Possible reasons for the failure of these trials could be the timing of the treatment and the dosing regimens chosen for the trials; further investigation in this area is warranted before dismissing the possibility of potential long-term benefits of anti-inflammatory drugs.

### Clinical and Epidemiological Evidence for Inflammation in PD

Over the past decade, a great wealth of new information has emerged to suggest that inflammation-derived oxidative stress and cytokine-dependent neurotoxicity are likely to contribute to nigrostriatal pathway degeneration and lead to idiopathic PD in humans. Postmortem analyses of brains from PD patients have confirmed the presence of inflammatory mediators in the area of substantia nigra (SN), where maximal destruction of vulnerable melanin-containing dopamine (DA)-producing neurons occurs in PD patients. Signs of inflammation included activated microglia and accumulation of cytokines (including TNF-$\alpha$, IL-1$\beta$, IL-6, and IFN-$\gamma$). The advent of technologies (such as positron emission tomography brain scans) that can be conducted in living humans confirms that patients with idiopathic PD have markedly elevated microglia activation in the pons, basal ganglia, striatum, and frontal and temporal cortical regions, irrespective of the number of years with the disease, compared to age-matched healthy controls. As stated earlier, activated microglia overproduce cytokines (including TNF-$\alpha$, IL-6, and IL-1$\beta$) which exert neurotoxic effects on DA neurons; and SN dopaminergic neurons may be uniquely vulnerable to neuroinflammatory insults that enhance oxidative stress. Elevated levels of cytokines may act in an autocrine manner to maintain activation of abundant numbers of microglia in the midbrain, potentiating inflammatory responses (autoamplification of ROS, NO, and superoxide radicals to form highly oxidizing peroxynitrite species) and creating an environment of oxidative stress which further hastens damage of DA neurons. Although the genes for various cytokines, chemokines, and acute-phase proteins have been surveyed and individual reports demonstrate that certain single nucleotide polymorphisms are overrepresented in certain cohorts of individuals affected with PD, most of these findings have not been replicated in independent studies. Meta-analyses of multiple such association studies are crucial to assess the overall genetic effect of cytokine gene polymorphisms on disease.

### Evidence for Oxidative Stress and Inflammation in Experimental Models of PD

A better understanding of the extent to which environmental factors that elicit chronic inflammation modify the known genetic risks for developing PD is needed. The two neurotoxins most widely used to induce nigral DA neuron loss in rodents are 1-methyl-4-phenyl-1,2,3,6-tetrahydropyridine (MPTP) and 6-hydroxydopamine (6-OHDA). These neurotoxin models are accompanied by a robust glial reaction. Studies with nonhuman primates exposed to MPTP indicate that the cycle of neuroinflammation triggered by these neurotoxins persists long after the initial insult abates and may contribute to the progressive degeneration of the nigrostriatal pathway. Although each neurotoxin may trigger different initial cascades of events, they all consistently involve oxidative stress as the critical mechanism that elicits DA neuron death. In addition, there are endogenous toxins that can contribute to nigral DA neuron loss and to the risk of developing PD. These include tetrahydroisoquinolines (TIQs), which are produced via the reaction of DA with aldehydes and act to decrease dopaminergic neurotransmission, generate ROS, and trigger apoptosis. Nitric oxide (NO) also contributes to DA neuron death through mechanisms that, while not completely understood, are likely to involve mediation of excitotoxicity, activation of poly(ADP ribose) polymerase-1 (PARP-1), DNA damage, activation of caspase-dependent and-independent cell death, and/or nitrosylation of proteins, including $\alpha$-synuclein and Parkin. Thus, general synthetic antioxidants as well as natural products with antioxidant properties (including green tea, omega oils, curcumin, and various varieties of berries) are being intensely investigated for their ability to provide DA neuroprotection in experimental models of PD and in the clinic.

Findings obtained from histopathologic, genetic, and pharmacologic studies in animals implicate a role for cytokine-driven inflammation and neurotoxicity in loss of nigral DA neurons induced by MPTP, 6-OHDA oxidative neurotoxin, and bacteriotoxin. Rapid increases in cytokine levels are detectable in rodent midbrain SN within hours of *in vivo* administration of 6-OHDA or MPTP. Pharmacologically, chronic infusion of various anti-inflammatory compounds, including cyclooxygenase-2 (COX-2)-selective nonsteroidal anti-inflammatory drugs (NSAIDs) or soluble TNF-selective inhibitors, rescues nigral DA neurons from progressive degeneration and death. The higher sensitivity of nigral DA neurons to injury induced by neuroinflammatory mediators may be secondary to reduction of endogenous antioxidant capacity (i.e., glutathione depletion). In summary, an overwhelming number of studies now indicate that chronic elevation of certain cytokines in the CNS can contribute to progressive degeneration of nigral DA-producing neurons. These findings raise the interesting possibility that environmental triggers initiate cytokine-driven neuroinflammation and contribute to the development of PD in humans.

### Anti-inflammatory Drugs and PD

Additional strong evidence that inflammatory processes modulate PD risk came from a prospective study of hospital workers; the study indicated that the incidence of sporadic PD among chronic users of over-the-counter NSAIDs was 46% lower than in age-matched non-NSAID users. NSAIDs scavenge free oxygen radicals and inhibit cyclooxygenase-2, which is known to participate in oxidation of DA. Although the specific NSAID target(s) that mediate(s) the effect associated with lower PD risk remain(s) unidentified, these findings suggest that neuroinflammatory processes contribute to DA neuron loss and development of PD in humans. In concert with findings derived from experimental models, these epidemiological findings raise the possibility that anti-inflammatory therapy could delay or prevent onset of PD.

Steroids have anti-inflammatory properties and have been shown to rescue DA neurons in rodent models of PD; however, because of their well-established side effects, they are not likely to be suitable for long-term use in humans. Clinical trials in PD patients using short-term administration of immunophilin ligands (which lack immunosuppressive properties of the parent compounds) have also had limited success. At the time of this writing, minocycline, the tetracycline derivative which readily crosses the BBB and is well-tolerated in humans, was under evaluation in phase III clinical trials to determine if it could alter long-term progression of PD and other neurodegenerative diseases. A reexamination of available classes of anti-inflammatory agents may be needed to identify other compounds that can cross the BBB. Alternatively, novel modes of therapeutic delivery for introduction of promising anti-inflammatory compounds that cannot readily cross the BBB will need to be developed. Time will tell if successful development and timely use of anti-inflammatory approaches can modify the progressive course of this disease and afford therapeutic benefit in patients with early signs of PD.

## Inflammation in Acute CNS Injury: Traumatic Brain Injury

Inflammation is a hallmark of all forms of acquired brain injury (i.e., traumatic brain injury (TBI), cerebral ischemia or stroke, and spinal cord injury). Worldwide, millions of people are treated each year for significant head injury; a substantial proportion die, and many more are disabled. The progressive tissue damage that occurs following TBI has been attributed to cerebral inflammation and apoptotic cell death. Strategies to inhibit one or both of these pathways are being investigated as potential therapies for TBI patients (see later), though current therapy for TBI remains entirely supportive. The rationale behind mechanism-based drug design is to provide some degree of neuroprotection by preventing injured or vulnerable nerve cells from dying and/or to stimulate regenerative processes that could lead to the restoration or the formation of new connections lost on account of the injury. However, in the case of TBI, although it is clear that microglia activation occurs, the role of microglia in limiting versus promoting injury has not been clearly established. To add to an already complex scenario, significant activation of astroglia also occurs in TBI and recent evidence suggests that astrocytes may also participate in the neuroinflammatory process by regulating expression of heme oxygenase-1 (HO-1) in microglia and limiting ROS/RNS production.

Although there are no clinical treatments yet available to enhance repair of the damaged brain after trauma, a number of potential therapies have been investigated with varying degrees of success. For many years, corticosteroids had been used to treat head injuries in the emergency room. This popular practice was based on the rationale that if these drugs could reduce mortality by 1–2% they would have a significant impact given the large numbers of people that incur TBI each year. However, the largest head injury trial conducted to date has provided conclusive data to indicate that global immunosuppression of innate immune responses with corticosteroids actually negatively affects outcome after TBI. In 2001, after a successful pilot study, the Medical Research Council of Great Britain funded the CRASH (Corticosteroid Randomization after Significant Head Injury) trial to examine the effect of corticosteroids on death and disability after injury. Data obtained for nearly 10 000 patients at 6 months postintervention showed that the risk of death was higher in the corticosteroid group than in the placebo group, as was the risk of death or severe disability, with no evidence that the effect of corticosteroids differed by injury severity or time since injury. On the basis of this definitive study, it was concluded that corticosteroids should not be used routinely in the treatment of head injury. Another large randomized, controlled trial using therapeutic hypothermia in the setting of moderate to severe TBI also failed to show an improvement. However, recent data suggest that one way that hypothermia may be helpful is when it is administered early enough to be an immune modulator. Based on this, there is currently another National Institutes of Health (NIH)-funded phase I clinical trial looking at the efficacy of hypothermia initiated early following TBI in children.

Given that TBI can induce progressive neurodegeneration in association with chronic inflammation and that chronic treatment with the nonsteroidal anti-inflammatory drug, ibuprofen, had been shown to improve functional and histopathologic outcome in mouse models of AD, ibuprofen was also investigated for its ability to improve long-term outcome following experimental TBI. Disappointingly, chronic ibuprofen administration was found to worsen cognitive outcome following TBI. These data may have important implications for TBI patients who are often prescribed NSAIDs for chronic pain.

Minocycline, a tetracycline derivative with anti-inflammatory properties, has been intensely studied and found to be therapeutically effective in various models of CNS injury and disease via mechanisms involving suppression of inflammation and apoptosis. Although individual studies have shown some positive effects with minocycline, further studies on minocycline in TBI are necessary in order to consider it as an effective therapy for brain-injured patients.

The neuroprotective properties of progesterone (PROG) and its enantiomers (ent-PROG) or its metabolite (allopregnanolone) have been investigated after TBI. Studies have shown that a single injection of progesterone can attenuate cerebral edema when administered during the first 24 h after TBI in rats, but this regimen did not always produce functional benefits. In contrast, steady-state progesterone treatment after TBI decreased edema and anxiety and increased activity, thus enhancing behavioral recovery. Therefore, chronic pharmacological administration of progesterone may prove to be more beneficial in translational and clinical testing compared to bolus injections over the same period of time. In addition, ent-PROG treatment has been shown to decrease cerebral edema, cell death mediators, inflammatory cytokines, and reactive gliosis, and to increase antioxidant activity in experimental models of TBI. Based on these compelling experimental data supporting the neuroprotective properties of progesterone, a phase II randomized, double-blind, placebo-controlled clinical trial was conducted at an urban level I trauma center to assess the safety and potential benefit of administering progesterone to patients with acute TBI. The study showed that patients with moderate TBI who survived their injury were more likely to have a moderate to good outcome if they received progesterone than were those who received the placebo.

The pro-apoptotic cytokine TNF as well as the impact of TNF neutralization has been extensively investigated following TBI in a number of different experimental models (e.g., percussion injury and controlled cortical injury) with a number of different outcomes, depending on the timing of the intervention. Overall, TNF was found to have dual effects: in certain instances it enhanced survival and in others it promoted apoptosis during secondary damage after TBI. These findings are not surprising given that TNF signaling activates numerous downstream pathways that have the capacity to differentially regulate activation of microglia – nuclear factor-kappa B (NF-$\kappa$B)- and c-Jun N-terminal kinase (JNK)/AP-1-dependent gene transcription. Multiple studies have suggested that the protective effects of hypothermia following TBI may be mediated via TNF-R1 activation of the JNK and caspase-3 pathways.

Last, it is worth noting that repetitive TBI has been linked to the development of AD. Development of diffuse $\beta$-amyloid (A$\beta$) deposits and tau pathology approximately 1–3 weeks after head trauma has been reported clinically, and head trauma is the environmental factor most consistently associated with AD or PD risk. The mechanisms underlying this process are not clear but are likely to be influenced by apolipoprotein E (apoE) gene status. The $\varepsilon$4 allele of the *APOE* gene (*APOE* $\varepsilon$4), which confers higher AD risk to individuals, has been associated with both the vascular risk factors and the consequences (cognitive impairment, dementia) of white-matter lesions (WMLs), suggesting that *APOE* $\varepsilon$4 carriers may have diminished capacity for neuronal repair after TBI. Further studies are warranted to clarify this relationship.

## Conclusions

Over the past decade, the concept of the brain as an immune-privileged organ has been rejected in light of unequivocal evidence that the permeability of the blood–brain barrier can be regulated under normal conditions and may become dysregulated in chronic and acute neurodegenerative conditions. Reactive astrocytes, activated microglial cells, and overproduction of inflammatory mediators orchestrate neuroinflammatory responses that either promote or compromise neuronal survival, depending on type, duration, and context. In general, acute inflammatory responses during trauma to the CNS involve the neuroprotective actions of microglia aimed at limiting injury and enhancing neuronal repair. When inflammation becomes chronic, microglial activities create an environment that often enhances oxidative stress in neurons, contributes to neuronal dysfunction, and hastens neurodegeneration. While there is no evidence to support a role for any specific inflammatory mediator in the etiology of any neurodegenerative disease, cytokines in the brain and/or cerebrospinal fluid become elevated in a wide range of CNS disorders, including ischemia, trauma, multiple sclerosis, AD, and PD, raising the possibility that they modify (i.e., accelerate) disease progression. Collectively, whether cytokine action has beneficial or harmful outcomes in the brain depends on the dynamics, cellular source, degree of compartmentalization of cytokine release, pathophysiological context, and presence of co-expressed factors. Nigral DA neurons may be uniquely vulnerable to neuroinflammatory insults that enhance oxidative stress; and over the past decade, a great wealth of new information has emerged to suggest that inflammation-derived oxidative stress and cytokine-dependent neurotoxicity may contribute to nigrostriatal pathway degeneration and increase the risk for development of idiopathic PD in humans. Following acute TBI, cerebral inflammation and apoptotic cell death contribute to progressive tissue damage and may increase the risk for development of AD by promoting diffuse deposits of A$\beta$. No highly effective clinical treatments are available to enhance repair of the damaged brain, and use of high-dose steroids is now contraindicated in the management of TBI. Further basic science and more clinical research are needed to establish a causal link between neuroinflammation and neurodegeneration. If one is found, timely use of anti-inflammatory drugs may prove effective in slowing down the progressive neuronal death in chronic neurodegenerative disease and acute CNS injury.

*See also:* Animal Models of Parkinson's Disease; Cell Replacement Therapy: Parkinson's Disease; Oxidative Damage in Neurodegeneration and Injury; Parkinson's Disease: Alpha-Synuclein and Neurodegeneration.

## Further Reading

Arai H, Furuya T, Mizuno Y, et al. (2006) Inflammation and infection in Parkinson's disease. *Histology and Histopathology* 21: 673–678.

Block ML and Hong JS (2005) Microglia and inflammation-mediated neurodegeneration: Multiple triggers with a common mechanism. *Progress in Neurobiology* 76: 77–98.

Chen H, Zhang SM, Hernan MA, et al. (2003) Nonsteroidal anti-inflammatory drugs and the risk of Parkinson disease. *Archives of Neurology* 60: 1059–1064.

Chen H, Jacobs E, Schwarzschild MA, et al. (2005) Nonsteroidal antiinflammatory drug use and the risk of Parkinson's disease. *Annals of Neurology* 59: 988–989.

Edwards P, Arango M, Balica L, et al. (2005) Final results of MRC CRASH, a randomised placebo-controlled trial of intravenous corticosteroid in adults with head injury-outcomes at 6 months. *Lancet* 365: 1957–1959.

Hald A and Lotharius J (2005) Oxidative stress and inflammation in Parkinson's disease: Is there a causal link? *Experimental Neurology* 193: 279–290.

Hirsch EC, Hunot S, and Hartmann A (2005) Neuroinflammatory processes in Parkinson's disease. *Parkinsonism & Related Disorders* 11(supplement 1): S9–S15.

Jellinger KA (2004) Traumatic brain injury as a risk factor for Alzheimer's disease. *Journal of Neurology, Neurosurgery & Psychiatry* 75: 511–512.

McGeer PL, Itagaki S, Boyes BE, et al. (1988) Reactive microglia are positive for HLA-DR in the substantia nigra of Parkinson's and Alzheimer's disease brains. *Neurology* 38: 1285–1291.

McGeer PL, Schwab C, Parent A, et al. (2003) Presence of reactive microglia in monkey substantia nigra years after 1-methyl-4-phenyl-1,2,3,6-tetrahydropyridine administration. *Annals of Neurology* 54: 599–604.

Minghetti L (2005) Role of inflammation in neurodegenerative diseases. *Current Opinion in Neurology* 18: 315–321.

Mrak RE and Griffin WS (2005) Glia and their cytokines in progression of neurodegeneration. *Neurobiology of Aging* 26: 349–354.

Sanchez-Pernaute R, Ferree A, Cooper O, et al. (2004) Selective COX-2 inhibition prevents progressive dopamine neuron degeneration in a rat model of Parkinson's disease. *Journal of Neuroinflammation* 1: 6.

Wersinger C and Sidhu A (2002) Inflammation and Parkinson's disease. *Current Drug Targets – Inflammation & Allergy* 1: 221–242.

## Relevant Websites

http://www.apdaparkinson.org – American Parkinson Disease Association Inc.

http://www.pdgene.org – Database for Parkinson's disease genetic association studies, Massachusetts General Hospital/Harvard Medical School, The Michael J. Fox Foundation and Alzheimer Research Forum.

http://www.braininjury.com – Medical, legal, and informational resources for persons dealing with traumatic brain injury.

http://www.parkinson.org – National Parkinson Foundation.

http://www.pdf.org – Parkinson's Disease Foundation.

http://www.braintrauma.org – The Brain Trauma Foundation.

http://www.michaeljfox.org – The Michael J. Fox Foundation for Parkinson's Research.

http://www.worldpdcongress.org – World Parkinson Congress.

# Cholinergic System Imaging in the Healthy Aging Process and Alzheimer Disease

**S J Teipel**, University of Rostock, Rostock, Germany
**L T Grinberg**, Faculdade de Medicina da Universidade de São Paulo and Instituto Israelita de Ensino e Pesquisa Albert Einstein, Sao Paulo, Brazil
**H Hampel**, Ludwig-Maximilian University, Munich, Germany; and Trinity College Dublin, The Adelaide and Meath Hospital, Incorporating the National Children's Hospital, Dublin, Republic of Ireland
**H Heinsen**, University Würzburg, Würzburg, Germany



## Introduction

The cholinergic system has received increasing attention from both basic scientists and clinical researchers after the observation that Alzheimer's disease (AD), the most frequent human neurodegenerative disorder, is specifically associated with an early and selective impairment of cholinergic transmission in the brain. Postmortem examinations showed severe neuronal loss in the basal forebrain cholinergic nuclei and a decrease of choline acetyltransferase (ChAT) and cholinesterase activity in the cerebral cortex in AD. Furthermore, anticholinergic drugs induce memory dysfunction in both animal models and humans. Experimental lesions of cholinergic nuclei impair memory function in monkeys. Even in the absence of AD-related neurodegeneration, decline of cholinergic function has been linked to specific cognitive impairment, including working memory, in healthy aging. It has been suggested that the effects of cholinergic lesions on memory are mediated by the primary effect of cholinergic transmission on attention and executive function. Age-associated alterations of the cholinergic system may underlie an increased risk of delirium in aging. Besides its role as a neurotransmitter in the adult brain, several lines of research suggest that acetylcholine (ACh) plays a critical role in cortical development and in establishing synaptic contacts in neuronal networks that will subserve complex cognitive functions in adulthood. However, the cerebral cortex continues to adapt to the environment throughout the entire life span. Cholinergic function plays a critical role in neuronal plasticity in the adult brain. Therefore, cholinergic impairment in AD not only leads to impaired neuronal transmission within the cerebral cortex but also compromises the ability of the brain to compensate for progressive neurodegeneration. Some of the vulnerability of the cholinergic system in aging or AD may be mediated by an amyloid-related decrease of target-specific nerve growth factor. Consistent with the evidence for cholinergic dysfunction, several double-blind placebo-controlled trials have suggested that enhancement of cholinergic function using (acetyl-) cholinesterase inhibitors attenuates cognitive decline in AD. The findings on cholinergic dysfunction, however, have been limited in regard to the time course and the differential involvement of systems and compartments of the cholinergic system. These limitations are partly because evidence was derived from postmortem studies that intrinsically forbid longitudinal observation and are biased toward severe end stages of disease. Therefore, advanced imaging technology was developed to determine the integrity of the cholinergic system *in vivo*. One line of research is based on radiolabeled ligands to specific components of cholinergic transmission together with positron emission tomography (PET). Another approach uses pharmacological intervention with cholinergic and anticholinergic drugs and functional neuroimaging, using PET and functional magnetic resonance imaging (fMRI). Finally, studies have used structural MRI to determine alterations of cholinergic nuclei in the human basal forebrain. Understanding the rationale, implications, and limitations of these approaches requires a basic understanding of the anatomy and biochemistry of cholinergic transmission in the human brain, outlined in the following paragraphs.

## The Cholinergic System

### Anatomy of the Cholinergic System

According to Mesulam, the cholinergic pathway "is likely to constitute the single most substantial regulatory afferent of the cerebral cortex." Cholinergic projections in the cerebral cortex are involved in arousal, maintenance of attention, and REM sleep. The main source of cholinergic projections to the cerebral cortex are the magnocellular neurons of the nucleus basalis Meynert in the substantia innominata which separates the globus pallidus from the ventral surface of the basal forebrain. Studies in monkeys suggest a corticotopic organization of basal forebrain cholinergic nuclei. The cholinergic nuclei of the basal forebrain can be divided into subpopulations and named by adapting Mesulam's nomenclature. Cell group Ch1 corresponds to the nucleus septi medialis and innervates the hippocampus, cingulate, and retrosplenial cortex. Ch2 corresponds to the vertical limb of the nucleus of the diagonal band of Broca. Both groups are located medially to the nucleus accumbens septi. Ch3 designates cells associated with the horizontal limb of the diagonal band nucleus and, together with Ch2, innervates the entorhinal and olfactory cortex. The Ch4 group contains the nucleus basalis neurons within the substantia innominata and is the most extensive cholinergic cell group. The Ch4 group can be divided into anterior (Ch4a), intermediate (Ch4i), and posterior (Ch4p) subpopulations. The anterior subpopulation can be divided into medial (Ch4am) and lateral (Ch4al) cell groups. Ch4a extends from the posterior end of the tuberculum olfactorium to the posterior end of the anterior commissure. Ch4i starts at the anterior tip of the thalamus and extends to the posterior end of the ansa peduncularis, Ch4p extends from the posterior end of the ansa peduncularis to the level of the mammillary bodies. In the human brain there are two major pathways from Ch4 neurons to the cortex: the medial pathway supplies the medial frontal cortex and cingulum, and the lateral pathway projects to the dorsal and lateral prefrontal cortex and the temporal, parietal, and occipital neocortex. A cluster of magnocellular cholinergic neurons located in the anterior part of the basal forebrain (ventrolaterally to the anterior commissure) has been named the nucleus subputaminalis (NSP), or Ayla's nucleus. It is interesting that the lateral subdivision of the NSP has been described only in the human brain, not yet in monkey brains, and has been linked to the cortical speech area. **Figures 1–6** illustrate the anatomy of the basal forebrain nuclei based on Gallocyanin-stained sections through the basal forebrain of two control cases, a 29-year-old man and a 66-year-old man.

Because all cholinergic nuclei are concentrated in a very small area of the brain, basal forebrain lesions will easily lead to cholinergic deficits. In the same line, anticholinergic effects of medication or other substances will easily lead to cholinergic deficits, particularly in aging and neurodegeneration, in which the cholinergic input into the cortex is already decreased or damaged.

### Cholinergic Neurotransmission

Because ACh is among the most biologically potent neurotransmitters of the brain, its synaptic concentration is highly regulated. ACh is synthesized through ChAT from choline and acetyl-Co-A (**Figure 7**). The cholinesterases (ChEs) inactivate ACh through hydrolyzation (**Figure 8**).

The primary enzyme is acetylcholinesterase (AChE), but ACh is also a substrate for butyrylcholinesterase. Inferences of the regional distribution of cholinergic projections based on ChE produce too widespread estimates. ChE is widely distributed across the brain, probably as a safeguard against ACh overload within the brain. The localization of cholinergic projections is more precisely inferred from the distribution of the ChAT. ACh acts on two classes of receptors that are characterized by the pharmacological activity of two alkaloids: nicotine and muscarine. The nicotinic ACh receptor is a ligand-gated ion channel eliciting a fast response whereas the muscarinic ACh receptor is G-protein-coupled and elicits a slow response. Both receptor classes exhibit distinct subtypes, are often co-expressed in the same synapse, and occur pre- and postsynaptically.

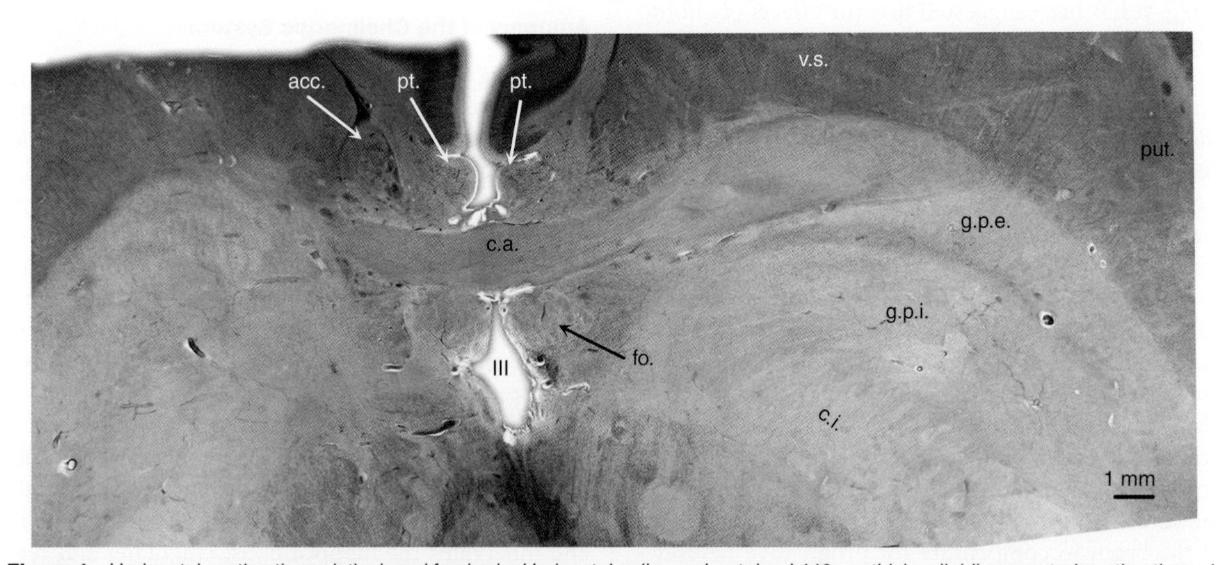


**Figure 1** Horizontal section through the basal forebrain. Horizontal gallocyanin-stained 440 μm-thick celloidin-mounted section through the fornix and dorsomedial parts of the anterior commissure of the right hemisphere of a 29-year-old man. Red line in **Figure 6** indicates plane of section in this figure. acc., nucleus accumbens.; c.a., anterior commissure; c.i., internal capsule; fo., fornix; g.p.e., external globus pallidus; g.p.i., internal globus pallidus; III, third ventricle; pt., paraterminal (subcallosal) gyrus; put., putamen; v.s., ventral striatum. Reproduced from Grinberg LT and Heinsen H (2007) *Dementia and Neuropsychologia* 1(2), with permission.

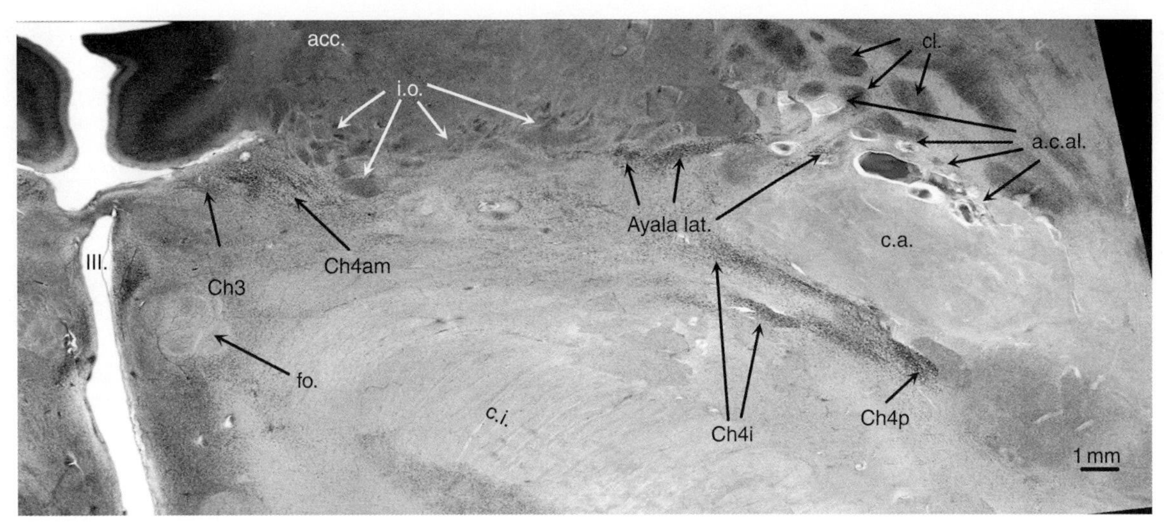


**Figure 2** Horizontal section through the basal forebrain. Horizontal gallocyanin-stained 440 μm-thick celloidin-mounted section 4.4 mm ventral to plane of section indicated in **Figure 1**, right hemisphere of a 29-year-old man. Plane of section is indicated in **Figure 6** by the white line. a.c.al., central anterolateral arteries; acc., nucleus accumbens.; Ayala lat., lateral or periputaminal part of Ayala's nucleus; c.a., anterior commissure; Ch4am, basal nucleus of Meynert, anteromedial part; Ch4i, basal nucleus of Meynert, intermediate part; Ch4p, basal nucleus of Meynert, posterior part; c.i., internal capsule; cl., claustrum; fo., fornix; III, third ventricle; i.o., olfactory islands. Reproduced from Grinberg LT and Heinsen H (2007) *Dementia and Neuropsychologia* 1(2), with permission.

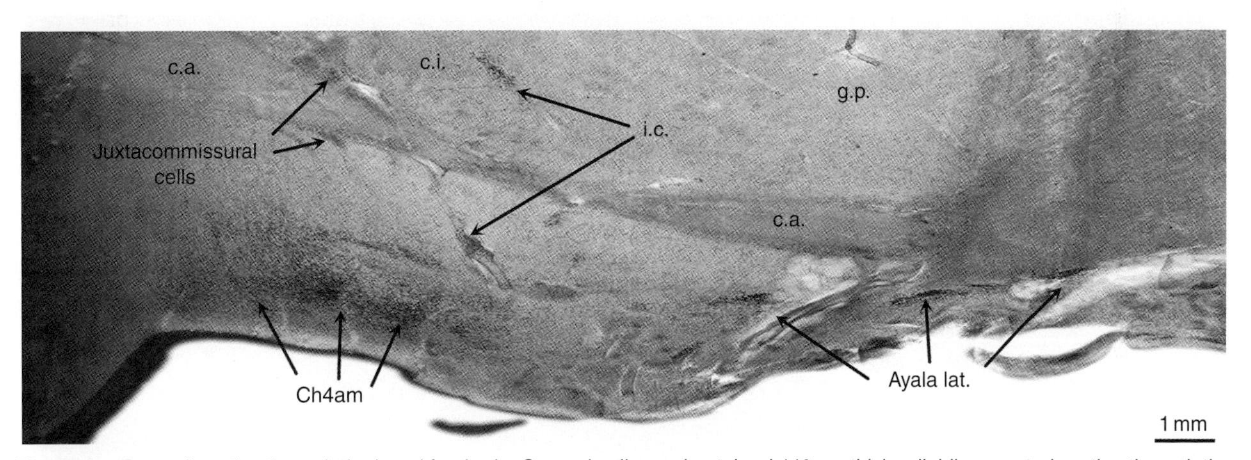


**Figure 3** Coronal section through the basal forebrain. Coronal gallocyanin-stained 440 μm-thick celloidin-mounted section through the fornix and dorsomedial parts of the anterior commissure of the right hemisphere of a 66-year-old man. Ayala lat., lateral or periputaminal part of Ayala's nucleus; c.a., anterior commissure; Ch4am, basal nucleus of Meynert, anteromedial part; c.i., internal capsule; g.p., globus pallidus; i.c., interstitial cells. Reproduced from Grinberg LT and Heinsen H (2007) *Dementia and Neuropsychologia* 1(2), with permission.

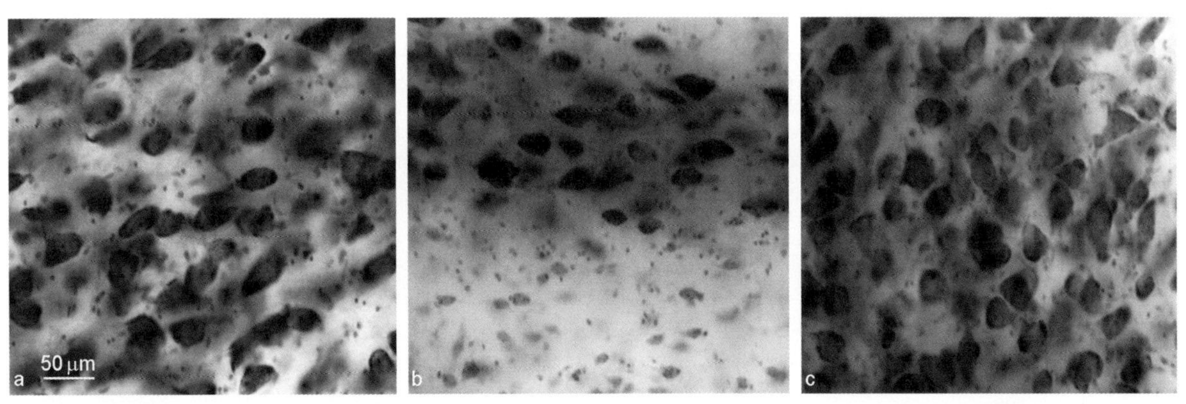


**Figure 4** Details from **Figures 1–3**. Microscopic view of (a) large chromophilic neurons of basal nucleus of Meynert, anteromedial part; (b) Ayala's lateral or periputaminal nucleus; and (c) basal nucleus of Meynert, posterior part (66-year-old man). Reproduced from Grinberg LT and Heinsen H (2007) *Dementia and Neuropsychologia* 1(2), with permission.

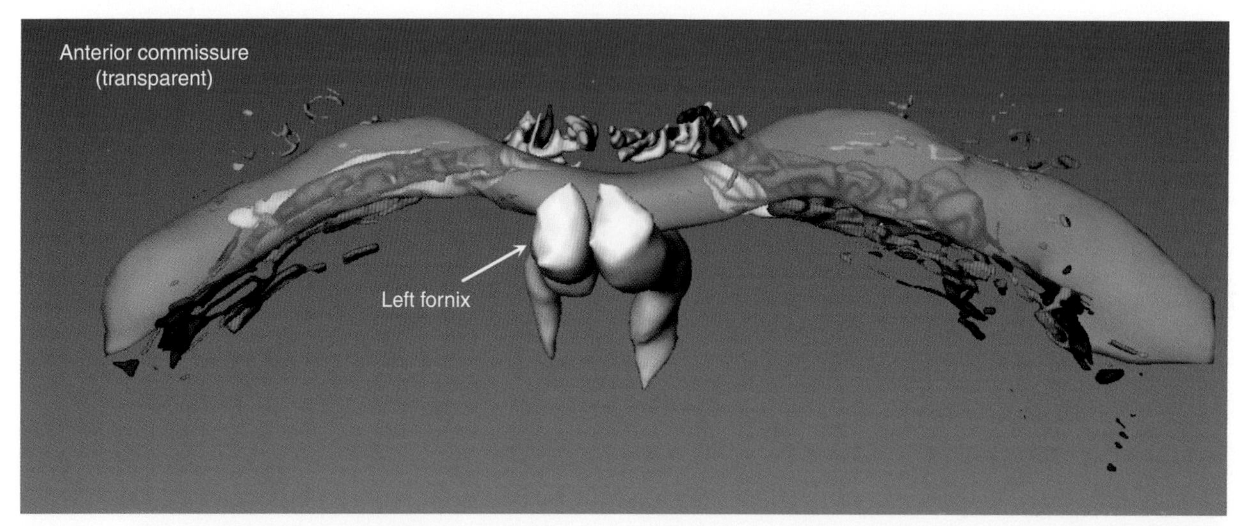


**Figure 5** Three-dimensional reconstruction of basal forebrain. Basal forebrain complex seen from dorsal after reconstruction with Amira software, 29-year-old man. Ch2, vertical limb of the nucleus of the diagonal band of Broca; Ch3, cells associated with the horizontal limb of the diagonal band nucleus; Ch4al, basal ncl. of Meynert, anterolateral part; Ch4am, basal ncl. of Meynert, anteromedial part; Ch4i, basal nucleus of Meynert, intermediate part; Ch4p, basal nucleus of Meynert, posterior part. Reproduced from Grinberg LT and Heinsen H (2007) *Dementia and Neuropsychologia* 1(2), with permission.

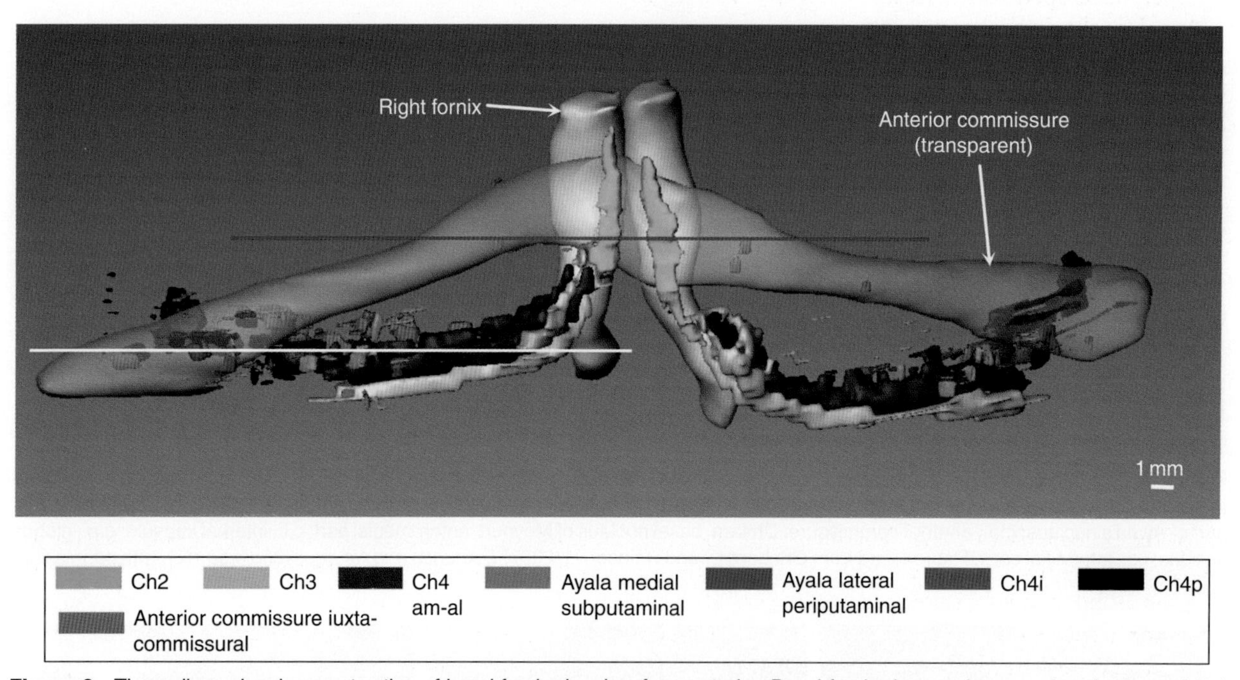


**Figure 6** Three-dimensional reconstruction of basal forebrain, view from anterior. Basal forebrain complex seen in a fronto-occipital perspective after reconstruction with Amira software, 29-year-old man. Ch2, vertical limb of the nucleus of the diagonal band of Broca; Ch3, cells associated with the horizontal limb of the diagonal band nucleus; Ch4al, basal ncl. of Meynert, anterolateral part; Ch4am, basal ncl. of Meynert, anteromedial part; Ch4i, basal nucleus of Meynert, intermediate part; Ch4p, basal nucleus of Meynert, posterior part. Reproduced from Grinberg LT and Heinsen H (2007) *Dementia and Neuropsychologia* 1(2), with permission.

$$\underset{\text{Choline}}{CH_3{-}\overset{CH_3}{\underset{CH_3}{N^+}}{-}CH_2{-}CH_2{-}OH} + \underset{\text{Acety1 CoA}}{CH_3{-}\overset{O}{\overset{\|}{C}}{-}S{-}CoA} \xrightleftharpoons[\text{acetyltransferase}]{\text{Choline}} \underset{\text{Acetylcholine}}{CH_3{-}\overset{CH_3}{\underset{CH_3}{N^+}}{-}CH_2{-}CH_2{-}O{-}\overset{O}{\overset{\|}{C}}{-}CH_3} + \underset{\text{Coenzyme A}}{HS{-}CoA}$$

**Figure 7** Acetylcholine synthesis. The reaction mediated by the choline acetyltransferase activity is shown.

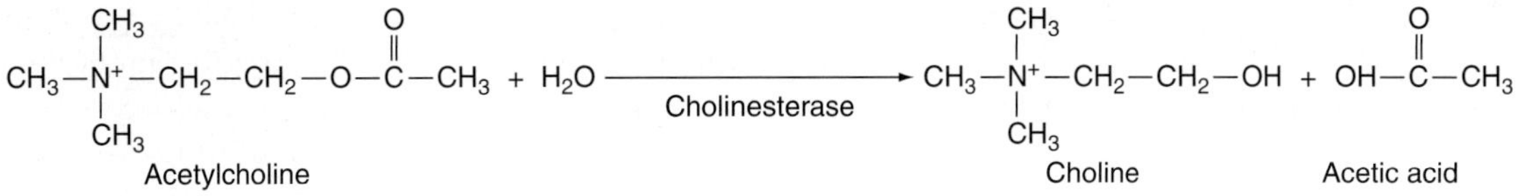

$$CH_3-N^+(CH_3)_2-CH_2-CH_2-O-C(=O)-CH_3 + H_2O \xrightarrow{\text{Cholinesterase}} CH_3-N^+(CH_3)_2-CH_2-CH_2-OH + OH-C(=O)-CH_3$$

Acetylcholine → Choline + Acetic acid

**Figure 8** Acetylcholine inactivation. The reaction mediated by acetylcholinesterase activity is shown.

## Targeting the Cholinergic System *In Vivo*

The short description of the anatomy and biochemistry of the cholinergic system in the brain reveals several potential targets for *in vivo* imaging of the integrity of the cholinergic system. Radiochemistry has developed $^{11}$C-labeled markers for nicotinic and muscarinic ACh receptors and for AChE that can be applied *in vivo* using PET. Cholinergic drugs together with cognitive tests during functional imaging using fMRI or PET have been used as a stress test to map the altered plasticity of the cholinergic system with aging and neurodegeneration. Finally, *in vivo* MRI has been developed to detect anatomical changes in the basal forebrain that may be related to the integrity of cholinergic neurons. The following paragraphs review the technical background of and findings obtained by these approaches.

### Pet Imaging of Cholinergic Receptor Binding and ChE Activity

**How can PET measure neuroreceptor binding and density *in vivo*?** In a neuroreceptor study with PET, the measured activity will always be a mixture of specifically bound, nonspecifically bound, and unbound radiotracer. Besides the binding kinetics and receptor distribution, peripheral clearance, regional cerebral blood flow (CBF), and transport across the blood–brain barrier will influence the profile of the time–activity curve. Therefore, the use of a simple quantitative measure, such as the ratio of a region of interest to a reference region (believed to represent nonspecific binding), will be inadequate to most receptor studies. Model-based approaches have been developed to relate the time–activity curve of a region of interest (the observed response function) to the arterial time–activity curve (the observed input function) through a defined model. Most of these models are based on the notion of compartments. In its most extended form, the compartment model encompasses three compartments, which are derived from a four-compartment model by collapsing two compartments into one (**Figure 9**).

A radiolabeled ligand is injected intravenously (compartment 1), resulting in an initial plasma concentration ($C_{Pl}$) that rapidly decreases because of

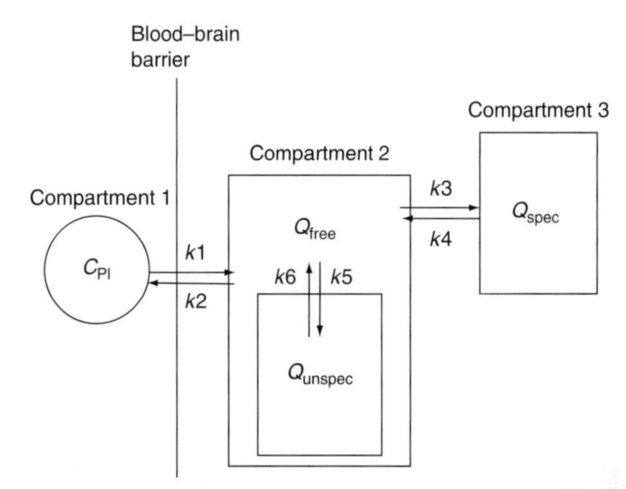


**Figure 9** Three-compartment model of ligand–receptor binding. Compartment 1 indicates plasma space; compartment 2, nonspecific binding space in tissue; and compartment 3, specific binding space. $C_{Pl}$ represents ligand concentration in plasma; $Q_{free}$ and $Q_{unspec}$, quantity of free exchangeable and nonspecifically bound ligand rapidly equilibrating with each other; and $Q_{spec}$, quantity of receptor-bound ligand. $k1$ to $k6$ are the rate constants for the ligand exchange between compartments in the specified directions. For further details, see text. Adapted with permission from *Brain Imaging: Applications in Psychiatry*, © 1989. American Psychiatric Publishing, Inc.

uptake into brain (and other tissues), resulting in an early equilibrium between vascular and extravascular brain compartments. The extravascular compartment is divided into three parts. The pool of unbound ligand (quantity of free ligand ($Q_{free}$)) is in rapid equilibrium with the nonspecific bound ligand pool (quantity of ligand reversibly bound to nonspecific proteins ($Q_{unspec}$)). Both compartments are considered a single exchangeable tissue pool (compartment 2). The quantity of free exchangeable ligand in compartment 2 ($Q_{free}$) tends to equilibrate with the quantity ($Q_{spec}$) of ligand specifically bound to neuroreceptor in compartment 3. The fractional rate constants $k_i$ govern the exchange of ligand between compartments in the specified direction, assumed to follow a first-order reaction kinetic.

Equilibrium is supposed to occur rapidly between the first (plasma) and second compartments (free and nonspecifically bound ligand). In contrast, the equilibration between the second and third compartments (specifically bound ligand) takes considerable time, depending on ligand–receptor affinity, and may even not occur within scanning time. There are essentially two models to determine neuroreceptor density and binding characteristics. The first measures receptor saturation with increasing ligand concentration and therefore requires equilibration of ligand between the second and third compartments during scanning time (steady-state model). The second measures the time course of specific brain uptake of ligand over time with different concentrations of an unlabeled receptor-blocking substance (kinetic model). This approach does not require equilibration between the nonspecific and specific compartments. Both models to assess neuroreceptor density and binding kinetics are based on several assumptions, some of which may be reasonable but may not be easily validated. Both approaches gave similar values for receptor density with different dopaminergic ligands (Wong: 17 pmol $g^{-1}$, Sedvall: 15 pmol $g^{-1}$). The results with respect to receptor binding affinity, however, are not directly comparable between both models, because different ligands are involved and different parameters are measured. In both approaches the tracer kinetic model requires an input function indicating delivery of tracer to the tissue. In some applications for receptor studies, an arterial input function can be replaced by a reference to tissue devoid of specific binding sites. Single scan acquisition allows for assessment of only one single parameter, whereas dynamic scanning to obtain the time course of tissue activity allows simultaneous estimation of multiple parameters.

Reliable quantification of receptor densities in AD using PET and single-photon emission computed tomography (SPECT) will require an appropriate model to rule out effects of cortical atrophy. Correction of PET data for cerebral atrophy based on corresponding high-resolution MRI scans has already been accomplished in a study on the μ-opioid receptor in the amygdala complex and may be routinely available in the future for other PET studies as well.

**Cholinergic Markers in PET** Table 1 gives an overview of the most important markers for the cholinergic system in SPECT and PET. Although a decreased nicotinic receptor density in AD has been established from neuropathological studies, studies of the cholinergic system can serve as an example of the pitfalls and controversies associated with *in vivo* receptor imaging using PET. Initial labeling of racemic nicotine with positron emitting $^{11}C$ in 1982 by Långström and thereafter of the nicotine (+)-(R)- and (−)-(S)-enantiomers was followed by attempts to visualize binding to the brain nicotinic ACh receptors with PET in monkeys and in humans. In AD, uptake of $^{11}C$-nicotine into frontal and temporal cortices was less than in healthy age-matched controls. These results are consistent with postmortem evidence of fewer nicotinic

**Table 1** PET radioligands for the cholinergic system

| *Compound* | *Label* | *Target* | *Main findings* |
|---|---|---|---|
| 2-[18F]fluoro-3-(2(S)-azetidinylmethoxy) pyridine (2-F-A-85380) | $^{18}F$ | $\alpha_4\beta_2$-nicotinic ACh-receptor (responsible for addicitive properties of nicotine?) | Applied to humans; no studies in AD so far reported. 88% occupancy of receptor after smoking of 1 cigarette (dependent smokers) |
| Nicotine | $^{11}C$ | nAChR | Unspecific binding, kinetic model suggests perfusion as main determinant of distribution |
| Norchlorofluoroepibatidine | $^{18}F$ | $\alpha_4\beta_2$-nicotinic AChR | Studies in baboons but not in humans because of toxicity of epibatidine |
| *N*-methyl-4-piperidyl benzilate (NMPB) | $^{11}C$ | mAChR | Reduced cortical uptake in AD |
| 3-R-quinuclidinyl-4-S-iodobenzilate (RS-IQNB) | $^{123}I$ | mAChR | Reduced uptake temporal and frontal in AD; SPECT marker |
| Tropanyl benzilate (TRB) | $^{11}C$ | mAChR | Slightly decreased binding in frontal cortex in aging |
| 3-(3-(3-Fluoropropyl)thio)-1,2,5-thiadiazol-4-yl)-1,2,5,6-tetrahydro-1-methylpyridine ([18F] FP-TZTP)FP-TZTP | $^{18}F$ | mAChR ($M_2$) | Greater receptor avidity in ApoE4 carriers, possibly reflecting lower occupancy by ACh |
| *N*-methyl-4-piperidyl acetate (MP4A) | $^{11}C$ | AChE | Reductions in AD |
| *N*-[$^{11}C$]methyl-pi-peridin-4-yl propionate ([$^{11}C$]PMP) | $^{11}C$ | AChE | Reductions in AD, correlations with cognitive performance |

ACh, acetylcholine; AChE, acetylcholinesterase; AD, Alzheimer's disease; ApoE4, apolipoprotein E; mAChR, muscarinic acetylcholine receptor; nAChR, nicotinic acetylcholine receptor; SPECT, single-photon emission computed tomography.

ACh binding sites in AD. However, the validity of nicotinic receptor imaging *in vivo* with $^{11}$C-nicotine was first questioned by Nybäck and co-workers. They showed in healthy participants that $^{11}$C-nicotine uptake did not decline after potential binding sites were blocked by unlabeled ligand, indicating no measurable impact of receptor availability on tracer uptake. The ratio between tracer uptake in temporal cortex and white matter was constant over time (compared with the previously described linearly ascending ratio between specific binding in the putamen and nonspecific binding in the cerebellum), which is not compatible with specific binding characteristics. Finally, the experimental data could be fitted with a two-compartment model, in which tracer uptake was determined solely by blood flow and blood–brain barrier permeation. A specific binding compartment was not required. Nybäck and colleagues concluded from these results that the regional reduction of $^{11}$C-nicotine uptake in AD reflects reduced CBF.

In the following years, new compounds were investigated for their potential to specifically detect nicotinic receptor binding *in vivo*. [$^{18}$F]norchlorofluoroepibatidine was successfully applied in baboon studies but because of toxicity could not be transferred to human studies. More-selective and less-toxic 3-pyridyl ether derivatives have been used successfully for development of a nicotinic ACh receptor (nAChR) ligand. The recently developed radioligands 2-[$^{18}$F]fluoro- and 6-[$^{18}$F]fluoro-3-(2(*S*)-azeti-dinylmethoxy) pyridine allow for visualization of brain $\alpha_4\beta_2$-nAChR with PET in humans. This nAChR subtype is expressed throughout the brain in pre- and postsynaptic locations and has been linked to the addicting properties of nicotine. *In vitro* studies on postmortem brain material using different ligands for autoradiography have obtained conflicting results with reductions of this specific nAChR in some but not all studies. The *in vivo* ligand A-85380 has shown its feasibility and validity in a familial form of frontal lobe epilepsy with mutations in the nAChR $\alpha 4$ or $\beta 2$ subunit but has not yet been applied to AD.

Several tracers have been applied to detect muscarinic AChRs (mAChRs). In postmortem studies these receptors have been found to be decreased with AD. In aging, autopsy studies have been inconsistent, with series with the greatest age ranges and participant numbers indicating more than 50% reductions in muscarinic receptor numbers in the striatum, hippocampus, and frontal cortex between ages 4 and 93 years. However, declines in cortical or hippocampal muscarinic receptors in series between the ages of 60 and 90 years were modest in some studies or not detectable in other studies. The first ligand to map mAChR *in vivo* in the human brain was 3-quinuclidinyl-4-iodobenzilate labeled with $^{123}$I ($^{123}$IQNB) in SPECT. In older normal adults, this tracer showed high radioligand uptake in the basal ganglia and occipital and insular cortex, and low uptake in thalamus and cerebellum. This regional pattern matches the relative distribution of mAChR in autopsy studies. Studies in AD showed a moderate decrease of tracer uptake in posterior temporal and frontal cortices. Unfortunately, the early uptake of $^{123}$IQNB is flow-dependent, so scans had to be performed 21 h after injection, making this probe difficult to handle in a clinical setting. Moreover, the spatial resolution of SPECT cameras is generally much lower than the resolution of PET, lowering the potential use of this approach for clinical application or research.

Binding of $^{11}$C-tropanyl benzilate (TRB) using PET was slightly but significantly decreased in the frontal cortex with aging. Binding showed a nonsignificant trend of increase in cerebellum and brain stem with aging. The authors interpreted these effects as indicating small decline of receptor density in frontal cortex in healthy aging. TRB has not yet been applied to studying patients with AD or other neurodegenerative disorders.

In one PET study using $^{11}$C-labeled *N*-methyl-4-piperidyl benzilate (NMPB), normal aging was associated with a reduction in muscarinic receptor binding in neocortical regions and thalamus. In AD patients, the three-compartment model appeared capable of dissociating changes in tracer transport from changes in receptor binding but suffered from statistical uncertainty, which required normalization to a reference region, therefore limiting its potential use in the study of neurodegenerative processes. After normalization, no regional changes in muscarinic receptor concentrations were observed in AD. These findings are in contrast to an earlier study combining measurement of mAChR with $^{11}$C-NMPB and $^{18}$fluorodeoxyglucose (FDG)-PET of cortical glucose consumption and $^{99}$Tc-hexamethylpropyleneamine oxime (HMPAO)-SPECT of CBF. In this study, AD patients had reduced $^{11}$C-NMPB uptake throughout the entire cortex relative to controls, which was more pronounced than reductions in CBF but less pronounced than changes in cortical glucose metabolism.

Application of M2 muscarinic receptor ligand 3-(3-(3-[$^{18}$F]Fluoropropyl)thio)-1,2,5-thiadiazol-4-yl)-1,2,5,6-tetrahydro-1-methylpyridine ([$^{18}$F]FP-TZTP) during PET revealed greater M2 receptor avidity in older adults carrying the apolipoprotein E4 (ApoE4) allele, a risk allele for sporadic AD. This alteration might be due to a decrease in synaptic ACh concentration through lower receptor occupancy. Both age and ApoE4 genotype independently led to a greater decrease in receptor avidity following the infusion of the AChE inhibitor physostigmin, probably reflecting the increased receptor occupancy by the endogenous ligand ACh. Other markers, such as N-$^{11}$C-methyl-3-piperidyl benzilate, have shown reduced mAChR binding in monkey cortical regions with aging but have not been applied to human studies so far.

In summary, in contrast to the nicotinic markers, a wide range of muscarinic markers are available with PET and SPECT, and these markers have been evaluated using kinetic modeling in aging and in AD. Results, however, are not absolutely consistent to date. Effects of aging on muscarinic receptor density appear to be modest at most, with the effects being concentrated in the frontal lobes. In AD, effects appear to be more widespread if they are detected at all. The most interesting finding to date seems to be the study using [18F]FP-TZTP with PET. This study showed an increased binding volume for the ligand, possibly reflecting decreased availability of endogenous ACh in at-risk groups of AD defined by the presence of the ApoE4 genotype.

A promising approach to quantify cholinergic impairment in AD was presented by Iyo and collaborators, who used an $^{11}$C-labeled substrate for AChE (*N*-methyl-4-piperidyl acetate (MP4A)) to measure catalytic activity of brain AChE with PET. This lipophilic tracer is taken up into the brain across the blood–brain barrier, where it is hydrolyzed by brain AChE to a hydrophilic metabolite that remains trapped within the brain. The ratio of metabolized to unmetabolized MP4A gives an estimate of AChE catalytic activity. Using this model in a pilot study with five AD patients and eight healthy age-matched controls, the investigators found AChE activity in AD reduced 20–25% in frontal and occipital cortex and 30–45% in temporal and parietal cortex. Subsequent studies using *N*-[$^{11}$C]methyl-pi-peridin-4-yl propionate as substrate for AChE confirmed an average decrease of 25–33% of mean cortical AChE activity in patients with moderate AD and 9–15% in mild AD. The extent of reduction in enzyme activity *in vivo* was much less than expected from postmortem findings. Still, the changes were more pronounced than were independently measured reductions of regional CBF. These data suggest that cholinergic denervation may be mild to modest in early stages of disease but appears to be independent of reduced CBF. Loss of cholinergic enzyme activity may precede metabolic and blood flow alterations in early AD, when severe functional impairment and final loss of neurons (both leading to decreased uptake of glucose) may not yet be markedly present. One study investigating cortical AChE activity and cognitive performance suggested that AChE activity is predominantly correlated with measures of working memory and attention, much less with measures of delayed short- and long-term memory. These data agree with experimental evidence that cholinergic transmission is predominantly involved in attention and executive function.

The hope that specific receptor imaging might improve clinical diagnosis and treatment of AD has not yet been fulfilled. The complexity of the methods, their limited availability, and inconsistency of results may be the major reasons why no studies have been reported so far using receptor PET as a diagnostic marker or secondary endpoint in pharmacological trials. Another reason may be the complex regulation of the cholinergic system. Surprisingly, an autopsy study based on the religious order population has shown an increase of ChAT activity in people with mild cognitive impairment whereas patients with AD showed a consistent reduction of cortical ChAT activity. In a subsequent discussion, many researchers argued that this finding may indicate that cholinergic lesions occur at a late stage in the development of the disease. A more differentiated discussion, however, has pointed out that some subregions of the basal forebrain cholinergic nuclei appear to be preserved until late stages of disease and might be the source of new cholinergic synapses into denervated brain areas. The stimulus for this sprouting of cholinergic axons might be the progressive loss of target neurons in the most vulnerable brain areas, such as the entorhinal cortex. These data, together with increased muscarinic receptor avidity reported in some studies of mild stages of AD, suggest a dynamic response of the cholinergic system to progressive neurodegeneration and decrease of ACh concentrations. These mechanisms most likely vary with the clinical stage of the disease and therefore make it difficult to identify patterns of receptor changes that are predictive of AD or respond specifically to treatment.

**Perspective of neuroreceptor imaging in Alzheimer's disease** Future studies using these techniques will have to address several issues. First, they need to more extensively study alterations in cholinergic transmission and ChE activity in preclinical stages and at-risk stages of AD. The finding of an upregulation of muscarinic receptor binding and ChAT activity in patients with mild cognitive impairment and even cognitively healthy people with ApoE4 genotype suggests preclinical compensatory processes whose breakdown may contribute to the clinical manifestation of dementia. Neuroreceptor imaging may contribute to a better understanding of these mechanisms. The binary discussion (early cholinergic involvement in AD as primary event vs. secondary involvement as consequence of degeneration of cortical input areas) may have to be replaced by a more differentiated view appreciating a stepwise involvement of some cholinergic nuclei and relative preservation of others that may partly compensate for the loss of cortical cholinergic innervation. Second, studies need to determine alterations in cholinergic binding over time to assess the potential use of these measures to map disease progression and treatment effects. Third, studies will have to consider that AD involves not only the cholinergic system but serotonergic, catecholaminergic, and glutamatergic transmission as well, any of which may contribute to both cognitive and behavioral symptoms of dementia.

### Functional Imaging of Cholinergic Treatment Effects

The short-acting AChE inhibitor (AChE-I) physostigmine increases working memory performance in young and older healthy participants and at the same time decreases regional CBF in prefrontal areas that are involved in task performance. These findings may indicate that in healthy adults, enhancement of cholinergic function can improve processing efficiency and thus reduce the effort required to perform a working memory task. A study in monkeys showed that application of an AChE-I could restore the reduced response of regional CBF to vibrotactile stimulation with aging. These data suggest that increased reaction time and reduced working memory performance in aging may be related to a functional decline of cholinergic transmission that can partly be restored by the application of AChE-Is. In agreement with monkey studies, two independent PET studies of the ChE-I donepezil showed a reduction of cortical AChE activity associated with neuropsychological effects of treatment in AD patients. These studies confirmed the notion that application of ChE-I inhibits cortical AChE and that this inhibition is associated with an attenuation of cognitive decline. Interpretation, however, is limited because these studies used an open-label design. Reductions ranged between 27% and 39% at therapeutic doses. The use of higher doses of the presently available ChE-I to reach a higher degree of AChE inhibition is limited because of side effects. Intravenous application of physostigmine decreased cholinesterase activity by 50%. Agents with a longer half-life than physostigmine's and higher inhibition potential than the presently available ChE-I approved for AD are needed to test whether even greater inhibition of ChE might lead to better clinical effects.

The ChE-I rivastigmine increased cortical metabolism in treatment responders compared with nonresponders and placebo-treated AD patients. In open-label studies, the ChE-Is reminyl and metrifonate increased metabolism in specific cortical regions in treatment responders compared with nonresponders. One study found relative preservation of cortical metabolism in temporoparietal and frontal association areas after 12 months of treatment with rivastigmine compared with untreated AD patients. These studies suggest that cholinergic treatment increases cortical metabolism in association with cognitive improvement through treatment. Only one of these studies, however, was placebo-controlled. All studies compared treatment responders with nonresponders. However, this approach has the disadvantage that it is difficult to disentangle treatment effects from other factors influencing cognition. Comparing responders with nonresponders shows the cortical substrate of cognitive change, which might only partly be related to treatment (since by definition both comparison groups are treated in the same way). In a recent study, the effect of donepezil on cortical metabolism was determined during rest and passive audiovisual stimulation in a double-blind crossover design. Effects were compared between donepezil and placebo-treated patients independent of clinical response. Resting state glucose metabolism was increased with donepezil in left prefrontal cortex and decreased in right hippocampus. Cortical response to activation was increased in right hippocampus with donepezil compared with placebo, suggesting that decreased hippocampus metabolism during rest with donepezil treatment was reversible and even led to an overshoot activation. The effects were spatially much more restricted than in earlier studies. This may be related to the use of a double-blind design and the fact that patients were not classified according to clinical response.

The first small-scale studies with functional MRI (fMRI) have shown effects of cholinergic treatment on regional activation during a memory task. During face encoding healthy older patients showed higher activity in the fusiform face area than did AD patients. After treatment with donepezil, AD patients showed fusiform activation similar to that of the controls. Exposure to the ChE-I galantamine produced increases in cortical activation during face recognition in patients with mild cognitive impairment. The available data on the effects of cholinergic treatment on the blood oxygen level-dependent response in fMRI are of limited value because there are no studies using a double-blind, placebo-controlled design.

### MRI Studies to Detect Atrophy of Cholinergic Nuclei

Hanyu and colleagues were the first to study atrophy of the substantia innominata as a surrogate measure of atrophy in nucleus basalis Meynert. They measured the thickness of the substantia innominata at its smallest extent in a coronal section through the anterior commissure. They found significant decline of the thickness of the substantia innominata in AD patients compared with controls. Moreover, they found that the thickness of the substantia innominata was inversely correlated with the clinical response to treatment with the ChE-I donepezil. Studying different types of dementia, they found that the reduction of thickness of the substantia innominata was more pronounced in patients with Lewy body dementia than in patients with AD and more pronounced in AD than in vascular dementia. The conclusiveness of the data, however, was limited because the two-dimensional measure of thickness was not accurately located within the substantia innominata and provided only a very rough estimate of atrophy. Recently an automated technique was proposed to determine signal changes within the substantia innominata on the basis of a proton density weighted scan through the basal forebrain. All MRI scans were normalized to a common standard space. Then a three-dimensional region of interest covering the entire extent of the substantia innominata was extracted. Differences in signal intensity across the entire substantia innominata were determined by means of voxelwise statistics. As the locations of nucleus basalis Meynert have not been mapped in MRI standard space before, postmortem MRI and histological sections of the brain of a nondemented older adult were used to transfer the locations of the nuclei of the nucleus basalis Meynert into standard space. These were used as anatomical reference for the signal changes in the substantia innominata from the *in vivo* data. Significant signal reduction was found in anterior lateral and intermediate nuclei of the nucleus basalis Meynert in AD. Moreover, signal reductions were correlated with regional distribution of cortical gray matter loss in AD, suggesting a somatotopic organization of cholinergic projections from the basal forebrain. Recently, a *z*-score approach was developed to determine signal reductions in single participants compared with a reference group of healthy controls. Significant reductions of signal intensity in the substantia innominata similar to those observed in the independent group comparisons were found.

Diffusion tensor imaging (DTI), another MRI modality, may become interesting in the near future for the detection of cholinergic abnormalities. It can be applied to determine the integrity of subcortical fiber tracts *in vivo*. The trajectories of cholinergic pathways in the human brain have been described. Using the Cholinergic Pathways Hyperintensities Scale with T2-weighted MRI scans, Behl and colleagues showed a better response to cholinergic treatment in those AD patients who had more lesions along cholinergic pathways compared with the patients with low cholinergic lesion load, even though overall load in white matter lesions was comparable between groups. It may be of interest in future studies to apply DTI together with basal forebrain scans to determine the selective loss of cholinergic projections from the basal forebrain to the cerebral cortex in AD. AD may serve as an ideal lesion model to study the neuroanatomy and neuropathology of the human cholinergic system. The advances in mapping and understanding alterations of the cholinergic system in AD can then be applied to other human diseases involving the cholinergic system and serve as target for animal models of cholinergic dysfunction.

## Imaging the Cholinergic System *In Vivo*: Future Perspectives

What will be the impact of all these techniques to describe cholinergic changes in AD and aging about 3 years ahead? To address this question three potential fields of application should be delineated: (1) clinical routine diagnostic and treatment, (2) clinical trials to establish new treatments (phase II trials), and (3) clinical neuroscience.

It may well be that none of the techniques described in this article will have a direct impact on clinical routine diagnostic or clinical evaluation of treatment efficacy in AD within the next 3 years. Before it is finally applied in clinical routine, a new diagnostic marker goes through three phases:

1. Establishment of the technical characteristics of the marker: reproducibility, effort of acquisition and analysis, analytical sensitivity, and specificity.
2. Establishment of the validity in selected samples: sensitivity and specificity.
3. Controlled diagnostic trial: application in an intent-to-diagnose design, in which patients are selected according to the indication of the diagnostic measure, not the presence or absence of the disease, so that positive and negative predictive values can be determined.

Most of the techniques to determine the functional or morphological integrity of the cholinergic system *in vivo* are presently in phase I: cholinergic receptor PET, fMRI of cholinergic treatment, and structural MRI of the cholinergic nuclei. Only FDG-PET in combination with cholinergic treatment is in phase II, with first clinical trials having applied PET as secondary outcome parameter, but these appear to be all small-scale monocenter studies in highly selected samples. Up to now, no multicenter phase III trial using PET has been undertaken to map the cholinergic system as diagnostic marker or secondary outcome parameter of treatment. One reason for this may be the high costs and limited availability even of FDG-PET for large-scale multicenter clinical trials.

The already completed European Network for Efficiency and Standardisation of Dementia Diagnosis (NEST-DD) and the recently launched US-based AD Neuroimaging Initiative (ADNI) both involve FDG-PET, although not in combination with cholinergic markers or treatment in a prospective design. Retrospective analysis of a subset of the data of the NEST-DD or ADNI may give some insight into the multicenter feasibility of this approach and the effect of cholinergic treatment on cortical metabolism in FDG-PET following a naturalistic design. There are plans now for multicenter molecular imaging studies that include receptor markers of the cholinergic system. Again, these studies will have to involve highly selected samples and therefore will not provide estimates of positive and negative predictive values of these techniques in a clinical population. Another reason these techniques will not likely be applied in clinical routine diagnostics in the near future is the limited budget of national health systems worldwide. From a socioeconomic point of view, the presently available treatment options for AD do not clearly justify the application of a very expensive methodology to give treatment to patients earlier. However, once these techniques have proven their potential to identify those patients who will have maximum benefit from available treatments or once disease-modifying treatments have become available, this situation will change. The costs of these techniques then will be paid off by the early application of disease-modifying treatments that decrease the costs of dementia care.

For the next years, the development of these techniques will probably have some impact on the development of new drugs. Presently, MRI-based volumetric markers are used as secondary outcome parameters in clinical trials on potential disease-modifying effects in AD. The regulatory authorities have indicated that they might accept biological markers as primary endpoints under certain conditions. It has to be shown that the biological marker is related to a clinically relevant outcome, reflects the supposed mechanism of action of the drug, and has high test–retest reliability. At some point receptor PET and imaging of cholinergic nuclei may be part of clinical trials to identify at-risk people and as a secondary endpoint for treatment. Receptor-PET will additionally play an important role in proof-of-concept studies. Receptor PET has already been applied in testing of new compounds and studying the mechanism of action of drugs, such as determining dopamine release and dopamine receptor antagonism of typical and atypical antipsychotics. Imaging cholinesterase inhibition, dynamic changes of nicotinic and muscarinic binding, and the functional consequence of cholinergic changes on brain reserve capacity will help evaluate the mechanism of action of presently available treatments, such as *N*-methyl-D-aspartate antagonists and ChE inhibitors. They may also help estimate the potential of new compounds to modulate the cholinergic system in the human brain *in vivo* before they enter into phase II and III trials.

The most important impact of these techniques in coming years will be their contribution to understanding of the organization and function of the cholinergic system in the human brain. Aging *per se* is associated with a decline of cholinergic markers in the brain. The total number of nucleus basalis Meynert neurons has been shown in the ninth decade to be 20–30% below that in newborns. If the reduction of neuron numbers is approximately linear with age, these data suggest that reduction would be about 3% per decade. However, within a narrow age range in adults (50–80 years), several studies could not detect significant reductions of neuron numbers, suggesting that the 30% reduction from newborns to centenarians may at least partly reflect developmental changes in childhood and adolescence rather than effects of aging. These findings of a limited reduction of cholinergic projections would fit well with the observation that, in contrast to a widely held belief, aging is not inevitably associated with dementia. It would be of high clinical relevance to confirm the postmortem data by a longitudinal imaging approach applying structural and functional markers of cholinergic projections and receptor availability. This would indicate that normal aging preserves a large enough capacity of the cholinergic system to help the brain adapt to environmental and intrinsic factors under normal conditions. In contrast, AD serves as a paradigm that stresses the cholinergic system and allows inferences about its normal function. The focus of research in the coming years will likely increasingly include the role of the cholinergic system in the maintenance of brain plasticity. Research will focus on one hand on the ability of the brain in healthy aging to respond to external and internal stimuli based on the preservation of a minimum of cholinergic function, and on the other hand on the failure of brain plasticity in the presence of severe cholinergic lesions such as in AD. A better understanding of the molecular basis of brain plasticity will help science not only understand the functional basis of learning in its broadest sense (adaption to a constantly changing environment) but also develop treatment paradigms that do not focus only on lowering the impact of pathological events but on increasing the ability of the brain to adapt to pathological changes.

*See also:* Aging of the Brain; Aging of the Brain and Alzheimer's Disease; Alzheimer's Disease: Neurodegeneration; Alzheimer's Disease: An Overview;

Brain Glucose Metabolism: Age, Alzheimer's Disease and ApoE Allele Effects; Cognition in Aging and Age-Related Disease; Functional Neuroimaging Studies of Aging.

## Further Reading

Arendt T, Bigl V, Arendt A, and Tennstedt A (1983) Loss of neurons in the nucleus basalis of Meynert in Alzheimer's disease, paralysis agitans and Korsakoff's disease. *Acta Neuropathologica* 61: 101–108.

Cullen KM and Halliday GM (1998) Neurofibrillary degeneration and cell loss in the nucleus basalis in comparison to cortical Alzheimer pathology. *Neurobiology of Aging* 19: 297–306.

Eckelman WC (2006) Imaging of muscarinic receptors in the central nervous system. *Current Pharmaceutical Design* 12: 3901–3913.

Gilmor ML, Erickson JD, Varoqui H, et al. (1999) Preservation of nucleus basalis neurons containing choline acetyltransferase and the vesicular acetylcholine transporter in the elderly with mild cognitive impairment and early Alzheimer's disease. *Journal of Comparative Neurology* 411: 693–704.

Heinsen H, Hampel H, and Teipel SJ (2006) Response to Boban et al: Computer-assisted 3D reconstruction of the nucleus basalis complex, including the nucleus subputaminalis (Letter to the editor). *Brain* 129(4): E43.

Holcomb HH, Links J, Smith C, and Wong D (1989) Positron emission tomography: Measuring the metabolic and neurochemical characteristics of the living human nervous system. In: Andreasen N (ed.) *Brain Imaging: Applications in Psychiatry*, pp. 235–370. Washington, DC: American Psychiatric Press.

Iyo M, Namba H, Fukushi K, et al. (1997) Measurement of acetylcholinesterase by positron emission tomography in the brains of healthy controls and patients with Alzheimer's disease. *Lancet* 349: 1805–1809.

Mesulam M (2004) The cholinergic lesion of Alzheimer's disease: Pivotal factor or side show? *Learning and Memory* 11: 43–49.

Mesulam MM, Mufson EJ, Levey AI, and Wainer BH (1983) Cholinergic innervation of cortex by the basal forebrain: Cytochemistry and cortical connections of the septal area, diagonal band nuclei, nucleus basalis (substantia innominata), and hypothalamus in the rhesus monkey. *Journal of Comparative Neurology* 214: 170–197.

Nybäck H, Halldin C, Åhlin A, Curvall M, and Eriksson L (1994) PET studies of the uptake of (S)- and (R)-11C-nicotine in the human brain: Difficulties in visualizing specific receptor binding in vivo. *Psychopharmacology* 115: 31–36.

Potkin SG, Anand R, Fleming K, et al. (2001) Brain metabolic and clinical effects of rivastigmine in Alzheimer's disease. *International Journal of Neuropsychopharmacology* 4: 223–230.

Selden NR, Gitelman DR, Salamon-Murayama N, Parrish TB, and Mesulam MM (1998) Trajectories of cholinergic pathways within the cerebral hemispheres of the human brain. *Brain* 121: 2249–2257.

Teipel SJ, Drzezga A, Bartenstein P, Moller HJ, Schwaiger M, and Hampel H (2006) Effects of donepezil on cortical metabolic response to activation during (18)FDG-PET in Alzheimer's disease: A double-blind cross-over trial. *Psychopharmacology (Berl)* 187: 86–94.

Teipel SJ, Flatz WH, Heinsen H, et al. (2005) Measurement of basal forebrain atrophy in Alzheimer's disease using MRI. *Brain* 128: 2626–2644.

# Programmed Cell Death

**R Oppenheim and C Milligan**, Wake Forest University School of Medicine, Winston-Salem, NC, USA



## Introduction

During the embryonic, fetal, larval, and early postnatal stages of development, there is a massive loss of undifferentiated and differentiating cells in most tissues and organs, including the central and peripheral nervous systems. This cell loss is a normal part of development, and it occurs by a highly regulated process known as programmed cell death (PCD). Developmental PCD is defined as the spatially and temporally reproducible, tissue and species-specific loss of cells whose occurrence serves diverse functions and is a process required for normal development. Perturbation of this normal process can be maladaptive, resulting in pathology.

## History

Although the occurrence of developmental PCD was first reported in the middle of the nineteenth century, it was not until the middle of the twentieth century that the occurrence and significance of PCD in the nervous system began to be appreciated by embryologists and developmental biologists. In a series of seminal papers by Viktor Hamburger and Rita Levi-Montelcini in the 1930s and 1940s, it was shown that the sensory and motor neurons in the spinal cord of the chick embryo are generated in excess during neurogenesis, followed by the PCD of approximately one-half of the original population. The period of cell loss was found to occur as sensory and motor neurons were establishing synaptic connections with their peripheral targets (e.g., muscle in the case of motor neurons). In a conceptual *tour de force*, Hamburger and Levi-Montalcini proposed that developing neurons compete for limiting amounts of target-derived survival-promoting signals (the winners survive and the losers undergo PCD) and in this way neurons are thought to optimize their innervation of targets by a process known as systems matching. It was this conceptual framework that led to the discovery of target-derived neurotrophic molecules and to the formulation of the neurotrophic theory that has been a major factor in fostering progress in this field for over 50 years. According to the neurotrophic theory, neurons that compete successfully for neurotrophic molecules avoid PCD, owing to the expression of survival promoting intracellular molecular genetic programs. The discovery of molecular genetic programs for both survival and death has revolutionized the study of PCD, resulting in the publication of thousands of articles each year since the early 1990s.

## Evolution

The occurrence of massive cell death during development is, on the one hand, counterintuitive, whereas, on the other hand, the fact of its existence in many vertebrate and invertebrate species argues for its significance as an adaptively significant and fundamental part of embryogenesis. The death of occasional cells during development is to be expected in biological systems in which accidental deleterious events may be lethal to individual cells. However, because the stereotyped death of large numbers of developing cells in all members of a species cannot be explained in this way, it raises two basic questions regarding the evolution of PCD: (1) because PCD occurs by a metabolically active, genetically regulated process (see the sections titled 'Molecular regulation of cell death and survival by neurotrophic factors' and 'Intracellular regulation of cell death'), how and why did the molecular mechanisms involved arise during evolution; and (2) what are the adaptive reasons for massive developmental cell death in the nervous system (see the section titled 'The functions of PCD in the nervous system')? A plausible answer to the first question is that PCD arose as a defense mechanism for eliminating abnormal cells that threatened the survival of the organism. According to this argument, the death-promoting intracellular machinery arose in host cells to defend against the spread of viral infection and, in response, viruses evolved survival-promoting genetic mechanisms to block or counter the host cell death-promoting defenses. Because these death- and survival-promoting mechanisms have been found to mediate PCD in a diversity of organisms (see the section titled 'Intracellular regulation of cell death'), they are considered to represent an evolutionarily conserved core molecular genetic program that has been coopted by the nervous system for regulating cell numbers.

## The Functions of PCD in the Nervous System

PCD in the nervous system involves both neurons and glia, and it occurs in the central and peripheral nervous system, the brain, and spinal cord and in virtually all neuronal and glial cell types. Accordingly, it is

**Table 1** Some possible functions of developmental PCD in the nervous system

| |
|---|
| Differential removal of cells in males and females (sexual dimorphisms) |
| Deletion of some of the progeny of a specific sublineage that are not needed |
| Negative selection of cells of an inappropriate phenotype |
| Pattern formation and morphogenesis |
| Deletion of cells that act as transient targets or that provide transient guidance cues for axon projections |
| Removal of cells and tissues that serve a transient physiological or behavioral function |
| Systems matching by creating optimal quantitative innervation between interconnected groups of neurons and between neurons and their targets (e.g., muscles, sensory receptors) |
| Systems matching between neurons and their glial partners by regulated glial PCD (e.g., Schwann cells and axons) |
| Error correction by the removal of ectopically positioned neurons or of neurons with misguided axons or inappropriate synaptic connections |
| Removal of damaged or harmful cells |
| Regulation of the size of mitotically active progenitor populations |
| Production of excess neurons may serve as an ontogenetic buffer for accommodating mutations that require changes in neuronal numbers in order to be evolutionary adaptive (e.g., increases or decreases in limb size may require less or more motor neuron death for optimal innervation) |
| Regulated survival of subpopulations of adult-generated neurons as a means of experience-dependent plasticity |

not surprising that the diversity of adaptive roles for PCD differs according to the animal species, cell type, nervous system region, and stage of development. A list of some of the most common reasons for PCD in the nervous system is provided in **Table 1**. Of these, the one that has received the most attention because of its central role in establishing optimal functional connectivity in neuronal circuits is systems matching (**Figure 1**). In birds and mammals, the evidence is compelling that for postmitotic neurons that are establishing interconnections with targets and afferents, systems matching, occurring in the framework of the neurotrophic hypothesis (see the section entitled 'Introduction'), is the primary reason for why a significant proportion of neurons (20–80%) undergo PCD. An analogous kind of systems matching adjusts myelinating glial cell numbers to the number of available axons.

## PCD by Autonomous versus Conditional Specification

PCD occurring during systems matching represents a paradigm example of the conditional specification of cell fate during development. Developmental biologists have identified two kinds of pathways that cells use for specifying their differentiated fate or phenotype. One pathway, autonomous specification, involves the differential segregation of cytoplasmic signals into daughter cells following mitosis. In this way, cells become different from one another by the presence or absence of these cytoplasmic signals with little if any contribution from signals from neighboring cells. The other pathway, conditional specification, requires signals from other cells (cell–cell interactions) to progressively restrict differentiation and determine whether a cell lives or dies. These cell–cell interactions can be of four types: (1) juxtacrine (direct cell–cell or cell–matrix contact); (2) autocrine (a secreted signal acts back on the same cell type from which the signal arose); (3) paracrine (a secreted diffusible signal from one cell that acts locally on a different cell type); and (4) endocrine (signaling via the bloodstream). In the developing vertebrate nervous system, neuronal and glial survival is largely dependent on conditional specification involving paracrine interactions that use neurotrophic factors (NTFs). For neurons, the most commonly used NTFs are members of three major families: (1) neurotrophins, (2) the glial cell line-derived family ligands (GFLs), and (3) ciliary NTFs. Each of these contains several distinct members that act preferentially on specific types of neurons via membrane-bound receptors. By contrast, the survival of glial cells depends on different families of trophic factors such as the neuregulins and insulin-like growth factors. Neurons and glial cells use paracrine signaling to promote survival by ligand–receptor interactions. However, there are some situations in which paracrine interactions can signal death versus survival.

## Molecular Regulation of Cell Death and Survival by NTFs

The dependence of neurons on NTFs for survival intuitively leads to the conclusion that the inactivation of NTF receptors, and the signal transduction associated with them, leads to the activation of cell death events (**Figure 2**). This has been best studied with neurotrophins and their receptors, notably the

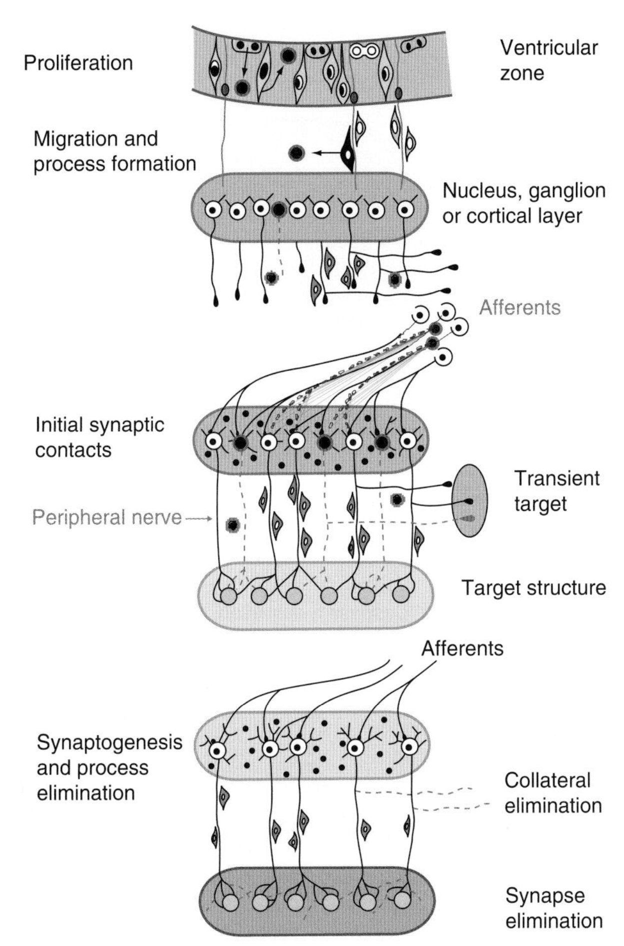


**Figure 1** Schematic illustration of some key steps in neuronal development. Neurons undergoing PCD (●) are observed during neurogenesis in the ventricular zone, during migration and while establishing synaptic contacts. Schwann cells in developing nerves also undergo PCD. ◊ represents peripheral glial (Schwann) cells; ● represents CNS glia (astrocytes and oligodendrocytes); ⊙ represents surviving, differentiating neurons (motor neurons) whose targets are skeletal muscle.

Trk receptors and p75 (**Table 2**). Activation of the trk receptors results in the activation of the phosphatidylinositol-3 kinase (PI3-K) and mitogen-activated protein kinase (MAPK) pathways. Akt, a substrate of PI3-K, has been reported to phosphorylate Bad, promoting its association with 14-3-3 and preventing the inactivation of Bcl-2 and Bcl-x. Activation of PI3-K/Akt, extracellular signal-regulated kinase-1/2 (ERK1/2), protein kinase (PK)C, or PKA promotes the serine phosphorylation of Bad and Bcl-2, and Gsk-3B phosphorylates and inactivates Bax – events that are associated with cell survival. The survival-promoting activity of phosphorylated Bcl-2, however, is controversial, being associated with motor neuron survival-induced trophic support, whereas in neurons treated with microtubule destabilizing agents it appears to promote death. The c-jun N-terminal kinase (JNK) pathway is reported to exhibit increased activity in neurons triggered to die. The activation of one of its substrates, c-jun, is also thought to play a role in mediating neuronal death. For example, the JNK activation of the BH3 proteins Bim and DP5 has been associated with neuronal death, and a JNK–p53–Bax pathway appears to be critical for death in specific cell types. Nonetheless, the JNK pathway has also been reported to be a critical mediator of survival-promoting events such as neurite outgrowth and may serve as a double-edged sword in neurons through changes in intracellular localization or specific activation or inactivation of individual isoforms.

The role of the p75 nerve growth factor (NGF) receptor in promoting neuronal survival or death is complicated. For example, complexes of p75 and the trk receptors bind neurotrophins and promote survival, while complexes of p75 and sortilin appear to promote binding of the proneurotrophins resulting in the death of neurons. Activation of p75 in the absence of Trk activation has most notably been associated with the death of neurons, but this appears to be developmentally regulated. The precise mechanism by which p75 promotes death is unclear. However, p75 is a member of the tumor necrosis factor receptor family, and another member of this family better known for its death promoting activity is Fas/CD95. Engagement of Fas by Fas ligand leads to the activation of caspase 8. This event is dependent on the formation of a death-induced signaling complex (DISC). Homotrimerization is the first step in this process. Engagement of Fas by its ligand can only occur when it is homotrimerized. This family of receptors is characterized by the presence of a death domain (DD) on the cytoplasmic region of the receptors. The Fas-associated death domain-containing protein (FADD) can then bind to Fas. In addition to the DD, FADD also contains a death effector domain (DED). Procaspase 8 binds the DED. The autolytic nature of caspases leads to active capsase 8 that can then go on to directly activate caspase 3 (in type 1 cells, e.g., thymocytes) or cleave Bid leading to changes at the mitochondria (in type 2 cells, e.g., hepatocytes and neurons) and cell death.

## Intracellular Regulation of Cell Death

Many of the intracellular mechanisms mediating neuronal cell death were investigated following initial work that suggested that new gene expression was required for the process. Horvitz and coworkers provided some of the first evidence that there was indeed a genetic component of PCD from their work in the 1980s with the free-living nematode *Caenorhabditis elegans*, although it was two decades later

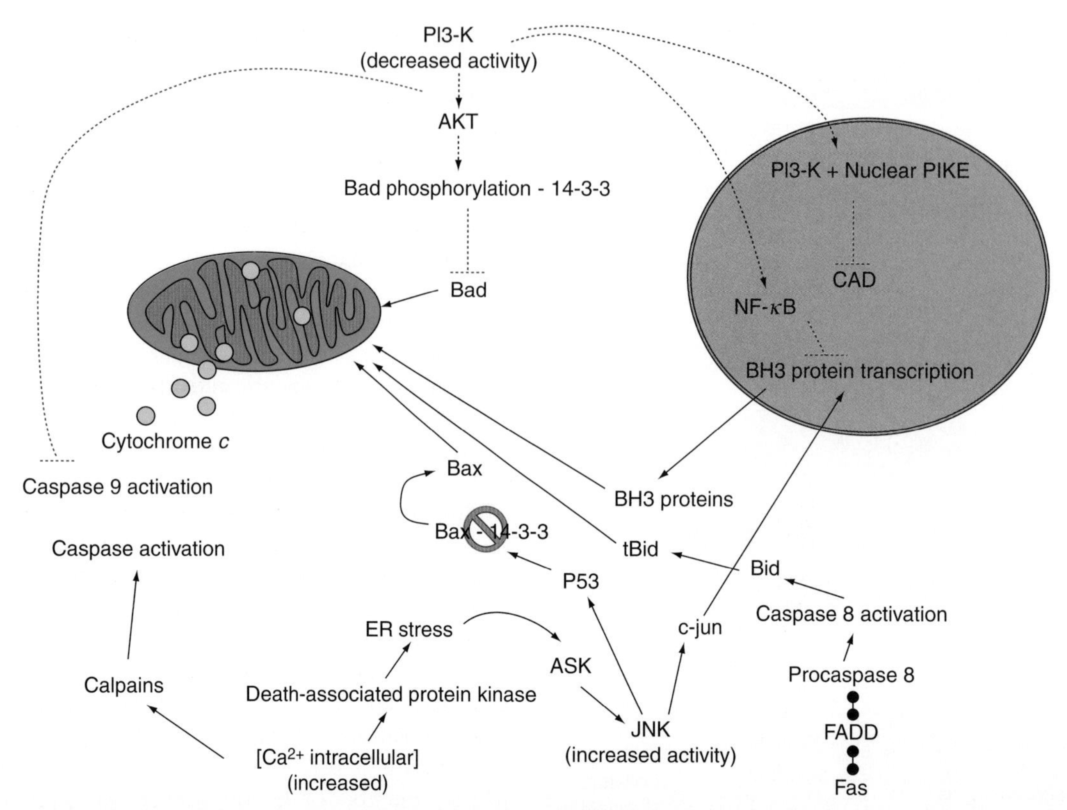


**Figure 2** Signals to die. Changes in signal transduction pathways, activation of receptors containing a DD, or changes in intracellular calcium concentrations have been identified as events that can lead to the activation of cell death-specific events. Loss of trophic support results in decreased activation of the phosphatidylinositol-3 kinase (PI3-K), thereby releasing inhibitory factors linked to this pathway. Alternatively, increases in intracellular calcium concentrations, activation of the FAS receptor, and increased c-jun terminal kinase activation are associated with the mitochondrial changes and caspase activation that occurs in cell death. ASK, apoptosis signal-regulating kinase; ER, endoplasmic reticulum; FADD, Fas-associated death domain-containing protein; JNK, c-jun N-terminal kinase; NF-κB, nuclear factor κB; PIKE, phosphatidylinositol-3 kinase enhancer.

that the significance of this work was fully appreciated (**Figure 3**). As these genes were identified, the sequence of events leading to the death of cells was pieced together.

The presence of Bcl-2 family proteins appears critical for mediating the survival or death of nervous system cells. During development, Bcl-2 expression is correlated with neuronal survival. With continued maturation and development, this expression declines while Bcl-x expression increases. The specific mechanisms responsible for this change in expression are currently not known. On the other hand, the expression of pro-death Bcl-2 family proteins such as Bid or Bax appears to be consistent throughout development, whereas the intracellular localization of these proteins appears to change in cells undergoing death. In healthy cells, Bax is localized more in the cytoplasm, whereas in dying cells the majority of Bax localizes to organelle membranes, including the mitochondria. The localization of Bax to the mitochondria corresponds to the release of cytochrome *c* into the cytoplasm. Once in the cytoplasm, cytochrome *c* binds Apaf-1, causing a conformational change in Apaf-1 to reveal a caspase recruitment (CARD) domain. In the presence of ATP, a heptamer of the cytochrome *c*–Apaf-1 complex is formed. Procaspase 9 has a high affinity for the CARD domain and localizes to the heptamer. This complex is referred to as the apoptosome. With the increased local concentration of procaspase 9, the autolytic property of caspases leads to the generation of active caspase 9. As an initiator caspase, active caspase 9 can cleave procaspase 3, resulting in the active form of this protease (**Figure 4**). The activation, inactivation, or destruction of specific substrates significantly contributes to the rapid degeneration of the cell. Caspase activity also leads to the activation of endonucleases that may play a role in the changes in nuclear morphologies that are observed in may cell types.

The removal of the dying cell is most likely just as critical an event in the cell death process as any of those already discussed. The event is accomplished by adjacent cells that can be nonprofessional phagocytes (e.g., Schwann cells) but more often involves

**Table 2** Specific molecules associated with cell death in neurons[a]

| | |
|---|---|
| Receptors | Fas |
| | P75 |
| | Trk receptors |
| Signal transduction pathways | PI3K–Akt |
| | ERK1,2 |
| | ERK 5 |
| | JNK |
| | PLC |
| | PKC |
| | PKA |
| Bcl-2 proteins | Bcl-2 |
| | Bcl-x |
| | Bid |
| | Bik |
| | Bag |
| | Bax |
| | Bad |
| | Dp5 |
| | Bim |
| | Puma |
| Mitochondrial proteins | Cytochrome *c* |
| | SMAC/Diablo |
| | XIAP |
| | AIF |
| | Endonuclease G Omi/HtrA2 |
| Caspases | |
| Initiator caspases | Caspase 8 |
| | Caspase 9 |
| Effector caspases | Caspase 2 |
| | Caspase 3 |
| Inflammatory caspases | Caspase 1 |
| Other proteins | Apaf-1 |
| | Amyloid precursor protein/ $\beta$-amyloid |
| | PARP |
| | p53 |
| | p35 |
| | Ubiquitin |
| | E2F |
| | Cyclin D1 |
| | Cyclin E |
| | Cyclin B |
| | Cdk4/6 |
| | Cdk2 |
| | Cdk1 |
| | Rb |
| | p130 |
| | HIF1-$\alpha$ |
| | CAD |
| | iCAD |

[a]AIF, apoptosis-inducing factor; CAD, caspase-activated DNase; HIF1-$\alpha$, hypoxia-inducing factor; iCAD, inhibitor of CAD; JNK, c-jun N-terminal kinase; PARP, poly(ADP-ribose) polymerase; PKA, protein kinase A; PKC, protein kinase C; PLC, phospholipase C; XIAP, x-linked inhibitor of apoptosis protein.

dedicated phagocytes such as tissue macrophage (microglia) or circulating monocytes. During central nervous system (CNS) development, resident microglia are not present in large numbers among dying cell populations. Intuitively, some signal must be sent by the dying cells to recruit the phagocytes into the area. One such chemotactic factor has been identified in nonneuronal cells, and interestingly, this factor is the phospholipid lysophosphatidylcholine. Caspase 3 activation appears to be necessary for the release of this factor. Once the phagocytic cells are in the region, they must be able to distinguish dying from healthy cells. Changes on the dying cell's surface include revealing of thrombospondin 1 binding sites, exposure of phosphatidyl serine, ATP-binding cassette (ABC-1) molecules, and carbohydrate changes. The phagocyte, in turn, expresses the phosphatidylserine receptor, as well as thrombospondin receptors that recognize the thrombospondin bound to the binding site on the dying cell, lectins that bind to the carbohydrate changes, and ABC-1 molecules that bind to like molecules. Phagocytosis occurs only when multiple changes are recognized, and this phagocytosis is limited and does not result in macrophage secretion of cytokines that would normally induce an inflammatory or immune response if foreign antigens were phagocytosed.

## Different Types of Neuronal Death

Many individuals studying cell death often rely on the critical papers of Currie, Kerr, and Wyllie to define and characterize the different types of cell death. These investigators described in detail two distinct morphological changes associated with cell death. An active process, apoptosis, was characterized by specific morphological changes that included condensation of nuclear chromatin, shrinking of the cytoplasm, and a breaking up of the cell into membrane-bound particles that were phagocytosed. Necrosis, on the other hand, was characterized by a swelling of the cytoplasm with eventual bursting of the cell (**Figure 5**). In this case, intracellular components are spilled into the extracellular space where they could initiate inflammatory and immune responses. However, electron microscopy studies of the developing nervous system indicated that neuronal death could not be so easily defined by only these two modes of death. During the naturally occurring death of some neurons, initial changes in dying cells are observed in the cytoplasm where there is an increase in the diameter of the cisternae of the rough endoplasmic reticulum (RER). Mitochondrial swelling was also observed, although it was not clear if these changes occur in the same cell. Nonetheless, during normal development initial changes are observed in the cytoplasm with little nuclear alterations in dying cells. The cell then appears to round up and break into pieces that are phagocytosed. Although cytoplasmic cell death appears to be more prominent

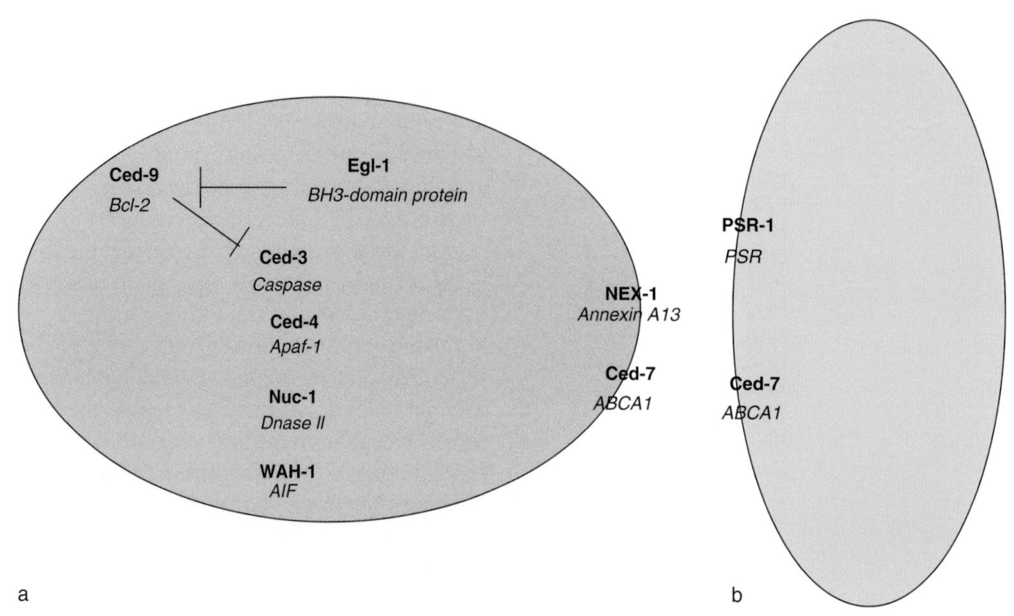


**Figure 3** Genes for the key regulators of cell death: (a) dying cell; (b) engulfing cell. Many of the key regulators of cell death are evolutionarily conserved. Many of the genes for these proteins were initially identified through genetic mutations in the free living nematode (bold type), with homologs later being identified in *Drosophila* and mammals (italics).

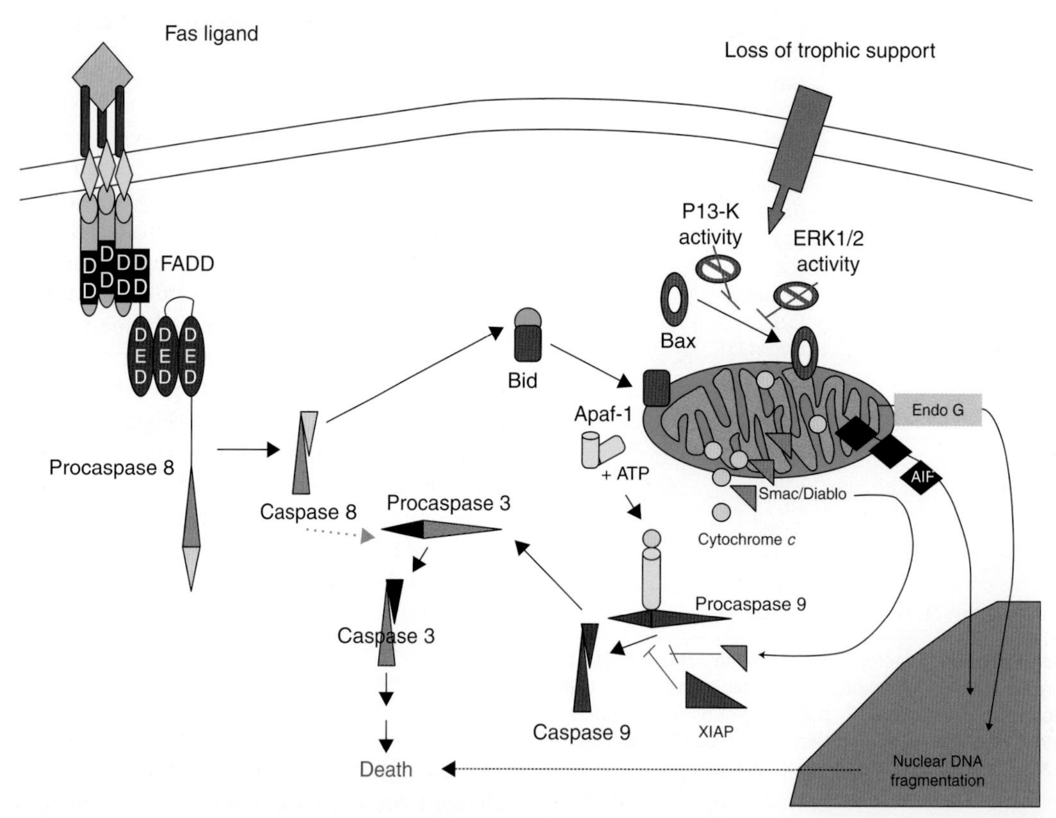


**Figure 4** Many of the critical components of neuronal death and their apparent sequence of activation during development. The mitochondria release many regulators that can lead to death with or without caspase activation. Neurons are considered type 2 cells where FAS engagement results in minimal caspase 8 activation, resulting in the cleavage of Bid and its subsequent movement to the mitochondria. The localization of Bax to the mitochondria appears to be a critical event in the death of neurons. AIF, apoptosis-inducing factor; ATP, adenosine triphosphate; DD, death domain; DED, death effector domain; ERK, extracellular signal-regulated kinase; FADD, Fas-associated death domain-containing protein; PI3-K, phosphatidylinositol-3 kinase; XIAP, x-linked inhibitor of apoptosis protein.

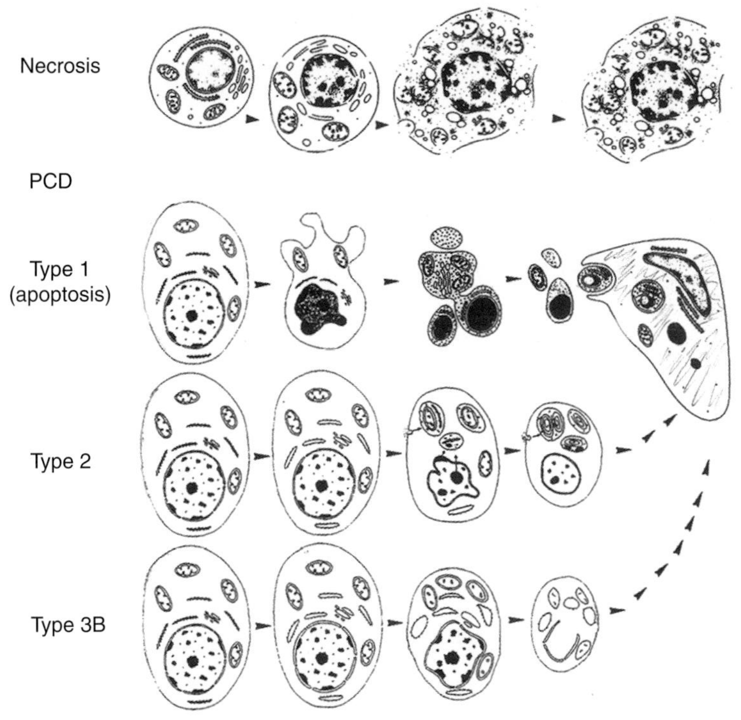


**Figure 5** Necrotic death and the three most common types of programmed cell death (PCD) observed in the nervous system. Type 1 meets the criteria for apoptosis. In PCD, phagocytes remove the resulting corpses, whereas necrosis in many cases results in inflammation.

during development, nuclear or apoptotic death also occurs. A third type of death is autophagy, characterized by the formation of numerous autophagic vacuoles. Nuclear changes may also occur in this type of death.

It is important to note that all three types of neuronal death appear to reflect a metabolically active process. Accordingly, it appears that a homogeneous mode of suicide does not occur but, rather, that processes leading to death appear to be context and developmentally dependent. One example is a neuronal population undergoing apoptosis when nuclear condensation is a prominent feature. On the other hand, in animals in which caspases have been inhibited or deleted, the same neuronal population undergoes a delayed death with major changes occurring in the cytoplasm.

## Pathological Neuronal Death

Many of the mechanisms that have been identified as playing a role in neuronal death during development are reported to be reactivated in the mature pathological nervous system. Alterations in the expression of Bcl-2 proteins are observed in animal models and in postmortem human tissue in Alzheimer's disease (AD), Parkinson's disease (PD), Huntington's chorea (HC), amyotrophic lateral sclerosis (ALS), and stroke. Caspase activation has also been reported in these and other disorders. Interestingly, although reports of apoptotic cells are present in these conditions, the majority of degenerating cells exhibit morphological changes reminiscent of cytoplasmic or autophagic death. This is most notable in the mutant SOD1 mouse model of ALS. During development, dying motor neurons exhibit many features of apoptosis, including possible Fas activation, Bax translocation, mitochondrial dysfunction, and caspase activation and condensation of nuclear chromatin. By contrast, in the adult spinal cord of SOD1 mice, degenerating motor neurons exhibit cytoplasmic vacuolization, mitochondrial dilation, and protein aggregation. Nonetheless, some of the cell death-associated events that occur during development are also observed. It is becoming increasingly recognized that neuronal dysfunction is most likely the event responsible for clinical symptoms in ALS and other neurodegenerative diseases. In experiments in which motor neuron death is inhibited in the SOD1 mouse, disease progression and survival of the animal are only very modestly affected; muscle denervation still occurs, and the animals die prematurely. Results such as this call into question the practical application of inhibiting cell death as a therapeutic approach for these disorders.

*See also:* Apoptosis in Neurodegenerative Disease; Autonomic and Enteric Nervous System: Apoptosis and Trophic Support During Development.

## Further Reading


Akhtar RS, Ness JM, and Roth KA (2004) Bcl-2 family regulation of neuronal development and neurodegeneration. *Biochimica et Biophysica Acta* 1644(2-3): 189–203.

Bronfman FC and Fainzilber M (2004) Multi-tasking by the p75 neurotrophin receptor: Sortilin things out? *EMBO Reports* 5(9): 867–871.

Brunet A, Datta SR, and Greenberg ME (2001) Transcription-dependent and-independent control of neuronal survival by the PI3K-Akt signaling pathway. *Current Opinion in Neurobiology* 11(3): 297–305.

Buss RR, Sun W, and Oppenheim RW (2006) Adaptive roles of programmed cell death during nervous system development. *Annual Review of Neuroscience* 29: 1–40.

Clarke PGH and Clarke S (1996) Nineteenth century research on naturally occurring cell death and related phenomena. *Anatomy and Embryology* 193: 81–99.

Fischer U, Janicke RU, and Schulze-Osthoff K (2003) Many cuts to ruin: A comprehensive update of caspase substrates. *Cell Death and Differentiation* 10(1): 76–100.

Greene LA, Biswas SC, and Liu DX (2004) Cell cycle molecules and vertebrate neuron death: E2F at the hub. *Cell Death and Differentiation* 11(1): 49–60.

Horvitz HR (2003) Worms, life, and death (Nobel lecture). *ChemBioChem* 4(8): 697–711.

Krantic S, Mechawar N, Reix S, and Quirion R (2005) Molecular basis of programmed cell death involved in neurodegeneration. *Trends in Neuroscience* 28(12): 670–676.

Lettre G and Hengartner MO (2006) Developmental apoptosis in *C. elegans*: A complex CEDnario. *Nature Reviews Molecular and Cell Biology* 7(2): 97–108.

Lindholm D and Arumae U (2004) Cell differentiation: Reciprocal regulation of Apaf-1 and the inhibitor of apoptosis proteins. *Journal of Cell Biology* 167(2): 193–195.

Morrison RS, Kinoshita Y, Johnson MD, Ghatan S, Ho JT, and Garden G (2002) Neuronal survival and cell death signaling pathways. *Advances in Experimental Medicine and Biology* 513: 41–86.

Oppenheim RW and Johnson JE (2002) Programmed cell death and neurotrophic factors. In: Zigmond MJ, et al. (eds.) *Fundamental Neuroscience*, pp. 499–532. New York: Academic Press.

Polster BM and Fiskum G (2004) Mitochondrial mechanisms of neural cell apoptosis. *Journal of Neurochemistry* 90(6): 1281–1289.

Ryan CA and Salvesen GS (2003) Caspases and neuronal development. *Biological Chemistry* 384(6): 855–861.

Sathasivam S and Shaw PJ (2005) Apoptosis in amyotrophic lateral sclerosis – What is the evidence? *Lancet Neurology* 4(8): 500–509.

Savill J and Fadok V (2000) Corpse clearance defines the meaning of cell death. *Nature* 407(6805): 784–788.

Stefanis L (2005) Caspase-dependent and- independent neuronal death: Two distinct pathways to neuronal injury. *Neuroscientist* 11(1): 50–62.

Wyllie AH, Kerr JF, and Currie AR (1980) Cell death. The significance of apoptosis. *International Review of Cytology 68: 251–306.*

Yuan J, Lipinski M, and Degterev A (2003) Diversity in the mechanisms of neuronal cell death. *Neuron* 40(2): 401–413.

# Apoptosis in Neurodegenerative Disease

**K A Roth and J J Shacka**, University of Alabama at Birmingham, Birmingham, AL, USA



## Apoptosis

In 1972, Kerr and colleagues coined the term apoptosis, which is Greek for 'falling off,' as in petals from a flower or leaves from a tree. This term was used to define a type of cell death characterized morphologically by the discrete budding of self-contained fragments from the cell membrane following a death stimulus and was distinct from necrotic (passive or accidental) cell death. Apoptosis, known also as type I cell death, is now defined by distinct morphological criteria, including pyknosis (condensation or reduction in size) of the cell and/or cell nucleus along with nuclear fragmentation and hyperchromatosis. The DNA of apoptotic nuclei is typically digested into 200–300 bp fragments which can be observed by enzymatic labeling of nicked DNA or by electrophoretic detection of DNA laddering. The plasma membrane of apoptotic cells remains relatively intact, with the dying cell budding off discrete packages of itself into 'apoptotic bodies' that are ultimately phagocytosed. This efficient and discrete phagocytosis in the absence of a ruptured plasma membrane generally limits the inflammatory response around an apoptotic cell and is unlike necrosis, which is characterized by a rupture of the plasma membrane and subsequent leakage of cell contents into the surrounding extracellular space with a resultant inflammatory response. In addition, unlike necrosis, apoptosis is an energy-dependent process and often requires new gene transcription and protein translation for cellular execution.

## Molecular Regulation of Apoptosis

Many seminal discoveries have been made in the past quarter century that have enhanced our understanding of cell death. Using the nematode *Caenorhabditis elegans* as a model, Ellis and Horvitz showed for the first time in the 1980s that apoptotic cell death was regulated by distinct genes. Subsequent cloning studies demonstrated that mammals also express homologs of these genes with similar function, although it has been found that the regulation of mammalian cell death is more complex than in the nematode. In 1984, a novel oncogene, B-cell lymphoma-2 (Bcl-2), was isolated and was found to have prosurvival or antiapoptotic functions. Mammalian cell apoptosis can be classified into distinct pathways, extrinsic (death-receptor mediated) or intrinsic (mitochondrial), and involves the participation of an extensive family of Bcl-2 family proteins (described below) and the activation of caspases (see **Figure 1**).

### Caspases

Caspases are cysteine-containing proteases that cleave substrates after aspartic acid residues and critically regulate both the transduction of apoptotic stimuli and the final execution of apoptotic death. There are 14 mammalian and 12 human caspases that can be divided into two basic groups, the interleukin-1$\beta$ converting enzyme (ICE)-like and the CED-3-like (named after the *C. elegans* homolog) families. The ICE-like caspases, which include caspase-1, caspase-4, caspase-5, and caspase-12, are involved in the proteolytic processing of cytokines and as such are related indirectly to neuron cell death as a function of modulating inflammatory responses. The CED-3-like caspases are directly involved in the apoptotic signaling cascade and can be further divided into two subgroups, the initiator caspases (caspase-2, caspase-8, caspase-9, and caspase-10) and the effector caspases (caspase-3, caspase-6, and caspase-7). Both initiator and effector caspases exist as inactive zymogens within the cell prior to receipt of a death stimulus. Initiator caspases are activated early in the apoptosis cascade by conformational change resulting from interaction with other apoptosis-associated proteins. Effector caspases are activated by the catalyzed cleavage from 'activated' initiator caspases. Both initiator and effector caspases have been shown to be proteolytically cleaved, but initiator caspases do not typically require cleavage for their activation. It is the selective cleavage of substrates by caspases which induces in large part the appearance of the morphological features of apoptosis.

### Bcl-2 Family

There are three major subgroups of proteins in the Bcl-2 family that are classified by composition of their Bcl-2 homology (BH) domains and by function (anti- vs. proapoptotic). It is widely believed that the relative 'ratio' of antiapoptotic versus proapoptotic Bcl-2 family members influences the response of neurons to apoptotic stress. Members of the antiapoptotic subgroup include Bcl-2, Bcl-$X_L$, Bcl-w, and Mcl-1 and possess four BH domains, except for Mcl-1, which lacks the BH4 domain. The importance of the BH1 and BH4 domains in the maintenance of antiapoptotic activity of Bcl-2 and Bcl-$X_L$ has been emphasized by molecular studies, where the targeted

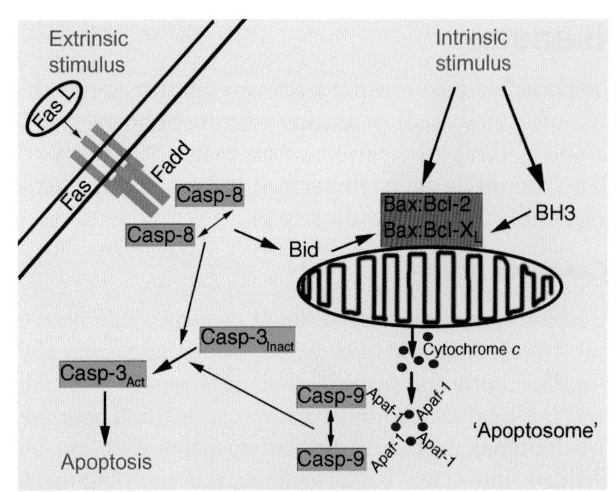


**Figure 1** Molecular regulation of cell death in mammals by the extrinsic (death receptor-mediated) versus intrinsic (mitochondrial) apoptotic pathways. Extracellular death receptor ligands (e.g., Fas ligand) bind to and activate their receptors (e.g., Fas), which in turn induces recruitment of adaptor proteins (Fadd) to the complex with subsequent recruitment, dimerization, and activation of caspase-8, an initiator caspase. Activation of caspase-8 in turn can directly cleave and activate caspase-3 resulting in apoptotic death. In addition, caspase-8 can cleave Bid, a proapoptotic, BH3-domain-only member of the Bcl-2 family, to truncated Bid which interacts with Bax at mitochondria and causes the subsequent activation of the intrinsic apoptotic pathway. Intrinsic death stimuli (e.g., DNA damage) can cause changes in expression and/or function of anti- and/or proapoptotic members of the Bcl-2 family. The ratio of anti- to proapoptotic Bcl-2 family members is thought to regulate the release of cytochrome *c* from mitochondria, which in turn binds with Apaf-1 and caspase-9 in a protein complex termed the 'apoptosome.' The binding of caspase-9 in the apoptosome induces its forced dimerization and subsequent activation, which in turn induces activation of caspase-3. The activation of caspase-3 and other effector caspases is responsible for the cleavage of downstream substrates that promotes in part the appearance of apoptotic morphology and ultimate cell death.

deletion of these regions has been shown to abrogate their antiapoptotic activity. Bcl-2 is expressed at high levels during nervous system development and remains high in the adult spinal cord and in sympathetic neurons, but decreases in relative expression in the adult brain. Upregulation of Bcl-2 has been documented in many neurodegenerative diseases, occurring possibly as an antiapoptotic neuron survival response or in reactive glial cell elements. Bcl-X$_L$ and Bcl-w are expressed at high levels in adult central nervous system (CNS) and have also been shown to be altered in neurodegenerative disease; in contrast, Mcl-1 is thought to play a more focused role in CNS development.

The proapoptotic, multidomain subgroup consists of Bax, Bcl-2-homologous antagonist/killer (Bak), Bok, and Bcl-X$_S$. These proteins share three homologous BH domains (BH1–3) except for Bcl-X$_S$, which lacks BH1 and BH2 domains but rather possesses a BH4 domain. Both Bax and Bak regulate the release of cytochrome *c* from mitochondria, a critical event in activation of the intrinsic apoptotic pathway. Upregulation of Bax has been documented in many neurodegenerative diseases. In contrast, while present in the CNS and similar in function to Bax, studies have indicated a limited role for Bak in promoting apoptotic neuron death. Bcl-X$_S$ is expressed at low levels in the mammalian nervous system and a regulatory role for Bok in CNS apoptotic death is yet to be described. The proapoptotic BH3 domain-only subgroup includes Bcl-2-associated death protein (Bad), Bid, Bcl-2 interacting mediator of cell death (Bim), Bcl-2 19 kDa interacting protein (BNip), death protein 5/harakiri (DP5/Hrk), Noxa, and p53-upregulated modulator of apoptosis (Puma). Although members of this subgroup are comparably the most diverse, their regulation of cell death depends critically upon the BH3 domain and the presence of Bax and/or Bak. The response of BH3 domain-only proteins to an apoptotic stress is regulated in part by distinct protein modifications, such as phosphorylation (Bim), dephosphorylation (Bad), induction of expression (Bim, DP5, Noxa, Puma), or cleavage (Bid). Altered regulation of BH3 domain-only proteins has been implicated in many neurodegenerative diseases, which are discussed below.

### Extrinsic versus Intrinsic Death Pathway

There are two distinct pathways of caspase-dependent cell death, the extrinsic or death receptor-mediated pathway and the intrinsic or mitochondrial-mediated pathway. The extrinsic pathway involves binding of a death receptor ligand (e.g., Fas ligand or TNF-$\alpha$) to a cell surface death receptor, which in turn stimulates receptor dimerization and the association of adaptor proteins to a newly formed complex that facilitates the association-induced activation of caspase-8. In the intrinsic pathway, a death stimulus regulates in part the levels and/or function of pro- versus antiapoptotic Bcl-2 family members, which induces the mitochondrial release of cytochrome *c* into the cytosol and in turn facilitates the energy-dependent activation of caspase-9 through formation of the 'apoptosome' complex with the protein Apaf-1. Both the intrinsic and extrinsic pathways converge at the level of effector caspase activation. Interestingly, cross-talk between these two pathways occurs via the cleavage of Bid by activated caspase-8 and the translocation of truncated Bid to the mitochondria, where it associates with the proapoptotic protein Bax to induce the release of cytochrome *c* and resultant activation of the intrinsic pathway.

## Apoptosis and Neurodegeneration

A variety of evidence for the occurrence of apoptotic death in human neurodegenerative disease has been cited to support the hypothesis that apoptosis-associated molecules such as Bcl-2 and caspase family members play a role in disease pathogenesis. Although the extent and quality of data vary for specific neurodegenerative diseases (see below), much of the data is correlational and fraught with technical and methodological obstacles inherent in the study of human postmortem brain tissue. Poorly validated antibodies to Bcl-2 and caspase family members and nonspecific 'markers' of apoptosis such as terminal deoxytransferase-mediated deoxyuridine triphosphate nick end labeling (TUNEL) have further complicated the analysis of the biological significance of apoptosis in neurodegenerative disease onset and progression. To date, specific mutations in apoptosis-associated genes have not been linked to any human neurodegenerative disease, raising concern about the primacy of altered apoptosis regulation in disease pathogenesis. Despite these caveats, it is clear that alterations in apoptosis-associated molecules occur in a variety of neurodegenerative diseases, that apoptotic death is one of several morphological forms of neuron death detected in human neurodegenerative disease, and that pharmacological inhibition of apoptotic death offers some hope for at least temporary preservation of neuron survival.

## Alzheimer's Disease

Alzheimer's disease (AD) is the most common human neurodegenerative disease and is the most prevalent form of dementia among elderly patients. Neuropathological hallmarks of AD include mature amyloid plaques and neurofibrillary tangles, which are considered the classical features of AD neuropathology and were used to identify the first case of AD more than 100 years ago. In addition to plaques and tangles, AD brains show decreased synapse density and profound neuron loss. To date, the majority of research efforts in AD has focused on mechanisms of plaque and tangle formation and their contribution(s) to neuron dysfunction and ultimate neuron death. Amyloid plaques are caused by the aberrant cleavage of the plasma membrane-bound amyloid precursor protein (APP), a soluble form of amyloid beta (A$\beta$), which ultimately deposits in the extracellular space in its fibrillar form. However, the formation of amyloid plaques correlates weakly at best with the symptoms and progression of AD. Neurofibrillary tangles are caused in part by the hyperphosphorylation of the cytoskeletal protein tau and more closely follow AD progression. Many studies have indicated that a loss in synapse density and neuron number correlate highly with dementia in AD, although the definitive causes of synapse and neuron loss in AD remain speculative. Whether neuron loss precipitates a decrease in synaptic integrity or if a decrease in neurotrophic support at the level of the synapse precipitates neuron death is not entirely clear in AD. Significant decreases in immunoreactivity for the synaptic protein synaptophysin have been detected in early AD prior to pronounced neuron loss, suggesting that neuronal dysfunction precedes neuron death. However, some brain regions, including the nucleus basalis of Meynert, which contains cell bodies of neurons which project to the cortex, show little neuron loss yet still exhibit AD neuropathology, suggesting that synapse function and cell death may not be directly related. Nevertheless, a better understanding of cell death pathways in AD combined with determining the events that trigger cell death in AD will hopefully generate more effective treatment strategies that preserve both neuron numbers and function.

A role for apoptosis in AD brain has been suggested previously by numerous studies. Techniques such as TUNEL staining have been developed that detect *in situ* the fragmentation of DNA, and studies of AD brain show a marked increase in numbers of such 'apoptotic' nuclei relative to control brains. However, subsequent studies have indicated that these detection methods are not specific for apoptotic nuclei, that is, necrotic nuclei have also been shown to exhibit fragmented DNA and are highly sensitive to fixation artifact. In addition, the large number of apoptotic neurons determined by these methods does not correlate well with the chronic neurodegenerative process of AD, which would predict low numbers of apoptotic neurons at any one time that are efficiently phagocytosed and degraded. Thus, despite findings of numerous studies, the morphological assessment of apoptotic neurons in AD remains inconclusive.

Many studies have indicated an altered regulation of Bcl-2 family members in AD brain. Both proapoptotic (Bak, Bad) and antiapoptotic (Bcl-2 and Bcl-$X_L$) Bcl-2 family members have been shown to be upregulated in AD brain. The increased expression of Bcl-2 and Bcl-$X_L$ may represent a chronic, antiapoptotic stress response of compromised neurons to compensate for increases in Bax. Based on these findings, a proposed therapeutic strategy in AD is the pharmacological delivery of agents that effectively increase the ratio of anti- to proapoptotic Bcl-2 family members in neurons. However, Bcl-2 family members are also expressed in glial cells, and increases in Bcl-2 have been shown to occur in reactive astrocytes in AD brain. Thus, expression data alone may provide an incomplete picture of the role of Bcl-2 family members and apoptosis in AD pathogenesis. Although they are clearly altered in AD brain, careful studies of Bcl-2 family members in the early stages of AD are required to differentiate possible etiologic from compensatory alterations in Bcl-2 family members specific for neurons.

*In vitro* models of AD have also implicated altered regulation of Bcl-2 family members. Treatment with A$\beta$, which causes the death of cultured neurons, induces the expression of proapoptotic Bcl-2 family members including Bax and DP5 while decreasing the expression of Bcl-2 and Bcl-w. In addition, the overexpression of Bcl-2, Bcl-$X_L$, or Bcl-w and the targeted deletion of Bax prevent A$\beta$-induced neuron death *in vitro*, which further implicates potential neuroprotective strategies *in vivo*. Future studies are necessary to validate the relative contribution of Bcl-2 family members for A$\beta$-induced neurotoxicity *in vivo*.

Caspase activation is a critical component of the apoptotic cascade and is largely responsible for the cytological changes that define apoptotic morphology. As such, many *in vivo* studies of human AD brain and *in vitro* models of AD have focused on levels and/or activation of caspases. An increased expression of caspase-1 has been reported in AD brain extracts, which may indicate activation of proinflammatory cytokines such as interleukin-1$\beta$ that is regulated by active caspase-1 and may lead to the indirect death of neurons in AD. Caspase-12 and caspase-4 are endoplasmic reticulum (ER)-associated proteins that mediate in part ER stress-induced apoptosis. While human isoforms of caspase-4 exist and have been shown along with caspase 12 to induce A$\beta$-induced neuron apoptosis, caspase-12 was shown recently not to be associated with ER stress-induced apoptosis in humans. In addition, caspase-4 has been shown to cleave APP and may thus be involved in the disease-associated formation of A$\beta$. Although little, if any, evidence currently exists for its activation in AD, future studies of apoptosis in AD brain warrant further investigation of caspase-4.

A clear role for caspase-2, caspase-8, or caspase-9 in AD has not yet emerged. Of the effector caspases, caspase-7 exists in very low levels and thus is unlikely to play a role in AD pathogenesis. Both caspase-6 and caspase-3 are abundant in human brain and several studies have suggested that caspase-6 may be involved in regulating both APP processing and apoptosis of human neurons. The majority of evidence for caspase activation in apoptosis has focused on caspase-3. Many studies have reported an increased expression and activation of caspase-3 in AD brain, but many conflicting reports have also reported little difference in levels or activity of caspase-3 in AD versus control brains. Caspase-3 has been shown to be activated in cultured neurons treated with A$\beta$; however, neurons deficient in caspase-3, while lacking apoptotic morphology, are not protected from the death-promoting effects of A$\beta$, which suggests that caspase-independent death pathways may also exist in AD (see below). While it remains entirely possible that caspase-dependent neuron death is important in AD, a comprehensive analysis of previous studies suggests a degree of caution in defining its relative contribution to AD pathogenesis. A potential disease-promoting role for nonlethal caspase-3 activation in AD neurons has also been proposed. Although cleaved 'activated' caspase-3 is typically associated with apoptotic cell death, cleaved caspase-3 immunoreactivity is consistently observed in AD neurons undergoing granulovacuolar degeneration, a nonapoptotic neurodegenerative process (**Figure 2**). The significance of this observation is currently unclear since granulovacuolar degeneration occurs in only a small fraction of degenerating neurons in the AD brain. If caspases are indeed involved in both sublethal and terminal events in AD neurons, caspase inhibitors may yet prove to be an effective therapy for preserving neuron function in AD.

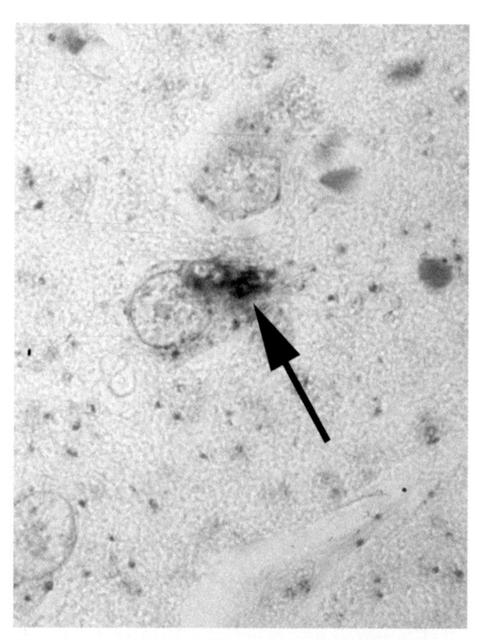

**Figure 2** Immunohistochemical detection of activated caspase-3 in AD brain shows strong labeling (dark deposit, indicated by an arrow) in hippocampal pyramidal neurons bearing features of granulovacuolar degeneration. Note that despite the presence of activated caspase-3 immunoreactivity, the nucleus appears intact and the neuron lacks apoptotic histological features.

## Parkinson's Disease

Parkinson's disease (PD), the second most common neurodegenerative disease after AD, is characterized by progressive motor impairment and is defined neuropathologically by the substantial loss of dopamine-containing neurons of the substantia nigra that project to the striatum along with the development of neuronal inclusions called Lewy bodies. Lewy bodies are cytoplasmic aggregates containing many proteins including α-synuclein, ubiquitin, and neurofilaments and occur in all affected brain regions of PD. The majority of PD cases are idiopathic, but approximately 5–10% of PD patients show familial patterns of inheritance, and to date several PD-linked gene mutations have been identified, such as those that encode for proteins including α-synuclein, parkin, and DJ-1.

The contribution of apoptosis in PD has been controversial both due to a limited number of reports suggesting alterations in apoptotic markers and morphology and to reports indicating little, if any, contribution of apoptosis to the neurodegenerative process in PD brain. This may be due to the chronic nature of the disease whereby only a few select neurons undergo apoptosis at any one time. Studies have indicated an upregulation in proapoptotic Bax and also antiapoptotic Bcl-2 and Bcl-$X_L$ in PD brain in comparison to age-matched controls. Increases in Fas ligand and activity of caspase-8 have been reported in PD brain, which may indicate the contribution of the extrinsic apoptotic death pathway. In addition, some studies have indicated an increase in the activity of caspase-3 in the PD substantia nigra along with apoptotic nuclear morphology, but it is not clear whether this is due to activation of the extrinsic and/or intrinsic apoptotic pathways or relates to disease pathogenesis.

Analysis of mutant PD genes, either in genetically modified mice or in cultured dopaminergic neurons, has determined that mutant α-synuclein or the targeted deletion of parkin induce apoptotic neuron death whereas overexpression of the wild-type genes prevent apoptosis. In addition, treatment with the dopaminergic neurotoxins such as 6-hydroxydopamine and 1-methyl, 4-phenyl, 1,2,3,6-tetrahydropyridine (MPTP) induce alterations in Bcl-2 family members, including increases in the expression of proapoptotic Bax and PUMA and decreases in expression of Bcl-2. MPTP-induced neurodegeneration can also be spared by the targeted deletion of Bax. Furthermore, the activation of both initiator (caspase-8 and caspase-9) and effector (caspase-3) caspases has been demonstrated in cultured dopaminergic neurons and in neurons of mice treated with MPTP. Thus, data from both *in vivo* and *in vitro* models of PD suggest a link between PD and apoptotic death and raise the possibility that inhibition of apoptotic death may prove beneficial in PD treatment strategies.

## Huntington's Disease

Huntington's disease (HD) induces chronic, progressive motor dysfunction and is caused by a mutation generating expanded CAG repeats in the *huntingtin* gene. In general, HD onset and prevalence is proportional to the number of CAG repeats. In the normal population, the number of CAG repeats varies from 6 to 35, but an increased risk is associated with between 36 and 41 CAG repeats and those with 42 or greater CAG repeats almost always manifest HD. The striatum is the principal area affected in HD with reports from postmortem studies demonstrating as much as 95% loss of striatal GABAergic, medium spiny neurons. In more severe cases of HD, other brain regions including the cortex, thalamus, globus pallidus, and hippocampus may also be affected.

Similar to AD and PD, there are conflicting reports of apoptosis in HD brain. Early reports of apoptosis in HD brain indicated increased TUNEL reactivity, but electrophoretic analysis of DNA fragments in HD brain did not reveal a characteristic apoptotic pattern, which possibly suggests a role for additional death processes, such as necrosis in HD brain. Regardless, there are scattered reports of caspase-8, caspase-9, and caspase-3 activation in HD brain, in addition to an increase in cytosolic cytochrome *c*, suggesting possible activation of both the extrinsic and intrinsic apoptotic pathways in HD. Increased levels of Bax and Bcl-2 have also been reported in HD brain specific to the caudate nucleus, which may be interpreted as an antiapoptotic survival response to an increase of Bax in stressed neurons. More studies are needed to determine the role of extrinsic versus intrinsic apoptotic death pathways in HD brain and the potential altered regulation of Bcl-2 family members in this devastating disease.

Animal models of HD, however, show more convincing evidence of a proapoptotic tone of Bcl-2 -family members in affected neuronal populations. One of the most well-studied mouse models of HD is the R6/2 transgenic mouse, which expresses only exon 1 of the *huntingtin* gene with 145 CAG repeats, shows a dramatic HD phenotype, and dies by 4 months of age. R6/2 mice show increases in mitochondrial Bim and Bax, a decrease in phosphorylated Bad, and activation of caspase-1 and caspase-3. Levels of Bcl-2 or Bcl-$X_L$ are unchanged in these mice, but the overexpression of Bcl-2 has been shown to slow disease progression. Treatment with 3-nitroprussic acid (3-NP), a chemical inhibitor of mitochondrial succinate dehydrogenase used to model HD, induces a selective striatal lesion when administered *in vivo* that increases the ratio of Bax/Bcl-2 and Bax/Bcl-$X_L$, thus shifting their balance to a proapoptotic state.

## Amyotrophic Lateral Sclerosis

Amyotrophic lateral sclerosis (ALS) is a fatal disease that affects one to two persons per 100 000 and is characterized by the progressive death of upper and lower motor neurons and eventual loss of motor function. Approximately 10% of ALS cases are familial, and of these 20% are linked to mutations in the gene encoding superoxide dismutase 1 (SOD1). It is well accepted that a percentage of dying motor neurons in ALS undergoes apoptosis, and apoptotic morphology has been identified in motor neurons of the human spinal cord and motor cortex. However, like most chronic neurodegenerative diseases, a major challenge in the study of apoptosis in ALS is the relative rapidity of the apoptotic process in the face of a disease with a slow time course, such that at any one time very few motor neurons are actively undergoing apoptosis. As such, there are reports indicating little, if any, evidence for increased apoptotic morphology in ALS cases when compared to control brains.

Studies of postmortem ALS spinal cord suggest a shift toward proapoptotic members of the Bcl-2 family, such as increased expression of Bax and DP5/Hrk and decreased expression of Bcl-2. Co-immunoprecipitation studies have also indicated an increase in Bax–Bax interactions and a decrease in Bax–Bcl-2 interactions in ALS tissue, which also suggests a shift to a proapoptotic state. A decrease in Bcl-2 may be of greater significance in ALS than in other neurodegenerative diseases since Bcl-2 remains high after neurodevelopment in the spinal cord relative to the brain, which suggests an important regulatory function of Bcl-2 in postmitotic spinal cord motor neurons. Analysis of symptomatic, transgenic mice with mutant SOD1 also shows decreased expression of Bcl-2 and Bcl-$X_L$ and increased expression of Bax and Bad. Truncated Bid is also evidenced in mutant SOD1 mice, which may suggest participation of the extrinsic apoptotic pathway in ALS.

*In vitro* studies of cultured motor neurons from mutant SOD1 mice indicate a role for Fas-mediated activation of caspase-8 and the extrinsic apoptotic pathway, but these same studies also report release of cytochrome *c* from mitochondria, suggesting activation of the intrinsic apoptotic pathway. Reports of altered regulation of Bax and Bcl-2 in human ALS would clearly support a role for the intrinsic apoptotic pathway in ALS. While there is no current evidence supporting the direct activation or altered expression of caspase-8 in human ALS, recent analysis of sera obtained from human ALS patients indicates an increase in antibodies to Fas that induced death of cultured motor neurons, which further suggests the contribution of the extrinsic apoptotic pathway. In addition to caspase-8, *in vitro* and animal models of ALS have shown activation of caspase-1, caspase-12, caspase-9, and caspase-3, while reports of caspase activation in human ALS tissue are limited to activation of caspase-1 and caspase-3. While good evidence exists supporting a role for apoptosis in ALS, more studies are needed to determine its relative contribution to the overall disease pathogenesis.

## Nonapoptotic Neurodegeneration

The reported inconsistencies for the role of apoptosis in neurodegenerative disease have led to an increasing number of studies investigating potential mechanisms of caspase-independent neuron death. Apoptosis-inducing factor (AIF) is a small protein that resides normally within the intermembrane space of mitochondria, and upon certain death stimuli translocates into the cytosol and ultimately the nucleus where it contributes to DNA fragmentation and chromatin condensation. The death-promoting effects of AIF have been shown to be independent of caspase activation, although the mitochondrial release of AIF has been shown to be Bax dependent. A role for AIF in cell death has been implicated in many neurodegenerative diseases including AD, PD, and ALS. Calpains are cytosolic proteases that are activated upon ER stress-induced intracellular calcium influx. While calpains have been shown to act downstream of caspases, many studies indicate caspase-independent, calpain-induced cell death. Calpain activation has been implicated in AD, PD, HD, and ALS. For instance, in an *in vitro* model of AD, the activation of calpains has been implicated in the A$\beta$-induced, caspase-independent degeneration of axons. $\alpha$-Synuclein has also been shown to be a substrate of activated calpain, which may lead to its aberrant deposition in PD.

The increased aggregation and deposition of specific cytoplasmic proteins in neurodegenerative diseases has implicated a role for disturbances in lysosomal function and/or type II autophagic cell death. Autophagy is a normal cellular process whereby vesicles shuttle cytoplasmic and organellar macronutrients through a series of fusion events to the lysosome for ultimate proteolytic degradation and recycling. Autophagic cell death is characterized by an aberrant accumulation of vacuoles that result either as a response to metabolic stress or to a disruption in lysosomal function, concomitant with morphological features of both apoptosis and necrosis. An accumulation of autophagic vacuoles has been clearly documented in AD, PD, and HD, which suggests that aberrant autophagic stress plays a significant role in the pathogenesis of these diseases.

Age-related lysosome disturbances may contribute to a compromise in protein degradation that allows for the accumulation of specific proteins in neurodegenerative disease. In addition, a compromise in the integrity of lysosome membranes may also induce leakage of lysosomal proteases, such as cathepsins, into the cytosol. Cathepsins have been shown to induce Bid- and Bax-dependent activation of the intrinsic apoptotic pathway and Bax-dependent release of AIF into the cytosol. In addition, cathepsins can also act as efficient effector proteases similar to that of effector caspases, thus creating another potential mechanism of caspase-independent death in neurodegenerative disease.

## Summary

Apoptotic death is a well-characterized and tightly orchestrated cell biological process whose involvement in normal brain development is unquestionable. Apoptotic neurons have been detected to varying degrees in virtually all human neurodegenerative diseases, and alterations in the expression of numerous apoptosis-associated genes and proteins have been widely reported. However, the significance of these findings is unclear, and separating correlation from causation has proven difficult. It remains possible that apoptosis-associated molecules and sublethal intracellular events contribute to neuron dysfunction early in disease pathogenesis, prior to significant loss of neurons, but definitive data on this conjecture are currently lacking for most neurodegenerative diseases. Neuroprotective agents that can inhibit both apoptotic and nonapoptotic neuron death may ultimately be used in conjunction with drugs designed to decrease disease-specific death stimuli, resulting in effective combination therapy to preserve neuron function in AD, PD, HD, ALS, and other neurodegenerative diseases.

*See also:* Alzheimer's Disease: Neurodegeneration; Animal Models of Motor and Sensory Neuron Disease; Animal Models of Alzheimer's Disease; Animal Models of Huntington's Disease; Autonomic and Enteric Nervous System: Apoptosis and Trophic Support During Development; Huntington's Disease: Neurodegeneration; Parkinson's Disease: Alpha-Synuclein and Neurodegeneration.

## Further Reading

Boatright KM and Salvesen GS (2004) Mechanisms of caspase activation. *Current Opinion in Cell Biology* 15: 725–731.

Clarke PG (1990) Developmental cell death. *Anatomy and Embryology* 181: 195–213.

Ellis HM and Horvitz HR (1986) Genetic control of programmed cell death in the nematode *C. elegans*. *Cell* 44: 817–829.

Hickey MA and Chesselet MF (2003) Apoptosis in Huntington's disease. *Progress in Neuro-Psychopharmacology and Biological Psychiatry* 27: 255–265.

Kerr JF, Wyllie AH, and Currie AR (1972) Apoptosis: A basic biological phenomenon with wide-ranging implications in tissue kinetics. *British Journal of Cancer* 26: 239–257.

Putcha GV, Harris CA, Moulder KL, et al. (2002) Intrinsic and extrinsic pathway signaling during neuronal apoptosis: Lessons from the analysis of mutant mice. *Journal of Cell Biology* 157: 441–453.

Roth KA (2001) Caspases, apoptosis and Alzheimer disease: Causation, correlation, and confusion. *Journal of Neuropathology and Experimental Neurology* 60: 829–838.

Sathasivam S, Ince PG, and Shaw PJ (2001) Apoptosis in amyotrophic lateral sclerosis: A review of the evidence. *Neuropathology and Applied Neurobiology* 27: 257–274.

Shacka JJ and Roth KA (2004) Regulation of neuronal cell death and neurodegeneration by members of the Bcl-2 family: Therapeutic implications. *Current Drug Targets – CNS and Neurological Disorders* 4: 25–39.

Tatton WG, Chalmers-Redman R, Brown D, and Tatton N (2003) Apoptosis in Parkinson's disease: Signals for neuronal degradation. *Annals of Neurology* 53: S61–S70.

# Autonomic and Enteric Nervous System: Apoptosis and Trophic Support during Development

**R O Heuckeroth**, Washington University School of Medicine, St. Louis, MO, USA



## Introduction

Programmed cell death (PCD) is an evolutionarily conserved process in multicellular organisms that is important for morphogenesis during development and for the maintenance of tissue homeostasis in organs with ongoing cell proliferation. PCD also occurs in many tissues in response to injury. Apoptotic cell death is one form of PCD that provides a mechanism to remove individual cells from tissues without damage to adjacent cells. Defects in molecular mechanisms that regulate cell death cause both cancer and abnormal development. Because many aspects of apoptosis have been reviewed over the past decade, this article focuses on the developmental role of apoptosis in the sympathetic, parasympathetic, and enteric nervous system and provides only a brief overview of apoptotic mechanisms.

## Apoptotic Cells and Molecular Mechanisms

Apoptosis is a form of cell death characterized by cell shrinkage, plasma membrane blebs, nuclear chromatin condensation, and DNA fragmentation. Phosphatidylserine, usually restricted to the inner face of the plasma membrane in live cells, also partially redistributes to the extracellular face of the plasma membrane. These changes largely result from activation of a set of cysteine-aspartyl-specific proteases called caspases. Caspases can be activated via either an extrinsic pathway or an intrinsic pathway. The extrinsic pathway is activated by binding of extracellular ligands to cell surface death receptors like Fas (Apo-1 or CD95), tumor necrosis factor receptor-1 (TNFR-1/p55/CD120a), interferon receptor, and TRAIL (TNF-related apoptosis-inducing ligand or Apo2-L) receptor. This binding leads to the recruitment of a variety of proteins that activate procaspase-8 or procaspase-10. In contrast, the intrinsic pathway is characterized by cytochrome c release from mitochondria. Cytochrome c in the cytoplasm induces heptamerization of Apaf-1, that then binds to and activates procaspase-9. These initiator caspases (-8, -9, and -10) activate the effector caspases (caspase-3 and caspase-7). The effector caspases have many substrates (estimated to include 0.5–5% of cellular proteins), induce DNA cleavage and mitochondrial permeabilization, and activate additional proteolytic cascades that eventually result in cell death.

## Apoptosis in the Nervous System and the Discovery of Neurotrophic Factors

PCD is common in most parts of the developing nervous system, where 20–80% of all neurons produced during embryogenesis die before adulthood. Typically, these cells are eliminated by PCD at the time that they would normally innervate their targets. In this setting, PCD is thought to occur to allow matching of the neuronal population to the size of the innervation target via target-derived, trophic factor-dependent cell survival. This mechanism also efficiently eliminates cells with abnormal migration or abnormal axon targeting because cells that fail to innervate appropriate targets do not receive adequate trophic factor. The observation that developing sympathetic and sensory neurons were dependent on exogenous trophic factor for survival led to the discovery of nerve growth factor (NGF) by Rita Levi-Montalcini and Stanley Cohen. This discovery prompted work that in turn resulted in the discovery of the related proteins brain-derived neurotrophic factor (BDNF), neurotrophin-3 (NT-3), and NT-4/5. An *in vitro* sympathetic neuron survival assay was also critical for discovery of neurturin (NRTN) and for the identification and characterization of the related proteins glial cell line-derived neurotrophic factor (GDNF), artemin (ARTN), and persephin (PSPN). Furthermore, molecular mechanisms of neonatal sympathetic neuron survival *in vitro* have been studied extensively and form the basis of much of what we know about neuronal apoptosis. In the rat, for example, superior cervical ganglion (SCG) sympathetic neurons are NGF dependent from embryonic day (E)16 to postnatal day (P) 7. NGF deprivation during this time results in PCD within 48 h. With this assay, investigators demonstrated that apoptosis is triggered after NGF withdrawal by increased c-Jun protein levels and c-Jun phosphorylation, increased translocation of the proapoptotic protein Bax from the cytosol to mitochondria, and Bax-induced cytochrome c release that causes apoptosis via the intrinsic pathway. Release of Smac/DIABLO and HtrA2/Omi from mitochondria is also important to relieve caspase inhibition by inhibitor of apoptosis proteins so that caspases can become activated.

## Lessons from Mutant Mice

The neurotrophic factors NGF, BDNF, NT-3, GDNF, NRTN, and ARTN are critical for the normal development and maintenance of the peripheral nervous system. Depending on the specific cell type and developmental time point, these factors determine nervous system structure and function via a number of distinct effects including preventing or in some cases promoting cell apoptosis. These proteins have other important roles as well:

- Promotion of cell survival
- Promotion of cell proliferation
- Support for neuronal differentiation
- Promotion of directed cell migration
- Promotion of neurite extension
- Axon guidance

Each of these actions is mediated by binding of trophic factors to specific cell surface receptors. There is also some ability for specific trophic factors to activate multiple receptors at least *in vitro*. Preferred receptor–ligand interactions, alternate receptor–ligand interactions, and a summary of interactions that are important *in vivo* are presented in **Figure 1**. Briefly, NGF, BDNF, and NT-3 bind to and activate the transmembrane tropomyosin-related kinase (Trk) receptors TrkA, TrkB, and TrkC, respectively. In addition, NT-4 activates TrkB, whereas NT-3 can directly activate TrkA and TrkB, but with lower efficiency than it activates TrkC. All these NTs also bind to the low-affinity receptor p75NTR. The interaction with p75NTR inhibits the activation of Trk receptors by their nonpreferred ligands and in some cases improves intracellular signaling. However, in the absence of specific Trk receptor activation, binding of proneurotrophins to p75NTR promotes cell death. For example, pro-BDNF binding to p75NTR on sympathetic neurons, which express TrkA but not TrkB, promotes cell death, but NGF promotes cell survival by activating both TrkA and p75NTR. This dual role for p75NTR in sympathetic neurons is thought to improve the specificity of axon targeting by eliminating cells whose axons innervate the wrong target and arrive at a source of BDNF but not of NGF. The proteins GDNF, NRTN, ARTN, and PSPN form a separate family of trophic factors called the GDNF-related ligands (GFLs). GFLs all activate the transmembrane tyrosine kinase Ret. Instead of binding directly to Ret, however, GFLs activate Ret by binding to a glycosylphosphatidylinositol-linked cell surface receptor (GFR$\alpha$1–4). Each GFL preferentially binds to a specific GFR$\alpha$ protein, with GDNF, NRTN, ARTN, and PSPN interacting best with GFR$\alpha$1, GFR$\alpha$2, GFR$\alpha$3, and GFR$\alpha$4, respectively. There is some *in vitro* cross-talk between GFLs and nonpreferred GFR$\alpha$ proteins, as indicated in **Figure 1**, but these interactions generally require higher trophic factor concentrations and do not appear to be physiologically important *in vivo*. However, receptor

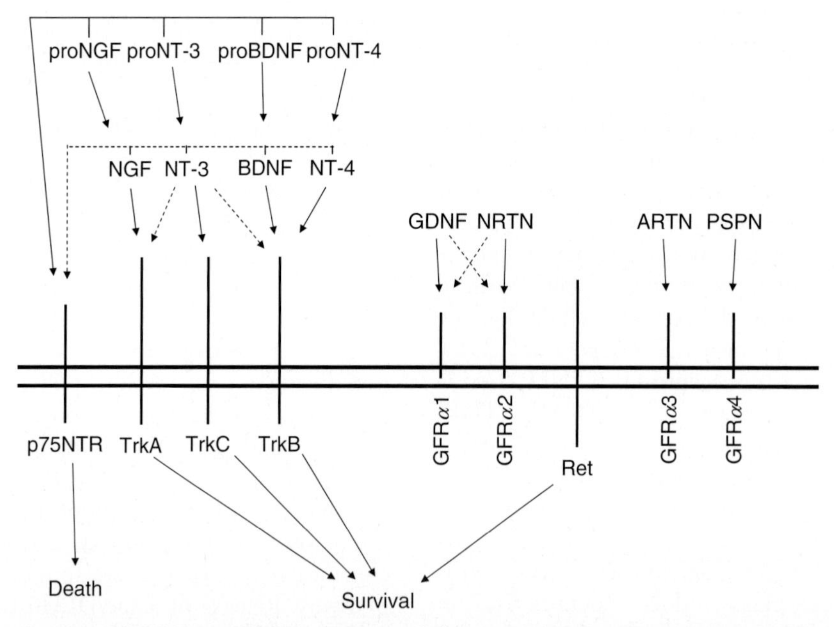


**Figure 1** Summary of neurotrophic factors and their receptors. Solid arrows show preferred receptor–ligand interactions. Dashed arrows show nonpreferred interactions between ligands and receptors. For neurturin (NRTN) and glial cell line-derived neurotrophic factor (GDNF), the nonpreferred interactions do not appear to be important *in vivo*. For neurturin (NT)-3, nonpreferred interactions are important *in vivo*. Interactions of neurotrophins with p75NTR are also likely to be physiologically relevant. ARTN, artemin; BDNF, brain-derived neurotrophic factor; GFR, growth factor receptor; NGF, nerve growth factor; PSPN, persephin.

cross-talk appears important in the pharmacological response to exogenously administered factors, presumably acting at supraphysiological levels.

Given the complexity of these systems, determining the relevance of specific trophic factors and their receptors for neuronal development and function has required a combination of *in vitro* studies and the analysis of mutant animals. While many conclusions can be drawn from this work, several important themes have emerged:

1. Patterns of trophic factor receptor and ligand expression correlate well with trophic factor dependence *in vivo* but incompletely predict function.
2. For many developing neurons, trophic factor dependence changes during development such that cells initially dependent on one factor become dependent on a different trophic factor at a later developmental stage.
3. Trophic factors may promote both survival and proliferation early in development or may direct precursor migration. At later stages, these factors may support survival (e.g., at the time of target innervation) or provide trophic support without being essential for survival.
4. Trophic factor dependence *in vitro* does not imply developmental stage-specific programmed cell death *in vivo*. Cell death *in vivo* requires both trophic factor dependence and trophic factor deficiency.
5. Trophic factor receptor expression also does not imply trophic factor-dependent survival since specific cell populations may respond to multiple trophic factors simultaneously.

A few examples will clarify the strategies employed to determine cell number in the autonomic and enteric nervous system (ENS). Effects of specific receptor and ligand mutations on sympathetic, parasympathetic, and enteric neurons are summarized in **Table 1**.

## Trk Receptors and Ligands Regulate Cell Death *In Vivo* in the Sympathetic Nervous System

The sympathetic chain in the mouse embryo arises at E11.5, and at that time, TrkC is expressed, but TrkA is not detectable. By E13.5, when the SCG forms, both TrkC and TrkA are detected in the SCG. TrkA expression becomes more robust by E15.5 and continues to be expressed at high levels throughout postnatal development. TrkC levels in the SCG fall significantly by P0, when only a few cells have detectable TrkC expression. In contrast, TrkB is not detected in sympathetic neurons at any developmental stage.

The pattern of TrkA receptor gene expression correlates well with the normal role of TrkA and NGF in development of the mouse sympathetic nervous system. In TrkA−/− mice, for example, neuron number is normal in the sympathetic chain at E11.5 and in the SCG and sympathetic chain at E13.5. By E15.5, there are 15% fewer SCG neurons in TrkA-deficient mice than wild-type (WT). By E17.5, TrkA−/− mice had 35% fewer SCG neurons than WT mice did, and almost all SCG neurons are lost in TrkA−/− mice by P9. At each of the periods investigated, the reduction in SCG neuron number in TrkA−/− mice was attributable to increased cell death. Similarly, in NGF-deficient mice, all SCG neurons are absent by P14, and increased cell death was observed at P3. These changes in neuron number demonstrate the critical role for NGF/TrkA signaling in sympathetic neuron survival and are consistent with results obtained in the 1960s with exposure to neutralizing antibodies to NGF. They also define a period of trophic factor-dependent cell death that correlates well with the timing of trophic factor dependence of SCG neurons in primary culture. Moreover, the striking similarity of the SCG phenotype for NGF−/− and TrkA−/− mice highlights the importance of this preferred ligand–receptor interaction.

In contrast to these results with TrkA−/− and NGF−/−mice, however, the effect of TrkC or NT-3 deficiency is less easily predicted based on gene expression patterns. For example, although TrkC is expressed in the sympathetic nervous system as early as E11.5, the SCG is normal at all developmental stages in TrkC−/− mice. This contrasts with the 50% reduction in SCG neurons found in NT-3-deficient mice. The time during development when NT-3 is required for sympathetic neuron survival, however, has been controversial, with an early report suggesting increased SCG precursor cell death between E11 and E17 and a later study using more animals failing to reproduce this observation and demonstrating increased programmed cell death after birth. Furthermore, injection of anti-NT-3 antibodies into neonatal rats caused a 60–80% loss of SCG neurons, suggesting a role for NT-3 in postnatal SCG survival. Together these observations demonstrate the complexity of determining how specific trophic factors regulate neuronal apoptosis, especially in cells with multiple trophic factor receptors. They further suggest that although NT-3 critically regulates SCG apoptosis, it may perform this function primarily via TrkA receptor activation.

The role TrkB and BDNF in SCG neuron survival is even more remarkable. Since TrkB is not expressed in the sympathetic nervous system, it might initially have been predicted that the SCG would be normal in

**Table 1** Neuron number in autonomic and enteric ganglia of mutant mice

| | *Sympathetic* | | *Parasympathetic* | | | | |
|---|---|---|---|---|---|---|---|
| | *SCG* | *Chain ganglia* | *Ciliary* | *Submandibutar* | *Sphenpalatine* | *Otic* | *Enteric* |
| TrkA KO | 100% loss by P9 | | | | | | |
| NGF KO | 100% loss by P7 | | | | | | |
| TrkB KO | Normal | | | | | | |
| BDNF KO | 36% increase | | | | | | |
| TrkC KO | Normal | | | | | | Selective neuron loss |
| NT-3 KO | 50% loss | | | | | | Selective neuron loss |
| p75NTR KO | Increased | | | | | | |
| Ret KO | Abnormal, but variable | Abnormal | 48% loss | 30–47% loss | 99–100% loss | 99–100% loss | 99% loss |
| GFRα1 KO | Normal | | | 33% loss | 99–100% loss | 99–100% loss | 99% loss |
| GDNF KO | 30% loss | | 40% loss | 36% loss | 99% loss | 86% loss | 99% loss |
| GFRα2 KO | Normal | | | 42% loss | Normal | 40% loss | |
| NRTN KO | Normal | | 50% loss | 45% loss | Normal | Normal | Normal |
| GFRα3 KO | Abnormal, but variable | Abnormal | | | | | |
| ARTN KO | Abnormal, but variable | Abnormal | | | | | Normal |

BDNF−/− mice. Remarkably, SCG neuron number is increased by 36% in BDNF−/−compared to WT animals at P15. This was hypothesized to occur because pro-BDNF promotes sympathetic neuron cell death by binding to p75NTR. Indeed, WT SCG neuron number decreases by 42% because of PCD between P0 and P23, but SCG neuron number increases between P0 and P23 in P75NTR−/− mice. These results suggest that both BDNF and p75NTR are required for naturally occurring cell death in the SCG of WT mice. Further support for this hypothesis is the observation that mice deficient in both p75NTR and TrkA have markedly reduced sympathetic neuron apoptosis compared to TrkA−/− animals. This mechanism allows for efficient elimination of cells with misguided axons that encounter pro-BDNF and reinforces the importance of TrkA activation for SCG survival.

## Ret Signaling Promotes Sympathetic Neuron Precursor Migration and Supports Axon Extension Required for Sympathetic Neurons to Innervate Targets and Encounter Trk Ligands

Ret activation via GFRα3 and ARTN is essential for normal sympathetic nervous system development. In this case, however, Ret signaling appears to be important for neuronal precursor migration and axon extension, but not directly for cell survival. This conclusion is based on the following observations. Ret−/− mice have smaller-than-normal SCG ganglia that appear in a more caudal position than in WT animals. Furthermore, sympathetic fiber innervation of targets in the nasal mucosa, eye, and skin of Ret−/− mice is almost completely absent, while sympathetic innervation density of the submandibular salivary gland is significantly reduced. In addition, there are defects all along the sympathetic chain in Ret−/− mice, including smaller-than-normal ganglia, abnormally located ganglia, and reduced target innervation. Detailed evaluation of the mechanism of these defects demonstrated that Ret−/− sympathetic neuron precursor lineage commitment appears normal, but differentiation of these cells is delayed, and neuronal cell death increases between E16.5 and P0 compared with WT mice. This increased cell death was particularly interesting since Ret is expressed in all sympathetic precursors at E11.5, but most of these cells lose Ret expression by E15.5. Furthermore, although increased cell death occurred in Ret−/− SCG, it did not occur preferentially in Ret-expressing cells.

Additional analysis demonstrated that the Ret ligand ARTN is expressed in blood vessels along the normal route of sympathetic axons where ARTN potently stimulates SCG axon growth and is a chemoattractant that directs axon pathfinding. It is interesting that ARTN- and GFRα3-deficient mice have variable defects in the SCG. Specifically, SCG size was near normal in ARTN−/− and GFRα3−/− mice if the ganglia were normal and in position and peripheral targets were innervated, but SCG size was markedly diminished and apoptosis was increased in mice with abnormally located SCG and reduced peripheral target innervation. Defects in ARTN−/− and GFRα3−/− mice were detected as early as E10.5, when sympathetic neuron precursors emerge from the neural crest, demonstrating an important role for Ret/ARTN/GFRα3 signaling in precursor migration and neurite extension. Overall, these data suggest that increased cell death in the sympathetic nervous system of Ret−/−, ARTN−/−, and GFRα3−/− mice, compared with WT mice, occurs because of failure to properly innervate targets and subsequent deficiency in target-derived trophic factor (i.e., NGF deficiency). The Ret ligand GDNF also appears to be important for SCG development; since GDNF−/− mice have been reported to have a 30% loss of SCG neurons. Although GDNF supports the survival of cultured SCG neurons, the precise role of GDNF for SCG neuron survival *in vivo* has not been defined. Furthermore, mice missing the preferred GDNF receptor GFRα1 have a normal SCG at P0, which suggests that GDNF effects on the SCG are mediated via an alternate signaling pathway. These analyses demonstrate the complexity of interpreting mouse phenotypes and primary culture data and the importance of careful mechanistic time-course studies to define the role of specific trophic factors and their receptors in neuron survival. They also demonstrate robustly that programmed cell death in response to trophic factor deficiency at the time of target innervation provides a powerful mechanism to ensure survival of only the sympathetic neurons that correctly innervate their targets.

## Programmed Cell Death in the Parasympathetic Nervous System

Cell death in parasympathetic neurons is much less well studied than in sympathetic neurons, but these cells do not appear to rely on neurotrophins (NGF, BDNF, or NT-3) for survival. Furthermore, although ciliary ganglion cell survival is supported by ciliary neurotrophic factor (CNTF) *in vitro*, CNTF does not appear to perform this function *in vivo* during normal development. Instead, Ret activation by GDNF and NRTN appears to be the most important determinant of parasympathetic neuron number. Unfortunately, a detailed analysis of cell apoptosis and proliferation within the parasympathetic nervous system of mutant mice has not been performed for most cell populations. Nonetheless, several important conclusions can be drawn about the role of GDNF and NRTN in parasympathetic neurons:

1. Different parasympathetic neuron populations respond differently to the loss of Ret signaling. For example, otic and sphenopalatine ganglia are essentially completely absent from Ret−/− and GDNF−/− mice, but the reduced cell number in the ciliary and submandibular ganglia of these mice is much less dramatic (**Table 1**).
2. Reduced number of ganglion cells does not imply increased apoptosis. This is demonstrated by analysis of the sphenopalatine ganglion in GDNF and GFRα1−/− mice. In these animals, the sphenopalatine ganglia are absent as early as E12.5. Furthermore, bromodeoxyuridine labeling in GDNF−/− mice demonstrated reduced proliferation of sphenopalatine ganglion cell precursors, but terminal transferase deoxyuridine triphosphate nick end labeling (TUNEL) analysis failed to demonstrate any increase in cell death. Thus, at least in the sphenopalatine ganglion, GDNF/Ret signaling is required for precursor proliferation and not for survival of mature neurons.
3. Reduced parasympathetic ganglion cell survival may result from abnormal ganglion cell precursor migration. This is illustrated by the otic ganglion in GDNF−/− mice. Like the sphenopalatine ganglion, most otic ganglion cells are absent in GDNF−/− animals. In this case, however, otic ganglion cell precursors migrate abnormally, and TUNEL staining demonstrated increased cell death in the abnormally migrating precursors.

Thus, while programmed cell death may be important for parasympathetic nervous system development, it is much less well understood than in the sympathetic nervous system. In particular, mechanisms for eliminating cells with abnormal axon targeting have not been well documented in the parasympathetic nervous system. Indeed, target innervation by parasympathetic sphenopalatine neurons is dramatically reduced in NRTN−/− and GFRα2−/− mice, but sphenopalatine ganglion numbers are normal, which suggests that these cells are not dependent on target-derived trophic factors for survival.

## Mechanisms Governing Neuron Number in the ENS

The ENS is a complex network of neurons and glia within the bowel wall that controls intestinal motility, responds to sensory stimuli, and regulates intestinal secretion and blood flow. To perform these functions, there are roughly as many neurons in the ENS as in the spinal cord, and the ENS comprises many distinct neuron subtypes that differ in function, transmitter phenotype, and pathways of axon pathfinding. Mechanisms of axon targeting in the ENS, however, and mechanisms to ensure that enteric neurons have correctly innervated their targets, are not yet understood. Furthermore, the ENS presents challenges that do not occur in other regions of the nervous system. This is especially true within the myenteric plexus since specific subtypes of myenteric neurons must extend their axons either orally (toward the mouth) or aborally (toward the end of the bowel) for the gut to function normally. Although there must be axon guidance cues present during development to direct these axons, adjacent regions of the bowel wall appear remarkably similar in the mature organism. Thus, unlike the sympathetic nervous system, where targets of innervation are far from the neuronal cell bodies and target-derived trophic factor dependence is an excellent mechanism for ensuring that only correctly targeted neurons survive, it is difficult to imagine how apoptosis could be used in the ENS to eliminate neurons whose axons project in the wrong direction. It is easier to imagine that apoptotic pathways could be important for enteric neurons projecting outside the muscular gut wall (e.g., to villi), but this has not yet been investigated.

With these ideas in mind, it is perhaps not surprising that apoptosis does not appear to occur within the developing or adult ENS of WT mice. This is not to suggest that enteric neurons are trophic factor-independent. Indeed, TrkA, TrkB, TrkC, p75NTR, Ret, GFRα1, and GFRα2 are all expressed within the ENS. Furthermore, enteric neurons undergo apoptotic cell death in primary culture when they are deprived of trophic factors and clearly respond to a variety of neuronal survival factors, including GDNF, NRTN, and NT-3, *in vitro*. In addition, Ret−/−, GFRα1−/−, and GDNF−/− mice miss essentially all enteric neurons from small bowel and colon. In fact, defective Ret signaling is the most commonly identified etiology of distal colon aganglionosis in humans (i.e., Hirschsprung disease). NT-3- and TrkC-deficient animals also have abnormal ENS development, with striking reduction in some subpopulations of enteric neurons. In contrast to these results, NRTN−/− and GFRα2−/− mice have a normal density of enteric neurons, but as in the parasympathetic nervous system, enteric neurons of NRTN−/− and GFRα2−/− mice are smaller than normal, with reduced neuronal projections, at least in some subtypes of neurons. Finally, apoptosis commonly occurs in ENS precursors of mice with a variety of mutations that cause intestinal aganglionosis, including in Ret−/−,

$Sox10^{Dom}/Sox10^{Dom}$, and Phox2b−/− animals. Thus, since ENS precursors depend on trophic factors for survival, but apoptosis does not appear to occur during normal ENS development, these factors must be produced in an adequate supply in WT animals to prevent programmed cell death. This further implies that ENS precursor proliferation must be carefully regulated to avoid producing more neurons than can be supported by available trophic factor. One way this occurs is that the availability of GDNF directly determines the rate of ENS precursor proliferation. Both increases and decreases in GDNF availability alter enteric neuron number via changes in precursor mitotic rates.

## Apoptosis in Neuronal Injury

In addition to the physiologic role of apoptosis during normal development, cellular injury may also cause cell death via apoptotic pathways. In the ENS, for example, apoptosis has been reported in age-related myenteric neuron loss, anti-HuD-associated paraneoplastic syndrome, colitis-induced neuronal injury, and diabetes-associated ENS injury. Furthermore, at least in mouse models of diabetes, apoptosis can be reduced by providing additional trophic factor (GDNF) *in vivo*. It is interesting that in contrast to the effect of diabetes on the ENS, sympathetic neuron cell death does not appear to occur in diabetic rats. Thus, once again, the importance of apoptotic pathways in the autonomic nervous system and the ENS is specific to the age of the animal, the apoptotic trigger, and the neuronal subtypes evaluated. Presumably, these differences in the extent of apoptosis in different regions of the nervous system and at different times during life reflect both the availability of trophic factors to support survival and the abundance of intracellular pro- and anti-apoptotic proteins.

## Summary

Programmed cell death is important for tissue morphogenesis during development and for the maintenance of tissue homeostasis during adult life. In part, apoptosis is valuable because it provides a way for the organism to specifically eliminate single cells that are no longer needed. This is important in the nervous system, where correct axon targeting and matching of the target size to the number of innervating nerve fibers is critical for function. Because both too much and too little cell death could be detrimental, carefully regulated intracellular and extracellular control mechanisms have been established to control PCD. Many of these mechanisms have been studied in detail in the sympathetic nervous system, where target-derived trophic factors are required for neuron survival during a defined developmental period when neurons are innervating their targets. Because trophic factors are active at the axon tip and are produced in the axon target this strategy effectively eliminates neurons that fail to innervate an appropriate target. Similar strategies are employed in most regions of the nervous system.

Mechanisms of PCD in the autonomic nervous system and the ENS highlight several important themes. First, for PCD to be useful for tissue morphogenesis, the axon tip and neuron cell body must be in different environments. For this reason, PCD tends to occur as neurons are innervating their final targets instead of when they first begin to extend axons. Because many parasympathetic neuron cell bodies are embedded in their targets and many enteric neurons project axons to targets in an environment similar to that around the cell body, it is more difficult to see how PCD could be used in the parasympathetic nervous systen and the ENS to ensure proper axon targeting. Indeed, it is difficult to find evidence that PCD occurs as a part of normal development in either the ENS or the parasympathetic nervous system. Second, for PCD to be effective, different targets need to produce different trophic factors, and subsets of neurons must respond to only a limited array of trophic factors. Even more effective would be a strategy to actively eliminate cells exposed to the 'wrong' trophic factor. This explains the variety of trophic factors and receptors present in the nervous system and the role of p75NTR to induce cell death in the absence of Trk receptor activation. Indeed, given the wide array of neuron subtypes and targets, it is remarkable that the nervous system can be established with so few trophic factors and receptors. It is probably for this reason that developing neurons may respond to a combination of trophic factors or change trophic factor dependence during development. Finally, most postmitotic neurons are needed throughout the life of the organism. For this reason, resistance to apoptosis in neurons that have correctly innervated their targets is important for longevity. In the nervous system, this is accomplished by limiting trophic factor-dependent cell survival to a small developmental period. It is important to note that this requires changes in the intracellular machinery for apoptosis once targets are innervated.

Tremendous advances have been made over the past decade in understanding the role of apoptosis in development and disease, but many challenges remain. In particular, it would be valuable to develop additional strategies to prevent neuronal cell death after injury. Ideally, these strategies should target specific cell populations since global inhibition of apoptosis is likely to be carcinogenic. Continued detailed analysis of the molecular mechanisms of apoptosis and survival in different defined cell populations is therefore critical to allow targeted therapy and reduce disease-related morbidity and mortality.

*See also:* Apoptosis in Neurodegenerative Disease; Neurotrophic Factor Therapy: GDNF and CNTF; Programmed Cell Death.

## Further Reading

Airaksinen MS and Saarma M (2002) The GDNF family: Signalling, biological functions, and therapeutic value. *Nature Reviews Neuroscience* 3(5): 383–394.

Baloh RH, Enomoto H, Johnson EMJ, and Milbrandt J (2000) The GDNF family ligands and receptors: Implications for neural development. *Current Opinion in Neurobiology* 10: 103–110.

Enomoto H, Crawford PA, Gorodinsky A, Heuckeroth RO, Johnson EM Jr, and Milbrandt J (2001) RET signaling is essential for migration, axonal growth, and axon guidance of developing sympathetic neurons. *Development* 128(20): 3963–3974.

Enomoto H, Heuckeroth RO, Golden JP, Johnson EM Jr, and Milbrandt J (2000) Development of cranial parasympathetic ganglia requires sequential actions of GDNF and neurturin. *Development* 127: 4877–4889.

Fagan AM, Zhang H, Landis S, Smeyne RJ, Silos-Santiago I, and Barbacid M (1996) TrkA, but not TrkC, receptors are essential for survival of sympathetic neurons *in vivo*. *Journal of Neuroscience* 16(19): 6208–18.

Francis N, Farinas I, Brennan C, et al. (1999) NT-3, like NGF, is required for survival of sympathetic neurons, but not their precursors. *Developmental Biology* 210(2): 411–427.

Gewies A (2003) ApoReview: Introduction to apoptosis. http://www.celldeath.de/encyclo/aporev/aporev.htm (accessed 3 August 2007).

Gianino S, Grider JR, Cresswell J, Enomoto H, and Heuckeroth RO (2003) GDNF availability determines enteric neuron number by controlling precursor proliferation. *Development* 130(10): 2187–2198.

Honma Y, Araki T, Gianino S, et al. (2002) Artemin is a vascular-derived neurotropic factor for developing sympathetic neurons. *Neuron* 35(2): 267–282.

Majdan M, Walsh GS, Aloyz R, and Miller FD (2001) TrkA mediates developmental sympathetic neuron survival in vivo by silencing an ongoing p75NTR-mediated death signal. *Journal of Cell Biology* 155(7): 1275–1285.

Newgreen D and Young HM (2002) Enteric nervous system: Development and developmental disturbances: Part 2. *Pediatric and Developmental Pathology* 5(4): 329–349.

Oppenheim RW (1991) Cell death during development of the nervous system. *Annual Review of Neuroscience* 14: 453–501.

Putcha GV, Harris CA, Moulder KL, Easton RM, Thompson CB, and Johnson EM Jr. (2002) Intrinsic and extrinsic pathway signaling during neuronal apoptosis: Lessons from the analysis of mutant mice. *Journal of Cell Biology* 157(3): 441–453.

Putcha GV and Johnson EM Jr. (2004) Men are but worms: Neuronal cell death in *C. elegans* and vertebrates. *Cell Death and Differentiation* 11(1): 38–48.

Reichardt LF (2006) Neurotrophin-regulated signalling pathways. *Philosophical Transactions of the Royal Society of London, Series B: Biological Sciences* 361(1473): 1545–1564.

Sastry PS and Rao KS (2000) Apoptosis and the nervous system. *Journal of Neurochemistry* 74(1): 1–20.

Wikipedia (2007) Apoptosis. http://en.wikipedia.org/wiki/Apoptosis (accessed 3 August 2007).

Yuan J and Yankner BA (2000) Apoptosis in the nervous system. *Nature* 407(6805): 802–809.

# Proteasome Role in Neurodegeneration

**L Korhonen, N Putkonen, and D Lindholm,**
Minerva Foundation Institute for Medical Research, Helsinki, Finland



## Introduction

Recent studies have shown the critical role of the proteasome system in different human neurological disorders. Herein, we review the current knowledge of the role of the proteasome in neuronal and axonal degeneration and discuss different ubiquitin E3 ligases, deubiquitinating enzymes (DUBs), and signaling pathways controlling the activity of the proteasome in these disorders.

## The Proteasome

Proteasomes are the main proteolytic machinery in cells and exhibit trypsin-like, chymotrypsin-like, and peptidyl-glutamyl peptide hydrolyzing enzymatic activity. The proteasomes are present as 26S and 20S particles in the cell. The 26S proteasome consists of a catalytic 20S core and two 19S regulatory subunits (19RS) at both ends of the 20S subunit forming a barrel-like structure. The 19RS consists of several proteins that mediate substrate recognition and delivery to the core complex for proteolytic degradation into short peptides. The 19RS also removes ubiquitin from proteins ensuring optimal levels of free ubiquitin in the cell. The 20S proteasome cleaves peptides and unfolded proteins without the regulatory unit. Apart from protein degradation, additional nonproteolytic functions have been ascribed to the proteasome. Thus, it was shown in *Saccharomyces cerevisiae* that the 19RS regulates gene transcription through an interaction with the activator protein, Gal4. In addition, 19RS promotes the Spt-Ada-Gcn5-acetyltransferase (SAGA) histone acetyltransferase complex to target promoters during gene activation.

In neurons as well as in other cells, the proteasomes are present in the nucleus, in the cell soma, as well as in the dendrites and the axons. Proteins destined for proteasomal degradation are marked with a chain of ubiquitin molecules in a three-step reaction catalyzed by the E1, E2, and E3 enzymes (see **Figure 1**). In mammalian cells, there is one E1 and a couple of E2's, but many more E3 ubiquitin ligases that can be divided into four major classes: the homologous to E6-AP C-terminus (HECT), Really Interesting New Gene (RING), plant-homeodomain (PHD), and UFD2-homology domain (U-box) families. The HECT domain ligases share a conserved C-terminal catalytic domain that forms a thioester bond with ubiquitin during the transfer from E2 to the substrate. The N-terminal regions of HECT domain ligases are involved in substrate specificity and recognition. The RING-finger ubiquitin ligases exist either as multisubunit complexes or as monomers and can interact directly with the E2 and the protein substrate. The PHD and U-box domain containing proteins, as well as the RING-finger E3 ligases, facilitate ubiquitin conjugation by acting as a bridge between the E2's and the protein substrates. In addition to the E3 ligases, the ubiquitin chain elongation factor E4 is involved in the efficient elongation of polyubiquitin chains. These proteins contain a protein motif known as the U-box and may act in concert with the E3 ligases. The C-terminus of Hsc0 interacting protein (CHIP) is a U-box containing protein important for the degradation of misfolded proteins, such as the Pael receptor in PD (see below).

Ubiquitin protein has seven lysine sites to which ubiquitin can be enzymatically linked. Attachment of multiple ubiquitin molecules to lysine 48 (called K48 ubiquitinylation) mediates mainly protein degradation, whereas K63 ubiquitinylation is involved in endocytosis, DNA repair, and in cell signaling. However, proteins can also become monoubiquitinated at different lysine sites as shown, for example, for receptor tyrosine kinases (RTKs), such as the epidermal growth factor receptor (EGFR) (see **Figure 1**).

The levels of ubiquitin are tightly regulated in the cells, among others by the action of the DUBs that remove and recycle ubiquitin from ubiquitin-conjugated proteins (see **Figure 1**). There are approximately 80 DUBs in the human genome that can be divided into five classes based on the ubiquitin protease domain: ubiquitin-specific proteases (USPs), ubiquitin C-terminal hydrolases (UCHs), Otubain proteases (OTUs), and Machado–Joseph disease proteases (MJDs) and JAMM (JAB1/MBN/Mov34). The latter is a metalloprotease whereas the others are cysteine proteases.

During neuronal differentiation the levels of ubiquitin and its conjugates, as well as protein ubiquitination and the activity of the proteasome, all increase suggesting an important function of these processes during development. In the mature nervous system, mutations in different components of the ubiquitin–proteasome system have been linked to the process of neurodegeneration, occurring in many human neurological disorders.

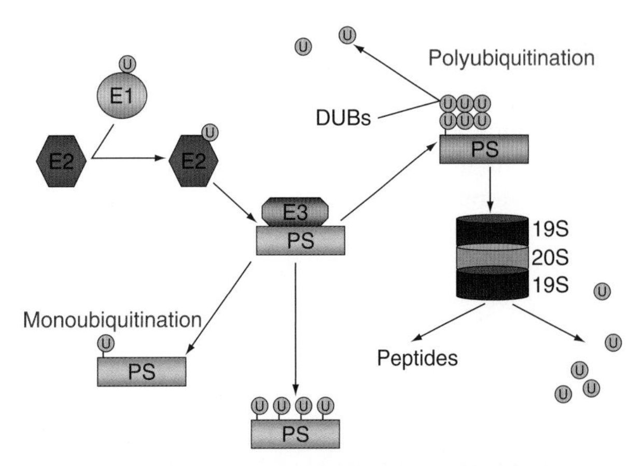


**Figure 1** Schematic view of the function of the ubiquitin–proteasome system (UPS) in protein turnover. Cellular proteins are ubiquitinated by the action of various enzymes acting in concert to add either poly- or monoubiquitinated species to the substrates. Polyubiquitinated proteins are usually doomed for degradation, while monoubiquitinated ones can have other destinies. The regulation of the UPS is complex and dysfunctions may occur in various human diseases. These aspects of the UPS are covered in great detail in recent reviews. U, ubiquitin; E1, ubiquitin-activating enzyme; E2, ubiquitin-conjugating enzyme; E3, ubiquitin ligase; 20S, catalytic core; 19S, regulatory particle; PS, protein substrate; DUB, deubiquitinating enzyme.

## Neuronal Degeneration and the Proteasome

Neurodegenerative diseases, such as Alzheimer's, Parkinson's, and Hungtinton's diseases (AD, PD, and HD), are characterized by the selective loss of neurons in specific brain areas. Distinct insoluble protein aggregates and inclusion bodies are found in the affected brain areas in form of tau-positive neurofibrillary tangles and amyloid $\beta$ (A$\beta$) peptide in the plaques in AD, $\alpha$-synuclein containing Lewy bodies in PD, and intranuclear inclusions of polyQ proteins in HD. The deposits contain misfolded proteins, molecular chaperones, and various components of the ubiquitin–proteasome system, such as ubiquitin, various E3 ligases, the 19RS and the 20S proteasome subunit. The accumulation of misfolded proteins in neurons can disturb cell metabolism in different ways and also cause dysfunction of the proteasome.

Normally in the cell, ubiquitin is generated from the ubiquitin precursor protein consisting of tandem ubiquitin molecules that are cleaved into monomeric ubiquitin by the UCH. However, a mutant form of ubiquitin (Ubiquitin$^{+1}$) has been identified in some neurodegenerative diseases. The ubiquitin$^{+1}$ is generated by a dinucleotide deletion in the mRNA encoding for the ubiquitin B precursor, resulting in a +1 frame shift in the C-terminus of the first encoded ubiquitin molecule. This molecular misreading leads to a 19-amino-acid-long extension of the protein that is resistant to the cleavage by the UCHL. Ubiquitin$^{+1}$ in high concentrations accumulates in neurons and glial cells and can inhibit the proteasome. Studies with human autopsy material and transgenic *ubiquitin*$^{+1}$ mice crossed with transgenic AD and other mouse models have shown that aberrant ubiquitin$^{+1}$ accumulates in AD and in some polyglutamine diseases. Interestingly, ubiquitin$^{+1}$ is not detected in synucleinopathies, such as in Lewy body disease and in multiple system atrophy.

Decreased proteasome activity has been reported in brain autopsy material from PD, AD, and HD patients. In sporadic PD, the 20S proteasome activity is impaired and various components, such as the 19RS and PA28 protein regulator, are expressed at lower levels compared to the controls. In addition, proteasome activity is decreased in neuronal cells after expression of the mutant forms of $\alpha$-synuclein and of other disease-causing proteins. Recent developments in fluorescent-based reporter systems monitoring proteasome activity *in vivo* has provided further evidence for an impaired proteasome function in neurodegenerative diseases. Treatment with chemical compounds inhibiting the proteasome have been reported to induce degeneration of dopaminergic neurons and cause PD-like symptoms in rats, as well as decrease the viability of nerve cells expressing the mutant amyloid precursor protein. In contrast, the proteasome inhibitors are shown to be partly neuroprotective in ischemic brain injuries, showing a complex relationship between the control of neuronal death and survival and the function of the proteasome.

Direct evidence for the role of the proteasome in neurodegeneration comes from studies of some familial and genetic forms of diseases. In PD, deletions and point mutations have been found in the protein Parkin in approximately 50% of patients with the autosomal recessive form of PD (AR-JP). Parkin is a 465-amino-acid-long protein with an N-terminal UBL domain and two C-terminal RING motifs that can function as an E3 ligase. Parkin is a component of the SCF-like E3 complex together with the F-box/WD repeats protein hSel-10 and Cullin. Mutations in Parkin are thought to reduce its ability to degrade substrates and decrease the overall proteasome activity in cells. Several substrates have also been identified for Parkin, such as the Parkin-associated endothelial-like (Pael) receptor, synphilin-1, and the Cell Division Control related protein (CDCrel-1). Pael-1 is a putative G-protein-coupled transmembrane polypeptide that can trigger the unfolded protein response in the endoplasmic reticulum, and Parkin can ubiquitinate and promote degradation of the insoluble receptor.

Synphilin-1 has an unknown function but is associated with $\alpha$-synuclein. CDCrel-1 is a member in the serpin family and has a role in synaptic vesicle dynamics. The mechanism by which loss-of-function of Parkin mutations causes PD is currently under intensive investigation.

Like Parkin, Dorfin is an E3 ligase with the ability to protect motor neurons against degeneration caused by mutant human copper/zinc superoxide dismutase 1 (SOD1) protein in amyotrophic lateral sclerosis (ALS). Dorfin ubiquitinates mutant SOD1 and reduces the occurrence of inclusion bodies of this protein in neurons. Dorfin also interacts with the AAA-ATPase valosine-containing protein (VCP) that can bind both the proteasome and polyubiquitinated proteins. The role of such shuttle proteins in different neurodegenerative diseases warrants more studies. Similarly, the precise role of Dorfin in sporadic ALS that represents the major form of the disease remains to be clarified.

For other diseases, it was shown that the ubiquitin E3 ligase, Ube3A, is mutated in Angelman's syndrome, characterized by mental retardation. Ube3A is crucial for long-term potentiation in mouse hippocampus, suggesting important protein targets in the nerve cells. In addition, it is known that anti-apoptotic proteins, such as the X chromosome-linked inhibitor of apoptosis protein (XIAP), exhibit ubiquitin ligase activity. XIAP is highly expressed in the nervous system and inhibits various caspases and further facilitates ubiquitinylation and degradation of pro-apoptotic proteins, such as Smac/DIABLO and Omi//HrtA2 released from the mitochondria during cell death. XIAP is an important regulator of neuronal cell death as shown in excitotoxicity, ischemia, and after nerve cell axotomy.

## Synaptic Dysfunction and the Proteasome

Apart from the ubiquitin E3 ligases, decreased activity of some DUBs has been linked to neuronal dysfunction. Usp14 is a USP acting on ubiquitinated target proteins, and mutations in the *usp14* gene cause ataxia in mice. This is the result of defects in synaptic transmission observed both at the neuromuscular junction and in hippocampal CA3–CA1 synapses, suggesting a presynaptic action of the Usp14 protein. Many of the DUBs are involved in the regulation of protein trafficking and endocytic pathways and are therefore expected to have important roles in vesicle transport and in the control of neuronal viability. The precise functions of the different DUBs in the nervous system and in neurological diseases remain to be studied.

Synapses undergo large structural changes during maturation and in response to neuronal activity. Recent studies have further shown that the proteasome is involved in the degradation of molecules belonging to the postsynaptic densities (PSDs) that contain various receptors and scaffolding proteins. In addition, synaptic activity influences the ubiquitination and turnover of a subset of PSD proteins, important for the control of synapse function, and maintenance. As shown by studies in models, systems such as *Aplysia* and *Drosophila* have shown that protein ubiquitination can influence synaptic development and plasticity. In the fruit fly, mutations in the *fat facets* (*faf*) gene, encoding a DUB enzyme, cause severe presynaptic overgrowth and defects in neurotransmitter release. Faf most probably acts by stabilizing the *liquid facets* (*laf*, *Drosophila* epsin) gene product, thereby influencing endocytosis and cell receptor signaling. Recent studies on the neuromuscular system in *Drosophila* revealed a relatively short time window for the degradation of synaptic proteins by the proteasome. Chemical compounds inhibiting the proteasome, as well as genetic manipulations in the fly, subsequently identified the *Drosophila* UNC protein, DUNC-13, as a selective target for the proteasome in the presynaptic terminal. Data showed that proteasome inhibition locally increases neurotransmitter release and the levels of DUNC-13, and enhances synaptic efficacy. Members of the UNC protein family are involved in synaptic vesicle priming and regulation of neurotransmitter release in different species, suggesting similar mechanism for the control of synaptic activity by the proteasome.

As indicated above, monoubiquitination regulates protein trafficking, involving endosomes and clathrin-mediated endocytosis. RTKs, such as the EGFR, are monoubiquitinated upon ligand binding. This leads to RTK internalization and degradation in lysosomes, thereby preventing receptor recycling to the plasma membrane. Similarly, several other receptors such as the neurotrophic factor receptors, TrkA and TrkB, are ubiquitinylated after receptor activation. Very recently, it was shown that the accumulation of mutant A$\beta$ peptide disturbs multivesicular body sorting by inhibiting the ubiquitin–proteasome system and thereby causes defects in receptor cycling. Current data suggest that receptor activation triggered by ubiquitinylation may serve as an important control step for signals mediating neuronal survival and degeneration.

Apart from this, several signaling pathways in the nervous system are regulated by the proteasome. In this context, the activation of the nuclear factor-kappa B (NF-$\kappa$B) pathway after proteasome degradation of the inhibitor-$\kappa$B protein mainly mediates neuronal survival. The nature of specific genes regulated by the different NF-$\kappa$B proteins in brain tissue, however, is only partly understood. In addition, the activities of the c-Jun-N-terminal kinase (JNK)-pathway can, depending on the cellular context, either increase nerve cell death or promote survival. The c-Jun-specific E3 ligase, Fbw7, is required for proper degradation of this protein, and mice deficient in Fbw7 exhibit tremor and loss of specific neuronal subtypes in the brain. The precise roles of the proteasome in controlling various signaling proteins and cascades in different neuronal populations afflicted by neurodegeneration need to be studied more in the future.

## Axonal Degeneration and the Proteasome

Apart from changes in the cell soma, various neurological diseases are often characterized by early disturbances in neuronal connectivity and in axonal functions. Particularly in AD, in ALS, and in multiple sclerosis, axons degenerate that leads to changes in neuronal signaling and cell communication. The axon and the synapse can also undergo degeneration due to genetic causes or disturbed metabolism by various chemicals and toxins, as well as following different types of nerve injuries. In these conditions, the neuronal soma will ultimately also degenerate by mechanisms involving disturbed axonal transport and signaling leading to a 'dying-back' disorder that follows the degeneration of the axon.

One model to study axonal reactions has been mechanical transections of a peripheral nerve in rodents that leads to a degeneration of the distal nerve stump. In these conditions, characteristic changes occur in the cellular composition and molecular interactions in the nerve following the injury, described under the term Wallerian degeneration. Studies of the mouse mutant (*WLd*$^s$) with a delay in Wallerian degeneration for about 2–3 weeks has helped to identify a gene product on chromosome 4 that is important for preservation of axonal integrity following the injury. The unique 85 kDa fusion protein in the *WLd*$^s$ mice is produced by gene rearrangements and consists of 70-amino-acid-long N-terminal portion of the ubiquitination factor E4b (UbE4b) connected by an 18-amino-acid stretch to the full-length nicotinamide mononucleotide adenylyl transferase (NMNAT) gene. Expression of the fusion protein in transgenic mice and using different vectors has been shown to delay axonal degeneration after traumatic and toxic injuries, as well as those due to genetic mutations.

The mechanisms by which WLd$^s$ protein affords neuroprotection and delays axonal degeneration has been intensively studied using both *in vitro* neuronal systems and transgenic mice. The *WLd*$^s$ mice protein in fusion gene is involved in the synthesis of nicotinamide adenine dinucletoide (NAD) that takes part in different cellular reactions. Increased levels of NAD have been shown to be neuroprotective acting either locally in the axons or indirectly in the nucleus by enhanced activity of the NAD-dependent deacetylase, SIRT1. SIRT1 in turn may induce transcription of genes involved in neuroprotection that then acts on the axons, and the effect of NAD can be blocked or enhanced using chemical compounds acting on SIRT1. In line with the presumed effect on gene transcription is the observation that the WLd$^s$ fusion protein is abundantly expressed in nuclei. However, studies on SIRT1 gene deleted mice as well as others suggest that other mechanisms and genes may also be activated in the *WLd*$^s$ mice.

Recent findings using transgenic mice indicate that apart from the NMNAT gene the N-terminal region in the *WLd*$^s$ gene is also important for counteracting axonal degeneration. The truncated UbE4b protein in this region is part of an E4 enzyme that can function in polyubiquitination and proteasome-mediated degradation of proteins. It may be that for full neuroprotection the whole molecule of WLd$^s$ is required. In the future, it is important to define the precise mechanisms and downstream gene products by which the WLd$^s$ protein can counteract axonal degeneration and increase neuronal survival in different systems.

Results obtained with the crush lesions in the optic nerve in rodents and with injured cultured neurons have shown that inhibition of the proteasome can retard the onset of axonal degeneration and specifically inhibit the fragmentation of microtubule that occurred early after the injury. Preserved axonal integrity was also observed after the expression of an ubiquitin protease that can reverse ubiquitination of protein substrates. These studies thus show that the local activity of proteasome is involved in axonal degeneration and certain drugs targeting this complex can inhibit this. Studies on the Gracile axonal dystrophy (*Gad*) mutant mice support further the involvement of the proteasome in axonal degeneration. Dysfunctional UCH-L1 leads to the degeneration of the gracile tract of the spinal cord and this protein can act both as a DUB and as an ubiquitin ligase. The exact mechanism causing axonal degeneration in the *Gad* mice is however so far unknown.

Studies in *Drosophila* have shown that the proteasome activity is involved in axonal degeneration of occurring during development. In axonal pruning, neuronal connections in the nervous system are refined and mature to meet requirements of functional activity. Interestingly, expression of an ubiquitin protease or a mutant form of the *Drosophila* E1 ubiquitin-activating enzyme, in the mushroom bodies, was found to inhibit pruning in the fly. This was also observed with mutations in two of the subunits in the 19S regulatory particle. These results indicate a crucial role for local proteasome activity in axon pruning and degeneration in *Drosophila*, supporting studies made on Wallerian degeneration in rodents. However, the identity of ubiquitination proteins as targets for the proteasome in the axons and in axonal degeneration remains to be studied.

In conclusion, available data support the view that the mechanisms underlying axonal degeneration and degeneration of the nerve cell body are inherently different. Death of the soma can occur through both caspase-dependent and caspase-independent mechanisms involving the mitochondria, whereas axonal degeneration occur likely via different mechanisms. However, both neuronal and axonal degeneration involves alterations in local activity of the proteasome including changes in ubiquitination of certain target proteins. As discussed above, data on the function of the WLd$^s$ fusion protein have shown that nuclear reactions with change in gene expression cause a delay in axonal degeneration and can contribute to nerve repair. Similar neuroprotective mechanisms with activation of specific gene programs also underlie the action of different neurotrophic factors that counteract neurodegeneration and promote neuronal survival.

## Conclusions

Ample evidence suggests that dysregulated proteasome activity can contribute to different neurodegenerative diseases involving both the cell soma and the axons. The discovery of the E3 ligase, Parkin, has given important insights into the role of proteasome in PD, and similar mechanisms or disturbances in protein turnover may prevail in other types of neurological disorders. Studies of the proteasome activity in axonal reactions and in the slow Wallerian degeneration mice (Wlds) have revealed mechanisms governing axonal degeneration and repair. A summary of the current knowledge about the involvement of the proteasome in axonal and nerve cell degeneration is given in **Table 1**. Increased insights into the function of the proteasome and its disturbances will deepen our understanding about the cellular mechanisms and proteins underlying different neurological disease and help us to design drugs to modify the cellular reactions involved and to counteract disease and axonal degeneration.

**Table 1** Summary of the evidence suggesting a role for the proteasome in neuronal and axonal degeneration

| *Defect in* | *Gene* | *Diseases/models* |
|---|---|---|
| *Ubiquitin gene* | | |
| | Ubiquitin$^{+1}$ | AD, HD: accumulation and cell death |
| *Ubiquitination enzymes and related proteins* | | |
| E3 ligases | UBE3A | Angelman syndrome |
| | Parkin | PD |
| | Dorfin | ALS (mouse model) |
| Ubiquitin hydrolase | UCH-L1 | PD, GAD |
| Ubiquitin proteases | usp14 | Ataxia (mouse model), synaptic defects |
| Fusion protein | NMNAT | Slow Wallerian degeneration (Wld$^S$): axonal degeneration |
| *Proteasome inhibition* | | |
| Synthetic inhibitors | | PD (rat model), cell models: decreased cell viability; axons: decreased axonal degeneration |
| Decreased proteasomal activity | | PD, HD (autopsy material, cell and mouse models) |
| *Aggregates* | | |
| PolyQ proteins | Different | HD, SCA, DRPLA, SBMA |
| Amyloid plaques | Aß | AD |
| Lewy bodies | α-Synuclein | PD |
| Bunina bodies | SOD-1 | ALS |
| *Others* | | |
| Axonal pruning | | *Drosophila* models with synthetic proteasome inhibitors |
| Signaling | | NF-κB |
| Anti-apoptotic proteins | XIAP | Smac/DIABLO |
| Receptor endocytosis | RTK, Glu6R<br>TrkB, TrkA | Receptor internalization |

SCA, spinocerebellar ataxia; DRPLA, Dentatorubral-Pallidoluysian atrophy; SBMA, spinal-bulbar muscular atrophy.

*See also:* Alzheimer's Disease: Neurodegeneration; Huntington's Disease: Neurodegeneration; Parkinson's Disease: Alpha-Synuclein and Neurodegeneration; Wallerian Degeneration.

## Further Reading

Conforti L, Fang G, Beirowski B, et al. (2007) NAD(+) and axon degeneration revisited: Nmnat1 cannot substitute for Wld(S) to delay Wallerian degeneration. *Cell Death Differentiation* 14: 116–127.

Fischer DF, DeVos RA, VanDijk R, et al. (2003) Disease-specific accumulation of mutant ubiquitin as a marker for proteasomal dysfunction in the brain. *FASEB Journal* 17: 2014–2024.

Glickman MH and Ciechanover A (2002) The ubiquitin–proteasome proteolytic pathway: Destruction for the sake of construction. *Physiological Reviews* 82: 373–428.

Hegde AN (2004) Ubiquitin–proteasome-mediated local protein degradation and synaptic plasticity. *Progress in Neurobiology* 73: 311–357.

Korhonen L and Lindholm D (2004) The ubiquitin proteasome system in synaptic and axonal degeneration: A new twist to an old cycle. *Journal of Cell Biology* 165: 27–30.

Lee D, Ezhkova E, Li B, Pattenden SG, Tansey WP, and Workman JL (2005) The proteasome regulatory particle alters the SAGA coactivator to enhance its interactions with transcriptional activators. *Cell* 123: 423–436.

Lindsten K, Menendez-Benito V, Masucci MG, and Dantuma NP (2003) A transgenic mouse model of the ubiquitin/proteasome system. *Nature Biotechnology* 21: 897–902.

Millard SM and Wood SA (2006) Riding the DUBway: Regulation of protein trafficking by deubiquitylating enzymes. *Journal of Cell Biology 173*: 463–468.

Moore DJ, West AB, Dawson VL, and Dawson TM (2005) Molecular pathophysiology of Parkinson's disease. *Annual Reviews of Neuroscience* 28: 57–87.

Nateri AS, Riera-Sans L, DaCosta C, and Behrens A (2004) The ubiquitin ligase SCFFbw7 antagonizes apoptotic JNK signaling. *Science* 303: 1374–1378.

Niwa J, Ishigaki S, Hishikawa N, et al. (2002) Dorfin ubiquitylates mutant SOD1 and prevents mutant SOD1-mediated neurotoxicity. *Journal of Biological Chemistry* 277: 36793–36798.

Wilson SM, Bhattacharyya B, Rachel RA, et al. (2002) Synaptic defects in ataxia mice result from a mutation in Usp14, encoding a ubiquitin-specific protease. *Nature Genetics* 32: 420–425.

Yi JJ and Ehlers MD (2005) Ubiquitin and protein turnover in synapse function. *Neuron* 47: 629–632.

# Genomics of Brain Aging: Nuclear and Mitochondrial Genomes

**A Prigione and G Cortopassi**, University of California at Davis, Davis, CA, USA
**C Ferrarese**, University of Milan-Bicocca, Monza, Italy



## Introduction

Over the past few years, significant advances in the study of brain genomics have been made due to the advent of DNA microarray technology, which has become the selected method for large-scale gene expression assessment. Gene microarrays provide a powerful new approach for addressing the complexity of brain function and aging-related processes, as they allow expression monitoring of thousand of genes simultaneously, permitting a more global analysis of genomic data.

The technique makes use of DNA sequences from a library of synthetic oligonucleotides; these are arrayed on a chip at a very high density. After extraction and labeling, nucleic acids from experimental samples are hybridized on the array chip and the amount of emitted fluorescence can be monitored, allowing a whole-genome expression analysis. The observations are not limited to single molecules, as the results can be integrated into biochemical pathways or classes of related genes using biostatistical tools. In fact, the significantly modified genes are automatically assigned to specific functional categories, such as molecular bioprocesses, chromosomes, cell compartments, and relevant pathways based on databased gene product functions (**Figure 1**).

Several recent endeavors have applied microarray technology to the study of the transcriptional modifications associated with the physiology and pathology of brain aging in both mammalian models and humans. Additionally, microarrays have been used to define the effects of mutations in the mitochondrial DNA (mtDNA) on nuclear gene expression. Mutations in the mitochondrial genome accumulate over time and mitochondria appear to play a central role in the process of brain aging. Herein we review the trends in gene regulation in the aging brain, the changes in mitochondrial mutations with age, and the direct consequences of mitochondrial mutations on the transcriptional profile of the nuclear genome.

## Gene Expression Profiles of the Aging Brain

The gift of long life is counterbalanced by the price of aging. Aging is a progressive and inevitable process that involves all the organs, and mental decline is a broadly experienced status consequent to physiological brain aging. The same changes experienced by brain cells during normal aging are exacerbated in vulnerable populations of neurons in neurodegenerative disorders. Neurodegenerative diseases, most frequently manifested in the form of Alzheimer's disease (AD) and Parkinson's disease (PD), affect an increasing proportion of the aging population and are considered as the pathological side effects of normal aging. Genetic, biological, and environmental factors all contribute to physiological and pathological brain aging, and thus understanding the transcriptional response in both cases might help in creating therapeutic strategies to fight pathology and in the definition of interventions to slow down physiological cognitive decline.

Several hypotheses have been proposed regarding the nature of neural death during aging, and the three main mechanisms involve oxidative stress, calcium dysregulation, and apoptosis. Oxidative stress occurs when the cellular redox homeostasis is imbalanced by an increase in pro-oxidant species, known as reactive oxygen species (ROS), without a corresponding increase in the protective antioxidant counterparts. ROS include a variety of diverse chemical species and, although partially generated by cytosolic enzyme systems such as NADPH oxidase or cytosolic reactions as dopamine catabolism, a vast majority of intracellular ROS are derived from the mitochondria. Mitochondrial ROS (superoxide anion, hydrogen peroxide, and hydroxyl radicals) are generated as by-products of mitochondrial oxidative phosphorylation (OXPHOS), which provides ATP via the reaction of hydrocarbons with oxygen. The free-radical theory of aging, formulated in the mid-1950s, hypothesizes that life span is limited by the accumulation of cellular damage from reaction with endogenous oxygen radicals. This endogenous formation of oxidative species may be spontaneous or triggered by specific external stimuli and can increase as humans age, eventually leading to irreversible and deleterious cellular modifications. Indeed, once ROS are generated, they can react with DNA, proteins, and lipids, and damage them. All organs and proteins can potentially be modified by oxidative stress, but certain tissues such as neurons and muscle may be particularly sensitive, as they are highly energy dependent. Energy-dependent tissues are in fact very sensitive to any mitochondrial alterations, the mitochondrion being the main cellular powerhouse and also the cellular compartment that has the greatest exposure to oxidative insults.

The 'calcium dyshomeostasis' hypothesis of brain aging proposes that early in the aging process, $Ca^{2+}$

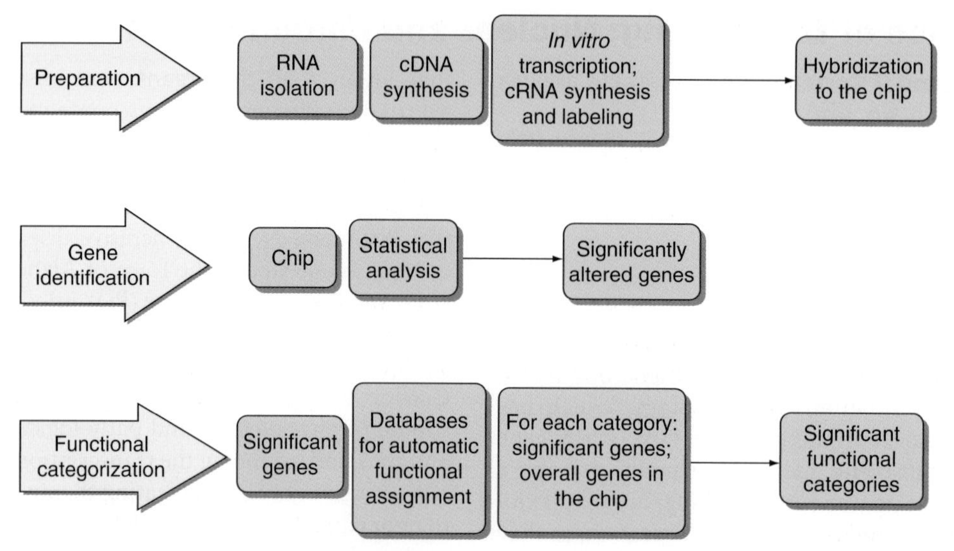


**Figure 1** A summary of basic steps in microarray technology.

balance in brain cells begins to be dysregulated, influencing multiple signaling pathways modulating cellular physiology, molecular functions, and cell structures of the normal brain. The same subtle age-dependent changes in $Ca^{2+}$ homeostasis may increase vulnerability to neurodegenerative diseases such as AD. Changes in free $Ca^{2+}$ concentrations in various subcellular compartments act as a universal signaling system and can be modified by uptake into the endoplasmic reticulum (ER) or into mitochondria, which can function as $Ca^{2+}$ reservoirs. Unbalanced increase of $Ca^{2+}$ influx into neurons can also be caused by excessive activation of glutamate receptors, a process known as glutamate excitotoxicity, which may result in dendritic damage and cell death. Glutamate is the main excitatory neurotransmitter in the central nervous system (CNS) and most neurons commonly receive synaptic glutamatergic input. Several molecular and cellular changes occurring during physiological aging and disease-specific abnormalities are known to increase neural vulnerability to excitotoxicity, and glutamatergic defects are thought to underlie the development of memory loss associated with aging.

Apoptosis is the most widely studied type of programmed cell death in the CNS. It can be triggered by several upstream factors, suggesting that, even if specific gene mutations give rise to different neurodegenerative disorders, there may be a convergence of disease- and age-related pathways on the same well-known cell death cascades. Apoptosis is regulated by cysteine proteases called caspases, which trigger cell death by degrading various structural proteins. Two main pathways are responsible for the activation of the caspases that execute the apoptotic death program. The extrinsic pathway involves cell-surface signaling complexes that activate caspase 8 and downstream caspases. The intrinsic, or mitochondrial, pathway controls activation of caspase 9 by regulating the release of cytochrome *c* from mitochondria. The latter appears to be predominant in the brain and, independently from the origin of the upstream activators, mitochondria, together with the ER, regulate and modulate the entire process, playing a role in integrating cellular stress signals.

Nonetheless, whether the mechanisms of cellular death involve ROS generation, $Ca^{2+}$ dysregulation, or activation of apoptotic factors, the study of nuclear profiling is of enormous help in understanding the common mechanisms upstream of neuronal death and the complex process of brain aging.

### Trends in Animal Models

Most genomic studies of brain aging have been performed in rodent models. Aging results in a gene expression profile indicative of oxidative stress, altered $Ca^{2+}$ regulation, apoptosis, and reduced mitochondrial function, suggesting that the three hypotheses regarding the nature of neural death may all be part of the same process, sharing mitochondrial dysfunction as a central common element. In addition, other relevant transcriptional modifications are related to inflammatory response, altered synaptic signaling, decreased protein turnover, and decreased levels of growth and trophic factors. Overall, upregulated genes can be assigned to the following categories: stress response, $Ca^{2+}$-related signal transduction, vesicle trafficking, myelin and lipid synthesis, inflammatory response, and glial function (**Table 1**).

The stress response category comprises genes involved in oxidative stress, heat shock response, lysosomal proteases, and apoptosis. These include ROS-activated elements, such as glutathione-*S*-transferase, several heat shock factors, components of the lysosomal proteolytic system, and injury or hypoxic stress-induced elements. The stress category thus appears to play an essential role in the aging process, although most likely a secondary one consequent to cellular insults, rather than being a primary step. The putative triggers could be calcium pathway dysfunctions as well as mitochondrial modifications. In fact, upregulation of transcripts encoding $Ca^{2+}$-regulating pathways has been detected, including modulators of $Ca^{2+}$-induced release and the voltage-dependent anion channel (VDAC) that regulates $Ca^{2+}$ flux into the mitochondria and plays a central part in apoptosis. Elevated intracellular $Ca^{2+}$ concentration can reduce neural activity through the activation of inhibitory $Ca^{2+}$-dependent conductances; amplified $Ca^{2+}$ release and influx in aged neurons may eventually lead to decreased neural responsiveness and synaptic plasticity.

Vesicle trafficking genes include those encoding SNARE (*N*-ethylmaleimide-sensitive factor attachment protein receptor) proteins that are involved in docking and fusion of synaptic vesicles to the active zone, as well as in the calcium-triggering step itself, as part of the exocytotic process at the presynaptic plasma membrane. Their increase has been proposed to be linked to the upregulation of myelin-related proteins, myelin-associated oligodendrocytic basic protein, and myelin-associated glycoprotein. The production and the secretion of myelin in oligondendrocytes may be increased in response to light-activated demyelinating processes occurring during aging. Demyelination and, most importantly, oxidative stress can then trigger a widespread neuroinflammatory response.

Neuroinflammation is mediated by microglial activation and cytokine release and can exacerbate the initial demyelination and redox status disequilibrium, creating a positive feedback. Activated transcripts included the microglia and macrophage migration factor, receptors that mediate the microglial activation pathway, inflammation markers, complement cascade genes, and major histocompatibility complex (MHC) class I antigen-presenting molecules. Both the widespread inflammatory response and the myelin production involve glial cells, and, as further confirmation, glial function genes appear to be upregulated in rodent brain aging.

Downregulated genes are mainly involved in mitochondrial function and energy metabolism, protein synthesis and turnover, synaptic plasticity, extracellular matrix (ECM) production, and neurotrophic factors. Downregulated mitochondrial transcripts include genes responsible for mitochondrial function, as well as electron transport chain genes. The overall inhibition of mitochondrial-related transcripts may be interpreted either as a primary event related to aging or, more likely, as the nuclear response to mitochondrial alterations which cause or are caused by oxidative stress. Mitochondrial dysfunction may be responsible, together with calcium dyshomeostasis, for evoking the cellular stress response; the downregulation of mitochondrial genes could then represent a nuclear attempt to shut down a process that is not working properly. The consequences would be deleterious for energy metabolism and would lead to metabolic and biosynthetic involutions. Accordingly, a decrease in protein biosynthesis and turnover is also observed.

**Table 1** Global view of nuclear transcriptional changes induced by brain aging (in mammalian models and in humans) and by human mtDNA mutations

| *Brain aging* | *mtDNA mutations* |
|---|---|
| Upregulated transcripts | Upregulated transcripts |
| Stress response | Stress response |
| $Ca^{2+}$-related signal transduction | $Ca^{2+}$-related signal transduction |
| Vesicle trafficking | Vesicle trafficking |
| Myelin and lipid synthesis | Cell cycle and cell proliferation |
| Inflammatory response | Glycolysis |
| Glial functions | |
| Downregulated transcripts | Downregulated transcripts |
| Mitochondrial function and energy metabolism | Mitochondrial function and energy metabolism |
| Protein biosynthesis | Protein biosynthesis |
| Protein turnover | Protein turnover (mtDNA deletion specific) |
| Synaptic plasticity | Synaptic transmission |
| Extracellular matrix production | Cell migration |
| Growth and neurotrophic factors | Oligodendrosis |

Transcripts encoding the main cellular pathway responsible for cytoplasmic protein degradation, the ubiquitin–proteasome system (UPS), are in fact mainly downregulated. Interestingly, UPS inhibition is known to cause accumulation of altered or misfolded proteins in some neurodegenerative diseases (e.g., PD and AD).

Downregulated activity-dependent synaptic/neurite plasticity genes comprise genes which have been previously linked to synaptogenesis, neurite remodeling, plasticity, or memory. Reduced synaptic signaling and plasticity in neurons have been linked to disruption of memory consolidation, which is a common feature of both physiological and pathological cognitive impairment. Moreover, some of these genes are also significant components of the ECM, and their downregulation may suggest a genomically mediated ECM erosion, which has been hypothesized to be related to the demyelination process. As a further confirmation, other ECM elements are also downregulated.

Inhibition of several genes associated with either trophic or growth functions has been detected. Lack of neurotrophic factors, which are involved in neural development and rescue from excitotoxic or metabolic insults, can thus exacerbate the whole process, leading aged brain cells to death. Transcripts comprise fibroblast growth factor, glial neurotrophic factor, and factors involved in neural migration.

Similar genomic transcriptional profiles have been identified in rodent models for AD, as multiple studies have demonstrated increased expression of inflammation-associated genes, apoptosis-related transcripts, and altered mitochondrial energy metabolism genes. Thus, common central elements for physiological and pathological aging appear to be calcium homeostasis, mitochondrial dysregulation, and stress- and inflammation-mediated neural damage. So far, the main mechanism that has proved effective in retarding the aging process in rodents appears to be caloric restriction (CR). CR may also extend the maximum life span of several nonmammalian models. Explanations involve decrease in ROS production, decreased body temperature associated with a hypometabolic state, hormone signaling changes, and alterations in gene expression and protein degradation. Interestingly, CR is found to selectively attenuate the age-associated induction of genes encoding for inflammatory and stress responses.

### Trends in Humans

The patterns of changes in gene expression observed in microarray experiments performed on aged human neurons have mainly reproduced the ones identified in rodents, helping to define a potential fingerprint of brain aging. The human transcriptional profiles are characterized by the induction of genes mediating stress response and repair functions, inflammatory and immune responses, myelination, and lipid metabolism. Transcripts reduced with aging include genes involved in signal transduction, synaptic function and plasticity, learning and memory processes, calcium homeostasis, protein turnover, and mitochondrial functions (**Table 1**).

Genomic studies of human pathological aging mainly have focused on AD. Two basic approaches are used: the comparison of AD-affected and AD-resistant brain tissues from the same individuals or the comparison of AD-affected brain structures of AD patients and age-matched controls. Although transcriptional changes of important elements of AD neuropathology have been identified, alterations shared with normal aging are also observed. Common trends are represented by the activation of apoptotic and neuroinflammatory signaling, stress response, and lipid metabolism, and the inhibition of the gene categories involved in synaptic plasticity and neurotrophic factors.

Taken together, common features of brain aging appear to be the repression of genes implicated in mitochondrial oxidative respiration, the decline in physiological neural activities, and inflammatory and glial reactions. In order to establish a cascade of events, some authors hypothesize that the primary dysfunctions occurring in aged neurons may be either calcium or mitochondrial dysfunctions, leading to cellular stress and genome damage, which in turn may compromise biosynthetic and metabolic pathways as well as systems that subserve synaptic functions and neural survival. Eventually, neural alterations might translate into widespread reactions involving glial responses, inflammation, and demyelination.

## Mitochondrial Mutations and Aging

Mitochondria, in addition to their well-known role in respiration and OXPHOS, are important sites for diverse cellular functions such as fatty acid oxidation and heme, urea, amino acid, and phospholipid biosyntheses. However, as explained previously, their relevance for the aging process consists in the central role they play in ROS generation, $Ca^{2+}$ signaling, and apoptosis.

Mitochondria, unlike other cellular compartments, possess an individual genome. mtDNA is a small 16.6 kb double-stranded circular DNA that encodes 13 respiratory chain polypeptides and 24 nucleic acids that are needed for intramitochondrial protein synthesis. mtDNA genes cannot be translated by use of the universal genetic code, hence the need for mtDNA to encode its own ribosomal RNA and transfer RNA genes. Replication is independent of the cell cycle and can occur even in nondividing cells. The rest of the respiratory chain elements and mitochondrial proteins (approximately 900 gene products) are all encoded by the nuclear genome. Nuclear DNA is also relevant for it maintenance, as it encodes the only DNA polymerase known to be targeted to mitochondria: DNA polymerase-$\gamma$ (POLG). POLG is responsible for all aspects of mtDNA synthesis, replication, and repair of mtDNA damage. It corrects mistakes made during replication by copying and proofreading the genome and it can participate in the resynthesis of DNA during the DNA repair processes. Overall, the cross-talk between nuclear and mitochondrial genomes plays an essential function in regulating mitochondrial activities.

mtDNA damage is a very likely event, given the proximity of mitochondria to an environment rich with reactive species, the lack of genomic introns and protective histones, and the presence of a single DNA polymerase. Consequently, mtDNA accumulates mutations more than 10 times faster than does the nuclear genome. Because multiple copies of mtDNA reside within each mitochondrion, and several hundred mitochondria are present within each cell, when a mutation occurs, the result is a dual population of mtDNAs – mutant and wild-type – coexisting within the same cell, a phenomenon called heteroplasmy. Moreover, mtDNA randomly segregates during mitosis, leading to different levels of heteroplasmy in different tissues. Eventually, when the amount of mutated mtDNA in a particular tissue reaches a threshold, pathological manifestations can become evident.

In the general population, mitochondrial genes in several nonreproductive (somatic) tissues accumulate mutations over time, and this event has been proposed to underlie the deleterious cellular modifications associated with aging. Somatic mtDNA mutations, in fact, may lead to energy-generation defects, increased number of harmful free radicals, or calcium dyshomeostasis, resulting in lowered cell and tissue functions characteristic of advanced age. Point mutations (single base-pair changes), deletions, and wider rearrangements of mtDNA can all occur and increase as humans age. These mutations are unevenly distributed and can accumulate clonally in certain areas, causing a mosaic pattern of respiratory chain deficiencies in tissues such as skeletal muscle, heart, and brain. A frequently observed mutation is a deletion of mtDNA of about 5 kb in length, the so-called common deletion, which preferentially occurs at much higher levels in nervous and muscle tissues. Deletions, accumulated exponentially with age, reach much lower concentrations in most tissues of healthy individuals than in patients affected by mitochondrial disorders such as Kearns–Sayre syndrome (KSS) or chronic progressive external opthalmoplegia (CPEO). However, deletions levels may come close to 5% in muscle tissue and to 50% in substantia nigra. These deletions occur as spontaneous errors in individual cells as a consequence of age, and rise in concentration over time, with each cell containing a different predominant deletion species. Age-dependent mtDNA deletions have also been linked with respiratory chain defects in muscle and substantia nigra.

More recently, it has been demonstrated that a defective POLG is correlated with an increased level of mtDNA mutations, comprising deletions and point mutations. Eliminating the proofreading activity, POLG becomes error prone, giving rise to increased somatic mtDNA mutations. Accumulation of mutations is associated with reduced life span and premature onset of aging-related phenotypes in mice models carrying defective POLG, demonstrating a causative link between mitochondrial mutations and aging in mammals. In POLG mutant mice, the increase in mutations in the mitochondrial genome is not associated with oxidative stress, but has been correlated with the induction of apoptotic markers. Accelerated development of aging through mtDNA mutations may thus occur even without redox imbalances. In the apoptotic pathway, mitochondrial dysfunction can lead to mitochondrial outer membrane permeabilization and subsequent release of cytochrome *c* in the cytoplasm, which in turn activates the key effector, protease caspase 3. However, the hypothesis of a contribution of oxidative DNA damage to aging is supported by other studies in humans and mice. In fact, the event leading to an error-prone POLG might be secondary to protein damage caused by ROS. However, the disruption of physiological mitochondrial activities appears to be a major element in the chain of events leading to aging.

Interestingly, CR, the only nutritional intervention that retards aging in animal models, has been demonstrated to reduce the mitochondrial-mediated apoptosis pathway and to delay the accumulation of mtDNA mutations in rodents. Indeed, CR reduces fiber loss and mitochondrial abnormalities in aged muscle and attenuates the increase in cytosolic cytochrome *c* and caspase 2 activity with age. As a therapeutic strategy, the possible application of CR to the human population would face ethical and practical problems, but the study of the specific mechanisms responsible for the antiaging influences of CR could lead to the development of pharmacological agents. Moreover, the demonstration that CR increases life span by decreasing the accumulation of mtDNA mutations and reducing mitochondrial apoptosis strengthens the concept that losses of mtDNA integrity and mitochondrial function are central factors in brain aging.

## Genomic Effects of Mitochondrial Mutations

Mutations in gene products expressed in the mitochondrion cause a nuclear transcriptional response that leads to neurological disease, such as Leber's hereditary optic neuropathy (LHON), neurogenic ataxia and retinitis pigmentosa (NARP), mitochondrial myopathy/encephalopathy/lactacidosis/stroke (MELAS), myoclonic epilepsy with ragged red fibers (MERRF), KSS, and CPEO. As indicated previously, the same mutations may play a causative role in the aging process as well. In order to assess mitochondrial disease nuclear genomic profiles, microarray analyses have been performed in different cell types. The results have demonstrated that a similar transcriptional profile is shared among different mitochondrial diseases, as if each mitochondrial disease and mitochondrial mutation could generate a common nuclear response (**Table 1**).

Upregulated transcripts encode for genes in the categories of stress response, calcium signaling regulation, cell cycle and cell proliferation, vesicle-mediated transport, and glycolysis. In the stress response category, ER- and unfolded protein response (UPR)-related transcripts are detected. This suggests that mitochondrial inhibition might be sufficient for activation of the UPR, which occurs when ER function is perturbed. Apoptosis is known to be stimulated in the presence of prolonged ER stress. Thus, mitochondrial mutations or other factors inducing a cellular stress response may eventually give rise to the mitochondrial apoptotic pathway. UPR induction may occur consequently to ER stress, which can be generated as a result of mitochondrial mutations, either by increased ROS production or by $Ca^{2+}$ signaling perturbations caused by mitochondrial bioenergetic defects. Upregulation of transcripts related to $Ca^{2+}$ modulation is in fact detected, including VDAC, which is thought to reside at the mitochondrial–ER interface and to control $Ca^{2+}$ transients. Moreover, it has been shown that depletion of mitochondrial membrane potential induces both the UPR genes and cell cycle genes. Accordingly, cell cycle and cell proliferation genes are induced. As observed for the brain aging genomic profiles, vesicle trafficking genes, protein transport genes, and ER-to-Golgi transport genes appear to be induced also as a consequence of mitochondrial mutations. However, the reason for this activation is still unclear. Glycolysis transcripts, such as aldolase C, are also upregulated, and their increase may be interpreted as the nuclear attempt to restore energetic metabolism functions.

Functional clusters inhibited by mitochondrial diseases include mitochondrial ATP synthesis, protein biosynthesis, synaptic transmission and vesicular secretion, cell migration, and oligodendrogenesis. Transcripts coding for mitochondrial proteins and specifically for ATP synthesis are downregulated. The data suggest that the nucleus senses the irreversible depletion of mtDNA-encoded mitochondrial subunits and transfer RNAs, and responds by downregulating the interacting subunits that would normally form a functional complex. The downregulation of nuclear-encoded mitochondrial ribosomal subunits and OXPHOS transcripts thus appears to reinforce the mitochondrial defect initiated by the mutations. As observed in aged brain transcriptomes, deleterious modifications of energy metabolism may lead to metabolic and biosynthetic involutions, as the functional category of protein biosynthesis appears to be inhibited. Moreover, members of the vesicular secretion pathway, or proteins that are transported by the vesicular secretion pathway, are also decreased. In addition to their function in vesicular secretion, some gene products of this category have also been shown to be involved in cellular migration, as cellular migration is dependent upon the process of vesicular secretion, both for secretion of cell migratory signals into the ECM as well as for the interpretation of them. Additionally, UPR and ER stress induction are known to inhibit protein synthesis and vesicular secretion. Oligodendrocyte-specific transcripts are found downregulated in mitochondrial diseases. Oligodendrocytes are especially dependent on lipid synthesis, cell-to-cell contact, synaptogenesis, and migration. In addition, the dihydroceramide desaturase transcript (DEGS1), which produces ceramide, the precursor of sphingomyelin, a major component of the myelin sheath, is also inhibited. These findings suggest that demyelination could be an important pathophysiologic feature of mitochondrial mutations, especially because myelin is transported by the vesicular secretion pathway.

Overall, microarray data show that there is a significantly shared transcriptomal imprint of mitochondrial mutations, which implicates specific pathophysiological pathways. The data support the hypothesis that mitochondrial mutations may preferentially causes ER stress, which in turn triggers the UPR and decreases vesicular secretion, resulting in dysfunctions of secreted proteins involved in synaptogenesis, neural cell migration, and oligodendrocyte function.

Interestingly, most of these transcriptional modifications are similar to the ones relative to brain aging, as if mtDNA mutations play a primary causative event in the aging process.

Recently published microarray data from our laboratory, conducted on several cell types harboring mtDNA deletions, show a prominent inhibition of transcripts encoding ubiquitin-mediated proteasome activity. Thus, the downregulation of the ubiquitin–proteasomal system appears to be a unique consequence of mtDNA deletions. As already described, mtDNA deletions are the most common type of mitochondrial mutations known to accumulate exponentially during aging and to preferentially occur at higher levels within the CNS, and in the substantia nigra in particular. Thus, the genomic signal of UPS inhibition characteristic of brain aging might be secondary to an increase in mtDNA deletions, and this may even be more dramatic for neurodegenerative diseases with UPS defects, such as PD, which is characterized by specific death of dopaminergic neurons within the substantia nigra.

## Conclusion

The field of aging research has been completely transformed in the past decade. Genomic profiles associated with brain aging became available due to the advent of microarray technology, and it is now widely accepted that the aging process is subject to regulation, like most biological processes. Mitochondrial dysfunctions were already known to play a relevant role because of the centrality of the organelle in apoptosis, $Ca^{2+}$ regulation, and oxidative stress, and because mtDNA mutations increase with age. In addition, the nuclear transcriptional profiles consequent to mtDNA mutations are showing multiple similarities with the ones relative to brain aging. Similar trends in activation and inhibition of functional categories can be observed. Overall, it is becoming more evident that mutations in the mitochondrial genome in aged neurons play a causative role rather than merely being correlative, and the concepts connected to the mitochondrial theory of aging are thus undergoing change.

*See also:* Alzheimer's Disease: Neurodegeneration; Gene Expression in Normal Aging Brain; Glial Cells: Microglia During Normal Brain Aging; Glial Cells: Astrocytes and Oligodendrocytes During Normal Brain Aging; Lipids and Membranes in Brain Aging; Rodent Aging; Synaptic Plasticity: Neuronogenesis and Stem Cells in Normal Brain Aging.

## Further Reading

Beal MF (2005) Mitochondria take center stage in aging and neurodegeneration. *Annals of Neurology* 58: 495–505.

Blalock EM, Chen KC, Stromberg AJ, et al. (2005) Harnessing the power of gene microarrays for the study of brain aging and Alzheimer's disease: Statistical reliability and functional correlation. *Ageing Research Reviews* 4: 481–512.

Blalock EM, Geddes JW, Chen KC, et al. (2004) Incipient Alzheimer's disease: Microarray correlation analyses reveal major transcriptional and tumor suppressor responses. *Proceedings of the National Academy of Sciences of the United States of America* 101: 2173–2178.

Cortopassi G, Danielson S, Alemi M, et al. (2006) Mitochondrial disease activates transcripts of the unfolded protein response and cell cycle and inhibits vesicular secretion and oligodendrocyte-specific transcripts. *Mitochondrion* 6(4): 161–175.

Cortopassi GA, Shibata D, Soong NW, et al. (1992) A pattern of accumulation of a somatic deletion of mitochondrial DNA in aging human tissues. *Proceedings of the National Academy of Sciences of the United States of America* 89: 7370–7374.

Dickey CA, Loring JF, Montgomery J, et al. (2003) Selectively reduced expression of synaptic plasticity-related genes in amyloid precursor protein + presenilin-1 transgenic mice. *Journal of Neuroscience* 23: 5219–5226.

DiMauro S and Schon EA (2003) Mitochondrial respiratory-chain diseases. *New England Journal of Medicine* 348: 2656–2668.

Finkel T and Holbrook NJ (2000) Oxidants, oxidative stress and the biology of ageing. *Nature* 408: 239–247.

Lee CK, Weindruch R, and Prolla TA (2000) Gene-expression profile of the ageing brain in mice. *Nature Genetics* 25: 294–297.

Loeb LA, Wallace DC, and Martin GM (2005) The mitochondrial theory of aging and its relationship to reactive oxygen species damage and somatic mtDNA mutations. *Proceedings of the National Academy of Sciences of the United States of America* 102: 18769–18770.

Lu T, Pan Y, Kao SY, et al. (2004) Gene regulation and DNA damage in the ageing human brain. *Nature* 429: 883–891.

Mattson MP and Magnus T (2006) Ageing and neuronal vulnerability. *Nature Reviews Neuroscience* 7: 278–294.

Trifunovic A, Wredenberg A, Falkenberg M, et al. (2004) Premature ageing in mice expressing defective mitochondrial DNA polymerase. *Nature* 429: 417–423.

## Relevant Websites

http://www.affymetrix.com – Affymetrix Corporation product/research Website.

http://www.alz.org – Alzheimer's Association (Chicago, IL, USA).

http://www.nia.nih.gov – National Institute on Aging (U.S. National Institutes of Health).

http://www.parkinson.org – National Parkinson Foundation, Inc. (Miami, FL, USA).

http://www.umdf.org – United Mitochondrial Disease Foundation (Pittsburgh, PA, USA).

# Gene Expression in Normal Aging Brain

**T E Morgan**, University of Southern California, Los Angeles, CA, USA



## Introduction

Genes, lifestyle, and life circumstances all contribute to aging. Impairments in cognitive and motor skills commonly occur with increasing age. These progressive changes in cognition are subtle but become apparent later in life. Aging is a major risk factor for numerous neurological disorders. Cognitive and motor impairments resulting from neurodegenerative diseases are distinct from age-related changes and are influenced by peripheral diseases, such as diabetes and cardiovascular disease. In fact, there are strong associations between cardiovascular disease, diabetes, and Alzheimer's disease (AD).

The complex process of brain aging involves multiple systems, cells types, and cellular pathways. As with aging in general, numerous mechanisms have been proposed to underlie brain aging, including instability of nuclear and mitochondrial genomes, neuroendocrine dysfunction, production of and damage from reactive oxygen species, altered calcium metabolism, and inflammation-mediated neuronal damage. The common thread woven through all of these mechanisms is changes in gene expression.

## Minimal Neuron Loss with Age

Prior to advances in stereological procedures for estimating neuron number it was widely believed that brain aging was accompanied by significant neuron death. Today, as verified by electron microscopy, stereologic counts, and gene expression data, brain aging is not due to loss of neurons. Instead, loss of synapses, morphologic changes in dendritic arbors and spines, and changes in myelination contribute to age-related cognitive and motor impairment. The causes of these changes are not known. Molecular studies with transgenic mouse models indicate that specific genetic manipulation can lead to defects in long-term potentiation (LTP) and learning in the absence of degenerative changes (e.g., *N*-methyl-D-aspartate receptor subunits and calcium- and calmodulin-dependent kinase II). These studies support the important role that specific genes have in brain aging.

## Gene Alleles: Single Nucleotide Polymorphisms

Brain aging is not only influenced by changes in gene expression but is also affected by the particular gene allele one inherits. Specific sequence variations (single-nucleotide polymorphism, or SNP), which make people more or less susceptible to certain diseases, have also been shown to influence age-related cognitive changes. An SNP in the coding or regulatory region can affect transcription, translation, and ultimately function. For example, in humans, there are three unique alleles for the apolipoprotein E (*APOE*) gene (designated *APOE* ε2, ε3, and ε4). *APOE* ε3 allele is most prevalent, followed by *APOE* ε4, with rare occurrence of *APOE* ε2. Genetic linkage studies indicate that the inheritance of *APOE* ε2 is associated with increased longevity. The neurotrophic activities of *APOE* are allele specific, with ε2 > ε3 > ε4. Furthermore, inheritance of *APOE* ε4 increases the risk for AD. Another example is brain-derived neurotrophic factor (BDNF), a neurotrophin that is essential for neuronal growth, survival, and function. In addition, BDNF is a critical factor in learning and memory. An SNP identified in the coding region of BDNF (Val66Met), which affects the pro-region of the protein, is associated with age-related cognitive deficits.

## Gene Expression

Most studies that examine age-related gene expression compare mRNA levels in brain tissue of different age groups. A number of important caveats must be acknowledged when interpreting these studies. First, gene expression is highly selective to different brain regions and specific cell types. Second, alterations of mRNA levels may represent changes in gene expression or mRNA stability or turnover. Third, changes in mRNA levels may not ultimately reflect translational or functional changes. Fourth, aging changes can be quite subtle, which requires confirmation of each potential age-regulated gene by multiple methods and labs, and even in a variety of species. Unfortunately, most studies do not look at the 'middle years' of a species' life span, and results are generally reported from only two age groups, young and old. Careful examination at multiple ages throughout the life span is necessary for identifying potential intervention strategies.

Age-related pathways that may contribute to aging are difficult to ascertain in humans. Since gene expression in the human brain can only be examined at autopsy, the changes measured may be causing aging or may be a consequence of aging. Due to this difficulty, animal models provide a reasonable approach to studying brain aging. In addition, specific pathways can be modeled *in vitro* but these studies are complicated by the fact that neurons are postmitotic.

Some of the first studies documenting changes in gene expression with age showed declines in specific pre- and postsynaptic markers which are often correlated with age-related decline of cognition and motor skills. Synaptophysin, a presynaptic vesicular protein that is often used as a marker for synaptic plasticity and integrity, decreases with increasing age in hippocampal and cortical regions. Specific subunits of *N*-methyl-D-aspartate (NMDA) receptors (e.g., NR1, NR2), which have critical roles in learning and memory, decrease with age. Also, dopamine receptor (D2R) levels decrease with increasing age. The decline in expression of these synaptic components effectively reduces synaptic transmission, which contributes to age-related cognitive and motor changes.

In addition to these neuronal changes, age-related gene expression also occurs in glia. Glial fibrillary acidic protein (GFAP) is an astrocyte-specific intermediate filament that consistently shows elevation in aging brain. Beginning at middle age, the increased expression of GFAP ranges from 1.5 to 4 times normal, depending on brain region. Widely accepted as an indicator of astrocyte activation, the consequence of the progressive increase during aging is not known. Recent data suggest that the age-related increase of GFAP may contribute to the loss of synaptic plasticity during aging. Astrocytes originating from aged rat brains possess elevated GFAP and are less able to support neurite outgrowth. Experimental manipulation of GFAP levels (increase with complementary DNA; decrease with small interfering RNA) demonstrates a direct correlation between GFAP levels and neurite outgrowth. This exciting line of discovery implicates astrocytes as a driver of age-related cognitive changes.

Myelin-associated oligodendrocyte basic protein (MOBP) is increased with age in oligodendrocytes. Although MOBP is the third most abundant protein in myelin, its exact physiological function is not known. Based on its localization to compact myelin, it is proposed to have a role in stabilization of the myelin. White-matter integrity declines with increasing age. In addition, a decline in the volume of brain white matter had been shown with magnetic resonance imaging (MRI). The impact of these age-related changes in the ultrastructure of the myelin sheath may have a profound effect on neuronal function.

Another glial gene that is increased with age is apolipoprotein D (ApoD). As with many of the age-regulated genes, the exact function of ApoD is not known. It is widely expressed by both astrocytes and oligodendrocytes. The ApoD protein can transport small hydrophobic ligands, including cholesterol, pregnenolone, and sterols, but it is not known if these are its physiological ligands. Importantly, the *Drosophila* homolog has been shown to increase stress resistance and extend life span. In studies of transgenic flies that overexpress the homolog of human ApoD, the protein named glial Lazarillo (GLaz), the flies were less susceptible to hyperoxic conditions, had improved motor and behavioral activities when subjected to sublethal oxygen exposure, and lived 29% longer. Therefore, the age-associated increase in ApoD expression may be a compensatory response to the age-related progressive increase in oxidative stress.

Genotypic changes in microglia are believed to underlie their activated phenotype which appears with increasing age. One indication of this is the increased expression of CD68 (the mouse homolog is macrosialin). CD68 is a lysosomal/endosomal-associated membrane protein with characteristics of scavenger receptors. Its function is not known but is believed to play a role in phagocytosis. Under normal conditions CD68 is predominantly found in microglia residing in white matter. Since oxidized lipids induce CD68 expression, it may be playing a role in age-related myelin changes.

Studies examining the expression of individual genes that change with age are vital to advancing our knowledge of the molecular mechanisms underlying aging; however, such studies do not address the polygenetic component of brain aging. This need is met by large-scale microarray analysis, which allows simultaneous examination of thousands of genes in a single sample. This type of analysis has been performed on brains from many mammalian species, including humans, chimpanzees, and rodents. There is significant change with age in a relatively small percentage (<5%) of the >20 000 genes examined. Although specific genes across all species may not be identical, classes of genes emerge (**Table 1**). The greatest percentage of genes that are upregulated consists of those involved in inflammatory processes and the stress response. Downregulated genes concentrate in synaptic function and plasticity. A comparison of gene classes in different species suggests that inflammatory and oxidative pathways are at the center of age-related cognitive and motor deficits.

An important group of inflammation-related genes that are increased in the aging brain includes components of the complement cascade. Resident brain cells produce each member of the complement cascade. Activation of the complement cascade initiates innate immune responses such as phagocytosis, chemotaxis, and cytolysis. The initial components, C1q, C3, and C4, are commonly shown to be increased in the aging brain, suggesting that activation of the complement cascade occurs during aging.

## Brain Aging across Species

A new informatics tool to compare microarray results from different specifies and/or different laboratories is the web database and analysis platform, Gene Aging Nexus (GAN). GAN is freely accessible online (see 'Relevant website' section), where new datasets can be uploaded, queries can be run on existing datasets, and extensive integrative analysis can be performed to identify differentially or co-expressed genes with functional annotation. The database currently has data for the aging brain in humans, mice, and rats. Seven genes are identified to be increased with age across these three species (**Table 2**).

This cross-species comparison identifies known age-changers such as *GFAP* and *APOD*, but also identifies age-regulated genes that have not been documented prior to the microarray analysis. For example, annexin A3 (gene *ANXA3*) increases in human, mouse, and rat brains. Annexin A3 is a member of the annexin family of proteins that have calcium-dependent phospholipid-binding activities. The function of annexin A3 is not known; it is expressed in activated microglia in response to neuron injury. Carbonic anhydrase IV (gene *CA4*) expression is decreased in the aged brain. Reduced levels of carbonic anhydrase IV may contribute to age-related declines in synaptic transmission, since this protein is responsible for buffering the alkaline pH shifts associated with neuronal activity.

It should be noted that some of the individual genes identified as age changers were not shown to be changed during aging in the microarray studies. Detection of individual genes is influenced by the specific set of oligonucleotide probes used on a particular microarray chip, which may explain these differences. Also, differences could arise due to the complex statistical considerations microarray analysis requires.

**Table 1** Functional gene classes showing consistent changes with brain aging

| *Age change* | *Functional class* | *Gene family* |
|---|---|---|
| Increase | Inflammatory and stress response | Complement subunits, heat shock proteins, cathepsins, inflammatory mediators |
| Decrease | Synaptic function and plasticity | Presynaptic and postsynaptic proteins, neurotransmitter receptors |

## Brain Aging and Insulin-Like Growth Factor

A discussion of aging is not complete without including insulin-like growth factor-1 (IGF-1). The importance of the insulin/IGF-1 pathway in aging and longevity is apparent from the vast experimental literature documenting a central role for IGF-1 pathways in regulating the life span of diverse organisms, including yeast, worms, flies, and mice. IGF-1 pathways orchestrate cell metabolism, growth, and survival. Systemic IGF-1 decreases with increasing age. IGF-1 crosses the blood–brain barrier and resident brain cells express IGF-1. Correlational and experimental observations suggest that IGF-1 has a generally positive influence on cognitive performance. At the cellular level, IGF-1 is neurotrophic and neuroprotective while participating in the stress response. The influence IGF-1 has on cellular activities predicts a critical role for IGF-1 pathways in brain aging.

**Table 2** Age-regulated genes common to human, rat, and mouse brain: fold change (*p*-value)

| *Human gene nomenclature/protein* | *Human* | *Mouse* | *Rat* |
|---|---|---|---|
| *ANXA3*/annexin A3 | +2.2 (0.006) | +1.4 (0.04) | +1.3 (0.002) |
| *APOD*/apolipoprotein D | +2.0 (0.000 05) | +2.6 (0.000 03) | +1.4 (0.003) |
| *CA4*/carbonic anhydrase IV | −1.5 (0.01) | −1.9 (0.009) | −1.3 (0.009) |
| *EGR2*/early growth response 2 | −1.6(0.04) | +1.6(0.04) | +2.8(0.000 9) |
| *GFAP*/glial fibrillary acidic protein | +1.5(0.01) | +2.1(0.000 1) | +1.7(0.05) |
| *MOBP*/myelin-associated oligodendrocyte basic protein | +2.1(0.03) | +1.8(0.007) | +1.6(0.02) |
| *QDPR*/quinoid dihydropteridine reductase | +1.4(0.04) | +1.3(0.04) | +1.6(0.000 003) |

Age-regulated genes identified from three brain aging datasets of gene expression microarray analysis in Gene Aging Nexus.
Human data from B Yankner, Harvard Medical School; mouse data from T Prolla/R Weindruch, University of Wisconsin-Madison; rat data from P Landfield, University of Kentucky.

## Regulation of Gene Expression

Gene expression is regulated through transcription, translation, and posttranslational means. Transcriptional factors, which are at the center of gene regulation during development, may also have a major role during aging. Generally, transcription factors exist in an inactive form and become active through phosphorylation, ligand binding, or removal of inhibitory subunits. Likely candidates involved in regulating age changes include activator protein-1 (AP-1), glucocorticoid receptor (GR), cyclic AMP response element-binding protein (CREB), peroxisome proliferator-activated receptor (PPAR), and nuclear factor-kappa B (NF-$\kappa$B). Each transcriptionally activates unique signal transduction pathways that are implicated in learning and memory. The promoter region of the age-regulated gene *GFAP* contains putative transcriptional binding sites for all of these transcription factors. Although transcriptional regulation of *GFAP* through these sites has been extensively documented, it is still unresolved which ones are responsible for increasing *GFAP* expression with age. Additional levels of regulation have recently been identified. These include the silencing RNAs, microRNAs, and chromatin acetylation pathways. The contribution each of these has in aging is not known.

## Attenuation of Brain Aging

Dietary restriction (DR; also called caloric restriction, or CR), which is achieved through restricted caloric intake, is the most powerful and general manipulation of aging processes in laboratory animals. In addition to extending life span, DR attenuates many of the age-related changes that occur in the brain. Depending on the brain region, 30–75% of the age-associated gene expression changes are either completely or partially prevented by DR. In the neocortex, DR prevents the age-related increased expression of 65% of those genes involved in the inflammatory response.

In addition to DR, specific dietary nutrients that possess anti-inflammatory and antioxidant properties have beneficial effects on brain aging. For example, supplementing rodent diets with strawberry, blueberry, or spinach extracts attenuates age-related motor and cognitive declines. Polyphenolics, such as the flavonoids found in these dietary supplements, are believed to underlie this beneficial effect. The omega-3 unsaturated fatty acids, such as eicosapentaenoic acid (EPA) and docosahexaenoic acid (DHA), improve cognitive performance. The phytoalexin, resveratrol, also has anti-aging properties. Since resveratrol is concentrated in the skin of red grapes, one has an additional reason to savor that glass of burgundy.

## Brain Aging and AD

Since aging is the primary risk factor for AD it should not be a surprise that age-regulated genes are also found in AD. In fact, inflammatory and oxidative pathways are implicated in AD progression. Activated astrocytes and microglia are integral components of AD-related pathologies. Furthermore, using transgenic mouse models of AD, experimental reduction in disease progression occurs with many of the methods shown to be beneficial for brain aging (e.g., DR, DHA, blueberry extract, and resveratrol).

## Technologic Advances

New technologies will soon advance our understanding of brain aging significantly. Microarray analysis is now available for screening microRNAs and SNPs. Laser capture microdissection (LCM) will allow researchers to perform molecular analysis of pure cells from specific microscopic regions of the brain; in addition, the synergism of molecular and computational biology disciplines is poised to make major contributions to the complex, multifaceted process of brain aging.

## Open Questions

Some of the critical issues that remain to be addressed include the unique roles that transcription, translation, and posttranslational modification (which currently is a major target of drug development) have in orchestrating age-related changes in motor and cognitive skills. The delineation between normal age-related changes and disease-related changes, such as those that occur in AD and Parkinson's disease, will identify unique molecular mechanisms underlying normal brain aging. Also, careful consideration must be given to the specific contributions made by vascular aging in the aging brain.

## Summary

Underlying normal age-related cognitive and motor deficits are gene expression changes that define the aging brain. The specific changes in gene expression that occur in each type of brain cell contribute to subtle dysfunctions in brain activities. A comparison of gene classes in the different species supports the belief that the aging brain experiences progressive increases of inflammatory and oxidative stress pathways, and that this results in the accumulation of altered or damaged proteins that interrupt normal synaptic functions. The current status of age-related gene expression supports an important role for glia in orchestrating these age changes in the brain.

*See also:* Aging of the Brain; Aging of the Brain and Alzheimer's Disease; Alzheimer's Disease: An Overview; Alzheimer's Disease: Transgenic Mouse Models; Alzheimer's Disease: MRI Studies; Animal Models of Alzheimer's Disease; Genomics of Brain Aging: Nuclear and Mitochondrial Genomes; Genomics of Brain Aging: Apolipoprotein E; Genomics of Brain Aging: Twin Studies.

## Further Reading

Blalock EM, Geddes JW, Chen KC, et al. (2004) Incipient Alzheimer's disease: Microarray correlation analyses reveal major transcriptional and tumor suppressor responses. *Proceedings of the National Academy of Sciences of the United States of America* 101: 2173–2178.

Lee CK, Weindruch R, and Prolla TA (2000) Gene-expression profile of the ageing brain in mice. *Nature Genetics* 25: 294–297.

Lu T, Pan Y, Kao SY, et al. (2004) Gene regulation and DNA damage in the ageing human brain. *Nature* 429: 883–891.

Morgan TE, Wong AM, and Finch CE (2006) Anti-inflammatory mechanisms of dietary restriction in slowing aging processes. In: Chanson P (ed.) *Endocrine Aspects of Successful Aging*, pp. 191–206. Berlin: Springer.

Pan F, Chiu CH, Pulapura S, et al. (2007) Gene Aging Nexus: A web database and data mining platform for microarray data on aging. *Nucleic Acids Research* 35: D756–D759.

## Relevant Website

http://gan.usc.edu – Gene Aging Nexus (online database of aging-related data developed at USC Davis School of Gerontology).

# Genomics of Brain Aging: Apolipoprotein E

**D K Lahiri and B Maloney**, Indiana University School of Medicine, Indianapolis, IN, USA



## Introduction

Dementia is rapidly becoming a common diagnosis in the current aging population, and the numbers are expected to rise exponentially in coming years. Alzheimer's disease (AD) alone now affects over 4.5 million people in the United States. In addition, millions more are currently affected by vascular dementia, Lewy body disease, and frontotemporal dementia (FTD). Therefore it is critical to understand potentially modifiable risk factors and early disease markers and the application of new disease-specific diagnostic tools and treatment modalities for these age-related dementias. In the present review, we focus on the current state of knowledge and directions of research on the role of apolipoprotein E (ApoE), encoded by the *APOE* gene, on the major dementias, particularly AD.

ApoE has a major role in cholesterol transport and plasma lipoprotein metabolism. It is also an important cholesterol transporter in the brain, being involved in neuronal repair, dendritic growth, maintenance of synaptic plasticity, and anti-inflammatory activities. Cholesterol regulates amyloid production and its deposition. ApoE is associated with both plaques and tangles, which are the two major pathological hallmarks of AD. Importantly, plaque formation seems to require the presence of ApoE. It is essential for amyloid deposition in the $\beta$-amyloid protein precursor (APP) Val717 $\rightarrow$ Phe (V717F) transgenic model of AD, suggesting that the role of ApoE in AD is to increase the yield and/or toxicity of amyloid plaques. *APOE* is known to have three alleles, $\varepsilon2$, $\varepsilon3$, and $\varepsilon4$. The risk of AD increases in accordance with genotype such that $\varepsilon4 > \varepsilon3 > \varepsilon2$. The $\varepsilon4$ allele has also been associated with anomic aphasia in FTD, while $\varepsilon2$ may be a risk factor for FTD in general, with the $\varepsilon2/\varepsilon4$ genotype conferring greatest risk for primary progressive aphasia.

However, multiple factors in addition to ApoE play an important role in neurological pathogenesis, making ApoE only one of several important contributors. In transgenic mice, for example, *APOE* $\varepsilon4$ expression causes neuropathology and behavioral deficits, but it is unclear whether the mechanism involves modulation of amyloid levels. ApoE is clearly an important and robust genetic risk factor for AD and other aging-related dementias. Perhaps the most important question in the area of age-related disorders concerns how ApoE accelerates dementia pathogenesis. In this regard, AD would serve as a model for ApoE's overall role, as this is currently the most extensively characterized relationship.

## ApoE Expression in Different Tissues and Cell Types and during Aging

ApoE mRNA is most abundant in the liver, followed by the brain, where it is synthesized and secreted primarily by astrocytes, oligodendrocytes, activated microglia, and ependymal layer cells. The synthesis and secretion of ApoE also occur in some human neurons, in which ApoE is induced by brain injury. Central nervous system (CNS) neurons express ApoE at lower levels than do astrocytes. APOE protein and mRNA are detected in cortical and hippocampal neurons in humans and in transgenic mice expressing human ApoE under the control of the human *APOE* gene promoter. ApoE is expressed in primary cultured human CNS neurons and in several human neuronal cell lines. Cerebrospinal fluid (CSF) ApoE is a component of CSF high-density lipoproteins (HDLs), the major lipoprotein type in the CSF.

*APOE* expression is upregulated in astrocytes during aging. Macrophages are the major source of *APOE* among various cell types studied. Two distal enhancers that specify *APOE* gene expression in macrophages were reported in transgenic mice with specific constructs of the human *APOE/APOC1/APOC1'/APOC4/APOC2* gene cluster. A multienhancer-1 (ME-l) element consists of a 620-bp sequence located 3.3 kb downstream of the *APOE* gene and the multienhancer-2 (ME-2) element comprises a 619-bp region located 15.9 kb downstream of the *APOE* gene and 5.9 kb downstream of the *APOC1* gene. Both ME-1 and ME-2 elements drive *APOE* gene expression in astrocytes of the olfactory bulb, cerebellum, and hippocampus in transgenic mice.

## ApoE and AD

Alzheimer's disease, the most common form of dementia among elderly, is a progressive, degenerative disorder of the central nervous system. The consequence of the disease is loss of memory; this and other cognitive and behavioral symptoms progressively impair function in the activities of daily living. Neurochemically, AD is characterized by selective neuronal cell death, the presence of amyloid deposits in the core of neuritic plaques, and the formation of neurofibrillary tangles in the brain of afflicted individuals. Plaque and tangles are deposited extraneuronally and intraneuronally, respectively. Notably, ApoE protein has been reported to co-localize with both of these neuropathological lesions of AD.

An increase in ApoE mRNA was reported in the hippocampus in AD and in rats after entorhinal cortex lesioning. In both AD and the lesion model, investigators observed a shift in the location of astrocytes containing prevalent ApoE mRNA, from the neuropil to regions with densely packed neurons. The increased abundance of ApoE mRNA in astrocytes close to neuron cell bodies could be indicative of lipid uptake in regions where neurons are degenerating or where synaptic remodeling is taking place.

## Protein Isoforms of ApoE and Their Role in AD

ApoE is a polymorphic 299-amino-acid protein with a molecular mass of 34 200 Da. The human *APOE* gene consists of at least three alleles ($\varepsilon2$, $\varepsilon3$, $\varepsilon4$) encoding three protein isoforms. The three common ApoE isoforms, $\varepsilon2$, $\varepsilon3$, and $\varepsilon4$, are all products of the same gene, with three alleles at a single gene locus. These isoforms differ from one another at residues 112 and 158. ApoE $\varepsilon3$ has Cys112 and Arg158, ApoE $\varepsilon4$ has arginine at both positions, and ApoE $\varepsilon2$ has cysteine. In almost all populations, the $\varepsilon3$ allele accounts for the vast majority of the *APOE* gene pool (typically ~70–80%) and the $\varepsilon4$ and $\varepsilon2$ alleles account for 10–15% and 5–10% of the population, respectively.

Among these isoforms, the $\varepsilon4$ allele of *APOE* constitutes a major susceptibility factor for the development of the familial and sporadic forms of late-onset AD. Individuals with one copy of *APOE* $\varepsilon4$ show a higher risk compared to those without the allele, and two copies of the *APOE* $\varepsilon4$ further increase the risk of AD. At the protein-binding level, ApoE binds $\beta$-amyloid (A$\beta$) peptide with high avidity, and the three isoforms differ in their binding properties, which may contribute to the isoform-specific effects of ApoE on A$\beta$ deposition. Animal studies suggest that ApoE participates in $\beta$-amyloid deposition in an isoform-specific manner. At drug treatment level, AD patients with ApoE $\varepsilon4$ responded poorly to treatment with tacrine compared to patients with ApoE $\varepsilon2$ and ApoE $\varepsilon3$.

### ApoE Isoforms and Familial AD-Linked Mutations

Familial AD (FAD) is caused by mutations in several genes, including *APP*, *PSEN1*, and *PSEN2*, which encode $\beta$-amyloid protein precursor, presenilin 1 (PS1), and presenilin 2 (PS2), respectively. Mutations in *APP*, *PSEN1*, and *PSEN2* account for only a small percentage of AD patients, even among those who have early age of onset and a family history of the disease. The role of ApoE isoforms in the pathogenesis of AD is visible by their interaction with FAD mutations, as well as in Down syndrome. In those cases when FAD is associated with a mutation in APP or increased load of APP, the presence of an *APOE* $\varepsilon4$ allele, in comparison to *APOE* $\varepsilon3$, significantly reduces the age of onset of AD. In contrast, most mutations in PS1 and PS2, which together account for the bulk of the known FAD families, are only weakly affected by the *APOE* $\varepsilon4$ allele. This finding cannot be explained by assuming a ceiling effect on the age of onset. It would appear that the *APOE* $\varepsilon4$ allele affects a particular pathway that is either removed or saturated by the PS1/2 mutations. It is then most likely that ApoE influences the pathogenesis of AD by altering the pathology, although cell culture studies indicate that there are no changes in A$\beta$ production because of co-expression of any of the three *APOE* alleles.

### Neurobiological Mechanisms: ApoE Isoform Status Influences Risk of AD by a Relatively Selective Effect on Episodic Memory

The neurobiological mechanisms linking the $\varepsilon4$ allele risk of AD are beginning to be understood. Previous histopathological studies report that the $\varepsilon4$ allele is associated with the number of neuritic plaques and neurofibrillary tangles, suggesting that the effect of the $\varepsilon4$ allele on disease may be mediated by an increase in the rate that AD pathology accumulates, but not all studies have found this association, and others suggest that the $\varepsilon4$ allele is related to other mechanisms, such as neural repair or survival. Data from the Religious Orders Study, an ongoing longitudinal clinical–pathologic study of aging and AD, were used to examine the pathologic mechanisms linking the $\varepsilon4$ allele to the development of AD. These studies suggest that *APOE* allele status influences risk of AD by a relatively selective effect on episodic memory. Other studies suggest that the presence of a single *APOE* $\varepsilon4$ allele is associated with an increased rate of hippocampal volume loss in healthy women in their sixth decade of life, but is not related to any detectable memory changes.

### Neuropathological Association and Role of *APOE* $\epsilon4$ Allele in the Clinical Manifestation of AD

Bennett and colleagues have tested the hypothesis that the $\varepsilon4$ allele is associated with the clinical manifestations of AD through an association with the pathologic hallmarks of disease among 128 deceased participants in the Religious Orders Study. The effect was present when measures of neuritic plaques, diffuse plaques, and neurofibrillary tangles were analyzed separately and when measures of five different cognitive systems (episodic memory, semantic memory, working memory, perceptual speed, and visuospatial ability) were analyzed independently. The ε4 allele is, therefore, associated with the clinical manifestations of AD through an association with the pathologic hallmarks of AD, rather than some other mechanism (e.g., direct effect on neuronal survival).

The relationships underlying neuropathology also exist in normal individuals. A study of Aβ42 and Aβ40 concentration in individuals without AD revealed a significant association between the *APOE* ε4 allele and greater decline in the amount of Aβ42 in cerebrospinal fluid than in non-*APOE* ε4 individuals. This sharper decrease in Aβ42 in CSF is likely due to greater precipitation of Aβ42 as amyloid plaque in brain parenchyma.

### ApoE Genotype/Age Interaction Affects Cerebral Glucose Metabolism

Positron emission tomography (PET) studies in healthy elders showed that the *APOE* ε4 allele is disruptive to the regional cerebral metabolic rate of glucose metabolism (rCMRGlu), perhaps via an interaction with the aging process. Another study assessed whether this interaction occurs in AD patients. ApoE groups were comparable for age, gender, age at onset, and disease duration. Correlations between age and rCMRGlu confirmed a negative relationship for both groups. For example, lower rCMRGlu was found within the frontal and cingulate areas for *APOE* ε4 carriers as compared with the noncarriers. Additionally, a significant ApoE/age interaction was detected in the frontal and anterior cingulate cortex, with the *APOE* ε4 carriers having a steeper regression slope with respect to the noncarriers. These results indicate that age-related regional rCMRGlu decreases within the frontal and anterior cingulate areas may be more severe in AD patients carrying the *APOE* ε4 allele.

### Increased ApoE mRNA Level in AD Subjects Carrying the *APOE* ε4 Allele

In addition to functional alterations associated with the ε4 isoform, gene expression of *APOE* has been assessed in the brains of AD subjects according to *APOE* genotype. Increased ApoE mRNA was reported in AD subjects carrying the ε4 allele. ApoE mRNA level was significantly higher in the AD group than in the control group ($p < 0.05$). ApoE mRNA level in the AD group with ε4 was also significantly higher than that in the control group with ε4 ($p < 0.05$). However, this is controversial. For example, in another study, levels of ApoD, but not ApoE, were reported to increase in the CSF and hippocampus of AD patients as a function of inheritance of the APOE ε4 allele.

### *APOE* ε4 Still May Not Be a Major Susceptibility Factor for All Racial and Ethnic Groups

Even considering all of the aforementioned evidence, the relationship of the ApoE isoform to AD risk is not necessarily clear-cut. To understand the role of the ε4 allele in different racial groups, Bennett and colleagues presented data from a geographically defined biracial population on the south side of Chicago. The effect of the ε4 allele on the risk of AD differed strongly between African-Americans and Caucasians. For example, among Caucasians the presence of the ε4 allele was associated with a 2.73-fold increase in risk, while among African-Americans there was no significant increase in risk. Other studies also report a weaker (or no) association of the ε4 allele among racial and ethnic minorities, although this remains controversial. Other genetic and environmental risk factors may be important for the development of AD among racial and ethnic minorities in particular.

## The Regulatory Region of the *APOE* Gene

In addition to the differential risk related to ε2/ε3/ε4 alleles, additional risk for AD may be related to transcriptional activity of the *APOE* gene. ApoE mRNA has been shown to be significantly more abundant in AD patients than in a control group, regardless of ApoE isoform genotype. In addition, reports have been made of correlation related to the presence of promoter single-nucleotide polymorphisms (SNPs), ApoE expression levels, and risk of AD.

The gene encoding ApoE in humans and in the most common model system (mice) resides in a well-conserved cluster of genes. The human *APOE* gene is on chromosome 19, while the mouse homolog, *Apoe*, is on mouse chromosome 7 (**Figure 1**). The genes for ApoE of mouse and human both consist of a 5′-untranslated region (5′-UTR) divided between two exons, three coding-sequence exons, and a terminator (**Figure 2**). All introns in the human gene are longer than those in the mouse.

### Structure of the 5′-Flanking Region of the Gene for ApoE

The proximal region of the ApoE gene promoter is highly conserved in the mouse, rat, and human; the relative position of the 'TATA box' and the two copies of the 'GC box' are identical. Such characterization of the ApoE promoter is important for understanding the mechanism of ApoE gene expression and its regulation.

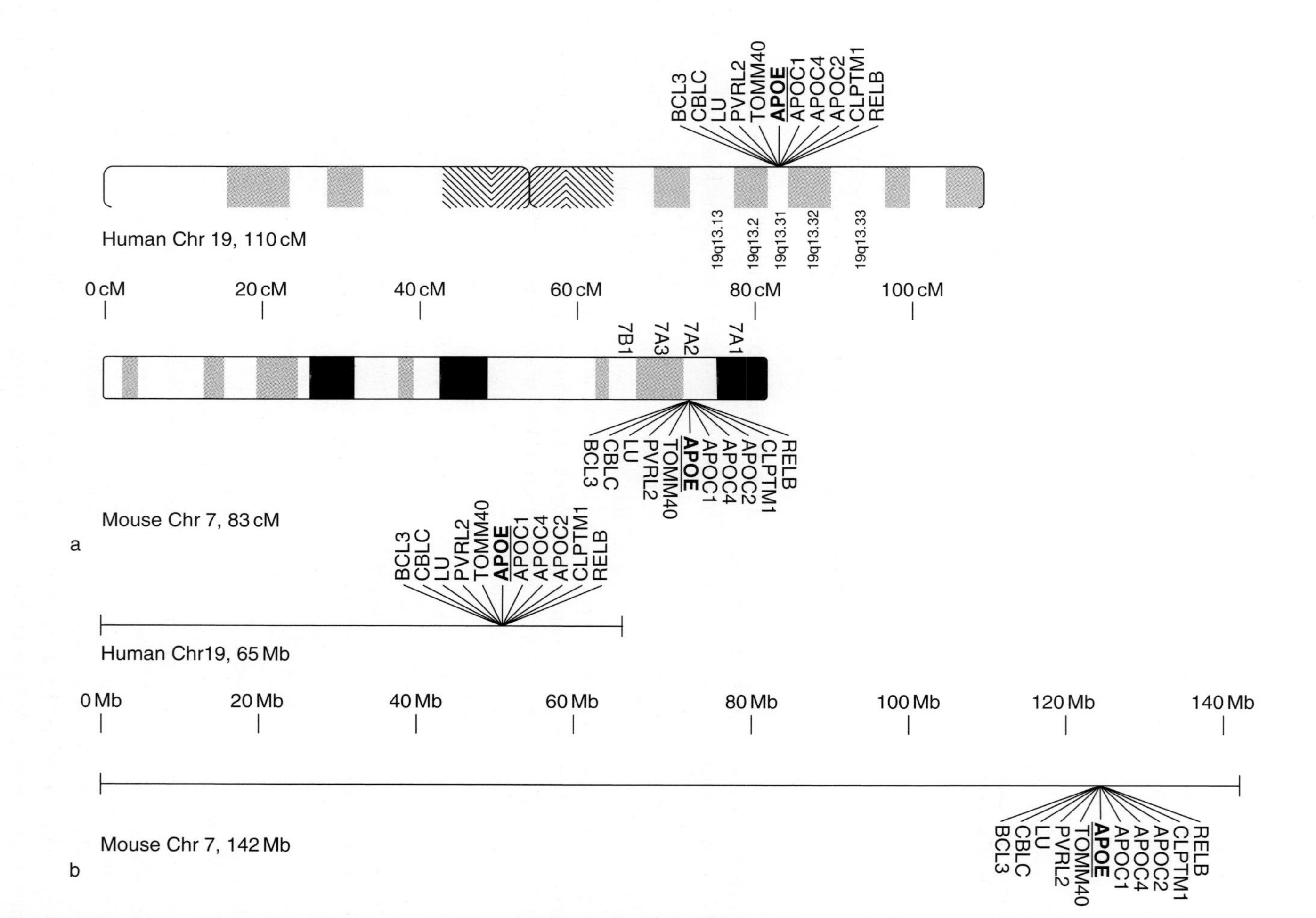


**Figure 1** Human and mouse chromosome regions near their respective (*APOE*, *Apoe*) genes encoding apolipoprotein E. Databases for human and mouse genomes were searched for genes in the region of each species' apolipoprotein E gene on the respective chromosomes. Homologous genes were mapped according to genetic (cM, centimorgan; 1 cM ~ 1 Mb) and physical (Mb, megabase pair) distances. Mouse maps are presented in 'reversed' orientation for clarity of comparison. (a) Genetic maps of human chromosome 19 and mouse chromosome 7. Maps are to the same centimorgan scale. (b) Physical maps of human chromosome 19 and mouse chromosome 7. Maps are to the same megabase pair scale.

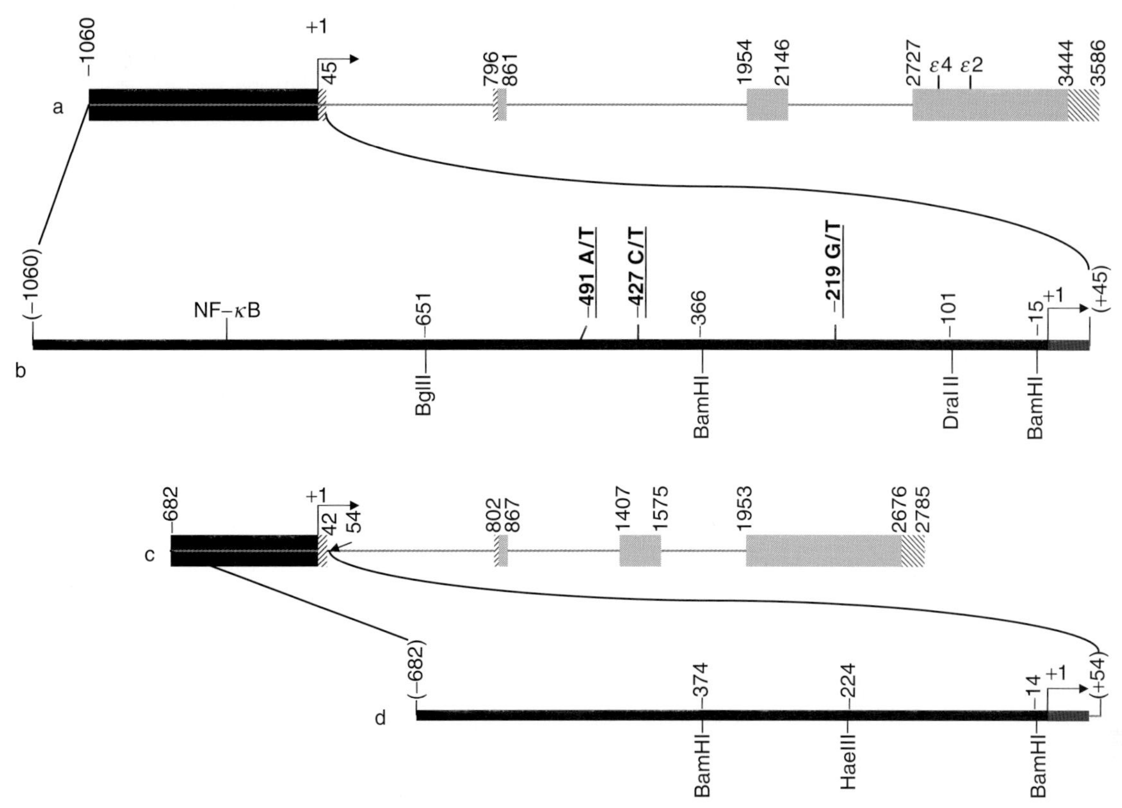


**Figure 2** Human (a, b) and mouse (c, d) genes that encode apolipoprotein E, and 5′-flanking region structures. (a) Map of the human *APOE* gene, including introns, exons, terminator region, and 5′-flanking region. The 5′-untranslated region is in blue, coding sequence exons are indicated in green, and the 3′-untranslated region/terminators are in red. Locations of the sites of the ε2 and ε4 alleles are indicated. (b) Map of the human *APOE* 5′-flanking region, including locations of the polymorphisms at −491, −427, and −219. (c) Map of the mouse *Apoe* gene, including introns, exons, terminator region, and 5′-flanking region. The 5′-untranslated region is in blue, coding sequence exons are indicated in green, and the 3′-untranslated region/terminators are in red. (d) Map of the mouse *Apoe* 5′-flanking region. Dashed vertical line indicates the 5′-untranslated region exon/intron border. Maps (a) and (c) are drawn on the same scale, and maps (b) and (d) are drawn on the same scale. See text for discussion of other sites. Adapted from Maloney B, Ge Y-W, Alley GM, and Lahiri DK (2007) Important differences between human and mouse APOE gene promoters with implications for Alzheimer's disease. *Journal of Neurochemistry* 103: 1237–1257.

Transcription control of the mouse *Apoe* (m-*Apoe*) gene has been studied. Promoter activity of a 729-bp 5′-flanking region, which is located 729-bp upstream from the translation initiation codon (ATG), has been thoroughly examined. The 729-bp region was cloned into the reporter chloramphenicol acetyl transferase (*CAT*) gene in a promoterless vector (pBLCAT3). Transfection studies of the m-*Apoe* promoter in rat glial C6 and neuronal PC12 cell lines suggest that the 729-bp *Apoe* region (from position −675 to +54) is functionally active in these cell types. The serial deletion analysis indicates that the 278-bp promoter region (from −224 to +54) has the highest, and the 68-bp region (from −14 to +54) the lowest, activity on the reporter gene in neuronal and astrocytic cell lines. These results indicate that the 151-bp region (from −374 to −224) has a negative regulatory effect on the reporter gene. DNA–protein interaction studies reveal that the 68-bp region binds a specific transcription factor(s) in PC12 nuclear extracts. These functional studies suggest that mouse *Apoe* gene can also be expressed in neuronal cells in addition to the astrocytic cells.

### Promoter Polymorphisms of the *APOE* Gene

In addition to the highly characterized isoform ε2/ε3/ε4 polymorphisms of *APOE*, three promoter SNPs have been reported with some level of associated AD risk. These polymorphisms reside at −491 (A/T), −427 (C/T), and −219 (G/T). Of the three, −491A/T has the most agreement in the literature as representing a risk factor (as −491A) for AD independent of *APOE* ε4, although some reports only indicate that it increases the effect of the ε4 allele and it has been reported that this polymorphism has no effect in some ethnic groups. In addition, it has been suggested that −491A is associated with increases in subjective memory loss. The −491T variant has been found to correlate with improved response to the drugs atorvastatin and benzafibrate. The −427C/T polymorphism has usually been reported to offer no significant association with AD. The −219G/T polymorphism has been shown to have no significant effect on age-related neurodegeneration, to be an independent risk factor (as −219T) for AD, or to only be a risk factor in association with *APOE* ε4. Interestingly, the −219T polymorphism has been suggested to have an association with coronary artery disease and open-angle glaucoma. The incompleteness of effect for these polymorphisms strongly indicates that environmental interaction plays an important role in any effects ApoE might have on aging and age-related pathology.

### Interaction of Environmental Factor(s) with the *APOE* Gene Promoter

The *APOE* promoter region contains important transcription factor sites through which the promoter interacts with various cellular factors. There are some functionally tested DNA sites in the *APOE* promoter for possible binding of interesting transcriptional factors. Nuclear factor-kappa B (NF-κB) and activator protein-2 (AP-2) sites present within the *APOE* promoter are mainly responsible for environmental factor-mediated *APOE* gene expression. DNA sequence analysis predicts the presence of putative sites for activator protein-1 (AP-1) and AP-2, and a functional site for the NF-κB transcription factor (TF) in the human *APOE* promoter. Gel electrophoretic mobility analysis has confirmed the presence of one predominant NF-κB site within the human *APOE* promoter. The significance of the presence of these important TF sites in the promoter is evident from a finding that cytokine/Aβ-induced glial activation, NF-κB, and ApoE play critical roles in neuroinflammatory process seen in AD. Notably, some of the aforementioned *APOE* TF sites are also present in the *APP* promoter region, suggesting a common regulatory pathway for both of these important genes in the pathogenesis of AD. Thus, these sites can potentially be effective drug targets to control the expression of the *APOE* gene.

In addition to acute environmental–gene promoter interactions, the possibility may exist that *APOE* gene expression may be subject to alteration in a LEARn (latent early-life associated regulation) fashion. LEARn begins by transient exposure of an organism to a stimulus such as lead (Pb), which alters DNA methylation and/or oxidation patterns. However, unlike conventional Pb-toxicity, there is no acute effect of the environmental exposure. Instead, the potential of gene expression is latently altered, only to be triggered later in life by a second event, such as the widespread changes in levels of transcription and inflammatory factors seen in the transition from maturity to old age. Mechanistically, a gene's primary DNA sequence would not be altered either by the original stimulus nor by the later trigger. Instead, methylation (or oxidation) would produce a latent 'somatic epitype' that would ultimately determine gene expression levels. The LEARn model provides a mechanism for explanation of the difficulty in tracking environment–gene interactions.

### Interaction of the Aβ Peptide with the *APOE* Promoter

Treating cultured cells with the Aβ peptide increased *APOE* promoter activity in a dose-dependent manner. Interestingly, Aβ-mediated *APOE* promoter activity is regulated by the NF-κB site, as evident from other functional studies, and this interaction can be altered by salicylate. Thus, the NF-κB site could potentially be a therapeutic target to control the increased *APOE* promoter activity seen in AD (**Figure 3**).

### Effects of Statins on *APOE* Gene Expression: Role in Regulating Inflammation and Cell Proliferation/Differentiation

The effects of statins have been tested on the activation of transcription factors, which are known to regulate inflammation and cell proliferation/differentiation. Simvastatin, atorvastatin, and lovastatin inhibited the binding of nuclear proteins to both the NF-κB and AP-1 DNA consensus oligonucleotides in human endothelial and vascular smooth muscle cells. Hydroxymethyl glutaryl coenzyme A (HMG-CoA) reductase inhibitors upregulate the inhibitory IκB-α protein levels in endothelial cells and decrease c-Jun mRNA expression in smooth muscle cells. Thus, HMG-CoA reductase inhibitors downregulate the activation of transcription factors NF-κB, AP-1, and hypoxia-inducible factor-1α. These results corroborate the concept that statins have anti-inflammatory and antiproliferative effects that are relevant in the treatment of atherosclerotic diseases. Current studies are directed to examine the role of those transcriptional factors' sites within the *APOE* promoter that would be affected by the HMG-CoA reductase inhibitors, which otherwise regulate inflammatory transcription factors that are involved in age-related disorders such as AD.

## Interaction of ApoE with Other Risk Factors: Oxidative Stress, Inflammation, and Cholesterol Transport

Besides genetic etiological effects on AD, ApoE interacts with other risk factors, such as oxidative and inflammatory processes, nutrition, and environment. ApoE has been shown to possess anti-inflammatory

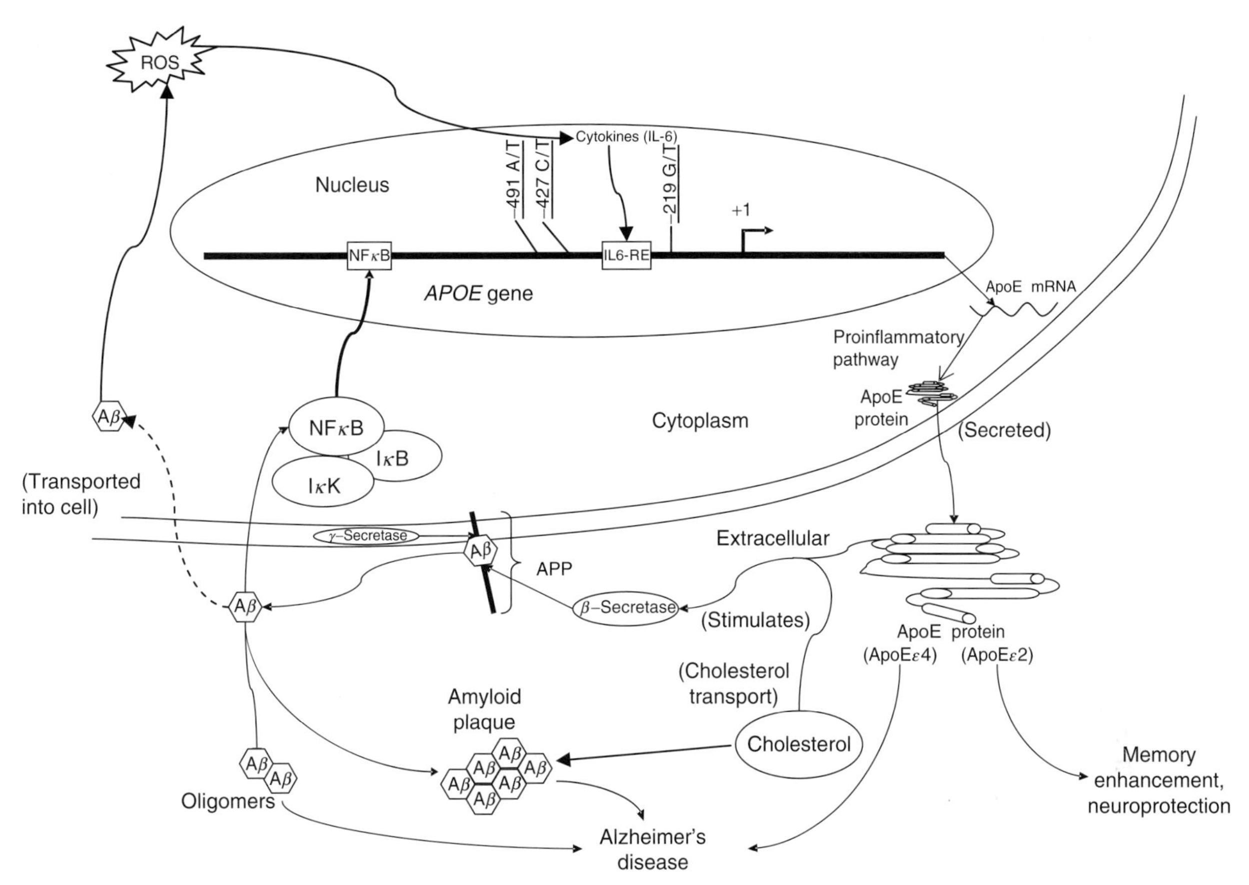


**Figure 3** Schematic of interactions of the apolipoprotein E (ApoE)-encoding *APOE* gene with β-amyloid (Aβ) and with environmentally driven risk factors that lead to Alzheimer's disease pathogenesis. The *APOE* regulatory region contains binding sites for several transcription factors, such as nuclear factor-kappa B (NF-κB), and inflammatory factors, such as interleukins (IL-6 RE; interleukin-6 response element), which modulate expression of the *APOE* gene. In addition, promoter single-nucleotide polymorphisms influence ApoE levels and risk for age-related neurodegenerative disorders such as Alzheimer's disease. Modulation of this regulatory region alters *APOE* transcription levels, thereby changing ApoE protein synthesis. ApoE acts as a cholesterol transporter protein and carries cholesterol into the brain. Cholesterol modulates β-amyloid protein precursor (APP) processing, as it increases the activity of one of the key secretase enzymes, β-secretase, which results in an increased Aβ production. Aβ plays several roles in the neuropathology: Aβ transported into the cell contributes to reactive oxidative species (ROS), inducing cytokine response. Oligomeric Aβ causes neurodegeneration, and excessive Aβ results in amyloid deposits, which also contain cholesterol and ApoE. Furthermore, Aβ activates NF-κB, which in turn goes to the nucleus and activates the *APOE* gene. In addition to this pathway, the *APOE* ε2 allele has a positive effect on memory and protective effect against Alzheimer's disease. However, allele ε4 may have a role in neurotoxicity and promotes fibrillogenesis, as deduced from cell culture and animal studies. All of these events lead to neurodegeneration and accelerate Aβ deposition. Thus, a vicious loop of interacting Aβ, *APOE*, cholesterol, and APP forms a cascade of events, culminating in Alzheimer's disease pathogenesis. From Lahiri DK, Sambamurti K, and Bennett DA (2004) Apolipoprotein gene and its interaction with the environmentally driven risk factors: Molecular, genetic and epidomiological studies with Alzheimer's disease. *Neurobiology of Aging* 25: 651–660.

properties in a variety of models. During normal aging, the brain undergoes several alterations, including reduced neuronal functioning and alterations in glia homeostasis. An increase in inflammatory signaling has also been reported in some studies of the aging brain, with inflammation potentially mediating age-related changes in the brain. Age-related alterations in ApoE expression have been shown to correlate with age-related changes in the cytokine interleukin-1β (IL-1β) in the brain of aging rats. In another study, ApoE expression was not altered during normal brain aging, but a relationship may exist between ApoE and IL-1β transcription in the cerebral cortex.

The role of cholesterol and oxidative stress is important in the pathogenesis of AD. Our current understanding of the role of oxidative stress in AD has resulted in the potentially beneficial use of vitamin E or selegiline, in delaying the clinical progression of functional loss and the time of nursing home placement. Cerebral oxidative damage is a feature of aging and increases in neurodegenerative diseases such as AD. In aged male and female mice, it has been shown that lack of ApoE is associated with oxidative damage in the brain, with results more pronounced for male mice, especially those fed a tocopherol-deficient diet, and that supplementation with tocopherol is not effective in reversing arachidonic acid oxidation in APOE(−/−) mice.

## Summary

Apolipoprotein E is encoded by a gene located on the long arm of human chromosome 19 (19q13). The protein is a vital part of cholesterol transport and functions in neurite outgrowth. ApoE mRNA and protein are found predominantly in astrocytes within the CNS. ApoE appears in three isoforms, ε2/ε3/ε4. The *APOE* ε4 allele is associated with the development of late-onset familial AD as well as late-onset sporadic AD. *APOE* ε4 has a gene dose effect on the risk and age of onset of AD. At the gene expression level, there is a report of a high expression of *APOE* mRNA in the brains with sporadic AD. Two promoter polymorphisms, −491A/T and −219G/C, have been repeatedly shown to influence AD risk. At the protein level, ApoE has been found to co-localize with one of the typical neuropathological lesions of AD, specifically the extracellular amyloid deposits.

Herein we have reviewed *APOE* as a risk factor for AD (as the most common form of aging-related neurodegeneration), both from the standpoint of the ε4 allele and potentially influential promoter polymorphisms, in general populations, among different racial and ethnic groups. The effect of the ε4 allele on annual decline in episodic memory is significantly stronger than is its effect on decline in other cognitive systems and at baseline, but there is an equal and opposite effect of the ε2 allele. Thus, the *APOE* allele status may influence risk of AD by a relatively selective effect on episodic memory. In addition, allele status alters cerebral glucose metabolism and influences age-related hippocampal volume loss.

*APOE* operates in age-related neurodegeneration, such as in AD, in a complex set of feedback loops that are influenced by both intrinsic and environmental factors (**Figure 3**). Briefly, the *APOE* gene regulatory region is a target for transcription factors such as NF-κB and inflammatory factors such as interleukins. Single-nucleotide polymorphisms in the promoter influence ApoE levels. Modulation of this regulatory region by differences in transcription factor/cytokine binding would alter *APOE* transcription levels, changing ApoE protein synthesis levels. Different responses to the same levels of transcription factors according to promoter polymorphism would also alter *APOE* mRNA levels. ApoE acts as a cholesterol transporter protein and carries cholesterol into the brain. Cholesterol modulates APP processing, since it increases the activity of one of the key secretase enzymes, β-secretase, which results in increased Aβ production. Aβ transported back into the cell would contribute to reactive oxidative species (ROS), inducing cytokine response. Oligomeric Aβ would cause gross neurodegeneration, and excess extra cellular Aβ accumulation results in amyloid deposits, which also contain cholesterol and ApoE. Furthermore, Aβ activates NF-κB, which in turn activates the *APOE* gene. Thus a vicious loop of interacting Aβ, *APOE*, cholesterol, and APP forms a cascade of events culminating in AD pathogenesis.

The proximal region of the *APOE* gene is highly conserved in the mouse, rat, and human; the relative positions of the 'TATA box' and the two copies of 'GC box' are identical. The human *APOE* promoter DNA contains putative sites for AP-1, AP-2, and NF-κB transcription factors. The NF-κB site interacts with Aβ, and this interaction can be blocked by salicylate. The promoter region also contains inflammatory response transcription factor sites such as IL-6 response element binding protein. Moreover, some of these transcription factor sites should provide suitable targets for drug development in treating devastating aging-related diseases such as AD.

*See also:* Aging of the Brain; Aging of the Brain and Alzheimer's Disease; Alzheimer's Disease: An Overview; Alzheimer's Disease: Molecular Genetics; Alzheimer's Disease: Transgenic Mouse Models; Alzheimer's Disease: MRI Studies; Animal Models of Alzheimer's Disease; Brain Glucose Metabolism: Age, Alzheimer's Disease and ApoE Allele Effects; Gene Expression in Normal Aging Brain; Genomics of Brain Aging: Twin Studies.

## Further Reading

Artiga MJ, Bullido MJ, Frank A, et al. (1998) Risk for Alzheimer's disease correlates with transcriptional activity of the APOE gene. *Human Molecular Genetics* 7: 1887–9182.

Bales KR, Du Y, Holtzman D, et al. (2000) Neuroinflammation and Alzheimer's disease: Critical roles for cytokine/Abeta-induced glial activation, NF-kappaB, and apolipoprotein E. *Neurobiology of Aging* 21: 427–432.

Biere AL, Ostaszewski B, Zhao H, et al. (1995) Co-expression of beta-amyloid precursor protein (betaAPP) and apolipoprotein E in cell culture: Analysis of betaAPP processing. *Neurobiology of Disease* 2: 177–187.

Corder EH, Saunders AM, Strittmatter WJ, et al. (1993) Gene dose of apolipoprotein E type 4 allele and the risk of Alzheimer's disease in late onset families. *Science* 261: 921–923.

Evans DA, Bennett DA, Wilson RS, et al. (2003) Incidence of Alzheimer disease in a biracial urban community: Relation to apolipoprotein E allele status. *Archives of Neurology* 60: 185–189.

Graff-Radford NR, Green RC, Go RC, et al. (2002) Association between apolipoprotein E genotype and Alzheimer disease in African American subjects. *Archives of Neurology* 59: 594–600.

Harwood DG, Barker WW, Ownby RL, et al. (2002) Apolipoprotein E polymorphism and cognitive impairment in a bi-ethnic community-dwelling elderly sample. *Alzheimer Disease & Associated Disorders* 16: 8–14.

Hebert LE, Scherr PA, Bienias JL, et al. (2003) Alzheimer disease in the US population: Prevalence estimates using the 2000 census. *Archives of Neurology* 60: 1119–1122.

Lahiri DK (2004) Apolipoprotein E as a target for developing new therapeutics for Alzheimer's disease: Characterization of the regulatory regions of the apolipoprotein gene. *Journal of Molecular Neuroscience* 23: 225–233.

Lahiri DK, Alley GM, Ge YW, et al. (2002) Functional characterization of the 5′-regulatory region of the murine apolipoprotein gene. *Annals of the New York Academy of Sciences* 973: 340–344.

Lahiri DK and Maloney B (2006) Genes are not our density: The somatic epitype bridges between the genotype and the phenotype. *Nature Reviews Neuroscience* 7: doi:10.1038/nrn2022-C1.

Lahiri DK, Maloney B, Basha MR, Ge Y-W, and Zawia NH (2007) How and when environmental agents and dietary factors affect the course of Alzheimer's disease: The 'LEARn' model (Latent Early Associated Regulation) may explain the triggering of AD. *Current Alzheimer Research* 4: 219–228.

Maloney B, Ge Y-W, Alley GM, and Lahiri DK (2007) Important differences between human and mouse APOE gene promoters with implications for Alzheimer's disease. *Journal of Neurochemistry* 103: 1237–1257.

Mayeux R, Small SA, Tang M, et al. (2001) Memory performance in healthy elderly without Alzheimer's disease: Effects of time and apolipoprotein-E. *Neurobiology of Aging* 22: 683–689.

Namba Y, Tomonoga M, Kawasaki E, et al. (1991) Apolipoprotein E immunoreactivity in cerebral amyloid deposits and neurofibrillary tangles in Alzheimer's disease and kuru plaque amyloid in Creutzfeldt–Jakob disease. *Brain Research* 541: 163–166.

Pitas RE, Ji ZS, Weisgraber KH, et al. (1998) Role of apolipoprotein E in modulating neurite outgrowth: Potential effect of intracellular apolipoprotein E. *Biochemical Society Transactions* 26: 257–262.

Refolo LM, Malester B, LaFrancois J, et al. (2000) Hypercholesterolemia accelerates the Alzheimer's amyloid pathology in a transgenic mouse model. *Neurobiology of Disease* 7: 321–331.

Saunders AM, Trowers MK, Shimkets RA, et al. (2000) The role of apolipoprotein E in Alzheimer's disease: Pharmacogenomic target selection. *Biochimica et Biophysica Acta* 1502: 85–94.

Slooter AJ, Houwing-Duistermaat JJ, van Harskamp F, et al. (1999) Apolipoprotein E genotype and progression of Alzheimer's disease: The Rotterdam Study. *Journal of Neurology* 246: 304–308.

Wang JC, Kwon MD, Shah P, et al. (2000) Effects of *APOE* genotype and promoter polymorphisms on risk of Alzheimer's disease. *Neurology* 55: 1644–1649.

Wilson RS, Schneider JA, Barnes LL, et al. (2002) Effect of apolipoprotein E ε4 allele status on decline in different cognitive systems and development of Alzheimer's disease over a 6-year period. *Archives of Neurology* 59: 1154–1160.

# Genomics of Brain Aging: Twin Studies

**C A Reynolds**, University of California at Riverside, Riverside, CA, USA



## The Twin Method

The study of adult twin siblings offers a distinctive perspective on brain aging. Monozygotic (MZ; identical) twins share 100% of their genes in common, while dizygotic (DZ; fraternal) twins share 50% of their segregating genes in common. This fortunate experiment of nature affords multiple avenues to consider the etiology of brain aging. For example, comparing the relative similarity of MZ twins to DZ twins allows one to consider the extent of genetic and environmental influences on individual variation in cognitive performance or dementia status. If MZ twins are more similar than are DZ twins in recall memory performance, for example, then genetic influences are implicated. For this same trait, if DZ twins are as similar to each other as are MZ twins, then shared environmental effects, such as family environments and school environments, are implied. To the extent that MZ twins are dissimilar (e.g., a correlation less than 1.0), then nonshared environmental influences are implied.

### Quantitative Traits

For quantitative traits, such as recall memory performance scores, covariances between twin pair scores are calculated for MZ pairs ($C_{mz}$) and for DZ pairs ($C_{dz}$). Next, a formal biometric model is fit to the covariances. The biometrical model quantifies understanding of all possible, though anonymous, genetic and environmental causes and shows how these factors contribute to sibling similarity. For example, the biometrical model could state that individual differences in recall memory performance ($V_{mem}$) are a function of genetic factors that act additively ($V_a$), plus rearing environment factors ($V_r$), plus unique person-specific (nonshared) experiences ($V_e$):

$$V_{mem} = V_a + V_r + V_e \qquad [1]$$

Next, the model specifies that two related individuals have similar recall memory scores due to the amount of genes shared, and the environment shared. For MZ twins reared together, their recall memory scores are similar, or co-vary, because the twins are genetically identical and they share a rearing environment:

$$C_{mz} = V_a + V_r \qquad [2]$$

For DZ twins reared together, their recall memory scores are similar because the twins share half of their segregating genes in common and share a rearing environment:

$$C_{dz} = 1/2V_a + V_r \qquad [3]$$

Heritability is often reported as a summary statistic and is a numerical ratio of genetic factors ($V_a$) over the total amount of individual differences in recall memory ($V_{mem}$). Thus, heritability ($h$) indicates the proportion of individual differences in recall memory scores that is due to genetic differences:

$$h^2 = V_a / V_{mem} \qquad [4]$$

Environmentality estimates are computed in similar fashion (e.g., shared environmentality is the ratio of $V_r/V_{mem}$). In longitudinal studies, there is increasing emphasis on unscaled genetic and environmental variance (i.e., $V_a$, $V_r$, $V_e$). For example, one can better judge whether both genetic and environmental influences are increasing or decreasing with age, a process that otherwise would be masked with ratio-based heritability and environmentality statistics. This knowledge is important in theorizing about potential gene action, specific environmental factors to consider, or longevity-related mechanisms whereby individuals with particular genotypes may die at earlier (or later) ages.

### Disease Traits

Disease traits are typically measured in terms of absence or presence (i.e., dichotomously), or otherwise as ordinal categories. Concordance rates are commonly reported for dichotomous traits and are compared between MZ and DZ twin pairs. Probandwise concordance is the simple proportion of co-twins of affected proband twins that exhibit the disease trait (e.g., Alzheimer's disease (AD)). In much the same way as for quantitative measures of pair similarity, the extent of differences between MZ and DZ concordance rates gives an indication of genetic influence. Furthermore, the extent to which the MZ concordance rate is less than 100% suggests the degree to which nonshared environmental influences, including random factors, are important.

Other measures of pair similarity may be reported for dichotomous or ordinal categorical traits, including tetrachoric and polychoric correlations, respectively. Herein, the assumptions are that there is an underlying continuum of liability or risk of the disease and that the observed proportion of those with the disease constitutes a threshold point on the normal curve where the disorder is manifested.

The interpretation of the correlations computed for twin pairs is that they reflect the magnitude of pair similarity for the underlying disease liability continuum. Prevalence rates increase concordance rates even when the underlying pair similarity for the disease liability is identical. Thus, tetrachoric correlations are preferred over concordance rates when the prevalence rates of the disease differ between groups included in the analysis, as can be the case for males versus females (e.g., AD). The biometrical model described in eqns [1]–[3] is applied to the tetrachoric or polychoric correlations in similar fashion.

### Assumptions of the Twin Method

Analytical requirements include the assumption of equal environments for twins-reared-together, as well as no assortative mating. The equal environments assumption (EEA) states that the shared environment (e.g., family environment, school environment) contributes equally to a trait in both MZ and DZ pairs. The vast majority of analyses which have tested for EEA violation across multiple psychological and psychiatric traits have indicated little support for EEA violation, thus providing sufficient confidence in the findings from studies of twins-reared-together, though it must be noted that some controversy remains over the adequacy of tests for violations of the EEA.

A second assumption is the absence of assortative mating – that is, spouses or mates do not select one another on the basis of the trait in question. There is ample evidence that spouses are moderately similar in terms of cognitive abilities, extant from the time of marriage. For aging-related cognitive traits (i.e., cognitive aging, AD), there are few data reported on spouse similarity. For now, it appears that the similarity between spouses for dementia is no greater than the similarity between any two people of the same age and sex.

### Co-Twin Control

In studies of discordant twins, the co-twin control method compares affected twins to their unaffected partners on a host of environmental factors. This design is advantageous to comparable case-control designs, as matching the discordant twins to their co-twin accounts for any confounding genetic of familial factors that might bear upon the disease trait. Indeed, discordant MZ twin pairs who share 100% of their genes in common are most valuable for full control of genetic and familial factors.

### Candidate Genes

Twin siblings who differ in terms of their recall memory scores, for example, and who differ in genotype at a particular gene locus (e.g., *APOE* ε3/ε4 vs. *APOE* ε3/ε3), give potential information about the association of the candidate gene with recall memory performance. True gene effects contribute to within-twin-pair differences, in the case of DZ twins, as well as to between-twin-pair differences for both MZ and DZ twins. Thus, formal tests of association in twins compare whether, according to genotype, within-pair differences and between-pair differences in trait scores are of the same magnitude. Twin designs have a distinct advantage over standard case-control designs, in which genetic association may be spurious if cases and controls originate from separate populations with differing gene frequencies (i.e., population stratification).

## Cognitive Aging and Dementia via Twin Models

### Twin Studies

Several research programs include adult twins as a focus in studies of cognitive abilities, most situated in the US (e.g., MTSADA, Minnesota Twin Study of Adult Development and Aging; NHLBI, National Heart, Lung, and Blood Institute Twin Study; New York Twin Study) or Scandinavia (LSADT, Longitudinal Study of Aging Danish Twins; Norwegian Twin Register; OCTO-Twin, Origins of Variance in the Oldest Old; SATSA, Swedish Adoption/Twin Study of Aging; the Gender Study of Opposite-Sex Twins), though others are located in Germany (German Observational Study of Adult Twins), Italy (Italian Twin Registry), Japan (Osaka/Kinki University Twin Study; OUATR, Osaka University Aged Twin Registry), and The Netherlands (Netherlands Twin Register). Among these research programs, the longitudinal studies, LSADT, NHLBI, OCTO-Twin, and SATSA have provided much of our knowledge about the etiology of cognitive decline. Summaries of findings from these studies are presented in the following sections.

Twin studies of cognitive impairment and dementia are fewer in number than are those addressing cognitive aging, and include two large-scale programs in the USA (NAS-NRC Registry, National Academy of Sciences – National Research Council Registry of World War II veteran twins; NHLBI), and several in Scandinavia (LSADT; Norwegian Twin Register; HARMONY, Study of Dementia in Swedish Twins).

### Contributions of Genes and Environments

**General cognitive abilities** Cross-sectional findings of adult twins indicate a strong heritability for general cognitive ability, or *g*, with a median value of 0.69 (see **Figure 1**). Therefore, 69% of the individual

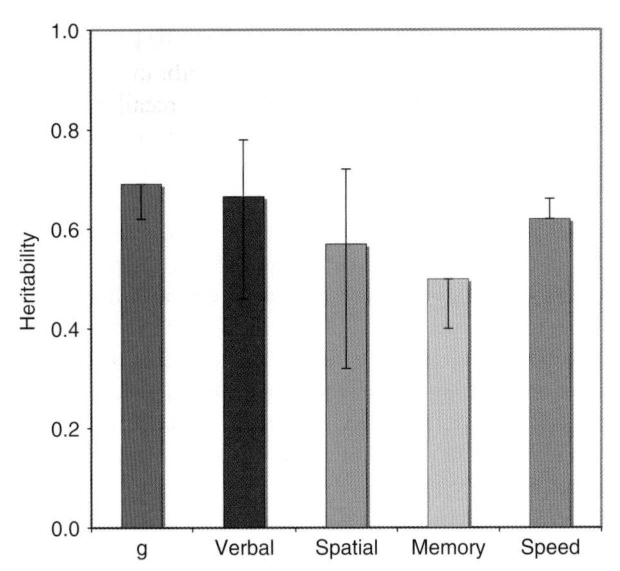


**Figure 1** Heritability estimates of cognitive abilities; *g* represents general cognitive ability. Note: Tops of solid colored bars indicate median heritability estimates from all adult twins studies that included older adult twins. Lower ends of 'error bars' indicate median old-old estimates and upper ends indicate young-old median.

differences in general cognitive ability results from genetic differences. Furthermore, the described studies find no evidence for shared environmental factors (e.g., rearing environment) affecting general cognitive abilities in adults, thus the remaining proportion of individual difference, 0.31, or 31%, is due to person-specific or nonshared environmental influences. Studies that report separate estimates for young-old adults versus old-old adults consistently suggest a waning of heritable influences for old-old adults, and concomitantly an increase in nonshared environmentality influences. The median heritability estimate for young-old twins is 0.69 versus 0.62 in the old-old (see error bars in **Figure 1**). Swedish twin studies, in particular SATSA and OCTO-Twin, suggest even more marked difference in heritability: as high as 0.80 in young-old twins and 0.62 in old-old twins. Thus, cross-sectional findings imply waning genetic influences in old-old age.

**Specific cognitive abilities** Specific ability domains, such as verbal, spatial, memory, and perceptual speed abilities, show varying patterns of the extent of genetic influences. The median heritability ranges from 0.50 for memory to 0.66 for verbal abilities (see **Figure 1**). Virtually all cross-sectional studies suggest that the remaining variation is largely due to nonshared environmental influences, with few hints of shared environmental effects on select cognitive measures within a domain. Furthermore, no convincing evidence for sex differences in heritability or for sex-limited genetic or environmental factors has been presented to date.

The median estimate of 0.66 for verbal ability across all studies is flanked by a median heritability of 0.78 in the young-old and a heritability of 0.46 in the old-old, again suggesting a similar pattern of waning genetic influences in old-old age. The most disparate pattern between young-old and old-old heritability estimates is observed for spatial ability (median of 0.72 in young-old and 0.32 in old-old), and the least is observed for perceptual speed (median of 0.66 in young-old and 0.62 in old-old). The estimates for the memory domain indicate a median heritability of 0.50 across all studies, and a median heritability of 0.40 in old-old versus 0.50 in young-old adult twins.

The memory domain is made up of multiple distinctive processes, including working memory, semantic memory, episodic memory, and long-term memory. Twin studies suggest higher heritability estimates for working memory measures than for episodic memory measures. Furthermore, the SATSA study indicates a lower heritability for working memory measures in young-old versus old-old (0.36 vs. 0.63) but a higher heritability for episodic memory measures in young-old versus old-old (0.50 vs. 0.33), as has been noted for the other cognitive domains.

Across domain, the differences in heritability and environmentality estimates among young-old and old-old suggest that new environmental factors may emerge in old-old adults, or perhaps that cohort differences are important. Longitudinal studies of twins best address these possible explanations and most directly assess the heritability of cognitive decline.

**Cognitive change** Longitudinal studies of twins have added to our understanding of the etiology of cognitive abilities and particularly of cognitive decline with age. **Figure 2(a)** illustrates the nature of the data analyzed in longitudinal studies of twins. Twins have their own individual cognitive test score trajectory that can be aligned to age of measurement. Furthermore, the trajectories within and between pairs are examined to establish the extent of environmental and genetic influences. The illustrative trajectories in **Figure 2(a)** suggest that cognitive performance (e.g., episodic memory ability) declines with age for all four twins. Furthermore, the extent of decline differs across individuals and within pairs, particularly for the DZ twin pair versus the MZ pair. Analyses of longitudinal twin data consist of fitting characteristic curves (e.g., linear, quadratic) to individual cognitive data and comparing twin similarity for curve parameters. By and large, the curves of MZ twin pairs are more similar than are those of DZ twin pairs, but

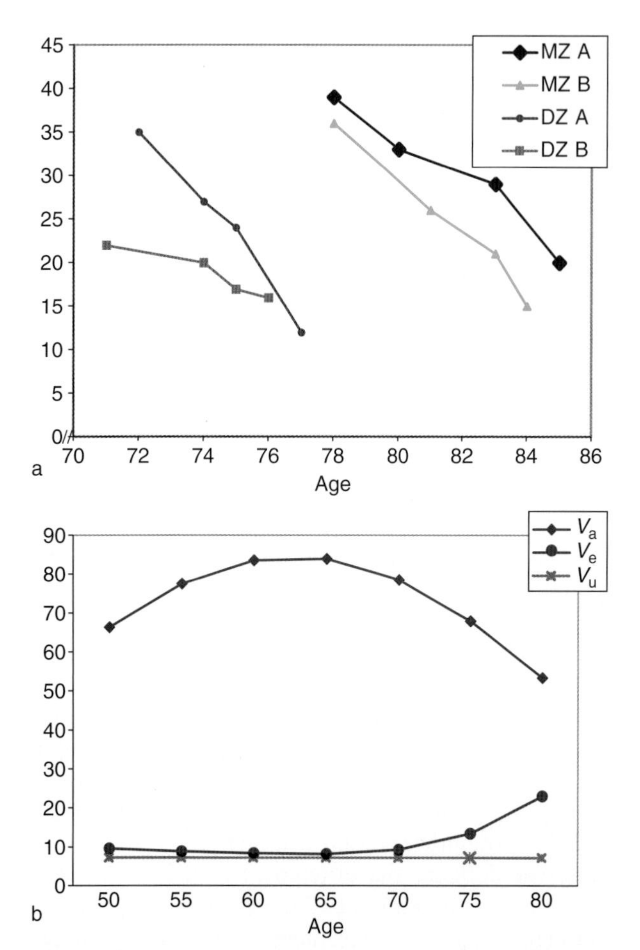


**Figure 2** Application of longitudinal twin models to longitudinal cognitive data: (a) illustrative data for monozygotic (MZ) and dizygotic (DZ) pairs of twins, ages 70–86 years; (b) actual results from the Swedish Adoption/Twin Study of Aging for general cognitive ability, *g*, for twins aged 50–80 years. $V_a$ represents additive genetic variance contributing to systematic longitudinal growth; $V_e$ represents person-specific (nonshared) environmental variance contributing to systematic longitudinal growth; $V_u$ represents occasion-specific variance (nongrowth-related variance). (b) Data from Reynolds CA, Finkel D, McArdle JJ, Gatz M, Berg S, and Pedersen NL (2005) Quantitative genetic analysis of latent growth curve models of cognitive abilities in adulthood. *Developmental Psychology* 41: 3–16; and adapted from Reynolds CA (in press) Genetic and environmental influences on cognitive change. In: Hofer SM and Alwin D (eds.) *The Handbook of Cogitive Aging. Interdisciplinary Perspective*. London: SAGE, with permission.

pairs become less similar to one another with age, particularly after age 65 years. Genetic factors influence average cognitive performance levels, as well as the shape of curve trajectories, though to a lesser degree. Furthermore, the course of genetic and environmental influences with age is consistent with cross-sectional studies, with the exception of memory; that is, there are waning genetic influences after age 65 years and increasing nonshared environmental influences. **Figure 2(b)** shows findings from the SATSA study for general cognitive ability, *g*, where only additive genetic factors ($V_a$) and nonshared environmental factors ($V_e$) contribute to individual differences in average performance and change. Genetic factors contribute the most to individual differences in *g* with age, but a decreasing trend is evident after age 65 years, when nonshared environmental factors assume greater importance. Thus, unique person-specific environmental factors may become amplified in late adulthood, or, alternatively, new environmental factors emerge. Likewise, the increase in nonshared environmental effects may suggest the possibility of gene–environment (G × E) interactions, as G × E effects become part of the nonshared environment estimate if present but not modeled directly.

The pattern of increasing unique person-specific environmental factors and decreasing role for genetic factors has been observed nearly uniformly across cognitive domains, with the exception of cognitive measures that tap memory. For example, for both working memory and episodic memory tasks, both genetic and unique person-specific environmental factors increase in importance with age, suggesting the amplification or emergence of new genetic and environmental factors in late adulthood. Thus, the search for measured genes and measured environments for cognitive change is of interest.

**Endophenotypes of cognitive aging** Locating relevant and heritable intermediary neural traits (i.e., endophenotypes) which relate to cognitive aging is important to understanding mechanisms underlying cognitive change. Studies of adult twins suggest that processing speed, brain volume, and electroencephalogram (EEG) measures may be reasonable endophenotypes in the search for measured genes, given their significant though typically modest to moderate relationships with measures of intellectual ability. Recently, regional brain volume metrics have been associated with cognitive performance in Dutch twins, as well as in aging male twins from the NHLBI study, in which a significant percentage of the relationship between lateral ventricular volumes and executive functioning performance was due to common genetic influences; other lobar brain volume–cognition relationships were associated through shared and nonshared environmental pathways. Furthermore, higher white-matter hyperintensities (WMHs) predict greater rates of decline in perceptual speed performance, and this association may be due largely to common genetic effects influencing both cognitive function and WMH.

**Cognitive functioning** The heritability for mental status is 0.49 across studies. However, mental status performance as assessed with telephone screening methods tends to yield lower heritability estimates than when assessed with in-person instruments (e.g., Mini-Mental Status Exam), which may in part reflect selection effects on health for those well enough to participate in person or by telephone. Furthermore, compared to cognitive performance measures or risk of AD, heritability of cognitive dysfunction is typically lower and may be the result of the diverse range of factors that lead to cognitive impairment, including health conditions and poor lifestyle habits, some of which show little evidence of genetic influences, such as cardiovascular disease.

**Dementia** Late-onset AD is due to multiple factors and is highly heritable, with MZ twins showing consistently higher concordance (and tetrachoric correlations) compared to DZ twins. **Figure 3** shows the median heritability for AD of 0.52 across all studies to date, with minimum and maximum values indicated by the error bars. The HARMONY study, at present the biggest population-based twin study of AD, estimates heritability to be as large as 79% for prevalent AD, while nonshared environmental effects account for the remaining variation in risk. Heritability of AD appears to be of similar magnitude for males and females. Vascular dementia, on the other hand, appears to be influenced completely by shared and nonshared environmental factors, given essentially equal concordance rates for MZ and DZ twins. Late-onset Parkinson's disease (PD), of which dementia sometimes features, appears to be mediated primarily by environmental factors, with nearly identical but low concordance rates in MZ and DZ twin pairs.

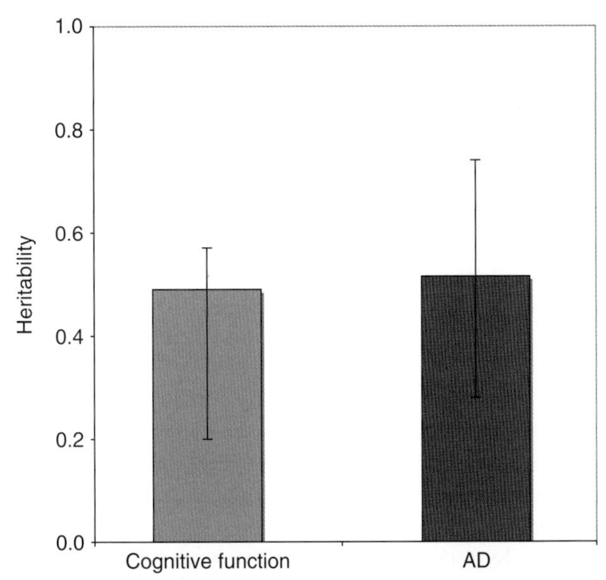


**Figure 3** Heritability estimates of cognitive functioning and Alzheimer's disease (AD). Note: Tops of solid colored bars indicate median heritability estimates from all adult twins studies. Lower ends of 'error bars' indicate lowest estimate and upper ends indicate highest estimate.

### Measured Genes

Twin sibling designs are valuable to address the presence of true gene-trait association. However, such studies have not been used to examine candidate association with cognitive aging traits with much frequency, or the pair aspect of the twins has not been used to full advantage. However, reports from twin studies are beginning to appear.

**Cognitive aging and decline** The *APOE* gene coding for the cholesterol transporter apolipoprotein E is considered a major genetic risk factor for late-onset AD. The weight of evidence for *APOE* and cognitive change in nondemented adults has been less consistently observed, in comparison. Twin association studies suggest an effect on memory-related tasks: the *APOE* $\varepsilon 4$ polymorphism has been associated with poorer recall memory, poorer working memory performance, and a steeper rate of change in working memory over age.

Apart from *APOE*, additional reports using twin models of association with cognitive aging or related traits are beginning to appear in the literature. For example, a recent SATSA study of twins suggests an association of the −1438 G/A [dbSNP: rs6311] variant of the *HTR2A* gene that codes for 5-hydroxytryptamine 2A (5-$HT_{2A}$; serotonin) receptors with differences in memory decline over time in late adulthood. Those with the common allele (G) exhibited higher performance on an episodic figure recognition task over time than did those with the rarer allele (A). These results extend a prior case-control study and are consistent with molecular and imaging studies that indicate decreases in serotonin 2A receptors with age in hippocampus and prefrontal cortex regions. Additional promising gene candidates have been reported in association studies for general intelligence in child and adult twins, including cholinergic muscarinic receptor 2 (*CHRM2*) gene polymorphisms. Finally, there are candidates for cognitive aging and decline that ought to be considered from a twin perspective, including the catechol O-methyltransferase (*COMT*) gene, which has been associated with performance and/or change in executive functioning and spatial performance in older adults.

**Cognitive functioning** *APOE* $\varepsilon 4$ status is related to an increased risk of mild cognitive impairment (MCI) according to analyses of twins as individuals from the

NHLBI study. Aging Danish DZ twins from the LSADT study were examined for four polymorphisms in the Werner syndrome *WRN* gene. In an analysis of all twins, without consideration of pair status, a positive association was found between the *Wrn-i1* gene variant located in a noncoding region of the *WRN* gene and cognitive functioning, a composite measure consisting of word fluency, word learning, and span memory performance, as well as self-rated health. Among DZ pairs who differed in genotype for the *WRN* gene variant, findings for the cognitive composite were consistent with the overall findings, but too few pairs led to lack of significance. Overall, the findings implicate *WRN* in cognitive functioning and health, and a further search of variants in regulatory regions is warranted.

**Dementia** *APOE* ε4 status predicts concordance for AD among twin pairs, as well as increased risk of AD among first-degree relatives. Several additional gene association studies have included twins in analyses of AD risk but have not considered AD risk association from a twin perspective *per se* to confirm associations. Beyond *APOE*, the most well-known and replicated genetic risk factors for late-onset AD, genes for insulin-degrading enzyme *(IDE)* and ATP-binding cassette A1 transporter (*ABCA1*), have been associated with increased AD risk or with earlier age of onset. Thus, a broader role of cholesterol metabolism on AD risk may be implied (e.g., *APOE*, *ABCA1*). Replication via within- and between-twin pair comparison is needed to confirm these and other associations suggested in the epidemiological literature.

### Measured Environmental Factors

Both positive and negative aspects of the environment appear to play an increasingly vital part in cognitive aging and risk of dementia. Epidemiological research points to low education, poor early-life socioeconomic factors, and head injury as risk factors, while protective factors may include physical and cognitive activity, antioxidant vitamin supplements, and nonsteroidal anti-inflammatory drugs (NSAIDs). Twin studies of measured environmental influences on cognitive aging or dementia have been conducted generally using two methods: by addressing the etiology of associations between complex traits to determine if their common variance is due to genetic or environmental pathways (e.g., education and cognitive functioning), or by comparing MZ twins discordant for AD or cognitive decline, for example, and examining differences in potential risk or protective variables (e.g., NSAID use).

**Cognitive aging and decline** Low education is a risk for compromised mental functioning. Theories have weighed in with biological (e.g., genetic factors influence both educational attainment and intelligence) or environmental explanations (e.g., cognitive reserve, or 'use-it-or-lose-it'). Twin studies suggest that education, cognitive ability level, and mental status performance are primarily associated due to genetic influences in common to all three, supportive of biological theories. However, as is discussed later, some weight should be accorded to environmental explanations, given the predictive value of pair differences in education and dementia risk in discordant twins.

Lobar brain-volume relationships with cognitive performance appear to be associated through environmental pathways in an analysis conducted on data from elderly male twins participating in the NHLBI twin study. Additional analyses further suggest that lower total brain volume is associated with a greater rate of decline in memory performance, perhaps influenced by common environmental factors.

**Dementia** Recent co-twin control studies of dementia-discordant twin pairs in the HARMONY study suggest that environmental factors may mediate dementia or AD risk, including occupational complexity, physical exercise, participation in leisure activities decades earlier, and tooth loss prior to age 35. Unaffected twins had a greater lifetime occupational complexity score, greater involvement in leisure activities 20 years prior to assessment, and were less likely to report tooth loss prior to age 35 years than was their AD affected twin partner. The reported tooth loss in earlier adulthood may reflect inflammation processes that result from poor health or nutrition factors earlier in life. Additional reports from the NAS–NRC Twin Registry and OCTO-Twin studies suggest that use of anti-inflammatory medications (NSAIDs, aspirin) may reduce risk of AD, replicating prior epidemiological studies.

In addition, co-twin-control reports from HARMONY indicate significant environmental factors important to diseases other than AD, or that are nonspecific to dementia type. Deficient physical exercise was more often observed for dementia-affected twins than for their co-twins, exclusive of AD. Risk of dementia of any type was increased in the case of lower educational attainment, and was evident to greater degree in dementia-affected twins than in their co-twins. The findings for education are supportive of environmental explanations, such as the cognitive reserve hypothesis, which posits that higher educational attainment may afford protection via greater synaptic density or cognitive compensation, despite neuropathologies, and thereby lower risk of, or delay clinically, meaningful cognitive impairments. Indeed, a host of potential neuroprotective or compensatory factors may coalesce in twins with higher education. In education-discordant twins, the twin with higher education reported greater participation in leisure activities and higher occupational status, for example.

The evidence from co-twin-control studies indicates intriguing possibilities that particular environmental factors may modify dementia or AD risk, though some factors appear to confer advantages or disadvantages in early or mid-adulthood, a time when individuals may not reflect on their future health.

### Gene–Environment Interplay

Environmentally mediated risk factors, as described earlier, may influence cognitive decline or risk of AD, and recent epidemiological literature suggests often this is dependent on genotype. For example, mid-adulthood hypertension or the occurrence of stroke or head injury, when coupled with *APOE* $\varepsilon 4$ carrier status, may lead to poorer cognitive performance or poorer cognitive recuperation, respectively. Viewing gene–environment interplay from a twin perspective could replicate and build upon existing epidemiologic research, but this is not often done, or the twin aspect is ignored. One twin approach to establish $G \times E$, while controlling for genetic similarity, is to compare MZ twins (e.g., on measures of cognitive decline), stratified by genotype, to examine whether pair differences are smaller or greater for pairs with particular genotypes. This approach would indicate whether individuals with certain genotypes are more or less 'sensitive' or responsive to environmental factors. The advantage of the twin $G \times E$ model is that background genetic and familial factors would be controlled.

## Conclusion

Twin studies provide a unique window on the etiology of brain aging. Identical twins are more similar or concordant than are fraternal twins for cognitive aging, cognitive dysfunction, and certain forms of dementia (i.e., AD). Other dementia- and aging-related traits show little evidence of genetic influences, including vascular dementia and PD, in which identical and fraternal twins are equally similar. Longitudinal twins studies of cognitive aging indicate that genetic influences decrease with age while person-specific environmental influences increase in importance. Co-twin-control studies suggest that individuals may be able to modify their risk of AD or dementia through environmental means, but that this may require effort even before midlife, before individuals may be cognizant of dementia risk. Epidemiological studies hint that the degree to which risk may be modifiable could be dependent on genotype. Twin approaches to gene–environment interplay should confirm and build upon epidemiological findings, given the strength of the twin design.

*See also:* Aging of the Brain and Alzheimer's Disease; Aging: Extracellular Space; Aging: Brain Potential Measures and Reaction Time Studies; Alzheimer's Disease: An Overview; Alzheimer's Disease: Molecular Genetics; Alzheimer's Disease: MRI Studies; Animal Models of Alzheimer's Disease; Brain Volume: Age-Related Changes; Synaptic Plasticity: Learning and Memory in Normal Aging.

## Further Reading

Bendixen MH, Nexo BA, Bohr VA, et al. (2004) A polymorphic marker in the first intron of the Werner gene associates with cognitive function in aged Danish twins. *Experimental Gerontology* 39: 1101–1107.

Boomsma D, Busjahn A, and Peltonen L (2002) Classical twin studies and beyond. *Nature Reviews Genetics* 3: 872–882.

Carmelli D, Reed T, and DeCarli C (2002) A bivariate genetic analysis of cerebral white matter hyperintensities and cognitive performance in elderly male twins. *Neurobiology of Aging* 23: 413–420.

Finkel D and Reynolds CA (in press) Behavioral genetic investigations of cognitive aging. In: Kim Y-K (ed.) *Handbook of Behavior Genetics*. New York: Springer.

Gatz M (2007) Genetics, dementia, and the elderly. *Current Directions in Psychological Science* 16(3): 123–127.

Gatz M, Prescott CA, and Pedersen NL (2006) Lifestyle risk and delaying factors. *Alzheimer Disease and Associated Disorders* 20: S84–S88.

Gatz M, Mortimer J, Fratiglioni L, et al. (2006) Potentially modifiable risk factors for dementia in identical twins. *Alzheimer's & Dementia: The Journal of the Alzheimer's Association* 2: 110–117.

McClearn GE, Johansson B, Berg S, et al. (1997) Substantial genetic influence on cognitive abilities in twins 80 or more years old. *Science* 276: 1560–1563.

Posthuma D and de Geus EJC (2006) Progress in the molecular-genetic study of intelligence. *Current Directions in Psychological Science* 15: 151–155.

Reynolds CA, Finkel D, McArdle JJ, et al. (2005) Quantitative genetic analysis of latent growth curve models of cognitive abilities in adulthood. *Developmental Psychology* 41: 3–16.

*Special Issue on Aging. Behavior Genetics* 33(2): 79–203.

*Special Issue on TwinRegisters, Across the Globe. Twin Research* 5 (5): v505.

Vogler GP (2006) Behavior genetics and aging. In: Birren JE and Schaire KW (eds.) *Handbook of the Psychology of Aging*, 6th edn., pp. 41–55. Amsterdam: Elsevier.

## Relevant Websites

http://www.sri.com – Genetic Epidemiology and Aging Research (NHLBI Twin Study), Stanford Research Institute.

http://www.meb.ki.se – Karolinska Institutet, Department of Medical Epidemiology and Biostatistics (Swedish Twin Registry, Including the HARMONY and SATSA Studies).

http://www.psych.umn.edu – Minnesota Twin Family Study (Study of Adult Development), University of Minnesota.

http://www.fds.duke.edu – Program on Population, Policy, and Aging, Duke University (Longitudinal Study of Aging Danish Twins).

http://www.iom.edu – Study of U.S. Veteran Twins, Institute of Medicine of the National Academies (NAS–NRC Twin Registry).

# Sensory Aging: Hearing

**M K Pichora-Fuller**, University of Toronto, Mississauga, ON, Canada
**E MacDonald**, University of Toronto, Toronto, ON, Canada



## Presbycusis

### Causes

The term presbycusis, or presbyacusis, refers to age-related hearing loss. There are many causes of hearing loss in older people, including environmental factors such as exposure to noise and ototoxic drugs, genetic factors, and generalized effects of aging such as cell damage and neural degeneration. Many age-related hearing impairments are types of sensorineural hearing loss involving damage to the cochlea, or inner ear. The hallmark of this damage is the elevation of the thresholds of audibility for high-frequency pure tones.

Our knowledge of cochlear pathology, its effects on perception, and many issues pertaining to treatment and rehabilitation can be generalized across the adult age range. In addition to types of cochlear damage that are common to adults across the age range, the peripheral and central auditory systems of older adults may also be damaged in ways that are not typical in younger adults. Furthermore, older adults often exhibit perceptual deficits that are disproportionate to the problems observed in younger adults with similar hearing thresholds. Importantly, studies of several species of mammals have demonstrated that hearing declines as a function of chronological age, even when all genetic and potentially damaging environmental effects have been carefully controlled.

The classification of subtypes of presbycusis, defined according to the particular structures of the auditory system that are affected by age, has continued to be refined for over four decades. Unfortunately, there is not a straightforward correspondence between damage to particular structures and perceptual deficits. Damage at multiple sites likely contributes to the differences in auditory processing that are observed between younger and older adults with similar hearing thresholds. Although high-frequency pure-tone threshold elevation is very common in older adults, the more age-specific structural and perceptual aspects of presbycusis have been the focus of most research conducted in the past decade.

### Prevalence

The prevalence of hearing loss increases with age and is greater in males than in females. The prevalence of presbycusis, based on audiometric pure-tone threshold data, ranges from almost 50% to over 80%, depending on the sample, age, and the definition of hearing loss. Some reports indicate that hearing loss is the third most common chronic disability in senior citizens. About three-quarters of those who have impaired hearing are over the age of 75 years.

Age-related hearing loss characterized by high-frequency threshold elevations can begin in the fourth decade of life. The degree of hearing loss and the range of frequencies affected continue to increase with age. The specific prevalence estimates based on pure-tone thresholds depend on the degree of loss and on which frequencies are used to determine the inclusion criteria. One explanation for the greater prevalence of hearing loss in males is that their cumulative exposure to noise, including industrial and military noise, is typically greater than that of females. Age cohorts may also differ in history of noise exposure associated with wars and changing industrial practices, as well as changes in longevity.

In industrialized countries, a century ago, fewer than 1 in 20 persons lived to age 65 years, but by 1980, there were 1 in 10, and by 2030, 1 in 5 persons will be over age 65 years. As the baby-boom generation enters retirement age and longevity increases, with the fastest growing age group now being those over the age of 85 years, because hearing loss increases with age, estimates of the prevalence of hearing loss increase for the coming decades.

In general, self-reported hearing loss yields prevalence estimates that are lower than those based on pure-tone thresholds measured objectively in the clinic. A typical finding is that by 65 years of age, over one-third of adults report hearing problems, and the prevalence of hearing loss has been reported to be as high as 80% for the frail elderly living in residential care facilities. The most obvious explanation for this apparent discrepancy in the estimates based on subjective and objective measures is that the effect of hearing loss on functioning in everyday life depends greatly on the individual's auditory ecology and not only on the degree of hearing impairment.

### Research Issues

Most early research on presbycusis was confounded by the high degree of correlation between high-frequency hearing loss and age. More recent studies of the effects of age on auditory perception have attempted to minimize this confound by matching younger and older adults as closely as possible for audiometric thresholds, by simulating the effects of hearing loss in younger listeners with normal hearing, or by amplifying sound so that it is equally audible to younger and older listeners. By controlling better for the effects of audibility it has been possible to isolate changes in auditory perception that are specific to aging.

Studies of nonhuman mammal species have enabled researchers to differentiate between the effects on the peripheral auditory system of genetic factors, environmental factors, and aging *per se*. A high degree of variability in hearing thresholds in quiet-reared gerbils and patterns of hearing loss in different strains of mice suggest that there is a strong genetic component in age-related hearing loss. A consistent finding is that heritability coefficients for age-related hearing loss in humans are strong (0.22–0.55), especially for mother/daughter and sister/sister relationships, with magnitudes similar to those reported for hypertension and hyperlipidemia. Whereas in studies of humans it is virtually impossible to eliminate the role of environmental factors that are known to cause hearing loss, it has been possible to rear animals in quiet and to control their diet and acoustical environment so that the effects of aging can be isolated.

Animal models have greatly advanced our understanding of age-related changes in peripheral auditory anatomy and physiology and how these changes relate to changes in pure-tone thresholds. Nevertheless, the auditory central nervous system, from the brain stem to the cortex, also undergoes significant age-related changes. It is more difficult to compare the nature and consequences of these changes across species, especially with regard to the perception of complex signals such as speech or music. Studies of the human central auditory system using electrophysiological techniques, such as evoked potentials and brain imaging, are beginning to address questions concerning age-related changes in the more cortical aspects of auditory perception.

## Anatomy and Physiology

The main structures of the peripheral auditory system are the outer, middle, and inner ear. The effects of aging on auditory perception result mainly from changes in the inner ear and the auditory nervous system.

### Outer Ear and Middle Ear

The outer ear, including the pinna, concha, and the ear canal to the ear drum, does not undergo any age-related change that significantly affects auditory perception in everyday life. Two outer ear conditions that may affect hearing testing are cerumen buildup and collapsing ear canals. Buildup of cerumen, or ear wax, may impede the perception of sound and possibly cause tinnitus, or ringing in the ears; however, cerumen can be safely removed by a qualified health professional. The cartilage of the outer ear loses rigidity, with the consequence that the ear canal may collapse if pressure is applied to the pinna. Collapsing canals occur in about a third of older adults. The occlusion produced when the ear canal collapses impedes the transmission of high-frequency sounds such that thresholds measured using circumaural earphones may over-estimate the degree of hearing loss. This testing artifact can be avoided by using insert earphones. A tester can easily identify a collapsing ear canal by pressing on the pinna with a finger and observing a change in the shape of the opening of the ear canal.

The middle ear cavity, from the ear drum to the inner ear, is normally air filled and contains a chain of three ossicles. The middle ear structures transduce the airborne acoustical sound vibrations arriving at the ear drum into mechanical vibrations that are relayed to the fluid-filled inner ear. Some older adults have hearing loss resulting from middle ear damage that developed earlier in their lifetime; however, the middle ear does not undergo any changes specifically related to age that significantly affect auditory perception in everyday life or during hearing testing.

### Inner Ear

The high-frequency threshold elevations commonly observed in older adults may be partially the result of environmental factors such as exposure to noise or ototoxic drugs, rather than age *per se*. Losses due to such environmental factors are characterized by damage to and loss of the outer hair cells of the cochlea, or inner ear. In the normal cochlea, the outer hair cells have a gain-control function that enables low-intensity sounds to be heard. The outer hair cells are motile and under efferent control. Their status is evaluated clinically by testing otoacoustic emissions. Importantly, there is less change in otoacoustic emissions with age than would be expected if outer hair cell damage were the only pathology contributing to the high-frequency hearing loss typical of presbycusis.

In addition to outer hair cell damage resulting from environmental factors rather than to age *per se*, there are two subtypes of presbycusis in which cochlear pathology is attributable specifically to aging. One subtype of presbycusis, sensory loss, characterized by atrophy or degeneration of the sensory receptor cells, the inner hair cells, and supporting cells, is now considered to be of only minor importance in the aged ear. A second subtype of presbycusis, metabolic loss, characterized by atrophy and degeneration of the lateral wall of the cochlea, especially the stria vascularis and spiral ligament, is now considered to be the predominant and most distinctive lesion of the aging ear.

The stria vascularis has a dense capillary network and an exceptionally high metabolic rate. There are significant losses of strial capillaries in aging animals. Age-related changes in the stria vascularis affect the basic mechanoelectric transduction process of the cochlea, the generation of electrochemical gradients, and the regulation of cochlear ionic homeostasis. Specifically, there is an age-related decrease in the endocochlear potential and in the gain of the cochlear amplifier. Importantly, metabolic presbycusis accounts for high-frequency threshold elevations, but is not observed in cases of noise-induced or ototoxic hearing loss. Furthermore, animal studies show that there can be at least temporary reversals of age-related threshold elevations on the order of 20–40 dB when the endocochlear potential is increased by the introduction of a direct current voltage into the scala media of the cochlea.

### Auditory Nerve

A third subtype of presbycusis, neural loss, is characterized by loss of spiral ganglion neurons and pronounced degeneration of the auditory nerve. In quiet-reared gerbils and other species, spiral ganglion cell density decreases and the average volume of the surviving cells also decreases in older compared to younger ears. In aging gerbils, reduced slopes of the input–output functions are observed for the compound action potential of the auditory nerve and for the auditory brain stem response. Such reduced amplitudes of physiological responses in aging ears to moderate to high-level signals, in cases in which there are only small changes in auditory thresholds, are consistent with reduced synchronization of neural firing. Importantly, these changes have been observed in the presence of intact inner hair cells, consistent with age-related primary degeneration of the auditory nerve, as opposed to degeneration secondary to loss of sensory receptor cells, as is observed in cases of noise-induced or ototoxic hearing loss.

## Perception

Auditory aging alters the perception of sound. The most commonly reported problem of older adults is difficulty understanding speech in noise. For the most part, the perceptual consequences of high-frequency hearing loss due to environmental factors are well known and will be the same for older adults as for younger adults. The perceptual consequences of the three subtypes of presbycusis are less well known, but seem to be distinctive. Importantly, older adults often find it more challenging than do their younger counterparts with similar audiograms when listening in complex acoustical environments, even though they have little or no problem listening in quiet situations. Furthermore, these problems can occur even if the audiometric thresholds of the older adults are within the normal range for speech or if sound is adequately amplified by hearing aids.

Psychoacoustic studies have documented changes in how older adults perceive the basic dimensions of sound. Studies have also documented age-related declines in speech perception and spoken language comprehension in more realistic, complex listening conditions. In humans, it has been difficult to distinguish the perceptual consequences of outer hair cell damage due to environmental factors from the perceptual consequences of the three subtypes of presbycusis, because most older adults have a combination of these pathologies. Insight into how the perceptual abilities of older adults resemble, or differ from, those of younger adults have been provided by studies in which the pure-tone thresholds of younger and older adults are matched as closely as possible, especially in the low-frequency range from 250 to 3000 Hz that is most important for speech perception.

### Psychoacoustics

**Absolute auditory thresholds** Compared to a population of human adults with normal hearing, the perceptual decline that defines the degree of hearing loss is the increase in absolute thresholds. The absolute threshold of audibility is the lowest intensity at which a sound is detectable. Clinically, these thresholds are measured for pure-tone frequencies at octave intervals between 125 and 8000 Hz and they are plotted on an audiogram in decibels referenced to standards based on population hearing level (HL) norms, or dB HL. The typical hearing loss associated with presbycusis is characterized by normal thresholds, less than or equal to 25 dB HL, in the low-frequency range, and a sloping high-frequency hearing loss with thresholds increasing progressively with frequency. **Figure 1** depicts the common progression of audiometric pure-tone threshold hearing loss with age for males and females.

An increase in high-frequency absolute thresholds, and thus a decline in sensitivity, can affect perception and function in everyday life; for example, some speech sounds, especially consonants, become inaudible. In addition, such hearing loss is usually

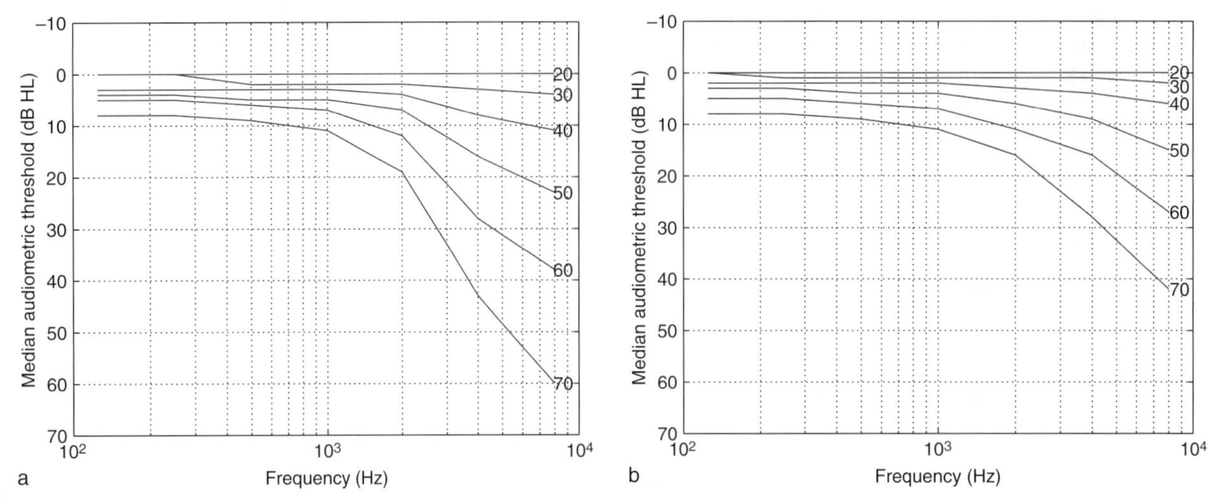


**Figure 1** Median thresholds in decibel 'hearing level' (dB HL; i.e., relative to normal hearing in 18-year-olds) as a function of age for males (a) and females (b), based on population data from the International Standards Organization (ISO 7029).

associated with alterations in the perception of the physical dimensions of suprathreshold sounds. The main physical dimensions of suprathreshold sound are frequency, intensity, and duration. In general, variations in the frequency of sound are perceived as variations in pitch, while variations in the intensity of sound are perceived as loudness. Older adults often exhibit changes in suprathreshold perception that are not predictable from their audiometric thresholds.

**Frequency selectivity and discrimination** The cochlea is tonotopically arranged such that different regions respond maximally to different frequencies (i.e., there is place coding of frequency). The characteristic frequency, or the frequency that elicits the maximum response, is highest in the basal region of the cochlea (nearest to the middle ear) and it decreases more or less logarithmically with distance along the cochlea, with the most apical region having the lowest characteristic frequency. The tuning of the cochlear response to different frequencies has been examined using psychoacoustic experiments in humans and physiological measurements in mammals. The frequency tuning of the cochlea is often characterized in terms of auditory filter properties. The auditory filters of listeners with outer hair cell loss are broader than are those of listeners with normal hearing. The degree to which the auditory filters are broadened is correlated with degree of audiometric hearing loss.

A perceptual consequence of broadened auditory filters, or decreased frequency selectivity, is an increased susceptibility to masking of one signal by a simultaneous competing signal, especially masking signals lower in frequency than the target signal. Listeners with outer hair cell loss often have difficulty understanding speech in noise because of their increased susceptibility to masking. The deleterious effects of the upward spread of masking continue to undermine speech perception in noise, especially when the noise is the speech of other talkers, even when hearing aids amplify sound sufficiently to restore the audibility, because both the target and any competing signals are amplified.

The effect of outer hair cell loss on ability to discriminate sequentially presented tones differing in frequency is less clear. While frequency discrimination is usually adversely affected, there is significant variability among individuals, and the size of the just-noticeable frequency difference is generally not well correlated with the degree of threshold hearing loss.

In general, older adults with normal audiograms for low-frequencies do not discriminate sequentially presented tones of different frequencies as well as younger adults do. Importantly, these age-related differences in frequency discrimination do not appear to vary with presentation level and they are larger for low-frequency pure tones than for high-frequency tones. The place coding of frequency is dominant for high frequencies while temporal coding is dominant at low frequencies. Temporal coding of low frequencies is possible because the timing of the firing of primary auditory neurons is phase-locked to the signal which causes the movement of the cochlear structures. Thus, the synchronized timing of the firing of auditory neurons provides a cue that can be used to code low frequencies. At higher frequencies, the speed of firing needed for this type of coding exceeds the limits of the neural response. The pattern of age-related declines in frequency discrimination is consistent with disruptions in temporal coding, rather than with the disruptions in place coding associated with outer hair cell damage that results in high-frequency threshold elevations.

**Loudness and intensity discrimination** Loudness recruitment is a phenomenon that is observed with almost all cases of outer hair cell loss. This phenomenon refers to an increased rate of growth in perceived loudness as the intensity level of a signal is increased. Consider an individual with a unilateral hearing loss – that is, one ear with normal thresholds and abnormally elevated thresholds in the other ear. While the range of audible sound levels will be different for both ears, the full range of perceived loudness will be the same (i.e., from barely audible to painfully loud). In the case of complete recruitment, once a tone becomes audible in the ear with elevated thresholds, the perceived loudness of that tone will grow faster there than in the normal ear as the level of the tone is increased, until the sound level is sufficiently high to produce the same perceived loudness in both ears. Overrecruitment or underrecruitment can also occur in some cases of cochlear hearing loss; respectively, these refer to the perceived loudness of a sound presented at a high level as being either greater or less in the impaired ear compared to the normal ear.

Typically, when tested at low intensity levels relative to their absolute thresholds, listeners with cochlear hearing loss who have recruitment discriminate smaller intensity differences compared to listeners with normal absolute thresholds. This is the basis of the behavioral tests sometimes used in the clinical evaluation of cochlear hearing loss characterized by recruitment. However, when tested at high intensity levels relative to their absolute thresholds, listeners with recruitment discriminate intensity differences that are similar, or sometimes larger, compared to young adults with normal audiograms.

Older adults with normal audiograms in the low-frequency range often exhibit larger intensity discrimination thresholds than do younger adults, with the largest age-related differences in intensity discrimination occurring for low-frequency tones. For high-frequency tones, loudness is mostly coded by the spread of activation along the length of the cochlea; that is, the overall rate of neural firing increases because activation occurs at more places in the cochlea. However, coding of loudness for low-frequency tones is influenced by temporal cues based on the number of neurons firing synchronously. The pattern of larger age-related threshold differences for low-frequency tones than for high-frequency tones is not consistent with the disruptions in place coding associated with outer hair cell damage resulting in high-frequency threshold elevations, but it is consistent with an age-related decline in neural synchrony.

**Temporal processing** The pattern of age-related differences in frequency and intensity discrimination for low frequencies is consistent with declines in auditory temporal processing. Age-related declines in temporal processing undermine the coding of the durational and transitional time-varying properties of the input signal and the extraction of signals from noise.

The gap detection threshold, the smallest gap between two sound markers that a listener can detect, is the most common measure of temporal resolution. Many older adults do not detect gaps until the gaps are significantly longer than those that can be detected by younger adults. Notably, age-related differences in gap detection thresholds are not associated with absolute threshold elevations. The effect of age on gap detection threshold is more pronounced when the markers surrounding the gap are shorter than 10 ms, when the gap is nearer to the onset or offset of the signal, and when the markers differ in frequency composition.

Older adults are also significantly worse than younger adults in discriminating the duration of two tones. The effect of age on duration discrimination is more pronounced when the target tone is embedded in a sequence of tones. There are also significant age-related declines in the discrimination and recall of the temporal order of short sequences of sounds, especially when the sounds are presented at fast rates.

Synchrony or periodicity coding is another aspect of auditory temporal processing that declines with age. Loss of neural synchrony is implicated by the pattern of age-related differences in frequency and intensity discrimination, as well as by the pattern of age-related differences in masking-level differences. Other evidence pointing to loss of periodicity coding in older adults is that they have more difficulty than do younger adults in detecting mistuned harmonics in complex sounds and in segregating concurrently presented sounds.

### Speech Processing

Like listeners of any age who have outer hair cell loss, audiometric thresholds explain nearly all of the problems encountered by older adults when they listen to speech in relatively quiet environments. However, specifically age-related deficits seem to be most apparent in difficult or complex auditory perception tasks, such as when speech is presented rapidly or in noisy or reverberant conditions, particularly in a background of interrupted or modulated noise, or when there is competing speech, even from a single talker.

Importantly, age-related deficits are much larger for temporally complex conditions compared to spectrally complex conditions. There is growing evidence that age-related declines in auditory temporal processing account for the disproportionate difficulties in understanding speech in noise that are experienced by older adults, including those for whom the audibility of speech is not a problem because they have normal audiometric thresholds throughout the speech frequency range up to 3000 Hz.

Declines in auditory temporal processing affect the ability of older adults to use different types of speech cues. At the sentence level, prosodic cues, such as syllable stress and rhythmic patterning, impart syntactic and affective information. At the word level, gaps and durations serve phonemic contrasts that enable word recognition (e.g., the stop consonant gap of [p] distinguishes the words split and slit). At the subphonemic level, periodicity cues derived from the voice fundamental frequency and harmonic structure contribute to voice quality and talker identification. Older adults exhibit deficits at all levels of speech processing that can be related to declines in auditory temporal processing.

### Interactions between Auditory and Cognitive Processing

There is a strong correlation between sensory and cognitive aging. The difficulties of older adults in understanding spoken language in complex listening conditions may be exacerbated by the effect of age on cognitive processing, including declines in working memory and attention. As the listening condition becomes more adverse, it becomes more cognitively demanding for listeners of any age to process heard information. However, older adults demonstrate preserved knowledge and better ability, compared younger adults, to use contextual cues to compensate for difficulty hearing in challenging listening conditions.

### Rehabilitation

In quiet, older adults with outer hair cell loss exhibit perceptual deficits resembling those of younger adults with comparable loss. Beyond the difficulties in speech perception explained by loss of audibility, many older adults also have problems attributable to subtypes of presbycusis involving declines in temporal processing. Currently, hearing aids are unable to correct for these declines; however, other assistive listening technology which enables the segregation of sound sources can reduce the challenge of listening in complex acoustical environments. Other forms of rehabilitation focus on training the older person and their communication partners to use strategies to avoid and cope with everyday listening challenges.

## Further Reading


Cilento B, Norton S, and Gates G (2003) The effects of aging and hearing loss on distortion-product otoacoustic emissions. *Otolaryngology, Head and Neck Surgery* 129: 382–389.

Davis AC (1995) *Hearing Impairment in Adults.* London: Whurr Publishers Ltd.

Frisina DR, Frisina RD, Snell KB, et al. (2001) Auditory temporal processing during aging. In: Hof PR and Mobbs CV (eds.) *Functional Neurobiology of Aging*, pp. 565–579. New York: Academic Press.

Gallacher J (2005) Hearing, cognitive impairment and aging: A critical review. *Reviews in Clinical Gerontology* 14: 199–209.

International Organization for Standardization (2000). *ISO 7029–2000: Acoustics – Statistical Distribution of Hearing Thresholds as a Function of Age,* 2nd edn. Geneva: International Organization for Standardization.

Kiessling J, Pichora-Fuller MK, Gatehouse S, et al. (2003) Candidature for and delivery of audiological services: Special needs of older people. *International Journal of Audiology* 42(supplement 2): 2S92–2S101.

Moore BCJ (1998) *Cochlear Hearing Loss.* London: Whurr Publishers Ltd.

Pichora-Fuller MK and Carson AJ (2000) Hearing health and the listening experiences of older communicators. In: Hummert ML and Nussbaum J (eds.) *Communication, Aging, and Health: Linking Research and Practice for Successful Aging*, pp. 43–74. Mahwah, NJ: Lawrence Erlbaum Associates.

Pichora-Fuller MK and Singh G (2006) Effects of age on auditory and cognitive processing: Implications for hearing aid fitting and audiological rehabilitation. *Trends in Amplification* 10: 29–59.

Schneider BA and Pichora-Fuller MK (2000) Implications of perceptual deterioration for cognitive aging research. In: Craik FIM and Salthouse TA (eds.) *The Handbook of Aging and Cognition,* 2nd edn., pp. 155–219. Mahwah, NJ: Lawrence Erlbaum Associates.

Vaughan N and Fausti S (eds.) (2006) The aging auditory system: Considerations for rehabilitation. Proceedings from the National Center for Rehabilitative Auditory Research (NCRAR), 2005. *Seminars in Hearing* 27(4): 213–352.

Willott JF (1991) *Aging and the Auditory System: Anatomy, Physiology, and Psychophysics.* San Diego: Singular.

# Sensory Aging: Vision

**J C Blanks and C K Dorey**, Florida Atlantic University, Boca Raton, FL, USA



## Introduction

Aging of the eye affects all structures, and changes in visual function accompany the aging process. The elderly often exhibit a decrease in visual acuity, ocular accommodation, and dark adaptation. Diseases can interfere with the aging process and cause blindness. Blindness among white and black persons in the US was caused by age-related macular degeneration (AMD; 46% and 4% of (cases); cataract (9% and 37%); glaucoma (6% and 26%); and diabetic retinopathy (5% and 7%). Vision loss is currently a major public health problem in the United States.

Approximately 3.3 million Americans age 40 years and older are blind or have low vision. With aging of the US population, the number of Americans with eye disease will increase such that by 2020 the number of people who are blind or have low vision is projected to increase substantially, to 5.5 million. Advanced age is the strongest risk factor for AMD, which is currently estimated to affect 20–25 million people in the industrialized world, causing severe vision loss in 8 million. With the aging of society, the number of people with AMD will triple in the next 40 years. This article discusses the normal pattern of aging in the choroid, retinal pigment epithelium (RPE), retina, vitreous humor, lens, and cornea (**Figure 1**). Diseases of aging associated with these structures are also covered.

## Outer Retina

The retina is divided into the light-detecting outer retina and the visual-processing inner retina (**Figure 2**). The outermost layer of the retina, composed entirely of rod and cone photoreceptors, is physiologically and functionally dependent on the neuroectoderm-derived RPE that covers the photoreceptors, and the choriocapillaris that delivers vitamin A, nutrients, and all oxygen utilized by the photoreceptors. The choriocapillaris is separated from the RPE and photoreceptors by Bruch's membrane, a thick avascular area of extracellular matrix (**Figure 3**). Gradual age-related changes in the RPE, Bruch's membrane, and choriocapillaris become quite marked at advanced age, and influence photoreceptor life and function.

It has been suggested that it would be productive to consider age-related degenerative diseases as a cluster of problems associated with poor aging, rather than as problems arising from disease-specific pathologies. AMD may well be an archetypical condition for exploration of this view. As the following discussion illustrates, reduced photoreceptor function in older eyes can derive from age-related changes in any or all of these functionally intertwined layers. Moreover, age-related changes observed in the RPE are known to induce pathology.

## Photoreceptors

### Aging Changes in Photoreceptors

Decline in visual acuity is especially prominent in low light. Human rods are lost more rapidly with age, declining $\sim$3% per decade, than are cones, which decline in number by 1.8% per decade. With increasing age, the diameter of rod outer segments and the optical density of rhodopsin increase, whereas the scotopic sensitivity of the rods, and the time needed to recover visual sensitivity following a bleach, both decline. Accumulated oxidative damage contributes to photoreceptor loss.

Photoreceptor outer segments contain photosensitive visual pigments (e.g., rhodopsin), which mediate light damage to the photoreceptors, and a high concentration of readily oxidized polyunsaturated fatty acids. This structure and the component environment (high concentration of oxygen and constant light exposure) result in a high rate of photooxidation in outer segment membranes. Photoreceptor function is maintained by rapid turnover of outer segments: new materials are added proximally, and the RPE ingests and digests the distal oxidized tips daily (**Figure 4**). Photoreceptor degeneration occurs if the RPE is unable to ingest or digest the photoreceptor tips – age-related changes in the RPE affect both processes.

Photoreceptors and the RPE are protected from damaging blue light by accumulation of the carotenoids lutein and zeaxanthin in photoreceptor axons and outer segments, and in axons in the inner plexiform layer of the central retina. The high carotenoid concentration in the retina creates a visible yellow region known as macular pigment.

### AMD-Related Changes in Photoreceptors

Soft drusen, a clinical hallmark of AMD, cause photoreceptor distortions (**Figure 5**). Large areas of soft drusen are associated with disturbances in photoreceptor physiology, including decreased sensitivity of rods and short-wavelength cones, delayed dark adaptation of rods, and, with longer duration of AMD,

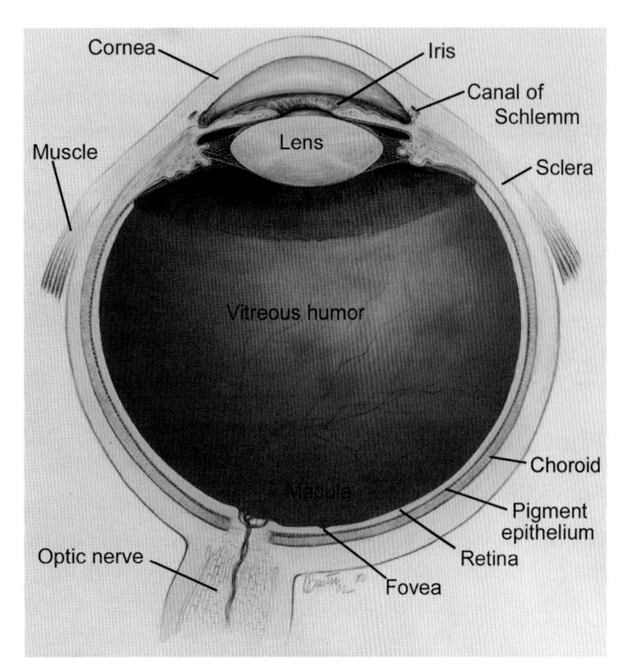


**Figure 1** Diagram of a human eye, illustrating the structures discussed in the text. Data from the National Eye Institute, National Institutes of Health (NEI/NIH) website, http://www.nei.nih.gov.

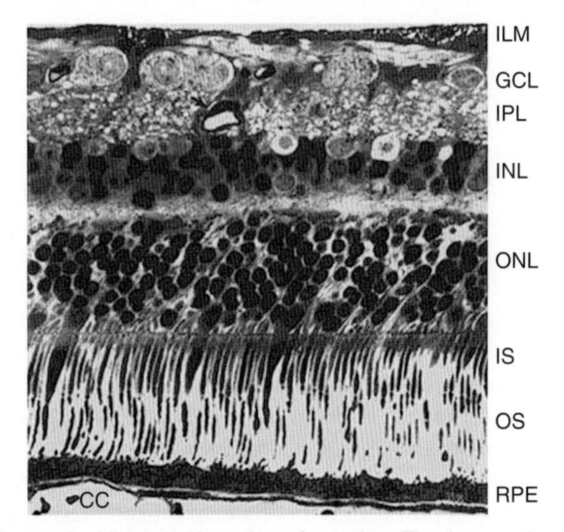


**Figure 2** Histological section of a retina. The inner retina is composed of neural cells and their processes, including the inner limiting membrane (ILM), ganglion cell layer (GCL), inner plexiform layer (IPL), and inner nuclear layer (INL). The outer retina contains only photoreceptor nuclei in the outer nuclear layer (ONL) and photoreceptor inner and outer segments (IS, OS) in the subretinal space. The photoreceptors are covered and maintained by the retinal pigment epithelium (RPE) and the vascular supply (choriocapillaris; CC). Modified from Kolb H, Fernandez E, and Nelson R, *WebVision: The Organization of the Retina and Visual System*, available online at http://webvision.med.utah.edu/.

deficiencies in color vision. Rod photoreceptors are lost preferentially and at an accelerated rate, leading to measurable loss of extrafoveal acuity, while retention of cones leaves central visual acuity unaffected.

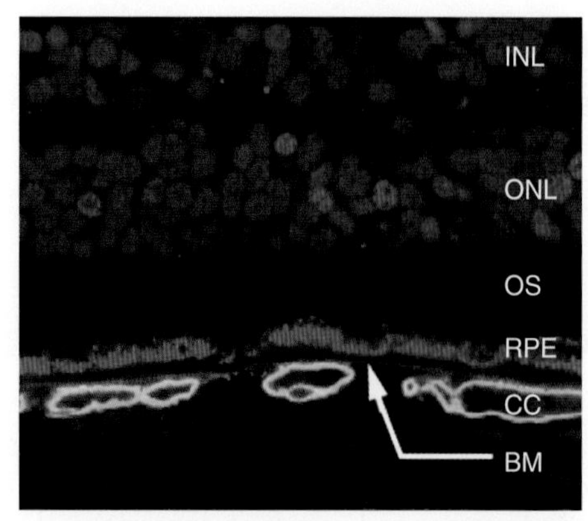


**Figure 3** Histological section of retina stained to reveal nuclei (red) in the inner nuclear layer (INL) and outer nuclear layer (ONL), and choriocapillaris (CC, yellow-green). Autofluorescence from retinal pigment epithelium (RPE) lipofuscin and Bruch's membrane (BM) are also shown in red. Choriocapillaris (CC) profiles contact BM; outer segments (OS) are not stained. Adapted from Yang Z, Alvarez BV, Chakarova C, et al. (2005) Mutant carbonic anhydrase 4 impairs pH regulation and causes retinal photoreceptor degeneration. *Human Molecular Genetics* 14: 255–265, with permission from Investigative Ophthalmology Visual Science.

Eventually in 'dry AMD,' visual field defects occur in macular areas of geographic atrophy, where both photoreceptors and RPE are missing. 'Wet AMD' – characterized by growth of choroidal neovascularization into Bruch's membrane and/or the photoreceptor layer – also causes marked loss of visual acuity. Both types can ultimately cause blindness.

## Retinal Pigment Epithelium

The RPE is a single layer of nondividing cells that extend processes across the subretinal space toward the photoreceptors. The basal surface of the RPE rests on the innermost layer of Bruch's membrane (see **Figure 2**). Age-related changes dramatically alter both structure and function of the RPE, impact photoreceptor function, and may contribute to the pathogenesis of AMD.

### RPE Functions Essential for Survival and Function of Photoreceptors

**Visual cycle** Photoreceptor function is dependent on constant delivery of vitamin A in the proper form. Retinol in the choriocapillaris is transported through Bruch's membrane to the basal RPE, transported into the RPE, isomerized to the required retinol isomer, and transported to the photoreceptors; after light absorption, it cycles back to the RPE.

**Absorption of short-wavelength light** The apical RPE contains numerous eumelanin granules. Melanin absorption of ultraviolet (UV) and blue light serves

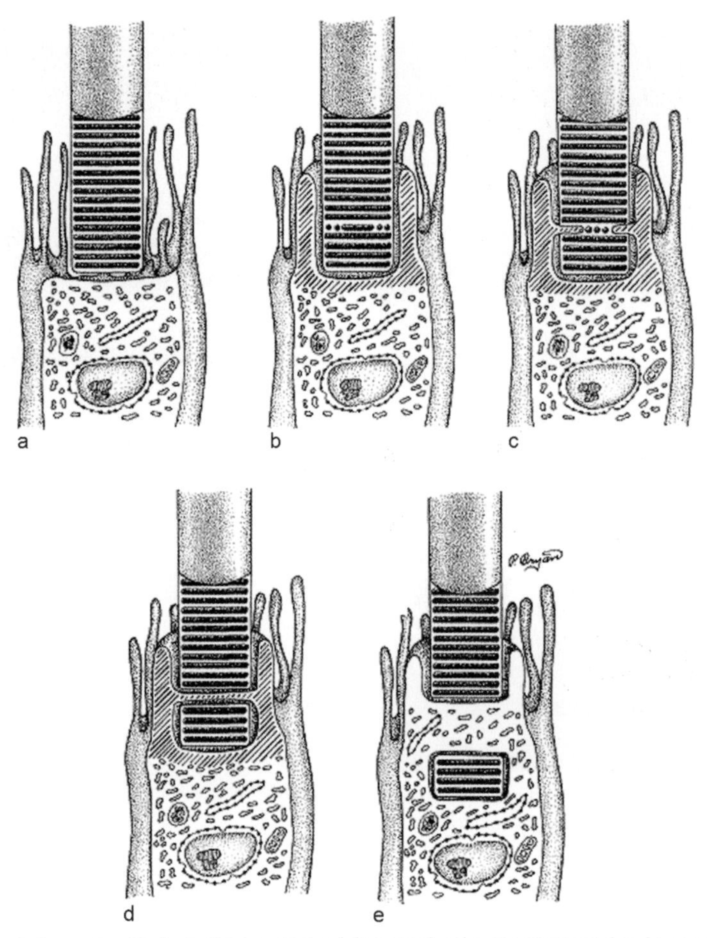


**Figure 4** Illustration of retinal pigment epithelium phagocytosis of damaged rod outer segment tips. In panel (e), lysosomal enzymes will enter the phagosome to digest the tip, leaving fluorescent residue. Adapted from Matsumoto B, Defoe DM, and Besharse JC (1987) Membrane turnover in rod photoreceptors: Ensheathment and phagocytosis of outer segment distal tips by pseudopodia of the retinal pigment epithelium. *Proceedings of the Royal Society of London, Series B: Biological Sciences* 230: 339–354, with permission from The Royal Society.

two purposes: reduced scatter of light back onto the photoreceptors improves visual acuity whereas reduced exposure to blue light reduces generation of singlet oxygen by A2E, an endogenous photosensitizer (see later).

**Blood–retinal barrier** The RPE serves as the interface between the photoreceptors and the rest of the body. Its tight junctions form the outer blood–retinal barrier, regulating movement of molecules and cells into the retina. Factors expressed by RPE cells modulate the activities of inflammatory cells that enter Bruch's membrane, and control their access to the retina. Risk for inflammatory damage to the retina is reduced by RPE inhibition of T cell proliferation and activation, and inhibition of complement cascades. However, when exposed to inflammatory cytokines, the RPE produces factors chemotactic for neutrophils and monocytes and expresses cellular adhesion molecules that facilitate migration of leukocytes across the RPE barrier.

### Age-Related Changes in the RPE

**Age-related cell loss** The RPE population declines about 3% per decade, losing 30% of RPE cells over a lifetime. As in aging neurons, RPE cells experience age-related loss in number and area of mitochondria, and loss of cristae. These losses occur earlier and are exaggerated in AMD. Taken together, these findings indicate that the ability of the RPE to execute its phagocytic, transport, and visual cycle functions will decline with age.

**Increased oxidative damage** A major change with advancing age of the RPE is decreasing ability to face oxidant challenge. Expression of antioxidant enzymes and levels of blue-light-absorbing melanin both decline. At the same time, the RPE accumulates a blue-light-absorbing photosensitizer. Increasing oxidative stress may contribute to the age-related increase in DNA mutations, and is strongly linked to the progression of AMD; a multicenter prospective

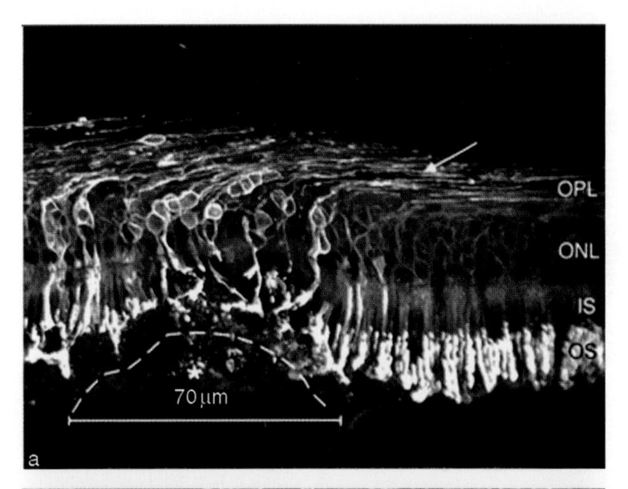


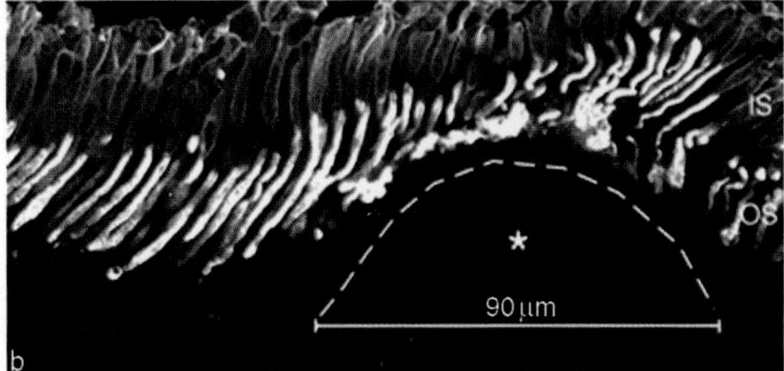


**Figure 5** Immunohistochemistry of rod (a) and cone (b) opsins. Drusen deposits (*; inside dashed lines) cause shortening of outer segments (OS) and relocation of opsin to (a, b) inner segments (IS), and (a) outer nuclear layers (ONL) and rod axons (arrow) in the outer plexiform layer (OPL). Reproduced from Johnson PT, Lewis GP, Talaga KC, et al. (2003) Drusen-associated degeneration in the retina. *Investigative Ophthalmology & Visual Science* 44: 4481–4488, with permission from Investigative Ophthalmology & Visual Science.

clinical trial has demonstrated that high doses of antioxidants (vitamin C, vitamin E, $\beta$-carotene, and zinc) slow progression of AMD.

Ingestion of oxidized outer segments probably impairs RPE digestion, since lysosomal enzyme activity declines in RPE fed oxidized lipid, and in other cell types fed oxidized protein. Moreover, exposure to oxidized membranes may help switch the RPE to a proinflammatory state: ingestion of oxidized outer segments triples RPE production of monocyte chemotactic protein-1, which would increase macrophages along Bruch's membrane.

**Lipofuscin accumulation** With advancing age, RPE lysosomes become engorged with lipofuscin (**Figure 6**), autofluorescent residue from the daily ingestion of oxidatively damaged tips of photoreceptor outer segments. The prominent fluorophore (A2E) is of particular interest, for it can be cytotoxic. The number of lipofuscin granules increases with age: by the ninth decade they occupy almost 20% of the free cytoplasmic space in the macular RPE cell.

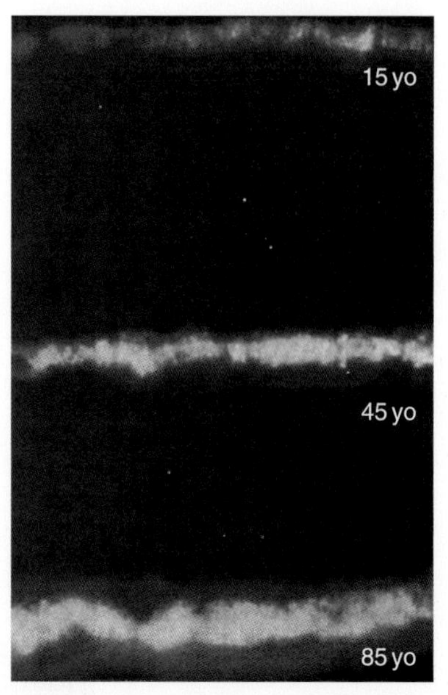


**Figure 6** Increase in lipofuscin autofluorescence in eyes from 15-year-old (15 yo), 45-year-old, and 85-year-old donors. Reproduced from Okubo A, Rosa RH Jr., Bunce CV, et al. (1999) The relationships of age changes in retinal pigment epithelium and Bruch's membrane. *Investigative Ophthalmology & Visual Science* 40: 443–449, with permission form Investigative Ophthalmology & Visual Science.

Lipofuscin levels are low in the fovea, increase across the macula, and decrease in the periphery. Regions with higher fundus autofluorescence (of RPE lipofuscin) exhibit delayed dark adaptation and decreased scotopic sensitivity, and are regions of preferential loss of rod photoreceptors in AMD.

**A2E: Fluorophore by-product of the visual cycle** A2E is a prominent lipofuscin fluorophore, composed of two molecules of vitamin A (retinol) and one molecule of ethanolamine. Upon absorption of a photon, visual pigments release highly reactive retinaldehyde. Under normal conditions, the photoreceptor-specific ATP binding cassette transporter (ABCA4) rapidly transports retinaldehyde across the disk membrane to the cytoplasm for reduction. When the retinaldehyde levels are sufficiently high to permit two retinaldehydes to sequentially interact with one phosphatidylethanolamine, an A2E precursor is formed. RPE digestion releases A2E. A2E formation is accelerated under conditions of high light irradiance, low antioxidant capacity in the photoreceptor, or when the ABAC4 activity is impaired by oxidative damage or inherited polymorphisms. A2E accumulates rapidly in Stargardt's retinal dystrophy (juvenile macular degeneration), caused by polymorphisms in ABCA4. The role of A2E is strongly implicated by evidence that mice lacking this enzyme accumulate high levels of A2E before photoreceptor atrophy.

**Toxicity of A2E** A2E may contribute to pathogenesis of AMD. RPE cells loaded with A2E *in vitro* have impaired lysosomal acidification and, consequently, inhibited lysosomal enzymes and impaired digestion of outer segments. Noxious A2E is also a photosensitizer in blue light, generating reactive singlet oxygen and lysing lysosomes. Released into the cytoplasm, A2E targets mitochondria, releases apoptosis-inducing proteins, and causes RPE apoptosis. The blue light hazard for RPE is increased by higher A2E levels and is decreased by absorption of blue light by lens brunescence, macular pigment optical density (determined by dietary lutein and zeaxanthin), and RPE melanin (which declines with age). Higher dietary, plasma, and retinal levels of lutein and zeaxanthin are associated with decreased risk for AMD; higher retinal zeaxanthin prevents light-induced photoreceptor death in quail.

#### The RPE in AMD

The functionally impaired RPE is unable to support the visual cycle – consequently, patients exhibit impaired dark adaptation. RPE changes contribute to the pathology seen in Bruch's membrane and the choriocapillaris. Foci of high lipofuscin (fundus hyperfluorescence) surrounding geographic atrophy (**Figure 7**) identify areas of subsequent photoreceptor and RPE apoptosis. The RPE in AMD has reduced ability to suppress inflammation, as described in the following sections.

## Bruch's Membrane and Choriocapillaris

The choriocapillaris is a bed of interconnected, highly fenestrated capillaries lying on the outer surface of Bruch's membrane (**Figure 8**). All peptide factors and waste products from the RPE and all nutrients from the choriocapillaris must pass through Bruch's membrane. Unfortunately, the aging Bruch's membrane accumulates debris that interferes with transport and induces inflammatory reactions. The RPE and choriocapillaris are functionally intertwined – death of either layer will ultimately cause attenuation or death of the other. The RPE produces angiogenic vascular endothelial growth factor (VEGF) and anti-angiogenic factors (e.g., pigment epithelial growth factor and endostatin) that maintain and regulate growth and status of the choriocapillaris.

#### Age-Related Changes in the Choriocapillaris

The choriocapillaris volume declines with advancing age. By age 90 years, almost half of the capillary

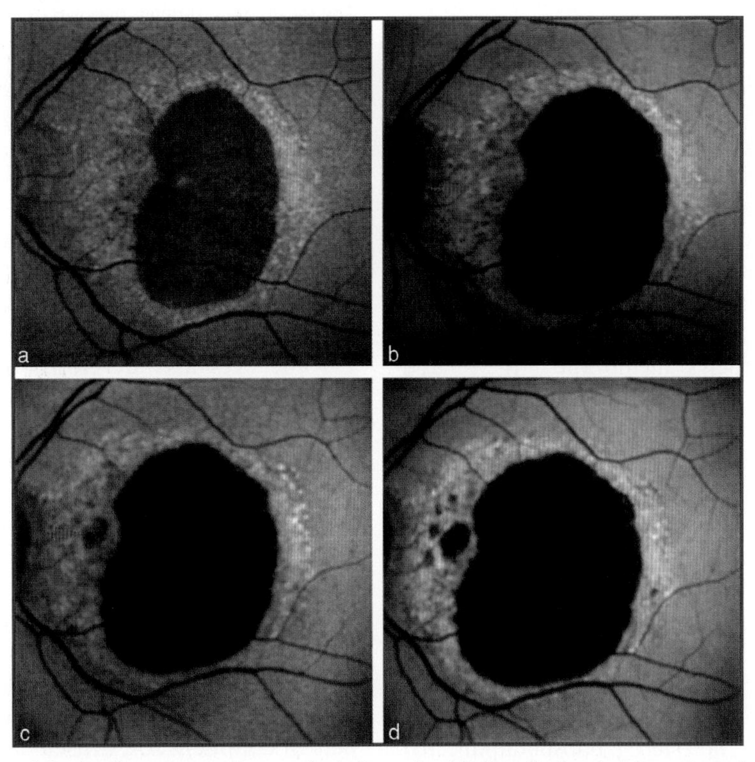


**Figure 7** Clinical images of fundus fluorescence from retinal pigment epithelium lipofuscin taken at yearly intervals (a–d). The dark areas are geographic atrophy, where both retinal pigment epithelium and photoreceptors have died. Atrophy expands in areas of higher lipofuscin; the bright area in b (arrow) became an atrophic area 1 year later (arrow in c). Reproduced from Holz FG, Bellman C, Staudt S, et al. (2001) Fundus autofluorescence and development of geographic atrophy in age-related macular degeneration. *Investigative Ophthalmology & Visual Science* 42: 1051–1056, with permission form Investigative Ophthalmology & Visual Science.

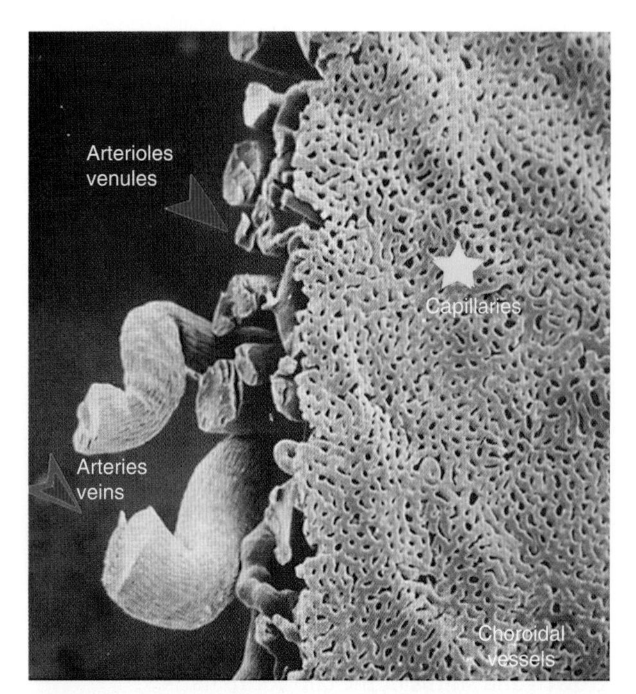


**Figure 8** The meshwork of the healthy choriocapillaris (star), and larger choroidal arteries and veins that supply it, are visible in this corrosion cast of a human choroid. Reproduced from Zhang HR (1994) Scanning electron-microscopic study of corrosion casts on retinal and choroidal angioarchitecture in man and animals. *Progress in Retinal & Eye Research* 13: 243–270, with permission from Investigative Ophthalmology & Visual Science.

profiles have been lost and their diameter has decreased by 30%. Consequent RPE hypoxia induces production of factors that can cause choroidal neovascularization in AMD. RPE production of antiangiogenic factors declines with aging and oxidative stress. At advanced age, the balance of angiogenic and antiangiogenic factors is tipped toward angiogenesis, increasing risk for choroidal neovascularization and visual loss due to retinal edema or scarring.

**Aging changes in Bruch's membrane** Age-related changes can be characterized as deposits which thicken Bruch's membrane and/or change in its biophysical properties. Deposits of amorphous material and long spacing collagen accrue above and below the RPE basal lamina, whereas vesicular and fibrillar deposits accumulate within Bruch's membrane. Focal exaggeration of these deposits under the basal lamina form a small lump that elevates RPE and is visible clinically as hard drusen in virtually all older eyes (**Figure 9(a)**). The thickening of Bruch's membrane is linearly correlated with lipofuscin accumulation. Lipids, cholesterol, phospholipids, triglycerides, and apolipoproteins accumulate throughout life. The hydraulic conductivity of Bruch's membrane declines rapidly, losing 70% by age 30 years; permeability to macromolecules declines 10% per decade.

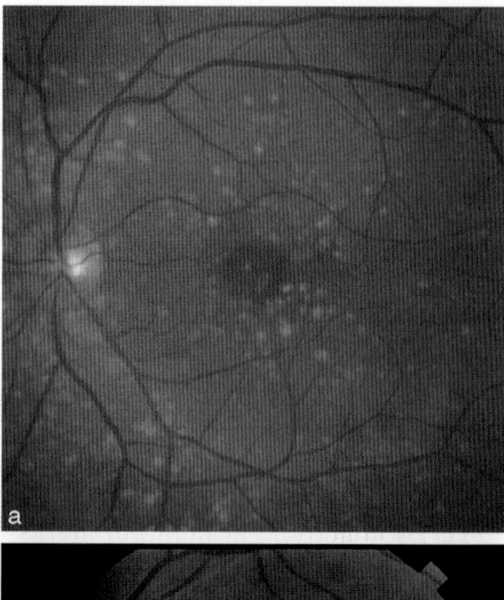


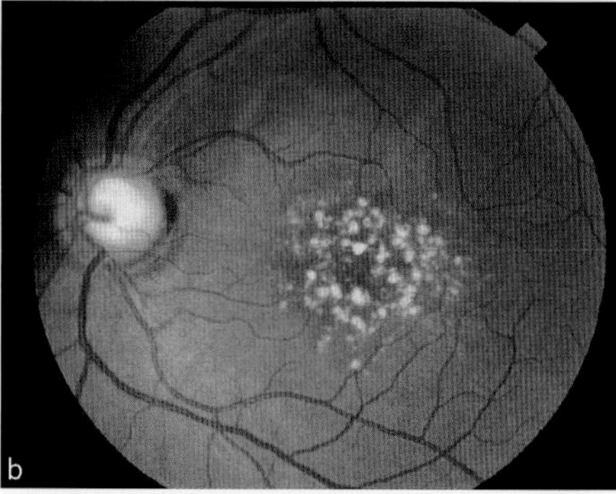


**Figure 9** Drusen in clinical photographs of the human macula. Scattered hard drusen deposits with distinct boundaries are seen in most older eyes (a), whereas the presence of soft indistinct drusen (b) is associated with decline in photoreceptor function, and high risk for age-related macular degeneration. (a) Reproduced from Galan A and Tavolato M (2004) Cataract surgery as risk factor for AMD progression. *Oculista Italiano* 138, with permission from La Oculista Italiano.; (b) Reproduced from Charles Retina Institute, http://www.charles-retina.com, with permission from Dr. Charles.

Reduced permeability to RPE factors could contribute to loss of the choriocapillaris.

Some eyes also develop larger soft drusen with indistinct borders; this feature is used to identify eyes at risk for AMD. Bruch's membrane in these eyes is thickened by ubiquitous layers of membranous debris and proteins; localized areas of thicker deposits form soft drusen where RPE displacement narrows the subretinal space, and distorts the regular array of the photoreceptors. Photoreceptors over drusen exhibit altered distribution of opsin, shortened outer segments, and altered synaptic terminals. Patients with maculopathy exhibit reduced color contrast sensitivity and decreased light sensitivity. Large numbers of confluent soft drusen deposits increase risk of progression to geographic atrophy or choroidal neovascularization.

The known components of hard and soft drusen are not different. Deposits include oxidatively damaged proteins, and serum and RPE proteins that regularly traverse Bruch's membrane. It is noteworthy that complement C5 and the membrane attack complex (MAC) (C5–C9) are found in all types of drusen.

## Inflammation in Drusen and AMD

Advanced glycation endproducts (AGEs) are cross-linked complexes of oxidized sugars, lipids, and proteins; they are considered a universal phenomenon of aging. Continuous deposits of AGEs are found in Bruch's membrane in AMD (**Figure 10**). Cellular responses to AGEs are mediated by cell surface receptors for advanced glycation endproducts (RAGEs) (**Figure 11**). Photoactivation of A2E induces RAGE expression in cultured RPE. RAGEs are expressed in the RPE over drusen, but not in RPE associated with normal extracellular matrix. RAGEs have been strongly implicated in the pathogenesis of complications of diabetes, atherosclerosis, and arthritis. Activation of RAGEs is associated with release of reactive oxygen species and increased expression of RAGEs, nuclear factor-kappa B (NF-$\kappa$B)-regulated inflammatory cytokines, chemotactic factors for inflammatory cells, adhesion molecules, and members of the complement cascade.

### Complement Activation

Activation of the complement cascade on cell surfaces can result in damage by the MAC, a membrane-piercing aggregate of the final five complement components (C5–C9). Activated complement components, including MAC, are routinely found in both hard and soft drusen (**Figure 12**). Cell responses to MAC include shedding of damaged membrane and/or production of regulators (such as fibronectin or clusterin) that impede formation of MAC. Vitronectin and clusterin are present in drusen and are produced by RPE cells (**Figure 12(c)**).

Despite evidence that complement activation occurs in drusen, only a fraction of people with drusen develop AMD. When complement C3b is deposited on cell surfaces, factor H prevents cell damage by binding and inactivating C3b. Recently, single polymorphisms in factor H were found in ~50% of patients with AMD; adding significant variants in other complement factors (component C2 regulator factor B) predicts up to 75% of AMD outcomes. Thus

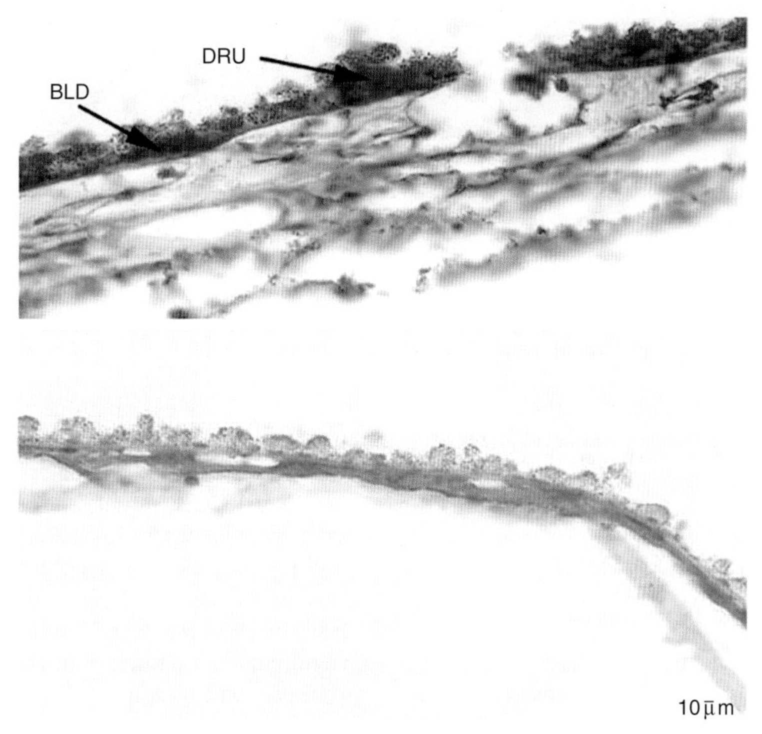


**Figure 10** Immunohistochemical localization (blue) of advanced glycation endproducts in druse (DRU) and thickened Bruch's membrane under the retinal pigment epithelium (brown) of an 82-year-old donor (BLD, basal laminar deposit). Reprinted from Farboud B, Aotaki-Keen A, Miyata T, et al. (1999) Development of a polyclonal antibody with broad epitope specificity for advanced glycation endproducts and localization of these epitopes in Bruch's membrane of the aging eye. *Molecular Vision* 5: 11, with permission from Molecular Vision.

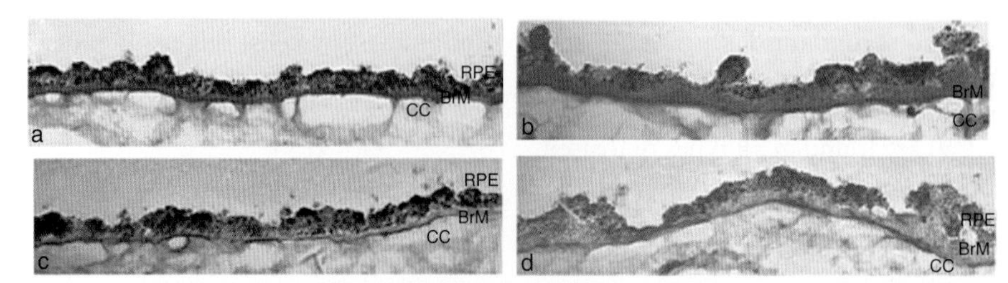


**Figure 11** Blue stain reveals immunohistochemical localization of the receptors for advanced glycation endproducts in an 85-year-old donor eye. The receptors are present in (a, c) normal retinal pigment epithelium (RPE) and Bruch's membrane (BrM) and in (b, d) thick basal deposits in Bruch's membrane. CC, choriocapillaris. Reproduced from Yamada Y, Ishibashi K, Bhutto IA, et al. (2006) The expression of advanced glycation endproduct receptors in RPE cells associated with basal deposits in human maculas. *Experimental Eye Research* 82: 840–848, with permission from Elsevier.

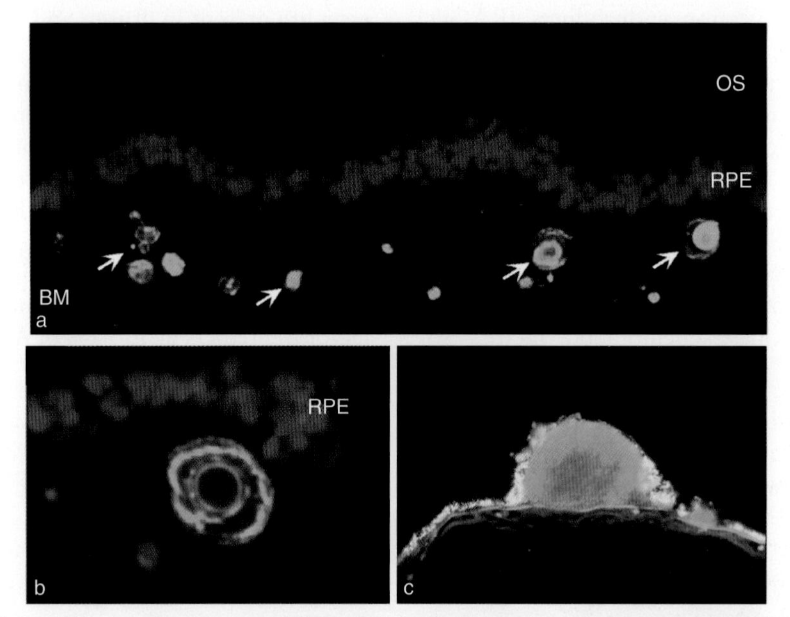


**Figure 12** (a, b) Activated C3b complement (green) is present in drusen under retinal pigment epithelium (RPE) with lipofuscin autofluorescence (red) (BM, Bruch's membrane; OS, outer segment). (c) RPE cells (yellow autofluorescence) secrete inhibitors of complement activation such as vitronectin (green), which is localized throughout Bruch's membrane and drusen. (a, b) Reprinted from Johnson LV, Leitner WP, Staples MK, et al. (2001) Complement activation and inflammatory processes in Drusen formation and age related macular degeneration. *Experimental Eye Research* 73: 887–896, with permission from Elsevier Science & Technology; (c) Reprinted from Hageman GS and Mullins RF (1999) Molecular composition of drusen as related to substructural phenotype. *Molecular Vision* 5: 28, with permission from Molecular Vision.

inability to prevent complement-mediated cell damage may be a key factor determining whether the inflammation observed in aging eyes will lead to macular degeneration.

## Normal Aging of the Inner Retina

### Age-Related Changes of the Inner Retina

While there is a paucity of studies analyzing changes in interneurons (bipolar, horizontal, and amacrine cells) in the retina with age, a number of histological studies show an inherent loss of retinal ganglion cells (RGCs) with age. RGCs, located in the innermost layer of the retina (adjacent to the inner limiting membrane and the vitreous humor; **Figure 13**) have large somas with dendritic processes that project into the inner synaptic layer, where they synapse with amacrine cells and bipolar cells. A single axon projects from each RGC and runs centripetally along the vitreous surface in the nerve fiber layer (NFL). These axons (approximately 1 million in total) produce the first action potentials in the visual system and carry information already processed by the retina to targets in the brain (primarily the lateral geniculate nucleus of the thalamus) for further processing. Although a larger percentage of RGCs is lost with age in the peripheral retina than in the central retina, the number of RGCs actually lost is similar, because of the higher density of RGCs in the central region. Therefore, a relatively constant number of RGCs die with age in different retinal locations. The

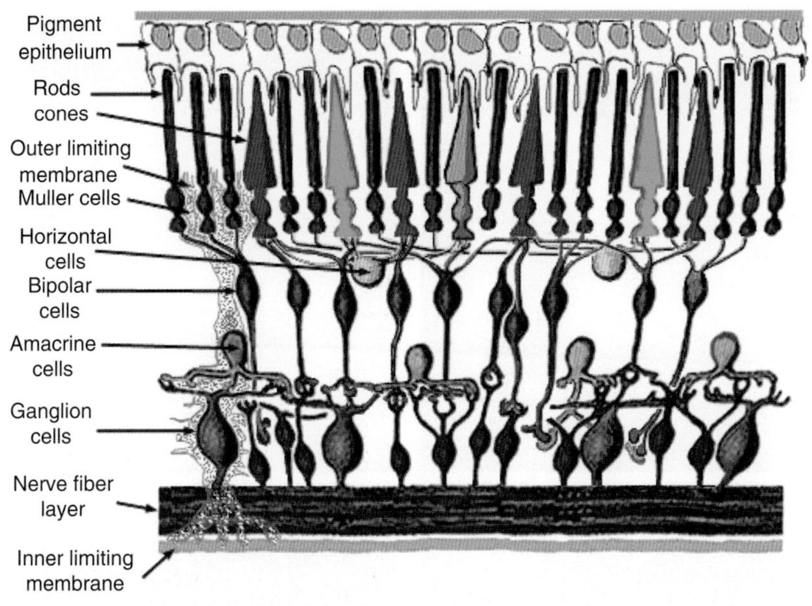


**Figure 13** Schematic illustrating the synaptic connections between the photoreceptors and horizontal and bipolar cells of the inner retina, and between bipolar cells and ganglion cells and amacrine cells. Note that the retinal pigment epithelium is at the top. From Kolb H, Fernandez E, and Nelson R, *WebVision: The Organization of the Retina and Visual System*, available online at http://webvision.med.utah.edu.

naturally occurring RGC loss with age means that about 5000 RGC axons (about 0.5%) die per year per eye. Noninvasive imaging techniques that measure the topography of the optic nerve head and the thickness of the NFL show at least a 20% thinning of the NFL with age. This thinning appears to be more pronounced in the superior quadrant. The reduction in the number of visual cells (photoreceptor cells and/or RGCs) in the retina and in higher visual centers is thought to be associated with the reduced visual acuity, decreased contrast sensitivity, and higher thresholds for light detection in visual field measurements that occur with age.

### Diseases of Aging Inner Retina

RGC loss may also play a role in visual dysfunction observed in neurodegenerative diseases. Parkinson's disease (PD) is characterized by a progressive loss of dopamine-containing neurons in the substantia nigra. There is also a progressive loss of dopaminergic cells in the retina and possibly within other areas of the visual system in PD. Selective spatial–temporal abnormalities in RGC function (arising from altered receptive field organization) may be due to the retinal dopamine deficiency.

Many patients with Alzheimer's disease (AD) also exhibit impairments of visual perception and eye movements. These defects can include anomalies in color vision, spatial contrast sensitivity, susceptibility to visual masks, electroretinographic and visual evoked potential abnormalities, and saccadic and eye tracking dysfunction. The precise basis for these deficits is unknown, but senile plaques, neurofibrillary tangles, and amyloid protein deposits are found in multiple regions of the brain that mediate vision, excluding the retina. Some of these visual impairments may also be due to retinal lesions, since RGC loss, reduction in the number of RGC axons in the optic nerve, and electroretinographic abnormalities have been reported in AD patients.

## Glaucoma

Glaucoma is an eye disease in which the normal fluid pressure inside the eye slowly rises, leading to vision loss or blindness. The disease is characterized by changes in the morphology of the optic disk, the pattern of visual field loss, and the progressive death of RGCs. RGC damage is usually not uniformly distributed throughout the eye. The uneven regional death of RGC axons in the NFL is clearly demonstrated in stained specimens of retinal whole mounts from glaucomatous animals (**Figure 14**). The principal treatment for glaucoma involved lowering the intraocular pressure (IOP). Increased IOP is directly related to the flow of aqueous fluid. Aqueous fluid is produced by the ciliary body and passes through the pupil into the anterior chamber of the eye (**Figure 15**). Aqueous fluid nourishes the lens and cornea and generally drains through the trabecular meshwork, a complicated arrangement of cells and extracellular 'beams' found at the junction of the iris, cornea, and sclera. This

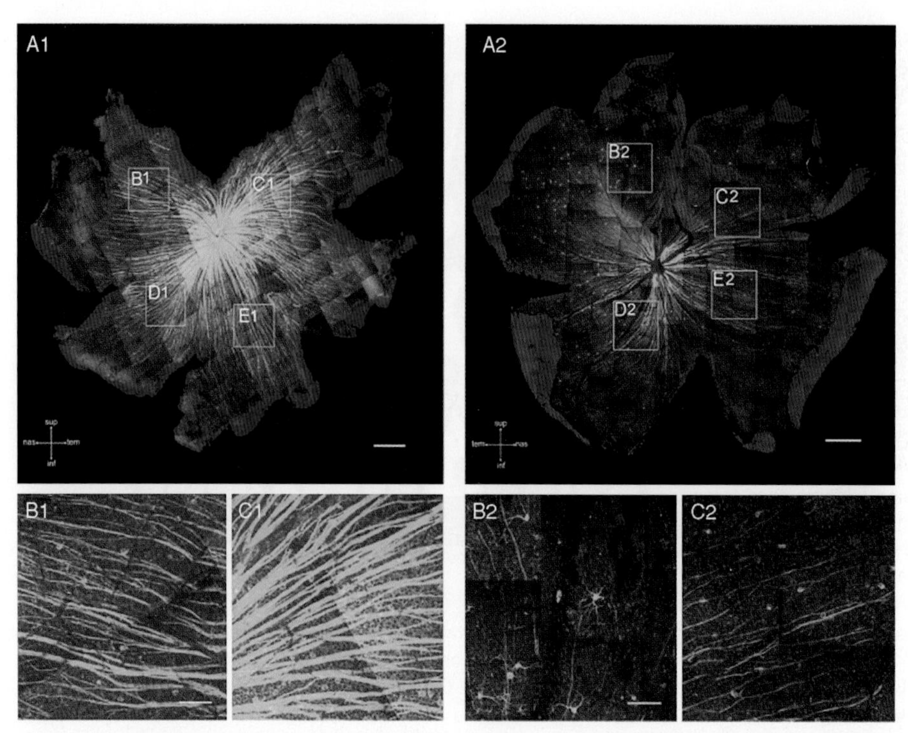


**Figure 14** Comparison of ganglion cell axons entering the optic disks of normal (A1; higher power views B1, C1) and glaucomatous (A2; higher power views B2, C2) retinas. More fibers are present in the central retina (C1, C2) than in the periphery (B1, B2) but numbers in both areas are diminished in glaucoma. (Higher power views of insets D1, D2, E1, and E2 not shown.) © Jakobs et al., 2005. Originally published in *The Journal of Cell Biology*, doi: 10.1083/jcb.200506099.

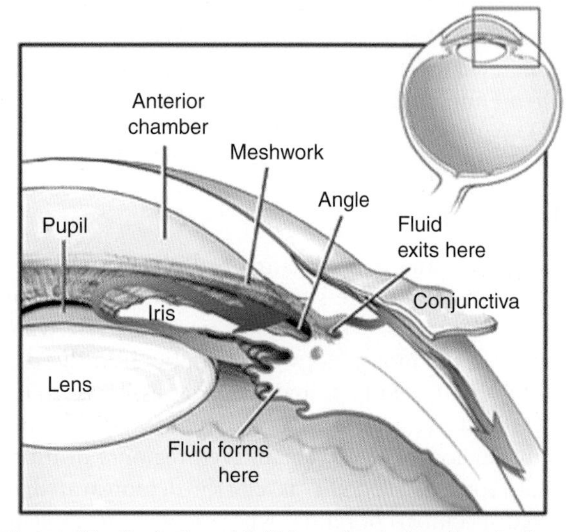


**Figure 15** Illustration of fluid formation in the ciliary body and movement around the iris to the anterior chamber, where fluid exits through the canal of Schlemm (seen in **Figure 1**). Impaired drainage raises intraocular pressure, causing glaucoma. Courtsey: National Eye Institute, National Institutes of Health.

outflow pathway maintains normal levels of aqueous humor outflow. When aqueous production exceeds outflow, the pressure within the eye rises.

The most common form of glaucoma is primary open angle glaucoma (POAG). It is a late-onset disease usually accompanied by elevated IOP resulting from pathology in the aqueous outflow pathway (the trabecular meshwork and Schlemm's canal). A combination of genetic and environmental factors is likely involved in the pathology of the outflow pathway. Oxidative stress may play an important role in the pathophysiology of POAG, since the cells of the trabecular meshwork are constantly exposed to an oxidative environment. Decreased antioxidant potential, increased expression of oxidative stress markers, and increased oxidative DNA damage have all been reported in glaucoma patients.

One or more of the following events may lead to the apoptosis of RGCs observed in POAG: elevated IOP, mechanical compression, ischemic changes, glutamate neurotoxicity, astrocytic reaction, autoimmune attack, neurotrophic deprivation, intracellular calcium toxicity, changes due to reactive oxygen species, and/or nitric oxide production. Several studies suggest that the region where the axons exit the eye, the lamina cribrosa, is the primary site of mechanical damage leading to RGC death. Most likely the compromised blood flow in the optic nerve noted in glaucoma patients would cause ischemic injury to RGCs. Therefore, it is possible that excitotoxic injury (routinely seen in ischemic injury) may play a major role in the pathogenesis of glaucoma.

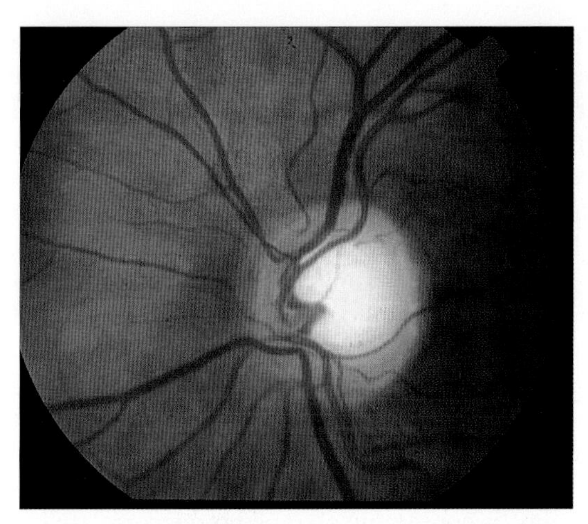

**Figure 16** Erosion of fibers in glaucoma causes the pale area around the optic disk (yellow) and cupping, where vessels curve downward toward the optic disk. Reproduced from the Southwestern Medical Center; image available online at http://www.swmed.edu/home_pages/ophth/images/dlc-fun8.jpg, with permission from UT Southwestern Medical Center.

Clinically, the loss of RGCs is monitored by University of Texas by biomicroscopy and is seen as a progressive loss of tissue in the optic nerve head. The optic nerve head in POAG has an unusual appearance on examination with an indirect ophthalmoscope – it appears to have a halo around the rim of the optic disk due to the massive loss of optic nerve fibers (**Figure 16**). Visual field testing is important in documenting the location and progression of RGC loss during the progression of the disease. A scotoma is an area of relative or complete blindness in an area of the visual field. Glaucoma invariably causes isolated scotomas in the visual field, such that the patient loses the ability to see one or more peripheral regions, while vision is completely normal in other areas of the eye. As the disease progresses, the field of vision gradually narrows and blindness can result.

#### Risk Factors for Primary Open Angle Glaucoma

Advanced age and increased IOP are fundamental risk factors for POAG. Other predisposing factors include mutations in specific genes, toxic effects, and mechanical damage due to elevated IOP. There is broad consensus that genetic risk factors play an important role this disease. The trabecular meshwork-induced glucocorticoid response (TIGR) gene, *MYOC*, for juvenile glaucoma (J-POAG) was identified by mapping and mutational analysis. This gene codes for myocillin, a protein which plays an important role in maintaining the structural integrity of the trabecular meshwork. A mutation in this gene is a high risk factor for the onset of POAG. Other strong risk factors include positive family history of glaucoma in a first-degree relative, higher incidence in African-Americans, and central corneal thickness less than 0.5 mm.

A possible diagnostic indicator for glaucoma may be the presence of the endothelial leukocyte adhesion molecule-1 (ELAM-1). ELAM-1, a member of the cell-adhesion family of molecules (CAMs), is present in trabecular meshwork cells from glaucomatous donor eyes, compared to normal eyes. This finding suggests that the activation of a tissue-specific stress response in the aqueous outflow pathway of the eye defines the glaucoma disease phenotype.

#### Treatments for Glaucoma

Current therapeutic strategies for glaucoma are aimed at reducing IOP by pharmacologically decreasing aqueous humor production or by microsurgically increasing the outflow capacity. However, lowering the IOP in glaucoma may not prevent further damage.

#### Neuroprotection or Rescue of RGCs

Knowledge of mechanisms of neuronal death of RGCs, and prevention, delay, or reversal of the processes, are now sufficient that glaucoma therapy may be designed to preserve RGCs and their axons. Neuroprotection is a therapeutic strategy to keep RGCs alive and functional. Neuroprotective strategies proven successful in animal models of glaucoma have included blocking retinal excitotoxicity mediated by glutamate and administration of neurotrophic factors (such as brain-derived neurotrophic factor) which maintain RGC survival. To date, none of the these strategies has worked in randomized controlled clinical trails in patients with glaucoma, but the use of neuroprotective agents to treat glaucoma still remains promising.

### Vitreous Humor

Eighty percent of the interior volume of the eye is made up of the vitreous humor. The primary structure of the vitreous is collagen and hyaluronic acid. In the vitreous, the predominant proteoglycans are chondroitin and heparan sulfate.

#### Age-Related Changes

The vitreous is primarily a collagen gel at 2 or 3 years of age, but the gel portion decreases and the liquid portion increases steadily throughout life. The continual increase in the liquid fraction of the vitreous suggests that there is a gradual change in the vitreous collagen. Data suggest that bundles of collagen form in the vitreous with aging.

The vitreous is attached to the posterior retina, but detaches with aging in 25–30% of the population. The process of posterior vitreal detachment occurs early in the second to third decades of life in people with high myopia, but typically occurs anywhere along the inner margin of the retina between the fourth and ninth decades of life. Detachment occurs suddenly, when fluid trapped in the middle of the vitreous humor floods into the interface between the vitreous base and the inner limiting membrane (**Figure 17**). Because the attachment over the macula is much firmer, the floating vitreous gel tugs on the retina. The process can cause the individual to see 'sparkles' or 'flashes' of light. When the vitreous base is firmly attached, the tugging on the retina can cause a tear in the retina, or a macular hole.

## Lens

### Normal Aging of the Lens

The lens is the optically clear structure located behind the iris and in front of the vitreous humor and retina (see **Figure 1**). The lens functions to focus light onto the retina. In order to form an image on the retina, the lens must maintain its shape, transparency, and refractive index. The lens has only one cell type, and these cells are precisely aligned to lead to the transparency of the lens. Young lens cells have organelles, but the organelles are destroyed during early development. What remains are cells made up of a dense solution of large proteins termed crystallins, specifically the $\alpha$-, $\beta$-, and $\lambda$-crystallins. The $\alpha$-crystallins act as molecular chaperones as well as structural proteins. As stated by Bloemendal and colleagues, "The absence of protein turnover in the lens and the ability of the lens to retain life-long transparency means that the crystallins must meet not only the requirement of solubility associated with high cellular concentration but that of longevity as well. Longevity, as it relates to proteins, is commonly assumed to be correlated with long-term retention of native structure, which in turn can be due to inherent thermodynamic stability, efficient capture and refolding of non-native protein by chaperones, or a combination of both."

The complexes formed by crystallin proteins make the cytoplasm optically homogeneous – that is, the refractive index does not change inside the cell or

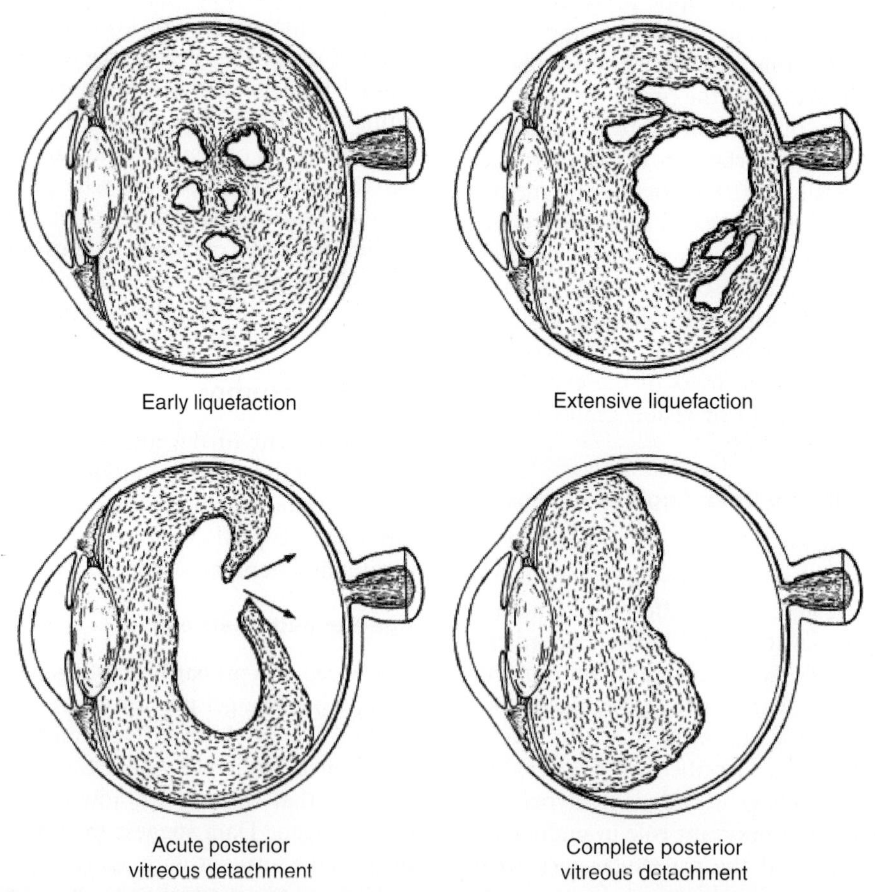


**Figure 17** Liquefaction of the vitreous during aging can cause complete or partial vitreous detachment and is a risk factor for retinal detachment and macular holes. Reproduced from Bishop PN (2000) Structural macromolecules and supramolecular organisation of the vitreous gel. *Progress in Retinal & Eye Research* 19: 323–344, with permission from Elsevier Science & Technology.

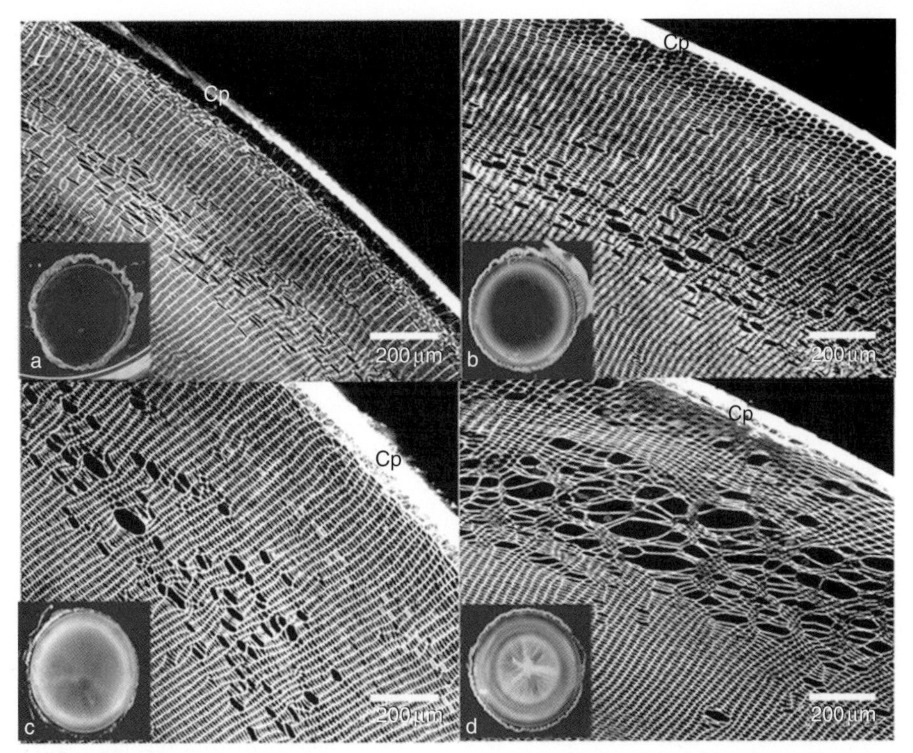


**Figure 18** The regularly arrayed fiber cells in a young lens (a; Cp, capsule) result in a transparent lens (inset, a). As the fiber cells become more disorganized with damage (b, c), there is a corresponding increase in the opacity of the lens, ultimately forming a full cataract (inset, d). Reproduced from Merriman-Smith BR, Krushinsky A, Kistler J, and Donaldson PJ (2003) Expression patterns for glucose transporters GLUT1 and GLUT3 in the normal rat lens and in models of diabetic cataract. *Investigative Ophthalmology & Visual Science* 44: 3458–3466, with permission form Investigative Ophthalmology & Visual Science.

from one cell to another. As the lens ages, new layers of fibers are added to the cortex, causing the lens nucleus to become compressed, harder, and more compact. Yellowing of the lens also occurs with age. The lens increases in bulk and mass, gaining weight steadily until about age 50 years. Some lenses remain clear with advancing age; but significant protein oxidation at the center of the lens causes cataracts. The key factor in preventing oxidation in the nucleus of the lens seems to be the concentration of nuclear glutathione (L-$\gamma$-glutamyl-L-cysteinylglycine; GSH). Nuclear sclerosis progresses slowly over the years, causing only a slight change in refraction, but these changes may not significantly affect vision. Further progression of nuclear sclerosis can lead to loss of color discrimination and also loss of distance vision.

### Diseases of Aging Lens

Any opacity of the lens or its capsule is called a cataract. Age-related cataracts are divided into nuclear, cortical, and posterior subcapsular types, depending on their location and clinical appearance. Lens crystallins accumulate damage with age (from ultraviolet light, oxidation, dehydration, or elevated blood sugar from diabetes), and may collapse into misfolded fibers. The misfolded fibers aggregate, producing heterogeneity in the protein concentration within the lens. This protein aggregation can lead to light scattering and to cataract formation (**Figure 18**). The most common treatment for cataracts is surgical removal.

It should be noted that oxidation is the hallmark of age-related nuclear (ARN) cataracts. Loss of protein sulfhydryl groups and the oxidation of methionine residues are progressive and increase as the cataract worsens. Therefore, the ARN cataract is associated with a loss of GSH in the center of the lens, and extensive modifications to the nuclear proteins, including coloration, oxidation, insolubilization, and cross-linkng.

### Risk Factors and Prevention

Cataracts are the most common cause of reversible vision loss worldwide. Age is by far the major risk factor for cataract formation. Ethnic variation has been reported with different cataract types. Genetic factors may account for as much as 50% of the severity of nuclear cataracts and may also be important in the development of cortical cataracts. Many reports suggest that women are at slightly higher risk than are men to develop cataracts. Other risk factors are of much less importance to cataractogenesis, but are ones that we may be able to control: cigarette smoking, exposure to sunlight, alcohol use, and low dietary levels of carotenoids or antioxidants. Epidemiological studies also suggest that exposure to ultraviolet B causes cortical cataract changes.

## Presbyopia

Presbyopia is the loss of accommodative amplitude which occurs with age. Around age 45 years, most people are unable to hold a book at the normal reading distance and still see the text clearly (i.e., the near point of accommodation is farther away than the normal reading or working distance). Presbyopia is largely thought to be associated with age-related changes in the lens. The many changes in the lens from birth onward (increase in diameter and thickness, volume, and mass; addition of new cells; changes in curvature) occur during the decline of accommodation that occurs with age. The link between lenticular changes and presbyopia is thought to be due to the increase in bulk and mass of the lens, such that the elastic forces are less effective in changing the curvature of the lens.

## Cornea

### Normal Aging of the Cornea

Research in aging of the cornea has focused on the inability of the corneal endothelium to replace cells that die with age. The corneal endothelium is a single layer of metabolically active cells that lines the entire posterior surface of the cornea. The aqueous humor in the anterior chamber of the eye is in direct contact with the corneal endothelium. The loss of the corneal endothelium during a life span is substantial: about half of the endothelium is lost. As endothelial cells die, neighboring endothelial cells expand and migrate in order to maintain a confluent layer of cells. Age-related changes in the tear film that bathes the cornea cause dry eye, a condition of irritating, burning, or itching eyes found in over 10 million people in the US.

In contrast to the endothelium, the corneal epithelium heals quickly and completely. The corneal epithelium replaces itself about every 10 days. It is believed that stem cells for the corneal epithelium are located in the limbus of the eye (the region between the transparent cornea and the sclera). If the limbal epithelium is damaged, then the epithelium is unable to repair itself. Corneal transplants were developed to replace damaged tissue with healthy limbal tissue.

## Further Reading


Algvere PV, Marshall J, and Seregard S (2006) Age-related maculopathy and the impact of blue light hazard. *Acta Ophthalmologica Scandanavica* 84: 4–15.

Asbell PA, Dualan I, Mindel J, et al. (2005) Age-related cataract. *Lancet* 365: 599–609.

Bloemendal H, de Jong W, Jaenicke R, et al. (2004) Ageing and vision: Structure, stability and function of lens crystallins. *Progress in Biophysics & Molecular Biology* 86: 407–485.

Bonnel S, Mohand-Said S, and Sahel JA (2003) The aging of the retina. *Experimental Gerontology* 38: 825–831.

Bourne WM and McLaren JW (2004) Clinical responses of the corneal endothelium. *Experimental Eye Research* 78: 561–572.

Budovsky A, Muradian KK, and Fraifeld VE (2006) From disease-oriented to aging/longevity-oriented studies. *Rejuvenation Research* 9: 207–210.

Congdon N, O'Colmain B, Klaver CC, et al. (2004) Causes and prevalence of visual impairment among adults in the United States. *Archives of Ophthalmology* 122: 477–485.

Gupta N and Yucel YH (2007) Glaucoma as a neurodegenerative disease. *Current Opinion in Ophthalmology* 18: 110–114.

Hageman GS, Anderson DH, Johnson LV, et al. (2005) A common haplotype in the complement regulatory gene factor H (HF1/CFH) predisposes individuals to age-related macular degeneration. *Proceedings of the National Academy of Sciences United States of America* 102: 7227–7232.

Harman A, Abrahams B, Moore S, and Hoskins R (2000) Neuronal density in the human retinal ganglion cell layer from 16–77 years. *Anatomical Record* 260: 124–131.

Jackson GR and Owsley C (2003) Visual dysfunction, neurodegenerative diseases, and aging. *Neurology Clinics* 21: 709–728.

Jackson GR, Owsley C, and Curcio CA (2002) Photoreceptor degeneration and dysfunction in aging and age-related maculopathy. *Ageing Research Reviews* 1: 381–396.

Quigley HA (2005) Glaucoma: Macrocosm to microcosm the Friedenwald lecture. *Investigative Ophthalmology & Visual Science* 46: 2662–2670.

Ramasamy R, Vannucci SJ, Yan SS, et al. (2005) Advanced glycation end products and RAGE: A common thread in aging, diabetes, neurodegeneration, and inflammation. *Glycobiology* 15: 16R–28R.

Ramrattan RS, van der Schaft TL, Mooy CM, et al. (1994) Morphometric analysis of Bruch's membrane, the choriocapillaris, and the choroid in aging. *Investigative Ophthalmology & Visual Science* 35: 2857–2864.

Sacca SC, Izzotti A, Rossi P, and Traverso C (2007) Glaucomatous outflow pathway and oxidative stress. *Experimental Eye Research* 84: 389–399.

Sparrow JR and Boulton M (2005) RPE lipofuscin and its role in retinal pathobiology. *Experimental Eye Research* 80: 595–606.

Stadtman ER (2004) Role of oxidant species in aging. *Current Medicinal Chemistry* 11: 1105–1112.

Truscott RJ (2005) Age-related nuclear cataract-oxidation is the key. *Experimental Eye Research* 80: 709–725.

Zarbin MA (2004) Current concepts in the pathogenesis of age-related macular degeneration. *Archives of Ophthalmology* 122: 598–614.


## Relevant Websites

http://www.afb.org – American Foundation for the Blind (link to CareerConnect > For Employers > Visual Impairment and

Your Workforce for information on normal changes in the aging eye).

http://webvision.med.utah.edu – John Moran Eye Center, University of Utah (WebVision: The Organization of the Retina and Visual System).

http://www.nei.nih.gov – National Eye Institute (U.S. National Institutes of Health).

http://www.stlukeseye.com – St. Lukes Cataract & Laser Institute (Florida, USA).

http://www.agingeye2.net – The Eye Digest.

# Sensory Aging: Chemical Senses

**R L Doty**, University of Pennsylvania, Philadelphia, PA, USA



## Introduction

The senses of taste and smell are critical for esthetics, discerning the flavor of foods and beverages, and detecting environmental toxins, including spoiled foods, leaking natural gas, polluted air, and smoke. Although age-related losses are present in all sensory modalities, the chemical senses – particularly olfaction – seem to be unduly influenced by aging. While approximately 2% of the American population has demonstrable olfactory loss before the age of 65 years, this percentage rises to 50% for those between the ages of 65 and 80 years, and to 75% for those above this age range. This likely explains why a disproportionate number of the elderly die from accidental gas poisoning, and why many elderly report that their food lacks flavor. The latter problem can lead to decreased motivation to eat, nutritional deficiencies, depression, and, in rare instances, death.

The age-related declines in taste and smell function are influenced by a number of factors, including gender. For example, in the case of odor identification ability, men exhibit a greater age-related decline in function than do women, a sex difference that increases in magnitude in the later years of life. The reasons why women maintain better function than men into older age are poorly understood, although estrogen may be involved. If this is the case, estrogen's organization effects on the brain may be largely responsible, since some chemosensory-related sex differences are present before puberty.

## Age-Related Alterations in Olfaction

The decline in olfactory function observed in the later years of life is, on average, quite large. This decrement is apparent on a wide variety of olfactory tests, including tests of odor detection, identification, discrimination, memory, and suprathreshold odor intensity perception. In addition, age-related declines have been noted in tests sensitive to chemically mediated somatosensory sensations derived from intranasal stimulation of the trigeminal cranial nerve (CN V), such as irritation, sharpness, coolness, warmth, and fullness. The changes in odor identification ability, as measured by the standardized University of Pennsylvania Smell Identification Test, and odor memory, as measured by a standardized Odor Memory Test, are shown in **Figures 1** and **2**, respectively. Such age-related changes are accompanied by alterations in brain activation, as indexed by functional magnetic resonance imaging (fMRI; **Figure 3**). The losses have been demonstrated in a number of different cultural groups, are much more pronounced than the influences of gender or smoking, and are present in the prepubescent period (**Figure 4**).

In general, the age-related decrease in sensitivity occurs for a spectrum of odorants, although some odors are more influenced than others, depending upon such factors as an individual's threshold to the odorant, whether multiple chemicals are involved in producing the odor, the number of receptor types activated by the odorant or odorant combination, and the nature of the psychophysical functions relating odorant concentration to perceived intensity. The tendency for threshold values of different odorants to be correlated suggests that a 'general olfactory acuity' factor exists, perhaps analogous to the general intelligence factor derived from items of intelligence tests. Interestingly, thresholds of a number of sensory modalities tend to be correlated with one another, as well as with several cognitive measures (e.g., verbal memory), implying that a 'general sensory acuity/cognitive factor' may also exist that subsumes intermodal correlations.

It is noteworthy that experience with odors can enhance olfactory sensitivity, suggesting the possibility that variation in age-related changes among people could reflect differences in prior experience with odorants. Improvement in olfactory function as a result of prior exposure is greater in women than in men, or in some cases may even be confined to women, implying that somehow the female brain is more plastic in this regard. The underlying physiological mechanisms regarding this phenomenon are poorly understood.

## Age-Related Alterations in Taste

Commonly, persons who complain of taste loss have, in fact, altered olfactory function. This is because flavor sensations, such as chocolate, mint, pizza, apple, strawberry, and so on, are dependent upon stimulation of the olfactory receptors via the retronasal route. Blocking the nose while eating, for example, greatly decreases or eliminates such flavor sensations, since active movement of air into the nasal pharynx from the oral cavity is significantly impeded.

The sensations mediated by the taste system are largely limited to such qualities as sweet, sour, bitter, salty, and metallic. Unlike olfaction, which is mediated by a single cranial nerve (CN I), the taste system

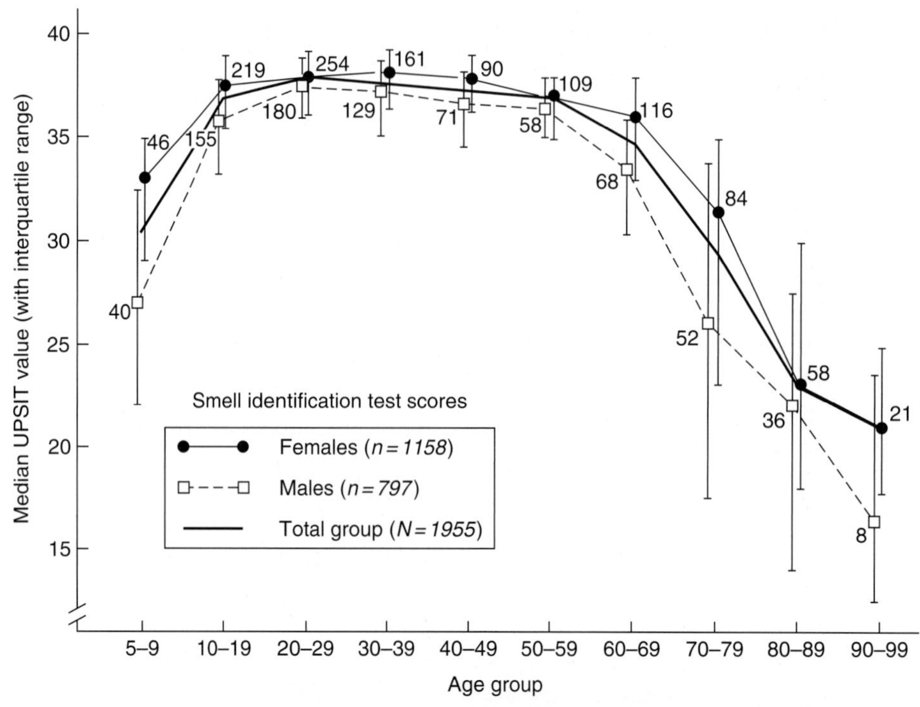


**Figure 1** Scores on the University of Pennsylvania Smell Identification Test (UPSIT) as a function of age in a large heterogeneous group of subjects. Numbers by data points indicate sample sizes. From Doty RL, Shaman P, Applebaum SL, et al. (1984). Smell identification ability: Changes with age. *Science* 226: 1441–1443. Reprinted with permission from AAAS.

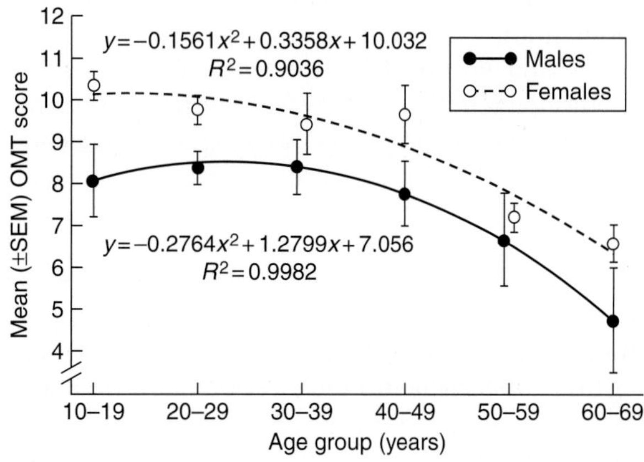


**Figure 2** Scores on the Odor Memory Test (OMT) as a function of sex and age. Data averaged across 10, 30, and 60 s delay intervals. From Choudhury ES, Moberg P, and Doty RL (2003) Influences of age and sex on a microencapsulated odor memory test. *Chemical Senses* 28: 799–805. Copyright 2003, Oxford University Press.

is subserved by a number of cranial nerves: the facial nerve (CN VII), the glossopharyngeal nerve (CN IX), and the vagus nerve (CN X). Like olfaction, age-related decreases in gustatory function are well documented. However, when the whole mouth is evaluated (e.g., by "sip and spit" testing), the magnitude of the age-related loss is not as marked as that seen for the sense of smell. This is not true when small regions of the tongue are evaluated, where marked age-related declines are observed.

Decreases in taste threshold sensitivity have been reported for such tastants as caffeine, citric acid, hydrochloric acid, magnesium sulfate, propylthiourea, quinine, sodium chloride, sucrose, tartaric acid, and a large number of amino acids. As with odorants, not all taste qualities appear to exhibit the same degree of age-related loss. Thus, sweet taste sensation appears to be less ravaged by age than are salty, bitter, and sour sensations. **Figure 5** shows the marked age-related decrement to NaCl seen in small regions of the tongue.

Changes in suprathreshold taste perception to chemicals as a result of aging have also been observed. For example, flatter functions relating tastant concentrations to perceived intensity have been reported in elderly persons relative to young adults, at least for some stimuli. However, the slope of such functions can be markedly altered by changes in intensity that occur at the extremes of the stimulus continuum, and may not reflect decreased sensitivity across the entire suprathreshold stimulus range. In a manner analogous to what occurs in olfaction, sex differences in taste function become more apparent in the later years.

Although the genetic influences on smell function are poorly established, a number for taste have been worked out. For example, the sensitivity to bitter-tasting agents is a heritable trait, presumably having

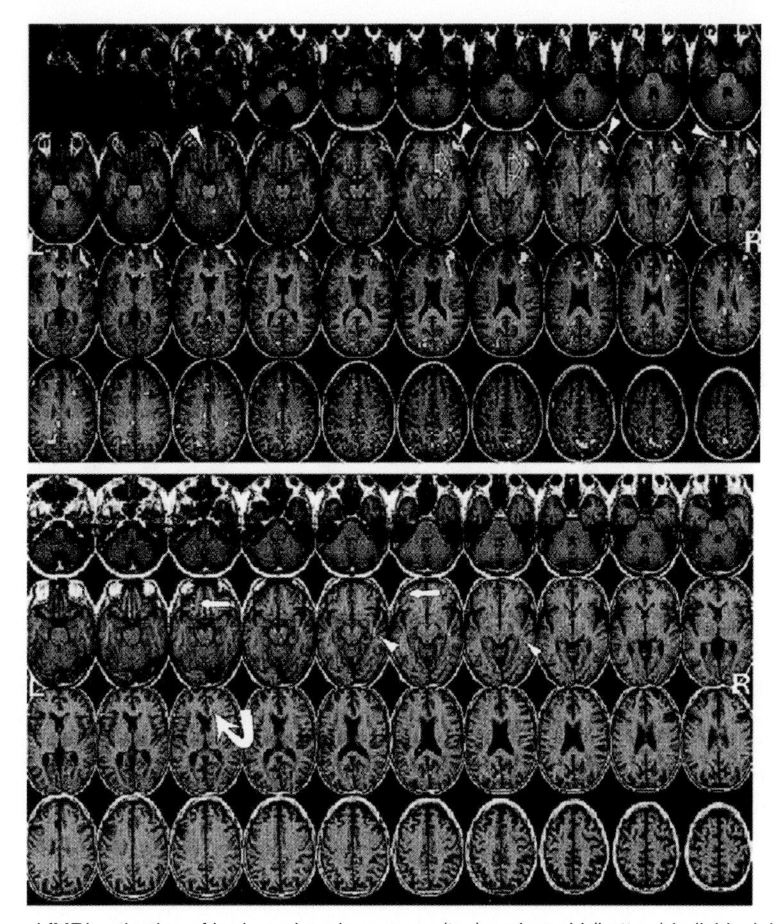

**Figure 3** Odorant-induced fMRI activation of brain regions in a young (top) and an old (bottom) individual. In the image of the young person's brain, note the strong activation in both inferior frontal lobes (arrowheads), on the right more than left. Perisylvian activation inferiorly on the right (open arrows) is also noteworthy. In the bottom image(older person), note minimal activation in the left inferior frontal (straight arrows), right perisylvian (arrowheads), and right cingulate (curved arrow) areas. From Yousem DM, Maldjian JA, Hummel T, et al. (1999) The effect of age on odor-stimulated function MR imaging. *American Journal of Neuroradiology* 20: 600–608. Copyright 1999, American Society of Neuroradiology.

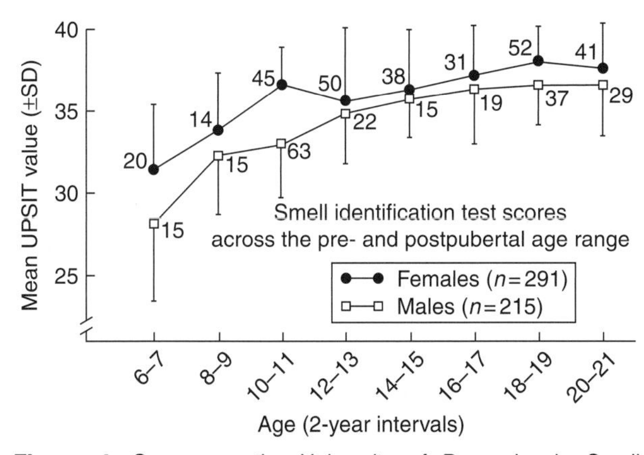


**Figure 4** Scores on the University of Pennsylvania Smell Identification Test (UPSIT) as a function of gender across the prepubertal, adolescent, and young adult years. Note the consistently better performance of women than of men. From Doty RL (1986) Gender and endocrine-related influences on human olfactory perception. In: Meiselman HL and Rivlin RS (eds.) *Clinical Measurement of Taste and Smell*, pp. 377–413. New York: Macmillan Publishing Co. Copyright 1986, Macmillan Publishing Company.

evolved to protect against ingestion of poisonous toxins, such as a number of bitter-tasting plant-derived alkaloids and glycosides. The tastants that have received the most experimental attention in this regard are phenylthiocarbamide (PTC) and 6-*n*-propylthiouracil (PROP). About three times more Europeans and North Americans of European descent are PTC 'nontasters' than are Asians, Africans, and Native Americans. Tasters and nontasters exhibit different food preferences, as well as different preferences for hot spices. Research from the 1930s suggested that sensitivity to PTC followed a simple Mendelian pattern, with tasters being homozygous or heterozygous for the dominant gene, and nontasters having two recessive genes. In the 1980s, alternative genetic models proved to be a better fit to the data (e.g., a two-locus model in which one locus controls PTC taste perception and the other controls more general taste acuity). Recently, genes have been identified that account, to a large degree, for bitter taste perception.

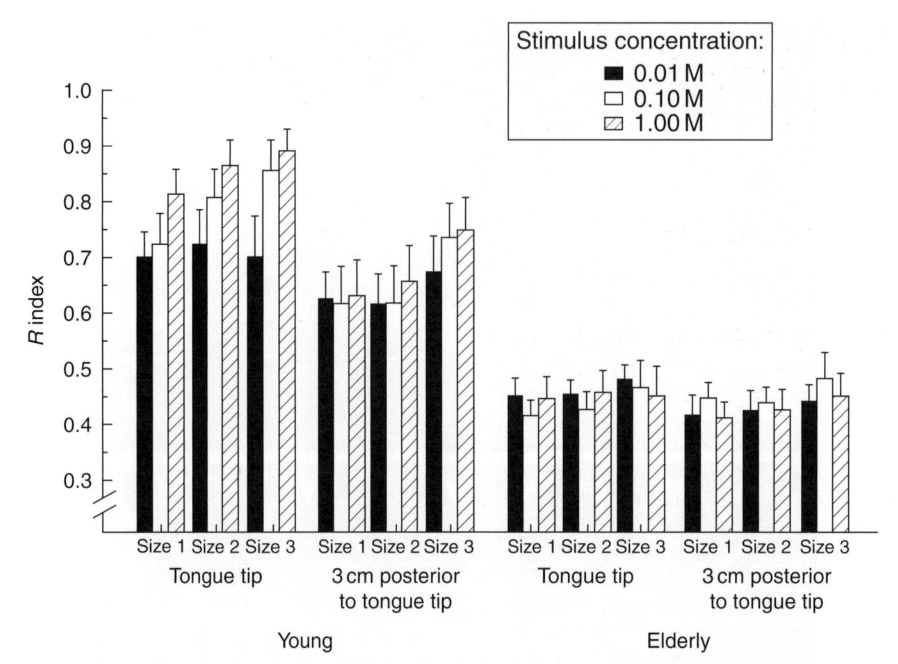


**Figure 5** Detection sensitivity to NaCl in two age groups for three stimulus concentrations at two tongue loci for three different sizes of tongue regions. Size 1 = 12.5 $mm^2$; size 2 = 25 $mm^2$; size 3 = 50 $mm^2$. Means reflect scores on a nonparametric signal detection measure of olfactory sensitivity. Error bars represent +1 SEM. From Matsuda T and Doty RL (1995) Regional taste sensitivity to NaCl: Relationship to subject age, tongue locus and area of stimulation. *Chemical Senses* 20: 283–290. Copyright 1995, Oxford University Press.

Despite evidence for strong heritability, sensitivity to PTC and PROP declines with age. In one study, for example, the percentage of a group of 378 women who perceived PROP as being strong dropped from over 70% between the ages of 18 and 29 years to approximately 50% between the ages of 60 and 70 years. Corresponding values for men in the same age ranges were 53% and 30%. The complexity of the inheritance of the PTC taster gene is illustrated by a community living in an isolated mountain region of Ecuador. In this group, PTC nontasters eat maize-derived bread which contains a goitrogen, and consistently develop large unsightly goiters. PTC tasters, on the other hand, avoid such bread and do not develop such goiters. However, the PTC nontasters exhibited greater resistance to malaria, providing an explanation as to why the nontasting genes remain in the gene pool.

## Alterations in Olfactory and Gustatory Function in Alzheimer's Disease

A number of age-related neurodegenerative diseases, most notably Alzheimer's disease (AD), exhibit marked olfactory loss relative to age-matched controls. Taste loss, while found in some studies, is much less marked. As with the influences of age, the olfactory decrements of AD are observed on numerous types of tests, and can be seen in functional imaging studies (**Figure 6**). A recent meta-analysis of 11 studies examining olfaction in AD found effect sizes ranging from 0.98 to 12.15 (median = 2.17), suggesting this is a robust phenomenon.

It has been argued that olfactory test scores are inversely related to the severity of dementia. However, such associations are weak. Moreover, they are enigmatic, given the influences of dementia on nonolfactory components of the tests. The AD-related olfactory dysfunction may reflect, to some degree, physiological changes associated with normal aging, although very early stage AD patients with mild dementia consistently score more poorly on olfactory tests than do age-matched controls. Recent studies suggest that smell loss may be occurring in the so-called preclinical stage of the disease process, and that it is more likely to occur in relatives than in nonrelatives of AD patients. In one study of 1604 nondemented community-dwelling senior citizens 65 years of age or older, a test of odor identification was a better predictor of future cognitive decline than was a global cognitive test. Persons who were anosmic and possessed at least one $\varepsilon 4$ allele of apolipoprotein E, a genetic risk factor for AD, had 4.9 times the risk of exhibiting cognitive decline than did persons with normal smell not possessing this allele. This is in contrast to the 1.23 times greater risk for cognitive decline in persons with normal smell who possessed this allele.

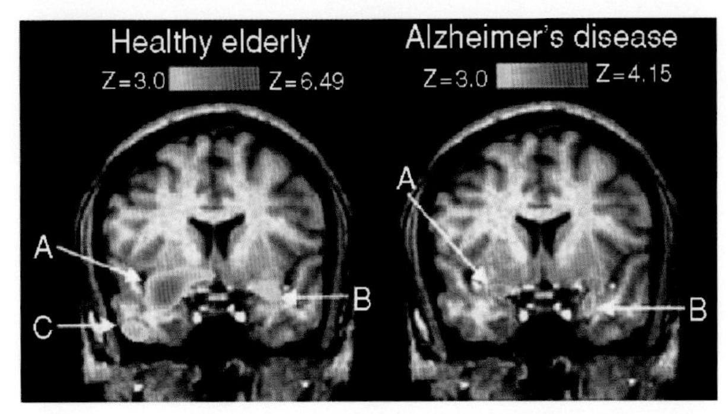


**Figure 6** Left: Whole brain subtraction of olfactory stimulation minus control stimulation in healthy elderly control participants, co-registered on a T1-weighted reference magnetic resonance image ($N=7$). Right and left are reversed. Regions of interest are as follows: (A) right frontotemporal junction (piriform area), (B) left piriform area, and (C) right anterior–ventral temporal lobe (Brodmann 20). Right: Same whole brain subtraction in patients with Alzheimer's disease. Regions of interest are as follows: (A) 7 mm anterior to the frontotemporal junction (piriform) region in the control participants' image and (B) left amygdala–uncus. From Kareken DA, Doty RL, Moberg PJ, et al. (2001) Olfactory-evoked regional cerebral blood flow in Alzheimer's disease. *Neuropsychology* 15: 18–29. Copyright © 2001, American Psychological Association.

## Causes of Olfactory Deficits in Aging and Alzheimer's Disease

The most common cause of age-related alterations in the ability to smell appears to be damage to the olfactory neuroepithelium from environmental insults (e.g., from viruses, bacteria, toxins, and pollutants). The dendritic knobs and the olfactory receptor-laden cilia are exposed rather directly to the external milieu, thereby being susceptible to nosogenic attack. In general, the olfactory neuroepithelium undergoes cumulative damage throughout life, exhibiting increasing numbers of islands of respiratory-like metaplasia soon after birth, decreased epithelial thickness, and decreased numbers of olfactory receptor cells. Some environments, such as cities or workplace environments polluted with airborne toxins, accelerate such pathology. Older persons may be more susceptible to epithelial damage than are younger ones, having a less resilient olfactory membrane due to such factors as reduced protein synthesis (as in hypothyroidism), loss of neurotrophic factors, altered vascularity and decreased intramucosal blood flow, altered nasal mucus viscosity, atrophy of secretory glands and lymphatics, and decreases in enzymes that deactivate xenobiotics within the olfactory mucosa. Some age-related deficits reflect subclinical neuropathology of the development of plaques and neurofibrillary tangles (NFTs) related to AD. Additionally, the number and cross-sectional area of the foramina of the cribriform plate are decreased in the elderly as a result of appositional bone growth, in effect pinching off the bundles of olfactory nerve axons as they traverse the foramina. In one study of this phenomenon, the mean area of the foramina was 47.3% less in men over the age of 50 years compared to those under the age of 50 years. The corresponding figure for women was 28.8%.

Despite the fact that olfactory receptor cells have some capacity to reconstitute themselves, this plasticity is compromised by age-related processes. Animal studies suggest that the ratio of dead or dying cells to the number of receptor cells increases with age, implying age-related decreases in mitotic activity. In older animals, epithelial repair is slower or nonexistent after damage by chemical agents such as zinc sulfate or methyl-formimino-methylester, unmasking an altered neurogenic process. Damage to the olfactory receptors leads to degenerative changes within the glomerular and mitral cell layers of the olfactory bulb. The number of mitral cells and glomeruli decreases steadily with age at an approximate rate of 10% per decade. By the ninth and tenth decades of life, less than 30% of these structures are present. *In vivo* quantitative magnetic resonance imaging studies confirm that the volume of the human olfactory bulbs and tracts decreases with age. Higher order structures associated with olfactory processing, such as the piriform cortex, show much less age-related change than do the olfactory bulb and epithelium. For example, in one study of rats, no significant age-related volumetric changes were found within cortical laminae Ia and Ib. The numerical and surface densities of the synaptic apposition zones in layer Ia, which are formed largely by mitral cell axons, were also not altered by age, although the proportion of layer Ia occupied by dendrites and spines was modestly decreased. An increase in the proportion of glial processes, but not in the proportion of axons and terminals, was also present.

The basis of the olfactory loss of patients with AD is poorly understood. No one has found within the olfactory epithelia of AD patients marker proteins or messenger RNAs that are specific to the disease. Nevertheless, tau-reactive and partially ubiquitin-reactive dystrophic neurites have been found in the lamina propria of AD patients. Both tau and ubiquitin contribute to the aggregation events associated with AD pathogenesis (e.g., NFTs). Recent behavioral studies in transgenic mice reveal that those that overexpress tau develop olfactory dysfunction, whereas those that overexpress $\beta$-amyloid, the major aberrant protein associated with neuritic plaques, do not. Interestingly, the $\varepsilon 4$ allele of apolipoprotein E is more than threefold greater in AD than in control olfactory receptor cells, and heat-shock proteins – proteins that are highly conserved and upregulated in cells after exposure to stressors – are markedly downregulated in the olfactory epithelia of AD patients. This conceivably increases the susceptibility of the receptor cells to damage from xenobiotic agents.

There is considerable evidence that the olfactory bulbs and tracts are damaged early in the AD process, although the most salient early signs of AD neuropathology appear within transentorhinal cortical regions. The pathology then seems to spread from these cortical areas to other brain regions. Postmortem studies report up to a 40% reduction in the cross-sectional area of the olfactory tracts of individuals with AD, as well as ~50% loss of myelinated axons in these tracts. The AD-related NFTs and neuropil treads and PD-related Lewy bodies have been found in all layers of the olfactory bulb, with the exception of the olfactory nerve cell axon layer. Interestingly, the olfactory bulbs of over 40% of nondemented persons 50 years of age and older exhibit NFTs, suggesting such individuals may be in the earliest stages of developing the ultimate disease phenotype. Central limbic structures that receive olfactory bulb projections preferentially exhibit relatively high numbers of NFTs and neuritic plaques, including the hippocampus, the periamygdaloid nucleus, the prepiriform and piriform cortices, and the entorhinal cortex.

Among the more interesting and controversial theories concerning the etiology of the olfactory dysfunction in AD and some other neurodegenerative disorders (e.g., Parkinson's disease), as well as the etiology of the diseases themselves, is the 'olfactory vector hypothesis.' One form of this hypothesis suggests that the olfactory loss results from a xenobiotic agent (e.g., virus, bacterium, prion, pesticide) that enters the brain from the nasal cavity via the olfactory fila. Since the time of Galen it was known that dyes can move from the nasal cavity into the cerebrospinal fluid (CSF), and as early as 1912 it was shown that the olfactory neuroepithelium is a major route of entry of the poliovirus into the brain. The latter observation led to the discovery that cauterizing the olfactory mucosa of monkeys with picric acid and other caustic chemicals protected them against contracting polio. This led, in the late 1930s, to public health initiatives to prophylactically spray the noses of Toronto school children with zinc sulfate during polio epidemics.

The central-to-peripheral direction of the development of the neuropathology of AD need not be in opposition to the olfactory vector hypothesis. Thus, a xenobiotic could be transported to brain regions that are more vulnerable to its effects and could initiate genetically determined pathology in the susceptible region. The possibility also exists that damage to the olfactory system, for whatever reason, leads to degenerative alterations in brain structures that ultimately produce the cognitive alterations associated with AD. Limited support for this understudied concept comes from the observation that older rats that have been anosmic for a period of time have considerable difficulty in learning an active avoidance learning task, unlike older rats of the same age whose anosmia is of recent origin.

## Causes of Taste Deficits in Aging and Alzheimer's Disease

Relative to olfaction, comparatively little research has been done to explain age- or AD-related alterations in gustatory function. As with other sensory systems, a relationship between the number of functioning receptor elements and the system's sensitivity is present. Thus, the perceived intensity of tastants presented to localized regions of the anterior tongue is correlated with the number of fungiform papillae and, hence, the number of taste buds within the stimulated regions. Individuals with higher densities of taste buds and fungiform papillae rate the intensity of a number of tastants as being 'more strong' than do individuals with lower densities of such structures. Age-related loss of taste buds has been found for human circumvallate papillae, although recent studies of rodent, monkey, and human tongues suggest that taste bud numbers on the fungiform papillae may be little influenced by age. For example, the average percentage of fungiform papillae-containing taste buds in Fischer 344 rats aged 4 to 6 months, 20 to 24 months, and 30 to 37 months was found in one study to be 99.6%, 99.3%, and 94.7%, respectively. A human study found no statistically meaningful relationship between age and taste bud densities on either the tip or the midregion of tongues from young adults (22–36 years, $N=5$), middle-aged adults (50–63 years, $N=7$), and old adults (70–90 years, $N=6$). This work, however, is limited by its small sample sizes and marked variability in the number of taste buds on the tongue at all ages.

Although taste bud numbers appear not to be markedly decreased in old rats, electrical responsiveness of the chorda tympani nerve, which innervates the anterior tongue, is decreased to a number of salts, acids, and sugars. Among the possible explanations of this phenomena are decreased intrinsic reactivity of taste buds to taste solutions, decreased multiple neural innervation of taste buds by some taste fibers, alterations in the general structure of the epithelium (which, e.g., might impair the movement of stimulus solution into the taste bud pore), and decreased taste nerve responsiveness, *per se*. It is also possible that some taste buds, although anatomically present, are not fully functional because of altered receptor cell turnover or related metabolic events. Since taste buds function in a complex ionic milieu and are bathed with saliva and other secretory products, changes in such products may also undergo age-related changes. There is suggestion that heightened taste threshold values and flattened suprathreshold psychophysical functions observed in many elderly persons reflect background tastes noticeable at low, but not at moderate or high, stimulus concentrations. Changes in both the neural and oral chemical milieu (e.g., salivary constituents) could contribute to this noise. Evidence that improved oral hygiene improves taste sensitivity in some elderly persons is in accord with this hypothesis.

Relatively few studies have examined the structure of the gustatory system of patients with AD. In one study, the innervation of the foliate and circumvallate taste buds was examined in five AD (three men, two women) and two control (both women) study participants, using antisera to protein gene product 9.5 (PGP 9.5), neuron-specific enolase (NSE), tyrosine hydroxylase (TH), dopamine $\beta$-hydroxylase (D$\beta$-H), and calcitonin gene-related peptide (CGRP). The mean number of PGP 9.5-immunoreactive intragemmal nerve fibers in taste buds of the foliate and circumvallate papillae was lower in the AD patients. Although very provocative, this study is limited by the small sample of individuals evaluated.

*See also:* Aging of the Brain; Aging of the Brain and Alzheimer's Disease; Alzheimer's Disease: MRI Studies.

## Further Reading

Arvidson K and Friberg U (1980) Human taste: Response and taste bud number in fungiform papillae. *Science* 209: 807–808.

Baker H and Genter MB (2003) The olfactory system and the nasal mucosa as portals of entry of viruses, drugs and other exogenous agents into the brain. In: Doty RL (ed.) *Handbook of Olfaction and Gustation*, pp. 549–573. New York: Marcel Dekker.

Bartoshuk LM, Rifkin B, Marks LE, et al. (1986) Taste and aging. *Journal of Gerontology* 41: 51–57.

Braak H, Braak E, Yilmazer D, et al. (1996) Pattern of brain destruction in Parkinson's and Alzheimer's diseases. *Journal of Neural Transmission* 103: 455–490.

Choudhury ES, Moberg P, and Doty RL (2003) Influences of age and sex on a microencapsulated odor memory test. *Chemical Senses* 28: 799–805.

Cowart BJ (1981) Development of taste perception in humans: Sensitivity and preference throughout the life span. *Psychological Bulletin* 90: 43–73.

Curcio CA, McNelly NA, and Hinds JW (1985) Aging in the rat olfactory system: relative stability of piriform cortex contrasts with changes in olfactory bulb and olfactory epithelium. *Journal of Comparative Neurology* 235: 519–528.

Dalton P, Doolittle N, and Breslin PA (2002) Gender-specific induction of enhanced sensitivity to odors. *Nature Neuroscience* 5: 199–200.

Doty RL (1986) Gender and endocrine-related influences on human olgactory perception. In: Meiselman HL and Rivlin RS (eds.) *Clinical Measurement of Taste and Smell*, pp. 377–413. New York: Macmillan Publishing.

Doty RL (2003) Odor perception in neurodegenerative diseases. In: Doty RL (ed.) *Handbook of Olfaction and Gustation*, pp. 479–501. New York: Marcel Dekker.

Doty RL, Shaman P, Applebaum SL, et al. (1984) Smell identification ability: Changes with age. *Science* 226: 1441–1443.

Ferreyra-Moyano H and Barragan E (1989) The olfactory system and Alzheimer's disease. *International Journal of Neuroscience* 49: 157–197.

Hinds JW and McNelly NA (1977) Aging of the rat olfactory bulb: Growth and atrophy of constituent layers and changes in size and number of mitral cells. *Journal of Comparative Neurology* 72: 345–367.

Kalmey JK, Thewissen JG, and Dluzen DE (1998) Age-related size reduction of foramina in the cribriform plate. *Anatomical Record* 251: 326–329.

Kareken DA, Doty RL, Moberg PJ, et al. (2001) Olfactory-evoked regional cerebral blood flow in Alzheimer's disease. *Neuropsychology* 15: 18–29.

Kishikawa M, Iseki M, Nishimura M, et al. (1990) A histopathological study on senile changes in the human olfactory bulb. *Acta Pathologica Japonica* 40: 255–260.

Langan MJ and Yearick ES (1976) The effects of improved oral hygiene on taste perception and nutrition of the elderly. *Journal of Gerontology* 31: 413–418.

Loo AT, Youngentob SL, Kent PF, et al. (1996) The aging olfactory epithelium: Neurogenesis, response to damage, and odorant-induced activity. *International Journal of Developmental Neuroscience* 14: 881–900.

Matsuda T and Doty RL (1995) Regional taste sensitivity to NaCl: Relationship to subject age, tongue locus and area of stimulation. *Chemical Senses* 20: 283–290.

McBride MR and Mistretta CM (1986) Taste responses from the chorda tympani nerve in young and old Fischer rats. *Journal of Gerontology* 41: 306–314.

Mesholam RI, Moberg PJ, Mahr RN, et al. (1998) Olfaction in neurodegenerative disease: A meta-analysis of olfactory functioning in Alzheimer's and Parkinson's diseases. *Archives of Neurology* 55: 84–90.

Miller IJ and Reedy FE Jr. (1990) Variations in human taste bud density and taste intensity perception. *Physiology & Behavior* 47: 1213–1219.

Mistretta CM (1984) Aging effects on anatomy and neurophysiology of taste and smell. *Gerodontology* 3: 131–136.

Murphy C and Gilmore MM (1989) Quality-specific effects of aging on the human taste system. *Perception & Psychophysics* 45: 121–128.

Nakashima T, Kimmelman CP, and Snow JB Jr. (1984) Structure of human fetal and adult olfactory neuroepithelium. *Archives of Otolaryngology* 110: 641–646.

Ohm TG and Braak H (1987) Olfactory bulb changes in Alzheimer's disease. *Acta Neuropathologica* 73: 365–369.

Rosli Y, Breckenridge LJ, and Smith RA (1999) An ultrastructural study of age-related changes in mouse olfactory epithelium. *Journal of Electron Microscopy* 48: 77–84.

Schiffman SS, Hornack K, and Reilly D (1979) Increased taste thresholds of amino acids with age. *American Journal of Clinical Nutrition* 32: 1622–1627.

Schultz EW and Gebhardt LP (1937) Zinc sulfate prophylaxis in poliomyelitis. *Journal of the American Medical Association* 108: 2182–2184.

Smith CG (1942) Age incident of atrophy of olfactory nerves in man. *Journal of Comparative Neurology* 77: 589–594.

Stevens JC and Cain WS (1986) Aging and the perception of nasal irritation. *Physiology & Behavior* 37: 323–328.

Weiffenbach JM, Cowart BJ, and Baum BJ (1986) Taste intensity perception in aging. *Journal of Gerontology* 41: 460–468.

Yamagishi M, Takami S, and Getchell TV (1995) Innervation in human taste buds and its decrease in Alzheimer's disease patients. *Acta Oto-Laryngologica* 115: 678–684.

Yousem DM, Maldjian JA, Hummel T, et al. (1999) The effect of age on odor-stimulated functional MR imaging. *American Journal of Neuroradiology* 20: 600–608.

# Neuromuscular Junction (NMJ): Aging

**G C Sieck and C B Mantilla**, Mayo Clinic College of Medicine, Rochester, MN, USA



The neuromuscular junction (NMJ) is the final effector for neural control of muscle contraction. Mammalian NMJs serve as the sole communicative link between a motor neuron and the muscle fibers it innervates, which together constitute a motor unit (**Figure 1**). Mechanical properties of motor units are critically important in determining the functional performance of a skeletal muscle across a range of physiological behaviors. In fact, motor units exhibit considerable diversity in terms of their morphological, mechanical, and fatigue properties, and recruitment of specific types of motor units is a major mechanism in neural control of muscle force generation. Despite the diversity among motor unit types, there is remarkable homogeneity in the mechanical and biochemical properties of muscle fibers comprising an individual motor unit. Thus, the properties of muscle fibers and motor neurons appear to be precisely matched to facilitate neuromotor control.

Synaptic plasticity is a hallmark of the nervous system's adaptive response to both intrinsic and extrinsic stimuli and forms the basis for changes in the efficacy of synaptic transmission. With aging, there are a number of changes that could affect plasticity of the NMJ: (1) neuromuscular activity is generally reduced with concomitant unloading of limb muscle fibers; (2) the number of motor neurons is reduced, resulting in denervation of some muscle fibers, which are subsequently reinnervated by the sprouting of axons from remaining motor neurons; and (3) muscle fiber size is reduced in a fiber-type-dependent manner. In the late 1940s, Donald Hebb introduced the concept of activity-dependent synaptic plasticity, which addresses the structural and functional changes that occur at synapses in response to altered use (either an increase or a decrease in activity). According to the Hebbian theory, synaptic efficacy is enhanced when the extent of correlation between pre- and postsynaptic activity increases. With an age-related decrease in neuromuscular activity, synaptic plasticity may occur, but only if there are alterations in the fidelity of neuromuscular transmission (i.e., the extent of correlation between pre- and postsynaptic activity). Clearly, conditions other than changes in activity can also drive synaptic plasticity. For example, with aging there is a loss of motor neurons, which leads to denervation. Following denervation, axonal regeneration and reinnervation of muscle fibers can occur, but this is affected by activity and trophic influences. With aging, there is also a decrease in muscle fiber size (i.e., age-related sarcopenia). A number of studies have reported changes in NMJ morphology and neuromuscular transmission in conditions in which muscle fiber size either increases or decreases, which may relate to common trophic influences. Each of these conditions, in which long-term plasticity of the NMJ occurs, varies with motor unit and/or muscle fiber type.

Although motor unit heterogeneity in mammalian skeletal muscles reflects a continuum of mechanical and biochemical properties of muscle fibers, motor units are generally categorized into different types based on mechanical and fatigue properties of motor unit fibers. Accordingly, motor units are classified into four types: (1) slow-twitch, fatigue-resistant (type S); (2) fast-twitch, fatigue-resistant (type FR); (3) fast-twitch, fatigue-intermediate (type FInt); and (4) fast-twitch, fatigable (type FF) (**Figure 2**). Muscle fiber type classification follows a corresponding scheme, whether based on histochemistry or myosin heavy chain (MHC) isoform expression. Thus, muscle fibers are classified into four types: type I ($MHC_{Slow}$), type IIa ($MHC_{2A}$), type IIx ($MHC_{2X}$), and type IIb ($MHC_{2B}$). The innervation ratio (i.e., the number of muscle fibers innervated by a motor neuron) varies across muscles, ranging from 10 or less in small intrinsic hand and eye muscles to several hundred in larger limb and postural muscles. Within a muscle, the innervation ratio is greater at type FInt and FF motor units compared to type S and FR units. In muscles of mixed fiber type composition, differences in fiber size also exist across motor unit types, with type IIx and/or IIb fibers having greater cross-sectional area than type I or IIa fibers. Across muscles the cross-sectional areas of a given fiber type may vary. For example, type I fibers in the rat soleus muscle are much larger than type I fibers in muscles with mixed fiber type composition such as the diaphragm. Although controversial, some studies suggest that the specific force (i.e., force per unit cross-sectional area) of type IIx and/or IIb fibers is greater than that of type I and IIa fibers. Together, the greater innervation ratio, larger fiber size, and greater specific force contribute to the greater forces generated by type FInt and FF motor units compared to type S and FR units. The recruitment of motor units is generally matched to these mechanical and fatigue properties, such that type S and FR motor units are recruited first, followed by type FInt and FF units. In fact, forces generated during most sustained motor behaviors require the recruitment of only type S and FR motor units, and only during high-force, short-duration motor behaviors would it be necessary to recruit more fatigable FInt and FF units.

## Aging Effects on Motor Units

The basis for classification of motor unit types does not change with aging. However, age-related changes can affect the proportion of motor unit types, their innervation ratio, and their mechanical properties. For example, with aging there is a decrease in the number of motor neurons innervating skeletal muscles, and this loss of motor neurons appears to be selective for FInt and FF motor unit types. Thus, the proportion of these motor unit types decreases with age. It is possible that the selective loss of more fatigable motor units reflects age-related changes in neuromuscular activity, but this has not been clearly established. With the loss of motor neurons, some type IIx and/or IIb fibers are denervated. As a result, remaining motor axons sprout and reinnervate these muscle fibers, leading to an increase in innervation ratio of the remaining motor units. In addition, the size of type IIx and/or IIb fibers decreases with age, possibly the result of motor neuron loss and denervation, decreased neuromuscular activity, unloading, or other underlying mechanisms. Finally, specific force also decreases with age, being more pronounced at type IIx and/or IIb fibers. Together, these age-related changes tend to decrease the diversity among motor unit types, with a lower proportion of FInt and FF motor unit types that generate greater forces. The age-related decrease in the force generated by FInt and FF motor units (due to decreased fiber cross-sectional area and specific force) may be partially offset by a concomitant increase in the force generated by type S and FR motor units (due to an increased innervation ratio). Certainly, these changes

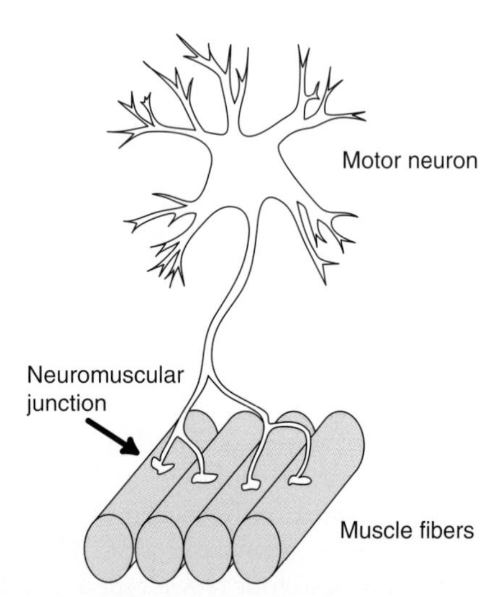


**Figure 1** Schematic representation of the elements of a motor unit. The neuromuscular junction (NMJ) serves as the sole communicative link between the motor neuron and the muscle fibers it innervates.

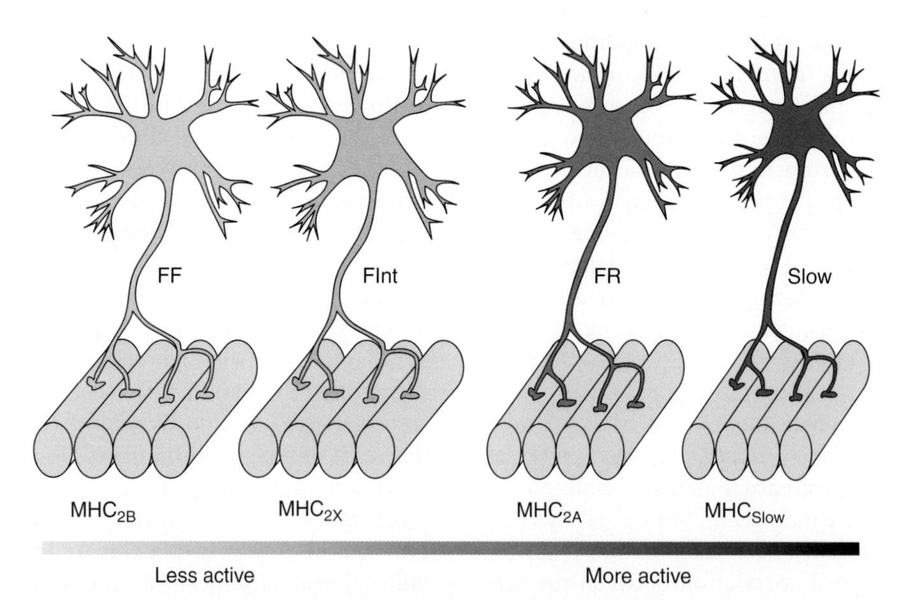


**Figure 2** Motor units are classified based on their mechanical and fatigue properties. Four types are commonly described: (1) slow-twitch, fatigue-resistant (type S); (2) fast-twitch, fatigue-resistant (type FR); (3) fast-twitch, fatigue-intermediate (type FInt); and (4) fast-twitch, fatigable (type FF). These generally correspond to the expression of specific myosin heavy chain (MHC) isoforms in the muscle fibers: type I, $MHC_{Slow}$; type IIa, $MHC_{2A}$; type IIx, $MHC_{2X}$; and type IIb, $MHC_{2B}$. Activation history of motor units also differs, with type S and FR motor units (innervating type I and IIa muscle fibers, respectively) being more frequently recruited than type FInt and FF motor units (innervating type IIx and/or IIb muscle fibers).

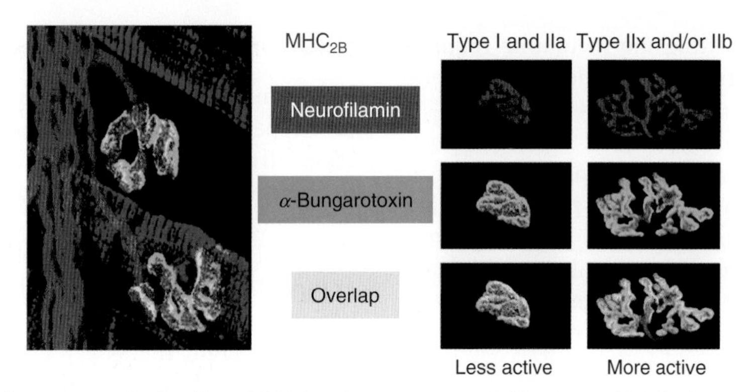


**Figure 3** The structure of neuromuscular junctions (NMJs) varies across muscle fiber types. Rat diaphragm muscle fibers were labeled with antibodies specific to the $MHC_{2B}$ isoform (blue). Note that other muscle fibers remain unlabeled and appear black. The motor nerve endings were labeled with antibodies specific to neurofilamin (red), whereas the motor end-plates were stained with rhodamine-conjugated α-bungarotoxin (green). Note the differences in size and complexity (number and length of branches) across fiber types, with NMJs present at type I or IIa fibers being smaller and less complex than those at type IIx and/or IIb fibers (magnification, ×400).

in mechanical properties might affect recruitment of motor units, particularly during high-force, short-duration motor behaviors. With aging, there appears to be a preservation of mechanical properties of motor units required for low-force, sustained motor behaviors. In some cases, the advantage of such preservation is quite obvious – for example, recruitment of type S and FR motor units in the diaphragm muscle to sustain ventilation or a similar recruitment of motor units in antigravity muscles to sustain posture.

## Structure of Neuromuscular Junctions

The structural properties of NMJs also vary considerably across different motor unit types (**Figure 3**). In rat muscles of mixed fiber type composition, NMJs at type I and IIa fibers are smaller, more compact, and show less complexity in branching patterns than NMJs at type IIx and/or IIb fibers. However, in the soleus muscle, NMJs are larger compared to NMJs at type IIx and/or IIb fibers in the extensor digitorum longus muscle, which is also a muscle with fairly homogeneous fiber type composition. It is possible that fiber size is an important determinant of NMJ size regardless of fiber type. Other factors, such as age, species, and activation history, may play important roles in determining NMJ size. Differences also exist across motor unit types in the ultrastructure of pre- and postsynaptic elements of NMJs. The presynaptic terminal surface area is greater at type IIx and/or IIb fibers compared to type I and IIa fibers. The density of active zones at presynaptic terminals does not appear to differ across fiber types; however, because of the difference in total surface area, the total number of active zones is greater at type IIx and/or IIb fibers compared to type I and IIa fibers. The number of docked vesicles at each active zone

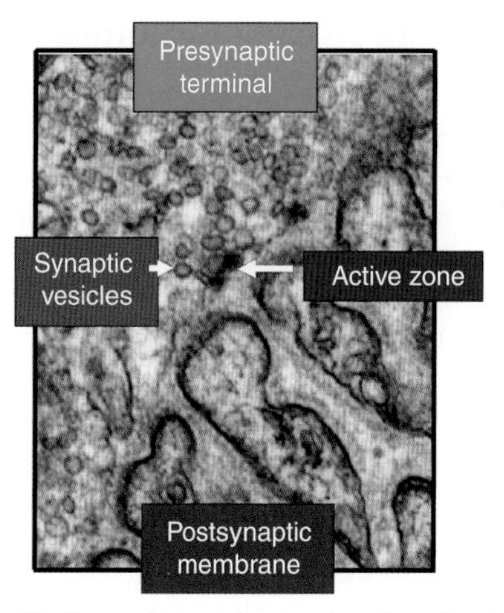


**Figure 4** Electron micrograph showing the ultrastructure of neuromuscular junctions.

(i.e., the readily releasable pool, defined morphologically as those vesicles that are fused to the terminal membrane and thus immediately available for release) is comparable across fiber types (**Figure 4**). However, given the greater number of active zones at type IIx and/or IIb fibers, the total number of synaptic vesicles in the readily releasable pool is greater at these fibers compared to type I and IIa fibers. Synaptic vesicles at presynaptic terminals are also separated into a pool of vesicles that are immediately adjacent to active zones (within 200 nm) and those that are more distant (reserve pool). The densities of synaptic vesicles in both the immediately adjacent pool and the reserve pool are greater at presynaptic terminals of type I and IIa fibers compared to type IIx and/or IIb fibers. Mitochondrial volume density is also greater at presynaptic terminals innervating type I and IIa fibers compared to those innervating type IIx and/or

IIb fibers, possibly reflecting the metabolic requirements of increased activation of these presynaptic terminals.

At motor end-plates of type I and IIa fibers, postsynaptic folding is less complex compared to those at IIx and/or IIb fibers. In addition, at type I and IIa fiber motor end-plates, cellular organelles including mitochondria, rough endoplasmic reticulum, free polysomes, and nuclei are frequently interposed between postsynaptic specializations and myofibrils. Indeed, muscle fiber types can be classified based on these morphological differences at the NMJ. There is no evidence for fiber type differences in the density of cholinergic receptors at the postsynaptic membrane. However, the density of sodium channels near the motor end-plate is much higher at type IIx and/or IIb fibers than at type I fibers.

## Aging Effects on NMJ Structure

Aging is associated with an increased number of nerve terminal branches at type IIx and/or IIb diaphragm muscle fibers and a greater fragmentation of nerve terminals. This has also been reported as an age-related increase in complexity of nerve terminal arborization at type IIx and/or IIb fibers. The increased number of terminal branches may reflect sprouting as motor neurons degenerate with aging. The increased branching and fragmentation (complexity) of presynaptic terminals with aging does not lead to an increase in the actual surface area of presynaptic terminals. Indeed, presynaptic terminal area has been reported to decrease with aging. Thus, although it has not been directly assessed, it may be that the total number of active zones at presynaptic terminals decreases with aging and, as a result, the number of synaptic vesicles in the readily releasable pool may decrease. Such changes may reflect alterations in the cyclical remodeling of NMJs reflecting outgrowth and retraction of pre- and postsynaptic elements. It is possible that trophic influences that affect this normal remodeling are altered with aging. Note that an increase in the size of NMJs does not necessarily correlate with improved synaptic efficacy. A possible reflection of instability of NMJ remodeling may be the increased expression of neural cell adhesion molecule that occurs with aging. There is an age-related gradual decrease in the number of cholinergic receptors at the motor end-plate and extrajunctional receptors appear. With aging, there is no change in the numbers of sodium channels at skeletal muscle fibers. However, it is not known whether the relative proportion of junctional, perijunctional, or extrajunctional channels changes with age.

## Neuromuscular Transmission

Neuromuscular transmission also varies across motor unit types as a result of differences in safety factor, especially during repeated activation. Safety factor is determined by the ratio of end-plate potential (EPP) amplitude to the threshold for muscle fiber action potential generation. The amplitude of EPPs is greater at type IIx and/or IIb fibers, whereas the threshold for action potential generation is lower at type I and IIa fibers. Overall, the safety factor for neuromuscular transmission during a single evoked response is higher at type IIx and/or IIb fibers compared to type I or IIa fibers. However, during repetitive stimulation the EPP amplitude progressively declines across all fiber types, but the decline is greater at type IIx and/or IIb fibers. Accordingly, type IIx and/or IIb fibers are more susceptible to neuromuscular transmission failure.

Across muscles, whether of mixed or homogeneous fiber type composition, the total number of vesicles in the readily releasable pool is greater at type IIx and/or IIb fibers compared to type I and IIa fibers. The total number of synaptic vesicles released during a single nerve impulse (i.e., quantal content) is higher at type IIx and/or IIb fibers than at type I and IIa fibers. However, there appears to be no fiber type difference in the probability of synaptic vesicle release during a single nerve impulse. With repetitive nerve stimulation, quantal content decreases across all fiber types. Initially, there is a rapid decline (within the first 500 ms) in quantal content that is dependent on the rate of stimulation. This rapid decline in quantal content reflects both a depletion of the readily releasable pool of synaptic vesicles and a decrease in the probability of release. Quantal content continues to decline during repetitive stimulation but at a much slower rate, reflecting a balance between depletion and repletion of the readily releasable pool of synaptic vesicles. Two mechanisms underlie replenishment of synaptic vesicles in the readily releasable pool: (1) recruitment from the immediately adjacent (or available) pool of vesicles and (2) recycling of released vesicles. As mentioned previously, the density of vesicles in the immediately available pool is greater at type I and IIa fibers compared to type IIx and/or IIb fibers. This is consistent with a higher rate of recruitment of vesicles into the readily releasable pool at type I and IIa fibers. It also appears that the extent of synaptic vesicle recycling (e.g., as measured by the uptake of styryl dyes such as FM4–64 or FM1–43) is greater at type I and IIa fibers compared to type IIx and/or IIb fibers. A greater extent of replenishment of the readily releasable pool at type I and IIa fibers during repetitive stimulation may underlie their reduced susceptibility to neuromuscular transmission failure. Clearly this is important since these motor units are more frequently recruited and are responsible for sustained motor behaviors (e.g., breathing and posture).

## Aging Effects on Neuromuscular Transmission

With aging, it has been reported that the safety factor for neuromuscular transmission decreases progressively. Unfortunately, there is little, if any, information regarding the effect of aging on safety factor across different fiber types. As mentioned previously, the safety factor for neuromuscular transmission depends on both EPP amplitude and the threshold for generating an action potential in muscle fibers. Quantal content and EPP amplitude decrease with aging, especially at type IIx and/or IIb fibers. However, the cross-sectional area of type IIx and/or IIb fibers also decreases with aging, which would be associated with an increase in input resistance and, thus, a lower threshold for action potential generation. Therefore, although it has not been directly examined, the safety factor for neuromuscular transmission at type IIx and/or IIb fibers may not be affected by aging. In contrast, the cross-sectional area of type I and IIa fibers changes very little with aging. Accordingly, the threshold for action potential generation would remain unaffected, and an age-related decrease in EPP amplitude would be associated with a decrease in safety factor at these fibers. Such an effect would be especially pronounced during the rapid decline in quantal content that occurs with repetitive stimulation.

## Conclusions

There is surprisingly little direct information about the effects of aging on the long-term plasticity of NMJs. Yet aging is clearly associated with changes that could affect plasticity of the NMJ. Physical activity is reduced with aging, and the resulting unloading of limb muscles clearly has an effect on NMJ structure and function that is more pronounced than in younger animals. There is a loss of motor neurons with aging, resulting in the functional denervation of muscle fibers, which are subsequently reinnervated by the sprouting of axons from remaining motor neurons. This process leads to loss of motor unit diversity, with greater loss of those motor units that are able to generate greater forces. Concomitantly, there is a reduction in muscle fiber size at these same motor units. Thus, with aging, there appears to be preservation of motor units required for low-force, sustained motor behaviors, which may be advantageous, for example, in the maintenance of adequate ventilation or posture.

## Further Reading


Balice-Gordon RJ (1997) Age-related changes in neuromuscular innervation. *Muscle & Nerve* 5: S83–S87.

Courtney J and Steinbach JH (1981) Age changes in neuromuscular junction morphology and acetylcholine receptor distribution on rat skeletal muscle fibres. *Journal of Physiology* 320: 435–447.

Fahim MA, Holley JA, and Robbins N (1983) Scanning and light microscopic study of age changes at a neuromuscular junction in the mouse. *Journal of Neurocytology* 12: 13–25.

Kelly SS and Robbins N (1987) Statistics of neuromuscular transmitter release in young and old mouse muscle. *Journal of Physiology* 385: 507–516.

Mantilla CB, Rowley KL, Fahim MA, Zhan WZ, and Sieck GC (2004) Synaptic vesicle cycling at type-identified diaphragm neuromuscular junctions. *Muscle & Nerve* 30: 774–783.

Prakash YS, Miller SM, Huang M, and Sieck GC (1996) Morphology of diaphragm neuromuscular junctions on different fibre types. *Journal of Neurocytology* 25: 88–100.

Prakash YS and Sieck GC (1998) Age-related remodeling of neuromuscular junctions on type-identified diaphragm fibers. *Muscle & Nerve* 21: 887–895.

Robbins N (1992) Compensatory plasticity of aging at the neuromuscular junction. *Experimental Gerontology* 27: 75–81.

Rosenheimer JL and Smith DO (1985) Differential changes in the endplate architecture of functionally diverse muscles during aging. *Journal of Neurophysiology* 53: 1567–1581.

Ruff RL (1996) Sodium channel slow inactivation and the distribution of sodium channels on skeletal muscle fibres enable the performance properties of different skeletal muscle fibre types. *Acta Physiologica Scandinavica* 156: 159–168.

Sieck GC and Fournier M (1989) Diaphragm motor unit recruitment during ventilatory and nonventilatory behaviors. *Journal of Applied Physiology* 66: 2539–2545.

Smith DO (1988) Muscle-specific decrease in presynaptic calcium dependence and clearance during neuromuscular transmission in aged rats. *Journal of Neurophysiology* 59: 1069–1082.

Smith DO and Weiler MH (1987) Acetylcholine metabolism and choline availability at the neuromuscular junction of mature adult and aged rats. *Journal of Physiology* 383: 693–709.

Wokke JH, Jennekens FG, van den Oord CJ, Veldman H, Smit LM, and Leppink GJ (1990) Morphological changes in the human end plate with age. *Journal of the Neurological Sciences* 95: 291–310.

Wood SJ and Slater CR (1997) Safety factor at the neuromuscular junction. *Journal of Physiology* 500: 165–176.

# Basal Ganglia: Functional Models of Normal and Disease States

**M A Kliem and T Wichmann**, Emory University, Atlanta, GA, USA



## Functional Anatomy of the Basal Ganglia

### General Circuit Overview

The term 'basal ganglia' encompasses the striatum, the external and internal segments of the globus pallidus (GPe and GPi, respectively), the pars compacta and the pars reticulata of the substantia nigra (SNc and SNr, respectively), and the subthalamic nucleus (STN). These structures are components of larger cortico–subcortical reentrant pathways (see **Figure 1**). Anatomical and physiological studies have shown that the basal ganglia, thalamus, and cortex are linked to form a system of segregated parallel circuits, which permit the basal ganglia to simultaneously participate in different functions, such as movement, reward, and procedural learning. Five major circuits have been identified, and were named after the presumed functions of the cortical areas they involve: a 'motor' circuit, centered on the precentral cortical motor fields; an 'oculomotor' circuit, originating from the frontal and supplementary eye fields; two prefrontal circuits, involving the dorsolateral prefrontal and lateral orbitofrontal cortices; and a 'limbic' circuit, centered on the anterior cingulate and medial orbitofrontal cortices. Each of these circuits may be composed of several subcircuits.

Although these circuits originate and terminate in different cortical areas, and occupy different domains in the basal ganglia and the thalamus, they share certain anatomical features, which are outlined in the simplified schemes shown in **Figures 1** and **2**. Each circuit arises from a separate cortical area and passes through different portions of the striatum, GPe, GPi, SNr, and STN, and parts of the ventrolateral, ventral anterior, and mediodorsal (VL, VA, and MD, respectively) thalamus. In each of these loops, the striatum and STN serve as 'input' stages of the basal ganglia, whereas GPi and SNr serve as 'output' stages, sending inhibitory projections to thalamus and brain stem. The thalamus projects back to the cortical areas of origin of the respective circuit to complete the reentrant loops.

### Motor Circuit

Of the different circuits passing through the basal ganglia, the motor circuit has received the most attention, because abnormalities in this circuit may play a major role in the development of the motor abnormalities in Parkinson's disease and other movement disorders. The motor circuit arises from the precentral motor fields and engages the putamen, the 'motor' portions of GPe, STN, and GPi/SNr, as well as parts of the VL thalamus. The circuit is organized into somatotopic regions that contain groups of neurons which discharge either in conjunction with the preparation for or the execution of limb movements. The motor circuit may also play a role in motor functions, ranging from the scaling of kinematic parameters such as the amplitude or velocity of movement, the initiation or execution of internally triggered movements, the sequencing of movements, and the switching between movements or motor programs, to a role in procedural learning.

### Direct and Indirect Basal Ganglia Pathways

Two major projection systems link the striatum to the output structures of the basal ganglia (**Figure 2**, left panel): a monosynaptic inhibitory ($\gamma$-aminobutyric acid (GABA)ergic) 'direct' pathway between striatum and GPi/SNr, and a polysynaptic 'indirect' pathway which involves GPe and STN. In the indirect pathway, the projections between the striatum and GPe, and between GPe and the STN are inhibitory (GABAergic), while the STN–GPi/SNr pathway is excitatory (glutamatergic). In addition, a reciprocal inhibitory connection between both segments of the globus pallidus which circumvents the STN is also considered as part of the indirect pathway.

The striatal neurons that give rise to the direct and indirect pathways differ in several ways which may strongly affect their functions. One difference is that in rodent experiments the two populations of striatal neurons appear to receive inputs from different populations of cortical neurons. Furthermore, direct pathway neurons receive stronger inputs from the intralaminar nuclei of the thalamus (the centromedian and parafascicular nucleus; CM/Pf), relative to indirect pathway neurons. Inputs from other thalamic nuclei (such as feedback circuits emanating from VA/VL) target direct and indirect neurons alike.

There are also biochemical differences between direct and indirect pathways. Although both groups of neurons are GABAergic, the direct pathway neurons express substance P as a cotransmitter, while indirect pathway neurons express enkephalin instead. Another important difference is that the activity of the direct and indirect pathways appears to be differentially modulated by dopamine, released from terminals of the nigrostriatal projection. Dopamine has been shown to inhibit the activity of striatal neurons

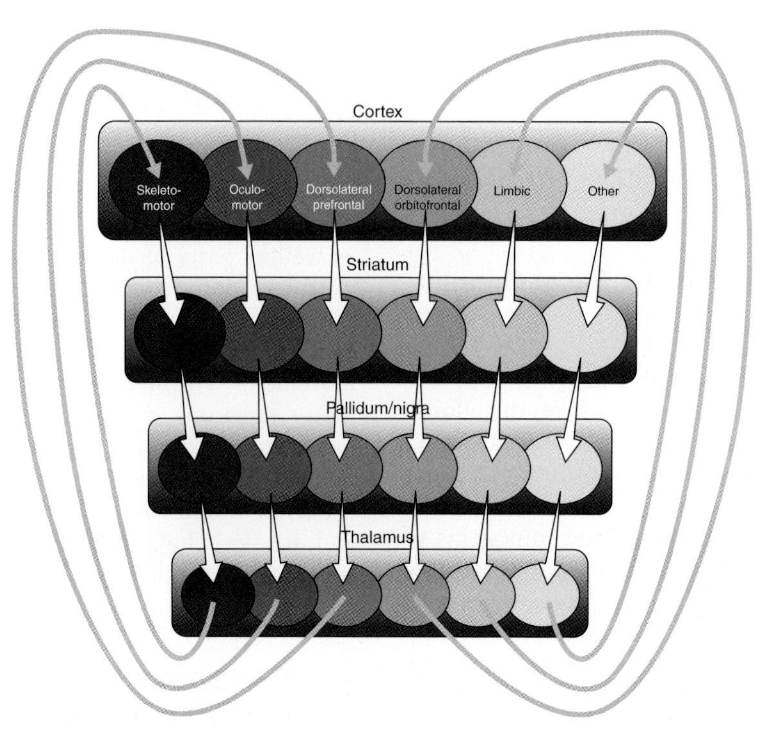


**Figure 1** Schematic view of the large-scale anatomic organization of related areas of cortex, basal ganglia, and thalamus. All three structures form reentrant circuits which occupy anatomically distinct areas.

that give rise to the indirect pathway via activation of dopamine D2 receptors, and to facilitate the activity of direct pathway neurons via activation of D1 receptors.

Activation of the direct pathway by cortical inputs would inhibit the activity of the output nuclei GPi and SNr, and thereby disinhibit thalamocortical projection neurons. In contrast, activation of striatal neurons that give rise to the indirect pathway would have a net excitatory effect on GPi/SNr activity and would thereby act to inhibit thalamocortical neurons (**Figure 2**, left panel). Dopamine may have a substantial effect on the balance of activity between direct and indirect pathways. Through its differential effects, exerted via activation of different dopamine receptor types, a reciprocal relationship may exist between striatal dopamine levels and basal ganglia output. Increased release of dopamine may activate the direct pathway and inactivate the indirect pathway so that basal ganglia output from GPi/SNr to the thalamus is reduced. Conversely, reduced release of dopamine would result in disinhibition of the indirect pathway and reduced facilitation of the direct pathway, leading to increased basal ganglia output. There is also evidence that dopamine may be released at extrastriatal sites, including the pallidum, STN, and SNr. For instance, dopamine released from local dendrites of SNc neurons may activate D1 receptors on terminals of the striatonigral projection. This may then result in increased local GABA release in the SNr and subsequent inhibition of the activity of SNr neurons.

The anatomic model described here can be applied to generate a description of the involvement of the basal ganglia in the control of movements. According to this model, movements are initiated at the cortical level. Cortical inputs to the basal ganglia may then transiently activate both direct and indirect pathways. A phasic increase of inhibition of GPi via the monosynaptic direct pathway may lead to a brief disinhibition of thalamocortical neurons, which may then allow increased activity in thalamic neurons projecting to specific cortical areas, thus facilitating the movement. Activation of the indirect pathway may increase GPi/SNr output and may thereby have the effect of inhibiting or stabilizing the activity in the thalamus and brain stem. This may suppress potential competing movements, or terminate ongoing movements.

While such models are attractive due to their simplicity, several new findings suggest that views based purely on the anatomic connections are not accurate. For instance, while in the older models the overall effect of phasic activation of corticostriatal projection neurons would entirely depend on the interplay between the direct and indirect pathways,

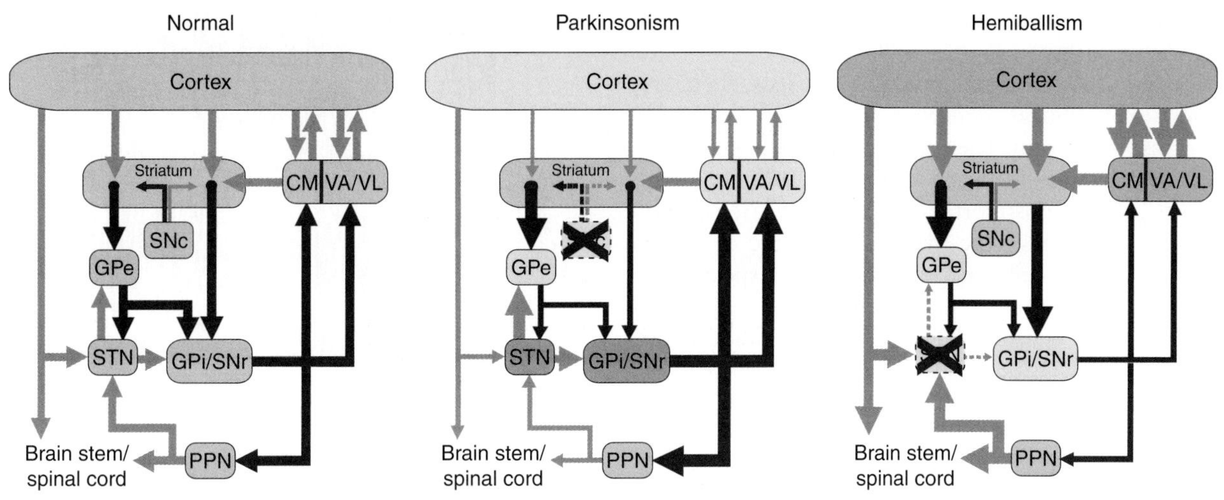


**Figure 2** Activity in the basal ganglia–thalamocortical circuitry (normal) and overall activity changes occurring in parkinsonism and hemiballism. The basal ganglia circuits include striatum, external and internal globus pallidal (GPe, GPi, respectively) segments, substantia nigra pars reticulata and pars compacta (SNr, SNc, respectively), and subthalamic nucleus (STN). Cortical input reaches the basal ganglia via the corticostriatal and the cortico–subthalamic connections. Basal ganglia output is directed toward the centromedian (CM) and ventral anterior/ventral lateral (VA/VL) nucleus of thalamus, as well as the pedunculopontine nucleus (PPN). Excitatory connections are shown in gray, inhibitory pathways in black. In the parkinsonian state, loss of dopamine in the striatum results in activity changes throughout the basal ganglia (represented by the thickness of the connecting arrows), leading to increased basal ganglia output to the thalamus. In hemiballism, damage to the STN results in reduced excitatory input to the basal ganglia output nuclei, leading to a cascade of other changes throughout the entire circuitry.

it now seems that basal ganglia output may also be shaped by nonlinear effects that go beyond straight excitation or inhibition. For example, recent studies of the interaction between GPe and STN have shown that release from GABAergic inhibition does not simply allow STN neurons to return to their baseline activity, but may also result in significant transient rebound excitation of such neurons. Similarly, glutamatergic inputs to basal ganglia neurons may not only lead to activation of recipient neurons, but may also alter firing patterns, such as burst firing or oscillatory discharge. The impact of these transient changes on the overall functioning of the basal ganglia is not clear at this time.

At a yet higher level of complexity, recent work has suggested that some of the basal ganglia functions can only be understood in terms of coordinated interactions of large neuronal ensembles in cortex and basal ganglia. For instance, there is emerging evidence that ensemble activity involving large populations of basal ganglia neurons may play a prominent role in the transmission and processing of cortical oscillatory activity.

## Pathophysiology of Motor Hypo- and Hyperkinetic Movement Disorders

Most movement disorders of basal ganglia origin are believed to arise from dysfunction of the 'motor' circuit, resulting in alterations of GPi and SNr output. Clinically, these disorders are dominated by either poverty of movement (hypokinetic disorders) or an excess of movement (hyperkinetic disorders). Hypokinetic disorders (e.g., Parkinson's disease) feature an impairment of movement initiation (akinesia) and reductions in the amplitude or velocity of voluntary movements (bradykinesia). By contrast, hyperkinetic disorders (e.g., chorea or ballism) are characterized by involuntary movements.

### Hypokinetic Disorders

The most common hypokinetic disorder is Parkinson's disease, which is characterized by akinesia, bradykinesia, a 5 Hz tremor at rest, and muscular rigidity. The most salient pathological sign is the finding of degeneration of the SNc neurons that give rise to the dopaminergic nigrostriatal projection. The earliest and most prominent dopamine loss is found in the motor portion of the striatum, the putamen. Studies of the neuronal activity in the basal ganglia in primates which were rendered parkinsonian by treatment with the dopaminergic neurotoxin 1-methyl-4-phenyl-1,2,3,6-tetrahydropyridine (MPTP) have suggested the pathophysiological changes that are shown in **Figure 2** (middle panel). According to this model, loss of dopamine in the striatum appears to have the opposite effect on the activity in the indirect and direct pathways, eventually leading to an overall increase in activation of GPi and SNr via the indirect pathway, and reduced inhibition of GPi and SNr via the direct pathway.

Perhaps related to striatal dopamine loss, there are also morphological changes in striatal neurons in the Parkinsonian brain. These changes may further influence the balance between direct and indirect pathways. Thus, in postmortem tissue from patients with Parkinson's disease, the overall size of the dendritic tree of striatal cells and the number of specialized dendritic membrane compartments (spines) on striatal projection neurons that receive cortical and other inputs were significantly reduced. Similar changes, particularly affecting the indirect pathway, have also been documented in rodent models of the disorder.

The activity of indirect pathway neurons downstream from the striatum is also altered. For instance, GPe activity is reduced in MPTP-treated animals, whereas discharge rates in STN and GPi/SNr are increased. An important piece of evidence in favor of the hypothesis that STN overactivity directly contributes to the development of parkinsonism is that lesions of this nucleus ameliorate all of the major parkinsonian motor signs.

Imaging studies of cortical function have shown that the activity changes in the basal ganglia in parkinsonism are accompanied by reduced activity in related cortical motor areas. In addition, there is a task-dependent increase of activation in some cortical areas which are normally not activated under these circumstances. The aberrant activation of these areas may be the product of compensatory mechanisms which may add to the inefficiencies of movement generation in parkinsonism.

In addition to the rate changes, a variety of changes in discharge patterns may also play a role, ranging from increased synchronization between neighboring neurons to oscillatory burst patterns (**Figure 3**). Recent reviews of the pathophysiology of parkinsonism have emphasized synchronized oscillations in the basal ganglia, based largely on recordings of local field potentials (LFPs) in patients with Parkinson's disease undergoing functional neurosurgery. These recordings have suggested that abnormal beta-range oscillations develop throughout the entire basal ganglia–thalamocortical circuitry, and these may play a major role in the production of parkinsonian signs. At the same time, 'prokinetic' normal oscillations in the gamma range appear to be suppressed in the parkinsonian state. Several mechanisms have been described that may account for oscillatory and non-oscillatory changes in phasic discharge properties, including pacemaker mechanisms that result from the interaction of different nuclei (for instance, GPe and STN), membrane properties of individual basal ganglia neurons that may predispose these cells to the development of recurring bursts, or altered cortical activity patterns that may directly drive basal ganglia neurons.

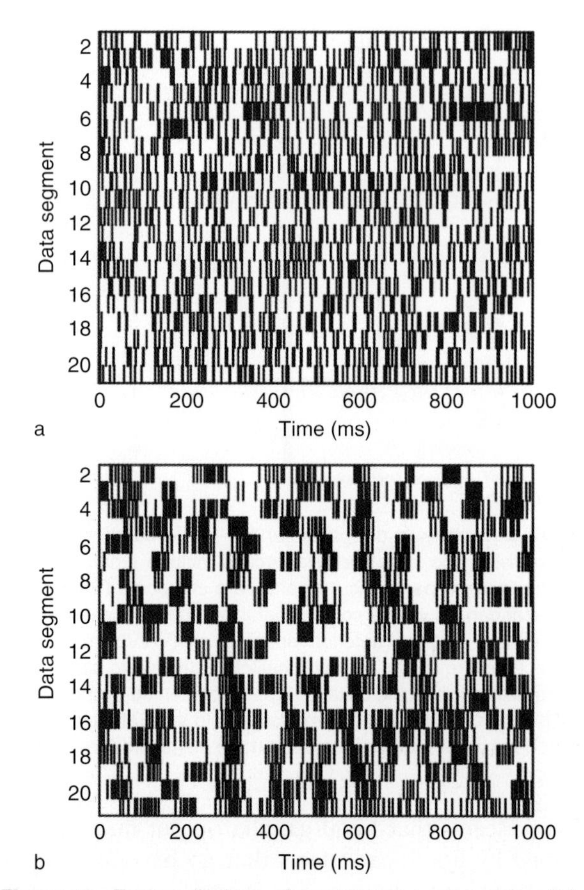


**Figure 3** Raster displays of spontaneous neuronal activity recorded in the internal segment of the globus pallidus in normal (a) and Parkinsonian (b) primates. Shown are 20 consecutive 1000 ms segments of data in each panel. Each of the short vertical lines in the figure represents a single action potential. In addition to an increase in activity, there are also obvious changes in the firing patterns, with a marked prominence of bursts and oscillatory discharge patterns in the parkinsonian state.

The link between abnormal network activity in the basal ganglia–thalamocortical circuitry and parkinsonian motor signs has not been firmly established. It appears that some aspects of akinesia and tremor are due to abnormal oscillatory bursting in the basal ganglia output nuclei. Bradykinesia may result from more specific disturbances of cortical processing by excessive phasic inhibition of thalamocortical circuits. Recent evidence indicates that damage to sites which are anatomically linked to the basal ganglia, but have historically not been considered to be part of these structures (in particular, the pedunculopontine nucleus), may also play a role in the development of akinesia.

Parkinsonism is also associated with nonmotor abnormalities, such as oculomotor disturbances, autonomic dysfunction, depression, anxiety, sleep disturbances, impaired visuospatial orientation, and cognitive abnormalities. It is likely that some of these impairments are due to abnormal discharge in nonmotor circuits of the basal ganglia, which may be affected by dopamine loss in much the same way as is the motor circuit. In addition, degeneration of neurons outside of the basal ganglia has been described to occur in Parkinson's disease. Thus, in addition to nigral damage, postmortem examinations of brains from parkinsonian patients show substantial damage in pigmented brain stem nuclei, the olfactory tubercle, and cortex.

### Hyperkinetic Disorders

In contrast to hypokinetic diseases, hyperkinetic disorders appear to be associated with abnormally low basal ganglia output from GPi and SNr. Reduced basal ganglia output may arise from a shift of the balance of activity between direct and indirect pathway activity toward the direct pathway. Thalamocortical neurons under such conditions may become more responsive to cortical inputs or exhibit an increased tendency to discharge spontaneously, thus leading to involuntary movements. The involuntary movements seen in ballism and Huntington's disease, as well as those induced by dopaminergic drugs, may all arise from this mechanism.

The term 'hemiballism' refers to involuntary proximal limb movements on one side of the body; these are most frequently due to damage restricted to the contralateral STN (see **Figure 2**, right panel). The STN damage, in turn, is thought to lower the STN's excitatory drive on GPi. These changes have been directly demonstrated in primate and rodent experiments, in which STN lesions result in a significant reduction of tonic discharge in GPi and a decrease in the phasic responses of GPi neurons to limb displacement.

In early stages of Huntington's disease, degeneration of striatal neurons that give rise to the indirect pathway is thought to result in excessive inhibition of STN neurons and abnormally low GPi activation. This, in turn, may result in proximal and distal involuntary movements. In later stages of the disease, the degeneration spreads to involve direct pathway neurons, and the associated phenotype changes in many patients are reminiscent of parkinsonism. Perhaps as a consequence of extensive neuronal degeneration of nonmotor circuits, most patients also develop cognitive and emotional disturbances.

Dystonia, a movement disorder which is often associated with basal ganglia lesions, is dominated by abnormal twisting movements and postures. The pathophysiology of dystonia is poorly understood, most likely due to the fact that dystonia can be a symptom of a variety of diseases with very different pathophysiologic backgrounds. Recordings in dystonic patients suggest that discharge rates in GPi are reduced, likely due to overactivity along the direct pathway. Similar to parkinsonism, however, the level of synchrony of spiking activity of neighboring neurons is increased in dystonia, and there is increased bursting throughout the basal ganglia. Recently, oscillatory LFP changes have also been identified in this condition.

### Surgical Treatments

As a result of the better understanding of the pathophysiologic abnormalities that may underlie movement disorders such as Parkinson's disease, neurosurgical interventions in the basal ganglia have been reintroduced as effective treatments for these conditions. Initially, neurosurgeons performed almost exclusively ablative procedures as treatments for movement disorders, including procedures such as pallidotomies or thalamotomies. In the meantime, however, the focus has shifted to deep brain stimulation therapy in which electrical leads are implanted into the basal ganglia or thalamus and are used for chronic high-frequency stimulation. Conceivably, both types of procedures work through elimination of disruptive basal ganglia outputs. While therapeutic lesion effects in Parkinson's disease are perhaps understandable in terms of a reversal of the described rate changes in the basal ganglia, deep brain stimulation may target predominately firing pattern abnormalities of the stimulated circuitry.

## Nonmotor Symptoms Related to Basal Ganglia Dysfunction

Pathology in the nonmotor circuits of the basal ganglia may be important not only for the development of some of the nonmotor signs accompanying Parkinson's disease and other movement disorders, but also for several unrelated behavioral and psychiatric disorders. Because anatomical and functional details regarding these circuits and the disease associated with their dysfunction are not completely worked out, they are only briefly mentioned here.

The dorsolateral prefrontal circuit originates in Brodmann areas 9 and 10 of the cortex, and its subcortical route involves the head of the caudate nucleus, the dorsomedial portion of GPi, and the rostral SNr, as well as portions of the VA and MD nuclei of the thalamus. This circuit has been implicated broadly in 'executive functions' – for instance, the organization of behavioral responses to complex problems or the use of verbal skills in problem solving. Damage to portions of this circuit is associated with a variety of behavioral abnormalities related to these cognitive functions.

The dorsolateral orbitofrontal circuit involves the lateral prefrontal cortex, the ventromedial caudate nucleus, and associated areas of GPi/SNr and thalamus. This circuit appears to play a role in the mediation of empathetic and socially appropriate behaviors. Consequently, damage to this area is associated with irritability, emotional lability, failure to respond to social cues, and lack of empathy.

The limbic circuit arises in the anterior cingulate gyrus of cortex and involves the ventral striatum (which also receives 'limbic' input from hippocampus, amygdala, and entorhinal cortices). The ventral striatum then projects to the ventral and rostromedial GPi and rostrodorsal SNr. From there the pathway continues to neurons in the paramedian portion of the MD nucleus of the thalamus, and back to the anterior cingulate cortex. This circuit may be important in motivated behavior and may convey reinforcing stimuli to diffuse areas of the basal ganglia and cortex.

In a general sense it appears that the psychiatric disorders resulting from dysfunction of the nonmotor circuits parallel those seen in the motor circuit. One of the best studied psychiatric disorders in this regard is obsessive–compulsive disorder. Functional imaging studies have demonstrated abnormalities in the activity of limbic basal ganglia–thalamocortical projection systems in patients with this disorder. Reminiscent of the situation in Parkinson's disease, patients with obsessive–compulsive disorder appear to access 'unusual' brain regions (such as portions of the temporal cortices) when they are confronted with stimuli that call for recruitment of corticostriatal systems.

Tourette syndrome, a disorder in which obsessive–compulsive symptoms occur in combination with vocal or motor tics, is likely associated with abnormalities in the limbic and motor basal ganglia–thalamocortical circuitry. Not unexpectedly, the motor symptoms in this disease are reflected in additional changes in brain activation, particularly in the sensorimotor cortices and supplementary motor area.

Finally, the basal ganglia may also be involved in the development of depression, even in the absence of motor dysfunction. Thus, positron emission tomography studies in patients with major depression have shown reduced metabolic activity or blood flow in cortical portions of the limbic system, especially the orbitofrontal cortex and prefrontal cortices, in conjunction with similarly reduced activity in related areas of the basal ganglia, particularly the ventral striatum and caudate nucleus. In addition, stimulation of the limbic portion of the STN, or the SNr in parkinsonian patients who are treated with deep brain stimulation, can induce dramatic mood changes, ranging from depression to mania.

## Conclusion

Our understanding of the role of the basal ganglia in normal movement remains very limited. Much more is known about abnormalities associated with basal ganglia disorders. By contributing to the reemergence of neurosurgical and other interventions as treatments for movement disorders, models of basal ganglia dysfunction have already had a major impact in the treatment of diseases such as Parkinson's disease or dystonia.

The models described herein continue to evolve. Gradually, the view of the basal ganglia as a set of nuclei by which cortical activity is simply relayed back to cortex is replaced by more general organizational schemes in which single-cell spiking and large-scale oscillatory and nonoscillatory activities of local or spatially distributed ensembles of neurons play a role.

*See also:* Animal Models of Parkinson's Disease; Cell Replacement Therapy: Parkinson's Disease; Deep Brain Stimulation and Movement Disorder Treatment; Parkinsonian Syndromes.

## Further Reading

DeLong MR (2000) The basal ganglia. In: Kandel ER, Schwartz JH, and Jessel TM (eds.) *Principles of Neural Science*, 4th edn., pp. 853–867. New York: McGraw-Hill.

Graybiel AM (2005) The basal ganglia: Learning new tricks and loving it. *Current Opinion in Neurobiology* 15: 638–644.

Hutchison WD, Dostrovsky JO, Walters JR, et al. (2004) Neuronal oscillations in the basal ganglia and movement disorders: Evidence from whole animal and human recordings. *Journal of Neuroscience* 24: 9240–9243.

Middleton FA and Strick PL (2000) Basal ganglia and cerebellar loops: Motor and cognitive circuits. *Brain Research Reviews* 31: 236–250.

Nutt JG and Wooten GF (2005) Clinical practice. Diagnosis and initial management of Parkinson's disease. *New England Journal of Medicine* 353: 1021–1027.

Smith Y, Raju DV, Pare JF, et al. (2004) The thalamostriatal system: A highly specific network of the basal ganglia circuitry. *Trends in Neuroscience* 27: 520–527.

# Aging and Memory in Animals

**P R Rapp**, Mount Sinai School of Medicine, New York, NY, USA



## Introduction

From a life-span perspective, aging comprises the late component of a genetically determined program of development, maturation, and senescence, interacting with a complex array of environmental factors. Commonly viewed as a process of deterioration, growing old is associated with sharply increased risk for many diseases and disabilities that compromise independent living, placing a heavy burden on families, caregivers, and society. Nonetheless, a majority of people successfully accommodate certain physical signs of aging, and in fortunate cases, old age can represent a rewarding period of new intellectual engagement, novel pursuits, and achievement. Such positive outcomes become increasingly unlikely in the face of failing cognitive function, and partly for this reason, disorders that lead to diminished mental capacity are among the most feared consequences of aging. The leading cause of dementia, Alzheimer's disease, ultimately results in a dense amnesia, gradually robbing patients of the lifetime of memories that define their personal history and identity. Even in the absence of disease, many people experience memory impairment that, although relatively mild, can cause considerable anxiety and compromise the quality of life. As the populations of industrialized countries rapidly age (**Figure 1**), we face a growing challenge of identifying ways to promote healthy cognitive aging and to maximize optimal functioning.

Research has illuminated many of the key features of age-related cognitive decline in humans; enabled by advances in *in vivo* brain imaging, it has begun to reveal how the neural systems organization of memory is altered. Studies of cognitive aging in humans are complicated by a variety of methodological factors, however, and they are limited by the range of applicable experimental approaches. A vexing issue is that individuals in preclinical stages of Alzheimer's disease and other disorders affecting cognition are difficult to identify with confidence. As a consequence, it is often unclear to what degree observed impairment is attributable to disease rather than normal nonpathological aging. Relating these deficits to underlying biological causes is also problematic, and although noninvasive imaging techniques continue to yield remarkable discoveries, defining the neurobiological mechanisms of cognitive aging requires experimental approaches not suitable for investigation in humans. Research in animal models has played a critical role in efforts to address these issues.

## Cognitive Aging – A Neuropsychological Framework

There is a broad consensus on a number of central concepts concerning the neuropsychology of normal memory in young adults. Rather than comprising a unitary, monolithic capacity, multiple forms of memory can be distinguished according to the qualities of the information remembered, endurance, flexibility, and a variety of other characteristics. A major advance in this field is the recognition that the existence of multiple forms of memory follows from the organization of brain systems that mediate these capacities. Converging evidence from rats, monkeys, and amnesic humans, for example, indicates that normal memory for the facts and events of daily life critically requires a collection of anatomically related structures in the medial temporal lobe, including the hippocampus and adjacent parahippocampal cortical areas. Damage to the prefrontal cortex, by comparison, largely spares this declarative or episodic type of memory and, instead, leads to deficits in a variety of memory-related processes, including the strategic use of remembered information when confronted with novel circumstances and the ability to recall the source from which information was acquired.

Informed by this background, research detailing the specific profile of behavioral deficits that emerge during aging can provide a window on the functional status of the neural systems that mediate memory and other cognitive processes. The Morris water maze has been used widely in studies of this sort in rats, partly because memory assessed by this procedure critically requires the functional integrity of the hippocampus. Across trials in this task, animals learn the spatial location of an escape platform hidden in a pool of clouded water, and navigation is guided by the memory of the platform position in relation to cues surrounding the maze. Young adult rats acquire this task rapidly, learning to swim directly to the escape platform from any point around the perimeter of the apparatus. By roughly 24 months of age, acquisition rates are impaired in a substantial proportion of rats, and by comparison with young adults, spatial searching is less tightly focused on the escape location during probe testing, when the platform is removed from the apparatus. There are a variety of reasons to presume that this pattern reflects a genuine impairment in learning and memory rather than the effects of

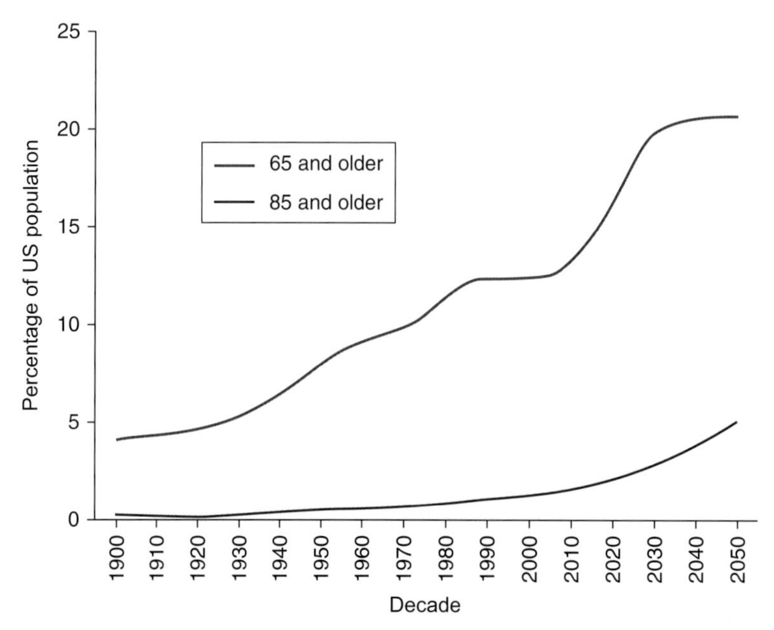


**Figure 1** Society is aging rapidly; the historical and projected percentages of the US population that is over 65 and 85 years. Results compiled from US Census Bureau figures.

aging on sensorimotor function, motivation, and other performance factors. In a variant of testing that shares many of the same performance demands as the hidden goal task, rats swim to a visible escape platform that either protrudes slightly above the water's surface or is marked with a prominent visual cue. Performance in this cue-approach version of the water maze is relatively insensitive to aging, suggesting that, in the absence of a requirement to learn and remember information about the spatial location of the escape platform, age-related deficits in swimming ability, the motivation to escape, and visual acuity are not sufficient to prevent performance comparable to that of young adults. Notably, a qualitatively similar pattern of impaired spatial learning, together with preserved cue-approach performance, is seen in rats following the disruption of hippocampal function.

Age-related impairment in learning and memory for spatial information is not peculiar to rats, and parallel deficits have been documented in both humans and monkeys. The hippocampus is not restricted to processing spatial information, however, and in efforts to define how aging influences medial temporal-lobe memory, research in monkeys has relied predominantly on other sorts of assessments. In a particularly widely used procedure, that is delayed nonmatching-to-sample (DNMS), trials consist of two phases; the presentation of a sample item followed by a recognition test. A retention interval is imposed between the two, and monkeys demonstrate recognition by choosing the novel object (or nonmatch) when it is presented together with the previously viewed sample. Training typically proceeds by testing subjects with a short retention interval of 8 or 10 s until they master the nonmatching rule and subsequently introducing more challenging memory delays, up to 10 min or longer.

Visual object recognition in young adult macaque monkeys declines gracefully over increased retention intervals, confirming that the task effectively taxes memory. The average life span in rhesus macaques is less than 25 years, and monkeys at this age and older display modest but reliable deficits in visual recognition. Specifically, whereas older subjects score as well as controls at relatively short delays, recognition accuracy falters at longer intervals. This profile is significant because it demonstrates that sensorimotor function, motivation, and attention in the aged monkey are sufficiently intact to support normal performance and that age-related DNMS deficits emerge specifically in relation to increased demands on memory. In addition, this delay-dependent pattern of impairment is qualitatively similar to the deficits observed in young monkeys with experimental damage involving the hippocampal region. Thus, across both rats and monkeys, behavioral studies converge, concluding that memory mediated by the hippocampal system is vulnerable to normal aging.

## Individual Variability in Cognitive Aging

An important insight from recent research is that the cognitive effects of aging vary considerably across individuals such that many experience little or no decline

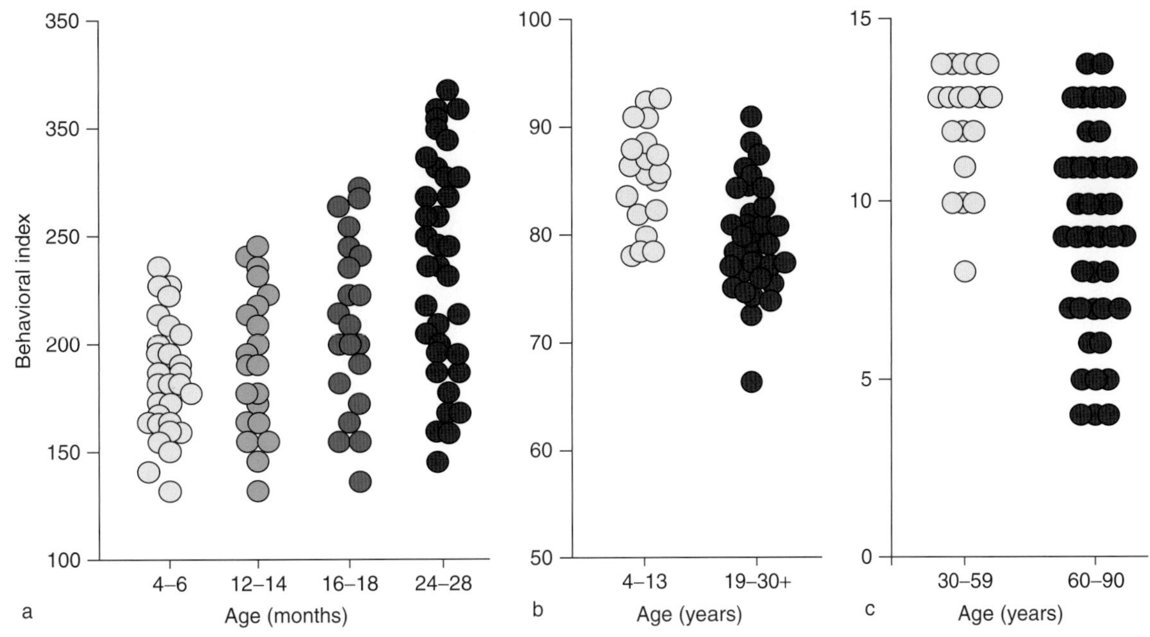


**Figure 2** The status of memory varies considerably across aged individuals: (a) spatial memory in rats; (b) object recognition memory in monkeys; (c) delayed recall in humans. For rats (a), low scores represent better memory; in monkeys (b) and humans (c), high values reflect better performance. Symbols signify scores for individual subjects. Note that in all species the status of memory among the aged is distributed across a broad range, from individuals that score on a par with the best young adults to other individuals that exhibit substantial impairment. (a) Data courtesy of M Gallagher, Johns Hopkins University; (b) data courtesy of P Rapp, Mount Sinai School of Medicine; (c) data courtesy of M Albert, Johns Hopkins School of Medicine.

but others develop substantial impairment (**Figure 2**). Individual variability is widely recognized as a prominent feature of human aging, but it has often been considered a nuisance factor, making it difficult to draw firm conclusions about the likely course of decline. In addition, impaired function in some aged people is likely a prodromal harbinger of disease, obscuring individual variability associated with the process of normal aging itself. Here again, studies in animal models that are spared the spontaneous, dementia diseases of human aging have proved key.

The status of spatial learning and memory varies substantially among aged rats and is continuously distributed from subjects that perform on par with the best young adults to other aged individuals that display marked impairment, outside the range of adult control values (**Figure 2(a)**). Findings from monkeys are similar, indicating that roughly 40% of aged subjects exhibit marked recognition memory deficits, relative to young animals, as assessed by DNMS (**Figure 2(b)**). These findings mirror observations in humans, establishing the important concept that, in the absence of frank disease, marked cognitive decline is not an inevitable consequence of growing older (**Figure 2(c)**). In addition, the recognition of individual differences in the cognitive outcome of aging has enabled a powerful approach for exploring the neurobiological basis of dysfunction. Whereas aging research traditionally has comprised comparisons between groups distinguished on the basis of chronological age, an alternative approach involves the analysis of age-matched groups that differ according to neuropsychological indices of aging. As described in subsequent sections, studies adopting this approach have substantially advanced progress toward defining the neural alterations that specifically distinguish aged individuals with cognitive deficits from others of the same age with preserved function.

## Neurophysiology of Cognitive Aging

Neuropsychological studies are an important source of evidence suggesting that memory mediated by the hippocampus is vulnerable to aging. Cognitive data alone, however, are a secondary proxy, and other sorts of approaches are needed to directly document the status of information processing in the aged hippocampus. An illuminating strategy used by a number of investigators involves recording the firing patterns of hippocampal neurons as freely moving rats navigate a maze apparatus. Under these conditions, neuronal activity in a large proportion of hippocampal pyramidal cells reliably and selectively increases when subjects move through a specific spatial location in the testing environment – a phenomenon referred to as place-field firing. Many features of place-field activity, including the incidence, selectivity, and specificity of location-related firing, appear normal in the aged hippocampus, even among rats with pronounced spatial learning and memory impairment. This sparing counts against the idea that deficits in memory result from generalized, nonspecific deterioration in the functional integrity of the hippocampus. In fact, recent evidence indicates that the influence of aging on hippocampal firing patterns is highly specific and coupled with the status of spatial memory measured behaviorally.

One means of defining the particular constellation of stimuli that controls place cell activity involves the use of probe conditions in which subsets of cues in the testing environment are manipulated independently. After mapping the place fields of neurons in a maze setting that is held constant, for example, the constellation of experimenter-defined tactile and odor cues on the surface of the apparatus can be rotated in the opposite direction of the extra-maze visual and auditory stimuli. Recorded under these conditions, a large percentage of neurons in the young adult hippocampus either stop firing or develop entirely new place fields, unrelated to the rotated configuration of either the intra- or extra-maze cues. Such remapping is thought to enable the flexibility of hippocampal information processing, providing the basis for encoding the relevant predictive relationships between familiar stimuli when they appear in novel circumstances. A similar incidence of flexible place-field remapping is observed in aged rats that perform normally on the hidden platform version of the Morris water maze. In contrast, aged rats with documented deficits in spatial learning and memory exhibit far less flexibility in place-field activity. Neuronal firing in this subpopulation appears abnormally rigid and disproportionately driven by the configuration of extra-maze cues in the testing environment relative to other available stimuli.

As a consequence of encoding a reduced scope of available information, it might be predicted that the aged hippocampus is prone to interference between environments that share a subset of cues. Consistent with this prediction, when aged rats repeatedly explore an environment containing cues that were also present in another familiar setting, place-field remapping occurs substantially more frequently than in young subjects (**Figure 3**). An orthogonal pattern of results is observed when unit activity is recorded in a highly familiar environment, followed by exploration of a novel setting, and subsequent return to the original environment. Place-field firing in memory-impaired aged rats initially persists across the familiar and novel conditions, and repeated exposure is needed to induce the remapping observed in young rats from the outset. Related findings also indicate that at least some features of age-related alteration in place-field firing are regionally selective across the principal cell layers of the hippocampus, appearing to be particularly pronounced in the CA3 field. Considered together, the picture emerges that prominent features of normal cognitive aging are coupled with reliable changes in key computational capacities of the hippocampus and that the specific nature of these alterations contributes to the spatial learning and memory impairment seen in a substantial proportion of aged individuals.

Although the precise cell biological basis of age-related change in hippocampal neurophysiology remains to be defined, recent research has considerably narrowed the range of plausible accounts. A particularly long-standing idea is that the complex constellation of neurobiological alterations that occurs during aging ultimately results in neuron death in vulnerable brain areas such as the hippocampus and that this loss of structural integrity is the proximal cause of cognitive aging. Enabled by the development and validation of efficient methods of morphometric quantification, however, studies taking advantage of these stereological tools have prompted a new consensus. Across all the animal models examined, including rats and monkeys, the findings consistently demonstrate that marked neuron loss in the hippocampus and associated components of the medial temporal-lobe memory system is not a necessary consequence of normal aging. Moreover, among investigations incorporating behavioral assessment prior to sacrifice and stereological analysis, age-related deficits in learning and memory frequently occur in the absence of detectable neuron loss in these structures (**Figure 4**). Alongside this preservation, synaptic connectivity that critically participates in hippocampal functional plasticity is compromised, and in some cases, these changes are correlated with individual variability in the cognitive outcome of aging.

Enduring changes in synaptic strength, mediated by long-term potentiation (LTP)-like mechanisms, are widely presumed to underlie the establishment of place-field firing and other electrophysiological correlates of behavior that are vulnerable to aging. Considerable attention has therefore focused on the possibility that deficits in LTP and other plasticity mechanisms might contribute to the cognitive outcome of aging. Many fundamental electrophysiological properties of neurons throughout the hippocampus are preserved during aging, including the resting membrane potential, the depolarization threshold for action potential firing, and, under maximal nonphysiological stimulation conditions, the induction of LTP. Studies using weaker stimulation, however, reveal that the threshold for LTP induction is increased in both the dentate gyrus

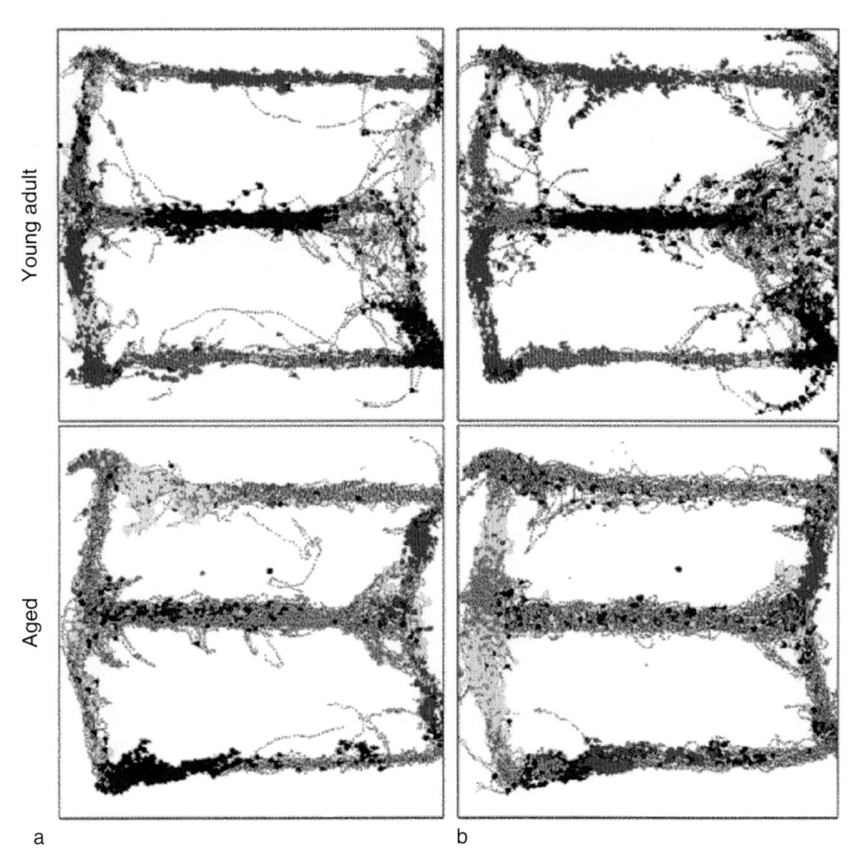


**Figure 3** Neuronal firing of multiple individual hippocampal pyramidal neurons as young and older rats explored a figure-8 maze on two occasions: (a) maze exploration A; (b) maze exploration B. Light gray lines illustrate the path subjects took as they navigated the apparatus; the locations at which neurons fired above background levels are represented as color-coded dots. As shown, the coding of spatial location is prone to disruption in the hippocampus of aged rats with memory deficits. Note that in the young rat, the distribution of place-field firing is highly consistent across the two bouts of exploration. In contrast, on roughly 30% of such occasions, the aged hippocampus exhibited completely different place-field mappings, as though the familiar test setting was represented as multiple environments. Reprinted by permission from Macmillan publishers Ltd: (*Nature*) (Barnes CA, Suster MS, Shen J, and McNaughton BL (1997) Multistability of cognitive maps in the hippocampus of old rats. *Nature* 388: 272–275), copyright (1997).

and CA1 fields of the aged hippocampus. The long-term maintenance of synaptic enhancement, which requires gene expression, protein synthesis, and synaptic remodeling, is also compromised and declines toward the baseline more rapidly in aged subjects than young adults. A significant challenge for future research is to identify the relevant linkages between the effects of aging documented at different levels of analysis (determining whether synaptic alterations in critical circuitry account for changes in the computational capacities of hippocampal neurons) and, ultimately, age-related deficits in learning and memory.

Although it seems reasonable to speculate that age-related changes in synaptic plasticity contribute to observed alterations in hippocampal place-field firing and spatial learning and memory impairment, emerging evidence reveals considerable complexity in the underlying mechanism of these effects. LTP in the CA1 field of the hippocampus reflects the operation of two components: the *N*-methyl-D-aspartate receptor (NMDAR)-dependent pathway and a second, less well-studied form mediated by voltage-dependent $Ca^{2+}$ channel (VDCC) activity. Using pharmacological manipulations to dissect the contribution of these components, current evidence indicates that, even under conditions in which the magnitude and endurance of compound LTP is similar, there is a shift toward a greater role of VDCC-mediated enhancement in the aged hippocampus compared to the young hippocampus. Related findings suggest that aging is also accompanied by a change in the bidirectional balance between synaptic enhancement and long-term depression (LTD). Prolonged low-frequency stimulation that is without effect in the young hippocampus produces an NMDAR-dependent LTD at the aged CA3 to CA1 synapses, and this vulnerability appears at least partly attributable to altered $Ca^{2+}$ signaling. Studies in behaviorally characterized rats extend these observations, suggesting that, whereas age-related change in NMDAR-mediated LTD is unrelated to individual

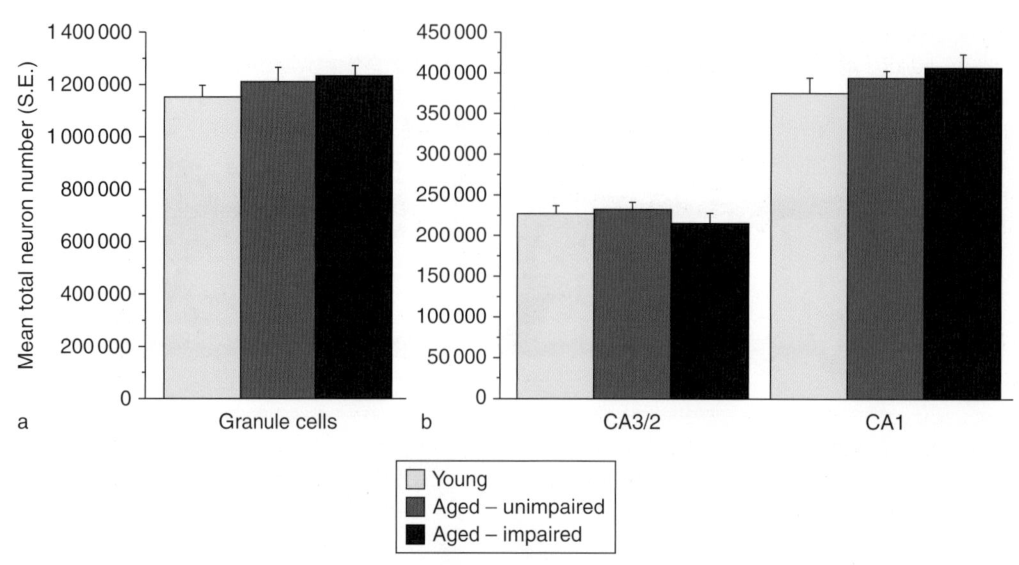


**Figure 4** Estimated total number (unilateral) of neurons in the principal cell fields of the hippocampus for young rats and aged (unimpaired and impaired) rats based on their performance in a water maze test of spatial learning and memory: (a) granule cells; (b) CA3/2 and CA1. The results illustrate that neuron number is stable as a function of chronological age and cognitive status. Memory impairment during normal aging does not require neuron death in the hippocampus. Reproduced from Rapp PR and Gallagher M (1996) Preserved neuron number in the hippocampus of aged rats with spatial learning deficits. *Proceedings of the National Academy of Sciences of the United States of America* 93: 9926–9930. Copyright (1996) National Academy of Sciences, US.

differences in the cognitive outcome of aging, NMDAR-independent LTD is increased in magnitude and correlated with preserved spatial memory in aged rats. Together, these findings indicate that the basis of age-related decline in hippocampal memory is likely to involve dynamic interactions between multiple plasticity pathways and that active compensatory mechanisms are available in support of optimal hippocampal function during aging. Defining and harnessing these mechanisms through pharmacological or other means are important goals of current research efforts.

## Horizons in Research on Cognitive Aging

### Aging beyond the Hippocampus

This article has centered on medial temporal-lobe memory as a particularly well-studied system in animal models of age-related cognitive decline. This focus, however, should not be taken to suggest that this form of memory is affected in isolation. Indeed it is entirely clear that other systems are also vulnerable, prominently including cognitive capacities supported by the prefrontal cortex. Relevant research in rats has begun to emerge only recently, in part because the normative functional organization of the prefrontal cortex has proved challenging to characterize. Nonetheless, comparable to findings in humans, aged rats exhibit deficits in attentional set shifting, or the ability to update and modify behavior in response to altered task contingencies. This capacity is a member of a broad domain, termed executive function, that mediates the strategic use and manipulation of memory in support of adaptive behavior. Parallel findings in monkeys demonstrate age-related impairment on a nonhuman primate adaptation of the Wisconsin Card Sorting Test, a widely used neuropsychological assessment of executive function in humans. Another, particularly well-documented signature of cognitive aging is that older monkeys display robust deficits relative to controls on delayed-response tests that measure the spatiotemporal organization of memory. This procedure challenges animals to remember and select the location of a reward presented prior to a delay interval of several seconds or more. Because each location is rewarded frequently over the course of the many trials that make up daily test sessions, the opportunity for interference is substantial, placing considerable demands on the temporal organization of memory to distinguish information relevant to the current trial from target locations used earlier. Delayed-response performance, like other measures of executive function, critically requires the functional integrity of the prefrontal cortex. Thus, taken together, the available neuropsychological evidence prompts the conclusion that cognitive capacities supported by the prefrontal cortex are vulnerable to normal aging. In addition, because these signatures of aging are largely unrelated to the status of medial temporal memory, the emerging consensus is that cognitive decline reflects the independent vulnerabilities of multiple memory-related brain structures rather than a process of widespread generalized deterioration.

### Intervention

Considerable attention in research on the biology of aging has been directed at efforts to extend the life span. Improving the quality of life is at least as important a goal, however, and in this regard, animal models can be valuable in informing research efforts to promote healthy cognitive aging. This perspective, emphasizing the importance of bringing into convergence evidence from multiple levels of analysis, from basic research to human clinical trials, is illustrated by a consideration of the results from the Women's Health Initiative (WHI). The WHI is among the largest double-blind, placebo-controlled clinical trials ever mounted in the United States and is aimed at defining the health and cognitive effects of hormone replacement therapy in postmenopausal women. A large body of basic research had provided ample reason to suspect that ovarian hormone replacement following menopause might offer significant benefit in terms of neuroprotection and cognitive function, and indeed, results from a number of relatively small observational studies in women supported this prediction. The surprising result of the WHI, in contrast, is that, in addition to significantly increasing the risk for a number negative health outcomes, including cancer and stroke, hormone replacement offered no benefit in terms of cognitive function. In fact, among the few who developed Alzheimer's disease, the incidence was nearly double in treated women compared with placebo controls.

Evidence from animal research suggests a number of factors that might contribute to these unexpected results. In rats, for example, the cognitive benefit of exogenous hormone replacement depends on initiating treatment soon after the onset of estrogen deficiency, and efficacy is compromised with delayed intervention. That there is a temporal window of opportunity for influencing cognitive function with ovarian hormone treatment is relevant to the WHI because, in that study, women averaged 65 years of age at the beginning of the trial, nearly 15 years older than the typical onset of menopause. The specific schedule of dosing may also be important, and adopting the standard clinical practice at the time, the regimen used in the WHI consisted of the chronic, daily administration of conjugated equine estrogen (CEE), with or without a synthetic progestin (medroxyprogesterone acetate (MPA)). Chronic, continuous treatment, however, may lead to the habituation or desensitization of the neurobiological responses that mediate the cognitive influence of estrogen. Supporting this speculation, acute dosing in rats is reportedly at least as, if not more, effective than continuous replacement. Findings in nonhuman primates are similar, demonstrating that a single injection of estradiol every 21 days in aged ovariectomized monkeys substantially reverses a key signature of cognitive aging relative to age-matched vehicle control values. The specific formulations used for replacement may also be critical, based on *in vitro* evidence that estradiol and progesterone produced by the ovaries can have very different, and sometimes opposite, cell biological effects compared to the CEE and MPA compounds typically prescribed in clinical practice. Although the precise influence of these and other variables remains to be defined, current research establishes the value of investigations in animal models for addressing these issues and of broader efforts aimed at the rational development of strategies to promote healthy cognitive aging.

*See also:* Aging of the Brain; Aging of the Brain and Alzheimer's Disease; Cognition in Aging and Age-Related Disease; Functional Neuroimaging Studies of Aging; Gene Expression in Normal Aging Brain; Lipids and Membranes in Brain Aging; Neuroendocrine Aging: Pituitary-Gonadal Axis in Males; Neuroendocrine Aging: Hypothalamic-Pituitary-Gonadal Axis in Women; Rodent Aging; Synaptic Plasticity: Learning and Memory in Normal Aging.

## Further Reading

Bachevalier J, Landis LS, Walker LC, et al. (1991) Aged monkeys exhibit behavioral deficits indicative of widespread cerebral dysfunction. *Neurobiology of Aging* 12: 99–111.

Barnes CA, Suster MS, Shen J, and McNaughton BL (1997) Multistability of cognitive maps in the hippocampus of old rats. *Nature* 388: 272–275.

Burke SN and Barnes CA (2006) Neural plasticity in the ageing brain. *Nature Reviews Neuroscience* 7: 30–40.

Gallagher M and Rapp PR (1997) The use of animal models to study the effects of aging on cognition. *Annual Review of Psychology* 48: 339–370.

Hedden T and Gabrieli JD (2004) Insights into the ageing mind: A view from cognitive neuroscience. *Nature Reviews Neuroscience* 5: 87–96.

Morrison JH and Hof PR (1997) Life and death of neurons in the aging brain. *Science* 278: 412–419.

Rapp PR and Gallagher M (1996) Preserved neuron number in the hippocampus of aged rats with spatial learning deficits.

*Proceedings of the National Academy of Sciences of the United States of America* 93: 9926–9930.
Sherwin BB (2006) Estrogen and cognitive aging in women. *Neuroscience* 138: 1021–1026.
Tanila H, Shapiro M, Gallagher M, and Eichenbaum H (1997) Brain aging: Changes in the nature of information coding by the hippocampus. *Journal of Neuroscience* 17: 5155–5166.

# Aging and Memory in Humans

**A M Brickman and Y Stern**, Columbia University, New York, NY, USA



## Memory Systems

Memory is the explicit or implicit recall of information encoded in the recent or distant past. Current conceptualizations of memory, however, do not view the construct as a unitary system but rather divide it into hierarchical taxonomic modules based on duration of retention and the type of information that is being retrieved. Among the more fully elucidated conceptualizations of memory systems is one characterized by Larry Squire and colleagues, in which long-term memory is divided into declarative and nondeclarative subcomponents. Declarative, or explicit, memory refers to the ability to consciously recall facts (semantic memory), events (episodic memory), or perceptual information (perceptual memory). Nondeclarative memory requires the implicit recall of information and is usually divided into procedural, priming, or simple conditioning paradigms. Information that is retained on the order of seconds or minutes is usually referred to as short-term memory and is thought to represent a memory system distinct from long-term memory. Working memory, which comprises short-term memory, refers to the short-term store required to perform certain mental operations during retention. The following sections examine the impact of normal aging on different types of memory, as well as some of the potential moderators and mediators of cognitive aging. The information presented is organized hierarchically, following the memory systems just outlined.

Increased age puts an individual at risk for the development of neurodegenerative disorders, such as Alzheimer's disease (AD). Central to the dementia syndrome that characterizes AD is the gradual and progressive loss of long-term memory functions. Although the vast majority of older adults do not develop dementia, most experience some degree of cognitive change. Following the elucidation of memory systems and their component parts in the cognitive and cognitive neuroscience literature, there has been a recent interest in the impact of age on the different memory systems, independent of the devastating effects of dementia. Among well-screened individuals who do not meet diagnostic criteria for dementia, both cross-sectional and longitudinal studies demonstrate that the different memory component systems do not uniformly age; rather they show differential vulnerability to aging effects.

## Long-Term Memory

### Declarative Memory

**Semantic memory** As noted, semantic memory refers the recall of general or factual knowledge. Older adults commonly complain of subjective semantic memory problems when, for example, they report difficulty recalling the names of common objects or other well-learned information. Yet, despite these subjective complaints, semantic memory is among the more stable memory systems across the adult life span. The construct can be operationally defined by requiring subjects to define words or provide the answers to factual questions (e.g., the capital of a certain country), such as on the Vocabulary and Information subtests of the Wechsler Adult Intelligence Scales. Semantic memory is often included as part of the definition of 'crystallized intelligence,' which reflects an accumulation of information acquired over time and that is relatively impermeable to the effects of normal aging or mild brain disease.

It is well established that semantic memory shows very little decline in normal aging. In fact, semantic knowledge accumulation and memory increase into the sixth and seventh decades of life and may show only a gradual decline afterward. Much of our understanding of the impact of age on semantic memory has come from large-scale longitudinal and cross-sectional studies of normal aging. For example, longitudinal data from the Canberra Study , which followed a random sample of adults over the age of 70 years, demonstrated that crystallized abilities remained stable over approximately 8 years. This pattern of stability was apparent when age-associated differences (i.e., cross-sectional analysis) were considered as well. Similarly, Denise Park and her colleagues measured knowledge-based verbal ability, including three semantic memory tasks tapping word knowledge, in a large sample of healthy adults ranging in age from 20 to 92 years, and found a gradual increase in performance across the age groups. This finding is again consistent with the idea that semantic knowledge accumulates across the life span with little or no deleterious effects of normal aging.

Although there is little cross-sectional or longitudinal evidence to suggest that semantic memory changes significantly with normal aging, why are subjective complaints of semantic recall so common among older adults? One phenomenon, termed 'tip-of-the-tongue' (TOT), may explain this occurrence. TOT refers to the common experience in which individuals have the feeling that they know the correct information (e.g., a person's name, a relatively low-frequency word), yet they are unable to recall it explicitly. The frequency of the TOT phenomenon increases with advancing age and may underlie the perceived difficulty with semantic recall.

**Episodic memory** The contrast between episodic memory and semantic memory was introduced by Endel Tulving in the early 1970s. Episodic memory refers to the explicit recollection of events, the 'what,' 'where,' and 'when' of information storage, and though it is conceptually distinct from semantic memory, the two memory systems interact. Episodic memory binds together items in semantic memory to form conceptually related time-based events. For example, the explicit recall of a learned story about a cowboy requires episodic memory for the story and semantic memory, or prior knowledge, of the items contained within the story.

Unlike semantic memory, episodic memory declines considerably with age. Older adults, for example, when prompted, have more difficulty recalling what they had for breakfast than do younger adults. The formal observation that episodic memory is affected by aging has existed for decades and has been documented numerous times.

Episodic memory is typically tested by requiring persons to learn information explicitly (e.g., a list or story) and recall it after a delay period. The three aspects of episodic memory include the encoding phase, the storage phase, and the retrieval of the encoded and stored information. These three phases show differential aging effects. Older adults' overall difficulty on tasks of episodic memory may be partially accounted for by a more shallow depth of encoding, as compared to younger adults – that is, older adults recall less information because of more limited processing of the initial study stimuli. This idea is supported by findings of greater age-related decline in acquisition or early retrieval of new information than in the degree of forgetting (i.e., the amount of information lost relative to the amount of information encoded). However, on free-recall tasks of list-learning paradigms, older adults recall fewer absolute words than do their younger adult counterparts; when given the correct stimulus in a recognition paradigm, older adults tend to incorrectly endorse more distractor stimuli, or foils.

Several observations about episodic memory and aging have emerged from the recent literature. First, the pattern of age-related differences in episodic memory appears to be similar across several modalities and domains, such as story recall, paired associate learning, face and word recall, and recognition paradigms of verbal and nonverbal information. Second, cross-sectional data suggest that age-associated episodic memory decline begins as early as age 20 years and decline linearly until about age 60, at which time there is a more precipitous decline. Third, whether the amount of interindividual variability increases systematically as a function of age is still somewhat unclear. Greater variability with aging would suggest that episodic memory decline might be a marker for insipient brain pathology, rather than a primary aspect of normal aging.

There are several competing theories postulating the potential mechanisms for age-associated declines in episodic memory. They can generally be divided into four areas, including an age-related failure of memory monitoring, or metamemory; age-associated decreases in the depth of initial encoding; age-related impairment in processing of contextual information; and age-associated decline in a number of processing resources. While these theories have not been fully substantiated empirically, the latter has received the greatest amount of support in the cognition literature. Proponents of a 'resource reduction hypothesis' argue that central to age-associated changes in episodic memory is a reduction in primary cognitive resources such as attention or working memory, or a reduction in the ability to engage due to an age-associated diminution in attentional inhibitory control. Others argue that age-associated memory decline is not due to a reduction in available attentional resources *per se*, but rather to age-related declines in perceptual processing speed.

**Source memory** Episodic memory comprises the information that is being recalled as well as the context in which the information was learned. This latter aspect is referred to as source memory, and there is increasing evidence that even when older adults successfully recall information, they may have difficulty recalling the source in which the information was acquired. The phenomenon has been demonstrated in the identification of the temporal context in which an item was learned, as well as the spatial and perceptual context. Studies by Daniel Schacter and colleagues required older adults to listen to different speakers read different blocks of declarative statements, and found that memory for the source of the declarative sentences was disproportionately worse than was memory for the statements. This general finding has been well replicated and has been extended to demonstrate age-associated impairment in both specific-source memory and partial-source memory. For example, older adults have difficulty, relative to younger adults, remembering which of four people spoke a word (i.e., specific-source memory) as well as remembering partial information about the person who spoke the word, such as his or her gender (i.e., partial-source memory).

### Nondeclarative Memory

Nondeclarative, or implicit, memory describes the memory system that allows learning outside of conscious awareness. It is generally divided into procedural memory and priming, each hypothesized to be mediated by a distinct neurobiological system. In general, nondeclarative systems are relatively spared across the adult life span, particularly compared to episodic memory, which shows the greatest aging effects.

**Procedural memory** Procedural or skill learning is one type of nondeclarative memory that refers to the nonconscious acquisition of motoric sequences. A common example of procedural memory is the process of learning how to drive an automobile. Initially, a novice driver needs to recollect consciously how to control each aspect of the automobile and the sequence in which to do so. With practice, the individual skill required to operate the vehicle enters into procedural memory, and control becomes relatively more automatic.

Little work has been conducted that explicitly examines the impact of chronological age on procedural memory, and results have been somewhat inconclusive; some studies show a decline with age and others show no apparent aging effect. A potential confounding factor in the examination of procedural memory among older adults is how much the experimental paradigm draws upon pure motoric speed and on other nonprocedural cognitive abilities, such as working memory, as well as whether the experimental paradigm addresses procedural learning versus long-term procedural memory. Studies that examine the effects of age on experts, such as pianists or typists, show a general trend of age-associated slowing, but a relative maintenance in measures of performance. Age-associated preservation of procedural memory may also be dependent on the amount of deliberate practice.

When considering age effects on procedural memory, it is important to dissociate performance time, learning, and memory for the task. For example, older adults tend to perform procedural memory tasks more slowly and learn procedural sequences at a slower rate, compared to younger adults. However, after initial acquisition, older adults will relearn a procedural memory task at rates similar to or more rapid than the rates at which younger adults relearn, even after a 2-year interval without practice. While procedural memory and aging remain somewhat understudied, there is some consensus that older adults have lasting preservation of procedural or motor memory. It is important, however, to distinguish between age-associated changes in motoric or perceptual processing speed and age-related changes in procedural memory; when accounting for aspects of the former, the latter appears to be relatively spared.

**Priming** Repetition priming is a special type of implicit memory that refers to the implicit impact that prior exposure to a stimulus has on later test performance. For example, an individual is more likely to complete the word stem STR__ with ONG (to form the word STRONG) than with EET (to form the word STREET) if he/she had previously studied the word STRONG. There is a long history of the examination of aging effects on priming in multiple modalities, with somewhat mixed results. In general, studies from the 1980s suggested that there are few differences between younger and older adults for priming across a number of tasks, such as word stem completion, picture naming, and word identification. There is evidence of a strong dissociation in the elderly between a preserved ability to perform implicit tasks and a deficit in performance on declarative tasks when compared to younger participants, although there are small, but reliable, age-associated decrements in priming for verbal abilities regardless of whether the dependent measure is accuracy or latency or whether the type of priming is item or associative.

Investigators such as John Rybash have distinguished among five types of priming, including (1) perceptual-item priming, (2) perceptual-associative priming, (3) conceptual-item priming, (4) conceptual-associative priming, and (5) perceptual-motor priming. Perceptual tasks are dependent upon implicit recall of properties of the stimuli presented during encoding, whereas conceptual tasks refer to priming for object meaning. There is some evidence that with normal aging there is relative preservation of performance on tasks that require perceptual priming and a relative decline in tasks that require conceptual priming. However, other investigators, such as Debra Fleischman and John Gabrieli, argue that studies that have shown no aging effect for priming may have had insufficient measurement reliability or power to detect age differences, and positive studies may have included a minority of individuals with insipient dementia. In contrast to Rybash, they argue that perceptual priming is vulnerable to aging effects, whereas conceptual priming remains relatively intact. Differences between these two conclusions may reflect differing positions on inclusion criteria for categories of priming.

For example, Fleischman and Gabrieli exclude all tasks that might include the processes that explicit memory retrieval tasks also engage. Further, they consider tasks such as category-exemplar generation to be conceptual, whereas Rybash considers these to be perceptual. In summary, although the literature is somewhat mixed, the effects of normal aging on repetition priming are of small magnitude or nonexistent, depending on modality.

## Short-Term Memory

In contrast to long-term memory, in which information is stored for minutes to years, short-term memory refers to holding information in conscious awareness for the duration of seconds. While short-term memory is qualitatively a type of memory distinct from long-term memory, the two systems interact. For example, in order for information to be stored in long-term memory, the information must first exist within short-term memory. The exact mechanisms by which short-term memory is transferred to long-term storage have yet to be fully elucidated, but the 'modal model' of information transfer, proposed by Atkinson and Shiffrin in the 1960s, provided a theoretical framework that has dominated the cognitive literature for decades. Briefly, information flows from the environment to primary sensory (perceptual) stores and then into a short-term store, which can include rehearsal, coding, or decision. The capacity of the short-term store is limited and information can leave the system (i.e., forgetting), lead to a response output, or enter long-term storage, which is fairly permanent. When information is drawn from long-term memory, it exists in short-term memory while in conscious awareness. In terms of the capacity of short-term memory, a heuristic offered by Miller in the 1950s is that the average short-term memory span for a person is seven, plus or minus two items, or chunks. Although this heuristic has remained popular, more recent studies suggest that short-term memory span is closer to four items, or is contingent upon an individual's processing speed.

The term 'short-term memory' refers to the passive short-term store of information, but speaks little to the process by which information is retained in the store. It is difficult to discuss the effects of age on short-term memory without consideration of these processes. The term 'working memory' is typically applied to the process by which information is held in short-term memory and refers to the cognitive manipulation of information that is contained within short-term memory. An influential model that attempts to explain how information is stored within short-term memory was proposed by Baddeley and Hitch, in which a supervisory 'central executive' controls the information flow among three 'slave systems,' including the phonological or articulatory loop, the visuospatial sketchpad, and the episodic buffer. The first two systems rehearse phonological or visual information, respectively, and the third system integrates information from the other two systems and interfaces with long-term memory. The central executive acts to coordinate the flow of information among the slave systems, to direct attention to appropriate internal and external stimuli, and to suppress irrelevant stimuli.

Working memory paradigms require individuals to perform mental operations on items held in conscious awareness, such as reordering words or numbers. While there is little evidence that short-term memory *per se* significantly declines with normal aging, there is ample evidence that working memory abilities do. Within working memory, efficacy of inhibition and smaller span are particularly vulnerable to the effects of aging.

Working memory in aging studies is often operationally defined by tasks such as letter rotation, reading span, computation span, and line span. Performance on these tasks tends to show a linear age-associated decline that is similar to that seen for tasks of long-term episodic memory and speed of processing. Some research suggests that age-associated decline in working memory abilities mediates age-associated decline in other memory and cognitive domains. Indeed, statistical control of performance on tasks of working memory often attenuates the observed age-associated decline on other cognitive tasks.

## Summary and Course

Normal age-associated memory decline is not uniform. Older adults evidence worse performance on long-term memory tasks compared to younger adults, but these differences are relatively greatest on tasks of declarative episodic memory. Semantic memory, on the other hand, remains relatively stable across the adult life span or may even increase as more semantic knowledge is accumulated with age. Similarly, working memory, or the manipulation of information that is held in conscious awareness, shows marked decline in with normal aging, and some theorists propose that working memory deficits mediate age-associated decline in other cognitive domains.

In terms of the course of memory changes across the adult life span, results from cross-sectional and longitudinal studies suggest that subtle memory changes can begin as early as the early or middle twenties and continue to decline linearly with age. Some authors distinguish between 'lifelong decline' and 'late-life decline.' Performance on tasks of episodic and working memory seems to begin to decline in the twenties and continues to decline linearly across the life span, which is supported by cross-sectional aging and cognition studies. Some longitudinal studies, however, suggest a curvilinear course of memory decline, with a more precipitous decline after about age 60 years, preceded by relatively little decline with age. Short-term memory store appears to remain relatively stable until about age 70, at which point it begins to drop, and, as noted, semantic abilities appear to remain relatively stable, at least until late life.

## Moderators and Mediators of Cognitive Aging

A consistent observation in the aging and cognition literature is a greater amount of variability in memory performance with increasing age. This finding is evident both when age is considered as a cross-sectional variable and when individual age trajectories are examined over time; some adults experience great amounts of cognitive decline while others experience relatively little. Increased variability suggests that there are moderators and mediators of cognitive aging. Two factors in particular, including cerebrovascular or cardiovascular risk and cognitive reserve, have received considerable attention as modulators that may account for some of the increased variability, and these are potential targets for intervention or prevention of cognitive decline.

Cerebrovascular or cardiovascular risk factors may mediate the relationship between chronological age and the neurobiological changes that underlie cognitive aging. Increased blood pressure is associated with reduced psychomotor speed, visuoconstruction ability, learning, memory, and executive functioning. Similarly, both insulin resistance and diabetes are associated with diminished cognitive abilities in later life. Hyperlipidemia has also been identified as a potential cerebrovascular risk factor that negatively impacts cognition in older adults. While it is still somewhat unclear by what mechanism cerebrovascular or cardiovascular risk factors impact cognition, it is likely that they interact with age to have a cumulative effect on cognition in later life. Furthermore, intervention studies that target the vascular system, such as cardiovascular fitness regimens, are associated with increased cognitive functioning among older adults.

Cognitive reserve is another factor that may account for the increased variability observed in cognitive aging. The cognitive reserve hypothesis stems from the observation that there is a disconnection between degree of brain pathology and its cognitive manifestation and suggests that the brain actively attempts to cope with brain pathology by using existing processing approaches or by recruiting compensatory networks. Although cognitive reserve is often applied to clinical entities, such as AD or stroke, it may be operative among the normal aged. That is, cognitive reserve may moderate the relationship between normal age-associated neurobiological changes and their cognitive outcome. Indeed, proxies for reserve, such as measures of education or literacy and IQ, are related to the degree of age-associated memory decline. Declarative memory scores of older adults with lower levels of literacy decline at a greater rate than do those with higher levels of literacy. Like vascular risk factors, interventions that target cognitive reserve may improve the course of cognitive aging.

Other factors, such as genes and nutritional exposure, could potentially impact the course of memory decline with normal aging. It is also important to note that the possibility of inclusion of individuals with insipient dementia may have contaminated some studies that attempted to elucidate the pattern of age-associated memory decline, and future studies of normal aging should focus on disentangling normal cognitive decline from pathological aging at the earliest point possible. With the powerful capabilities of neuroimaging techniques, we can begin to understand the complex relationships among normal aging, neurobiology, and cognition while defining potentially modifiable moderators and mediators of cognitive aging.

*See also:* Aging of the Brain; Aging of the Brain and Alzheimer's Disease; Aging and Memory in Animals; Cognition in Aging and Age-Related Disease; Episodic Memory; Functional Neuroimaging Studies of Aging; Lipids and Membranes in Brain Aging; Metal Accumulation During Aging; Short Term and Working Memory.

## Further Reading

Atkinson RC and Shiffrin RM (1968) Human memory: A proposed system and its control processes. In: Spence KW and Spence JT (eds.) *The Psychology of Learning and Motivation: Advances in Research and Theory,* vol. 2, pp. 89–195. London: Academic Press.

Baddeley AD (2000) The episodic buffer: A new component of working memory? *Trends in Cognitive Science* 4: 417–423.

Christensen H (2001) What cognitive changes can be expected with normal ageing? *Australian and New Zealand Journal of Psychiatry* 35: 768–775.

Cowan N (2001) The magical number 4 in short-term memory: A reconsideration of mental storage capacity. *Behavioral and Brain Sciences* 24: 87–114; discussion 114–185.

Craik FIM and Salthouse TA (eds.) (2000) *The Handbook of Aging and Cognition,* 2nd edn. Mahwah, NJ: Lawrence Erlbaum.

Fleischman DA and Gabrieli JD (1998) Repetition priming in normal aging and Alzheimer's disease: A review of findings and theories. *Psychology and Aging* 13: 88–119.

Grady CL and Craik FI (2000) Changes in memory processing with age. *Current Opinion in Neurobiology* 10: 224–231.

Hedden T and Gabrieli JD (2004) Insights into the ageing mind: A view from cognitive neuroscience. *Nature Reviews Neuroscience* 5: 87–96.

Krampe RT and Ericsson KA (1996) Maintaining excellence: Deliberate practice and elite performance in young and older pianists. *Journal of Experimental Psychology: General* 125(4): 331–359.

LaVoie D and Light LL (1994) Adult age differences in repetition priming: A meta-analysis. *Psychology & Aging* 9: 539–553.

Light LL (1991) Memory and aging: Four hypotheses in search of data. *Annual Review of Psychology* 42: 333–376.

Park DC, Smith AD, Lautenschlager G, et al. (1996) Mediators of long-term memory performance across the life span. *Psychology & Aging* 11: 621–637.

Rybash JM (1996) Implicit memory and aging: A cognitive neuropsychological perspective. *Developmental Neuropsychology* 12: 127–179.

Salthouse TA (1993) Speed mediation of adult age differences in cognition. *Developmental Psychology* 29: 722–738.

Schacter DL, Kaszniak AW, Kihlstrom JF, et al. (1991) The relation between source memory and aging. *Psychology & Aging* 6: 559–568.

Small SA, Stern Y, Tang M, et al. (1999) Selective decline in memory function among healthy elderly. *Neurology* 52: 1392–1396.

Squire LR (2004) Memory systems of the brain: A brief history and current perspective. *Neurobiology of Learning and Memory* 82: 171–177.

Stern Y (2002) What is cognitive reserve? Theory and research application of the reserve concept. *Journal of the International Neuropsychological Society* 8: 448–460.

Stern Y (ed.) (2006) *Cognitive Reserve: Theory and Applications.* New York: Taylor & Francis.

Tulving E (2002) Episodic memory: From mind to brain. *Annual Review of Psychology* 53: 1–25.

# Cognition in Aging and Age-Related Disease

**E A Kensinger**, Boston College, Chestnut Hill, MA, USA



## Introduction

As the average age of the population increases, there is growing interest in understanding the cognitive and neural changes that accompany aging. It is now clear that significant cognitive decline is not an inevitable consequence of advancing age. This realization has spurred researchers to examine what separates high-performing older adults from lower-performing older adults and to investigate how the changes with successful aging differ from those that result from age-related disease. Data acquired using behavioral testing, functional neuroimaging, and structural neuroimaging are beginning to inform these issues, although a number of questions remain.

## Cognitive Declines with Healthy Aging

Not all cognitive domains are affected equally by age, and not all cognitive processes show age-related decline. If an older adult were asked to list the cognitive declines that have been most notable to him or her, it is likely that at least one of the following would make the list: problems paying attention to relevant information and ignoring irrelevant information in his or her environment, word-finding difficulties, or problems remembering the context in which information was learned.

Much of the research within the field of cognitive aging is focused on understanding whether the pattern of age-related cognitive decline can be explained by domain-general (or 'core') cognitive changes or whether changes in domain-specific processes are required to describe the data.

### Domain-General Theories of Cognitive Aging

Domain-general theories of aging are based on the hypothesis that there is a shared ability that cross-cuts all of the tasks on which older adults are impaired: these theories suggest that although aging affects a range of cognitive functions, there is one central or core deficit underlying the myriad changes. This article focuses on three core deficits that have been proposed to explain the pattern of age-related declines: changes in sensory perception, changes in inhibitory ability, and changes in speed of processing.

**Sensory deficits** The sensory deficit hypothesis of aging proposes that the cognitive changes with aging may be attributed to changes in sensation (i.e., deficits in vision and hearing). Indeed, aging is associated with dramatic sensory declines: by the eighth decade of life, the majority of older adults have significant hearing loss and a reduced ability to discriminate colors and luminance. Support for the hypothesis that these sensory deficits may underlie cognitive changes has come from two lines of research. First, across a wide range of cognitive tasks, older adults' performance correlates strongly with their sensory abilities. Second, in young adults, cognitive impairments can arise when the to-be-processed stimuli are degraded. For example, when asked to remember words pronounced against a noisy background, or when asked to match digits and symbols written in low-contrast font, young adults perform comparably to older adults. Thus, it is plausible that age-related deficits on many cognitive tasks stem, at least in part, from reductions in sensory processing. For example, changes in audition may result in older adults' slowed performance on tasks requiring auditory processing and could explain older adults' poorer memory for auditory information. It is also possible, however, that in at least some instances the correlation derives from a common influence underlying both the sensory and cognitive changes. For example, individuals who have greater brain atrophy or dysfunction may be more likely to have both sensory deficits and cognitive impairments. Regardless of the precise mechanism through which sensory deficits relate to cognitive ones, evidence of significant sensory changes with age underscores the importance of modifying testing procedures to minimize the influence of sensory deficits on older adults' cognitive test performance (e.g., using louder or higher-contrast stimuli).

**Inhibition** Hasher, Zacks, and colleagues have proposed that older adults' cognitive deficits may relate to their inability to ignore irrelevant information in the environment while focusing attention on goal-relevant information. If an older adult is seated at a restaurant that has many tables in close proximity to one another, he or she may have difficulty paying attention to the conversation at his or her table while ignoring the conversations at nearby tables. In the laboratory, inhibitory deficits can result in responses to previous (but not current) targets. For example, after reading the sentence "Before going to bed, please turn off the STOVE," older adults will be more likely than young adults to believe that the target word was LIGHT rather than STOVE. The age-related increase in this type of error is thought to occur because older adults have a hard time inhibiting the strong association present in the 'garden path sentence' (e.g., between the idea of going to bed and the idea of turning off the light). Inhibitory deficits also frequently emerge when older adults are required to task-switch or to set-shift. On these tasks, older adults must first pay attention to one aspect of a stimulus (e.g., match items based on their shape) and then another (e.g., match items based on their color). Older adults often have a harder time than young adults when they must ignore the previously relevant dimension (e.g., the shape of the items) and instead focus their attention on another stimulus attribute (e.g., the color of the items).

These data clearly indicate that inhibitory deficits can occur on a range of tasks requiring the ability to selectively attend to information in the environment or to inhibit a strong association or response. However, inhibitory deficits may impair performance not only on tasks that directly assess inhibitory ability, but also on assessments of working memory capacity: if older adults have a hard time distinguishing relevant from irrelevant information, this likely means that they store task-irrelevant information, reducing the storage capacity available for task-relevant information.

**Speed of processing** Older adults have a slower speed of processing than young adults. This slowed processing is noted at the motor level, but it also is apparent at a cognitive level. For example, older adults will tend to be slightly slower than young adults when they must slam on the brakes at a red light; this slowing may primarily be due to motor changes, because the association of red = stop remains very strong with aging. The reaction time differences between young and older adults will be exaggerated, however, if older adults must decide whether to slam on the brakes or to hit the gas as they approach a light that has just turned yellow. This additional slowing likely results because of the increased cognitive processing that must occur before the appropriate action can be selected.

Salthouse and colleagues have suggested that this decline in processing speed may underlie the age-related changes in cognitive function. It is apparent how slowed speed of processing could be detrimental to performance on any type of timed task. Importantly, however, a slower speed of processing could also manifest itself on nontimed tasks. For example, imagine that I read aloud the following arithmetic problem, and ask you to solve it in your head, with no time limit: "Jimmy walks up to a store counter with three packs of gum, each costing 50 cents. He gives the sales clerk $5. Because the clerk is out of dollar bills, she gives Jimmy his change in quarters. How many quarters does Jimmy receive from the sales clerk?" On the face of it, this is a task of working memory ability (the ability to store various pieces of information and to update the information as you work through the problem) that might be thought of as being independent from a measure of speed of processing, because there is no time limit for solving the task. However, if it takes someone a little longer to process the phrase "Jimmy walks up to a store counter with three packs of gum," it is possible that they will have a harder time attending to the phrase "costing 50 cents." Similarly, if it takes someone longer to multiply 3 by 50, it is possible that by the time that calculation is completed they will have forgotten the amount of money that Jimmy gave to the clerk. In other words, cognitive performance can suffer because the slowed mental operations cannot be carried out within the necessary time frame, and because the increased time between mental operations can make it more difficult to access previously processed information. Thus, a slower speed of processing may lead to a poorer encoding of information and a reduced ability to store information. In support of the hypothesis that processing speed changes may underlie much of the cognitive decline with aging, controlling for speed of processing often eliminates age differences on cognitive tasks, and longitudinal studies have shown a strong relation between changes in speed of processing and changes in performance on a large number of cognitive tasks.

### Domain-Specific Theories of Cognitive Aging

In contrast to the domain-general theories of cognitive aging, domain-specific theories propose that some age-related declines may not be explained by core deficits that affect all aspects of cognition, but rather by changes that have a larger impact on one area of cognition than on another.

**Word-finding difficulties and transmission deficits** Older adults often have difficulties retrieving the appropriate name for a person, place, or thing. These word-finding problems can be manifest in various ways: excessive use of pronouns (due to difficulty generating the proper nouns), decreased accuracy and increased reaction time when asked to name items, and increased tip-of-the-tongue experiences. The tip-of-the-tongue state occurs when a person has access to a word's meaning, but not the phonological features of the word. The word seems just out of reach. Older adults tend to have more tip-of-the-tongue experiences than young adults, particularly for proper names, and the accuracy of the phonological information available during a tip-of-the-tongue state (e.g., the first letter of the word, the number of syllables) tends to be more accurate for young adults than for older adults.

Burke, MacKay, and colleagues have suggested that these word-finding difficulties result from the fact that, with age, the links connecting one unit to another within the memory system become weaker. Thus, more links must be active in order for older adults to generate the correct name for an object or a person. This transmission deficit will mean that older adults will be relatively good at generating words when there are lots of links converging onto the word but will show larger impairments when trying to generate words that have fewer associated links. Most everyday objects have many semantic associations (e.g., individuals know that apples are red, of waxy texture, round, etc.). This convergence onto the word 'apple' (referred to by Burke and colleagues as summation of priming) makes it relatively easy to generate the name of the object. In contrast, proper names (with the exception of nicknames) are arbitrary and do not benefit from the same summation of priming (e.g., there is nothing that is required for someone to be a Jane vs. a Linda). Consistent with a transmission deficit, older adults remain relatively good at generating words of everyday objects for which they know a lot of semantic information but show larger impairments when asked to generate proper nouns.

**Contextual memory and associative binding deficits**
Episodic memory can be thought of as including memory for two different types of information: memory for the item previously encountered (which typically can be based on familiarity) and memory for the broader context in which that item was encountered (which requires recollection). For example, when I pass by someone on the street, I might recognize that I have seen the person before (item memory). I also may remember that I met the person at a recent conference (an item–context association) or that the person is my colleague's husband (an item–item association).

Older adults remain quite good at using familiarity to recognize previously encountered people or items. However, older adults are particularly impaired at using recollection to remember the contextual details of an event. They seem to have difficulties binding together multiple event details into one cohesive memory. These deficits arise when trying to remember both item–item associations and item–context associations ('source' memory). Naveh-Benjamin and colleagues have proposed that this associative memory deficit underlies older adults' episodic memory difficulties.

Two broad types of memorial deficits may underlie this decreased ability to remember item–item or item–context associations. First, older adults have difficulties initiating effective encoding 'strategies' that would promote memory for the associative details of an experience (as proposed by Craik, Jennings, and colleagues). When they are given a strategy to use as they learn information (e.g., if they are asked to tell a story that binds the item to its context), older adults often perform as well as young adults on tasks requiring associative or contextual memory. Thus, at least some of the age-related deficits in remembering contextual details seem to result from deficits at encoding. Second, older adults seem to have difficulties either in forming a long-lasting 'bond' between an item and its context, or in retrieving that bound representation. Thus, given retrieval support (e.g., an untimed recognition task), older adults will tend to perform better than when given little retrieval support (e.g., a recall task).

## Preserved Cognitive Function with Healthy Aging

Although the previous sections have focused on age-related declines, aging is not associated with across-the-board deficits in cognitive. In fact, some aspects of cognition remain markedly stable, or even improve, as individuals age.

### Crystallized Intelligence

Crystallized intelligence refers to the ability to retrieve and use information that has been acquired throughout a lifetime. It often is contrasted with fluid intelligence, the ability to store and manipulate new information. As discussed earlier, fluid intelligence processes tend to be disrupted by healthy aging. Crystallized intelligence, in contrast, remains stable across the life span. Thus, older adults are very good (often better than young adults) at defining words, answering questions that rely on general world knowledge (e.g., "Who wrote the 'Star Spangled Banner'?"), detecting spelling errors, or carrying out skills related to jobs that they have held for many years.

### Emotion Regulation

Another important area of preservation (or enhancement) is within the realm of emotion regulation. After about the age of 60, the ability to regulate emotion seems to start to improve. Thus, older adults show lower rates of depression than young adults. Compared to young adults, their good moods last longer, and they are able to rebound more quickly from negative mood states. Older adults seem to focus more on positive information in their environment and to choose activities (e.g., spending time with close family or friends) based on their potential for emotional fulfillment.

At least some aspects of memory for emotional information also seem to be preserved with aging. While older adults tend to show overall poorer memory than young adults on a variety of tasks requiring retrieval of contextual information, studies of 'flashbulb memories' (memories for a public event that was highly surprising and emotional) have suggested that older adults, like young adults, are more likely to remember contextual details if the event contains emotional relevance than if it does not. The emotional memory benefit for older adults may be particularly pronounced for positive information, although a number of studies have suggested that older adults receive a memory boost when asked to remember negative experiences as well.

## Neural Changes with Healthy Aging

Although healthy aging is associated with brain changes, not all regions are affected equally. To date, the vast majority of studies investigating the neural changes with healthy aging have used structural magnetic resonance imaging (MRI) (allowing examination of the volume of various brain regions) and functional MRI (fMRI)or positron emission tomography (PET) (allowing indirect measurements of neural activity as individuals are performing a task). These studies have suggested that the largest changes in structure and function occur in the prefrontal cortex and in the medial temporal lobe, while other cortical and subcortical regions remain relatively preserved across the life span. Current research also is focused on examining how the connections between regions are affected by aging. Diffusion tensor imaging studies are being conducted to examine age-related changes in white matter tracks, and structural equation modeling of fMRI and PET data is being used to investigate age-related changes in the functional connections between brain regions. These methods may provide new insights into the neural changes that mediate older adults' cognitive decline.

### Changes in Prefrontal Cortex

The prefrontal cortex shows notable changes with aging. At a structural level, there is evidence of atrophy, both in the gray matter and in the white matter. The gray matter declines may reflect reductions in the number of cells (due to cell death) or may be a sign of neuronal shrinkage. The white matter changes reflect axonal abnormalities and may result in slowed neurotransmission. It is plausible that these white matter changes may mediate the cognitive slowing that accompanies healthy aging.

Prefrontal function also is altered with aging. Across a range of working memory and episodic memory tasks, older adults seem to show a different pattern of prefrontal activity than young adults, with reduced activity in some prefrontal regions and increased activity in other regions. As noted by Cabeza and colleagues, particularly in the prefrontal cortex, older adults often show a hemispheric asymmetry reduction. In other words, on tasks leading to unilateral prefrontal activity in young adults, older adults will tend to show bilateral recruitment. It is not yet clear whether this bilateral recruitment reflects compensatory activation, or whether it is a result of pathological changes (e.g., hemispheric release from inhibition). Some of the strongest evidence in favor of the compensatory view has come from studies comparing the pattern of neural recruitment in older adults who perform a task as well as young adults (high performers) and in older adults who perform the task more poorly (low performers). These studies have found that the high performers tend to recruit the prefrontal cortex bilaterally, whereas the low performers show unilateral prefrontal activity. To the extent that the prefrontal activity underlies the initiation of goal-relevant task strategies, it would make sense that the older adults who recruit additional prefrontal regions would be those who would be best able to perform the tasks. Nevertheless, future studies are required to clarify whether this compensatory hypothesis can account for all of the findings of bilateral prefrontal recruitment with age.

### Medial Temporal Lobe Changes

The hippocampus proper is the other region that shows large age-related change. While there are structural changes in the hippocampus, it is not clear whether there is significant cell loss in the hippocampus with aging, or whether the structural changes are related more to neuronal atrophy (shrinkage). There also is ambiguity regarding the regions of the hippocampus that are most affected by aging. High-resolution MRI, allowing the distinction of the various hippocampal subfields (CA1, CA3, dentate gyrus, subiculum) may allow better assessment of the structure (and function) of these regions.

Functionally, the hippocampus tends to be under-recruited by older adults during both the encoding and the retrieval phases of recollective or associative memory tasks, and these functional changes often correlate with the older adults' reduced performance on the tasks. Given the critical role of the hippocampus in forming vivid and detailed memories, it makes sense that the functional and structural changes in this region would correspond with older adults' difficulties in remembering the context in which information was learned.

### Changes in Emotion Processing Regions

The fact that older adults show improved emotion regulation, and a memory enhancement for many types of emotional information, is consistent with the neural evidence indicating that emotion processing regions (particularly the amygdala and orbitofrontal cortex) are relatively spared in healthy aging. The amygdala shows minimal atrophy with healthy aging; its atrophy is on par with the decline in whole-brain volume (with a 1–3% reduction in volume every decade). Similarly, the orbitofrontal cortex seems to undergo little volumetric decline with age, particularly as compared to other regions of the prefrontal cortex.

## Mild Cognitive Impairment

As discussed previously, some cognitive impairment is a natural part of the aging process. For some adults, however, advancing age is associated with fairly severe impairments in recent memory. Although the deficits do not impair their ability to function in daily life (and thus they do not meet the criteria for dementia), these individuals have cognitive impairments that exceed those that typically accompany healthy or successful aging. 'Benign senescent forgetfulness' was the first term used (by Kral and colleagues) to describe these individuals' impairments, although there were no strict diagnostic criteria associated with the concept. 'Age-associated memory impairment' (proposed by Crook and colleagues) was the first attempt at a standardized definition, requiring an individual to have subjective memory complaints and to perform at least 1 standard deviation below the mean for young adults on a standardized memory task. This concept was criticized by a number of researchers, who believed that the concept was too restrictive. Mild cognitive impairment (MCI) (defined by the Mayo Clinic Alzheimer's Disease Research Center) is the most recent in a series of attempts to characterize these individuals who straddle the boundary between healthy aging and dementia. A diagnosis of MCI requires subjective memory complaints and impairment in one area of cognition (scores must be more than 1.5 standard deviations below age-scaled norms), but with deficits not severe enough to interfere with activities of daily living or to result in a diagnosis of dementia.

There has been tremendous interest in defining this group of individuals, because as treatments that slow or reverse the development of Alzheimer's disease (AD) become available, it will be critical to have a method for diagnosing individuals at risk for, or in the prodromal stages of, the disease. Individuals with MCI seem to be an excellent population to be the targets of such treatments, because they are at increased risk for development of dementia, and of AD in particular. In fact, by some estimates, the vast majority of patients with MCI will eventually meet the diagnostic criteria for dementia.

The link between MCI and AD is supported not only by the high conversion rate, but also by the overlapping neuropathological and genetic features. Like AD patients, those with MCI have significant structural and functional changes in the medial temporal lobe. They also have alterations in the concentration of amyloid beta protein, the protein associated with neuritic plaque formation in AD. Moreover, the $\varepsilon 4$ allele of the apolipoprotein E, associated with an increased risk of developing AD, is overrepresented among individuals with MCI.

## Alzheimer's Disease

AD was first described by Alois Alzheimer in 1907. In the landmark publication, Alzheimer reported a case study of a woman with severe psychiatric symptoms and memory deficits. An autopsy conducted upon her death revealed a large quantity of intracellular neuritic plaques and extracellular neurofibrillary tangles, now recognized as the pathologic hallmarks of AD.

### Cognitive Changes in AD

Although dementia (a loss of intellectual function severe enough to interfere with daily activities) can result from a variety of etiologies, AD is by far the most common cause, accounting for an estimated two-thirds of all cases of dementia. Because AD can only be confirmed at autopsy, its clinical diagnosis must be an exclusionary one. Thus, the clinical profile of AD requires memory impairment plus decline in one other area of cognition (language, motor function, attention, executive function, personality, or object recognition). The deficits must have a gradual onset, and they must progress continually and irreversibly. When these criteria are met, a diagnosis of 'probable' AD is given. When made by a trained clinician, this diagnosis will be accurate in the vast majority (80–90%) of cases.

**Episodic memory** In contrast to healthy older adults, who remain able to successfully remember previously encountered information (though perhaps not the context in which it was encountered), the most notable deficit for patients with mild AD is an inability to remember information encountered in the recent past. This deficit extends across different types of encoding tasks (e.g., incidental or intentional; deep or shallow processing) and exists regardless of the stimulus materials (e.g., pictures, words, faces, autobiographical events, emotional stimuli) or the task's retrieval demands (e.g., recall, forced-choice recognition, yes–no recognition). In fact, deficits in episodic memory tend to be the best way of distinguishing people with AD from healthy older adults.

**Semantic memory** In contrast to episodic memory, semantic memory (general world knowledge) is relatively spared with mild AD. As the disease progresses, however, significant semantic deficits arise. The deficits are particularly pronounced on word-finding tasks, with the extent of such deficits being useful for tracking the severity of AD. Current research is examining the extent to which the breakdown in semantic memory is due to changes in the structure of the memory networks (i.e., to a problem with the storage of such knowledge) or to difficulties retrieving the stored information (i.e., a problem with access).

**Working memory and executive function** In addition to deficits in long-term memory, Alzheimer's patients also show deficits in the online processing of information (working memory). Deficits are particularly pronounced on tasks requiring dual-task performance, suggesting that a primary deficit in AD may be in executive functions, the ability to flexibly shift attention and to attend to goal-relevant information.

## Neural Changes in AD

Although plaques and tangles often are apparent throughout the brain in the later stages of AD, early in the course of the disease, the medial temporal lobe regions are those most affected. Even early in the disease, the hippocampal formation shows marked atrophy, and a volumetric reduction in the entorhinal cortex (which serves as the site of input to the hippocampus) is one of the best indicators that an individual has early AD. Given the essential role of the hippocampus for memory formation and retrieval (as demonstrated by the link between hippocampal damage and amnesia), it makes sense that patients with mild AD would be best identified by their difficulties remembering recently learned information (as discussed previously).

The amygdala is another region of the medial temporal lobe that is affected early in the disease process. This region, through its interactions with other medial temporal lobe structures, is thought to be essential for enhancing individuals' memories for highly emotional events. It is likely because of the amygdalar damage that patients with AD do not show a memory benefit for emotional information.

In addition to the medial temporal lobe changes, the nucleus basalis also shows significant cell loss in mild AD. This region of the ventral forebrain contains many of the brain's cholinergic neurons; thus, the damage to this region impedes cholinergic neurotransmission. Many of the first approved therapies for AD have been aimed at increasing the amount of acetylcholine in the brain, by blocking the function of acetylcholine esterase (the enzyme that breaks down the neurotransmitter acetylcholine). The minimal effectiveness of these acetylcholine esterase inhibitors in affecting cognitive function in AD patients suggests that the acetylcholine deficiency is not the only cause of the cognitive dysfunction in AD (and, indeed, at least by late-stage AD, there is marked depletion of other neurotransmitters, including norepinephrine, dopamine, and serotonin). The underwhelming effect of the cholinesterase inhibitors has led to continued research on alternate therapeutic options for AD, which may prove to be more effective in altering the progression of the disease.

### Neural Changes in Later-Stage AD

In later stages of the disease, there is increased atrophy throughout the medial temporal lobe, and the cellular abnormalities also become apparent in the frontal lobe and throughout the temporal lobe. Because the frontal lobe is essential for higher-level executive functions, and the temporal neocortex is critical for the ability to retrieve semantic information, the advancing neuropathology in these regions clearly is linked to the cognitive deficits that arise in moderate AD. By the very late stages of AD, neuropathology is abundant throughout the cortical and subcortical structures, even within primary sensory regions (e.g., auditory, visual, and motor cortex).

## Individual Differences in Aging

Unlike many diseases or illnesses that can be linked to a specific cause, MCI and AD appear to arise due to a combination of many factors, both environmental and biological. Research has begun to elucidate some of the traits that can increase the likelihood of disease development ('risk factors' such as prior head injury, mutations in the apolipoprotein, presenilin-1, and presenilin-2 genes, or having the $\varepsilon 4$ allele of apolipoprotein E) or that seem to reduce the probability of disease ('protective factors' such as high education level, intake of antioxidants, or having the $\varepsilon 2$ allele of apolipoprotein E). Research continues to investigate the reliability of these factors in predicting disease development and to examine whether any of the protective factors can alter disease progression once the pathological hallmarks of disease are present.

More broadly, researchers are beginning to recognize the necessity of taking an individual-differences approach to understanding the cognitive and neural changes that accompany aging. Research is now exploring the potential differences between high-performing and low-performing older adults. It is plausible that a better understanding of what differentiates the most successful agers from those who age less gracefully will lead to ideas for behavioral or neurobiological interventions that can boost the performance of individuals who experience significant age-related cognitive decline.

*See also:* Aging of the Brain; Aging of the Brain and Alzheimer's Disease; Aging and Memory in Humans; Alzheimer's Disease: MRI Studies; Episodic Memory.

## Further Reading

Burke DM and Mackay DG (1997) Memory, language, and ageing. *Philosophical Transactions of the Royal Society of London, B Series (Biological Sciences)* 352: 1845–1856.

Cabeza R (2002) Hemispheric asymmetry reduction in older adults: The HAROLD model. *Psychology and Aging* 17: 85–100.

De Toledo-Morrell L, Goncharova I, Dickerson B, Wilson RS, and Bennett DA (2000) From healthy aging to early Alzheimer's disease: *In vivo* detection of entorhinal cortex atrophy. *Annals of the New York Academy of Sciences* 911: 240–253.

Grady CL and Craik FIM (2000) Changes in memory processing with age. *Current Opinion in Neurobiology* 10: 224–231.

Kensinger EA (2006) Remembering emotional information: Effects of aging and Alzheimer's disease. In: Welsh EM (ed.) *Frontiers in Alzheimer's Disease Research*. Hauppauge, NY: Nova Science Publishers.

Li KZH and Lindenberger U (2002) Relations between aging sensory/sensorimotor and cognitive functions. *Neuroscience and Biobehavioral Reviews* 26: 777–783.

Light L (1992) The organization of memory in old age. In: Craik FIM and Salthouse TA (eds.) *The Handbook of Aging and Cognition*, pp. 111–165. Hillsdale, NJ: Erlbaum.

Mather M and Carstensen LL (2005) Aging and motivated cognition: The positivity effect in attention and memory. *Trends in Cognitive Sciences* 9: 296–502.

Park DC and Gutchess AH (2002) Aging, cognition, and culture: A neuroscientific perspective. *Neuroscience and Biobehavioral Reviews* 26: 859–867.

Salthouse TA (1996) The processing-speed theory of adult age differences in cognition. *Psychological Review* 103: 403–428.

# Declarative Memory System: Anatomy

**W A Suzuki**, New York University, New York, NY, USA



## Introduction

Think for a moment about all the memories you conjure up in a typical day. Perhaps something in today's newspaper reminds you of a recent debate with a friend, a TV program brings to mind a memorable vacation, or a perfume sampled at the department store recalls a favorite aunt. Our ability to retrieve the factual and event-based information from our past is called declarative memory. More formally, declarative memory is defined as the facts and events from our own personal histories that can be consciously brought to mind. We form and retrieve declarative memories every hour of the day. Without a functioning declarative memory system, as is the case in amnesia, we would be unable to hold down a job, keep track of our daily appointments, or find our way around. Declarative memory not only plays a critical part of our daily lives, but the facts and events in our declarative memory help shape our personalities and define who we are.

Surprisingly, the first systematic studies of human memory did not begin until the late 1800s. In 1881, the French psychologist Theodule Ribot published *Les Maladies de la Mémoire* (Diseases of Memory). In it, Ribot provided the first systematic description of human amnesic patients, noting that when brain injury causes memory impairment, recent memories are more susceptible to disruption than remote memories. This principle has become known as Ribot's Law. In 1885, Hermann Ebbinghaus published *Uber das Gedachtnis* (On Memory), in which he was the first to apply systematic experimental approaches to the study of normal human memory. For example, he showed that some memories are only short-lived while others are long lasting, and the amount of repetition can influence the duration of the memory. In 1890, the famous American psychologist William James published *Principles of Psychology* in which he described distinctions between what he called primary memory, defined as information that forms the focus of current attention, and secondary memory, defined as memory that persists even after it has dropped out of consciousness.

By the beginning of the twentieth century a major debate in the field concerned whether memory function was localized to a particular brain area or was distributed throughout the brain. The results from early experimental studies were mixed. However, in the 1920s Karl Lashley, a well-known psychologist at Harvard University, performed a series of influential experiments that addressed this debate. He trained rats to run in a simple spatial maze and then examined the effect of lesions to different parts of their cortex. He found that the larger the cortical lesion, the more severe the memory impairment on the rat's maze performance. He referred to these findings as the Law of Mass Action and concluded that memory was distributed over widespread cortical areas. This view dominated the field for more than 25 years, until findings from a single landmark study transformed our understanding of the localization of memory in the brain.

In 1957, Scoville and Milner described the effects of experimental bilateral lesions of the medial temporal lobe in 10 patients. These surgeries were done in an attempt to relieve a variety of psychiatric or neurological conditions, including schizophrenia, manic depressive psychosis, and, in one case, epilepsy. One striking and completely unexpected result of these experimental procedures was a profound memory loss that appeared to be correlated to the amount of bilateral hippocampal damage sustained by the patients. This memory loss was most obvious and easiest to study in patient H.M., who underwent the surgery in an effort to relieve severe and intractable epilepsy. Other patients also exhibited memory loss, but were more difficult to study because of their persistent psychiatric conditions. H.M., who received one of the largest medial temporal lobe resections, was described as being able to recall nothing of recent events of his life, though his childhood memories remained intact. Moreover, while H.M.'s memory impairment was profound, affecting memory for all sensory modalities tested (i.e., visual, auditory, somatosensory, and even olfactory memory), his general cognitive and intellectual abilities were spared.

This landmark study demonstrated several fundamental principles of memory organization in the brain. First, it showed definitively that memory could be localized to a particular brain area, namely the medial temporal lobe. Second, it demonstrated that memory could be studied independently of other general cognitive functions. Third, it led the way to more recent demonstrations that the medial temporal lobe has a critical and selective role in establishing declarative memory for facts and events. Fourth, this report was the catalyst for the ensuing experimental studies focused on defining more precisely the neuroanatomical basis of declarative memory. The following sections describe findings from the series of experimental lesion studies and neuroanatomical studies in animals that, cumulatively, led to the identification of the subset of medial temporal lobe structures important for declarative memory.

## The Amygdala, Hippocampus, and Development of an Animal Model of Human Amnesia

The amygdala and the hippocampus are two prominent structures that sit just deep to the cortex of the medial temporal lobe. The amygdala is a rounded almond-shaped structure that is situated in the anterior portion of the medial temporal lobe behind the cortex of the temporal pole. The hippocampus is an irregular tube-shaped structure that is located just posterior to the amygdala. The noted neurosurgeon Dr. William Scoville, who performed the surgery on patient H.M., described the medial temporal lobe resection as removing 8 cm of the medial temporal lobe tissue, including the temporal pole, amygdala, and approximately two-thirds of the anterior–posterior length of the hippocampus (**Figure 1**). Scoville and Milner initially hypothesized that bilateral damage to the hippocampus was responsible for the severe memory impairment, but because H.M.'s resection also included bilateral damage to the amygdala, they could not rule out the possibility that combined damage to both the hippocampus and the amygdala was collectively responsible for the deficit.

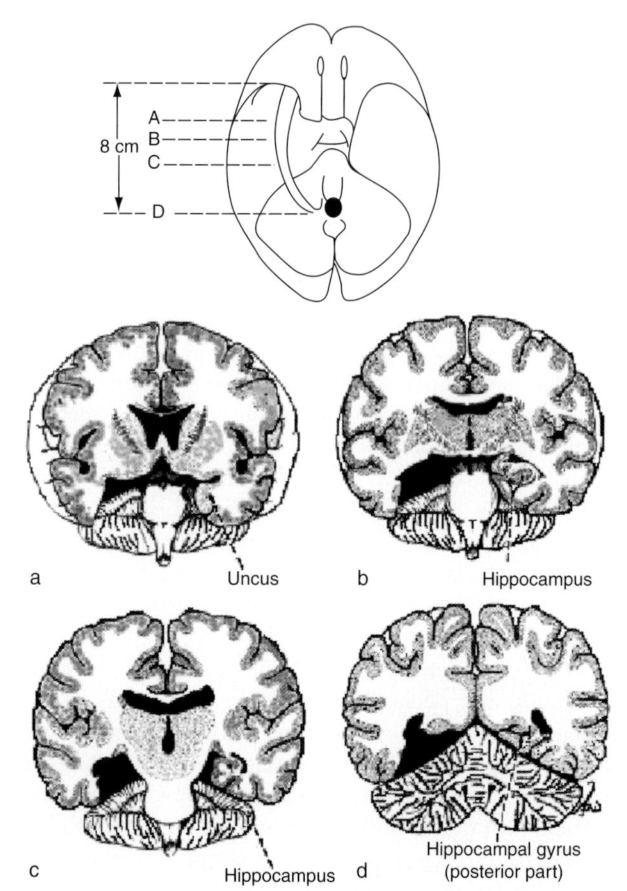


**Figure 1** This is a reproduction of the original from the 1957 work of Scoville and Milner, illustrating Scoville's estimate of H.M.'s medial temporal lobe resection. The drawing at the top shows a ventral view of the human brain, indicating the estimated boundaries of the lesion. (a through d) Coronal sections arranged from anterior (a) to posterior (d), showing the estimated extent of the lesion. In these coronal sections, the left side shows the estimated amount of tissue removed while the right side is shown intact in order to label the different target brain structures. Reproduced from Scoville WB and Milner B (1957) Loss of recent memory after bilateral hippocampal lesions. *Journal of Neurology, Neurosurgery & Psychiatry* 20: 11–21, with permission from the BMJ Publishing Group.

To address the question of which medial temporal lobe structure or structures underlie declarative memory, experimentalists attempted to develop an animal 'model' of human amnesia. Nonhuman primates have been particularly valuable animal model systems because their general brain organization, intelligence, and dominant visual system mirror our own. The strategy was to determine if bilateral medial temporal lobe lesions in an animal model could reproduce the profound memory impairment seen in patient H.M. If the deficit could indeed be reproduced, this animal model then becomes a very useful tool with which to examine the effect of damage to specific structures or combinations of structures on memory. In 1978, Mishkin was the first to successfully develop an animal model of human amnesia in the monkey. This study focused on recognition memory, defined as the ability to determine that a given stimulus has been seen before. Recognition memory is a form of declarative memory known to be impaired in human amnesic patients. In his seminal study, Mishkin examined the effect on recognition memory of combined damage to the hippocampus and the amygdala, made to resemble the damage sustained by patient H.M. Animals with separate lesions of either the hippocampus or the amygdala were also tested. All of the lesions included some inadvertent damage to the surrounding cortical areas. Recognition memory was tested using a delayed nonmatching to sample task (**Figure 2(a)**). In this now widely used task, animals are first shown either a single sample stimulus or a series of to-be-remembered sample stimuli. After a variable delay interval, animals are presented with a sample stimulus and a novel stimulus. They must learn to choose the novel, or 'nonmatching,' stimulus to receive a reward. Mishkin's study showed convincingly that combined, but not separate, lesions of the hippocampus and amygdala produced a severe recognition memory deficit (**Figure 2(b)**). These results suggested that it was combined damage to

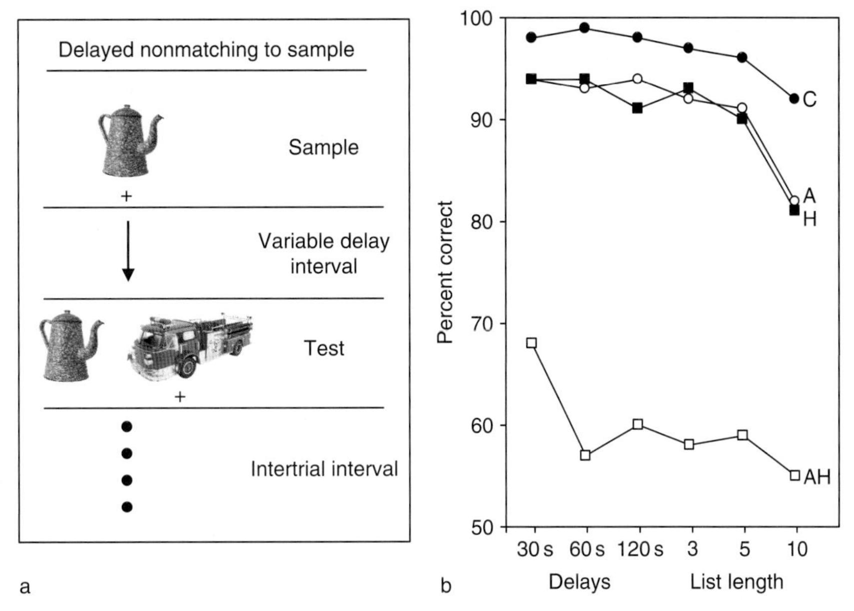


**Figure 2** (a) This panel shows a schematic diagram of the delayed nonmatching to sample task, a benchmark task used for studying recognition memory in monkeys. Animals are first presented with a rewarded (+) 'sample' stimulus over a central food well. After displacing the sample object and retrieving the reward, there is a variable delay interval (typically between 8 s and 10 min), after which time the sample stimulus and a novel stimulus are presented over two lateral food wells. The animals learn to choose the novel, or 'nonmatching,' test stimulus to receive a reward (+). After the intertrial interval, the sequence is repeated using a novel pair of objects. The objects used in this task are typically 'junk' three-dimensional objects, including small toys, containers, or other similarly sized objects. (b) Illustration of the performance of groups of monkeys with combined damage to the amygdala and hippocampus (AH lesion group), or separate amygdala (A group) or hippocampal (H group) lesions on the delayed nonmatching to sample task. Delay interval indicates the time delay between the sample and the test phase of the task. List length refers to a condition where lists of 3, 5, or 10 sample stimuli were presented before the test phase. This manipulation increases memory demand by increasing interference between the sample stimuli. Note the profound memory deficit of the AH group even on the shortest delay interval used. Reprinted by permission from Macmillan Publishers Ltd: *Nature* (Mishkin M (1978) Memory in monkeys severely impaired by combined but not by separate removal of amygdala and hippocampus. *Nature* 273: 297–298.), © 1978.

the hippocampus and amygdala that was responsible for H.M.'s severe memory deficit.

Other scientists replicated the severe memory impairment observed in monkeys following large medial temporal lobe lesions (**Figure 4(a)**). This lesion has been referred to by Zola-Morgan and his colleagues as the H+A+ lesion, where H refers to hippocampal damage and A to amygdala damage, with the + symbols indicating inadvertent damage to the surrounding cortex made to access the underlying hippocampus and amygdala (**Figure 3**). Additional studies showed that the memory impairment exhibited by the H+A+ group paralleled key features of human amnesia. Specifically, it was shown that the impairment extended to both the visual and the somatosensory modalities, that it was long lasting, and that the memory impairment was selective for only certain forms of memory while other forms of memory known to be spared in human amnesia were intact.

Findings from human neuropsychological studies reported at about the same time that the effects of H+A+ lesions being characterized showed that bilateral damage limited to the hippocampus resulted in significant memory impairment. However, the memory impairment was not as severe as that of patient H.M. To confirm these observations in human patients experimentally, Zola-Morgan and his colleagues tested a group of monkeys with bilateral lesions limited to the hippocampus and the adjacent cortex (H+ group) on the delayed nonmatching to sample task. The H+ lesion group exhibited significant memory impairment on this task, but similar to the observations in human patients, the deficit was not as severe as the deficit seen following large H+A+ lesions (**Figure 4(a)**). Note that this finding differed from Mishkin's report of no memory impairment following hippocampal lesions (H lesion in **Figure 2(b)**, which also included damage to the surrounding cortex), most likely because that study used the relatively less sensitive prelesion training strategy while the H+ group was tested using a more sensitive postlesion training strategy.

These findings suggested that additional damage sustained by the H+A+ group, but not the H+ group, must be underlying this more severe deficit. Based on the original report by Mishkin, the first

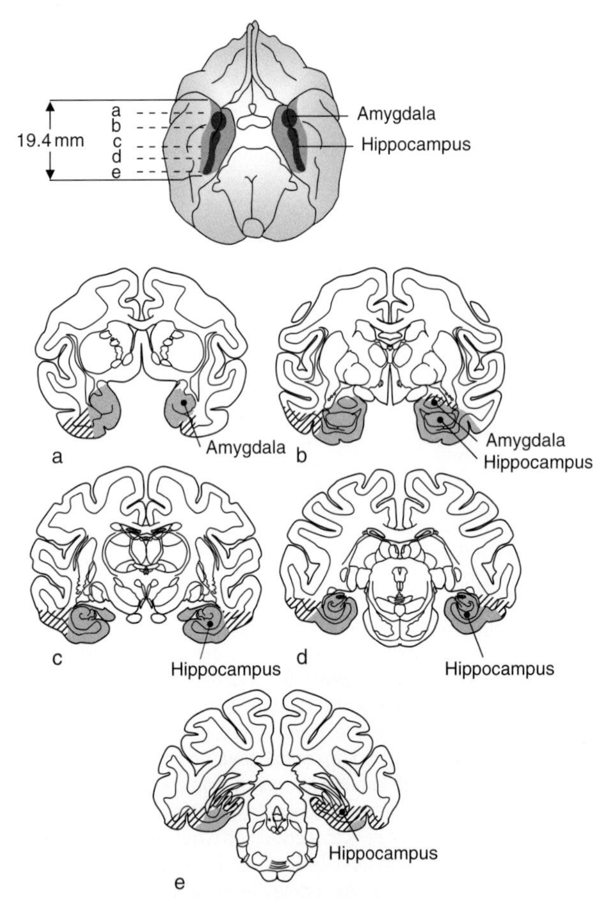


**Figure 3** Schematic illustration of the extent of damage sustained in Zola-Morgan's H+A+ lesion group. Shown on the top is a ventral view of a macaque monkey brain. Shaded areas indicate the approximate location of the intended lesion. Also shown are the relative locations of the amygdala and hippocampus, which are situated just below the cortex of the ventral medial temporal lobe. Drawings (a)–(e) show coronal sections through the lesion site, illustrating the largest (stippled) and smallest (gray area) extent of damage. Note the similarities between the lesion illustrated here and the illustration of H.M.'s presumed lesion in **Figure 1**. Also note the similarity between the intended lesion of the AH group of Mishkin (**Figure 2(b)**) and the H+A+ group of Zola-Morgan and colleagues. The major difference was simply the nomenclature used by the two groups. Adapted from Zola-Morgan S and Squire LR (1985) Medial temporal lesions in monkeys impair memory on a variety of tasks sensitive to human amnesia. *Behavioral Neuroscience* 99: 22–34.

obvious candidate was the additional damage to the amygdala. To test this idea, Zola-Morgan and colleagues developed techniques that would allow selective damage of the amygdala without involvement of the surrounding cortex. In contrast to the prediction from Mishkin's study, Zola-Morgan showed that animals with bilateral lesions limited to the amygdala performed completely normally on the delayed nonmatching to sample task (**Figure 4(b)**). Even more convincingly, it was shown that the addition of a selective amygdala lesion to a H+ lesion (H+A lesion group) did not exacerbate the memory deficit observed following H+ lesions alone. These findings showed definitively that the additional damage of the amygdala was, in fact, not contributing to the severe memory impairment of the H+A+ group. This left the possibility that the inadvertent damage to the surrounding cortex was underlying the more severe memory deficit of the H+A+ group. This possibility was particularly surprising because ever since the original description of H.M., attention had focused almost exclusively on the likely role of the hippocampus and amygdala in memory. Cortical damage was acknowledged (**Figures 1** and **3**), but generally considered unrelated to the memory impairment. Two additional pieces of evidence emerging at the time further strengthened the idea that the surrounding cortical areas may be important for declarative memory.

### The Entorhinal Cortex

The first piece of evidence came from anatomical studies examining the connections of the monkey entorhinal cortex, a brain area strongly associated with the hippocampus. In the early 1900s the famous Spanish neuroanatomist Ramón y Cajal first noted the massive projections from layers II and III of the entorhinal cortex to the dentate gyrus, part of the hippocampal region. Based on this strong anatomical projection, Cajal speculated that the functions of the entorhinal cortex and hippocampus would also be strongly linked. To emphasize this strong anatomical link, the term 'hippocampal formation' is often used to refer to the hippocampal region, including the dentate gyrus, hippocampus proper, and subicular complex, together with the entorhinal cortex.

**Boundaries and connections** In the monkey, the entorhinal cortex is situated medial to the rhinal sulcus on the ventromedial aspect of the anterior temporal lobe (**Figures 5** and **7**). Important early anatomical studies by Van Hoesen and colleagues described some of the cortical inputs to the monkey entorhinal cortex. However, the full complement of anatomical projections to this area was not known. To address this question, Insausti, Amaral, and colleagues analyzed a series of retrograde tracer injections made throughout the monkey entorhinal cortex. Retrograde tracers are transported from the axons that terminate in the injection site to the cell bodies of origin, where the tracer can be visualized. Plotting the locations of retrogradely labeled cell bodies resulting from a given injection site identifies all the brain areas that project to that site, while the number of labeled cells in each brain area provides an indication of the relative strength of that projection.

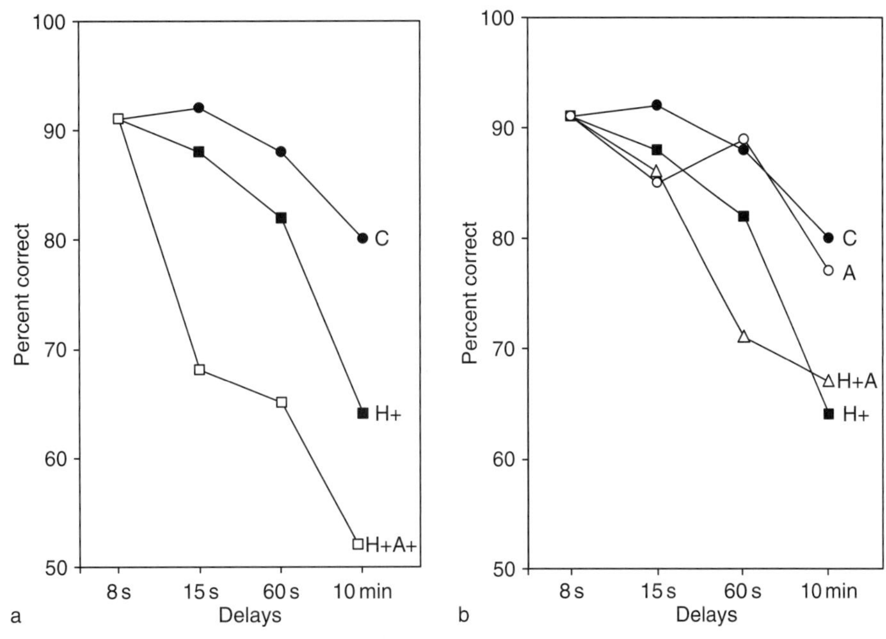


**Figure 4** (a) This graph illustrates the memory deficit exhibited by the H+ and H+A+ group on the delayed nonmatching to sample task. Note that the H+ group exhibited a significant but less severe memory impairment compared to the H+A+ group. For both groups, the most severe impairment is seen at the longest delay intervals. (b) This graph illustrates the effect of selective amygdala damage on the delayed nonmatching to sample task. This graph makes two key points: first, that bilateral lesions limited to the amygdala that do not involve the surrounding cortex (A group) do not cause memory impairment, and second, that the addition of an A lesion to an H+ group (H+A) does not exacerbate the memory deficit of the original H+ group. This study showed definitively that the amygdala was not contributing to the severe visual recognition memory deficit exhibited by the original H+A+ group. C, control group. Adapted from Zola-Morgan S, Squire LR, and Amaral DG (1989) Lesions of the amygdala that spare adjacent cortical regions do not impair memory or exacerbate the impairment following lesions of the hippocampal formation. *Journal of Neuroscience* 9: 1922–1936; and Zola-Morgan S, Squire LR, Amaral DG, and Suzuki WA (1989) Lesions of perirhinal and parahippocampal cortex that spare the amygdala and hippocampal formation produce severe memory impairment. *Journal of Neuroscience* 9: 4355–4370.

**Figure 5** illustrates the cortical connections of the monkey entorhinal cortex. A major finding was that the entorhinal cortex receives almost two-thirds of its direct cortical inputs from the surrounding perirhinal and parahippocampal cortices. Earlier anatomical studies suggested that these areas are higher-order association areas that receive convergent inputs from multiple sensory modalities. Other direct inputs to the entorhinal cortex include olfactory and somatosensory inputs from the piriform cortex and insula, respectively, spatial inputs from the retrosplenial cortex, auditory inputs from the superior temporal gyrus, and multimodal sensory inputs from the orbitofrontal cortex. The discovery that the major cortical inputs to the entorhinal cortex originate in the surrounding perirhinal and parahippocampal cortices suggested that these adjacent cortical areas were more intimately linked to the hippocampal formation than previously appreciated.

The second critical piece of evidence highlighting the possible role of the perirhinal and parahippocampal cortices in memory was a reexamination of the extent of the lesion in Zola-Morgan's original H+A+ group. This reexamination revealed that not only was the entorhinal cortex removed in this group, but also a much larger extent of both the perirhinal and the parahippocampal cortices was damaged than had originally been appreciated. These findings, together with the newly described connections of the entorhinal cortex, strongly suggested that the extra damage to the surrounding perirhinal and parahippocampal cortices was underlying the more severe memory impairment in the H+A+ group relative to the H+ group.

### The Perirhinal and Parahippocampal Cortices

**Lesion studies** To test the hypothesis that the perirhinal and parahippocampal cortices contribute to memory, Zola-Morgan and colleagues examined the effects of bilateral lesions limited to these areas that spared the adjacent and underlying entorhinal cortex, hippocampus, and amygdala (PRPH lesion group) on a battery of memory tasks, including the delayed nonmatching to sample task. Consistent with this hypothesis, PRPH lesions produced a severe recognition memory impairment that was similar in severity to the deficit exhibited by the original H+A+ group (**Figure 6**). Like human amnesia, the impairment of

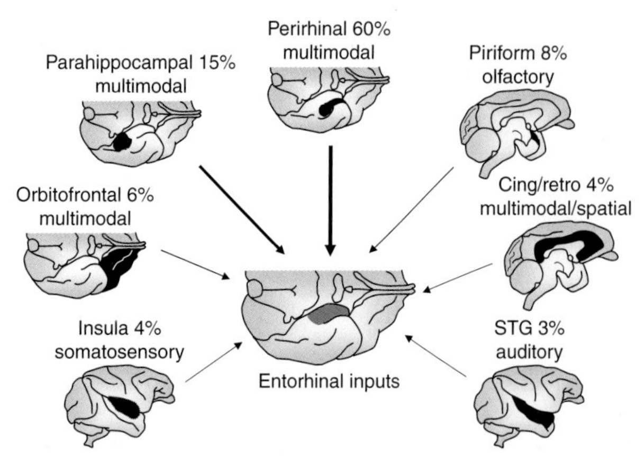


**Figure 5** Illustration of the relative strength of the cortical inputs to the monkey entorhinal cortex. These data were taken from one representative case of retrograde tracer injection in the entorhinal cortex (Case IM-6 of Insausti and colleagues). The entorhinal cortex receives nearly two-thirds of its direct cortical input from the adjacent perirhinal and parahippocampal cortices. Other inputs include olfactory inputs, somatosensory inputs as well as strong multimodal input from the orbitofrontal cortex, and spatial input from the retrosplenial cortex as well as auditory input from the superior temporal gyrus (STG). Reproduced from Suzuki WA (1996) Neuroanatomy of the monkey entorhinal, perirhinal and parahippocampal cortices: Organization of cortical inputs and interconnections with amygdala and striatum. *Seminars in the Neurosciences* 8: 3–12.

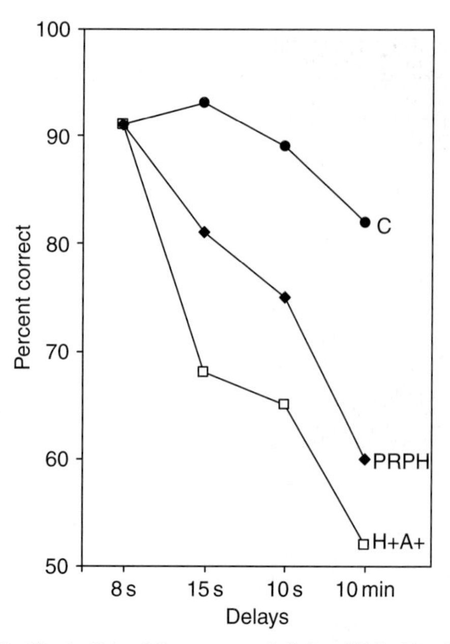


**Figure 6** Illustration of the severe deficit exhibited by the perirhinal and parahippocampal (PRPH) group on the delayed nonmatching to sample task. The performance of the H+A+ group is shown for comparison. Adapted from Zola-Morgan S, Squire LR, Amaral DG, and Suzuki WA (1989) Lesions of the perirhinal and parahippocampal cortex that spare the amygdala and hippocampal formation produce severe memory impairment. *Journal of Neuroscience* 9: 4355–4370.

the PRPH lesion group encompassed multiple sensory modalities and was long lasting. Findings from numerous other studies confirmed that damage including the perirhinal and parahippocampal cortices resulted in severe memory impairment. These lesion findings provided definitive evidence that the perirhinal and parahippocampal cortices contributed to memory. Subsequent neuroanatomical studies focused on these areas helped clarify how these areas contribute to memory.

**Boundaries and connections** A major factor that slowed progress in identifying the memory functions of the perirhinal and parahippocampal cortices was the fact that at the time, very little was known about these cortical medial temporal lobe areas. For example, both the boundaries and the nomenclature applied to these areas had been highly inconsistent, making it difficult to know where one area ended and the next one began. Amaral and colleagues performed a preliminary reevaluation of the boundaries of these areas as part of their description of the cortical inputs to the entorhinal cortex. The current boundaries of the perirhinal and parahippocampal cortex are informed by data from tract tracing studies, cytoarchitectonic analyses of the cell staining and density patterns, as well as chemoarchitectonic analyses of neuropeptide staining patterns. **Figure** 7 illustrates the boundaries of the perirhinal and parahippocampal cortices on a ventral view of the monkey brain as well as in coronal sections. By comparing the boundaries of the perirhinal and parahippocampal cortices shown in the coronal sections in **Figure** 7 to the boundaries of the H+A+ lesion shown in the coronal sections of **Figure** 3, one can gain an appreciation for how much of the perirhinal and parahippocampal cortex was damaged in the original H+A+ group.

What are the connections of the perirhinal and parahippocampal cortices? Early anatomical studies suggested that both the perirhinal and parahippocampal cortices received convergent input from areas involved in processing multiple sensory modalities. But like the entorhinal cortex, it was unclear if all of the inputs to the perirhinal and parahippocampal cortices had been identified. To address this question, a series of neuroanatomical studies were undertaken to define the full complement of cortical inputs to these areas. The results from a series of retrograde tracer injections made throughout the perirhinal and parahippocampal cortices revealed two major findings. First, these areas receive a massive convergence of input from widespread areas, including visual, somatosensory, and auditory association areas, as well as higher order multimodal association areas, including the frontal lobe and the dorsal bank of the

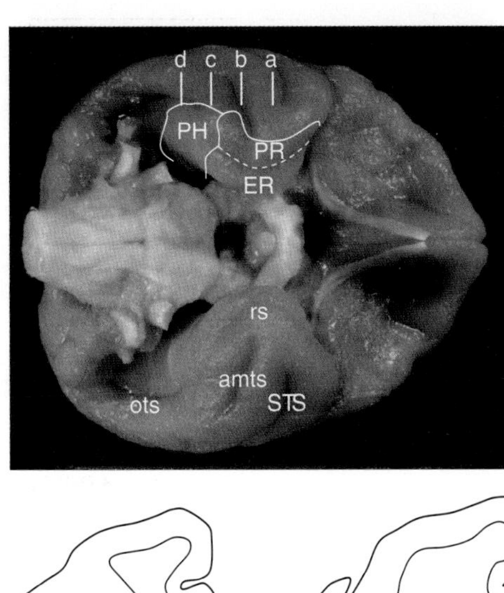


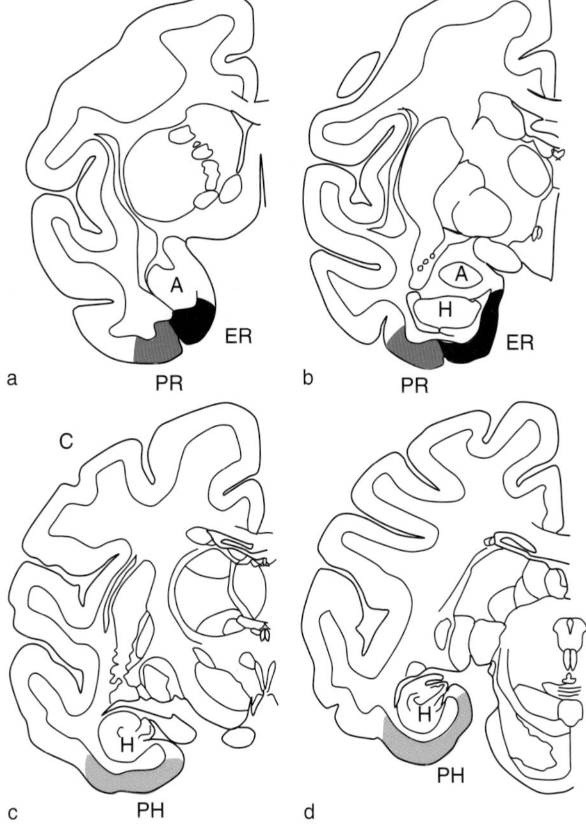


**Figure 7** (Top) Photograph of the ventral view of a macaque monkey brain, where the boundaries of the entorhinal (ER), perirhinal (PR), and parahippocampal (PH) cortices are shown on the right hemisphere and the major sulci are labeled on the left hemisphere. The dashed line on the right hemisphere shows the location of the rhinal sulcus, which forms the boundary between the entorhinal cortex medially and the perirhinal cortex more laterally. Drawings (a)–(d) illustrate individual coronal sections through the perirhinal and parahippocampal cortices. The locations of these coronal sections are indicated by vertical lines in the ventral view. Note that by comparing this figure with **Figure 3**, one can gain an appreciation of the extent of the perirhinal and parahippocampal cortices damaged in the H+A+ group. A, amygdala; amts, anterior middle temporal sulcus; H, hippocampus; ots, occiptotemporal sulcus; rs, rhinal sulcus; STS, superior temporal sulcus.

superior temporal sulcus. Second, the perirhinal and parahippocampal cortices receive distinctly different sets of cortical inputs, suggesting potentially different roles in memory.

The major input to the perirhinal cortex (**Figure 8**) arises from the adjacent visual area TE, a major component of the 'ventral visual pathway' important for processing visual object information. The second strongest input to the perirhinal cortex originates in the parahippocampal cortex, suggesting strong cross-talk between the two areas. Other inputs-include multimodal sensory input from the dorsal bank of the superior temporal sulcus and orbitofrontal cortex as well as somatosensory input from the insular cortex. These projections are largely reciprocal. The powerful inputs from visual area TE, taken together with the devastating effects of lesions, including the perirhinal cortex on the delayed nonmatching to sample task, suggest that perirhinal cortex is particularly important for visual object memory.

Like the perirhinal cortex, the parahippocampal cortex also receives strong visual input, but this input comes from more posterior visual areas TEO and V4 (**Figure 9**). In contrast to the perirhinal cortex, with its prominent inputs from the highest order visual area TE, the parahippocampal cortex receives more prominent input from the so-called dorsal visual processing pathway, important for processing visual–spatial information. These spatial inputs arise from the retrosplenial cortex, posterior parietal cortex, dorsal bank of the superior temporal gyrus, and dorsolateral regions of the prefrontal cortex. These findings suggest that the parahippocampal cortex may be particularly important for spatial memory.

## Neuroanatomy of Declarative Memory

**Figure 10** illustrates the connections of the key medial temporal lobe structures important for declarative memory. These areas include the hippocampus, together with the surrounding and strongly interconnected entorhinal, perirhinal, and parahippocampal cortices. The connections of these areas exhibit a hierarchical organization, with the hippocampus acting as the ultimate site of information convergence. The hippocampus receives its major input and provides it major output to the entorhinal cortex, which receives its major cortical inputs and outputs from the surrounding perirhinal and parahippocampal cortices. The latter two areas act as a gateway of sensory information flow between the hippocampal formation, on the one hand, and widespread areas of neocortex, on the other.

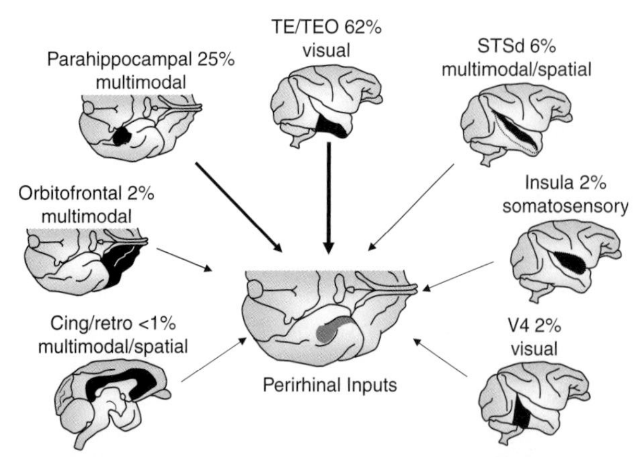


**Figure 8** Schematic illustration of the relative strength of the cortical inputs to the perirhinal cortex. These data were taken from one representative case of retrograde tracer injection in the perirhinal cortex (Case M3–90 of Suzuki and Amaral). The strongest input to the perirhinal cortex arises from visual area TE and TEO, important for processing visual object information. The parahippocampal cortex also provides a very strong input to the perirhinal cortex. Other inputs include somatosensory input from the insula as well as multimodal input from the dorsal bank of the superior temporal sulcus (STSd) and orbitofrontal cortex. Thus, while the perirhinal cortex is a site of multimodal sensory convergence, its major input originates in visual areas TE and TEO. Reproduced from Suzuki WA (1996) Neuroanatomy of the monkey entorhinal, perirhinal and parahippocampal cortices: Organization of cortical inputs and interconnections with amygdala and striatum. *Seminars in the Neurosciences* 8: 3–12.

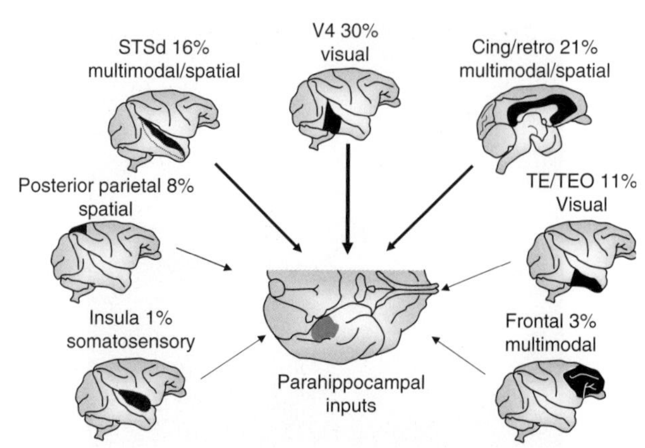


**Figure 9** Schematic illustration of the relative strength of the cortical inputs to the parahippocampal cortex. The percentages shown here are taken from one prototypical case of retrograde trace injection in the parahippocampal cortex (Case M2–90 of Suzuki and Amaral). Like the perirhinal cortex, the parahippocampal cortex also receives strong visual input, but this input arises from more posterior visual area V4. Relative to the perirhinal cortex, the parahippocampal cortex receives disproportionately more direct input from the so-called dorsal visual processing stream, important for processing spatial information. These inputs arise from the retrosplenial cortex, the dorsal bank of the superior temporal sulcus, the posterior parietal lobe, and the dorsolateral prefrontal cortex. Reproduced from Suzuki WA (1996) Neuroanatomy of the monkey entorhinal, perirhinal and parahippocampal cortices: Organization of cortical inputs and interconnections with amygdala and striatum. *Seminars in the Neurosciences* 8: 3–12.

Understanding the anatomical connections of these medial temporal lobe areas provides the foundation for elucidating the role of these structures in declarative memory. For example, these anatomical connections suggest that the severe memory impairment exhibited by the PRPH group is not due exclusively to damage of the perirhinal and parahippocampal cortices, but also to the fact that the PRPH lesion cuts off virtually all of the cortical interconnections of both the entorhinal cortex and the hippocampus (**Figure 10**). Importantly, numerous studies have confirmed that damage limited to the hippocampal formation both in humans and in monkeys also produce significant, though less severe, memory deficits than the deficits seen following damage to the surrounding cortex. The powerful and reciprocal interconnections between these medial temporal lobe structures suggest that functional interactions between these areas may also be important for memory. This idea is supported by findings showing that the extent of medial temporal lobe damage correlates with the severity of impairment on the delayed nonmatching to sample task. However, the differences in the connections of these areas, particularly between the perirhinal and parahippocampal cortex, suggest that these cortical areas may also participate in different forms of memory, with perirhinal cortex important for object memory and the parahippocampal cortex important for spatial memory. This idea is also supported by experimental findings in monkeys.

The lesion and anatomical studies in the monkey model system also make strong prediction with respect to the actual brain damage sustained by patient H.M. Specifically, it is clear that the impairment in patient H.M. is substantially more severe than the memory impairment in patients with damage limited to the hippocampal formation bilaterally. Findings from animal studies predict that that patient H.M.'s lesion involves the hippocampus together with the surrounding cortex. Unfortunately, for many years the precise extent of H.M.'s lesion could not be confirmed with structural magnetic resonance imaging (MRI) scans because it was thought that the surgical clips left in place after his original surgery were not compatible with imaging. However, it was later determined that the clips would not pose a danger to H.M., and in 1997 Corkin and colleagues brought the study of the neuroanatomy of declarative memory full circle when they published a historic MRI analysis of H.M.'s brain lesion. This study confirmed that H.M. sustained bilateral damage to the amygdala as well as approximately the anterior

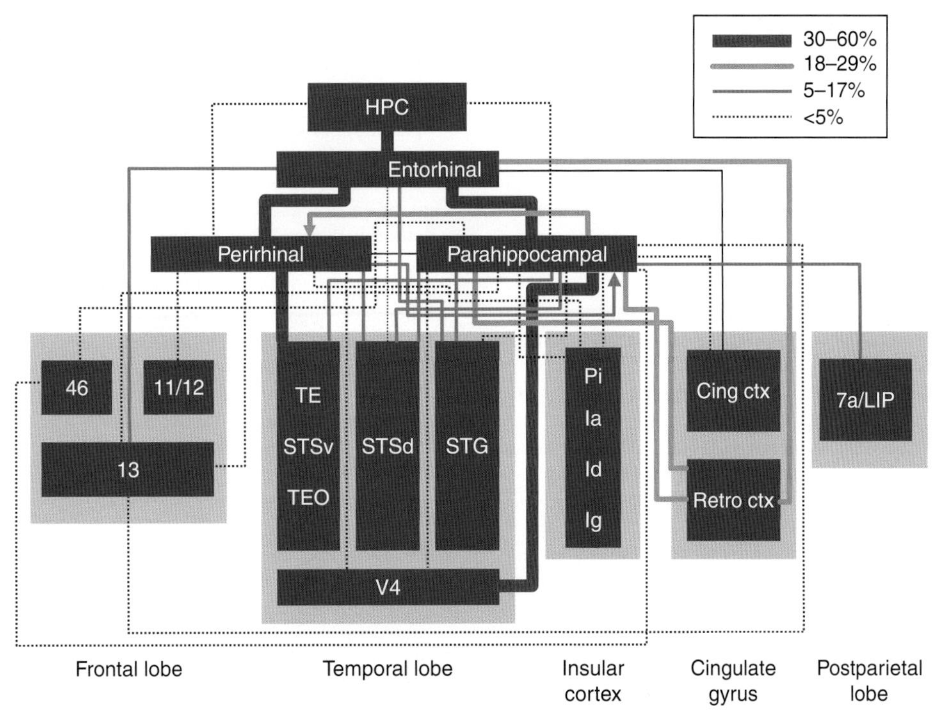


**Figure 10** Schematic illustration of the connections of the hippocampus and the entorhinal, perirhinal, and parahippocampal cortices. The relative strength of the connections estimated from retrograde tracer experiments is illustrated by the color and thickness of the interconnecting lines (see key in upper right). Note the hierarchical organization, with the hippocampus acting as the pinnacle of the hierarchy. The hippocampus receives its major inputs and outputs from the entorhinal cortex, which receives its major inputs from the perirhinal and parahippocampal cortices. These latter two cortical areas act as the major gateway of sensory information flow between widespread areas of the neocortical mantel, on the one hand, and the hippocampal formation, on the other. Note the differences in the patterns of inputs between the perirhinal and parahippocampal cortices. These anatomical connections provide a framework for understanding the mnemonic functions of the medial temporal lobe. HPC, hippocampus; Ia, agranular portion of the insula; Id, dysgranular portion of the insula; Ig, granular portion of the insula; LIP, lateral intraparietal area; STG, superior temporal gyrus; STSd, dorsal bank of the superior temporal sulcus. Reproduced from Suzuki WA and Amaral DG (1994) Perirhinal and parahippocampal cortices of the macaque monkey: Cortical afferents. *Journal of Comparative Neurology* 350: 497–533.

half of the hippocampus. Moreover, consistent with the predictions from the animal studies, the MRI scans confirmed that the resection also involved all of the entorhinal cortex and parts of the perirhinal cortex bilaterally, though the parahippocampal cortex was spared. These findings, taken together with studies of other well-characterized amnesic patients with MRI or histological confirmation of the medial temporal lobe damage, support the idea that results from experimental animal studies are directly applicable to humans. These parallel findings reinforce the value of animal models in elucidating the principles of memory organization in the brain.

The discussion here has emphasized the lesion and anatomy studies in monkeys. However, important studies in the rodent model system have also provided fundamental contributions to our understanding of the medial temporal lobe areas important for declarative memory. Taken together, these experimental studies have not only identified the structures important for declarative memory, but have also laid the groundwork for the next generation of questions concerning how each of these medial temporal lobe areas contribute and/or interact to acquire new declarative memories. These studies also form the foundation for clinical strategies to treat various neurological conditions wherein declarative memory is impaired, including both normal human aging and Alzheimer's disease. Progress in the basic research of the brain basis of declarative memory will be critical in addressing these pressing clinical issues.

*See also:* Episodic Memory.

## Further Reading

Amaral DG, Insausti R, and Cowan WM (1987) The entorhinal cortex of the monkey: I. Cytoarchitectonic organization. *Journal of Comparative Neurology* 264: 326–355.

Corkin S, Amaral DG, GilbertoGonzalez R, et al. (1997) H.M.'s medial temporal lobe lesion: Findings from magnetic resonance imaging. *Journal of Neuroscience* 17: 3964–3979.

Insausti R, Amaral DG, and Cowan WM (1987) The entorhinal cortex of the monkey: II. Cortical afferents. *Journal of Comparative Neurology* 264: 356–395.

Lavenex P and Amaral DG (2000) Hippocampal-neocortical interaction: A hierarchy of associativity. *Hippocampus* 10: 420–430.

Mishkin M (1978) Memory in monkeys severely impaired by combined but not by separate removal of amygdala and hippocampus. *Nature* 273: 297–298.

Mishkin M and Murray EA (1994) Stimulus recognition. *Current Opinion in Neurobiology* 4: 200–206.

Scoville WB and Milner B (1957) Loss of recent memory after bilateral hippocampal lesions. *Journal of Neurology, Neurosurgery & Psychiatry* 20: 11–21.

Squire LR and Zola-Morgan S (1991) The medial temporal lobe memory system. *Science* 253: 1380–1386.

Stefanacci L, Buffalo EA, Schmolck H, et al. (2000) Profound amnesia after damage to the medial temporal lobe: A neuroanatomical and neuropsychological profile of patient E.P. *Journal of Neuroscience* 20: 7024–7036.

Suzuki WA (1996) Neuroanatomy of the monkey entorhinal, perirhinal and parahippocampal cortices: Organization of cortical inputs and interconnections with amygdala and striatum. *Seminars in the Neurosciences* 8: 3–12.

Suzuki WA and Amaral DG (1994) Perirhinal and parahippocampal cortices of the macaque monkey: Cortical afferents. *Journal of Comparative Neurology* 350: 497–533.

Suzuki WA and Amaral DG (2003) Where are the perirhinal and parahippocampal cortices? A historical overview of the nomenclature and boundaries applied to the primate medial temporal lobe. *Neuroscience* 120: 893–906.

VanHoesen GW and Pandya DN (1975) Some connections of the entorhinal (area 28) and perirhinal (area 35) cortices of the rhesus monkey. III. Efferent connections. *Brain Research* 95: 48–67.

Zola-Morgan S and Squire LR (1985) Medial temporal lesions in monkeys impair memory on a variety of tasks sensitive to human amnesia. *Behavioral Neuroscience* 99: 22–34.

Zola-Morgan S and Squire LR (1993) Neuroanatomy of memory. *Annual Review of Neuroscience* 16: 547–563.

Zola-Morgan S, Squire LR, Amaral DG, and Suzuki WA (1989) Lesions of the amygdala that spare adjacent cortical regions do not impair memory or exacerbate the impairment following lesions of the hippocampal formation. *Journal of Neuroscience* 9: 1922–1936.

Zola-Morgan S, Squire LR, Amaral DG, and Suzuki WA (1989) Lesions of perirhinal and parahippocampal cortex that spare the amygdala and hippocampal formation produce severe memory impairment. *Journal of Neuroscience* 9: 4355–4370.

# Episodic Memory

**M E Wheeler and E J Ploran**, University of Pittsburgh, Pittsburgh, PA, USA



Of all human capabilities, the one central to our development of a unique sense of personal identity is episodic memory. The ability to form and retrieve memories of our life experiences provides an opportunity to identify those events as our own and embed the memories in a malleable framework of related experiences. This process in turn influences our interpretation, and therefore learning, of new experiences. Indeed, people who have lost the ability to access their past episodic memories have sometimes reported feeling a loss of self identity. Over the past 50 years, significant advances in the philosophical, theoretical, and empirical study of memory have begun to shed light on the nature of memory and the intricate cognitive and neural mechanisms involved in remembering past episodes.

## What Is Episodic Memory?

Episodic memories are consciously perceived memories of personal events or episodes. One defining feature of episodic memories is that they are the personal memories. Among other types of information, they include where you were, what you were doing, whom you were with, what you saw, how you felt, and what you were thinking during a particular event. A second defining feature is that episodic memories are accompanied by a conscious awareness of the temporal context in which the events occurred. In bringing to mind events from childhood, there is an awareness that the events occurred before the present time, and perhaps an awareness of precisely when they occurred.

Episodic memories represent one type of declarative memory – memories which are accessible to consciousness and which can be readily communicated or declared to others. Nondeclarative memories, such as learned skills, are memories that are not directly available to consciousness but nevertheless influence behavior. Semantic memories are memories of factual knowledge and represent a second type of declarative memory. Declarative memory is dependent upon the integrity of medial temporal lobe (MTL) structures including hippocampus, entorhinal and perirhinal cortex, and parahippocampal gyrus, and also on the fornix and anterior and mediodorsal nuclei in the thalamus. Significant damage to these structures produces amnesia and consequently an impaired ability to form and retrieve declarative memories. By some accounts, episodic and semantic memories arise from different memory, and by extension, brain systems. These accounts are supported in part by behavioral dissociations and in part by evidence that some patients are impaired on tests of episodic but not semantic memory while other patients are impaired on tests of semantic but not episodic memory. Further support for different memory systems has come from neuroimaging studies which demonstrate dissociations among brain regions associated with episodic and semantic memory tasks. By other accounts, episodic and semantic memories differ according to the presence or absence of personal spatial and temporal context, but are otherwise similar. When the initial context has been lost or is inaccessible, or the current situation does not encourage conscious recollection, the result is retrieval of semantic knowledge. By this account, the difference is due to the different processing demands of a given situation rather than differences in the underlying memory systems. These accounts are not necessarily mutually exclusive.

A distinction is often drawn between memories of personal events accompanied by subjective awareness, and memories of personal facts related to those events. For example, you may remember purchasing your first car, bringing back to mind where you found it, how it appeared, and how the seller talked. However, some of these remembered details can also be retrieved as facts. You know that you bought your car from an individual rather than a dealer. You know that it was blue. And you know that the seller was a man. Another distinction is sometimes made between episodic memories pertaining to a single event (recollections) and those pertaining to repeated similar episodes (generic personal memories). Autobiographical memory is often used synonymously with episodic memory. However, in some instances, the two terms are used to denote memories of life experiences and laboratory events, respectively.

## Reconstructing Memories Using Retrieval Cues

Prevailing frameworks of episodic memory describe remembering as a dynamic process during which an active model or simulation of an event is constructed from memory traces. A memory trace, also called an engram, is a neural change reflecting the thoughts, emotions, and sensory features of the original experience. The construction account of remembering rejects the proposition that memory traces are replayed like a song or a movie. Instead, memory cues are used to create a mental simulation of an event by drawing upon available memories and filling in the missing pieces with educated guesses. Ulric Neisser compares this process to that of assembling a dinosaur from mere bone chips. The reconstruction process begins with a retrieval cue, and subsequent retrieval is influenced by an individual's past experiences, his or her current state of mind, and the contextual demands in which remembering occurs. For example, accessing the same memory trace on two different occasions may produce qualitatively different recollections if the first occasion is immediately following a transcontinental flight and the second is after a full night of rest. The German biologist Richard Semon coined the term ecphory to describe this synergistic interaction of the original memory trace and the current retrieval cue to form the episodic memory experience. Because of an inherent dependence on the retrieval cue, memories entering consciousness need not be veridical representations of the mnemonic content of a stored engram. By this view, only the products of the interpretive interaction of current intent and past experience are available to awareness.

The effectiveness of a retrieval cue depends in large part upon the degree of similarity between the cue and the original event. The more similar the relationship, the more likely it is that the retrieval cue will lead to ecphory. Therefore, the way in which events are interpreted at encoding influences how they can later be accessed and reconstructed. Endel Tulving refers to this relationship as the encoding specificity principle. Daniel Schacter provides a memorable example of this principle by describing an experiment carried out by the artist Sophie Calle at The Museum of Modern Art. In the experiment, museum staff were asked to recollect as much detail as they could about various paintings. The level of detail provided by the staff differed widely, from just a few words to detailed sentences. Calle found that the degree and quality of detail was related to the type of work carried out by the staff member. Staff in charge of art placement provided structural descriptions of the paintings, while those in charge of procuring the art provided more detailed accounts and interpretations of the artwork. The differing experiences of the museum staff influenced how they thought about the artwork, and this in turn influenced the types of memory they were later able to retrieve. According to the encoding specificity principle, the cues that will best evoke episodic memories of the paintings will be those that most closely match the initial processing.

## Components of Remembering

Episodic remembering is typically perceived as a unitary phenomenon. A memory comes to mind, sometimes easily and sometimes after some effort, but the details of the process are often unavailable to superficial introspection. There is an abundance of research demonstrating, however, that remembering depends on a variety of memory-specific and domain-general cognitive abilities, including among others, formation of search strategies, memory-based decision making, and postretrieval assessment. The most convincing evidence for separable components in human episodic memory has come from studies that produce a dissociation, either by identifying patients in whom lesions to different parts of the brain produce different types of memory deficit, or by identifying brain areas in which neural activity modulates according to variations in one component differently than the other components.

### Strategic Processing

At times, memories of past events come to mind rather easily, seeming to pop into awareness after an encounter with a retrieval cue. At other times, a feeling of knowing can lead to a more strategic search through memory for the appropriate information. Strategic search involves generating retrieval cues from memory and using those cues to guide further searches. Models of strategic processing in remembering typically describe three basic stages, including initial identification of the desired information (target), a search for information and a matching process in which retrieved information is compared with the target, and an assessment or monitoring of the degree of accuracy. Neuropsychological studies of patients with lesions in the prefrontal cortex (PFC) have provided valuable insight into the role of frontal cortex in memory. Frontal patients are readily able to encode and retrieve memories. In fact, they tend to perform at levels approaching that of control subjects on simple item recognition tests that can be accomplished using nonrecollective memory (i.e., familiarity). However, patients with prefrontal damage perform significantly more poorly than control subjects on memory tests that require strategic processing, such as source memory tests of cued recall, free recall, and recognition. Source memory tests require the ability to identify the appropriate learning context, such as the time (was this item studied in the first or second study session?), place (was this item studied in the laboratory or in the classroom?), or feature (was this item presented in a male or female voice?). In these frontal patients, memory deficits are due not to problems forming memories (as in amnesia) but instead are related to impairments of organization of new memories, strategic access to memory contexts, and the supervisory control required to maintain information in working memory and assess retrieval effectiveness. As a result, frontal patients can experience difficulty under situations that require flexible processing, such those required by source memory tasks. This difficulty is sometimes observed in the form of confabulations – the generation of false and contradictory statements.

Results from neuroimaging studies are consistent with the neuropsychological findings. Because the need for strategic search typically increases as retrieval difficulty increases, retrieval difficulty can be manipulated in order to examine the neural substrates of strategic processing in memory. For example, difficulty can be manipulated by varying encoding strength so that weakly encoded items are more difficult to retrieve than strongly encoded items, or by varying the amount of detail required so that retrieving more detail is more difficult. These studies demonstrate that frontal lobe regions near the anterior extent of the left inferior frontal gyrus (aIFG), left anterior prefrontal cortex (aPFC), and bilateral posterior middle frontal gyrus (pMFG) consistently increase activity as retrieval difficulty increases. There is also evidence for functional specialization within this set of regions. pMFG activity tends to also modulate according to the type of information retrieved, with a preference toward processing of verbal stimuli in the left hemisphere and nonverbal stimuli in the right. Other brain regions also appear to play an important role in strategic processing. Bilateral anterior insula/frontal operculum (AI/FO), the anterior cingulate cortex (ACC), and basal ganglia (bilateral caudate nucleus, putamen) also tend to increase activity as retrieval difficulty increases. While there has been relatively little attention given to the role of AI/FO and basal ganglia in these studies, the ACC has received quite a bit of attention. Current models suggest that ACC is involved in reward-based decisions, motor control, error detection, emotional self-control, problem solving, and/or conflict monitoring. The demand for all of these functions conceivably increases as retrieval difficulty increases, and therefore often it is a particular challenge to clearly identify the source of signal modulations.

### Contents of Remembering

There is a host of evidence from neuropsychological, single-unit physiology, event-related potential (ERP), and neuroimaging studies supporting the view that regions of the nervous system that initially process an event (i.e., data-processing regions) are reactivated when the event is later remembered. For example, recalling your car's seller involves the coordinated recruitment of regions of the brain that initially processed the appearance of his face, the color of the car, the sound of his voice, and your thoughts about the situation. Work on human and nonhuman primates is particularly relevant to episodic remembering and mental imagery, but extends also to semantic knowledge. The findings indicate that memory traces are distributed in the nervous system as a function of the type of material encoded. For example, patients with focal lesions can exhibit impaired memory for visual objects, yet show preserved memory for spatial relationships. Neuroimaging experiments have shown that data-processing regions that are active during initial encoding of stimuli are, to some extent, reactivated during later remembering. It is yet to be determined how such a coordinated reactivation is organized and controlled, but prominent theories propose an interaction of frontoparietal attentional control processes and a medial temporal associational mechanism that provides an index of stored memory traces.

An unresolved issue concerning how memory contents are reconstructed is related to the degree to which data-processing centers are reactivated. In the visual system, for example, the most basic aspects of visual stimuli are processed in early stages, with complexity and receptive field size increasing as information is transmitted to later stages. Thus, while fundamental stimulus features such as line segments located in discrete parts of the visual field are processed in early stages of the visual processing hierarchy, more complex features that are less dependent upon location in the visual field, such as object form, are processed in later stages. Evidence that visual memories are reconstructed from the bottom up arises from studies that show involvement of early and late regions during visual retrieval. By this view, remembering the appearance of your car involves reconstructing the visual image from basic featural components up to the whole object. Other studies, however, indicate that visual memories are reconstructed by recruiting the levels most specialized to process the desired information. That is, remembering the general appearance of your car is accomplished by recruitment of higher-level object-processing regions, while remembering the color is accomplished by recruitment of color-processing regions.

### Retrieval Success (Ecphory)

Episodic memory retrieval is associated with a conscious awareness of reexperiencing an event that occurred in the past. As noted previously, it has been convincingly demonstrated that the MTL plays a critical role in retrieval success. Recent studies using ERP and neuroimaging techniques have identified a set of neo- and subcortical areas that also play a role in retrieval success. ERP studies of recognition memory were the first to note a distinct set of frontal and parietal signals related to the successful identification of previously studied items. This old/new, or retrieval-success effect, has been further studied in studies using ERP, positron emission tomography (PET), and functional magnetic resonance imaging (fMRI). The ubiquitous finding across studies is that recognizing old versus new items modulates activity in left PFC, as well as lateral and medial posterior parietal cortex. In PET and fMRI studies, frontal activations have been identified in bilateral dorsolateral PFC, medial PFC, and frontal polar cortex. Parietal activations have been consistently found in the anterior intraparietal sulcus and lateral parietal cortex near the supramarginal gyrus. While the findings are significant because they guide future research, the functional role of each of these regions has not been elucidated by the old/new manipulation.

In dual-process accounts of memory, recognition can arise from the fundamentally different memory processes of recollection and familiarity. Recollection is the ability to retrieve episodic memories, whereas familiarity is a strength-based signal that information is old. Familiarity is not supported by recollective evidence. Source memory tests, and the remember/know test developed by Endel Tulving, have been used to successfully dissociate these processes and further characterize the role of retrieval-success regions. Importantly, activity in the hippocampus has been consistently tied to memory decisions based on recollection, but not to those based on familiarity/knowing in the absence of recollection. In contrast, a related region of MTL near the rhinal cortex has been associated more with familiarity-based decisions. The neuroimaging results are consistent with an extensive animal literature that supports a role of hippocampus in recollection, and perirhinal cortex in nonrecollective recognition decisions. However, as there is also evidence that patients with lesions restricted to hippocampus proper are impaired on recognition memory, it is not currently known precisely how different MTL structures contribute to memory.

Somewhat surprisingly, ERP and fMRI studies suggest that parietal regions involved in old/new decisions are differentially related to whether those decisions were based on recollection or familiarity/knowing. ERP studies tend to indicate that parietal signals are related to recollection rather than familiarity. This general finding is supported by fMRI such that lateral parietal cortex near the supramarginal gyrus has been shown to respond preferentially to decisions based on recollection. The supramarginal regions have also been reported in studies evaluating temporal judgments and temporal orienting of attention, suggesting that they could be involved in providing temporal context during remembering, allowing us to mentally travel through time. Functional MRI studies also suggest that slightly more medial parietal cortex, near the anterior intraparietal sulcus (particularly in the left hemisphere), is similarly responsive to decisions based on recollection and familiarity. Interestingly, the literature on patients with posterior parietal lesions offers no supporting evidence for parietal involvement in long-term memory processes. Instead, parietal lesions tend to disrupt visuospatial and visuomotor abilities. There is evidence, however, that lesions affecting the angular gyrus in the left hemisphere impair short-term memory. In addition, studies of epileptic patients indicate that seizures originating in posterior parietal cortex can be associated with déjà vu or a disturbance in consciousness. This observation is broadly consistent with neuroimaging experiments that have found posterior parietal activity related to temporal attention. Thus, posterior parietal cortex may play a role in establishing temporal context during remembering.

## Retrieval Mode

What would life be like if every stimulus evoked a memory of a past experience? As noted previously, ecphory occurs when there is an interaction between a retrieval cue and a stored memory trace. A retrieval cue can be any stimulus in the internal or external environment, including a word spoken by a companion, the sight of a flower, or a thought about an aching joint. Most stimuli we encounter, however, do not evoke episodic memories. In fact, because there would be little opportunity to adequately process and interpret relevant stimuli, it could be quite detrimental to basic survival if every stimulus cued an episodic memory.

At the same time, it is possible to allow stimuli to serve as cues to remember past events. One could freely associate by allowing each thought entering awareness to serve as a cue to remember a past experience. Or one could look at items in the room and think about how they were obtained. In both cases, the thoughts and objects do not typically evoke episodic memories. However, this simple suggestion could have suddenly turned an object on your desk into a cue that sparked a memory about where you purchased it. The all-or-none state in which stimulus cues serve to evoke memories is known as retrieval mode.

Studies using neuroimaging, ERPs, and neuropsychological techniques suggest that the most anterior portions of the PFC are important for engaging in retrieval mode. A large number of neuroimaging studies using PET have consistently found right aPFC activity during episodic memory tests. aPFC was not found to modulate depending upon whether retrieval is successful or unsuccessful, indicating that it is not important for retrieving individual items from memory. Recent studies using fMRI and ERP add some insight into the role of aPFC by showing that right aPFC is active in a sustained manner, throughout the testing session and independently of the transient trial responses. aPFC could therefore be important for maintaining task-level rather than trial-level processes, playing a role in maintaining retrieval mode.

## Associative Nature of Memory

Some theories of episodic memory consider the connections between the facts of the memory and the context in which they were encoded (i.e., time, place, emotional state), as well as the connections between items presented simultaneously. Since we are constantly presented with many items at once, it may not be the case that each memory trace is completely separate from the others; it is possible that each memory item can be related to many other items for organizational and efficiency reasons. Imagine if each chapter in this encyclopedia had to describe each piece of the brain it references and how certain processes work. This would add length and complexity to every chapter. Instead, the use of linking each chapter to related items allows each chapter to describe a specific piece while the reader can reference related pieces as necessary to form a whole. In this way, each related memory item can potentially serve as a recollective cue for the others in the organized group. While this can be true for the network of declarative facts alone, it can also connect declarative facts to the contextual items that allow for episodic memory. For instance, in the first car example given above, perhaps you remember that the salesperson was nervous because he had a very clammy handshake. This would be an example of the connection between a declarative fact (the seller was nervous) and a contextual fact (your disgust at his handshake) to form an episodic memory.

Several computational models attempt to use an associative system to represent the process of memory. One example is Lynne Reder's source activation confusion (SAC) model of memory, which suggests that people create different nodes of information at the time of encoding. One type of node contains the basic semantic knowledge of the item, while another type contains the contextual elements of the encoding episode. A memory within the SAC model can have many nodes and associations between these nodes, as well as common nodes that are associated with many memories (e.g., all the words studied during in an experiment might be connected to a context node encompassing the experiment room, computer, lighting, etc). A person experiences an episodic recollection when not only the item node is reinstated during recall, but also nodes containing information about the context. It is the activation of the association between the item and context nodes that makes the difference between remembering an event episodically versus simply a feeling of familiarity. It has also been suggested that a lack of association, or binding, of elements into memory can cause a lack of episodic experience at a later time. Other models that focus on the interconnected nature of memory include the search of associative memory (SAM) model and versions based on parallel distributed processing (PDP).

Neuroimaging studies have sought to investigate the neural mechanisms behind the creation of associations among items to be remembered. These studies involve presentation of multiple stimuli at one time and participants either study the items singly or as a group. The hypothesis is that studying each item singly does not prompt the creation of associations with the other items, while studying the items as a group does. These studies have shown that the hippocampus and parahippocampal gyrus (particularly on the right) are more active during associative encoding than single item encoding. Other neuroimaging studies have focused on the encoding and retrieval of contextual information about the remembered items. A variety of paradigms can be used to test recollection of contextual information, and often include a question at test regarding the temporal order of items, list membership, or color/font/spatial location at study. These studies have also found activation in the hippocampus and parahippocampal gyrus. Other areas of interest include bilateral dorsolateral PFC and parts of parietal cortex, which may be involved in the more specific contextual memories of spatial locations.

*See also:* Declarative Memory System: Anatomy.

## Further Reading

Aggleton J and Brown M (1999) Episodic memory, amnesia, and the hippocampal–anterior thalamic axis. *Behavioral and Brain Sciences* 22: 425–489.

Brewer W (1996) What is recollective memory? In: Rubin D (ed.) *Remembering Our Past: Studies in Autobiographical Memory*, pp. 19–66. New York: Cambridge University Press.
Buckner RL (2003) Functional–anatomic correlates of control processes in memory. *Journal of Neuroscience* 23: 3999–4004.
Buckner RL and Wheeler ME (2001) The cognitive neuroscience of remembering. *Nature Reviews Neuroscience* 2: 624–634.
Burgess PW and Shallice T (1996) Confabulation and the control of recollection. *Memory* 4: 359–411.
Moscovitch M, Rosenbaum RS, Gilboa A, et al. (2005) Functional neuroanatomy of remote episodic, semantic and spatial memory: A unified account based on multiple trace theory. *Journal of Anatomy* 207: 35–66.
Schacter DL (1996) *Searching for Memory.* New York: BasicBooks.
Shimamura AP (2002) Memory retrieval and executive control processes. In: Stuss D and Knight R (eds.) *Principles of Frontal Lobe Function*, pp. 210–220. Oxford: Oxford University Press.
Squire L (1987) *Memory and Brain.* New York: Oxford University Press.
Squire L (1992) Memory and the hippocampus: A synthesis from findings with rats, monkeys, and humans. *Psychological Review* 99: 195–231.
Tulving E (1983) *Elements of Episodic Memory.* Oxford: Clarendon Press.
Tulving E and Craik F (eds.) (2000) *The Oxford Handbook of Memory.* Oxford: Oxford University Press.
Velanova K, Jacoby LL, Wheeler ME, McAvoy MP, Petersen SE, and Buckner RL (2003) Functional–anatomic correlates of sustained and transient processing components engaged during controlled retrieval. *Journal of Neuroscience* 23: 8460–8470.
Wheeler MA, Stuss DT, and Tulving E (1997) Toward a theory of episodic memory: The frontal lobes and autonoetic consciousness. *Psychological Bulletin* 121: 331–354.
Wheeler ME, Petersen SE, and Buckner RL (2000) Memory's echo: Vivid remembering reactivates sensory-specific cortex. *Proceedings of the National Academy of Sciences of the United States of America* 97: 11125–11129.

# Semantic Memory

**A Martin**, National Institute of Mental Health, Bethesda, MD, USA

Published by Elsevier Ltd.

## Introduction

'Semantic memory' refers to a major division of long-term memory that includes knowledge of facts, events, ideas, and concepts. As such, it covers a vast terrain, ranging from information about historical and scientific facts, details of public events, and mathematical equations to the information that allows us to identify objects and understand the meaning of words. Semantic memory can be distinguished from episodic/autobiographical memory by an absence of temporal and spatial details about the context of learning. In relation to episodic memory, semantic memory is considered to be both a phylogenically and an ontologically older system. Although many animals, especially mammals and birds, acquire information about the world, they are assumed to lack the neural machinery to consciously recollect detailed episodes of their past. Finally, although retrieval of semantic memory often requires explicit, conscious mediation, the organization of semantic memory can also be revealed via implicit tasks such as semantic priming.

## Semantic Memory and the Medial Temporal Lobe Memory System

Studies of patients with amnesia due to damage to the medial temporal lobes have established three broadly agreed on facts about the functional neuroanatomy of semantic memory. First, like episodic memory, acquisition of semantic memories is dependent on medial temporal lobe structures, including the hippocampal region (CA fields, dentate gyrus, and subiculum) and surrounding neocortex (parahippocampal, entorhinal, and perirhinal cortices). Damage to these structures results in deficient acquisition of new facts and public events and the extent of this deficit is roughly equivalent to the deficit for acquiring personal information about day-to-day occurrences. However, despite broad agreement that acquiring semantic memories requires medial temporal lobe structures, there is disagreement concerning the role of the hippocampal region. One position, championed by Larry Squire and colleagues, holds that the hippocampus is necessary for acquiring semantic information. In contrast, others have argued that acquisition of semantic memories can be accomplished by the surrounding neocortical structures alone; participation of the hippocampus is not necessary. Recent studies have favored the hippocampal position by showing that carefully selected patients with damage limited to the hippocampus are impaired in learning semantic information, and that the impairment is equivalent to their episodic memory deficit. One potentially important caveat to this claim comes from studies of individuals who have sustained damage to the hippocampus at birth or during early childhood. These cases of developmental amnesia have disproportionately better semantic than episodic memories, suggesting that the hippocampal region may not be necessary for acquiring semantic information. Reconciliation of this issue will depend on direct comparison of adult onset and developmental amnesias with regard to extent of medial temporal lobe damage and its behavioral consequences.

The second major finding established by studies of amnesic patients is that the medial temporal lobe structures have a time-limited role in the retrieval of semantic memories. Retrieval of information about public events shows a temporally graded pattern with increasing accuracy for events further in time from the onset of the amnesia. Conceptual information about the meaning of objects and words acquired many years prior to the amnesia onset remains intact as assessed by both explicit and implicit tasks. Third, semantic memories of all types are stored in the cerebral cortex. In amnesic patients, impairments in semantic memory for information acquired prior to amnesia onset are directly related to the extent of damage to cortex outside the medial temporal lobes.

## Cortical Lesions and the Breakdown of Semantic Memory

Studies of semantic memory in amnesia have concentrated largely on measures of public event knowledge. The reason for this is that these tasks allow memory performance to be assessed for events known to have occurred either prior to, or after, amnesia onset. These measures also allow performance to be evaluated for events that occurred at different times prior to amnesia onset to determine if the memory impairment shows a temporal gradient – a critical issue for evaluating theories of memory consolidation. Because these patients have either no or, more commonly, limited damage to regions outside the medial temporal lobes, they are not informative about how semantic information is organized in the cerebral cortex. To address this issue, investigators have turned to patients with relatively focal lesions compromising different cortical areas. In contrast to the studies of amnesic patients, these studies have focused predominantly on measures designed to probe knowledge of object concepts.

### Object Concepts

An object concept refers to the representation (i.e., information stored in memory) of an object category (a class of objects in the external world). Concepts are central to all aspects of cognition; they are the glue that holds cognition together. The primary function of a concept is to allow us to quickly draw inferences about an object's properties. That is, identifying an object as, for example, a 'hammer' means that we know that this is an object used to pound nails, so that we do not have to rediscover this property each time the object is encountered. Object concepts are hierarchically organized, with the broadest knowledge represented at the superordinate level, more specific knowledge at an intermediary level referred to as the 'basic level,' and the most specific information at the subordinate level. For example, 'dog' is a basic-level category that belongs to the superordinate categories 'animals' and 'living things,' and has subordinate categories such as 'poodle' and 'collie.' As established by Eleanor Rosch and others in the 1970s and 1980s, the basic level has a privileged status. It is the level used nearly exclusively to name objects (e.g., 'dog' rather than 'poodle'). It is also the level at which we are fastest to verify category membership (i.e., we are faster to verify that a picture is a 'dog' rather than an 'animal' or a 'poodle'). It is also the level at which subordinate category members share the most properties (e.g., collies and poodles have similar shapes and patterns of movement). Finally, the basic level is the easiest level at which to form a mental image (you can easily imagine an elephant but not an 'animal'). Studies of patients with cortical damage have documented the neurobiological reality of this hierarchical scheme and the central role of the basic level for representing objects in the human brain.

### Semantic Dementia and the General Disorders of Semantic Memory

Several neurological conditions can result in a relatively global or general disorder of conceptual knowledge. These disorders are considered general in the sense that they cut across multiple category boundaries; they are not category specific. Many of these patients suffer from a progressive neurological disorder of unknown etiology referred to as semantic dementia (SD). General disorders of semantic memory are also prominent in patients with Alzheimer's disease (who, compared to SD patients, typically have a greater episodic memory impairment), and can also occur following left hemisphere stroke, prominently involving the left temporal lobe. The defining characteristics of this disorder, initially described by Elizabeth Warrington and colleagues in the mid-1970s, are deficits on measures designed to probe knowledge of objects and their associated properties. These deficits include impaired object naming (with errors typically consisting of semantic errors – retrieving the name of another basic-level object from the same category, or retrieval of a superordinate category name), impaired generation of the names of objects within a superordinate category, and an inability to retrieve information about object properties – including sensory-based information (shape, color) and functional information (motor-based properties related to the object's customary use, but this may include other kinds of information not directly related to sensory or motor properties). The impairment is not limited to stimuli presented in a single modality, like vision, but rather extends to all tasks probing object knowledge regardless of stimulus presentation modality (visual, auditory, tactile) or format (words, pictures). The semantic deficit is hierarchical in the sense that broad levels of knowledge are often preserved, while specific information is impaired. Thus, these patients can sort objects into superordinate categories, having, for example, no difficulty indicating which are animals, which are tools, which are foods, and the like. The difficulty is manifest as a problem distinguishing among the basic-level objects as revealed by impaired performance on measures of naming and object property knowledge.

Recent studies have expanded our understanding of SD in two important ways: one is related to location of neuropathology, the other to functional characteristics of the disorder. The initial neuropathological and imaging studies of SD indicated prominent atrophy of the temporal lobes, especially to the anterolateral sector of the left temporal lobe, including the temporal polar cortex inferior and middle temporal gyri, and the most anterior extent of the fusiform gyrus. However, recent advances in neuroimaging that allow for direct and detailed comparison of brain morphology in SD patients relative to healthy controls have shown that the atrophy extends more posteriorally along the temporal lobe than previously appreciated. In fact, the amount of atrophy in ventral occipitotemporal cortex, including the posterior portion of the fusiform gyrus, has been reported to be as strongly related to the semantic impairment in SD as is atrophy in the most anterior regions of the temporal lobes.

The other major advance in our understanding of SD is that it is not as global a conceptual disorder as initially thought. Rather, certain domains of knowledge may be preserved, and the pattern of impaired and preserved knowledge appears to be related to the locus of pathology. Specifically, left-sided atrophy seems to impair information about all object categories except person-specific knowledge (i.e., information about famous people), which, in turn, is associated with involvement of the right anterior temporal lobes. Also relatively spared is knowledge of number and mathematical concepts, and information about motor actions needed to use familiar tools. Both of these knowledge domains are associated with left posterior parietal cortex.

### Category-Specific Disorders of Semantic Memory

Although reports of cases of relatively circumscribed knowledge disorders date back over 100 years, the modern era of the study of category-specific disorders began in the early 1980s with the seminal reports of Warrington and colleagues. Category-specific disorders have the same functional characteristics as does SD, except that the impairment is largely limited to members of a single superordinate object category. For example, a patient with a category-specific disorder for 'animals' will have greater difficulty naming and retrieving information about members of this superordinate category relative to members of other superordinate categories (e.g., tools, furniture, flowers). Similar to patients with SD, patients with category-specific disorders have difficulty distinguishing among basic-level objects (e.g., between dog, cat, horse), thereby suggesting a loss or degradation of information that uniquely distinguishes among members of the superordinate category (e.g., four-legged animals).

Although a variety of different types of category-specific disorders have been reported (e.g., for fruits and vegetables), most common have been reports of patients with relatively greater knowledge deficits for animate entities (especially animals) than for a variety of inanimate object categories. While less common, other patients show the opposite dissociation of a greater impairment for inanimate manmade objects (including common tools) than for animals and other living things.

**Models of category-specific disorders** Two major theoretical positions have been advanced to explain these disorders. Following the explanation posited by Warrington for her initial cases, most current investigators assume that category-specific deficits are a direct consequence of an object property-based organization of conceptual knowledge, an idea that was prominent in the writings of Karl Wernicke, Sigmund Freud, and other behavioral neurologists during the late nineteenth and early twentieth centuries. The central idea is that object knowledge is organized in the brain by sensory (e.g., form, motion, color, smell, taste) and motor properties associated with the object's use, and in some models, by other functional/verbally mediated properties, such as where an object is typically found. In this view, category-specific semantic disorders occur when a lesion disrupts information about a particular property or set of properties critical for defining, and for distinguishing among, category members. Thus damage to regions that store information about object form, and form-related properties like color and texture, will produce a disorder for animals. This is because visual appearance is assumed to be a critical property for defining animals, and because the distinction between different animals is assumed to be heavily dependent on knowing about subtle differences in their visual forms. A critical prediction of sensory/motor-based models is that the lesion should affect knowledge of all object categories with this characteristic, not only animals. In a similar fashion, damage to regions that store information about how an object is used should produce a category-specific disorder for tools, and all other categories of objects defined by how they are manipulated.

The alternative to these property-based theories is the domain-specific view championed most recently by Alfonso Caramazza and colleagues. On this account, our evolutionary history provides the major constraint on the organization of conceptual knowledge in the brain. Specifically, the theory proposes that selection pressures have resulted in dedicated neural machinery for solving, quickly and efficiently, computationally complex survival problems. Likely candidate domains offered are animals, conspecifics, plant life, and possibly tools. Property-based and category-based accounts are not mutually exclusive. For example, it is certainly possible that concepts are organized by domains of knowledge, implemented in the brain by large-scale property-based systems. Much of the functional neuroimaging evidence (to be discussed later) is consistent with this view.

**Functional neuroanatomy of category-specific disorders** There is considerable variability in the location of lesions associated with category-specific disorders for animate and inanimate entities. Nevertheless, some general tendencies can be observed. In particular, category-specific knowledge disorders for animals are disproportionately associated with damage to the temporal lobes. The most common etiology is herpes simplex encephalitis, a viral condition with a predilection for attacking anteromedial and inferior (ventral) temporal cortices. Category-specific knowledge disorders for animals also have been reported following focal, ischemic lesions to the more posterior regions of ventral temporal cortex, including the fusiform gyrus. In contrast, category-specific knowledge disorders for tools and other manmade objects have been most commonly associated with focal damage to lateral frontal and parietal cortices of the left hemisphere. However, it is important to stress that the lesions in patients presenting with category-specific knowledge disorders are often large and show considerable variability in their location from one patient to another. As a result, these cases have been relatively uninformative for questions concerning the organization of object memories in cerebral cortex. In contrast, recent functional neuroimaging studies of the intact human brain have begun to shed some light on this thorny issue.

## Organization of Conceptual Knowledge: Neuroimaging Evidence

The functional neuroanatomy of semantic memory has been explored using a wide variety of experimental approaches and paradigms. The overarching goal of these studies has been to identify brain regions that show a heightened response during a conceptual or semantic processing task (e.g., is the object a living thing?) versus performing a nonsemantic, but equally difficult, task (e.g., does the object's name contain the letter 'p'?) with the same set of stimuli. Assigning a particular brain region a role in semantic processing is further strengthened when the activity is associated with semantic task performance, regardless of physical differences in the stimuli used to denote the objects (i.e., visual or auditory, pictures or words). As a result, activity in these brain regions can be linked to conceptual processes and not features of stimulus input modality or format.

Studies based on this type of approach have consistently isolated two regions, the left ventrolateral prefrontal cortex (VLPFC) and the ventral and lateral regions of the temporal lobes (typically stronger in the left than in the right hemisphere). Detailed evaluations of left VLPFC have shown that this region is critically involved in the top-down control of semantic memory. Specifically, left VLPFC is responsible for guiding retrieval and postretrieval selection of conceptual information stored in posterior temporal and perhaps other cortical regions. This neuroimaging finding is consistent with studies of patients with left inferior frontal lesions that have word-retrieval difficulties, but retain concept knowledge of the words they have difficulty retrieving. In contrast, a large body of evidence has linked activity in posterior regions of the temporal lobes to memory representations, especially concerning conceptual representations of specific properties associated with concrete objects.

### Object Knowledge Is Organized by Sensory and Motor-Based Properties

One source of evidence concerning the organization of object property information comes from property production tasks. In these tasks, individuals are presented with an object, denoted either by a picture or the object's name, and are required to generate a word denoting a specific property associated with that object. Several studies have shown that generating an action associate (e.g., saying 'pull' in response to a child's wagon) elicits heightened activity in premotor cortex as well as in a posterior region of the left lateral temporal cortex, centered on the middle temporal gyrus (pMTG) just anterior to primary visual motion-sensitive cortex. In contrast, relative to action word generation, color word generation (e.g., saying 'red' for the child's wagon) activates the fusiform gyrus, anterior to regions associated with passive perception of color. Subsequent studies have shown that tasks requiring finer perceptual discrimination of colors, or perceiving more complex patterns of object motion, elicit activity in the same temporal lobe regions active when color or action knowledge is retrieved. The findings from these and similar studies suggest two important points. First, that information about different types of object-associated properties (e.g., color, action) is represented in different anatomical regions. Second, these regions directly overlap with sites that mediate perception of these object properties. A number of studies have replicated these two basic points using property verification tasks requiring subjects to answer yes/no questions about the relation between a particular object and property. These studies have linked retrieval of information about visual, auditory, and tactile properties to regions associated with processing within each of these sensory modalities. Similarly, retrieval of object use-associated information has been linked to motor regions active when objects are actually manipulated. These studies suggest that the same neural systems are involved, at least in part, in perceiving, acting on, and knowing about specific object properties.

### Neural Networks for Animate Entities and Tools

A large number of studies have addressed questions concerning how information about different object categories is represented in the brain. Many of these studies were motivated by the neuropsychological evidence for category-specific disorders discussed in the preceding sections. As a result, these studies have concentrated on the neural systems for perceiving and knowing about two broad domains: animate agents (living things that move on their own), and tools (manmade objects with a close association between their function and motor movements associated with their use). These studies have provided evidence for four major points about the organization of conceptual knowledge. First, information about a specific object category is not represented in a single cortical region, but rather is represented by a network of discrete regions that may be widely distributed throughout the brain. Second, the informational contents of these regions are related to specific properties associated with the object. Third, some of these property-based regions are automatically active when objects are identified, thus suggesting that object perception is associated with the automatic retrieval of a limited set of associated properties that may be necessary and sufficient to identify that object. These properties prominently include information about what the object looks like, how it moves, and, for a tool, its use. Fourth, this object property-based information is stored in sensory and motor systems active when that information was acquired.

Conceptual representations of animate entities have been most strongly linked to two regions of temporal lobe cortex, the lateral region of the fusiform gyrus, located on the ventral surface of the temporal lobe (including the fusiform face area; FFA), and the posterior region of the superior temporal sulcus (pSTS), located on the lateral surface of the temporal lobe (**Figure 1**). Activation of these regions is typically stronger in the right, rather than in the left, hemisphere. In contrast, conceptual representations of tools have been linked to four regions. Included in this network are the medial portion of the fusiform gyrus, pMTG (as discussed previously in relation to generating action words), and posterior parietal and ventral premotor cortices. Activity in these regions is typically stronger in the left, rather than in the right, hemisphere (**Figure 1**).

Activity in these networks transcends stimulus features, thus strengthening the claim that these regions are involved in conceptual processing. For example, category-related activity in the lateral region of the fusiform gyrus has been shown not only in response to pictures of people and animals, but also to the written and heard names of animals, human voices, animal-associated sounds (like the moo of a cow), and when simply imagining faces. In addition, category-related activity in the lateral fusiform gyrus has been found in response to degraded and abstract visual

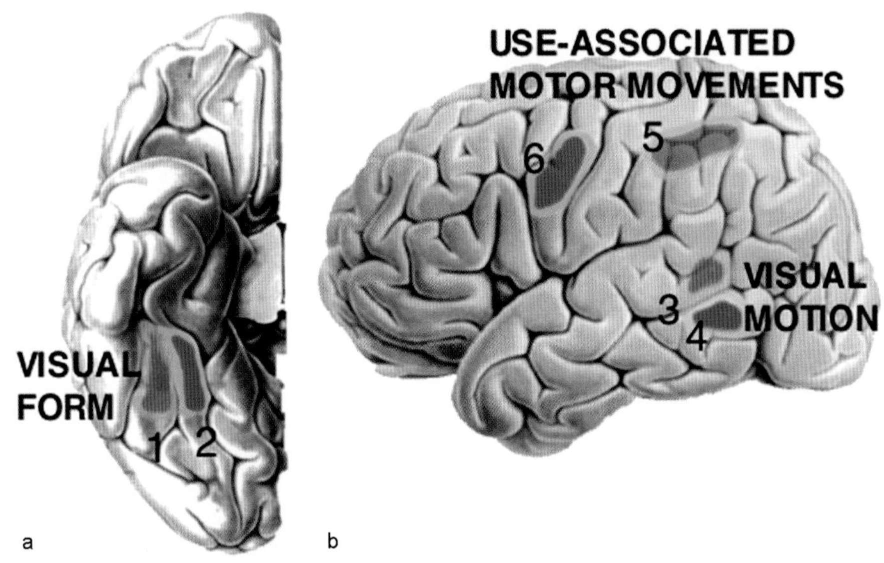


**Figure 1** Schematic illustration of location of regions showing category-related activity for animate entities (red) and tools (blue). (a) Ventral view of the right hemisphere, showing relative location of regions assumed to represent visual form and form-related properties such as color and texture of animate entities (1, lateral region of the fusiform gyrus, including, but not limited to the fusiform face area) and tools (2, medial region of the fusiform gyrus). (b) Lateral view of the left hemisphere, showing relative location of regions assumed to represent biological motion (3, posterior region of the superior temporal sulcus) and rigid motion vectors typical of tools (4, posterior region of the middle temporal gyrus). Also shown are the relative locations of the posterior parietal (5, typically centered on the intraparietal sulcus) and ventral premotor (6) regions of the left hemisphere assumed to represent information about the motor movements associated with using tools.

stimuli such as humanlike stick figures, point light displays interpreted as human figures in motion, degraded pictures of objects when misinterpreted as faces, and when simple geometric shapes in motion are interpreted as depicting social interactions. Similarly, category-related activity in the more medial region of the fusiform gyrus has been shown not only in response to pictures of tools, but also to the written and heard names of tools and to tool-associated sounds (like the banging of a hammer). In addition, category-related activity in the medial fusiform gyrus has been observed in response to degraded and abstract visual stimuli such as point light displays interpreted as depicting tools in motion, and when simple geometric shapes in motion are interpreted as depicting mechanical interactions. Taken together, these findings indicate that the critical determinant of differential activity in these regions is the conceptual interpretation assigned to a stimulus (e.g., as a person, animal, or tool), not the physical characteristics of the stimulus, *per se*.

### Linking Category-Related Representations to Sensory and Motor Properties

Neuroimaging studies also have provided evidence about the functional role of regions showing category-related activity. The two regions in the posterior region of the temporal lobe that show a stronger response to animate rather than inanimate objects – the lateral region of the fusiform gyrus and pSTS – have been linked to representing the visual form and the flexible, fully articulated, patterns of motion associated with animate objects, respectively. These findings are consistent with a large body of neurophysiological data from studies of monkeys showing that neurons in inferior temporal cortex and STS are differentially tuned to visual form and biological motion, respectively. In contrast, the two regions in the posterior region of the human temporal lobe that show a stronger response to tools than to animate objects – the medial region of the fusiform gyrus and pMTG – have been linked to representing the visual form and rigid, unarticulated patterns of motion typically associated with manipulable man-made objects.

Other studies have provided evidence linking the representation of tools to two regions of the dorsal action-processing stream – the left posterior parietal cortex and left ventral premotor cortex. These findings are consistent with data from monkey neurophysiology showing that neurons in ventral premotor and parietal cortices respond both when monkeys grasp objects and when they see objects that they have had experience manipulating. Along with pMTG, functional brain imaging studies in humans have shown that these dorsal regions are active when individuals retrieve information about the functional properties of objects, suggesting that information about how objects are used is stored in these regions. Support for this claim comes from recent neuropsychological investigations linking impaired knowledge of tools and their associated actions to damage to either left pMTG, posterior parietal, or premotor cortices.

Finally, recent studies have shown that the category-related, property-based neural systems discussed here are not only active when objects are encoded into memory, but also that this activity is reinstated prior to recalling that information at a later time. These findings underscore the central role that these systems play in memory encoding, storage, and retrieval.

*See also:* Aging and Memory in Humans; Declarative Memory System: Anatomy.

## Further Reading

Caramazza A and Shelton JR (1998) Domain-specific knowledge systems in the brain: The animate–inanimate distinction. *Journal of Cognitive Neuroscience* 10: 1–34.

Damasio H, Tranel D, Grabowski T, et al. (2004) Neural systems behind word and concept retrieval. *Cognition* 92: 179–229.

Forde EME and Humphreys GW (2002) *Category Specificity in the Brain and Mind.* Hove and New York: Psychology Press.

Garrard P and Hodges JR (1999) Semantic dementia: Implications for the neural basis of language and meaning. *Aphasiology* 13: 609–623.

Manns JR, Hopkins RO, and Squire LR (2003) Semantic memory and the human hippocampus. *Neuron* 38: 127–133.

Martin A (2007) The representation of object concepts in the brain. *Annual Review of Psychology* 58: 25–45.

Martin A and Caramazza A (eds.) (2003) *The Organization of Conceptual Knowledge in the Brain: Neuropsychological and Neuroimaging Perspectives.* Hove and New York: Psychology Press.

Murphy GL (2002) *The Big Book of Concepts.* Cambridge, MA: The MIT Press.

Polyn SM, Natu VS, Cohen JD, et al. (2005) Category-specific cortical activity precedes retrieval during memory search. *Science* 310: 1963–1966.

Vargha-Khadem F, Gadian DG, Watkins KE, et al. (1997) Differential effects of early hippocampal pathology on episodic and semantic memory. *Science* 277: 376–380.

Warrington EK (1975) The selective impairment of semantic memory. *Quarterly Journal of Experimental Psychology* 27: 635–657.

Warrington EK and Shallice T (1984) Category specific semantic impairments. *Brain* 107: 829–854.

# Short Term and Working Memory

**B R Postle**, University of Wisconsin – Madison, Madison, WI, USA
**T Pasternak**, University of Rochester, Rochester, NY, USA



## Introduction

### Definition

Short-term memory (STM) refers to the active retention (for humans, the 'keeping in mind') of information when it is not accessible from the environment. Working memory can be thought of as 'STM+,' the '+' referring to the ability to manipulate or otherwise transform this information, to protect it in the face of interference, and to use it in the service of such high-level behaviors as planning, reasoning, and problem solving. STM and working memory are of central importance to the study of high-level cognition because they are believed to be critical contributors to such essential cognitive functions and properties as language comprehension, learning, planning, reasoning, and general fluid intelligence.

### Historical Backdrop

The modern study of working memory began with the work of Jacobsen, who, in the 1930s, demonstrated that large bilateral lesions of the prefrontal cortex (PFC) in the monkey produced profound deficits in spatial working memory, and thus Jacobsen ascribed to the PFC a function he termed "immediate memory." A subsequent study, however, revealed that this deficit could be erased by darkening the testing cage and thus reducing interference during the memory delay. These results highlighted a critical question, still the subject of an active debate, which is whether PFC plays a critical role in storage in addition to its role in behavioral control.

The early 1970s witnessed two developments that were seminal in shaping contemporary conceptions of working memory. The first was the observation made by Joaquin Fuster that individual neurons in PFC of monkeys, trained to compare visual stimuli separated by a memory delay, exhibit sustained activity throughout that delay. This observation suggested a neural correlate of two potent ideas from physiological psychology – that of a PFC-dependent immediate memory and that of a reverberatory mechanism for "a transient 'memory'" proposed by psychologist Donald Hebb. The second development, which occurred in the field of human cognitive psychology, was the multiple component model of working memory proposed by Alan Baddeley and Graham Hitch. In its initial instantiation this model comprised two independent buffers for the storage of verbal and of visuospatial information, and a central executive to control attention and to manage information in the buffers. Prompted by these two developments, the neuroscientific and the psychological study of working memory each proceeded along parallel, but largely independent, paths until the late 1980s, when a third important advance occurred.

The third advance was the conceptual integration of the neuroscientific and psychological traditions of working memory research, proposed by Patricia Goldman-Rakic, that the sustained delay-period neural activity in PFC and the storage buffers of the multiple-component model of Baddeley and Hitch were cross-species manifestations of the same fundamental mental phenomenon. This association between prefrontal cortex and working memory has been very influential in systems and cognitive neuroscience.

### Current State of Working Memory Research

A growing body of evidence provided by behavioral, physiological, and neuroimaging studies indicates that information about sensory stimuli may be stored in a segregated, feature-selective manner, and that the relevant cortical regions include relatively early stages of sensory cortical processing. The principle emerging from this work is that the same brain regions that are responsible for the precise sensory encoding of information also contribute to its short-term retention. In the remaining portions of this article we briefly describe these recent advances, with the emphasis on working memory for fundamental dimensions of sensory stimuli.

## Visual Working Memory

In the laboratory, the ability to briefly retain visual information can be measured with delayed discrimination tasks. In the simplest version of such tasks, individuals discriminate between two stimuli, the sample and the test, separated by a temporal delay of various durations, and report whether and how the test differs from the previously seen sample. The results of such experiments have revealed that fundamental stimulus features, such as orientation, contrast, size, or speed, can be faithfully preserved for many seconds, although the duration of this preservation often differs for different features. For example, stimulus size or orientation can be retained accurately longer than can luminance contrast. This difference suggests that different stimulus attributes may be retained by separate, feature-selective mechanisms. More direct support for feature-selective storage mechanisms comes from studies that have used interference consisting of an irrelevant stimulus (a 'memory mask') introduced during the delay separating S1 and S2. With this procedure, determination of the parameters that maximize the interfering effects of the mask can provide insights into the nature of the remembered stimulus and thus into the mechanisms involved in its short-term retention. For example, the memory for spatial frequency of gratings (size) can be disrupted only if the masking stimulus is of a different spatial frequency, irrespective of its orientation. This selective interference suggests that information about spatial frequencies may be preserved by mechanisms that are narrowly tuned for that stimulus attribute, and closely associated with its processing, but distinct from mechanisms concerned with stimulus orientation. Similar specialization is seen within the domain of visual motion, in that stimulus speed can be preserved by the mechanisms that are relatively narrowly tuned for speed, independently of stimulus direction.

Representation of visual motion in memory appears to be localized in space in a manner consistent with properties of neurons in cortical area MT, a region specialized in processing of visual motion, thereby supporting the idea that neurons processing visual motion may also be involved in its short-term storage. Overall, behavioral studies of mechanisms that preserve basic attributes of visual stimuli can be characterized as narrowly tuned, spatially localized filters, supporting a model of working memory that involves the contribution of sensory cortical areas.

A close link between visual processing and storage is also suggested by the consequences of cortical damage. For example, damage to motion processing area MT in monkeys results in deficits in remembering motion direction, but only when the remembered stimuli require area MT for their encoding (**Figure 1**). Similarly, damage to inferotemporal cortex, associated with processing of complex shapes, results in deficits in remembering such shapes. Humans with focal lesions in the occipitotemporal cortical area, implicated in processing of specific visual features, also have difficulties remembering these features.

The activity of visual cortical neurons during working memory tasks also attests to their role in sensory maintenance. Such tasks similar to the delayed discrimination tasks described earlier, require the comparison of two stimuli separated by a memory delay. When such stimuli consist of moving random-dot stimuli and the monkeys are required to identify and remember the dot direction, the activity of MT neurons during the memory delay shows selectivity reflecting the remembered direction. This activity is pronounced early in memory delay but weakens toward its end and does not correlate with the animals' decision, suggesting that additional neural mechanisms are needed to account for the maintenance of motion signals. However, at the time of the comparison and decision, these neurons are strongly affected by the remembered direction, suggesting that throughout the task they have access to the remembered direction, and thus are likely to participate in the circuitry subserving storage. Similar behavior has been observed in inferotemporal (IT) cortex, associated with the processing of complex shapes, in that stimulus-selective activity occurs not only in response to specific shapes but also during the memory delay. During the comparison phase of the task, these neurons, like neurons in area MT, reflect the remembered shape. This type of activity is consistent with the possibility that visual cortical neurons actively participate the circuitry subserving storage of signals they process, although the nature of this participation awaits further study.

Functional imaging studies of visual working memory indicate that regions of the visual system are differentially recruited by working memory tasks, depending on the stimulus dimension that must be remembered. For example, working memory for the location versus the identity of visual stimuli, such as color patches and geometric shapes, recruits domain-specific memory-related activity in posterior cortical regions associated with the 'where' and 'what' visual streams, respectively. For location memory this includes dorsal occipital cortex, the intraparietal sulcus, and superior parietal lobule, whereas for object identity it includes ventral temporal and occipital cortex. Location memory tasks also typically recruit frontal regions associated with oculumotor control – the frontal eye fields and the supplementary eye fields.

Within the ventral streams, delay-period activity is segregated in a category-specific manner. For example, multiple images of faces and of naturalistic scenes can be presented as samples, with a postsample cue indicating which is the to-be-remembered category. When individuals are cued to remember the faces, delay-period activity is selectively elevated in the fusiform face area (FFA). When they are cued to remember scenes, delay-period activity is selectively elevated in the parahippocampal place area (PPA). Another study of working memory for faces featured three delay periods interposed between the presentation of the first and second, second and third, and third and fourth stimuli. The logic was that the multiple distracting events in this task might serve to 'weed out' from the first delay period activity that was not involved directly in storage, because only regions

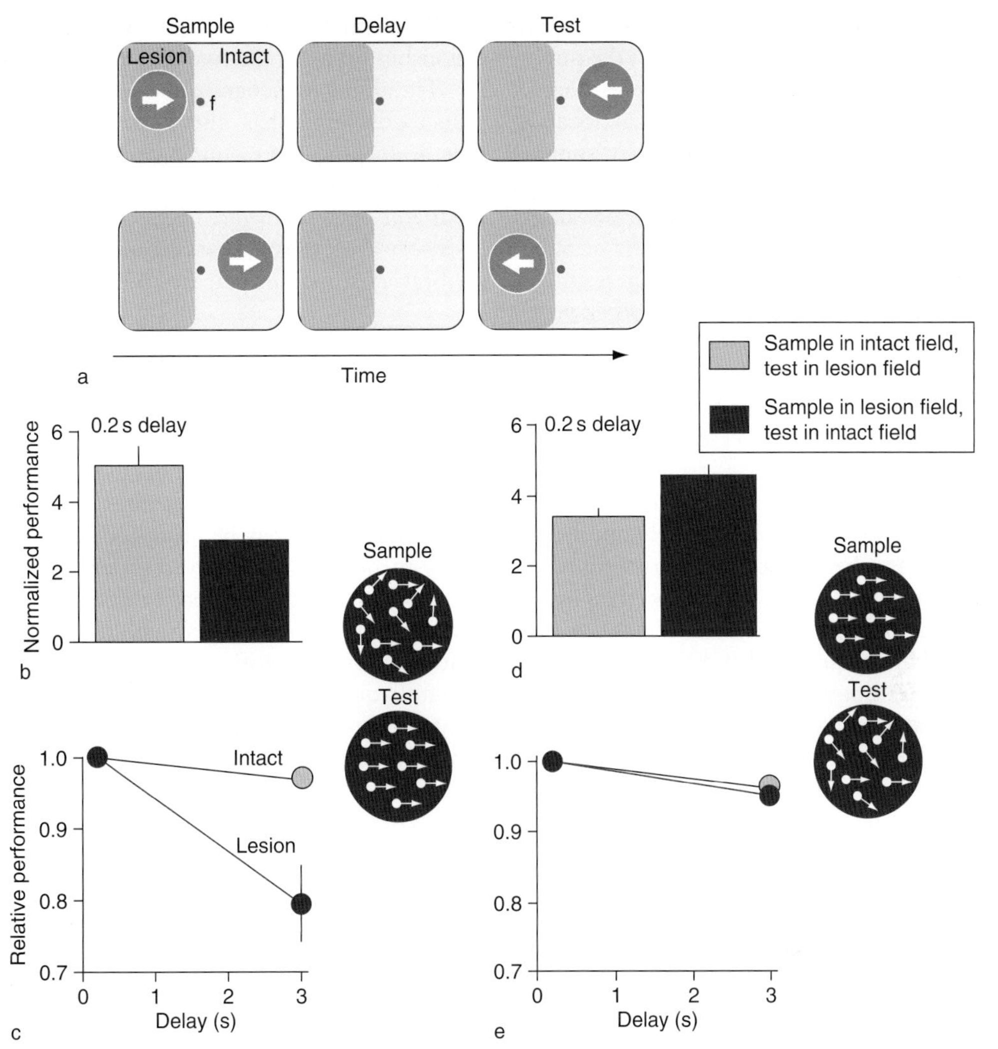


**Figure 1** The role of cortical areas processing visual motion in remembering motion direction. (a) Monkeys with unilateral lesions of motion processing areas MT/MST performed a task requiring integration of complex motion and remembering its direction. In this task, sample and test stimuli were separated by a delay and positioned in opposite hemifields, so that one was placed in the lesioned and the other in the corresponding location in the intact hemifield. The monkeys reported whether sample and test moved in the same or in different directions by pressing one of two response buttons. During each trial the monkeys fixated a small target at the center of the display (blue dot) while attending to moving random dots (indicated by arrows) presented in the periphery. (b, d) Normalized direction range thresholds (1/(360-range threshold)/360) measured when the delay between sample and test was minimal (0.2 s). The sample or the test stimulus was composed of dots moving in a range of directions, while the other stimulus contained only coherent motion (stimulus configurations shown to the right of each set of plots). Performance was decreased whenever the stimulus containing noncoherent motion, and thus requiring integration, was placed in the lesioned field, demonstrating the importance of areas MT/MST for motion integration. (c, e) Effect of memory delay on performance for two direction range tasks. Performance was measured with both stimuli placed in the intact (gray circles) or in the lesioned hemifields (blue circles), either by varying the range of directions in the sample, while the test moved coherently (left plots), or by varying the range of directions in the test, while the sample moved coherently (right plots). Thresholds were normalized to the data measured at a 0.2 s delay. Error bars are SEM. A delay-specific deficit was present only when the remembered stimulus (sample) contained a broad range of directions and required integration. This result demonstrates that stimulus conditions requiring motion integration depend on intact areas MT/MST. However, coherently moving random dots can be discriminated and remembered at a normal level even in the absence of areas MT/MST. This shows that a cortical area involved in processing of sensory signals is also involved in their storage. Adapted from Bisley JW and Pasternak T (2000) The multiple roles of visual cortical areas MT/MST in remembering the direction of visual motion. *Cerebral Cortex* 10: 1053–1065.

with activity necessary for retaining the information to the end of the trial would maintain their activity across distracting stimuli. The results revealed that the posterior fusiform gyrus was the only region that retained the relevant signal during the last delay, immediately preceding the decision. Still other studies of working memory for faces have varied memory load (i.e., the number of items to be remembered), on the logic that storage-related activity is sensitive to variations in load. These studies confirmed the importance of posterior fusiform gyrus in short-term storage.

Studies of working memory for spatial location have also addressed the mechanisms that support storage. One of these is attention-based rehearsal, a mechanism hypothesized to contribute to the short-term retention of locations via covert shifts of attention to the to-be-remembered location. These studies presented sample stimuli in one or the other visual hemifield while individuals fixated a central spot. Delay-period activity was greater in the hemisphere contralateral to the target location, an effect comparable to what one sees in studies of attention. This lateralized delay-period bias was strongest in extrastriate regions, decreased in magnitude across the parietal cortex, and was no longer reliable in frontal cortex. A second mechanism for the short-term retention of location information is prospective motor coding – the formulation, and then retention, of a motor plan for the acquisition of a target with, say, a saccade or a grasp. Electrophysiology studies in monkeys and neuroimaging studies in humans that encourage a prospective strategy localize this activity to frontal oculomotor regions, to prefrontal cortex, and to the caudate nucleus.

## Tactile Working Memory

As with visual stimuli, tactile stimuli can be faithfully represented in working memory. Delayed discrimination of vibration stimuli can be performed with delays of many seconds, although the accuracy of this discrimination is maximal at short delays and decreases rapidly during the first 5 s of the delay. At longer delays, however, performance does not continue to deteriorate, suggesting that a two-stage memory process might be involved. Delayed discrimination of vibration stimuli can be disrupted by the application of transcranial magnetic stimulation (TMS) to human primary somatosensory cortex (area S1) during the initial portion of the delay period. This effect not only implicates area S1 in storage of vibration information, but also demonstrates the vulnerability of the storage mechanism early in the delay, an effect also observed in studies of working memory for visual motion.

In the awake behaving monkey, electrophysiological evidence for delay-period activity in S1 is mixed, with some laboratories reporting evidence in favor of it and others failing to find it. In area S2, however, there is no such equivocality, with strong evidence for stimulus frequency-specific activity persisting far into the delay period. Furthermore, responses of S2 neurons to the test stimulus contain information about the remembered stimulus, thereby reflecting the relationship between the two stimuli.

The ability to recognize and remember objects based on tactile input has also been studied with human neuroimaging. For example, positron emission tomography (PET) studies have revealed activation in the parietal operculum (area S2), an associative somatosensory area, during working memory tasks involving vibratory stimuli or palpated wire forms.

## Auditory Working Memory

Although humans discriminate both pitch and loudness of sounds with high accuracy and remember these dimensions for many seconds, memory for pitch and loudness decline at different rates, suggesting that the two dimensions may be processed separately in auditory memory. The observed differences in the precision of memory for intensity and pitch parallel the findings in vision for contrast and spatial frequency (see earlier). As with vision, the use of interfering stimuli during the memory delay has revealed the nature of representation of auditory stimuli in memory. For example, memory for pitch can be disturbed by distractor tones, but only if these tones are within a narrow range of frequencies, relative to the frequency of the remembered stimulus, supporting the existence of separate pitch memory modules. Furthermore, sound frequency is likely to be stored separately from its location, suggesting that auditory memory obeys the same patterns as does auditory perception for physical parameters of the remembered stimuli.

In the auditory cortex, neuronal activity during the delay period of a delayed two-tone discrimination carries information about the frequency of the sample tone. In addition, the response to the second tone depends on whether its frequency matches that of the first, implicating neurons in auditory cortex in the circuitry subserving working memory for tone frequency. Participation of auditory cortex in auditory working memory is also supported by studies of the human using magnetoencephalography (MEG).

Memory-related activity has also been seen in associative auditory cortex of humans during tasks requiring memory for differences in pitch and tonal sequences; the right superior temporal lobe was associated with tonal sequences, and pitch judgments were associated with increased right frontal lobe activation. Interestingly, the auditory cortex appears to be maximally active early in the delay, whereas regions in the supramarginal gyrus and parts of the cerebellum are activated later, suggesting that for audition, too, working memory may be accomplished in multiple stages.

The involvement of regions processing auditory information in working memory is also supported by selective deficits in patients with damage to auditory associative cortex in the right temporal lobe. These patients are impaired primarily on tone discrimination with long, distracted delays.

## Multiple Encoding in Working Memory

It seems unlikely that the representational bases of working memory are limited to the domain in which stimuli are perceived. In addition to the multiple serially ordered stages in working memory for visual, tactile, and auditory domains, sensory information can also be represented in parallel in multiple formats. Thus, as reviewed earlier, working memory for location can be accomplished via retrospective sensory memory-based mechanism, as well as a prospective motor planning mechanism. Analogously, there is growing evidence that working memory for visual identity recruits verbal (and, perhaps, semantic) representations in addition to the ventral stream visual mechanisms already discussed.

## Working Memory and Prefrontal Cortex

For many years PFC has been strongly associated with the storage of sensory information in working memory. Consistent with this idea are the facts that PFC is directly interconnected with cortical areas processing fundamental sensory dimensions, and that PFC neurons often display stimulus-related activity during the memory delay. However, a growing body of evidence suggests that PFC is more likely to play a key role in directing attention to behaviorally relevant sensory signals and in making decisions concerning these signals, than in directly supporting their retention in short-term and working memory. Consistent with this view is the fact that extensive damage to prefrontal cortex does not eliminate the ability to perform simple short-term memory tasks, such as the digit span and spatial span variants of immediate serial recall. The limited effect of such lesions is apparent on delay tasks if during the memory period there is no interference and no requirement to keep track of multiple items. Also consistent with this view is the finding that although the delay activity of PFC neurons recorded in monkeys performing a delayed direction discrimination task (see earlier) was direction-specific, it did not predict decisions about stimulus direction that they made at the end of the trial. The contribution of these neurons to the task became apparent only after the end of the memory delay, when the monkeys compared test direction with that of the remembered sample. Furthermore, direction selectivity in PFC neurons, prevalent during the task, was greatly reduced when the monkeys were not required to use the information about stimulus direction. This work showed that PFC neurons are capable of gating sensory signals according to their behavioral relevance, it but does not support the key role for PFC neurons in the maintenance of these signals. Other studies, designed to isolate the contributions of attention and response selection from those of sensory storage, indicate a stronger role for the former in PFC.

In the human neuroimaging literature there is some debate about stimulus domain- or category-specific segregation of delay-period activity in PFC – for example, comparable patterns of delay-period activity during tasks involving memory for location versus memory for objects or memory for faces versus memory for scenes. The same is true for tasks that involve haptic object encoding versus those that involve visual object encoding. Similarly, in tasks requiring cross-modal integration, individual PFC neurons can be active during the memory delay following the presentation of different stimulus modalities. These examples suggest that PFC assists early-level sensory processing regions to form supramodal mental representations of objects and/or of task contingencies when the relevant stimuli belong to more than one sensory domain.

Another strategy for evaluating the neural basis of sensory storage is to require the retention of information across multiple delay periods interrupted with intervening distracting stimuli. This approach has revealed that neurons in the monkey PFC, but not IT cortex, could sustain a representation of the sample stimulus across multiple delays. Robust sample-specific activity across multiple delay periods is not limited to PFC, however, and has also been observed in other regions, including the temporal pole and the entorhinal cortex of the medial temporal lobe. Functional magnetic resonance imaging (fMRI) studies in humans remembering faces showed that only posterior fusiform gyrus retains sustained activity across all three delay periods, although many regions displayed activity during the first delay period. An analogous result was seen in the individuals remembering spatial locations, with distraction-spanning delay-period activity distributed across multiple areas, including intraparietal sulcus, superior parietal lobule, frontal eye fields, and supplementary eye fields in frontal cortex (**Figure 2**). Finally, repetitive TMS (rTMS) applied to PFC has failed to disrupt working memory performance, but has altered performance when applied to parietal cortex.

Although the weight of evidence is inconsistent with a storage role for PFC, virtually every electrophysiological and neuroimaging study of primate

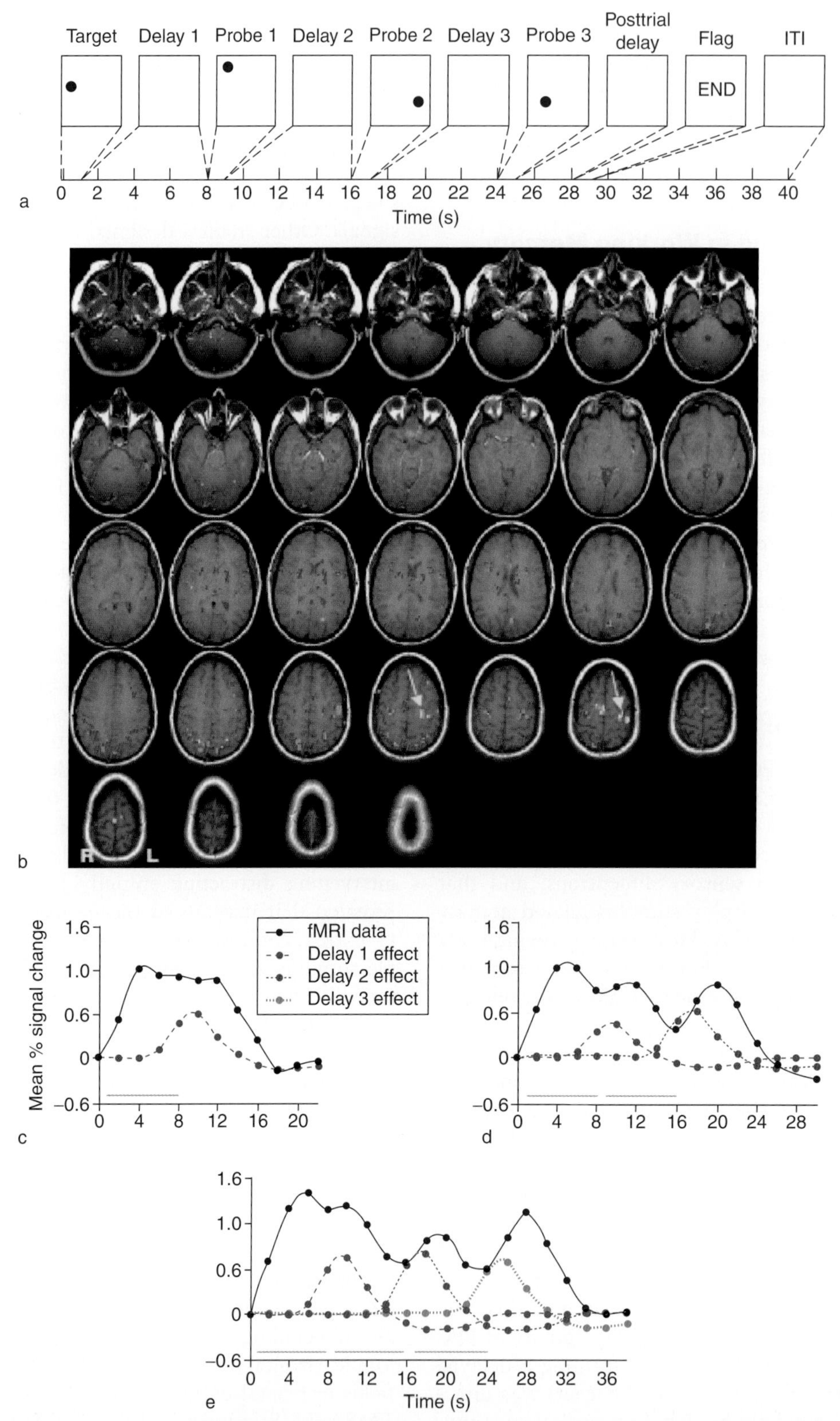


**Figure 2** A functional magnetic resonance imaging experiment demonstrating sustained memory-related activity for location across multiple delay periods. (a) Behavioral task. Individuals view and encode the target location, then, after a 7 s delay, indicate with a button press whether the probe does or does not appear in the same location. One-third of the trials end after probe 1. On two- and three-delay trials, the offset of probe 1 is followed by another delay period, after which individuals evaluate the location of probe 2 with respect to the target. On one- and two-delay trials, the 'END' message appears at times 12 and 20 s, respectively. (b) Results from a single representative individual. Voxels in red showed sustained delay-period activity for delay 1 only. Voxels in blue are the subset of voxels

working memory finds delay-period activity in this region of the brain. Thus, it is important to address the role of this activity. If this activity does not represent the storage of information, what are alternative explanations of its function? One possibility is that PFC plays an important role in the control of potentially disrupting effects of distraction and interference during working memory tasks. Lesion, electrophysiology, and fMRI studies have all provided evidence that dorsolateral PFC can accomplish this by controlling the gain of activity in sensory processing regions, such that the delay-period processing of potentially distracting stimuli is suppressed. Analogously, a region of ventrolateral PFC has been implicated in the control of proactive interference, the deleterious effect of previous mental activity on current task performance. In addition, an important function of delay activity in PFC may be the attentional selection of task-relevant information, as well as in planning for the response. This is consistent with the well-characterized role for PFC in the biasing of stimulus–response circuits so that novel or less salient behaviors can be favored over well-learned associations and behavioral routines. Finally, the potential role of PFC in the integration of cognitive and motivational factors is supported by the modulation of delay-period activity of its neurons during spatial memory tasks by the type of the anticipated reward.

## Conclusion

The emerging picture of the neural basis of working memory is of a class of behaviors that does not depend on one or more functionally specialized regions. Rather, working memory is supported by the coordinated activity of circuits responsible for the sensory processing of the critical information, and, to varying degrees, those that control the flexible allocation of attention and the selection of task-relevant behavior.

## Further Reading

Baddeley A (1992) Working memory. *Science* 255: 556–559.

Bisley JW and Pasternak T (2000) The multiple roles of visual cortical areas MT/MST in remembering the direction of visual motion. *Cerebral Cortex* 10: 1053–1065.

Cowan N (2005) *Working Memory Capacity*. Hove, UK: Psychology Press.

Fuster J (1995) *Memory in the Cerebral Cortex*. Cambridge, MA: MIT Press.

Goldman-Rakic PS (1995) Cellular basis of working memory. *Neuron* 14: 477–485.

Hebb D (1949) *Organization of Behavior*. New York: Wiley.

Miller EK and Cohen JD (2001) An integrative theory of prefrontal cortex function. *Annual Review of Neuroscience* 24: 167–202.

Pasternak T and Greenlee M (2005) Working memory in primate sensory systems. *Nature Reviews Neuroscience* 6: 97–107.

Petrides M (2000) Dissociable roles of mid-dorsolateral prefrontal and anterior inferotemporal cortex in visual working memory. *Journal of Neuroscience* 20: 7496–7503.

Postle BR (2006) Distraction-spanning sustained activity during delayed recognition of locations. *NeuroImage* 30: 950–962.

Postle BR (2006) Working memory as an emergent property of the mind and brain. *Neuroscience* 139: 23–38.

Postle BR and Hamidi M (in press) Nonvisual codes and nonvisual brain areas support visual working memory. *Cerebral Cortex-Advance Access*. published online on December 5, 2006.

Romo R and Salinas E (2003) Flutter discrimination: Neural codes, perception, memory and decision making. *Nature Reviews Neuroscience* 4: 203–218.

Todd J and Marois R (2005) Posterior parietal cortex activity predicts individual differences in visual short-term memory capacity. *Cognitive, Affective & Behavioral Neuroscience* 5(2): 144–155.

Zaksas D and Pasternak T (2006) Directional signals in the prefrontal cortex and area MT during a working memory for visual motion task. *Journal of Neuroscience* 26: 11726–11742.

with delay-period activity sustained across delay 1 and delay 2. Voxels in yellow are the still smaller subset with delay-period activity sustained across all three delays. Note that although many regions, including the prefrontal cortex, show delay-period activity during delay 1, only dorsal-stream parietal and frontal oculomotor regions sustain this delay-period activity across all three delays. Arrows highlight the voxels from left frontal eye field (the activity of which is shown in panels c, d, and e). (c) Activity from three-delay voxels in the left frontal eye field, averaged across one-delay trials. 'Delay 1 effect' reflects the estimated magnitude of delay 1 activity. Gray bar along the horizontal axis indicates the duration of the delay period. (d) Activity from these same frontal eye field voxels averaged across two-delay trials. Graphical conventions are the same as in panel b. (e) Activity from these same frontal eye field voxels averaged across three-delay trials. Graphical conventions are the same as in panel c. Adapted from Postle BR (2006) Distraction-spanning sustained activity during delayed recognition of locations. *NeuroImage* 30: 950–962.

# Language in Aged Persons

**E A L Stine-Morrow and M C Shake**, University of Illinois at Urbana–Champaign, Champaign, IL, USA



## Introduction

Communicative competence is essential for effective functioning through adulthood. The ability to understand and remember language depends on a coordinated array of processing components that translate an orthographic or acoustic signal (i.e., printed symbols and speech sounds, respectively) into a representation of the meaning of written text or speech that changes dynamically as the signal unfolds in real time. Language production can be thought of as a reverse process, requiring message formulation from which a surface form is generated. Aging brings both growth (e.g., knowledge) and decline (e.g., speed of processing). As a result, there are a variety of changes in how language is processed, both normatively and in what is required for successful performance. In this article, we consider the nature of these changes. We first review prominent theories of cognitive aging and then present a conceptual model of language processing. We then consider how language comprehension, memory, and production change with age, taking into account data from behavioral paradigms, event-related potentials, and imaging.

## Theories of Cognitive Aging

Although there are a number of competing theories for cognitive and brain mechanisms that underlie age-related changes in cognition, there is broad agreement that at a coarse level, age effects on cognition can be characterized as a result of two competing forces. On the one hand, senescence drives a decline in mental mechanics, the speed and accuracy with which elementary information processing components can be completed. On the other hand, there is accumulating evidence that the brain has immense potential for plasticity into late life so that experience-based growth in knowledge systems, skill, and expertise offers potential for growth. Language processing can potentially depend on both of these systems.

There are a variety of cognitive theories characterizing the specific nature of declines in mechanics with age, all of which have influenced both behavioral and neuroscience approaches to the study of language. The slowing hypothesis suggests that aging brings a systematic decrease in the speed with which mental operations can be performed. The working memory (WM) hypothesis posits that aging is associated with a decline in the capacity to perform basic processing operations and store their products (as might be required, for example, as one listens to a lecture and tries to construct an understanding of what is currently being discussed and integrate it with what has come before). The inhibition deficit (ID) hypothesis focuses on the important role that inhibitory function plays in effective language processing by suppressing irrelevant environmental input, inappropriate (or no longer appropriate) features of meaning, and incorrect interpretations. According to the ID account, inhibitory function is reduced with age, thus compromising comprehension by allowing irrelevant (or incorrect) information to be incorporated into the language representation – and consuming working memory capacity that might otherwise be used for effective processing. Finally, drawing on the WM hypothesis, the effortfulness hypothesis identifies sensory loss as critical to understanding the aging language system, not simply as a direct influence on the quality of signal, but as an influence on central resources that are strained by the attempt to interpret the muddy signal.

At the same time, aging may bring growth in vocabulary and some aspects of verbal ability, with evidence suggesting that such growth depends on habitual engagement with literacy-based activities. Particularized knowledge systems and expertise develop differentially as a consequence of occupational and leisure pursuits. Well-developed procedural skills, such as reading, can be well maintained into late life. Evidence suggests that language processing may impact the trajectory of mental mechanics as well, with several demonstrations of enhanced executive function among fluent bilinguals, who habitually manage two language systems.

Patterns of growth and decline are also evident in the aging brain, with selective changes in both brain structure and function. Consistent with the notion of age-related declines in controlled attention, as suggested by WM and ID accounts, the prefrontal cortex is the brain region most vulnerable to deterioration. Both evoked potential and imaging data very often show more diffuse activation in the aging brain, a pattern that has been explained alternately in terms of dedifferentiation (effectively, inhibition failure) and as compensatory recruitment.

## The Nature of Language Processing

In both reading and speech understanding, the signal unfolds linearly in time. However, the structure of thought arising from this process is not linear and certainly represents information beyond the verbatim input. The task faced by theories of language comprehension, then, is to explain how this mental representation is constructed and updated over time during comprehension, as well as later when certain tasks require retrieval or use of the text representation. Contemporary models of language comprehension distinguish among distinct processes that operate in concert to construct different facets of the language representation. At the surface level, individual lexical items (words) are encoded from the orthographic or acoustic signal, and their meanings are activated. The stream is parsed into syntactically coherent segments (e.g., clauses and sentences) that establish the thematic roles (functions such as agent and object) for the constituents of the sentence. The semantic representation can be described in terms of integrated ideas (or propositions) that establish relationships among concepts given by the text, a representation called the textbase. Knowledge plays a role in facilitating integration, enabling elaborative inference, and evoking a simulation of the situation suggested by the text. Production can be described as the reverse of this process. To express a thought, a syntactic structure has to be formulated, particular lexical items must be selected to fill thematic roles (a process called lemma selection), and the phonology must be assembled (phonological encoding).

## Age-Related Change in Language Processing

### Language Comprehension and Memory

Consistent with the divergent age trajectories of mechanics and knowledge-based processes, aging appears to have the strongest impact on resource-consuming aspects of language processing.

**Word processing** Vocabulary often shows an increase with age, particularly among those who read regularly. Word recognition appears to be highly resilient through late life. For example, word frequency effects (i.e., faster processing for more familiar words) in reading and word naming are typically at least as large for older adults as they are for young. Sublexical features (e.g., neighborhood density), however, may have a smaller effect on processing time on older readers, suggesting that accumulated experience with literacy may increase the efficiency of orthographic decoding.

By contrast, declines in auditory processing can make spoken word recognition more demanding so that more acoustic information is needed to isolate the lexical item. Such effects may not merely disrupt encoding of the surface form but also tax working memory resources that would otherwise be used to construct a representation of the text's meaning. For example, when elders with normal or impaired hearing listen to a word list and are interrupted periodically to report the last word heard, they may show negligible differences. However, if asked to report the last three words, hearing-impaired elders will likely show deficits. The explanation for such a provocative finding is that the hearing-impaired elders overcome a sensory loss at some attentional cost so as to exert a toll on semantic and elaborative processes that enhance memory. Presumably, the same mechanisms would operate in ordinary language processing.

At the same time, there is evidence that older adults can take differential advantage of context in the recognition of both spoken and written words, especially in noisy environments. Semantic processes at the lexical level also appear to be largely preserved. Semantic priming effects (i.e., facilitation in word processing by prior exposure to a related word) are typically at least as large among older adults as among the young. Also, in the arena of neurocognitive function, evoked potentials show similar lexical effects for young and old – a reduced N400 component for related words relative to unrelated controls. One area of difficulty that older adults may have in word processing is in deriving the meaning of novel lexical items from context, with research showing that older adults are likely to infer more generalized and imprecise meanings relative to the young – a difference that can be largely accounted for in terms of working memory deficits.

**Syntactic processing** Syntax, or the set of rules directing appropriate formation of grammatical sentences, guides the near instantaneous and incremental processing of words into coherent chunks as we encounter them, a phenomenon often termed 'parsing.' Paradigms in neuroscience have shed some light on the influence of aging on brain activity during syntactic parsing. Most of this research indicates that initial parsing processes are fundamentally preserved with aging. Studies using functional magnetic resonance imaging (fMRI), for example, show that for simple sentences, the left peri-sylvian areas responsible for sentence processing produce similar patterns of activation for young and old. Also, attempts to understand ungrammatical sentences produce similar effects among young and old in evoked potentials – for example, exaggerated left anterior negativity to number violations or an exaggerated P600 to phrase structure violations. Other research with evoked potentials suggests a qualitative shift in brain region recruitment, with brain potentials to syntactic violations becoming less asymmetric and more frontal with age. Despite substantial preservation of certain aspects of syntactic processing, complex syntax can especially compromise the older adult's ability to interpret a sentence's meaning and remember it, a finding often attributed to processes downstream from the initial parse. Such effects may be due to cognitive constraints imposed by age-related declines in the working memory resources needed to retain the products of parsing in memory. For example, behavioral and imaging studies comparing the effects of syntactically simpler subject-relative sentences (e.g., 'The dog that chased the boy is friendly') to more complex object-relative sentences (e.g., 'The dog that the boy chased is friendly') suggest that age differences emerge primarily for more complex constructions. As syntactic structure becomes more complex, older adults are more likely to show comprehension errors (e.g., was it the boy who chased the dog or vice versa?) than younger adults and may show relatively less activation in the left peri-sylvian sentence processing region. However, variability among older adults is high, so there is typically a subset of elders who perform at least as well as the young. These high-comprehension older adults appear to achieve good performance in part by upregulating additional brain regions associated with verbal working memory (e.g., left dorsal portions of the inferior frontal cortex and right temporal–parietal regions).

**Textbase processing** Older adults typically show poorer memory for the content expressed directly by the text. Online measures of reading (e.g., reading time and probed recognition) suggest that the fundamental mechanisms of the system used to construct an integrated representation of ideas are preserved. However, processes used to construct the textbase (e.g., to instantiate and integrate concepts in the text) are among the most resource-consuming of those required in language understanding and are hence the most vulnerable to aging. When reading is self-paced, older adults require more time for effective propositional encoding (e.g., as indexed by effective reading time, the time allocated per idea unit recalled). In listening, when the pace is controlled by the speaker, older adults may have particular difficulty in understanding and retaining the information, especially as propositional density (ideas expressed per word) is increased or in noisy environments. Older adults appear to have no difficulty drawing anaphoric inference (i.e., correctly identifying the referent when the pronoun is used to refer to a noun that was introduced earlier) over short distances, but they may have difficulty when the pronoun and referent are separated by intervening text so that the referent must be retained in working memory. Similarly, older adults may have difficulty in reactivating nouns during sentence comprehension in the way that younger adults do, which may compromise the ability to represent a coherent meaning. For example, younger adults trying to understand an object-relative construction, as described previously (e.g., 'The $\text{dog}_i$ that the boy chased $t_i$ is friendly') will reactivate the object (e.g., dog) of the matrix clause following the verb (i.e., at $t_i$, the gap in the phrase structure that leaves a trace of the constituent moved in the relativized construction) and actually show facilitated processing of the object at the gap site. Older adults are less likely to reactivate the object as the noun-gap distance is increased (e.g., 'The $\text{dog}_i$ wearing the blue bandana that the boy chased $t_i$ is friendly').

The general impression from the behavioral data that the semantic (textbase) representation is more fragmented and less distinctive among older adults is reinforced by evidence from evoked potentials. Semantic anomaly (e.g., 'The dog wearing the blue bandana is a tree') exaggerates the N400 component of the evoked potential for the anomalous word (in this case, 'tree') for both younger and older adults; however, the brain response to this anomaly is less pronounced among older adults. Such findings suggest that the encoding of the sentence context up to the anomalous word was not as distinctive among the older adults so that the perception of anomaly was not as strong when the implausible word was encountered. Interestingly, there is no evidence of a delayed response to the anomaly with age; rather, the data are more consistent with a reduced level of response. Contrary to early claims in the literature that aging brings no semantic deficits, it appears that difficulty in integrating concepts can lead to an impoverished semantic representation.

Ironically, older adults may ordinarily allocate less attention to textbase processing. This is demonstrated in reading time for naturalistic text, in which regression analysis is used to isolate the effects of textbase processes. For example, with lexical properties such as length and frequency controlled, reading time increases when new concepts are introduced or at the ends of syntactic constituents, which are sites at which readers pause briefly to integrate concepts into a coherent meaning. These data show that older readers may allocate less time to these processes. However, analogous to fMRI findings with sentence processing, older adults who show high levels of language memory performance often overallocate attention to these processing components or, as described later, may shift allocation to process more holistic features of the discourse. Some older adults do show very good language comprehension and memory, with data from behavioral studies and neuroimaging suggesting that compensatory recruitment of additional resources underlies their successful performance.

Less is known about the factors that engender this recruitment. However, age effects on language performance are often moderated by verbal ability: more verbal older adults are more likely to show event-related potential patterns similar to young, allocate resources to construct the textbase, and show relatively good textbase memory. To the extent that literacy activities are practiced across the life span, growth in knowledge-based systems may support language processing at different levels.

**Situation model** Aside from deriving ideas directly from the text, language understanding also involves elaboration on these ideas based on existing knowledge. Some theories focus on the perceptual quality of this level of representation, which gives rise to a perceptual simulation of the events described by the text. For example, in narrative understanding, readers track the goals and emotional reactions of characters, as well as their movement through space and time. Behavioral methods to study this level of representation include probe recognition for objects in the narrative as well as reading time, both of which show subtle effects of situation model processing. Readers are slower to verify the existence (in the narrative world) of objects that are spatially distant from the protagonist relative to those that are nearby. Readers also slow down when new characters are introduced or when there is a spatial discontinuity (e.g., the locus of narrative events shifts from the village to the castle) or a temporal discontinuity (e.g., 'The next day...'). When the text describes a goal to be achieved (e.g., Susan intends to buy her mother a purse for her birthday), the goal is activated in memory until it is achieved so that concepts related to the goal are more quickly verified as long as the goal is open (e.g., the purse will be more quickly verified if Susan cannot find the purse when she goes to the store relative to a condition in which she succeeds in securing the gift). To the extent that these paradigms have been used to explore adult age differences in situation model processing, there has been very little evidence of developmental differences in situation construction and updating.

Outside of narrative comprehension, it appears that older adults also show the perceptual symbols effect, suggesting that fundamental processes of perceptual simulation are preserved. For example, adults are faster to verify that a pictured object was described in a sentence if the picture of the object represents its shape implied from the sentence (e.g., in encountering, 'She watched the egg frying in the pan,' readers are faster at verifying that the sentence described an egg if the drawing depicts a cracked egg with yolk and white relative to a picture of an egg in its shell, with the reverse effect for 'She watched the egg boiling in the pot'). According to perceptual symbol theory, response time is faster when the picture matches the shape implied by the sentence because sentence processing evokes a perceptual simulation of the events described so that the implied shape is more available when the picture is presented. This perceptual symbols effect is at least as large in the elderly as it is among the young.

Older adults may particularly rely on the situation model to support textbase processing. For example, in ambiguous text (e.g., 'The strength and flexibility of this equipment is remarkable. Not everyone is capable of using it even though most try at one point or another...'), older readers take differential advantage of titles (e.g., Driving a Car) that disambiguate the meaning to facilitate processing. Because the title renders the situation instantly transparent, both younger and older adults are more efficient in reading when it is available; however, older adults show this effect to a larger degree.

To the extent that the hallmark of situation model processing is an integration of textbase content with knowledge, one might expect that older adults would be particularly adept at inferential processing; however, this is not always the case. Although older adults are more likely to draw elaborative inference (e.g., in recall, to annotate their recollections with personal experiences or related information learned in another context), if inference is constrained so that it requires retrieval of textbase content, age deficits are the norm.

**Discourse structures and context** Beyond sentence processing, different genres of text have characteristic forms. For example, narratives typically begin by introducing a setting and characters and proceed to describe a series of episodes in which goals or problems are introduced to be resolved, and so on. Expository texts have certain characteristic forms of argumentation (e.g., problem–solution and thesis–evidence). Older readers generally appear to track these larger discourse structures in the same way as the young. For example, scrambling a narrative so that the canonical form is disrupted will impair recall similarly for young and old. However, other research suggests that older readers may have more difficulty in remembering narratives relative to matched expository text, presumably because of the demands of establishing and tracking complex plot structures. For example, older adults are more likely to have difficulty when two plots are interwoven.

**Recap** Collectively, these findings are generally consistent with what would be expected given declines in mechanics accompanied by experience-based growth in knowledge. The most resource-intensive aspects of language processing requiring the associationist connection of concepts in propositional encoding and storage in working memory show the largest age deficits. At the same time, sensory deficits may indirectly tax working memory by diverting resources to interpret muddy input. However, knowledge-based processes are largely preserved and may support these other processes.

### Language Production

In contrast to comprehension, production shows more pronounced differences as a function of age at almost every level of analysis. In this domain, methodological approaches are almost exclusively behavioral, and even among behavioral measures, the focus has been on word production and on comparing global measures of spontaneous production (e.g., diaries) because of the problems associated with constraining production in a way that can be measured (although there are interesting recent innovations).

Word production stands in sharp contrast to receptive word processes in showing age differences. Picture naming is slower and likely to produce more errors among older adults. As do the young, older adults show effects of name agreement and word frequency on naming latency, suggesting that processes of lemma selection and phonological encoding are intact. Older adults may show relatively faster response time when name agreement is high, suggesting perhaps enhanced effects of knowledge on lemma selection. Aging is also likely to bring an increase in tip-of-the-tongue experiences, that feeling of knowing the word one wants to say (e.g., this experience is evoked among some people by asking them the name of a person who collects stamps) but not quite being able to come up with it (e.g., philatelist). Interestingly, spelling errors also increase with age. This asymmetry between comprehension and production has been explained in terms of transmission deficits between the phonological and semantic networks. It is assumed that connections among lexical and phonological units generally decrease with increased age and disuse. Although the semantic representation is rich in connections, phonological units are constrained (e.g., there are rich and variable ways to encode the concept of chocolate, but if one cannot activate /ch/, production is necessarily disrupted).

Cross-sectional studies of spontaneous production show that older adults generally produce shorter sentences with fewer complex syntactic constructions (e.g., embeddings and left-branching constructions). Diary studies have shown longitudinal change toward simpler syntactic forms and reduced information content as a function of normal aging, with declines exaggerated among those later diagnosed with Alzheimer's disease.

In constrained production (e.g., construct a novel sentence from a given set of target words), age differences may be negligible when the number of target words to be accommodated in the production is small; however, as the number of targets is increased, older adults are more likely to become dysfluent, presumably because of the greater demands on syntactic construction and proposition assembly. Another form of constrained production that has been explored with aging is to create a sentence to describe a picture. In this paradigm, too, older speakers are likely to produce dysfluencies. The monitoring of eye movements has been used to examine the extent to which older adults might plan their utterances ahead to enhance fluency, but evidence on this issue is not yet clear.

Finally, there have been several demonstrations in the contexts of laboratory tasks (e.g., text recall) that older speakers may be more likely to produce off-target speech. Whereas some have shown this to be related to independent estimates of inhibition and have labeled this phenomenon verbosity, other research has demonstrated that such age differences in such elaboration can be exaggerated for personal topics, suggesting that such patterns may be related to social demands rather than to individual differences in cognitive function. This issue remains to be resolved.

## Conclusion

There appears to be a two-way street between the multidirectional change in adult cognition and language processing. Age-related declines in sensory processes and cognitive capacity impact the ability to derive meaning from text and spoken language, especially as syntactic structure becomes more complex or speech rate is increased. Age-related declines also impact name retrieval and the syntactic complexity of production. Knowledge-driven processes can offset these declines to a large degree so that in many arenas, aging individuals will show better performance than their younger counterparts. At the same time, language processing (e.g., habitual reading, or learning a second language) appears to exercise the cognitive system so as to contribute to cognitive vitality.

*See also:* Aging of the Brain; Cognition in Aging and Age-Related Disease; Dementia and Language.

## Further Reading

Bialystok E, Craik FIM, and Freedman M (2007) Bilingualism as a protection against the onset of symptoms of dementia. *Neuropsycholgia* 45: 459–464.

Federmeier KD, McLennan DB, De Ochoa E, and Kutas M (2002) The impact of semantic memory organization and sentence context information on spoken language processing by younger and older adults: An ERP study. *Psychophysiology* 39: 133–146.

Federmeier KD, Van Petten C, Schwarz TJ, and Kutas M (2003) Sounds, words, sentences: Age-related changes across levels of language processing. *Psychology and Aging* 18: 858–872.

Griffin ZM and Spieler DH (2006) Observing the what and when of language production for different age groups by monitoring speakers' eye movements. *Brain and Language* 99: 272–288.

Johnson RE (2003) Aging and the remembering of text. *Developmental Review* 23: 261–346.

Kemper S and Mitzner TL (2001) Language production and comprehension. In: Birren JE and Schaie KW (eds.) *Handbook of the Psychology of Aging,* 5th edn., pp. 378–398. New York: Academic Press.

Kemper S, Thompson M, and Marquis J (2001) Longitudinal change in language production: Effects of aging and dementia on grammatical complexity and propositional content. *Psychology and Aging* 16: 600–614.

Meyer BJF and Pollard CK (2006) Applied learning and aging: A closer look at reading. In: Birren JE and Schaie KW (eds.) *Handbook of the Psychology of Aging,* 6th edn. New York: Elsevier.

Radvansky GA, Zwaan RA, Curiel JM, and Copeland DE (2001) Situation models and aging. *Psychology and Aging* 16: 145–160.

Stine-Morrow EAL (2007) The Dumbledore hypothesis of cognitive aging. *Current Directions in Psychological Science* 16: 289–293.

Stine-Morrow EAL, Miller LMS, Gagne DD, and Hertzog C (in press) Self-regulated reading in adulthood. *Psychology and Aging.*

Stine-Morrow EAL, Miller LMS, and Hertzog C (2006) Aging and self-regulated language processing. *Psychological Bulletin* 132: 582–606.

Thornton R and Light LL (2005) Language comprehension and production in normal aging. In: Birren JE and Schaie KW (eds.) *Handbook of the Psychology of Aging,* 6th edn. New York: Academic Press.

Wingfield A and Grossman M (2006) Language and the aging brain: Patterns of neural compensation revealed by functional brain imaging. *Journal of Neurophysiology* 96: 2830–2839.

Wingfield A and Stine-Morrow EAL (2000) Language and speech. In: Craik FIM and Salthouse TA (eds.) *The Handbook of Aging and Cognition,* 2nd edn., pp. 359–416. Mahwah, NJ: Erlbaum.

Wingfield A, Tun PA, and McCoy SL (2005) Hearing loss in older adults: What it is and how it interacts with cognitive performance. *Current Directions in Psychological Science* 14: 144–148.

# Dementia and Language

**S Kemper**, University of Kansas, Lawrence, KS, USA
**L J P Altmann**, University of Florida, Gainesville, FL, USA



## Effects of Dementia on Language

Language production begins with the formulation of a message and includes steps of discourse planning, lexical selection, and syntactic encoding. Language comprehension involves the analysis of discourse, syntactic, and lexical representations. Because of the many levels of analysis and the necessity of integrating information from all levels, both aspects of language use are compromised to some extent in all types of dementia beyond the small but significant effects of normal aging. Errors can arise at any stage of production or comprehension as a result of attentional lapses, processing limitations, or execution problems. For example, normative aging processes may exacerbate production or comprehension problems due to general slowing of cognitive processes, reductions of working memory capacity, or a breakdown of inhibitory processes. In dementia, these deficits are exacerbated by the breakdown of cortical connectivity within and between language processing regions, depending on the locus and severity of damage. This article examines what is currently known about the way dementia affects language use and attempts to distinguish pathological changes from normative age-related changes to language.

One issue has tended to dominate theoretical discussions of the effects of dementia on language: the modularity of phonological, syntactic, semantic, and/or discourse processes. Modularity, or autonomy, theory holds that language production or comprehension involves a sequence of processing stages or levels. Typically, four properties are ascribed to this sequence of processing stages: (1) each stage corresponds to a distinct linguistic rule system; (2) each stage involves a discrete representational system; (3) processing flows unidirectionally through the sequence of stages; and (4) each stage may be characterized by independent constraints on the rules, representations, or processing operations at that stage. Most models assume that there are a number of autonomous levels of processing, including a phonological level, which specifies the sequence of sounds composing each word; a morphological level, at which part-of-speech information is determined; a syntactic level, in which individual words are formed into phrases and clauses; a semantic level, at which meaning relations within and between words and phrases are determined; and a discourse, or pragmatic level of message structure and function. The degree of interactivity among these levels of representation and the possibility of bidirectional flow of information between levels are at the center of this controversy. Specific forms of dementia can attack any or all these levels, perhaps reflecting the mapping of linguistic functions onto cortical or subcortical structures.

A variety of specific tests for the effects of dementia on language are commonly included as part of the assessment of dementia, including tests of verbal fluency, picture naming, and immediate and delayed recall of word lists and short passages. For example, the Mini-Mental State Examination includes tests of sentence production, object naming, immediate and delayed verbal memory, and the ability to follow verbal commands, in addition to tests of orientation, attention, and constructional praxis. Psycholinguists and neuropsychologists have assessed the effects of dementia on various language-related tasks, such as picture naming, word-picture and sentence-picture verification, metalinguistic judgments, and reading and listening comprehension.

Many linguistic tests are relatively complex, involving a variety of component processes that may be affected by normal aging or other age-related conditions; some of these tests are often used to assess other cognitive processes, such as executive function. One example is verbal fluency. Tests of verbal fluency, such as the Controlled Oral Word Association Test, allow individuals a set amount of time, often 1 min, to generate words in response to a cue, such as words beginning with a target letter or phoneme (e.g., words beginning with an 'f') or members of a semantic category (e.g., the names of animals). Because the task draws on both semantic memory and executive function, deficits are found in nearly all dementias. For example, it is well established that individuals with Parkinsonian dementia and Alzheimer's dementia have verbal fluency deficits, particularly on tests involving retrieval of members of semantic categories. These deficits in individuals with Parkinsonian dementia have typically been attributed to nonsemantic problems associated with executive function, such as problems of inhibitory control, whereas similar deficits in individuals with Alzheimer's dementia have been attributable to a breakdown of semantic memory or semantic retrieval processes. Some attempts have been made to develop separate semantic and executive-function components of verbal fluency performance by tracking the time course of responses, requiring responses to alternate or switch between two categories, and varying other task parameters.

## Differential Effects

Dementias differentially affect language processes, primarily because of the location and scope of damage from the particular disease. Thus, some symptoms may be similar to those resulting from stroke affecting the same cortical region. However, dementias are degenerative, so the initial effects often are much milder than those resulting from stroke, more difficult to detect in early stages, and progressive. Alzheimer's dementia has been studied in some depth, whereas research into other dementias has been more limited.

### Alzheimer's Dementia

Early in the disease, Alzheimer's dementia primarily affects posterior association cortex and only rarely extends to the perisylvian region. Consequently, phonology and phonetics, the sounds of words and the ability to articulate them, are typically affected only late in the disease, if at all. Similarly, morphological processing (e.g., appropriate use of verb tenses and sentence structure) is also considered a primarily perisylvian function that is usually preserved during early stages of the disease, although grammatical errors increase in frequency as the disease progresses.

In contrast, Alzheimer's dementia has a more significant effect on other aspects of language that rely on semantic knowledge (i.e., word meanings) and conceptual knowledge (i.e., knowledge about items and their attributes) and on the ability to integrate the meanings of several words. Often an early sign of Alzheimer's dementia is a marked increase in word-finding problems during conversation, as individuals experience frequent 'tip of the tongue' events, substitute generic references (e.g., "that thing"), and rely on metalinguistic references (e.g., "oh, you know who I mean"). At diagnosis, most Alzheimer's dementia patients score below normal on picture-naming tests, and this impairment increases over time. Errors are typically substitutions of words from the same semantic category, (e.g., saying 'horse' for 'giraffe') or category names (e.g., 'animal' or 'tool'), although other types of errors, such as visual errors and failures to respond, also occur. However, the structure of semantic knowledge remains intact until quite late in the disease; thus, patients have little difficulty sorting pictures by semantic category or verifying the category membership of an item. Performance on word–picture matching tasks in which the individual must identify a spoken or written target within an array of pictures is also more preserved than picture naming, particularly if there is little visual similarity among distracter pictures. Visual-perceptual difficulties may also contribute to the difficulties of Alzheimer's dementia patients in such tasks.

Extensive research has suggested that word-finding problems in conversation and picture-naming tasks are due to both a degradation of the meanings of words and difficulty accessing word meanings. Preserved partial meanings are evident in 'circumlocutions'; for example, someone unable to recall 'horse' may say, "Oh, you ride it. You put a leather thing on its back and you ride it. They race them too." The disease appears to impair the ability to focus word searches, resulting in activation of conceptually similar words (e.g., horse, deer, cow, donkey). These alternatives may compete for production, contributing to a breakdown in word retrieval. As the disease progresses, word meanings themselves appear to deteriorate. Even so, individual words, particularly high-frequency, highly familiar words, may be preserved, and comprehension is somewhat spared relative to production throughout the disease process.

Oral reading is preserved in mild Alzheimer's dementia, although comprehension of written sentences and texts is impaired. Thus, they may be able to read connected text that they cannot understand. In contrast, writing is impaired relatively early in Alzheimer's dementia and is characterized by simplified grammatical structures, loss of content, and regularized spelling.

Although the speech of individuals with mild or moderate Alzheimer's dementia is often described as 'empty speech,' it is quite distinct from the empty speech produced by individuals with fluent aphasia following stroke. Individuals with Alzheimer's dementia do not produce strings of words with little relevance to context, perseverate, or use ungrammatical phrases or 'word salad' jargon until extremely late in the disease, if at all. Instead, the conversational speech of individuals with Alzheimer's dementia is characterized by the overuse of general nouns (e.g., 'thing,' 'stuff') and verbs (e.g., 'get,' 'do') and by an overreliance on pronouns. These phenomena decrease the amount of information conveyed and can make following their conversations difficult. Even so, in Alzheimer's dementia, the fundamental building blocks of sentences are relatively intact; sentences contain noun and verb phrases and may also include complex grammatical constructions. Thus, conversational speech is often relatively coherent until moderate and late stages of the disease.

Understanding conversational speech requires listeners to combine the meanings of many words, often across sentence boundaries; keep these in memory; and integrate the new information with past knowledge. Thus, it is an extremely demanding task. Patients with Alzheimer's dementia often miss a good deal of the information in a conversation; however, they may understand the gist of a conversation by relying on contextual cues. As the disease progresses, the appropriateness of language declines as behavior becomes increasingly disinhibited. Even mild-mannered Alzheimer's dementia patients may become verbally abusive. Typically, speech output also declines as the need to communicate dwindles. Gradually, conversational speech disappears, eventually ending in mutism.

### Parkinsonian Dementia

Parkinson's disease primarily affects the ability to smoothly coordinate and perform speech, resulting in slow, dysarthric articulation in combination with a soft, breathy voice. The ability to produce consonants that require tongue movements, such as /t/, /d/, /l/, /n/, /k/, and /g/, is most impaired. Individuals with Parkinsonian disease may evidence 'articulatory undershoot,' in which the tongue does not quite reach the appropriate spot in the mouth or does not achieve full closure to block the flow of air when necessary. Further, if they do achieve adequate closure, they may either release it too fast or maintain it too long, decreasing intelligibility. Speech rate varies among patients; some may speak faster than normal, others at a normal rate. Often, there is a lack of awareness of these deficits; individuals with Parkinson's disease may feel that they are shouting when they are barely audible to others.

Recent research has revealed that other aspects of language use are also impaired in Parkinsonian dementia, suggesting either that subcortical structures play a role in language use or that the reduced dopamine production in Parkinson's disease impairs frontal lobe function, secondarily affecting language use. Converging evidence for the latter argument comes from findings that individuals with Parkinsonian dementia often score poorly on tests of immediate memory, executive function, and processing speed, all of which have been found to be associated with performance on language tasks in healthy older adults. Consequently, performance on all language tasks that rely on memory or executive function, such as reading comprehension and the production of complex sentences, is likely to be impaired. Micrographia, extremely small writing, is characteristic of Parkinson's disease. However, there are no studies on the content of writing in Parkinsonian dementia and very few on reading connected text or producing connected speech.

### Frontotemporal Dementias

Frontotemporal dementia (FTD) is usually described as having three variants that differ in their behavioral manifestation depending on the locus of damage. These include primary nonfluent aphasia, semantic dementia, and frontal-variant FTD. In contrast to Alzheimer's disease, damage in FTD is relatively focal; consequently, episodic memory, spatial cognition, and orientation are usually preserved until late in the disease. Frontal-variant FTD manifests primarily as social and behavioral impairments. Language is relatively unimpaired in this group, although there are mild deficits affecting category and letter fluency and a very mild impairment of picture naming. Most language impairments in this subgroup seem to be related primarily to deficits in executive function.

**Primary nonfluent aphasia** Primary nonfluent aphasia, also known as primary progressive aphasia, is characterized by increasing difficulty programming speech movements, which leads to severely impaired spoken output, or dysarthria, from early in the disease. Speech becomes slower and more effortful over time and is marked by long word-finding pauses, phonemic paraphasias, and distortions. Intonation patterns also deteriorate, possibly because of the increasing effort needed to produce speech or the decreasing length of utterances.

Word-finding difficulties are often among the initial complaints of individuals with primary nonfluent aphasia; these difficulties are evident in both picture naming and spontaneous speech and become increasingly severe over time. As mentioned above, the difficulty seems to be in determining and producing the sounds of the words. In contrast, comprehension of single words as measured by word–picture matching tests is largely intact until late in the disease. Thus, in individuals with primary nonfluent aphasia (about 30% at the mild stage of the disease), there is a large discrepancy between the ability to comprehend words and the ability to name pictures, and this discrepancy increases as the disease progresses. Based on these findings, several researchers consider that the primary impairment in primary nonfluent aphasia affects the phonological output representations.

Primary nonfluent aphasia may affect different parts of speech in different ways; phonemic paraphasias may occur primarily with nouns, and semantic paraphasias primarily with verbs. There is mounting evidence that in some patients, primary nonfluent aphasia degrades the semantic or conceptual representations of verbs and that this degradation, in turn, disrupts sentence production. As in acquired aphasias, verb impairments are often accompanied by impaired use of verb endings and difficulties comprehending complex or reversible sentences. Some individuals with primary nonfluent aphasia have been described as producing agrammatic speech. Very little is currently known about how primary nonfluent aphasia affects written production or the comprehension of sentences or discourse in either spoken or written modalities.

**Semantic dementia** Semantic dementia, also called progressive fluent aphasia, results when FTD primarily affects the anterior inferior temporal lobe. The defining feature in semantic dementia is a progressive loss of semantic and conceptual knowledge. Thus, semantic dementia affects performance on all tasks that require the access to or manipulation of word meanings, leading to extremely poor picture naming and severely impaired verbal fluency. In contrast, sentence grammar and discourse are relatively preserved although disrupted by the prevalence of word-finding difficulties.

Because the disorder impairs conceptual knowledge as well as semantic knowledge, individuals with semantic dementia may not be able to demonstrate how to use common items, such as a hammer or comb. Visual-perceptual aspects of meaning (e.g., shape and relative size) can be especially impaired, particularly knowledge of biological categories, such as fruits, vegetables, or animals. Comprehension of sentences is also impaired, although the patients may be able to abstract a 'gist' meaning based on whatever partial semantic information is available to them. Due to the deterioration of conceptual knowledge, reading comprehension and written communication are also impaired. Reading aloud may resemble surface alexia, in which irregularly spelled words are pronounced as they are written. Similarly, writing may also be impaired, with regularized spelling and low information content.

**Other dementias: vascular dementia and Lewy body disease** The effects of other dementias on language have not been well studied. Two of the more common types of dementia on which there is some language research are vascular dementia and Lewy body disease. Vascular dementia is difficult to differentiate from Alzheimer's dementia without reliance on neuroimaging. Both syndromes show impairments in word finding, picture naming, and letter and category fluency. However, individuals with vascular dementia are particularly poor at letter fluency relative to category fluency, and the reverse is true for individuals with Alzheimer's dementia. Individuals with vascular dementia also outperform those with Alzheimer's dementia on verbal memory tasks, such as free verbal recall, delayed verbal recall, and recognition memory of verbal material. Both vascular dementia and Alzheimer's dementia patients show a range of attention, executive function, and working memory impairments. Thus, the cognitive and linguistic characteristics of vascular dementia and Alzheimer's dementia are very similar, and to date, there are no 'cut-off' scores available to distinguish between these groups.

Lewy body disease primarily affects visual-spatial skills and the extrapyramidal systems rather than language, although it frequently accompanies both Parkinson's disease and Alzheimer's dementia. While those with Lewy body disease and Alzheimer's dementia are similar with respect to performance on picture-naming and word fluency tasks, attention and executive function are usually more impaired in Lewy body disease. On the other hand, individuals with Lewy body disease typically have better verbal memory, especially with respect to delayed recall and recognition memory, than do individuals with Alzheimer's dementia. Individuals with Lewy body disease also exhibit profound visuospatial and visuoconstructional impairments very early in the disease; for example, they are often unable to copy the overlapping pentagons in the Mini-Mental State Examination, even at an early stage of the disease. The spontaneous speech of the two groups differs qualitatively. Conversational speech in Alzheimer's dementia is characterized by a lack of content and relatively intact grammar; conversational speech in Lewy body disease is marked by confabulation.

Unfortunately, few studies have examined sentence and discourse level language production or comprehension in either spoken or written modalities in individuals with vascular dementia or Lewy body disease.

## Interactions and Interventions

Family members and other caregivers of individuals with dementia typically cite communication problems as a major source of caregiver stress. Breakdowns in communication, whether from word-finding problems, dysarthria, or repetitive questions, may lead to other problem behaviors, such as agitation and aggression, and may impair quality of care. Little systematic research has assessed intervention and training programs targeted at alleviating communication problems with dementing individuals – largely as a result of the perception that, because these are degenerative conditions, language treatment would be a waste of resources. Two studies of semantic training in semantic dementia have shown immediate gains in naming ability but no maintenance of these gains at follow-up. At least one study has shown maintenance of treatment gains in picture naming in Alzheimer's disease 2 months after treatment. However, most intervention programs target caregivers, emphasizing the use of functional communication strategies, such as making eye contact, repeating key words, using yes–no questions and lexical prompts, and providing memory books and other conversational stimuli. Other programs have targeted the development of techniques for minimizing repetitive questions and other disruptive behaviors.

A special speech register, sometimes termed 'elderspeak,' is often recommended as a way to enhance or facilitate communicating with older adults, especially those with dementia. Elderspeak has been characterized as involving a simplified speech register with exaggerated pitch and intonation, simplified grammar, limited vocabulary, and slow rate of delivery. Many of the characteristics of elderspeak, such as its slow rate and simplified syntax and vocabulary, may be helpful for ensuring comprehension in impaired populations, although there has been little research assessing its effectiveness. On the other hand, certain characteristics of elderspeak, such as the use of exaggerated prosody, collective pronouns (e.g., saying 'we' instead of 'you'), diminutives (e.g., 'Chrissy' instead of Chris or Mrs. Smith), and terms of endearment (e.g., 'honey' and 'darling'), resemble another register of speech, 'baby talk.' These characteristics are often perceived as demeaning or controlling. Alternative strategies for enhancing communication with adults with dementia must be developed by combining clinical observation with empirical research.

## Implications and Conclusions

There has been considerable interest in using linguistic measures to predict risk, aid in diagnosis, and track the progression of dementia, particularly Alzheimer's dementia. Performance on tests of verbal memory, such as immediate and delayed word list recall, as well as tests of verbal fluency, appear to be particularly sensitive indicators of risk for dementia and its onset and seem to covary with other known risk factors, such as apolipoprotein E genotype. However, deficits in verbal fluency are not specific to Alzheimer's disease, as discussed above. The relationship between linguistic ability and the risk for Alzheimer's disease and longevity has been the focus of the Nun Study, an ongoing longitudinal, epidemiological study of aging. Low linguistic ability in young adulthood, determined from an analysis of language samples written by the nuns at age 18 to 32 years, was associated with increased risk for poor performance on cognitive and memory tests in late adulthood, increased neuropathology characteristic of Alzheimer's disease, and increased all-cause mortality among participants. These findings suggest that linguistic ability may offer a general measure of cognitive and neurological development. Low linguistic ability in young adulthood may reflect suboptimal neurocognitive development, which in turn may increase or reflect susceptibility to Alzheimer's and other dementing diseases.

*See also:* Aging of the Brain and Alzheimer's Disease; Aging and Memory in Humans; Alzheimer's Disease: An Overview; Cognition in Aging and Age-Related Disease; Language in Aged Persons; Semantic Memory.

## Further Reading

Altmann LJP, Kempler D, and Andersen ES (2001) Speech production in Alzheimer's disease: Re-evaluating morphosyntactic preservation. *Journal of Speech, Language, and Hearing Research* 44: 1069–1082.

Bayles KA and Kaszniak AW (1987) *Communication and Cognition in Normal Aging and Dementia.* Boston: College-Hill.

Grossman M (1999) Sentence processing in Parkinson's disease. *Brain and Cognition* 40: 387–413.

Grossman M and Ash S (2004) Primary progressive aphasia: A review. *Neurocase* 10: 3–18.

Hamilton HE (1994) *Conversations with an Alzheimer's Patient.* Cambridge, UK: Cambridge University Press.

Kemper S, Thompson M, and Marquis J (2001) Longitudinal change in language production: Effects of aging and dementia on grammatical complexity and propositional content. *Psychology and Aging* 16: 600–614.

Ripich DN (1994) Functional communication with AD patients: A caregiver training program. *Alzheimer's Disease and Associated Disorders* 8: 95–109.

Rogers TT, Ivaniou AD, Patterson K, and Hodges JR (2006) Semantic memory in Alzheimer's disease and the frontotemporal dementias: A longitudinal study of 236 patients. *Neuropsychology* 20: 319–335.

Ryan EB, Giles H, Bartolucci G, and Henwood K (1986) Psycholinguistic and social psychological components of communication by and with the elderly. *Language and Communication* 6: 1–24.

Savundranayagam MU, Hummert ML, and Montgomery RJV (2005) Investigating the effects of communication problems on caregiver burden. *Journal of Gerontology: Social Sciences* 60: S48–S55.

Snowdon DA, Kemper SJ, Mortimer JA, Greiner LH, Wekstein DR, and Markesbery WR (1996) Cognitive ability in early life and cognitive function and Alzheimer's disease in late life: Findings from the Nun Study. *Journal of the American Medical Association* 275: 528–532.
Williams KN (2006) Improving outcomes of nursing home interactions. *Research in Nursing and Health* 29: 121–133.

## Relevant Websites

http://www.alz.org – Alzheimer's Association research page.
http://www.alzforum.org – Alzheimer Research Forum.
http://www.alzheimers-research.org – Alzheimer's Research Foundation.

# Aging: Brain Potential Measures and Reaction Time Studies

**E Golob**, Tulane University, New Orleans, LA, USA
**H Pratt**, Technion – Israel Institute of Technology, Technion City, Israel
**A Starr**, University of California at Irvine, Irvine, CA, USA



## Brain and Cognitive Aging

Normal aging affects brain functions in a variety of ways. Some regions, such as prefrontal cortex, medial temporal lobe, and neuromodulatory systems (cholinergic, noradrenergic, and serotonergic), show marked age-related changes, whereas other regions, such as primary sensory and motor regions, show comparatively little change during normal aging. Age-related changes in memory function are also selective, which is related to regional differences in the impact of brain aging. In general, age-related declines are most evident in certain types of explicit memory tasks that require conscious retrieval of remembered information, but they are small or absent for many implicit memory tasks. For example, episodic memory is one type of explicit memory, and it is defined as memory for specific events and their context, such as remembering what one did last weekend. In contrast, implicit memory can be expressed even if a person does not consciously remember a specific learning episode or cannot articulate what was learned. An example of implicit memory is the gradual development of proficiency in various kinds of motor skills, such as riding a bicycle. Because the adverse effects of aging are most evident for explicit memory, this article focuses on event-related potential (ERP) and electroencephalogram (EEG) findings while subjects perform explicit memory tasks. For most studies of human aging, the comparison group of young subjects consists of young adult college students, typically between the ages of 18 and 25 years. Depending on the study, older subjects can range in age from the 60s, termed 'young-old,' to older than 90 years, termed the 'oldest-old.'

## EEG and ERP methods

The EEG is a measure of continuous variations in voltage over time from electrodes placed on the scalp. It was first developed in the 1930s by Hans Berger, and it was one of the first methods for non-invasively measuring brain activity in humans. Event-related potentials can be used to define neural activity measured by the EEG that is time-locked to an event, such as the onset of a stimulus, the onset of a behavioral response, or their correlation with cognitive processing. An example of an ERP following presentation of an auditory stimulus is shown in **Figure 1(a)**. Depending on the component of interest, the EEG waveforms associated with tens to thousands of events (stimuli or behavioral responses) are averaged together to yield the ERP. The main ERP measures employed are peak amplitude and latency of individual components. Amplitude is conventionally defined as the voltage difference of a peak along the waveform relative to a baseline period, such as the mean amplitude during the 100 ms before stimulus presentation (**Figure 1(b)**). Latency is defined as the time between stimulus onset and the peak of a given component.

## Processing Speed and Reaction Time

A basic observation of cognitive aging is that performance of a wide variety of reaction time tasks declines during the course of normal aging, even when subjects are in excellent health. This is often depicted using a Brinley plot, whereby reaction times on multiple tasks in young subjects are arranged in ascending order on the *x*-axis and plotted against the respective reaction times of older subjects performing the same tasks on the *y*-axis. Reaction times in young and older subjects show a linear relationship, with a slope $>1$ indicating larger absolute age effects on reaction time on tasks having longer overall reaction times.

Large-scale cross-sectional and longitudinal studies suggest that a general factor can account for most of the typical cognitive changes associated with normal aging ($>90\%$ of variance). Theorists have proposed that the general factor may indicate fluid intelligence or a 'resource' such as processing speed. Processing speed as a general factor is supported by observations that age differences in reaction time can predict age differences on more complicated tests, such as episodic memory or verbal fluency. Thus, slowing of reaction time appears to reflect a fundamental aspect of cognitive aging. As a practical matter, when comparing reaction times among young and older subjects performing two or more tasks, it is often worthwhile not only to express the absolute differences between tasks for each group but also to include a relative measure, such as percentage increase from task A to B in each group, to take into account age-related slowing of overall reaction time.

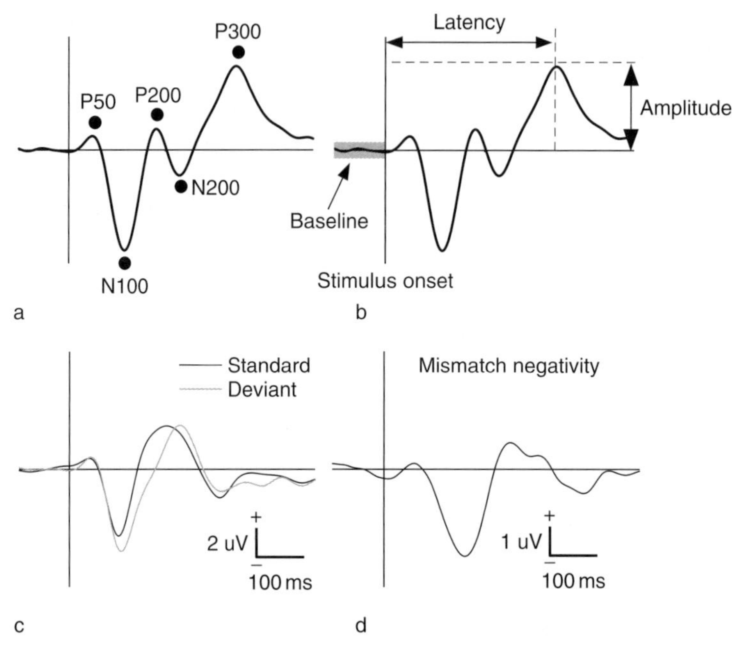


**Figure 1** Examples of event-related potential components and measurement of amplitude and latency. (a) Example of auditory event-related potentials elicited by infrequent target stimuli during a target detection task in young subjects. The vertical line indicates stimulus onset. Voltage is plotted as a function of time, from a 100 ms prestimulus baseline to 500 ms after stimulus onset. Components are named by convention using polarity ('P' or 'N') followed by approximate latency. Thus, the 'P50' is a positive component having a latency of ~50 ms. The component labeled 'P300' refers to the P300b (or P3b) component. A different component called the P3a is elicited by distractor stimuli, often a novel sound other than the task-relevant target or nontarget stimuli. (b) Illustration of amplitude and latency measures of the P300 component. (c) Example of the mismatch negativity. Subjects passively listened to standard (90%) and deviant (10%) tone stimuli. Event-related potentials to standard and deviant stimuli are plotted together. Note that potentials to deviants are more negative than standards from ~100 to 240 ms after stimulus presentation, indicating the mismatch negativity. (d) Example of mismatch negativity difference waveform. To better visualize the mismatch negativity, the event-related potential waveform at each time point for standards can be subtracted from the deviant waveform. Note the differences in scale for (d) vs. (a)–(c). To be consistent with event-related potential examples in (a)–(c), negativity is plotted downward. However, in many publications the mismatch negativity is shown with negative polarity plotted upwards.

## Sensory Memory

Sensory memory is a relatively automatic form of memory and has a duration of several seconds. In the auditory modality sensory memory is important for the perception of speech and various aspects of auditory scene perception. The mismatch negativity (MMN) is an ERP that has been extensively used to study sensory memory (**Figures 1(c)** and **1(d)**). The MMN is elicited by a stimulus that deviates from a previously established pattern of stimuli presented sequentially. The rationale for its use is that sensory memory can be probed by first establishing a standard sensory memory representation and then presenting a stimulus that is predicted to differ from the sensory memory representation in some respect. If the MMN is elicited, it can be inferred that the deviant aspect of the stimulus was not a property of the standard stimulus representation (mismatch). It is a way to infer memory representations by defining the 'limits' of the sensory memory representation, after which a stimulus is considered different.

Studies of the MMN in normal aging consistently report that when the deviant stimulus differs from the standard in terms of acoustic properties such as frequency or intensity, there are little or no age differences. However, if deviant stimuli differ in terms of timing, such as stimulus duration or interstimulus interval, then age differences are often evident, with smaller MMN amplitudes in older compared to young subjects.

## Working Memory

Working memory is the process of maintaining and manipulating information that is relevant to performing a task for short periods of time (seconds to minutes). A typical example of maintaining information in working memory is remembering the digits of an unfamiliar phone number prior to placing the call. An example of manipulation is the addition of an area code before the phone number in working memory and reordering the digits to correctly dial the number.

Working memory is frequently studied using variations of a paradigm developed by Saul Sternberg, which provides the ability to document both behavioral and brain activity changes accompanying working memory retrieval. In the Sternberg task, subjects perform a series of working memory trials. In each trial a set of items, typically numbers or letters, are memorized in sequence. After a delay of a few seconds (typically 2–4 s), retrieval is assessed by presenting a probe item. Subjects then determine whether or not the probe was a member of the current memory set, and press one of two buttons to indicate their decision. When a range of memory loads are tested (from one to approximately six to eight items), reaction time shows a positive linear relationship with memory load, with longer reaction times to probes when more items are memorized. Thus, the Sternberg task is a useful tool to define systematically the impact of memory load on reaction time and ERP measures during encoding (memorized items) and memory retrieval (probe), which is important because limited capacity is a defining feature of working memory.

The slope of the memory load versus reaction time function is usually steeper in older subjects than in young subjects, indicating slower retrieval speed. Age effects are sometimes smaller when subjects are rigorously screened to exclude all but the healthiest older subjects or use young-older people (in their 50s), but retrieval speed is still slower in older groups relative to young adults in their 20s. Although not always examined, slower retrieval speed can persist even when there are controls for overall differences in reaction time, if present.

Most ERP studies using the Sternberg paradigm have focused on a prolonged late positive waveform present from approximately 200 to 800 ms that is elicited by probes. The most prominent component is often called the 'late positive wave' or 'P300'; it is similar to the P300 shown in **Figure 1** in response to targets, but it is longer in duration. Detailed analyses indicate that probes elicit a set of components comprising the late positive waveform and likely correlate with somewhat different cognitive processes during memory retrieval. Control tasks show that the late positive wave is not present when the same stimuli are presented as in the working memory task, but probes do not require comparison to memory. In young subjects, increased memory load is associated with increases in P300 latency and decreases in P300 amplitude. Older subjects show a similar pattern of results as a function of memory load, with overall increases in P300 latency and a weaker correlation between P300 latency and reaction time.

In auditory versions of the Sternberg task, a component called the 'sustained frontal negativity' is elicited by memorized items and probes when environmental sounds are used, especially when most probes match memorized items. It is unclear if the sustained frontal negativity is present for other modalities. The amplitude of the sustained frontal negativity is reduced in older subjects relative to young subjects. It has been suggested that a smaller sustained frontal negativity may be associated with age-related changes in prefrontal cortex function during working memory.

Studies have examined potentials that occur before the P300/late positive wave and sustained frontal negativity. The purpose of examining shorter latency components is to determine if modality-specific components associated with sensory cortical activity are also modulated by working memory demands. In the auditory modality, amplitude of the N100 component is affected by memory load during both encoding of items and retrieval, with linear reductions in amplitude as memory load is increased. Amplitude of the N100 to probes is also modulated by the details of encoding, such as the order of item presentation, and is not affected by memory load when memorized items are presented in the visual modality. Both results are consistent with psychological theories emphasizing a linkage between memory retrieval and specific processes during memory encoding, such that memory retrieval may reinstate aspects of perceptual processing.

Compared to young subjects, older subjects exhibit two main differences in auditory potentials in the Sternberg task. During encoding, young subjects have shorter N100 latencies with increasing memory load, whereas N100 latency in older subjects is invariant across memory loads. The second main age difference is that probe N100 amplitudes to memorized items show increases with greater memory load in older subjects, but the opposite is seen in the young as amplitudes decrease with memory load. In addition, N100 amplitudes to probes that were not in the memorized list also exhibit smaller changes in N100 amplitude as a function of memory load relative to young subjects. The significance of age-related changes in auditory potentials is unclear, but it has been proposed that changes in prefrontal cortex function may be relevant because prefrontal cortex can regulate the responsiveness of sensory cortex.

## Declarative: Episodic Memory

ERP and EEG studies of episodic memory typically employ a recognition format instead of having subjects freely recall the memorized material. In recognition memory tests, the experimenter presents stimulus items to which the subject has to make a judgment, such as whether or not the item was previously memorized. A recognition format is preferred for ERP and EEG studies because the ability to precisely control stimulus timing is necessary to define neural responses, such as ERPs, that are time-locked to stimulus presentation. The Sternberg task described previously is also a recognition task because subjects judge whether the probe was a memorized item.

Target detection, or 'oddball,' tasks, first described by Samuel Sutton and colleagues in 1965, can be considered a simple recognition memory test. Subjects are presented a sequence of stimuli and are instructed to detect the infrequent occurrence of a predefined target stimulus. Subjects usually either press a button to each target or count the number of targets presented in the sequence. An example of ERPs recorded during a target detection task is shown in the top of **Figure 1**. Both target and nontarget stimuli elicit sensory potentials, such as P50, N100, and P200, in response to auditory stimuli. However, targets also elicit two additional components that are small or absent compared to nontargets: the N200 and P300 complex. N200 and P300 are associated with the active identification of targets because they are not present during passive listening tasks, such as those used to elicit the mismatch negativity. When the responses to nontargets are analyzed in detail as a function of the stimulus sequence history, a positive component at 250 ms can be seen when the likelihood for a target to appear is high. Thus, each stimulus in the train appears to be processed slightly differently depending on the expectations of the listener.

Many studies have shown that P300 changes substantially during normal development. P300 latency is shortest in early adulthood (teens) and has a gradual, progressive, increase in latency across age cohorts until at least the 10th decade of life (**Figure 2**). As a group, P300 latency to auditory stimuli for subjects in their 80s is approximately 100 ms longer than that of young adults in their early 20s (300 vs. 400 ms). The same pattern of age differences is observed for P300 to visual stimuli, which has somewhat longer latencies (by approximately 60 ms for all age groups) relative to auditory stimuli. When the distribution of voltage across the entire scalp at the time of the P300 peak is examined with an array of electrodes, the focus of maximum amplitude is more anterior in older relative to younger subjects. Age differences in P300 latency are probably not attributable to differences in sensory function because stimuli are easily distinguishable by all subjects, and age differences in P300 latency are not affected by manipulations of stimulus discriminability. Response preparation is also unlikely because, unlike most reaction time tasks, reaction times to targets show little change as a function of age in healthy subjects and P300 latency shows little or no change with response demands. Thus, although the precise cognitive correlates of P300 latency are uncertain, age differences are likely associated with memory-dependent stimulus classification processes.

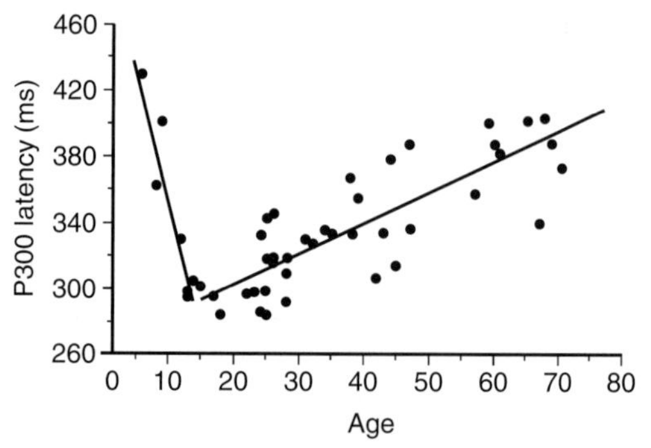


**Figure 2** P300 latency as a function of age. Linear decreases in P300 latency are seen during childhood, with shortest latencies during late adolescence. Gradual increases in P300 latency are then observed throughout the adult years. An auditory target detection task was used to assess P300 latency to infrequent targets (20% stimuli) that differed from nontargets according to pitch. Circles indicate measures from individual subjects. Note that 'P300' refers to the P300b component. Data replotted from Goodin DS, Squires KC, Henderson BH, and Starr A (1978) Age-related variations in evoked potentials to auditory stimuli in normal human subjects. *Electroencephalography and Clinical Neurophysiology* 44(4): 447–458.

In most episodic memory studies subjects are requested to remember multiple items rather than just one target item as in target detection. Two main paradigms are used extensively in episodic memory studies using ERPs: study/test and continuous recognition. In the study/test paradigm, items are encoded during a study period and memory is later assessed during a test period. One advantage of the study/test paradigm is that it permits separate assessment of potentials during encoding and retrieval. In contrast, in continuous recognition tasks, subjects are presented a sequence of items and each item is typically judged as either old or new. Thus, encoding and retrieval processes are operative during each item presentation.

Presentation of items in both paradigms elicits positive slow wave potentials. The slow waves begin as early as ~250 ms and last for hundreds of milliseconds depending on the component and task. During retrieval in both study/test and continuous recognition previously studied items elicit slow waves that are more positive relative to new items – a phenomenon called the 'ERP old/new effect.' In the study phase of study/test tasks positive slow waves following item presentation are also present. Moreover, items that are later remembered, via either free recall or correct recognition, elicit more positive slow waves than items that are later not remembered. This finding is called the 'subsequent memory effect.' Here, we focus on episodic memory retrieval because only a few studies have examined subsequent memory effects as a function of aging, and the subsequent memory effect is sometimes not observed in young subjects.

There are two main positive slow waves associated with old/new effects in study/test paradigms: one that is largest over left parietal recording sites and another that is maximal over right frontal areas. The slow waves are also distinguished by onset time and duration, with the left parietal slow wave having an earlier onset and shorter duration compared to the right frontal slow wave. The two slow waves can be dissociated by manipulating task demands. The left parietal slow wave is associated with retrieval of item information, such as old versus new, whereas the right frontal slow wave is associated with item information and contextual information. Item information is specific knowledge presented during the study session, whereas contextual information reflects other aspects of the learning situation that are either not necessarily relevant to the memory test or are common to multiple items. For example, subjects may memorize words presented on two separate lists and be tested later using item recognition for the words (old/new words). In this example, the actual words on the lists would be considered item information, whereas remembering which of the two lists contained a given word would be considered contextual information.

The old/new effect at left parietal sites changes little with aging, suggesting that certain retrieval processes associated with item recognition may not change substantially during aging. In contrast, the right frontal slow wave does exhibit age differences. Although additional study is needed, the right frontal slow wave is usually attenuated in older subjects relative to young subjects when judgments of context are assessed. Age differences in right frontal slow wave amplitude are accompanied by behavioral differences, with less accurate judgments of item context in older subjects. Thus, behavioral and ERPs studies are consistent in showing that normal aging is associated with small differences in recognition memory for items, whereas substantial age differences are evident when memory for context is assessed.

## Conclusions

This summary of findings on memory and aging using reaction time and electrophysiological measures underscores the benefit of combining behavioral and physiological tools. Behavioral measures reflect the overt product of cognitive and response selection. Different components of electrophysiological responses reflect specific aspects of covert brain processing. Thus, ERPs and EEG provide a tool to define covert memory processes and compare these memory processes as a function of normal aging.

*See also:* Aging of the Brain; Aging: Extracellular Space; Brain Volume: Age-Related Changes; Cognition in Aging and Age-Related Disease; Functional Neuroimaging Studies of Aging.

## Further Reading

Allison T, Hume AL, Wood CC, and Goff WR (1984) Developmental and aging changes in somatosensory, auditory and visual evoked potentials. *Electroencephalography and Clinical Neurophysiology* 58(1): 14–24.

Friedman D (2000) Event-related brain potential investigations of memory and aging. *Biological Psychology* 54(1–3): 175–206.

Goodin DS, Squires KC, Henderson BH, and Starr A (1978) Age-related variations in evoked potentials to auditory stimuli in normal human subjects. *Electroencephalography and Clinical Neurophysiology* 44(4): 447–458.

Iragui VJ, Kutas M, Mitchiner MR, and Hillyard SA (1993) Effects of aging on event-related brain potentials and reaction times in an auditory oddball task. *Psychophysiology* 30(1): 10–22.

Naatanen R (1992) *Attention and Brain Function.* Hillsdale, NJ: Lawrence Erlbaum.

Sternberg S (1966) High-speed scanning in human memory. *Science* 153(736): 652–654.

Sutton S, Braren M, Zubin J, and John ER (1965) Evoked-potential correlates of stimulus uncertainty. *Science* 150(700): 1187–1188.

# Neuroendocrine Aging: Pituitary Metabolism

**W E Sonntag**, University of Oklahoma Health Science Center, Oklahoma City, OK, USA
**C B Herenu and R G Goya**, National University of La Plata, La Plata, Argentina



## Overview

A substantial volume of empirical and scientific evidence has accumulated that demonstrates that biological aging is associated with functional deficits at the cellular, tissue, organ, and system levels. Several theories have been proposed to explain these changes as well as the increased risk of disease with age. However, to date, no single explanation satisfactorily accounts for the physiological and pathological events related to biological aging, and it is apparent that several factors interact, resulting in the decline in tissue function that creates diversity in the aging phenotype. Current concepts of aging encompass genetic, free radical, somatic mutation, DNA repair, and cellular theories, all of which are not mutually exclusive. Although there is evidence to support a relationship between these theories of aging and end-of-life pathologies and/or life span, there is no overwhelming evidence that the progressive development of the aging phenotype can be adequately explained by these classical theories.

The neuroendocrine theory of aging evolved from observations that animals exhibit predictable and progressive impairments in a number of physiological processes over time, including, but not limited to, decreases in reproductive and immune function, decreases in muscle mass and function, increased adipose tissue mass, and decreases in glucose utilization and cognitive function. The role of the neuroendocrine system in these processes is related to the observation that many of the hormones regulated by the neuroendocrine system have an important trophic and integrative role in maintaining tissue function, withdrawal of hormonal support mimics some of the phenotypes observed in aging animals and humans, and hormone replacement therapy in many cases improves function in older individuals and, in some unique cases, life span.

The neuroendocrine system includes the hypothalamus and associated brain structures as well as the pituitary gland. This system includes neurotransmitters and neuropeptides within the brain that regulate hypothalamic-releasing and -inhibiting hormones secreted into hypophyseal portal blood that reaches the pituitary gland. These hormones influence the secretion of anterior pituitary hormones into the bloodstream and subsequently regulate tissue function. The posterior pituitary also is an important part of the neuroendocrine system but, in contrast to the anterior pituitary gland, is composed of long axons from specific hypothalamic nuclei. The hypothalamus and pituitary gland have the capacity to detect neural activity and/or humoral secretions from target tissues and adjust activity to maintain an optimal internal environment or 'milieu' for tissue function. It is well established that the neuroendocrine system has a critical role in regulating tissue growth and metabolism through the release of growth hormone (GH) and thyroid stimulating hormone (TSH); reproductive function through the release of luteinizing hormone, follicle stimulating hormone, prolactin, and oxytocin; and plasma electrolytes and responses to stress through secretion of vasopressin and adrenocorticotropin, respectively. In addition, the hypothalamus has an important role in the integration of parasympathetic and sympathetic nervous system activity and can thereby influence a wide variety of functions, including heart rate, blood pressure, vascular responses, and glucose metabolism, among others. Specific nuclei within the hypothalamus also regulate biological rhythms. More recently, the regulation of food intake, fat metabolism, and body composition has been shown to be regulated through the hypothalamus by its response to leptin, cholecystokinin, and ghrelin via the synthesis of neuropeptide Y and other orexigenic peptides. Unfortunately, the categorization of hormones and their primary functions noted above is an overly simplistic and only partial view of the neuroendocrine system, and it is now known that critical interactions occur between hormones that contribute to the regulation of cellular function. Because many of the early events of aging include alteration in systems regulated by the neuroendocrine axis, it was proposed (and subsequent studies supported the conclusion) that age-dependent alterations in the neuroendocrine system result in a progressive series of events that are manifest as biological aging.

Although the etiology of the age-related changes in the neuroendocrine system is unknown, it has been proposed that cellular and molecular events in specific subpopulations of neurons within the hypothalamus and brain and/or supporting structures are a contributing factor in the dysregulation of this system. The cause of the specific perturbations may be related to genetic errors or increased free radicals that lead to progressive aberrations in tissue function. As a result, the contribution of neuroendocrine theories to aging is unique compared with other theories of aging in that alterations in this system are not considered the primary causative factor in biological aging but rather are important mediators of aging initiated by cellular changes in specific subpopulations of neurons or systems that closely interact with hypothalamic neurons. In this article, the major alterations within the neuroendocrine axis are discussed, with special emphasis on the regulation of GH, insulin-like growth factor 1 (IGF-1), TSH, and prolactin since these hormones mediate or have the potential to mediate some of more important physiological changes that occur with age. Insulin has been included in this review because of its importance for biological aging, homology to IGF-1, and potential interactions with several neuroendocrine systems.

## GH and IGF-1

### Invertebrate Studies

Results from studies conducted over the past several years have revealed that life span in invertebrates can be substantially increased by disrupting the insulin/IGF-1 pathway. In *Caenorhabditis elegans*, insulin and IGF-1 share a common signaling pathway, and the protein receptor encoded by the daf-2 gene binds to a circulating insulin-like substrate initiating a cascade of events, including activation of age-1, a homolog of mammalian phosphoinositide-3-OH kinase. This kinase, in turn, activates the Akt-encoded protein kinase B (PKB). PKB is a serine/threonine kinase that blocks the function of a forkhead/winged helix transcription factor (daf-16), at least in part, by preventing its translocation to the nucleus.

Disruption of the daf-2 pathway (including mutations in daf-2, age-1, or Akt genes) in *C. elegans* extends life span from 30% to 100%. This appears to be the result of the release of the Akt/PKB 'brake' on daf-16, which is the primary target of the daf-2 pathway. Although the specific mechanisms of the increased life span are not well understood, it is known that daf-16 regulates gene expression that is essential to environmental stress resistance, development, and dauer formation.

In addition to the studies in *C. elegans*, results indicate that disruption of the insulin/IGF-1 signaling pathway in flies and yeast also increases life span. These effects on life span in multiple species raise the possibility that these signaling molecules regulate a conserved pathway that influences longevity in mammals. Although there are promising data in this regard, these issues are more complex in mammals because IGF-1 and insulin regulate two independent but interacting signaling pathways and appropriate models to address these questions are not readily available.

### Mammalian Studies: Background

Our understanding of GH and IGF-1 regulation and actions in mammals began in the early part of this century. It was known that a substance present in blood promotes body growth, but it was only after the isolation of bovine GH from the pituitary gland in 1945 that the biological effects of the hormone became evident. It was subsequently shown that this protein hormone stimulated cellular amino acid uptake, DNA, RNA, and protein synthesis and has a role in cellular division and hypertrophy.

Plasma levels of GH reveal a pattern of discrete pulses that increase after the onset of sleep. Although the precise function of this ultradian pattern remains unknown, the pulsatile release of GH has been confirmed in every species examined to date and is closely related to the biological actions of the hormone. GH in adolescents and young adults is characterized by relatively low-amplitude pulses throughout the day and a large pulse after the onset of sleep. In male rats, GH is characterized by an ultradian rhythm with high amplitude secretory pulses every 3–3.5 h; between pulses, GH decreases to almost undetectable levels. This secretory pattern is regulated by two different hormones released from the hypothalamus: GH-releasing hormone (GHRH) increases GH release, and somatostatin inhibits GH release. Both hormones are secreted in a phasic manner, with GHRH contributing to high-amplitude GH pulses and somatostatin to trough levels. The dynamic interrelationship between these hypothalamic hormones is necessary for pulsatile secretion. A number of other factors either within the hypothalamus or in plasma contribute to the regulation of GH release, either by acting directly on the pituitary gland or by regulating hypothalamic somatostatin and GHRH secretion. Both GH and IGF-1 inhibit GH secretion in a typical feedback relationship: They increase somatostatin and decrease GH RH release from the hypothalamus, resulting in a suppression of GH release.

More recently, there has been increased interest in small peptides that stimulate the release of GH. One of these peptides, ghrelin, is produced in the stomach and gastrointestinal tract. Although it does not interact with the GHRH receptor, it has been shown to act synergistically with GHRH to regulate both GH release and food intake. This peptide is particularly important from a clinical standpoint since oral administration of peptide and nonpeptide analogs of ghrelin increase GH secretion.

Although GH is one of the most potent anabolic factors in the body, most of the anabolic actions of the hormone are regulated through the secretion of IGF-1. IGF-1 binds with high affinity to the type 1

IGF receptor found in tissues throughout the body. The IGF-1 receptor shares 50% amino sequence homology with the insulin receptor, and competitive binding and affinity cross-linking studies have demonstrated that IGF-1 binds to the insulin receptor with 100 times lower affinity than to the IGF-1 receptor. IGF-1 is synthesized mainly in the liver under the regulation of GH but is also synthesized in smaller quantities in almost all tissues. Although regulation of local or 'paracrine' IGF-1 activity is poorly understood, it represents an important source of the hormone that also appears to be regulated by GH.

IGF-1 secreted from the liver circulates in the blood, either free or bound to specific binding proteins that prolong the half-life of the peptide. At present, at least six binding proteins have been identified and constitute an intricate transport system for the IGFs that regulate their availability to specific tissues. It is now clear that IGF binding proteins are critical regulators of IGF-1 activity yet may have actions independent of IGF-1.

The actions of GH are not entirely mediated through IGF-1, and in muscle and adipose tissue, GH can have direct effects. Specifically, GH decreases insulin sensitivity (see the section titled 'Insulin and glucose regulation') by direct effects on skeletal muscle and increases release of free fatty acids (and decreases fat mass) by increasing the activity of hormone-sensitive lipase in adipose tissue. These findings have led to the concept that the major anabolic actions of GH are mediated through IGF-1 (through either endocrine or paracrine regulation), whereas the effects of GH on free fatty acid levels are mediated by direct actions on muscle and adipocytes (**Figure 1**).

### Effects of Age on the GH/IGF-1 Axis

Alterations in the regulation of GH have an important role in the physiological and biochemical changes normally associated with aging. Early studies in humans indicated that the amplitude of GH pulses and the rise in the concentrations of GH after insulin-induced hypoglycemia were blunted in older individuals. Subsequent studies demonstrated a prominent decrease in the amplitude of GH pulses in older animals although there were no changes in either the basal concentrations or the ultradian rhythm of GH. The age-related decline in GH results in a progressive reduction in plasma concentrations of IGF-1 in both animals and humans. Subsequent studies in various strains of rats and mice, nonhuman primates, and humans consistently confirmed the decline in GH and IGF-1 concentrations with age and suggested that changes in these hormones are a robust marker of biological aging in mammalian species.

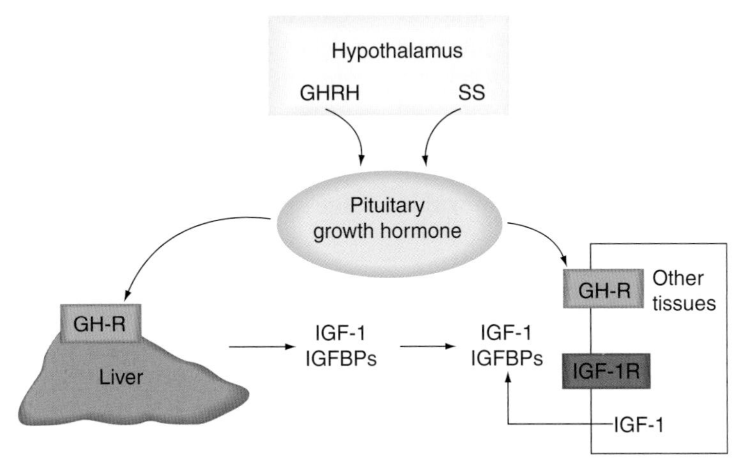


**Figure 1** Growth hormone (GH), produced in the anterior pituitary, is modulated by two hypothalamic hormones: growth hormone-releasing hormone (GHRH), which stimulates both the synthesis and secretion of GH, and somatostatin (SS), which inhibits GH release in response to GHRH. GH also feeds back to inhibit GHRH secretion and probably has a direct inhibitory effect on secretion from the somatotroph (GH-producing cells). In mammals, GH is secreted in pulsatile bursts from the anterior pituitary gland, a pattern that is necessary to achieve full biological activity. GH binds with high affinity to its receptor (GH-R), found in tissues throughout the body, and activation of this receptor stimulates the synthesis and secretion of insulin-like growth factor 1 (IGF-1). Although 90% of circulating IGF-1 is synthesized and secreted by the liver, many types of cells, including some found in the brain and vasculature, are capable of IGF-1 production. Binding of the hormone to the IGF-1 receptor (IGF-1R) causes potent mitogenic effects, including increases in DNA, RNA, and protein synthesis. Although heterogeneity exists in the processing of IGF-1 messenger RNA, these transcripts appear to produce a single peptide that is homologous to the structure of proinsulin. Blood and tissue levels as well as activity of the peptide are regulated by IGF-1 binding proteins (IGFBPs). Although it was initially proposed that all the actions of GH were mediated through IGF-1, data from several studies support direct roles for GH in the regulation of lipolysis and insulin sensitivity that are independent of IGF-1. Reproduced from Carter CS, Ramsey MM, and Sonntag WE (2002) A critical analysis of the role of growth hormone and IGF-1 in aging and lifespan. *Trends in Genetics* 18: 295–301, with permission from Elsevier.

Restoration of GH and IGF-1 levels with purified GH preparations reverses the age-related decline in cellular protein synthetic capacity in both skeletal and cardiac muscle. Although these studies did not address the question of whether tissue response to GH diminished with age, they clearly indicated that (1) cellular protein synthesis decreases in an environment of reduced GH and IGF-1 levels and (2) GH and/or IGF-1 replacement has the capacity to attenuate a classical symptom of aging, that is, the decrease in cellular protein synthesis in older people. Furthermore, these studies were supported by clinical observations that GH replacement to humans for 6 months reversed the decline in bone mass, the negative nitrogen balance (an indicator of decreased whole-body protein synthesis), and the increase in fat mass characteristic of aging in humans. Subsequent reports demonstrated that GH or IGF-1 administration could partially reverse the decline in immune function; increase the expression of aortic elastin, lean body mass, skin thickness, and vertebral bone density; and improve cognitive function in older animals and humans. Although the beneficial effects of GH administration have been confirmed repeatedly, deleterious side effects of GH replacement severely limit its usefulness as a treatment to modulate physiological changes with age. For example, GH increases cartilage growth, contributes to carpal tunnel syndrome, impairs insulin sensitivity, increases blood glucose, and in several cases has been found to increase neoplastic disease. In fact, elevated IGF-1 levels are a risk factor for breast, gastrointestinal, prostate, and lung cancer. These studies provide compelling data suggesting that the development of the aging phenotype results, at least in part, from dysfunction within the neuroendocrine axis and, more specifically, deficiencies in the secretion of GH and IGF-1. Therefore, it is evident that decreases in GH and IGF-1 have clinical significance and are a contributing factor in the genesis of tissue dysfunction in older animals and humans.

### Actions of GH and IGF-1 in Older Animals

Although GH and IGF-1 are well known for their ability to increase cellular protein synthesis, muscle and bone mass, and immune function and to decrease fat mass, recent data suggest that the age-related decrease in these hormones has an important role in the genesis of brain aging. Studies indicate that IGF-1 reverses the age-related decline in cerebral microvascular density and blood flow and that IGF-1 crosses the blood–brain barrier and enhances learning and memory, long-term potentiation, glucose metabolism, synaptic density, neurogenesis, and neuronal survival and reduces oxidative stress in older animals. These results and recently published reports have led to the hypothesis that age-related deficiency in circulating GH and IGF-1 leads to vascular insufficiency and deficits in synaptic function and cell replacement in the central nervous system and are a contributing factor in the genesis of brain aging (**Figure 2**).

### Animal Models of GH/IGF-1 Excess and Deficiency

Our understanding of the effects of GH/IGF-1 deficiency and replacement in aging humans and animal models is complicated by results of studies with transgenic and mutant animals. Transgenic mice with varying levels of GH overexpression demonstrate an inverse relationship between GH expression and life span. The reduction in life span observed with

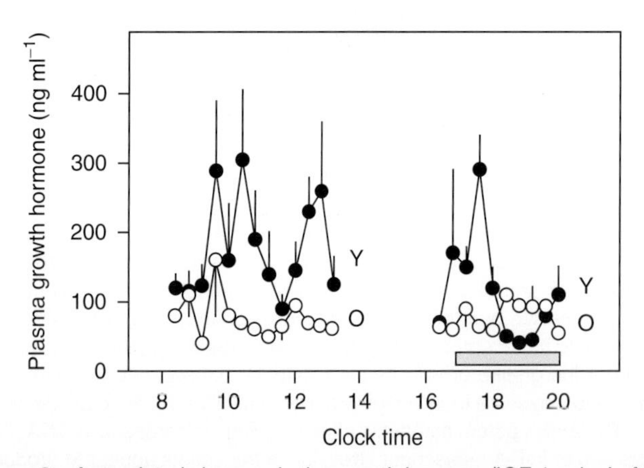


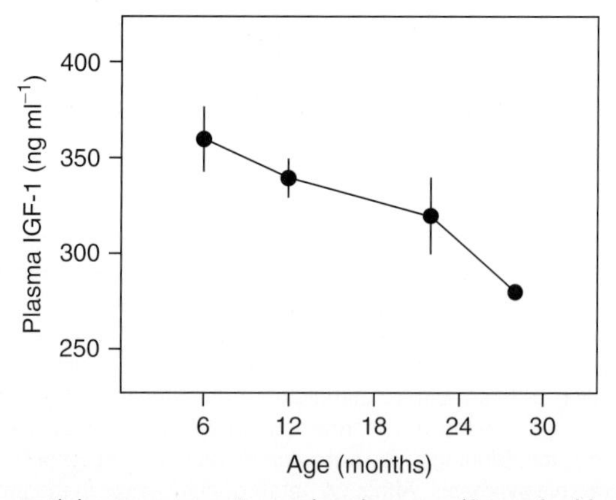


**Figure 2** Age-related changes in the growth hormone/IGF-1 axis. Left, growth hormone secretory pulses in young (6-month-old; solid circles; Y) and old (20-month-old; open circles; O) Sprague–Dawley male rats. Serial samples of blood were removed at 20-min intervals from conscious, freely moving animals and analyzed by radioimmunoassay. Hatched area represents the dark phase of the light/dark cycle. Right, age-related changes in plasma IGF-1. Data represent mean ± SE. Reproduced from Carter CS, Ramsey MM, and Sonntag WE (2002) A critical analysis of the role of growth hormone and IGF-1 in aging and lifespan. *Trends in Genetics* 18: 295–301, with permission from Elsevier.

overexpression of GH is generally attributed to increased pathologies, primarily renal lesions and a very high frequency of hepatocellular tumors, including adenoma and carcinoma. While the reduced life span observed in models of GH overexpression appears to support the hypothesis that GH reduces life span, it is important to note that GH levels in these overexpression models are 1000–10 000-fold greater than those found under physiological conditions. Therefore, the distinction between the potential of pharmacological doses of GH to increase pathology and reduce life span has to be considered separately from the physiological changes in growth that occur during the normal life span of most mammalian models. Therefore, it remains unclear whether transgenic models of GH overexpression provide relevant information on the role of GH and IGF-1 in normal life span regulation (**Figure 3**).

Perhaps the most commonly used animal models to investigate the effects of GH and IGF-1 are those that show deficiency in GH expression or have mutations in the GH/IGF-1 axis. Some of these models include those with deficiencies in transcription factors necessary for the generation of GH, TSH, and prolactin-producing cells in the pituitary gland (e.g., Snell and Ames dwarf mice), and others have selective impairments in the GH/IGF-1 axis (e.g., $GHRHR^{-}$/$GHRHR^{-}$, GHRKO, and $IGFR^{+/-}$ mice). Despite the limitations of the models with multiple hormonal deficiencies, even models with specific deficiencies in GH and IGF-1 during early development result in impaired development of related endocrine systems, and alterations in corticosterone, insulin, glucose, and thyroid hormones are routinely observed. As a result, data from these models must be interpreted cautiously since they demonstrate the potential of alterations in GH/IGF-1 early during development to affect developmental processes that regulate life span but cannot address whether these effects are mediated directly by the GH/IGF-1 actions on target tissues or are mediated indirectly via the developmental other endocrine systems dependent on GH and IGF-1. The distinction is important to resolve for the direction of future scientific studies.

The marked contrasts noted in the studies described above related to the beneficial effects of GH replacement on function in humans and animal models and the contrasting studies demonstrating that deficiencies in GH and IGF-1 increase life span represent the crux of the current issue in the GH/IGF-1 field. Recent studies have attempted to circumvent the developmental effects of GH/IGF-1 deficiency by developing a model of adult-onset GH/IGF-1 deficiency. In this model, a mutant animal that exhibits reduced levels of GH around puberty is administered GH replacement to increase circulating IGF-1 and ensure the development of other endocrine systems until adulthood. The GH injections are then terminated and the animal allowed to age normally, creating the model of adult-onset GH deficiency (AO-GHD), and these animals are compared to dwarf (GH-deficient) or wild-type (normal) animals. In these studies, no differences in life span were evident between normal and GH-deficient animals, but it is interesting that AO-GHD animals lived 15% longer than either group. End-of-life pathology indicated that both AO-GHD and GH-deficient animals had less kidney pathology and neoplastic disease compared with wild-type animals but died primarily from cardiac thrombus and intracerebral hemorrhage. The increased life span in the AO-GHD animals resulted from a corresponding delay in the appearance of cardiac thrombus and intracerebral hemorrhage. This study provided compelling data that GH and IGF-1 levels around puberty have an important role in determining life span, that life span and the factors that determine life span may be a separate entity from age-related pathology, and finally that functional changes with age are separate from life span.

The effects of GH and IGF-1 on aging and life span highlight several unresolved issues in the field

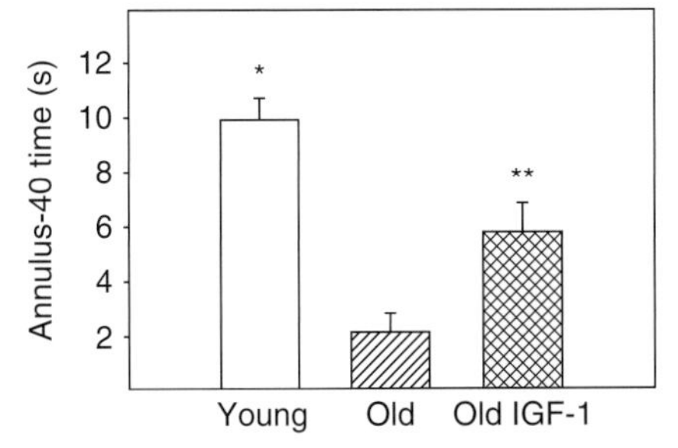


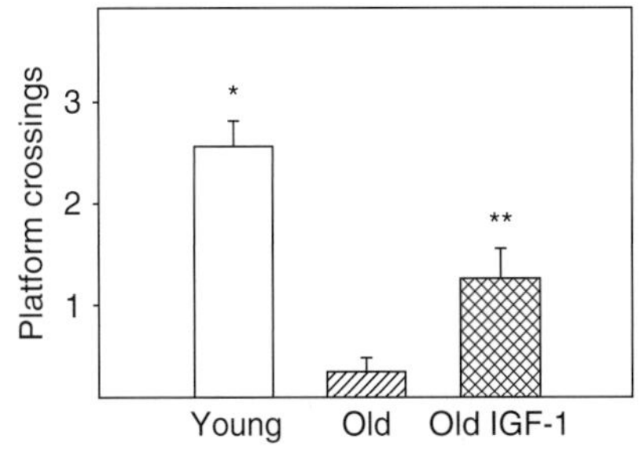


**Figure 3** Probe trial in the Morris water maze assessed spatial reference memory. Performance in both Annulus-40 time (left) and platform crossings (right) that are measures of memory declined with age ($p < 0.001$ comparing young and old control (old) animals). IGF-1 treatment for 28 days reversed this decline in both measures (*$p < 0.05$ comparing young to old and old IGF–1; **$p < 0.01$ in both measures comparing old control (old) and old IGF-1 treated animals). Data are shown as mean ± SE (unpublished data).

of gerontology. Deletion or mutation of a single gene, as occurs in transgenic or mutant animals, can result in developmental anomalies, but consequences for data interpretation remain uncertain. Most investigators recognize the importance of the age-related decline in tissue function, pathology, and life span; however, few studies of life span include either functional or pathological analyses. Until gerontologists agree on standards for the conduct of aging studies designed to assess the effects of these interventions on aging, such issues will remain unresolved (**Table 1**).

## Insulin and Glucose Regulation

Since the initial studies demonstrating that suppression of signaling in the insulin/IGF-1 pathway in *C. elegans* increases life span, there has been much interest in insulin signaling in mammalian models and its potential contribution to life span. In contrast to the common insulin/IGF-1 pathway in invertebrates, separate insulin and IGF-1 signaling pathways have evolved in mammals.

In humans, the regulation of blood glucose and insulin secretion is perhaps one of the most important areas of endocrinology because pertubations in this pathways are closely associated with cardiovascular disease and many other pathological conditions that increase with age. The postprandial rise in insulin secretion in mammals results from an increase in blood glucose that affects the pancreatic $\beta$ cell. Also, glucagon-like peptide-1 release from the intestine sensitizes the $\beta$ cells, resulting in optimal insulin output into the circulation. Although insulin action is a more complicated process than described here, its primary action is to stimulate glycogen synthesis in liver and, after binding to its receptor on muscle and adipose tissue, to activate glucose transporter-4 proteins in the membrane of the cell through a complicated signal transduction mechanism. The result is glucose uptake in these tissues and a reduction in blood glucose levels. In nonobese older study participants, there are minor, albeit significant, changes in both insulin secretion and glucose tolerance tests with age. These changes alone are not sufficient to result in pathology. However, in obese study participants regardless of age, basal glucose levels rise, resulting in increased pathology. The interaction of body composition (adipose tissue mass) and glucose levels has been widely recognized and results in insulin receptor insensitivity, potentially from a rise in cytokines (e.g., tumor necrosis factor-$\alpha$ and interleukin-6) and other hormones (e.g., leptin) secreted by adipose tissue, that interferes with insulin signaling mechanisms designed to lower blood glucose. In addition, recent studies indicate that adipose-like tissue accumulates within skeletal muscle and is an important factor in the increased incidence of disability with age. Whether these changes in adipose tissue are important corollary events or actually cause the increase in disability remains to be determined.

Unfortunately, it is unclear whether alterations in insulin levels or insulin signaling directly contribute to the development of the aging phenotype. Certainly, one might expect that obesity and the pathologies associated with obesity synergize with the pathologies of aging to shorten life span. But the interactions between these pathways are complex and remain underinvestigated. In this regard, there are data to

**Table 1** End-of-life pathology in a model of adult-onset growth hormone deficiency[a]

| *Pathology* | *Heterozygous* | *GHD* | *AO-GHD* |
|---|---|---|---|
| Tumor bearing (%) | 88 | 73 ($p = 0.34$) | 63 ($p = 0.04$) |
| Tumor burden (mean) | $1.3 \pm 0.2$ | $0.92 \pm 0.2$ ($p =$ ns) | $1.0 \pm 0.2$ ($p =$ ns) |
| Fatal tumor (%) | 57 | 36 ($p = 0.21$) | 31 ($p = 0.11$) |
| Fatal pituitary adenoma (%) | 39 | 21 ($p = 0.30$) | 13 ($p = 0.08$) |
| Grade of pituitary adenoma | $1.57 \pm 0.3$ | $1.69 \pm 0.4$ ($p =$ ns) | $1.19 \pm 0.3$ ($p =$ ns) |
| Fatal chronic nephropathy (%) | 74 | 0 ($p < 0.0001$) | 0 ($p < 0.0001$) |
| Severity of nephropathy | $4.0 \pm 0.3$ | $0.8 \pm 0.2$ ($p < 0.001$) | $1.0 \pm 0.2$ ($p < 0.001$) |
| Intracranial hemorrhage/thrombus (%) | 17 | 50 ($p = 0.06$) | 44 ($p = 0.14$) |
| Age of occurrence of hemorrhage/thrombus (months) | 31.9 | 30.7 | 34.1 |
| Disease burden (total) | $6.6 \pm 0.5$ | $5.0 \pm 0.5$ ($p = 0.009$) | $5.5 \pm 0.4$ ($p = 0.067$) |

[a]End-of-life pathology was determined in the following experimental groups: (1) heterozygous (normal) male Lewis rats injected with saline from 4 to 14 weeks of age (normal size, normal GH/IGF-1 levels); (2) GH-deficient (GHD) dwarf Lewis males administered saline from 4 to 14 weeks of age (dwarf, GHD); and (3) GHD dwarf Lewis males administered porcine growth hormone from 4 to 14 weeks of age (Adult-onset GHD (AO-GHD)). AO-GHD animals exhibited a 12–15% increase in life span compared with either heterozygous or GHD animals ($p < 0.05$). ns, not significant.

suggest that a knockout of the insulin receptor specifically in adipose tissue increases life span in mice, but no age-related pathological analyses or performance measures were done, and therefore this remains an important area for future research.

## Endocrine Effects of Moderate Caloric Restriction: Interactions with Insulin and IGF-1

Despite the absence of changes in insulin and glucose levels in nonobese humans with age, there is important information to suggest that the insulin and IGF-1 pathways may have important roles in determination of life span in invertebrate and vertebrate subjects. Consistently, moderate caloric restriction (60% of *ad libitum* feeding) has been shown to increase life span in rodent models. Although there are data to suggest that the decrease in nutrients alters several cellular metabolic pathways (e.g., SirT1, PGC1$\alpha$) that may contribute to regulation of life span, alterations in caloric intake have important effects on several hormonal systems that may affect pathology and life span. For example, reduction in calories increases GH, cortisol, and catecholamine secretion and reduces circulating insulin and IGF-1 levels. These changes are a normal physiological response designed to access nutrients that have been stored during periods of nutrient excess. The result of these changes is an increase in gluconeogenesis, glycogenolysis, ketogenesis, and fatty acid metabolism and a decrease in insulin sensitivity and cellular protein synthesis, all of which are designed to maintain blood glucose and other nutrients in the bloodstream. The endocrine changes that result from a reduction in caloric intake appear to have a direct effect on life span and age-related pathology in small animals. For example, chemical-induced bladder cancer is markedly reduced in caloric restricted compared with *ad libitum* animals, and replacement of IGF-1 to calorie restricted animals increases the incidence of this pathology. These data suggest that the decrease in IGF-1 that occurs during caloric restriction is an important part of the protective mechanism against chemical-induced cancer and provides indirect evidence that chronically lower levels of IGF-1 may protect animals from other types of neoplastic diseases that increase with age. This conclusion is supported by data from many transgenic and mutant animal strains and studies using the model of AO-GHD that show protection against dimethylbenz[a]anthracene-induced mammary cancer. In the studies using the AO-GHD model, GH replacement resulted in a dose-related rise in mammary cancer. Since many mouse strains are highly susceptible to neoplastic disease, any intervention that delays or prevents this pathology will naturally increase life span. Although it is unlikely that the increased life span from moderate caloric restriction results from reduced levels of IGF-1 and/or insulin alone, the relationship between IGF-1 and neoplastic disease is compelling, and further research to understand the specific hormones that may mediate the actions of moderate caloric restriction is required (**Figure 4**).

## Prolactin

### Basic Regulation and Actions

Prolactin secretion from the pituitary lactotroph in humans and animal models is regulated by hypothalamic dopamine secretion into the hypothalamic-pituitary portal vessels. The dopaminergic neurons of the rat hypothalamus are grouped into two main areas, $A_{12}$ and $A_{14}$. The dopaminergic perikarya of the $A_{12}$ area are located in the arcuate nucleus, with their axons projecting into the external zone of the median eminence. There, dopamine is released from the axon terminals into the portal vessels of the median eminence and travels to the anterior pituitary lobe. The $A_{14}$ dopamine neurons are primarily located within the paraventricular and periventricular

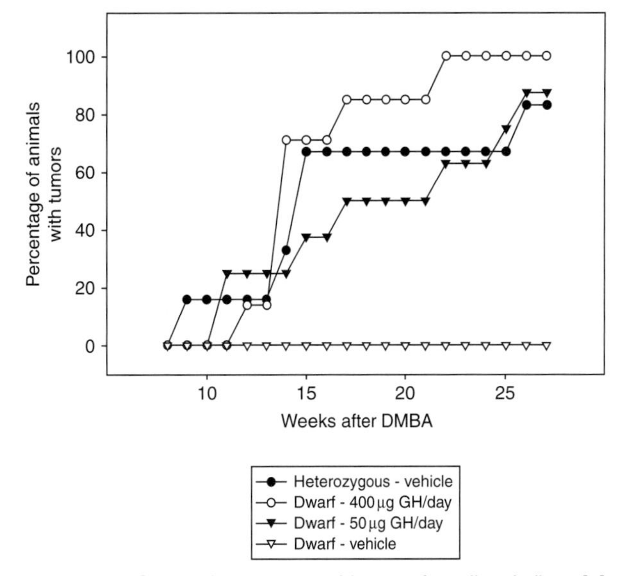


**Figure 4** Cumulative tumor incidence after dimethylbenz[a] anthracene (DMBA) injection in growth hormone-deficient dwarf animals and heterozygous (normal) control female rats. At 27 weeks after DMBA injection, 80% of heterozygous animals exhibited mammary tumors, while dwarf animals administered saline, 50 μg, or 400 μg porcine growth hormone per day demonstrated 0%, 83%, and 100% tumor incidence, respectively. Reproduced from Ramsey MM, Ingram RL, Cashion AB, et al. (2002) Growth harmone-deficient dwarf animals are resistant to dimethylbenzanthracine (DMBA)-induced mammary carcinogenesis. *Endocrinology* 143: 4139–4142, copyright 2002, The Endocrine Society.

hypothalamic nuclei and project their axons through the internal portion of the median eminence to terminate on both the neural and intermediate pituitary lobes. The $A_{12}$ area and its corresponding axon terminals constitute the tuberoinfundibular dopaminergic (TIDA) system, whereas the $A_{14}$ area and its fibers are known as the periventricular dopaminergic system. Both systems regulate lactotrophs exerting a tonic inhibitory control on prolactin secretion and lactotroph proliferation.

Plasma concentrations of prolactin are lowest during waking hours in humans and highest during sleep. The variable patterns of secretion are dependent on gender, environmental conditions including stress, reproductive status (including lactation), estrous and menstrual cycles, pregnancy, and other factors. Prolactin participates in multiple roles throughout the body although its primary actions are related to reproduction and maternal behavior. In some mammals, prolactin is important for the secretory activity of the corpus luteum. Within the mammary gland, prolactin induces growth and synthesis of milk (e.g., uptake of some amino acids; synthesis of the milk proteins casein and $\alpha$-lactalbumin; uptake of glucose, lactose, and fats) and maintains milk secretion. Also, prolactin has a role in the immune system, stimulating mitogenesis in normal lymphocytes. Prolactin also appears to have a role in the regulation of solute and water transport across mammalian cell membranes. In keeping with its lactogenic properties, it affects water transport across amniotic membranes and acts as an osmoregulatory hormone.

### Prolactin Secretion and Aging

The dopaminergic neurons of the central nervous system are among the cells most susceptible to damage produced by the aging process. For example, Parkinson's disease is a human condition characterized by a degeneration of nigrostriatal dopamine neurons. This disease affects 0.1–0.3% of the population and is the most conspicuous reflection of the vulnerability of dopamine neurons to age. In female rodents, degeneration and loss of TIDA neurons during normal aging is associated with progressive hyperprolactinemia and the development of pituitary prolactinomas. It is interesting that patients with Parkinson's disease usually reveal functional alterations within the hypothalamoprolactin axis.

Female rats ranging in age from 6 to 32 months have been studied to ascertain the impact of aging on the number and morphology of dopamine neurons of the $A_{12}$ and $A_{14}$ areas as well as on serum prolactin levels. In these studies, a progressive loss of both TIDA and periventricular dopaminergic neurons was evident, and these changes became more conspicuous at extreme ages. It is interesting that in old female rats, restoration of hypothalamic IGF-1 using recombinant adenoviral vectors that express rat IGF-1 was highly effective in restoring hypothalamic dopamine function, suggesting important interactions between the age-related decrease in IGF-1, the deterioration of dopamine activity, and the age-related rise in prolactin levels.

In addition to IGF-1, hypothalamic lactotrophs are controlled by a number of factors. For example, estrogen has been demonstrated to have an important effect on lactotrophs and to stimulate the development of tumorigenesis within the pituitary. Although there are data demonstrating that estrogen acts on dopaminergic cells within the hypothalamus, several investigators have proposed a lactotroph, folliculostellate cell interaction as a potential mechanism for the role of estrogen in the genesis of pituitary prolactinomas. Under basal conditions, prolactin cells are under a tonic inhibitory autofeedback of transforming growth factor-$\beta$ (TGF$\beta$)-1 acting on the TGF$\beta$II receptor. When estrogen levels rise, lactotroph cells are stimulated to release high levels of TGF$\beta$3, which then acts on TGF$\beta$II receptors on folliculostellate cells, inducing the production and release of basic fibroblast growth factor, which in turn acts on the lactotroph to stimulate both cell proliferation and prolactin secretion. These novel interactions of estrogen on the pituitary gland may interact synergistically with the reduced levels of dopamine to produce the pronounced rise in pituitary microadenomas and frank tumors that occurs with age.

## TSH: Tri-iodothyronine, Thyroxine, and Thyroid Function

Cellular metabolism is regulated by the secretion of TSH release from the anterior pituitary gland. This hormone enters the circulation and stimulates the thyroid gland to produce both tri-iodothyronine ($T_3$) and thyroxine ($T_4$). These hormones circulate bound to thyroid binding globulin and after transport are taken up by the cell. $T_3$ is the more potent thyroid hormone, but $T_4$ can be converted to $T_3$ with the enzyme deiodinase. Levels of these hormones regulate basal metabolic activity throughout the body. $T_3$ and $T_4$ are controlled by negative feedback regulation, primarily at the level of the anterior pituitary gland, by suppression of TSH release although the hypothalamus also produces thyroid-releasing hormone (TRH) that can stimulate TSH release. TRH has only a minor role in the regulation of TSH release under normal conditions, but in response to cold stress or other challenges, TRH exerts an important influence. In addition to prominent effects on cellular metabolism and temperature regulation, thyroid hormones have a permissive role in the regulation of many organs and tissues, including reproductive function, bone metabolism, and glucose metabolism. Therefore, changes in the secretion or response to thyroid hormones would be expected to have broad-ranging effects on the health of the organism.

Apart from the appearance of specific thyroid diseases that increase with age, only subtle changes in the levels of thyroid hormones seem to occur in otherwise healthy older animals or humans. Investigators have reported a reduction in deiodinase activity that leads to reduced levels of $T_3$ within target organs, but the significance of these changes for the development of physiological changes in tissue function have not been well defined. Similarly, it is of interest that numerous animal models that exhibit increased life span also exhibit a decrease in thyroid hormone levels, but the significance of these changes remains to be determined.

## Conclusions

There are now convincing data that alterations within the neuroendocrine system contribute to both the physiological and pathological manifestations of aging. Decreases in GH and IGF-1, for example, are important factors in the accumulation of body fat mass, the reduction in cellular protein synthesis, and impairments in some aspects of immune function with age. The effects of moderate caloric restriction are mediated, at least in part, through physiological changes affected by the neuroendocrine system. These studies, together with results from transgenic and mutant animals, provide compelling data of an important link between the neuroendocrine axis and the development of the aging phenotype. Nevertheless, controversies remain that can be resolved only by the development of transgenic animal models permitting close regulation of both GH and IGF-1 and assessing function, pathology, and life span.

Although the specific etiology of the age-related neuroendocrine dysfunctions is unknown, increases in oxidative damage in specific hypothalamic nuclei, impairments in neurotransmitter levels and turnover, and diminished plasticity within hypothalamic nuclei are potential mechanisms. An alternative explanation for the alterations within the neuroendocrine system is that an active regulation of hypothalamic function exists throughout the life span. Decreases in GH and IGF-1, for example, are initiated immediately after puberty and progress with age. In fact, age-related changes in the regulation of this system may be consistent with a model of antagonistic pleiotropy, a theory on the evolution of aging that suggests that the expression of particular genes is beneficial early in life but becomes detrimental as the organism ages. Certainly, the beneficial actions of GH and IGF-1 on muscle mass and organ development early in life would be expected to increase general fitness as well as reproductive fitness, whereas continued high levels may increase pathology and limit life span. Thus, the age-related decrease in GH and IGF-1 as well as other hormones regulated by the neuroendocrine axis may be a physiologically regulated, adaptive process rather than the result of damage to specific hypothalamic nuclei. The potential significance of such changes for this and other systems regulated by the neuroendocrine axis awaits further investigation.

*See also:* Neuroendocrine Aging: Pituitary-Gonadal Axis in Males; Neuroendocrine Aging: Hypothalamic-Pituitary-Gonadal Axis in Women; Neuroendocrine Aging: Pituitary-Adrenal Axis.

## Further Reading

Bartke A (2005) Minireview: Role of the growth hormone/insulin-like growth factor system in mammalian aging. *Endocrinology* 146: 3718–3723.

Bartke A, Masternak MM, Al-Regaiey KA, and Bonkowski MS (2007) Effects of dietary restriction on the expression of insulin-signaling-related genes in long-lived mutant mice. *Interdisciplinary Topics in Gerontology* 35: 69–82.

Carter CS, Ramsey MM, and Sonntag WE (2002) A critical analysis of the role of growth hormone and IGF-1 in aging and lifespan. *Trends in Genetics* 18: 295–301.

Goya RG, Sarkar DK, Brown OA, and Herenu CB (2004) Potential of gene therapy for the treatment of pituitary tumors. *Current Gene Therapy* 4: 79–87.

Harper JM, Durkee SJ, Dysko RC, Austad SN, and Miller RA (2006) Genetic modulation of hormone levels and life span in hybrids between laboratory and wild-derived mice. *Journals of Gerontology. Series A, Biological Sciences and Medical Sciences* 61: 1019–1029.

Herenu CB, Brown OA, Sosa YE, et al. (2006) The neuroendocrine system as a model to evaluate experimental gene therapy. *Current Gene Therapy* 6: 125–129.

Herenu CB, Cristina C, Rimoldi OJ, et al. (2007) Restorative effect of insulin-like growth factor-I gene therapy in the hypothalamus of senile rats with dopaminergic dysfunction. *Gene Therapy* 14: 237–245.

Kenyon C (2005) The plasticity of aging: Insights from long-lived mutants. *Cell* 120: 449–460.

Ramsey MM, Ingram RL, Cashion AB, et al. (2002) Growth harmone-deficient dwarf animals are resistant to dimethylbenzanthricine (DMBA)-induced mammary carcinogenesis. *Endocrinology* 143: 4139–42.

Sarkar DK (2006) Genesis of prolactinomas: Studies using estrogen-treated animals. *Frontiers in Hormone Research* 35: 32–49.

Sonntag WE, Carter CS, Ikeno Y, et al. (2005) Adult-onset growth hormone and insulin-like growth factor I deficiency reduces neoplastic disease, modifies age-related pathology, and increases life span. *Endocrinology* 146: 2920–2932.

Sonntag WE, Ramsey M, and Carter CS (2005) Growth hormone and insulin-like growth factor-1 (IGF-1) and their influence on cognitive aging. *Ageing Research Reviews* 4: 195–212.

# Neuroendocrine Aging: Pituitary-Adrenal Axis

**E R de Kloet**, LACDR, Leiden University Medical Center, Leiden, The Netherlands



## Introduction

Aging is characterized by decreased ability to maintain homeostasis and consequent less efficient adaptation to change. The impact of this age-related breakdown of adaptation is far more disruptive when integrative homeostatic communication systems are also affected. These systems include the sympathetic and parasympathetic branches of the autonomic nervous system and the neuroendocrine hypothalamic–pituitary–adrenal (HPA) axis (**Figure 1**). The amine, neuropeptide, and steroid components of these systems signal specific information to ensure coordination of brain and body functions during adaptation. Such coordination involves the control of emotions and cognitive performance as well as energy metabolism, cardiovascular, inflammatory, and immune regulation. The question addressed here is whether aberrant HPA activity is the cause or the consequence of signs of aging and age-related pathology. The focus is on the glucocorticoids, which operate in a bimodal fashion crucial for defense and recovery of homeostasis.

## HPA Axis and Glucocorticoids

### Glucocorticoid Cascade Hypothesis

The HPA axis promotes an organisms' ability to cope with environmental stressors and facilitates behavioral adaptation. At senescence, coping and adaptation usually become less efficient and as a consequence HPA reactions may become inadequate, excessive, or prolonged. In particular, the excessive and prolonged secretion of glucocorticoids has received attention. Elevated glucocorticoid levels imply that the cost to defend homeostasis has become too high. The outcome is enhanced wear and tear, a process termed 'allostatic load' by Bruce McEwen. This notion materialized in Robert Sapolsky's glucocorticoid cascade hypothesis, which focused on the damaging outcome of chronic glucocorticoid overexposure during the aging process.

While acute rises in glucocorticoid levels are essential for recovery, adaptation, and survival, prolonged elevated glucocorticoid levels may indeed become damaging to brain and body. It is thought that such elevated hormone levels are primarily caused by slowing down of cognitive operations and less efficient homeostatic control mechanisms. In the brain, the damage particularly affects the hippocampus, which is an important glucocorticoid target and is crucial for learning and memory processes. Because of ensuing feedback resistance, glucocorticoid levels further rise. Hence a feed-forward cascade develops, leading ultimately to degeneration and death. This appealing hypothesis clearly has supporting evidence: for instance, rainbow trout swimming upriver for reproductive purposes eventually die of cushingoid symptoms.

### HPA Axis in Aging Rodents

In support of the glucocorticoid cascade hypothesis, excessive HPA axis activation in aging has been reported. Sprague–Dawley, Long–Evans, Fisher, and Wistar rats generally show prolonged stress-induced corticosterone levels during aging. Several scenarios have linked rises in corticosterone level to aging. First, stress and hypercorticism have been proposed to promote cognitive aging. Second, aging may cause initial hypercorticism, which then, in a feed-forward cascade of increasing feedback resistance, would further promote the aging process. Third, stress, together with hypercorticism, would amplify the individual variation in cognitive aging. However, hypercorticism during aging appears not at all to be a common phenomenon.

In aged Brown Norway rats, exposure to a novel environment or a conditioned emotional stressor produces excessive adrenocorticotropic hormone (ACTH) peak values, while corticosterone responses are not different from those of young controls, indicating an efficient feedback operation. The relatively low corticosterone concentration is partly compensated by reduced corticosteroid-binding globulin (CBG) capacity, yielding an increased free fraction of nonbound corticosterone. Hence, Brown Norway rats show adrenocortical deficiency during aging, suggesting that healthy aging and low corticosterone levels coincide. This demonstrates, however, that hypercorticism is not an inevitable consequence of aging.

The profound differences in 50% survival age between different rat strains raise the question of the appropriate rodent model for aging research. The criteria for the 'ideal' model include an almost rectangular survival curve and thus a high estimated 50% survival age. Included also is the occurrence of age-related, multiple pathologies rather than inbred-strain specific diseases (e.g., the high prevalence of testis- or pituitary-tumors in F344 or Wistar rats). The animals used in many aging studies in fact either fail to meet these criteria or should be considered middle-aged rather than old. The Brown Norway strain does

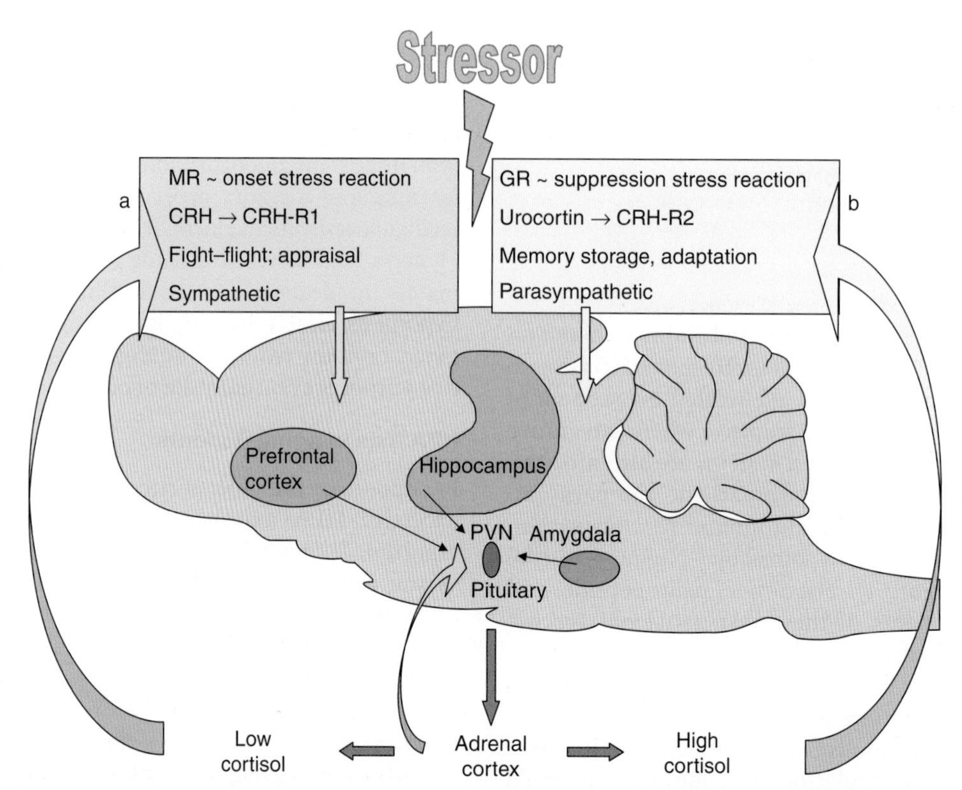


**Figure 1** Bimodal organization of neuroendocrine stress response. Stressful stimuli, through afferent pathways, stimulate corticotropin-releasing hormone (CRH) and vasopressin neurons in the paraventricular nucleus (PVN), thus orchestrating behavioral, autonomous, and neuroendocrine response patterns. Physical stressors (e.g., pain, blood loss, heat, and cold) directly stimulate the PVN via ascending aminergic pathways. Psychological stressors require processing in limbic–cortical brain structures (e.g., hippocampus, amygdala, and prefrontal cortex) that communicate transsynaptically with the PVN. The neuroendocrine response pattern enhances the secretory bursts of cortisol (primates) or corticosterone (rodents) from the adrenal cortex. The corticosteroids feed back on the brain circuits that have produced their own secretion and prevent them from overexcitation. These actions exerted by the steroids are mediated by (a) mineralocorticoid receptors (MR) and (b) glucocorticoid receptors (GR). Under psychosocial stress conditions, MRs regulate the onset of the stress response, which – through CRH receptor 1 (CRH-R1) – determines the extent of CRH action on hypothalamic–adrenal–pituitary activation, sympathetic nervous system responses, and flight–flight behavior. After stress progressively activates GRs, the rise in corticosterone levels facilitates adaptation, promotes memory storage, and hence prepares for coping with future events. CRH-R2 activation by urocortins and parasympathetic nervous activity is thought to synergize with GRs to promote recovery. MRs and GRs operate in complementary fashion through a (non)genomic mechanism to facilitate defense and recovery of homeostasis; imbalance is thought to promote signs of aging and age-related pathology. From de Kloet ER, de Rijk RH, and Meijer OC (2007) Is there an imbalanced response of mineralocorticoid and glucocorticoid receptors in depression? *Nature Clinical Practice Endocrinology & Metabolism* 3: 168–179.

meet these criteria and is, as such, generally accepted as a good rodent model to study aging.

Another important consideration in these types of studies is the individual variation within a genetically homogeneous cohort of animals – variation that becomes much more pronounced at old age. With respect to many behavioral and neurobiological parameters, such as memory performance, individual animals can often be classified in impaired, intermediate, and nonimpaired subgroups. In aged Long–Evans or Brown Norway rats, approximately 30% will show excellent memory performance, whereas the rest will be partially or severely impaired; in Long–Evans rats, the extent of impairment appears related to the persistence in corticosterone secretion after stress.

### HPA Axis Pulsatility

Glucocorticoids are secreted in a circadian pattern, with peak levels before the active period, which, for the nocturnal rat, is in early evening. However, hormones are secreted in distinct pulses, as is well documented for the gonadotropin-releasing hormone and growth hormone systems. Also, the HPA axis displays an ultradian rhythm and produces a complex pattern of approximately 1-h-long glucocorticoid secretory pulses. The variation in amplitude of the glucocorticoid pulses produces the circadian profile.

A stressor initiates an HPA pulse. The pulse frequency and amplitude increase, but it makes a difference at what phase of the ultradian rhythm the stressor is experienced. As shown in Stafford Lightman's laboratory, a stressor at the ascending phase of the pulse results in a profound glucocorticoid response, while at the descending phase a refractory period exists, leading to strongly attenuated glucocorticoid secretion. This raises the possibility that modulation of glucocorticoid pulsatility serves as means of communication. Frequency encoding by glucocorticoid hormones implies that the pattern of hormone secretion matters, rather than absolute hormone concentration. It implies also that glucocorticoid pulses throughout the circadian cycle likely are required for synchronization and coordination of daily and sleep-related events.

Glucocorticoid pulsatility has an immediate consequences for HPA aging studies. This is one may wonder about the relationship between cortisol changes observed after a stressor if disparate pulsatility patterns among individuals are grouped. In some rat strains, aging is characterized by a prolonged response to stress. Taking the ultradian rhythm into account, this apparent persistence may be due to desynchronization of the pulsatile corticosterone pattern among individuals: the pulse may be delayed in some and advanced in others. Such disordered pulsatile patterns may appear seemingly prolonged after grouping the data of each individual. Hence the pulsatile pattern of corticosterone needs to be examined during the aging process.

### Glucocorticoid Feedback

The recognition of pulsatile patterns of hormone secretion casts an entirely new light on the classical view of regulation and function of the HPA axis. The pulse generator is not known, but probably resides proximal to or within the hypothalamic corticotropin-releasing hormone (CRH) neurons. Furthermore, hormone pulses can be amplified and dampened at each level of the HPA axis. Central modulation occurs via afferent inputs that are involved in processing of stressful information. In this respect, excitatory afferents from the hippocampus and prefrontal cortex that process contextual information innervate a $\gamma$-aminobutyric acid (GABA)ergic inhibitory neuronal network controlling the CRH neurons. Emotional arousal through nervous input from the amygdala excites the paraventricular nucleus (PVN), as do the monosynaptic ascending aminergic pathways from the brain stem. How these inputs are integrated and translated in a CRH pulse is not known; current evidence suggests that a stressor advances the onset of the pulse and can enhance pulse frequency and amplitude.

Glucocorticoids feed back precisely in those cells and circuits that initially activated the HPA pulse generator, but in what modes does the feedback action operate? The answer to this question has been guided by the discovery of the receptors for the glucocorticoids. Two types of receptors have been identified: mineralocorticoid receptors (MRs) and the glucocorticoid receptors (GRs), which together control a spectrum of actions over time domains ranging from seconds to hours. The GRs prevail in the PVN and pituitary primary feedback sites, where they exert negative control over the stress-induced HPA pulse. The limbic–midbrain afferents express both MRs and GRs that modulate the input secondarily in either feed-forward or feedback mode, depending on time and context, and thus are capable of modulating the HPA pulse either way. The MR is promiscuous and binds aldosterone as well as corticosterone and cortisol, but is exposed to much larger amounts of the latter steroid. This is because the limbic brain lacks the enzyme 11$\beta$-hydroxysteroid dehydrogenase type 2 (11-HSD-2), which confers aldosterone selectivity to MRs in kidney cells. Instead, limbic neurons express 11-HSD-1, a reductase that regenerates bioactive glucocorticoids.

Glucocorticoid action mediated by limbic MRs and GRs modulates higher brain functions involved in behavioral and physiological adaptations to the stressor. GRs become substantially occupied at glucocorticoid peaks during ultradian pulses and after stress, while MRs are substantially occupied already at the troughs. The MR- and GR-mediated glucocorticoid actions act in concert with the other stress signaling systems. **Figure 1** depicts the bimodal nature of the stress system, with CRH/CRH receptor 1 (CRH-R1) and corticosterone/cortisol through MRs controlling the onset of the pulse, and possibly urocortin/CRH-R2, with GRs facilitating the termination of the pulse. This dualism in the stress system presents a new working model to examine the HPA axis during the aging process.

## Brain Corticosteroid Receptors

### Corticosteroid Receptor Balance and Aging

Crucial for stress and aging are two fundamental observations. First, MRs and GRs act as transcription factors and they are co-localized in limbic neurons, but have distinctly different properties. MRs have a tenfold higher affinity than GRs do for cortisol/corticosterone, and both act through transactivation at overlapping but distinct hormone response elements (HREs). Additionally, GRs are capable of interacting with common transcription factors such as activator protein-1 (AP-1) and nuclear factor-kappa B (NF-$\kappa$B), with rapid effects on gene transcription by transrepression. Thus, in response to a brief corticosterone pulse, waves of primary and secondary gene responses occur and initially act through predominant transrepression, probably dampening the initial stress reactions. Then transactivation prevails to exert recovery of homeostasis and adaptation.

Second, it was recently discovered by Henk Karst and Marian Joëls that the very same nuclear MRs also mediate nongenomic actions in the same neurons. Glucocorticoid pulses were found to rapidly enhance excitatory glutamate transmission in the hippocampus through nongenomic MRs; CRH release from the PVN was found to be suppressed through membrane GR-like receptors. Thus, it is very likely that these nongenomic effects have an important control function in activation and termination of the HPA pulse.

Previously, the slower genomic MR actions in the hippocampus appeared important for maintaining stability and integrity of limbic neuronal circuits. Situations with mostly nuclear MRs, but few GRs, activated are associated with small calcium currents, reduced spike frequency accommodation, stable responses to repeated stimulation of glutamatergic pathways, and relatively small responses to biogenic amines. Nuclear MR activation thus seems to guarantee a stable background of neuronal firing. Activation of co-localized nuclear GRs (and MRs) with higher glucocorticoid concentrations was found to result in enhanced calcium influx, stronger spike frequency accommodation, changes in ion channel conductances, and marked responses to biogenic amines such as serotonin. Thus, nuclear MRs also control, on another time scale, the onset of the stress reactions, and GRs control recovery and adaptation.

How this complementary function of membrane and nuclear MRs and GRs accommodates glucocorticoid pulsatility needs to be investigated. One testable prediction is that membrane MRs control the ascending, and membrane GRs control the descending, arm of the pulse. The genomic MRs promote long-term stability and defense of homeostasis, while genomic GRs facilitate recovery after a stressful challenge and help to prepare for the future. These functions are indicated for MRs as being proactive, determining the future state versus the reactive mode mediated by GRs, referring to the consequence of a stress-induced glucocorticoid pulse.

The MR/GR balance hypothesis predicts that, upon imbalance, threats to homeostasis are less well communicated. At a certain threshold, such hampered communication may lead to a condition of neuroendocrine dysregulation and impaired behavioral adaptation, which potentially can aggravate age-related deterioration and promote susceptibility to age-related disease to which an individual is predisposed. Aging, therefore, is characterized by a changing MR/GR balance and enhanced HPA lability.

### Modulation of Brain Corticosteroid Receptors and HPA Axis: Implications for Aging

Interestingly, individual differences in cognitive impairment that appear at old age are subject to manipulation by early life events, such as handling of newborns and depriving newborns of maternal contact. In rats, maternal deprivation or postnatal handling during the first 2 weeks of life can alter stress responsiveness permanently. Aged Long–Evans animals handled postnatally showed an attenuated stress-induced rise in corticosterone levels and are cognitively less impaired than are animals not handled as infants. Handled Long–Evans animals show a higher expression of hippocampal GRs, an effect that is established during the handling procedure by an epigenetic mechanism acting in GR promoter regions.

The outcome of maternal deprivation depends on the age at which separation occurred and duration, and strain and gender are also important variables. Deprivation of maternal contact in Brown Norway rats for 24 h at postnatal day 3 was recently shown to cause a profound midlife surge in stress-induced corticosterone output, but senescent animals and their control, mother-reared littermates displayed comparable attenuated corticosterone responses. It was expected that the early adverse experience would cause cognitive decline in all of the aged rats. This was, however, not observed: rather, maternal deprivation amplified individual differences in old age, and gave rise to a complete shift in the percentages of animals classified as impaired, partially impaired, or nonimpaired, on the basis of their performance in the Morris water maze. Most senescent rats that were mother-reared as pups showed average partially impaired performance, but senescent rats subjected to maternal deprivation were either fully impaired or nonimpaired, at the expense of the average (**Figure 2**).

Whether such a previous 'midlife' surge in HPA activity is of relevance for the subsequent aging trajectory has been tested in various ways. In the classical Phil Landfield experiments, Fisher rats adrenalectomized at midlife and provided with low amounts of corticosterone, sufficient to occupy MRs, showed at senescence less age-related decline in various behavioral and neuroanatomical markers than did their untreated controls. Likewise, in Jonathan Seckl's laboratory, mutant mice in which the 11-HSD-1 gene was deleted, preventing local reactivation of corticosterone, age-related learning impairments were ameliorated. Alternatively, rats receiving high doses of corticosterone during midlife deteriorated cognitively toward old age. However, Carmen Sandi showed that such

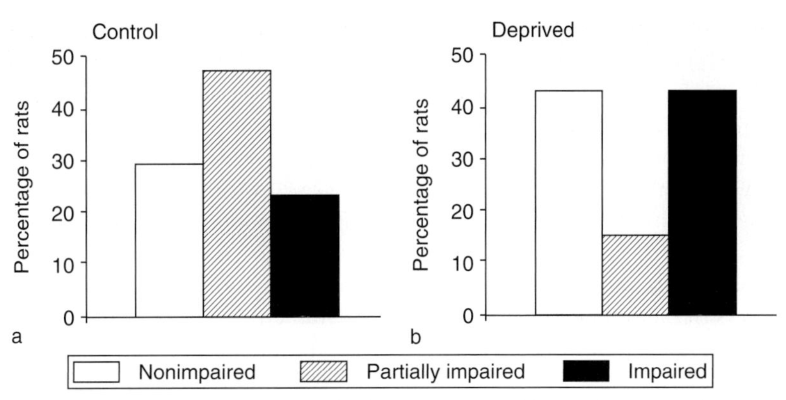


**Figure 2** Adverse early life experience amplifies individual differences in cognitive performance. At senescence, the distribution of the quality of spatial performance differs between (a) control ($n = 34$) and (b) maternally deprived ($n = 47$) animals, yielding a bell-shaped vs. a U-shaped distribution, respectively. Spatial learning of maternally deprived rats is impaired, at the expense of average performance (i.e., partially impaired). From de Kloet ER and Oitzl MS (2003) Who cares for a stressed brain? The mother, the kid or both. *Neurobiology of Aging* 24: S61–S65.

corticosterone-induced decline depends on the context in which corticosterone operates. In these experiments, Wistar rats were distinguished at midlife by high or low reaction to novelty exposure and were subsequently exposed for 28 days to stressors. Interestingly, both high and low responders showed enhanced corticosterone reactivity at senescence, but only the high novelty responders deteriorated in cognitive performance at senescence. Which factor determines this context-dependent action of corticosterone is not known.

A consistent finding has been the downregulation of hippocampal MRs in aged animals, irrespective of whether these were rodents, dogs, or primates. Strategies to enhance MR expression indeed were able to enhance cognitive performance. One example is conferred by nature: the successfully aging Brown Norway rat strain has a gain-of-function mutation in the MR. Another example is provided by tricyclic antidepressants, which in the senescent animal had no effect on cognitive performance. If, however, treatment started at midlife and was maintained until senescence, the antidepressant ameliorated cognitive impairment in the aged rats, while the drug induced MRs and downregulated corticosterone secretion. Collectively, these data suggest that the outcome of midlife corticosterone for the aging process depends on genetic and environmental contexts, in a fashion involving the corticosteroid receptors.

## Glucocorticoids and Aging: Mechanisms

### Age-Related Changes during Hypercorticism: The Relevance of Context

The relevance of context in corticosteroid action is provided by studies using a certified strategy for longevity: caloric restriction (CR). It has been known for a long time that reduction of caloric intake by at least 20% slows down the aging process and extends the life span of rats and other organisms. The aged CR rats have improved cognitive performance in a maze test, long-term potentiation (LTP) is facilitated, and neurogenesis is enhanced. However, this condition of energy deprivation also results in high circulating glucocorticoid levels. How is it possible that in some conditions chronically elevated glucocorticoids cause damage, but in others, such as in CR, elevated glucocorticoid levels exert protection? According to Caleb Finch, this paradox may be explained by the following processes: CR promotes (1) maintenance of microvasculature integrity, (2) attenuation of oxidative damage, (3) enhanced synthesis of growth factors, and (4) attenuation of glial cell proliferation. It is thought that these effects exerted by CR override the negative effects of elevated glucocorticoids. It might be that CR imposes a different context for corticosterone, resulting in suppression of oxidative damage and glial cell activation; the factors determining this context are not known.

### Possible Mechanisms Underlying Glucocorticoid-Related Damage

Excessive glucocorticoid exposure is associated with age-related impairment in memory performance, but what is the underlying mechanism? One factor is the overall increase in calcium currents, which has a number of consequences, including suppressed synaptic efficacy, as seen from impaired LTP. Glial cell proliferation and function are affected and dendritic branching in the CA3 pyramidal cell field is suppressed. Other factors related to energy metabolism, cell birth, viability, recovery, and growth also could be implicated in corticosteroid-dependent degenerative changes accompanying the aging process.

Cortisol can suppress hippocampal glucose metabolism, which is known to reflect learning and memory capacity. Positron emission tomography (PET) measurements of cerebral glucose metabolism in elderly persons indeed show in cortical and hippocampal areas a generalized hypometabolism that precedes subsequent hippocampal volume reductions and reduced performance in long-term memory tasks. The beneficial effects of glucose, and also of insulin administration, on aspects of rat and human cognitive performance are of interest, although variable results have been reported.

Brain-derived neurotrophic factor (BDNF) and nerve growth factor (NGF), and their receptors, may be implicated, since age-associated spatial learning impairments in rats can be reversed by administration of some but not all growth factors. Most attention is on the age-related atrophy of cholinergic neurons innervating hippocampus and neocortex. Recently, in rhesus monkeys, learning impairment as well as the associated atrophy of subcortical cholinergic neurons could even be reversed using neurotrophin gene therapy. This suggests that growth factors hold some potential for treating age-related changes.

Another possibility is hippocampal dentate gyrus neurogenesis, which persists from early life but slows down at old age. It is thought that neurogenesis allows the dentate gyrus to continuously rejuvenate its neuronal population in order to adapt to change. Newborn cells become functional after 2–3 weeks as judged from increases in synaptic plasticity and LTP. Stress and elevated corticosterone concentrations suppress the rate of neuronal birth in young, adult, and aged rats. Alternatively, high corticosterone levels – as a result of calorie restriction or increased physical activity through voluntary running, and also estrogen treatment, enriched environmental housing, or training in a hippocampal learning task – increase neurogenesis into old age. Further studies are needed to explore the significance of neurogenesis for aging in these various contexts.

## HPA Axis, Aging, and Age-Related Pathology in Humans

Similar to that in rodents, increased HPA activity is not a universal feature of human aging. A flattened circadian rhythm in circulating cortisol is the most consistent finding. In support of enhanced HPA activity, increased numbers of CRH/vasopressin-expressing neurons were found in the PVN of aging males. Stress responsiveness is affected by aging, as a (stressful) public speech task revealed that the subgroup with an early (anticipatory) increase in cortisol in response to the stimulus was impaired in declarative memory, whereas elderly subjects without this early rise in cortisol showed no memory deficits. These data suggest that some of the results from rat studies can be extrapolated to the human situation. However, the design of the experiments and the nature of the stressors are for obvious reasons quite different.

New information also suggests that corticosteroid receptor polymorphisms are associated with aging. In one study from Steven Lamberts' group, the ER22/23EK polymorphism, conferring a relative resistance to the GRs, was found associated with a healthy metabolic profile. The GR variant (frequency 8.9% in adults) had a significantly higher frequency in the older population. The MR Il80V polymorphism discovered by Roel de Rijk shows loss of function, and, as a result, carriers of this gene variant (frequency, 14%) display strongly enhanced neuroendocrine and autonomic responses in the Trier social stress test. Aged individuals carrying this gene variant showed a higher prevalence of depressive symptoms accumulating in the aging population.

In a prospective study, cortisol appeared related to cognitive function. Sonia Lupien observed that only aged individuals with rising cortisol levels over the 4-year period were correlated with reductions in explicit memory and attention tasks. In parallel, a reduction of 14% in hippocampal volume occurred. These volume reductions are already apparent in elderly individuals who display only mild cognitive impairment preceded by glucose hypometabolism. Similar changes in cognitive performance and hippocampal volume have been observed after exogenous glucocorticoids and in Cushing's disease. Significant correlations have, furthermore, been found between hippocampal size and delayed, but not primary or immediate, memory performance, independent of gender, generalized cerebral atrophy, or age. The extent of these early atrophic changes of the hippocampus was found to be a risk factor for accelerated memory dysfunction in normal aging.

Enhanced HPA activity, cognitive decline, and hippocampal volume reduction strongly depend on the genotype of a patient. These measures are reduced in carriers of the apolipoprotein E4 allele, which is a risk factor for age-related cognitive deficits and Alzheimer's disease. In general, basal and stress-induced HPA activation are further enhanced by Alzheimer's disease pathology. This raises the question whether Alzheimer's disease hippocampal damage is the cause or the consequence of excessive glucocorticoids. Since Alzheimer's disease pathology is absent in patients treated with corticosteroids or in severely depressed patients suffering from hypercortisim, it has been argued by Dick Swaab that these hormones are obviously not the trigger. Rather, aberrant HPA activity is considered of relevance in the variable onset and progression of Alzheimer's disease.

## Future Directions

A fundamental question in neuroendocrine aging is to what extent the HPA axis contributes to physical and behavioral signs of aging and age-related pathology. This question has received renewed interest because HPA hormones, particularly the glucocorticoids, have been found to be released in pulsatile fashion, to coordinate cell and organ function. Moreover, at the tissue level, the glucocorticoids operate through complementary MR- and GR-mediated actions in the limbic brain. While MRs are involved in stability and the onset of the stress response, recovery and adaptation after stress are mediated by GRs. In balance, MRs and GRs are crucial for homeostasis and health; imbalance and consequent HPA lability are characteristic features of aging.

A major issue is how glucocorticoids should be positioned in the current theories of aging. The thrifty genotype (James Neel) and phenotype (David Barker) hypotheses of aging hinge on nutrition and metabolism. The underlying genes and physiology do not act by themselves, but need to be regulated. Glucocorticoids are essential for this purpose, but how glucocorticoid action can change from protective to damaging is not known. Also, Caleb Finch's 'gero-inflammatory manifold,' favoring systemic and local inflammations as common denominators in age-related degenerations, has an obvious link to glucocorticoids, but glucocorticoid's place in the sequence of events is virtually unknown. Whether susceptibility genes and adverse life experiences aggravate age-related processes unidirectionally in every individual, or act on a common unknown substrate, amplifying individual differences *per se*, is an unresolved issue as well.

Thus, the issue on aberrant HPA axis and glucocorticoids, as either the cause or the consequence of aging processes, is unresolved. Yet, the significance of HPA pulsatility and MR/GR responsiveness provides ample opportunity for aging research. The current evidence suggests that MR/GR imbalance compromises the ability of organisms to defend and to restore homeostasis. Such imbalance may result in HPA dysregulation enhancing vulnerability through the very same genes that are otherwise essential for maintaining homeostasis, resilience, and health. The identity of these genes is not known.

*See also:* Aging of the Brain; Neuroendocrine Aging: Pituitary Metabolism; Neuroendocrine Aging: Pituitary-Gonadal Axis in Males; Neuroendocrine Aging: Hypothalamic-Pituitary-Gonadal Axis in Women; Synaptic Plasticity: Learning and Memory in Normal Aging.

## Further Reading

Buckley TM and Schatzberg AF (2005) Aging and the role of the HPA axis and rhythm in sleep and memory consolidation. *American Journal of Geriatric Psychiatry* 13: 344–352.

de Kloet ER, de Rijik RH, and Meijer OC (2007) Is there an imbalanced response of mineralocorticoid and glucocorticoid receptors in depression? *Nature Clinical Practice Endocrinology & Metabolism* 3: 168–179.

de Kloet ER, Joëls M, and Holsboer F (2005) Stress and the brain: from adaptation to disease. *Nature Reviews Neuroscience* 6: 463–475.

de Kloet ER and Oitzlr MS (2003) Who cares for a stressed brain? The mother, the kid or both. *Neurobiology of Aging* 24: S61–S65.

de Kloet ER, Vreugdenhil E, Oitzl MS, et al. (1998) Brain corticosteroid receptor balance in health and disease. *Endocrine Reviews* 19: 269–301.

de Nicola AF, Saravia FE, Beauquis J, et al. (2006) Estrogens and neuroendocrine hypothalamic–pituitary–adrenal axis function. *Frontiers of Hormone Research* 35: 157–168.

Finch CE and Crimmins EM (2004) Inflammatory exposure and historical changes in human life-spans. *Science* 305: 1730–1739.

Finch CE and Stanford CB (2004) Meat adaptive genes and the evolution of slower aging in humans. *Quarterly Review of Biology* 79: 3–50.

Karst H, Berger S, Tronche F, et al. (2005) Mineralocorticoid receptors are indispensable for nongenomic modulation of hippocampal glutamate transmission by corticosterone. *Proceedings of the National Academy of Sciences of the United States of America* 102: 19204–19207.

Landfield PW (1978) Hippocampal aging and adrenocorticoids: Quantitative correlations. *Science* 202: 1098–1102.

Lucassen PJ and De Kloet ER (2001) Glucocorticoids and the aging brain. In: Hof PR and Mobbs CV (eds.) *Functional Neurobiology of Aging*, pp. 883–903. San Diego: Academic Press.

McEwen BS (2003) Interacting mediators of allostasis and allostatic load: Towards an understanding of resilience in aging. *Metabolism* 52: 10–16.

Meaney MJ and Szyf M (2005) Maternal care as a model for experience-dependent chromatin plasticity? *Trends in Neurosciences* 28: 456–463.

Oitzl MS, Workel JO, Fluttert M, et al. (2000) Maternal deprivation affects behaviour from youth to senescence: Amplification of individual differences in spatial learning and memory in senescent Brown Norway rats. *European Journal of Neuroscience* 12: 3771–3780.

Patel NV and Finch CE (2002) The glucocorticoid paradox of caloric restriction in slowing brain aging. *Neurobiology of Aging* 23: 707–717.

Sapolsky RM (1992) *Stress, the Aging Brain and the Mechanism of Neuron Death*. Cambridge, MA: MIT Press.

Young EA, Abelson J, and Lightman SL (2004) Cortisol pulsatility and its role in stress regulation and health. *Frontiers in Neuroendocrinology* 25: 69–76.

## Relevant Website

http://www.lifespannetwork.nl – LifeSpan Network (a consortium of European laboratories studying mechanisms of aging and development).

# Neuroendocrine Aging: Pituitary–Gonadal Axis in Males

**P Y Liu, P Takahashi, and A Nehra**, Mayo Clinic, Rochester, MN, USA
**S M Pincus**, Guilford, CT, USA
**D M Keenan**, University of Virginia, Charlottesville, VA, USA
**J D Veldhuis**, Mayo Clinic, Rochester, MN, USA



## Overview

The human hypothalamic–pituitary–testicular axis maintains testosterone production and spermatogenesis through an intricate feedforward and feedback system involving the intermittent exchange of neurohormonal signals. The chief components of this ensemble are hypothalamic gonadotropin-releasing hormone (GnRH), pituitary luteinizing hormone (LH), and testicular sex steroids (testosterone and estradiol). Epidemiological data and clinical experiments jointly indicate that healthy older men have lower bioavailable and free testosterone concentrations, higher sex hormone-binding globulin (SHBG) concentrations, and decreased daily testosterone secretion rates, compared with young individuals. Mechanistic analyses predict each of impaired GnRH outflow, blunted LH-stimulated testosterone secretion, and reduced negative feedback by testosterone in the older male. Aging also disrupts quantifiable synchrony among sleep stage, the secretion of LH, testosterone, prolactin, and follicle-stimulating hormone (FSH), and the oscillations in nocturnal penile tumescence, denoting erosion of coordinate central neurohormonal outflow. Prolactin, estradiol, and inhibin B concentrations fall and FSH concentrations rise with age. How age affects spermatogenesis in healthy populations is not known. In summary, aging in men is marked by multisite regulatory failure within the gonadal axis. Further studies are required to define the tempo, variability, reversibility, and import of such interlinked deficits, and to delineate safe and effective interventional strategies to obviate aging-related frailty.

The human hypothalamic–pituitary–testicular axis maintains a eugonadal state via reciprocal feedforward and feedback connections. Multiple changes in this network occur with age, resulting in a gradual (1–2% $yr^{-1}$) decline in testosterone availability. Whether aging also decreases spermatogenesis is not established due to the inherent difficulty in obtaining semen from representative populations. Less testosterone production may contribute to aging-associated increases in visceral fat, insulin resistance, falls, and fractures, and decreases in bone mineral density, muscle mass and strength, physical performance, perceived quality of life, libido, and potentia. Mechanistic inquiries dissecting the bases for testosterone depletion with age should facilitate the development and optimization of preventive and therapeutic strategies. Such strategies must carefully balance potential risks of androgen-induced prostatic disease, cardiovascular disease, erythrocytosis, and/or obstructive sleep apnea against psychological and physical benefits.

## Rate of Evolution and Extent of Relative Hypoandrogenemia in Older Men

Testosterone depletion in aging is an increasingly pertinent clinical issue, given epidemiological data linking hypoandrogenemia with sarcopenia, osteopenia, visceral adiposity, insulin resistance, reduced physical stamina, sexual dysfunction, impaired quality of life, and (possibly) depressive mood and waning cognitive function. Impoverished testosterone availability in older men has been affirmed by (1) longitudinal investigations in healthy cohorts, (2) meta-analysis of cross-sectional data, and (3) direct sampling of the spermatic vein. In general, bioavailable (non-SHBG-bound) and free testosterone concentrations fall more markedly than do total values, as illustrated for cross-sectional data in **Figure 1**. Total testosterone concentrations are partially maintained by a concomitant increase in SHBG concentrations in aging men. A 15-year prospective analysis in the United States (New Mexico) reported that total testosterone concentrations fall by approximately 0.39 $nmol\,l^{-1}\,yr^{-1}$ in men after the age of 60 years; the Massachusetts Male Aging Cohort studies estimated a 0.8–1.3% annual decrement in bioavailable testosterone concentrations and the Baltimore Longitudinal Study of Aging predicted an annual decline of 4.9 pmol testosterone per nanomole SHBG. In the last study, the age-related prevalence of hypogonadal testosterone:SHBG ratios, judged by young-adult normative criteria, exceeded 20%, 30%, and 50% at ages 60, 70, and 80 years, respectively. Thus, reduced systemic testosterone availability is especially significant in societies with high longevity.

Comorbidity in aged individuals can markedly exacerbate androgen deficiency. Comorbid risk factors include chronic systemic disease, acute illness, stress, weight loss, hospitalization, institutionalization, and the use of certain drugs. The degree to which illness-induced androgen decrements are reversible in various settings is not yet well articulated. Whether certain comorbidities represent indications for testosterone replacement also requires further investigation.

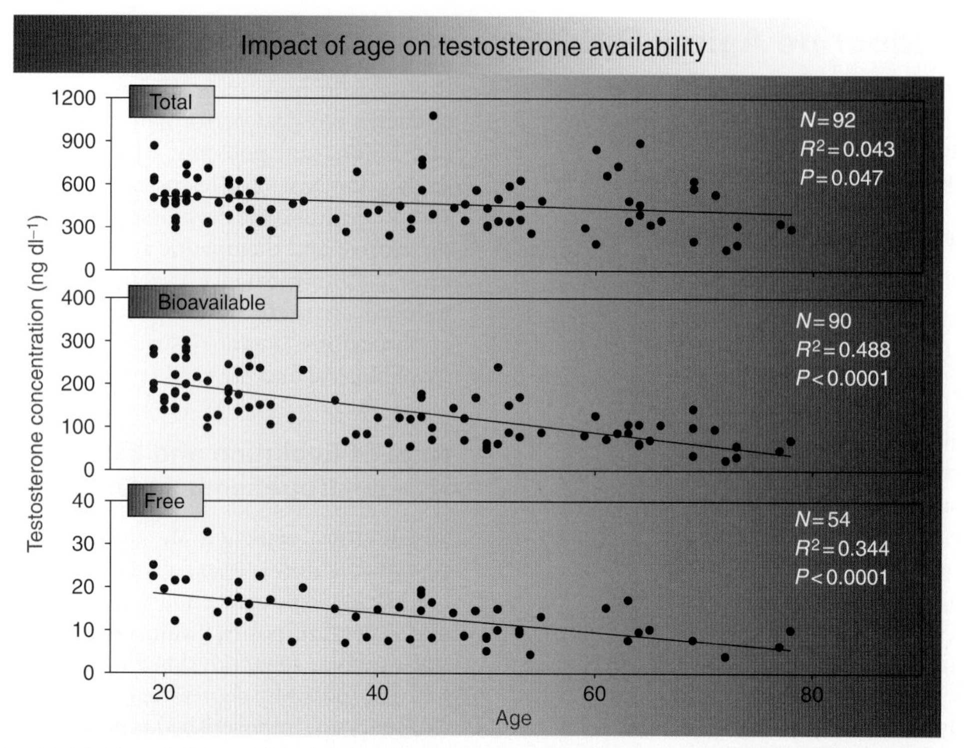


**Figure 1** Plots of total, bioavailable, and free (unbound) testosterone concentrations regressed on age in healthy volunteers from Olmsted County, Minnesota, USA. The $R^2$ values denote the fraction of variability accounted for by age.

## Mechanistic Bases of Testosterone Depletion

The fundamental mechanisms that cause hypoandrogenemia in the aging male remain unknown. However, recent clinical investigations have disclosed multiple alterations in the gonadal axis in older men: (1) lower amplitude LH pulses, suggesting reduced drive by GnRH; (2) more frequent LH pulses and less orderly patterns of LH release, consistent with diminished negative feedback; (3) preserved LH and heightened FSH responses to exogenous GnRH pulses, indicating lack of primary pituitary failure; (4) reduced pulsatile and thereby total daily testosterone secretion; and (5) impaired testosterone responses to elevated endogenous LH concentrations (stimulated by flutamide, tamoxifen, GnRH, or anastrozole) and to infused pulses of recombinant human LH. The collective findings indicate that relative androgen deficiency in the older male reflects multisite failure of the GnRH–LH–testosterone axis.

An investigative challenge arises in verifying specific mechanisms that mediate gradual aging-related adaptations in the human gonadal axis. A recent strategy is to exploit an ensemble regulatory concept of reciprocal interactions among GnRH, LH, and testosterone (**Figure 2**). An integrative biomathematical approach allows one to formalize and quantify underlying feedback and feedforward interactions objectively. The analytical rationale is to reconstruct aging-related adaptations among all three primary signals – GnRH, LH and testosterone – rather than evaluate any one signal in isolation. According to ensemble analyses, a parsimonious explication of collective clinical data is that aging (1) attenuates hypothalamic GnRH outflow, (2) impairs testicular responsiveness to LH pulses, and (3) decreases androgenic negative feedback. Longitudinal studies will be required to establish the relative importance of individual regulatory deficits in healthy aging (**Figure 3**).

## Putative Pathophysiological Mechanisms in the Aging Male

### Hypothalamic GnRH Deficiency in Older Men

Indirect clinical observations support the hypothesis that hypothalamic GnRH outflow is reduced in older men (**Table 1**). In particular, aging individuals maintain lower LH pulse amplitudes (due to a reduced mass of LH secreted per burst) and more irregular LH secretory patterns (increased approximate entropy), yet respond normally to exogenous GnRH pulses. Although such observations are congruent with decreased endogenous GnRH drive, hypothalamic GnRH secretion cannot be monitored directly

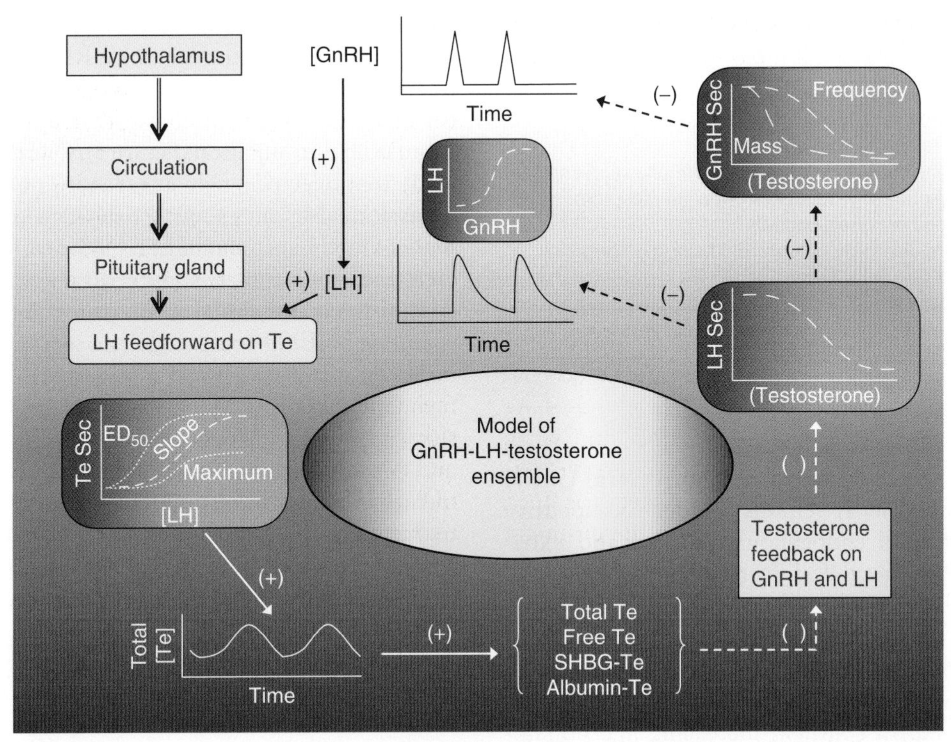


**Figure 2** Simplified structure of the male gonadotropin-releasing hormone (GnRH), luteinizing hormone (LH), and testosterone (Te) (i.e., hypothalamic–pituitary–Leydig cell) axis with key feedback (–, inhibitory) and feedforward (+, stimulatory) interactions. Smooth curves denote *in vivo* dose–response interfaces mediating GnRH-induced LH secretion (top middle), LH-stimulated testosterone secretion (Te Sec; bottom left; SHBG, sex hormone-binding globulin), and testosterone-enforced negative feedback on the pituitary gland (right middle) and hypothalamus (rightmost top). Possible adaptations in LH → Te feedforward sensitivity (slope), potency (one-half-maximal stimulus concentration, or $ED_{50}$), and efficacy (maximal responsiveness) are illustrated (left middle).

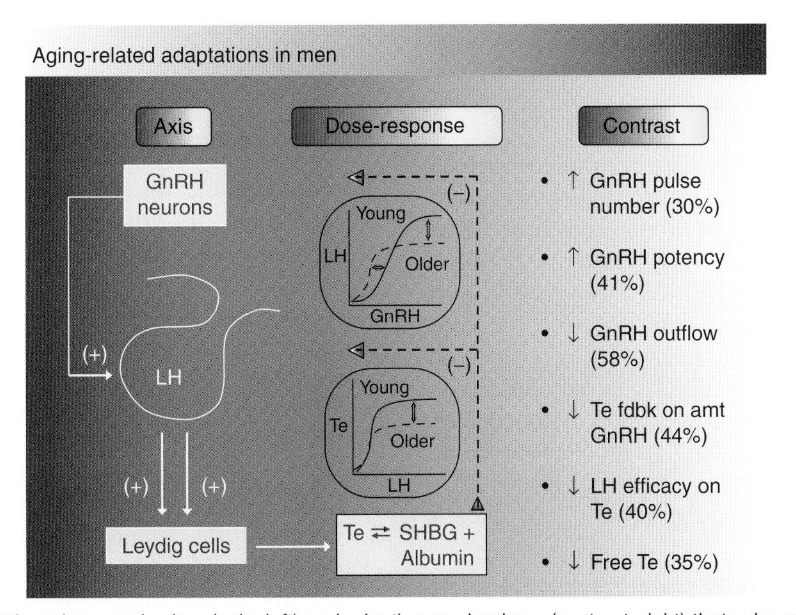


**Figure 3** Schema of inferred anatomic sites (axis; left) and adaptive mechanisms (contrast; right) that subserve relative testosterone (Te) deficiency in healthy older men (age >60 years), compared with young (age <35 years) men (GnRH, gonadotropin-releasing hormone; LH, luteinizing hormone; SBHG, sex hormone-binding globulin). Arrows define stimulation (+) and inhibition (–). The two encircled insets (dose–response; center) depict age-related contrasts in GnRH → LH drive (top center) and LH → Te feedforward (bottom center).

**Table 1** Indirect evidence for impaired gonadotropin-releasing hormone drive in aging men

| |
|---|
| Low-amplitude spontaneous luteinizing hormone pulses |
| Increased or normal potency of exogenous gonadotropin-releasing hormone pulses |
| Blunted unleashing of pulsatile luteinizing hormone secretion after blockade of sex hormone feedback or opiate receptors |
| Normal luteinizing hormone bioactivity |
| Quantifiably more disorderly patterns of luteinizing hormone release |
| Asynchrony of central neurohormone outflow: luteinizing hormone, follicle-stimulating hormone, prolactin, sleep stage, and nocturnal penile tumescence |
| Mathematical predictions based on an ensemble concept |

in the human. One means to estimate central GnRH outflow noninvasively entails graded competitive inhibition of both endogenously driven and exogenous GnRH-induced LH secretion. Inhibition can be enforced by escalating doses of a potent and selective GnRH-receptor antagonist (such as ganirelix). In this paradigm, age does not affect LH responses to an exogenous GnRH stimulus, whether or not ganirelix is administered. In contrast, increasing age potentiates suppression of endogenously driven LH pulses by any given blood concentration of the receptor antagonist. Given that GnRH and ganirelix compete for the same pituitary receptor, greater inhibition of LH pulses by the receptor antagonist in older men points to less hypothalamic GnRH outflow to drive gonadotrope cells. This important inference is confirmed analytically using a model in which one assumes that (1) unobserved hypothalamic GnRH concentrations stimulate LH secretion via a positive logistic function and (2) testosterone inhibits GnRH outflow (release and action) via a negative exponential function.

Laboratory experiments have also delineated significant (but not exclusive) central hypogonadotropism associated with hypoandrogenemia in the aged male rodent. Relevant evidence includes (1) reduced *in vivo* LH pulse amplitude, (2) diminished *in vitro* GnRH secretion by hypothalamic tissue, (3) attenuated castration-induced LH release, (4) normal LH responses to injected GnRH, (5) altered GnRH neuronal synaptology, and (6) restoration of sexual activity in the aged rat following hypothalamic neuronal transplantation. Nonetheless, potential species differences limit facile extrapolation of these data to humans.

### Enhanced Gonadotrope Secretory Responsiveness to Small Amounts of GnRH

In principle, smaller LH pulses in older men could reflect diminished hypothalamic GnRH secretion and/or decreased pituitary sensitivity to available GnRH. In the latter context, initial studies using a single supramaximal dose of GnRH were contradictory. Recent investigations incorporating physiological doses of GnRH disclosed that aging potentiates submaximal stimulation (**Figure 3**). One dose–response study that injected single intravenous pulses of GnRH in random order on a single morning estimated 1.6-fold higher GnRH-stimulated LH, FSH, and free $\alpha$-subunit secretion in aging men. Another dose–response analysis administered individual intravenous doses of GnRH spanning a 1000-fold range on separate, randomly ordered mornings. This stratagem disclosed that older men maintain nearly twofold greater gonadotrope sensitivity to GnRH (steeper slope of the dose–response) and twofold greater GnRH potency (50% lower dose of GnRH required to stimulate LH secretion one half-maximally) than do young counterparts, but have comparable maximal LH secretion (denoting age-equivalent GnRH efficacy). Heightened acute gonadotrope responsiveness to submaximal amounts of GnRH (2.5, 10, and 25 ng $kg^{-1}$) in older men could reflect upregulation of GnRH receptors, augmentation of pituitary LH stores, and/or feedback withdrawal due to lower concentrations of total, bioavailable, or free testosterone. The last consideration has been affirmed analytically. A third investigation infused pulses of GnRH (100 ng $kg^{-1}$) or saline intravenously every 90 min for 14 days by portable infusion pump in young and older men. The GnRH clamp increased pulsatile LH secretion equivalently in the two age groups, indicating comparable gonadotrope responsivity. However, bioavailable and free testosterone concentrations remained 50% lower in elderly men, suggesting reduced bioactivity of secreted LH or impaired testicular steroidogenesis.

### Impaired Leydig Cell Steroidogenesis in Older Men

Administration of the LH surrogate, human chorionic gonadotropin (hCG), fails to stimulate maximal young-adult testosterone concentrations in many, but not all, aging men. Impaired testis responsivity in this setting is difficult to interpret definitively, because the hCG stimulus is not physiological. Three concerns exist: (1) the half-life of hCG is 20–30 h (compared with 1 h for LH), thus enforcing nonpulsatile drive, (2) hCG rapidly downregulates Leydig cell steroidogenesis *in vivo* and *in vitro*, and (3) conventional hCG doses exceed the lutropic activity of an LH pulse by >300-fold, thus testing only maximal testis responses. The last issue is relevant, because low testosterone production occurs in older men when LH concentrations are normal or only minimally elevated.

A near-physiological protocol to examine Leydig cell responsiveness involves intravenous infusion of pulses of recombinant human LH during acute suppression of

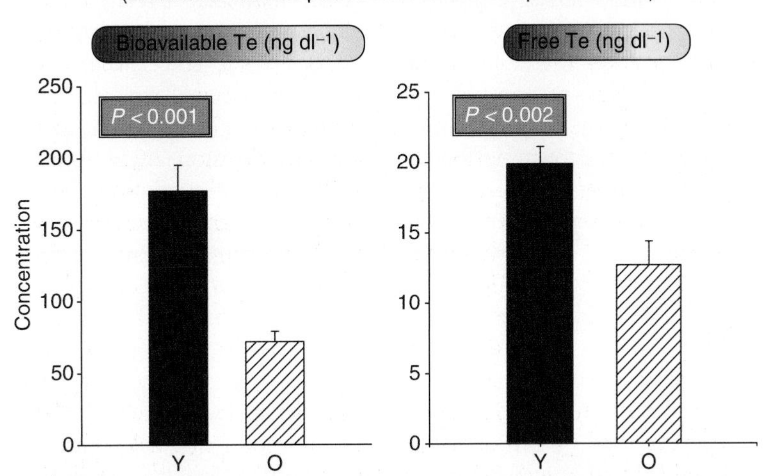


**Figure 4** Pulses of recombinant human luteinizing hormone (LH) after gonadotropin-releasing hormone (GnRH) receptor blockade fail to normalize bioavailable (left) and free (right) testosterone (Te) concentrations in older men. Adapted from Liu PY, Takahashi PY, Roebuck PD, et al. (2005) Aging in healthy men impairs recombinant human luteinizing hormone (LH)-stimulated testosterone secretion monitored under a two-day intravenous pulsatile LH clamp. *Journal of Clinical Endocrinology & Metabolism* 90(10): 5544–5550. Copyright 2005, The Endocrine Society.

LH secretion with a GnRH-receptor antagonist. Application of this paradigm for 48 h unmasked 50% lower exogenous LH-stimulated bioavailable and free testosterone concentrations in older, as compared to young, men (**Figure 4**). Whether more extended delivery of normal-amplitude LH pulses would restore young-adultlike testosterone secretion in older individuals is not known. Although experiments in the aged Brown Norway rat suggest a fixed steroidogenic defect in Leydig cells, the exact relevance of the animal model to the human is not certain.

### Aging Restricts Feedback by Endogenous Testosterone on GnRH and LH Secretion

Clinical studies have reported both greater and less reduction of LH concentrations following supplementation with testosterone or 5α-dihydrotestosterone in older men, as compared to young men. Discrepancies could relate to nonphysiological routes, amounts, preparations and patterns of androgen administration, confounding effects of endogenous testosterone, dissimilar study populations, and/or single measurements of LH concentrations. On the other hand, aging appears to mute negative feedback by endogenous testosterone consistently. The latter inference applies to comparisons made by cross-correlation, approximate entropy, and ensemble-model analyses of frequently sampled LH and testosterone concentrations. Feedback attenuation is not due solely to systemic androgen depletion in older individuals, because feedback strength decreases per unit testosterone concentration (**Figure 5**). The molecular mechanisms that mediate impaired feedback are not known. However, histochemical analyses have quantitated reduced androgen-receptor expression in the brain and pituitary gland of the aged rat and in genital fibroblasts of the older human.

Reduced negative feedback by androgens could contribute to the higher frequency, lower amplitude, and more irregular LH secretion patterns that typify aging individuals. In fact, blocking testosterone or estradiol synthesis with steroidogenic enzyme inhibitors elicits rapid and irregular LH pulses even in healthy young men. The sex steroid-depleting regimens fail to increase LH secretory burst size equally in older and young men. Therefore, smaller LH secretory bursts in aging individuals are not attributable to hypoandrogenemia or hypoestrogenemia *per se*. On the other hand, more frequent LH secretory bursts and less orderly LH secretion patterns may reflect both reduced feedback efficacy and reduced testosterone availability for feedback. Verification of relative age- and androgen-dependent altered LH secretion would require experimentally clamping systemic testosterone availability in different age groups.

## FSH and Inhibin B

Hypothalamic–pituitary–gonadal control of FSH secretion is less well understood because (1) feedforward relationships among FSH, testosterone, estradiol, and the 90-day spermatogenic cycle in the human are complex; (2) sperm production and

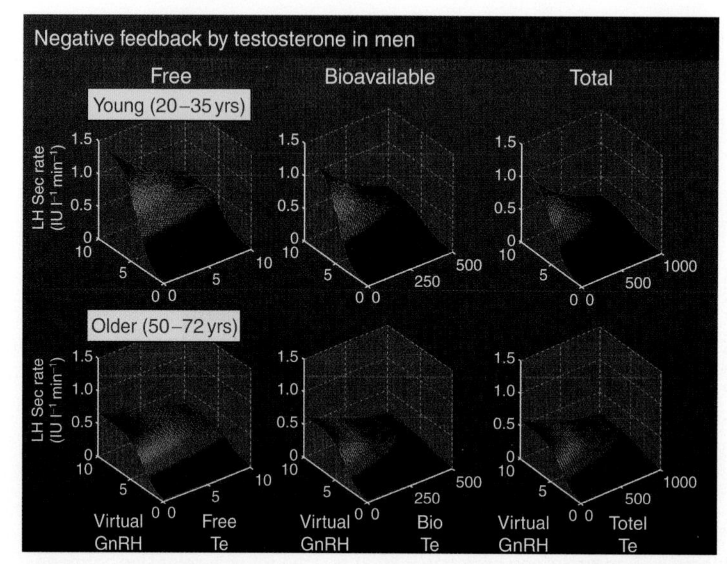


**Figure 5** Analytically reconstructed surfaces depicting virtual (calculated) gonadotropin-releasing hormone (GnRH) feedforward and free testosterone (Te) feedback onto luteinizing hormone (LH) secretion in men in the indicated age cohorts. Each surface interlinks pulsatile LH secretion (vertical axis) with estimated GnRH drive (oblique axis) and Te feedback (horizontal axis). Te feedback is calculated for each of free (left), bioavailable (middle), and total (right) Te concentrations. Aging is marked in three ways: by decreased maximal LH secretion, defined at maximal virtual GnRH outflow at zero (extrapolated) Te concentrations (vertical axis intercept of the surface); by less Te feedback on LH secretion at any given GnRH signal strength (right-to-left downslope at top of surface due to increasing free Te concentrations); and by greater potency of virtual GnRH (leftward shift in GnRH → LH dose–response function at any given Te concentration). Adapted from Keenan DM, Takahashi PY, Liu PY, et al. (2006) An ensemble model of the male gonadal axis: Illustrative application in aging men. *Endocrinology* 147(6): 2817–2828, with permission. Copyright 2006, The Endocrine Society.

quality are difficult to monitor serially; (3) FSH secretion is restrained by all three of testosterone, estradiol, and inhibin B; (4) both GnRH and activin appear to maintain FSH secretion in animals; (5) recombinant human activin and inhibin B are not available for clinical studies; and (6) the prolonged half-life of FSH makes reliable pulsatility analysis more difficult (**Figure 6**). Available clinical studies show that greater age is associated with higher serum FSH concentrations but lower inhibin B (especially before age 35 years) and bioavailable sex-steroid concentrations, and loss of coordinate LH–FSH secretion. Although estradiol, testosterone, and nonaromatizable androgens inhibit FSH secretion in older men, how age influences sex-steroid feedback potency and efficacy remains unknown.

## Prolactin

Disorders associated with hyperprolactinemia are associated with suppressed LH pulsatility and reduced sex-steroid production. How hyperprolactinemia represses GnRH secretion is not established in humans. Data derived from animal models are species dependent. For example, prolactin administration suppresses LH

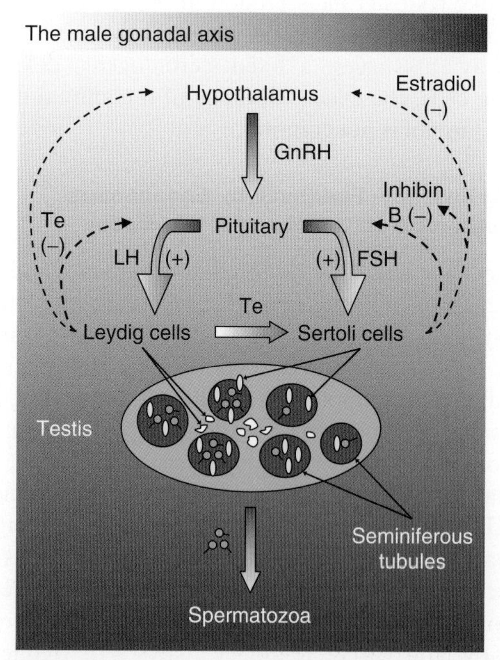


**Figure 6** Overall gonadal axis, including the gonadotropin-releasing hormone (GnRH)–luteinizing hormone (LH)–testosterone (Te) component (left) and the seminiferous tubular compartment with GnRH, follicle-stimulating hormone (FSH), estradiol, and inhibin B (right).

pulse frequency and amplitude in sheep and castrated rats. On the other hand, a prolactin receptor-knockout mouse strain had normal concentrations of LH, FSH, and testosterone, with preserved fertility and testis morphology. Lactotrope cell number does not change, but pulsatile prolactin secretion declines with age in humans. Prolactin is not known to regulate testis function directly in men, unlike in the rodent.

## Neurohormonal Network Regulation in Aging Men

Available analyses indicate that aging in men impairs hypothalamic GnRH outflow, LH-stimulated Leydig cell steroidogenesis, and testosterone-dependent negative feedback (**Figure** 3). How this tripartite conclusion is reached requires explanation. Strong interdependencies among GnRH, LH, and testosterone secretion signify the operation of networklike (ensemble) control of the gonadal axis. In ensemble systems, no one signal acts alone or can be viewed alone. Thus, to ensure valid mechanistic interpretation, all three signals must be analyzed together without perturbing their feedback or feedforward connections. The approximate entropy (ApEn) measure provides one means to do this. ApEn is a model-free statistic that quantifies the overall impact of regulatory inputs within an interlinked system. The endpoint is the relative regularity, orderliness, or consistency of recurring subpatterns in serial measurements. Thus, ApEn does not replace, but instead complements, the analysis of pulses or circadian rhythms. ApEn analyses unveil more irregular (less orderly) patterns of both LH and testosterone secretion with increasing age. In contrast, FSH release is quite irregular at all ages.

As an extension of the ApEn metric, the cross-ApEn statistic quantifies the joint pattern synchrony of two coupled processes, specifically paired signals. Unlike the linear cross-correlation coefficient, cross-ApEn provides a probabilistic measure of pattern similarities (reproducibility) independently of time lag. Cross-ApEn analyses unmask prominent age-related deterioration of LH's coupling with each of testosterone, FSH, and prolactin secretion, sleep-stage transitions, and nocturnal penile tumescence (NPT) oscillations. Directional cross-ApEn comparisons indicate that older individuals have symmetric erosion of LH → testosterone feedforward and testosterone → LH feedback linkages (**Figure** 7). Thus, aging alters ensemble control among LH, FSH, prolactin and testosterone secretion as well as sleep and autonomic nervous system outflow, which govern NPT.

Investigation of the male gonadal axis is enhanced by biomathematical formalism, which encapsulates nonlinear, dose-responsive, and time-delayed feedforward and feedback coupling among GnRH, LH, and testosterone. One analytical construct allows one to estimate endogenous dose–response properties noninvasively

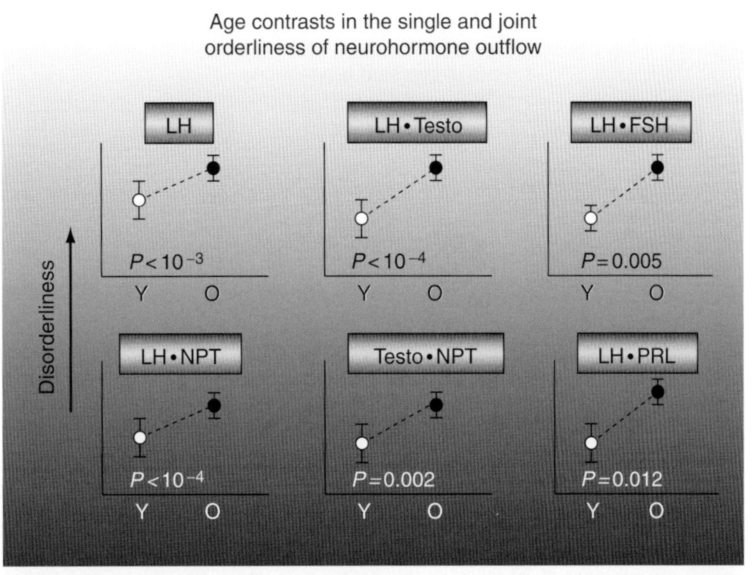


**Figure 7** Disruption of central neurohormone outflow in the aging male quantified by the cross-approximate entropy (ApEn) statistic, which quantifies loss of orderliness. Cross-ApEn values for each pair of neurohormone outflow measures are elevated in older men, compared with those in young men, signifying deterioration of joint synchrony (LH, luteinizing hormone; Testo, testosterone; NPT, nocturnal penile tumescence; FSH, follicle-stimulating hormone; PRL, prolactin). Greater asynchrony of paired signals in turn denotes disruption of the physiological pathways connecting the signals. Adapted from Veldhuis JD, Iranmanesh A, Godschalk M, et al. (2000) Older men manifest multifold synchrony disruption of reproductive neurohormone outflow. *Journal of Clinical Endocrinology & Metabolism* 85(4): 1477–1486, with permission. Copyright 2000, The Endocrine Society.

without the injection of GnRH, LH, or testosterone or disruption of feedback linkages. Experimental validation required repetitive direct sampling of cavernous sinus and internal jugular venous concentrations of GnRH, LH, and testosterone in the awake stallion and ram, and peripheral LH and spermatic vein testosterone concentrations in the human. **Figure 8** illustrates the results of noninvasive analytical reconstruction of feedforward dose–response functions linking LH concentrations and testosterone secretion rates, using only paired systemic measurements of the two hormones in 15 older and 15 young men. The methodology is illustrated schematically in **Figure 9.** This approach predicts that aging selectively impairs

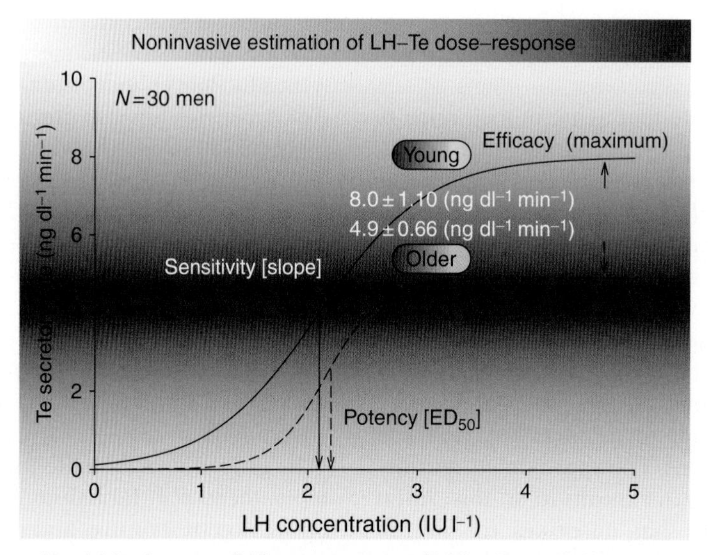


**Figure 8** Analytically estimated luteinizing hormone (LH) → testosterone (Te) feedforward dose–response functions in young ($N = 15$) and older ($N = 15$) men. LH efficacy is defined as asymptotically maximal Te secretion; potency is defined as [$ED_{50}$], the half-maximally effective stimulatory concentration (dose) of LH, and sensitivity is defined as the maximal positive slope. Only LH efficacy differed by age in these analyses. Adapted from Keenan DM and Veldhuis JD (2004) Divergent gonadotropin–gonadal dose–responsive coupling in healthy young and aging men. *American Journal of Physiology – Regulatory, Integrative and Comparative Physiology* 286(2): R381–R389.

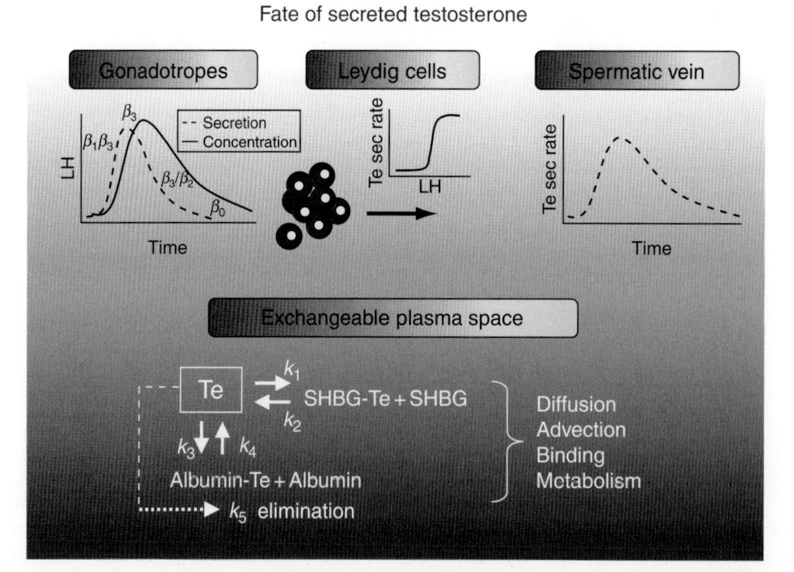


**Figure 9** Model of feedforward mechanism by brain gonadotropin-releasing hormone onto pituitary gonadotrope cells to induce a burst of luteinizing hormone (LH) secretion (top left). LH concentrations drive gonadal testosterone secretion (Te Sec) via a monotonic dose–response function (top middle). Burstlike Te secretion over time (top right) is the output of the simultaneously calculated LH → Te feedforward function. Testosterone entering the blood has the five possible fates, denoted as $k_1$–$k_5$, representing random diffusion, linear advection in blood, binding to albumin or sex hormone-binding globulin (SHBG), and irreversible removal or chemical metabolism.

**Table 2** Model-simulated implications of selected regulatory deficits in (aging) male gonadal axis

| *Deficit examined*[a] | *Predicted responses* | | | *Orderliness or regularity* |
|---|---|---|---|---|
| | *Available testosterone* | *Luteinizing hormone pulses* | | |
| | | *Amplitude* | *Frequency* | |
| (A) ↓ GnRH outflow | ↓ | ↓ | | ↑ |
| (B) ↓ LH → Te drive | ↓ | ↑ | ↑ | ↓ |
| (C) ↓ Te feedback action | ↑ | ↑ | ↑ | ↓ |
| Combined A, B, and C[a] | ↓ | ↓ | ↑ | ↓ |

[a]Mimics observed features in older men.
GnRH, gonadotropin-releasing hormone; LH, luteinizing hormone; Te, testosterone.

endogenous LH efficacy; that is, aging reduces maximally stimulated testosterone secretion without altering LH potency or testis sensitivity to LH. Randomly ordered dose-varying intravenous infusions of recombinant LH during GnRH-receptor blockade will be needed to establish such inferences empirically.

Computer-assisted simulations confer predictive insights. Recent model-based simulations of gonadal axis adaptations in the older male illustrate that (1) isolated attenuation of central GnRH drive forecasts lower LH pulse amplitude and attendant hypoandrogenemia, but not higher LH pulse frequency or greater irregularity; (2) exclusive impairment of Leydig cell steroidogenesis presages lower testosterone concentrations, higher LH pulse frequency, and greater disorderliness of LH secretion patterns, but not smaller LH pulses; (3) pure restriction of negative feedback projects normal or elevated testosterone concentrations with elevated LH pulse frequency, amplitude, and irregularity; and (4) combined reductions in GnRH outflow, LH-stimulated testosterone secretion, and negative feedback by testosterone predict all four of hypoandrogenemia, attenuated LH pulse amplitude, accelerated LH pulse frequency, and disorderly LH release. This tetrad is characteristic of older men. **Table 2** summarizes these concepts.Accordingly, from an objective modeling perspective, no single regulatory defect can explicate the neuroendocrine phenotype observed in healthy older men.

## Summary

Androgen deprivation adversely impacts libido and sexual potency, psychological well being, mood and (possibly) cognition, exercise tolerance, muscle mass, bone mineral density, intraabdominal adiposity, and lipid and carbohydrate metabolism. Clinical investigations in older men support triple impairment of hypothalamic GnRH secretion, pulsatile LH-stimulated testosterone synthesis, and testosterone-mediated negative feedback. Understanding the primary mechanistic bases of testosterone depletion in aging men should foster novel therapeutic strategies to maintain anabolism, enhance quality of life, and minimize physical frailty.

*See also:* Gene Expression in Normal Aging Brain; Neuroendocrine Aging: Pituitary Metabolism; Neuroendocrine Aging: Hypothalamic-Pituitary-Gonadal Axis in Women; Neuroendocrine Aging: Pituitary-Adrenal Axis; Sleep and Circadian Rhythm Disorders in Human Aging and Dementia.

## Further Reading

Feldman HA, Longcope C, Derby CA, et al. (2002) Age trends in the level of serum testosterone and other hormones in middle-aged men: Longitudinal results from the Massachusetts male aging study. *Journal of Clinical Endocrinology & Metabolism* 87: 589–598.

Foresta C, Bordon P, Rossato M, et al. (1997) Specific linkages among luteinizing hormone, follicle stimulating hormone, and testosterone release in the peripheral blood and human spermatic vein: Evidence for both positive (feed-forward) and negative (feedback) within-axis regulation. *Journal of Clinical Endocrinology & Metabolism* 82: 3040–3046.

Harman SM, Metter EJ, Tobin JD, et al. (2005) Longitudinal effects of aging on serum total and free testosterone levels in healthy men. Baltimore Longitudinal Study of Aging. *Journal of Clinical Endocrinology & Metabolism* 86: 724–731.

Keenan DM, Alexander S, Irvine CH, et al. (2004) Reconstruction of *in vivo* time-evolving neuroendocrine dose-response properties unveils admixed deterministic and stochastic elements. *Proceedings of the National Academy of Sciences of the United States of America* 101: 6740–6745.

Keenan DM, Takahashi PY, Liu PY, et al. (2006) An ensemble model of the male gonadal axis: Illustrative application in aging men. *Endocrinology* 147: 2817–2828.

Keenan DM and Veldhuis JD (2001) Hypothesis testing of the aging male gonadal axis via a biomathematical construct. *American Journal of Physiology – Regulatory, Integrative and Comparative Physiology* 280: R1755–R1771.

Liu PY, Swerdloff RS, and Veldhuis JD (2004) The rationale, efficacy and safety of androgen therapy in older men: Future research and current practice recommendations. *Journal of Clinical Endocrinology & Metabolism* 89: 4789–4796.

Liu PY, Takahashi PY, Roebuck PD, et al. (2005) Aging in healthy men impairs recombinant human luteinizing hormone (LH)-stimulated testosterone secretion monitored under a two-day intravenous pulsatile LH clamp. *Journal of Clinical Endocrinology & Metabolism* 90: 5544–5550.

Longcope C (1973) The effect of human chorionic gonadotropin on plasma steroid levels in young and old men. *Steroids* 21: 583–592.

Mahmoud AM, Goemaere S, DeBacquer D, et al. (2000) Serum inhibin B levels in community-dwelling elderly men. *Clinical Endocrinology (Oxford)* 53: 139–140.

Morley JE, Kaiser FE, Perry HM, 3rd, et al. (1997) Longitudinal changes in testosterone, luteinizing hormone, and follicle-stimulating hormone in healthy older men. *Metabolism: Clinical and Experimental.* 46: 410–413.

Mulligan T, Iranmanesh A, Kerzner R, et al. (1999) Two-week pulsatile gonadotropin releasing hormone infusion unmasks dual (hypothalamic and Leydig cell) defects in the healthy aging male gonadotropic axis. *European Journal of Endocrinology* 141: 257–266.

Pincus SM, Mulligan T, Iranmanesh A, et al. (1996) Older males secrete luteinizing hormone and testosterone more irregularly, and jointly more asynchronously, than younger males. *Proceedings of the National Academy of Sciences of the United States of America* 93: 14100–14105.

van den Beld AW, De Jong FH, Grobbee DE, et al. (2000) Measures of bioavailable serum testosterone and estradiol and their relationships with muscle strength, bone density, and body composition in elderly men. *Journal of Clinical Endocrinology & Metabolism* 85: 3276–3282.

Veldhuis JD, Iranmanesh A, and Keenan DM (2004) An ensemble perspective of aging-related hypoandrogenemia in men. In: Winters SJ (ed.) *Male Hypogonadism: Basic, Clinical, and Theoretical Principles*, pp. 261–264. Totowa, NJ: Humana Press.

Zmuda JM, Cauley JA, Kriska A, et al. (1997) Longitudinal relation between endogenous testosterone and cardiovascular disease risk factors in middle-aged men. A 13-year follow-up of former Multiple Risk Factor Intervention Trial participants. *American Journal of Epidemiology* 146: 609–617.

# Neuroendocrine Aging: Hypothalamic–Pituitary–Gonadal Axis in Women

**N E Rance**, University of Arizona College of Medicine, Tucson, AZ, USA



## Introduction

The loss of female reproductive function in midlife constitutes the most dramatic and consistent change during aging in humans. Because menopause occurs early, at a mean age of 51 years, the average woman will spend approximately one-third of her life span in the postmenopausal phase. The hallmark of menopause is ovarian failure, which leads to a reduction in circulating blood levels of estrogen and progesterone to castrate levels. In turn, the loss of ovarian hormones leads to hypersecretion of pituitary gonadotropins and the clinical symptoms experienced by postmenopausal women, including hot flushes, vaginal dryness, and mood changes. Postmenopausal women also have an increased risk for osteoporosis and cardiovascular disease, which may reflect the effects of estrogen withdrawal on age-related diseases. Thus, the repercussions of menopause can impact a women's health for a significant portion of her life span.

## Overview of the Hypothalamic–Pituitary–Ovarian Axis in Young Women

Reproduction is controlled via reciprocal interactions among the hypothalamus, anterior pituitary gland, and ovaries. Neurons in the medial basal hypothalamus that synthesize gonadotropin-releasing hormone (GnRH) project to fenestrated capillaries in the median eminence. In the median eminence, GnRH is secreted from axon terminals into hypophyseal portal vessels, a capillary system that carries blood to the pituitary gland. At the level of the anterior pituitary gland, GnRH stimulates the production and release of the gonadotropins, luteinizing hormone (LH) and follicle-stimulating hormone (FSH), into the systemic circulation. LH and FSH target the ovaries via the systemic circulation to regulate folliculogenesis, ovulation, and the secretion of steroid and glycoprotein hormones. In turn, the ovarian steroids, estrogen and progesterone, exert complicated feedback actions on both the hypothalamus and the pituitary. The ovary also secretes glycoprotein hormones (such as inhibin and activin) that modulate FSH secretion at the level of the anterior pituitary gland. The timing of the menstrual cycle is dependent upon the pattern of ovarian hormone secretion. In addition, multiple neural inputs to hypothalamic GnRH neurons convey information on the status of physiologic systems to optimize the timing for reproduction.

GnRH is secreted in a pulsatile manner, leading to short-duration peaks of LH in peripheral plasma. Studies in the rhesus monkeys have delineated the medial basal hypothalamus as the control center for pulsatile LH secretion. The medial basal hypothalamus is the site of multiunit electrical activity (the GnRH pulse generator) that is synchronized with pulses of LH in the peripheral circulation. Neurons with high levels of GnRH gene expression are scattered within the primate medial basal hypothalamus, further evidence of its importance in the control of the reproductive axis. The frequency and amplitude of GnRH pulses vary over the menstrual cycle, and the pattern of GnRH pulsatility plays an important role in pituitary responsiveness. For example, continuous (rather than pulsatile) administration of GnRH (or one of its analogs) actually suppresses LH secretion. Moreover, the pattern of LH and FSH pulses in the systemic circulation influences ovarian follicle development.

## Ovarian Aging and the Menopausal Transition

Ovarian aging is a critical determinant of menopause. At the time of birth, there are approximately 1 million primordial follicles. The absolute number of ovarian follicles declines exponentially until approximately 37 years of age, at which time follicular loss accelerates (**Figure 1**). At the time of the perimenopausal period, the follicle population has been reduced to less than 1000, and after menopause the ovary is virtually depleted of follicles. Fertility begins to decline in the third and fourth decades of life and this process accelerates after 35 years of age.

A staging system for the phases of reproductive life in women is shown in **Figure 2**. The staging system uses the final menstrual period as a reference point. In the reproductive phase, menstrual cycles are regular, except for a variable time immediately following puberty. The reproductive phase is followed by the menopausal transition, heralded by the onset of cycles of variable lengths. It is important to note, however, that alterations in hormone secretion occur quite early, even before the onset of irregular cycles.

The first endocrine marker of reproductive aging consists of an elevation of serum FSH early in the follicular phase. The rise in serum FSH has been termed monotropic because it is unaccompanied by an elevation in serum LH. The monotropic increase in FSH secretion in the late reproductive stage is inversely correlated with declining serum levels of inhibin B (**Figure 3**). Inhibin B is secreted by granulosa cells in developing antral follicles. Lower levels of inhibin B may reflect a diminished number of primordial follicles recruited into the developing pool or decreased numbers of granulosa cells for each follicle. Loss of the restraining effect of inhibin B on the anterior pituitary gland results in the reciprocal increase in serum FSH. Early development of the dominant follicle, secondary to increased FSH, may lead to shortening of the follicular phase and marked increases in estrogen secretion.

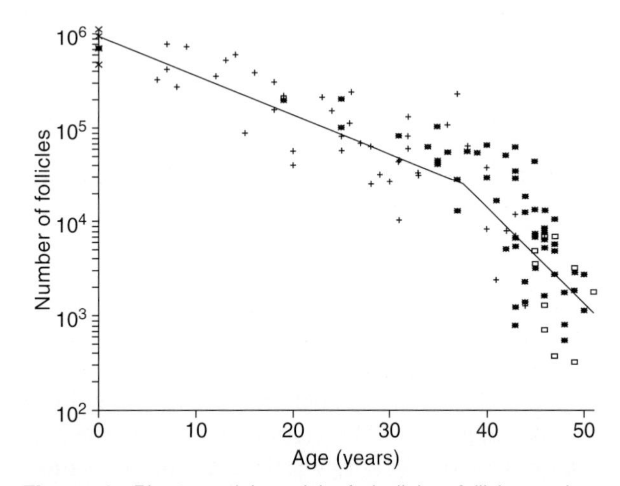


**Figure 1** Biexponential model of declining follicle numbers in pairs of human ovaries, from neonatal age to 51 years old. Each symbol represents one person. The different symbols represent data obtained from several studies. Reprinted from Faddy MJ, Gosden RG, Gougeon A, et al. (1992) Accelerated disappearance of ovarian follicles in mid-life: Implications for forecasting menopause. *Human Reproduction* 7: 1342–1346, with permission.

The menopause transition is a prolonged period of time characterized by irregular cycle lengths and periods of amenorrhea (**Figure 2**). During the early menopausal transition, there may be elevated levels of FSH throughout the cycle and lower levels of inhibin A, inhibin B, and progesterone. Serum estrogen levels of the early menopausal transition are either preserved or increased. In the late menopausal transition, frequent anovulatory cycles occur, with substantial variability in hormone levels. The multiinstitutional Study of Women's Health across the Nation (SWAN) has classified the anovulatory cycles into three types based on daily urine hormone levels: (1) cycles in which estrogen fails to induce an LH surge, (2) cycles with an LH surge but disrupted or dysfunctional corpus luteum formation, and (3) cycles with tonically elevated LH and FSH but with no evidence of a rise in estrogen. The variable cycles in the late menopausal transition lead to markedly fluctuating levels of ovarian hormones. The fluctuating levels of estrogen in particular may account for many of the symptoms experienced at this stage, including breast tenderness, menorrhagia, hot flushes, sleep disturbances, and mood changes.

The monotropic rise in FSH secretion has been postulated to produce the accelerated phase of follicular

Final menstrual period (FMP)

| Stages: | −5 | −4 | ϖ3 | −2 | −1 | 0 | +1 | +2 |
|---|---|---|---|---|---|---|---|---|
| Terminology: | Reproductive | | | Menopausal transition | | | Postmenopause | |
| | Early | Peak | Late | Early | Late* | | Early* | Late |
| | | | | Perimenopause | | | | |
| Duration of stage: | Variable | | | Variable | | (a) 1 yr | (b) 4 yrs | Until demise |
| Menstrual cycles: | Variable to regular | Regular | | Variable cycle length (>7 days different from normal) | ≥2 skipped cycles and an interval of amenorrhea (≥60 days) | Amen × 12 mos | None | |
| Endocrine: | Normal FSH | | ↑ FSH | ↑ FSH | | | ↑ FSH | |

*Stages most likely to be characterized by vasomotor symptoms ↑ = Elevated

**Figure 2** The Stages of Reproductive Aging Workshop staging system, showing endocrine levels of follicle-stimulating hormone (FSH) at different stages. Reprinted from Soules MR, Sherman S, Parrott E, et al. (2001) Executive summary: Stages of Reproductive Aging Workshop (STRAW). *Fertility and Sterility* 76: 874–878, with permission.

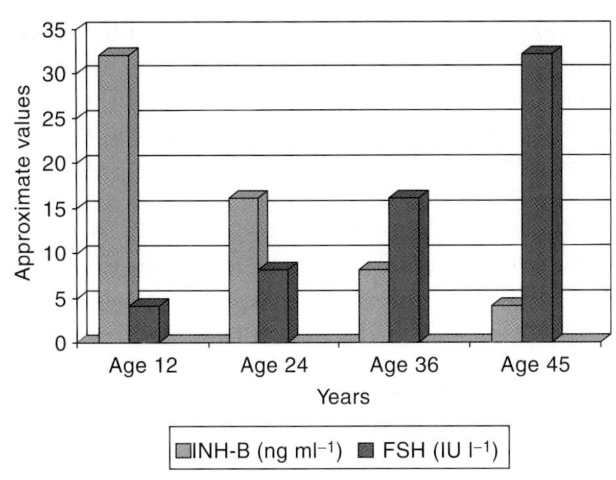


**Figure 3** As ovarian reserve declines throughout reproductive life, inhibin B (INH-B), a product of small, growing follicles, decreases. Concomitant with this relative loss of INH-B, follicular-phase follicle-stimulating hormone (FSH) increases in a reciprocal fashion. Reprinted from *American Journal of Medicine*, 118 (supplement 12B), Santoro N (2005) The menopausal transition 8–13, with permission from Elsevier.

decline that occurs after the age of 37 years. Higher levels of FSH (secondary to decreased inhibin B) could recruit increased numbers of follicles into the growing pool. Once this happens, follicles must either ovulate or become atretic. An abnormal FSH signal could also be detrimental to ovarian function independent from its effects on follicular recruitment. The steady loss of ovarian follicles in late reproductive life could reach a critical threshold and initiate a cascade of events leading ultimately to complete ovarian failure.

A hypothalamic contribution to the acceleration of follicular loss has also been proposed. For example, in experimental animals, altered patterns of LH pulsatility (secondary to GnRH pulses) are detrimental to ovarian function. Moreover, reducing the frequency of GnRH pulses can produce isolated FSH secretion in experimental animals, providing a potential mechanism whereby an altered hypothalamic signal could contribute to the monotropic rise in FSH and the subsequent acceleration of follicular decline. The slowing of LH pulses in older, regularly cycling women, however, has only been demonstrated in one study, and this finding has not yet been replicated. Therefore, the proposal that hypothalamic factors trigger the ovarian failure of menopause is controversial.

## Effects of Ovarian Failure on the Hypothalamic–Pituitary Axis

The hormonal milieu of the postmenopausal phase of life is characterized by profound ovarian hormone deficiency. There is a marked increase in serum gonadotropins in postmenopausal women, similar to the hormone profile seen in young, oophorectomized women. The postmenopausal increase in serum gonadotropins is secondary to both loss of steroid negative feedback and removal of the restraining action of inhibin on FSH secretion from the pituitary gland. In addition, the loss of ovarian hormones shifts the composition of the gonodotropins to more acidic isoforms. The shift of gonadotropin isoforms contributes to elevated levels of LH and FSH in the peripheral circulation by retarding their clearance. Administration of gonadal steroids still suppresses the high levels of serum gonadotropins in elderly women, indicating that the postmenopausal hypothalamus remains sensitive to negative feedback effects. Overall, many aspects of hypothalamic and pituitary function remain intact in postmenopausal women, in contrast to the complete failure of the ovaries.

Based on the central role of GnRH in the reproductive axis, research has focused on the status of GnRH neurons in postmenopausal women. Because direct measurements of GnRH secretion require invasive procedures, information on the function of these neurons has been obtained through indirect methods. For example, assessment of GnRH pulses can be inferred by examination of pulses of the free $\alpha$ subunit (FAS) of the pituitary glycoprotein hormones in peripheral plasma. This method was developed because measurements of LH pulses are complicated by the prolonged half-life of LH and the rapid frequency of pulses in postmenopausal women. The quantity of GnRH secretion can also be estimated by pharmacological methods such as competitive receptor blockage. Another approach to studying GnRH neurons is to quantify relative levels of gene expression in postmortem tissues using *in situ* hybridization histochemistry. This method is well suited for the investigation of human autopsy material due to the relative stability of mRNA species with prolonged postmortem intervals. A further level of anatomical resolution is provided by the localization of mRNA species at the cellular level within tissue sections.

The function of GnRH neurons in postmenopausal women appears to be dominated by the removal of steroid negative feedback. Studies have shown that negative feedback effects of estrogen and progesterone on LH secretion are maintained in postmenopausal women. The hypothalamic content of the GnRH peptide decreases in postmenopausal women, similar to response in young women who have had their ovaries removed. Therefore, it has been hypothesized that the decreased GnRH content in postmenopausal women is secondary to increased

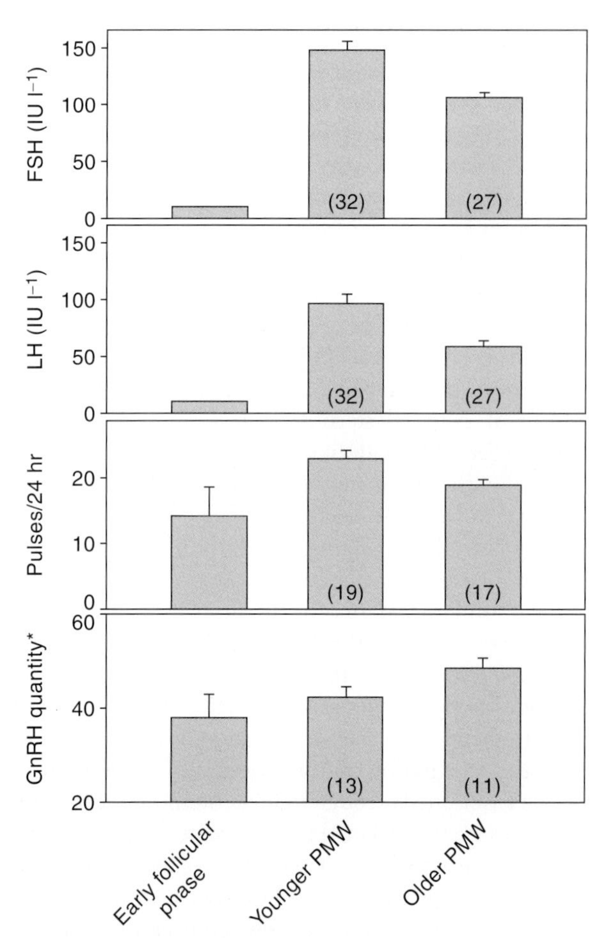


**Figure 4** Studies in younger (age 45–55 years) and older (age 70–80 years) postmenopausal women (PMW) indicate that follicle-stimulating hormone (FSH) and luteinizing hormone (LH) levels increase after menopause, but decrease with aging. Gonadotropin-releasing hormone (GnRH) pulse frequency, assessed by pulsatile secretion of the free $\alpha$ subunit, is also increased in relation to the early follicular phase, but is not different from the frequency in the late follicular phase of midcycle surge in healthy women. After menopause, pulse frequency decreases with aging. The quantity of GnRH (100% inhibition of LH in response to submaximal GnRH receptor blockade) increases with aging. The number of individuals contributing to each bar is indicated in parentheses. Reprinted from Hall JE (2004) Neuroendocrine physiology of the early and late menopause. *Endocrinology and Metabolism Clinics of North America* 33: 637–659, with permission.

secretion into the portal vessels. As would be expected with the removal of steroid negative feedback, GnRH pulse frequency (assessed by peripheral FAS pulses) is increased in young postmenopausal women compared with premenopausal women in the early follicular phase (**Figure 4**). Moreover, the quantity of GnRH secreted (estimated by competitive receptor blockage) is increased in young postmenopausal women and continues to increase with age. These findings agree with the demonstration of increased GnRH gene expression in the medial basal hypothalamus of postmenopausal women. Taken together, these studies provide evidence that GnRH neurons in postmenopausal women, instead of age-related signs of degeneration, exhibit compensatory responses due to removal of ovarian hormones.

## Effects of Menopause on Neuronal Morphology and Neuropeptide Gene Expression in the Human Hypothalamus

More than four decades ago, Sheehan and Kovacs described differences in hypothalamic neuronal morphology in premenopausal and postmenopausal women. In postmenopausal women, pronounced enlargement of neurons was observed in the infundibular nucleus of the medial basal hypothalamus. The cellular hypertrophy was characterized by increased Nissl substance (indicative of increased protein synthesis) and enlarged nuclei and nucleoli (**Figure 5**). Based on similar findings in other hypogonadal conditions, the authors proposed that neuronal hypertrophy in the infundibular nucleus was secondary to loss of ovarian function.

Subsequent studies using *in situ* hybridization have characterized the phenotype of the hypertrophied neurons and have shown numerous alterations in hypothalamic neuropeptide gene expression in comparing tissues of premenopausal and postmenopausal women. The most dramatic change in gene expression occurs in the subpopulation of infundibular neurons that undergoes neuronal hypertrophy. These neurons co-express estrogen receptor-$\alpha$ and two tachykinin neuropeptides, neurokinin B (NKB) and substance P (SP). In addition to the marked increase in cell size, the number of neurons expressing NKB and SP gene transcripts in the infundibular nucleus increases by more than sixfold, indicating a marked increase in tachykinin gene expression. GnRH gene expression is also elevated in postmenopausal women, but in a separate subpopulation of neurons within the medial basal hypothalamus. The elevation in GnRH mRNA provides evidence of increased hypothalamic GnRH secretion contributing to the rise in plasma gonadotropins in postmenopausal women.

In contrast to elevated NKB and GnRH gene expression, the number of neurons expressing proopiomelanocortin (POMC) gene transcripts in the infundibular nucleus is significantly reduced in older women. Moreover, the number of neurons expressing POMC mRNA is inversely correlated with age and the expression of neuropeptide Y mRNA. Numerous studies in experimental animals have revealed that POMC and neuropeptide Y neurons in the arcuate nucleus (the rodent homolog of the infundibular nucleus) are essential components of hypothalamic

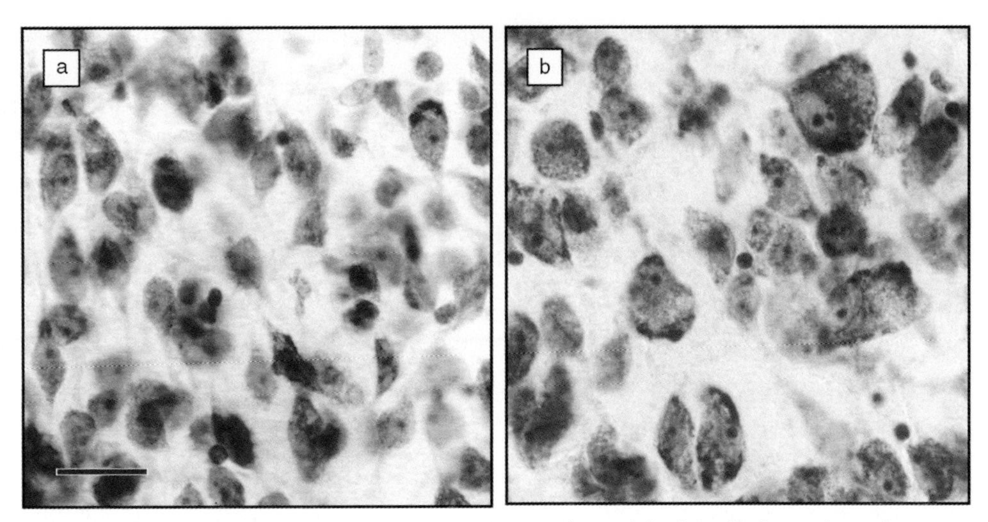


**Figure 5** Representative photomicrographs of cresyl violet-stained sections of the infundibular nucleus of young, premenopausal (a) woman and an older, postmenopausal (b) woman. In the section from the older woman, note the considerably enlarged neurons, with both increased nuclear size and increased Nissl substance. Scale bar = 25 μm (a, b). Reprinted from Abel TW and Rance NE (2000) Stereologic study of the hypothalamic infundibular nucleus in young and older women. *Journal of Comparative Neurology* 424: 679–688, with permission.

regulatory circuits controlling body weight and reproduction. Arcuate POMC and neuropeptide Y neurons express leptin receptors and exert opposing actions on feeding behavior and energy balance. Experimental paradigms in laboratory animals in which nutritional status or serum leptin is altered produce reciprocal changes in POMC and neuropeptide Y gene transcripts. Thus, the finding of a reciprocal relationship between POMC and neuropeptide Y gene expression in the human arcuate (infundibular) nucleus suggests that this relationship is conserved among species.

A major limitation in interpreting changes in gene expression in premenopausal and postmenopausal women is the confounding influence of two variables, age and ovarian status. To control for the effects of aging, young cynomolgus monkeys have been used as a model to determine the effects of ovariectomy and steroid hormone replacement on hypothalamic neuropeptide gene expression (**Table 1**). Importantly, ovariectomy of young monkeys induces both hypertrophy and increased NKB gene expression in the infundibular nucleus. Similarly, GnRH gene expression in the medial basal hypothalamus is increased by ovariectomy of young cynomolgus monkeys. Conversely, estrogen replacement of young ovariectomized monkeys decreases NKB and GnRH gene transcripts. These data provide strong support for the hypothesis that increased NKB and GnRH gene expression observed in the hypothalamus of postmenopausal women is secondary to ovarian failure and not due to age *per se*. In contrast to the marked effect of ovarian hormones on NKB gene expression, ovariectomy or steroid replacement had no effect on POMC or neuropeptide Y gene expression in the primate model. These data suggest that factors other than ovarian failure may be more important in influencing this peptide system in postmenopausal women.

The association of increased NKB gene expression with estrogen withdrawal and gonadotropin hypersecretion suggests that these neurons participate in the hypothalamic circuitry regulating steroid negative feedback. This hypothesis is supported by the demonstration of estrogen receptor-$\alpha$ within the hypertrophied neurons and the location of these neurons within the primate control center for reproduction. Studies in experimental animals have also provided strong support for this hypothesis. Expression of NKB gene transcripts is increased by ovariectomy and reduced by estrogen replacement in a variety of species. NKB gene expression in the arcuate nucleus varies with the estrous cycle, and these neurons are sexually dimorphic. Moreover, anatomic studies have also revealed a close relationship between NKB and GnRH neurons. In the rat median eminence, there is close apposition of NKB and GnRH fibers at a site where GnRH fibers express the receptor for NKB, the neurokinin-3 receptor. Finally, LH secretion is modulated by central injections of a neurokinin-3 receptor agonist. Thus, multiple lines of evidence support the hypothesis that the hypertrophied neurons in the infundibular nucleus of postmenopausal women play a role in the regulation of steroid negative feedback on gonadotropin secretion.

**Table 1** Effects of ovariectomy and estrogen replacement on neuropeptide gene expression in the medial basal hypothalamus of young cynomolgus monkeys

| *Neuron phenotype* | *Changes in postmenopausal women* | *Effects of ovariectomy of cynomolgus monkeys* | *Effects of estrogen replacement in ovariectomized monkeys* |
|---|---|---|---|
| NKB | Hypertrophy, ↑ mRNA | Hypertrophy, ↑ mRNA | ↓ mRNA below levels of detection |
| GnRH | ↑ mRNA | ↑ mRNA | ↓ mRNA |
| POMC | ↓ mRNA | No change | No change |
| NPY | ↑ mRNA | No change | Not tested |

NKB, neurokinin B; GnRH, gonadotropin-releasing hormone; POMC, proopiomelanocortin; NPY, neuropeptide Y.

## Aging of the Reproductive Neuroendocrine Axis, Independent of the Effects of Gonadal Status

As chronological age advances, hypothalamic aging and secondary effects from the aging of other neuroendocrine systems become increasingly important. Age-associated changes occur in the reproductive neuroendocrine axis independently from those caused by ovarian failure. For example, plasma LH is lower in postmenopausal women (between 51 and 70 years old), compared to women (between 25 and 40 years old) with premature ovarian failure. A consistent decline in serum LH has also been observed in association with aging after the menopause. Moreover, some studies have reported an age-associated decline in the frequency of LH pulses, although these studies are confounded by the prolonged half-life of LH after the menopause.

Recent studies by Hall and colleagues have provided a detailed picture of the aging of the hypothalamic–pituitary axis by comparing hormone levels in postmenopausal women who are 45–55 years of age and 70–80 years of age. Using a sampling interval of 5 min and measurements of the gonadotropin free $\alpha$ subunit, a reduction in GnRH pulse frequency was observed in the older postmenopausal women. Because the GnRH pulse generator resides within the hypothalamus, these studies provide evidence of hypothalamic neuroendocrine aging that is independent of the changes in gonadal function.

Despite slower GnRH pulses and reduced serum LH, older postmenopausal women retain the capacity to decrease GnRH secretion in response to steroid feedback. Interestingly, indirect pharmacological measurements provide evidence that GnRH secretion increases with age in postmenopausal women. The age-related rise in GnRH secretion is consistent with the demonstration of increased GnRH secretion documented by push–pull perfusion in older rhesus monkeys. Moreover, the reciprocal changes in POMC and neuropeptide Y (NPY) gene expression that occur in the medial basal hypothalamus of older women also appear to be independent of gonadal status. It is hypothesized that loss of an inhibitory effect of POMC, combined with stimulation by NPY, could explain why GnRH secretion could continue to rise, even after the menopausal transition.

## Summary

The hallmark of reproductive aging in women is degeneration of ovarian follicles, which accelerates around the age of 37 years. An early sign of reproductive aging is a rise in serum FSH, which is inversely correlated with decreased inhibin B secretion from developing antral follicles. After a transition phase characterized by irregular cycles and fluctuating hormone levels, the final menstrual period marks the cessation of cycles. In the postmenopausal period, the reproductive neuroendocrine axis is dominated by the withdrawal of ovarian hormones. Age-associated changes in the hypothalamic pituitary axis also occur and are independent of gonadal status.

*See also:* Hormones and Memory; Neuroendocrine Aging: Pituitary Metabolism; Neuroendocrine Aging: Pituitary-Gonadal Axis in Males; Neuroendocrine Aging: Pituitary-Adrenal Axis.

## Further Reading

Abel TW and Rance NE (2000) Stereologic study of the hypothalamic infundibular nucleus in young and older women. *Journal of Comparative Neurology* 424: 679–688.

Escobar CM, Krajewski SJ, Sandoval-Guzmán T, et al. (2004) Neuropeptide Y gene expression is increased in the infundibular nucleus of postmenopausal women. *Journal of Clinical Endocrinology & Metabolism* 89: 2338–2343.

Faddy MJ, Gosden RG, Gougeon A, et al. (1992) Accelerated disappearance of ovarian follicles in mid-life: Implications for forecasting menopause. *Human Reproduction* 7: 1342–1346.

Gill S, Lavoie HB, Bo-Abbas Y, et al. (2002) Negative feedback effects of gonadal steroids are preserved with aging in postmenopausal women. *Journal of Clinical Endocrinology & Metabolism* 87: 2297–2302.

Gill S, Sharpless JL, Rado K, et al. (2002) Evidence that GnRH decreases with gonadal steroid feedback but increases with age in postmenopausal women. *Journal of Clinical Endocrinology & Metabolism* 87: 2290–2296.

Gore AC, Windsor-Engnell BM, and Terasawa E (2004) Menopausal increases in pulsatile gonadotropin-releasing hormone release in a nonhuman primate (*Macaca mulatta*). *Endocrinology* 145: 4653–4659.

Hale GE and Burger HG (2005) Perimenopausal reproductive endocrinology. *Endocrinology and Metabolism Clinics of North America* 34: 907–922, ix.

Hall JE (2004) Neuroendocrine physiology of the early and late menopause. *Endocrinology and Metabolism Clinics of North America* 33: 637–659.

Hall JE, Lavoie HB, Marsh EE, et al. (2000) Decrease in gonadotropin-releasing hormone (GnRH) pulse frequency with aging in postmenopausal women. *Journal of Clinical Endocrinology & Metabolism* 85: 1794–1800.

Knobil E (1980) The neuroendocrine control of the menstrual cycle. *Recent Progress in Hormone Research* 36: 53–88.

Rance NE and Young WS, III (1991) Hypertrophy and increased gene expression of neurons containing neurokinin-B and substance-P messenger ribonucleic acids in the hypothalami of postmenopausal women. *Endocrinology* 128: 2239–2247.

Sandoval-Guzmán T, Stalcup ST, Krajewski SJ, et al. (2004) Effects of ovariectomy on the neuroendocrine axes regulating reproduction and energy balance in young cynomolgus macaques. *Journal of Neuroendocrinology* 16: 146–153.

Santoro N (2005) The menopausal transition. *American Journal of Medicine* 118(supplement 12B): 8–13.

Soules MR, Sherman S, Parrott E, et al. (2001) Executive summary: Stages of Reproductive Aging Workshop (STRAW). *Fertility and Sterility* 76: 874–878.

Weiss G, Skurnick JH, Goldsmith LT, et al. (2004) Menopause and hypothalamic–pituitary sensitivity to estrogen. *JAMA: Journal of the American Medical Association* 292: 2991–2996.

# Neuroimmune System: Aging

**E Bellavista and C Franceschi**, University of Bologna, Bologna, Italy



## The Immune-Neuroendocrine System: The Evolutionary Perspective

The immune system, consisting of innate and adaptive branches, acts to recognize and cope with a variety of stressors, including physical, emotional, chemical, and biological agents. The evolutionary perspective indicates that this system is a part of a larger network involving the neuroendocrine circuit, termed the immune-neuroendocrine suprasystem, which is able to integrate immunity, stress response, and inflammation. The major characteristic is that this suprasystem is conserved with variations and increased complexity, from invertebrates to vertebrates, and a pool of well-known molecules (pro-opiomelanocortin-derived peptides, cytokines, biogenic amines, glucocorticoids, and nitric oxide) are used for intra- and intersystem communication. The main cellular actor of this network is the macrophage, which produces all of the previously mentioned mediators.

In a number of low eukaryotes, such as yeast and worm, the ability to cope with stressors is provided by a defense network including superoxide dismutase and catalase (free radical scavenger enzymes), heat shock proteins and other chaperones, DNA repair genes, and intracellular signaling molecules (e.g., H-RAS). A study on *Caenorhabditis elegans*, a favorite model for the genetic investigation of the physiology and developmental biology of multicellular organisms, showed that the resistance to pathogen is mediated by a Toll/interleukin-1 receptor domain adaptor protein and a conserved p38 mitogen-activated protein kinase family (PMK)-1 pathway. Moreover, this circuit is involved in mediating resistance to osmotic stress, in addition to mediating immune signaling.

In invertebrates (mollusks, insects, and annelids), immune and neuroendocrine functions are performed by a cell type called immunocyte, the ancestor of macrophage, that can be considered an 'immune-mobile brain.' This phagocyting cell expresses molecules such as adrenocorticotropin hormone (ACTH), corticotropin-releasing hormone (CRH), interleukin-like IL-1$\alpha$, -1$\beta$, -2, and -6, and tumor necrosis factor (TNF)-$\alpha$ that are able to affect migration and increase phagocyte activity against pathogens. Moreover, immunocytes express a NO synthase-like molecule producing NO, which provokes the clumping of bacteria around them and their subsequent killing. Pathogen recognition in *Drosophila melanogaster* is mediated by different molecules, including a member of the immunoglobulin superfamily which is diversified by alternate splicing. This molecule is called Dscam, the homolog of Down syndrome cell adhesion molecule, and its isoforms are expressed in immune tissues as membrane and soluble proteins.

During evolution, the old molecular and cellular effectors merged with the new sophisticated recognition system, the adaptive clonotypical immunity, to face the increasing complexity of the organisms, which required a more sophisticated type of circuitry and recognition. Specialized function for the defense of the organism occurred in jawed fish, in which the thymus, the B and T lymphocytes, and the major histocompatibility complex (MHC) appeared in order to improve recognition of external molecules, on the one hand, and avoid self-destruction, on the other hand. The thymus, a neuroendocrine organ, was built up with cellular and molecular components present in invertebrates, and it is able to perform a local immune and stress response. Accordingly, the appearance of a new organ such as the thymus can be envisaged as an example of tinkering – that is, the combinatorial reutilization of preexisting materials, a strategy largely used in evolution.

In mammalian, immune and neuroendocrine responses are bidirectionally regulated through more specialized tissues and organs, where neuropeptides and cytokines represent the signaling molecules responsible for chemical information. In particular, in the central nervous system (CNS), the hypothalamus integrates information from physical (e.g., trauma and infection) and emotional stimuli (e.g., bereavement) and, in response to stress, releases CRH from the paraventricular nucleus, which acts on the pituitary gland to induce the release of ACTH to stimulate the adrenal gland and the production of glucocorticoids (cortisol) and dehydroepiandrosterone (DHEA) from the cortex. The adrenal gland is also regulated by the sympathetic nervous system, secreting catecholamines in response to acetylcholine release from the fibers innervating adrenal medulla. Catecholamines and glucocorticoids mediate the 'fight-or-flight' response to stress but at the same time induce suppression of the majority of immune functions. The principal mechanism of glucocorticoids is the downregulation of pro-inflammatory cytokine synthesis and the inhibition of enzymes such as cyclooxegenase-2 (COX-2). Moreover, they affect cell-mediated immune function, suppressing natural killer (NK) cell cytotoxicity and neutrophil chemotaxis. By contrast, DHEA is primarily immune enhancing and counterbalances the potentially detrimental effect of long-term stimulation of the hypothalamic–pituitary–adrenal (HPA) gland axis. In particular, DHEA improves NK cell number and cytotoxicity and induces superoxide generation in neutrophils. As an example of this connection between the CNS and the immune system, we studied the effect of an emotional stimuli, such as bereavement, on the immune system and found functional alterations of immune parameters, such as responsiveness of peripheral blood lymphocytes to mitogens. In particular, despite a normal number of circulating lymphocyte subsets, the functional activity of NK cells was markedly reduced. Moreover, changes in the intracellular concentration of $\beta$-endorphin in peripheral blood mononuclear cells correlated with anxiety and depression scores. Other distress-generating events (e.g., divorce, caring for a relative affected by Alzheimer's disease, and unemployment) are associated with alterations in both cellular and humoral immune response and enhanced risk of developing infectious disease.

Evidence also supports the idea that information can flow from the immune system to the CNS – for example, the innervation of the immune system and organs, the production of neuropeptides by cells of the immune system, and the presence of receptors for both cytokines, as well as neurotransmitters on immune and CNS cells. An example of such communication is that, upon peritoneal infection with bacteria, resident macrophages produce IL-1, which acts on the vagus nerve to cause behavioral changes and illness symptoms, and on hypothalamus and pituitary gland to produce CRH and ACTH. Additionally, lymphocytes produce $\beta$-endorphin to modulate pain sensations, acting on peripheral sensory nerves. Moreover, evidence from our laboratory indicates that a pure immunological stimulus is able to impact on the brain inducing neuroendocrine changes and to modify cognitive performance. Indeed, healthy volunteers were vaccinated against hepatitis B and tested during a 6-month period on a simple reaction times task and the Stroop task. Although we did not find improvement in the general efficiency of the CNS as indexed by simple task reaction, when we considered higher cognitive function (Stroop task), the performance of vaccinated subjects was faster and more accurate. Additionally, these effects were not acute, being evident 4 weeks after antigenic stimulation, and they persisted even after 25 weeks.

Thus, the available data in the literature suggest that a bidirectional circuitry is present between the immune–endocrine and the nervous system, where both systems continuously alert and inform each other.

## Immunosenescence

The aging of the immune system (immunosenescence) is a recent phenomenon related to the linear improvement in survival and life span not predicted by evolutionary force. Indeed, the immune system has probably been selected to serve individuals living until reproduction, as supported by the thymic ontogenesis and involution. Nowadays, the immune system supports the soma of individuals living 80–120 years; during this increased life span, elderly people have to cope with long-lasting antigenic burden (clinical and subclinical infection, as well as continuous exposure to other types of antigens), which causes a chronic activation of the immune system. In order to adapt to and counteract the effect of chronic stress, the immune system undergoes reshaping and remodeling, in which some parameters decrease with age while others increase or remain unchanged. The adaptive immunity is the most affected by aging, being characterized by expanded clones of memory and effector T cells, mostly prominent in $CD8^+$ and specific for viral antigens. In particular, we showed that Epstein–Barr virus (EBV) and cytomegalovirus (CMV) infections are responsible for the expansion of memory $CD8^+$ T cell clones specific for a limited number of viral peptide epitopes which accumulate in the immune system of older people, thus filling the 'immunological space.' We have shown that $CD4^+$ T cells are also affected by viral chronic infections such as CMV. Indeed, when cells from $CMV^+$ $CD4^+$ T cell are stimulated *in vitro* with specific peptides of the CMV, they produce seven or eight times more inflammatory cytokines such as interferon-$\gamma$ (IFN-$\gamma$) and TNF-$\alpha$, thus contributing to the inflammatory status present in elderly people. Moreover, aging is associated with a shrinkage of T cell repertoire and a decrease in thymic emigrant T cells and virgin T cells. The result is the filling of the immunological space (the physical space in which all possible interactions among immune cells and/or cell subsets must occur) with indolent cells with memory markers which could exert a suppressor/negative effect on other bystander T cells, including the few naive (virgin or antigen-nonexperienced) cells. Considering the innate immunity, different studies have shown that age-associated alterations are also present in this branch of immunity, but overall it is well preserved. Our attention was particularly focused on NK cells, macrophages, and cytokine circuits. NK cells are large granular lymphocytes expressing surface markers such as CD16 and CD56 and lacking the CD3 T cell receptor. They display two alternative mechanisms of cytotoxicity activity: the MHC unrestricted cytotoxicity against neoplastic and virus-infected cells and antibody-dependent cell-mediated cytotoxicity. In elderly people, an increase in the absolute and relative number of NK cells, with a mature phenotype ($CD56^{dim}$), is seen, with no variation in their killing activity as a whole, despite the presence of subtle impairment of killing activity on a per cell basis. Moreover, a feedback loop between NK cells and macrophages has been described. Since NK cells secrete factors such as IFN-$\gamma$ which activate macrophages to eliminate pathogens, some of the age-associated macrophage dysfunctions may be secondary to alteration in cytokine production of NK cells. Additionally, an age-related reduction in hydrogen peroxide and nitric oxide synthesis by macrophage has been described, paralleled by an induction of suppressive substances (e.g., prostaglandin E) which have an inhibitory effect on dendritic cells, the main antigen presenting cells. *In vitro* studies on cytokine production by peripheral mononuclear cells from healthy young and aged subjects identified a complex scenario. Significant increases in levels of IL-6, TNF-$\alpha$, and IL-1$\beta$ were found in mitogen-stimulated cultures from aged donors despite a similar amount of cytokines in unstimulated cultures from young and aged subjects. Moreover, chemokines such as MCP-1 and RANTES, soluble TNF-$\alpha$ receptor I and II, and IL-1 receptor antagonist increased in the elderly. For one of the most important proinflammatory cytokines, IL-6, synthesis in macrophage cells is controlled by the CRH–DHEA circuit. Whereas production of cortisol remains constant or even increases with age, levels of DHEA decrease gradually from the third decade, reaching levels that are only 10–20% of their maximum by the eight decade; aged-related decrease of DHEA correlates with IL-6 overproduction. The elderly thus gradually approach a status of relative glucocorticoid excess with age, with a potential negative impact on baseline immune function and an exaggerated response to stressors. The correlation between DHEA and IL-6 has been reported in several studies. In particular, it has been demonstrated that in normal subjects, serum IL-6 levels increased with age, whereas serum DHEA concentrations decreased. Furthermore, serum IL-6 levels were inversely correlated with serum DHEA levels. In addition, in *in vitro* experiments, DHEA inhibited the lipopolysaccharide-stimulated IL-6 production from monocytes and peripheral blood mononucleated cells, even in a narrow concentration range.

## Inflamm-Aging

As previously described, the major consequence of chronic exposure to antigenic stress is the progressive activation of macrophages and related cells in most organs and tissues of the body, leading to a progressive chronic, low-grade proinflammatory status which is the principal characteristic of the aging process. We called this phenomenon 'inflamm-aging,' and we postulated that it contributes substantially to the pathogenesis of all major age-related diseases, such as cardiovascular diseases, diabetes type 2, dementia, and cancer, and also to complex pathological conditions very common and characteristic of elderly people, such as frailty and sarcopenia. Thus, directly or indirectly, inflamm-aging is a major determinant of morbidity and mortality in the elderly. However, inflammatory status *per se* is compatible with extreme longevity in good health, and paradoxically proinflammatory characteristics have been documented in healthy centenarians, suggesting that inflammation is a physiologic phenomenon and a beneficial response to cope with chronic stressors. We also found evidence that centenarians activate a strong and effective anti-inflammatory response mediated by cytokines such as IL-10 and T cell transforming growth factor-$\beta$ capable of counteracting their inflammatory status. Moreover, the rate at which the threshold of proinflammatory status is reached and over which disease/disabilities emerge and the individual's ability to adapt to stressors can be considered a complex trait with high interindividual variability and a strong genetic component. We propose that the persistence over time of inflammatory stimuli, such as CMV chronic infection, represents the biological background favoring susceptibility to a variety of diseases. The absence of robust gene variants and/or the presence of frail gene variants is necessary to develop overt organ-specific, age-related disease. We predicted that the concomitant presence of high inflammatory status and inherited variants of frailty genes or the absence of robust genes can explain why some individuals, but not others, are more susceptible to certain types of age-related disorders.

## Inflammation and the Neuroendocrine System

### Major Depression

The proinflammatory status in elderly people is associated with several inflammatory-based pathologies, such us cardiovascular disease, sarcopenia, and Alzheimer's disease. Moreover, studies suggest that proinflammatory cytokines play an important role in the pathophysiology of major depression. This psychiatric disorder, at a neurobiological level, is associated with alterations in neurotransmitter function involving serotonin, norepinephrine, and dopamine. Hypersecretion of CRH and alteration of its regulation by glucocorticoids are also described. Regarding the emerging role of the immunological pathway in major depression, several studies have shown that patients with this pathology, who are otherwise medically healthy, have increased serum and plasma concentrations of IL-6 and/or C-reactive protein and elevated IL-1$\beta$ and TNF-$\alpha$ in peripheral blood and cerebrospinal fluid. Additionally, increased levels of chemokines and adhesion molecules, such as E-selectin, have been described. A positive correlation between plasma concentrations of various inflammatory mediators and depressive symptom severity has been reported. Finally, a nascent literature suggests that functional allelic variants of genes for IL-1$\beta$ and TNF-$\alpha$ increase the risk for depression and are associated with reduced responsiveness to antidepressant therapy.

Given the primary factors that are associated with the pathophysiology of depression, it is likely that peripherally released inflammatory cytokines can access the brain. Because of the large size of cytokines and their resultant inability to readily penetrate the blood–brain barrier, attention has focused on routes by which they access the brain, including through leaks in the blood–brain barrier, the binding to specific transport, or the activation of vagal afferent fibers which transmit cytokine signals to specific brain nuclei. Once in the brain, there is a CNS cytokine network made up of cells that not only produce cytokines and express cytokine receptors but also amplify cytokine signals, which in turn can have profound effects on neurotransmitter and CRH function, as well as behavior. These cytokine-induced behavioral changes are associated with alterations in the metabolism of serotonin and dopamine in brain regions essential for the regulation of emotion and psychomotor function. Moreover, these cytokines induce the activation of the HPA axis hormone, as well as CRH in both the hypothalamus and the amygdala, regions that have an important role in fear and anxiety. Finally, NF-$\kappa$B induction in the brain may also contribute to alterations in neuronal growth and survival, especially through the induction of nitric oxide production and, ultimately, oxidative stress. The discovery that depression shares common features with inflammation and stress system activation led to a novel perspective to treatment. Indeed, patients with a history of nonresponse to antidepressants have been found to exhibit increased plasma concentration of IL-6 compared to treatment-responsive patients. Moreover, antidepressants may only address select symptoms (e.g., depressed mood and anxiety), whereas others are relatively unaffected. Thus, inhibition of proinflammatory cytokine signaling represents a viable strategy for the treatment of depression, especially in patients with evidence of increased inflammatory activity who might be less likely to respond to convectional agents.

### Anorexia Nervosa

Prolonged elevation of proinflammatory cytokines may induce a cascade of biochemical effects in the CNS, involving secretion of neurotransmitters, neuropeptides, neurohormones, and peripheral hormones, which may be etiopathogenetic for the development of anorexia nervosa. Indeed, in experimental animal models, IL-1$\beta$, IL-6, and TNF-$\alpha$ introduced directly in the CNS induced anorexia, fatigue, apathy, depressed mood, and impaired secretion of catecholamines, serotonin, and HPA axis hormones. Although there are no reports regarding the cerebrospinal fluid concentration of such cytokines in patients affected by anorexia nervosa, we performed a study on a group of anorectic subjects (restricted and bingeing–purging) measuring plasma concentrations of IL-1$\beta$, IL-6, TNF-$\alpha$, and their soluble receptors. Surprisingly, we found that the concentrations of the three cytokines were normal in our patients, as well as the concentrations of the TNF-$\alpha$ soluble receptors. Interestingly, the levels of the IL-6 soluble receptor were lower than normal in our patients. Since there was a positive correlation between the levels of cytokine and its soluble receptors, the low level of receptors contrasts with the normal levels of IL-6. The biological significance of the soluble IL-6 receptor is not clear, but an impairment of the proinflammatory cytokine pathway might be involved in the development of anorexia nervosa.

### Alzheimer's Disease

One of the most important age-associated pathologies is Alzheimer's disease (AD), a heterogeneous and progressive neurodegenerative form of dementia. Neuropathological hallmarks of AD are neuronal and synapsis loss, extracellular $\beta$-amyloid deposition (neuritic/senile plaques), and intracellular deposition of hyperphosphorylated tau-based paired helical filaments in hippocampus, temporal and frontal cortex. Inflammation clearly occurs in this pathologically vulnerable region of the AD brain, where the principal proinflammatory stimuli are highly insoluble A$\beta$42 and A$\beta$40. Indeed, these peptides are recognized by astrocytes and microglia, which in turn produce many cytokines such as IL-1, IL-6, and TNF-$\alpha$. In particular, TNF-$\alpha$ acts as a proapoptotic agent, which contributes to neuronal loss in AD. Moreover, the senile plaques are potent activators of the complement cascade; during this process, small fragments, known as anaphylotoxins, are produced and stimulate inflammation. Chemokines and acute phase proteins are also co-localized as secondary components in neuritic plaques and are overproduced in brain of the AD patient. Other important mediators of inflammation in AD are prostaglandins, a group of fatty acids derived from the precursor arachidonic acid, with COX catalyzing the initial step. Both the constitutive COX-1 and the inducible COX-2 mRNA, as well as their protein products, are upregulated in affected brain areas. COX-1 is mainly expressed in microglia, whereas COX-2 is abundantly expressed in pyramidal neurons and seems to play a role in synaptic plasticity. Epidemiological studies have shown that levels of proinflammatory mediators, particularly cytokines such as IL-6, IL-10, and TNF-$\alpha$, in blood and brain may be under genetic control and the risk of AD is influenced by several polymorphisms in the noncoding region of these genes. Thus, the overall chances of an individual developing AD might be profoundly affected by a susceptible proinflammatory profile, reflecting the combined influence of inheriting multiple high-risk alleles. The identification of such a profile might in the future lead to strategies for therapeutic intervention in the very early stages of the disorders. Indeed, one possible strategy to suppress inflammatory response would be the administration of nonsteroidal anti-inflammatory drugs (NSAIDs). Epidemiological investigations indicate that the use of NSAIDs decreases the risk of AD by approximately 50% and delay the onset of AD by 5–7 years. (R)-flurbiprofen, the R enantiomer of racemic flurbiprofen, is under license for the potential treatment of AD, modulating the gamma secretase and the production of A$\beta$42. Indeed, in a murine model of AD, it lowered the level of this toxic peptide, and chronic administration reduced brain amyloid pathology and prevented defects in learning and memory. In a phase II clinical trial in AD, patients demonstrated a benefit in cognitive and behavioral performance. Two pivotal phase III clinical trials in AD are under way.

### Frailty Syndrome

Another aged-related disorder, involving innate immunity and inflammation, is frailty and sarcopenia. Frailty has been defined as a decline in lean body mass and decreased muscle strength, endurance, balance, and walking performance, accompanied by a high risk of disability, falls, hospitalization, and mortality. This syndrome reflects a profound metabolic imbalance due to overproduction of catabolic cytokines (likely TNF-$\alpha$) and the diminished availability of anabolic hormones such as insulin-like growth factor-1 (IGF-1), resulting from aging and the presence of associated chronic conditions. Sarcopenia, the loss of muscle mass and strength that occurs with normal aging, is a central part of the frailty syndrome, probably due the decline of neural, hormonal, and environmental trophic signals to muscle. The most

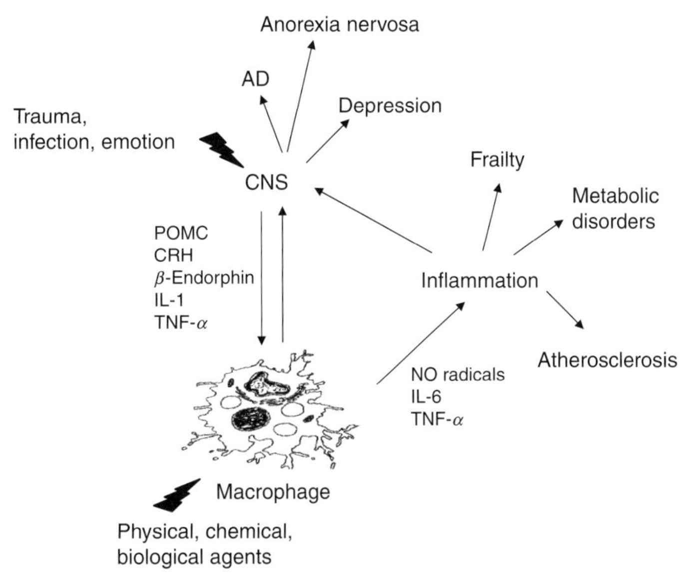


**Figure 1** The central role of the macrophage in stress response, inflammation, and age-related diseases. The immune and neuroendocrine systems share common mediators to exchange information and they are affected by different stressors. The macrophage is the most important actor in integrating information, and its chronic activation in aging supports the development of inflammatory diseases.

important endogenous cause may be the irreversible age-related loss of α-motor units in the central nervous system. TNF-α has effects that may contribute directly to sarcopenia, including increased basal energy expenditure, and anorexia, and it is associated with wasting/cachexia in chronic inflammation disorders. In the CNS, TNF-α can drive neuronal apoptosis, as previously described for AD. The role of IL-6 in sarcopenia is not clear, even though the effect of IL-6 in the regulation of IGF-1 action has been widely documented. To better understand the involvement of this circuit in this complex condition, we evaluated the relationship of plasma concentrations of IGF-1 and IL-6 with muscle function in apopulation-based sample of elderly people. We found that higher IL-6 and lower IGF-1 plasma levels are associated with lower muscle strength and power, suggesting a multifactorial relationship between these two mediators. Thus, insulin resistance in the elderly and loss of insulin anabolic action on muscle are factors favoring the onset and progression of this pathology.

## Conclusion

In this article, we discussed the most important link between the immune system and the neuroendocrinal circuit. These three systems form a complex network, conserved during evolution, in which cytokines and neuropeptides are the mediators. The most important actor which regulates this network is the macrophage, a very ancestral cell that is able to integrate immunity, stress response, and inflammation (**Figure 1**). Aging is characterized by a proinflammatory status which supports the most important aged-related pathologies.

*See also:* Neuroendocrine Aging: Pituitary Metabolism; Neuroendocrine Aging: Pituitary-Gonadal Axis in Males; Neuroendocrine Aging: Hypothalamic-Pituitary-Gonadal Axis in Women; Neuroendocrine Aging: Pituitary-Adrenal Axis; Thermoregulation: Autonomic, Age-Related Changes.

## Further Reading

Blalock JB (2005) The immune system as the sixth sense. *Journal of Internal Medicine* 257: 126–138.

Butcher SK and Lord JM (2004) Stress responses and innate immunity: Aging as contributory factor. *Aging Cell* 4: 151–160.

De Martinis M, Franceschi C, Monti D, and Ginaldi L (2005) Inflamm-aging and lifelong antigenic load as major determinants of ageing rate and longevity. *FEBS Letters* 579: 2035–2039.

Franceschi C and Bonafè M (2003) Centenarians as model of healthy aging. *Biochemical Society Transactions* 31: 457–461.

Franceschi C, Bonafè M, and Valensin S (2000) Human immunosenescence: The prevailing of innate immunity, the failing of clonotypic immunity, and the filling of immunological space. *Vaccine* 18: 1717–1720.

Franceschi C, Bonafè M, Valensin S, et al. (2000) Inflamm-aging. An evolutionary perspective of immunosenescence. *Annals of the New York Academy of Sciences* 908: 208–218.

Franceschi C, Valensin S, Bonafè M, et al. (2000) The network and the remodelling theories of aging: Historical background and new perspectives. *Experimental Gerontology* 35: 879–896.

Glaser R and Kiecolt-Glaser JK (2005) Stress-induced immune dysfunction: Implications for health. *Nature Reviews Immunology* 5: 243–251.

Harding G, Mak YT, Evans B, et al. (2006) The effects of dexamethasone and dehydroepiandrosterone (DHEA) on cytokines and receptor expression in a human osteoblastic cell line: Potential steroid-sparing role for DHEA. *Cytokine* 36(1–2): 57–68.

Jacob F (1977) Evolution and tinkering. *Science* 196(4295): 1161–1166.

Licastro F, Candore G, Lio D, et al. (2005) Innate immunity and inflammation in ageing: A key role for understanding age-related diseases. *Immunity & Aging* 18: 2–8.

Ottaviani E and Franceschi C (1996) The neuroimmunology of stress from invertebrates to man. *Progress in Neurobiology* 48(4–5): 421–440.

Ottaviani E and Franceschi C (1997) The invertebrate phagocytic immunocyte: Clues to a common evolution of immune and neuroendocrine systems. *Immunology Today* 18(4): 169–174.

Ottaviani E, Franchini A, and Franceschi C (1997) Pro-opiomelanocortin-derived peptides, cytokines, and nitric oxide in immune responses and stress: An evolutionary approach. *International Review of Cytology* 170: 79–141.

Ottaviani E, Valensin S, and Franceschi C (1998) The neuroimmunological interface in an evolutionary perspective: The dynamic relationship between effector and recognition system. *Frontiers in Bioscience* 3: 431–435.

Pawelec G, Solana R, Remarque E, and Mariani M (1998) Impact of aging on innate immunity. *Journal of Leukocyte Biology* 64: 703–712.

# Autonomic Neuroplasticity and Aging

**T Cowen**, University College London, London, UK



## Introduction

Autonomic neurons, in common with neurons from all parts of the nervous system, undergo extensive and continual adaptation in response to intrinsic and extrinsic stimuli. The nature of neural plasticity and the factors that influence it vary throughout life. Broadly, these changes fall into three stages: development, maturation and adulthood, and old age. During development, neurons undergo differentiation toward their adult phenotype. This process includes acquiring appropriate morphology, expression of neurotransmitters and receptors, and patterns of activity and signaling pathways adapted to particular functions and connectivity. Toward the end of this period, usually during the perinatal period in mammals, neurons that have not made appropriate connections with their target tissues (which may be other neurons or effector organs such as muscles and glands) die by programmed cell death. During maturation and adulthood, plasticity takes different forms. The broad outline of the phenotype acquired during development – morphology, neurotransmitter, and other functional characteristics – remains stable. However, neurons retain responsiveness to altered demand. During adulthood and aging, many autonomic neurons retain a remarkable capacity to adapt to changing circumstance. However, a minority of cells become subject to selective vulnerability, undergo atrophy, and may die, probably contributing to the functional losses of old age which include deficits in autonomic control of homeostasis. The molecular mechanisms that underlie autonomic neural plasticity are beginning to be understood. A recurring theme throughout the life span is the interaction between neurons and their target tissues, which has profound effects on many aspects of neural plasticity.

## Plasticity in the Developing Autonomic Nervous System

### Neuronal Growth

During embryonic development, differentiating neurons of the autonomic nervous system (ANS) migrate to appropriate peripheral locations, such as the pre- and paravertebral autonomic ganglia and the enteric nerve plexuses of the gut. From these locations, the neurons extend dendrites to form synapses with spinal autonomic neurons of the intermediolateral column. They also extend axons toward their peripheral targets which may be smooth muscle effectors in organs such as blood vessels, gut, or iris, or glands such as salivary, pineal, or sweat glands, or glands of the gastrointestinal tract. Guidance cues for migratory neurons and growing neurites are provided by bound molecules of the extracellular matrix (ECM) as well as by diffusible factors. In recent years, it has become increasingly clear that guidance, like other epigenetic processes in the nervous system, is regulated by a combination of positive and negative influences. In the case of neural guidance, bound and diffusible factors may act as chemoattractants or chemorepellents. Chemoattractants include the ECM glycoprotein laminin and the diffusible neurotrophic and other growth factors, while chemorepellents include members of the semaphorin family, netrins, and ECM glycoproteins. Some but not all of these molecules have been investigated in relation to autonomic neurons. The combination of attractive and repulsive signals ensures that growing neurites follow appropriate pathways, including established nerve fascicles, grow toward appropriate targets, and avoid inappropriate ones.

Around the time that neurons approach their target areas, the peripheral tissues start to synthesize a locally specific cocktail of factors, which attract the growing axons and stimulate synaptogenesis. Bound factors such as laminin stimulate growth of autonomic neurites, for example, through contact between receptors on axon terminals and laminin located in glial or smooth muscle basal laminae. Laminin binds to integrin receptors of the tyrosine kinase family, which initiate neuronal growth and survival signals through ERK/MAP-kinases and PI3-kinase. Diffusible growth-promoting factors produced by targets include the neurotrophin family of growth factors, nerve growth factor (NGF), and neurotrophin-3 (NT3) having the most clearly defined functions in the ANS. These molecules generate growth responses in sympathetic and other peripheral neurons by activating the Trk family of tyrosine kinase receptors, of which TrkA and TrkC are the receptors principally responsible for mediating the effects of NGF and NT3, respectively. The contribution of the nonspecific neurotrophin receptor, p75, remains controversial. Some suggest a synergistic role with Trk receptors in promoting growth and survival; others argue for an inhibitory role for p75 in neuronal growth and survival, probably acting in situations where Trk's are absent. A characteristic of these signaling systems is that they often act in combination, with bound and diffusible factors either working synergistically or providing a balance of positive and negative influences.

### Locally Specific Phenotype

Target specificity of innervation is achieved by secretion by target tissues of locally specific combinations of factors, which stimulate particular patterns of neuronal growth as well as neuronal synthesis of particular neuropeptides. Growth of large, complex sympathetic neurons (supplying, e.g., the iris where fast physiological reflexes are required) correlates with high levels of synthesis of NGF. In contrast, smaller sympathetic neurons supplying sparse innervation to blood vessels (where much slower, less specific neuromuscular responses are required) are associated with low levels of NGF synthesis in the vessel wall (**Figure 1**). Target-derived factors also ensure the adoption by the neurons of a suitable neurotransmitter phenotype The mechanism underlying establishment of neurotransmitter phenotype has been investigated widely, particularly in the minority of sympathetic nerves of the ANS which utilize acetylcholine as a neurotransmitter. Target-derived molecules are clearly involved in this process; however, their identity remains elusive and may include the cytokines ciliary neurotrophic factor, leukemia inhibitory factor, or cardiotrophin-1 and/or the neurotrophic factor, glial cell-derived neurotrophic factor (GDNF).

### Neuronal Survival

The nervous system at large generates more neurons during embryogenesis than are required in maturity.

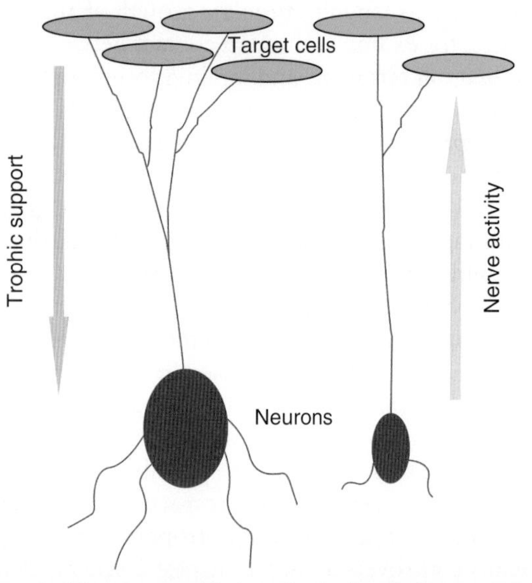


**Figure 1** The neurotrophic hypothesis. Target tissues support neuronal growth, survival, and phenotype through their production of neurotrophic and other factors which are taken up by neurons and retrogradely transported to the cell body where they initiate gene transcription. Neurons regulate the supply of these factors through their activity. Targets producing high levels of neurotrophic factors become innervated by large, complex neurons; conversely, targets producing lower levels of these factors become innervated by smaller ones. These different phenotypes are likely to be functionally adaptive. However, they appear to have pleiotropic effects on the survival trajectories of these neurons in old age (see text).

Fine-tuning of the numbers of surviving neurons to the physical size and physiological demands of peripheral organs occurs by ensuring that the cocktail of diffusible growth factors produced by the target tissue provides sufficient but not superfluous support for the innervating neurons. Neurons that reach their target compete for this scarce resource by growing axons bearing receptors which uptake the appropriate factors, generating further growth and survival signals. Those that fail to take up a sufficient supply undergo programmed cell death or apoptosis. This process was extensively investigated in sympathetic neurons of the ANS and only later understood to be important in many areas of the nervous system. This model of developmental regulation of neuron survival has become encapsulated in the 'neurotrophic theory' (**Figure 1**). An interesting recent addition to this concept concerns the emerging evidence that physiological activity of the target, which is at least partly a result of the activity of the innervating neurons, can itself stimulate synthesis or secretion of growth factors. This important observation suggests that neurons have the capacity to regulate their target-derived trophic support.

Similar groups of molecules contribute to the survival as well as to the growth of ANS neurons. Thus, target-derived NGF and NT3 signaling through Trk receptors regulate survival of numbers of sympathetic neurons appropriate to the particular target tissues during development while the survival of gut neurons is regulated partly by laminin, and subsequently by GDNF and NT3. Pruning of neuron numbers takes place at different periods of perinatal life depending on the target area. In general, it starts in the few days before birth and continues for a variable number of weeks after birth. The signaling pathways mediating developmental survival are similar to those driving neurite growth, that is, tyrosine kinase signaling through ERK/MAP-kinase and PI3-kinase. It seems most likely that downregulation of these pathways (i.e., by subthreshold levels of the ligand neurotrophins), particularly of PI3-kinase, is the key signal for neuronal death. The death signal involves a pro-apoptotic shift in the downstream effectors of tyrosine kinase signaling, which include expression of the BCL-2 and caspase families of proteins. Knockout studies in mice indicate that the p75 receptor, signaling through NFκB, is also involved in the developmental regulation of cell death in NGF-dependent autonomic and other neurons.

## Plasticity in the Adult and Aging ANS

### Neuronal Growth

Adult neurons retain the capacity to respond sensitively to extrinsic growth signals. Body organs continue to grow beyond the period of naturally occurring postnatal neuronal cell death. Existing and relatively stable populations of neurons are therefore required to expand their axonal and dendritic connections by continued growth and synaptogenesis. *In vitro* studies have shown that adult neurons are at least as sensitive as early postnatal neurons in terms of their growth responsiveness to neurotrophins. Indeed, even those neurons that survive into old age retain growth responsiveness to neurotrophins; strong nerve sprouting has been observed *in vivo* in response to treatment with neurotrophic factors. However, there is an age-related reduction in the dose-response curve *in vivo* and *in vitro*.

That plasticity is a physiological feature of adult autonomic neurons is supported by several lines of evidence. Pregnancy and the estrous cycle are accompanied by substantial cyclical changes in the autonomic neurons innervating the uterus. Sympathetic nerve fibers undergo atrophy in late estrus and pregnancy, followed by regeneration. Pregnancy-induced neuronal atrophy may protect the fetus against vasoconstriction caused by homeostatic responses such as fight and flight. Multiple pregnancies may result in the death of some of the innervating sympathetic neurons. Transplantation studies and experimental enlargement of organs demonstrate further that neurons, particularly autonomic neurons, retain an extraordinary ability to expand their field of innervation. Partial obstruction of the bladder can result in a roughly tenfold increase in muscle mass to which the innervating autonomic neurons respond with an increase in size of around 80%. Removal of the obstruction results in reversion of neuron size to control values over a few weeks. Comparable manipulations in the gut and iris – both autonomically innervated tissues – have yielded similar results.

The same neurotrophic factors, which regulate growth of neurons during development, are retained by the adult target tissues, generally at unchanged levels. In the case of adult sympathetic neurons, these factors include NGF and NT3. Of these, NGF produces the substantially stronger growth response, although both factors are present in many of the target tissues of sympathetic neurons. TrkA and TrkC receptors are present on adult sympathetic neurons to mediate the growth responses. The increased target area resulting from the physiological or experimental expansion of organs described above is therefore likely to induce nerve growth by increasing the availability of neurotrophins in the vicinity of existing nerve terminals. The role of the nonspecific p75 receptor remains as controversial in the adult nervous system as it is at earlier stages of development. Some argue that p75 inhibits neuronal growth in adult rodent neurons while others consider that p75 supports the trophic role of NGF by heterodimerization with Trk. These apparently contradictory roles for this receptor may result from different expression ratios of p75 and Trk in different neurons and under different circumstances. Local availability of neurotrophins can affect the expression of both p75 and Trk receptors in the adult nervous system, thus providing a further level at which nerve growth can be regulated.

The signaling pathways involved in adult neuron growth are less well defined than those regulating survival. While inhibition of tyrosine kinase signaling largely inhibits extension of neurites from adult sympathetic neurons *in vitro*, there is less conclusive definition of the downstream pathways involved. This may be because neurite extension involves integrin-as well as neurotrophin-mediated activation of tyrosine kinase signaling, with resulting convergent activation of ERK/MAP-kinase and PI3-kinase.

During aging, some autonomic neurons undergo atrophy (shrinkage of their dendritic and axonal arbors and cell soma), while others continue to grow into old age. This phenomenon has been called 'selective vulnerability' and affects sympathetic neurons projecting to vascular targets and sweat and pineal glands. In contrast, neurons projecting to iris and submandibular gland show no evidence of atrophy. To the extent that the subpopulations of neurons innervating each target tissue exhibit specific patterns of growth or atrophy during aging, selective vulnerability involves interactions between particular target tissues and neurons. However, it has not proved possible to demonstrate that reduced levels of NGF availability from targets causes age-related atrophy of fibers.

### Locally Specific Phenotype

Adult neurons retain locally specific morphology and neurochemical phenotype. Thus, subpopulations of adult sympathetic neurons, retrogradely traced from particular target tissues, exhibit different sizes which appear to relate to physiological function in the same way that developing neurons adapt to the physiological function of the end organ. In addition, adult neurons continue to express locally specific patterns of neuropeptides and transmitters. For example, neuropeptide Y expression is highest in superior cervical ganglion (SCG) neurons projecting to iris, lower in those projecting to the middle cerebral artery, and lowest in neurons projecting to the submandibular gland. There is evidence that these patterns of neurotransmitter expression remain plastic in adulthood. Transplantation in adult rats of sweat glands (normally innervated by cholinergic, vasoactive intestinal peptide (VIP)-expressing sympathetic neurons) into contact with catecholaminergic sympathetic neurons induces a shift to a cholinergic, VIP-expressing phenotype in those neurons, which extend processes to innervate the transplanted tissue. There is no obvious loss of this form of plasticity in old age.

### Neuronal Survival in Adulthood

While neurons retain growth responsiveness to extrinsic factors in adulthood and even in old age, the factors that determine survival, at least of sympathetic and other (sensory) peripheral neurons, appear to change dramatically during maturation and in aging. Between birth and 12 weeks of age, rodent sympathetic neurons, from being dependent for their survival on an extrinsic source of neurotrophin, become almost completely independent and able to survive in dissociated cell culture for many days and even weeks under serum- and NGF-free conditions (**Figure 2**). The loss of survival dependence on target-derived signals may be seen as an adaptation allowing adult neurons a degree of resistance against an axonal injury that separates the neuron from its target, thus allowing the neuron a period of respite in which to initiate regeneration.

The principal signaling pathway that mediates neuronal survival independence, as this phenomenon is called, is PI3-kinase. Pharmacological and/or genetic inhibition of this pathway at different points leads to the rapid death of cultured adult sympathetic neurons; inhibition of ERK/MAP-kinase is without significant effect. This suggests strongly that while neurons lose their dependence on signaling through Trk receptors to regulate their survival, the downstream PI3-kinase pathway remains a key determinant of neuronal resistance to cell death. It is feasible that autonomic and perhaps other neurons protect their survival in adulthood by cell-autonomous activation of PI3-kinase in a way that resembles a mechanism by which cancer cells avoid apoptosis.

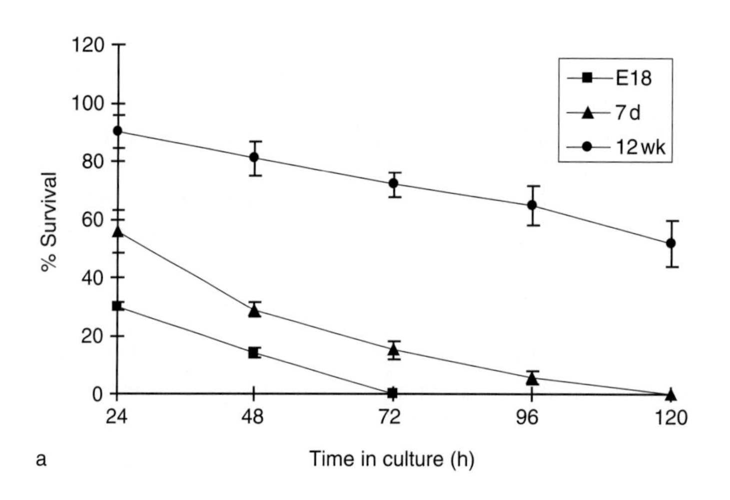


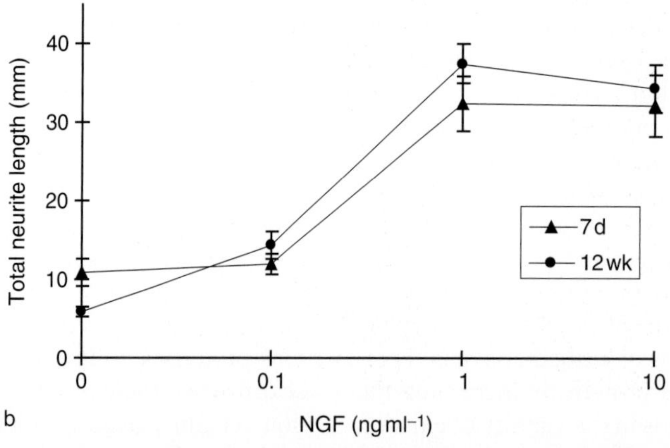


**Figure 2** Differential regulation of neuronal growth and survival during maturation. During postnatal maturation, sympathetic neurons lose their survival dependence on NGF. Growth, however, remains NGF dependent. (a) Sympathetic neurons become independent of NGF for their survival between embryonic day 18 (E18) and 12 weeks of age, as demonstrated by a dramatic increase in survival with age in dissociated, pure neuron cell cultures maintained over several days in the absence of serum and NGF. Note that by 12 weeks, approximately 60% of neurons survive for longer than 5 days *in vitro*, whereas no E18 neurons survive longer than 3 days. (b) Postnatal sympathetic neurons require NGF to initiate growth and to extend neurites *in vitro*. Note that neurons fail to grow in the absence of NGF; dose responsiveness is similar in 12-week-old and 7-day-old neurons.

It has recently become clear that production of neuronal progenitors, at least in the central nervous system (CNS), continues throughout life under the influence of a range of epigenetic factors, including physical exercise, brain activity, and injury. It is not clear, however, whether the ANS can continue to generate neuronal precursors in adulthood or whether plasticity mechanisms similar to those seen in development and maturation regulate the differentiation, survival, and integration of progenitor cells into the functioning adult nervous system.

### Neuronal Survival in Old Age

Selective vulnerability to neuron cell death is a well-established characteristic of the aging nervous system as a whole, as well as of the ANS. The groups of neurons affected are similar to those which exhibit age-related atrophy (see the earlier section on neuronal growth). Among sympathetic neurons, only certain populations undergo cell death, while many survive into extreme old age. Roughly 50% of vascular-projecting neurons in the SCG are lost in aging rats, while neurons projecting to the iris are less significantly reduced in numbers. Autonomic neurons from the myenteric plexus of the gut are also vulnerable to age-related cell death, with roughly 50% of neurons lost from for small and large intestine of aging rats and guinea pigs (**Figure 3**). There is limited but less secure evidence that similar patterns of myenteric neuron loss occur in humans, at least from the large intestine.

In line with evidence that adult neurons become relatively independent of target-derived trophic support, it is probable that locally selective patterns of neuron death during aging are not the result of

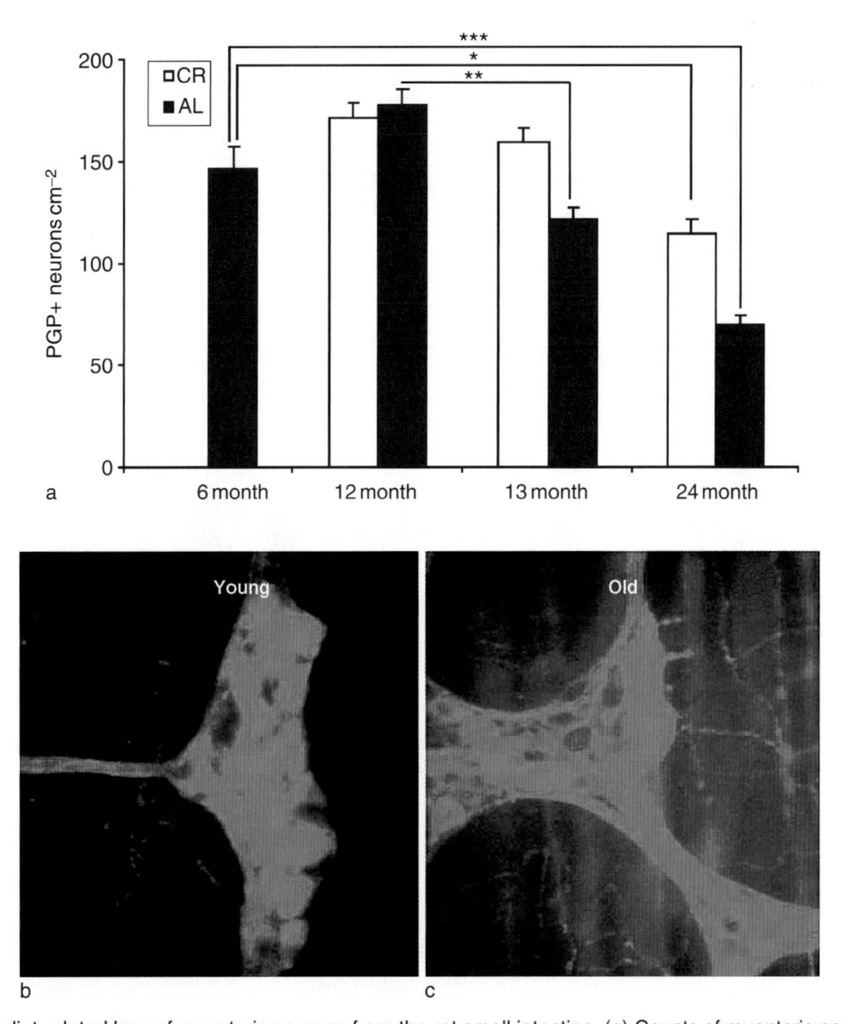


**Figure 3** Age- and diet-related loss of myenteric neurons from the rat small intestine. (a) Counts of myenteric neurons using the pan-neuronal marker, PGP 9.5. The histogram shows significant losses of total numbers of neurons, commencing at 12–13 months of age in *ad libitum*-fed (AL) animals, which reaches a peak of 51% at 24 months. There is a much smaller but statistically significant ($p < 0.05$) reduction seen in animals of the same age (24 months) but fed a calorie-restricted (CR) diet, demonstrating substantial protection by CR against age- and diet-related loss of myenteric neurons. Statistically significant differences are indicated by $^{*}p < 0.05$; $^{**}p < 0.01$; $^{***}p < 0.001$. (b, c) Photomicrographs of young (6 months; b) and old (24 months AL; c) PGP-immunostained whole-mount preparations of rat myenteric plexus. Note the large spaces in the old ganglia indicating where neuron cell loss has occurred.

reduced availability of neurotrophins. Studies of different target areas show no evidence of altered levels of NGF or NT3 protein or mRNA in targets of aging sympathetic neurons, mirroring similar observations in targets of vulnerable NGF-dependent neurons in the CNS. The search for causes of age-related neurodegeneration is a high priority in aging research. A number of avenues are being explored including whether expression of the p75 or Trk receptors are selectively altered in subgroups of aging sympathetic neurons vulnerable to age-related cell loss. In the absence of changes in levels of extrinsic factors, these are likely to be cell-autonomously regulated. Subpopulations of neurons that exhibit age-related cell loss project to target areas that throughout life express relatively low levels of neurotrophic factors; conversely neurons that survive successfully into old age tend to project to targets expressing high levels of neurotrophic factors (see the above discussion on locally specific phenotype). These differential levels of neurotrophin synthesis are retained throughout life and are reflected in altered patterns of p75 and Trk receptor expression, which correlate directly with neurotrophin availability. It therefore seems likely that growth factor signaling is set during maturation at levels that are adapted to particular physiological requirements of subgroups of neurons. This maturational setting affects the capacity of neurons to survive into old age.

Oxidative and other forms of free radical damage are heavily implicated in neurodegenerative disease as well as in normal aging. It is now known that neurotrophic factors can influence antioxidant defense in neurons and that free radical levels increase in neurons, such as those of the myenteric plexus, prior to the onset of age-related cell loss. A working hypothesis, then, is that differential, lifelong patterns of neurotrophin signaling are established during early postnatal life, and become more or less fixed in maturity. These differential and adaptive levels of signaling then determine, among other things, neuronal capacity to resist prolonged exposure to free radical damage. The neurons with low levels of signaling succumb first (**Figure 4**).

An alternative way in which neurotrophin signaling might be altered in adulthood and old age concerns the recent discovery that the precursor pro-form of NGF, previously thought to be expressed within target cells prior to its cleavage by enzymes of the proconvertase family, is widely distributed in the adult nervous system, including in nonneural targets of autonomic nerves. ProNGF levels are elevated in the pregnant uterus, in association with the atrophy and degeneration of the innervating sympathetic neurons (see the earlier section on neuronal growth). ProNGF is also the predominant form of NGF present in the Alzheimer's diseased brain and in the spinal cord after injury – two situations known to

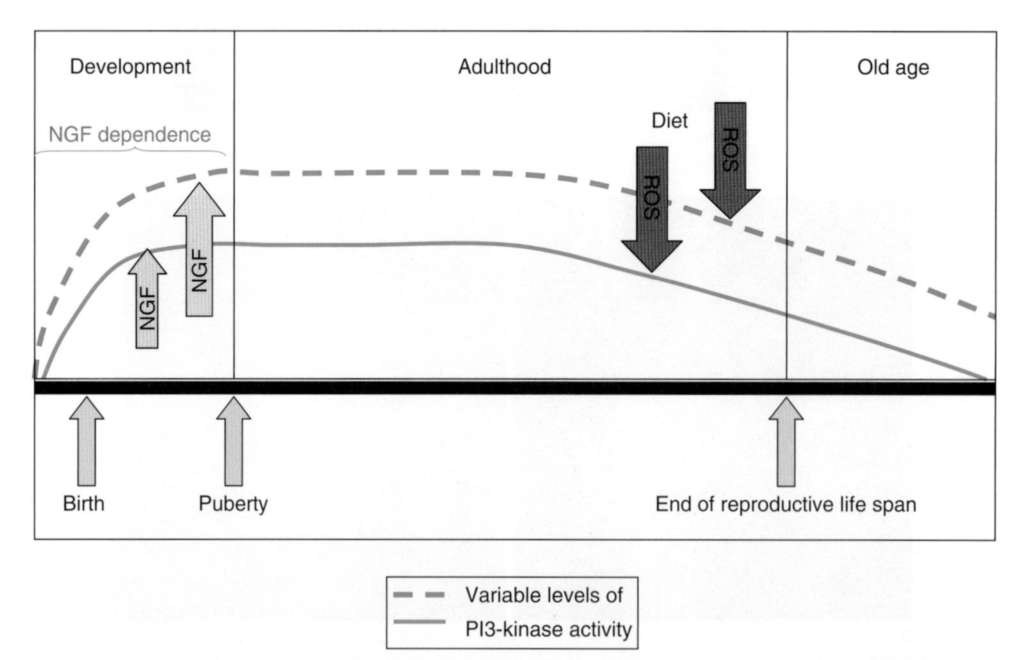


**Figure 4** Model illustrating mechanisms regulating lifelong survival of autonomic neurons. During perinatal life, autonomic neurons become increasingly independent of target-derived NGF for their survival as a result of upregulation of intraneuronal PI3-kinase activity. The degree of upregulation is variable (see dashed and solid green lines), locally specific and adaptive, being set initially by the levels of the target-derived NGF (large and small arrows). Later in life (around puberty and after), PI3-kinase activity is maintained independently of NGF availability. Toward the end of the reproductive life span, exposure to reactive oxygen species (ROS) downregulates PI3-kinase activity, increasing vulnerability to age-related neuron cell death. AL feeding (see **Figure 3**) increases exposure to ROS and thereby accentuates this process, while CR provides protection.

be associated with widespread neuron cell death. ProNGF has been shown to induce apoptotic cell death in PC12 cells as well as SCG neurons from young animals. ProNGF signals via the receptor sortilin (first discovered as a member of the Vsp10 family of sorting receptors and later identified as the receptor for neurotensin) acting in combination with p75. At present, there is little evidence from the aging nervous system that proNGF contributes to age-related cell death. However, the sortilin receptor is expressed in the adult (including human) brain; therefore, this possibility cannot be ruled out.

### Extrinsic Factors Affecting Neuron Survival

Factors that influence life span have been shown, unsurprisingly, to also affect neuron survival. The classical extender of life span in rodents, caloric restriction (CR), has remarkable effects in rescuing myenteric neurons of rats from age-related cell death. CR also reduces free radical levels in neurons, and appears to achieve this by influencing neurotrophic factor-mediated antioxidant defense. Exercise and the environmental stimuli to which laboratory rodents are exposed (including group vs single animal housing) may also affect health, life span, and, perhaps, cell survival. Further studies are required to quantify these phenomena and to explore the underlying mechanisms.

## Conclusions

Autonomic neuronal plasticity is regulated by core mechanisms which are set up during development, modified and consolidated during maturation, and maintained into adulthood and old age. These mechanisms, which are adaptive, may have deleterious pleiotropic consequences for some groups of aging neurons, making them vulnerable to neurodegeneration. Neurodegeneration probably occurs initially as loss of terminal nerve fibers (axons and dendrites) and shrinkage of the cell soma and may progress ultimately to cell death. Interactions between neurons and their target tissues, including the provision of nerve growth and survival promoting factors by targets and the related expression of appropriate receptors in neurons, comprise an intimate and essential interaction at the core of plasticity responses throughout life. Intraneuronal signaling pathways have specific roles in regulating plasticity, including growth and survival responses, and, probably, maintenance of neurotransmitter phenotype and antioxidant defense. Extrinsic factors, including diet and other behaviorally regulated phenomena, have major effects on neural plasticity, probably by acting on intraneuronal signaling pathways.

*See also:* Autonomic and Enteric Nervous System: Apoptosis and Trophic Support During Development.

## Further Reading

Anderson DJ (2000) Genes, lineages and the neural crest: A speculative review. *Philosophical Transactions of the Royal Society of London B: Biological Sciences* 355: 953–964.

Brauer MM, Shockley KP, Chavez R, Richeri A, Cowen T, and Crutcher KA (2000) The role of NGF in pregnancy-induced degeneration and regeneration of sympathetic nerves in the guinea pig uterus. *Journal of the Autonomic Nervous System* 79: 19–27.

Bruno MA and Cuello AC (2006) Activity-dependent release of precursor nerve growth factor, conversion to mature nerve growth factor, and its degradation by a protease cascade. *Proceedings of the National Academy of Sciences of the United States of America* 103: 6735–6740.

Cowen T (2002) Selective vulnerability in adult and ageing mammalian neurons. *Autonomic Neuroscience* 96: 20–24.

Cowen T and Gavazzi I (1998) Plasticity in adult and ageing sympathetic neurons. *Progress in Neurobiology* 54: 249–288.

Dontchev VD and Letourneau PC (2003) Growth cones integrate signaling from multiple guidance cues. *Journal of Histochemistry and Cytochemistry* 51: 435–444.

Fahnestock M, Michalski B, Xu B, and Coughlin MD (2001) The precursor pro-nerve growth factor is the predominant form of nerve growth factor in brain and is increased in Alzheimer's disease. *Molecular and Cellular Neuroscience* 18: 210–220.

Gabella G, Berggren T, and Uvelius B (1992) Hypertrophy and reversal of hypertrophy in rat pelvic ganglion neurons. *Journal of Neurocytology* 21: 649–662.

Gershon MD (1998) V. Genes, lineages, and tissue interactions in the development of the enteric nervous system. *American Journal of Physiology* 275: G869–G873.

Glebova NO and Ginty DD (2005) Growth and survival signals controlling sympathetic nervous system development. *Annual Review of Neuroscience* 28: 191–222.

Miller FD and Kaplan DR (2001) Neurotrophin signalling pathways regulating neuronal apoptosis. *Cellular and Molecular Life Sciences* 58: 1045–1053.

Nykjaer A, Lee R, Teng KK, et al. (2004) Sortilin is essential for proNGF-induced neuronal cell death. *Nature* 427: 843–848.

Orike N, Middleton G, Buchman VL, Cowen T, and Davies AM (2001) Role of PI 3-kinase, Akt and Bcl-2-related proteins in sustaining the survival of neurotrophic factor-independent adult sympathetic neurons. *Journal of Cell Biology* 154: 995–1005.

Thrasivoulou C, Soubeyre V, Ridha H, et al. (2006) Reactive oxygen species, dietary restriction and neurotrophic factors in age-related loss of myenteric neurons. *Aging Cell* 5: 247–257.

Xu XM, Fisher DA, Zhou L, et al. (2000) The transmembrane protein semaphorin 6A repels embryonic sympathetic axons. *Journal of Neuroscience* 20: 2638–2648.

# Thermoregulation: Autonomic, Age-Related Changes

**R B McDonald**, University of California, Davis, CA, USA



## Introduction

The inability to appropriately thermoregulate during exposure to heat and cold has been generally accepted as an age-related change in humans. Epidemiological observations of increased mortality during severe cold and hot weather also emphasize a possible age-related attenuation in thermoregulation. It remains unclear, however, the extent to which biological factors impact clinical hyper- and hypothermia in senescent humans. A majority of case studies have found that diminished thermoregulatory capacity in the senescent human reflects more closely socioeconomic factors such as insufficient living area heating or cooling, malnutrition, and lack of adequate health care. Nonetheless, experimental evidence in humans and laboratory rodents has reported senescent-related alteration in autonomic mechanisms that regulate body temperature. In this article, we will briefly explore possible affects of biological aging on the autonomic control of thermoregulation.

## 'Problems' of Interpretation of Aging Research

### Definition of Aging

Research into the mechanisms underlying the aging process, including altered autonomic regulation of body temperature, poses experimental design concerns that do not often occur in other disciplines. Lack of a generally accepted definition of aging should be viewed as primary among these concerns. The difficulty in defining aging arises from an inability to determine when aging begins. Biological aging should include all events occurring to a particular system throughout the life span. Evolutionary, genetic, and developmental data have clearly established a link between development and the rate of aging. Although some investigations have begun to approach aging from this broad perspective, such experimental designs are difficult to construct and restrict the investigator to observational studies. Narrowing the definition of biological aging to a postreproductive period in which biological deterioration can be clearly identified increases mechanistic observations. However, since narrow definitions of aging do not include early life events, the true cause and effect relationship may be missed. Nonetheless, Dr. Edward Masoro's definition of biological aging has provided a good guide for most aging research:

> Aging can be defined as deteriorative changes with time during post-maturational life that underlie and increase vulnerability to challenges, thereby decreasing the ability of the organism to survive. (Masoro EJ, 1995: 3–21)

### Designs in Research on Thermoregulation and Aging

Appropriately interpreting results from investigations evaluating the biology of senescence can be difficult for the nonspecialist. The difficulty in interpretation arises from the fact that extrinsic and behavioral factors such as physical activity level, acclimation, and disease have significant biological aging effects that, in turn, can affect thermoregulation. Cross-sectional designs, investigations that compare different age cohorts at one point in time, are the most common but least powerful design used in senescence research. The length of the human life span precludes, for the most part, longitudinal design except by large multiinstitutional programs (e.g., the Baltimore Longitudinal Study), and these investigations are limited to observational evaluation.

Cross-sectional designs can be problematic because selecting the appropriate older age group that represents accurately the 'normal' senescence process cannot be easily achieved; we return to the question, 'when does aging begin?' This problem reflects that the beginning of senescence in humans is highly variable, may not affect all physiological systems equally, and does not correlate with chronological age. For example, all 70-year-old humans will have some functional loss in at least one physiological system, but not all 70-year-old people will experience decrements in the same physiological system. Investigations using cross-sectional designs will control for individualized senescence by including an older age group with a wide range of ages (e.g., 65–80 years). Such a wide age range will often result in significant error about the mean as compared to the variation seen in younger groups.

Rats and mice have been the overwhelming animals used to model humans in senescence-related research evaluating possible alterations of thermoregulation. Their short life span (2–4 years) provides the investigator an opportunity for mechanistic study using powerful longitudinal designs. However, application of results to humans may be limited due the differences in effector organs responsible for heat dissipation and

**Table 1** Tissues and organs of heat dissipation and production in humans, rats, and mice

| *Species* | *Insulation* | *Heat dissipation (hot environment)* | *Heat production (cold environment)* |
|---|---|---|---|
| Human | Subcutaneous fat | Convection resulting from the transfer of heat from blood to skin throughout body<br>Evaporation resulting from increased sweat gland function | Shivering via increased skeletal muscle metabolism |
| Rats and mice | Fur and air gap between skin and muscle | Convection resulting from the transfer of heat from blood to skin limited to tail | Nonshivering thermogenesis in brown adipose tissue<br>Shivering via increased skeletal muscle metabolism |

production (**Table 1**). It remains unclear as to how or if neural feedback, from the different sensor systems and effector organs of heat production and dissipation in rodents versus humans, may impact thermoregulation during senescence.

### Disease and Its Effect on Thermoregulation during Senescence

The separation between disease and the biological process of aging presents yet another issue that must be considered when investigating senescent-related effects on thermoregulation. Clearly, older individuals are more likely to have disease than are younger individuals. It can be difficult to determine if the finding of altered thermoregulation reflects biological aging *per se* or reflects the outcome of a disease: the discussion herein relies on data from investigations in which a clear attempt has been made to control for disease.

### Acclimation and the Sedentary Life Style

The ability to thermoregulate appropriately in hot and cold environments depends, in part, on one's acclimation to the ambient conditions. Well-acclimatized individuals will tolerate deviations from thermoneutrality better than will the nonacclimated. Moreover, a sedentary lifestyle in which individuals live at ambient temperatures close to thermoneutrality will reduce sweat gland density, capillary density of skin, and muscle mass, all outcomes that are independent of age and all of which impact thermoregulatory capacity.

## Physiology of Thermoregulation

### Hypothalamic Regulation of Body Temperature

Humans and rodents maintain a constant body temperature by balancing heat production with heat loss. The equilibrium point of heat production and loss, thermoneutrality, occurs at an ambient temperature of approximately 21 °C in the human and 23–25 °C in rats and mice, and does not appear to change with age. Exposure to environmental temperatures above or below thermoneutrality will result in various physiological responses that work to keep a stable core body temperature (38 °C humans, 38–39 °C mice and rats). Communication between neural centers within the hypothalamus and peripheral sensors and effector organs regulate, almost exclusively, the processes that maintain stable body temperature in homeotherms (**Figure 1**). The preoptic area of the hypothalamus contains heat- and cold-sensitive neurons that respond to changes in blood temperature, although heat-sensitive neurons far outnumber cold-sensitive sensors. Thus, detection of increased body temperature occurs, for the most part, within the hypothalamus. Detection of a cold environment occurs via temperature sensors within the skin; these sensors send afferent signals to the posterior hypothalamus at or near the level of the mammillary bodies (cold sensors in the skin outnumber heat sensors 10 to 1). The posterior hypothalamus also serves to coordinate the temperature-sensing 'data' from both the preoptic area and the periphery and, in turn, takes appropriate action through efferent signals.

### Neural Regulation of Skin Blood Flow and Sweat Glands

Changes in skin blood flow, comprising the initial physiological response to increased or decreased ambient temperatures, are regulated by sympathetic vasodilation and vasoconstriction mechanisms. An increase in core temperature causes the release of vasoconstrictor tone, resulting in increased blood flow. If convective heat losses resulting from relaxation of the vasoconstrictor tone continue to be insufficient to cool the core, cholinergic active vasodilation causes further increase in skin blood flow that, in turn, increases convective heat loss. Sweat glands are innervated by cholinergic fibers that secrete

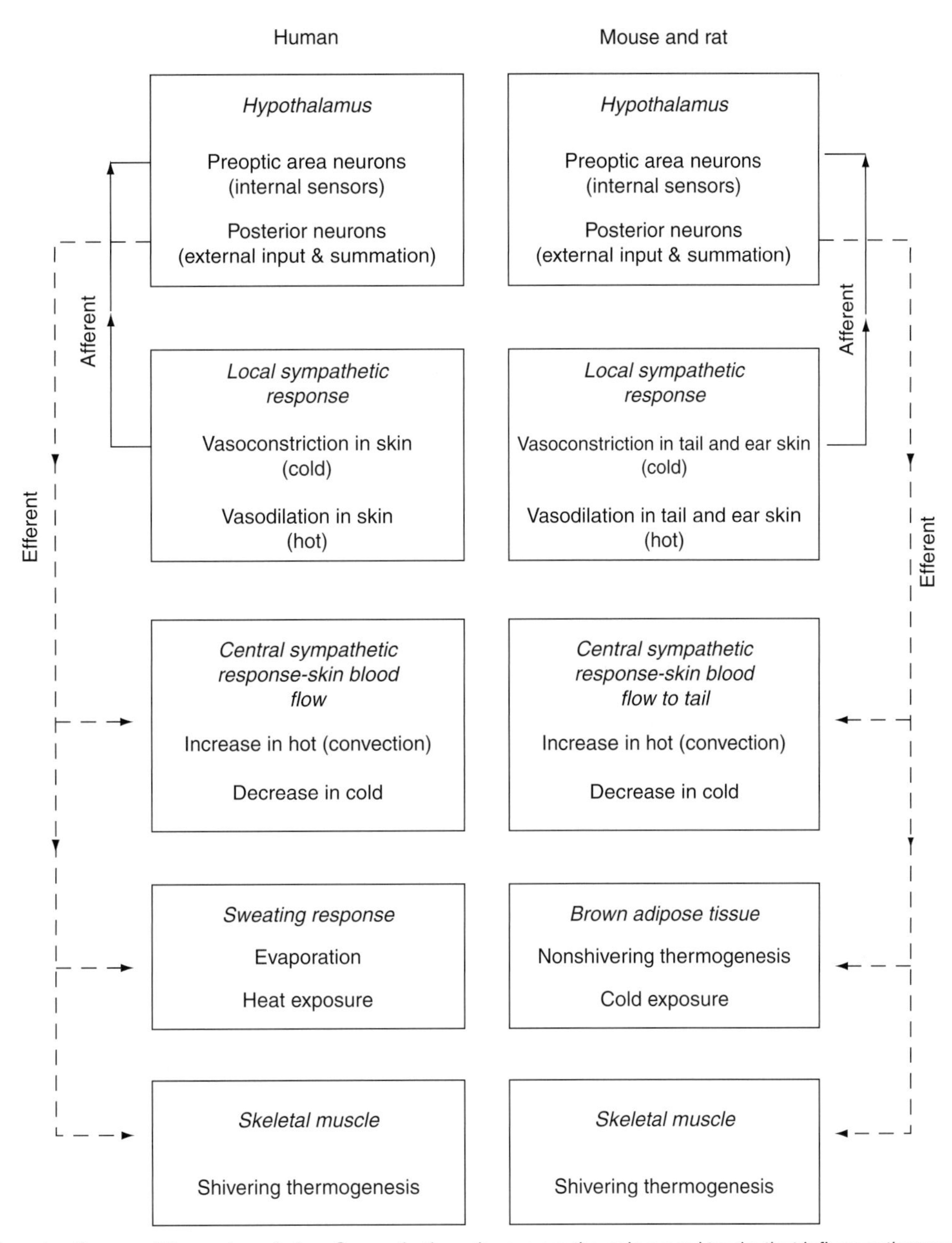


**Figure 1** Neural pathways of thermal regulation. Sympathetic pathways are the sole neural tracks that influence thermoregulation; the parasympathetic system does not exert a significant effect on thermoregulation. Vasoconstriction and initiation of nonshivering thermogenesis occur in response to the release of norepinephrine attaching to $\alpha$-adrenergic receptors in blood vessels and to $\alpha$- and $\beta$-adrenergic receptors in brown adipose tissue. Vasodilation in the human is caused by epinephrine exciting the $\beta$-adrenergic receptor; acetylcholine causes vasodilation in rats (muscarinic receptors). Sweat glands are innervated by cholinergic nerve fibers that secrete acetylcholine, but the glands are also responsive to epinephrine secreted by the adrenal gland.

acetylcholine and are stimulated simultaneously with active vasodilation. Decreased skin and core temperatures result in activation of norepinephrine-mediated vasoconstriction beyond the normal vasoconstriction tone. Various other peptides, such as neuropeptide Y, can be co-released with norepinephrine and may also have a significant effect on vasoconstriction.

## Effect of Senescence on the Response to Heat and Cold Exposure in Humans

### Changes to Core Temperature

Mean core temperatures of humans exposed to cold or heat decrease (cold) or increase (hot) slightly more in senescent versus adult individuals. The differences in core temperature are modest (no more than 0.1–0.2 °C difference) and are not generally considered physiologically important. Physical activity and acclimation to the ambient temperature improve the core temperature of older individuals, suggesting that behavior rather than aging *per se* may be of greater importance as the cause of senescent-related alterations in core temperature. The experimental evidence does not support the suggestion that the older population (defined here as those over 65 years of age) has a higher risk of developing clinical hypo- or hyperthermia, as compared to younger populations. However, recent investigations have found that senescent-related alterations in core temperature leading to thermal sickness are progressive in the aged cohort. That is, the most significant alteration in the loss or gain in body heat, as measured by core temperature, appears to be limited to the oldest old (i.e., individuals in their ninth and tenth decades of life).

### Autonomic Response

**The onset of vasoconstriction and vasodilation during thermal stress** While changes in core temperature in response to heat and cold stress do not appear to be a general hallmark of senescence, there is agreement that the autonomic response may be altered. For example, the time to onset of sweating during passive heat exposure occurs significantly later in older versus younger individuals and is independent of the anatomical location (**Figure 2**). Once sweating has begun, however, the rate and amount appear to be similar in both young and old individuals. Moreover, there does not appear to be a significant difference in sweat glad density or skin temperature threshold that initiates sweating between healthy senescent and younger individuals. Together, lack of differences in sweat rate, sweat glad density, and thresholds suggests that the delay in sweating during passive heat stress reflects some alteration at the interface between efferent neural signaling and the effector organ.

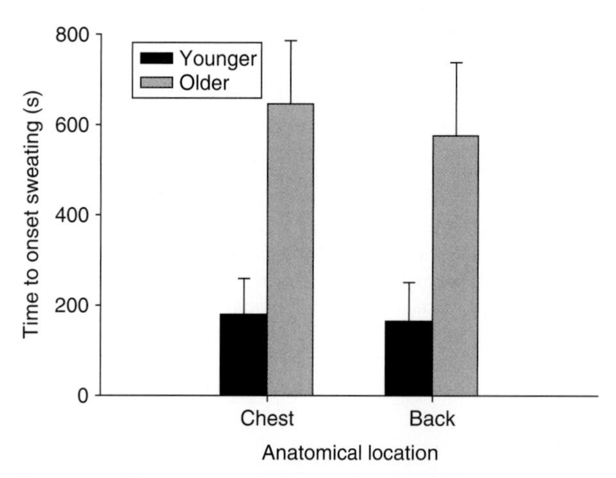


**Figure 2** Time to onset of sweating in younger (ages 20–24 years; $n=7$) and older (64–76 years; $n=7$) men exposed to heat (35 °C) and a relative humidity of 45%. The lower leg of each person was also placed in a water bath (42 °C) during the passive heat exposure. With kind permission from Springer Science + Business Media: Inoue Y, Shibasaki M, Ueda H, et al. (1999) Mechanisms underlying the age-related decrement in the human sweating response. *European Journal of Applied Physiology* 79(2): 121–126.

Differences observed in the onset of vasoconstriction during cold exposure are consistent with senescent-related attenuation in the autonomic response within the efferent pathways during a thermal challenge. That is, the cold-induced decrease in skin blood flow occurs much later in the senescent individual as compared to a younger adult. However, an older individual's capacity to sense a drop in ambient temperature, as determined by subjective rating scales, does not differ from that of younger persons. The ability of the elderly to accurately sense a decrease in ambient temperature corresponds to the decrease in skin temperature. If the afferent signal to the hypothalamus was disrupted, these two measures – the sensing of cold and drop in skin temperature – should be discontinuous. Therefore, altered skin blood flow concomitant with appropriate thermal senescing suggests that the neural mechanisms that inform the thermoregulatory center of a change in ambient temperature – the afferent pathways – function normally. The delay in cold-induced vasoconstriction during senescence implies that the disruption in the autonomic response more likely reflects changes to efferent pathways. It remains unclear, however, if the senescent-related alteration in autonomic response to heat and cold stress occurs at the level of the hypothalamus or the peripheral effector organs.

**Senescence and central-mediated autonomic response to thermal stress** Evaluating the central-mediated (hypothalamic) autonomic response to thermal stress in humans is difficult, and most conclusions on a central-mediated role are based on circumstantial rather than direct experimental evidence. For example, the circadian rhythm of body temperature has been used as an indicator of hypothalamic function, as the suprachiasmatic nucleus regulates exclusively this 'biological clock.' Several investigations have found a senescent-related decrease in the amplitude of the circadian rhythm of body temperature and sleep, a finding consistent with alterations in hypothalamic regulation. However, senescent-related changes in the circadian rhythm of body temperature and sleep are almost always associated with an underlying disease, a sedentary life style, and/or psychological conditions that disrupt normal patterns of daily life. Investigations controlling for these confounding variables have found that alterations in the circadian rhythm of body temperature and sleep are limited, for the most part, to the oldest individuals in the senescent group, a conclusion similar to that suggested for alterations in core temperature.

**Senescence, skin blood flow, and vasodilation response to heat** Several investigations have demonstrated that skin blood flow of elderly persons at rest and during heat stress does not increase to the same degree as that observed in younger individuals. Initially, senescent-related changes to the skin architecture, including diminished arteriole and capillary density, were suggested to be the primary cause of altered skin blood flow. Investigations comparing different anatomical sites for a senescent-related attenuation of skin blood flow have shown that the largest alterations occur at sites normally exposed to the environment. For example, heat-induced increase in forearm-skin blood flow occurs significantly later and is less robust than that observed from sites on the chest and back. These data imply that senescent-related alterations to skin blood flow caused by changes in the skin structure reflect more closely photoaging than they do the biological process of senescence.

Nonetheless, attenuated heat-induced skin blood flow of older individuals has also been reported at anatomical sites typically not exposed to the sun (i.e., chest and thigh). The question, then, becomes 'does the blunted heat-induced skin blood flow observed in the elderly reflect a decrease in vasodilation or the failure to release vasoconstriction tone?' Evaluation of decreased vasodilation versus sustained vasoconstriction tone can be achieved by comparing skin treated with bretylium tosylate, an agent that blocks vasoconstriction, to nontreated skin. These investigations have consistently shown no effect on skin blood flow by the blocking agents. That is, reduced heat-stress blood flow in the senescent individual appears to reflect altered active vasodilation.

The mechanisms underlying the altered vasodilation of the elderly during heat exposure remains to be elucidated. There is agreement that aged vessels are less responsive to neural transmitters, have fewer receptors, and less ligand binding to the receptor, but these effects have not been tested directly during heat exposure and also appear to be influenced significantly by underlying disease. Findings of a diminished level of neural density in senescent skin have been inconsistent. Increases, decreases, and no difference in neural density of aging skin have been reported. This inconsistency most likely arises from different anatomical sites sampled and the disease states of the individuals.

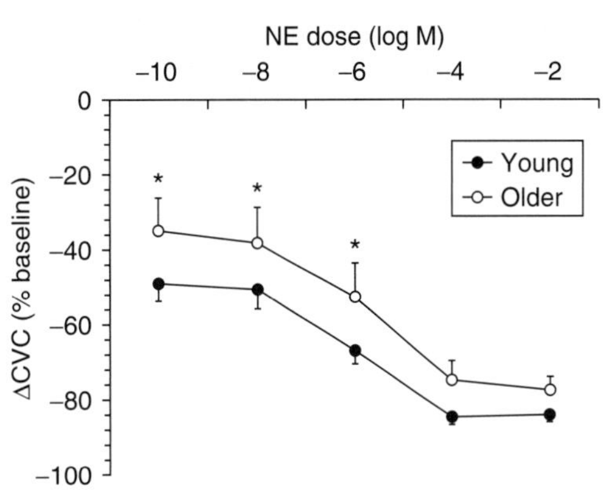


**Figure 3** Mean cutaneous vasoconstriction (CVC) responses to sequential norepinephrine (NE) infusions in young ($n=11$) and older ($n=11$) individuals. Reproduced from Thompson CS, Holowatz LA, and Kenney WL (2005) Cutaneous vasoconstrictor responses to norepinephrine are attenuated in older humans. *American Journal of Physiology – Regulatory, Integrative and Comparative Physiology* 288(5): R1108–R1113, used with permission from the American Physiological Society.

**Senescence, skin blood flow, and the vasoconstriction response to cold** The decrease in skin blood flow during local and systemic cooling of healthy senescent individuals appears to be blunted compared to younger persons. Altered senescent-related skin blood flow during cold exposure reflects diminished vasoconstriction and appears to be related to both a decreased effect by norepinephrine (**Figure 3**) and a loss of contribution to vasoconstriction from co-released transmitters. Cold-induced vasoconstriction has been shown to be enhanced by neurotransmitters co-released with norepinephrine. When yohimbine and propranolol are used to antagonize norepinephrine's effects during local cold exposure, vasoconstriction decreases, but remains robust in the younger individuals, indicating contribution by co-released neurotransmitters. However, yohimbine and propranolol completely block vasoconstriction in the elderly. That is, co-released peptides, such as neural peptide Y (NPY), must be compensating for the diminished effect of norepinephrine in the younger vessel, yet there is no compensation in the older vessel. While some evidence suggests that the co-released neural peptide may be NPY, the precise identity of the agent remains to be elucidated.

### Summary of the Effects of Senescence on the Response to Heat and Cold Exposure in Humans

Senescent-related alterations to core temperature during thermal stress are minimal. Reports of a greater risk of hypo- and hyperthermia in the elderly have now been judged to reflect more closely factors such as disease, physical activity level, psychological impairment, and socioeconomic level, rather than biological aging *per se*. Sympathetic regulation of vasoconstriction and vasodilation appears to decline with age and reflects a general desensitivity to neural transmitters. Vasoconstriction and vasodilation can be improved by acclimation to the ambient temperature and increased physical activity, suggesting that behavior may also a have a significant role in the maintenance of sympathetic regulation. Recent investigations have suggested that if diminished thermoregulatory capacity occurs as the result of biological aging, the effects are progressive within the senescent cohort. That is, altered thermoregulation is most pronounced in people nearing the end of their natural life span (i.e., the ninth and tenth decades of life).

## Effect of Senescence on the Response to Cold Exposure in Laboratory Rodents

The use of laboratory rodents, particularly rats and mice, to evaluate the affect of senescence on thermal stress has been, for the most part, limited to investigations describing changes to cold exposure. Young and old rodents do not tolerate heat exposure greater that 2–3 °C above normal body temperature (38–39 °C) and mortality rates increase significantly after 2 h. Thus, the discussion presented here is limited to the autonomic response observed in cold exposure of senescent mice and rats.

### Response of Core Temperature to Cold Exposure

Observations of the effect of advanced age on cold-exposed core temperature in rats and mice have not been consistent. The differences in mean values for cold-exposed core temperature, when present, are modest, and the variations about the mean are considerably greater than those seen in younger animals. If an age-related decrease in cold-exposed temperature exists in animals, the decline appears to be more pronounced in males than in females (**Figure 4**). An inability to maintain core temperature during senescence does not correlate to a decline in the mean cold exposure oxygen consumption, a finding that indicates appropriate shivering response. Small and nonsignificant declines in muscle mass among 27-month-old versus 12- and 5-month-old rats are consistent with adequate shivering during senescence. Therefore, if senescence does cause altered cold-exposed core temperature in rodents, then the changes must be due to increased heat loss – that is, altered vasoconstriction, or alterations to brown adipose tissue (BAT) nonshivering thermogenesis (NST).

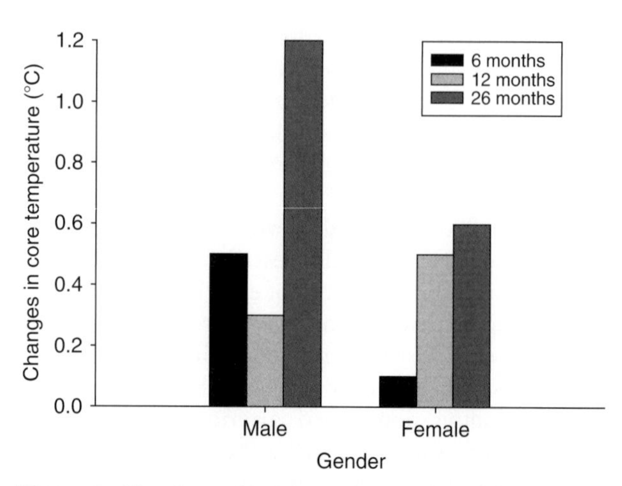


**Figure 4** The change in core temperature in male and female Fischer 344 rats after 4h of exposure to 6 °C. Resting core temperature did not differ between 6-, 12-, and 26-month-old groups. Data adapted from Gabaldón AM, Gavel DA, Hamilton JS, et al. (2003) Norepinephrine release in brown adipose tissue remains robust in cold-exposed senescent Fischer 344 rats. *American Journal of Physiology – Regulatory, Integrative and Comparative Physiology* 285(1): R91–R98, with permission.

### Autonomic Response to Cold Exposure

**Nonshivering thermogenesis: A mechanism of heat production in rodents** Brown adipose tissue is the principal anatomic location of NST. The contribution of NST to overall heat production during cold exposure in the rat has been estimated to be as high as 30–40% of the total body oxygen consumption. Nonshivering thermogenesis in BAT is initiated by release of norepinephrine from terminal endings of sympathetic neurons (**Figure 5**). Norepinephrine binds primarily to $\beta_3$-adrenergic receptor, and the transduction of this signal is mediated through a second-messenger pathway involving adenylyl cyclase. The oxidation of fatty acids, the primary fuel substrate for BAT, is uncoupled from the electron transfer system by a novel protein, uncoupling protein 1 (UCP1), so that heat, rather than ATP, is the metabolic end product.

**Senescence and nonshivering thermogenesis** BAT mass declines significantly with age. However, the decrease in BAT mass with chronological age most likely reflects long-term disuse of this tissue, because animals are typically housed at thermoneutral temperatures. This suggestion is consistent with the observation that cold exposure of old rats at 10 °C for five consecutive days significantly increases brown fat mass and BAT cell size.

Brown fat thermogenesis potential as measured by UCP1 concentration does not change with age. Therefore, any senescence-related attenuation of

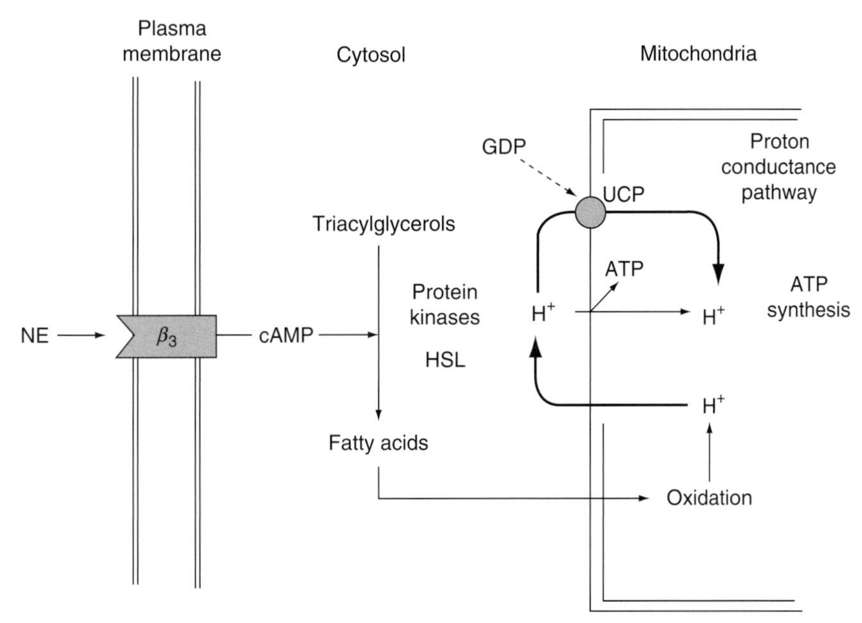


**Figure 5** Schematic description of the influence of norepinephrine (NE) on the uncoupling of oxidative phosphorylation from electron transport that leads to nonshivering thermogenesis. $\beta_3$, $\beta_3$-adrenergic receptor; HSL, hormone-sensitive lipase; GDP, guanosine diphosphate; UCP, uncoupling protein.

NST more likely reflects alteration in sympathetic activity. This suggestion seems unlikely, however, as senescence has been associated with an increase rather than a decrease in sympathetic activity. Elevated plasma norepinephrine levels have been observed in several studies, and neural recordings of sympathetic signaling to brown fat are greater in 30- versus 5-month-old mice exposed to several hours of cold. Finally, brown fat tissue norepinephrine release measured during cold exposure remains robust in old rats (**Figure 6**). The preponderance of evidence indicates that neural signaling to aging brown fat is intact in the senescent animals.

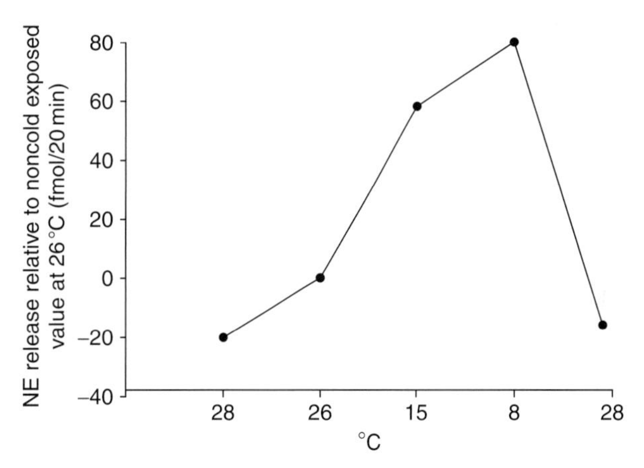


**Figure 6** Net brown adipose tissue norepinephrine (NE) release in male 26-month-old rats during cold exposure. Values are expressed relative to the non-cold-exposed period of 26 °C. Data adapted from Gabaldón AM, Gavel DA, Hamilton JS, et al. (2003) Norepinephrine release in brown adipose tissue remains robust in cold-exposed senescent Fischer 344 rats. *American Journal of Physiology – Regulatory, Integrative and Comparative Physiology* 285(1): R91–R98, with permission.

**Vascular response to cold exposure in rats** Rodents rely on sympathetic-mediated changes in vessel diameter to regulate blood flow during exposure to cold. However, vasoconstriction is limited to the exposed skin of the tail. Vasodilation occurs in organs (such as skeletal muscle and BAT) that rely on an increase in local blood flow to enhance tissue oxygen demands of cold-exposed energy production. Indirect measurement of blood flow using radiolabeled microspheres has demonstrated that older rats are capable of increased vasodilation during cold exposure to the same degree as younger animals (**Table 2**). Vasoconstriction, as measured by heat loss index of the tail, also declines to the same degree in young and old rats. Vasoconstriction and vasodilation during cold exposure appear to respond appropriately in aged animals.

## Summary of Autonomic Reponses to Cold in Rodents

The decline in mean core temperature in old versus young rats is modest, is greater in males than in females, and supports the results observed in humans. Vasoconstriction and vasodilation respond appropriately during cold exposure, and brown fat thermogenesis does not appear to be altered. These results

**Table 2** Tissue masses and resting (23 °C) and cold-exposed (6 °C) regional blood flow of brown adipose tissue and soleus muscles in rats

| *Rat tissue/age* | *Mass (g)* | *Blood flow (ml min$^{-1}$ g$^{-2}$)* | | |
|---|---|---|---|---|
| | | *Resting* | *Cold (2 h)* | *Cold (4 h)* |
| Brown adipose tissue | | | | |
| 12 months old | 0.44 ± 0.03 | 0.74 ± 0.19 | 5.41 ± 0.41 | 2.67 ± 0.24 |
| 26 months old | 0.44 ± 0.04 | 0.63 ± 0.21 | 5.52 ± 0.64 | 4.46 ± 0.59 |
| Soleus muscle | | | | |
| 12 months old | 0.12 ± 0.01 | 0.63 ± 0.07 | 1.25 ± 0.12 | 1.09 ± 0.22 |
| 24 months old | 0.13 ± 0.01 | 0.72 ± 0.18 | 1.09 ± 0.11 | 1.21 ± 0.21 |

Data adapted from McDonald RB, Hamilton JS, Stern JS, et al. (1989) Regional blood flow of exercise-trained younger and older cold-exposed rats. *American Journal of Physiology – Regulatory, Integrative and Comparative Physiology* 256(5): R1069–R1075, with permission.

suggest that autonomic control of thermoregulation in aged rats remains intact. It must be kept in mind that conclusions suggesting little or no change in thermoregulation of the aged rodent are based on the group means. Significant declines in core temperature and alteration in autonomic control of cold-induced thermoregulation can be identified in individual animals.

## Response to Thermostress and the Rate of Aging

The evidence from studies in humans and rats suggests that senescence-related alterations in autonomic control of thermoregulation may be limited to individuals nearing the end of their natural life span. Moreover, investigations evaluating possible senescent-related alterations to heat and cold exposure in both humans and rodents have shown considerable inconsistency of results and high levels of variation about the mean for the outcome variable in old versus young groups. Investigations in both humans and animals report that core temperatures for some individuals do not change, whereas core temperatures for other individuals of identical age decrease (cold) or increase (hot) significantly. Together, these finding would suggest that distinct cohorts exhibiting different senescent-related outcomes may exist within the population of individuals considered old.

### Cold-Exposed Core Temperature and Weight Loss

Our laboratory began investigating the possibility of multiple cohorts within the aged population when, by serendipity, we noticed that the old animals exhibiting decreasing core temperature during cold exposure were also the animals with the lowest body weights. Retrospective analysis in our longitudinal studies revealed that a drop in core temperature

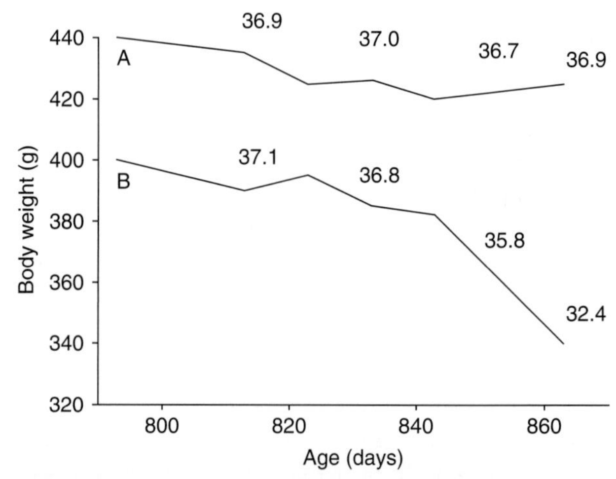


**Figure 7** Relationship between body weight and intraperitoneal temperatures during cold exposure in two aged-matched male Fischer 344 rats. Line A is a weight-stable animal that lived to an age of 911 days. The rat represented by line B died at 870 days. Values above lines are the 4 h intraperitoneal temperatures (°C) of rats exposed to 6 °C. Data adapted from McDonald RB, Florez-Duquet M, Murtagh-Mark CA, et al. (1996) Relationship between cold-induced thermoregulation and spontaneous rapid body weight loss of aging F344 rats. *American Journal of Physiology – Regulatory, Integrative and Comparative Physiology* 271(5): R1115–R1122, with permission.

corresponded to weight loss and that the rats with altered core temperature died within days or a few weeks after the onset of weight loss. This suggested that alterations in autonomic regulation of body temperature decline only near the end of the life span.

We formally tested this hypothesis by measuring body weight daily and performing cold-exposure trials (6 °C for 4 h) on an every-other-week basis starting at 20 months of age (mean life span of Fischer 344 rats is 24–25 months). We found that cold-exposed core temperature in all old animals declined only after the onset of body weight loss and declining food intake (**Figure 7**). No decline in core temperature in senescent versus presenescent rats occurred during the weight-stable phase. The decline in core temperature during the weight-loss phase was rapid, and the several animals became hypothermic, with body temperature below 33 °C, within 2 h of cold exposure. Most important, however, was our finding that the start of the terminal phase and decline of core temperature occurred in all animals tested and at various ages between 22 and 32 months. That is, loss in thermoregulatory capacity was not related to chronological age.

**Hypothalamic function, thermoregulation, and the terminal phase** The rapid and spontaneous loss in weight and food intake decline, combined with the dramatic alterations in cold-exposed thermoregulation, suggested a disruption in a central-mediated mechanism. Experiments were initiated to test the functionality and responsiveness of the hypothalamus, the primary center of thermoregulation. As a first step in evaluating hypothalamic function during the terminal phase, endogenous circadian rhythm of body temperature was followed longitudinally starting at 20 months of age. Continuous measurement of core temperature under constant darkness was performed by radiotransmitters implanted in the visceral cavity. Disruption of endogenous phase, rhythm amplitude, and period coincided with the start of body weight and food intake decline and occurred in all animals (**Figure 8**). Moreover, 24 h mean body temperature, a highly stable variable throughout the majority of the life span, abruptly declined at the start of weight loss and continued until death. These results indicated strongly that neural centers within the hypothalamus were disrupted during the terminal phase and that it was only during the rapid and spontaneous loss in weight that differences in autonomic function in thermoregulation were noted.

The rapid decline in core temperature leading to hypothermia in the senescent rats precludes experiments using the cold-exposure paradigm. We confirmed, however, that loss in mean body temperature coincided precisely with diminished food intake, a physiological system also highly regulated by the hypothalamus. Thus, we are able to use the decline in food intake as a surrogate model for temperature regulation and track possible alteration in hypothalamic function. To this end, several experiments have been completed evaluating the role of NPY, a potent stimulator of food intake, in hypothalamic regulation.

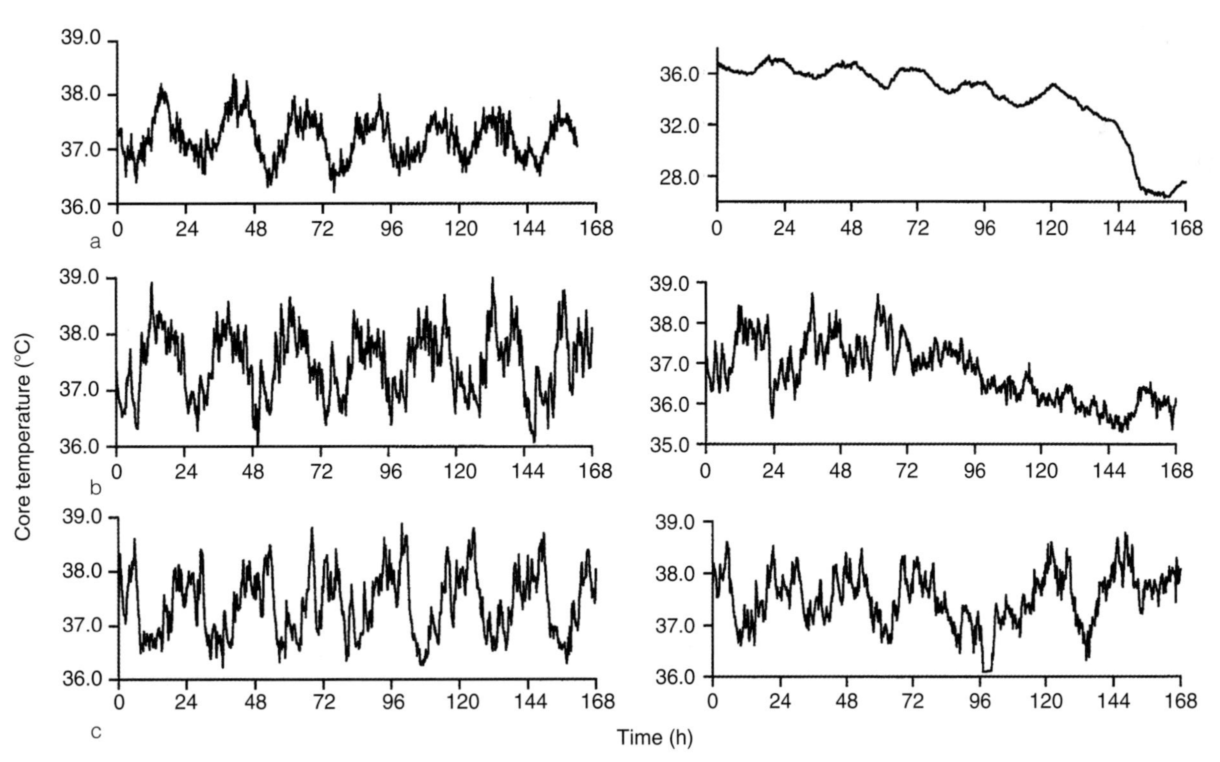


**Figure 8** Seven days of core temperature data from three representative old male F344 rats (a–c) during analysis periods of presenescence (left) and senescence (right). Note that the temperature scale in the senescence period for rat a is different from that for rats b and c. Reproduced from McDonald RB, Hoban-Higgins TM, Ruhe RC, et al. (1999) Alterations in endogenous circadian rhythm of core temperature in senescent Fischer 344 rats. *American Journal of Physiology – Regulatory, Integrative and Comparative Physiology* 276(3): R824–R830, used with permission from the American Physiological Society.

The infusion of NPY directly into the paraventricular nucleus of the hypothalamus does not stimulate food intake in the senescent rat to the same degree as that observed during the weight-stable period. The lack of an effect of NPY on stimulating food intake in the senescent rat does not reflect declining receptor number, fewer NPY-sensitive neural cells, or alterations to pathways of peptides that are co-released with NPY – for example, $\gamma$-aminobutyric acid (GABA).

We have recently completed preliminary work describing gene expression within the hypothalamus between weight-stable and non-weight-stable old rats using microarrays. We find that differences in DNA expression are limited to only a few genes (3% of those tested, a finding constant with other microarray analyses in aging research) and that the genes with the greatest difference in expression were limited to specific types of cell functions. The majority of expression difference occurred in genes coding for ion channel proteins. A potassium channel had the highest degree of difference. The other genes were involved in signal transduction and nerve growth and maintenance factors. We did not find significant difference in any genes that coded for specific neural transmitters involved in thermoregulation or food intake. Our results indicate that the loss in function in the hypothalamus that may impact thermoregulation occurs at the level of cell processes common to all physiological systems and is not limited to any particular system.

## Conclusion

The data from both humans and laboratory animals are consistent in that autonomic control of thermoregulation is well maintained through out most of the life span, including during a significant portion of old age. While there appear to be some alterations to autonomic mechanisms that can be directly related to biological aging, these changes tend to be modest and do not significantly affect core temperature. The current data suggest that many of the changes occurring to sympathetic-mediated vasoconstriction, vasodilation, and NST during old age can be prevented or reversed through acclimation and changes in life style.

Nonetheless, there does appear to be a period near the end of the life span in which alterations to autonomic regulation of body temperature declines. Evaluations of the vasoconstriction and vasodilation mechanisms in humans have found that heat conservation and convective heat loss do result in changes to core temperature in the oldest old. These observations in humans have been confirmed in rodents, in which losses in several autonomic mechanisms of thermoregulation coincide with a terminal phase of life characterized by declining body weight, food intake, and mean body temperature.

*See also:* Neuroimmune System: Aging; Cognitive Aging in Nonhuman Primates.

## Further Reading

Coppola JD, Horwitz BA, Hamilton JS, et al. (2005) Reduced feeding response to muscimol and neuropeptide Y in senescent F344 rats. *American Journal of Physiology – Regulatory, Integrative and Comparative Physiology* 288(6): R1492–R1498.

Florez-Duquet M and McDonald RB (1998) Cold-induced thermoregulation and biological aging. *Physiological Reviews* 78: 339–358.

Gabaldón AM, Gavel DA, Hamilton JS, et al. (2003) Norepinephrine release in brown adipose tissue remains robust in cold-exposed senescent Fischer 344 rats. *American Journal of Physiology – Regulatory, Integrative and Comparative Physiology* 285(1): R91–R98.

Horwitz BA, Gabaldón AM, and McDonald RB (2001) Thermoregulation during aging. In: Hoff PR and Mobbs CV (eds.) *Functional Neurobiology of Aging*, pp. 839–854. San Diego: Academic Press.

Inoue Y, Shibasaki M, Ueda H, et al. (1999) Mechanisms underlying the age-related decrement in the human sweating response. *European Journal of Applied Physiology* 79: 121–126.

Inoue Y, Kuwahara T, and Araki T (2005) Maturation- and aging-related changes in heat loss effector function. *Journal of Physiological Anthropology and Applied Human Science* 23: 289–294.

Kenney WL and Munce TA (2003) Invite review: Aging and human temperature regulation. *Journal of Applied Physiology* 95: 2598–2603.

Masoro EJ (1995) Aging: Current concepts. In: Masoro EJ (ed.) *Handbook of Physiology: Aging*, pp. 3–21. New York: Oxford University Press.

McDonald RB, Florez-Duquet M, Murtagh-Mark CA, et al. (1996) Relationship between cold-induced thermoregulation and spontaneous rapid body weight loss of aging F344 rats. *American Journal of Physiology – Regulatory, Integrative and Comparative Physiology* 271(5): R1115–R1122.

McDonald RB, Hoban-Higgins TM, Ruhe RC, et al. (1999) Alterations in endogenous circadian rhythm of core temperature in senescent Fischer 344 rats. *American Journal of Physiology – Regulatory, Integrative and Comparative Physiology* 276(3): R824–R830.

Pierzga JM, Frymoyer A, and Kenney WL (2003) Delayed distribution of active vasodilation and altered vascular conductance in aged skin. *Journal of Applied Physiology* 94: 1045–1053.

Thompson CS, Holowatz LA, and Kenney WL (2005) Cutaneous vasoconstrictor responses to norepinephrine are attenuated in older humans. *American Journal of Physiology – Regulatory, Integrative and Comparative Physiology* 288(5): R1108–R1113.

# Sleep and Circadian Rhythm Disorders in Human Aging and Dementia

**M V Vitiello**, University of Washington, Seattle, WA, USA



## General Introduction

In 2007, approximately 12% of the United States population consisted of people 65 years of age or older. By 2030, this proportion will rise to 20%. Because the older portion of the population is increasing twice as fast as other age groups, the number of people 65 year of age or older in the United States will effectively double, to 72 million, by 2030.

Alzheimer's disease (AD) is a progressive neurodegenerative disorder, common in older adults, that accounts for approximately two-thirds of all dementias worldwide. In the United States, it has recently been estimated that between 2 and 4 million older adults have AD. Further estimates suggest that this number is likely to quadruple over the next 50 years.

Among older adults, one of the major changes that commonly accompany the aging process is an often profound disruption of an individual's daily sleep–wake cycle. This disruption is often further exacerbated in older individuals suffering from neurodegenerative disorders, such as AD. Here we review what is known about sleep and circadian rhythms in normal aging and in AD.

## Normal Aging

As many as 50% of older individuals complain about sleep problems, including disturbed or light sleep, frequent awakenings, early morning awakenings, and undesired daytime sleepiness. Such disturbances are often associated with impaired daytime function and compromised quality of life. The most striking change in the sleep in older adults is the frequent interruption of nighttime sleep by periods of wakefulness, possibly the result of an age-dependent intrinsic lightening of sleep homeostatic processes. Further, older adults are more easily aroused from nighttime sleep by auditory stimuli, suggesting that their sleep may be more easily disrupted by exogenous stimuli. Both of these changes are indicative of impaired sleep maintenance and depth and contribute to the characterization of the sleep of older adults as lighter, or more fragile, than that of younger adults.

These age-associated increases of nighttime wakefulness are mirrored by increases in daytime fatigue, excessive daytime sleepiness, and increased likelihood of napping or falling asleep during the day. Aging is also associated with a tendency to both fall asleep and awaken earlier. That is, a tendency for older individuals to be larks rather than owls. Older individuals also tend to be less tolerant of phase shifts in time of the sleep–wake schedule such as those produced by jet lag and shift work. These changes suggest an age-related breakdown of the normal adult circadian sleep/wake rhythm.

Polysomnographic (PSG) studies of sleep architecture have consistently demonstrated four age-related changes in nighttime sleep quality: reductions in total sleep time, reductions in sleep efficiency, reductions in slow wave sleep, and increases in waking after sleep onset. Nevertheless, age-related changes in numerous other PSG sleep measures remain unclear. Sleep onset latency, stages 1 and 2, and rapid-eye movement (REM) have all been reported to change with age, whereas other studies have observed no such age-related changes.

It is important to note that even carefully screened older adults who do not complain of sleep disturbance and who have minimal medical burdens show the changes described when compared to younger adults. This suggests that at least some of the sleep disturbance seen in older adults is part of the aging process *per se*, apparently independent of any medical or psychiatric illnesses or primary sleep disorders and often referred to as age-related sleep change. As to whether this age-related decline in the ability to generate sleep equates with a decreased need for sleep in the later years of the human life span remains unclear. Nevertheless, the available scientific evidence suggests that it is important to remember that as individuals age, they make need to modify their expectations about the duration and quality of their nighttime sleep.

### Sleep in Normal Aging

Clearly sleep changes with advancing age, but an important question remains: exactly when during the adult life span do these changes occur? It has typically been assumed that the age-related sleep changes that characterize the sleep of older adults begin to appear in early adulthood and progress steadily across the full continuum of adult human life span, including the older adult years, so that, for example, the sleep of the typical 75-year-old was assumed to be worse than that of the typical 65-year-old.

However, the results of a recent and comprehensive meta-analysis of objective sleep measures across the human life span demonstrated that the bulk of the age-related changes seen in adult sleep patterns do not occur continuously across the human adult life span but, rather, occur between early adulthood, beginning at age 19, and that after age 60 changes in sleep macroarchitecture becomes effectively asymptote, declining only minimally from age 60 to age 102. These sleep changes across adult life-span are summarized in **Table 1**. When the full adult life span is examined (2391 adults, ages 19–102 years), such analyses confirmed the four most consistently reported age-related changes in PSG studies of sleep macroarchitecture: decreases in total sleep time (TST), decreases in sleep efficiency (SE), decreases in slow wave sleep (SWS), and increases in waking after sleep onset (WASO). Also confirmed were less consistently reported age-related sleep changes; increases in stage 1 (S1) and stage 2 (S2) sleep and decreases in REM sleep were all observed in their meta-analyses. Conversely, age-related changes in both sleep latency (SLAT) and REM latency (REMLAT) were minimal.

It is crucial to note that all these significant age changes in objectively assessed sleep architecture were found only when the full adult life span of 19–102 years was examined. When the sleep of only older adults (1142 adults, ages 60–102 years) was examined, the meta-analyses demonstrated that no sleep measures showed any significant age-related change within the older adult portion of the study sample, with the single exception of SE, which declined significantly but at a modest rate of ~3% per decade.

**Table 1** Summary of significant findings from meta-analyses examining the association between various sleep measures and age

| | *Adults 19–102 years*[a] | *Older adults 60–102 years*[b] |
|---|---|---|
| Total sleep time | ⇓ | ⇔ |
| Sleep latency | ⇔ | ⇔ |
| WASO | ⇑ | ⇔ |
| Sleep efficiency | ⇓ | ⇓ |
| Stage 1 (%) | ⇑ | ⇔ |
| Stage 2 (%) | ⇑ | ⇔ |
| SWS (%) | ⇓ | ⇔ |
| REM (%) | ⇓ | ⇔ |
| REM latency | ⇔ | ⇔ |

[a]*n* = 2391 adults.
[b]*n* = 1142 adults.
From Ohayon MM, Carskadon M, Guilleminault C, and Vitiello MV (2004) Meta-analysis of quantitative sleep parameters from childhood to old age in healthy individuals: Developing normative sleep values across the human lifespan. *Sleep* 27(7): 1255–1273.
⇓, decreased; ⇑, increased; ⇔, unchanged; REM, rapid-eye movement sleep; SWS, slow wave sleep; WASO, waking after sleep onset.

These findings were comparable for both older men and older women.

These findings seem counterintuitive and fly against our commonly held concepts of sleep changes with aging. However, it is important to remember that Ohayon and colleagues used very rigorous selection criteria, such that the subjects involved in the studies analyzed were not representative of the older population *per se* but, rather, were in excellent health and more likely represented individuals who were optimally or successfully aging. That is, these findings represent normative age-related changes in sleep architecture and not average sleep changes of the entire aging population. When various medical and psychiatric comorbidities that typically accompany the aging process were controlled for and optimal aging was examined, the bulk of age-related sleep changes occurred in early and middle adulthood (ages 19–60 years); after age 60, assuming the individual remained in good health, further age-related sleep changes were likely to be, at most, modest. Conversely, if such comorbidities were present, these minimal age-related sleep changes might well be exacerbated, consistent with the level of sleep-related complaints seen in the general aging population.

## Circadian Rhythms in Normal Aging

Not only does the quality of sleep change across the human life span, but the timing of when that sleep occurs also apparently changes with aging. Circadian rhythms are those that occur within a period of 24 h (from the Greek about (*circa*) a day (*dies*)), such as the adult human sleep–wake cycle. Interestingly, as with sleep, there is a considerable disparity in the conventional wisdom concerning circadian rhythms and aging and which evidence is supports or does not support these conventionally held beliefs. The conventional wisdom regarding what happens to human circadian processes with aging holds that (1) the circadian amplitude is reduced; (2) the circadian phase is advanced (i.e., the circadian rhythm moves earlier relative to the environment); (3) the circadian free-running period (tau) is shortened; and (4) the ability to tolerate rapid phase shifts, such as those experienced in shift work or rapid transmeridian travel (jet lag), declines.

However, the available evidence convincingly supports only two of these beliefs: that older people tend to have earlier circadian phases, with a corresponding tendency to go to bed and arise from bed earlier than younger adults; and that they have more trouble than younger adults adjusting to the rapid phase shifts due to shift work and jet lag, at least in terms of sleep quality, subjective complaint, and performance measures. Interestingly, the data in support of diminished circadian amplitudes and shortened circadian taus in healthy older adults are, at best, equivocal. Again, it is important to note that these circadian changes are those seen in normal, or more accurately normative, aging, in which age-related health comorbidities have been carefully screened out.

### Excessive Daytime Sleepiness and Napping in Normal Aging

Two other commonly held assumptions about sleep and aging are that older adults typically report more excessive daytime sleepiness (EDS) and nap more than younger adults. Few community-based epidemiological studies have reported the prevalence of regular napping and its association with sleep complaints and other mental and physical health problems, especially in relation EDS. Somewhat counterintuitively, although epidemiological studies typically show a significant increase in the prevalence of regular napping with advancing age, EDS does not demonstrate a similar increase in prevalence among older adults. Consistent with the findings already described for nighttime sleep measures, very recent findings indicate that the presence of comorbidities (medical illness, depression, etc.) is highly associated with the likelihood of an older adult reporting regular napping or EDS. That is, healthy older adults, even those complaining of significant nighttime sleep disturbance, are much less likely to report regular napping or EDS than their more health-burdened cohorts.

There is considerable debate as to whether regular napping among older adults, particularly those in good health, may be beneficial to daytime wakefulness or perhaps detrimental to their nighttime sleep propensity. Studies of naps in healthy noncomplaining older adults demonstrated that napping had only a mild to moderate impact on nighttime sleep quality and may even result in improved postnap cognitive performance. However, these results should be interpreted with caution because, again, it is important to emphasize that the subjects involved were carefully screened healthy older adults. It is unclear if similar results would be obtained, for example, with a sample of older insomniacs.

### Causes of Sleep and Circadian Disturbances in Older Adults

It is often repeated that epidemiological studies report that as much as 50% of older adults complain of significant, chronic sleep disturbance. However, it is important to keep in mind that the other 50% of older adults do not so complain. As previously reviewed, the sleep of very healthy older adults changes only very slowly across the later human life span. Conversely it has also been shown that older adults who do not complain of any sleep problems nevertheless have objective sleep quality that is markedly compromised (e.g., less TST, less SE, less SWS, and more WASO) compared to that of healthy, non-sleep-complaining, younger adults. Nevertheless, there are many factors in addition to the age-related sleep and circadian changes already discussed that contribute to the significant sleep disturbance reported by half of all older adults. These include (1) medical and psychiatric comorbidities and their treatments, such as cardiovascular disease, arthritis, depression, or AD and many of the drugs used to treat them; (2) primary sleep disorders, many of which, such as sleep apnea, restless legs syndrome, and REM behavior disorder, tend to occur with increasing frequency in older adults; (3) the many behavioral, environmental, and social factors, often collectively referred to as sleep hygiene, that have the potential to maximize or compromise an individual's sleep quality; and (4) some combination of these factors. A more comprehensive list of such contributing factors with specific examples appears in **Table 2**.

**Table 2** Causes of poor sleep in older adults

| *Cause/problem* | *Examples* |
|---|---|
| Physiological | Age-related sleep change |
| | Age-related circadian rhythm change (phase advance) |
| Medical illnesses | Arthritis or other conditions causing chronic or intermittent pain |
| | Chronic cardiac or pulmonary disease |
| | Gastro-esophageal reflux disorder |
| Psychiatric illnesses | Depression |
| | Alzheimer's disease |
| Medications | Diuretics (leading to nocturnal awakenings) |
| | Inappropriate use of OTC medications |
| Primary sleep disorders | Sleep disordered breathing (sleep apnea) |
| | Restless legs syndrome |
| | REM behavior disorder |
| Behavioral/social | Retirement/lifestyle change reducing need for regular bed and rise times |
| | Death of a family member or friend |
| | Inappropriate use of social drugs |
| | Transmeridian air travel |
| | Napping |
| Environmental | Bedroom environment (e.g., ambient noise, temperature, light, bedding) |
| | Moving to a new home or downsizing to a smaller space or a retirement community or related facility |
| | Institutionalization |

OTC, over-the-counter; REM, rapid-eye movement.
Reproduced from Vitiello MV (2006) Sleep in normal aging. *Sleep Medicine Clinics* 1(2): 171–176, with permission from Elsevier.

## Alzheimer's Dementia

Both normal aging and AD are associated with disturbances in the daily sleep–wake cycle, although the disturbances in AD are typically much more severe. Within the AD population itself, significant sleep disturbances are quite common, affecting as much as half of community- and clinic-based samples.

For AD patients, sleep disturbance adds an additional burden to the compromised function and quality of life directly attributable to dementia. For AD caregivers, disturbances in patients' sleep and nighttime behavior, particularly the reduced nighttime sleep time, increased nighttime wakefulness, and wandering that commonly require considerable caregiver attention, with a subsequent chronic sleep loss for the caregivers themselves, are a significant source of physical and psychological burden and are often cited as one of the principal reasons for a family's decision to institutionalize a demented person.

### The Biological Bases of Sleep Disturbances in AD

The sleep disturbances that accompany early-stage or mild dementia are remarkable in that they appear as exacerbations of the sleep changes found in normal aging rather than as unique disease-related phenomena. The sleep of AD patients is marked by an increased duration and frequency of awakenings, decreased SWS and REM sleep, and more daytime napping. Damage to neuronal pathways that initiate and maintain sleep is the most likely cause of the acceleration of these age-related sleep changes in AD patients. These compromised neural structures, including the suprachiasmatic nucleus (SCN) of the hypothalamus and neuronal pathways originating in subcortical regions, also regulate arousal and the sleep–wake cycle. Recent work suggested that at least some of the sleep disruption seen in AD many be an inherent trait, probably linked to apolipoprotein E (APO E) status, such that AD patients positive for the APO E $\varepsilon4$ allele show greater sleep disruption over time.

There is considerable evidence that sleep disturbance grows more severe with increasing severity of AD. However, the moderate, or intermediate, stages of the disease are typically when most other behavioral disturbances occur with peak frequency. The breakdown of the basic circadian sleep–wake rhythm of dementia patients can be severe and in extreme cases may lead to an extremely fragmented sleep–wake pattern or nearly complete day–night sleep pattern reversals. In end-stage AD, patients may appear to sleep throughout most of the day and night, awakening only for brief periods. However, as of this time there have been no prospective longitudinal studies of sleep in AD that would inform an accurate understanding of both the course and the individual differences of these sleep disturbances. This remains an important gap in both the biology and therapeutics of sleep disturbances in demented patients.

The current standard of care for treating cognitive disorders in AD is the use of oral acetylcholinesterase inhibitors. These drugs may improve sleep patterns in some AD patients, particularly by increasing REM sleep. Because they act by enhancing cholinergic transmission in the brain, the involvement of both forebrain and brain stem cholinergic nuclei in regulating sleep–wake cycles and arousal forms the rationale for expecting both positive and negative sleep effects with use of these agents.

### Circadian Rhythm Disorder in AD

AD patients commonly manifest a breakdown of their circadian sleep–wake rhythm. Termed irregular sleep–wake rhythm disorder (ISWRD), this perturbation of circadian rhythmicity is commonly associated with neurological impairment in older adults, such as the deficits experienced by AD patients. ISWRD is characterized by the relative absence of a circadian pattern in the sleep–wake cycle. Total sleep time can well be comparatively normal, but, instead of being consolidated into a single long relatively continuous major sleep period, sleep times are shortened and, in extreme cases, relatively brief sleep periods are distributed almost randomly throughout the day and night. The cause, or more likely the causes, of this association are unknown, but AD-related damage to the SCN, locus of the circadian pacemaker, is clearly implicated as an important, if not a major, etiological factor. Experimental ablation of the SCN in animals produces a loss of circadian rhythmicity that strongly resembles the sleep–wake pattern typically seen in older adults with dementing disorders, particularly in the later stages of the dementia. However, it is important to note that clinical studies, the bulk of which have been carried out in older adults with dementia (particularly AD), have rarely used formal sleep diagnostic criteria for ISWRD.

Although complaints of nighttime sleep fragmentation and daytime napping have been consistently reported to increase with age, the weight of the available evidence strongly indicates that, although the prevalence of ISWRD increases with age, this increase is secondary to the increased prevalence of associated medical disorders with advancing age. Age is apparently not an independent risk factor for ISWRD; rather, it is probably the age-related medical comorbidities which constitute the main risk factors for sleep pathology, including both increasing disturbances in sleep architecture and the increase in prevalence of ISWR with advancing age. ISWRD is particularly associated with neurological impairment in older adults, most significantly AD.

Although there is no direct evidence for a genetic basis for ISWRD, there are several lines of evidence that suggest that the sleep disturbance seen in AD is at least partially based on genetic factors. Actigraphic studies of AD patients have demonstrated longitudinal deterioration of sleep quality, and most of this longitudinal variance in sleep appears to be related to an inherent trait of the individual patient. This suggests that genetic factors may help determine the ultimate course and level of sleep deterioration seen in a given AD patient, a hypothesis consistent with research indicating that much circadian variation in physiological systems is controlled by a limited number of similar genes across species. Regardless, the tentative relationship of genetic factors as a risk for ISWRD remains to be fully explored.

Older adults are exposed to reduced light levels in their daily lives relative to younger individuals. This reduction may be exacerbated by disorders of vision, which are common in older adults and which many further attenuate the impact of ambient light on the SCN. Finally, the SCN itself is adversely impacted by age. The impact of each of these factors is magnified in patients with AD, including those who are community dwelling, who have been shown to be exposed to less light than age-matched healthy normal controls. Furthermore, the retina and the optic nerve can both be compromised in AD. Older adults, especially those who are institutionalized, are exposed to less intense light than younger people, but questions remains as to the importance of reduced light exposure as an etiological factor in ISWRD.

### Other Causes of Sleep and Circadian Disturbances in AD

Population-based studies examining the causes, incidence, and persistence of sleep disturbances in AD patients are lacking; consequently, little is known about the risk factors for their development. Therefore, we must look to the literature concerning sleep disturbance in the nondemented elderly and factor this with clinical assessments of individual patients to make informed inferences about the AD population. Certainly the same neurodegenerative mechanisms that result in the evolving cognitive deficits seen in AD are a potential primary cause of the sleep disruption seen in this population. However, it is important to recognize that sleep can be disrupted for many other reasons as well. The likely major causes of sleep disruption in AD, as with normal aging, include (1) physiological changes that arise as part of normal nonpathological aging; (2) sleep problems due to comorbid physical or mental health conditions, including AD, and possibly their treatments; (3) primary sleep disorders; (4) poor sleep hygiene or sleep-related habits; or (5) commonly, some combination of these factors.

Studies of patients living in institutional settings provide much of the available information about sleep and circadian disturbance in cognitively impaired patients. Environmental factors that promote circadian dysregulation in dementia patients living in such settings include light, noise, activity schedules, and the needs of staff. Ambient light levels are typically too low in many congregate care facilities to support natural light-dependent internal rhythms, and noisy conditions, especially during the night, are both common and inimical to sleep. The staffing schedules and timing of specific activities in many facilities that care for demented people may be driven less by the needs of the patients than by compliance with federal and state requirements governing nursing-home operations. Regulatory requirements in general fail to incorporate many of the positive evidence-based practices found to be beneficial for demented residents, focusing more attention on feeding and bathing schedules, injury prevention, and detection of medical problems than on sleep and other issues related to quality of life.

## Conclusion

Previously it was assumed that sleep and circadian rhythms decline inexorably with advancing age. However, recent research shows that, at least for health older adults, most of the age changes in sleep that have typically been thought to progress steadily across the adult human life span actually occur before age 60 and that adults over age 60 who remain healthy can expect their sleep quality to remain relatively stable as they age. Similarly age-related changes in circadian changes, again in older adults free of comorbid illnesses, are not as marked as was previously assumed. However, even healthy older adults tend to have earlier circadian phases, with a corresponding tendency to go to bed and arise from bed earlier than younger adults and to experience more trouble than younger adults adjusting to the rapid phase shifts such as those experienced in shift work and jet lag, at least in terms of sleep quality, subjective complaint, and performance measures.

Although sleep quality and circadian rhythms are apparently relatively well preserved in healthy older adults, the negative impact of age-related comorbid illnesses on sleep and circadian rhythms is also clear. AD is a classic example of this negative impact, with AD patients typically demonstrating progressively fragmented sleep and irregular circadian rhythms with increasing severity of the disease.

Nevertheless, there is a growing body of evidence that the sleep quality and circadian rhythmicity of AD patients can, like those of other less-than-healthy older adults, be improved through nonpharmacological treatment approaches, whether the AD patients are community dwelling or institutionalized. Such groundbreaking work offers a firm foundation for future efforts to improve the sleep and related circadian rhythms of older adults, regardless of their health status.

*See also:* Aging of the Brain and Alzheimer's Disease; Alzheimer's Disease: An Overview; Alzheimer's Disease: Molecular Genetics; Alzheimer's Disease: Transgenic Mouse Models; Animal Models of Alzheimer's Disease; Brain Glucose Metabolism: Age, Alzheimer's Disease and ApoE Allele Effects.

## Further Reading

Alessi CA, Martin JL, Webber AP, Cynthia Kim E, Harker JO, and Josephson KR (2005) Randomized, controlled trial of a nonpharmacological intervention to improve abnormal sleep/wake patterns in nursing home residents. *Journal of the American Geriatrics Society* 53: 803–810.

Alessi CA and Schnelle JF (2000) Approches to sleep disorders in the nursing home setting. *Sleep Medicine Reviews* 4(1): 45–56.

Ancoli-Israel S, Gehrman P, Martin JL, et al. (2003) Increased light exposure consolidates sleep and strengthens circadian rhythms in severe Alzheimer's disease patients. *Behavioral Sleep Medicine* 1: 22.

Bonnefond A, Harma M, Hakola T, Sallinen M, Kandolin I, and Virkkala J (2006) Interaction of age with shift-related sleep-wakefulness, sleepiness, performance, and social life. *Experimental Aging Research* 32(2): 185–208.

Campbell SS, Murphy PJ, and Stauble TN (2005) Effects of a nap on nighttime sleep and waking function in older subjects. *Journal of the American Geriatrics Society* 53: 48–53.

Cajochen C, Munch M, Knoblauch V, Blatter K, and Wirz-Justice A (2006) Age-related changes in the circadian and homeostatic regulation of human sleep. *Chronobiology International* 23(1–2): 461–474.

Dijk DJ, Duffy JF, and Czeisler CA (2001) Age-related increase in awakenings: Impaired consolidation of nonREM sleep at all circadian phases. *Sleep* 24(5): 565–577.

Dijk DJ, Duffy JF, Riel E, Shanahan TL, and Czeisler CA (1999) Ageing and the circadian and homeostatic regulation of human sleep during forced desynchrony of rest, melatonin and temperature rhythms. *Journal of Physiology* 516: 611–627.

Duffy JF, Zeitzer JM, Rimmer DW, Klerman EB, Dijk DJ, and Czeisler CA (2002) Peak of circadian melatonin rhythm occurs later within the sleep of older subjects. *American Journal of Physiology, Endocrinology and Metabolism* 282(2): E297–E303.

Foley DJ, Ancoli-Israel S, Britz P, and Walsh J (2004) Sleep disturbances and chronic disease in older adults: Results of the 2003 National Sleep Foundation Sleep in America Survey. *Journal of Psychosomatic Research* 56: 497–502.

Foley DJ, Monjan AA, Brown SL, Simonsick EM, Wallace RB, and Blazer DG (1995) Sleep complaints among older persons: An epidemiological study of three communities. *Sleep* 18: 425–432.

Foley DJ, Monjan AA, Simonsick EM, Wallace RB, and Blazer DG (1999) Incidence and remission of insomnia among elderly adults: An epidemiologic study of 6,800 persons over three years. *Sleep* 22(supplement 2): S366–S372.

Foley DJ, Vitiello MV, Bliwise DL, Ancoli-Israel S, Monjan AA, and Walsh JK (2007) Frequent napping is associated with excessive daytime sleepiness, depression, pain and nocturia in older adults: Findings from the national sleep foundation "2003 sleep in America" poll. *American Journal of Geriatric Psychiatry* 15: 344–350.

Hoffman MA and Swaab DF (2006) Living by the clock: The circadian pacemaker in older people. *Ageing Research Review* 5(1): 33–51.

Martin JL, Webber AP, Alam T, Harker JO, Josephson KR, and Alessi CA (2006) Daytime sleeping, sleep disturbance, and circadian rhythms in the nursing home. *American Journal of Geriatric Psychiatry* 4: 121–129.

McCurry SM, Gibbons LE, Logsdon RG, Vitiello MV, and Teri L (2005) Nighttime insomnia treatment and education for Alzheimer's disease: A randomized, controlled trial. *Journal of the American Geriatrics Society* 53: 793–802.

McCurry SM, Logsdon RG, Teri L, and Vitiello MV (2007) Evidence-based psychological treatments for older adult sleep disorders. *Psychology and Aging* 22(1): 18–27.

Monk TH (2005) Aging human circadian rhythms: Conventional wisdom may not always be right. *Journal of Biological Rhythm* 20(4): 366–374.

Monk TH, Buysse DJ, Carrier J, Billy BD, and Rose LR (2001) Effects of afternoon "siesta" naps on sleep, alertness, performance, and circadian rhythms in the elderly. *Sleep* 24(6): 680–687.

Ohayon MM, Carskadon MA, Guilleminault C, and Vitiello MV (2004) Meta-analysis of quantitative sleep parameters from childhood to old age in healthy individuals: Developing normative sleep values across the human lifespan. *Sleep* 27(7): 1255–1273.

Rybarczyk B, Stepanski E, Fogg L, Lopez M, Barry P, and Davis A (2005) A placebo-controlled test of cognitive behavioral therapy for comorbid insomnia in older adults. *Journal of Consulting Clinical Psychology* 73(6): 1164–1174.

Van Someren EJ (2000) Circadian and sleep disturbances in the elderly. *Experimental Gerontology* 35(9–10): 1229–1237.

Van Someren EJ, Riemersma RF, and Swaab DF (2002) Functional plasticity of the circadian timing system in old age: Light exposure. *Progress in Brain Research* 138: 205–231.

Vitiello MV (2000) Effective treatment of sleep disturbances in older adults. *Clinical Cornerstone* 2(5): 16–27.

Vitiello MV (2006) Sleep in normal aging. Sleep Medicine Clinics 1(2): 171–176.

Vitiello MV and Borson S (2001) Sleep disturbances in patients with Alzheimer's disease: Epidemiology, pathophysiology and management. *CNS Drugs* 15(10): 777–796.

Vitiello MV, Larsen LH, and Moe KE (2004) Age-related sleep change: Gender and estrogen effects on the subjective-objective sleep quality relationships of healthy, non-complaining older men and women. *Journal of Psychosomatic Research* 56(5): 503–510.

Vitiello MV, Moe KE, and Prinz PN (2002) Sleep complaints co-segregate with illness older adults: Clinical research informed by and informing epidemiological studies of sleep. *Journal of Psychosomatic Research* 53: 555–559.

Vitiello MV, Prinz PN, Williams DE, Frommlet MS, and Ries RK (1990) Sleep disturbances in patients with mild-stage Alzheimer's disease. *Journal of Gerontology* 45(4): M131–M138.

Wu YH and Swaab DF (2005) The human pineal gland and melatonin in aging and Alzheimer's disease. *Journal of Pineal Research* 38: 145–152.

Yesavage JA, Friedman L, Kraemer H, et al. (2004) Sleep/wake disruption in Alzheimer's disease: APOE status and longitudinal course. *Journal of Geriatric Psychiatry and Neurology* 17(1): 20–24.

Yesavage JA, Taylor JL, Kraemer H, Noda A, Friedman L, and Tinklenberg JR (2002) Sleep/wake cycle disturbance in Alzheimer's disease: How much is due to an inherent trait? *International Psychogeriatrics* 14(1): 73–81.

# Sleep in Aging

**M Rissling and S Ancoli-Israel**, University of California at San Diego, San Diego, CA, USA



## Background

Sleep is one of many biological processes that invariably undergoes change with age. Older adults may experience an increase in sleep disturbance as the homeostatic and circadian rhythm processes regulating the sleep cycle become more fragile. Sleep is also affected by the many complications associated with physical and mental illness that often accompany aging. It is important to understand the distinction between expected changes in sleep associated with aging and those likely to arise from pathology unrelated to aging *per se*. This article reviews both.

## Changes in Sleep with Age

It is generally well accepted that aging results in an overall decrease in slow-wave sleep (SWS). The amplitude of slow-wave delta waves is attenuated, and the percentage of SWS observed is markedly decreased. This process is gradual and begins in young adulthood, at around 20 years of age, but stabilizes by about age 60, with no further reduction in SWS. There is evidence of a gender difference, with men showing greater decreases in SWS than women. Latency to rapid eye movement (REM) sleep is also decreased in the older adult and is considered to be a direct consequence of the circadian phase advancement that occurs with aging.

The largest overall change observed in older adults is an increase in stage shifts, with more time spent awake in bed following sleep onset. The disruptions are often characterized by early morning awakenings and an inability to fall back to sleep and may be due to a number of sleep disturbances (such as insomnia or other specific sleep disorders) rather than to aging. The result is marked decreases in sleep efficiency (defined by the percentage of sleep that occurs while in bed) and consequently increases in daytime sleepiness. Stage 1 sleep is slightly increased in association with such arousals; stage 2 sleep appears to be more or less constant throughout the life span.

The need for sleep, or the sleep drive, however, does not diminish with age. Rather, it is the ability to satisfy the sleep drive that becomes impaired. Therefore, although the need for sleep remains the same, nighttime sleep is curtailed, with the remaining sleep drive redistributed to the daytime. The decreased ability to maintain nighttime sleep is most often attributed to changes in circadian rhythms, insomnia comorbid with medical or psychiatric illnesses and medication use, and the presence of other specific sleep disorders, such as periodic limb movements in sleep or restless leg syndrome (RLS), REM behavior disorder, and sleep-disordered breathing.

### Changes in Circadian Rhythm

Circadian rhythms, such as hormone secretion, body temperature, and the sleep–wake cycle, fluctuate every 24 h. In humans, the sleep–wake cycle is regulated by the combined effect of an endogenous pacemaker and exogenous stimuli. The hypothalamic suprachiasmatic nucleus (SCN) is considered the endogenous clock of the brain and is believed to be the primary mediator of circadian rhythms. Circadian rhythms maintained by the SCN are naturally entrained by exogenous stimuli, or zeitgebers (time cues). The most significant zeitgeber is light, but other factors, such as physical activity, the timing of meals, and social interactions, have also been shown to be important regulators.

As people age, their circadian rhythms become weaker, desynchronize, and lose amplitude. It is hypothesized that the progressive deterioration of the SCN and its subsequent weakened functioning contribute to the disruption of circadian rhythms in older adults. The most common desynchronized pattern often seen in older adults is an advancement of the sleep phase.

Older adults with advancement of the sleep phase have increased numbers of early morning awakenings and difficulty maintaining sleep. They typically experience an increased drive for sleep between 7 p.m. and 9 p.m. and a drive to wake up between 3 a.m. and 5 a.m. Despite being sleepy earlier in the evening, they will often try to stay up later because of societal pressure, resulting in fewer hours spent in bed sleeping. The decreased total sleep time at night then results in daytime sleepiness and napping. Some older adults take inadvertent evening naps in front of the television or while reading early in the evening and then have difficulty falling asleep when they go to bed later in the evening; yet they still wake up early in the morning. This pattern gets interpreted as a complaint of insomnia, that is, difficulty falling asleep, difficulty staying asleep, and daytime sleepiness.

### Treatment of Circadian Rhythm Disturbances

The most common treatment for circadian rhythm disturbances is bright light therapy. An advanced circadian rhythm can be delayed and the sleep–wake cycle strengthened with evening light exposure. As the sun is the strongest source of bright light, older adults with advanced rhythms should spend more time outdoors during the late afternoon or early evening and avoid bright light in the morning hours (as morning light would advance their rhythms even more). Studies have also shown that exposure to artificial light via a bright light box in the early evening has a similar effect and can improve sleep continuity in both healthy and institutionalized older adults. Adherence to a regular sleep schedule will promote a more robust sleep–wake cycle. Physical activity has also been shown to improve nighttime sleep quality. Such therapies are without adverse effects, and each has been shown to improve mood, performance, and quality of life and to reduce fatigue.

While melatonin has been shown to be safe, the efficacy for circadian rhythm disturbances has not been established. There is little consensus on the recommended dose or timing of administration, and melatonin purity and composition are currently not regulated in the United States by the Food and Drug Administration (FDA).

## Insomnia

Insomnia is defined as difficulty falling asleep or staying asleep or nonrestorative sleep which results in daytime consequences. Insomnia is found in 30–50% of those more than 60 years old. Gender differences exist as well, with older women being more likely to complain of insomnia than older men. In older adults, sleep maintenance insomnia (an inability to maintain sleep throughout the night) and early morning or terminal insomnia (awakening early in the morning, with difficulty returning to sleep) are the most common complaints. Insomnia in the older adult is most often comorbid with medical and psychiatric illness or the medications used to treat those illnesses. Insomnia may also be a complaint in patients with specific sleep disorders, such as RLS or periodic limb movements in sleep.

### Consequences of Insomnia

The consequences of disturbed sleep are well known and may include difficulty sustaining attention, slowed response time, difficulty with memory, and decreased performance. These consequences can be more pronounced in older adults and may even be misinterpreted as symptoms of dementia. Insomnia disrupts the sleep of 64% of bed partners of both healthy and demented older adults and 84% of caregivers (i.e., who may sleep in separate rooms). Nighttime restlessness secondary to insomnia is often cited as the primary reason for institutionalizing a demented relative.

Untreated insomnia is associated with a nearly 500% increase in the general risk for serious accidents and injuries, including car accidents due to daytime sleepiness and hip fractures due to falls. In the older adult, insomnia is most often comorbid with medical and psychiatric disorders and the medications used to treat them.

### Medical and Psychiatric Illness and Insomnia

Older adults are the most susceptible to comorbid insomnia as they are more likely to have multiple medical conditions. Data from the National Sleep Foundation 2003 poll on sleep in older adults showed that those with three or four medical conditions, including cardiac disease, pulmonary disease, stroke, and depression, reported significantly more sleep complaints than did those with none, one, or two conditions.

Medical and psychiatric illnesses most commonly associated with insomnia include pain resulting from osteoarthritis or malignancies, chronic obstructive pulmonary disease, congestive heart failure, nocturia (e.g., due to enlarged prostate) and incontinence, neurological deficits (e.g., stroke, Parkinson's disease, and dementia), gastroesophageal reflux, depression, and anxiety disorder. Difficulty falling asleep has a prevalence rate of 31% for patients with osteoarthritis and 66% for those with chronic pain conditions. In addition, 81% of arthritis patients, 85% of chronic pain patients, and 33% of diabetes patients report difficulty staying asleep.

Two additional conditions that are associated with aging and contribute to poor sleep are nocturia and menopause. Nocturia or frequent voluntary voiding of urine during the night and nocturnal polyuria, the production of an abnormally large volume of urine during sleep, are common causes of awakenings and disruption of nighttime sleep. Although nocturia is often perceived as causing nocturnal awakenings, it has been hypothesized that the fragmented sleep experienced by the older adult leads to more awareness of the need to void. The circadian rhythm of urine excretion is also altered with age such that there is a fall of renal concentrating ability, sodium conservation, and secretion of renin–angiotensin–aldosterone, and arginine vasopressin secretion and an increase of atrial natriuretic hormone. The bladder's ability to store urine decreases with age, and age-related lower urinary tract problems, such as detrusor overactivity, can affect frequency and urgency.

Menopause is associated with hormonal, physiologic, and psychological changes that affect sleep and may play a critical role in modulating both the presence and the degree of sleep disruption. Menopausal vasomotor symptoms or hot flashes are known to lower sleep efficiency and cause multiple arousals during the night. Insomnia is a common symptom of menopause and may be partially related to psychological factors such as depression and anxiety.

In fact, insomnia can be prognostic or indicative of depression. While depressed mood may predict insomnia, unresolved insomnia also places adults at greater risk for developing depression, anxiety, and substance abuse. Studies in younger adults have suggested that treating the insomnia may improve depression.

### Medication Use and Insomnia

Alerting or stimulating drugs, such as central nervous system stimulants, beta-blockers, bronchodilators, calcium channel blockers, corticosteroids, decongestants, stimulating antidepressants, stimulating antihistamines, and thyroid hormones are known to cause difficultly falling asleep when taken late in the day. Sedating drugs, such as sedative antihistamines, antihypertensives, and sedative antidepressants, may cause excessive daytime sleepiness when taken early in the day and contribute to daytime napping behavior and consequently delay nighttime sleep onset. Adjusting the dose or the time of day of administration often improves the insomnia.

### Treatment of Insomnia

Effective treatments for insomnia in older adults include cognitive behavioral therapy (CBT) and pharmacological therapy.

**CBT** CBT is a combination of cognitive restructuring and behavioral modification for insomnia. The cognitive portion deals with misconceptions or unrealistic expectations about sleep (i.e., absolute requirement of 8 h or more of sleep), amplification of the consequences of not getting enough sleep, incorrect thinking about the cause of insomnia, and misunderstandings about healthy sleep practices. The behavioral component of CBT involves a combination of sleep restriction therapy, stimulus control therapy, and good sleep hygiene practices. **Tables 1–3** list the rules for each of these, respectively.

A recent National Institutes of Health (NIH) State-of-the-Science Conference on Insomnia concluded that CBT is as effective as prescription medications for the treatment of chronic insomnia and that there are indications that the beneficial effects of CBT may last well beyond the termination of treatment. If pharmacologic treatment is initiated, it is most effective when combined with CBT.

**Table 1** Stimulus control therapy

| |
|---|
| Stimulus control is a behavioral therapy that seeks to remove the association of sleeplessness with the bed. The following steps reflect the principles of stimulus control: |
| Step 1. Go to bed only when sleepy. |
| Step 2. Use the bed only for sleeping – do not read, watch television, or eat in bed. |
| Step 3. If unable to fall asleep within about 20 min (don't watch the clock), move to another room. Stay up until really sleepy. The goal is to associate the bed with falling sleep quickly. |
| Step 4. If you are awake at night for more than about 20 min, get out of bed. |
| Step 5. Repeat Step 4 as often as necessary. |
| Step 6. Awaken at the same time every morning regardless of total sleep time. |
| Step 7. Avoid naps. |

From Bootzin RR, Epstein D, and Wood JM (1991) Stimulus control instructions. In: Hauri PJ (ed.) *Case Studies in Insomnia*, pp. 19–28. New York: Plenum; Bootzin RR and Epstein D (2000) Stimulus control. In: Lichstein KL and Morin CM (eds.) *Treatment of Late-Life Insomnia*, pp. 167–184. Thousand Oaks, CA: Sage and Morin C (2005) Psychological and behavioral treatments for primary insomnia. In: Kryger MH, Roth T, and Dement WC (eds.) *Principles and Practice of Sleep Medicine*, 4th edn., pp. 726–748. Philadelphia: Elsevier/Saunders.

**Table 2** Sleep restriction therapy

| |
|---|
| Sleep restriction therapy that seeks to keep patients from spending excessive time in bed. The following steps reflect the principles of sleep restriction therapy: |
| Step 1: Cut bedtime to actual amount patient reports sleeping but not less than 4 h per night. |
| Step 2: Prohibit sleep outside these hours. |
| Step 3: Get up at the same time each morning. |
| Step 4: Have patient report daily the amount of sleep obtained. |
| Step 5: Compute sleep efficiency (SE); based on moving average of five nights; when SE is >85%, increase bedtime by15 min. |
| Step 6: Compute sleep efficiency (SE); when SE is > 80%, increase time in bed by going to bed 15 min earlier. Continue to get up at same time each morning. |

From Glovinsky PB and Spielman AJ (1991) Sleep restriction therapy. In: Hauri PJ (ed.) *Case Studies in Insomnia*, pp. 49–63. New York: Plenum; Morin C (2005) Psychological and behavioral treatments for primary insomnia. In: Kryger MH, Roth T, and Dement WC (eds.) *Principles and Practice of Sleep Medicine*, 4th edn., pp. 726–748. Philadelphia: Elsevier/Saunders; and Spielman AJ, Saskin P, and Thorpy MJ (1987) Treatment of chronic insomnia by restriction of time in bed. *Sleep* 10: 45–56.

**Pharmacological therapy** There are several classes of medications used to treat insomnia in older adults, in particular benzodiazepine receptor agonists (benzodiazepines and nonbenzodiazepines), melatonin receptor agonists, antidepressants, antipsychotics, and anticonvulsants. In addition, over-the-counter

**Table 3** Sleep hygiene rules for older adults

1. Maintain a consistent bedtime and uptime.
2. Use the bed only for sleep and satisfying sex. Avoid watching television, eating, or discussing stressful issues while in bed.
3. Minimize noise and temperature extremes.
4. Keep the bedroom environment as dark as possible. Minimize the exposure to bright light if you get out of bed during the night. Use the dimmest night-light possible.
5. Avoid naps if you have difficulty falling asleep at night. If must nap, restrict napping to 30 min or less in the early afternoon.
6. Avoid nicotine near bedtime and during night awakenings.
7. Avoid all forms of caffeine for 4–6 h before bedtime.
8. Although alcohol may help you fall asleep, metabolism of alcohol will cause awakenings.
9. Avoid eating heavy meals and large amounts of protein close to bedtime. A light meal of carbohydrates or dairy products (particularly milk) is preferable.
10. Avoid vigorous exercise just before bedtime.
11. Avoid having pets in bed at nighttime.

From Morin C (2005) Psychological and behavioral treatments for primary insomnia. In: Kryger MH, Roth T, and Dement WC (eds.) *Principles and Practice of Sleep Medicine*, 4th edn., pp. 726–748. Philadelphia: Elsevier/Saunders; and Hauri PJ (1991) Sleep hygiene, relaxation therapy, and cognitive interventions. In: Hauri PJ (ed.) *Case Studies in Insomnia*, pp. 65–86. New York: Plenum.

**Table 4** Does and half-life of US Food and Drug Administration-approved agents for insomnia

| *Generic agent (trade name)* | *Dose (mg)*[a] | *Half-life (h)* |
|---|---|---|
| Benzodiazepine receptor agonists: benzodiazepines | | |
| Flurazepam HCL (Dalmane) | 15, 30 | 47–100 |
| Quazepam (Doral) | 7.5, 15 | 39–73 |
| Estazolam (ProSom) | 0.5, 1, 2 | 10–24 |
| Temazepam (Restoril) | 7.5, 15, 30 | 3.5–18.4 |
| Triazolam (Halcion) | 0.125, 0.25 | 1.5–5.5 |
| Benzodiazepine receptor agonists: nonbenzodiazepines | | |
| Eszopiclone (Lunesta) | 1, 2, 3 | 6 |
| Zaleplon (Sonata) | 5, 10, 20 | 1 |
| Zolpidem (Ambien) | 5, 10 | 1.4–4.5 |
| Zolpidem MR (Ambien CR) | 6.25, 12.5 | 2.8–2.9 |
| Melatonin receptor agonist | | |
| Ramelteon (Rozerem) | 8 | 2–5 |

[a]The lowest dose is generally recommended as the starting dose for older people.
*Physicians' Desk Reference* (1991, 1999, and 2004).
FDA (2004) Lunesta (eszopiclone) Tablets: 1 mg, 2 mg, 3 mg. Retrieved January 24, 2008, from www.fda.gov/cder/foi/label/2004/021476lbl.pdf.
From Consensus conference (1984) *JAMA* 251: 2410–2414.

**Table 5** Indications for the newer nonbenzodiazepine and melatonin receptor agonist agents

| *Agent* | *Indicated for sleep onset* | *Indicated for sleep maintenance* | *Required time in bed (h)* |
|---|---|---|---|
| Eszopiclone[a] | X | X | 8 |
| Zaleplon[b] | X | | 4 |
| Zolpidem[c] | X | | 7–8 |
| Zolpidem MR[d] | X | X | 7–8 |
| Ramelteon[e] | X | | No minimum time |

[a]From Lunesta (eszopiclone) prescribing information. Sepracor.
[b]From SONATA (zaleplon) prescribing information. King Pharmaceuticals, Inc.
[c]From Ambien (zolpidem tartrate) prescribing information. Sanofi-Synthelabo Inc.
[d]From Ambien CR (zolpidem tartrate extended-release tablets) prescribing information. Sanofi-Synthelabo Inc.
[e]From Rozerem (ramelteon) prescribing information. Takeda Pharmaceutical Company Inc.

medications such as antihistamines (diphrenhydramine) and nutritional or herbal supplements are often used. The NIH conference concluded that there is no systematic evidence for the effectiveness of antidepressants, antipsychotics, anticonvulsants, antihistamines, or herbal preparations in the treatment of insomnia. The panel also expressed significant concerns about the risks associated with the use of these medications, particularly in older adults.

The NIH panel also concluded that the benzodiazepine receptor agonists and in particular the newer nonbenzodiazepines have been shown to be the most effective and safest of the sedative hypnotics (at the time of the conference, the melatonin receptor agonist remelteon had not yet been approved by the FDA). The key to successful pharmacologic treatment of insomnia lies in choosing the sedative or hypnotic with the characteristics that best fit the patient's complaints. (It is important to note that only ten agents have been approved by the FDA for the treatment of insomnia. It is important to note that there are only 10 agents approved by the FDA for the treatment of insomnia, listed in **Table 4** with the recommended dose and half-life. **Table 5** lists the indications for the five newer nonbenzodiazepine receptor agonists and the melatonin receptor agonist.)

If sedatives or hypnotics are presecribed, the potential side effects must be taken into account. The administration of long-acting benzodiazepines can cause adverse daytime effects such as excessive daytime sleepiness and poor motor coordination, which may lead to injuries. In older adults, the risk of falls, cognitive impairment, and respiratory depression are of particular concern, although recent data suggest that insomnia *per se*, and not hypnotics, might increase the risk of falls. Chronic use of long-acting benzodiazepines can lead to tolerance and withdrawal symptoms if they are abruptly discontinued. In addition, the potential for exacerbating coexisting medical conditions such as severe hepatic or renal disorders exists when some of these medications are used. The newer nonbenzodiazepine agents have been shown to be effective with a lower propensity for causing clinical residual effects, withdrawal, dependence, or tolerance.

## Specific Sleep Disorders

There are also specific sleep disorders which are quite common in the older adult and which contribute to poor sleep. These would include sleep-disordered breathing (SDB), RLS and periodic limb movements in sleep, and REM behavior disorder.

### SDB

SDB, also called sleep apnea, is a disorder in which the patient periodically stops breathing during sleep. Each apnea (complete cessation in respiration) or hypopnea (partial decrease in respiration) lasts a minimum of 10 s and generally ends with an arousal. This results in fragmented sleep and nocturnal hypoxemia. The total number of apneas plus hypopneas per hour of sleep is called the apnea–hypopnea index (AHI) or respiratory disturbance index (RDI). Depending on the laboratory, an AHI or RDI greater than or equal to 5–10 is required for the diagnosis of SDB.

The prevalence of SDB in older adults has been estimated to be between 30% and 60%, substantially higher than in younger adults. Cross-sectional studies have shown that there is a small increase in prevalence of SDB with increasing 10-year age groups. In the longest longitudinal study, older adults followed for 18 years showed changes in AHI only when associated with changes in body mass index.

There are two main types of SDB. The first, central sleep apnea, occurs when the respiratory centers in the brain shut down during sleep, and there is no attempt to breathe. Central sleep apnea is most common in congestive heart failure and is often associated with Cheyne–Stokes breathing. The second, obstructive sleep apnea (OSA), is the more common and occurs when the muscles of the airway collapse. The two main symptoms of OSA are loud snoring and excessive daytime sleepiness (EDS). EDS, which results from the recurrent nighttime arousals and sleep fragmentation, may manifest as unintentional napping or falling asleep at inappropriate times during the day. EDS can cause reduced vigilance and is associated with cognitive deficits, which may be particularly serious in older adults who may already have some cognitive impairment at baseline.

In addition to the cognitive deficits that may occur as a result of SDB, there is some evidence that many of the progressive dementias (e.g., Alzheimer's and Parkinson's disease) involve degeneration in areas of the brain stem responsible for the regulation of respiration and other autonomic functions relevant to sleep maintenance, thus placing the patient at increased risk for SDB.

Patients with SDB are at greater risk for hypertension, cardiac arrhythmias, congestive heart failure, myocardial infarction, and stroke. Most of the research, however, has not focused on older adults, so the relationship between SDB and some of these various morbidities remains unknown in the older population. Research that has targeted older adults has reported an association between SDB and the risk of developing cardiovascular disease, including coronary artery disease and stroke.

**Treatment of SDB** Treatment of SDB in older adults should be guided not only by the severity of the SDB but by the significance of the patient's symptoms. Patients with more severe SDB (AHI > 20) deserve a trial of treatment. For those with milder SDB (AHI < 20), treatment should be considered if comorbid conditions such as hypertension, cognitive dysfunction, or EDS are present. Treatment should never be withheld because of age or because of assumed noncompliance.

The gold standard for treating SDB is continuous positive airway pressure (CPAP), a device which provides continuous positive pressure via the nasal passages or oral airway, thus creating a pneumatic splint to keep the airway open during inspiration. If used appropriately, CPAP has been shown to be safe and efficacious. Alternative treatments include oral appliances, weight loss (although most older adults with SDB are not obese), and surgery; however, no other treatments have been shown to be as effective as CPAP.

**Significance of SDB in development** The presentation of sleep apnea in the older adult is not necessarily the same as in the younger adult. This has led to the idea that OSA in older adults might be a different disorder from OSA in younger adults. Two questions that arise are whether OSA is pathological in older adults and whether OSA in the older patient should be treated, and if so, when?

From a clinical standpoint, does it matter what is causing SDB in the older adult? Does it matter if the SDB in the older adult is the same as or different from the SDB seen in the younger adult? Probably not. What matters is whether SDB in the older adult has negative consequences. Most studies suggest that SDB does not increase the risk of mortality in the older adult, but that older adults with SDB are excessively sleepy and that SDB may contribute to decreased quality of life, increased cognitive impairment, and greater risk of nocturia, hypertension, and cardiovascular disease. In addition, older adults with both insomnia and SDB have more functional impairment than those with just insomnia.

Any patient, no matter the age, who presents with symptoms of impairment as well as with a history of traffic accidents and repeated falls, should be evaluated for a sleep disturbance. Treatment of a symptomatic older patient should not be withheld on the basis of age, as the consequence of SDB can be just as serious in older as in younger patients. The bottom line is that if the sleep apnea is associated with clinical symptoms (e.g., hypertension, cognitive dysfunction, nocturia, high levels of sleep-disordered breathing, cardiac disease), then it should be treated, regardless of the age of the patient.

### Periodic Limb Movements and RLS

The condition known as periodic limb movements in sleep (PLMS) is characterized by repetitive leg movements during sleep, occurring every 20–40 s and recurring several hundred times over the course of the night, each causing a brief awakening, which results in sleep fragmentation. Patients with PLMS may complain of EDS and/or insomnia. The number of limb movements per hour of sleep is called the periodic limb movement index, with ten or more limb movements per hour needed for a clinical diagnosis. A diagnosis of PLMS therefore must be made with an overnight sleep study that records leg movements. The etiology of PLMS is unknown. PLMS can be seen in patients with fibromyalgia and in conjunction with other primary sleep disorders, including SDB and narcolepsy.

While the prevalence of PLMS in adults is estimated to be 5–6%, the rate increases dramatically with age, with reported prevalence rates of up to 45% in community-dwelling people over the age of 65.

RLS, a condition strongly linked to PLMS, is characterized by leg dysesthesia, often described as 'a creepy crawling' or 'restless' sensation, occurring in the relaxed awake state. The uncomfortable sensations can be relieved only by movement. The diagnosis of RLS is made on the basis of history alone, often with the question, When you try to relax in the evening or sleep at night, do you ever have unpleasant, restless feelings in your legs that can be relieved by walking or movement? Patients may have no knowledge that they kick, so interviewing the patient's bed partner may be helpful in elucidating the history of both RLS and PLMS. Patients with symptoms of RLS should also be assessed for anemia, uremia, and peripheral neuropathy as each of these conditions can cause or exacerbate RLS.

**Treatment of PLMS and RLS** RLS and PLMS are treated primarily with dopamine agonists, which reduce the number of kicks and associated arousals. At this time, only ropinirole and pramipexole have been FDA-approved for the treatment of RLS.

### REM Behavior Disorder

REM behavior disorder (RBD) is characterized by the intermittent absence of normal skeletal muscle atonia during REM sleep, resulting in patients' acting out their dreams. RBD generally occurs during the last third of the night, when REM sleep is more common. The movements during RBD may be violent and potentially harmful to patient and bed partner. Vivid dreams, consistent with the patient's aggressive and/or violent behavior, may be recalled on waking.

The prevalence of RBD is unknown, but it has been shown to be most common in older men. The etiology of RBD is also unknown, but there is a strong association between idiopathic RBD and neurodegenerative diseases, including Parkinson's disease, multiple system atrophy, and Lewy bodies dementia, with the RBD often preceding the other disorders by years. Data suggest that close to half of RBD patients develop Parkinson's disease within several years of the RBD diagnosis. The onset of acute RBD has been associated with stress disorders and the use of monoamine oxidase inhibitors and fluoxetine as well as withdrawal of REM-suppressing agents (e.g., alcohol, tricyclic antidepressants, amphetamines, and cocaine).

**Treatment of RBD** Treatment of RBD with clonazepam, a long-acting benzodiazepine, results in partial or complete cessation of abnormal nocturnal motor movements in 90% of patients. However, because of the long half-life, many patients report residual sleepiness in the morning and during the day. Recently, melatonin has also been shown to be an effective treatment. In addition to the pharmacologic treatment, other aspects of treatment should include good sleep hygiene and education for the patient and bed partner. The bedroom should be made safer by removing heavy or breakable or potentially injurious objects from the bed's vicinity. Heavy curtains should be placed on bedroom windows and doors, and windows should be locked at night. Finally, to avoid falling out of bed, patients may consider sleeping on a mattress on the floor.

## Conclusion

The ability to have restful sleep decreases with age but is likely independent of age itself. Medical and psychiatric disorders, medication, circadian rhythm changes, and primary sleep disorders (such as SDB,

RLS or PLMS, and RBD) contribute to older adults' decreased ability to maintain a full night's sleep. Each of these conditions is treatable, and health care professionals should never withhold treatment because of age or because of assumed noncompliance.

*See also:* Sleep and Circadian Rhythm Disorders in Human Aging and Dementia.

## Further Reading


Ancoli-Israel S (2005) Sleep and aging: Prevalence of disturbed sleep and treatment considerations in older adults. *Journal of Clinical Psychiatry* 66: 24–30.

Ancoli-Israel S (2007) Guest editorial: Sleep apnea in older adults: Is it real and should age be the determining factor in the treatment decision matrix? *Sleep Medicine Review* 11: 83–85.

Ancoli-Israel S, Kripke DF, Klauber MR, et al. (1991) Sleep disordered breathing in community-dwelling elderly. *Sleep* 14(6): 486–495.

Bliwise DL (1993) Sleep in normal aging and dementia. *Sleep* 16(1): 40–81.

Bootzin RR and Epstein D (2000) Stimulus control. In: Lichstein KL and Morin CM (eds.) *Treatment of Late-Life Insomnia*, pp. 167–184. Thousand Oaks, CA: Sage.

Bootzin RR, Epstein D, and Wood JM (1991) Stimulus control instructions. In: Hauri PJ (ed.) *Case Studies in Insomnia*, pp. 19–28. New York: Plenum Press.

Edinger JD, Heolscher TJ, Marsh GR, Lipper S, and Ionescu-Pioggia M (1992) A cognitive-behavioral therapy for sleep-maintenance insomnia in older adults. *Psychological Aging* 7: 282–289.

Foley DJ, Monjan A, Simonsick EM, Wallace RB, and Blazer DG (1999) Incidence and remission of insomnia among elderly adults: An epidemiologic study of 6,800 persons over three years. *Sleep* 22: S366–S372.

Glass J, Lanctot KL, Herrmann N, Sproule BA, and Busto UE (2005) Sedative hypnotics in older people with insomnia: Meta-analysis of risks and benefits. *British Medical Journal* 331: 1169.

Glovinsky PB and Spielman AJ (1991) Sleep restriction therapy. In: Hauri PJ (ed.) *Case Studies in Insomnia*, pp. 49–63. New York: Plenum.

Hauri PJ (1991) Sleep hygiene, relaxation therapy, and cognitive interventions. In: Hauri PJ (ed.) *Case Studies in Insomnia*, pp. 65–86. New York: Plenum.

Lichstein KL and Reidel BW (1994) Behavioral assessment and treatment of insomnia: A review with an emphasis on clinical application. *Behavior Therapy* 25: 659–688.

Martin JL and Ancoli-Israel S (2003) Insomnia in older adults. In: Szuba MP, Kloss JD, and Dinges DF (eds.) *Insomnia: Principles and Management*, pp. 136–154. Cambridge, UK: Cambridge University Press.

Morin C (2005) Psychological and behavioral treatments for primary insomnia. In: Kryger MH, Roth T, and Dement WC (eds.) *Principles and Practice of Sleep Medicine*, 4th edn., pp. 726–748. Philadelphia: Elsevier/Saunders.

Morin CM, Colecchi C, Stone J, Sood R, and Brink D (1999) Behavioral and pharmacological therapies for late life insomnia. *Journal of the American Medical Association* 281: 991–999.

National Institutes of Health (2005) State-of-the-Science conference statement on manifestations and management of chronic insomnia in adults. *Sleep* 28(9): 1049–1057.

Spielman AJ, Saskin P, and Thorpy MJ (1987) Treatment of chronic insomnia by restriction of time in bed. *Sleep* 10: 45–56.

Van Someren EJ (2000) Circadian rhythms and sleep in human aging. *Chronobiology International* 17(3): 233–243.

Vitiello MV (2000) Effective treatment of sleep disturbances in older adults. *Clinical Cornerstone* 2(5): 16–47.


## Relevant Websites

http://www.aasmnet.org – American Academy of Sleep Medicine.
http://www.sleepfoundation.org – National Sleep Foundation.
http://www.sleepresearchsociety.org – Sleep Research Society.

# Erectile Dysfunction

**S H Tariq and J E Morley**, St. Louis University and Veterans Affairs Medical Center, St. Louis, MO, USA



Erectile dysfunction (ED) is defined as an inability to achieve or maintain an erection sufficient for satisfactory sexual performance. ED is estimated to occur in 10–20 million men, which increases to 30 million if partial ED is included. ED can occur in men of all ages, but it is more common in the elderly and is not a part of normal aging.

## Prevalence

The prevalence of ED varies. The reported prevalence of ED is 5% in 40-year-olds, which triples to 15% at the age of 70. The occurrence of ED increases with age: it occurs in 27% of men at age 70 years, 55% at age 75 years, and 75% at age 80 years. The prevalence of ED is 50% in patients with renal disease and it increases to 75% in those who are on dialysis. ED develops in approximately 50% of patients with diabetes within 10 years of diagnosis.

ED is a major health concern, strongly associated with increasing age. Men with ED have impaired quality of life compared with unaffected individuals. Since sexual activity contributes significantly to quality of life, patients expect physicians to address such an important issue.

## Physiology of Penile Erection and Changes with Aging

Sexual activity consists of libido and potency. Libido is the sexual desire/drive, thoughts and fantasies, satisfaction, and pleasure. Potency is the ability to attain and maintain an erection and to ejaculate. There are three pathways for penile erection, which are summarized in **Table 1**. Masters and Johnson described changes in the four stages of the sexual response cycle – excitement, plateau, orgasm, and resolution – with aging and these are outlined in **Table 2**.

## Causes of ED

The etiology of ED varies significantly depending on the clinic population, as shown in **Table 3**. Etiologic causes for ED include vascular, endocrine (hypogonadism, thyroid problems, diabetes, and hyperprolactinemia), neurologic, urologic, psychogenic, medication, and alcoholism. Depression as a cause of ED is more common in those younger than 60 years of age. Similarly, anxiety and marital discord are more common in the population group younger than 60 years of age.

## Medical Conditions Associated with ED

The most common medical conditions associated with ED are hypertension and coronary artery disease. Fifty-seven percent of patients who underwent coronary artery surgery had ED. Other conditions associated with ED include diabetes, smoking, and alcohol use. An increase in total cholesterol has been associated with ED.

## Evaluation

A detailed history is absolutely essential to differentiate difficulties with ED from a decrease in libido. **Table 4** highlights some of the important areas that are essential in evaluating ED. A history from a partner will help uncover some of the problems with their sex life. The risk factors for ED during history include vascular disease, hormonal problems (thyroid and diabetes), neurological dysfunction, medication, past surgery, or trauma; psychological and social factors, including an evaluation for depression, should be assessed. A number of medications are associated with an increase or decrease in libido, and these are shown in **Table 5**. Since depression can contribute to ED, a screening assessment tool such as the Beck Depression Inventory or Yesavage Geriatric Scale may be used. A number of screening tools are also available to screen patients for androgen deficiency, such as the Saint Louis University Androgen Deficiency in Aging Males (ADAM), which is shown in **Table 6**. A positive answer to question 1 or 7 or any other three questions is considered to indicate a high likelihood of low testosterone. The ADAM questionnaire has a sensitivity of 88% and a specificity of 60%. There is a clear improvement in the ADAM questionnaire after treatment with testosterone.

Physical examination should include assessment of secondary sexual characteristics along with a detailed examination searching for vascular and neurological problems. Local penile shaft examination is also important to rule out Peyronie's disease.

Laboratory investigations such as diabetes, renal functions, lipid profile, thyroid function tests, prolactin levels, and penile brachial pressure index will help search for the etiology of ED. The most important test for determining a decrease in sex drive or libido is to check total testosterone and bioavailable testosterone

**Table 1** Physiology of penile erection

- Stimulation of the genitalia, in which sensation in the sensory receptors of the penile shaft and glans penis is carried to the spinal cord by afferent nerves. The sensory perception is then passed on to the higher centers, and efferent impulses leave the spinal cord via the sacral parasympathetic center to the pelvic plexus and the cavernous nerves into the corpus cavernosum.
- Psychogenic stimulation that occurs through visual or auditory stimuli. This is more complex and less understood.
- Penile erection occurs during the rapid eye movement stage of sleep. The physiology is not known.
- During erection the parasympathetic nervous system causes vasodilatation, increased blood supply and pressure to the cavernous spaces, and venous occlusion.
- Flaccid state is achieved by the release of catecholamines, which causes contraction of the smooth muscles around the sinusoids and arterioles, expelling the blood out of the sinusoidal spaces and decreasing blood flow, resulting in a flaccid state.
- Neurotransmitters such as nitric oxide, prostaglandins $E_2$ and $E_1$, vasoactive intestinal peptide, neuropeptide Y, and acetylcholine cause smooth muscle relaxation and penile erection. Smooth muscle contraction is caused by prostaglandin $F_{2\alpha}$, substance P, histamine, and norepinephrine.

**Table 2** Changes with aging

- The excitement phase shows a delay in erection, decrease in erectile strength as well as decrease in scrotal congestion, and testicular elevation with advanced age.
- The plateau phase in prolonged along with a decrease in preejaculatory. The orgasmic phase is shorter, with a decrease in the ejaculatory force. The time for ejaculation is longer compared to that of young adults.
- In the resolution phase, there is rapid detumescence and rapid testicular descent.
- The refractory period is also prolonged with age.
- The penis becomes less sensitive to stimulation with age and it takes longer to achieve erection. The erection becomes more dependent on physical stimulation of the penis and less responsive to visual, nongenital, and psychologic stimulation.

**Table 3** Causes of erectile dysfunction

| | *Type of clinic* | | | | | | | | |
|---|---|---|---|---|---|---|---|---|---|
| | *Endocrine* | *Urology* | *Sex therapist* | *Family practice* | *Medical outpatient* | *Outpatient clinic* | *Urology*[a] | *Outpatient* | *Geriatric clinic* |
| No. of patients | 105 | 165 | 101 | 154 | 120 | 93 | 31 | 121 | 297 |
| Mean age (years) | | | | | | 61 | 53 | 68 | 66 |
| Etiology (%) | | | | | | | | | |
| Endocrine | 44 | 6 | 45 | 24 | 29 | 35 | | | 54 |
| Hypogonadism | 26 | 6 | 39 | 23 | 19 | 9.6 | 19 | 2.6 | 43–81 |
| Thyroid | 3 | | 4 | | 4 | 2 | | | |
| Hyperprolactinoma | 8 | | 2 | | 6 | 4 | | | 0.3 |
| Diabetes mellitus | 7 | 19 | 2 | 20 | 9 | 22 | | | 27–29 |
| Vascular | | 8 | 1 | 37 | | 37.6 | 68 | 21 | 39–55 |
| Neurogenic | | 4 | 2 | 5 | 7 | 17 | 26 | 17[b]/11[c] | 3–10 |
| Surgical/trauma | | 8 | | | | | | | |
| Urologic | | 8 | | 49 | 6 | 19 | 3 | 1.3 | 7–12 |
| Alcoholism | 3 | 0.6 | | 7 | 7 | 20 | | | 12–14 |
| Systemic disease | 6 | | 2 | 5 | 5 | | | | |
| Medication | 7 | 2 | 0 | 12 | 25 | 67 | | 4 | 58 |
| Psychiatric | 14 | | | | | | | | 5–18 |
| Psychogenic | 14 | 51 | 19 | 60 | 14 | 15 | 19 | 9.2 | 3–9 |
| Unknown | 19 | 0.6 | | 0.7 | | | 4 | | |

[a]All diabetic male patients.
[b]Diabetic neuropathy.
[c]Nondiabetic neuropathy.

in older adults (non-sex hormone-bound testosterone). Cross-sectional studies have demonstrated that levels of testosterone decline with age, although the Baltimore Longitudinal Aging Study failed to find a decrease in testosterone levels in males. The decline in testosterone is 100 ng/dl per decade. A number of studies have reported a decline in bioavailable testosterone with aging. A decline in bioavailable testosterone is

**Table 4** Evaluation of erectile dysfunction

| |
|---|
| History |
| Review of medications |
| Screening for depression |
| Screening for hypothyroidism |
| Screening questionnaires for ED/hypogonadism |
| Physical examination |
| Laboratory tests |
| Renal function |
| Liver function tests |
| Complete blood count |
| Prostatic serum antigen |
| Total or free or bioavailable testosterone |

**Table 5** Drugs that possibly affect sexual desire in men

| |
|---|
| Drugs that possibly decrease sexual desire |
| Antihistamine/barbiturates/benzodiazepines (sedative effect) |
| Tricyclic antidepressant |
| Diuretics |
| Spironolactone/hydrochlorothiazide/acetazolamide/methazolamide |
| Antihypertension |
| Reserpine/propranolol/clonidine/alpha-methyldopa/prazosin |
| Cimetidine |
| Hormones |
| Estrogen/medroxyprogesterone |
| Clofibrate |
| Digoxin |
| Lithium |
| Drugs that possibly increase sexual desire |
| Androgens (in androgen-deficit states) |
| Baclofen (possible anti-anxiety effect) |
| Benzodiazepines (anti-anxiety effect) |
| Haloperidol (indirect effect due to improved sense of well-being) |

**Table 6** St. Louis University ADAM Questionnaire

1. Do you have a decrease in libido (sex drive)?
2. Do you have a lack of energy?
3. Do you have a decrease in strength and/or endurance?
4. Have you lost height?
5. Have you noticed a decreased 'enjoyment of life'?
6. Are you sad and/or grumpy?
7. Are you erections less strong?
8. Have you noted a recent deterioration in your ability to play sports?
9. Are you falling asleep after dinner?
10. Has there been a recent deterioration in your work performance?

seen in 50% of men by 50 years of age. Luteinizing hormone levels fail to increase appropriately in response to a decrease in testosterone levels with age, suggesting failure of the hypothalamic–pituitary unit (i.e., secondary hypogonadism).

## Treatment

Current treatments for ED include oral medications, intracavernosal injections, testosterone therapy, vacuum pumps, penile prostheses, and vascular surgery. First-line therapy includes oral drug therapy, the use of vacuum devices, and, less frequently, sex therapy. Referral to a specialist may be required upon request of the patient if initial therapy failed or a more comprehensive evaluation is needed. Patients with a suspected psychogenic etiology should be considered for sexual counseling or psychiatric referral. Involvement of the partner in the diagnostic and therapeutic plans from the beginning is crucial. If a particular drug is related to ED, then that drug should be stopped and a different class of drug should be used.

## Oral Therapy

### Phosphodiesterase-5 inhibitors

Currently, there are three drugs available in the United States and all inhibit phosphodiesterase-5 (PDE-5), which is responsible for the degradation of cyclic guanosine monophosphate (cGMP) in the corpus cavernosum. When the nitric oxide and cGMP pathway is activated during sexual stimulation, inhibition of cGMP by PDE-5 inhibitors results in increased levels of cGMP in the corpus cavernosum and increased relaxation of corpus cavernosal smooth muscle cells in response to sexual stimulation. This causes an increase in penile blood flow and erection. Thus, sexual stimulation is required and PDE-5 inhibitors help maintain it.

Sildenafil is well absorbed during fasting, and the plasma concentrations are maximal within 30–120 min. It is mainly excreted by the liver, and its half-life is approximately 4 h. The starting dose is 50 mg taken 1 h before sexual activity, not more than once per day. The quality of life consequences of ED include depression, anxiety, and loss of self-esteem. One study determined correlations between erectile function (EF), intercourse success, and emotional well-being measured with the Self-Esteem and Relationship (SEAR) questionnaire in men treated with sildenafil citrate for ED and stratified by age (<50 years, 50–65 years, and >65 years). For the overall population, there was significant improvement ($p < 0.0001$) in the Self-Esteem subscale and intercourse success rate. Secondary outcomes (i.e., EF domain of the International Index of Erectile Function and event log frequency of erection hard enough for sexual intercourse and of ejaculation/orgasm) also improved. All ten correlations were positive in men of all ages. The most common adverse events were mild to moderate headache (12%), vasodilatation (7%), and rhinitis (4%). Sildenafil improved EF and there was an increased intercourse success rate, which correlated positively with improvement in SEAR measures of self-esteem and sexual relationship.

In hypogonadal patients with ED, androgen supplementation is first-line therapy. If there is no response to androgen alone or sildenafil alone, combined use may improve EF and enhance the therapeutic effect of PDE-5 inhibitors.

Tadalafil has a half-life of 17.5 h. At a dose of 20 mg, tadalafil enabled 73–80% of sexual intercourse attempts to be completed successfully with appropriate sexual stimulation between 30 min and 36 h after dosing. Tadalafil is well tolerated. Side effects are headache and dyspepsia, which are mostly mild or moderate and transient. There were no significant laboratory abnormalities or electrocardiogram changes. It is preferred over sildenafil by men in the treatment of ED, secondary to its longer duration of action, especially by those who do not have any risk factors for developing chest pain. In a small number of patients refractory to sildenafil therapy, a trial of vardenafil might be helpful.

Vardenafil is well tolerated. Side effects are headache, cutaneous flushing, dyspepsia, and rhinitis. Vardenafil successfully treated ED after nerve sparing radical retropubic prostatectomy.

All PDE-5 inhibitors are absolutely contraindicated with the use of nitrates and are hazardous in patients with positive stress test not on nitrites; patients on a complicated, multidrug, antihypertensive program; patients with congestive heart failure and borderline low-volume status; and patients on drugs which prolong the half-life of these medications.

### Phentolamine

Phentolamine is an $\alpha$-adrenergic blocker with poor efficacy in improving ED but may be used with other agents. A prospective, placebo-controlled, double-blind study (40 patients) showed that phentolamine had a success rate of 20.0%, 30.0%, and 36.7% at doses of 20, 40, 60 mg, respectively, in the treatment of ED compared to placebo. The mean age was 48 years for those who responded to phentolamine compared to 52.2 years for the nonresponders. An open-label, multicenter study compared the efficacy and safety of oral sildenafil and phentolamine in men with ED. Patients received sildenafil (25–100 mg, $n = 123$) or phentolamine (40 mg, $n = 119$) for 8 weeks, and efficacy was assessed. Approximately twice as many men receiving sildenafil had successful attempts at sexual intercourse, improved erections, and improved ability to have sexual intercourse compared with men receiving phentolamine. The adverse events included rhinitis, headache, tachycardia, and nausea, with a higher frequency reported in patients receiving phentolamine than sildenafil (41% vs. 33%), with the exception of headache, which was reported more frequently in sildenafil users. Overall, sildenafil was more effective and appeared to be better tolerated than phentolamine for the treatment of ED.

### Yohimbine

Yohimbine is an $\alpha_2$-adrenergic receptor antagonist; it acts at the serotonergic and adrenergic receptors in brain centers associated with libido and penile erection. Its exact mechanism of action is unknown. Yohimbine has been shown to be relatively effective in psychogenic ED, although other studies have shown no statistical significant difference from placebo. Yohimbine carries a risk of hypertension, palpitation, fine tremor, and anxiety. Yohimbine has had questionable effects in men with organic ED. Of a total of 18 nonsmoking men, 50% were successful in completing intercourse in more than 75% of attempts. Yohimbine could be effective therapy to treat organic ED in some men.

### Apomorphine

Apomorphine is a sublingual formulation and is now undergoing clinical trials. It acts centrally on dopamine $D_2$ receptors of the hypothalamus and activates pathways that involve nitric oxide and oxytocin signals. It is generally well tolerated at doses of 2 or 3 mg, and it is not an opioid. An observational cohort study was conducted in England between October 2001 and December 2002. The study cohort comprised 11 185 patients, 99.3% of whom were men, with a median age of 61 years. The study concluded that the most frequently reported prescribing indication was ED and the most frequently reported reason for stopping apomorphine was that it was "not effective." However, in a small percentage of patients apomorphine has been effective. Headache was the most commonly reported adverse drug reaction and the most frequently reported clinical condition occurring in the first month.

### Medicated Urethral System for Erection

The medicated urethral system for erection (MUSE) uses prostaglandin $E_1$, which is an endogenous unsaturated 20-carbon fatty acid derivative of arachidonic acid, and alprostadil is a more stable, synthetic form of prostaglandin $E_1$. Prostaglandin $E_1$ causes smooth muscle relaxation and vasodilatation by acting on adenylate cyclase to increase the intracellular cyclic adenosine monophosphate concentration. The first application is usually at a dose of 50 μg in the physician's office because of the potential complications. The dose can be titrated (250–1000 μg), depending on the erectile response. Alprostadil is well tolerated and effectively restores the capacity for erections and sexual intercourse. A restrictive band can be placed at the base of the penis to increase efficacy, especially if the phenomena of pelvic steal syndrome is present. The advantages of transurethral therapy include local application, minimal systemic effects, and the rarity of drug interactions. Side effects include penile pain, local discomfort, hypotension, dizziness, and syncope. Trauma, if it occurs, consists of a superficial urethral abrasion, no urethral stricture, penile fibrosis, or priapism.

## Intracavernous Injection Therapy

The most commonly used intracavernous drugs in the United States are alprostadil and a combination of papaverine, phentolamine, and alprostadil. Intracavernous injection of alprostadil is an effective therapy with tolerable side effects. It has a success rate of 60–70% in the treatment of vasculogenic and 100% in neurogenic ED. It is contraindicated in sickle cell anemia, schizophrenia or other psychiatric disease, or severe venous leakage. Pain is reported to occur in 13–33%, fibrotic complication in 1–57%, and priapism in 5–23% of patients. If priapism does occur, the patient should be informed to take oral benadryl and, if it is not effective, should then seek immediate access for medical treatment. Injection therapy should be avoided in patients who are on chronic warfarin therapy because it increases the risk of hematomas and bleeding.

Alprostadil is the only agent approved in the United States. It is given in doses of 5–20 μg. It is indicated if oral therapy fails or is not suitable; for diabetics, especially those on insulin; for spinal cord injury; and for radical prostatectomy.

Papaverine is a nonspecific PDE inhibitor. The usual dose is 15–60 mg. It is highly effective (up to 80%) in men with psychogenic and neurogenic ED, but it is less effective when ED is of vascular etiology (36–50%). Advantages include its low cost and stability at room temperature. Its major drawbacks are priapism, corporal fibrosis, and occasional increases in serum aminotransferase concentrations.

## Testosterone Therapies

Testosterone replacement can be achieved in several ways: oral and injectable therapies, transscrotal and transdermal patches, subcutaneous gels, buccal preparations, and selective androgen receptor molecule (no effects on prostate). Some preparations are under development, such as sublingual and inhalation forms.

### Effects of Testosterone Replacement

**Coronary artery disease** Studies have shown that low testosterone levels are associated with both coronary artery disease and the degree of atherosclerosis. Testosterone dilates the brachial artery and therefore increases blood flow, decreases ST depression during exercise stress test, and reduces or has no effect on myocardial infarction. Testosterone replacement therapies have not demonstrated an increased incidence of cardiovascular disease or events.

**Lipid profiles** Testosterone decreases cholesterol and low-density lipoprotein cholesterol levels and high-density lipoprotein (HDL) cholesterol. A meta-analysis of testosterone replacement showed that HDL levels were reduced in three studies and unchanged in 15 studies. Five studies showed a reduction in total cholesterol, two showed an increase, and 12 were unchanged. Low-density cholesterol was either unchanged or reduced in 14 of the 15 studies.

**Libido and ED** Epidemiologic studies have shown that decreased sexual activity and libido are associated with a decrease in free or bioavailable testosterone levels. These studies suggested that aging was a more important contributor than androgen levels in sexual behavior in healthy older men. Testosterone replacement improves libido and quality of erection in hypogonadal men. Sildenafil is reported to improve erections in older men. It is reported that certain men with low testosterone do not respond to sildenafil alone unless their testosterone is replaced. This seems to be because testosterone is essential for the synthesis of nitric oxide synthase.

**Body composition and frailty** Decrease in muscle mass and strength is associated with aging. Resistance training restores the muscle mass and strength. Excessive loss of muscle in a person results in sarcopenia and increases the risk of becoming frail. Studies on testosterone replacement have clearly shown that it increases the muscle mass, and this is reported even in older men who are not hypogonadal. The increased strength is reported in the upper extremities in hypogonadal men, but studies have failed to show increased strength in men who are not hypogonadal. 5α-Dihydrotestosterone has been shown to increase knee flexion strength. Testosterone replacement is also beneficial in improving the functional index measure during rehabilitation following hospitalization.

Leptin levels increase with age in males, which has been shown to be associated with decreases in testosterone. Testosterone replacement decreases fat mass, with subcutaneous adiposity being reduced to a greater extent than abdominal visceral fat.

**Behavioral effects** Epidemiologists have shown a strong correlation between low testosterone and cognitive decline with aging. Low testosterone at middle age is predictive of developing Alzheimer's disease. Testosterone replacement improves working memory, spatial memory, and trail making B; effects on verbal memory are controversial and in some studies testosterone replacement has failed to alter cognition.

The SAMP8 mouse has low testosterone levels. Testosterone replacement in this mouse model of Alzheimer's disease improves learning and memory. Testosterone treatment decreased amyloid precursor protein. Patients with Alzheimer's disease have low testosterone levels in brain tissue compared with controls. Testosterone replacement has been shown to produce small improvements in cognition in people with Alzheimer's disease.

Dysphoria has been related to low levels of testosterone. Testosterone replacement did improve depression in one study, but other studies have failed to show the same results.

**Bone** Testosterone is converted by the process of aromatization to estradiol, which acts on bone. Testosterone increases bone mineral density by acting on the osteoblasts and thus increases bone mineral density at the hip and lumbar spine. There are no studies examining the effect of testosterone on hip fractures.

**Polycythemia** Lower testosterone levels are associated with anemia in men. A rise in hematocrit is generally beneficial in anemia; elevation above the normal range may have grave consequences resulting in coronary, cerebrovascular, or peripheral vascular circulation. Injections increased hematocrit by 44%, transdermal nonscrotal patch by 15%, and scrotal patch by 5%. There is a direct relationship between testosterone dosage and the incidence of erythrocytosis. Erythrocytosis was reported in 2.8% of men on 5 mg of nonscrotal patches and 11% with 50 mg $day^{-1}$ and 18% with 100 mg $day^{-1}$ gel, respectively.

**Benign prostatic hypertrophy** Benign prostatic hypertrophy (BPH) is induced by androgens and it decreases with decreases in serum testosterone. However, a number of studies have failed to show that testosterone replacement exacerbates the voiding symptoms of BPH or causes increased urinary retention compared to placebo.

**Prostate cancer** Case reports have suggested that testosterone replacement therapy may convert occult cancer into a clinically apparent lesion. Prospective studies demonstrate a low frequency of prostate cancer in association with testosterone replacement therapy. The prevalence rate for prostate cancer is similar to that of the general population, as evident by prospective trials of testosterone replacement therapy showing prostate cancer in 1.1% of men followed for 6–36 months. There is some concern that the underlying prevalence of occult prostate cancer in men with low testosterone levels appears to be substantial. Despite decades of research, there is no compelling evidence that testosterone has a causative role in prostate cancer. Other side effects include gynecomastia, water retention, hypertension, and sleep apnea.

## External Devices

A vacuum device is an external cylinder placed over the penis to allow air to be pumped out, resulting in engorgement of the penis with blood, creating the erection. A constriction band is fitted at the base of the penis to maintain the erect state. It is a noninvasive approach. Vacuum devices are suitable for a range of patients with chronic or occasional ED. They can be used after the removal of infected or defective penile prosthesis and they may prevent scar contraction. Disadvantages are discomfort, poor erection quality, and refusal of this mechanical therapy. Overall satisfaction has been reported to be between 70 and 94%. Caution should be exercised for patients using anticoagulation therapy and those with bleeding disorders. Vacuum device therapy might be a useful option in patients for whom other therapies are contraindicated, such as patients with heart disease.

## Surgery

Penile prostheses are indicated after failure of other therapies for men of middle or old age. A psychiatrist or psychologist should participate in the preoperative evaluation of patients with psychogenic ED and determine the need for the prosthesis. Advantages are ease of use, reliability, and normal appearance and sensation. It requires destruction of the corpus cavernosum, which prevents pharmacological treatment later. Arterial revascularization may be indicated in young individuals with normal corporeal venous function who have arteriogenic ED secondary to trauma. Venous ligation for veno-occlusive dysfunction was not effective and therefore was abandoned.

## Further Reading

Bakhshi V, Elliott M, Gentili A, Godschalk M, and Mulligan T (2000) Testosterone improves rehabilitation outcomes in ill older men. *Journal of the American Geriatrics Society* 48: 550–553.

Barrett-Connor E, Goodman-Gruen D, and Patay B (1999) Endogenous sex hormones and cognitive function in older men. *Journal of Clinical Endocrinology & Metabolism* 84: 3681–3685.

Collins WE, Mckendry JBR, Silverman M, et al. (1983) Multidisciplinary survey of erectile impotence. *Canadian Medical Association Journal* 128: 1393–1399.

English KM, Mandour O, and Steeds RP (2000) Men with coronary artery disease have lower levels of androgens than men with normal coronary angiograms. *European Heart Journal* 21: 890–894.

Kaiser FE, Viosca SP, Morley JE, et al. (1988) Impotence and aging: Clinical and hormonal factors. *Journal of the American Geriatrics Society* 36: 511–519.

Kinsey AC, Pomeroy WB, and Martin CE (1948) *Sexual Behavior in the Human Male*. Philadelphia, PA: Saunders.

Morley JE, Charlton E, Patrick P, et al. (2000) Validation of a screening questionnaire for androgen deficiency in aging males. *Metabolism* 49: 1239–1242.

Morley JE, Perry HM 3rd, Kaiser FE, et al. (1993) Effects of testosterone replacement therapy in old hypogonadal males: A preliminary study. *Journal of the American Geriatrics Society* 41: 149–152.

NIH Consensus Development Panel on Impotence (1993) Impotence. *Journal of the American Medical Association* 270: 83–90.

Sih R, Morley JE, Kaiser FE, et al. (1997) Testosterone replacement in older hypogonadal men: A 12-month randomized controlled trial. *Journal of Clinical Endocrinology & Metabolism* 82: 1661–1667.

Slag MF, Morley JE, Elson MK, et al. (1983) Impotence in medical clinic outpatients. *Journal of the American Medical Association* 249: 1736–1740.

Tariq SH, Haleem U, Omran ML, et al. (2003) Erectile dysfunction: Etiology and treatment in young and old patients. *Clinics in Geriatric Medicine* 19: 539–551.

Tariq SH, Haren MT, Kim MJ, and Morley JE (2005) Andropause: Is the emperor wearing any clothes? *Reviews in Endocrine and Metabolic Disorders* 6: 77–84.

Wei M, Macera CA, Davis DR, et al. (1994) Total cholesterol and high density lipoprotein cholesterol as important predictors of erectile dysfunction. *American Journal of Epidemiology* 140: 930–937.

Wittert GA, Chapman I, Haren M, et al. (2003) Oral testosterone supplementation increases muscle and decreases fat mass in healthy elderly males with low-normal gonadal status. *Journals of Gerontology, Series A: Biological Sciences and Medical Sciences* 58: 618–625.

# Alzheimer's Disease: An Overview

**P I Moreira**, University of Coimbra, Coimbra, Portugal
**X Zhu and M A Smith**, Case Western Reserve University, Cleveland, OH, USA
**G Perry**, University of Texas at San Antonio, San Antonio, TX, USA



## Major Neuropathologic Hallmarks

The distinctive brain lesions, senile plaques (SPs) and neurofibrillary tangles (NFTs), used by Alois Alzheimer together with the clinical deficits to describe Alzheimer's disease (AD), are still used today as the defining features for diagnosis (**Figure 1**). In addition to these striking changes, there is variable cerebral cortical atrophy, particularly of the temporal and frontal lobes, and associated ventricular dilation, both of which are consequences of the neuron loss and astrocyte proliferation in affected regions.

Both NFTs and SPs are found in normal aged persons, but it is their quantitative increase that defines the pathologic diagnosis of AD. NFTs consist of abnormal 20 nm helical filaments with an 80 nm half-periodicity, termed paired helical filaments, and 12–15 nm straight filaments (**Figure 2**). NFTs usually occur in large numbers in the brain of a patient with AD, particularly in the entorhinal cortex; hippocampus; amygdala; association cortices of the frontal, temporal, and parietal lobes; and certain subcortical nuclei that project to these regions. The subunit protein of the paired helical filaments is the microtubule-associated protein tau $\tau$. Paired helical filaments are not limited to the tangles found in the neuronal cell bodies but also occur in smaller bundles in many of the dystrophic neurites present around the amyloid plaques. Biochemical studies reveal that the tau protein present in paired helical filaments is a hyperphosphorylated, insoluble form of this normally highly soluble protein. The insoluble tau aggregates in the tangles are often complexed with ubiquitin, a feature they share with numerous other intraneuronal protein inclusions in etiologically diverse disorders, such as Parkinson's disease and diffuse Lewy body dementia. If this complexing with ubiquitin represents an attempt by the neuron to mark the altered tau proteins for degradation by the proteasome, it seems to be largely without benefit for the patient.

The amyloid-$\beta$ (A$\beta$) peptides of SPs consist of 7–10 nm helical filaments (**Figure 2**) that share with paired helical filaments the ability to bind the dye Congo red and appear birefringent green when viewed under cross-polarized light, a property of $\beta$-pleated molecular sheets. SPs (or neuritc plaques) contain extracellular deposits of the A$\beta$42/40 (explained below) surrounded by dystrophic neurites (axons and dendrites), activated microglia, and reactive astrocytes. A large portion of the A$\beta$ in these neuritic plaques is in the form of insoluble amyloid fibrils, but these are intermixed with a poorly defined array of nonfibrillar forms of the protein. Once protein sequencing established that A$\beta$ was the subunit of fibrillar amyloid, A$\beta$ immunohistochemistry revealed several deposits in brains of patients with AD that lacked the dystrophic neurites and altered glia that characterize the neuritic plaques. Such plain A$\beta$ deposits, referred to as 'diffuse' plaques, exist mostly in a nonfibrillar (that is, 'preamyloid') form. The diffuse deposits are composed of A$\beta$42, which is far more prone to aggregation than the slightly shorter and less hydrophobic A$\beta$40. In healthy individuals, A$\beta$40 and A$\beta$42 make up 90% and about 10%, respectively, of the A$\beta$ peptides that are normally produced by brain cells throughout life. Note that A$\beta$ plaques do not occur simply in these two extreme forms (diffuse and neuritic) but rather as a continuum in which mixtures of nonfibrillar and fibrillar forms of the peptide can be associated with varying degrees of surrounding neuritic and glial alteration.

## Neurotransmitter Deficits

Since the demonstration of a 90% drop in cortical choline acetyltransferase activity, there has been considerable interest in understanding whether neurotransmitter replacement therapy would be beneficial in the treatment of AD. The 'cholinergic hypothesis' states that decreased cholinergic transmission plays a major role in the expression of cognitive, functional, and possibly behavioral symptoms in AD. The cholinergic hypothesis rests on pathological, biochemical, and pharmacological observations. Cholinergic neurons in the ventral forebrain are depleted; many of those that remain contain NFTs. As a result of these pathological changes, there are decreases in biochemical indices of cholinergic function in neocortex and hippocampus that correlate with dementia severity. The hypothesis is further supported by an extensive literature from pharmacological studies using cholinergic agonists and antagonists, ablative lesions of the cholinergic pathway, and transgenic animal models that emphasize the close connection between cognition and the cholinergic neurotransmission. It has also been proposed that the cholinergic deficit plays a role not only in the cognitive symptoms but also in the behavioral changes observed in AD. A limitation of the cholinergic hypothesis is the lack of cholinergic deficit observed in early stages of AD or in patients with mild cognitive impairment.

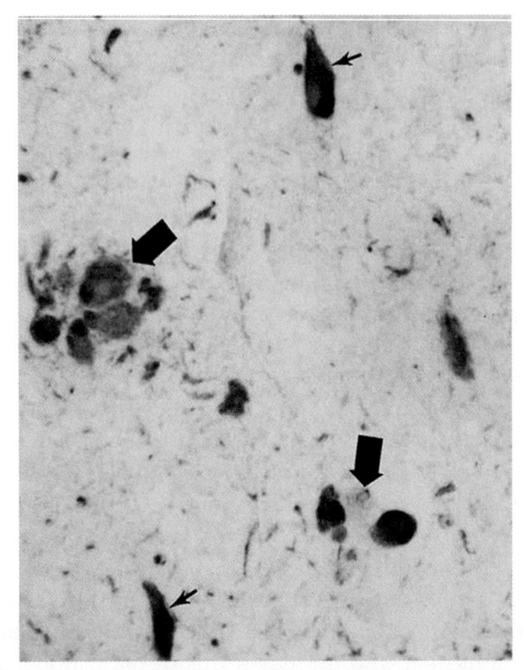

**Figure 1** Characteristic pathologic lesions of Alzheimer's disease, neurofibrillary tangles (small arrows), and senile plaques (large arrows) are readily immunostained with antibodies to tau. (×760)

Other neurotransmitters may also be important in AD. For example, levels of three important neurotransmitters, serotonin, somatostatin, and noradrenaline, are reduced in the brains of some Alzheimer's patients. It has been suggested that these abnormalities are related to sensory disturbances and aggressive behavior. However, most neurotransmitter research in dementia continues to focus on acetylcholine because of its close ties to memory and reasoning.

## Genetics

Autosomal dominant inheritance of mutations in the amyloid-$\beta$ protein precursor (A$\beta$PP), PSEN1, and PSEN2 genes localized in chromosomes 21, 14, and 1, respectively, are responsible for familial early-onset AD (FAD). PSEN1 and PSEN2 encode for homologous polytopic membrane proteins, termed presenilin 1 and 2 (PS1 and PS2). To date, 15 missense mutations in A$\beta$PP, 199 in PS1, and 11 in PS2 have been reported to cause FAD. FAD mutations in PSEN1 cause most aggressive forms of AD, in some cases with onset younger than 30 years. Individuals with trisomy 21 (Down syndrome) have an extra copy of the A$\beta$PP gene and develop AD pathology as early as 20 years of age. Even though FAD-linked mutations in A$\beta$PP account for less than 5% of total AD cases, these autosomal dominant mutations are highly penetrant, and the clinical and pathologic symptoms of individuals with FAD mutations are nearly identical to those of patients

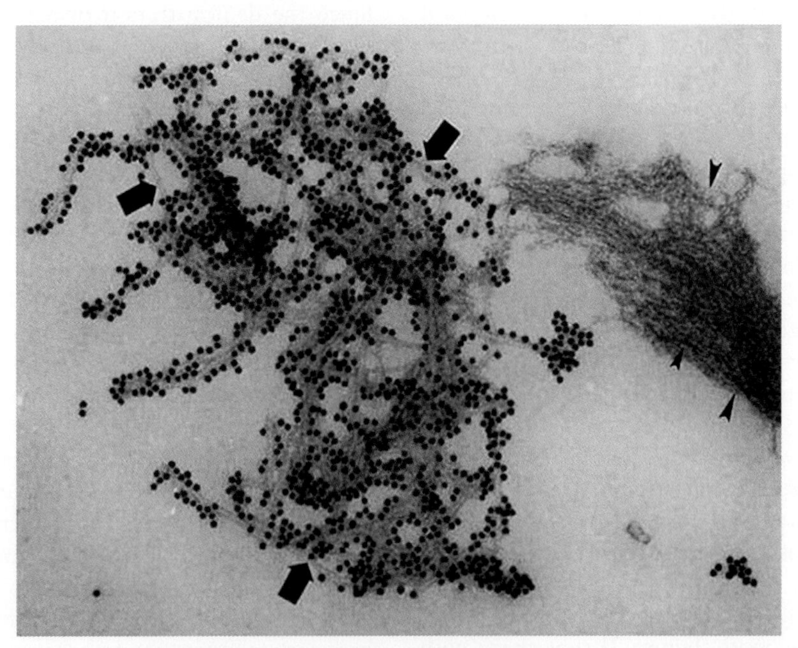

**Figure 2** Structural and antigenic differences between paired helical filaments (large arrows) and amyloid-$\beta$ (A$\beta$) filaments (arrowheads) are readily apparent in this negatively stained preparation. Heavy subunit of neurofilaments is localized by colloidal gold to paired helical filaments but not to A$\beta$ filaments. (×105 000)

with late-onset sporadic AD. Biochemically, FAD-associated mutants in PSEN1 and PSEN2 lead to selective increase in the levels of Aβ42 species, which readily aggregate *in vitro* and are the initial Aβ species deposited in the brains of individuals with AD and Down syndrome. FAD-linked mutations in AβPP either increase Aβ42 or overall Aβ production or generate highly fibrillogenic Aβ variants. In addition to early-onset FAD mutations, it was also discovered that the presence of ε4 allele of the apolipoprotein E (APOE) gene is a risk factor for familial late-onset AD. APOE plays a vital role in the metabolism and clearance of Aβ along with α2-macroglobulin (α2M) and low-density lipoprotein receptor. The biochemical outcome of harboring the APOE ε4 allele includes increased Aβ aggregation and decreased Aβ clearance.

The number of genes involving distinct proteins suggests that the pathologic and clinical entity we call AD is not unique to abnormalities in a single gene locus or etiology. Significantly, the vast majority of AD is not linked to any established genetic abnormality.

## Aβ and Its Protein Precursor

Understanding of the pathogenesis of Aβ deposition was greatly advanced by the sequencing of Aβ by G Glenner and C Wong in 1983. Subsequent cloning showed that Aβ is a 39–42-amino-acid fragment of a larger 695–770-amino-acid, membrane-spanning glycoprotein termed Aβ precursor protein (AβPP). The gene for AβPP resides as a single copy on the long arm of chromosome 21 and forms the locus for some cases of familial AD.

Aβ is proteolytically released from a large type 1 membrane glycoprotein of unknown function, the AβPP, via sequential cleavages by two aspartyl proteases, referred to as the β- and γ-secretases. The Aβ region of AβPP comprises the 28 residues just outside the single transmembrane domain, plus the first 12–14 residues of that buried domain. On this basis, Aβ was originally assumed to arise only under pathological circumstances, in that the second cleavage was thought to require some kind of prior membrane disruption to allow access of γ-secretase and a water molecule to the otherwise intramembranous region. This concept was disproved in 1992, when Aβ was shown to be constitutively released from AβPP and secreted by mammalian cells throughout life and thus occur normally in plasma and cerebrospinal fluid. This discovery enabled the dynamic study of Aβ production in cell culture and animal models, including examination of the effects of AD-causing genetic mutations. Moreover, high-throughput screening could now be conducted on cultured cells to identify Aβ-lowering compounds and determine their mechanism. Most AβPP molecules that undergo secretory processing are cleaved by α-secretase, rather than β-secretase, near the middle of the Aβ region. This releases the large, soluble ectodomain (APPs-α) into the medium and allows the resultant 83-residue, membrane-retained, C-terminal fragment (C83) to be cleaved by γ-secretase, generating the small p3 peptide. α-secretase acts on AβPP molecules at the cell surface although some processing also occurs in intracellular secretory compartments. The precise subcellular loci of the β- and γ-secretase cleavages are unclear but likely include early, recycling endosomes. The functional consequences of the proteolytic processing of AβPP remain ill-defined. The current leading hypothesis is that cleavage by α-secretase followed by γ-secretase enables the release of the AβPP intracellular domain (AICD) into the nucleus, where it may participate in transcriptional signaling. The APPs-α derivative secreted as a result of this processing appears to have distinct extracellular functions. For example, those APPs-α isoforms that contain an alternatively spliced Kunitz protease inhibitor domain function as serine protease inhibitors, including by inhibiting of Factor XIa in the coagulation cascade.

The importance of AβPP in the primary etiology of AD is fairly well established, with FAD associated with a number of mutations in the AβPP gene on chromosome 21. Some of the mutations in AβPP leading to AD have been related to AβPP processing, yielding more Aβ or the longer form of Aβ, which has a greater propensity to form Aβ fibrils. Additional support for the importance of AβPP in the pathogenesis of AD comes from Down's syndrome, in which an extra copy of chromosome 21 leads at midlife to a spectrum of pathologic changes similar to those found in AD.

## Cytoskeletal Abnormalities

A fundamental process in the pathogenesis of AD is the breakdown of the cytoskeleton. The main component of NFTs is the paired helical filaments (PHF), which are mainly comprised of the protein tau in an abnormally phosphorylated status. Tau filaments accumulate in dystrophic neurites as fine neuropil threads or as bundles of PHF in neuronal bodies forming the NFTs, which become extracellular ghost tangles after the death of the neuron. The severity of dementia has been correlated with accumulation of NFTs in different brain regions, while with SP, such correlation has not been demonstrated. In AD, tau binds with lower affinity to microtubules, and it self-aggregates into aberrant structures, probably helped by other molecules. As a consequence of hyperphosphorylation, tau shows a loss of microtubule-binding capacity and is accumulated in neuronal bodies.

The finding that neurons are primarily responsible for AβPP production and also contain NFTs highlights a key issue in AD, the relationship of Aβ to NFTs. The application of monoclonal antibodies, immunoelectron microscopy, and antibody affinity techniques has shown that microtubule-associated protein tau and the heavy subunit of neurofilaments are major components of NFTs. Recent studies demonstrated a direct high-affinity interaction between tau and AβPP. More studies are required to understand the pathologic significance of tau-AβPP interaction in Aβ deposition.

Sparing solubility and heterogeneity of enriched fractions have hampered efforts to define NFTs quantitatively, and there is now a considerable effort to understand how the two identified posttranslational modifications of NFTs, phosphorylation and glycation, mediate the transformation of soluble tau into insoluble paired helical filaments. Indeed, although increased phosphorylation would lead to microtubule instability – a key hallmark of AD – phosphorylation does not mediate paired helical filament insolubility. Conversely, oxidation of tau confers the same solubility properties as paired helical filaments. Therefore, it is likely that oxidative cross-links play a role in NFT insolubility. Significantly, oxidation is one of the earliest changes in the disease that leads to neuronal dysfunction. It is not surprising, therefore, that therapeutic efforts to reduce oxidative stress slow disease progression and/or decrease the incidence of the disease.

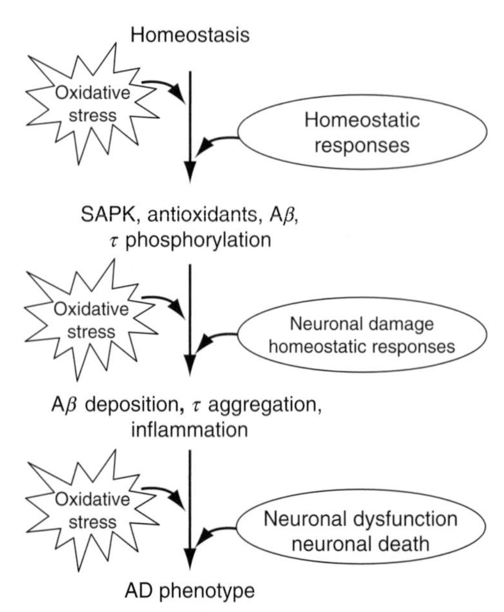


**Figure 3** The sequence of events in Alzheimer's disease (AD). Oxidative stress, the earliest event occurring in AD, initially leads neurons to activate neuronal defenses, including stress-activated protein kinases (SAPK), tau (τ) phosphorylation, and amyloid-β (Aβ), to maintain homeostatic balance. However, given the chronic and insidious nature of oxidative stress, progression of the disease through Aβ deposition, tau aggregation, and inflammation overwhelms initial compensatory mechanisms and culminates in neuronal dysfunction and death, that is, the AD phenotype.

## Oxidative Stress in AD

It has been shown that increased oxidative damage is a prominent and early feature of vulnerable neurons in AD. Over the past decade, an oxidative stress-related modification of macromolecules has been described in association with the susceptible neurons of AD. Such modifications include advanced glycation end products, nitration, lipid peroxidation adduction products, carbonyl-modified neurofilament protein, and free carbonyls, as well as glycation and glycoxidation products. Levels of these markers are initially elevated following some unknown triggering neuronal event, but these levels soon decrease as the disease progresses to advanced AD. Together these findings suggest that increased oxidative damage is not the terminal sequelae of the disease but instead plays an initial role. They also suggest that damage does not mark further destruction by reactive species and is instead marked by a broad array of increased cellular defenses. It can be argued that in AD, these defenses are responsible for the reduction of damage if we view AD in isolation. However, when seen in the context of other conditions in which reactive oxygen and nitrogen species are involved and damage is either limited or absent, such as Parkinson's disease, this result raises the question of whether oxidative damage noted in AD may be better thought of as homeostatic, that is, that oxidative damage could initiate signal transduction pathways to manipulate cellular responses to stress, which are characterized by increased levels of reactive oxygen and nitrogen species. Furthermore, there is evidence that in the first stage of AD development, Aβ deposition and hyperphosphorylated tau function as compensatory responses and downstream adaptations to ensure that neuronal cells do not succumb to oxidative damage (**Figure 3**).

## Conclusions

The numerous genetic abnormalities as well as the high prevalence of sporadic disease suggest that AD has a defined pathogenesis stemming from numerous etiologies. Consequently, AD is a syndrome now defined by the most striking aspects of that pathogenesis, that is, NFTs and SP. Since oxidative damage is the earliest described cytopathological abnormality that occurs in vulnerable brain regions and selective neuronal populations, it is extremely important to decipher the causes and consequences of neuronal oxidative stress. Clarification of mechanistic viewpoints concerning AD pathophysiology may provide insights into efficacious therapeutics. In particular, investigation must uncover what the relationships are between increased oxidative stress and other facets of the disease such as regionally selective neuronal degeneration, A$\beta$ deposition as SP, and tau phosphorylation and aggregation as NFTs.

*See also:* Aging of the Brain and Alzheimer's Disease; Alzheimer's Disease: Neurodegeneration; Alzheimer's Disease: Molecular Genetics; Alzheimer's Disease: Transgenic Mouse Models; Alzheimer's Disease: MRI Studies; Animal Models of Alzheimer's Disease; Axonal Transport and Alzheimer's Disease; Brain Glucose Metabolism: Age, Alzheimer's Disease and ApoE Allele Effects.

## Further Reading


Glenner GG and Wong CW (1984) Alzheimer's disease: Initial report of the purification and characterization of a novel cerebrovascular amyloid protein. *Biochemical and Biophysical Research Communications* 120: 885–890.

Hendrie HC (1998) Epidemiology of dementia and Alzheimer's disease. *American Journal of Geriatric Psychiatry* 6: S3–S18.

Hernandez F, Engel T, Gomez-Ramos A, et al. (2005) Characterization of Alzheimer paired helical filaments by electron microscopy. *Microscopy Research and Technique* 67: 121–125.

Katzman R (1993) Clinical and epidemiological aspects of Alzheimer disease. *Clinical Neuroscience* 1: 165–170.

Khachaturian ZS (1985) Diagnosis of Alzheimer's disease. *Archives of Neurology* 42: 1097–1105.

Khachaturian ZS and Mesulam M-M (eds.) (2000) *Special Issue: Alzheimer's disease: A Compendium of Current Theories. Annals of the New York Academy of Sciences* 924.

Levy-Lahad E, Wasco W, Poorkaj P, et al. (1995) Candidate gene for the chromosome 1 familial Alzheimer's disease locus. *Science* 269: 973–977.

Moreira PI, Honda K, Zhu X, et al. (2006) Brain and brawn: Parallels in oxidative strength. *Neurology* 66: S97–S101.

Perry G (ed.) (1993) Alzheimer's Disease. *Clinical Neuroscience* 1: 163–224.

Perry G and Smith MA (1993) Senile plaques and neurofibrillary tangles: What role do they play in Alzheimer disease? *Clinical Neuroscience* 1: 199–203.

Post SG (2000) *The Moral Challenge of Alzheimer Disease: Ethical Issues from Diagnosis to Dying*, 2nd edn. Baltimore: Johns Hopkins University Press.

Roses AD (1994) Apolipoprotein E affects the rate of Alzheimer disease expression: $\beta$-Amyloid burden is a secondary consequence dependent on ApoE genotype and duration of disease. *Journal of Neuropathology and Experimental Neurology* 53: 429–437.

Selkoe DJ (1994) Normal and abnormal biology of the $\beta$-amyloid precursor protein. *Annual Review of Neuroscience* 17: 489–517.

Selkoe DJ (1994) Alzheimer's disease: A central role for amyloid. *Journal of Neuropathology and Experimental Neurology* 53: 438–447.

Selkoe DJ (2004) American College of Physicians, and American Physiological Society. Alzheimer disease: Mechanistic understanding predicts novel therapies. *Annals of Internal Medicine* 140: 627–638.

Sherrington R, Rogaev EI, Liang Y, et al. (1995) Cloning of a gene bearing missense mutations in early-onset familial Alzheimer's disease. *Nature* 375: 754–760.

Smith MA, Sayre LM, Monnier VM, et al. (1995) Radical ageing in Alzheimer's disease. *Trends in Neurosciences* 18: 172–176.

Smith MA, Siedlak SL, Richey PL, et al. (1995) Tau protein directly interacts with the amyloid $\beta$-protein precursor: Implications for Alzheimer's disease. *Nature Medicine* 1: 365–369.

Tanzi RE and Bertram L (2005) Twenty years of the Alzheimer's disease amyloid hypothesis: A genetic perspective. *Cell* 20: 545–555.


## Relevant Websites

http://www.alz.org – Alzheimer Association.

http://www.alzforum.org – Alzheimer Research Forum.

http://www.molgen.ua.ac.be – Molecular Genetics Department, VIB, The Flanders Institute for Biotechnology.

http://www.nih.gov/nia – National Institute on Aging.

# Alzheimer's Disease: Neurodegeneration

**N J Cairns**, Washington University School of Medicine, St. Louis, MO, USA



## Neuropathology

Alzheimer's disease (AD) is clinically, neuropathologically, and genetically heterogeneous. Subgroups include familial and sporadic forms; there are also phenotypic differences according to the gene and the defect within a single gene. Not infrequently, atypical cases may be seen, and AD may be found in combination with another neurodegenerative disease, most frequently dementia with Lewy bodies, which may also contribute to the cognitive deficits.

Patients with AD often die of bronchopneumonia and there is often little pathology outside the central nervous system (CNS). Depending on the stage of disease, the brain may appear unremarkable to the naked eye or may be grossly atrophic (**Figure 1**). Brain weight is typically reduced, often to less than 1000 g, from the average of 1250–1400 g. The atrophy is usually symmetrical and affects the frontal, temporal, and parietal lobes, with relative sparing of the sensorimotor cortices and occipital lobe, although all cortical areas may be affected in the most severe cases. Senile plaques and neurofibrillary tangles are frequently found with other lesions which are not specific to AD, as they are found in the aged brain, including granulovacuolar degeneration, Hirano bodies, gliosis, synaptic and neuronal loss, and white-matter and vascular changes.

The senile or neuritic plaque is one of the signature lesions of the AD brain (**Figure 2**). These complex extracellular structures range in size from 50 to 200 μm and are readily visualized by silver impregnation methods. In research settings and diagnostic laboratories, immunohistochemistry may be available, and these structures can be detected by antibodies raised against specific epitopes of the pathological protein (A$\beta$). Immunohistochemistry generally reveals much more extensive pathology than that seen by traditional silver and other staining methods and is now the method of choice for detecting the molecular pathology of most neurodegenerative diseases, not just of AD. The classical senile plaque consists of an amyloid core with a ring or crown, as seen in cross section with the light microscope, of argyrophilic axonal and dendritic processes, amyloid fibrils, astrocytic processes, and microglial cells. The neuritic processes of the senile plaque are often dystrophic and contain abnormal paired helical filaments (PHFs) made up largely of hyperphosphorylated tau protein. The amyloid is composed of 5–10 nm filaments of amyloid $\beta$-protein (A$\beta$), a 39- to 43-amino-acid (4 kDa) protein, a cleavage product of a transmembranous amyloid precursor protein (APP).

A$\beta$ immunohistochemistry reveals a wider spectrum of plaque types than is seen with traditional staining methods. A$\beta$ deposits include subpial, diffuse, ring-with-core, compact, vascular, and dyshoric deposits (**Figure 3**). A$\beta$ plaques may be found throughout the brain, including the neocortex, amygdala, hippocampus, striatum, pallidum, nucleus basalis of Meynert, thalamus, midbrain, medulla oblongata, cerebellar cortex, and spinal cord. Immunohistochemistry using antibodies that recognize a full-length A$\beta_{1-42(43)}$ or a truncated A$\beta_{1-40}$ detects most A$\beta$ species in the AD brain. The predominant species in sporadic AD is A$\beta_{1-42(43)}$, which is present in most plaques types; A$\beta_{1-42(43)}$ is found only in diffuse plaques, indicating that this species may be more toxic. This widespread deposition of A$\beta$ protein is evidence that AD is an A$\beta$-amyloidosis of the CNS. *In vitro* and *in vivo* models also provide evidence that soluble A$\beta$ oligomers may also be injurious to neurons and disrupt synaptic function (**Figures 4–6**).

The second histological signature lesion of AD is the neurofibrillary tangle (NFT; **Figures** 7 and **8**). The NFT is not specific to AD; NFTs occur in aging and other neurodegenerative diseases, including Down syndrome, dementia pugilistica, postencephalitic parkinsonism, amyotrophic lateral sclerosis–parkinsonism–dementia complex of Guam, subacute sclerosing panencephalitis, dementia with tangles with and without calcification, and myotonic dystrophy. NFTs also are characteristic of frontotemporal lobar degeneration (FTLD) with tauopathy, which includes corticobasal degeneration, progressive supranuclear palsy, and argyrophilic grain disease, but their ultrastructure and tau isoform patterns, as demonstrated by immunohistochemistry or biochemistry, differ from those seen in AD and the aging brain.

In AD, as in the aged brain of cognitively normal individuals, NFTs are typically numerous in the medial temporal lobe, but in AD their distribution is much more widespread and may involve neocortex and subcortical nuclei. NFTs are intraneuronal inclusions composed of cytoskeletal components, mainly tau, and their shape is determined by the type of neuron in which they develop – for example, in pyramidal neurons they are flame-shaped, while in subcortical neurons they may be globose in appearance. Mature NFTs are

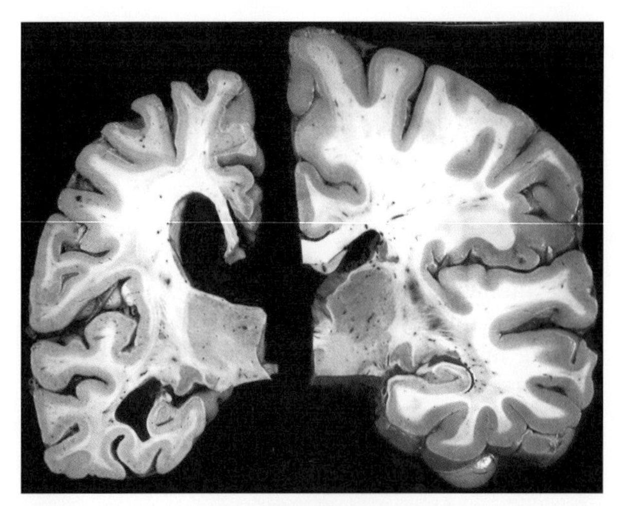

**Figure 1** On the left, a coronal slice of the hemi-brain of a 70-year-old patient with severe Alzheimer's disease, showing atrophy: the lateral ventricle is enlarged with rounding of its angle, several gyri are narrowed, the hippocampus is small, and the lateral fissure is widened. On the right, a slice of the right hemi-brain of a normal age-matched individual.

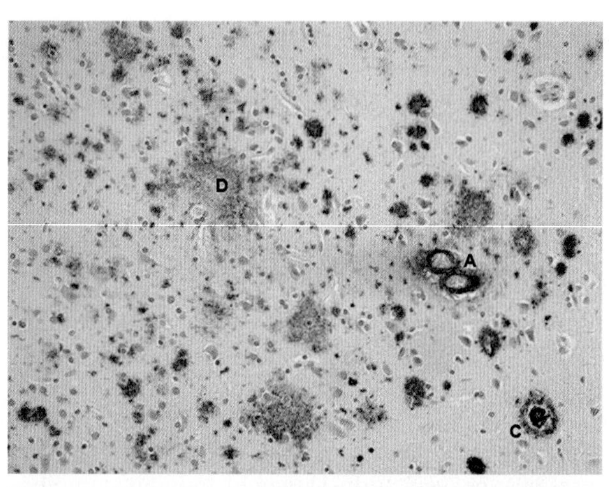


**Figure 3** Extracellular amyloid $\beta$-protein deposits are present through all layers of the neocortex and have variable morphology: a diffuse plaque (D), a ring-with-core plaque (C), and cerebral amyloid angiopathy (A). Amyloid $\beta$-protein (10D5 antibody) immunohistochemistry.

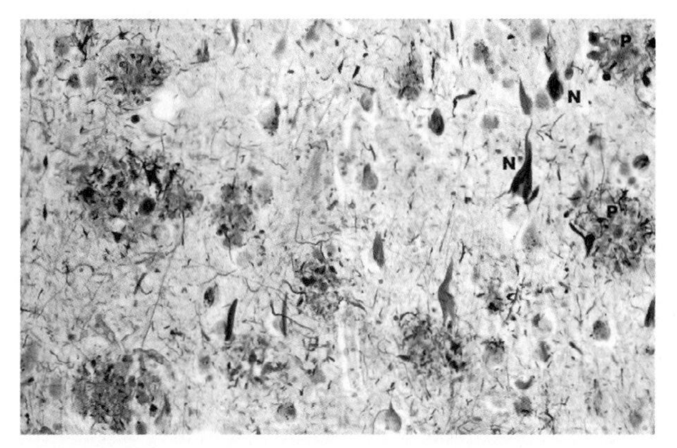


**Figure 2** Neurofibrillary tangles (N) and neuritic plaques (P) in the hippocampus. Modified Bielschowsky silver impregnation.

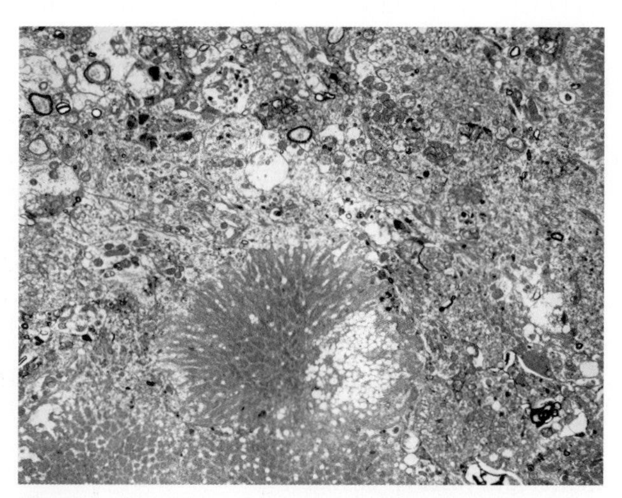

**Figure 4** A low-power electron micrograph revealing a compact senile plaque.

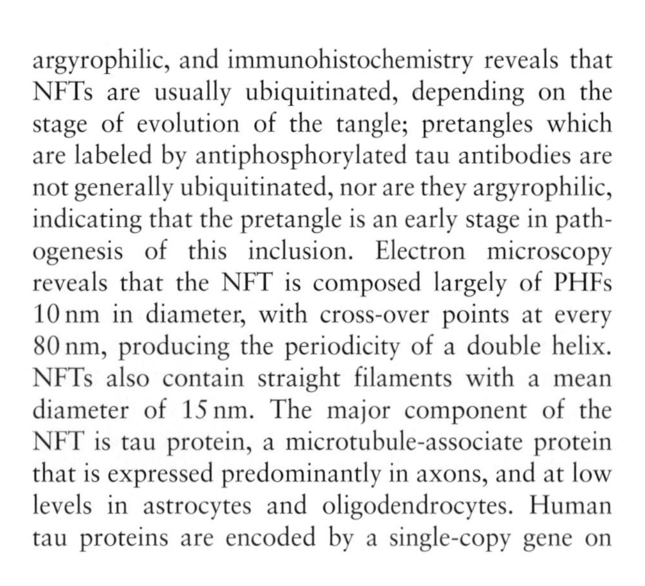

argyrophilic, and immunohistochemistry reveals that NFTs are usually ubiquitinated, depending on the stage of evolution of the tangle; pretangles which are labeled by antiphosphorylated tau antibodies are not generally ubiquitinated, nor are they argyrophilic, indicating that the pretangle is an early stage in pathogenesis of this inclusion. Electron microscopy reveals that the NFT is composed largely of PHFs 10 nm in diameter, with cross-over points at every 80 nm, producing the periodicity of a double helix. NFTs also contain straight filaments with a mean diameter of 15 nm. The major component of the NFT is tau protein, a microtubule-associate protein that is expressed predominantly in axons, and at low levels in astrocytes and oligodendrocytes. Human tau proteins are encoded by a single-copy gene on chromosome 17q21. In adult human brain, alternative splicing of exons 2, 3, and 10 generates six tau isoforms, ranging from 352 to 441 amino acids in length, which differ by the presence of either three or four microtubule binding repeats (3R tau or 4R tau, respectively), of 31 or 32 amino acids each. Additionally, alternative splicing of exons 2 and 3 leads to the absence (0N) or presence of inserted sequences of 29 (1N) or 58 (2N) amino acids in the N-terminal third of the molecule. In the adult human brain, the ratio of 3R:4R tau isoforms is approximately 1:1, while in other FTLDs with tauopathy there is a preponderance of 3R tau (Pick's disease), 4R tau (corticobasal degeneration, progressive supranuclear palsy, and agyrophilic grain disease), or 3R, 4R, or 3R and 4R tau, as seen in cases of FTLD with

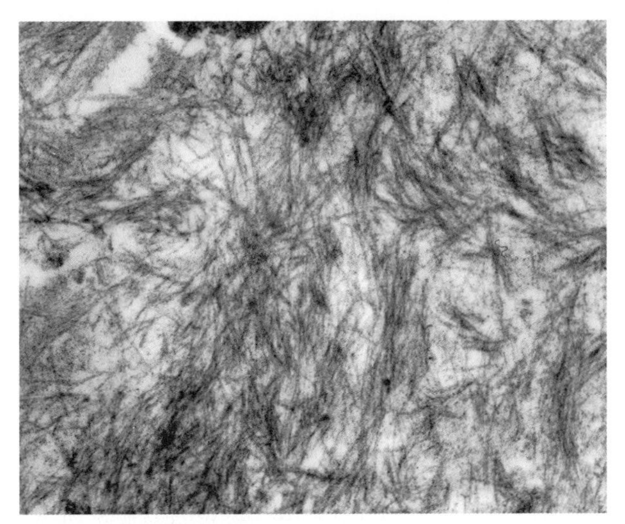

**Figure 5** A high-power electron micrograph of loosely aggregated amyloid filaments of the corona of an amyloid $\beta$-protein plaque.

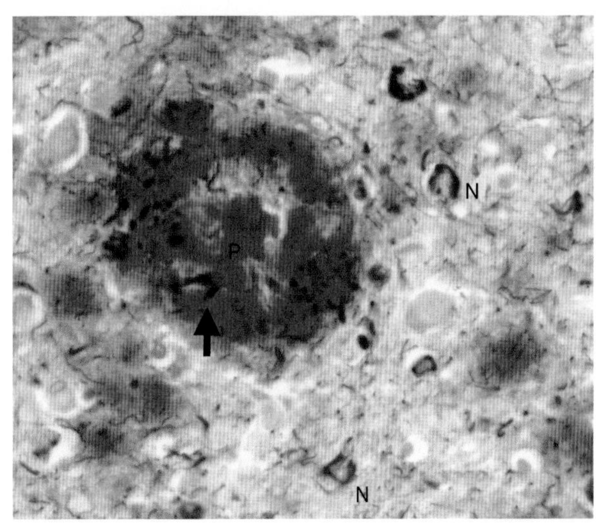


**Figure 7** An amyloid $\beta$-protein plaque (P) and neurofibrillary tangles (N) in the superior temporal lobe of an Alzheimer's disease brain. The plaque contains tau-immunoreactive dystrophic neurites (arrow). Tau (black, PHF1 antibody) and amyloid $\beta$-protein (red, 10D5 antibody) immunohistochemistry.

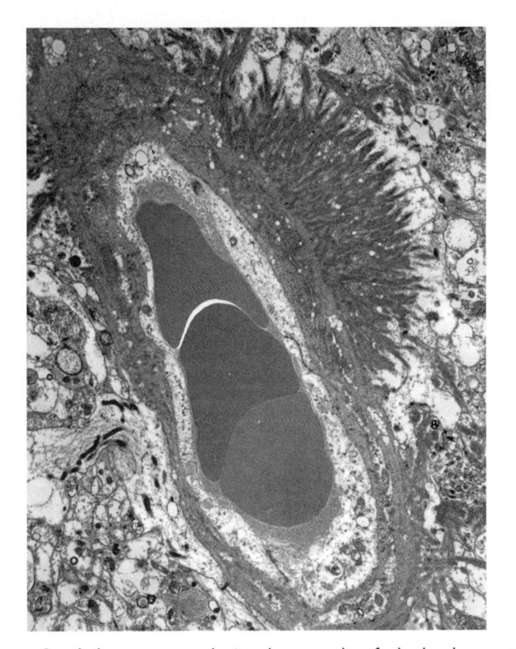

**Figure 6** A low-power photomicrograph of dyshoric cerebral amyloid angiopathy with amyloid fibrils extending from the vessel wall and radiating into the surrounding brain parenchyma.

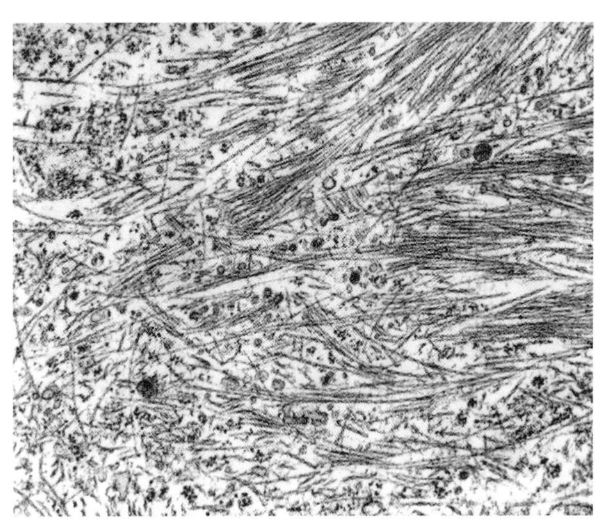

**Figure 8** A high-power electron micrograph revealing paired helical filaments of a neurofibrillary tangle. Micrograph kindly provided by the late Professor LW Duchen, University of London, UK.

*microtubule-associated protein tau (MAPT)* gene mutation.

Tau binds to and stabilizes microtubules (MTs) and promotes MT polymerization. The function of tau as an MT binding protein is regulated by phosphorylation. Phosphorylation at approximately 30 of these sites has been identified in normal tau protein. Several protein kinases and protein phosphatases have been implicated in regulating the phosphorylation state and thus the function of tau. The phosphorylation sites are clustered in regions flanking the MT binding repeats, and increasing tau phosphorylation at multiple sites negatively regulates MT binding. All six isoforms of tau in AD are hyperphosphorylated, and phosphorylation-specific anti-tau antibodies may be used to identify three sites of tau pathology: the NFT, dystrophic neurites of senile plaques, and neuropil threads. In both sporadic and familial AD, tau is hyperphosphorylated, and it is this 'abnormal' tau that is the principal component of the filamentous aggregates in neurons and one of the pathological hallmarks of this disorder.

## The A$\beta$ Cascade Hypothesis

The initial observation that A$\beta$ deposits are present in the blood vessels of young patients with Down syndrome (DS) was followed by the discovery that A$\beta$ plaques and NFTs are present in adult DS and AD. These studies led to the hypothesis that a gene encoding A$\beta$ was located on the long arm of chromosome 21. Subsequently, mutations in the APP gene were identified in a family with early-onset AD, and then additional mutations were reported. These neuropathological, biochemical, and genetic findings led to the hypothesis that increased A$\beta$ production leads to a catastrophic cascade, including synaptic alterations, fibrilization, microglial and astrocytic activation, abnormal phosphorylation of tau proteins to form oligomers and the PHFs of NFTs, progressive synaptic and neuronal loss, loss of multiple neurotransmitters (especially acetylcholine), and ultimately to dementia (**Figure 9**).

The generation of A$\beta$ peptides is an example of regulated intramembrane proteolysis. APP processing is regulated by either of two membrane-bound proteases: $\alpha$-secretase (a type of metalloproteinase) and $\beta$-secretase, also called $\beta$-site APP-cleaving enzyme (BACE). The membrane-associated fragment caused by BACE cleavage may undergo further cleavage by the $\beta$-secretase complex, an aspartyl protease composed of presenilin-1 (PS1) or PS2, nicastrin, APH1 (anterior pharynx defective-1), and an anti-presenilin protein enhancer (PEN2). All four proteins are necessary and sufficient to form the $\gamma$-secretase complex. Further evidence for the importance of A$\beta$ oligomerization in AD comes from the observation that mutations in PS1 and PS2, components of the $\gamma$-secretase complex, lead to enhanced production of A$\beta_{1-42}$. Neuropathological studies confirm that

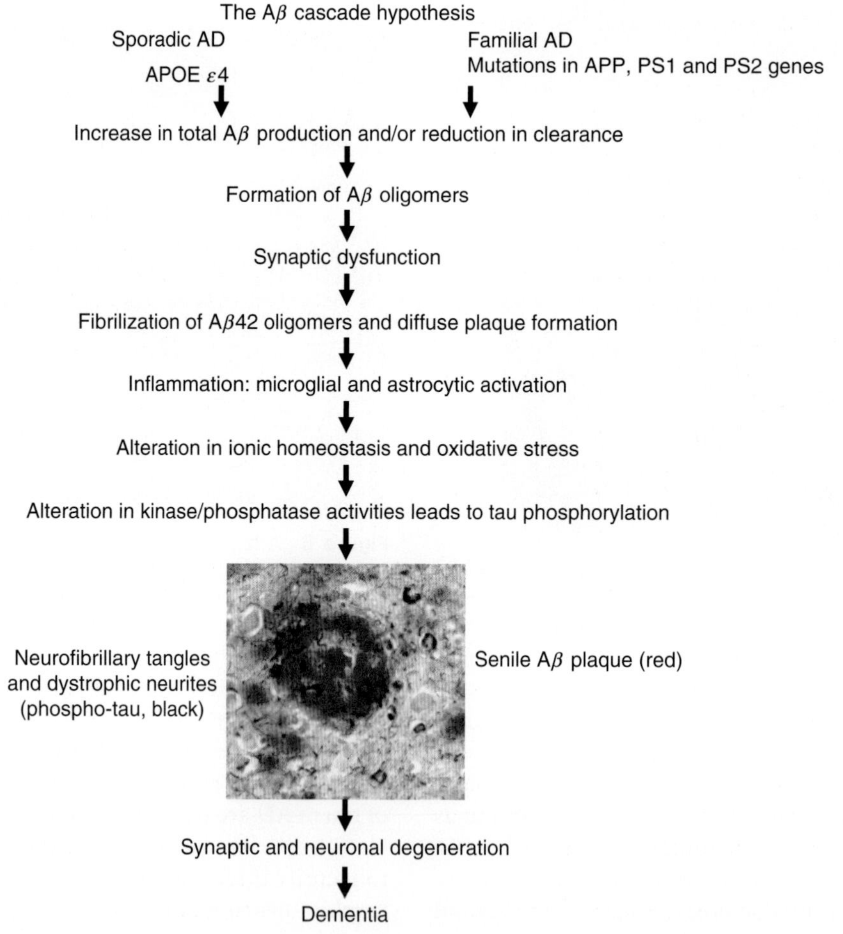


**Figure 9** The amyloid $\beta$-protein (A$\beta$) cascade hypothesis of Alzheimer's disease (AD). APOE, apolipoprotein E; APP, amyloid precursor protein; PS1, PS2, presenilins.

$A\beta_{1-42}$ is present in diffuse plaques, an early stage in senile plaque pathogenesis. The molecular dissection of the $A\beta$ cascade has generated several potential targets for therapeutic intervention: inhibitors of $\beta$- and $\gamma$-secretases that generate $A\beta$, statins, $A\beta$ vaccines, and nonsteroidal anti-inflammatory drugs (NSAIDs) are some of the classes of compounds targeting $A\beta$-induced neurodegeneration.

## The Tau Hypothesis

Several sporadic and familial neurodegenerative disorders, including AD, that are characterized clinically by dementia and/or motor dysfunction are characterized pathologically by abnormal intracellular accumulations of the microtubule-associated protein tau (MAPT); these disorders are collectively called tauopathies. The progressive accumulation of filamentous tau inclusions in the absence of other disease-specific neuropathological abnormalities provides evidence implicating tau dysfunction in disease onset and progression. However, the discovery of pathogenic *MAPT* gene mutations in the heterogeneous group of disorders known as frontotemporal lobar degeneration (FTLD) with *MAPT* mutation, also called frontotemporal dementia with parkinsonism linked to chromosome 17 (FTDP-17), provided confirmation of the central role of tau abnormalities in the etiology of neurodegenerative tauopathies. These findings have opened up novel areas for investigation into the mechanisms of tau dysfunction and the relationship of tau abnormalities to brain degeneration.

Several transgenic models of tau pathology have been generated by overexpressing human tau proteins in mice. However, these mice are either asymptomatic or develop pathology that is localized to the spinal cord and/or lacks many of the key features of tau-based disorders. In contrast, the introduction of the P301L mutation led to the development of transgenic mice that develop age- and gene dose-dependent accumulation of tau tangles in the brain and spinal cord, with associated nerve cell loss and gliosis as well as behavioral abnormalities. Similar to human disease, the tau aggregates are composed of only mutant human tau, further implicating the P301L change in promoting the selective aggregation of mutant tau. Other systems have also been developed to model various aspects of human tauopathies, including a transgenic mouse overexpressing the shortest human tau isoform; this mouse acquired age-dependent taupathology similar to that seen in FTLD with *MAPT* mutation. Overexpression of either wild-type or mutant tau in *Drosophila melanogaster* demonstrates features of tauopathy, including adult-onset progressive neurodegeneration with accumulation of abnormal tau. However, the neurodegeneration in this model occurred in the absence of NFT formation. More recent studies have demonstrated NFT-like pathology when tau is co-expressed with *shaggy*, a homolog of glycogen synthase 3-kinase (GSK3), an enzyme implicated in tau phosphorylation. Neurodegeneration and defective neurotransmission have also been demonstrated in *MAPT* transgenic *Caenorhabditis elegans*. In this model, panneuronal expression of normal and mutant tau resulted in altered behavior, accumulation of insoluble phosphorylated tau, age-dependent loss of axons and neurons, and structural damage to axonal tracts. These models recapitulate various features of the tauopathies and highlight targets for disease-modifying therapies, not only for AD but also for related tauopathies. Thus, microtubule-stabilizing drugs might ameliorate the sequestration of tau into NFTs, and drugs such as LiCl, in an animal model of tauopathy, that impede the abnormal phosphorylation of tau by inhibiting GSK3 indicate a fruitful area for drug discovery.

## Timeline of Neurodegenerative Changes in AD

Many nondemented older individuals have few or no AD-type changes in the brain. However, about 30% of cognitively normal individuals over the age of 75 years have both diffuse and neuritic plaques throughout the neocortex, identical to those seen in individuals with symptomatic AD (**Figure 10**). The change in the brain that corresponds most closely to the clinical symptoms of those with very mild dementia (Clinical Dementia Rating = 0.5) is neuronal loss in the medial temporal lobe, including the entorhinal cortex and hippocampus; cognitively normal individuals, in contrast, have minimal or no loss of neurons from these areas. Thus, there is a 'preclinical' stage of AD in which there are sufficient plaques in neocortical areas for a neuropathological diagnosis of AD, but there is not sufficient synaptic and neuronal loss to produce cognitive symptoms. In familial cases, misfolding of $A\beta$ may start at a very young age, and only after two to three decades may result in the neuropathology of AD. Research efforts are now focusing on this preclinical stage of AD, which may extend years or even decades before the onset of clinical symptoms. The development of antecedent biomarkers that detect asymptomatic AD is a priority; efforts include clinical, cognitive, neuropsychological, structural and functional imaging, genetic, and proteomic studies. Tau and $A\beta$ protein levels in cerebrospinal fluid are potentially additional markers of preclinical

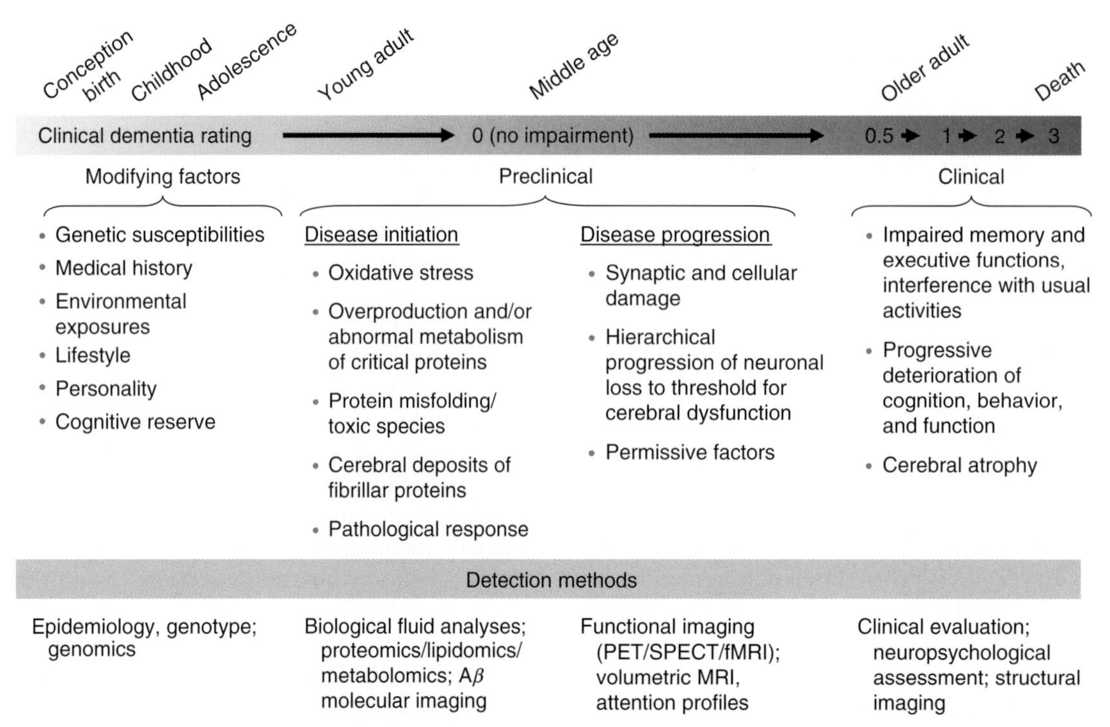


**Figure 10** Hypothetical timeline for Alzheimer's disease. PET, positron emission tomography; SPECT, single-photon computed tomography; fMRI, functional magnetic resonance imaging.

AD, as is the A$\beta$ imaging agent, Pittsburgh compound B (PIB). It is likely that a combination of biomarkers will provide greater diagnostic accuracy than is possible using any single analyte or measure.

## Conclusions

The accumulations of filamentous A$\beta$ species in the extracellular space, and of abnormally phosphorylated tau proteins within neurons, are the signature lesions of AD. Protein misfolding, fibrilization, and aggregate formation are not unique to AD, but are common features of a wide variety of sporadic and familial neurodegenerative disorders (synucleinopathies, tauopathies, trinucleotide repeat disease, and TDP-43 proteinopathies). These diseases are distinguished by the distinct topographic and cell type-specific distribution of inclusions. The biochemical and ultrastructural characteristics of the inclusions also reveal a significant phenotypic overlap. The discovery that multiple mutations in *APP*, *PS1*, and *PS2* genes lead to abnormal protein aggregation in AD demonstrates that neuronal A$\beta$ mismetabolism is sufficient to produce neurodegenerative disease. Experimental evidence indicates that mutations lead to specific alterations in expression, function, and biochemistry of A$\beta$ and tau proteins. The realization that AD pathology is present many years prior to the onset of clinical symptoms has led to dramatic progress in the search for antecedent biomarkers which herald disease onset. These new markers, once validated, will facilitate the identification of individuals at the preclinical or early stage of AD, and these individuals stand to benefit greatly from emerging therapeutic interventions.

*See also:* Aging of the Brain and Alzheimer's Disease; Alzheimer's Disease: An Overview; Alzheimer's Disease: Molecular Genetics; Alzheimer's Disease: Transgenic Mouse Models; Alzheimer's Disease: MRI Studies; Axonal Transport and Alzheimer's Disease; Axonal Transport and Neurodegenerative Diseases; Brain Glucose Metabolism: Age, Alzheimer's Disease and ApoE Allele Effects; Oxidative Damage in Neurodegeneration and Injury; Transgenic Models of Neurodegenerative Disease.

## Further Reading

Braak H, Alafuzoff I, Arzberger T, et al. (2006) Staging of Alzheimer disease-associated neurofibrillary pathology using paraffin sections and immunocytochemistry. *Acta Neuropathologica* 112: 389–404.

Cairns NJ, Lee VM-Y, and Trojanowski JQ (2004) The cytoskeleton in neurodegenerative diseases. *Journal of Pathology* 204: 438–449.

Csernansky JG, Wang L, Swank J, et al. (2005) Preclinical detection of Alzheimer's disease: Hippocampal shape and volume predict dementia onset in the elderly. *NeuroImage* 25: 783–792.

Forman MS, Trojanowski JQ, and Lee VM-Y (2004) Neurodegenerative diseases: A decade of discoveries paves the way for therapeutic breakthroughs. *Nature Medicine* 10: 1055–1063.
Goedert M, Spillantini MG, Cairns NJ, et al. (1992) Tau proteins of Alzheimer paired helical filaments: Abnormal phosphorylation of all six brain isoforms. *Neuron* 8: 159–168.
Haass C and Selkoe DJ (2007) Soluble protein oligomers in neurodegeneration: Lessons from the Alzheimer's amyloid beta-peptide. *Nature Reviews – Molecular Cell Biology* 8: 101–112.
McKhann GM, Albert MS, Grossman M, et al. (2001) Clinical and pathological diagnosis of frontotemporal dementia: Report of the Work Group on Frontotemporal Dementia and Pick's Disease. *Archives of Neurology* 58: 1803–1809.
Mirra SS, Heyman A, McKeel D, et al. (1991) The Consortium to Establish a Registry for Alzheimer's Disease (CERAD). Part II. Standardization of the neuropathologic assessment of Alzheimer's disease. *Neurology* 41: 479–486.
Mirra SS and Hyman BT (2002) Ageing and dementia. In: Graham DI and Lantos PL (eds.) *Greenfield's Neuropathology*, 7th edn., pp. 195–271. London: Arnold.
Price JL, Davis PB, Morris JC, et al. (1991) The distribution of tangles, plaques and related immunohistochemical markers in healthy aging and Alzheimer's disease. *Neurobiology of Aging* 12: 295–312.
Price JL and Morris JC (1999) Tangles and plaques in nondemented aging and "preclinical" Alzheimer's disease. *Annals of Neurology* 45: 358–368.

## Relevant Website

http://www.alzforum.org – Alzheimer Research Forum.

# Axonal Transport and Alzheimer's Disease

**L S B Goldstein**, University of California at San Diego, La Jolla, CA, USA



## Basic Features of Alzheimer's Disease

The initiating symptoms or presenting symptoms of Alzheimer's disease (AD) are well known. Problems with memory, language, and cognition are all characteristic features of the behavioral symptoms that signal the initiation of this terrible neurodegenerative disorder. Most AD occurs relatively late in life, generally after 65 years of age. It afflicts 10% of people over the age of 65 and 50% of people over the age of 85. The time from diagnosis to fatality is highly variable, ranging anywhere from a few years to over a decade or more. The tragedy of AD is that not only are there substantial deficits in cognition accompanied by neurodegenerative changes in the brain, there are presently no effective drugs and no effective treatments, and the disease progresses inexorably in its course to end stage, in which a persistent vegetative state is common.

The pathological hallmarks in the brains of people who have had AD are well recognized in postmortem autopsy material. These features include amyloid beta (A$\beta$) plaques and neurofibrillary tangles. Amyloid plaques consist of aggregated A$\beta$ peptides derived by proteolytic processing of the amyloid precursor protein (**Figure 1**). Amyloid plaques are generally associated with dystrophic neurites, which can be abnormal axons or dendrites that are characterized by substantial swelling and distension, and accumulations of vesicles and organelles. Dystrophic neurites surround and invade the plaques, but are also in regions of diseased brains that lack amyloid plaques. Neurofibrillary tangles are aggregates of a protein called tau that normally binds to and stabilizes microtubules, and which, following hyperphosphorylation and conformational change, forms characteristic paired helical filaments. Although amyloid plaques and neurofibrillary tangles are the hallmark diagnostic pathological features, their correlation with cognitive characteristics of AD is not consistent and is controversial. Many researchers have argued that loss of synapses in key areas of the diseased brain is the earliest pathological feature of AD that is most highly correlated with cognitive defects. Neuronal death is an invariant though relatively late feature of AD, as is loss of axons and white matter. The loss of axons, white matter, and synapses could be a consequence of early changes in axonal transport in AD.

Although most AD is late onset, there are well-recognized hereditary forms (familial AD, or FAD) that can begin as early as the fourth decade of life and are highly virulent in their course. Mutations in the gene encoding amyloid precursor protein (APP) and mutations in the genes encoding presenilin (PS) proteins are the major recognized causes of FAD. Duplications of the APP gene can also give rise to mid- to late-onset AD. Two other major risk factors are known. One major genetic risk factor is the allelic state in the gene (*APOE*) encoding apolipoprotein E (ApoE), which has a role in cholesterol trafficking. While the $\varepsilon 2$ variant of the *APOE* gene is protective, the $\varepsilon 4$ variant predisposes to late-onset AD. Another major risk factor is traumatic brain injury. How these risk factors lead to AD is far from clear, and indeed, what molecular defects and pathways lead to any form of AD remain controversial.

The dominant hypothesis proposed to account for the molecular and cellular changes found in AD is the amyloid cascade hypothesis, which proposes that AD is caused by toxic A$\beta_{40}$ or A$\beta_{42}$ peptide fragments liberated by proteolysis of APP (**Figure 2**). In this hypothesis, the hereditary FAD mutations are proposed to generate excess toxic forms of A$\beta$, and apolipoprotein E variants are proposed to affect the rates of turnover of these peptides. Most important, the amyloid cascade hypothesis and its major variants indicate that the toxic A$\beta$ peptides initiate a biochemical and cellular cascade of pathology leading to synaptic loss, neurofibrillary tangles, and ultimately neuronal death. While there is considerable evidence for the amyloid cascade hypothesis, principally in the human APP FAD mutations, questions remain, and there are a number of alternative hypotheses that have been proposed over the years. Another class of hypothesis suggests that early deficits in axonal transport either cause, or contribute substantively to, the development of AD.

## Axonal Transport Is Disrupted in AD

There has been considerable evidence for some time that axonal transport is disrupted at late stages of AD. For example, although basal fore-brain cholinergic neurons exhibit phenotypes that suggest that they lack retrograde neurotrophic input from nerve growth factor, diseased brains appear to retain high levels of

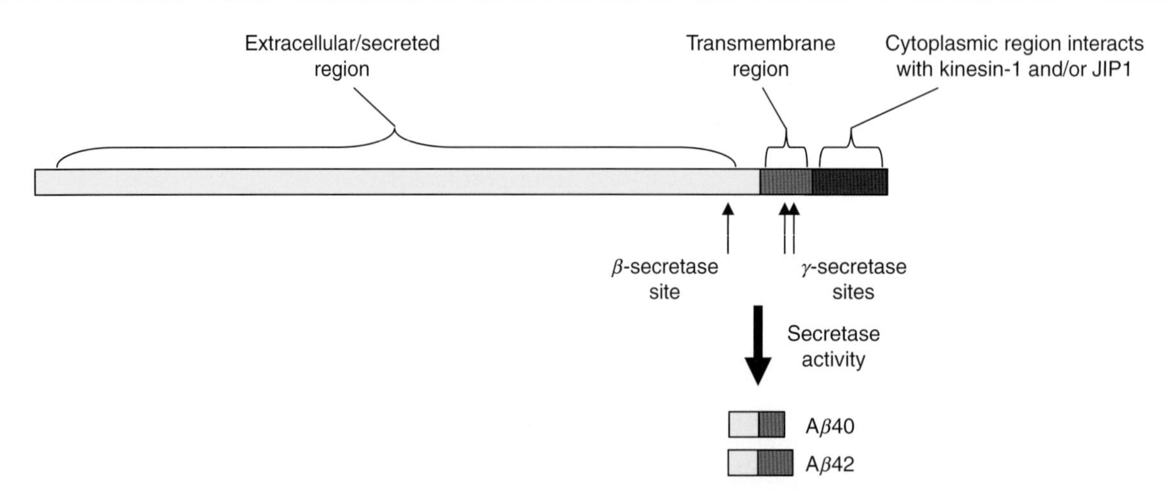


**Figure 1** Schematic diagram of amyloid precursor protein. The cytoplasmic region is thought to interact with the axonal transport machinery either directly via binding to the kinesin light chain subunit of kinesin-1, or indirectly via binding to the Jun N-terminal kinase-interacting protein 1, JIP1. Sites for proteolytic activity by $\beta$- and $\gamma$-secretases are shown, which together liberate amyloid beta (A$\beta$) peptides that are either 40 or 42 amino acids in length.

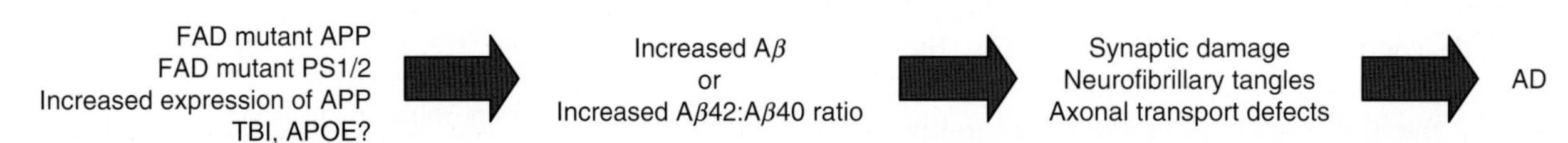


**Figure 2** A standard view of the amyloid cascade hypothesis in which the various mutations, genetic variants, and environmental insults induce a series of molecular events caused by increasing levels of toxic human amyloid beta (A$\beta_{40}$ or A$\beta_{42}$), culminating in Alzheimer's disease (AD). FAD, familial Alzheimer's disease; APP, amyloid precursor protein; PS1/2, presenilins 1 and 2; TBI, traumatic brain injury; APOE, apolipoprotein E.

nerve growth factor in the cortex where basal forebrain cholinergic neurons ordinarily connect to cells that secrete nerve growth factor (**Figure 3**). The levels of cholinergic enzymes in the cortex appear also to be reduced in AD, even though basal forebrain cholinergic neurons appear to express normal levels of cholinergic markers even after the onset of AD.Finally, there are reports that axonal transport is defective in neurons taken from brains of AD patients postmortem, and there are long-standing observations of destruction of the microtubule cytoskeleton in axons in diseased neurons. These disruptions have for some time been thought to be related to the hyperphosphorylation and aggregation of tau protein into neurofibrillary tangles, although what events are causative and what events are consequences of each other remain unclear.

The observation that axonal transport is defective at late stages of AD has led a number for workers over the years to propose that axonal transport defects might be early, possibly causative, events in AD. However, evidence for early defects in axonal transport in AD was lacking until recently. Recent work has provided evidence for early changes in axonal transport in postmortem AD brains and in animal models of AD in which APP is overexpressed. One striking example is the finding of dystrophic axons early in AD and in AD models prior to the formation of amyloid plaques, and thus not associated with, or in the vicinity of, amyloid plaques. These dystrophic axons exhibit signs of axonal transport failure, including massive accumulations of vesicles and organelles in axons that appear to be distended by these accumulations. This type of phenotype is characteristic of mutations that disrupt the axonal transport machinery directly in *Drosophila* and in the mouse. In this context, there is some evidence that deficits in axonal transport can induce formation of, or enhance accumulation of, A$\beta$. A final line of evidence that raises the possibility that axonal transport deficits are relevant to the initiation of AD is the body of evidence that suggests that there are age-dependent changes and reductions in microtubules in axons. Observed changes include accumulation of APP and other transported material in axons, increases in axonal dystrophies, and some direct evidence that transport itself is reduced in aging neurons relative to younger neurons.

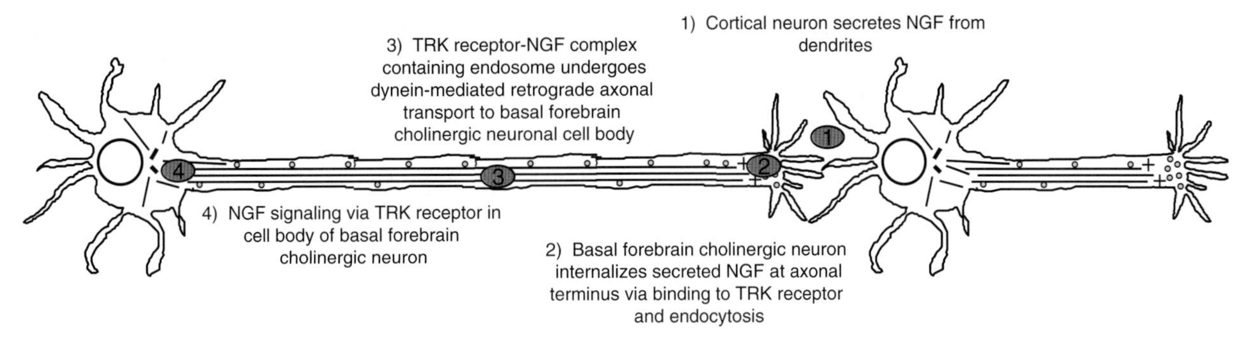


**Figure 3** Pathway of nerve growth factor (NGF) signaling from cortical neurons to target cholinergic neurons in the basal forebrain. This transport and signaling pathway appears to be defective in Alzheimer's disease. Secretion of NGF by cortical neurons (1) is followed by internalization at presynaptic endings of basal forebrain cholinergic neurons (2). Retrograde transport of vesicles containing NGF and activated TrkA in postendosomal carriers (3) culminates in arrival at the basal forebrain cholinergic neuron cell body and signaling (4).

## Roles and Effects of AD Risk Factors and Genes in Axonal Transport

Several reports suggest that APP and its proteolytic products may play normal roles in, and/or be able to interfere with, axonal transport. For example, deletion of the APP gene in *Drosophila* and in the mouse appears to generate mild defects in axonal transport. Overexpression of APP and its relatives dramatically interferes with axonal transport in *Drosophila* and in the mouse, and even simple duplication of the APP gene in the mouse appears to significantly inhibit some forms of axonal transport – in particular, retrograde transport of the nerve growth factor (NGF)–neurotrophin receptor signaling complex to the basal forebrain. Although this could be the result of, and often has been interpreted as such, an excess of toxic human A$\beta$, strong evidence in *Drosophila* and in the mouse suggests that even if human A$\beta$ is toxic to axonal transport, other factors, perhaps the C-terminus of the APP protein, may be bigger contributors to these defects.

The mechanism of APP-induced defects in axonal transport remains unknown. One suggestion, based on biochemical evidence from a number of groups, is that the cytoplasmic tail of APP can form either a direct or indirect interaction with the kinesin light chain subunits of conventional kinesin, kinesin-1. If indirect, this interaction may occur via formation of a ternary complex with the Jun kinase scaffolding protein called JIP1 (c-Jun N-terminal kinase-interacting protein 1). There is also evidence from the squid and *Drosophila* that the C-terminus of APP interacts with the axonal transport machinery and is required for interference with axonal transport in overexpression situations. This C-terminus is also required to generate movements of APP in *Drosophila* neurons and in squid axoplasm. These findings are consistent with the proposal that overexpression of APP poisons axonal transport by competing for either JIP1 or kinesin light chain, which are needed for axonal transport of other materials (**Figure 4**). Although several groups have provided evidence in favor of this hypothesis, other groups do not agree. Strong additional evidence in support of this idea comes from the observation that reductions in kinesin-1 subunits in *Drosophila* or in the mouse enhance transport defects caused by overexpression of APP.

There is considerable evidence that FAD mutations and deletion mutations in the gene encoding presenilin 1 may also interfere with axonal transport. One obvious mechanism that is consistent with the amyloid cascade hypothesis is that presenilins might cause axonal transport defects by enhancing or altering A$\beta$ production in neurons, which is then toxic to axonal transport. However, the presenilin mutations that cause familial forms of AD in humans can poison axonal transport of APP, and perhaps other proteins, in the mouse in the absence of human APP. These findings, combined with considerable evidence that kinesin-1 is required for APP transport in axons and with evidence that presenilin and $\beta$-secretase and related proteins may be in a common axonal vesicle with APP suggest that these processes may be tied together in a way that does not rely on APP processing to generate A$\beta$. For example, one possible mechanism of presenilin influence on the axonal transport machinery may be via control of kinesin light chain phosphorylation by glycogen synthase-3$\beta$ (GSK-3$\beta$) or other kinases. A related and surprising finding is that overexpression of $\beta$-secretase, which is another protease that may generate A$\beta$ from APP, can generate defects in axonal transport of APP, and may shift the cellular sites of A$\beta$ production and secretion. These ideas have been hotly debated in the literature; further evidence is required to establish or refute these proposals.

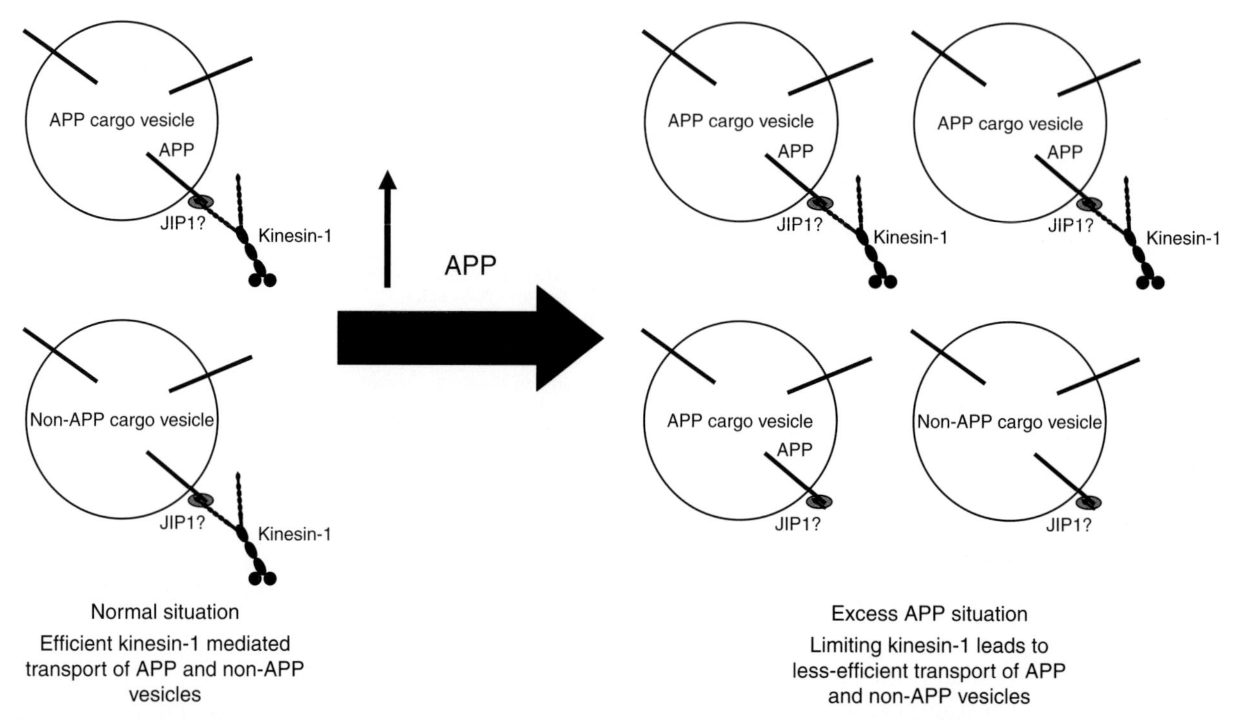


**Figure 4** A possible mechanism by which excess amyloid precursor protein (APP) poisons axonal transport. In the normal situation, there is enough membrane-bound kinesin-1 to move APP and non-APP vesicular and organellar cargoes. In the presence of excess APP, competition for limiting kinesin-1 may occur so that some vesicles (intermittently) lack kinesin-1 and can stall in the axon, leading to axonal blockages and transport deficits. Although this schematic shows kinesin-1 as the limiting protein, other proteins or adapters, such as Jun N-terminal kinase-interacting protein 1 (JIP1), could be the vulnerable limiting component.

Another possible link of axonal transport to the pathogenesis of AD is via tau protein. This protein, which aggregates into neurofibrillary tangles, appears to play a normal role in the control of vesicle transport along microtubules, perhaps most prominently for vesicles transported by kinesin-1. There is considerable evidence that tau protein, which binds to microtubules, may directly regulate the movement of vesicles by kinesin-1 along microtubules. This function, in addition to the potential role of tau in initiating the polymerization of microtubules or in enhancing the stability of microtubules, may play important roles in the effects of tau protein and its mutants or of inappropriate phosphorylation variants on the axonal transport machinery. There is also evidence that overexpression of tau protein and its disease-causing mutations poison axonal transport directly.

Another important risk factor for AD is traumatic brain injury. There is evidence that injury to the brain induces so-called diffuse axonal injury, among other consequences. Diffuse axonal injury is of particular interest because it is effectively injury to the axons, which has been reported to result in large accumulations of APP and A$\beta$ peptides at injury sites in axons. The possibility that axonal injuries cause biochemical changes in the axonal transport machinery is likely, given recent evidence that injury to axons induces activation of the damage-signaling kinase, Jun kinase (JNK). JNK is known to phosphorylate tau protein, and has been implicated in some aspects of axonal transport regulation. Thus traumatic brain injury may exert its influence on axonal transport defects via JNK activation and phosphorylation of targets such as tau and perhaps the kinesin heavy chain subunit of kinesin-1.

Another possible risk factor for AD is cholesterol. It is known that genotypic variants at the apolipoprotein E gene in humans have a large influence on the probability of developing late-onset AD. The $\varepsilon 4$ variant confers the highest risk and the $\varepsilon 2$ variant is protective. There is accumulating evidence that cholesterol and phospholipids might have important regulatory effects, not only on vesicular traffic, but also on kinesin-mediated axonal transport. One important kinesin family that is known to be essential for the transport of synaptic vesicle proteins is kinesin-3. This type of kinesin motor can bind to phospholipids in membrane bilayers, and has been reported to be activated by lipid clustering in vesicular membranes via oligomerization of the motor protein. Since cholesterol may have an impact on the ability of phospholipids to cluster, such a mechanism might affect transport directly. Another possible linkage of cholesterol to AD comes from the pediatric disease called Neimann–Pick

C type 1 (NPC1), which is a disorder of cholesterol transport and metabolism caused by defects in the *NPC1* gene. Such mutations appear to interfere with cholesterol trafficking in nonneuronal cells, possibly via the endocytic pathway. It is less clear what NPC1 mutations do in neurons, but it is clear that the mutations induce significant neurodegeneration in patients who have them, and ultimately lead to death. The striking clue is that these mutations appear to generate excess A$\beta$ peptides and to induce the aberrantly phosphorylated forms of tau protein that lead to the formation of neurofibrillary tangles. Further work will be necessary to establish the precise molecular link between cholesterol trafficking and AD. It is intriguing, however, that defects in movement of cholesterol and related components in axons may be an important part of the problem.

Finally, A$\beta$ peptide has been reported by a number of groups to induce phenotypes in cultured neurons that mimic those seen in neurons bearing axonal transport mutations. One group has reported that this may be a result of the effects of A$\beta$ peptides on the state of microtubules; this influence appears to be mediated by extracellular A$\beta$ peptides. Further work is necessary to clarify this association.

## Role of Axonal Transport Failure in AD

The key issue is whether axonal transport defects are an early or late event in AD, and whether they are a consequence of A$\beta$ toxicity, or a stimulus for A$\beta$ generation or turnover, or both. Although further work is clearly required to resolve these issues, a testable type of hypothesis, presented in **Figure 5**, encapsulates some key features. Key elements are that axonal transport defects can induce synaptic defects directly, activate kinases that may induce tau hyperphosphorylation, and instigate or enhance A$\beta$

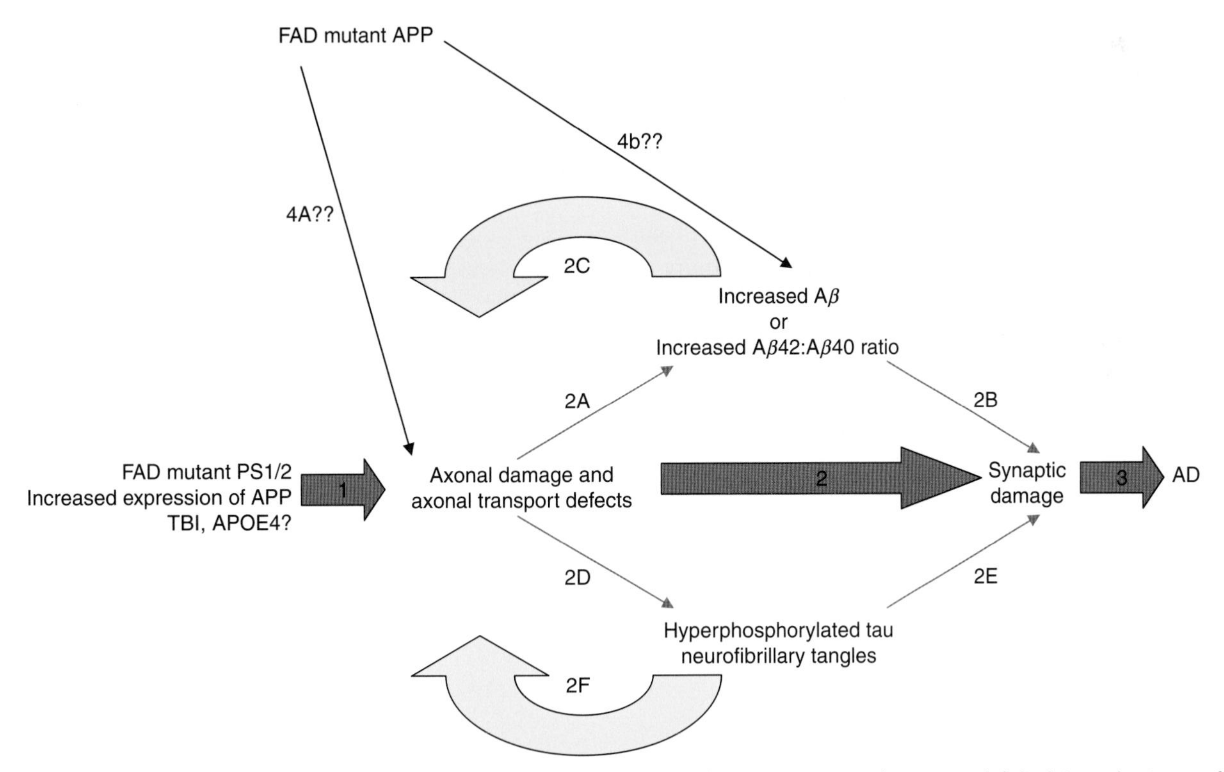


**Figure 5** One possible alternative to the amyloid cascade hypothesis that incorporates axonal transport deficits into early stages of Alzheimer's disease (AD) causation or progression. (1) In familial Alzheimer's disease (FAD), mutant presenilins (PS1/2), excess amyloid precursor protein (APP), traumatic brain injury (TBI), and apolipoprotein E $\varepsilon$4 (APOE4) are suggested to initiate neuronal damage by interfering with axonal transport and/or causing axonal damage. (2) Axonal damage and axonal transport deficits may cause synaptic damage directly, or indirectly by the indicated alternative pathways. Axonal damage might lead to increased amyloid beta or increased ratio of the two types (A$\beta_{40}$, A$\beta_{42}$) (2A), which may cause synaptic damage directly (2B), or may lead to further axonal damage and transport defects in an autocatalytic loop (2C). Axonal damage may also lead to activation of Jun N-terminal kinase, which could initiate hyperphosphorylation of tau protein and formation of neurofibrillary tangles (2D), which may cause synaptic damage (2E) and cause further defects in axonal transport and axonal damage (2F) in an autocatalytic loop. Synaptic damage caused by all of these insults may be the primary neuronal defect causing cognitive symptoms in AD (3). How FAD mutant APP might initiate these proposed events is unknown, but could happen either by directly interfering with axonal transport or causing axonal damage (4A) or by increasing A$\beta$ formation (4B).

formation, which may have its own toxic consequences. Unique aspects of this class of hypothesis include an autocatalytic spiral of events and multiple entry points, including mutation and environmental insults. In at least one form of hereditary AD, it can be argued strongly that this hypothesis may have validity in view of evidence that moderate increases in APP expression can induce significant axonal transport deficits in the absence of human A$\beta$. Although these ideas focus on axons, it is likely that comparable phenomena and behaviors exist in dendrites, for which less information is currently available.

## Conclusion and Perspectives

The fundamental problem in evaluating whether defects in axonal transport can induce or significantly accelerate early stages of AD derives from the nature of the experimental systems. Only humans develop true AD, but axonal transport and biochemical defects induced by agents and mutations that can cause AD can only be rigorously studied at present in animal models of the disease. Further resolution of these issues will require noninvasive imaging suitable for studying axonal transport in living human patients, and the development of true human neuronal models of disease, perhaps using human embryonic or other types of stem cells. Finally, it is likely that truly effective therapies for AD will require a detailed and mechanistic understanding of the events leading from causative insults to the biochemistry of neuronal function and viability; axonal transport appears to be an important process in disease progression, and its role needs to be rigorously evaluated.

*See also:* Aging of the Brain and Alzheimer's Disease; Animal Models of Alzheimer's Disease; Axonal Transport and Neurodegenerative Diseases; Stroke: Neonate vs. Adult.

## Further Reading


Goldstein LS (2003) Do disorders of movement cause movement disorders and dementia? *Neuron* 40: 415–425.

King ME, Kan HM, Baas PW, et al. (2006) Tau-dependent microtubule disassembly initiated by prefibrillar beta-amyloid. *Journal of Cell Biology* 175: 541–546.

Lee EB, Zhang B, Liu K, et al. (2005) BACE overexpression alters the subcellular processing of APP and inhibits Abeta deposition *in vivo*. *Journal of Cell Biology* 168: 291–302.

Muresan Z and Muresan V (2005) c-Jun $NH_2$-terminal kinase-interacting protein-3 facilitates phosphorylation and controls localization of amyloid-beta precursor protein. *Journal of Neuroscience* 25: 3741–3751.

Pigino G, Morfini G, Pelsman A, et al. (2003) Alzheimer's presenilin 1 mutations impair kinesin-based axonal transport. *Journal of Neuroscience* 23: 4499–4508.

Satpute-Krishnan P, DeGiorgis JA, Conley MP, et al. (2006) A peptide zipcode sufficient for anterograde transport within amyloid precursor protein. *Proceedings of the National Academy of Sciences of the United States of America* 103: 16532–16537.

Shen J and Kelleher RJ, III (2007) The presenilin hypothesis of Alzheimer's disease: Evidence for a loss-of-function pathogenic mechanism. *Proceedings of the National Academy of Sciences of the United States of America* 104: 403–409.

Stokin GB and Goldstein LS (2006) Axonal transport and Alzheimer's disease. *Annual Review of Biochemistry* 75: 607–627.

Stokin GB, Lillo C, Falzone TL, et al. (2005) Axonopathy and transport deficits early in the pathogenesis of Alzheimer's disease. *Science* 307: 1282–1288.

Terry RD (1996) The pathogenesis of Alzheimer disease: An alternative to the amyloid hypothesis. *Journal of Neuropathology & Experimental Neurology* 55: 1023–1025.

# Aging of the Brain and Alzheimer's Disease

**D L Price, A V Savonenko, M Albert, J C Troncoso, and P C Wong**, The Johns Hopkins University, Baltimore, MD, USA



## Introduction

Over the past decade, major advances have been made in understanding the causes and mechanisms of age-associated alterations in memory and cognition, including mild cognitive impairment (MCI) and Alzheimer's disease (AD), which is the most common cause of dementia in the elderly. For many reasons (prevalence, lack of mechanism-based treatments, cost of care, and impact on individuals and caregivers), AD is one of the most challenging illnesses in medicine. However, extraordinary progress has been made in deliniating the pathology, biochemistry, and neurobiology of the disease; the clinical and pathological bases of MCI and AD; the utility of new diagnostic approaches and outcome measures; and the value of transgenic models of the genetic forms of familial AD and gene-targeted models which discloses mechanism-based targets for therapy of AD. These discoveries are leading to new treatments for this type of dementia.

## Cognitive and Memory Impairments in the Elderly

Although many older individuals remain intellectually intact, a significant number of the elderly show declines in memory and cognitive abilities. These alterations are usually mild, occur relatively late in life, and involve speed of learning, complex problem solving, ability to retain large amounts of new information, and visuospatial skills. Vocabulary, information storing, and comprehension skills often remain relatively stable into old age. Some older individuals have mild cognitive impairments (MCIs) in which memory loss exceeds that expected for age, but they do not meet criteria for AD. However, among individuals with MCIs, there may be considerable heterogeneity (i.e., interindividual variability in the rate and severity of progression). The variability of alterations in memory with age makes it sometimes difficult to determine, after a single examination, whether an individual with mild age-associated memory impairments will remain relatively stable or will progress to severe dementia. The most accurate clinical approach to this problem is to assess cognitive abilities and memory performance repeatedly in the same patient over a period of time. Several neuropsychological tests are valuable in predicting AD: delayed recall tasks (recently learned information), as assessed by the California Verbal Test and Wechsler Memory Scale, and measures of executive functions (ability to organize and plan), as assessed by the Trail-Making Test and Self-Ordering Test. On the basis of clinical assessments correlated with postmortem examinations of the brain, MCI is now regarded as a transitional state between the normal functional state of older persons and early AD. Imaging studies have predictive value regarding the progression to AD.

## Alzheimer's Disease: Clinical Features, Diagnostic Studies, Neuropathology and Biochemistry, and Current Treatments

AD is the most common cause of senile dementia, a term that refers to a syndrome occurring in older individuals that results in memory loss and cognitive impairments of sufficient severity to interfere with social, occupational, and personal functions. This type of dementia affects more than 4 million people in the United States. Because of increased life expectancy and the postwar baby boom, this population is the fastest growing segment of society. During the next 25 years, the number of people with AD in the United States will triple, as will the cost of care and treatment.

A majority of individuals with sporadic AD exhibit the first clinical signs during their seventh decade. However, some people develop disease in midlife; in these cases, a family history of the illness is more likely. In both the sporadic and familial forms of AD, affected individuals show difficulties with memory, problem solving, executive functions for language, calculation, visuospatial perceptions, judgment, and behavior; mental functions and activities of daily living are increasingly impaired. Some patients develop psychotic symptoms, such as hallucinations and delusions. In the late stages, individuals are mute, incontinent, bedridden, and usually die of intercurrent medical illnesses.

Other causes of dementia syndromes include cerebrovascular disease, alone or in combination with AD; Lewy body dementia; Parkinson's disease; alcoholism; drug intoxications; infections, such as with human immunodeficiency virus (HIV) (i.e., acquired immunodeficiency syndrome (AIDS)) and syphilis; brain tumors; vitamin deficiencies (e.g., $B_{12}$); thyroid disease; and a variety of other metabolic disorders. Because some of these disorders are treatable, it is important for physicians to exclude these entities before making a diagnosis of possible AD. At present, except for brain biopsy, it is not possible to definitively establish the diagnosis of early AD in living humans, and it is extremely important for the physician to exclude other causes of dementia syndromes, because some of these illnesses respond to specific treatments. Imaging studies are of great value in exclusions and are of increasing utility in supporting the diagnosis of AD. At present, the only available treatment for AD is for symptoms.

As mentioned previously, to establish a diagnosis of AD, clinicians rely on histories from patients and informants; physical, neurological, and psychiatric examinations; neuropsychologic testing; laboratory studies, including examinations of cerebrospinal fluid (CSF); and a variety of laboratory tests, including neuroimaging studies. The clinical profile, in concert with a variety of imaging and laboratory assessments (levels of CSF $\beta$-amyloid (A$\beta$) decrease and tau proteins increase), allows the clinician to make a diagnosis of possible or probable AD. Several imaging strategies and studies of biomarkers, particularly of CSF, are useful laboratory approaches. In early AD, magnetic resonance imaging (MRI) often discloses atrophy of specific regions of the brain, while positron emission tomography (PET) with [$^{18}$F]deoxyglucose, or single-photon emission computerized tomography (SPECT), demonstrates decreased glucose utilization and early reductions in regional blood flow in the parietal and temporal lobes. PET, following administration of brain penetrant $^{11}$C-labeled Pittsburgh compound B (PIB), which binds to A$\beta$ with high affinity, discloses patterns which are thought to reflect the A$\beta$ burden in the brain. In concert, these various approaches as applied to patients should increase the accuracy of diagnosis in earlier stages of disease and allow assessments of the efficacies of new antiamyloid therapeutics. The presence of the apolipoprotein E (apoE) $\varepsilon$4 allele confers risk in late-onset disease (see later); apoE genotyping is a useful research tool, but it is not helpful for routine diagnostic purposes in individual patients.

AD is the result of abnormalities associated with dysfunction and death of specific populations of neurons, particularly those cells in neural systems participating in memory and cognitive functions. Abnormalities in the brain selectively involve neurons in the neocortex, entorhinal area, hippocampus, amygdala, nucleus basalis, anterior thalamus, and several brain stem monoaminergic nuclei (in particular the locus coeruleus and raphe complex). Dysfunction of neurons in these brain regions and circuits is reflected by the presence of cytoskeletal abnormalities (neurofibrillary tangles (NFTs)) in these cells, the presence of neuritic plaques in brain regions receiving inputs from these nerve cells, and reductions in transmitter markers of these neurons in their target fields. The cellular pathology is characterized by the presence of intracellular and extracellular protein or peptide aggregates: phosphorylated tau is assembled into the paired helical filaments (PHFs) within NFTs and in swollen neurites (around plaques); and A$\beta$ peptides exist in extracellular $\beta$-pleated sheet conformations assembled into oligomers, which are the principal constituent of amyloid plaques. Ultimately, affected neurons die; often 'tombstone' tangles, amyloid deposits, and glial reactions remain to indicate sites of neurodegeneration.

The clinical manifestations of amnestic mild cognitive impairment (aMCI) and early Alzheimer's disease (eAD) result from abnormalities occurring among populations of neurons in neural systems/brain regions essential for memory, learning, and cognitive performance. Damaged circuits include those in the basal forebrain cholinergic system, the amygdala, the hippocampus, the entorhinal and limbic cortex, and the neocortex. In a recent study, the character, abundance and distribution of the lesions (i.e., diffuse plaques, neuritic plaques, and tangles) were correlated with clinical signs in cognitively characterized controls, individuals with aMCI, or cases of eAD. No differences were observed in the number of diffuse plaques among the groups. In persons with aMCI, tangles were significantly increased in regions of the ventral medial temporal lobe as compared to controls; individuals with eAD showed greater numbers of NFTs and neuritic plaques in both frontal and temporal regions. Individuals with aMCI exhibited increased numbers of neuritic plaques in neocortical regions as compared to controls, but not as compared to cases of eAD. Memory deficits correlated most closely with the abundance of NFTs in CA1 of the hippocampus and in the entorhinal cortex, a finding which leads to the interpretation that tangles were more important than amyloid deposition in the progression from normal function to aMCI to eAD, and that tangles in the medial temporal lobe play a key role in memory declines in aMCI. Data from multiple studies are interpreted to indicate that aMCI reflects a transitional state in the evolution of AD. Because the regional distributions of NFTs correlate most closely with the degree of clinical impairments from aged healthy controls to individuals with aMCI to cases of AD, the spread of NFTs beyond the medial temporal lobe is thought to be linked to the development of dementia.

As mentioned previously, clinical signs reflect cellular abnormalities within specific neural circuits responsible for memory and cognition. Affected neurons exhibit conformationally altered isoforms of tau comprising the PHFs in NFTs, neurites, and neuropil threads. Axonal pathologies include varicosities and terminal clubs, the latter representing disconnected synapses and transected axons (also observed in aged, memory-impaired rhesus monkeys with A$\beta$ deposits, and in some transgenic monkey models of A$\beta$ amyloidosis). The A$\beta$-containing neuritic plaques are sites of synaptic disconnection in regions receiving inputs from diseased populations of neurons; associated with these lesions are decrements in generic and transmitter-specific synaptic markers in the target fields of degenerating nerve cells and local astroglial and microglial responses (particularly associated with plaques). Ultimately, there is evidence of death of neurons, possibly by apoptosis. Thus, the clinical manifestations of aMCI and AD reflect a progressive disruption of synaptic communications involving subsets of neural circuits, first associated with degeneration of axon terminals and, later, death of neurons.

In one hypothetical model mechanistically linking A$\beta$ and phosphorylated tau, A$\beta$42 species liberated at terminals oligomerize to form A$\beta$ assemblies or A$\beta$-derived diffusible ligands (ADDLs), which are linked to synaptic damage. Subsequently, not yet identified retrograde signals, which originate at disconnected terminals, trigger the activation of kinases (or the inhibition of phosphatases) in cell bodies. The phosphorylation of tau at a variety of sites is associated with conformational changes in this protein and the formation of PHFs and, eventually, NFTs. Secondary disturbances in axonal transport can, in turn, compromise the functions and viability of neurons. Eventually, affected nerve cells die and extracellular tangles remain as markers of the nerve cells destroyed by disease.

These cellular abnormalities have profound clinical consequences. Abnormalities that damage the entorhinal cortex, hippocampus, and other circuits in the medial temporal cortex are presumed to be critical for memory impairments. Higher cognitive deficits, such as disturbances in executive functions for language, calculation, problem solving, and judgment, are believed to be related to pathology in the neocortex. Alterations in the basal forebrain cholinergic system may contribute to memory difficulties and attention deficits. The behavioral and emotional disturbances presumably may reflect involvement of the limbic cortex, amygdala, thalamus, and monoaminergic systems.

At present no cure exists for this devastating illness. Available treatments for AD target symptoms, and many present-day therapies focus on treating associated conditions such as depression, agitation, sleep disorders, hallucinations, and delusions. One therapeutic approach has been to try to support the functions of the basal forebrain cholinergic system, which is one of the circuits severely damaged in AD. Several strategies have been developed to influence this neurotransmitter system. Unfortunately, precursor loading (i.e., choline, lecithin) and muscarinic agonist approaches to improve cholinergic functions did not prove effective. Several acetylcholinesterase inhibitors, which prolong the half-life of the transmitter in the synaptic cleft, have been approved for the treatment of AD in the United States. These drugs are sometimes associated with side effects and they have, at best, a very modest effect on cognitive functions and the ability to perform activities of daily living. Clinical trials have tested the efficacies of anti-inflammatory compounds, estrogens, plant extracts, and antioxidants without great benefit. In an antiexcitotoxic approach, memantine, an *N*-methyl-D-aspartate (NMDA) receptor antagonist, has been suggested to reduce the rate of clinical deterioration of patients with moderate to severe AD.

## Genetic Causes and Risk Factors for AD

Established genetic risk factors implicated in AD include: mutations in the *amyloid beta (A4) precursor protein* gene (*APP*; chromosome 21), mutations in *presenilin 1* (*PSEN1*; chromosome 14) and *presenilin 2* (*PSEN2*; chromosome 1), and the presence of a risk susceptibility allele ($\varepsilon$4) of *APOE* (chromosome 19), which predisposes to later onset AD and some cases of late-onset familial AD. Autosomal dominant mutations in *APP*, *PSEN1*, or *PSEN2* usually cause disease earlier than occurs in sporadic cases, with the majority of mutations in *APP*, *PSEN1*, and *PSEN2* influencing $\beta$-site APP cleaving enzyme 1 (BACE1) or $\gamma$-secretase cleavages of APP to increase the levels of all A$\beta$ species or the relative amounts of toxic A$\beta$42.

A member of the *APP* gene family (including also *APLP1* and *APLP2*), *APP* encodes a type I transmembrane protein of unknown function which is abundant in the nervous system, rich in neurons, and transported rapidly anterograde in axons to terminals. In the central nervous system (CNS), BACE1 cleaves at the +1 and +11 sites; subsequently, the $\gamma$-secretase complex cleaves at a variety of sites (see later), which generate the N- and C-termini of A$\beta$ peptides, respectively. A variety of *APP* mutations alter the processing of APP and influence increase in the production of A$\beta$ peptides or the amounts of the more toxic A$\beta$42. The APP Swedish double mutation (*swe*) enhances manyfold the BACE1 cleavage at the N-terminus of A$\beta$ (+1 site), resulting in substantial elevations in levels of all A$\beta$ peptides. $APP_{717}$ mutations influence $\gamma$-secretase cleavage to increase secretion of A$\beta$42, which is the most toxic peptide. In contrast, other mutations may promote local fibril formation and vascular amyloidosis. Moreover, individuals with increased *APP* gene dosage are prone to AD; familial duplications of *APP* or trisomy 21 (Down syndrome) are associated with extra copies of *APP* and develop AD pathology relatively early in life.

*PSEN1* and *PSEN2* encode two highly homologous and conserved 43–50 kDa multipass transmembrane proteins that are involved in Notch1 signaling pathways critical for cell fate decisions. Presenilins are endoproteolytically cleaved by a 'presenilinase' to form an N-terminal ~28 kDa fragment and a C-terminal ~18 kDa fragment; both fragments are critical components of the $\gamma$-secretase complex. Nearly 50% of early-onset cases of familial AD are linked to >90 different mutations in *PSEN1*. A relatively small number of *PSEN2* mutations also cause autosomal dominant familial AD. The majority of abnormalities in *PSEN* genes are missense mutations that enhance $\gamma$-secretase activities and increase the levels of the A$\beta$42 peptides.

## Biochemistry of Amyloidosis: APP and the Secretases

APP is cleaved by $\beta$- and $\gamma$-secretases, releasing the ectodomain of APP (APPs), liberating a cytosolic fragment termed APP intracellular domain (AICD), and generating several species of A$\beta$ peptides. In the central (but not peripheral) nervous system, A$\beta$ peptides are generated by sequential endoproteolytic cleavages by BACE1 (at the A$\beta$ +1 and +11 sites) to generate APP-$\beta$ C-terminal fragments (APP-$\beta$CTFs) and by the $\gamma$-secretase complex (at several sites, varying from A$\beta$ 36 to 38, 40, 42, and 43) to form A$\beta$ species peptides. The intramembranous cleavages of APP-$\beta$CTF by $\gamma$-secretase releases an AICD, which can form a complex with Fe65, a nuclear adaptor protein; Fe65 and A$\beta$ or Fe65 alone (in a novel conformation) can gain access to the nucleus to influence gene transcription, a signaling mechanism analogous to that occurring in the Notch1 pathway (next to the Notch intracellular domain (NICD)). It has been speculated that the AICD signaling pathway may play a role in learning and memory. In other cells in other organs, APP is cleaved endoproteolytically within the A$\beta$ sequence through alternative, nonamyloidogenic pathways: $\alpha$-secretase (TNF-$\alpha$ converting enzyme, or TACE) cleaves between residues 16 and 17; BACE2 cleaves between residues 19 and 20 and 20 and 21. These cleavages, which occur in nonneural tissues, preclude the formation of A$\beta$ peptides and serve to protect these cells/organs from A$\beta$ amyloidosis.

BACE1, encoded by a gene on chromosome 11, is a transmembrane aspartyl protease that preferentially cleaves APP at the +1 and +11 sites of A$\beta$ in APP and is critical for the generation of A$\beta$. BACE1 is present in the brain but is virtually undetectable in nonneural tissues; BACE1-specific immunoreactivities are readily localized in the hippocampus. Importantly, as compared to wild-type APP, BACE1 cleaves APP (swe) cleaved approximately 100-fold more efficiently at the +1 site, resulting in a greater increase in BACE1 cleavage products (elevating cell A$\beta$ species) in the presence of this mutation. Significantly, *BACE1*-deficient mice show no overt developmental phenotype, and in cultures of $BACE1^{-/-}$ neurons, the secretion of A$\beta1_{40/42}$ and A$\beta11_{40/42}$ is abolished. Thus BACE1 is the principal neuronal $\beta$-secretase is responsible for the penultimate A$\beta$ cleavage. In contrast, BACE2, which makes antiamyloidogenic cleavages at sites +19/+20 of A$\beta$, acts more like $\alpha$-secretase; it does not appear to play a significant role in APP processing in neurons.

$\gamma$-Secretase, essential for the regulated intramembranous proteolysis of a variety of transmembrane proteins, is a multiprotein catalytic complex that includes PSEN1 and PSEN2 (described earlier); nicastrin (Nct), a type I transmembrane glycoprotein; and Aph-1 (anterior pharynx defective) and Pen-2 (presenilin enhancer), two multipass transmembrane proteins. Presenilin contains aspartyl residues that play roles in these intramembranous cleavages; substitutions at D257 in transmembrane (TM) segment 6 and at D385 in TM segment 7 reduce A$\beta$ secretion and cleavage of Notch1 *in vitro*. The functions of the various $\gamma$-secretase proteins and their interactions in the complex are not yet fully defined. In one model, Aph-1 and Nct form a precomplex which interacts with presenilin; subsequently, Pen-2 enters the complex, where it is critical for the 'presenilinase' cleavage of presenilin into two fragments. In concert, this complex is responsible for $\gamma$-secretase cleavages of APP, Notch, and a variety of other transmembrane proteins. With regard to APP, this complex makes the final cleavage that generates a peptide (especially A$\beta$) and liberates AICD. Many familial AD-linked *PSEN* mutations promote the cleavages that lead to formation of A$\beta$42.

## Animal Models of Aging and AD

Both spontaneously occurring and genetically engineered diseases in animals have been used to model features of AD.

For example, an age-related spontaneous disorder occurs in elderly rhesus monkeys (*Macaca mulatta*). With an estimated life span of more than 35 years (×2.8, for equivalent human age), these rhesus monkeys show cognitive and memory deficits that appear at the end of the second and at the beginning of the third decades of life. These animals develop virtually all of the brain abnormalities (amyloid deposits, neuritic plaques, scattered tangles, and modest reductions in transmitter markers) that are observed in older humans and, to a greater degree, in cases of AD. Genetically engineered models relevant to AD are discussed in the following section.

## Genetic Models of A$\beta$ Amyloidosis

In mice, expression of *APPswe* or $APP_{717}$ (with or without mutant *PSEN1)* leads to an A$\beta$ amyloidosis in the CNS. Mutant *APP/PSEN1* mice develop accelerated disease secondary to increased levels of A$\beta$ (particularly A$\beta$42) associated with the presence of diffuse A$\beta$ deposits and neuritic plaques in the hippocampus and cortex. Levels of A$\beta$ peptides, particularly A$\beta$42, increase in brain with age, and oligomeric species (ADDLs, A$\beta$*56, etc.) appear in the CNS.

Over time, mice carrying mutant transgenes exhibit A$\beta$ deposits, swollen neurites in proximity to these deposits, and neuritic plaques associated with glial responses. Some lines of mice show evidence of amyloid in vessels (congophilic angiopathy). In forebrain regions, the density of synaptic terminals and levels of several neurotransmitter markers are reduced. In some settings, there are deficiencies in synaptic transmission. Moreover, some lines of mice show evidence of degeneration of subsets of neurons.

Behavioral studies of lines of transgenic mice, including those generated by Dr. David Borchelt, disclose deficits in spatial reference memory (Morris water maze task) and episodic-like memory (repeated reversal and radial water maze tasks). At 6 months of age, *APPswe/PSEN1ΔE9* mice develop plaques, but all genotypes are indistinguishable from nontransgenic animals in all cognitive measures. However, in 18-month-old cohorts, *APPswe/PSEN1ΔE9* mice perform all cognitive tasks less well than do mice of all other genotypes. In these animals, amyloid burdens are high; decreases are detectable in levels of several neurotransmitter markers. The strongest relationships exist between deficits in episodic-like memory tasks and total A$\beta$ loads in the brain. Collectively, these studies suggest that, in *APPswe/PSEN1ΔE9* mice, some form of A$\beta$ (ultimately associated with amyloid deposition) can disrupt circuits critical for memory, particularly episodic-like memory. Some of these impairments have been linked to the presence of A$\beta$ oligomers (see later) and can be reversed by antibody-mediated reductions of levels of brain A$\beta$. These mice do not recapitulate the complete phenotype of AD, but they are very useful subjects for research designed to examine behavioral consequences of A$\beta$ amyloidosis in the CNS, to delineate disease mechanisms, and to test novel therapies.

Over the past decade, a variety of A$\beta$ species, including monomers, oligomers, A$\beta$-derived diffusible ligands, structural assemblies, and amyloid deposits in neuritic plaques, have been suggested to play important roles in impairing synaptic communication. For example, in one study, when naturally secreted A$\beta$ peptides were injected into the ventricular system of rats, long-term potentiation (LTP) was inhibited in the hippocampus; the activity of the peptide was completely blocked by the injection of a monoclonal A$\beta$ antibody, a finding consistent with the concept that oligomers are the toxic moiety and that they are both necessary and sufficient to perturb learned behavior. Active immunization was less effective in rescuing function and correlated most closely with the levels of antibodies recognizing oligomers. More recently, studies of Tg2576 mice suggested that extracellular accumulations of a 56 kDa soluble amyloid assembly, termed A$\beta$*56, purified from the brains of memory-impaired mice, can interfere with memory when administered to young rats.

## Models of Tau Abnormalities

The paucity of tau abnormalities in various lines of mutant *APP* mice may be related to differences in tau isoforms expressed in this species, as compared to humans. Early efforts to express mutant *tau* transgenes in mice did not lead to striking clinical phenotypes or pathology. More recently, mice overexpressing *tau* show clinical signs, attributed to degeneration of motor axons. When prion or Thy1 promoters are used to drive $tau_{P301L}$ (a mutation linked to autosomal dominant frontotemporal dementia with parkinsonism), some brain and spinal cord neurons develop tangles. Mice expressing $APPswe/tau_{P301L}$ exhibit enhanced tangle-like pathology in limbic system and olfactory cortex. Moreover, injection of A$\beta$42 fibrils into specific brain regions of $tau_{P301L}$ mice increases the number of tangles in those neurons projecting to sites of A$\beta$ injection. A triple transgenic mouse (3×Tg-AD), created by microinjecting *APPswe* and $tau_{P301L}$ into single cells derived from monozygous $PSEN1_{M146V}$ knockin mice, develop age-related plaques and tangles as well as deficits in LTP which appear to antedate overt pathology. However, mice bearing both mutant *tau* and *APP* (or *APP/PSEN1*) or mutant *tau* mice injected with A$\beta$ may not be ideal models of familial AD because the presence of the *tau* mutation alone is associated with the development of tangles and disease.

## Targeting of Genes in the Amyloidogenic Pathway

To begin to understand the functions of some of the proteins thought to play roles in AD, investigators have targeted a variety of genes encoding BACE1, PSEN1, Nct, and Aph-1.

### *BACE1*$^{-/-}$ Mice

These animals mate successfully and exhibit no overt pathology. *BACE1*$^{-/-}$ neurons do not cleave at the +1 and +11 sites of A$\beta$, and the production of A$\beta$ peptides is abolished, establishing that BACE1 is the neuronal $\beta$-secretase required to generate the N-termini of A$\beta$. However, *BACE1*$^{-/-}$ mice show altered performance on some tests of cognition and emotion (see later); the former deficits can be rescued by overexpression of *APP* transgenes.

### *PSEN1*$^{-/-}$ Mice

These embryos develop severe abnormalities of the axial skeleton, ribs, and spinal ganglia, a lethal outcome which resembles a partial *Notch1*$^{-/-}$ phenotype. *PSEN1*$^{-/-}$ cells show decreased levels of secretion of A$\beta$ related to the fact that PSEN1 (along with Nct, Aph-1, and Pen-2) is a component of the $\gamma$-secretase complex that carries out the final (S3) intramembranous cleavage of Notch1. Without $\gamma$-secretase cleavage, the NICD is not released from the plasma membrane and cannot reach the nucleus to provide a signal to initiate transcriptional processes essential for cell fate decisions. Significantly, conditional *PSEN1/2* targeted mice show impairments in memory and synaptic plasticity in the hippocampus, raising the question (posed particularly effectively Dr. Jie Shen and colleagues) as to the roles of loss of presenilin function in neurodegeneration and AD.

### *Nct*$^{-/-}$ Mice

These embryos die early and exhibit several patterning defects, including abnormal segmentation of somites; this phenotype closely resembles that seen in *Notch1*$^{-/-}$ and *PSEN1/2*$^{-/-}$ embryos. Importantly, *Nct*$^{-/-}$ cells do not secrete A$\beta$ peptides, whereas *NctT*$^{+/-}$ cells show reduction of ~50%. The failure of *NctT*$^{-/-}$ cells to generate A$\beta$ peptides is accompanied by accumulation of APP C-terminal fragments. Importantly, *Nct*$^{+/-}$ mice develop tumors of the skin, presumably related to reduced levels of signaling by Notch1, which appears to act as a tumor suppressor in the skin.

### *Aph-1a*$^{-/-}$ Mice

Three murine *Aph-1* alleles (*Aph-1a*, *Aph-1b*, and *Aph-1c*) encode four distinct Aph-1 isoforms: Aph-1aL and Aph-1aS (derived from differential splicing of *Aph-1a*), Aph-1b, and Aph-1c. *Aph-1a*$^{-/-}$ embryos show patterning defects that resemble, but are not identical to, those of *Notch1*, *Nct*, or *PSEN* null embryos. Moreover, in *Aph-1a*$^{-/-}$-derived cells, the levels of Nct, presenilin fragments, and Pen-2 are decreased. There is an associated reduction in levels of the high-molecular-weight $\gamma$-secretase complex and a decrease in secretion of A$\beta$. In *Aph-1a*$^{-/-}$ cells, other mammalian Aph-1 isoforms can restore the levels of Nct, presenilin, and Pen-2.

## Experimental Therapeutics

The availability of models of amyloidogenesis provides opportunities to ablate or knock down genes, to modulate cleavages, and to influence clearance; such experiments have set the stage for testing the influences of A$\beta$ production, APP cleavage patterns, peptide neurotoxicity, and promotion of clearance and/or degradation of A$\beta$. Because it is not possible to discuss all experimental treatment in mouse models of A$\beta$ amyloidosis, we focus on selected studies that illustrate the experimental strategies directed at specific therapeutic targets that we predict will provide mechanism-based therapeutic benefits to patients with AD.

#### Reductions in BACE1 Activity

Deletion of *BACE1* in *APPswe/PSEN1ΔE9* mice prevents both A$\beta$ deposition and age-associated cognitive abnormalities that occur in this transgenic model. The *BACE1*$^{-/-}$ *APPswe/PSEN1ΔE9* mice do not develop the A$\beta$ deposits or the age-associated abnormalities in working memory that occur in the *APPswe/PSEN1ΔE9* model of A$\beta$ amyloidosis. Similarly, *BACE1*$^{-/-}$ Tg2576 mice show rescue from age-dependent memory deficits and physiological abnormalities. Moreover, A$\beta$ deposits are sensitive to *BACE1* dosage and can be efficiently cleared from regions of the CNS when *BACE1* is silenced at these sites. Inhibitors of $\beta$-secretase, conjugated to carrier peptides, can inhibit *in vitro* and *in vivo* (following intraperitoneal injection into Tg2576 mice). Conditional expression systems or RNA interference (RNAi) silencing allow investigators to examine the pathogenesis of diseases and to assess the degrees of reversibility of the disease processes. Clearly, BACE1 is an attractive therapeutic target. However,

$BACE1^{-/-}$ mice manifest alterations in both hippocampal synaptic plasticity and in performance on tests of cognition and emotion; the memory deficits, but not emotional alterations, in $BACE1^{-/-}$ mice are prevented by co-expression of *APPswe/PSEN1ΔE9* transgenes, suggesting that APP processing influences cognition/memory and that the other potential substrates of BACE1 may play roles in neural circuits related to emotion. BACE1 and APP processing pathways appear to be critical for cognitive, emotional, and synaptic functions. Moreover, BACE1 cleaves reuregulin (NRC), an axonal signal for myelination; inhibition of β-secretase activity is an exciting therapeutic opportunity, but future studies should be alert to potential mechanism-based side effects that may occur with strong inhibition of the enzyme.

### Inhibition of γ-Secretase Activity

Both genetic and pharmaceutical lowering of γ-secretase activity decreases production of Aβ in cell-free and cell-based systems and reduces levels of Aβ mutant mice with Aβ amyloidosis. Thus, γ-secretase activity is a significant target for therapy. Treated mice show reduced levels of Aβ and amyloid plaques. However, γ-secretase activity is also essential for processing of Notch, which is critical for lineage specification and cell growth during embryonic development. Significantly, inhibitors of γ-secretase activity reduce production of Aβ, but can also have profound effects on T and B cell development and on the appearance of intestinal mucosa (proliferation of goblet cells, increased mucin in gut lumen, and crypt necrosis). Significantly, $Nct^{+/-}$*APPswe/PSEN1ΔE9* mice develop skin tumors, presumably, in part, because reduced γ-secretase acts, via Notch signaling, as a tumor suppressor in skin. Clinicians carrying out trials of this class of inhibitor will have to be alert to several potential adverse events associated with reduced activity of this enzyme complex.

### γ-Secretase Modulation by Nonsteroidal Anti-inflammatory Compounds

Retrospective epidemiological studies suggested that significant exposure to nonsteroidal anti-inflammatory drugs (NSAIDs) reduces risk of AD, an outcome initially interpreted as related to suppression of the well-documented inflammatory process occurring in brains of cases of AD. However, more recent studies indicate that a subset of NSAID compounds in this class can modulate secretase cleavages (to shorter, less toxic Aβ species) without altering Notch or other APP processing outcomes. Short-term treatment of mutant mice appears to have some benefit in terms of lowering Aβ and plaque pathology. This strategy is now being evaluated in drug trials in humans.

### Aβ Immunotherapy

Multiple lines of evidence, including studies of lesions of entorhinal cortex or perforant pathway (by lesioning cell bodies or axons/terminals transporting APP to terminals, respectively), indicate that removing the source of Aβ significantly reduces Aβ in target fields. Similarly, increasing local levels of degrading enzymes, including insulin-degrading enzymes (IDEs) and neprolysin (NEP), can cleave Aβ and can reduce levels of the Aβ peptide.

To date, the most exciting findings regarding clearance of Aβ are derived from studies using active and passive Aβ immunotherapy. In treatment trials in mutant mice and in rodents, both Aβ immunization (with Freund's adjuvant) and passive transfer of Aβ antibodies reduce levels of Aβ and the level of plaque burden. Although, the mechanisms of enhanced clearance are not certain, at least two not mutually exclusive hypotheses have been suggested: (1) a small amount of Aβ antibody reaches the brain, binds to Aβ peptides, promotes the disassembly of fibrils, and, via the Fc antibody domain, encourages activated microglia to enter the affected regions and remove Aβ; and (2) serum antibodies serve as a 'sink' to draw the amyloid peptides from the brain into the circulation, thus changing the equilibrium of Aβ in different compartments and promoting removal of Aβ from the brain. Whatever the mechanism, Aβ immunotherapy in mutant mice is successful in partially clearing Aβ, in attenuating learning and behavioral deficits in several cohorts of mutant *APP* mice, and in partially reducing tau abnormalities in the triple transgenic mice. However, several studies have documented that immunotherapy may be associated with brain hemorrhages related to congophilic angiopathy. In these settings, the presence of amyloid presumably can weaken vascular walls; potentially, removal of some intramural vascular amyloid could lead to rupture of damaged vessels and bleeding.

Although mutant mice that received immunotherapy were not reported to develop evidence of meningoencephalitis, a subset of patients in a clinical trial did manifest these problems. Individuals receiving vaccinations with preaggregated Aβ and an adjuvant (followed by a booster) develop antibodies that recognize Aβ in the brain and vessels. Unfortunately, although phase I trials with Aβ peptide and adjuvant vaccination were not associated with any adverse events, phase II trials detected complications (meningoencephalitis) in a subset of patients and were suspended. The pathology in the index case, consistent with T cell meningitis, was interpreted to show some clearance of Aβ deposits, but some regions contained a relatively high density of tangles, neuropil threads, and vascular amyloid. A$\beta$ immunoreactivity was sometimes associated with microglia, and T cells were conspicuous in the subarachnoid space and around some vessels. In another case, there was significant reduction in amyloid deposits in the absence of clinical evidence of encephalitis. The trial was stopped. Assessment of cognitive functions in a small subset of patients (30) who received vaccination and booster immunizations disclosed that patients who generated A$\beta$ antibodies (as measured by a new A$\beta$ assay) had a slower decline in several functional measures. The events occurring in this subset of patients illustrate the challenges of extrapolating outcomes in mutant mice to human trials. Investigators continue to pursue the passive immunization approaches and are attempting to make new antigen/adjuvant formulations that do not stimulate T-cell-mediated immunologic attack.

## Conclusions

Over many years, investigators have more accurately defined mild cognitive impairment and early AD, have developed diagnostic approaches, and have clarified the character and stages of pathology and related the findings to clinical signs. They have greatly enhanced our understanding of the mechanisms underlying the biochemistry of A$\beta$ plaques and tau-related pathology. Following leads from human autopsy studies and from investigations of *in vitro* and *in vivo* models, investigators are now on the threshold of implementing novel treatments based on an understanding of the neurobiology, neuropathology, biochemistry, and genetics of AD. Moreover, a variety of tools, including amyloid imaging and measure of A$\beta$ flux between various compartments, are now available to assess efficacy of treatment. It is anticipated that exciting discoveries over the next few years will lead to the design of new mechanism-based therapies that can be tested in models, and that these approaches will be introduced into the clinic for the benefit of patients with this devastating illness.

*See also:* Aging of the Brain; Alzheimer's Disease: Neurodegeneration; Alzheimer's Disease: An Overview; Alzheimer's Disease: Transgenic Mouse Models; Animal Models of Alzheimer's Disease; Axonal Transport and Alzheimer's Disease; Brain Glucose Metabolism: Age, Alzheimer's Disease and ApoE Allele Effects; Cholinergic System Imaging in the Healthy Aging Process and Alzheimer Disease; Dementia and Language.

## Further Reading

Bard F, Cannon C, Barbour R, et al. (2000) Peripherally administered antibodies against amyloid $\beta$-peptide enter the central nervous system and reduce pathology in a mouse model of Alzheimer disease. *Nature Medicine* 6(8): 916–919.

Barten DM, Guss VL, Corsa JA, et al. (2005) Dynamics of $\beta$-amyloid reductions in brain, cerebrospinal fluid, and plasma of $\beta$-amyloid precursor protein transgenic mice treated with a $\gamma$-secretase inhibitor. *Journal of Pharmacology and Experimental Therapeutics* 312(2): 635–643.

Borchelt DR, Ratovitski T, VanLare J, et al. (1997) Accelerated amyloid deposition in the brains of transgenic mice coexpressing mutant presenilin 1 and amyloid precursor proteins. *Neuron* 19(4): 939–945.

Braak H and Braak E (1994) Pathology of Alzheimer's disease. In: Calne DB (ed.) *Neurodegenerative Diseases*, pp. 585–613. Philadelphia: W.B. Saunders.

Braak H and Braak E (1991) Neuropathological staging of Alzheimer-related changes. *Acta Neuropathologica* 82: 239–259.

Buxbaum JD, Thinakaran G, Koliatsos V, et al. (1998) Alzheimer amyloid protein precursor in the rat hippocampus: Transport and processing through the perforant path. *Journal of Neuroscience* 18(23): 9629–9637.

Cai H, Wang Y, McCarthy D, et al. (2001) BACE1 is the major $\beta$-secretase for generation of A$\beta$ peptides by neurons. *Nature Neuroscience* 4(3): 233–234.

Cao X and Sudhof TC (2001) A transcriptionally [correction of transcriptively] active complex of APP with Fe65 and histone acetyltransferase Tip60. *Science* 293(5527): 115–120.

Cleary JP, Walsh DM, Hofmeister JJ, et al. (2005) Natural oligomers of the amyloid-$\beta$ protein specifically disrupt cognitive function. *Nature Neuroscience* 8(1): 79–84.

Cummings JL (2004) Alzheimer's disease. *New England Journal of Medicine* 351(1): 56–67.

Davies P and Maloney AJF (1976) Selective loss of central cholinergic neurons in Alzheimer's disease. *Lancet* 2: 1403.

De Strooper B, Saftig P, and Craessaerts K (1998) Deficiency of presenilin-1 inhibits the normal cleavage of amyloid precursor protein. *Nature* 391(6665): 387–390.

DeMattos RB, Bales KR, Cummins DJ, et al. (2001) Peripheral anti-A$\beta$ antibody alters CNS and plasma A$\beta$ clearance and decreases brain A$\beta$ burden in a mouse model of Alzheimer's disease. *Proceedings of the National Academy of Sciences of the United States of America* 98: 8850–8855.

DeMattos RB, Bales KR, Cummins DJ, et al. (2002) Brain to plasma amyloid-$\beta$ efflux: A measure of brain amyloid burden in a mouse model of Alzheimer's disease. *Science* 295: 2264–2267.

Farzan M, Schnitzler CE, Vasilieva N, et al. (2000) BACE2, a $\beta$-secretase homolog, cleaves at the $\beta$ site and within the amyloid-$\beta$ region of the amyloid-$\beta$ precursor protein. *Proceedings of the National Academy of Sciences of the United States of America* 97: 9712–9717.

Ghiso J and Wisniewski T (2004) An animal model of vascular amyloidosis. *Nature Neuroscience* 7(9): 902–904.

Goedert M and Spillantini M (2006) Neurodegenerative alpha-synucleinopathies and tauopathies. In: Siegel GJ, Albers RW, Brady S, and Price DL (eds.) *Basic Neurochemistry: Molecular, Cellular, and Medical Aspects,* 7th edn., pp. 745–759. Boston, MA: Elsevier Academic Press.

Hock C, Konietzko U, Streffer JR, et al. (2003) Antibodies against $\beta$-amyloid slow cognitive decline in Alzheimer's disease. *Neuron* 38(4): 547–554.

Jankowsky JL, Fadale DJ, Anderson J, et al. (2004) Mutant presenilins specifically elevate the levels of the 42 residue $\beta$-amyloid peptide *in vivo*: Evidence for augmentation of a 42-specific γ-secretase. *Human Molecular Genetics* 13(2): 159–170.

Jicha GA, Parisi JE, Dickson DW, et al. (2006) Neuropathologic outcome of mild cognitive impairment following progression to clinical dementia. *Archives of Neurology* 63(5): 674–681.

Killiany RJ, Hyman BT, Gomez-Isla T, et al. (2002) MRI measures of entorhinal cortex vs hippocampus in preclinical AD. *Neurology* 58(8): 1188–1196.

Klunk WE, Engler H, Nordberg A, et al. (2004) Imaging brain amyloid in Alzheimer's disease using the novel positron emission tomography tracer, Pittsburgh compound-B. *Annals of Neurology* 55: 1–14.

Klyubin I, Walsh DM, Lemere CA, et al. (2005) Amyloid β protein immunotherapy neutralizes Aβ oligomers that disrupt synaptic plasticity *in vivo*. *Nature Medicine* 11(5): 556–561.

Laird FM, Cai HS, Savonenko AV, et al. (2005) BACE1, a major determinant of selective vulnerability of the brain to Aβ amyloidogenesis is essential for cognitive, emotional and synaptic functions. *Journal of Neuroscience* 25(50): 11693–11709.

Lazarov O, Morfini GA, Lee EB, et al. (2005) Axonal transport, amyloid precursor protein, kinesin-1, and the processing apparatus: Revisited. *Journal of Neuroscience* 25(9): 2386–2395.

Lesne S, Koh MT, Kotilinek L, et al. (2006) A specific amyloid-β protein assembly in the brain impairs memory. *Nature* 440 (7082): 352–357.

Lewis J, Dickson DW, Lin W, et al. (2001) Enhanced neurofibrillary degeneration in transgenic mice expressing mutant tau and APP. *Science* 293: 1487–1491.

Li T, Ma G, Cai H, et al. (2003) Nicastrin is required for assembly of Presenilin/γ-secretase complexes to mediate notch signaling and for processing and trafficking of β-amyloid precursor protein in mammals. *Journal of Neuroscience* 23(8): 3272–3277.

Ma G, Li T, Price DL, et al. (2005) APH-1a is the principal mammalian APH-1 isoform present in γ-secretase complexes during embryonic development. *Journal of Neuroscience* 25(1): 192–198.

Markesbery WR, Schmitt FA, Kryscio RJ, et al. (2006) Neuropathologic substrate of mild cognitive impairment. *Archives of Neurology* 63(1): 38–46.

Martin LJ, Pardo CA, Cork LC, et al. (1994) Synaptic pathology and glial responses to neuronal injury precede the formation of senile plaques and amyloid deposits in the aging cerebral cortex. *American Journal of Pathology* 145(6): 1358–1381.

Morris JC, Storandt M, Miller JP, et al. (2001) Mild cognitive impairment represents early-stage Alzheimer disease. *Archives of Neurology* 58(3): 397–405.

Morris R and Mucke L (2006) Alzheimer's disease: A needle from the haystack. *Nature* 440(7082): 284–285.

Nestor PJ, Scheltens P, and Hodges JR (2004) Advances in the early detection of Alzheimer's disease. *Nature Medicine* 10(supplement): S34–S41.

Nicoll JA, Wilkinson D, Holmes C, et al. (2003) Neuropathology of human Alzheimer disease after immunization with amyloid-β peptide: A case report. *Nature Medicine* 9(4): 448–452.

Oddo S, Caccamo A, Shepherd JD, et al. (2003) Triple-transgenic model of Alzheimer's disease with plaques and tangles: Intracellular Aβ and synaptic dysfunction. *Neuron* 39(3): 409–421.

Petersen RC (2003) Mild cognitive impairment clinical trials. *Nature Reviews Drug Discovery* 2(8): 646–653.

Price DL, Tanzi RE, Borchelt DR, et al. (1998) Alzheimer's disease: Genetic studies and transgenic models. *Annual Review of Genetics* 32: 461–493.

Price JL and Morris JC (1999) Tangles and plaques in nondemented aging and "preclinical" Alzheimer's disease. *Annals of Neurology* 45(3): 358–368.

Rogaev EI, Sherrington R, Rogaeva EA, et al. (1995) Familial Alzheimer's disease in kindreds with missense mutations in a gene on chromosome 1 related to the Alzheimer's disease type 3 gene. *Nature* 376: 775–778.

Rovelet-Lecrux A, Hannequin D, Raux G, et al. (2006) APP locus duplication causes autosomal dominant early-onset Alzheimer disease with cerebral amyloid angiopathy. *Nature Genetics* 38(1): 24–26.

Savonenko AV, Laird FM, Troncoso JC, et al. (2005) Role of Alzheimer's disease models in designing and testing experimental therapeutics. *Drug Discovery Today* 2(4): 305–312.

Savonenko A, Xu GM, Melnikova T, et al. (2005) Episodic-like memory deficits in the APPswe/PS1dE9 mouse model of Alzheimer's disease: Relationships to β-amyloid deposition and neurotransmitter abnormalities. *Neurobiology of Disease* 18(3): 602–617.

Schenk D, Hagen M, and Seubert P (2004) Current progress in β-amyloid immunotherapy. *Current Opinion in Immunology* 16(5): 599–606.

Selkoe D and Kopan R (2003) Notch and presenilin: Regulated intramembrane proteolysis links development and degeneration. *Annual Review of Neuroscience* 26: 565–597.

Sheng JG, Price DL, and Koliatsos VE (2002) Disruption of corticocortical connections ameliorates amyloid burden in terminal fields in a transgenic model of Aβ amyloidosis. *Journal of Neuroscience* 22(22): 9794–9799.

Sisodia SS, Koo EH, Beyreuther KT, et al. (1990) Evidence that β-amyloid protein in Alzheimer's disease is not derived by normal processing. *Science* 248: 492–495.

Vassar R, Bennett BD, Babu-Khan S, et al. (1999) β-Secretase cleavage of Alzheimer's amyloid precursor protein by the transmembrane aspartic protease BACE. *Science* 286: 735–741.

Weggen S, Eriksen JL, Das P, et al. (2001) A subset of NSAIDs lower amyloidogenic Aβ42 independently of cyclooxygenase activity. *Nature* 414: 212–216.

Wolfe MS, Xia W, Ostaszewski BL, et al. (1999) Two transmembrane aspartates in presenilin-1 required for presenilin endoproteolysis and γ-secretase activity. *Nature* 398(6727): 513–517.

Wong PC, Li T, and Price DL (2005) Neurobiology of Alzheimer's disease. In: Siegel GJ, Albers RW, Brady S, and Price DL (eds.) *Basic Neurochemistry: Molecular, Cellular, and Medical Aspects*, 7th edn., pp. 781–790. Boston, MA: Elsevier Academic Press.

Wong GT, Manfra D, Poulet FM, et al. (2004) Chronic treatment with the γ-secretase inhibitor LY-411,575 inhibits β-amyloid peptide production and alters lymphopoiesis and intestinal cell differentiation. *Journal of Biological Chemistry* 279(13): 12876–12882.

Wong PC, Price DL, and Cai H (2001) The brain's susceptibility to amyloid plaques. *Science* 293: 1434–1435.

# Alzheimer's Disease: Molecular Genetics

**R Sherrington**, Unité de Recherche en Neuroscience, Ste-Foy, QC, Canada
**P H St. George-Hyslop**, University of Toronto, Toronto, ON, Canada



## Introduction

Alzheimer's disease (AD) is the most common form of degenerative dementia of the human central nervous system. It is defined clinically as a progressive loss of cognitive function with the onset of a slowly progressive impairment of memory during mid- to late adult life. The neuropathologic hallmarks of this disease include amyloid deposits, neurofibrillary tangles, astrocytic gliosis, and reductions in the number of neurons and synapses in many areas of the brain, but especially from the cerebral cortex and the hippocampus.

The etiology of AD is complex. Multiple epidemiologic studies have proposed a number of potential risk factors, including environmental (head trauma, smoking, and exposure to heavy metals such as aluminum), sociologic (depression and level of education), biologic (increasing age, hyperthyroidism, late maternal age), and family history (of Down syndrome or AD).

The repeated observation in multiple epidemiologic surveys that a positive family history is a strong risk factor for AD clearly suggest that genetic factors play a role in this disease. A simple autosomal dominant inheritance with age-dependent penetrance has been observed in many pedigrees, especially those with early age of onset. However, in many families multiply affected by AD, the pattern of inheritance is unclear and could fit one of several genetic models, including a high-frequency, low-penetrance single gene disorder or a multifactorial model in which several genes or nongenetic factors interact. Using the powerful tools of genetic linkage analysis, significant progress has been made in unraveling the genetic etiology. Four loci that play a role in the genetic susceptibility of AD have been identified, namely, genes on chromosomes 21 ($\beta$-amyloid precursor protein ($\beta$APP)), 19 (apolipoprotein E (APOE)), 14 (presenilin 1 (PS1)), and 1 (presenilin 2 (PS2)).

## $\beta$-Amyloid Precursor Protein

The $\beta$APP gene was the first gene found to bear mutations capable of causing early-onset familial AD (FAD). The $\beta$APP gene on chromosome 21 encodes an alternatively spliced single spanning transmembrane protein. The longer isoform, APP770 (770 amino acids), contains two exons, which encode a Kunitz protease inhibitor (KPI) domain and a 19-amino acid sequence homologous to the ox-2 antigen. The KPI-containing isoforms are expressed in most tissues. In contrast, the shorter isoform APP695 (695 amino acids), without the KPI domain, is predominantly expressed in the brain. The $\beta$APP protein undergoes a series of endoproteolytic cleavages that give rise to A$\beta$ peptide, a 4 kDa peptide resulting from proteolytic cleavages of the full-length protein from residues 672–711 that spans the last 28 residues of the extracellular domain and the first 14 residues of the transmembrane domain. One of these cleavages, which results from the action of a putative membrane-associated $\alpha$-secretase, liberates the extracellular N-terminus of $\beta$APP and is thus a nonamyloidogenic pathway, because this cleavage precludes the formation of A$\beta$ peptide. The other cleavage pathway, which occurs in part in the endosomal-lysosomal compartment, involves the recently identified $\beta$-secretase (BACE) and the putative $\gamma$-secretases, which give rise to a series of peptides that contain the 40–42 amino acid A$\beta$ peptide. A$\beta$ peptides ending at residue 42 or 43 (long-tailed A$\beta$) are thought to be more fibrillogenic and more neurotoxic than A$\beta$ ending at residue 40, which is the predominant isoform produced during normal metabolism of $\beta$APP. The activity of these secretases, and especially g-secretase, giving rise to the more fibrillogenic and potentially neurotoxic long-tailed Ab1–42, appears to play an important role in the pathogenesis of AD. It has been suggested that processing of $\beta$APP into A$\beta$ and the subsequent accumulation of A$\beta$ in the brain are central events in the pathogenesis of AD in both genetic and nongenetic forms. Although A$\beta$ peptides are secreted by cells under normal physiologic conditions, how they induce neurodegeneration in AD is still unclear.

Five mutations in the $\beta$APP gene (APP V7171, APP V717G, APP V717F, APP KM670/671NL, and APP A692G) can cause early-onset FAD. The mutations occur within or near the Ab peptide. Mutations in this gene account for only a small fraction of early-onset FAD cases (less than 0.1%). The function of the $\beta$APP protein is unknown, although roles in cell adhesion, synaptic growth, and neural repair have been proposed. It has also been suggested that the $\beta$APP may function as a G-coupled receptor linked through an interaction in the C-terminus to $G_o$ protein.

## The APOE Gene

The APOE gene is located on chromosome 19q13 and encodes the major apolipoprotein expressed in the central nervous system. APOE plays an important role in triglyceride-rich lipoprotein metabolism and regulation of cholesterol. It has also been suggested that APOE plays a role in repair, growth, and maintenance of myelin and axonal membranes. Studies implicating the APOE gene as an AD susceptibility locus derived from several intersecting lines of investigation (linkage analysis on late-onset AD families, APOE immunoreactivity in senile plaques and NFTs of patient with AD, and APOE binding to A$\beta$). Association studies revealed that inheritance of the APOE e4 allele (Cys112Arg) is associated with a copy number-dependent increase in risk of both sporadic and late-onset FAD and with a decrease in the age of onset in late-onset AD. Conversely, the e2 allele confers a protective effect against late-onset AD. However, it seems that the inheritance of the e4 allele (APOE4) may be neither necessary nor sufficient by itself to cause AD, and the predictive value of the APOE4 in presymptomatic people is unclear. The mechanism by which the inheritance of the APOE4 increases risk of AD and how this might be addressed therapeutically are still unclear. Moreover, it has been recently reported that APOE represents only 7–9% of the total variation in age at onset of AD, indicating that other loci may have greater importance than does APOE.

## The PS Genes

Genetic linkage studies, using large numbers of pedigrees with early-onset autosomal dominant FAD, mapped a common locus (AD3) to chromosome 14q24.3. The AD3 locus is associated with a particularly aggressive form of early-onset (between 30 and 60 years of age) AD and accounts for up to 50% of early-onset FAD cases. Subsequent positional studies led to the isolation of a novel gene termed PS1. Mutational analysis of the PS1 gene have identified at least 70 different predominantly missense mutations in residues that are highly conserved in evolution. The PS1 is predicted to be an integral membrane protein with at least seven hydrophobic membrane-spanning domains (transmembrane domains (TM)) and two acidic hydrophilic domains located at the N-terminus and between TM6 and TM7. The amino acid sequence of the PS1 protein shows a weak homology to the *C. elegans* SPE-4 protein, which is thought to be involved in membrane budding and fusion events in a membrane-bound cytoplasmic organelle derived from the spermatocytes' Golgi network. Subsequently, a strong homology was found between PS1 and SEL-12, another *C. elegans* protein. SEL-12 facilitates signaling mechanisms in the LIN-12/Notch family of intercellular receptors, which are responsible for direct signal transmission from the cell surface to the nucleus during intercellular signal transduction specifying cell fate. These later data suggest that PS1 may have a role in the vertebrate Notch signaling pathway. This is further supported by the fact that PS1 knockout mice show developmental abnormalities similar to those of mice with targeted knockouts of the murine Notch1 gene.

Following the isolation of the PS1 gene causing the AD3 subtype of AD, a second gene, PS2, mapped to chromosome 1, was discovered on the basis of its strong nucleotide and amino acid sequence homology to PS1. The structural organization of the PS2 is very similar to that of PS1, including the transmembrane and acidic hydrophilic domains. These observations suggest that the PS1 and the PS2 genes are members of a gene family. Mutational analysis of the PS2 gene led to the discovery of six different mutations. The phenotype of these families shows a later onset than the phenotype associated with mutations in PS1 or $\beta$APP genes (onset 50–70 years).

All the PS1 or PS2 mutations lead to significantly increased levels of secreted A$\beta$42 compared with wild-type controls. These observations suggest a dominant gain of function for the mutant PSs because even though the endogenous wild-type alleles are present, the level of A$\beta$42 is increased in cells expressing mutant cDNAs. How mutant presenilins alter APP processing is unknown. It has been proposed that PS1 (and probably PS2) is part of a complex that includes the recently discovered nicastrin, involved in the proteolytic cleavage of the C-terminal fragment of APP (mediated by $\gamma$-secretase). Other functions in intracellular signaling, suppression of apoptosis, and protein/membrane trafficking have been suggested for both PS1 and PS2.

## Other AD Genes

Although four different genetic loci associated with inherited susceptibility to AD ($\beta$APP, APOE, PS1, and PS2) have been identified, they account for only about half of the genetic forms of AD. Thus at least one and possibly several other genes remain to be identified. Several polymorphisms in different genes have been associated with AD. Recent findings suggest the presence of new AD genes on chromosomes 9, 10, and 12. However, the lack of replication of these associations in different AD pedigrees makes these results unclear at the moment.

## Animal Models

Several transgenic murine model lines have been created using a variety of $\beta$APP constructs and have met with varying degrees of success. The most successful model to date has been a transgenic line in which a mutant human $\beta$APP minigene has been placed under the control of the platelet-derived growth factor receptor beta subunit promoter. This latter model demonstrates abundant A$\beta$ deposition and synaptic pathology. Targeted null mutation (knockout) of the murine $\beta$APP gene was created but is not very illuminating because it leads only to subtle phenotypes including minor weight loss, decreased locomotor activity, abnormal forelimb motor activity, and minor nonspecific degrees of reactive gliosis in the cortex.

Transgenic mice overexpressing wild-type or mutant PSs have been created. All mutants show an increase in A$\beta$42 production in the brain compared with wild-type. Despite the A$\beta$42 accumulation in the brain, these PS transgenic mice have not shown any major neuronal death. However, these mice showed reduced spontaneous alternation performance in a Y maze. PS1 knockout mice have also been generated. These knockout mice developed deformations of the caudal axial skeleton, hemorrhages in the central nervous system, and impairment in neurogenesis and usually did not survive beyond the first day after birth. These studies show that, at least, PS1 is required for proper formation of the axial skeleton and is involved in normal neurogenesis and survival of progenitor cells and neurons in specific brain subregions.

*See also:* Aging of the Brain; Aging of the Brain and Alzheimer's Disease; Alzheimer's Disease: An Overview; Alzheimer's Disease: Transgenic Mouse Models; Dementia and Language; Gene Expression in Normal Aging Brain; Genomics of Brain Aging: Apolipoprotein E; Genomics of Brain Aging: Twin Studies.

## Further Reading

Bertram L and Tanzi RE (2001) Dancing in the dark? The status of late-onset Alzheimer's disease genetics. *Journal of Molecular Neuroscience* 17(2): 127–136.

Farrer LA, Myers RH, Connor L, et al. (1991) Segregation analysis reveals evidence of a major gene for Alzheimer disease. *American Journal of Human Genetics* 48: 1026–1033.

Games D, Adams D, Alessandrini A, et al. (1995) Alzheimer-type neuropathology in transgenic mice overexpressing V717F beta-amyloid precursor protein. *Nature* 373: 523–527.

Goate AM, Chartier-Harlin MC, Mullan M, et al. (1991) Segregation of a missense mutation in the amyloid precursor protein gene with familial Alzheimer disease. *Nature* 349: 704–706.

Katzman R (1986) Alzheimer's disease. *New England Journal of Medicine* 314(15): 964–973.

Katzman R and Kawas C (1994) The epidemiology of dementia and Alzheimer disease. In: Terry RD, Katzman R, and Hick KL (eds.) *Alzheimer Disease*, pp. 105–122. New York: Raven Press.

Prince JA, Feuk L, Sawyer SL, et al. (2001) Lack of replication of association findings in complex disease: An analysis of 15 polymorphisms in prior candidate genes for sporadic Alzheimer's disease. *European Journal of Human Genetics* 9(6): 437–444.

Rogaev EI, Sherrington R, Rogaeva EA, et al. (1995) Familial Alzheimer's disease in kindreds with missense mutations in a novel gene on chromosome I related to the Alzheimer's disease type 3 gene. *Nature* 376: 775–778.

Saunders A, Strittmatter WJ, Schmechel S, et al. (1993) Association of apolipoprotein E allele e4 with late-onset familial and sporadic Alzheimer disease. *Neurology* 43: 1467–1472.

Selkoe DJ (1994) Normal and abnormal biology of $\beta$-amyloid precursor protein. *Annual Review of Neuroscience* 17: 489–517.

Selkoe DJ (2001) Alzheimer's disease: Genes, proteins, and therapy. *Physiological Reviews* 81(2): 741–766.

Sherrington R, Rogaev EI, Liang Y, et al. (1995) Cloning of a gene bearing missense mutations in early-onset familial Alzheimer's disease. *Nature* 375: 754–760.

Warwick DE, Payami H, Nemens EJ, et al. (2000) The number of trait loci in late-onset Alzheimer disease. *American Journal of Human Genetics* 66(1): 196–204.

## Relevant Website

http://www.alzforum.org – Alzheimer Research Forum.

# Alzheimer's Disease: MRI Studies

**P M Thompson and A W Toga**, UCLA School of Medicine, Los Angeles, CA, USA



## Introduction

Alzheimer's disease (AD) is the leading cause of senile dementia, affecting 10% of those over age 65. The disease causes irreversible memory loss, behavioral and cognitive decline, personality changes, and a decreasing ability to cope with everyday life. Ironically, up to 30 years elapse between the onset of the cellular pathology that causes AD (amyloid plaques and neurofibrillary tangles in the brain) and the clinical changes that lead to diagnosis. To help understand how the disease emerges and progresses, imaging technology can be applied that is safe, repeatable, and widely available. Magnetic resonance imaging (MRI) scans of AD patients reveal profound anatomic changes: severe cortical and hippocampal atrophy, sulcal and ventricular enlargement, and reduced gray matter and white matter volume. These changes occur in a distinct spatial and temporal sequence, and correlate with cognitive and metabolic decline. If patients are scanned repeatedly with MRI as their disease progresses, dynamic maps can be reconstructed that reveal a shifting pattern of cortical changes. This spreading cortical atrophy mirrors the spread of the underlying pathology (as defined by tangle and amyloid plaque deposition). Repeat MRI scanning can monitor disease progression in individual patients and can evaluate how drugs oppose these changes. It can also clarify how anatomical deficits link with cognitive and behavioral deficits as they emerge in individuals and populations.

## Impact

AD is a severe and growing public health crisis. The incidence of AD doubles every 5 years after age 60. It afflicts 1% of those aged 60 to 64 and 30% to 40% of those over 85. Without a cure, the number of AD victims will rise from 2.0 to 3.5 million now to an estimated 10 to 14 million by 2030. A number of promising AD treatments are now being developed. These range from acetylcholinesterase inhibitors, which ballast neurotransmitter function, to experimental vaccines, which directly attack the amyloid plaques that are a key element of AD pathology. Most therapeutic trials of new drugs in AD rely primarily on cognitive tests to determine efficacy. Neuroimaging, however, can be extremely beneficial in this research. It supplies a variety of biological markers that measure disease progression. With novel brain mapping techniques, the disease can be tracked as it spreads in the living brain. In Alzheimer's patients, MRI scans show prominent hippocampal atrophy. Diffuse tissue loss is also found in the medial temporal lobes. These deficits are progressive, and their magnitude correlates with cognitive decline. Because it is vital to detect the disease early, there is great interest in developing MRI measures that predict imminent transition to dementia, in healthy elderly subjects. For example, significant hippocampal volume deficits are found in subjects with mild memory impairments, who do not yet have dementia. These deficits shown on MRI can also help to predict how soon an elderly individual will develop AD. Reliable MRI predictors are especially valuable, as cholinergic drugs are most effective in the mildest phases, when widespread neuronal loss has not yet occurred.

Several neuroimaging measures can be used to characterize dementia. For instance, MRI scans can assess the integrity of medial temporal lobe structures involved in memory, such as the entorhinal cortex and hippocampus. The region and rate of atrophic brain changes can be measured as the disease progresses, as can the profile of cortical thinning and gray matter loss. The required three-dimensional MRI scans can be performed in approximately 10 min, on a conventional 1.5-Tesla scanner. Although MRI is not routinely used to diagnose AD, new techniques in brain image analysis can be applied to MRI scans to reveal how the disease emerges, and track how medications affect the disease process. MRI scans are often used in dementia research to (1) screen at-risk populations to find anatomical measures that might help predict each individual's likelihood of developing AD, (2) discriminate AD from normal aging and other dementias (such as frontotemporal and Lewy body dementias, which have distinct anatomic patterns), and (3) monitor disease progress and therapeutic response, gauging the effectiveness of drug treatments.

## Brain Tissue Loss and Cognitive Decline

In the 1990s, MRI research in dementia focused on measuring medial temporal lobe structures. This was because AD pathology typically starts in the temporal cortex adjacent to the entorhinal cortex and quickly spreads to the entorhinal cortex before involving the hippocampus. This temporal lobe pathology persists for several years before spreading cortically to engulf the rest of the temporal, frontal, and parietal lobes. A more recent trend in dementia research has been to move from cross-sectional studies to dynamic measures. Serial MRI scans (acquired from the same patients repeatedly over time) can provide much greater power to detect pathological atrophy, because they provide a baseline reference point to calculate change. Fox et al. found that AD patients lose brain tissue at a faster 'rate' than age-matched controls. Evaluated with MRI for 5–8 years, AD patients lost brain tissue at a median rate of 2.20% per year (range, 0.82–4.19) versus 0.24% per year in controls (range, −0.35–0.64). These rates correlated with the rate of cognitive decline, reflected by worsening performance on the Mini Mental Status Examination (MMSE). In a recent 52 week clinical trial of milameline (a muscarinic receptor agonist), Jack et al. noted that hippocampal volume, measured with MRI, also tracked cognitive decline. Perhaps the most prominent sign of AD seen on an MRI scan is that the lateral ventricles are often greatly enlarged. Bradley et al. measured the ratio of the ventricular volume to the total brain volume (the ventricle-to-brain ratio (VBR) in 39 elderly subjects scanned with serial MRIs over 3 to 6 month intervals. The VBR rate of change was 15.6% ± 2.8% (mean ± SD) per year for AD patients compared with 4.3% ± 1.1% per year in controls ($P < 0.001$). VBR did not separate groups when measured at only a single time point, supporting the value of longitudinal assessments. Power calculations revealed that 135 subjects would be needed in each arm of a placebo-controlled clinical trial if this measure of AD progression were to detect a 20% reduction in the excess rate of atrophy over 6 months, with 90% power.

## Gray Matter Deficits

Brain changes in AD can also be visualized using three-dimensional maps. **Figure 1** shows the spatial pattern of cortical gray matter loss in mild to moderate AD. This type of image is a composite map; it results from a sequence of image processing steps that compare scans of AD patients with matched healthy elderly subjects. With image analysis techniques, three-dimensional brain MRI scans can be split up

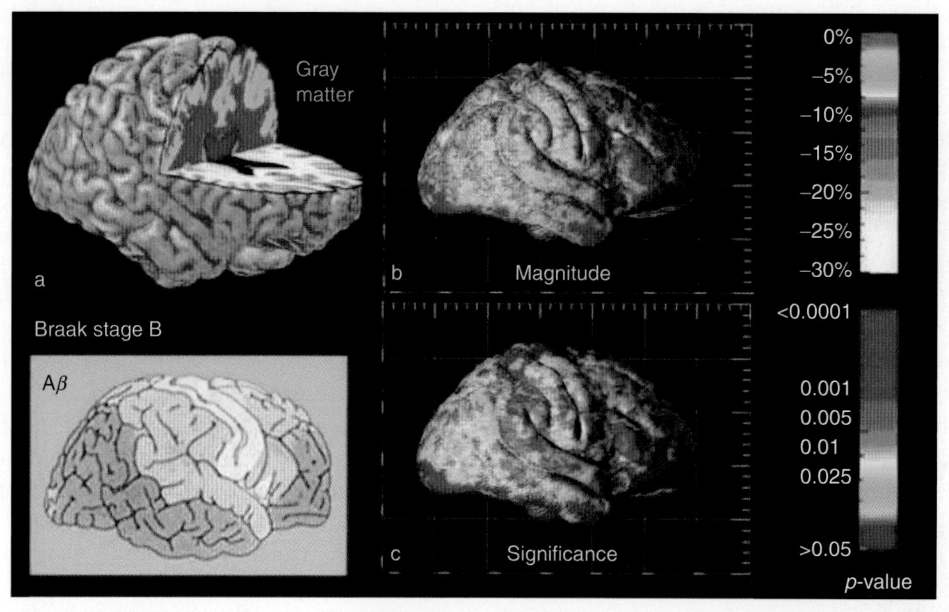


**Figure 1** Gray matter deficits in early AD. Here the local amount of cortical gray matter (green colors, (a)) is compared across 26 patients with mild to moderate AD (age, 75.8 ± 1.7 years; MMSE score, 20.0 ± 0.9) and 20 matched elderly controls (age, 72.4 ± 1.3 years). At this stage of AD, 30% of the cortical gray matter has been lost in the temporoparietal regions (b). (c) Statistical significance of these deficits. The pattern of temporal lobe gray matter loss, seen on MRI, spatially matches the pattern of beta-amyloid (A$\beta$) deposition seen postmortem. The inset panel (Braak stage B) is adapted from data reported by Braak and Braak (1997). It shows regions with minimal (white), moderate (orange), and severe (red) A$\beta$ deposition. Amyloid deposition and gray matter loss may not be synchronized, so these maps may represent different stages of AD; however, there is a clear spatial agreement in the severity of the deficits, between MRI and A$\beta$ maps. Primary sensorimotor regions (white in the amyloid map), and the superior temporal gyri (blue colors in (c)) are spared relative to other temporal lobe gyri. These MRI patterns have been replicated in independent studies by Thompson PM, Mega MS, Woods RP, et al. (2001) Cortical change in Alzheimer's disease detected with a disease-specific population-based brain atlas. *Cerebral Cortex* 11: 1–16, Baron JC, Chetelat G, Desgranges B, et al. (2001) *In vivo* mapping of gray matter loss with voxel-based morphometry in mild Alzheimer's disease. *Neuroimage* 14: 298–309, and O'Brien JT, Paling S, Barber R, et al. (2001) Progressive brain atrophy on serial MRI in dementia with Lewy bodies, AD, and vascular dementia. *Neurology* 56: 1386–1388.

into regions representing gray matter, white matter, and cerebrospinal fluid. A measure is then computed that is related to the thickness of the cortical gray matter at each cortical location. Computer analyses then compare the amount of gray matter at each cortical location across subjects, while adjusting for potentially confounding factors, such as age and gender effects, and gyral patterning differences. Differences can be visualized locally in the form of color-coded statistical maps. These show how much gray matter volume is reduced in AD patients relative to healthy controls in each cortical region

## Maps of Disease Progression

MRI scanning also reveals a dynamically spreading wave of gray matter loss in the brains of patients with AD. With novel brain mapping methods, the loss pattern can be visualized as it spreads over time from temporal and limbic cortices into frontal and occipital brain regions, sparing sensorimotor cortices. These shifting deficits are correlated with cognitive decline. As shown in **Figure 2**, cortical atrophy occurs in a well-defined sequence as the disease progresses, mirroring the temporal sequence of beta-amyloid (A$\beta$) and neurofibrillary tangle accumulation observed at autopsy. The trajectory of deficits also matches the sequence of metabolic decline typically observed with positron emission tomography.

To map AD progression, advancing deficits can be visualized as dynamic video maps that change over time, which distinguish different phases of AD and differentiate AD from normal aging. Frontal brain regions, spared early in the disease, show pervasive deficits later (>15% loss). Local gray matter loss rates ($5.3 \pm 2.3\%$ per year in AD vs. $0.9 \pm 0.9\%$ per year in controls) are faster in the left hemisphere than the right, at least at this stage of AD. Transient

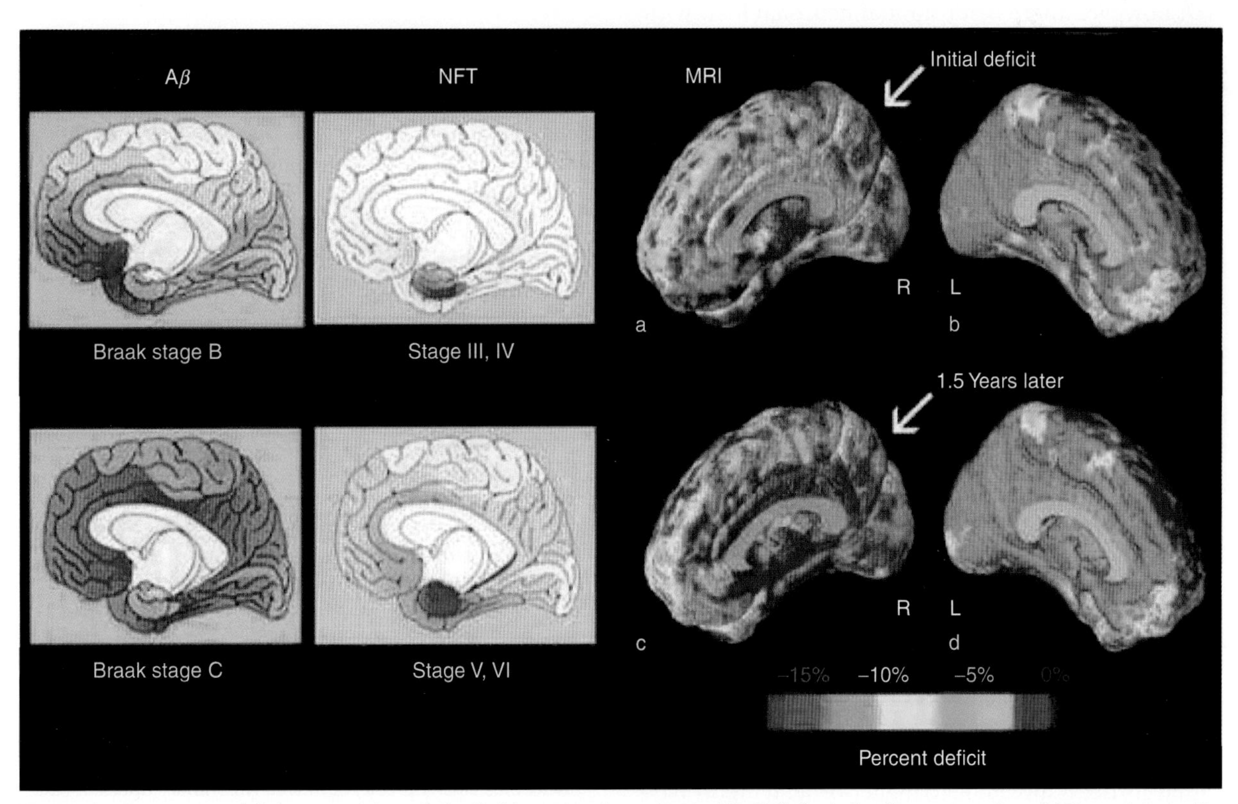


**Figure 2** Gray matter deficits spread through the limbic system in moderate AD. Deficits during the progression of AD are detected by comparing average profiles of gray matter volumes between 12 AD patients (age, $68.4 \pm 1.9$ years) and 14 elderly matched controls (age, $71.4 \pm 0.9$ years). Colors show the average percent loss of gray matter relative to the control average. Profound loss engulfs the left medial wall (>15% (b), (d)). On the right, however, the deficits in temporoparietal and entorhinal territory (a) spread forward into the cingulate gyrus 1.5 years later (c), after a 5-point drop in average MMSE test scores. Limbic and frontal zones are prominently divided, with different degrees of impairment (c). MRI-based changes, observed in living patients, agree strongly with the spatial progression of A$\beta$ and neurofibrillary tangle (NFT) pathology observed postmortem (Braak stages B and C and III to VI. NFT accumulation is minimal in sensory and motor cortices, but occurs preferentially in entorhinal pyramidal cells, the limbic periallocortex (layers II/IV), the hippocampus/amygdala and subiculum, the basal forebrain cholinergic systems and subsequently in temporoparietal and frontal association cortices (layers III/V). Left four panels adapted from Braak H and Braak E (1997) Staging of Alzheimer-related cortial destruction. *International Psychogeriatrics* 9(supplement 1): 257–271; discussion 269–272.

barriers to disease progression also appear. A frontal band (0–5% loss) is sharply delimited from the limbic and temporoparietal regions that show severest deficits in AD (>15% loss). This pattern is consistent with the hypothesis that AD pathology spreads centrifugally from limbic/paralimbic to higher-order association cortices. This degenerative sequence, observed as it develops in living patients, provides a quantitative, dynamic visualization of cortical atrophic rates in dementia, over a period of cognitive decline lasting 1.5 years.

## What Is Gray Matter Atrophy?

Gray matter atrophy observed on MRI is linked with cognitive decline in AD and is attributable to several cellular processes. In healthy aging, age-related neuronal loss does not occur in most neocortical regions and appears specific to the frontal cortex and some hippocampal regions (e.g., CAI and the subiculum). In AD, however, there is substantial neuronal loss, with severe early losses in layer II of the entorhinal cortex.

## A$\beta$ and Neurofibrillary Tangle Maps

MRI-based maps of cortical atrophy agree strongly with postmortem maps of A$\beta$ deposition (A$\beta$, **Figure 2**). A$\beta$ is an insoluble protein that is a key feature of Alzheimer pathology. The spatial congruence of these two maps supports the hypothesis that A$\beta$ deposition may participate in the cascade of events that leads to regional gray matter atrophy and neuronal cell loss. In both maps, primary sensorimotor cortices are relatively spared until late in the disease, and the superior temporal gyrus is less affected than other temporal lobe gyri. In early AD, intraneuronal filamentous deposits, or neurofibrillary tangles (NFTs), also accumulate within neurons. These deposits are composed of hyperphosphory lated tau protein. This cellular pathology disrupts axonal transport and induces widespread metabolic decline; it eventually leads to neuronal loss, observed as gross atrophy on MRI. Braak and Braak noted on autopsy that NFT distribution was initially restricted to entorhinal cortices, spreading to higher-order temporoparietal association cortices, then frontal, and ultimately primary sensory and visual areas. MRI scans suggest that a similar wave of cortical atrophy can be mapped in patients while they are alive. This provides a biological marker of disease progression that can monitor the effects of therapy.

It remains a mystery why brain changes in AD occur in this sequence. Braak and Braak suggested that the atrophic trajectory in AD is somewhat the reverse of the sequence in which cortical areas are myelinated during development. For example, primary sensory regions myelinate first and degenerate last, and temporal regions mature last but degenerate first in AD. This palindromic sequence is largely supported by the pattern of cortical changes observed on MRI. The selective vulnerability of specific cortical systems in AD may relate to differences in cellular maturational rates and/or plasticity. The most plastic systems may also be most vulnerable to AD.

## Conclusion

MRI scans can measure brain change in AD, mapping the disease process in detail. MRI measures of disease progression, along with measures of genetic risk and abnormalities in specific neuropsychological tests, now provide key quantitative predictors to monitor brain degeneration and gauge how well it is decelerated or delayed in clinical trials.

*See also:* Aging of the Brain and Alzheimer's Disease; Alzheimer's Disease: Molecular Genetics; Axonal Transport and Alzheimer's Disease; Brain Glucose Metabolism: Age, Alzheimer's Disease and ApoE Allele Effects; Functional Neuroimaging Studies of Aging.

## Further Reading

Arnold SE, Hyman BT, Flory J, et al. (1991) The topographical and neuroanatomical distribution of neurofibrillary tangles and neuritic plaques in the cerebral cortex of patients with Alzheimer's disease. *Cerebral Cortex* 1: 103–116.

Baron JC, Chetelat G, Desgranges B, et al. (2001) *In vivo* mapping of gray matter loss with voxel-based morphometry in mild Alzheimer's disease. *Neuroimage* 14: 298–309.

Braak H and Braak E (1997) Staging of Alzheimer-related cortical destruction. *International Psychogeriatrics* 9(supplement 1): 257–261; discussion 269–272.

Bradley KM, Bydder GM, Budge MM, et al. (2002) Serial brain MRI at 3–6 month intervals as a surrogate marker for Alzheimer's disease. *British Journal of Radiology* 75: 506–513.

Chetelat G and Baron JC (2003) Early diagnosis of Alzheimer's disease: Contribution of structural neuroimaging. *Neuroimage* 18: 525–541.

Convit A, de Asis J, de Leon MJ, et al. (2000) Atrophy of the medial occipitotemporal, inferior, and middle temporal gyri in non-demented elderly predict decline to Alzheimer's disease. *Neurobiology of Aging* 21: 19–26.

Fox NC, Crum WR, Scahill RI, et al. (2001) Imaging of onset and progression of Alzheimer's disease with voxel-compression mapping of serial magnetic resonance images. *Lancet* 358: 201–205.

Frisoni GB, Laakso MP, Beltramello A, et al. (1999) Hippocampal and entorhinal cortex atrophy in frontotemporal dementia and Alzheimer's disease. *Neurology* 52: 91–100.

Gomez-Isla T, Price JL, McKeel DW Jr., et al. (1996) Profound loss of layer II entorhinal cortex neurons occurs in very mild Alzheimer's disease. *Journal of Neuroscience* 16: 4491–4500.

Jack CR Jr., Slomkowski M, Gracon S, et al. (2003) MRI as a biomarker of disease progression in a therapeutic trial of milameline for AD. *Neurology* 60: 253–260.

Jobst KA, Smith AD, Szatmari M, et al. (1994) Rapidly progressing atrophy of medial temporal lobe in Alzheimer's disease. *Lancet* 343: 829–830.

Kaye J, Moore M, Kerr D, et al. (1999) The rate of brain volume loss accelerates as Alzheimer's disease progresses from a presymptomatic phase to frank dementia. *Neurology* 52(supplement): A569–A570.

Laakso MP, Lehtovirta M, Partanen K, et al. (2000) Hippocampus in AD: A 3-year follow-up MRI study. *Biological Psychiatry* 47: 557–561.

Malmgren R (2000) Epidemiology of aging. In: Coffey CE and Cummings JL (eds.) *Textbook of Geriatric Neuropsychiatry*, pp. 17–31. Washington, DC: American Psychiatric Press.

Mesulam MM (2000) A plasticity-based theory of the pathogenesis of Alzheimer's disease. *Annals of the New York Academy of Sciences* 924: 42–52.

Morrison JH and Hof PR (1997) Life and death of neurons in the aging brain. *Science* 278: 412–419.

O'Brien JT, Paling S, Barber R, et al. (2001) Progressive brain atrophy on serial MRI in dementia with Lewy bodies, AD, and vascular dementia. *Neurology* 56: 1386–1388.

Peters A, Morrison JH, Rosene DL, and Hyman BT (1998) Feature article: Are neurons lost from the primate cerebral cortex during normal aging? *Cerebral Cortex* 8: 295–300.

Scahill RI, Schott JM, Stevens JM, et al. (2002) Mapping the evolution of regional atrophy in Alzheimer's disease: Unbiased analysis of fluid-registered serial MRI. *Proceedings of the National Academy of Sciences of the United States of America* 99: 4703–4707.

Thal DR, Rub U, Orantes M, and Braak H (2002) Phases of A-beta deposition in the human brain and its relevance for the development of AD. *Neurology* 58: 1791–1800.

Thompson PM, Hayashi KM, de Zubicaray G, et al. (2003) Dynamics of gray matter loss in Alzheimer's disease. *Journal of Neuroscience* 23: 994–1005.

Thompson PM, Mega MS, Woods RP, et al. (2001) Cortical change in Alzheimer's disease detected with a disease-specific population-based brain atlas. *Cerebral Cortex* 11: 1–16.

# Brain Glucose Metabolism: Age, Alzheimer's Disease, and ApoE Allele Effects

**A Pupi, M T R De Cristofaro, B Nacmias, and S Sorbi**, University of Florence, Florence, Italy
**L Mosconi**, New York University School of Medicine, New York, NY, USA



## Technical Issues in Brain Glucose Metabolic Imaging in Aging and Degeneration

The cerebral metabolic rate of glucose (CMRglc) can be measured *in vivo* using positron emission tomography (PET) imaging with [$^{18}$F]-2D-fluoro-deoxyglucose (FDG) as the tracer. Before we present FDG-PET findings in aging individuals and Alzheimer's disease (AD) patients, some technical issues relevant to the use of PET in these populations should be described.

First, FDG is taken up in the brain at a gray matter: white matter ratio of 5:1. The cortical gray matter ribbon thickness varies between 1 and 4.5 mm, with an average of approximately 2.5 mm, which is well below the 4.25 mm spatial resolution of current state-of-the-art PET scanners. Therefore, the FDG-PET signal in the cortex spills out to the surrounding low-activity regions (white matter and cerebrospinal fluid, CSF). This effect is called partial volume effect (PVE). PVE can be thought of as a blurring of higher-activity gray matter structures with the surrounding lower-activity spaces. PVE generates an artifactual reduction of FDG-PET activity as the cortex gets thinner (**Figure 1**). This is relevant to elderly subjects and AD patients, whose brains may show cortical atrophy (i.e., cortical thinning due to neuronal loss) and ventricular enlargement. It is possible to correct for PVE using mathematical models and co-registration with magnetic resonance imaging (MRI), which offers a structural image of the distribution of gray and white matter and the CSF. Partial volume correction adjusts FDG uptake values as a function of the underlying tissue volume, therefore increasing CMRglc values in atrophic regions, and provides a measure of metabolic activity per gram of brain tissue. Despite the technical importance of PVE, several FDG-PET studies have shown that CMRglc reductions in AD patients compared to controls remain significant after PVE correction and therefore reflect true reductions of CMRglc.

Second, AD is associated with a phenomenon called selective neuronal vulnerability, in which specific regions are vulnerable to neurodegeneration and others are spared. CMRglc reductions are found in the brain regions most affected by AD pathology (i.e., amyloid-beta and tau pathology, and neuronal loss), for example, the medial temporal lobes (MTLs), including the hippocampus (HIP), entorhinal cortex (EC) and transentorhinal cortex, and parahippocampal gyrus; the posterior cingulate cortex (PCC); the parieto-temporal cortex (PTC); and the frontal association cortex (FC). Traditionally, FDG-PET scans in aging and AD were sampled using manually drawn regions of interest (ROI) of the brain structures known to be selectively involved, either on PET-co-registered MRI or directly on PET. However, the method is subject to high inter- and intra-operator variability. To overcome some of these issues, several methods of automatic ROI extraction and cross-modality co-registration have been developed, and their application guarantees a better reproducibility and higher anatomical accuracy of the results.

A totally different approach is based on voxel-based analysis (VBA), a family of techniques that analyzes brain images at the single voxel (the image-unitary 3D element) level to identify regions showing significant CMRglc reductions between groups of subjects by means of intrasubject or intersubject statistical comparisons and generation of statistical parametric maps. Because human brains differ in size, shape, and orientation, VBA methods adequately modify these features by virtue of warping and spatial normalization algorithms so that all images conform to a standardized brain shape in the stereotactic space (i.e., a template image) (**Figure 2**). After spatial normalization, statistical analysis is performed on a voxel-by-voxel basis, correcting for confounding variables known to affect CMRglc (e.g., age, gender, and education) by standard multiple linear regressions (**Figure 3**). Among these methods, the most widely used for FDG-PET analysis in dementia are Statistical Parametrical Mapping (SPM) (Wellcome Department of Neurology, London, UK) and NEUROSTAT (University of Washington, Seattle, WA, USA). Moreover, diagnostic VBA methods have also been developed to perform an automated comparison of a single patient's scan to a database of reference FDG-PET scans to facilitate clinical decision (see the sections titled 'Brain glucose metabolic alterations: operating characteristics in Alzheimer's disease and mild cognitive impairment' and 'FDG-PET: Work in progress').

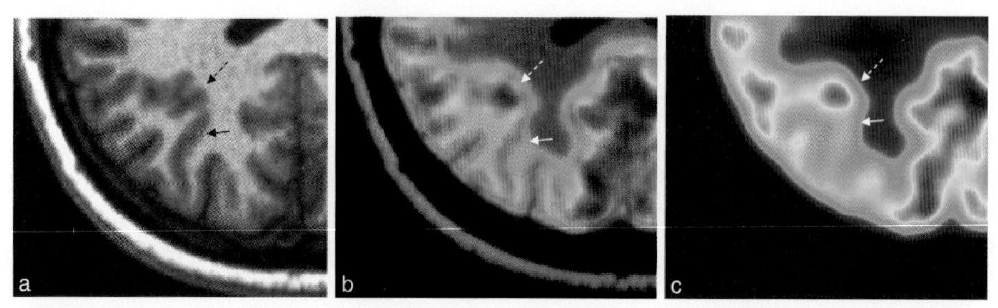


**Figure 1** Visual depiction of partial volume effect in the temporal–occipital region of a representative middle-age normal subject: (a) MRI; (b) co-registered MRI and FDG-PET; (c) FDG-PET. FDG uptake is represented on a green to red color scale, with red indicating high FDG uptake (i.e., high local metabolic activity). Among other factors, visualization of FDG uptake is related to the thickness of the cortex. Thinner cortical gyri show lower signal intensities (solid arrows) than do thicker cortical gyri or areas of convergence of different gyri (dotted arrows), which show higher metabolic activity. CMRglc may, in fact, be comparable in the two regions, and the apparent difference in intensity depends on the response of the PET scanner to the different thickness of the activity source (the cortex). CMRglc, cerebral metabolic rate of glucose; FDG, [$^{18}$F]-fluoro-2-deoxyglucose; MRI, magnetic resonance imaging; PET, positron emission tomography.

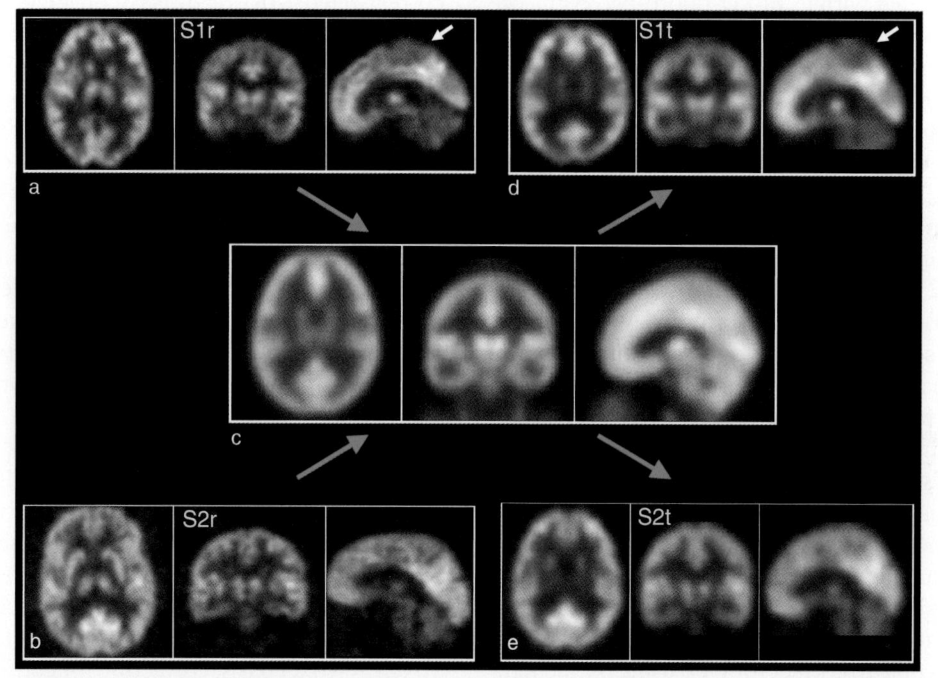


**Figure 2** Spatial normalization with voxel-based analysis (VBA): (a–b) raw images from two subjects, S1 and S2; (c) FDG-PET template from the Montreal Neurological Institute; (d–e) normalized images using SPM2. Spatial normalization is the key point to apply VBA. In (a) and (b), the two brains differ in their size, shape, and orientation. The VBA software contains a mathematical algorithm (usually very complex) which resizes, reshapes, and reorients the brain, minimizing, as the cost function, the differences using a FDG-PET template (c). After normalization (d–e), the subjects' brain images have the same size, shape, and orientation, but the metabolic differences are maintained (white plain arrow in (a) and (d), showing S1's brain cortical region with low metabolism). FDG, [$^{18}$F]-fluoro-2-deoxyglucose; PET, positron emission tomography; S1r, subject 1 raw images; S1t, subject 1 transformed images; S2r, subject 2 raw images; S2t, subject 2 transformed images; SPM, Statistical Parametrical Mapping.

## Molecular and Cellular Factors Involved with Glucose Hypometabolism in the Aging Brain and Alzheimer's Dementia

There seems to be very little or no relationship between CMRglc decreases and the density of pathological hallmarks of AD: amyloid-beta plaques and neurofibrillary tangles. Neuronal loss is thought to be the most likely cause of decreased FDG uptake, although CMRglc decrease seems to be an early event in AD, possibly occurring prior to neuronal loss. Moreover, there is a significant CMRglc decrease during normal aging, whereas age-related neuronal loss is minimal.

Another molecular factor that has been implicated in aging and AD, and their related CMRglc reductions,

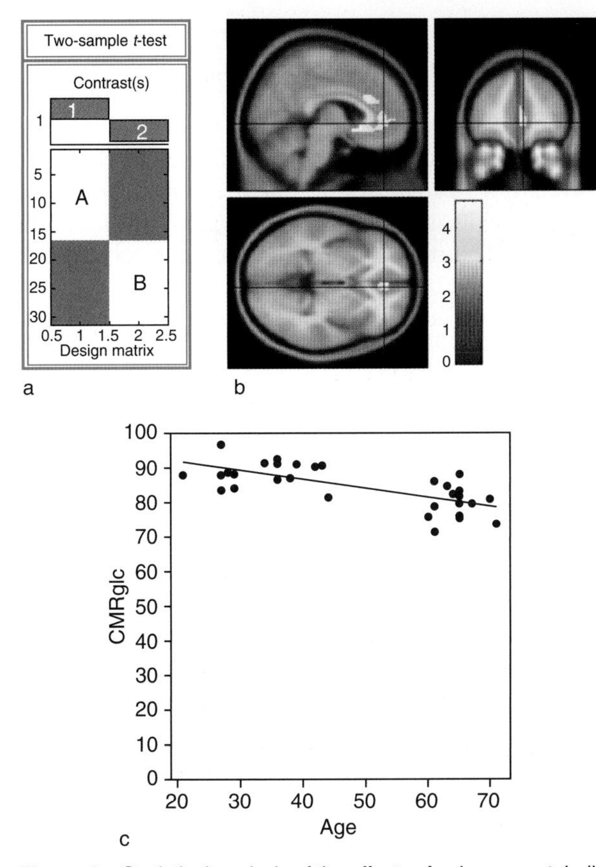


**Figure 3** Statistical analysis of the effects of aging on metabolic activity: (a) the basic components of the analysis, the proposed contrast and the groups; (b) the region(s) in which the metabolic activity is significantly different in accord with the proposed contrast; (c) CMRglc in each subject's ACC, normalized to the mean cortical value, plotted versus the subject's age. The second phase of voxel-based analysis (VBA) is the statistical analysis, performed after spatial normalization. This figure presents a basic SPM analysis. The example is a comparison of two groups of normal subjects, a group A of 16 subjects with mean age $34.3 \pm 6.1$ (range 21–44 years) and a group B of 15 subjects with mean age $64.9 \pm 3.0$ (range 60–71 years). The analysis is a *t*-test. The contrast applied in the study is shown at the top of (a) by the two small gray labels (1 and 2). In the lower part of (a), the two large white columns (A and B) indicate the subjects of the two groups. The two white columns indicate the number of subjects in each group (*y*-axis). The two gray labels indicate the direction of the contrast. In this study, the regions searched for are those in which the metabolism is higher in the younger group than in the older one (upward direction of the gray label 1 and downward for label 2). In (b), the regions in which the metabolic activity is significantly different in accord with the proposed contrast are shown in a scale which reports the value of *t* (0–5 in the color bar). In this example, the anterior cingulate cortex (ACC) fulfills the contrast and is colored yellow. The colored region is represented on a magnetic resonance template. As shown in (c), the slope of the regression line is 6.9% per decade. CMRglc, cerebral metabolic rate of glucose.

is oxidative stress (OS). OS is the result of a complex of damaging events correlated with the energetic production in neurons and astrocytes, and its disruptive effects may be differently expressed depending on individual brain compensatory capacity, which in turn is determined by genetic, environmental, and lifestyle factors. Over time, OS can impair cellular energy production at the mitochondrial level. Such impairment has been supposed to be the primary cause of CMRglc decreases on FDG-PET, a hypothesis contradicted by the known activity of allosteric (due to feedback activity) mechanisms which are activated by cellular energy demand and increase cellular glucose intake. Actually, AD neurons *in vitro* show an increased, not a decreased, glycolytic consumption to compensate for decreased energy production.

Synapse dysfunction and loss are instead currently seen as the most probable cause of CMRglc reduction in aging and AD. First, synapse molecular impairment, structural changes, and loss are early events in AD, as well as, to a lesser extent, in normal aging and senescence. Hippocampal synapse loss and dysfunction have been detected before neurofibrillary tangles in human tau transgenic mice. Second, there are sound observations linking local glucose intake with the activity of glutamatergic synapses, which constitute 80% of total brain synapses. A decline in synaptic function decreases local glutamate release and may therefore reduce glucose intake. Third, synaptic loss, like FDG uptake, closely correlates with cognitive deficits in AD. Finally, animal studies with synaptophisin (i.e., a presynaptic vesicle membrane protein used as a marker of synaptic density) have shown a significant positive correlation between synapse density and regional FDG uptake. Overall, these findings strongly support the hypothesis that resting-state CMRglc measures on FDG-PET reflect integrated synaptic activity and suggest that FDG-PET imaging may be ideal tool for investigating changes in brain function in aging and AD.

Several other factors may be implicated in FDG-PET changes in aging and AD. For example, amyloid-beta is deposited in the walls of cerebral arterioles in normal individuals and AD patients, causing vascular damage that may in turn impair glucose supply. Moreover, although it is generally accepted that FDG-PET measures are an index of neuronal activity, it is still debated whether and to what extent astrocytes play a role in the observed signal. Astrocytes, or neuroglial cells, are hypothesized to be responsible for the anaerobic component of the glycolytic path, converting glucose into lactate, which is then shuttled into neurons for further metabolism. This hypothesis is currently being tested. In this case, FDG uptake in the brain is limited to astrocytes, and FDG-PET images represent astrocytic, instead of neuronal, glucose metabolic rate. The observation of a tenfold prevalence of glial cells with respect to neurons in the brain and the poor understanding of nerve versus glial energy metabolism need further investigation.

## Brain Glucose Metabolism Changes with Aging

Aging is the most important risk factor for late-onset sporadic AD (LOSAD). Therefore, aging has been a point of interest since the beginning of FDG-PET studies in AD. Early FDG-PET studies that visually compared healthy elderly with young or middle-age normal subjects failed to show significant FDG-PET changes associated with age. The negative reports are probably due to the fact that age-related CMRglc changes are subtle and difficult to identify by visual inspection (**Figure 4**) on low-resolution PET scanners, such as those used in the early years. More recently, with improvements in PET spatial resolution and the use of more sophisticated analytical methods like VBA, age-related CMRglc changes have been detected. Cross-sectional FDG-PET studies have shown that normal elderly subjects, compared to young controls, show CMRglc decreases in the association neocortex, particularly in the FC and the anterior cingulate cortex (ACC) (**Figure 3**), and also in the PTC regions that are usually hypometabolic in clinical AD. On the other hand, the PCC, which is usually hypometabolic in AD, is apparently not affected by aging. These regional changes are paralleled by global CMRglc reductions of approximately 6% per decade (**Figure 3**), as demonstrated by quantitative CMRglc measurements. Interestingly, this rate of CMRglc decrease is consistent with the estimated rate of decrease in synaptic density measured postmortem in healthy elderly brains. This correspondence further suggests that FDG-PET measures and synaptic density are strongly related.

Longitudinal FDG-PET studies that monitored changes in cognitive performance in normal healthy subjects showed that CMRglc changes in the PTC correlate with cognitive decline over time. Moreover, there is evidence that CMRglc reductions in the EC and HIP among normal elderly accurately predict future decline from normal cognition to mild cognitive impairment (MCI), often a prodrome to AD, and dementia. These studies were based on the detection of group differences and did not examine whether individual patterns of baseline FDG-PET alterations may be predictive of clinical outcome in individual subjects. Nonetheless, longitudinal FDG-PET studies of normal elderly individuals followed for 6–14 years with serial FDG-PET exams have shown that the rate of CMRglc decline per year is highly predictive of clinical outcome. Longitudinally, in stable normal individuals the rate of hippocampal CMRglc reductions was less than 1% per year, whereas patients who declined to MCI and AD had significantly higher rates of CMRglc reductions of 2.4% and 4.4% per year, respectively. A similar effect was also found in the PCC in patients who declined to AD.

Overall, these data show that age-related CMRglc changes are found also in the brain regions showing CMRglc deficits in AD and that metabolic differences between healthy brain aging and AD may be more quantitative than topographic in nature (i.e., the magnitude rather than the site of CMRglc decrease may be important). Important issues stem

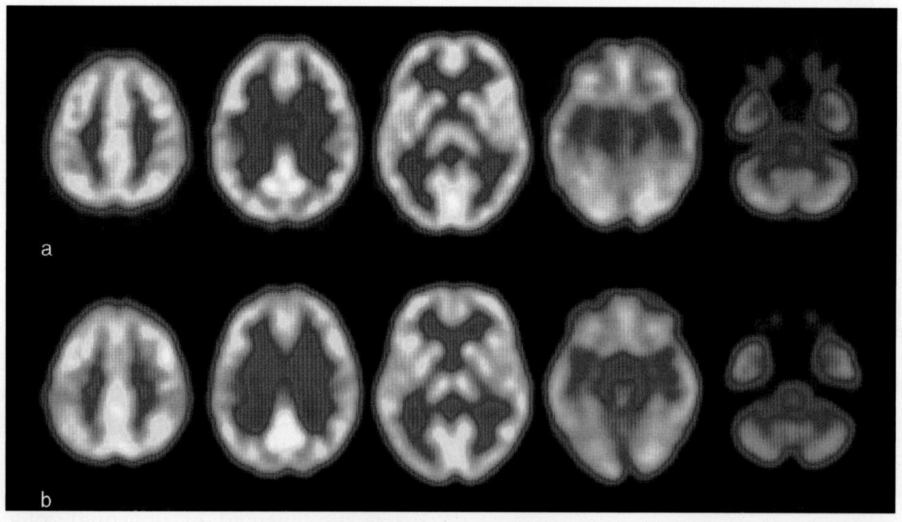


**Figure 4** FDG-PET changes in the brain with age: (a) young control (35 years old); (b) elderly control (65 years old). Axial FDG-PET images of the two subjects are shown from vertex (left) to cerebellum (right). The color scale is from blue (low activity) to white (high activity.) A visual search for differences between two subjects is not able to find differences. The brains were spatially normalized with SPM2. FDG, [$^{18}$F]-fluoro-2-deoxyglucose; PET, positron emission tomography; SPM, Statistical Parametrical Mapping.

from this consideration. First, deviations from age-associated CMRglc declines and CMRglc declines related to MCI and AD may be determined, on an individual basis, by the imbalance between damaging events and protective capacity, a theory currently at the basis of the pathogenesis of AD. Second, in operative terms, it is necessary to take into account the effects of age in any clinical FDG-PET studies aimed at identifying AD as opposed to normal aging.

## Brain Glucose Metabolic Alterations in AD: Pathophysiology

In the early 1980s, the clinical use of FDG-PET was restricted to epilepsy and brain tumors. But its potential in the study of degenerative brain diseases became clear as soon as an AD-specific distribution pattern of the CMRglc deficits was discovered.

Afterward, a more systematic approach to the examination of AD was made possible by the development and application of automated image analysis procedures, which enriched the FDG-PET evaluation of more objective criteria based on the statistical detection of CMRglc differences between AD and normal subjects. AD patients exhibit early CMRglc reductions in the PCC and in the PTC (**Figure 5**), followed by FC hypometabolism found in severe dementia. Quantitative measurements of CMRglc with FDG-PET have also shown that a widespread metabolic reduction underlies regional changes, partially sparing the basal ganglia and cerebellum. These signs are like a signature of AD in the brain and may

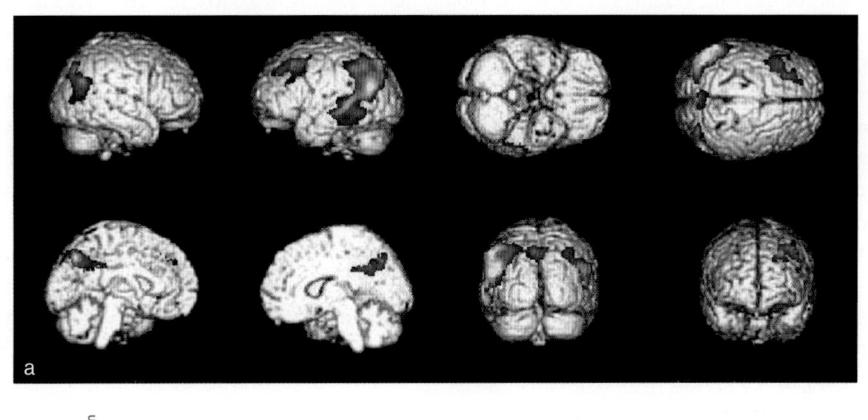


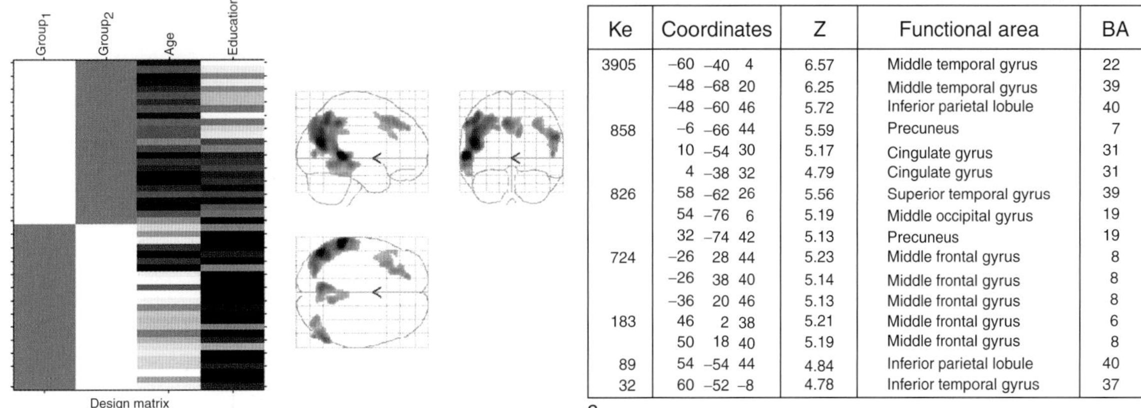


| Ke | Coordinates | Z | Functional area | BA |
|---|---|---|---|---|
| 3905 | −60 −40 4 | 6.57 | Middle temporal gyrus | 22 |
| | −48 −68 20 | 6.25 | Middle temporal gyrus | 39 |
| | −48 −60 46 | 5.72 | Inferior parietal lobule | 40 |
| 858 | −6 −66 44 | 5.59 | Precuneus | 7 |
| | 10 −54 30 | 5.17 | Cingulate gyrus | 31 |
| | 4 −38 32 | 4.79 | Cingulate gyrus | 31 |
| 826 | 58 −62 26 | 5.56 | Superior temporal gyrus | 39 |
| | 54 −76 6 | 5.19 | Middle occipital gyrus | 19 |
| | 32 −74 42 | 5.13 | Precuneus | 19 |
| 724 | −26 28 44 | 5.23 | Middle frontal gyrus | 8 |
| | −26 38 40 | 5.14 | Middle frontal gyrus | 8 |
| | −36 20 46 | 5.13 | Middle frontal gyrus | 8 |
| 183 | 46 2 38 | 5.21 | Middle frontal gyrus | 6 |
| | 50 18 40 | 5.19 | Middle frontal gyrus | 8 |
| 89 | 54 −54 44 | 4.84 | Inferior parietal lobule | 40 |
| 32 | 60 −52 −8 | 4.78 | Inferior temporal gyrus | 37 |

**Figure 5** Population study comparing brains of normal controls (NCs) and Alzheimer's disease patients (ADs), with age and education as confounding covariates: (a) significantly hypometabolic regions in ADs are depicted in red-to-yellow scale on a magnetic resonance template; (b) design matrix with age and education as confounding covariates and the classical SPM glass-brain presentation; (c) results table. Twenty-five NCs (age 61.3 ± 5.2 years, education 11.2 ± 4.1 years) ($Group_1$) and 25 ADs (age 72.2 ± 9.4 years, education 6.7 ± 3.2 years) ($Group_2$) were compared using SPM. In (a), the parietal and temporal regions are hypometabolic bilaterally. Metabolic involvement of the frontal lobe is shown on the left hemisphere. In the medial surface of the brain, severe hypometabolism in the posterior cingulate cortex is bilaterally shown. Lower left panel: In (c), the characteristics of each brain region are reported. Ke reports the extension (in voxels) of the regions. The column 'coordinates' reports (in millimeters) the peaks in the region, localized with *x* (left to right side, respectively, negative to positive values), *y* (back to front, respectively, negative to positive values), and *z* (above to below the intercommisural plane, respectively, negative to positive values) in Talairach space. The third and fourth columns are the names of the functional areas and the Broadman areas (BAs).

be used to diagnose and to differentiate AD from other dementias. In addition to differential diagnosis, the magnitude and extent of the CMRglc reductions on FDG-PET predict with high accuracy clinical progression and correlate with disease severity.

Despite the well-known early pathological involvement of the MTL in AD, only recently did clinical FDG-PET studies demonstrate CMRglc reductions in the MTL as an additional specific and consistent sign of early and advanced AD. This finding is in agreement with the finding of MTL synapse loss at a very early stage of AD. In contrast to the MTL, the PTC is affected by pathology in a later phase of AD, although PTC CMRglc decline is a relatively early event which does not seem to strictly correlate with local pathology. It has been suggested that PTC CMRglc reductions may be a distant effect of deafferentation of this region from the MTL, due to the disruption of neurons connecting the MTL to the PTC. Lesional studies in experimental animals demonstrated that the interruption of HIP and EC input/output connections can produce CMRglc decreases in the PTC. Moreover, a functional (metabolic) connectivity has been shown between the EC and PCC with FDG-PET in normal subjects and AD patients. In this model, damage to the MTL is a likely source of remote neocortical CMRglc impairment, or a sort of metabolic diaschisis due to the interruption of MTL inputs to the associative cortex. Later on, axonal transport dysfunction could participate in further decreasing cortical metabolic activity in the PTC and induce pathological changes such as synaptic and neuronal loss.

### Brain Glucose Metabolic Alterations: Operating Characteristics in AD and Mild Cognitive Impairment

The clinical diagnosis of AD may be difficult to make at the early stages of disease when symptoms are not fully expressed and are hard to differentiate from physiological age-related declines. There is a great need for an accurate and early AD diagnosis because several clinical entities exist that share similar symptoms with AD but that, unlike AD, are potentially treatable (e.g., dementia syndrome of depression) or have a different response to specific treatments (e.g., frontotemporal dementia, FTD, and anticholinesterase inhibitors). The increasing availability of medications for symptomatic treatment of mild to moderate AD, which may be most effective in the early phases of disease, and the potential for protective/reparative treatments for preclinical AD phases, namely MCI, represent an additional reasons for developing early image diagnostic tools.

Several studies have put forth FDG-PET as a candidate tool for the preclinical detection of dementia. The largest published study with pathologically verified cases showed that bilateral PTC hypometabolism on FDG-PET has a sensitivity of 94% and a specificity of 73% in distinguishing AD from normal aging and other neurodegenerative diseases (**Figures 6–8**). Therefore, the presence of PTC CMRglc alterations appears to be relatively specific to AD. In the same study, the presence of PTC hypometabolism was shown to have prognostic value in predicting the progressive course of dementia with 93% sensitivity and 76% specificity. In 2004, these results prompted US Medicare to approve the reimbursement of FDG-PET scans performed in patients with documented cognitive decline of at least 6 months and a recently established diagnosis of dementia that meets diagnostic criteria for both AD and FTD.

For clinical and strategic reasons, it is also important to develop early diagnostic procedures to forecast on an individual basis who among MCI subjects will eventually decline to AD. Recently, FDG-PET capability to predict a decline from MCI to AD has been tested by several research groups. A clear sign predictive of decline has not been found in the general population of MCI patients. By restricting the evaluation to amnestic MCI patients, who show higher rates of decline to AD (up to 50% in 3 years compared to other MCI subtypes), it appeared clear that CMRglc reductions in the PCC and PTC are present in the future decliners at least 1–3 years prior to the clinical diagnosis of AD, with prediction accuracies between 75% and 90%.

FDG-PET can attain diagnostic power in MCI when, in addition to the typical cortical pattern already described, it is used to also evaluate the MTL metabolism, which is more visible with the use of oblique axial projections (**Figures 9** and **10**). When the severity of hypometabolism of the PTC and PCC is associated with that of the MTL, FDG-PET is able to differentiate MCI patients from normal subjects with a 85% sensitivity and 71% specificity and from AD patients with a 100% sensitivity and 77% specificity.

### FDG-PET: Work in Progress

The most common approach to the diagnosis of AD with FDG-PET is by visual inspection, but visual analysis requires expert examiners. To obviate the need for expert readers, several automated diagnostic tools have been developed. They are based on the availability of a normal FDG-PET database and of VBA software. Thereafter, either the statistical parametric images are reviewed by the reader or the algorithm itself expresses a probabilistic attribution of the subject to a class of disease (AD, AD Lewy body variant, or FTD). The diagnostic performances of

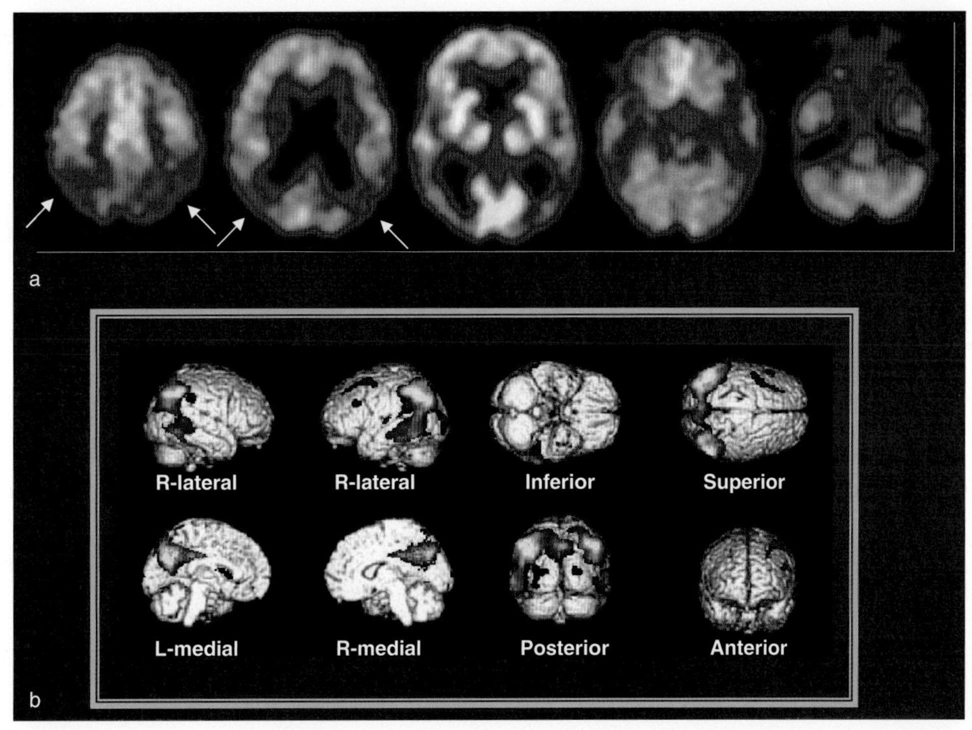


**Figure 6** FDG-PET brain pattern in a patient affected by Alzheimer's disease (AD): (a) axial FDG-PET images; (b) automated analysis in which the classic SPM display shows the hypometabolic regions in comparison with a database of normal controls. The patient is a woman 63 years old, with AD onset at 57 years old, MMSE 17. The images in (a) show bilateral parieto-temporal hypometabolism (white arrows). Metabolism is relatively high in the occipital cortex and in the basal ganglia. The display in (b) confirms the presence of symmetrical parieto-temporal and posterior cingulate cortex hypometabolism. FDG, [$^{18}$F]-fluoro-2-deoxyglucose; L, left; MMSE, Mini-Mental State Examination; PET, positron emission tomography; R, right; SPM, Statistical Parametrical Mapping.

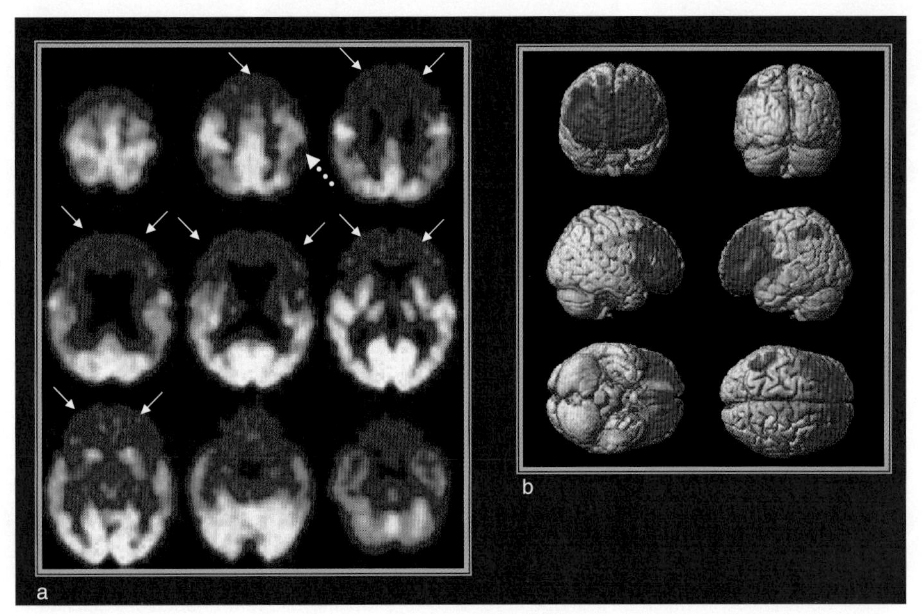


**Figure 7** FDG-PET brain pattern in a patient with frontal dementia: (a) axial FDG-PET images; (b) automated analysis in which the classic SPM display shows the hypometabolic regions in comparison with a database of normal controls. The images in (a) show bilateral frontal hypometabolism (white plain arrows). The display in (b) confirms the presence of symmetrical frontal hypometabolism. Parietal hypometabolism on the left side might be due to sulcal enlargement (dotted arrow). FDG, [$^{18}$F]-fluoro-2-deoxyglucose; PET, positron emission tomography; SPM, Statistical Parametrical Mapping.

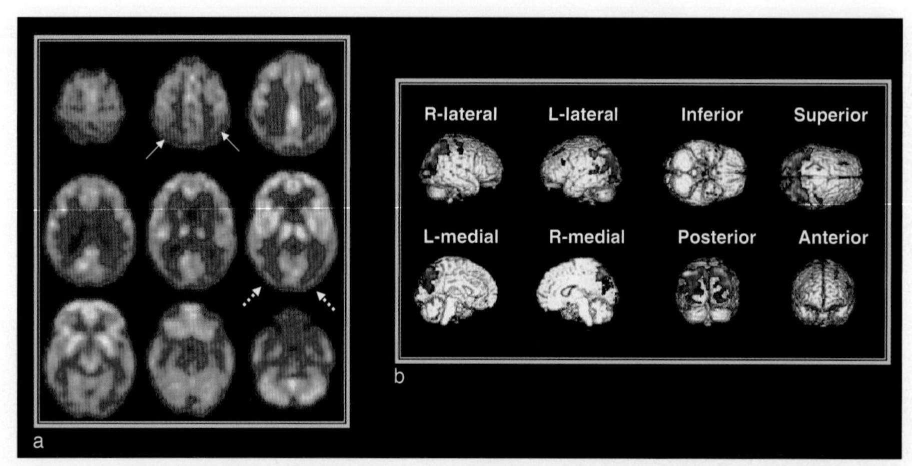


**Figure 8** FDG-PET brain pattern in a patient with AD Lewy body variant. (a) axial FDG-PET images; (b) automated analysis in which the classic SPM display shows the hypometabolic regions in comparison with a database of normal controls. The images in (a) show bilateral frontal and parietal hypometabolism (white arrows). In addition, occipital hypometabolism is shown (white dotted arrows). The display in (b) confirms the presence of symmetrical parieto-temporal hypometabolism and the involvement of occipital cortex. FDG, [$^{18}$F]-fluoro-2-deoxyglucose; PET, positron emission tomography; SPM, Statistical Parametrical Mapping.

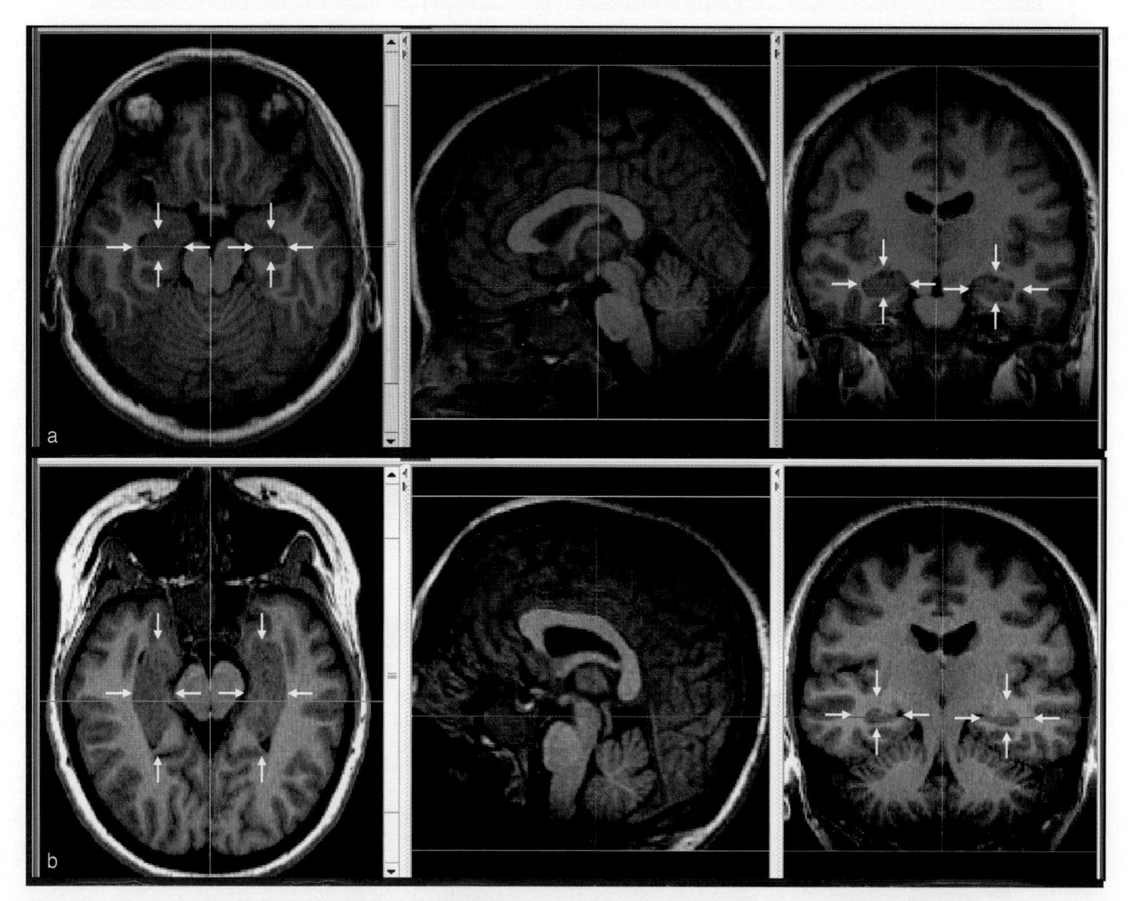


**Figure 9** Hippocampal visualization in which MRI images are used to show the different orientations applied at brain volume for examining cortical metabolism and hippocampal metabolism: (a) orientation commonly used with functional images; (b) orientation used for visualizing hippocampal bodies. In (a), the horizontal plane is oriented so as to include anterior and posterior commissurare (intercommisural plane). This is called bicommisural or fronto-occipital orientation. With this orientation, hippocampal bodies, surrounded by white arrows, are cut out of their major axis plane. In (b), the sagittal image (middle) clearly shows the different horizontal plane. The hippocampal bodies are cut in the plane (white arrows). Images are prepared with MIPAV. MRI, magnetic resonance imaging.

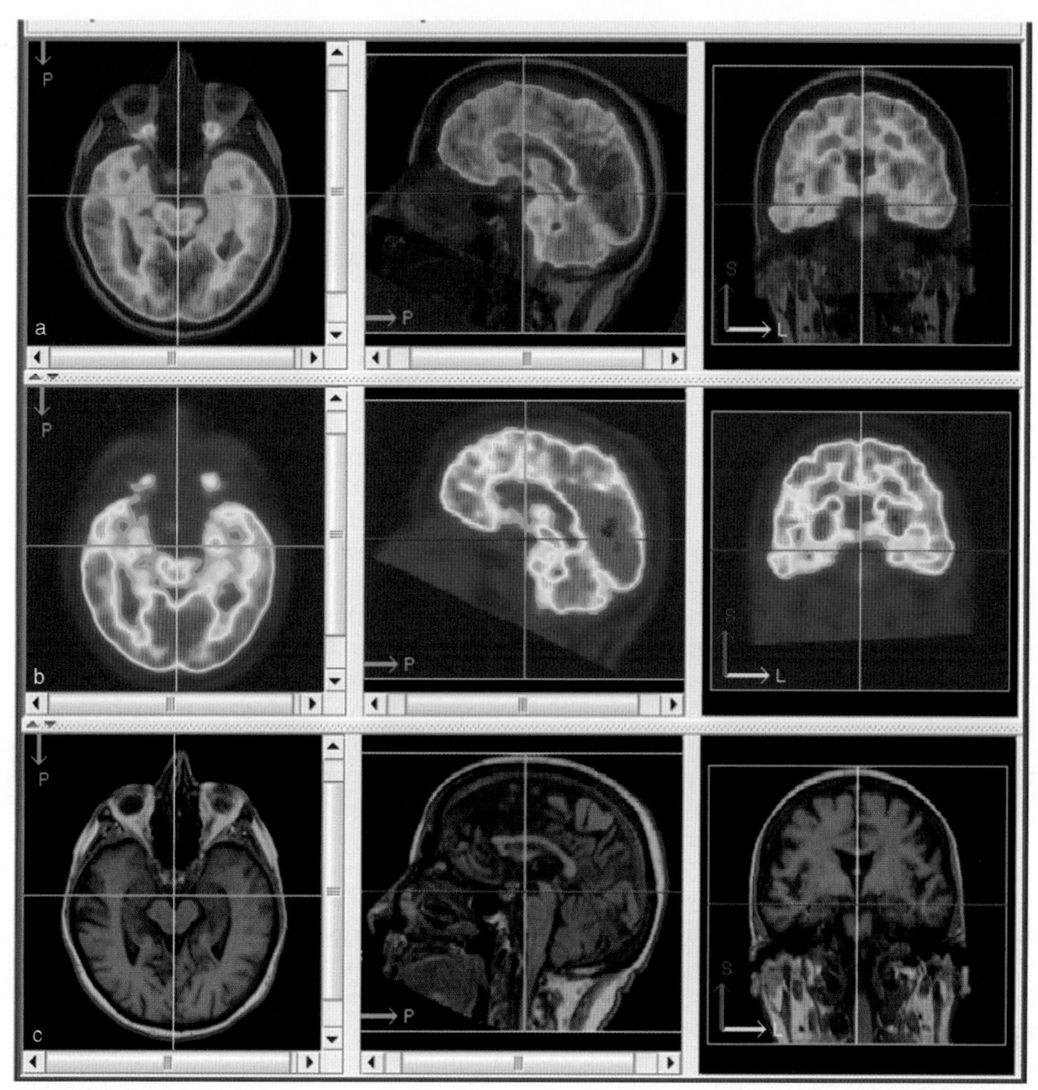


**Figure 10** Co-registered MRI FDG-PET images of a normal subject, 30 years of age: (a) MRI and FDG-PET fused images; (b) FDG-PET images; (c) MRI images. Hippocampal metabolism is usually lower than cortical metabolism. Images are prepared with MIPAV. FDG, [$^{18}$F]-fluoro-2-deoxyglucose; L, lateral; MRI, magnetic resonance imaging; P, parietal; PET, positron emission tomography.

the automatic diagnostic systems are comparable with that of visual inspection by experts. We caution the reader that the use of automatic diagnostic systems for AD has not been approved by regulatory institutions and that clinical application is not recommended. Examples of the application of VBA to single-case studies are given in **Figures 6–8**.

FDG-PET has also been proposed as a surrogate endpoint in AD clinical trials. A surrogate endpoint is a biological marker, laboratory measurement, or physical sign used as a substitute for a clinically meaningful endpoint that measures directly how a patient feels, functions, or survives. Using FDG-PET as surrogate endpoint in clinical trials in AD could substantially reduce the sample size necessary to detect a treatment effect, the trial duration, and the expense. Other analyses have shown that the use of FDG-PET as surrogate endpoint after a treatment to slow or halt the progression of AD could allow a tenfold reduction in the number of subjects needed to enroll in a clinical trial. FDG-PET is not currently recommended as a surrogate endpoint in phase III drug trials in AD because it has not been demonstrated that halting or hindering CMRglc reductions corresponds to improving the patient's health.

## Genetic Factors in Relation with Brain Metabolism Changes in Aging, Mild Cognitive Impairment, and AD

Correlative studies of CMRglc changes and genetic factors have been performed in normal individuals and in AD and MCI patients. The interest of these studies is twofold: to discover in genetically predisposed subjects the earliest metabolic signs of AD and to better investigate the cellular and molecular factors of regional metabolic reduction.

Few cross-sectional studies with FDG-PET have been done in subjects carrying autosomal dominant mutations which determine an early-onset familial AD manifestation (EOFAD). These genes are *APP, PS1*, and *PS2*, and their proteins are involved in the production and metabolism of amyloid precursor protein. The importance of these studies is that they may be performed long before the emergence of cognitive clinical symptoms and may help to better define the metabolic continuum among normal brain, aging, senescence, MCI, and AD. In these studies, the same regions affected by CMRglc decrease in MCI and AD (i.e., the PTC and HIP) were already hypometabolic more than a decade before subject's expected time of overt clinical dementia (**Figure 11**). Despite possible biological differences with nonautosomal AD, the finding of this specific endophenotype in autosomal EOFAD subjects, although cognitively normal, strengthens the importance of the CMRglc impairment in the PTC and HIP as a research and diagnostic marker of the evolution from normal brain function to dementia; also it may be proof of a direct relationship between the amyloid metabolism and CMRglc decrease, an issue currently still largely debated.

FDG-PET changes in relation with the major genetic risk factor of AD, the ε4 allele of apolipoprotein E (ApoE) gene, have been investigated in AD patients at the onset of cognitive symptoms. The AD-specific hypometabolic pattern does not differ among the different ApoE genotype groups (with 0, 1, or 2 ε4 alleles). Anyway, when researchers compared the AD genetically different groups directly, hypometabolic differences were evident within the brain areas adjacent to those typically hypometabolic in AD, in a ε4-related fashion, suggesting that the metabolic damage of AD might be exacerbated by the dose of the ε4 allele. This effect of ApoE ε4 on CMRglc is consistent with the less efficient protective role of ApoE ε4 protein toward synapses in A$\beta$ mouse models, as well as with the possibility that the ApoE ε4 protein directly damages the synapses, the main factor of FDG uptake in the brain. Moreover, the dose-dependent relationship between the number of copies of ε4 and the severity (extension) of CMRglc reduction in PTC is consistent with the known relationship of the ε4 dose and the age of onset of AD, such that ε4/ε4 subjects have an earlier onset than do heterozygous ε4 subjects.

The growing interest in the association of CMRglc and genetic factors in MCI patients derives from the compelling need to select the patients for protective/reparative interventions. Indeed, a genetic predisposition due to ApoE ε4 allele by itself is a predictor of conversion from MCI to AD with a good sensitivity (75%) but poor specificity (50%). When the genetic predisposition is considered together with FDG-PET hypometabolism in specific areas (the ACC and inferior FC), the capacity to predict conversion attains a sensitivity of 75% and a specificity of 89%.

An interesting issue is the study of brain metabolism in normal aging in association with the ApoE ε4 allele in order to select as early as possible the at-risk subjects for treatment and for protection against AD. In middle-age and older healthy subjects with normal

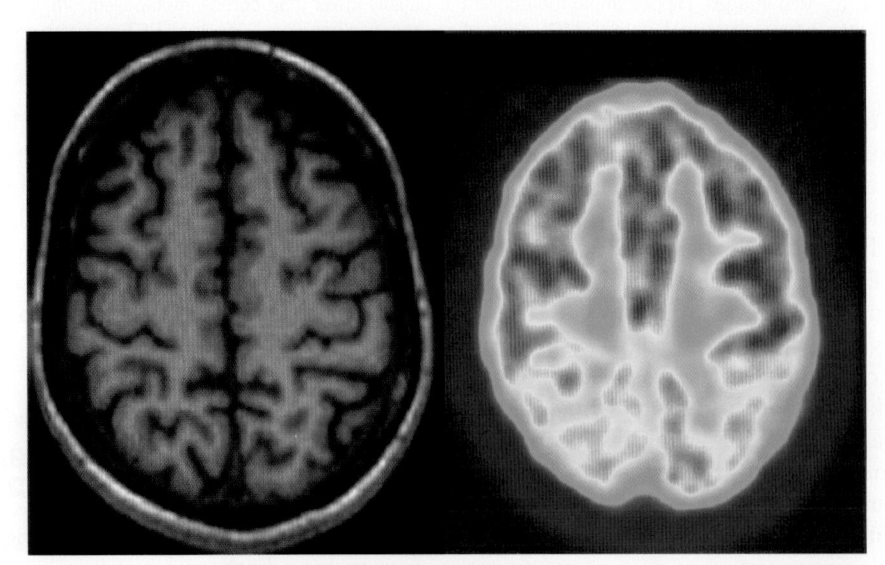

**Figure 11** Metabolic impairment in a preclinical AD genetic case. Metabolic reduction in temporoparietal cortex in a subject 35 years old with a genetic mutation PS1 Leu392Val. At the time of this FDG-PET study (right) the subject was cognitively normal (MMSE 30). MRI-coregistered image (left) does not show any atrophy in the corresponding regions.

memory performance, a single copy of the ApoE ε4 allele was associated with lowered PTC and PCC metabolism, which predicted the degree of cognitive decline after 2 years of longitudinal follow-up. This does not occur in the absence of the ApoE ε4 risk factor, representing possible proof that PTC hypometabolism might be a consequence of a genetic impairment of brain cells and, thus, might be causative of (instead deriving from) the cognitive impairment.

Although the ApoE ε4 allele by itself may account for a large fraction of genetic risk of developing AD, the role of several other susceptibility genes is suspected. Therefore, studies in AD patients have focused on LOFAD patients in which these unknown AD susceptibility genes may act in concert with epigenetic events to determine AD. In these patients (approximately one-third of all late-onset AD), there is a reduction of CMRglc in the PCC, HIP complex, and occipital regions compared with LOSAD patients, balanced for the presence of the ApoE ε4 allele. This finding suggests that these unknown susceptibility genetic factors may participate in the protective/reparative cellular functions against OS because they decrease CMRglc in brain regions in which a reduced capacity to compensate for stress and damages at the cellular level is known to be present.

*See also:* Aging of the Brain and Alzheimer's Disease; Alzheimer's Disease: An Overview; Alzheimer's Disease: MRI Studies; Axonal Transport and Alzheimer's Disease; Cholinergic System Imaging in the Healthy Aging Process and Alzheimer Disease; Genomics of Brain Aging: Apolipoprotein E.

## Further Reading

Mosconi L, Brys M, Glodzik-Sobanska L, et al. (2007) Early detection of Alzheimer's disease using neuroimaging. *Experimental Gerontology* 42: 129–138.

Reiman EM (2007) Linking brain imaging and genomics in the study of Alzheimer's disease and aging. *Annals of the New York Academy of Sciences* 1097: 94–113.

Silverman DH, Small GW, Chang CY, et al. (2001) Positron emission tomography in evaluation of dementia: Regional brain metabolism and long-term outcome. *Journal of the American Medical Association* 286: 2120–2127.

## Relevant Websites

http://www.loni.ucla.edu – Alzheimer's Disease Neuroimaging Initiative (ADNI) research.

http://www.fda.gov – Food and Drug Administration, Peripheral and Central Nervous System Drugs Advisory Committee.

http://www.cms.hhs.gov – Medicare expansion of the coverage of FDG-PET for the diagnosis of Alzheimer's disease.

http://mipav.cit.nih.gov – MIPAV (medical image processing, analysis, and visualization) software.

http://imaging.mrc-cbu.cam.ac.uk – Montreal Neurological Institute.

http://www.uni-koeln.de – Network for Efficiency and Standardisation of Dementia Diagnosis (NEST-DD).

www.fil.ion.ucl.ac.uk – Statistical Parametrical Mapping (SPM) website.

http://128.95.65.28 – University of Washington NEUROSTAT website.

# Parkinsonian Syndromes

**C Schwarz**, New York Presbyterian Hospital – Weill Cornell Medical Center, New York, NY, USA
**C Henchcliffe**, Weill Medical College of Cornell University, New York, NY, USA



## Introduction and Clinical Features

Parkinsonian syndromes occur as part of multiple underlying pathologies that involve the striatum (**Table 1**). This article focuses on Parkinson's disease (PD), which is the most common cause of parkinsonism, accounting for approximately 80% of cases. PD is the second most common neurodegenerative disease after Alzheimer's disease. Most cases of PD occur between ages 40 and 70 years, with the peak of onset in the sixth decade. In the United States, the prevalence is 1% by age 65 years and 3.5% by age 85 years. PD is suggested to arise due to both environmental and genetic causes. In approximately 90% of cases, no apparent genetic defect can be detected, and it is referred to as idiopathic. The disease is progressive, and treatment is mainly symptomatic. Clinically, PD is characterized by cardinal motor features that are caused by a loss of dopaminergic neurons in the substantia nigra, which project to the striatum. They include tremor at rest with a frequency of 4–7 Hz, lead-pipe rigidity of the extremities on examination, bradykinesia and akinesia (slowness and paucity or absence of movement), and postural instability later in the course of disease. Patients suffer from hypophonia (decreased voice volume), dysphagia, drooling, hypomimia (loss of facial expression or 'masked facies'), micrographia (decreased size of handwriting), flexed posture, shuffling gait, and freezing (motor block, often affecting gait). The onset is gradual and usually asymmetric. Dystonia may be associated with PD early in a minority of patients or, more often, later in the course of the disease. PD patients also suffer from a variety of nonmotor symptoms, including cognitive and behavioral changes, depression and anxiety, olfactory loss, sleep disorders, and autonomic symptoms. These are related to neurodegeneration in brain regions other than the substantia nigra and may precede motor symptoms. The longitudinal course of PD is followed clinically using the Unified Parkinson Disease Rating Scale, which consists of ratings for mentation, behavior and mood, the ability to conduct activities of daily living, and motor performance.

The previously mentioned features of PD are found in varying combinations in other parkinsonian syndromes (**Table 1**), raising a diagnostic challenge in some individuals. The presence of atypical features, however, may raise suspicion. Progressive supranuclear palsy (PSP) typically manifests with symmetric parkinsonism, impaired oculomotor function (particularly vertical gaze), prominent axial rigidity, and early falls. The cortical involvement of corticobasal-ganglionic degeneration may lead to apraxia, alien limb, cortical sensory impairment, and myoclonus, in the setting of asymmetric parkinsonism. Multiple system atrophy (MSA) may comprise prominent parkinsonism (striatonigral degeneration or MSA-P); cerebellar signs and symptoms (olivopontocerebellar atrophy or MSA-C); or autonomic dysfunction, including orthostatic hypotension and syncope, urinary symptoms, constipation, and sexual dysfunction (Shy–Drager syndrome or MSA-A). These Parkinson's-plus syndromes typically display more rapid progression and little, if any, response to dopamine replacement. The presence of dementia early in disease should raise suspicion of dementia with Lewy bodies (DLB), typically associated with hallucinations and fluctuating clinical symptoms. However, parkinsonism is also a feature of Alzheimer's disease and frontotemporal dementia in some patients, and tau mutations on chromosome 17 lead to autosomal dominant inherited frontotemporal dementia with parkinsonism (FTDP-17).

## Neuroimaging

In general, patients with parkinsonian syndromes undergo conventional magnetic resonance imaging (MRI) to rule out any structural abnormalities that may lead to parkinsonism, such as tumor, stroke or infection of the basal ganglia, or hydrocephalus. The neurodegenerative diseases under discussion in this article typically have subtle or no abnormalities revealed by MRI. However, in some with MSA, T2-weighted images demonstrate hyperintense signals along the lateral aspect of the putamina, the 'hot cross bun' sign due to degeneration of neural pathways in the pons, and/or cerebellar atrophy. Positron emission tomography (PET) and single photon emission computerized tomography (SPECT) allow imaging of basal ganglia and nigrostriatal tract integrity using appropriate radioligands. Fluorodopa PET has revealed progressive presynaptic dopaminergic nigrostriatal damage in PD patients: while not a satisfactory measure of progression, it may help to assess nigrostriatal tract integrity in atypical cases. Fluorodeoxyglucose PET shows normal or increased metabolism in the striatum in PD, whereas in Parkinson's-plus syndromes there may be decreased striatal metabolism. Currently, however, PET and SPECT are little used in clinical management of PD and other parkinsonian syndromes.

**Table 1** Differential diagnosis for parkinsonian syndromes

| |
|---|
| Parkinson's disease (sporadic, familial) |
| Secondary parkinsonism |
| Toxic: MPTP, manganese, copper, cyanide |
| Trauma |
| Tumor |
| Vascular |
| Postencephalitic |
| Hypoxia |
| Carbon monoxide |
| Normal pressure hydrocephalus |
| Drug-induced: dopamine antagonists and depletors |
| Hemiatrophy–hemiparkinsonism |
| Multisystem degeneration (Parkinson's-plus syndromes) |
| Multiple asystem atrophy (MSA) |
| MSA-P (striatonigral degeneration) |
| MSA-A (Shy–Drager) |
| MSA-C (sporadic olivopontocerebellar atrophy) |
| Progressive supranuclear palsy |
| Corticobasal ganglionic degeneration |
| Progressive pallidal atrophy |
| Lytico–Bodig (Parkinson–dementia–ALS complex of Guam) |
| Dementia associated: dementia with Lewy bodies, frontotemporal dementia, Alzheimer's disease |
| Hereditary neurodegenerative diseases |
| Lubag |
| Dopa-responsive dystonia |
| Olivopontocerebellar and spinocerebellar atrophy |
| Huntington's disease |
| Mitochondrial myopathies |
| Neuroacanthocytosis |
| Wilson's disease |
| Neurodegeneration with brain iron accumulation type1 |

## Pathology

The disease is characterized by loss of dopaminergic neurons in the substantia nigra pars compacta (SNpc), which affects nigrostriatal dopaminergic pathways leading to striatal dopamine (DA) deficiency. Depigmentation of the SNpc in autopsy tissue is caused by loss of these neurons, which contain neuromelanin. Findings indicate a more widespread pathology with neurodegeneration also involving the olfactory bulb, the dorsal motor nucleus of the vagus nerve, the nucleus basalis of Meynert, the superior cervical ganglia, the mesenteric plexus, and the locus coeruleus (**Table 2**). Neurodegeneration in these areas likely precedes neuronal death in the SNpc and accounts for some of the nonmotor symptoms in PD. Clinical symptoms of PD emerge when approximately 80% of striatal dopamine and 50% of nigral dopaminergic neurons are lost. Thus, early diagnosis remains a challenge.

A pathological hallmark of PD is the formation of intraneuronal cytoplasmic inclusion bodies, termed Lewy bodies (**Figure 1**). The major components of Lewy bodies are $\alpha$-synuclein and ubiquitin. Lewy bodies occur not only in the SN but also in other areas involved in neurodegeneration mentioned previously. The relationship between Lewy bodies and cell death remains to be determined, and of note, they are not found in PD arising from parkin mutations nor in all cases of PD associated with LRRK2 mutations.

## Etiology and Pathogenesis

Multiple lines of evidence point to a combination of environmental factors and genetic mutations or susceptibility factors as the cause of PD, and their relative contributions vary between individuals.

**Table 2** Proposed pathologic stages of Parkinson's disease (Braak classification)

| *Stage* | *Location* | *Structures affected* |
|---|---|---|
| 1 | Medulla oblongata | Dorsal nucleus of vagal and glossopharyngeal nerves and/or intermediate reticular zone; olfactory bulb and anterior olfactory nucleus |
| 2 | Medulla oblongata and pontine tegmentum | Pathology of stage 1 plus caudal raphe nuclei, gigantocellular reticular nucleus, and coeruleus–subcoeruleus complex |
| 3 | Midbrain | Pathology of stages 1 and 2 plus midbrain, especially in the pars compacta of the substantia nigra |
| 4 | Basal prosencephalon and mesocortex | Pathology of stages 1–3 plus prosencephalic and cortical involvement confined to the temporal mesocortex (transentorhinal region) and allocortex (CA2 plexus) |
| 5 | Neocortex | Pathology of stages 1–4 plus higher order sensory association areas of neocortex and prefrontal neocortex |
| 6 | Neocortex | Pathology of stages 1–5 plus sensory association areas of the neocortex and premotor areas; occasionally mild changes in primary sensory areas and the primary motor field |

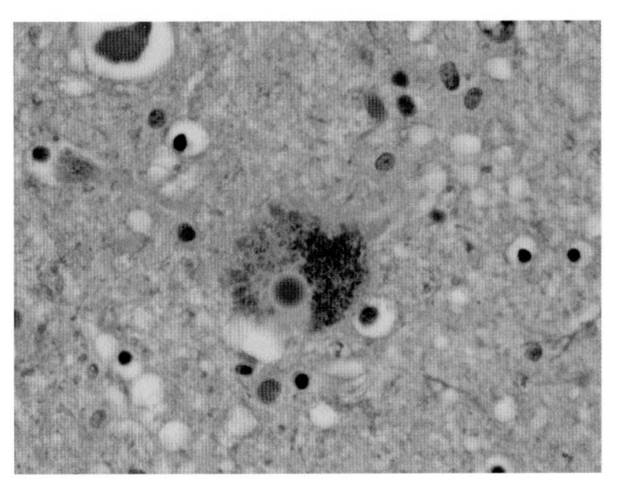

**Figure 1** Photomicrograph of a hematoxylin–eosin-stained section taken from the substantia nigra pars compacta at autopsy of a PD patient. A surviving dopaminergic neuron in the center of the field contains a cytoplasmic Lewy body, with a central eosinophilic dense core surrounded by a halo. Also in the cytoplasm are numerous neuromelanin granules (brown pigment). Cortical Lewy bodies are less well-defined in structure. Courtesy of Ehud Lavi, MD, Department of Pathology, Weill Medical College of Cornell University.

Several studies have examined the role of environmental factors in the pathogenesis of PD. These have identified a number of risk factors, including rural living, pesticide exposure, well water consumption, as well as mining and welding. Exposure to manganese, a component of herbicides, pesticides, and welding fumes, produces a parkinsonian syndrome and leads to damage of the globus pallidus with sparing of the substantia nigra. Exposure to the pesticide rotenone has been shown to reproduce neuropathological features of PD with cytoplasmic inclusions that contain ubiquitin and α-synuclein in rats. Ingestion of annonacin in the Caribbean is associated with parkinsonism in humans and induces nigral and striatal neurodegeneration in rats. Exposure of humans to 1-methyl-4-phenyl-1,2,3,6-tetrahydropyridine (MPTP) results in an acute parkinsonian syndrome. These findings have been used to create animal models of PD by administering a variety of the previously mentioned toxins to different species. In contrast, other environmental factors, such as cigarette smoking, alcohol, and caffeine, are associated with a lower risk for PD, although the reason for this negative association remains to be elucidated.

## PD Genetics: Implications for PD Pathogenesis

In the past few years, various gene mutations have been identified in familial PD (**Table 3**), beginning with the discovery of α-synuclein mutations in the large Contursi kindred in 1997. Even though only approximately 10% of PD cases are clearly hereditary, the study of genetic forms of the disease offers valuable insights into the pathogenesis of PD (**Table 3**). Aided by the discovery of specific genes mutated in familial PD, genetic animal models in different species have been generated to address the previous questions and to test pharmacological interventions that could be applied to nonfamilial forms of PD as well. The discovery of mutations in several PD genes involved in the function of the ubiquitin–proteasome system provides evidence that dysfunction of this pathway leads to abnormal protein aggregation, Lewy body accumulation, and neuronal death in PD. Study of normal and mutant function of several PD genes also strongly implicates a role for mitochondrial dysfunction and oxidative stress in PD pathogenesis. In addition, oxidative stress or environmental toxins most likely contribute to proteasomal dysfunction.

α-Synuclein (PARK1) was the first gene to be identified in hereditary PD, but mutations in this gene affect only a very small number of families. α-Synuclein is a 140-amino acid presynaptic protein that is involved in vesicle trafficking and neurotransmitter release, and it is a major component of Lewy bodies in PD and DLB. Three missense mutations and several genomic multiplications of the α-synuclein gene have been identified in hereditary PD. However, promoter polymorphisms may also be associated with sporadic PD. Mutant or overexpressed α-synuclein is misfolded and aggregates within the cell, thus impairing degradation by the ubiquitin–proteasome system. Transgenic mouse lines have been generated, including α-synuclein knockouts, and lines overexpressing human α-synuclein under the control of various promoters. The behavioral phenotype, such as decreased locomotor activity and increased righting time on an inverted screen, correlates with the level of overexpression of the protein. However, only a few of the lines have dopaminergic neurodegeneration in the striatum, which limits their usefulness as PD models. In contrast, injection of viral vectors containing human α-synuclein into adult rat SN leads to formation of α-synuclein-positive, Lewy body-like inclusions and dopaminergic neuronal cell death. A *Drosophila* model of PD overexpressing wild-type or mutant α-synuclein has been developed. Dopamine neuron loss in these flies was associated with Lewy body-like inclusions and behavioral changes that were corrected by administration of dopamine agonists and levodopa.

Mutations in the parkin gene (PARK2) cause juvenile autosomal recessive parkinsonism. They are found at high frequency in early onset PD (18% sporadic early

**Table 3** Parkinson's disease genetics

| *Gene* | *Inheritance* | *Function* | *Observations* |
|---|---|---|---|
| α-Synuclein (PARK1, -4) | Dominant | ? | Aggregation into oligomers or fibrils<br>Substrate for ubiquitin–proteasomal system<br>Knockout mutant: ↑ resistance to MPTP (mice)<br>Mutation: A53T overexpression; abnormal mitochondria (mice) |
| Parkin (PARK2) | Recessive | Ubiquitin E3 ligase | Ubiquitin–proteasomal system impairment<br>Mitochondrial outer membrane association (partial)<br>Mutation: abnormal mitochondria, ↑ sensitivity to oxidative stress (*Drosophila*) |
| UCH-L1 (PARK5) | Recessive | Ubiquitin C-terminal esterase L1 | Ubiquitin–proteasomal system impairment<br>Requires confirmation as PD gene in further families |
| PINK1 (PARK6) | Recessive | Mitochondrial serine–threonine kinase | Mitochondrial membrane localization<br>siRNA: ↑ sensitivity to MPP+, rotenone<br>Mutation: abnormal mitochondria, ↑ sensitivity to oxidative stress (*Drosophila*) |
| DJ-1 (PARK7) | Recessive | Oxidative stress sensor, chaperone | Oxidative stress causes re-localization to mitochondria (matrix/ intermembrane space)<br>Protects against oxidative stress<br>siRNA: ↑ sensitivity to oxidative stress (*Drosophila*)<br>Mutation: ↑ sensitivity to rotenone, paraquat, hydrogen peroxide (*Drosophila*); ↑ sensitivity to oxidative stress (mice) |
| LRRK2 (PARK8) | Dominant | Serine–threonine kinase | GTPase or MAPKKK dysregulation |
| Lysosomal ATPase (PARK9) | Recessive | ? | Responsible for Kufor–Rakeb syndrome of parkinsonism with dementia |

onset and 49% familial autosomal recessive early onset). Heterozygous mutations in the parkin gene might constitute a risk factor in sporadic PD, and polymorphisms in the parkin gene appear to confer an increased risk toward the development of idiopathic PD. Parkin is a ubiquitin E3 ligase that is involved in the ubiquitination of proteins prior to degradation by the proteasome. Mutations in parkin therefore interfere with normal clearing of proteins by the cell and lead to accumulation of parkin substrates, such as α-synuclein, Pael-R, synphilin 1, and CDCrel-1. A role for parkin in mitochondrial function has also been suggested. Parkin knockout mice, however, do not display accumulation of parkin substrates and there is no loss of dopaminergic neurons. These mice do show altered dopaminergic neurotransmission as well as behavioral changes but are not regarded as a good model for parkinsonism. Ubiquitin–proteasomal system dysfunction is further implicated by the finding of ubiquitin C-terminal hydrolase L1 (UCH-L1) mutation in a family with autosomal recessive PD, but its role in PD remains to be confirmed with additional families.

PINK1 (PTEN-induced kinase1) is another cause of autosomal recessive PD, although far less commonly than parkin. It localizes to the mitochondria and possibly exerts a protective effect on the cell.

Deletions or point mutations in the DJ-1 gene (PARK7) cause rare autosomal recessive early onset PD. DJ-1 is a multifunctional oncogene product, a regulatory subunit of RNA binding protein, and has been implicated in the cellular response to oxidative stress. DJ-1 knockout mice show an increased vulnerability to oxidative stress and decreased dopamine release from striatal neurons but no loss of dopaminergic neurons. These mice have decreased locomotor activity and increased sensitivity to the neurotoxin MPTP.

Leucine-rich repeat kinase 2 (LRRK2) is increasingly recognized as an important cause of autosomal dominant PD in particular populations. Its function remains to be elucidated, but its structure comprises a Ras-GTPase domain and a MAPKKK domain, thus implicating dysfunction of multiple cell signaling pathways.

Several important questions remain in the pathogenesis of genetic PD, such as the basis for selective neuronal vulnerability and the exact mechanism of cell death. It is unclear why there is a delayed clinical presentation in adulthood in the genetic forms of PD, even though the gene is expressed at or before birth. It is very likely that there are molecular changes in the affected cells before expression of the phenotype, which cannot yet be studied in patients.

## Neurotoxin-Based Animal Models of PD

Nongenetic, neurotoxin-based animal models of PD have played a critical role in understanding the pathways that lead to selective degeneration of dopaminergic neurons in sporadic PD. Toxin-induced animal models are based on the assumption that dopaminergic neurons follow a common pathway of cell death regardless of the initial stimulus. The toxins used are all thought to provoke the formation of reactive oxygen species. Elucidating the cascade that leads to cell death in these models might contribute to a better understanding, identify novel molecular pathways that might be targeted to prevent degeneration, and might lead to the discovery of novel treatment strategies. Moreover, they may prove valuable for preclinical testing of novel treatments that modulate the ubiquitin–proteasome pathway or mitochondrial function.

Exposure of humans and nonhuman primates to MPTP, a by-product of chemical synthesis of a synthetic heroin analog, results in an acute parkinsonian syndrome that recapitulates many clinical features of PD. Its metabolite, 1-methyl-4-pyridinium (MPP+), binds to the dopamine transporter and thus accumulates in dopaminergic neurons. It inhibits NADH dehydrogenase and NADH CoQ reductase of complex I of the mitochondrial electron transport chain (ETC), resulting in cell death of dopaminergic neurons of the SNpc.

Local stereotactic injection of the neurotoxin 6-hydroxydopamine (6-OHDA) into the SN or striatum of animals induces selective degeneration of the nigrostriatal dopamine system, resulting from preferential uptake by dopamine and noradrenergic transporters. Animals with 6-OHDA-induced lesions undergo extensive loss of DA neurons in the basal ganglia, with corresponding neurological impairments including akinesia, dyskinesias in response to dopaminergic treatment, and improved motor function after the administration of dopaminergic agents. However, 6-OHDA injections do not produce Lewy body-like inclusions.

The herbicide, paraquat, and the pesticide, rotenone, are also used to produce toxic animal models of PD. Intravenous infusion of rotenone, which binds to and inhibits complex I of the mitochondrial respiratory chain, produces dopaminergic degeneration in the SN and induces α-synuclein-positive Lewy body-like inclusions in rats. Animals treated with rotenone develop bradykinesia and abnormal posturing. Paraquat shows structural similarity to MPP+. When infused into mice, it leads to dopaminergic degeneration in the SN and formation of α-synuclein-positive inclusions.

## Oxidative Stress and Mitochondrial Dysfunction in PD Pathogenesis

There is mounting evidence supporting mitochondrial dysfunction, either primarily or secondarily, in the PD pathogenetic process, and several PD genes are suggested to be involved in mitochondrial function and oxidative stress (**Table 3**). The mitochondrial ETC is a major source of highly reactive free radicals, particularly when impaired, that may oxidatively damage neighboring molecules. Moreover, impaired mitochondrial function increases susceptibility to oxidative stress.

Complex I is the largest of the ETC complexes, and it is the major site of superoxide production in the ETC. Its activity in PD patients is decreased in the SN and other brain regions, and cell cybrids (fusion of platelets isolated from PD patients with cell lines deficient in mitochondria) can have a complex I deficiency. Moreover, higher levels of acquired mitochondrial DNA deletions have been demonstrated in autopsy brain tissue in PD compared with non-PD controls. Finally, several studies have suggested that specific clusters of mitochondrial DNA haplogroups may decrease the risk of PD. Postmortem analysis of SN tissue has provided direct evidence of oxidative stress in PD, with increased levels of malondialdehyde and cholesterol lipid hydroperoxides and others. Moreover, oxidative damage impairs the ubiquitination and degradation of proteins mediated by the ubiquitin–proteasomal system, suggesting a link between oxidative damage and characteristic abnormal protein aggregates. In addition to mitochondrial ETC dysfunction, impaired antioxidant mechanisms and increased exposure to environmental or endogenous sources of oxidative stress may all contribute to increased oxidative stress in PD. Glutathione, an important antioxidant, is decreased in the SN by approximately 50% in PD patients. Iron can act as a catalyst in the formation of hydroxyl radicals from hydrogen peroxide. Neuromelanin, present in some dopamine neurons, chelates iron but is present at decreased levels in PD substantia nigra. Several studies have demonstrated increased iron levels in the brain of PD patients, as well as decreased amounts of ferritin, which inhibits iron's catalytic properties. Therefore, the potential use of iron chelators, combined with monoamine oxidase (MAO) inhibition, is being pursued.

## Current Treatments for PD

Pharmacologic treatment is symptomatic. The mainstay of PD treatment is the DA precursor L-DOPA (L-3,4-dihydroxyphenylalanine). L-DOPA provides adequate control of akinesia and tremor for several years, but over time treatment is complicated by the development of motor fluctuations and dyskinesias, requiring more advanced therapeutic strategies. Many clinicians therefore prefer to use alternative therapies, such as dopamine agonists or MAO-B inhibitors, when possible, early in the disease. These may also be used as adjunctive therapies in later disease, along with catechol-O-methyltransferase inhibitors such as entacapone, which inhibit levodopa metabolism and help decrease motor fluctuations. Other agents in use are amantadine, with multiple mechanisms of action including NMDA receptor blockade, and anticholinergic agents, which are helpful for tremor. Side effect profiles of these, however, are often limiting. Cholinesterase inhibitors seem to have mild benefit for cognitive dysfunction, and various antidepressants are under investigation for PD-associated depression.

Deep brain stimulation (DBS) is a surgical approach to treating advanced PD in which electrodes are implanted into specific brain regions. Continuous high-frequency stimulation is then delivered via an externally programmable pulse generator. Stimulation of either the subthalamic nucleus (STN) or the internal segment of the globus pallidus (GPi) relieves bradykinesia, rigidity and tremor, as well as drug-induced dyskinesias in PD patients. DBS of the GPi and STN can reduce the need for dopaminergic medications and alleviate motor fluctuations. Stimulation of the ventral intermediate nucleus of the thalamus can reduce tremor associated with PD, but since other symptoms usually become more prominent over time, this target is seldom used. Although the exact mechanism of action remains to be determined, it is thought that DBS modulates a motor circuit formed by the basal ganglia, thalamus, and cortex that is dysfunctional in PD. Dopamine depletion in the basal ganglia induces changes in electrical activity in this circuit, resulting in increased output from the GPi and STN. DBS appears to act in restoring electrical imbalance in the affected pathways. Indications for DBS in PD are intractable tremor, drug-related motor fluctuations, and dyskinesias, with a good previous clinical response to medical treatment with L-DOPA being the best prognostic indicator for success of DBS. Unfortunately, 'midline symptoms,' such as postural instability, and nonmotor symptoms, such as cognitive decline, may continue to worsen after DBS.

## New Directions in PD Therapy

A great deal of research is dedicated toward identifying neuroprotective agents to prevent or treat PD. Several trials have investigated the neuroprotective properties of selegiline, a selective, irreversible inhibitor of MAO-B, which inhibits dopamine metabolism. Its metabolite, desmethylselegiline, has been shown to protect animals against MPTP toxicity by inducing transcriptional changes leading to upregulation of antiapoptotic and downregulation of proapoptotic genes. In PD patients, selegiline significantly delayed the need for levodopa compared to placebo, but this was most likely due to symptomatic effect, and any potential neuroprotective activity remains highly controversial. The Food and Drug Administration has approved rasagiline (*N*-propargyl-1R-aminoindan), a novel, highly potent irreversible MAO-B inhibitor for the treatment of PD. Studies have shown a neuroprotective effect in cell culture via activation of the antiapoptotic protein Bcl-2 and downregulation of the proapoptotic Bax family of proteins by its propargylamine moiety. A delayed-start design phase III clinical trial demonstrated greater benefit in subjects taking rasagiline early, suggesting a possible neuroprotective effect that remains to be clarified and is currently under investigation.

Based on increasing understanding of underlying pathogenetic mechanisms of PD, avenues to retard disease progression that include modulation of mitochondrial function, oxidative stress, and ubiquitin–proteasomal system function are being evaluated. One such promising compound is coenzyme Q10 (CoQ10), an electron acceptor for mitochondrial complex I, shown to be decreased in mitochondria of PD patients. Treatment of PD patients with CoQ10 leads to an increase of platelet mitochondrial complex I and III activity. In a phase II clinical trial, high-dose CoQ10 was well tolerated and it may slow clinical progression in PD patients. High-dose CoQ10 is under further evaluation in a large phase III clinical trial.

Restorative therapy is a future goal for treatment, potentially allowing arrest or reversal of disease progression. Transplantation of fetal and stem cells has attracted considerable attention since it might lead to a more physiologic replacement of dopamine, potentially avoiding the motor complications associated with long-term medical treatment. Based on evidence from animal models of PD that transplantation of fetal dopaminergic neurons can relieve PD symptoms, PD patients have received bilateral intrastriatal transplants of fetal mesencephalic cells, with demonstrated dopaminergic cell survival. However, development of 'runaway dyskinesias' – that is, dyskinesias persisting after overnight withdrawal of dopaminergic medication – indicates the need for more sophisticated understanding and modulation of transplanted cell function. Embryonic stem cells are pluripotent cells that can be differentiated into dopaminergic neurons. Transplantation of these cells into the striatum of MPTP-lesioned monkeys and rats after 6-OHDA administration leads to behavioral improvement. No clinical trials have been conducted in PD patients.

Neuroregeneration by infusion of growth factors in a site-specific manner would potentially avoid side effects resulting from activation of other neuronal circuits. Neurotrophic factors, which undergo retrograde axonal transport over intact nigrostriatal axons, have been infused into the striatum of PD patients.

A phase I trial continuously infused glial cell line-derived neurotrophic factor (GDNF) into the putamen of PD patients, which resulted in improvement of certain aspects of motor function with reduction of medication-induced dyskinesias. However, a subsequent randomized controlled study found no detectable clinical benefit from infused intraputamenal GDNF.

Finally, considerable interest has focused on the possibility of using viral vectors to deliver genes to the central nervous system. Lentiviral delivery of GDNF by injecting lenti-GDNF into striatum and substantia nigra of the MPTP primate model of PD reversed functional deficits. Clinical trials of gene therapy in PD patients are under way using adeno-associated virus to deliver glutamic acid decarboxylase to the STN and also aromatic L-amino acid decarboxylase and the trophic factor neurturin to the striatum.

*See also:* Animal Models of Parkinson's Disease; Cell Replacement Therapy: Parkinson's Disease; Deep Brain Stimulation and Parkinson's Disease; Deep Brain Stimulation and Movement Disorder Treatment; Parkinson's Disease: Alpha-Synuclein and Neurodegeneration; Transgenic Models of Neurodegenerative Disease.

## Further Reading

Au WL, Adams JR, Troiano A, et al. (2006) Neuroimaging in Parkinson's disease. *Journal of Neural Transmission Supplement* 70: 241–248.

Beal MF (2005) Mitochondria take center stage in aging and neurodegeneration. *Annals of Neurology* 58: 495–505.

Betchen SA and Kaplitt M (2003) Future and current surgical therapies in Parkinson's disease. *Current Opinion in Neurology* 16: 487–493.

Braak H, Del Tredici K, Rub U, et al. (2003) Staging of brain pathology related to sporadic Parkinson's disease. *Neurobiology of Aging* 24: 197–211.

Chaudhuri KR, Healy DG, and Shapira AH (2006) Non-motor symptoms of Parkinson's disease: Diagnosis and management. *Lancet Neurology* 5: 235–245.

de Lau LM and Breteler MM (2006) Epidemiology of Parkinson's disease. *Lancet Neurology* 5: 525–535.

Fahn S (2003) Description of Parkinson's disease as a clinical syndrome. *Annals of the New York Academy of Science* 991: 1–14.

Farrer MJ (2006) Genetics of Parkinson disease: Paradigm shifts and future prospects. *Nature Reviews Genetics* 7: 306–318.

Hardy J, Cai H, Cookson MR, et al. (2006) Genetics of Parkinson's disease and parkinsonism. *Annals of Neurology* 60: 389–398.

Lim KL, Dawson VL, and Dawson TM (2003) The cast of molecular characters in Parkinson's disease: Felons, conspirators, and suspects. *Annals of the New York Academy of Sciences* 991: 80–92.

Waters CH (2006) *Diagnosis and Management of Parkinson's Disease.* Caddo, OK: Professional Communications.

## Relevant Websites

http://www.MichaelJFox.org – Michael J. Fox Foundation for Parkinson's Research.

http://www.ninds.nih.gov – National Institute of Neurological Disorders and Stroke.

http://www.parkinson.org – National Parkinson Foundation.

http://www.pdf.org – Parkinson's Disease Foundation.

http://www.psp.org – Society for Progressive Supranuclear Palsy.

http://www.nlm.nih.gov – U.S. National Library of Medicine.

http://www.wemove.org – Worldwide Education and Awareness for Movement Disorders.

# Parkinson's Disease: Alpha-Synuclein and Neurodegeneration

**M Goedert**, Medical Research Council Laboratory of Molecular Biology, Cambridge, UK



## Introduction

Parkinson's disease (PD) is the most common movement disorder. It affects on the order of 1–2% of the general population older than 65. Clinical manifestations include bradykinesia, increased muscle tone, resting tremor, and postural imbalance. Neuropathologially, PD is defined by the loss of dopaminergic nerve cells in the substantia nigra and the presence there of abnormal proteinaceous inclusions known as Lewy bodies. Similar inclusions are also seen in nerve cell processes, where they are called Lewy neurites. The clinical symptoms of PD become manifest when, as a result of dopaminergic nerve cell loss in the substantia nigra, striatal dopamine levels are reduced by more than 70%. In addition, nerve cell loss and Lewy body pathology are found in a number of other brain regions, such as the dorsal motor nucleus of the vagus nerve, the nucleus basalis of Meynert, and some autonomic ganglia. Lewy bodies and Lewy neurites also constitute the defining neuropathological characteristics of Lewy body dementias (encompassing dementia with Lewy bodies and Parkinson's disease with dementia), common late-life dementias that exist in a pure form or overlap with the neuropathological characteristics of Alzheimer's disease (AD). Unlike PD, Lewy body dementias are characterized by the presence of a large number of Lewy bodies and Lewy neurites in cortical brain areas. Ultrastructurally, Lewy bodies and Lewy neurites consist of abnormal filamentous material. Even though Friedrich Lewy described the Lewy body as long ago as 1912, its molecular composition remained unknown until 1997.

A new direction to research on the etiology and pathogenesis of PD and Lewy body dementias was imparted by the twin discoveries that a missense mutation in the *α-synuclein* gene is a rare genetic cause of PD and that α-synuclein is the main component of Lewy bodies and Lewy neurites in PD, dementia with Lewy bodies, and several other diseases. Subsequently, the filamentous glial and neuronal inclusions of multiple system atrophy (MSA), also known as Papp–Lantos bodies, were found to be made of α-synuclein, revealing an unexpected molecular link with Lewy body diseases. These findings have placed dysfunction of α-synuclein at the center of PD and several atypical parkinsonian disorders (**Table 1**).

## The Synuclein Family

Synucleins are abundant brain proteins whose physiological functions are only poorly understood. The human synuclein family consists of three members – α-synuclein, β-synuclein, and γ-synuclein – which range from 127 to 140 amino acids in length and are 55–62% identical in sequence, with a similar domain organization. The N-terminal half of each protein is taken up by imperfect 11-amino acid repeats that bear the consensus sequence KTKEGV. Individual repeats are separated by an interrepeat region of five to eight amino acids. The repeats are followed by a hydrophobic intermediate region and a negatively charged C-terminal domain. By immunohistochemistry, α- and β-synuclein are concentrated in nerve terminals, with little staining of somata and dendrites. Ultrastructurally, they are found in proximity to synaptic vesicles. In contrast, γ-synuclein is present throughout nerve cells in many brain regions. In rat, α-synuclein is most abundant throughout telencephalon and diencephalon, with lower levels in more caudal regions. Beta-synuclein is distributed fairly evenly throughout the central nervous system, whereas γ-synuclein is most abundant in midbrain, pons, and spinal cord, with much lower levels in forebrain areas.

Synucleins are natively unfolded proteins with little secondary structure and have been identified only in vertebrates. Mammals and birds have three synuclein genes; a fourth gene (called γ2-synuclein) is present in teleost fish. Experimental studies have shown that α-synuclein can bind to lipid membranes through its N-terminal repeats, indicating that it may be a lipid-binding protein. It adopts structures rich in α-helical character on binding to synthetic lipid membranes containing acidic phospholipids. In presence of the lipid mimetic sodium dodecyl sulphate, this conformation is taken up by amino acids 1–98 and consists of two α-helical regions (residues 1–42 and 45–98) that are interrupted by a break of two amino acids (residues 43 and 44). Residues 99–140 are unstructured. In cell lines and primary neurons treated with high fatty acid concentrations, α-synuclein was found to accumulate on phospholipid monolayers surrounding triglyceride-rich droplets. Beta-synuclein binds in a similar way, but γ-synuclein fails to bind to lipid droplets and remains cytosolic. Accordingly,

**Table 1** α-Synucleinopathies

| |
|---|
| Idiopathic Parkinson's disease |
| Dementia with Lewy bodies |
| Pure autonomic failure |
| Lewy body dysphagia |
| Inherited Lewy body diseases |
| Incidental Lewy body disease |
| Rapid eye movement sleep behavioral disorder |
| Multiple system atrophy |

α-synuclein binds to fatty acids *in vitro*, albeit with a much lower affinity than physiological fatty acid-binding proteins. Both α- and β-synuclein have been shown to inhibit phospholipase D2. This isoform of phospholipase D localizes to the plasma membrane, where it may be involved in signal-induced cytoskeletal regulation and endocytosis. A fraction of α-synuclein in normal brain is phosphorylated at S129. Casein kinase-1 and casein kinase-2 phosphorylate S129 of α-synuclein *in vitro*, as do several G-protein-coupled receptor kinases.

Inactivation of the *α-synuclein* gene by homologous recombination does not lead to a neurological phenotype, with the mice being largely normal. A loss of function of α-synuclein is therefore unlikely to account for its role in neurodegeneration. Analysis of mice lacking β- or γ-synuclein has similarly failed to reveal any gross abnormalities. The same is true of mice double knockout for α- and β-synuclein. The most compelling evidence for a physiological function comes from experiments showing that increased expression of α-synuclein rescues mice lacking cysteine string protein α (CSPα), a presynaptic molecular chaperone that aids in the folding and refolding of synaptic proteins critical for neurotransmitter release and vesicle recycling. CSPα-deficient mice are normal at birth but develop progressive neurodegeneration that causes defects in synaptic transmission after 2–3 weeks of age and is lethal after 1–4 months. Deletion of CSPα inhibits soluble NSF (N-ethylmaleimide-sensitive factor) attachment protein receptor (SNARE) protein assembly, and transgenic α-synuclein ameliorates this inhibition in a phospholipid binding-dependent manner. Alpha-synuclein may therefore play a physiological role in maintaining synaptic integrity.

## Parkinson's Disease and Other Lewy Body Diseases

### *Alpha-Synuclein* Mutations Cause Familial PD

In 1990, an autosomal dominantly inherited form of PD was described in an Italian-American kindred (the Contursi kindred). It was the first familial form of disease in which Lewy bodies were shown to be

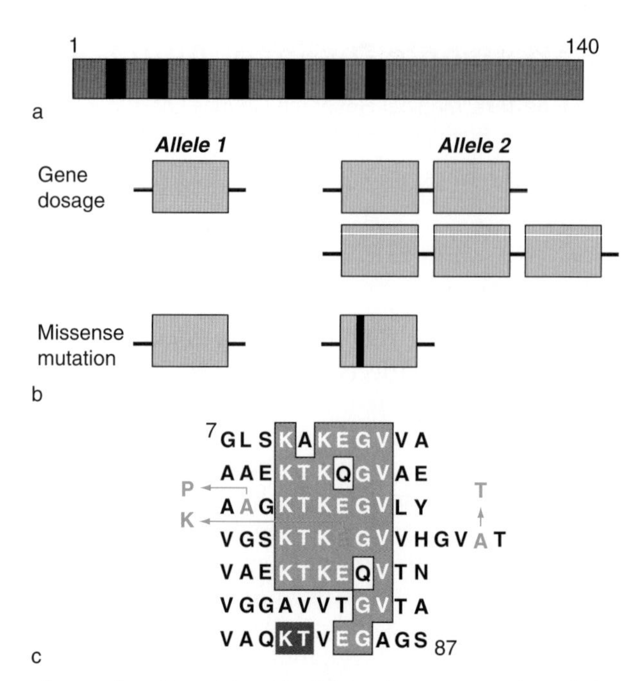


**Figure 1** α-Synuclein. (a) Schematic representation of the 140-amino-acid protein, with the N-terminal, lipid-binding repeats shown as black bars. (b) Duplications and triplications of the *α-synuclein* gene and missense mutations (black box) in the *α-synuclein* gene cause inherited forms of Parkinson's disease/Lewy body dementia. (c) Three missense mutations in *α-synuclein* are shown (in blue).

present. In 1997, a missense mutation (A53T) in the *α-synuclein* gene was identified as the cause of disease in the Contursi kindred (**Figure 1**). Subsequently, a second mutation (A30P) was identified in a German family with PD. More recently, a third mutation (E46K) was described in a Spanish family with PD and Lewy body dementia. All three missense mutations are located in the repeat region of α-synuclein.

In addition to changes in primary structure, overexpression of wild-type α-synuclein has also been identified as a cause of inherited PD and Lewy body dementia. Thus, duplications and triplications of the *α-synuclein* gene are responsible for autosomal dominantly inherited forms of disease with abundant and widespread Lewy body pathology. Triplications lead to early onset, rapidly progressive parkinsonism dementia with autonomic dysfunction. Duplication mutations result in disease that is more closely related to PD, with later onset parkinsonism, slower progression, and without autonomic dysfunction. In addition, polymorphic variations in the 5′ noncoding region of the *α-synuclein* gene have been found to be associated with idiopathic PD in some studies. They probably influence the level of α-synuclein expression. Variation in the expression levels of α-synuclein may be an important risk factor for the development of Lewy body diseases.

**Lewy Body Filaments Are Made of $\alpha$-Synuclein**

Shortly after the identification of the genetic defect responsible for PD in the Contursi kindred, Lewy bodies and Lewy neurites in the substantia nigra from patients with sporadic PD were shown to be strongly immunoreactive for α-synuclein (**Figure 2**). Subsequently, Lewy body filaments were found to be decorated by antibodies directed against α-synuclein. The same was true of the Lewy body pathology of dementia with Lewy bodies.

Filaments associated with PD and dementia with Lewy bodies are unbranched, with a length of 200–600 nm and a width of 5–10 nm. Full-length α-synuclein is present, with its N- and C-termini being exposed on the filament surface (**Figure 3**). The core of the filament extends over a stretch of about 70 amino acids that overlaps almost entirely with the lipid-binding region of α-synuclein. Of the three human synucleins, only α-synuclein is associated with the filamentous inclusions of Lewy body diseases. Prior to this work, ubiquitin staining was the most sensitive marker of Lewy body pathology. By the use of double-labeling, the number of α-synuclein-positive structures was found to be greater than that stained for ubiquitin, suggesting that ubiquitination of α-synuclein occurs after assembly. Phosphorylation and nitration are two additional posttranslational modifications of filamentous α-synuclein. Phosphorylation at S129 has been documented in Lewy bodies and Lewy neurites, and it has been suggested that it may trigger filament assembly. However, it remains to be determined whether phosphorylation at S129 occurs before or after filament assembly in the human brain. Nitration of filamentous α-synuclein was found using antibodies specific for nitrated tyrosine residues. As for ubiquitin, staining for nitrotyrosine was less extensive than staining for α-synuclein, suggesting that nitration of α-synuclein occurs after its assembly into filaments.

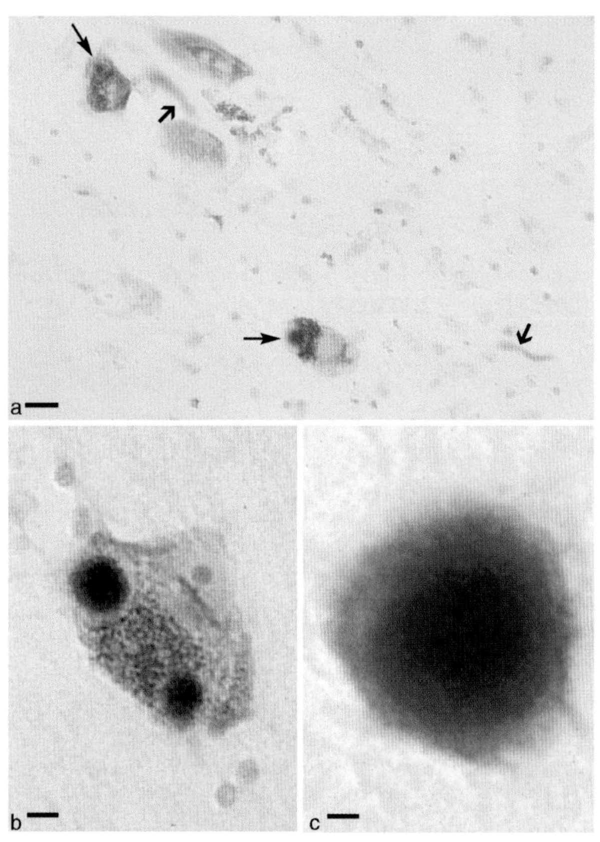


**Figure 2** Substantia nigra from patients with Parkinson's disease immunostained for α-synuclein. (a) Two pigmented nerve cells (large arrows), each containing an α-synuclein-positive Lewy body. Lewy neurites (small arrows) are also immunopositive. (b) Pigmented nerve cell with two α-synuclein-positive Lewy bodies. (c) α-Synuclein-positive extracellular Lewy body. Scale bar = 20 μm (a), 8 μm (b), 4 μm (c).

Lewy body pathology is also the defining feature of several other, rarer diseases, such as Lewy body dysphagia and pure autonomic failure. In these diseases, Lewy bodies and neurites are largely limited to the enteric and peripheral nervous systems. In PD, Lewy body pathology is also present in the enteric and autonomic nervous systems.

Incidental Lewy body disease describes the presence of small numbers of Lewy bodies and Lewy neurites at autopsy. It is observed in 5–10% of the general population older than 60 and is believed to represent a preclinical form of Lewy body disease. In cases with incidental Lewy body disease, the first α-synuclein-positive structures in the brain form in the dorsal motor nucleus of the vagus nerve, the intermediate reticular zone, the olfactory bulb, and the anterior olfactory nucleus. Rapid eye movement sleep behavioral disorder and olfactory dysfunction can be the clinical correlates of these abnormalities. The pathology ascends from vulnerable regions in the medulla oblongata to the pontine tegmentum, midbrain, basal forebrain, and cerebral cortex. Alpha-synuclein deposits may form even earlier in the enteric nervous system, the peripheral nervous system, and layer I of the dorsal horn of the spinal cord, suggesting that Lewy body diseases may begin outside the brain. Autonomic dysfunction and pain can be the clinical correlates of these abnormalities. The first deposits develop in the form of Lewy neurites, suggesting that the filamentous assembly of α-synuclein in axons may precede assembly in cell bodies and dendrites. Incidental Lewy body disease may be at one end of the spectrum of Lewy body diseases, with Lewy body dementias at the other end, and with Lewy body dysphagia, pure autonomic failure, and PD in between. These findings underscore the fact that there is more to PD than the mere degeneration of dopaminergic cells in the substantia nigra.

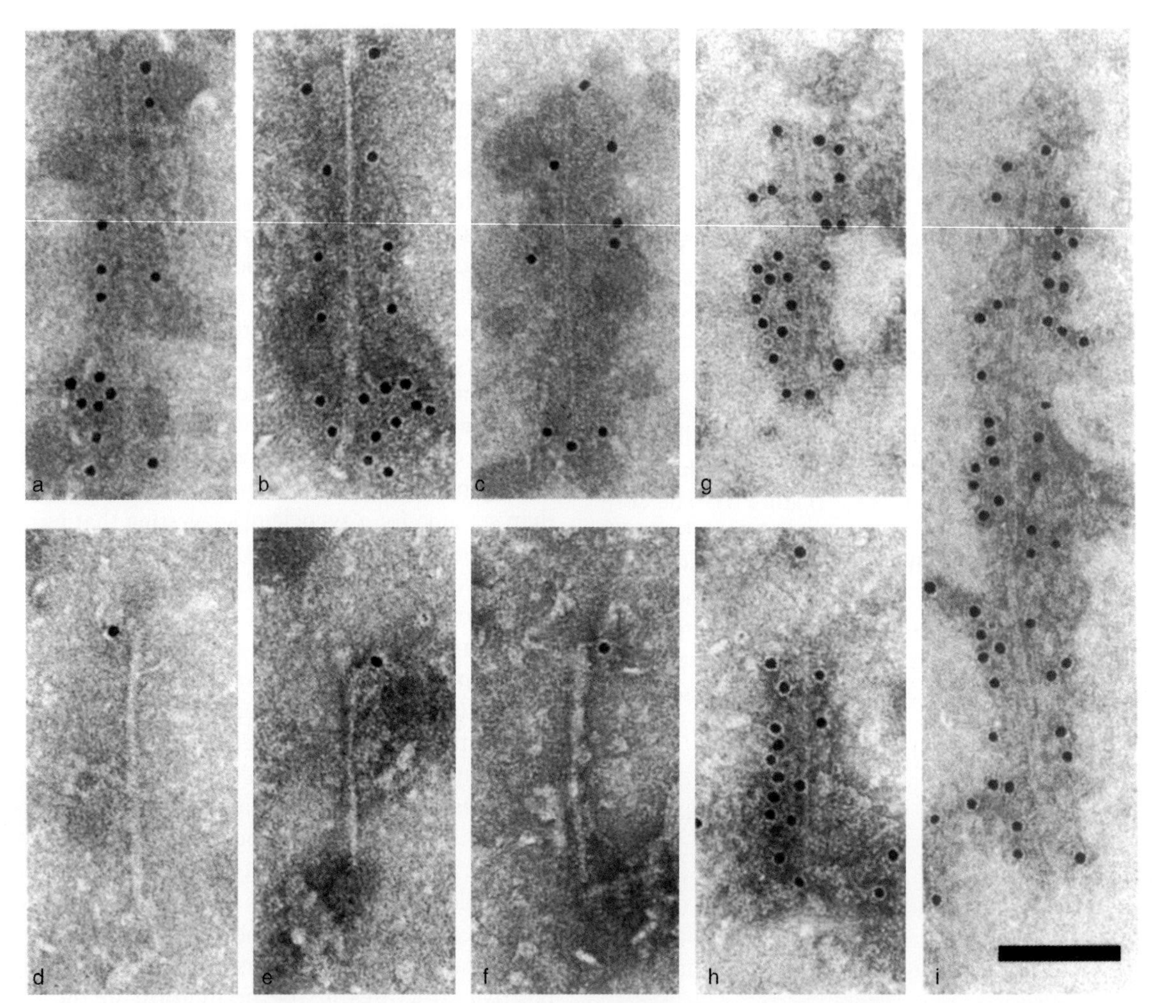


**Figure 3** Decoration of filaments from substantia nigra of Parkinson's disease brain with α-synuclein antibodies. (a–c) Antibody recognizing the C-terminus of α-synuclein. (d–f) Antibody recognizing the N-terminal region of α-synuclein. (g–i) Antibody recognizing the N-terminus of α-synuclein. The gold particles indicating positive staining appear as black dots. Scale bar = 100 nm.

Besides amyloid-β and tau deposits, Lewy body pathology is also present in some cases of familial AD, indicating that mutations in the amyloid precursor protein and the presenilin genes can lead to dysfunction of α-synuclein. Similarly, Lewy bodies and Lewy neurites are present in some cases of sporadic AD. Conversely, plaques and tangles are often found in Lewy body dementias. The amygdala is most consistently affected, suggesting that α-synuclein pathology may start there. These findings in AD define a type of α-synucleinopathy that is distinct from the one described above.

## MSA

Filamentous α-synuclein deposits characterize this atypical parkinsonian disorder, but Lewy body pathology is not generally observed. Instead, glial cytoplasmic inclusions (Papp–Lantos bodies), which consist of filamentous aggregates, are the defining neuropathological feature of MSA. They are found mostly in the cytoplasm and, to a lesser extent, the nucleus of oligodendrocytes. Inclusions are also observed in the cytoplasm and nucleus of some nerve cells and in neuropil threads. The principal brain regions affected are the substantia nigra, striatum, locus coeruleus, pontine nuclei, inferior olives, cerebellum, and spinal cord. Typically, nerve cell loss and gliosis are observed. The formation of glial cytoplasmic inclusions may be the primary lesion that will eventually compromise nerve cell function and viability.

Glial cytoplasmic inclusions are strongly immunoreactive for α-synuclein, and filaments isolated from the brains of patients with MSA are labeled by α-synuclein antibodies. Assembled α-synuclein is nitrated and phosphorylated at S129, and the number of α-synuclein-positive structures exceeds that stained by antiubiquitin antibodies, confirming that the accumulation of α-synuclein precedes ubiquitination. Filament morphologies and their staining characteristics were found to be similar but not identical to those of filaments extracted from the brains of patients with PD and dementia with Lewy bodies. This work revealed an unexpected molecular link between MSA and Lewy body diseases. The main difference is that in MSA, most of the α-synuclein pathology is found in glial cells, whereas in Lewy body diseases, most of the pathology is in nerve cells.

## Synthetic α-Synuclein Filaments

Recombinant α-synuclein assembles into filaments that share many of the morphological characteristics of the disease filaments. Assembly is a nucleation-dependent process that occurs through the N-terminal repeats of α-synuclein. The C-terminal region, in contrast, is inhibitory. Assembly is accompanied by the transition from random coil to a β-pleated sheet. By electron diffraction and solid-state nuclear magnetic resonance spectroscopy, filaments assembled from recombinant wild-type human α-synuclein show a conformation characteristic of amyloid fibers, with evidence for the existence of filaments with different morphologies. Assembly of monomeric α-synuclein into filaments probably passes through an oligomeric stage that after prolonged incubation results in the formation of elongated protofibrils that disappear on filament formation. Oligomeric α-synuclein may be toxic to nerve cells. Under the conditions of these experiments, β- and γ-synucleins failed to assemble, consistent with their absence from the filamentous lesions of the human diseases. Both β- and γ-synucleins have been shown to inhibit the assembly of α-synuclein *in vitro*. Inhibition of α-synuclein assembly by a number of small organic molecules has also been described.

Mutations E46K and A53T increase the rate of assembly of α-synuclein, indicating that this may be their primary effect. The mechanism of action of the A30P mutation is less clear. One study reported that it causes the accumulation of oligomeric, nonfibrillar α-synuclein, implying that this may be a toxic species. Mutation A30P reduces the binding of α-synuclein to rat brain vesicles, whereas mutation E46K increases lipid binding. Mutation A53T has no significant effect on lipid binding. Many factors have been shown to influence the rate and/or extent of α-synuclein assembly *in vitro*. They include nitration, oxidation, dopamine, cytochrome C, polyamines, some pesticides, several di- and trivalent cations, and sulphated glycosaminoglycans. However, the relevance, if any, of these factors for the transition from soluble to filamentous α-synuclein in nerve cells in the brain remains to be established.

## Animal Models of Human α-Synucleinopathies

### Rodents and Primates

Mice expressing wild-type and mutant human α-synuclein in nerve cells or glial cells have been shown to develop numerous α-synuclein-immunoreactive cell bodies and processes. However, filament formation and nerve cell loss were a less consistent feature. One study has documented the presence of α-synuclein filaments in brain and spinal cord of mice transgenic for A53T α-synuclein. The formation of inclusions correlated with the appearance of a severe movement disorder, suggesting a possible cause and effect relationship. A minority of inclusions was ubiquitin immunoreactive. Signs of Wallerian degeneration were in evidence in ventral roots, but motor neuron numbers were not reduced. A major difference with PD was the absence of significant pathology in dopaminergic cells of the substantia nigra. However, a more recent study of transgenic mice expressing a C-terminally truncated form of α-synuclein that fibrillizes more readily than the full-length protein *in vitro* has shown some evidence of filament formation and neurodegeneration in substantia nigra. Expression of high levels of human α-synuclein in oligodendroglia recapitulates features of MSA, including the accumulation of filamentous human α-synuclein aggregates in oligodendrocytes linked to their degeneration. In addition, mouse α-synuclein also accumulated in axons in association with neuronal cell loss and motor impairments.

Viral vector-mediated gene transfer differs from standard transgenic approaches by being targeted to a defined region of the central nervous system and by being inducible at any point during the life of the animal. Adeno-associated and lentiviral vectors have been used to express human wild-type and mutant α-synuclein in rodent and primate substantia nigra. Lewy body-like inclusions formed, and a significant proportion of nerve cells degenerated. In the rat, about a quarter of animals were impaired in spontaneous and drug-induced behavior. In the marmoset monkey, expression of human A53T α-synuclein in the substantia nigra resulted in the formation of inclusions and nerve cell death in most (but not all) animals. Fewer inclusions and less nerve cell death were observed on expression of wild-type human α-synuclein.

A neurotoxin model of α-synuclein pathology has been developed in the rat through the chronic administration of the pesticide rotenone, a high-affinity inhibitor of complex I. The rats developed a progressive degeneration of nigrostriatal neurons and Lewy body-like inclusions that were immunoreactive for α-synuclein and ubiquitin. Behaviorally, they showed bradykinesia, postural instability, and resting tremor. Under these conditions, the inhibition of complex I was only partial, indicating that a bioenergetic defect with adenosine triphosphate (ATP) depletion was probably not involved. Instead, oxidative damage might have played a role since partial inhibition of complex I by rotenone stimulates the production of reactive oxygen species. It would therefore seem that oxidative stress can lead to the assembly of α-synuclein into filaments.

### Flies, Worms, and Yeast

One of the first reports to describe the overexpression of human α-synuclein made use of *Drosophila melanogaster*, an organism without synucleins. Expression in nerve cells resulted in the formation of filamentous Lewy body-like inclusions and an age-dependent loss of some dopaminergic nerve cells. The inclusions were α-synuclein- and ubiquitin-positive. The ensuing motor defect was reversed by L-dopa and dopamine agonists. Toxicity of α-synuclein in this system depended on aggregation and required phosphorylation of S129. Chaperones modulated these effects. Thus, co-expression of heat shock proteins reduced the toxicity of α-synuclein, whereas a reduction in endogenous chaperones exacerbated nerve cell loss. Overexpression of wild-type and mutant human α-synuclein in *Caenorhabditis elegans* resulted in dopaminergic nerve cell loss and motor deficits, but in the apparent absence of α-synuclein fibrils.

When expressed in the yeast *Saccharomyces cerevisiae*, human wild-type and A53T α-synucleins were found to localize at the plasma membrane, whereas the A30P mutant had a cytosolic distribution, consistent with lipid binding *in vitro*. Unlike the A30P mutant, wild-type and A53T α-synucleins were toxic to yeast cells. A genome-wide screen to identify genetic enhancers of α-synuclein toxicity identified genes involved in lipid metabolism and vesicle-mediated transport as the predominant functional class. It has been suggested that toxicity of overexpressed α-synuclein in yeast may not require filament formation but may result from the coating of the plasma membrane and internal membranes via amphipathic α-helix formation, resulting in the disruption of normal membrane processes, including vesicle trafficking. Expression of α-synuclein has been shown to block selectively endoplasmic reticulum to Golgi transport, causing endoplasmic reticulum stress. The ensuing toxicity could be suppressed by overexpression of the small GTPase Ypt1/Rab1. It is interesting that overexpression of the latter also rescued nerve cell loss in invertebrate models of α-synuclein-induced neurodegeneration.

## Outlook

The work of the past 10 years has shown conclusively that a neurodegenerative pathway leading from soluble to insoluble, filamentous α-synuclein is central to Lewy body diseases and MSA. The development of experimental models of α-synucleinopathies has opened the way to the identification of the detailed mechanisms by which the formation of inclusions can cause disease. These model systems have also made it possible to identify disease modifiers that may lead to the development of mechanism-based therapies for these diseases. At a conceptual level, it will be important to understand whether α-synuclein has a role to play in inherited parkinsonian disorders caused by mutations in other genes. They include autosomal-recessive juvenile forms of parkinsonism caused by mutations in the *Parkin*, *DJ-1*, and *PINK-1* genes and the dominantly inherited diseases caused by mutations in the lysosomal type 5 adenosine triphosphatase gene (*ATP13A2*) and the leucine-rich repeat kinase-2 gene (*LRRK2*). The latter is particularly interesting since activating missense mutations in the kinase domain of LRRK2 are a relatively common cause of Lewy body PD, demonstrating the existence of a pathway leading from activation of LRRK2 to aggregation of α-synuclein. The molecular dissection of this pathway is likely to shed new light on the understanding of Lewy body diseases.

*See also:* Animal Models of Parkinson's Disease; Basal Ganglia: Functional Models of Normal and Disease States; Cell Replacement Therapy: Parkinson's Disease; Deep Brain Stimulation and Parkinson's Disease; Oxidative Damage in Neurodegeneration and Injury; Parkinsonian Syndromes.

## Further Reading

Baba M, Nakajo S, Tu PS, et al. (1998) Aggregation of α-synuclein in Lewy bodies of sporadic Parkinson's disease and dementia with Lewy bodies. *American Journal of Pathology* 152: 879–884.

Betarbet R, Sherer TB, MacKenzie G, et al. (2000) Chronic systemic pesticide exposure reproduces features of Parkinson's disease. *Nature Neuroscience* 3: 1301–1306.

Braak H, Del Tredici K, Rüb U, et al. (2003) Staging of brain pathology related to sporadic Parkinson's disease. *Neurobiology of Aging* 24: 197–211.

Chandra S, Gallardo G, Fernández-Chacón R, Schlüter OM, and Südhof TC (2005) α-Synuclein cooperates with CSPα in preventing neurodegeneration. *Cell* 123: 383–396.

Chartier-Harlin MC, Kachergus J, Roumier C, et al. (2004) Alpha-synuclein locus duplication as a cause of familial Parkinson's disease. *Lancet* 364: 1167–1169.

Feany MB and Bender WW (2000) A *Drosophila* model of Parkinson's disease. *Nature* 404: 394–398.

Fujiwara H, Hasegawa M, Dohmae N, et al. (2002) α-Synuclein is phosphorylated in synucleinopathy lesions. *Nature Cell Biology* 4: 160–164.

Goedert M (2001) Alpha-synuclein and neurodegenerative diseases. *Nature Reviews Neuroscience* 2: 492–501.

Ibanez P, Bonnet AM, Debarges B, et al. (2004) Causal relation between alpha-synuclein gene duplication and familial Parkinson's disease. *Lancet* 364: 1169–1171.

Krüger R, Kuhn W, Müller T, et al. (1998) Ala30Pro mutation in the gene encoding α-synuclein in Parkinson's disease. *Nature Genetics* 2: 106–108.

Lee VMY and Trojanowski JQ (2006) Mechanisms of Parkinson's disease linked to pathological α-synuclein: New targets for drug discovery. *Neuron* 52: 33–38.

Lewy F (1912) Paralysis agitans. In: Lewandowski M and Abelsdorff G (eds.) *Handbuch der Neurologie, vol. 3*, pp. 920–933. Berlin: Springer.

Paisán-Ruiz C, Jain S, Whitney Evans E, et al. (2004) Cloning of the gene containing mutations that cause *PARK8*-linked Parkinson's disease. *Neuron* 44: 595–600.

Papp MI, Kahn JE, and Lantos PL (1989) Glial cytoplasmic inclusions in the CNS of patients with multiple system atrophy. *Journal of Neurological Sciences* 94: 79–100.

Polymeropoulos MH, Lavedan C, Leroy E, et al. (1997) Mutation in the α-synuclein gene identified in families with Parkinson's disease. *Science* 276: 2045–2047.

Singleton AB, Farrer M, Johnson J, et al. (2003) α-Synuclein locus triplication causes Parkinson's disease. *Science* 302: 841.

Spillantini MG, Crowther RA, Jakes R, Hasegawa M, and Goedert M (1998) α-Synuclein in filamentous inclusions of Lewy bodies from Parkinson's disease and dementia with Lewy bodies. *Proceedings of the National Academy of Sciences of the United States of America* 95: 6469–6473.

Spillantini MG, Schmidt ML, Lee VMY, et al. (1997) α-Synuclein in Lewy bodies. *Nature* 388: 839–840.

Zarranz JJ, Alegre J, Gomez-Esteban JC, et al. (2004) The new mutation, E46K, of α-synuclein causes Parkinson and Lewy body dementia. *Annals of Neurology* 55: 164–173.

Zimprich A, Biskup S, Leitner P, et al. (2004) Mutations in *LRRK2* cause autosomal-dominant parkinsonism with pleomorphic morphology. *Neuron* 44: 601–607.

# Triplicate Repeats: Huntington's disease

**J-H J Cha and K B Kegel**, MassGeneral Institute for Neurodegenerative Disease, Charlestown, MA, USA



## Introduction

In 1872, the physician George Huntington described the neurologic disorder that has come to bear his name. This disorder was originally called Huntington's chorea, from the Greek word for dance, referring to the dancelike adventitious movements that are the hallmark of the disorder. Since chorea is not a universal feature, especially in the juvenile forms of the disease, the term Huntington's disease (HD) is now used. Huntington's original report noted the familial inheritance, the adult onset, the constellation of emotional, cognitive, and motor symptoms, and the inexorably progressive and ultimately fatal nature of the disease. Since that time, Huntington's disease has been a paradigmatic neurodegenerative disorder, with advances in HD at the forefront of molecular genetics, of ethical issues around genetic testing, as well as of many important concepts in modern neuroscience.

## Huntington's Disease Gene

Recombinant DNA techniques became available in the late 1970s, and their power was brought to bear on the search for the Huntington's disease gene. In 1983, using anonymous DNA markers spread over the human genome, David Housman at the Massachusetts Institute of Technology and James Gusella at Massachusetts General Hospital used DNA samples from the many members of a large Venezuelan family to localize the *HD* gene to the long arm of chromosome 4. In 1993, the Huntington's Disease Collaborative Research Group, a consortium of gene hunter laboratories from around the world, isolated a gene originally named *IT15* ('interesting transcript 15'); it was subsequently renamed *HD*.

The defect in *HD* was found to reside in an unstable region of the gene, in that a cytosine–adenine–guanosine (CAG) trinucleotide repeat present in normal alleles was abnormally expanded in the alleles of HD patients. CAG is the codon for glutamine, and this trinucleotide repeat gives rise to a polyglutamine moiety within the huntingtin protein. Normal huntingtin alleles contain from 6 to 35 CAG repeats, giving rise to 6–35 glutamines in the mature protein, and variation occurs in the normal human population. Patients with Huntington's disease invariably have alleles with more than 35 repeats, and greater than 40 repeats invariably gives rise to HD. There is a 'gray area,' between 35 and 39 repeats, where some uncertainty exists. Some patients with up to 39 repeats have lived into their 70s without developing overt signs of HD, although age of onset as late as 80 years has been described.

Age of onset correlates negatively with repeat length, although the correlation is strongest for high CAG repeat numbers. That is, although repeat numbers of greater than 70 invariably produce juvenile onset of HD, more common repeat numbers – for example, in the 40–45 range – produce a varied age of onset. Neuropathologic grade also varies with CAG number, with the most damage seen in brains with the highest CAG repeats. However, CAG repeat length accounts for only about 50% of the variability of age of onset, suggesting that there are other genetic or environmental influences on age of onset.

HD thus belongs to a novel family of neurologic diseases, the trinucleotide repeat disorders. Certain other diseases, such as myotonic dystrophy, fragile X syndrome, and Friedreich's ataxia, are characterized by trinucleotide repeat expansions in noncoding regions of the genes. In contrast, HD belongs to a family of disorders characterized by expansion of a CAG trinucleotide motif within the coding region of the gene. The CAG repeat disorders, or polyglutamine disorders, display common features, including dominant inheritance, adult onset, threshold levels of approximately 37 repeats, and progressive neurodegeneration (**Table 1**).

## Clinical Features

Huntington's disease is typically an adult-onset disorder characterized by insidious onset of both neurologic and psychiatric symptoms. In the United States, approximately 25 000 persons are affected by HD (about 10 per 100 000 population) and approximately 150 000 persons are at 50% risk for the disease by virtue of having an affected parent. Symptoms usually begin in the mid-30s to mid-40s, although disease onset can range from as young as 2 years or as old as 80 years. Initial symptoms include personality change and the gradual appearance of small involuntary movements. Symptoms progress, with chorea becoming more obvious and incapacitating. Over years, motor symptoms worsen such that walking

**Table 1** CAG trinucleotide repeat disorders

| *Disease* | *Protein* | *Inheritance* | *Normal repeat number* | *Range of CAG repeats in disease* |
|---|---|---|---|---|
| Spinobulbar muscular atrophy (Kennedy's disease) | Androgen receptor | X-linked dominant | 11–33 | 40–62 |
| Huntington's disease | Huntingtin | Autosomal dominant | 1–39 | 36–121 |
| Spinocerebellar ataxia, type 1 (SCA1) | Ataxin-1 | Autosomal dominant | <29–36 | 43–60+ |
| Spinocerebellar ataxia, type 2 (SCA2) | Ataxin-2 | Autosomal dominant | 15–24 | 35–59 |
| Spinocerebellar ataxia, type 3 (SCA3; Machado–Joseph disease) | Ataxin-3 | Autosomal dominant | 13–36 | 62–82 |
| Spinocerebellar ataxia, type 6 (SCA6) | $a_{1a}$ subunit of voltage-dependent calcium channel | Autosomal dominant | 4–16 | 21–27 |
| Spinocerebellar ataxia, type 7 (SCA7) | Ataxin-7 | Autosomal dominant | 7–17 | 38–130 |
| Dentatorubral-pallidoluysian atrophy (DRPLA) | Atrophin-1 | Autosomal dominant | 7–25 | 49–85 |
| Spinocerebellar ataxia, type 17 (SCA17) | TATA-binding protein (TBP) | Autosomal dominant | <42 | ≥43 |

becomes more difficult, as do speaking and eating. Weight loss is common, partially due to the extra energy required for adventitious movements but also due to increased resting basal energy expenditure. Most HD patients eventually succumb to aspiration pneumonia, due to swallowing difficulties.

About 10% of cases start before the age of 20 years. The juvenile form (Westphal variant) is more parkinsonian in nature. Rather than chorea, the prominent features are bradykinesia, rigidity, and tremor. Seizures can be present in juvenile patients. Juvenile-onset HD usually results from paternal genetic transmission. For individuals who develop symptoms before age 10 years, more than 90% have an affected father. This tendency for 'anticipation' – younger age of onset in successive generations – is especially pronounced in cases of paternal transmission. This phenomenon is not completely understood but may be the result of further CAG expansion during spermatogenesis. Interestingly, both the juvenile-onset and adult-onset phenotypes can be present within the same family.

New mutations have been described, usually from a parent carrying an 'intermediate allele.' Homozygotes for the *HD* gene have also been described. The clinical symptomatology manifested by these very few patients appears to be no worse than that manifested by patients carrying only a single *HD* gene. Thus, HD appears to be a truly genetically dominant disease.

## Pathology

Neuropathologically, HD is characterized by gross atrophy of the basal ganglia, specifically the striatum (caudate-putamen), although pathologic changes have also been described in cortex, thalamus, and subthalamic nucleus. Atrophy and gliosis of the caudate nucleus and putamen are progressive and marked. In juvenile-onset cases, cell loss tends to be more pronounced, and can also be seen in the cerebellum. Within the striatum, HD differentially affects subpopulations of neurons, with projection neurons preferentially being lost rather than interneurons. The $\gamma$-aminobutyric acid (GABA)-containing medium spiny neurons, which are the output projection neurons, comprise approximately 90% of striatal neurons. HD seems to spare the interneurons while devastating the projection neurons. Consistent with a loss of projection neurons was the early finding that GABA levels were markedly reduced in the caudate-putamen of HD patients.

Of the two populations of striatal projection neurons, the neurons of the indirect pathway (i.e., enkephalin/GABA-containing neurons) are affected first, thus providing an anatomical explanation for the increased movement which is the hallmark of HD. Consistent with this observation is the fact that in early-grade HD cases, markers for this population of striatal neurons are decreased, including dopamine D2 receptors, adenosine $A_{2a}$ receptors, and enkephalin. In early HD, thus, the indirect pathway is predominantly disrupted, and basal ganglia circuitry models predict an overall increase in movement, manifested as chorea and ballism (**Figure 1**). In later stages of adult HD, both populations of striatal projection neurons are affected, with concomitant loss of markers of the direct pathway (i.e., substance P/dynorphin/GABA-containing neurons), including dopamine D1 receptors and substance P. The functional correlate of degeneration of both direct and indirect pathways is a rigid bradykinetic state. In

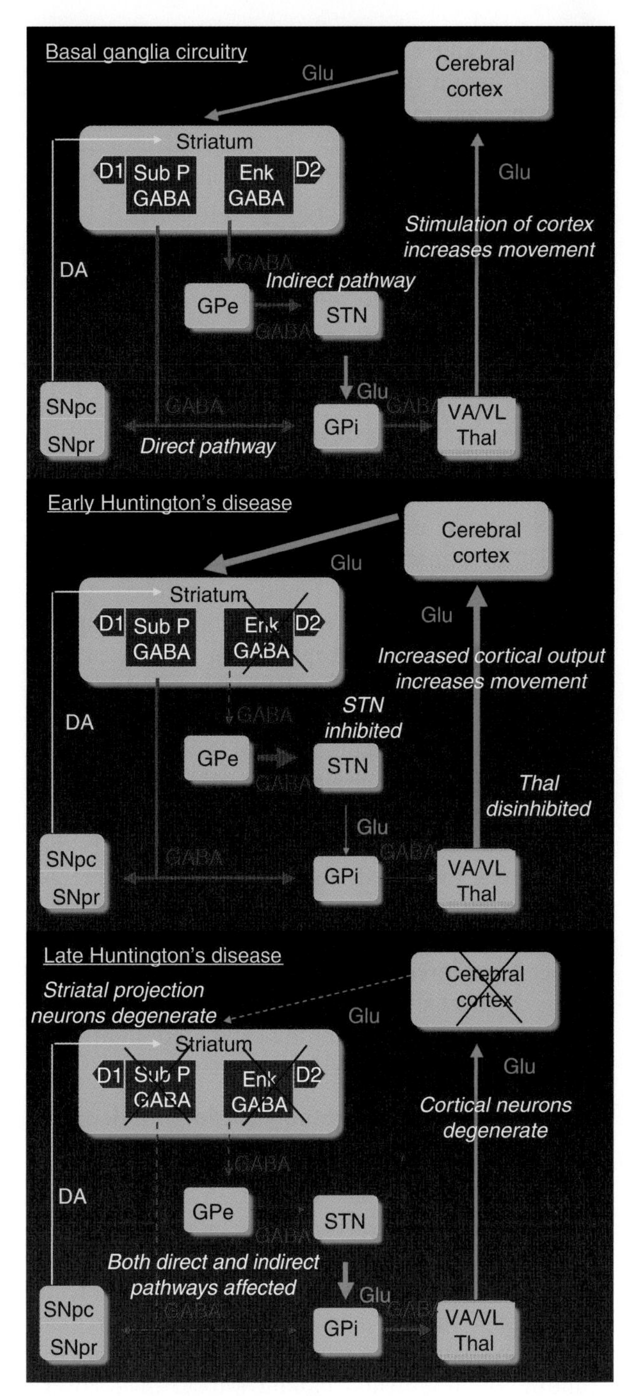


**Figure 1** Basal ganglia circuitry models of early and late Huntington's disease. Glu, glutamate; D1/D2, dopamine D1/D2 receptors; Sub P, substance P; Enk, enkephalin; GABA, $\gamma$-aminobutyric acid; DA, dopamine; VA/VL Thal, ventral anterior/ventral lateral thalamic nuclei; STN, subthalamic nucleus.

juvenile HD cases, which resemble Parkinson's disease, degeneration of both direct and indirect pathway striatal neurons is seen concurrently.

The mode by which neurons die in HD is unknown. One theory is that death in neurons occurs through apoptosis – that is, activation of a programmed cell death pathway. DNA fragmentation characteristic of apoptosis has been described in HD brain. However, it is not yet clear whether apoptosis is the primary pathologic event, or merely the final mode of exit for neurons which have been long compromised by some other process.

## Theories of HD Pathogenesis: Excitotoxicity and Mitochondrial Dysfunction

Prior to the discovery of the *HD* gene, the leading hypotheses concerning the pathogenesis of HD implicated either excitotoxicity or energetic dysfunction. Evidence for both theories has been demonstrated in both human and animal model studies of HD. Discovery of the *HD* gene in 1993 did not reveal the pathogenetic mechanism of HD. The novel protein huntingtin had no clear relationship to excitatory amino acid neurotransmission, nor to cellular or mitochondrial energetics.

Excitotoxicity is the process in which neuronal cells die as a result of excessive excitatory amino acid neurotransmission. Excitatory amino acids, especially glutamate, have been postulated to play a role in the pathogenesis of HD because intrastriatal injections of excitatory amino acid receptor agonists, particularly agonists acting at the *N*-methyl-D-aspartate subtype of glutamate receptor, reproduce the neuropathological features of HD. Excessive glutamate has been postulated to kill neurons in a number of neurological disorders, including hypoxia–ischemia, head trauma, epilepsy, schizophrenia, and neurodegenerative disorders such as Alzheimer's disease, Parkinson's disease, and amyotrophic lateral sclerosis. Postmortem analyses of HD brains demonstrate reduced glutamate receptors that might represent an effort by cells to reduce glutamate receptor activation and subsequent calcium influx, although this decrease in receptors may also reflect generalized neuronal loss.

Mitochondrial dysfunction has also been implicated as a pathologic mechanism in HD, potentially rendering cells vulnerable to normal ambient levels of extracellular glutamate. Positron emission tomography studies have demonstrated reduced striatal glucose metabolism in HD patients. Magnetic resonance imaging (MRI) spectroscopy shows increased levels of striatal glutamate/glutamine and lactate in HD patients, suggesting that glutamatergic function and abnormalities in energy metabolism may combine to produce pathology. Defects in mitochondrial metabolism have been demonstrated in HD brain tissue, especially in complex II–complex III activity. Further, intrastriatal injection of the complex II inhibitor malonate produces lesions reminiscent of HD. Systemic administration of another complex II inhibitor, 3-nitropropionic acid (3-NP), produces strikingly focal striatal lesions in rodents and primates, also reminiscent of HD striatal pathology. Currently, clinical trials are exploring the effects of coenzyme $Q_{10}$, a compound that boosts mitochondrial energetics. In addition, creatine, which is important for storage of high-energy phosphate (creatine phosphate) and subsequent production of ATP, has been shown to extend life in transgenic HD mice, and is being explored as a potential therapy for human HD patients.

## Huntingtin

*HD* encodes a novel protein called huntingtin. The huntingtin gene comprises 67 exons, encoding a 3144-amino-acid protein with an expected molecular mass of 348 kDa. Expectations were that the distribution of huntingtin would mirror the characteristic distribution of neuropathological damage. However, huntingtin mRNA is expressed in almost all tissues of the body and homogenously throughout the brain. No difference in expression is seen between HD and control cases. Immunohistochemical studies confirm the widespread neuronal distribution of the huntingtin protein.

Huntingtin plays an important role in early development, since mice with targeted knockout of the huntingtin gene are embryonic lethal. Humans with Wolf–Hirshhorn syndrome, who have a partial deletion of chromosome 4 and as a result have only one copy of the *HD* gene, do not have the a neurologic disease identical to HD. Therefore, having a single copy of a normal *HD* allele does not produce the HD phenotype and is sufficient for normal embryogenesis in humans. The mutation in huntingtin is thought to confer at least one gain of function. Mutant huntingtin possesses at least certain of the functional properties of wild-type huntingtin, in that mutant huntingtin can rescue the embryonic lethal phenotype seen in huntingtin-null knockout mice.

The huntingtin protein contains 'HEAT repeats,' which correspond to tandemly arranged curlicue-like structures that appear to serve as flexible scaffolding on which other components can assemble (the acronym 'HEAT' derives from names of four proteins found to contain the repeats). Huntingtin has been shown to have numerous protein-binding partners, and many of these interactors have differential affinity for the normal and polyglutamine-expanded versions of huntingtin. Numerous cytoskeletal proteins have been identified as huntingtin interactors. In addition, huntingtin binds to a number of transcriptionally active molecules, raising the possibility that mutant huntingtin may perturb the normal transcriptional activities of neurons. In addition to these interactions, nuclear localization and nuclear export domains have been identified within the huntingtin molecule, suggesting that huntingtin may be shuttled in and out of the nucleus.

The normal function of huntingtin is not completely known, although it has been proposed to act as a molecular scaffold at multiple locations within cells. Huntingtin is a soluble protein that associates with membranes. It can be transported by fast axonal transport, and may be involved in endocytosis and membrane recycling. Aside from functioning as a molecular scaffold, a number of other normal functions have been proposed for normal versions of huntingtin. Wild-type (normal) huntingtin has been found to have antiapoptotic properties. In addition, a permissive role in brain-derived neurotrophic factor (BDNF) transcription has been proposed. Wild-type huntingtin has been found to interact with the transcriptional repressor RE1-silencing transcription factor (REST) in the cytoplasm. In the absence of cytoplasmic sequestration of REST by wild-type huntingtin, REST is free to enter the nucleus, where it serves as a negative regulator or BDNF transcription.

The polyglutamine moiety has been shown to assume a $\beta$-pleated sheet conformation when the number of residues exceeds 35. Current opinion is that such a 'toxic fold' leads to aberrant binding interactions with other proteins, including with other copies of misfolded huntingtin. Consequently, one area that has been explored is whether chaperone proteins, such as the heat shock proteins, could mitigate toxicity by keeping polyglutamine-expanded huntingtin from assuming a misfolded conformation.

### Posttranslational and Proteolytic Modifications of Huntingtin

Several posttranslational modifications of the huntingtin protein have been identified, including phosphorylation, sumoylation, and palmitoylation. Sumoylation (so named for 'small ubiquitin-like modifier' proteins) of the first 17 amino acids of the N-terminal regions of huntingtin can promote nuclear localization of exon 1 huntingtin fragments. Palmitoylation may govern huntingtin's interaction with membranous structures such as the Golgi or synaptic vesicles. It is likely that these posttranslational modifications alter huntingtin's intracellular distribution as well as its stability and degradation.

At least five different proteolytic mechanisms have been identified for huntingtin: cleavage by the apoptotic proteases known as caspases, cleavage by the calcium-dependent protease calpain, degradation by the ubiquitin–proteasome system, cleavage by an aspartyl protease, and autophagic digestion in the lysosome. There has been much focus on proteolytic processing of huntingtin, since truncated fragments of huntingtin – specifically, those fragments containing the N-terminal portion of the molecule – appear to be especially toxic in a number of experimental paradigms. Compared to full-length versions of huntingtin, N-terminal fragments enter the nucleus more easily, form aggregates more readily, and induce cellular toxicity more readily. Thus, preventing certain cleavage events has been an attractive strategy by which to mitigate the toxic effects of huntingtin fragments.

### Polyglutamines and Aggregates

The polyglutamine moiety that is expanded in mutant forms of the huntingtin protein gives rise to the neuropathologic changes observed in HD. A transgenic mouse model expressing only exon 1 of the huntingtin gene (the portion containing the polyglutamine moiety) develops an abnormal neurologic phenotype. The fact that such a striking phenotype could be produced in an animal expressing only exon 1 of the *HD* gene added weight to the early 'toxic fragment' hypothesis. One of the striking findings which emerged out of the findings with transgenic animals is the observation of novel inclusions in neuronal nuclei, although huntingtin normally exists as a predominantly cytoplasmic protein. This remarkable finding prompted reexamination of human biopsy and necropsy material, and abnormal huntingtin- and ubiquitin-positive nuclear and cytoplasmic aggregates were also found. Although abnormal aggregates are a striking feature of human HD and transgenic mouse HD models, recent studies have raised controversy as to the importance of inclusions. In cell culture models, the presence of inclusions was not correlated with neuronal death. Other models have suggested that inclusions may actually serve as a protective function.

Several theories have developed concerning the gain-of-function which emerges when the length of the polyglutamine moiety extends into the pathologic range. One theory proposes that polyglutamine moieties serve as a substrate for the enzyme transglutaminase. Nuclear inclusions in HD are reminiscent of protein deposits that are pathologic hallmarks of other neurodegenerative diseases; an emerging theme is that neurodegenerative diseases may be disorders of altered protein folding. Polyglutamine moieties are found within other proteins, including transcription factors. Many studies suggest that polyglutamine aggregates may serve as sinks to sequester transcription factors or to deplete proteins involved in vesicle movement. Aggregates present in the cytoplasm may also physically block vesicles, preventing transport within axons and dendrites.

Huntingtin aggregates are dynamic. In *in vitro* systems, one can visualize the appearance and growth of aggregates. In a transgenic mouse model that expressed a portion of the huntingtin protein under the control of an inducible promoter, stopping expression of the mutant transgene led to regression of the size of intranuclear aggregates, attesting to the recuperative ability of neurons with aggregates.

## Pathogenesis in HD

### Transcriptional Dysregulation

Human HD brains show selective downregulation of neurotransmitters and neurotransmitter receptors, a finding that has been recapitulated in transgenic mouse models. Gene expression studies using DNA microarrays have confirmed widespread alteration of mRNA levels of genes involved in neurotransmission, calcium homeostasis, and intracellular signaling. While transcriptional dysregulation has been acknowledged as an important pathogenic feature in HD, the exact mechanisms have not been elucidated.

Normal huntingtin is found predominantly outside the nucleus, whereas mutant versions, especially N-terminal fragments, of huntingtin tend to localize in the nucleus, where they can form visible aggregates. Mutant huntingtin has been shown to bind directly to a number of transcription factors, including Sp1, p53, cAMP response element-binding protein (CREB)-binding protein (CBP), and nuclear corepressor (NCoR), raising the possibility that mutant huntingtin could interfere with the normal function of these factors. Transcriptional dysregulation in HD likely involves histone proteins and chromatin remodeling, as histone deactylase inhibitors, which promote histone acetylation, have been shown to have beneficial effects in a number of animal and cellular models of HD.

### Other Theories of Pathogenesis

Numerous other ideas have been advanced to account for HD pathogenesis, including synaptic dysfunction, disruption of axonal transport, apoptosis, inappropriate cell cycling, calcium dysregulation, oxidative stress, proteasome dysfunction, dysfunction of ion channels, endocrine dysregulation, disrupted mitochondrial biogenesis, altered lipid metabolism, loss of BDNF function, and altered intracellular second messenger signaling. A major challenge within the field is to prioritize the most relevant pathologic processes, while accounting for all of the observed abnormalities.

## Genetic Testing

Isolation of the *HD* mutation made direct DNA testing available. While HD genetic testing is now commercially available, testing should be approached judiciously. Genetic testing is performed in two contexts. Symptomatic gene tests are performed when a patient is manifesting symptoms that could be consistent with HD. For persons who are at risk for HD by virtue of having an affected relative, presymptomatic testing can be helpful in planning life events, such as whether to get married or have children. Caution should be exercised, however, as having a positive gene test result can have many unanticipated negative consequences. Persons undergoing presymptomatic testing should undergo detailed counseling with a genetic counselor. Preimplantation gestational testing, which can determine the gene status of a fetus, is available in some centers.

## Therapeutics

There are currently no effective therapies for preventing the onset or for slowing the progression of HD. Current therapies are symptomatic, and include the use of neuroleptics to decrease chorea and the use of psychotropic medications to address depression, obsessive–compulsive symptoms, or psychosis. In addition, speech therapy and physical therapy are useful in addressing the swallowing and walking difficulties that many HD patients experience. An important principle in HD treatment is that the brain is constantly degenerating, and thus the response to medications changes over time. Therapy must therefore be flexible, and constant reevaluation is required. In addition, because of altered by neurotransmitter receptors, paradoxical reactions to medications can occur. Careful empiric trials must be performed with each patient. Only one medication should be altered at a time.

Some efforts at neural transplantation have been made. Transplantation of embryonic stem cells as replacement therapy is currently being evaluated, but remains experimental at this time. Other surgical therapies, such as ablative approaches or deep brain stimulation, have not been extensively tried.

The pathologic mechanisms underlying HD are not yet completely understood; effective therapeutics depends on a more clear elucidation of pathogenic pathways. The Huntington Study Group, a multicenter academic consortium of HD clinicians, has conducted numerous clinical trials for therapeutic compounds, and many are currently under way. Further therapies will no doubt arise as the molecular pathogenesis HD is further delineated.

*See also:* Axonal Transport and Huntington's Disease; Huntington's Disease: Neurodegeneration; Protein Folding and the Role of Chaperone Proteins in Neurodegenerative Disease; Transgenic Models of Neurodegenerative Disease.

## Further Reading

Bates G, Harper P, and Jones L (eds.) (2002) *Huntington's Disease.* New York: Oxford University Press.

Cha J-HJ (2000) Transcriptional dysregulation in Huntington's disease. *Trends in Neurosciences* 23: 387–392.

Everett CM and Wood NW (2004) Trinucleotide repeats and neurodegenerative disease. *Brain* 127: 2385–2405.

Ferrante RJ, Kowall NW, Beal MF, et al. (1985) Selective sparing of a class of striatal neurons in Huntington's disease. *Science* 230: 561–564.

Friedlander RM (2003) Apoptosis and caspases in neurodegenerative diseases. *New England Journal of Medicine* 348: 1365–1375.

Gusella JF, Wexler NS, Conneally PM, et al. (1983) A polymorphic DNA marker genetically linked to Huntington's disease. *Nature* 306: 234–238.

Harjes P and Wanker EE (2003) The hunt for huntingtin function: Interaction partners tell many different stories. *Trends in Biochemical Sciences* 28: 425–433.

Huntington's Disease Collaborative Research Group (1993) A novel gene containing a trinucleotide repeat that is unstable in Huntington's disease chromosomes. *Cell* 72: 971–983.

Leegwater-Kim J and Cha J-HJ (2004) The paradigm of Huntington's disease: Therapeutic opportunities in neurodegeneration. *NeuroRx* 1: 128–138.

Li SH and Li XJ (2004) Huntingtin–protein interactions and the pathogenesis of Huntington's disease. *Trends in Genetics* 20: 146–154.

Mangiarini L, Sathasivam K, Seller M, et al. (1996) Exon 1 of the *hd* gene with an expanded CAG repeat is sufficient to cause a progressive neurological phenotype in transgenic mice. *Cell* 87: 493–506.

Vonsattel JP and DiFiglia M (1998) Huntington disease. *Journal of Neuropathology & Experimental Neurology* 57: 369–384.

Vonsattel JP, Myers RH, Stevens TJ, et al. (1985) Neuropathological classification of Huntington's disease. *Journal of Neuropathology & Experimental Neurology* 44: 559–577.

## Relevant Websites

http://www.hdfoundation.org – Hereditary Disease Foundation.
http://www.hdsa.org – Huntington's Disease Society of America.
http://www.huntington-study-group.org – Huntington Study Group.
http://www.wemove.org – Worldwide Education and Awareness for Movement Disorders.

# Huntington's Disease: Neurodegeneration

**M F Chesselet, M A Hickey, and C Zhu**, The David Geffen School of Medicine at UCLA, Los Angeles, CA, USA
**M S Levine**, Semel Neuroscience Institute, The David Geffen School of Medicine at UCLA, Los Angeles, CA, USA



## Huntington's Disease

Huntington's disease (HD) is characterized by progressive motor dysfunction, cognitive decline, and psychiatric disturbances. The mutation causing the disease is an unstable expansion in the number of CAG repeats in exon 1 of the gene that encodes huntingtin (**Figure 1**). Whereas the normal number of CAG repeats varies from 6 to 35, repeat lengths of 36–41 lead to an increased risk for the disease, and greater lengths cause HD with full penetrance. Although there is a direct correlation between repeat length and age of onset, with juvenile forms being usually caused by repeat lengths of over 60, there is considerable variation in age at onset for the same number of CAG repeats in the mid-range of the expansions.

## Neurodegeneration in HD

Neuronal degeneration is the pathological signature of HD. Progressive striatal and cortical atrophy can be detected by brain imaging *in vivo* and is striking at autopsy (**Figure 2**). The most clearly affected neuronal population is the medium-sized spiny GABAergic efferent neurons of the striatum, which comprise 95% of neurons in this region. Neuronal loss in the striatum proceeds in a medial to lateral gradient that led to the classification of the disease progression into four stages or grades. In contrast to efferent neurons, striatal interneurons are relatively spared, although whether this extends to all classes of interneurons remains a matter of debate. Although pathology is mostly restricted to the striatum and cortex at early stages, other brain areas such as the globus pallidus, hippocampus, thalamus, and cerebellum may be affected by grades 3 and 4. The involvement of the hippocampus and the cerebellum is particularly noticeable in cases of juvenile HD. Thus, an important point about neurodegeneration in HD is that although striatal efferent neurons are clearly more vulnerable than other neuronal types, this 'selective vulnerability' is not absolute, and greater CAG repeat burden seems to lead to the involvement of a wider range of neurons.

Ultrastructural studies in postmortem human brains, although notoriously difficult, revealed a loss of dendrites, dendritic spines, and synaptic connections in surviving neurons, with indentations of the nuclear membrane, disorganization of the nucleolus, and depletion of the ribosomes of the rough endoplasmic reticulum. Increased spine density and curved, elongated dendrites have also been observed, suggesting proliferative changes in spared neurons. These observations indicate that even surviving neurons are affected by the disease process, and these morphological changes may be related to neuronal dysfunction which occurs in humans long before evidence for neuronal degeneration. Similarly, decreases in neurotransmitter peptides in the axons of striatal efferent neurons have been detected in the few cases of presymptomatic carriers examined at autopsy. Finally, recent imaging studies revealed that alterations in white matter tracts could precede neuronal loss, as assessed by brain atrophy. It is striking, however, that surviving neurons even in advanced grades have high levels of the corresponding mRNA and that degenerating neurons appear to be intermingled with spared ones.

Neuropathological studies in humans have not fully resolved the mechanisms of neuronal cell death occurring. It has been suggested that apoptosis is the primary mechanism of cell death in HD. Evidence, however, remains inconclusive. Indeed, although there is no doubt that DNA nick labeling techniques revealed DNA damage in HD tissue, examination of these DNA fragments by gel electrophoresis did not reveal the characteristic apoptosis profile. This is not incompatible with the histological data as TUNEL or other DNA damage stains are by no means specific for apoptotic cell death. Nevertheless, a number of factors may hinder the detection of apoptotic cell death in postmortem brain, such as postmortem delay, the transient nature of apoptosis, or a low quantity of DNA fragments being present in tissue. Morphological changes are the most reliable method of detecting apoptosis, but they are difficult to assess in postmortem human tissue. Interestingly, caspase and calpain family members are increased in HD tissue. However, the extensive and early astrocytosis observed in HD brains argues against an apoptotic mechanism.

In conclusion, neurodegeneration is a critical aspect of HD but its mechanism remains uncertain. It affects primarily but not exclusively the striatum, and proceeds according to a well-characterized gradient

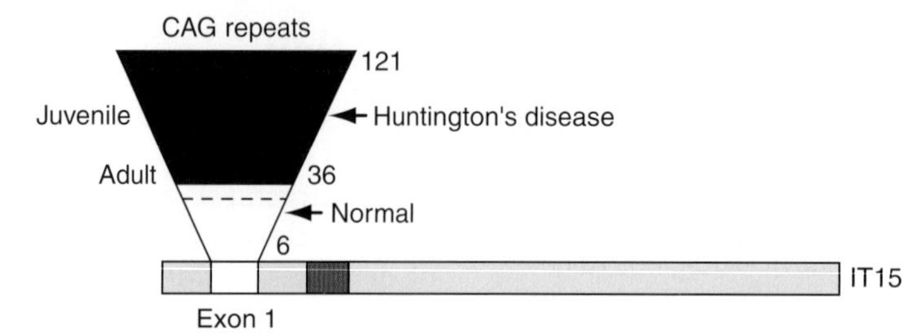


**Figure 1** Cartoon of the gene encoding human huntingtin. In the normal population, the CAG repeats in exon 1 range from 6 to 35. CAG repeat lengths of 42 and above cause HD while repeat lengths between 36 and 41 increase the risk of the disease.

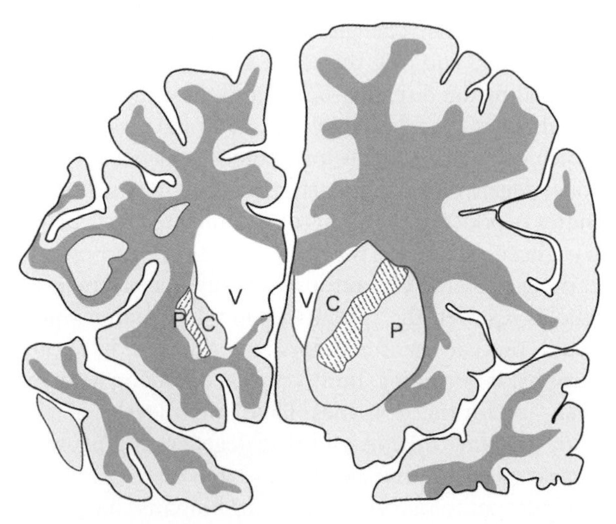


**Figure 2** Representation of degeneration in a human HD brain (left) compared with a normal brain (right). Representations are of coronal sections at the level of frontal lobe. Note the profound atrophy of the caudate nucleus, putamen and the cortex, and the ventricular enlargement in the HD brain. C, caudate nucleus; P, putamen; V, lateral ventricle. Adapted from Hickey MA and Chesselet MF (2003) The use of transgenic and knock-in mice to study Huntington's disease. *Cytogenetics and Genome Research* 100(1–4): 276–286, with permission from S. Karger AG, Basel.

in this region, where it mostly affects efferent neurons. Neuronal loss can be followed during the patient's life by assessing regional atrophy with structural brain imaging techniques but a period of neuronal dysfunction is likely to precede cell death, which appears to be stochastic even in the most affected regions. These observations point to the importance of understanding the early mechanisms of neuronal dysfunction caused by the HD mutation, rather than cell death, in order to identify most effective therapeutic targets.

## Neurodegeneration in *In Vitro* Models of HD

Since the identification of the HD-causing mutation in 1993, much effort has been devoted to reproducing HD in cellular and animal models. Expression of mutated huntingtin in yeast and mammalian cells leads to neurodegeneration under certain experimental conditions. In general, neurodegeneration is promoted by smaller fragments of mutated huntingtin compared to the full-length protein, suggesting that protein cleavage is an essential step in the pathophysiology of the disease. *In vitro* studies have revealed an association between cell death and the interaction of mutated huntingtin with transcriptional coactivators and transcriptional dysregulation. They also challenged the idea that neurodegeneration is directly linked to the formation of insoluble aggregates of mutated huntingtin. Indeed, aggregates can be neuroprotective despite frequent correlative evidence that neurodegeneration parallels the formation of aggregates. Several *in vitro* studies have suggested that mutated huntingtin can induce cell death by triggering apoptotic mechanisms, with evidence of DNA fragmentation, caspase activity, and cytochrome *c* release. However, the relevance of these mechanisms to the disease situation remains unclear, due to the usually high levels of mutant protein in cell lines.

## Neurodegeneration in *In Vivo* Models of HD

The first example of neurodegeneration in an animal model of HD was obtained in *Drosophila*. Expression of mutated huntingtin (exons 1–3 and part of 4) under the eye-specific expression construct pGMR led to progressive degeneration of ommatidia with reduced age of onset of degeneration and increased nuclear accumulation with increased polyglutamine length. Similarly, neurodegeneration was observed in *Caenorhabditis elegans*. These models were instrumental in demonstrating the ability of the chaperones to prevent HD-induced neurodegeneration. They also played a key role in furthering our understanding of mutation-induced mechanisms of neurodegeneration (see below).

Paradoxically, neurodegeneration has proved to be challenging to demonstrate in genetic mammalian models of HD. Prior to the identification of the HD mutation, neurodegeneration was created in the rodent or monkey striatum by local or peripheral injections of the neurotoxins quinolinic acid or 3-nitropropionic acid. The relevance of these models to HD was inferred from the similarity of their effects to the pattern of neuronal death observed in HD, with preferential loss of efferent neurons and relative sparing of interneurons. These models pointed to the potential role of excitotoxicity for inducing neurodegeneration in HD, an idea that remains timely (see below). Since the discovery of the HD-causing mutation, genetic mammalian models are a more direct route to understand HD pathophysiology, and many different mouse and even rat models have been created.

The first published transgenic mouse model of HD, the R6 line generated by the laboratory of Gillian Bates, remains the best characterized. The R6/2 line expresses exon 1 of the HD gene with approximately 145 CAG repeats. These mice have severe motor anomalies, cognitive decline, weight loss, and seizures; they die around 4 months of age. Levels of mRNA and protein for neuronal molecules present in striatal medium-sized spiny neurons are decreased in a 4-week-old mice, but evidence of neurodegeneration is only obtained late in disease development (**Figure 3**). This observation indicates that, similar to humans, neurological symptoms precede neurodegeneration in mice, indicating a protracted period of neuronal dysfunction before cell death occurs. Similar observations were made in a number of other mouse models of HD, including both transgenics, and mice with extended CAG repeats inserted in their own HD gene (knock-in mice). Most of these mice show brain atrophy at least at late stages of the disease, and some present evidence of 'dark cells' that are interpreted as degenerating neurons. Few models, however, show a decrease in the number of neurons in the striatum when examined with stereological methods.

Contrary to expectation, work in mice did not unequivocally clarify the mechanism of cell death induced by the HD mutation. In R6/2 mice, 'dark' neurons were found in the striatum and had dense nuclear condensation but no evidence of apoptotic bodies. The absence of DNA fragmentation with either TUNEL staining or *in situ* nick translation does not support an apoptotic mechanism, but some authors did report DNA fragmentation at the very late stages of disease. Also arguing against an apoptotic mechanism, the levels of proteins involved in apoptotic pathways were not changed in R6/2 mice, although increases in these proteins were reported in another R6 line, R6/1, which also showed 'dark' neurons. Increases in apoptotic proteins, TUNEL staining, and apoptotic profiles were reported in another full-length transgenic line, which is no longer in existence. It should be noted that these differences among transgenic models parallel the severity of the phenotype, with the mice that exhibit a more prolonged life and milder phenotype showing more evidence for apoptosis than mice with a more severe course of disease. A recent study proposed that mutated huntingtin causes neurodegeneration through a novel mechanism of cell death related to reduction in Pol II-dependent transcription. The cell death was prolonged, morphologically unlike apoptosis, necrosis or autophagy, and modulated by isoforms of yes-associated protein (YAP), and these isoforms reduced degeneration in an HD *Drosophila* model.

Although most knock-in and transgenic models show little cell loss, viral vectors encoding long

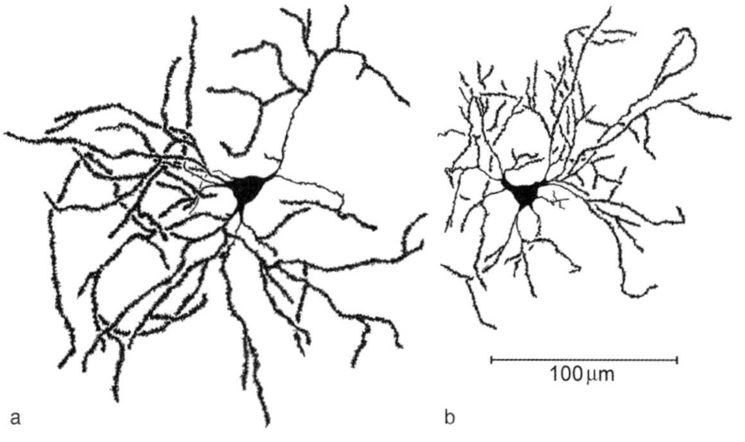


**Figure 3** Camera lucida drawings of Golgi impregnated medium spiny neurons in wild-type (a) and R6/2 transgenic HD (b) mice at 12 weeks of age. Note the loss of spines and reduction in the diameter in the dendritic field in the R6/2 transgenic neuron. Adapted from Klapstein GJ, Fisher RS, Zanjani H, et al. (2001) Electrophysiological and morphological changes in striatal spiny neurons in R6/2 Huntington's disease transgenic mice. *Journal of Neurophysiology* 86(6): 2667–2677, with permission from the American Physiological Society.

CAG repeats induce apoptotic cell death *in vivo* when introduced into the striatum. This may be due to the very high expression levels of the abnormal polyglutamine tract in this model.

## Pathophysiological Mechanisms of Neurodegeneration in HD

As indicated earlier, the precise mechanism of cell death in HD remains unclear. However, much information has accumulated in the last 13 years on the various ways mutated huntingtin alters cellular function. Evidence that the lack of huntingtin is embryonic lethal and that mutated huntingtin can rescue this phenotype suggests that the mutation causes an abnormal gain of function. What is not excluded, however, is that the detrimental effects of this gain of function are compounded by a loss of function of the normal protein, in particular its role in stimulating the production of brain-derived neurotrophic factor (BDNF), an essential growth factor for striatal medium-sized spiny neurons.

A compelling abnormal property of mutated huntingtin is its ability to form insoluble aggregates both in the cytoplasm and in the nucleus. These aggregates are usually detected with antibodies against the N-terminal portion of the protein (which contains the polyglutamine expansion), but they can also contain full-length huntingtin. Aggregates, which range in size from microaggregates to large nuclear inclusions, were described in human brain and later found to contain huntingtin. After their detection in the R6/2 line, they were described in all animal models of HD (**Figure 4**), as well as in many cellular models. Normal huntingtin does not form aggregates, although it can be recruited into aggregates of mutated protein. Aggregates are present in all diseases of CAG repeat expansion and can be formed by long polyglutamine repeats, either isolated or inserted in a nonrelevant gene.

Despite their obvious association with pathology, aggregates may not be pathological themselves. *In vitro* studies have shown that they do not always cause cell death or they can even be neuroprotective. A lack of causal relationship between aggregates and neuronal dysfunction has been demonstrated in several mouse models where a careful analysis has shown that behavioral deficits precede detectable aggregates, and in a recent description, mice expressing exon 1 and 2 of mutant huntingtin show widespread aggregate development, but no atrophy or motor impairments. Even in postmortem human brains, aggregate location does not correlate with neurodegeneration. Finally, several compounds have beneficial effects in

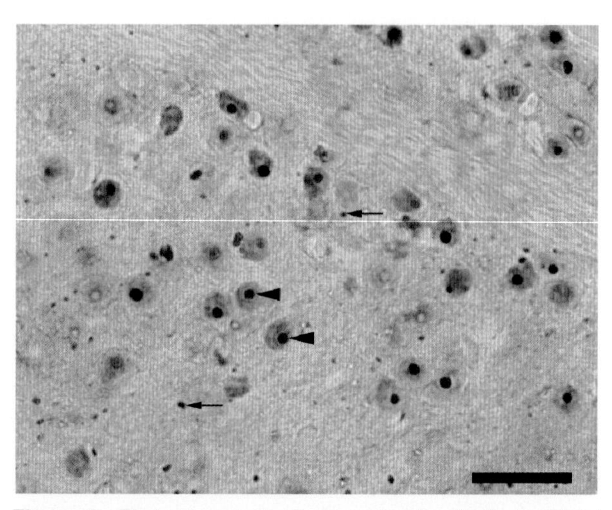

**Figure 4** Photomicrograph of aggregated huntingtin in striatal neurons in knock-in (CAG 140) mouse brain at 13 months of age (Zhu and Chesselet, unpublished). Mutant huntingtin forms nuclear inclusions (arrowhead) and neuropil aggregates (arrows) in striatal neurons. Immunohistochemistry with polyclonal EM48 (Gift from Dr. X.-J. Li, Emory University). Scale bar = 20 μm.

models of HD without causing a parallel decrease in aggregates.

Although huntingtin aggregates do not appear to directly cause neurodegeneration, they could contribute to pathology in different ways. Aggregated huntingtin may interfere with axonal transport, block proteasomal degradation of other proteins, and sequester proteins that are important for cell functioning. Indeed, transcriptional dysregulation appears to be a critical pathophysiological mechanism in HD and histone deacetylase inhibitors, which reverse this deficit, have beneficial effects in cellular and animal models. Interactions with transcription coactivators, however, may depend on huntingtin misfolding and/or truncation and do not always parallel aggregate formation.

The form of huntingtin that initiates the aggregation process (or is pathological in itself) is not fully identified. As indicated earlier, truncated huntingtin is more toxic than the full-length protein, suggesting that protein cleavage is important. Critical proteases that have come to light include several caspases and calpains. Indeed, recent data show that preventing cleavage of mutant huntingtin at the caspase 6 cleavage site (as opposed to the caspase 3 site) protects transgenic mice from neurodegeneration and motor impairments.

Whatever the pathophysiological mechanisms leading to neurodegeneration in HD, a main question that remains is the relative selective vulnerability of certain cell types despite the widespread expression of huntingtin. Recent studies in mice with restricted expression of the mutated protein suggest that the process may be initiated not in the striatum but in cortical interneurons, which in turn affect corticostriatal projections and ultimately result in the death of striatal neurons. A role for glutamatergic cortical inputs in HD was suspected since the experiments with quinolinic acid mentioned earlier. For example, cellular pathology caused by mutant huntingtin in presynaptic corticostriatal terminals might result in an increased release of glutamate. Alternatively, impaired clearance of glutamate from the synaptic cleft might increase glutamatergic neurotransmission. In both cases, striatal excitotoxicity could occur. Intracerebral microdialysis has shown that depolarizing concentrations of potassium chloride increase the extracellular concentrations of glutamate substantially more in HD mutant than in control mice, and, in particular, the glial glutamate transporter is downregulated before any evidence of neurodegeneration. This indicates that there is an impairment in glutamate transport and glutamate–glutamine cycling, and suggests that a defect in astrocytic glutamate uptake could contribute to the phenotype.

One of the first indications of electrophysiological changes in the corticostriatal pathway was the observation that the stimulus intensity necessary to evoke an excitatory postsynaptic potential (EPSP) in medium-sized spiny neurons was significantly increased in symptomatic HD transgenic mice. Another observation was that synaptic responses in transgenic mice displayed slower rise and incomplete decay of the EPSP. This phenomenon was hypothesized to indicate a larger contribution of a slower kinetic current, such as the one mediated by activation of *N*-methyl-D-aspartate (NMDA) receptors. This idea is supported by the occurrence of enhanced NMDA receptor-mediated responses. In addition to alterations in evoked synaptic responses, both transient and progressive changes in spontaneous synaptic currents occur. Spontaneous excitatory postsynaptic currents show a progressive reduction in frequency that becomes more evident as the HD phenotype becomes more severe. This effect was interpreted as a progressive disconnection between the striatum and its cortical inputs. There is also a transient expression of complex, large amplitude synaptic currents that coincide temporally with the onset of the behavioral phenotype. These large currents probably reflect dysregulation of glutamate release and/or an increase in cortical synchronization.

Changes in the corticostriatal pathway and glutamate receptors alone are probably insufficient to produce neurodegeneration. However, numerous membrane properties of striatal medium-sized spiny neurons are also altered. One of the earliest and most consistent alterations in the basic membrane properties is an increase in input resistance. This increase probably reflects loss of conductive membrane channels due to morphological changes such as reduced membrane area, possibly as a consequence of the loss of spines. Consistent with this observation, cell capacitance is significantly reduced in symptomatic animals. The increase in membrane input resistance could also be due to alterations in the number and/or properties of $K^+$ channels. As a consequence, many medium-sized spiny neurons have a depolarized resting membrane potential and are less able to repolarize. These alterations are particularly relevant because membrane depolarization can remove the $Mg^{2+}$ block of the NMDA receptor, causing neurons to become more depolarized when glutamatergic inputs are activated. These neurons will stay in a depolarized state for longer periods of time, thus predisposing them to excitotoxicity. Other voltage-gated conductances may also be affected as alterations in firing patterns occur. For example, there is a reduction in high voltage-activated $Ca^{2+}$ conductances in symptomatic HD mice but an increase in these currents in presymptomatic mice. Taken together, the complex set of alterations in voltage-gated conductances of medium-sized spiny neurons will produce dysfunctional cells that have altered responses to inputs. Early increases in $Ca^{2+}$ conductances will predispose cells to become more easily depolarized while subsequent decreases might be protective. Early increases in input resistance will decrease electrotonic decay of conductances allowing peripheral inputs to depolarize over longer distances while specific changes in $K^+$ channel function can also predispose neurons to remain depolarized and more excitable. In particular, decreased inward rectification could amplify excitatory inputs. These alterations emphasize that striatal medium-sized spiny neurons are dysfunctional and may be primed to be affected by abnormal cortical inputs. Thus, selective vulnerability in HD may result from a combination of changes in intrinsic properties of neurons and extrinsic inputs.

## Conclusion

Thirteen years after the identification of the mutation that causes HD, much has been learned about the disease but the exact mechanisms leading to neurodegeneration remain unclear. Although cell loss remains a hallmark of advanced HD, extensive evidence that neuronal dysfunction precedes cell death has shifted emphasis from preventing neurodegeneration to maintaining neuronal health. Even apoptotic mechanisms may contribute to pathology by cleaving the mutant protein into more toxic forms rather than leading directly to cell death. Current data point to a number of avenues for therapeutic development that are being actively pursued. Some successes have been obtained in animal models but the challenge remains to develop compounds that are safe enough for long-term use and target the most critical mechanisms rather than peripheral or secondary effects of the mutation.

*See also:* Animal Models of Huntington's Disease; Apoptosis in Neurodegenerative Disease; Axonal Transport and Huntington's Disease; Axonal Transport and Neurodegenerative Diseases; Oxidative Damage in Neurodegeneration and Injury; Transgenic Models of Neurodegenerative Disease; Triplicate Repeats: Huntington's disease.

## Further Reading

Arrasate M, Mitra S, Schweitzer ES, Segal MR, and Finkbeiner S (2004) Inclusion body formation reduces levels of mutant huntingtin and the risk of neuronal death. *Nature* 431(7010): 805–810.

Brouillet E, Conde F, Beal MF, and Hantraye P (1999) Replicating Huntington's disease phenotype in experimental animals. *Progress in Neurobiology* 59(5): 427–468.

Cattaneo E, Zuccato C, and Tartari M (2005) Normal huntingtin function: An alternative approach to Huntington's disease. *Nature Reviews Neuroscience* 6(12): 919–930.

Cha JH (2000) Transcriptional dysregulation in Huntington's disease. *Trends in Neurosciences* 23(9): 387–392.

DiFiglia M (1990) Excitotoxic injury of the neostriatum: A model for Huntington's disease. *Trends in Neurosciences* 13(7): 286–289.

Harper PS (1996) *Huntington's Disease.* London: Saunders.

Hickey MA and Chesselet MF (2003) Apoptosis in Huntington's disease. *Progress in Neuropsychopharmacology and Biological Psychiatry* 27(2): 255–265.

Hickey MA and Chesselet MF (2003) The use of transgenic and knock-in mice to study Huntington's disease. *Cytogenetics and Genome Research* 100(1–4): 276–286.

Klapstein GJ, Fisher RS, Zanjani H, et al. (2001) Electrophysiological and morphological changes in striatal spiny neurons in R6/2 Huntington's disease transgenic mice. *Journal of Neurophysiology* 86(6): 2667–2677.

Levine MS, Cepeda C, Hickey MA, Fleming SM, and Chesselet MF (2004) Genetic mouse models of Huntington's and Parkinson's diseases: Illuminating but imperfect. *Trends in Neurosciences* 27(11): 691–697.

Mangiarini L, Sathasivam K, Seller M, et al. (1996) Exon 1 of the HD gene with an expanded CAG repeat is sufficient to cause a progressive neurological phenotype in transgenic mice. *Cell* 87 (3): 493–506.

Menalled LB (2005) Knock-in mouse models of Huntington's disease. *NeuroRx* 2(3): 465–470.

Menalled LB and Chesselet MF (2002) Mouse models of Huntington's disease. *Trends in Pharmacological Sciences* 23(1): 32–39.

Saudou F, Finkbeiner S, Devys D, and Greenberg ME (1998) Huntingtin acts in the nucleus to induce apoptosis but death does not correlate with the formation of intranuclear inclusions. *Cell* 95(1): 55–66.

Vonsattel JP and DiFiglia M (1998) Huntington disease. *Journal of Neuropathology and Experimental Neurology* 57(5): 369–384.

# Axonal Transport and Huntington's Disease

**F Saudou and S Humbert**, Institut Curie and CNRS UMR 146, Orsay, France



## Introduction

Huntington's disease (HD) is a dominantly inherited neurodegenerative disease caused by abnormal polyglutamine (polyQ) expansion in a specific protein. HD is characterized by uncontrolled movements (chorea), personality changes, and dementia. Patients typically die 10–20 years after the appearance of the first clinical symptoms. In HD, the specific dysfunction and death of neurons in the striatum and cerebral cortex has been reported. The mutated protein is huntingtin, a large ubiquitous protein, the function of which is not fully understood. However, there is evidence that huntingtin is an antiapoptotic protein. Indeed, it is necessary for development since knockout mice die soon after gastrulation and conditional knockout mice, in which huntingtin is selectively turned off in brain and testis during adulthood, show a progressive degenerative neuronal phenotype and sterility. Moreover, overexpression of normal huntingtin protects striatal cells from a variety of apoptotic stimuli. Huntingtin becomes toxic when it contains an abnormal polyQ expansion. PolyQ-huntingtin induces the formation of neuritic and intranuclear inclusions, neuronal dysfunction, and, finally, neuronal death. The precise mechanisms underlying these phenomena are not well understood, and it is not known how increased neuronal death in the brain relates to huntingtin function and dysfunction.

Evidence suggests that huntingtin is involved in intracellular transport and that axonal transport is impaired in HD. Herein, we review studies demonstrating that huntingtin belongs to a protein complex that controls intracellular transport and that the alteration of this complex may participate in the pathogenesis of HD.

## Intracellular Transport and Neurons

Neurons are highly polarized cells composed of a cell body and an extensive network of cell processes: dendrites, and a single axon. The axonal extension can reach up to 1 m in length. Most protein synthesis is restricted to the cell body. Therefore, active transport is required to supply the axon with newly synthesized materials. Conversely, external signals have to be transmitted to the cell body from the cell's extremities, where the signaling cascade is activated. Here also, active transport of receptor–ligand complexes is required for a quick and appropriate response. Due to the length of their processes, transport efficiency is of particular importance in neurons.

To allow efficient communication between cell bodies and axon termini, molecular motor proteins continuously shuttle vesicles and organelles along the microtubules and actin filaments that make up the cellular cytoskeleton. This transport process mostly involves microtubules, which are polarized structures with 'plus' (cell extremity) and 'minus' (cell body) ends. Molecular motors are considered to be unidirectional: Dynein complexes are connected to retrograde transport (from plus to minus end), whereas kinesins are connected to anterograde transport (from cell body to plasma membrane).

These two classes of motor complexes are required for efficient transport in neurons. Disruption of these complexes has dramatic consequences on neurons and leads to severe diseases.

## Huntingtin and Intracellular Dynamics

There is growing evidence to suggest that huntingtin plays a role in intracellular dynamics. Cytoplasmic huntingtin co-localizes with microtubules and interacts with $\beta$-tubulin. Huntingtin is also found in neurites co-localizing with vesicles. In addition to this specific distribution, huntingtin interacts with many proteins that are implicated in intracellular trafficking. The best-described example is the case of huntingtin-associated protein 1 (HAP1), which is discussed later. Huntingtin also interacts with huntingtin-interacting protein 1 (HIP1), which binds to the clathrin light chain and regulates clathrin assembly, linking huntingtin to the control of endocytosis. A role of huntingtin in the control of endocytosis is further supported by the observation that huntingtin bound to the HAP40 protein acts as an effector of the small guanosine triphosphatase Rab5. In the pathological situation, a reduction of endosomal motility and endocytic activity in HD fibroblasts and mutant cells is observed. Finally, huntingtin associates with PACSIN1, a protein that is part of the endocytic machinery, which could act as a negative regulator.

In addition to regulating endocytosis at the plasma membrane, huntingtin may also regulate secretion and trafficking of proteins or organelles from the Golgi region. Indeed, HIP14, another huntingtin interacting protein that contains an ankyrin domain, is found at the Golgi and at vesicles. This protein is a palmitoyl transferase involved in the sorting of proteins from the Golgi region. Also, huntingtin binds optineurin, a protein associated with Rab8 and myosin VI. This complex is found at the Golgi and in vesicles. The depletion of optineurin leads to the disruption of the Golgi ribbon structure and to the reduction in the transport of VSV-G from the Golgi complex to the plasma membrane. Finally, two other proteins that are dysregulated in HD participate in the sorting of organelles from the Golgi complex. The DNAJ-containing HSJ1b protein, possibly by regulating clathrin uncoating, promotes the processing of vesicles that contain the brain-derived neurotrophic factor (BDNF) from the Golgi to the cytoplasm. HSJ1b protein colocalizes with the transglutaminase 2 (TGase 2) enzyme that has a negative role in BDNF sorting from the Golgi. Interestingly, TGase 2 is inhibited by cystamine and by its reduced form, cysteamine, resulting in an increase in the sorting of BDNF-containing vesicles from the Golgi and in the release of this trophic factor from the cell. Treating HD mice with cysteamine results in an increase in BDNF release in the brain and in the reduction of the loss of striatal neurons in HD mutant mice. That a compound such as cysteamine, which is a Food and Drug Administration-approved drug, is neuroprotective by enhancing the intracellular processing of vesicles suggests that strategies aiming at restoring the defects in intracellular trafficking of organelles are of therapeutic interest.

## Huntingtin Associates with the Molecular Motor Complex

The identification of HAP1 and its characterization with respect to huntingtin led to the hypothesis that both proteins may have an important function in the axonal transport machinery. Indeed, huntingtin and HAP1 are transported both anterogradely and retrogradely in rat sciatic nerves. Both proteins are associated with synaptic vesicles and localize with microtubules in neurites. The discovery in 1997 that HAP1 associates with the p150$^{Glued}$ subunit of dynactin further supported this possibility. Since then, a growing body of evidence has suggested that huntingtin has a major function in the control of axonal transport and that this function requires the HAP1 protein. Indeed, HAP1 binds not only to the p150$^{Glued}$ subunit of dynactin but also to the kinesin light chain 2, a subunit of the kinesin-1 complex. In agreement, depletion of HAP1 reduces kinesin-dependent transport of amyloid precursor protein vesicles. In addition to the indirect association of huntingtin to dynactin through HAP1, huntingtin also interacts with the dynein light chain of the dynein complex. Finally, huntingtin and HAP1 comigrate in the same fractions after sucrose gradient fractionation, further supporting the idea that huntingtin belongs to the molecular motor complex dynein/dynactin/kinesin-1.

## Huntingtin Stimulates Axonal Transport

In addition to the previously discussed biochemical evidence, videomicroscopy approaches in cells and in *Drosophila* have further established a functional role of both HAP1 and huntingtin in axonal transport. In *Drosophila*, reduction in huntingtin protein level by the RNAi approach results in axonal transport defects. Furthermore, huntingtin directly promotes intracellular transport of vesicles along microtubules in neuroblastoma and striatal cells. Indeed, using BDNF as a marker of intracellular trafficking in cells, Gauthier and collaborators showed that expression of huntingtin enhances the velocity of BDNF-containing vesicles while reducing the percentage of time spent pausing. In support of a positive role of huntingtin in stimulating axonal transport of vesicles that contain BDNF, downregulation of huntingtin by RNAi approaches resulted in a decrease in the velocity of moving vesicles and an increase in the percentage of time spent pausing. BDNF was chosen because it is an important factor in HD. It is produced in the cortex and is transported to the striatum, the major site of degeneration in HD, where it supports neuronal differentiation and survival. It inhibits polyQ–huntingtin-induced neuronal death and its level is abnormally low in HD patients. BDNF is synthesized from the large precursor protein pre-pro-BDNF that is proteolytically processed and moves through the Golgi apparatus to the trans-Golgi network, where it is packaged into vesicles. BDNF-containing vesicles are then transported along microtubules (MTs) to the plasma membrane and subsequently released through the regulated secretory pathway. BDNF-containing vesicles are immunopositive for the classical markers of secretion and their activity-dependent release requires an intact MT network because it is blocked by nocodazole, an MT depolymerizing agent.

The huntingtin-dependent transport of BDNF vesicles is bidirectional because huntingtin stimulates both anterograde and retrograde transport in axons. It requires the HAP1 protein. Indeed, short N-terminal fragments of huntingtin that do not contain the HAP1 interacting region are unable to stimulate intracellular transport. Also, BDNF transport is reduced after downregulation of HAP1 protein and, in these conditions, huntingtin is unable to enhance BDNF trafficking. The HAP1-dependent transport is not restricted to BDNF vesicles since the movement of amyloid precursor protein vesicles is also reduced by decreasing HAP1 levels. Since it is known that the movement of amyloid precursor protein vesicles is kinesin dependent, one can infer an important role of huntingtin and HAP1 in both anterograde and retrograde transport.

Although these studies revealed a role for huntingtin in axonal transport, the extend to which axonal transport depends on huntingtin remains unknown. Huntingtin stimulates the dynamic of BDNF-containing vesicles; however, whether other types of vesicles are regulated by huntingtin remains to be established. The observations that transport of amyloid precursor protein vesicles depends on HAP1 and that downregulation of huntingtin alters the general axonal transport in *Drosophila* strongly suggest that huntingtin regulates the transport of other small vesicles.

Milton, a *Drosophila* ortholog of HAP1, participates in the axonal transport of mitochondria, thus raising the possibility that huntingtin and HAP1 could also regulate the transport of mitochondria. However, the velocity of mitochondria is not regulated by huntingtin, and whereas HAP1 overexpression leads to the redistribution of BDNF vesicles in cells, it has no effect on mitochondria. Conversely, Milton, although known to redistribute mitochondria in cells, has no effect on BDNF vesicles. Therefore, HAP1 and Milton show specificity in the type of cargoes they are transporting.

## HD Pathological Situation and Axonal Transport Defects

Several studies indicate an alteration of axonal transport in the polyQ situation. However, depending on the stage of the disease, different mechanisms may be involved.

In striatal cells and in neurons, whereas normal full-length huntingtin stimulates BDNF transport, full-length huntingtin with an abnormally expanded polyQ expansion has lost this stimulatory ability. Also, a reduction in transport of BDNF is observed in striatal cells derived from knockin mice in which a CAG expansion has been inserted in the endogenous mouse huntingtin gene. These mutant striatal cells express full-length huntingtin at endogenous levels. Similarly, polyQ-containing polypeptides encompassing amino acids 1–548 of huntingtin inhibit fast axonal transport in isolated axoplams without the formation of detectable morphological aggregates. Thus, polyQ expansion of huntingtin results in a loss of huntingtin function in transport. This phenotype is mediated by HAP1, which forms a protein complex by binding to the $p150^{Glued}$ subunit of dynactin and simultaneously binding to a domain located in the N-terminal region of huntingtin. The formation of this complex subsequently stimulates microtubule-based transport. In contrast, when huntingtin contains an abnormal polyQ expansion, it interacts more strongly with HAP1 and $p150^{Glued}$, leading to the detachment of the molecular motors from the microtubules and thus to less efficient transport of BDNF vesicles.

The presence of an abnormal polyQ expansion in huntingtin leads to a reduced anterograde and retrograde transport of BDNF-containing vesicles. In cortical neurons that deliver BDNF to striatal neurons, the reduced transport of BDNF in the pathological situation has a direct consequence on the ability of these cells to release BDNF, rendering striatal neurons particularly vulnerable. In agreement, cortical ablation of BDNF in mice induces striatal dendrite deficits followed by neuronal loss. Once released, BDNF binds to its receptor TrkB; the BDNF–TrkB complex is endocytosed in striatal neurons and transported retrogradely to the nucleus, where it activates downstream signaling pathways. The consequences of axonal transport defects in these neurons remain to be established.

A critical step in HD pathogenesis is the cleavage of polyQ–huntingtin into N-terminal fragments containing the polyQ expansion. These fragments translocate into the nucleus and induce the formation of aggregates. In the nucleus, polyQ–huntingtin fragments cause neuronal death by a gain-of-function mechanism, leading to the dysregulation of transcriptional activity. In addition, N-terminal huntingtin fragments and their aggregates accumulate in axonal processes and terminals. What are the consequences of these aggregates on axonal transport? In *Drosophila*, N-terminal huntingtin polypeptide fragments containing the polyQ expansion accumulate in axonal inclusions and cause axonal trafficking defects. This subsequently induces neuronal death and organismal death. Furthermore, mutant huntingtin aggregates alter vesicular and mitochondrial transport in mammalian neurons. These aggregation-induced defects in transport may be due to a physical blockage of vesicles but may also involve the sequestration by mutant huntingtin aggregates of motor proteins (particularly $p150^{Glued}$ and kinesin heavy chain) from other cargoes and pathways. Interestingly, this aggregation-induced alteration of general transport appears to be independent of the protein context since it has also been reported to occur in other polyQ disorders.

In summary, in early stages of the disease, the role of huntingtin in transport is lost following the alteration of the huntingtin (soluble form)/HAP1 interaction. In later stages, polyQ–huntingtin forms neuritic aggregates that contribute to a trafficking defect by a gain-of-function mechanism.

## Disruption of Axonal Transport and Neurodegenerative Disorders

Transport defects have also been described in other polyQ neurodegenerative disorders. The polyQ-containing androgen receptor, the mutation of which is responsible for spinobulbar muscular atrophy, inhibits fast axonal transport. Non-polyQ neurodegenerative diseases have also been linked to transport failure. In Alzheimer's disease (AD), mutation of the amyloid precursor protein, responsible for familial forms of AD, disrupts axonal transport. Similarly, overexpression of tau, a neuronal microtubule-associated protein that accumulates in AD, inhibits organelle trafficking. Several different neurodegenerative disorders are caused by mutation within motor proteins. For instance, Charcot–Marie–Tooth disease type 2 is caused by a loss-of-function mutation in the kinesin KIF1B, and hereditary spastic paraplegia type 10 is caused by a missense mutation in KIF5A. Finally, a mutation in the $p150^{Glued}$ subunit of dynactin has been identified in a family with an autosomal dominant form of lower motor neuron disease. Furthermore, inhibition of dynein-mediated axonal transport causes neurodegeneration in mouse models.

## Conclusion

Slowing of transport might be a general phenomenon in neurodegenerative diseases. It can be either causative or contributory, depending on the disease. Neurons are highly susceptible to dysregulation of transport. Indeed, because transport in neurons occurs over very long distances, a decrease in intracellular trafficking has dramatic consequences on their viability. The development of accurate models of intracellular transport in neurons should therefore help researchers understand the molecular mechanisms underlying these diseases.

Axonal transport is a promising target for designing and testing treatments to slow or block neurodegeneration. In the case of HD, compounds that enhance transport or rescue huntingtin dysfunction might be of therapeutic interest since huntingtin directly controls transport of the pro-survival factor BDNF. This is of utmost importance because no treatment currently exists for this devastating disorder.

*See also:* Axonal Transport and Neurodegenerative Diseases; Huntington's Disease: Neurodegeneration; Triplicate Repeats: Huntington's disease.

## Further Reading

Altar CA, Cai N, Bliven T, et al. (1997) Anterograde transport of brain-derived neurotrophic factor and its role in the brain. *Nature* 389: 856–860.

Borrell-Pages M, Canals JM, Cordelieres FP, et al. (2006) Cystamine and cysteamine increase brain levels of BDNF in Huntington disease via HSJ1b and transglutaminase. *Journal of Clinical Investigation* 116: 1410–1424.

Cattaneo E, Zuccato C, and Tartari M (2005) Normal huntingtin function: An alternative approach to Huntington's disease. *Nature Reviews Neuroscience* 6: 919–930.

Chang DT, Rintoul GL, Pandipati S, and Reynolds IJ (2006) Mutant huntingtin aggregates impair mitochondrial movement and trafficking in cortical neurons. *Neurobiology of Disease* 22: 388–400.

Gauthier LR, Charrin BC, Borrell-Pages M, et al. (2004) Huntingtin controls neurotrophic support and survival of neurons by enhancing BDNF vesicular transport along microtubules. *Cell* 118: 127–138.

Gunawardena S, Her LS, Brusch RG, et al. (2003) Disruption of axonal transport by loss of huntingtin or expression of pathogenic polyQ proteins in *Drosophila*. *Neuron* 40: 25–40.

Hafezparast M, Klocke R, Ruhrberg C, et al. (2003) Mutations in dynein link motor neuron degeneration to defects in retrograde transport. *Science* 300: 808–812.

Heerssen HM, Pazyra MF, and Segal RA (2004) Dynein motors transport activated Trks to promote survival of target-dependent neurons. *Nature Neuroscience* 7: 596–604.

Hirokawa N and Takemura R (2005) Molecular motors and mechanisms of directional transport in neurons. *Nature Reviews Neuroscience* 6: 201–214.

LaMonte BH, Wallace KE, Holloway BA, et al. (2002) Disruption of dynein/dynactin inhibits axonal transport in motor neurons causing late-onset progressive degeneration. *Neuron* 34: 715–727.

Lee WC, Yoshihara M, and Littleton JT (2004) Cytoplasmic aggregates trap polyglutamine-containing proteins and block axonal transport in a *Drosophila* model of Huntington's disease. *Proceedings of the National Academy of Sciences of the United States of America* 101: 3224–3229.

Li SH, Gutekunst CA, Hersch SM, and Li XJ (1998) Interaction of huntingtin-associated protein with dynactin P150Glued. *Journal of Neuroscience* 18: 1261–1269.

Li XJ and Li SH (2005) HAP1 and intracellular trafficking. *Trends in Pharmacological Sciences* 26: 1–3.

Pal A, Severin F, Lommer B, Shevchenko A, and Zerial M (2006) Huntingtin–HAP40 complex is a novel Rab5 effector that regulates early endosome motility and is upregulated in Huntington's disease. *Journal of Cell Biology* 172: 605–618.

Piccioni F, Pinton P, Simeoni S, et al. (2002) Androgen receptor with elongated polyglutamine tract forms aggregates that alter axonal trafficking and mitochondrial distribution in motor neuronal processes. *FASEB Journal* 16: 1418–1420.

Sahlender DA, Roberts RC, Arden SD, et al. (2005) Optineurin links myosin VI to the Golgi complex and is involved in Golgi organization and exocytosis. *Journal of Cell Biology* 169: 285–295.

Saudou F, Finkbeiner S, Devys D, and Greenberg ME (1998) Huntingtin acts in the nucleus to induce apoptosis but death does not correlate with the formation of intranuclear inclusions. *Cell* 95: 55–66.

Szebenyi G, Morfini GA, Babcock A, et al. (2003) Neuropathogenic forms of huntingtin and androgen receptor inhibit fast axonal transport. *Neuron* 40: 41–52.

Trushina E, Dyer RB, Badger JD 2nd, et al. (2004) Mutant huntingtin impairs axonal trafficking in mammalian neurons *in vivo* and *in vitro*. *Molecular and Cellular Biology* 24: 8195–8209.

Yanai A, Huang K, Kang R, et al. (2006) Palmitoylation of huntingtin by HIP14 is essential for its trafficking and function. *Nature Neuroscience* 9: 824–831.

# Vascular Issues in Neurodegeneration and Injury

**L Longhi, E R Zanier, V Conte, and N Stocchetti**, University of Milan, Milan, Italy
**T K McIntosh**, Media NeuroConsultants, Inc., Media, PA, USA



## Introduction

Traumatic brain injury (TBI) is clinically classified as either focal or diffuse. Focal TBI most often involves contusion and/or lacerations accompanied by intraparenchymal hematomas. Diffuse brain swelling, vascular alterations, ischemic brain injury, and diffuse axonal injury (DAI) are frequent components of the diffuse TBI spectrum. The sequelae of TBI may be further divided into (1) primary injury (immediate, nonreversible, and dependent upon the physical factors involved in the injury) and (2) secondary or delayed injury (progressing from hours to months following injury and likely to be partially reversible). This secondary injury cascade is complex and poorly understood and includes the breakdown of the blood–brain barrier (BBB), edema formation, vasospasm, ionic dysregulation, impairment and/or uncoupling of energy metabolism, changes in intracranial pressure (ICP) and/or cerebral perfusion pressure (CPP), inflammation, expression of both pathogenic and protective genes and proteins, and activation and/or release of autodestructive factors. Since the pathological sequelae of TBI are multifactorial in nature, an improved understanding of the secondary injury cascade offers a unique opportunity for the development of targeted therapeutic interventions to attenuate cellular damage and improve functional recovery. One area of intense interest and research concerns the posttraumatic alterations in cerebrovascular function and their role in mediating delayed cell death and/or neurologic dysfunction. Here we review the current state of knowledge concerning the importance of vascular changes following TBI.

## Pathology of Vascular Disruption Following TBI

Following TBI, mechanical deformation because of compression, shear, and tension can cause a disruption of the blood vessel wall. Depending on the anatomic position of the injured vessel, hemorrhage can occur in the epidural, subdural, subarachnoid, and/or intraparenchymal compartments. Intracranial hematomas are an extremely important lesion that can complicate 25–45% of severe TBI cases, 3–12% of moderate TBI cases, and approximately 0.2% of mild TBI cases.

### Epidural Hematoma

Epidural hematoma (EDH) is caused by the vascular disruption of meningeal arteries, meningeal veins, and/or dural sinuses before they enter into the space between the brain and the dura mater. In the majority of cases, EDH occurs as a consequence of skull fracture and stripping of the endosteum and can therefore be considered a complication of fracture and not related to head acceleration (see **Table 1**).

In 50% of human TBIs, the most common location of an EDH is the squamosal portion of the temporal bone. In this area, the middle meningeal artery is partially embedded in the inner table of the skull and is therefore relatively fixed to the bone and more vulnerable to injury. As the hematoma enlarges, it strips the dura from the skull and forms a circumscribed ovoid mass with progressive compression of brain. Picard and colleagues have shown that craniotomy for prompt evacuation of an isolated acute EDH is one of the most cost-effective surgical procedures with excellent results in term of outcome. Postmortem analysis and experimental studies have shown that neuropathological damage associated with EDH is the result of massive ischemic injury associated with brain compression and low CPP.

### Subdural Hematoma

Subdural hematoma (SDH) occurs in the space located between the meningeal layer of the dura and the underlying arachnoid and is often associated with a hemorrhagic contusion, intracerebral hematoma breaking through the arachnoid, the rupture of one or more bridging veins, or the laceration of a cortical artery or vein (see **Table 1**). The cerebral convexity in the parietal region is the most common location of SDHs. Most SDHs are caused by motor vehicle-related accidents (MVAs), falls, and assaults. During falls, angular acceleration is often high and of short duration. Bridging veins (those portions of the superficial cerebral veins that cross the subdural space to reach venous sinuses) are relatively straight as they cross the subarachnoid space and are therefore highly vulnerable to sudden angular acceleration forces. In elderly patients because of brain atrophy, the bridging veins can be stretched, leading to an increased incidence of SDHs. Infants are also susceptible to SDHs because their bridging veins are thin-walled. In younger adults, MVAs are responsible for 56% of SDHs, while only

**Table 1** Pathology of vascular disruption following traumatic brain injury (TBI)

| *Pathology* | *TBI severity* | *Incidence (%)* | *Reference*[a] |
|---|---|---|---|
| Extradural hematoma | Mild/moderate/severe | 3 | Bullock et al. (2006) |
| | Severe | 9 | Bullock et al. (2006) |
| Subdural hematoma | Mild/moderate/severe | 11 | Servadei et al. (2000) |
| | Severe | 12–29 | Bullock et al. (2006) |
| Subarachnoid hemorrhage | Mild/moderate/severe | 40 | Servadei et al. (2002) |
| Intraparenchymal hematoma | Mild/moderate/severe | 15 | McCormick and Post (1996) |
| Delayed traumatic intracerebral hematoma | Moderate/severe | 3–7.5 | Bullock et al. (2006) |

[a]From Bullock MR, Chesnut R, Ghajar J, et al. (2006) Surgical management of acute epidural hematomas. *Neurosurgery* 58: S7–S15; Bullock MR, Chesnut R, Ghajar J, et al. (2006) Surgical management of acute subdural hematomas. *Neurosurgery* 58: S16–S24; Bullock MR, Chesnut R, Ghajar J, et al. (2006) Surgical management of traumatic parenchymal lesions. *Neurosurgery* 58: S25–S46; Servadei F, Nasi MT, and Giuliani G (2000) CT prognostic factors in acute subdural haematomas: The value of the 'worst' CT scan. *British Journal of Neurosurgery* 14: 110–116; Servadei F, Murray G, and Teasdale G (2002) Traumatic subarachnoid hemorrhage: Demographic and clinical study of 750 patients from the European Brain Injury Consortium Survey of Head Injuries. *Neurosurgery* 50: 261–269; McCormick PC and Post KD (1996) Trigeminal neurinomas. In: *Neurosurgery*, 2nd edn., pp. 1545–1552. New York: McGraw-Hill.

12% are caused by falls. The overall mortality rate following acute traumatic SDH ranges from 40% to 75%. If the mass lesion is evacuated early, the prognosis depends primarily on the severity of the underlying injury. Mass effect with hemispheric compression and ischemia is an important mechanism of brain damage following a SDH.

### Subarachnoid Hemorrhage

Traumatic subarachnoid hemorrhage (SAH) is caused by the bleeding of surface blood vessels into the subarachnoid space or by the rupture of bridging veins or traumatic aneurysm (see **Table 1**). Typical locations for SAH are the interpeduncular cistern and the Sylvian fissure. In a retrospective study conducted on 141 patients, Chieregato and colleagues reported that the best predictive factors for poor outcome were subarachnoid blood at admission, assessed by computed tomography (CT) scan, together with the Glasgow Coma Scale (GCS) score. However, in a multicenter prospective study conducted on 1005 patients with moderate or severe TBI, Servadei and colleagues observed that patients with SAH exhibit worse CT scan features, such as subdural hematomas and parenchymal damage, with higher risk for evolving contusion, suggesting that SAH may be an indicator of initial injury severity. Furthermore, progression of contusions usually occurs in locations in which cortical SAH is observed, suggesting that contusions are part of a similar process and that subarachnoid blood is probably an early sign of cortical microbleeding.

### Intraparenchymal Hematoma

Traumatic intraparenchymal hematoma (IPH) is caused by the bleeding of an intracerebral or subarachnoid vessel. Most IPHs develop as a complication of a contusion; although they can also be absent from the first CT scan, they may develop later. Delayed traumatic intracerebral hematoma (DTICH) usually occurs in an area of radiologically normal brain in patients with otherwise pathological CT scans. However, DTICH can also occur in areas of contusion on initial CT scan or may appear following resection of surgical brain lesions or resuscitation. The occurrence of traumatic IPHs is five times more common in patients older than 45 years, compared to younger patients, suggesting that age-related changes in the brain parenchyma (such as atrophy) and vasculature may facilitate the development of IPHs following TBI (see **Table 1**). The most common locations for traumatic IPHs include the frontal and temporal regions, where the brain can impact the roughened inner surfaces of the skull (i.e., orbital roof or sphenoid ridge).

A study by Chang and colleagues showed that over 60% of lesions remain unchanged within the first 3 days of injury, and nearly 40% progress to larger lesions, with 15% becoming larger than 5 $cm^3$. A majority of studies on traumatic IPH are observational and comparisons of outcome between surgical and nonsurgical groups are difficult. However, current evidence indicates that an IPH that is accompanied by progressive neurological dysfunction, refractory intracranial hypertension, or radiological signs of mass effect may be associated with a poor outcome if left untreated.

## Alterations in Cerebral Blood Flow Following TBI

Brain glycogen reserves are extremely limited, and since the brain has no significant storage capacity for glucose, it is highly dependent on blood flow. Inadequate substrate delivery and cerebral blood flow (CBF) are therefore among the most important factors contributing to mortality and morbidity following TBI. A consistent finding reported in the experimental TBI literature is that CBF is significantly reduced acutely following impact. The degree and time window of this reduction are related to the injury model, the species, and the methodology used to measure CBF.

### Ischemia

Measurement of CBF following TBI can identify hypoperfusion. However, because of metabolic derangement, true ischemia in these patients can only be defined by a CBF that is inappropriately low for tissue metabolism.

The classical definition of ischemia has the following implications:

- Reduction of CBF to less than 18 ml per 100 g tissue $min^{-1}$ (coupled with increase in cerebral oxygen extraction fraction/high arteriojugular oxygen content difference ($AJDO_2$) values).
- Reduction of ATP (otherwise termed 'energy failure').
- Increase in anaerobic glucose metabolism, thereby producing lactate.

TBI in rodents reduces cortical CBF distal to the impact by 50%, while CBF in the injury site is reduced by about 95%. Additional experimental studies have reported reduction in ATP content in traumatized cortical regions and elevated lactate levels after severe TBI. In the clinical setting, ischemic lesions, most commonly seen in the hippocampus and basal ganglia, have been detected in the majority (90%) of fatal TBI. In nonfatal TBI, profound reductions of cerebral CBF (CBF $<$18 ml per 100 g tissue $min^{-1}$) have also been shown in more than 30% of patients during the early (4–12 h) period following TBI. However, high $AJDO_2$ values (above 9 ml/100 ml) have been identified in only a minority of these patients, indicating a coupled CBF and metabolic depression. Since marked heterogeneity of CBF and metabolic patterns in the traumatically injured brain have been demonstrated, focal ischemia could remain undetected by global measurements such as $AJDO_2$. Recent $^{15}$O-positron emission tomography studies exploring regional pathophysiology measuring $O_2$ extraction fraction have shown that within 24 h a small but significant volume of brain tissue is at risk of ischemic damage. Other clinical studies, however, have reported that ischemic levels of CBF are only observed in irreversibly damaged tissue. Therefore the importance of ischemia in delayed neuronal injury following nonfatal TBI remains controversial. In an assessment of the similarities and differences between ischemia and TBI, Bramlett and colleagues have shown that mechanisms that cause cells to die are quite similar, but different for dysfunctional cells that survive the initial insult. The mechanisms underlying cellular vulnerability in ischemia and TBI may therefore be distinct.

### Uncoupling of Cerebral Blood Flow and Metabolism

Under normal conditions, CBF is coupled to metabolism whereby cortical activation is followed by CBF increase. Following TBI, several studies have shown that this CBF/metabolism coupling can be lost when acute glucose metabolism increase is associated with a reduction in CBF (**Figure 1**, upper panel). Similarly, human brain PET studies have shown that around cerebral contusions and beneath intracranial hematomas there is an area of increased glucose metabolism together with decreased CBF (**Figure 1**, lower panel). Therefore, the ability of the injured brain to respond adequately to subsequent perturbations that may exacerbate the imbalance between $O_2$/nutrient delivery and consumption is extremely limited. The concept of the vulnerability of the traumatized brain to secondary insults has been investigated by Jenkins and colleagues, who were the first to show that although neither mild TBI nor brief (6 min) bilateral carotid artery occlusion resulted in detectable hippocampal morphological damage/death, the same severity of incomplete ischemia resulted in profound CA1 hippocampal neuronal death when induced 1 or 24 h following a mild TBI. Therefore in the presence of secondary sublethal challenges (i.e., increased ICP, reduced CPP, fever, seizures), injured cells may undergo irreversible damage.

### Posttraumatic Vasospasm

Posttraumatic vasospasm occurs in approximately 25–40% of patients following severe TBI and is associated with the presence of traumatic SAH. In a study comparing 99 patients with posttraumatic SAH to 114 patients with aneurismal SAH, a clear difference was found in distribution, time course of CBF, and neurological changes in the two conditions. Furthermore, the distribution of the low-density areas on the CT scans did not correlate with vascular territories in traumatic SAH patients, suggesting that vasospasm following TBI does not contribute to secondary deterioration. Therefore the importance of monitoring TBI patients for vasospasm is not currently clear.

## Molecular Mediators of Vascular Damage after TBI

In addition to CBF changes (*vide supra*), post-TBI changes include morphologic damage to endothelial cells and perivascular cells, increased BBB permeability,

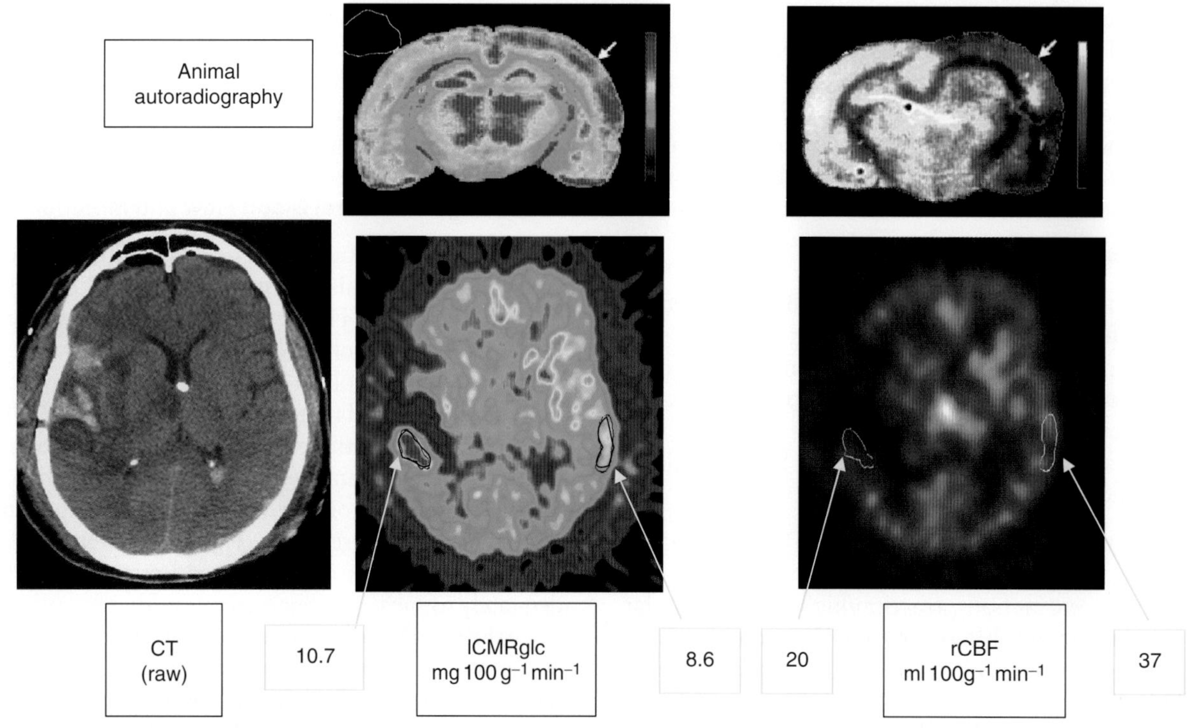


**Figure 1** Mismatch of local cerebral metabolic rate for glucose (lCMRglc) and regional cerebral blood flow (rCBF) following traumatic brain injury. (Top) 1 h following experimental traumatic brain injury, the rat brain exhibits evidence of local increase in lCMRglc, utilizing [$^{14}$C]deoxyglucose autoradiography (left), associated with a reduction of local cerebral blood flow, utilizing [$^{14}$C]iodoantipyrine autoradiography (right). (Bottom) In brain scans of a person with severe traumatic brain injury, confirmation of a similar mismatch of lCMRglc and rCBF in the tissue beneath a contusion. The computed tomography (CT) scan shows the contusion and the pericontusional tissue (left panel). The pericontusional rim (yellow arrows, middle panel) exhibits an increase of lCMRglc measured with $^{18}$F-labeled deoxyglucose positron emission tomography (PET scanning), associated with a reduction of rCBF, measured with $^{15}O\text{-}H_2O$ PET techniques (yellow arrows, right panel). Reproduced by permission of David A Houda, UCLA Brain Injury Research Center.

reduction or abolition of vasodilatory and vasoconstriction responses to changes in mean arterial pressure (MAP), changes in arterial tension of $O_2$ and $CO_2$ and blood viscosity as well as in cerebral metabolic activity. It is likely that some aspects of cerebrovascular dysfunction following TBI contribute to the increased sensitivity of the injured brain to secondary hypoxia and hypotension and to the increased mortality and morbidity that occurs due to secondary insults after TBI in humans.

The cerebral circulation is influenced by a variety of vasodilators and vasoconstrictors, and the observed reduction in CBF after TBI may be due to an imbalance in the levels/activity of vasoconstrictors and vasodilators and/or to local microthrombosis.

## Calcium

An increase in brain calcium ($Ca^{2+}$) flux has been documented following several experimental models of TBI. $Ca^{2+}$ entry into neurons may occur through membrane-associated receptors and voltage-sensitive $Ca^{2+}$ channels (VSCCs). Antagonism of these channels using various $Ca^{2+}$ channel blockers has been shown to be associated with attenuation of regional CBF reductions, brain edema, and motor and cognitive deficits following experimental TBI. However, when nimodipine, which blocks L-type VSCCs, was administered 24 h postinjury to severely head-injured patients, no significant clinical benefit was observed, except in a subgroup of patients with SAHs.

## Prostanoid, Eicosanoid, and Reactive Oxygen Species

Posttraumatic elevation of intracellular free $Ca^{2+}$ may precipitate an attack on the cellular membrane by activating phospholipases $A_2$ ($PLA_2$) and C (PLC), resulting in the release of free fatty acids (FFAs) and diacylglycerol (DAG). Increases in brain FFAs can lead to secondary damage through BBB breakdown and formation of cerebral edema, and, indirectly, via the generation of prostaglandins, thromboxanes, and leukotrienes (cyclooxygenase and lipooxygenase products from arachidonic acid), which may affect CBF and BBB permeability. **Figure 2** shows the pathways activated from arachidonic acid and the effect on cerebral circulation of the different products.

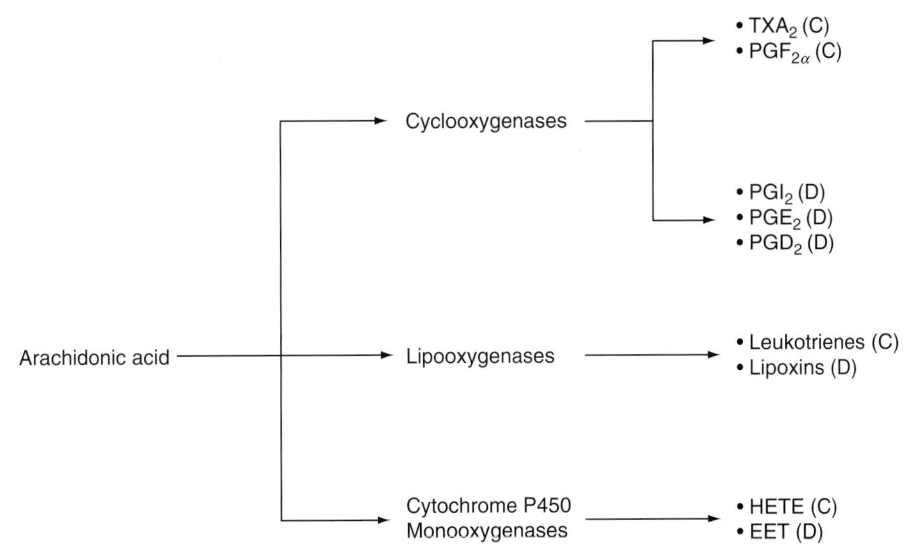


**Figure 2** Pathways of eicosanoid formation and effects on cerebral circulation. $TXA_2$, thromboxane $A_2$; $PGF_{2\alpha}$, $PGI_2$, $PGE_2$, $PGD_2$, prostaglandins; EET, epoxyeicosatrienoic acid; HETE, hydroxyeicosatetraneoic acid; C, vasoconstrictor; D, vasodilator.

Cyclooxygenases 1 (constitutive isoform) and 2 (inducible) have been shown to increase following experimental and clinical TBI as early as 3 h postinjury, together with increases in plasma and tissue levels of prostaglandins ($PGF_{2\alpha}$, $PGE_2$), prostacyclin, and thromboxane, with an imbalance toward vasocontrictors. Administration of cyclooxygenase (COX) inhibitors has resulted in mixed results, and additional studies are warranted to better understand the effects of COX inhibition on cerebrovasculature and neuroinflammation following TBI. Prostacyclin has also been shown to reduce cortical cell death and to improve compromised perfusion caused by posttraumatic vasoconstriction and microthrombosis following experimental TBI.

In addition, during the process of prostaglandin synthesis from arachidonic acid, reactive oxygen species (ROS) are generated by the mitochondria due to $Ca^{2+}$-mediated disorganization of the inner mitochondrial membrane, with subsequent disruption of the electron transport chain. Reactive oxygen species formation leads to peroxidative destruction of the cell membrane, proteins, and DNA. Under physiological conditions, ROS are scavenged by endogenous antioxidants, including superoxide dismutase (SOD), glutathione peroxidase, and catalase. Since the amount of ROS production exceeds the scavenging capacity of the cells, increasing the endogenous antioxidative mechanisms may have therapeutic value in TBI.

## Adenosine

Adenosine is a major regulator of the cerebral circulation known to cause vasodilation acting on two receptors located on the vascular endothelium and smooth muscle. An increase in cerebral metabolic activity and/or a mismatch between energy production and oxygen consumption results in a rise in adenosine concentration, which in turn can dilate the cerebral arterioles and increase local CBF. Following TBI in rats, interstitial adenosine has been shown to increase in the contusional/pericontusional area and, in humans, in the cerebrospinal fluid (CSF) of patients following TBI. Furthermore, secondary episodes of global brain hypoxia further increase CSF adenosine, suggesting a role during periods of secondary insults following TBI. Agonists of adenosine have been shown to attenuate cerebral hypoperfusion and posttraumatic behavioral deficits following experimental TBI in rats, suggesting that the neuroprotection obtained by adenosine receptors pathways might be due to an increase in CBF.

## Nitric Oxide

Nitric oxide (NO) is a vasodilator that has been suggested to play a role in damage, recovery, or both, after TBI. Three different systems are reportedly involved in NO synthesis: endothelial, neuronal, and inducible NO synthase (eNOS, nNOS, and iNOS). lIt has been observed that cerebral tissue NO levels increase transiently and then decrease within 15–30 min following experimental TBI.

When expression/levels of NO synthases are evaluated following TBI, it has been shown that brain tissue nNOS activity and NO levels increase markedly but briefly after TBI, and then decrease below preinjury levels for hours to days. Inducible NOS activity in the brain increases hours to days after TBI and remains elevated up to a week after injury.

In the cerebral vasculature, eNOS immunoreactivity and protein expression increase after TBI. Since CBF might be linked to NO levels, several studies have evaluated the hypothesis that increasing NO levels (or NO synthases activity) results in increased CBF and attenuated damage following TBI. The NOS substrate L-arginine improves CBF and attenuates histological damage after experimental TBI. Nonselective blockade of NOS has also been shown to be either detrimental or without influence on outcome following TBI in rats.

To clarify the beneficial or detrimental effects of iNOS-related NO production, studies employing genetically engineered mice and/or pharmacological inhibition have been performed: mice deficient in iNOS show increased cognitive deficits 15–18 days after TBI, suggesting a protective role of iNOS. This protective role of iNOS was confirmed in a rat model of TBI in which administration of iNOS inhibitors aminoguanidine and L-*N*-iminoethyl-lysine resulted in reduced learning capacity and greater hippocampal cell loss at 3 weeks after injury. It appears possible, therefore, that iNOS activity may exert detrimental effects in the acute postinjury period but can show beneficial effects in the chronic phase. Further work needs to be directed at clarifying the exact role of the different isoforms of NOS to draw a more comprehensive picture of the role of NO following TBI.

#### Endothelin-1

Endothelin (ET) is a 21-amino-acid peptide with potent vasoconstrictive activity. The ET family comprises ET-1, ET-2, and ET-3. Endothelin is produced in endothelial cells, as well as in several tissues, including the brain, where ET-1 has been found in neurons, glial cells, and choroid plexus cells. Synthesis of ET-1 is stimulated by thrombin, cytokines, transforming growth factor-$\beta$, vasopressin, angiotensin II, and oxyhemoglobin, while it is inhibited by NO. Endothelins acts on two receptors, $ET_A$ and $ET_B$, which have been observed in vascular smooth muscle cells (leading to vasoconstriction through an increase in $Ca^{2+}$) and endothelial cells (leading to vasodilation through the release of vasodilatory COX products and NO). Following experimental TBI, an increase in ET-1 mRNA expression and protein level has been observed in the brain and CSF, and this is associated with reduced CPP. A relationship has been observed between ET and NO (vasoconstriction and vasodilation) and inhibition of iNOS using antisense iNOS oligodeoxynucleotides following TBI in rodents; this interaction results in reduced synthesis of iNOS and a five- to sixfold increase in ET-1 concentration, both of which are associated with a marked reduction in CBF. In contrast, ET-1 antagonists have been associated with restored vasodilatory capacity of the cerebrovascular system following TBI, suggesting that ET-1 might be a target for therapeutic intervention.

#### Neuropeptide Y

Neuropeptide Y is a transmitter of the cerebrovascular sympathetic nerves known to induce vasoconstriction and reduction of CBF. It acts on a family of six different receptors localized in the smooth muscle cells of pial arteries and arterioles. In cerebral arteries, the vasoconstriction mediated by neuropeptide Y is mediated by changes in smooth muscle membrane potential and $Ca^{2+}$ increase (either via increased influx or increased release from intracellular stores); this vasoconstriction can be blocked by $Ca^{2+}$ channel blockers such as nimodipine, nifedipine, and verapamil. Following TBI in rodents, concentrations of neuropeptide Y are increased in the pericontusional cortex. In addition, neuropeptide Y-immunoreactive fibers are observed in the injured cortex and ipsilateral hippocampus, suggesting that an increase in cerebral neuropeptide Y concentrations after brain injury may be involved in the pathogenesis of reduced CBF.

#### Endothelium-Dependent Hyperpolarizing Factor

Recently, an additional endothelial-derived vasodilatory pathway that is different from NO and/or cyclooxygenase-derived products has been identified; this pathway is mediated by endothelium-derived hyperpolarizing factor (EDHF), which acts to hyperpolarize the vascular smooth muscle and activate potassium ($K^+$) channels on the vascular smooth muscle. Possible candidates of EDHF include epoxides, anandamide, $K^+$ ion, and an electrical coupling mediated by myoendothelial gap junctions. Following TBI in rodents, EDHF pathways are activated and the resultant vasodilation can be inhibited by charybdotoxin, an inhibitor of large-conductance $Ca^{2+}$-sensitive $K^+$ channels, suggestive of a compensatory role of EDHF during conditions characterized by compromised NO. Further work will determine whether the observed phenomenon of EDHF potentiation might be a protective mechanism initiated following cerebrovascular injury.

## Genetic Influences/Susceptibility to Vascular Changes after TBI

#### Apolipoprotein E

Lipoproteins (lipid transporters) are macromolecules that contain lipids and proteins, known as apolipoproteins, including apolipoprotein E (ApoE), which plays a major role in the metabolism of triglyceride lipoproteins (chylomicrons, chylomicrons remnants, very low-density and intermediate-density lipoproteins) and high-density lipoproteins (HDLs), and in the redistribution of lipids locally among cells. The brain is the second most abundant site of ApoE mRNA synthesis, which occurs primarily in astrocytes, but also in neurons and activated microglia. ApoE is also synthesized in macrophages, where it appears to play a role in modulating cholesterol accumulation. ApoE is the only lipoprotein present in the CSF and may function in redistributing lipids among CNS cells for normal lipid homeostasis, repairing injured neurons, maintaining synaptodendritic connections, and scavenging toxins. Polymorphisms of ApoE can occur as a result of mutation, and can be defined as a genetic locus where the most common allele occurs with a frequency of less than 100% in the general population. The most common polymorphisms are single-nucleotide polymorphisms (SNPs), resulting from a single base variation. The ApoE gene on chromosome 19 has three common alleles ($\varepsilon2$, $\varepsilon3$, and $\varepsilon4$), producing corresponding isoforms of the protein: ApoE2 (Cys112, Cys158), ApoE3 (Arg112, Cys158), and ApoE4 (Arg112, Arg158). These three isoforms differ in their abilities to accomplish critical tasks in the maintenance and repair of neurons.

Contusions in head-injured patients with ApoE4 might be more severe as a result of relatively deficient clotting mechanisms. Prolonged clotting times were also identified in ApoE4 carriers after stroke, providing further evidence that ApoE genotype is of relevance to the coagulation cascade. Animal models have provided further information about ApoE mechanisms and the response of the brain to injury. In normal rodents, ApoE immunoreactivity is found in glia, mostly astrocytes, and is not observed in neurons. After ischemic brain injury, the distribution changes and ApoE is increased in reactive astrocytes and neuropils, but later also in degenerating neuronal processes. This increase in ApoE levels has been interpreted as an attempt of the brain to protect itself. ApoE-deficient mice were found to have a greater extent of neuronal damage than did wild-type mice after TBI and ischemia, which was ameliorated by continuous intracerebral infusion of ApoE. However, different genotypes show specific abilities in the response to brain injury. Vascular factors such as thrombin, released during TBI, may also play an important role in the upregulation of ApoE levels. The increase in ApoE levels after hemorrhagic stroke or TBI may initiate a cascade of events leading to amyloid protein-lipoprotein ($A\beta$) deposition, neuroinflammation, and cognitive deficits. Increased neuroinflammation in ApoE4 carriers may alter the neurological outcome after TBI (see **Table 2**). Although ApoE4-associated vascular pathology may influence the outcome in survivors, it seems unlikely to significantly increase the probability of a fatal outcome after TBI, since ApoE4 carriers are equally represented in fatal cases (35%) and in nonfatal

**Table 2** Outcome of traumatic brain injury (TBI)

| *Outcome measure* | *Main results of TBI pathology* | *Reference*[a] |
|---|---|---|
| Recovery of consciousness | $\varepsilon4$ found more frequently in patients who did not recover consciousness at 1 year | Sorbi et al. (1995) |
| Glasgow Outcome Scale | $\varepsilon4$ associated with unfavorable outcome at 6 months | Teasdale et al. (1997) |
| Functional and cognitive assessment | $\varepsilon4$ associated with poorer clinical outcome at 6–8 months | Friedman et al. (1999) |
| Functional independence measures | $\varepsilon4$ associated with poorer rehabilitation at 6 months | Lichtman et al. (2000) |
| Learning/fluency test | $\varepsilon4$ had poorer memory outcome at 6 months | Crawford et al. (2002) |
| Glasgow Outcome Scale | $\varepsilon4$ associated with unfavorable outcome at 6 months | Chiang et al. (2003) |
| Neuropsychological tests | $\varepsilon4$ bearers had worse postinjury performance | Sundström et al. (2004) |
| Hematoma volume/Glasgow Outcome Scale | $\varepsilon4$ associated with larger hematoma volume and poorer outcome at 6 months | Liaquat et al. (2002) |
| Glasgow Outcome Scale $\pm$ posttraumatic seizures | $\varepsilon4$ associated with increased risk of seizures, independent of outcome at 6 months | Diaz-Arrastia et al. (2003) |
| Confusion index/ischemic damage | $\varepsilon4$ associated with greater incidence of moderate/severe contusions and severe ischemic damage at autopsy | Smith et al. (2006) |
| Cerebral amyloid angiopathy | $\varepsilon4$ associated with cerebral amyloid angiopathy at autopsy | Leclercq et al. (2005) |

[a]See full entries in 'Further reading' section.

cases (33%). Cerebral amyloid angiopathy in fatal cases of TBI is considerably more likely to occur in those with a genetic predisposition conferred by possession of an ApoE ε4 allele. Clinical studies of TBI have shown that possession of the ε4 allele of the apolipoprotein E gene is associated with a relatively poor outcome. Teasdale and colleagues demonstrated that 57% of ApoE ε4 carriers had an unfavorable outcome following TBI (defined as dead, in a vegetative state, or with severe disability), compared with 27% of noncarriers of ApoE ε4. Friedman and colleagues reported that only 3.7% of patients with the ε4 allele had a good recovery, compared with 31% of patients without the allele. It has been suggested that ApoE ε4 promotes production of Aβ aggregates, cytoskeleton alterations, and oxidative damage, while also impairing the ability of the brain to repair and regenerate (see **Figure 3**).

## Summary

Traumatic brain damage is a result of an immediate mechanical disruption of brain tissue, (including blood vessels), and delayed physiological, molecular, and cellular events/cascades that are potentially amenable to postinjury therapeutic intervention because of their progressive nature, from minutes to months after the initial trauma. Vascular pathology and alterations in CBF are major contributors to mortality and morbidity following TBI. Traumatic brain injury results in CBF reduction both in the experimental and clinical setting, and the reduced CBF in patients has been shown to render the brain vulnerable to secondary insults and has been associated with an unfavorable neurological outcome. The mechanisms responsible for the observed reductions in regional CBF (rCBF) following TBI have not been fully elucidated. During normal physiological conditions, CBF and metabolism are coupled, in contrast to alterations following TBI, when an uncoupling between CBF (reduced) and glucose metabolism (increased) is often observed. In addition, TBI causes an impairment in cerebral autoregulation, whereby patients are not able to tolerate hypoperfusion and remain vulnerable to hypotension and hypoxia. Since TBI is often followed by an upregulation and release of several vasoactive mediators which act in a complex network, it has been hypothesized that posttraumatic alterations in CBF might be the result of an imbalance between molecules with vasoconstrictive and vasodilatory properties. Alterations in $Ca^{2+}$, nitric oxide, endothelin, prostanoids, adenosine, neuropeptide Y, reactive oxygen species, and endothelium-derived hyperpolarizing factor have been observed following TBI and have been linked with the observed CBF alterations. In addition, patient genotype appears

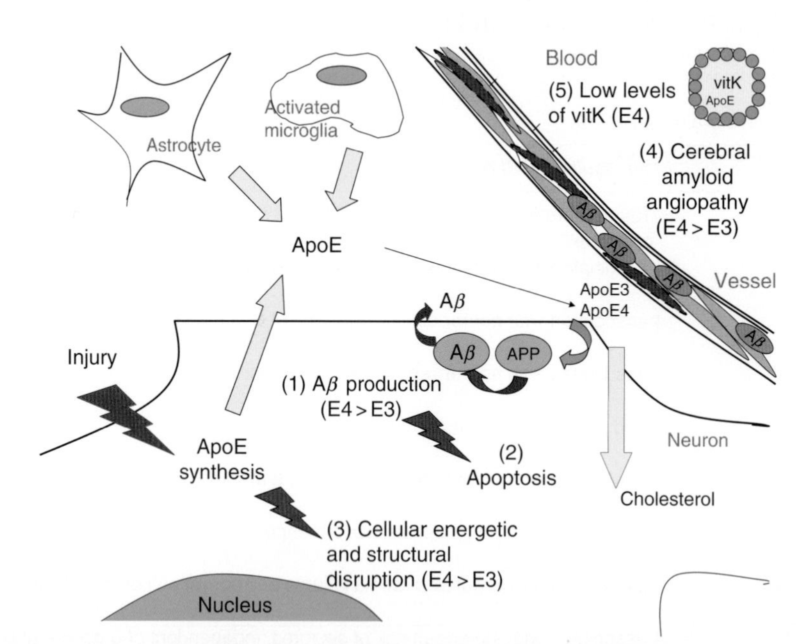


**Figure 3** Roles of ApoE in lipid redistribution among cells in the central nervous system and ApoE isoform-specific differences in neuropathology. ApoE is synthesized by astrocytes, activated microglia, and neurons. Potential detrimental roles for ApoE4 include (1) enhanced amyloid β-protein (Aβ) production, (2) potentiation of $A\beta_{1-42}$-induced lysosomal leakage and apoptosis, (3) enhanced neuron-specific proteolysis, resulting in translocation of neurotoxic ApoE4 fragments in the cytosol, where they are associated with cytoskeletal disruption and mitochondrial dysfunction (APP, amyloid precursor protein), (4) increased deposition of amyloid in cerebral cortical and leptomeningeal blood vessels, with consequent thickening of the walls and degeneration of the smooth muscle cells in the media, and (5) decreased levels of plasma vitamin K (vitK), which is a fat-soluble vitamin involved in clotting mechanisms.

capable of influencing both the vulnerability of the brain to injury and the posttraumatic damaging/protective response. Although laboratory investigations in which a single pathway has been modified have shown promising results, further work is warranted to better understand the molecular pathways associated with posttraumatic alterations in CBF/metabolism and to draw a comprehensive picture that will permit the successful translation of therapeutic strategies from animals models to the clinical setting.

## Further Reading


Bramlett HM and Dietrich WD (2004) Pathophysiology of cerebral ischemia and brain trauma: Similarities and differences. *Journal of Cerebral Blood Flow and Metabolism* 24: 133–150.

Bullock MR, Chesnut R, Ghajar J, et al. (2006) Surgical management of acute epidural hematomas. *Neurosurgery* 58: S7–S15.

Bullock MR, Chesnut R, Ghajar J, et al. (2006) Surgical management of acute subdural hematomas. *Neurosurgery* 58: S16–S24.

Bullock MR, Chesnut R, Ghajar J, et al. (2006) Surgical management of traumatic parenchymal lesions. *Neurosurgery* 58: S25–S46.

Chiang MF, Chang JG, Hu CJ, et al. (2003) Association between apolipoprotein E genotype and outcome of traumatic brain injury. *Acta Neurochirurgica* 145: 649–654.

Crawford FC, Vanderploeg RD, Freeman MJ, et al. (2002) APOE genotype influences acquisition and recall following traumatic brain injury. *Neurology* 58: 1115–1118.

DeWitt DS and Prough DS (2003) Traumatic cerebral vascular injury: The effects of concussive brain injury on the cerebral vasculature. *Journal of Neurotrauma* 20: 795–825.

Diaz-Arrastia R, Gong Y, Fair S, et al. (2003) Increased risk of late posttraumatic seizures associated with inheritance of APOE. *Archives of Neurology* 60: 818–822.

Friedman G, Froom P, Sazbon L, et al. (1999) Apolipoprotein E-ε4 genotype predicts a poor outcome in survivors of traumatic brain injury. *Neurology* 52: 244–248.

Golding EM (2002) Sequelae following traumatic brain injury. The cerebrovascular perspective. *Brain Research Reviews* 38: 377–388.

Hurley SD, Olschowka JA, and O'Banion MK (2002) Cyclooxygenase inhibition as a strategy to ameliorate brain injury. *Journal of Neurotrauma* 19: 1–15.

Leclercq PD, Murray LS, Smith C, et al. (2005) Cerebral amyloid angiopathy in traumatic brain injury: Association with apolipoprotein E genotype. *Journal of Neurology, Neurosurgery & Psychiatry* 74: 827a–828a.

Liaquat I, Dunn LT, Nicoll JA, et al. (2002) effect of apolipoprotein E genotype on hematoma volume after trauma. *Journal of Neurosurgery* 96: 90–96.

Lichtman S, Seliger G, Tycki B, et al. (2000) Apolipoprotein E and functional recovery from brain injury. *Neurology* 54: 2082–2088.

Mahley RW, Weisgraber KH, and Huang Y (2006) Apolipoprotein E4: A causative factor and therapeutic target in neuropathology, including Alzheimer's disease. *Proceedings of the National Academy of Sciences of the United States of America* 103(15): 5644–5651.

McCormick PC and Post KD (1996) Trigeminal neurinomas. In: *Neurosurgery*, 2nd edn., pp. 1545–1552. New York: McGraw-Hill.

Nortje J and Menon DK (2004) Traumatic brain injury: Physiology, mechanisms, and outcome. *Current Opinion in Neurology* 17: 711–718.

Smith C, Graham DI, Murray LS, et al. (2006) Association of APOE ε4 and cerebrovascular pathology in traumatic brain injury. *Journal of Neurology, Neurosurgery & Psychiatry* 77: 363–366.

Servadei F, Nasi MT, and Giuliani G (2000) CT prognostic factors in acute subdural haematomas: The value of the 'worst' CT scan. *British Journal of Neurosurgery* 14: 110–116.

Servadei F, Murray G, and Teasdale G (2002) Traumatic subarachnoid hemorrhage: Demographic and clinical study of 750 patients from the European Brain Injury Consortium Survey of Head Injuries. *Neurosurgery* 50: 261–269.

Sorbi S, Nacimas B, Piacentini S, et al. (1995) ApoE as a prognostic factor for post-traumatic coma. *Nature Medicine* 1: 135–137.

Sundström A, Marklund P, Nilsson L-G, et al. (2004) APOE influences on neuropsychological function after mild head injury: Within-person comparisons. *Neurology* 62: 1963–1966.

Teasdale GM, Nicoll JAR, Murray G, et al. (1997) Association of apolipoprotein E polymorphism with outcome after head injury. Early report. *The Lancet* 350: 1069–1071.

Waters RJ and Nicoll JA (2005) Genetic influences on outcome following acute neurological insults. *Current Opinion in Critical Care* 11: 105–110.

# Stroke: Injury Mechanisms

**A B Singhal and E H Lo**, Harvard Medical School, Boston, MA, USA
**J M Ren and S P Finklestein**, Biotrofix, Inc., Needham, MA, USA



## Introduction

Stroke is a clinical syndrome of sudden-onset neurological deficits resulting from vascular etiology. Ischemic stroke accounts for 80% of all strokes; hemorrhagic stroke accounts for the remaining 20%. Brain injury after stroke results from sudden cessation of blood flow, leading to impaired delivery of oxygen, glucose, and other nutrients to brain cells, triggering a cascade of ultimately lethal cellular events.

Several mechanisms of stroke-induced cell death have been delineated, the most important of which include glutamate excitotoxicity, acidotoxicity, oxidative/nitrative stress, apoptosis, inflammation, and peri-infarct depolarization. These fundamental pathways are interdependent, overlapping, and evolve over minutes, hours, or days to mediate injury in neuronal cells, glial cells, and vascular structures. The individual contribution of individual mechanisms toward the net stroke-related injury (**Figure 1**) is dependent on the degree and duration of ischemia, the timing of reperfusion, the type of brain tissue affected, physiological variables such as brain temperature, and genetic influences. For example, excitotoxicity has a dominant role in the 'core' of the ischemic territory where the reduction in blood flow is severe (less than 20% of baseline, typically 10–12 ml 100 $g^{-1}$ $min^{-2}$). In peripheral zones termed the 'ischemic penumbra,' where the blood flow reduction is abnormal but not as severe (20–60 ml 100 $g^{-1}min^{-2}$), cell death occurs relatively slowly via mechanisms such as inflammation, apoptosis, and peri-infarct depolarization. In these penumbral regions, brain electrical activity ceases, but the cells remain viable for an extended time period, constituting a target for stroke therapy.

In addition to the severity of ischemia, stroke-related injury is dependent on the type of brain tissue involved. Gray matter and white matter have different susceptibilities to ischemic injury. Among neurons, CA1 hippocampal pyramidal neurons, cortical projection neurons in layer 3, neurons in dorsolateral striatum, and cerebellar Purkinje cells are particularly susceptible. Among cell types comprising the white matter, oligodendrocytes are more vulnerable than are astroglial or endothelial cells. White-matter ischemia is typically severe, with rapid cell swelling and tissue edema because there is little collateral blood supply in deep white matter. Cell death mechanisms such as excitotoxicity have a different pathophysiology in white-matter ischemia, as compared to gray-matter ischemia. It is important to note that 'global' cerebral ischemia (which occurs after cardiac arrest) and 'focal' cerebral ischemia (which occurs due to thrombosis or embolism to a brain artery) result in different histological and behavioral outcomes and have different molecular mechanisms and treatments.

While restoration of blood flow can salvage penumbral brain regions, pharmacological interruption of one or more pathways of cell death can also mitigate stroke-related brain injury. Indeed, a major goal of uncovering pathways of cellular injury is to uncover targets for treatment. Numerous drugs have proved effective in reducing infarct size and functional deficits in animal stroke models. Although no neuroprotective drug to date has proved efficacious in humans, much knowledge has been gained. There is now a better understanding of factors such as the therapeutic time window and optimal patient selection, which are critical for success. Rapid advances in neuroimaging have made it possible to estimate the degree of ischemia and quantify penumbral tissue within minutes after symptom onset. These advances raise hope that successful stroke neuroprotection is imminent. In the following sections we review the major mechanisms of neuronal injury after stroke, emphasizing targets for intervention.

## Excitotoxicity and Ionic Imbalance

Glutamate, the main excitatory neurotransmitter, promotes the activity of virtually every neuron within the brain and spinal cord. It is synthesized within neurons, stored in presynaptic vesicles, released in a calcium-dependent manner to bind with specific receptors, and inactivated after release by high-affinity reuptake mechanisms. Glutamate flux is tightly regulated by its receptors and transporters. There are two categories of glutamate receptors: ionotropic N-methyl-D-aspartate (NMDA), α-amino-3-hydroxy-5-methyl-4-isoxalleprionate (AMPA), and kainate receptors, and metabotropic glutamate receptors (mGluRs). In the context of stroke-related injury, NMDAR and AMPAR activation mechanisms have been extensively studied.

Stroke-induced reductions in oxygen and glucose supply lead to the failure of energy-dependent processes such as that involving $Na^{+}$, $K^{+}$-ATPase,

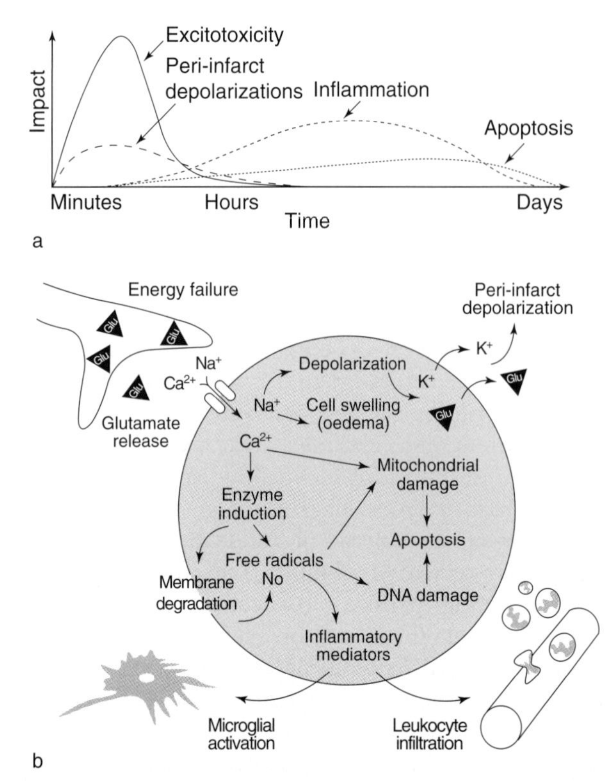


**Figure 1** Contribution of individual mechanisms toward the net stroke-related injury.

processes which normally maintain cellular homeostasis. This causes an abrupt cessation in the exchange of $Na^+$, $K^+$, and other ions across cell membranes; release of excitatory neurotransmitters such as glutamate; and inhibition of glutamate-reuptake mechanisms. Excessive glutamate binding to ionotropic NMDA and AMPA receptors promotes the deleterious influx of calcium into the cytoplasm and mitochondria, with activation of downstream phospholipases and proteases that degrade membranes and proteins.

$Ca^{2+}$ influx leads to activation of intracellular proteases such as calpains, which are involved in pro-death pathways such as those involving free radical-mediated mitochondrial damage, activation of Bad and subsequent release of cytochrome *c*, and cleavage of the $Na^+/Ca^{2+}$ exchanger, leading to further $Ca^{2+}$ influx. Calpains activate several caspases that in turn promote apoptosis, demonstrating the overlap between excitotoxic and apoptotic cell death mechanisms. Glutamate excitotoxicity and calcium influx also directly or indirectly influence intracellular signaling pathways. Three mitogen-activated protein kinase (MAPK) signaling pathways have been characterized: the extracellular signal-related kinase (ERK) MAPK pathway, the p38 MAPK pathway, and the c-jun N-terminal kinase (JNK) MAPK pathway. Growth factors, cytokines, and free radicals released after stroke all stimulate the MAPK pathways, providing further evidence for the overlap between the different fundamental pathways. ERK pathway activation is known to be neuroprotective; however, emerging data show that the ERK pathway is involved in mitochondrial cytochrome *c* release, activates caspase-3, and interacts with mGluRs to make them more susceptible to excitotoxic injury.

Because the endoplasmic reticulum (ER) participates in redistributing $Ca^{2+}$ ions within the cells, $Ca^{2+}$ influx also contributes to ER stress, which further exacerbates cell death. With regard to the mitochondria, $Ca^{2+}$ overload leads to the opening of the nonspecific mitochondrial permeability transition pore, which allows protons and other small molecules to cross the mitochondrial membrane, resulting in ATP depletion and cell death via necrosis or apoptosis. Whether necrosis or apoptosis dominates is determined by the duration of pore opening. If the pore reseals quickly, ATP production may be rescued, resulting in apoptotic death via cytochrome *c* release rather than necrotic cell death. In addition to the aforementioned processes, glutamate-dependent processes include the activation of transcription factors cAMP response element-binding protein (CREB) and nuclear factor-kappa B (NF-$\kappa$B), trophic factor signaling, and regulation of Bcl-2. Ionotropic glutamate receptors also promote ionic homeostasis perturbations that play a critical role in cerebral ischemia. Glutamate receptor activation is not uniformly deleterious, since different subtypes of receptors have prosurvival and proapoptotic effects. As explained next, this leads to a new level of complexity in finding suitable pharmaceutical agents for stroke.

Pharmaceutical agents that block the glutamate, NMDA, AMPA, and metabotropic subfamily receptors have been shown to attenuate infarct volumes in animal stroke models. However, these approaches were effective within a narrow 1 h time window, and a short therapeutic time window might explain the failure of several glutamate antagonists to show efficacy in clinical trials. Drugs that have failed clinical testing include lubeluzole, the competitive NMDA-antagonist selfotel, and several noncompetitive NMDA antagonists (dextrorphan, gavestinel, aptiganel, and eliprodil). The failure of these drugs is also attributable to drug toxicity, possibly due to inhibition of excitatory neurotransmission, and the fact that the NMDA receptor has variable functions depending on its location (synaptic or extrasynaptic), so that nonselective blockade of NMDA receptors may not be fruitful. For example, the activation of synaptic NMDA receptors promotes cell survival via mechanisms such as inactivation of the proapoptotic protein Bad, and the activation of CREB and expression of its target genes such as brain-derived neurotrophic factor (BDNF). Activation of extrasynaptic NMDA receptors causes inactivation of the CREB pathway and counters the increase in BDNF expression, leading to a proapoptotic effect. Therefore, the challenge for the future is to uncouple the essential excitatory neurotransmitter functions from deleterious effects – for example, by blocking downstream neurotoxic signaling pathways without blocking the receptors.

## Acidotoxicity

As compared to cell types in other organs, neurons are more vulnerable to ischemic injury. Experimental research has shown that the metabolic acidosis accompanying stroke can exacerbate neuronal death via excitotoxic mechanisms that are distinct from glutamate- and $Ca^{2+}$-mediated excitotoxicity. Neuronal acid-sensing ion channels (ASICs) have recently been characterized in the brain as ligand-gated channels belonging to the epithelial $Na^{+}$ channel superfamily. ASICs have at least six subunits, which form voltage-insensitive, amiloride-sensitive channels on the surface membrane. In contrast to other neurotransmitter-receptor channels that are inactivated with acidosis, these multimeric channels become activated when pH falls below 7.0, and their activation is half-maximum at a pH of 6.2, which is consistently achieved within ischemic regions. ASIC activation is also increased by arachidonic acid and lactic acid production, as well as membrane stretching and decreased extracellular $Ca^{2+}$ (all processes that routinely occur after stroke). The major role of ASICs is to gate amiloride-sensitive $Na^{+}$ currents. However, some subunits facilitate a glutamate-independent influx of $Ca^{2+}$ in the presence of acidosis. In experiments utilizing cell culture, patch clamp, and *in vivo* rodent stroke models, the tarantula toxin PcTX1 (an ASIC1a-specific antagonist) and amiloride have been shown to inhibit excitotoxic $Ca^{2+}$ currents via the ASIC1a ligand-gated ion channel. Moreover, ASIC1a channel inhibition has been documented to decrease infarct volumes by over 60%, acid-mediated toxicity is shown to be absent in mice lacking the ASIC1 gene, and ASIC transfected cells show increases in intracellular $Ca^{2+}$ following acidosis. These and other data indicate that ASIC1 causes a significant acid-dependent increase in excitotoxic cell death and that the ASIC1a subunit represents a promising stroke therapeutic target.

## Apoptosis

Apoptosis is an evolutionarily conserved process of programmed cell death. It is characterized histologically by nuclear and cytoplasmic condensation, nuclear fragmentation, and cell shrinkage. Apoptotic cells are terminal deoxynucleotidyl transferase-mediated dUTP nick-end labeling (TUNEL)-positive cells that exhibit DNA laddering. Cell type, age, and brain location render them more or less resistant to apoptosis or necrosis. Since apoptotic pathways require energy in the form of ATP, apoptosis predominantly occurs in the ischemic penumbra that sustains milder injury, rather than in the core, where ATP levels are rapidly depleted. Nevertheless, ionic imbalances can also trigger apoptosis under certain conditions, suggesting a close mechanistic interrelationship between glutamate-mediated excitotoxicity and apoptosis. The normal human brain expresses caspases 1, 3, 8, and 9, apoptosis protease-activating factor-1 (APAF-1), death receptors, P53, and a number of Bcl-2 family members, all of which are implicated in apoptosis.

Caspases are protein-cleaving enzymes (zymogens) that belong to a family of cysteine aspartases constitutively expressed in adult and newborn brain cells, particularly neurons. Apoptosis occurs via caspase-dependent as well as caspase-independent mechanisms. Caspases are sequentially cleaved and activated via two pathways: an extrinsic pathway dependent on death receptors, and an intrinsic mitochondrial pathway. With regard to the extrinsic pathway, several apoptotic triggers of have been identified. These include oxygen free radicals, Bid cleavage, death receptor ligation, DNA damage, and lysosomal protease activation. The tumor necrosis factor (TNF) superfamily of death receptors, most importantly Fas, powerfully regulates upstream caspase processes. Emerging data suggest that the cell nucleus is involved in releasing signals for apoptosis, and that the mitochondrion plays a central role in mediating apoptosis. At least four mitochondrial molecules mediate downstream cell death pathways: cytochrome *c*, secondary mitochondria-derived activator of caspase (Smac/Diablo), apoptosis-inducing factor (AIF), and endonuclease G. While cytochrome *c* and Smac/Diablo mediate caspase-dependent apoptosis, AIF and endonuclease G mediate caspase-independent apoptosis. Cytochrome *c* binds to Apaf-1, which combines with procaspase 9 to form the 'apoptosome' that activates caspase 9 in the presence of dATP. In turn, caspase 9 activates caspase 3. It is important to note that formation of the apoptosome complex is inhibited by Bcl-2, which becomes upregulated very early after stroke. Smac/Diablo binds to inhibitors of activated caspases and causes further caspase activation. After activation, caspases 3 and 7 (the executioner caspases) attack and degrade various substrate proteins, ultimately leading to DNA fragmentation and irreversible cell death. Over 30 proteins can be cleaved, including the DNA-repairing enzyme poly(ADP-ribose) polymerase (PARP), the cytoskeletal protein gelsolin, actin, presenilins, and other caspases. Caspase 3, the most prominent cysteine protease, is present in the infarct core as well as in the penumbra and is cleaved acutely after stroke. Subsequent caspase cleavage occurring after hours or days leads to delayed cell death.

Caspase-independent apoptosis has been shown to develop in cultured neurons by NMDA-receptor induced activation of PARP-1, and in fibroblasts exposed to oxidative stress. AIF is implicated as a key molecule in this cascade. AIF is released from mitochondria after PARP-1 activation, then relocates to the nucleus, where it binds DNA, promotes chromatin condensation, and leads to cell death via incompletely understood mechanisms. The exact significance of caspase-independent apoptosis remains to be determined.

Overexpression of Bcl-2, deletions of genes for Bid or caspase-3, and the use of caspase-3 inhibitors, peptide inhibitors, and antisense oligonucleotides that suppress the expression and activity of apoptosis-promoting genes have been shown to reduce infarct volume or decrease neurological deficits after stroke. A single intracerebroventricular injection of zDEVD-FMK, a relatively selective caspase-3 inhibitor, is shown to be neuroprotective when administered up to 9 h after transient focal stroke. However, caspase inhibitors do not reduce infarct size in all brain ischemia models. This variable effect may be attributable to different severities of ischemia, to limited potency or inability of caspase inhibitors to cross the blood–brain barrier, to the relatively minor impact of apoptosis as compared to excitotoxicity or other mechanisms on stroke outcome, and to upregulation of caspase-independent or redundant cell death pathways.

## Oxidative and Nitrosamine Stress

Free radicals subserve an important role in virtually all fundamental cell death pathways. Under resting conditions, mitochondria produce reactive oxygen species, such as superoxide and hydroxyl radicals, during electron transport. Oxidative and nitrosamine stresses are routinely modulated by endogenous antioxidant enzyme systems such as superoxide dismutase (SOD) and the nitric oxide synthase (NOS) family. Moreover, toxic free radicals are normally scavenged by SOD, catalane, and glutathione (GSH), and antioxidant vitamins (e.g., $\alpha$-tocopherol and ascorbic acid). Ischemia-induced accumulation of intracellular $Ca^{2+}$, $Na^{+}$, and ADP stimulates excessive mitochondrial oxygen free radical production, which overwhelms endogenous scavenging mechanisms. Reoxygenation during reperfusion provides large amounts of oxygen as a substrate for numerous enzymatic oxidation reactions in the cytosolic compartments, subcellular organelles, and mitochondria. There is a burst of oxygen free radicals from enzymatic conversions, including the cyclooxygenase-dependent conversion of arachidonic acid to prostanoids and degradation of hypoxanthine, the superoxide-mediated reduction of iron, and the generation of nitric oxide. Substantial amounts of free radicals are also generated during the inflammatory response after ischemia.

Oxygen radicals and oxidative stress facilitate mitochondrial transition pore (MTP) formation, which dissipates the proton-motive force required for oxidative phosphorylation and ATP generation. As a result, mitochondria release apoptosis-related proteins and other constituents within the inner and outer mitochondrial membranes. Free radical-mediated mechanisms signal the release of mitochondrial cytochrome *c* by mechanisms probably related to Bcl-2 and translocation of Bax. After its release, cytochrome *c* binds to Apaf-1 followed by caspase-9 to form the apoptosome complex that subsequently activates caspase-3 and other caspases. Activated caspase-3 goes on to cleave several nuclear DNA repair enzymes, culminating in apoptosis. Another major signaling pathway of oxygen radicals involves NF-$\kappa$B, an important transcription factor. NF-$\kappa$B activation results in its translocation into the nucleus and binding to the NF-$\kappa$B site of inducible genes such as cyclooxygenase-2 (COX-2), inducible nitric oxide synthase (iNOS), matrix metalloproteinases (MMPs), intercellular adhesion molecules (ICAM-1), and cytokines. Furthermore, after ischemia, the activation of NMDA receptors and formation of oxygen free radicals and nitric oxide by neuronal nitric oxide directly signal the mitochondrial release of cytochrome *c* or formation of peroxynitrite. Subsequent hydroxyl radical production results in lipid peroxidation and directly damages proteins, carbohydrates, and DNA, ultimately leading to necrotic cell death.

The significant contribution of free radical-mediated injury is apparent from studies showing reduced stroke-related tissue injury in mice overexpressing SOD, and in mice with reduced expression of the neuronal and inducible NOS isoforms. Treatment with free radical scavengers such as *N-tert*-butyl-$\alpha$-phenylnitrone ($\alpha$-PBN) is also shown to attenuate stroke-related injury. However, in clinical trials, free radical scavengers such as NXY-059 have failed to show any benefit.

## Inflammation

Inflammation is intricately related to atherosclerosis and stroke onset, as well as subsequent stroke-related tissue damage. Elevated stroke risk has been linked to high blood levels of inflammatory markers, such as C-reactive protein, interleukin-6 (IL-6), TNF-$\alpha$, and soluble intercellular adhesion molecule (sICAM). Arterial thrombosis associated with plaque ulceration is triggered by multiple processes involving endothelial activation and proinflammatory and prothrombotic interactions between the vessel wall and circulating blood elements. Ischemic stroke-related brain injury itself triggers inflammatory cascades within the parenchyma that further amplify tissue damage. Within a few hours after the onset of ischemia, the expression of several cytokines (including TNF-$\alpha$ and IL-1$\beta$) is upregulated in brain. By 24 h, brain microglia become activated. Adhesion molecules (including ICAM, and P- and E-selectins) are expressed in vascular endothelial cells, facilitating the attraction, rolling, and entering of blood-borne inflammatory cells into the ischemic brain – first neutrophils, at 24–48 h after ischemia, followed by lymphocytes and macrophages, peaking at 5–7 days after ischemia. Lymphocytes express inflammatory cytokines and chemokines (including TNF-$\alpha$ and IL-8), while macrophages participate in the scavenging of necrotic cell debris. The enzymes iNOS and COX-2 are also upregulated after cerebral ischemia, contributing to the formation of nitric oxide and oxygen free radicals.

There is evidence that the inflammatory cascade occurring after cerebral ischemia contributes to acute and delayed cell death. Blocking inflammatory cell infiltration or blocking or deleting the genes for various cytokines, adhesion molecules, or other inflammatory mediators reduces cell death in animal models of cerebral ischemia. On the other hand, infiltration of blood-borne cells following ischemia likely includes hematopoietic and other stemlike and progenitor cells, which may play a role in the repair of brain injury following stroke.

## Peri-infarct Depolarizations

Research in mouse, rat, and cat stroke models has shown that neurons and astrocytes in ischemic penumbral regions exhibit waves of spreading depression, detected as spreading negative slow-voltage variations. These waves, termed peri-infarct depolarizations (PIDs), exacerbate tissue damage due to the increased energy requirements for reestablishing ionic equilibrium in metabolically compromised ischemic tissues. In the initial 2–6 h after experimental stroke, PIDs result in a stepwise increase of core-infarcted tissue into adjacent penumbral regions. The frequency and total duration of spreading depression correlate with infarct size. Recent evidence suggests that PIDs contribute to the expansion of the infarct core throughout the period of infarct maturation. The significance of PIDs to human stroke pathophysiology is unclear; however, a recent study of patients with subarachnoid hemorrhage showed strong correlation between periods of spreading electrocorticographic depression and delayed ischemic neurological deficits and brain infarcts. These results suggest that spreading depolarizations with prolonged depressions may be a promising target for stroke treatment with pharmaceutical agents such as NMDA antagonists, or with physiological approaches such as hypothermia and hyperoxia.

## Contemporary Views on Stroke Pathophysiology: A Clinical Perspective

While each step in every cell death pathway provides an opportunity for pharmaceutical intervention, no drug has proved effective in clinical trials. Agents that have failed clinical testing include the lipid-peroxidation inhibitor tirilazad mesylate, the ICAM-1 antibody enlimomab, the calcium channel blocker nimodipine, the $\gamma$-aminobutyric acid (GABA) agonist clomethiazole, the glutamate antagonist and sodium channel blocker lubeluzole, the competitive NMDA antagonist selfotel, and noncompetitive NMDA antagonists such as dextrorphan, gavestinel, aptiganel, and eliprodil. The lack of efficacy can be related to diverse factors, including inadequate preclinical studies, inclusion of patients without evidence for salvageable tissue on brain imaging, and deficiencies in clinical trial methodology, particularly the use of crude 'global' outcome measures rather than outcome measures designed to assess specific modalities of neurological function, which may be more sensitive. From a basic science standpoint, the failure of clinical trials has exposed the shortcomings of traditional approaches that relied on targeting a single cell death pathway.

The overlapping features and interactions between pathways are now better appreciated, and newer approaches have focused on combining or adding drugs in series so as to target distinct pathways during the evolution of ischemic injury. Neuroprotective combinations that have shown success in animal models include the coadministration of an NMDA receptor antagonist with GABA receptor agonists, citicoline, free radical scavengers, cyclohexamide, and caspase inhibitors. While growth factors such as basic fibroblast growth factor (bFGF) have shown promise in promoting recovery after stroke, recent studies indicate that growth factors also exert powerful neuroprotective effects via downstream mechanisms, including upregulation of free radical-scavenging enzymes and antiapoptotic proteins. Therefore, it is not surprising that synergy is observed when combining citicoline with bFGF, and that caspase inhibitors given with bFGF or an NMDA receptor antagonist extend the therapeutic window with lower effective doses. Since early reperfusion with tissue plasminogen activator (tPA) has proved efficacy, attention is also being paid to neuroprotective strategies that decrease reperfusion injury, reduce postischemic hemorrhage, and inhibit downstream targets in cell death cascades. In animal models, thrombolysis has shown synergistic or additive effects with oxygen radical scavengers, AMPA receptor antagonists, NMDA blockers, MMP inhibitors, citicoline, topiramate, antileukocytic adhesion antibodies, and antithrombotics. Clinical trials have reported the feasibility and safety of tPA used in combination with clomethiazole or lubeluzole. However, in recent major clinical trials, the combination of the free radical scavenger NXY-059 with tPA showed no benefit.

Given the importance of targeting multiple pathways, more attention is being paid to physiological approaches such as modulation of brain temperature, optimizing blood glucose levels, improving brain oxygenation, and enhancing cerebral blood flow. While less specific, these approaches offer potential to interrupt nearly all deleterious biochemical and molecular pathways. For example, a reduction in brain temperature reduces cellular physiological processes and metabolic demands, as well as slows the rate of pathological processes such as lipid peroxidation and reduces the activity of certain cysteine or serine proteases. Hyperoxia favorably alters the levels of glutamate, lactate, Bcl-2, manganese SOD, and COX-2, and inhibits cell death mechanisms such as apoptosis. These physiological interventions are now being carefully tested in ongoing clinical trials.

Finally, it is now recognized that stroke-related injury is not simply restricted to cellular elements, but also involves the endothelial and vascular smooth muscle cells, the astro- and microglia, and the associated tissue matrix proteins – the integrated concept of the 'neurovascular unit.' Endothelial/astrocyte/matrix interactions are considered vital in protecting the integrity of brain tissue, and it is recognized that disruption of the neurovascular matrix, which includes basement membrane components such as type IV collagen, heparan sulfate proteoglycan, laminin, and fibronectin, upsets cell matrix and cell–cell signaling mechanisms that maintain neurovascular homeostasis. Emerging data show important linkages between tPA, MMPs, and brain edema and hemorrhage. The potential efficacy of MMP inhibitors in mitigating ischemic damage after stroke is being actively investigated.

## Summary

The current pathophysiological understanding of stroke is substantially based on experimental studies. Brain injury after cerebral ischemia results from a complex signaling cascade that evolves in a spatiotemporal pattern and differentially affects gray and white matter. Excitotoxicity leads to acute cellular necrosis predominantly in the core of the infarction. The surrounding ischemic penumbra suffers milder insults, with excitotoxic and inflammatory mechanisms leading to delayed cell death showing characteristics of apoptosis. Research to date has led to an improved understanding of the importance, timing, and functional relationships of individual mechanisms, and to a better understanding of the differences between humans and the animal models in which these cell death mechanisms were discovered. Despite advances in the knowledge of individual injury mechanisms and despite intense efforts to translate this knowledge to effective stroke treatments, no therapeutic strategy has proved successful. The traditional 'neurocentric' approach is quickly yielding to a modern strategy of targeting neurons and glial cells, in particular astrocytes, as well as the neurovascular unit. It is anticipated that advances in basic science research and molecular neuroimaging, combined with advances in the fields of genomics and proteomics – as well as refinements in patient selection, improved methods of drug delivery, use of more clinically relevant animal stroke models, and use of combination therapies that target the entire neurovascular unit – will soon make stroke neuroprotection an achievable goal.

*See also:* Animal Models of Stroke; Inflammation in Neurodegenerative Disease and Injury; Oxidative Damage in Neurodegeneration and Injury; Stroke: Neonate vs. Adult; Vascular Issues in Neurodegeneration and Injury.

## Further Reading

Barone FC and Feuerstein GZ (1999) Inflammatory mediators and stroke: New opportunities for novel therapeutics. *Journal of Cerebral Blood Flow & Metabolism* 19: 819–834.

Chan PH (2001) Reactive oxygen radicals in signaling and damage in the ischemic brain. *Journal of Cerebral Blood Flow & Metabolism* 21(1): 2–14.

Cramer SC, Koroshetz WJ, and Finklestein SP (2007) The case for modality-specific outcome measures in clinical trials of stroke recovery-promoting agents. *Stroke* 38: 1393–1395.

Dirnagl U, Iadecola C, and Moskowitz MA (1999) Pathobiology of ischaemic stroke: An integrated view. *Trends in Neuroscience* 22(9): 391–397.

Endres M and Dirnagl U (2002) Ischemia and stroke. *Advances in Experimental Medicine & Biology* 513: 455–473.

Fisher M (2004) The ischemic penumbra: Identification, evolution and treatment concepts. *Cerebrovascular Disease* 17(supplement 1): 1–6.

Gidday JM (2006) Cerebral preconditioning and ischaemic tolerance. *Nature Reviews Neuroscience* 7(6): 437–448.

Graham SH and Chen J (2001) Programmed cell death in cerebral ischemia. *Journal of Cerebral Blood Flow & Metabolism* 21: 99–109.

Hossmann KA (1996) Periinfarct depolarizations. *Cerebrovascular and Brain Metabolism Reviews* 8: 195–208.

Iadecola C, Cho S, Feuerstein GZ, et al. (2004) Cerebral ischemia and inflammation. In: Mohr JP, Choi DW, Grotta JC, et al. (eds.) *Stroke: Pathophysiology, Diagnosis, and Management*, pp. 883–893. Edinburgh: Churchill Livingstone.

Kroemer G and Reed JC (2000) Mitochondrial control of cell death. *Nature Medicine* 6: 513–519.

Lo EH, Dalkara T, and Moskowitz MA (2003) Mechanisms, challenges and opportunities in stroke. *Nature Reviews Neuroscience* 4(5): 399–415.

Lo EH, Moskowitz MA, and Jacobs T (2005) Exciting, radical, suicidal: How brain cells die after stroke. Review. *Stroke* 36(2): 189–192.

Mergenthaler P, Dirnagl U, and Meisel A (2004) Pathophysiology of stroke: Lessons from animal models. *Metabolic Brain Disease* 19(3–4): 151–167.

Mitsios N, Gaffney J, Kumar P, et al. (2006) Pathophysiology of acute ischaemic stroke: An analysis of common signaling mechanisms and identification of new molecular targets. *Pathobiology* 73(4): 159–175.

Nedergaard M and Dirnagl U (2005) Role of glial cells in cerebral ischemia. *Glia* 50(4): 281–286.

Petty MA and Wettstein JG (1999) White matter ischaemia. *Brain Research – Brain Research Reviews* 31: 58–64.

Ren JM and Finklestein SP (2005) Growth factor treatment of stroke. Review. *Current Drug Targets – CNS & Neurological Disorders* 4(2): 121–125.

Singhal AB, Lo EH, Dalkara T, and Moskowitz MA (2005) Advances in stroke neuroprotection: Hyperoxia and beyond. *Neuroimaging Clinics of North America* 15(3): 697–720, xii–xiii.

Syntichaki P and Tavernarakis N (2003) The biochemistry of neuronal necrosis: Rogue biology? *Nature Reviews Neuroscience* 4: 672–684.

Zukin RS, Jover T, Yokota H, et al. (2004) Molecular and cellular mechanisms of ischemia-induced cell death. In: Mohr JP, Choi DW, Grotta JC, et al. (eds.) *Stroke: Pathophysiology, Diagnosis, and Management*, pp. 829–854. Edinburgh: Churchill Livingstone.

# Stroke: Neonate vs. Adult

**A M Comi and M V Johnston**, Kennedy Krieger Institute, Johns Hopkins University, Baltimore, MD, USA



## Definition

Stroke is defined as brain injury occurring as a result of insufficient glucose and oxygen delivery to meet the needs of the involved brain tissue. Ischemic stroke is defined as brain injury resulting from blockage or other impairment of arterial blood delivery. The causes, molecular pathways, available treatments, and clinical outcomes are age dependant. This article focuses on contrasting neonatal arterial ischemic stroke with that in adults (see **Table 1**). Venous and hemorrhagic causes are not included. Neonatal arterial ischemic stroke is defined in this article as an arterial stroke presenting either (1) in the term infant during the first month of life, or (2) in the first year of life with early handedness and magnetic resonance imaging (MRI) evidence of a stroke presumed to have occurred in the neonatal period.

## Epidemiology

Overall, stroke is the third leading cause of death and the leading cause of disability in the United States. Although the incidence is highest in the elderly, about one-fifth of all strokes occur in individuals younger than 65 years. Recently, data have demonstrated that neonatal stroke occurs in approximately 1 in 4000 term live births. About 33% of children with cerebral palsy have the hemiplegic form affecting the motor functions of one side of the body. In many children, the hemiplegic cerebral palsy results from a stroke in the pre-, peri-, or postnatal period. Overall, the costs of stroke are projected to be in the tens of billions of dollars per year. Stroke in the developing brain is also an important cause of neurologic impairment that often persists into adulthood.

## Presentation

When recognized in the neonatal period, the most common presentation of stroke is seizures. Other less common presenting signs are lethargy or asymmetric limb movements. Other evidence of encephalopathy may also be seen, including poor suck, apnea, and hypotonia. In addition, pediatric patients with stroke often present with seizures. Children with ischemic middle cerebral artery stroke who present with seizures may be at increased risk for poor neurologic and functional outcome. When neonatal stroke presents later in the first year of life, the most common presentation is early handedness or other evidence of hemiparesis, with or without other developmental delays.

Stroke in adults is much less likely (compared with neonates) to present with acute seizures. Adult stroke is much more likely to present with the acute onset of a neurologic deficit such as a motor paresis (weakness of a limb), aphasia (inability to speak), alexia (inability to read), or apraxia (the inability to perform a previously learned motor task despite normal strength and coordination). The precise deficit in adult stroke is dependent on the location of the stroke and severity of the ischemic insult. Elegant localization of the stroke in an adult is often possible based on the constellation of neurologic deficits. This is usually not the case in neonates. Small vessel ischemia may also present with slowly evolving dementia and/or motor impairments in older adults.

## Location, Etiologies, Workup, and Management

The most common site of neonatal arterial stroke is in the distribution of the middle cerebral artery and on the left side. Less commonly, neonatal ischemic stroke may occur in a watershed distribution. Stroke in the adult is more variable and depends on the age and risk factors of the individual, and whether it is secondary to large or small vessel disease or embolism.

In elderly adults, the most common causes of ischemic stroke are cardiovascular, resulting in thrombotic, lacunar, or embolic strokes that may involve the cortical, subcortical, or brain stem structures. Diagnostic workup often includes an MRI (including diffusion and perfusion) of the brain, carotid ultrasound, and cardiac echo, depending on the location and type of stroke. In adults younger than 65 years, underlying causes of ischemic stroke are generally more diverse and include thrombophilias, infection, dissection, autoimmune and inflammatory causes, and malignancy. Workup in these younger adults therefore more often includes testing such as magnetic resonance angiogram (MRA) or computerized tomography angiogram (CTA) of the neck, a thrombophilia workup, a lumbar puncture, and testing for sources of autoimmunity or cancer.

Ischemic stroke in the neonate deserves a workup as well, although a specific etiology is less often found. Inherited thrombophilias may be sought, although a recent study of candidate gene polymorphisms in genes regulating thrombosis and thrombolysis did

**Table 1** Differences between neonatal and adult stroke

| | *Neonate* | *Adult* |
|---|---|---|
| Frequently presents with seizures | Yes | No |
| Frequently presents with focal deficits | No | Yes |
| Sex-related differences | Yes, but different from adults | Yes, but different from neonates |
| Prolonged prominent apoptosis after stroke | Yes | Not so much |
| Oxidative reserves | Less | More |
| Specific treatment available | No | Yes: tissue plasminogen activator (TPA) in some cases |
| Recurrence risk | Very low | Generally much higher |

not identify any variant allele more common in newborns with stroke than in the normal controls. An MRI/MRA with diffusion imaging is needed to determine the location and nature of the stroke and extent of brain damage. Infection should be excluded and an echocardiogram considered as well. Maternal medical issues and placental pathology also have important roles in the etiology of neonatal stroke.

Recurrence risk for ischemic stroke is related to the underlying etiology and ongoing risk factors for stroke. The recurrence risk for neonatal stroke overall is very low. It is likely that in most neonates the stroke occurs as a result of a convergence of factors related to the process of transitioning from the intranatal state to the postnatal state. As a result, these children are not likely to experience recurrence of stroke. It is not clear to what extent this recurrence risk is modified by the existence of one or more thrombophilias. Major congenital cardiovascular anomalies do carry a substantial risk for stroke recurrence. In young adults, the recurrence risk depends on whether the underlying cause is ongoing but in general is higher than in neonates. Ischemic stroke in the elderly is most often caused by multivessel disease, and overall recurrence risk is high.

The acute management of ischemic stroke in adults includes supportive measures such as preventing fevers, managing intracranial pressure and edema, and ventilation if required. Specific treatment is available in the form of tissue plasminogen activator (TPA) given either peripherally or intravascularly to break up the clot producing the ischemia. TPA is limited to the subset of patients presenting in the first few hours after stroke with no contraindication, but in this setting can be very helpful in reducing the morbidity from stroke. Aspirin is another antithrombotic treatment shown to be beneficial for the acute treatment of stroke in adults. No randomized clinical trials have addressed the acute management of perinatal stroke. Supportive care is also provided to neonates after a stroke, but anticoagulation is generally not used except for treatment of a known cardioembolic stroke.

## Gender Differences

Clinical ischemic stroke has long been recognized to be sexually dimorphic in adults; most studies have shown that while women have lower stroke incidence relative to men, early outcome may be more favorable in men. This difference has been largely attributed to a protective estrogen effect in women; the risk in postmenopausal women equals that in age-matched men. Until recently, little data were available from clinical trials in adults regarding gender and response to treatment. Female sex is a mildly unfavorable prognostic factor in rehabilitation outcome after stroke.

In the pediatric population, neonatal arterial ischemic strokes are more commonly diagnosed in boys. Boys are at higher risk for all stroke types than girls. In addition, male gender may be a prognostic factor for poor outcome after stroke. Little attention, however, has been given to whether or how clinical interventions for neonatal or pediatric stroke should be tailored for patient gender.

Very recent studies suggest that sex may influence the response to treatment of ischemic brain injury. The impact of animal sex on susceptibility to brain injury has recently been examined *in vitro*. Du and colleagues studied sex-segregated neuronal cultures and found that XX and XY cultures responded differently to cytotoxic and pharmacological manipulation. They reported that XY neurons were more sensitive to nitrosative stress and excitotoxicity than XX neurons. After $ONOO^-$ treatment, XY neurons demonstrated apoptosis-inducing factor (AIF)-mediated programmed cell death, but XX neurons underwent predominantly caspase-dependent apoptosis. Sex-related differences in response to hypoxia-ischemia have also been shown experimentally *in vivo*. Results of experiments with hypoxia-ischemia in 7-day-old mice showed that knocking out Parp-1 protected males but not females. These sex-related differences could be related to differences in expression of genes known to be on the X chromosome, such as AIF or X-linked inhibitor of apoptosis (XIAP). Poly(ADP-ribose) polymerase-1 (PARP-1) is a nuclear enzyme activated by DNA strand breaks that participates in DNA repair and replication. On the other hand, a similar experiment in adult Parp-1 knockout mice demonstrated that female adults sustained worse injury rather than less injury. Recently, Zhu et al. showed that in P9 C57/BL6J mice, neuronal death after hypoxic-ischemic injury proceeded predominantly via an AIF-mediated pathway in males versus a caspase-dependent pathway in females. These findings support the conclusion that the pathways mediating ischemic brain injury and neuroprotection may differ in males and females. This sex-related difference is relevant to both adults and neonates, but manifests differently in these two age groups.

## Mechanisms of Neuronal Injury

Both adult and neonatal animal models for stroke have been developed, and the basic pathways involved are similar. Broadly speaking, a core of severely injured tissue is formed at the epicenter of the ischemia that develops into pan-necrosis. Surrounding this core is an area of underperfused tissue, the penumbra. The penumbral tissue is at risk of becoming included in the infarcted core. Cerebral ischemia triggers the release of the excitatory neurotransmitter glutamate, and the energy uncoupling produced by ischemia prevents the active removal of glutamate from the synaptic cleft. Excess glutamate overstimulates postsynaptic receptors, thereby opening the associated ion channels. This allows sodium and calcium ions to enter the cell, while potassium flows outward. Excessive calcium influx initiates a cascade of energy-consuming and ultimately lethal intracellular events.

However, *in vivo* models have revealed important differences in the immature brain as compared to the adult brain exposed to ischemic injury. In the neonatal animal, apoptosis evolves over the week following the initial ischemic insult or longer. This contrasts with a much shorter period of apoptosis following ischemic injury in the adult brain. A range of morphologies from necrosis to apoptosis is more common in the neonatal brain. The oxidative reserve of the neonatal brain, in terms of its high oxygen demand and low concentration of antioxidants, is also less compared to that in the mature animal. This in conjunction with the high concentrations of unsaturated fatty acids and bioavailable iron suggests that the neonatal brain is more susceptible to oxidative damage compared with the mature brain. In the neonate, activation of the $Cl^-$ permeable $\gamma$-aminobutyric acid type A ($GABA_A$) receptors is excitatory and has been linked to a higher propensity for seizures. The immature brain also differs greatly from the mature brain in terms of the ion channel subunits expressed, leading to important differences in excitation and patterns of regional vulnerability to ischemic injury. These important differences make it essential that developmentally appropriate models are used for mechanistic and preclinical studies and emphasize the need for clinical trials in neonates as well as adults with stroke.

Seizures induced by global hypoxia, global ischemia, or hypoxic-ischemic injury have been noted in adult and immature rats. Studies examining interactions between seizures and hypoxic-ischemic brain injury in developing rats have produced various results. Status epilepticus induced by the GABA antagonist bicuculline after hypoxia-ischemia did not exacerbate brain damage in postnatal day 7 (P7) rats. On the other hand, in P10 rats subjected to unilateral carotid ligation and brief hypoxia, seizures induced by kainate significantly exacerbated brain injury. It is controversial and difficult to determine, however, whether seizures occur only secondary to the injury or in fact contribute to the ischemic brain injury.

## Outcome

Outcome in neonatal stroke is variable. Although the majority of children survive their stroke, about 75% have sequelae including cerebral palsy. In one study, 46% of children followed after neonatal stroke developed cerebral palsy. The assessment of motor movements in the weeks following birth can be useful in the early identification of hemiparesis. A delayed presentation is associated with the developmental of cerebral palsy, as are the radiologic findings of large stroke size and injury to Broca's area, the internal capsule, Wernicke's area, or the basal ganglia. Other long-term complications include epilepsy and learning disabilities. Outcome in adults depends on the age, associated medical comorbidities, type of stroke, and recurrence risk of the individual involved but is generally less good than in neonates. Nevertheless, the potential for lifelong neurologic impairments makes stroke in the neonate an important national health issue.

## Prevention

Prevention of stroke in adults is aimed at decreasing or treating the risk factors for stroke such as hypertension or hypercholesterolemia, eliminating sources of embolic clots such as the heart or carotid artery, and anticoagulation. In contrast, prevention of neonatal stroke risk factors (such as preeclampsia, extracorporeal membrane oxygenation (ECMO), maternal diabetes, chorioamnionitis, placental pathology, trauma, infection, or polycythemia) is often currently not possible and recurrence risk is low in most cases. No randomized clinical trials have been done to address the issue of primary or secondary prevention in neonatal stroke. Maternal thromboprophylaxis for maternal thrombophilia remains speculative, and aspirin use is generally not recommended after neonatal stroke.

## Future Directions

More rigorous preclinical study and clinical trial design may produce future success in neuroprotection. Other promising future approaches include enhancing the endogenous regenerative response with growth factors. A recent pilot trial with granulocyte colony-stimulating factor (GCSF), given to adults after stroke, reported that CD34+ cells were mobilized and the treatment was safe and well tolerated. Recent genetic studies have identified loci that confer susceptibility to stroke and have future application in the management of ischemic stroke patients. Advances in the treatment of stroke in adults will require careful study in developmental models and neonatal trials before they can be applied in neonates.

*See also:* Animal Models of Stroke; Stroke: Injury Mechanisms.

## Further Reading

Ashwal S, Russman BS, Blasco PA, et al. (2004) Practice parameter: Diagnostic assessment of the child with cerebral palsy: Report of the quality standards subcommittee of the American Academy of Neurology and the Practice Committee of the Child Neurology Society. *Neurology* 62: 851–863.

Chalmers EA (2005) Perinatal stroke-risk factors and management. *British Journal of Haematology* 130: 333–343.

Du L, Bayir H, Lai Y, et al. (2004) Innate gender-based proclivity in response to cytotoxicity and programmed cell death pathway. *Journal of Biological Chemistry* 279: 38563–38570.

Dzhala VI, Talos DM, Sdrulla DA, et al. (2005) NKCC1 transporter facilitates seizures in the developing brain. *Nature Medicine* 11(11): 1205–1213.

Ferriero DM (2004) Neonatal brain injury. *New England Journal of Medicine* 351: 1985–1995.

Hagberg B, Hagberg G, Beckung E, and Uvebrant P (2001) Changing panorama of cerebral palsy in Sweden. VIII: Prevalence and origin in the birth year period 1991–94. *Acta Paediatrica* 90: 271–277.

Hagberg H, Wilson MA, Matsushita H, et al. (2004) PARP-1 gene disruption in mice preferentially protects males from perinatal brain injury. *Journal of Neurochemistry* 90: 1068–1075.

Hunt RW and Inder TE (2006) Perinatal and ischaemic stroke: A review. *Thrombosis Research* 118: 39–48.

Jensen FE (2002) The role of glutamate receptor maturation in perinatal seizures and brain injury. *International Journal of Developmental Neuroscience* 20: 339–347.

Lee JL, Croen LA, Lindan C, et al. (2005) Predictors of outcome in perinatal arterial stroke: A population-based study. *Annals of Neurology* 58: 303–308.

Lynch JK, Hirtz DG, de Veber G, and Nelson KB (2002) Report of the National Institute of Neurological Disorders and Stroke workshop on perinatal and childhood stroke. *Pediatrics* 109: 116–123.

Miller SP, Wu YW, Lee J, et al. (2006) Candidate gene polymorphisms do not differ between newborns with stroke and normal controls. *Stroke* 37: 2678–2683.

Nakajima W, Ishida A, Lange MS, et al. (2000) Apoptosis has a prolonged role in the neurodegeneration after hypoxic ischemia in the newborn rat. *Journal of Neuroscience* 20: 7994–8004.

Paolucci S, Bragoni M, Coiro P, et al. (2006) Is sex a prognostic factor in stroke rehabilitation? A matched comparison. *Stroke* 37: 2989–2994.

Rice JE III, Vannucci RC, and Brierley JB (1981) The influence of immaturity on hypoxic-ischemic brain damage in the rat. *Annals of Neurology* 9: 131–141.

Sprigg N, Bath PM, Zhao L, et al. (2006) Granulocyte-colony-stimulating factor mobilizes bone marrow cells in patients with subacute ischemic stroke. *Stroke* 37: 2979–2983.

Towfighi J, Mauger D, Vannucci RC, and Vannucci SJ (1997) Influence of age on the cerebral lesions in an immature rat model of cerebral hypoxia-ischemia: A light microscopic study. *Brain Research Developmental Brain Research* 100: 149–160.

Wang MM (2006) Genetics of ischemic stroke: Future clinical applications. *Seminars in Neurology* 26(5): 523–530.

Wu YW, Lynch JK, and Nelson KB (2005) Perinatal arterial stroke: Understanding mechanisms and outcomes. *Seminars in Neurology* 25(4): 424–434.

Zhu C, Xu F, Wang X, et al. (2006) Different apoptotic mechanisms are activated in male and female brains after neonatal hypoxia-ischemia. *Journal of Neurochemistry* 96(4): 1016–1027.

# Aging: Invertebrate Models of Normal Brain Aging

**M Artal-Sanz, K Troulinaki, and N Tavernarakis**, Foundation for Research and Technology, Heraklion, Crete, Greece



## Introduction

The free-living soil nematode *Caenorhabditis elegans* has been extensively used for studying the genetic regulation of aging, in part because of its short life span and genetic homogeneity. In the nematode, neuroendocrine signaling, nutritional sensing, and mitochondrial functions have been shown to play important roles in the determination of life span. However, aging in *C. elegans* is mainly controlled by a neuroendocrine system, the DAF-2/insulin signaling pathway, which also regulates the life span of flies and mammals, indicating that this pathway is a universal longevity regulator.

The insulin signaling pathway in *C. elegans* was first genetically identified for its effects on dauer larva formation (DAF). Dauer is an alternative developmental stage induced by harsh environmental conditions such as starvation, high population density, or high temperature. Under normal conditions *C. elegans* develops to reproductive adulthood through four larval stages (L1–L4), in 3 days. However, when conditions are adverse, larvae arrest development at the second molt, to enter the dauer stage. Dauers do not feed, are resistant to stress, and can survive up to several months. Dauer larvae are considered to be nonaging because postdauer life span is not affected by the duration of the dauer stage. In addition to insulin signaling, another neuronal pathway that regulates the choice between reproductive growth and dauer entry is the DAF-7 transforming growth factor-$\beta$ (TGF-$\beta$) pathway.

The DAF-2 insulin/insulin-like growth factor-1 (IGF-1) receptor pathway is required for reproductive growth and metabolism, as well as for normal life span. *C. elegans* has a single transmembrane insulin receptor kinase, DAF-2. Upon ligand binding to DAF-2, the kinase domain of the receptor phosphorylates and activates AGE-1, which is a phosphatidylinositol 3-kinase (PI3K). Activated AGE-1 PI3K generates 3-phosphoinositides (PtdIns-3,4-P2 and PtdIns-3,4,5-P3), which are second messengers required for activation of downstream kinases. Downstream kinases include pyruvate dehydrogenase kinase and serine/threonine kinases (PDK-1, AKT-1, and AKT-2), which are protein kinase B (PKB) proteins. These protein kinases regulate the forkhead (FOXO) transcription factor DAF-16, which translocates to the nucleus depending on its phosphorylation level. Phosphorylated DAF-16 remains inactive in the cytoplasm, while upon dephosphorylation it enters the nucleus and exerts its effects on transcription. Thus, the insulin signaling pathway functions to block the nuclear localization of DAF-16. An antagonist of the DAF-2/AGE-1 signaling pathway is the DAF-18 phosphatase and tensin homolog (PTEN) lipid phosphatase (**Figure 1**).

## Neuronal Insulin-Like Signaling

Mutations in *daf-2*, *age-1*, or in other genes positively regulated by *daf-2* result in constitutive developmental arrest at the dauer stage. Reducing the activity of genes that antagonize insulin signaling, such as *daf-18* PTEN or *daf-16* FOXO, suppresses the dauer-constitutive phenotype of insulin signaling mutants. While severe mutations in the insulin pathway induce dauer arrest, a milder reduction of insulin signaling results in longer life span. For example, when temperature-sensitive *daf-2* and *age-1* mutants are grown at the permissive temperature until past the dauer arrest decision point, and then shifted to higher temperatures, *daf-2* and *age-1* mutants show increased life span that is dependent on DAF-16. DAF-16 is the main downstream target and major effector of DAF-2/insulin-like signaling regulating *C. elegans* life span. Signaling via the DAF-2/insulin-like receptor antagonizes the FOXO transcription factor DAF-16 by promoting its phosphorylation. Similar modulation of insulin-like signaling pathways in the fruit fly and mouse also modify life span.

The insulin/IGF-1 signaling pathway was first linked to aging in *C. elegans* when mutations in *daf-2* were found to double the life span of the worm. Subsequent investigations, aimed at identifying in which specific cells insulin signaling controls animal aging, supported a prominent role for the nervous system. In these studies, the life span of mosaic animals that had lost *daf-2* activity in different cell lineages was examined. After fertilization, the first cell division produces the AB and the $P_1$ cells. The AB descendants produce most of the neurons, the hypodermis, the pharynx, the excretory glands, and the vulva, while the $P_1$ descendants give rise to the muscles, the germ line, the somatic gonad, the hypodermal cells, a few pharyngeal cells, and a few neurons. AB mosaics, which had lost *daf-2* activity in the AB lineage but were *daf-2*(+) in the $P_1$ lineage, lived twice as long

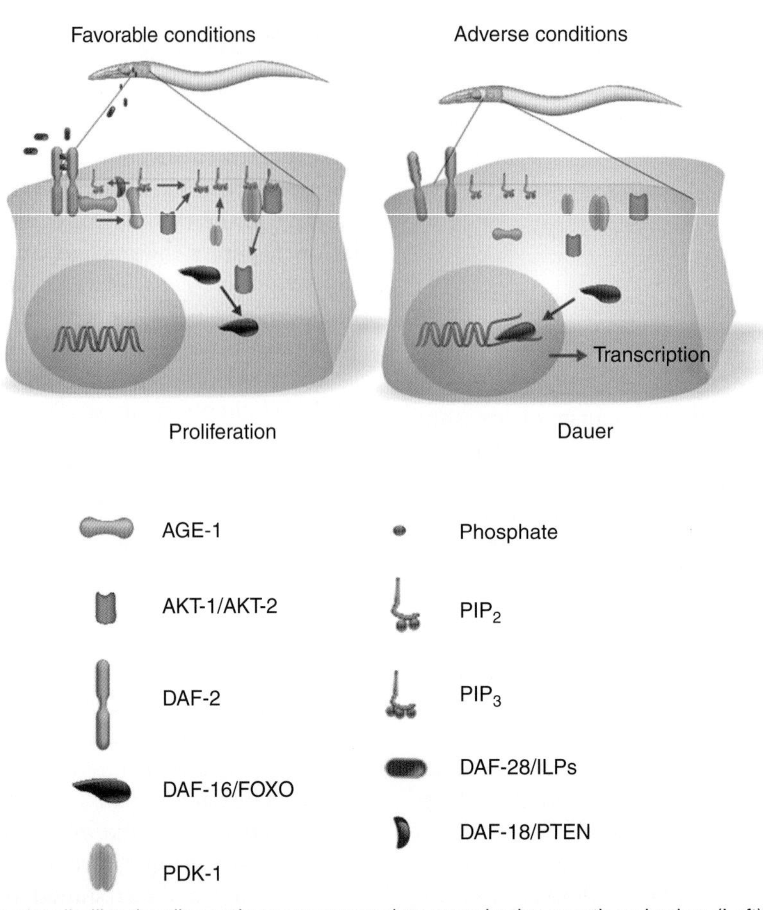


**Figure 1** The *C. elegans* insulin-like signaling pathway genes regulate reproductive growth and aging. (Left): Under favorable growing conditions, insulin-like peptides (ILPs) are produced from sensory neurons, to promote the reproductive mode. Binding of ILPs to the DAF-2/insulin receptor results in the phosphorylation of AGE-1/PI3K. Activated AGE-1 generates the phosphoinositides required for the activation of downstream kinases (PDK-1 and AKT-1). This conserved signaling cascade phosphorylates the transcription factor DAF-16/FOXO, preventing its nuclear localization. Retention of DAF-16 in the cytoplasm leads to normal reproductive growth and aging. (Right): Under unfavorable conditions (e.g., crowding or starvation), insulin signaling is inhibited, resulting in the nuclear localization of DAF-16. In the nucleus, DAF-16 regulates the transcription of genes that induce dauer entry and extend life span. Protein nomenclature: AGE, advanced glycosylation end product; AKT, serine/threonine kinase; DAF, abnormal dauer formation; FOXO, forkhead transcription factors, group O; PDK, pyruvate dehydrogenase kinase; PIP, phosphoinositol phosphate; PTEN, phosphatase and tensin homolog. Illustration: Liesbeth de Jong.

as wild-type animals. Although genetic mosaic analyses of *daf-2* support the interpretation that DAF-2 signaling from the nervous system controls longevity, these experiments did not assign longevity control by *daf-2* to specific cell types. In a complementary approach, cell-type-restricted promoters were used to drive the expression of *daf-2* and *age-1* cDNAs in *daf-2* and *age-1* mutants, respectively. Transgenic expression of *daf-2* and *age-1* in neurons suppressed the life span extension phenotype of the corresponding *daf-2* and *age-1* mutants. Life span extension is not rescued when insulin signaling is restored in muscles or in intestinal cells. However, tissue-specific expression, genetic mosaic analysis, and RNA interference (RNAi) experiments indicate that *daf-16* FOXO activity in neurons accounts for not more than 20% of the longevity seen in *daf-16*(−); *daf-2*(−) double mutant animals. Instead, intestinal expression of *daf-16* is sufficient to extend the life span of these animals by 50–60%. These findings indicate that an intricate signaling network regulates aging in *C. elegans* and that neuronal insulin-like signaling controls life span by producing downstream signals that control aging of nonneuronal target tissues.

## DAF-2 Insulin Receptor Function in the Nervous System

The *C. elegans* genome encodes more than 30 insulin-like ligands that might mediate input to the *daf-2* pathway through environmental cues, such as nutritional

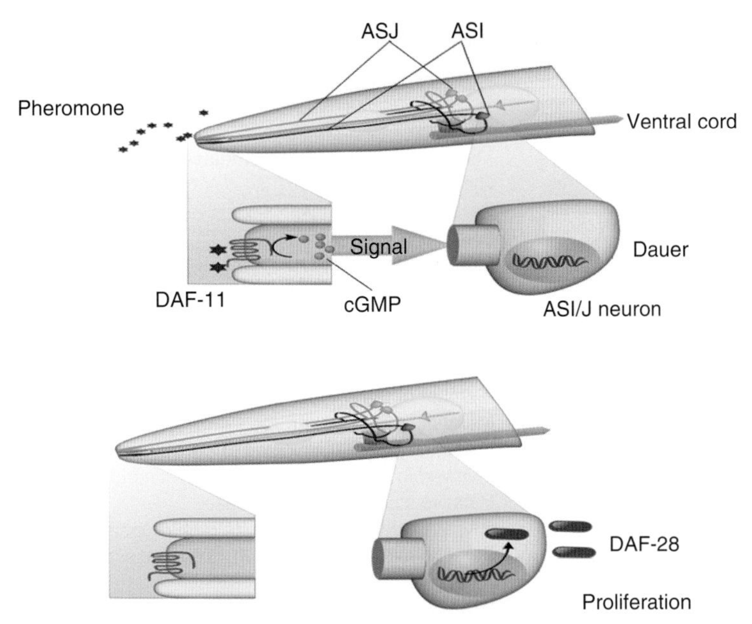


**Figure 2** Sensory neurons couple environmental cues to the production of ILPs, such as DAF-28. In the ciliated sensory neurons ASJ and ASI, DAF-11/guanylyl cyclase transduces dauer pheromone signals to cyclic guanosine monophosphate (cGMP) gated ion channels to induce dauer arrest (upper panel). At low pheromone concentration, DAF-28 is produced to activate DAF-2/insulin signaling and reproductive growth (lower panel). Illustration: Liesbeth de Jong.

status or growth conditions. These insulin genes are mainly expressed in neurons, although they are also found in the intestine, epidermis, muscle, and gonad. Some of the insulin-like peptides (ILPs) have been shown to influence longevity. Thus, neuroendocrine control of aging may entail environmental cues that influence neuronal production of insulin-like ligands. For example, *daf-28* encodes an insulin-like protein, which when mutated causes dauer arrest and downregulation of DAF-2 signaling. *daf-28* is expressed in two sensory neurons (ASI and ASJ) that regulate dauer arrest. The presence of dauer pheromone is sensed by DAF-11/guanylyl cyclase, which in turn downregulates *daf-28*. Conversely, in the absence of dauer pheromone, DAF-28 is produced to induce reproductive growth (**Figure 2**). Although the dauer pheromone does not appear to influence aging, ILPs can act as either agonists or antagonists on DAF-2 to regulate metabolism, reproductive growth, and life span.

In support of this notion, mutations in two genes involved in $Ca^{2+}$-regulated secretion (*unc-64*, encoding syntaxin, and *unc-31*, encoding calcium-dependent activator protein for secretion, or CAPS), result in an increased life span that is dependent on *daf-16*. *unc-31* is expressed exclusively in neurons, and although *unc-64* is expressed in many secretory tissues, including the nervous system and the intestine, it is the function of *unc-64* in neurons that influences aging. In mammals, insulin secretion by $\beta$ cells in the pancreas is a $Ca^{2+}$-regulated process. Therefore, a possible explanation is that the life span extension of *unc-64* and *unc-31* mutants is due to decreased secretion of a DAF-2 insulin-like ligand. Alternatively, *unc-64* and *unc-31* could regulate neurotransmitter input to other insulin-producing cells.

It has also been suggested that oxidative damage to neurons may be a primary determinant of life span. Loss of DAF-2 activity results in the activation of the FOXO transcription factor DAF-16, which controls the expression of antioxidant enzymes, such as superoxide dismutase (SOD) and catalase. Increased expression of these and possibly other free-radical-scavenging enzymes may protect neurons from oxidative damage. Thus, neuronal *daf-2* signaling might regulate animal life span by controlling the integrity of specific neurons that secrete neuroendocrine signals that might regulate the life span of target tissues. In support of this hypothesis, overexpression of Cu/Zn SOD exclusively in *Drosophila* motor neurons extends life span.

## DAF-16 Targets

An important question that arises is how the insulin/IGF-1 pathway ultimately regulates aging. As noted earlier, the DAF-16/FOXO transcription factor is a downstream effector of insulin signaling and regulates a wide range of physiological responses by altering the expression of genes involved in metabolism and energy generation, as well as antimicrobial and cellular stress response genes (**Table 1**).

**Table 1** Proteins implicated in neuroendocrine control of *C. elegans* life span, encoded by genes under control of the DAF-16/FOXO, transcription factor

| *Protein* | *Brief description* |
|---|---|
| SOD-3 | Manganese superoxide dismutase |
| CTL-1 | Cytosolic catalase |
| CTL-2 | Peroxisomal catalase |
| GST-4 | Glutathione-*S*-transferase |
| MTL-1 | Metallothionein-related cadmium-binding protein |
| OLD-1 | Putative receptor tyrosine kinase |
| SCL-1 | Secreted protein with sperm-coating protein (SCP) domain |
| LYS-7 | Lysozyme |
| LYS-8 | Lysozyme |
| DOD-1 | Member of the cytochrome P450 family |
| DOD-13 | Member of the cytochrome P450 family |
| DOD-16 | Cytochrome P450, involved in the oxidation of arachidonic acid to eicosanoids |
| FAT-7 | Involved in the biosynthesis of polyunsaturated fatty acid |
| DOD-11 | Putative alcohol dehydrogenase |
| GPD-2 | Glyceraldehydes-3 phosphate dehydrogenase |
| DAO-3 | Putative tetrahydrofolate dehydrogenase/cyclohydrolase |
| DOD-14 | Alcohol dehydrogenase |
| DOD-15 | UDP-glucuronosyl and UDP-glucosyl transferases |
| GCY-6 | Putative guanylyl cyclase |
| GCY-18 | Guanylate cyclase catalytic domain |
| VIT-2 | Vitellogenin |
| VIT-5 | Vitellogenin |
| PES-2 | Unknown function, putative role in ubiquitin-mediated protein degradation |
| PEP-2 | Oligopeptide transporter |
| HSP16 | Heat shock protein family |
| SIP-1 | Heat shock protein |
| INS-7 | Insulin/insulin-like growth factor-1 peptide |
| PNK-1 | Pantothenate kinase |
| MRP-5 | Adenosine triphosphate-binding protein, member of the subfamily C transporters |

SOD is one of the most effective intracellular enzymatic antioxidants. This antioxidant enzyme catalyzes the dismutation of superoxide anions ($O_2^-$) to oxygen ($O_2$) and to the less reactive species, hydrogen peroxide ($H_2O_2$). *sod-3* encodes one of two manganese-containing SODs. It is localized in the mitochondrial matrix and is abundant in neural tissues. *sod-3* was one of the first recognized targets of DAF-16. *sod-3* mRNA is increased in *daf-2* mutants and undetectable in the absence of DAF-16; microarray experiments show that expression of this gene is at least tenfold higher in wild-type worms compared to DAF-16-deficient mutants. Moreover, SOD-3 is upregulated in long-lived strains, such as *age-1* and *daf-2* mutants, and in response to exogenously imposed oxidative stress in a DAF-16-dependent manner. Therefore, increased detoxification from damaging free radicals by this enzyme may contribute to life span extension.

DNA microarray analysis shows that *daf-16* affects the expression of stress response genes, the products of which directly influence aging. For example, expression of the catalase genes *ctl-1* and *ctl-2*, the glutathione-*S*-transferase gene *gst-4*, and the small heat shock protein genes increases when the activity of *daf-2* is reduced, whereas the expression of these genes decreases with the reduction of *daf-16* activity. All of these genes function to promote longevity, probably by preventing or repairing oxidative and other forms of macromolecular damage.

The gene *mtl-1*, which encodes the basally expressed form of metallothionein, is another known target of DAF-16. Metallothioneins are small cysteine-rich, metal-binding proteins protecting cells against heavy metal toxicity and reactive oxide species (ROS)-associated damage. They are induced under a variety of stress conditions, such as conditions involving metal ions, inflammation, glucocorticoids, or oxidative stress. The expression of metallothionein genes is increased in *daf-2* mutants, compared to wild-type animals. Elevated levels of these proteins as well as antioxidant enzymes are expected to decrease ROS-associated damage and hence extend life span.

An additional downstream signaling factor that is regulated by insulin/IGF-1 signaling and promotes longevity is OLD-1, a transmembrane tyrosine kinase. OLD-1 expression is increased in long-lived *daf-2* and *age-1* mutants. This increase is dependent on DAF-16. Moreover, OLD-1 is necessary for the increased longevity of *daf-2* and *age-1* mutants and its overexpression increases life span and stress resistance. *old-1* mutations render animals more sensitive to ultraviolet light, starvation, and heat stress. These results point to a positive regulatory role for OLD-1 in life span and stress resistance.

The gene *scl-1* (which encodes sperm-coating protein (SCP)-like extracellular protein) is another target of DAF-16 that is essential for life span extension. SCL-1 has an SCP domain and is homologous to the mammalian cysteine-rich secretory protein (CRISP) family. These proteins enter the secretory pathway and are either released or anchored extracellularly by a transmembrane domain or a glycophosphatidylinositol (GPI) anchor. Expression of *scl-1* is elevated in long-lived *daf-2* and *age-1* mutants and is required for their extension of life span, since downregulation of *scl-1* reduces both life span and stress resistance of these animals. However, *scl-1* is not expressed in *daf-16* mutants. *scl-1* expression correlates with dauer morphogenesis; *scl-1* expression increases (fivefold) as worms enter dauer morphogenesis and decreases (four- to sevenfold) during dauer exit. This implicates *scl-1* in both aging and dauer formation. CRISP family proteins are involved in host defense systems in various organisms, and several functions have been postulated for SCP domain proteins, including cell adhesion, ligand function, protease inhibition, and other enzymatic activities. SCP proteins function in a variety of biological processes that involve signaling, which can be reconciled with SLP-1 being a secreted protein. However, the biochemical function of SCL-1 remains unknown.

DNA microarray analysis has also revealed a number of other *daf-2*/*daf-16*-regulated genes with substantial effects on life span. These include antimicrobial genes encoding lysozymes (*lys-7* and *lys-8*). Lysozymes are upregulated in *daf-2* mutants, and RNAi with these genes suppresses longevity of *daf-2* mutants. Consistently, *daf-2* mutants are resistant to bacterial pathogens. Other genes induced in *daf-2* and repressed in *daf-16* mutants encode proteins potentially involved in the synthesis of steroid or lipid-soluble hormones (e.g., cytochrome P450s, estradiol dehydrogenases, esterases, alcohol/short-chain dehydrogenases, and UDP-glucuronosyltransferases), and genes involved in fatty acid desaturation.

Genes with expression decreased in *daf-2* mutants and increased in *daf-16* mutants include those (*gcy-6* and *gcy-18*) encoding two receptor guanylate cyclases that are expressed in neurons. Further, RNAi with these genes prolongs life span, indicating a role for insulin signaling in sensing the environment. Other genes in this group include vitellogenin genes (*vit-2* and *vit-5*; yolk protein/apolipoprotein-like) and some genes for proteases and metabolic enzymes, such as amino- and carboxypeptidases; also included are *pes-2* (encoding a protein associated with ubiquitin-mediated protein degradation) and an amino oxidase gene, an aminoacylase gene, and an oligopeptide transporter (*pep-2*) gene. Inhibition of several of these genes results in life span extension, leading to the suggestion that the life span extension of *daf-2* mutants may involve reduced turnover of specific proteins.

Studies utilizing computational tools have also identified putative DAF-16 transcriptional targets. The genomes of both *C. elegans* and *Drosophila* were surveyed for genes with DAF-16 binding sites. Orthologous genes identified were further examined in wild-type and *daf-2* mutants animals; *pnk-1* and *mrp-5* were among several putative targets. *pnk-1* encodes a pantothenate kinase that is involved in the biosynthesis of coenzyme A, which is key to fat metabolism. This gene is upregulated in *daf-2* mutants, and RNAi inactivation results in a dramatic decrease of life span of both wild-type and *daf-2* mutant worms. *daf-2* mutants have increased fat storage, which may reflect *pnk-1* upregulation. *mrp-5* encodes an adenosine triphosphate-binding cassette C transporter. Proteins of this family modulate the secretion of insulin and participate in the transport of glutathione and nucleoside analogs. Inactivation of this gene by RNAi extends life span. It has been proposed that MRP-5 could affect aging by regulating the secretion of insulin or the transport of glutathione, an enzyme required for the antioxidant defense of the organism.

It is known that both heat shock and reduced insulin signaling trigger the nuclear localization of DAF-16, where it promotes the expression of small heat shock protein (*shsp*) genes (e.g., *hsp-16.1*, *hsp-16.49*, *hsp-12.6*, and *sip-1*). The *C. elegans* transcription factor, heat-shock factor-1 (HSF-1), is also required for the life span extension, as observed in *daf-2* mutants. HSF-1 acts in response to heat stress and reduced insulin signaling to activate the expression of *shsp* genes, together with DAF-16. Interestingly, overexpression of the gene encoding HSP70F increases longevity, similar to overexpression of the *shsp* gene *hsp-16*. Heat shock proteins are involved in reparation of misfolded or damaged proteins and are essential for recovery of cells after heat treatment. This indicates that protein misfolding and aggregation are important factors in aging.

In addition to inducing the expression of genes involved in several processes, DAF-16 can also act as a transcriptional repressor. For example, DAF-16 inhibits the expression of *ins-7*. This gene encodes an insulin/IGF-1 peptide. Its expression is repressed in animals with reduced DAF-2 activity and is elevated in animals with reduced DAF-16 activity. RNAi for *ins-7* increases the life span of wild-type animals and the frequency of dauer formation. Thus, INS-7 behaves as a putative DAF-2 agonist. It is hypothesized that when DAF-2 is active, it inhibits the activity of DAF-16, which allows *ins-7* to be expressed. The production of INS-7 leads to further activation of DAF-2. However, when DAF-2 activity is reduced, DAF-16 is activated and inhibits the expression of *ins-7*.

## Sensory Input and Neuroendocrine Signaling

*C. elegans* senses environmental cues through ciliated sensory neurons. Large numbers of genes required for the development and function of *C. elegans* sensory neurons have been identified. In addition to disrupting sensory neuron function, mutations in some of these genes increase life span.

Amphids are gustatory and olfactory neurons located at the *C. elegans* head. Gustatory neurons sense dauer pheromone, food, and amino acids in the environment. Olfactory neurons are responsible for sensing food-derived substances and volatile chemicals. To define which of these cells are involved in the regulation of life span, individual cells were ablated by a focused laser microbeam. Ablation of gustatory neurons revealed that only a specific subset (ASI and ASG, and not ADF, ASJ, and ASK ) may influence life span. This was also the case for the olfactory neurons (AWA and AWC), since only the ablation of AWA extended life span, whereas, ablation of AWC had no effect. This was further confirmed by using mutants with specific defects in these neurons. Combined ablation of gustatory and olfactory neurons results in greater longevity, compared to ablation of either gustatory or olfactory neurons alone, suggesting that these neurons function in distinct pathways to control life span. Many mutations abrogating sensory neurons, including putative chemosensory receptors, extend the life span of *C. elegans*, largely in a *daf-16*-dependent manner. The life span extension caused by gustatory neurons depends on *daf-16*, indicating that these neurons might modulate the *daf-2* pathway. However, in the case of olfactory neurons, effects on life span are only partly dependent on *daf-16*. Many sensory neurons in *C. elegans* produce ILPs. Therefore, it is plausible that perception affects life span by influencing the activity of the insulin signaling pathway. However, these mutations show complicated interactions with the insulin/IGF-1 pathway. Double mutants did not live longer than the *daf-2* single mutant; instead their life span was shorter. This may indicate that sensory control of life span is only partially dependent on the *daf-2* pathway. Likewise, some gustatory and olfactory neurons enhance, whereas others reduce, longevity. Therefore, environmental signals that affect life span may be relayed by a *daf-2*-independent pathway.

A possible additional mechanism for influencing life span by sensory inputs is the regulation of lipid accumulation. *C. elegans* mutant strains with defects in neuroendocrine signaling show increased fat accumulation and extended life span. This is the case for *daf-2* insulin receptor mutants, the tryptophan hydroxylase *tph-1* serotonin defective mutant, and the *tub-1* (tubby ortholog) mutant. In *C. elegans*, *tub-1* is expressed in sensory neurons and *tph-1* is expressed in serotonergic neurons. In addition, a genome-wide RNAi screen identified several genes involved in food sensation and neuroendocrine signaling that, when knocked down, resulted in aberrant fat accumulation. These include genes for glutamate and dopamine receptors, as well as chemoreceptor and olfactory receptor genes. Consistently, *C. elegans* mutants with either structural or functional defects in nine specific ciliated neurons show increased fatty acid accumulation in the intestine. Interestingly, some mutations that increase lipid accumulation also lengthen life span. Thus, ciliated neurons may sense environmental cues and express neuropeptides and insulin ligands to regulate metabolism and life span. However, while the insulin signaling pathway is a major regulator of *C. elegans* fat storage (*daf-2* mutants show increased lipid accumulation), fat accumulation in these sensory mutants is independent of DAF-16/FOXO.

## Other Neuroendocrine Mechanisms

In a chemical screen aimed at identifying drugs that delay aging, anticonvulsant medications were found to extend worm life span Anticonvulsants modulate neural activity in mammals and act presynaptically to modulate neuromuscular activity in the worm. Interestingly, anticonvulsants significantly increase the life span of *daf-2* loss-of-function mutants as well as that of *daf-16* mutants. This finding suggests that neural activity regulates aging by an additional mechanism independent of insulin-like signaling Similarly, these compounds further increased the life span of animals with mutations in genes important for the function of sensory neurons (*osm-3* and *tax-4*) and neurotransmission (*unc-31*, *unc-64*, and *aex-3*), highlighting the intricacy of the neuronal pathways that might be involved in the regulation of nematode aging.

The reproductive system also regulates aging in *C. elegans*. Ablation of the two germ line precursor cells, as well as mutations that reduce germ line proliferation, remarkably extend the *C. elegans* life span. This life span extension requires the presence of the somatic gonad, since ablation of both germ line and somatic gonad has no effect on life span. The life span extension of germ-line-ablated animals depends on DAF-16/FOXO and may be mediated hormonally, since it also depends on DAF-12, a nuclear hormone receptor, and DAF-9, a cytochrome P450 involved in the production of steroid hormones (3-keto sterols) that function as DAF-12 ligands. Another gene required for the increased life span associated with germ line loss is *kri-1*. The *kri-1* gene encodes a conserved protein with ankyrin repeats and is expressed in pharynx and intestine. Germ line ablation results in the nuclear localization of DAF-16 in the intestine (the worm's adipose tissue), and intestinal expression is sufficient to account for the observed longevity. KRI-1 and to a lesser extent DAF-12 and DAF-9 are required for the DAF-16 nuclear localization in the intestine of germ-line-defective animals. However, although loss of DAF-2 receptor activity promotes the nuclear localization of DAF-16 in many tissues, including the intestine, DAF-16 localization is not dependent on KRI-1, DAF-12, or DAF-9. Moreover, *kri-1* RNAi completely suppresses the life span extension of germ line mutants, while it has no significant effect on wild-type or *daf-2* mutants, indicating that modulation of life span by KRI-1 is specific to the reproductive-signaling pathway. Thus, the role of lipophilic hormones and KRI-1 on the nuclear localization of DAF-16 is germ line specific and independent of insulin signaling. Therefore, the germ line might possess a specific endocrine system to influence aging.

Nevertheless, neuronal and gonadal endocrine signaling mechanisms appear to interact in a complex manner. Somatic gonad ablation prevents the life span extension of germ-line-ablated wild-type animals, but it does not completely prevent life span extension in animals that lack olfactory neurons. Similarly, germ line ablation further extends the life span of *daf-2* mutants independently of whether or not the somatic gonad is present.

## Concluding Remarks

The nervous system performs the task of sensing and integrating environmental cues into coordinated physiological responses that will ensure maximal survival and reproductive fitness. In *C. elegans*, food availability, temperature, and a secreted pheromone are some of the sensory inputs that regulate the decision of entering the metabolically active reproductive mode or shifting to the nonreproducing, nonfeeding dauer larva stage, with large amounts of stored fat.

Despite its apparent simplicity, *C. elegans* has a surprisingly sophisticated neuroendocrine system that regulates development, metabolism, and life span. Both, insulin-like and TGF-$\beta$ signaling pathways act in parallel to regulate development and metabolism, with the insulin-like signaling pathway playing a major role in life span regulation. Importantly, the regulation of life span by insulin/IGF signaling is conserved across taxa, and reduction of insulin signaling has been shown to extend life span in worms, flies, and mammals. Similarly, the physiological processes involved in the aging process also appear to be conserved. For example, signals from the reproductive system also influence life span in mammals, and dietary restriction has been shown to extend life span in a wide variety of organisms. Likewise, sensory perception could also regulate life span in higher organisms, since blocking the sense of taste reduces insulin secretion in mammals, and the smell of food increases insulin levels in humans.

How physiological processes are coordinated by neuroendocrine signaling to meet the biological demands of an organism is still not completely understood. More than 30 ILPs are encoded in the *C. elegans* genome, some of which are agonists and others of which are antagonists. Some have been shown to influence aging, but many remain to be functionally characterized. It is of fundamental importance to understand which cells or tissues emit or receive signals to coordinate the aging process at the level of the whole organism. *C. elegans* has been instrumental for the discovery of conserved molecular pathways regulating aging. Its relatively short life span and its amenability for genetic and molecular analyses make it an ideal organism to pursue these studies further, aiming to ultimately understand why and how animals age.

*See also:* Aging of the Brain; Gene Expression in Normal Aging Brain; Neuroendocrine Aging: Pituitary Metabolism.

## Further Reading

Aamodt E (2006) *The Neurobiology of Caenorhabditis elegans*. Amsterdam; Boston: Elsevier/Academic Press.

Ailion M, Inoue T, Weaver CI, et al. (1999) Neurosecretory control of aging in *Caenorhabditis elegans*. *Proceedings of the National Academy of Sciences of the United States of America* 96: 7394–7397.

Alcedo J and Kenyon C (2004) Regulation of *Caenorhabditis elegans* longevity by specific gustatory and olfactory neurons. *Neuron* 41: 45–55.

Antebi A (2004) Long life: A matter of taste (and smell). *Neuron* 41: 1–3.

Boulianne GL (2001) Neuronal regulation of lifespan: Clues from flies and worms. *Mechanisms of Ageing and Development* 122: 883–894.

Braeckman BP, Houthoofd K, and Vanfleteren JR (2001) Insulin-like signaling, metabolism, stress resistance and aging in *Caenorhabditis elegans*. *Mechanisms of Ageing and Development* 122: 673–693.

Evason K, Huang C, Yamben I, et al. (2005) Anticonvulsant medications extend worm life-span. *Science* 307: 258–262.

Finch CE and Ruvkun G (2001) The genetics of aging. *Annual Review of Genomics and Human Genetics* 2: 435–462.

Hekimi S and Guarente L (2003) Genetics and the specificity of the aging process. *Science* 299: 1351–1354.

Kenyon C (2005) The plasticity of aging: Insights from long-lived mutants. *Cell* 120: 449–460.

Libina N, Berman JR, and Kenyon C (2003) Tissue-specific activities of *Caenorhabditis elegans* DAF-16 in the regulation of lifespan. *Cell* 115: 489–502.

Mak HY, Nelson LS, Basson M, et al. (2006) Polygenic control of *Caenorhabditis elegans* fat storage. *Nature Genetics* 38: 363–368.

Murphy CT, McCarroll SA, Bargmann CI, et al. (2003) Genes that act downstream of DAF-16 to influence the lifespan of *Caenorhabditis elegans*. *Nature* 424: 277–283.

Schreibman MP and Scanes CG (1989) *Development, Maturation, and Senescence of Neuroendocrine Systems: A Comparative Approach*. San Diego: Academic Press.

Tatar M, Bartke A, and Antebi A (2003) The endocrine regulation of aging by insulin-like signals. *Science* 299: 1346–1351.

Thomas JH (1999) Lifespan. The effects of sensory deprivation. *Nature* 402: 740–741.

Walker DW, McColl G, Jenkins NL, et al. (2000) Evolution of lifespan in *Caenorhabditis elegans*. *Nature* 405: 296–297.

Wolkow CA (2002) Life span: Getting the signal from the nervous system. *Trends in Neurosciences* 25: 212–216.

Wood WB (1988) *The Nematode Caenorhabditis elegans*. Cold Spring Harbor, NY: Cold Spring Harbor Laboratory.

Yeoman MS and Faragher RG (2001) Aging and the nervous system: Insights from studies on invertebrates. *Biogerontology* 2: 85–97.


## Relevant Websites

http://www.arclab.org – Aging Research Centre.

http://www.afar.org – American Federation for Aging Research.

http://www.ncoa.org – National Council on Aging.

http://www.nia.nih.gov – National Institute on Aging (U.S. National Institutes of Health).

http://sageke.sciencemag.org – Science of Aging Knowledge Environment, an interdisciplinary repository of issues related to aging (American Association for the Advancement of Science).

http://www.geron.org – The Gerontological Society of America.

# Cognitive Aging in Nonhuman Primates

**M G Baxter**, Oxford University, Oxford, UK



## Introduction

Aged monkeys provide an extremely useful model system for understanding both the nature of normal cognitive aging in humans, and the neurobiological basis of these cognitive impairments, due to their well-characterized behavioral repertoires, similarity in brain structure to humans, and lack of age-related neuropathological conditions like Alzheimer's disease. There is a striking similarity between the profile of cognitive impairment in aged nonhuman primates and that of aged humans that are not affected by age-related neuropathological conditions like Alzheimer's disease.

Five general domains of cognitive function can be identified that have been examined in aged nonhuman primates: visual recognition memory, spatial memory, stimulus-reward associative learning, relational memory, and attentional processing/executive function. The general approach to be taken with each domain of cognitive function is to consider the evidence for impairments in a particular domain in aged nonhuman primates, and then to review the neuropsychological evidence from young animals linking each domain to the function of particular neural circuits. Much research in this area has been guided by the thesis that a particular behavioral task, known to be sensitive to damage to a particular neural system in young monkeys, should provide a functional index of the integrity of that neural system in aging. Accordingly, the approach taken in this article is task-oriented, within the broad category of each particular domain of cognitive function.

The vast majority of studies of cognitive aging in nonhuman primates have examined rhesus monkeys (*Macaca mulatta*) or other species of macaques. This is advantageous for the current discussion, because most neuropsychological studies in nonhuman primates have also been done in macaques, providing a ready comparison with the effects of aging on cognition. An approximate ratio of 1:3 can be used to relate the rhesus monkey age to the human age, that is, a 20-year-old rhesus monkey is approximately equivalent to a 60-year-old human. Assessments of cognitive aging in other species of monkeys, as well as in great apes, are much more limited in scope. To the extent that other primate species may be advantageous for modeling particular aspects of the aging process – for example, neuroendocrine and reproductive aging, or amyloid deposition (which occurs naturally in vervet monkeys), an important direction for future work is to adapt behavioral testing paradigms that have been developed for macaque monkeys for the testing of other species.

When data are available, the relationship of performance on different tasks by the same monkeys will be discussed. Identifying the dissociations between impairments in cognitive function has been the most productive strategy to understand the organization of cognitive systems. If impairments in two different tasks tend to be correlated across a group of monkeys, a common neural substrate for that impairment might be hypothesized. In contrast, if a group of aged monkeys is impaired on one task but performs normally on another, or if impairment on two different tasks is unrelated, it might be supposed that different neural systems might underlie the different impairments. Also, when data are available, the apparent time-course of age-related cognitive decline will be discussed, although these observations are usually based on cross-sectional comparisons of different groups of aged monkeys.

The overall picture of cognitive aging in nonhuman primates is that functional impairments are seen in many different cognitive domains, and these impairments do not resemble the effects of a single brain lesion or disruption to a particular neurotransmitter system. Moreover, marked individual differences are observed such that on most tasks it is possible to identify aged monkeys that are as proficient as young monkeys, whereas other aged monkeys will demonstrate severe impairments. This pattern resembles cognitive aging in other mammalian species, including rodents and humans. But nonhuman primates make several unique contributions to the study of the neuroscience of aging. Because their cortical development is much greater than that of rodents, particularly in terms of the development of the prefrontal cortex, they are extremely valuable as models for age-related changes in the neurobiology of the prefrontal cortex and how these changes affect cognition. Like female humans, female rhesus monkeys have a menstrual cycle and undergo menopause, and therefore, they are extremely valuable for understanding the relationship between reproductive aging and cognitive aging.

## Visual Recognition Memory

Behavioral tasks to measure stimulus recognition, literally 'knowing the stimulus again,' were developed as part of an effort to develop an animal model of human amnesia. In terms of the overall neuropsychological profile of the aged monkey, deficits in spatial working memory are much more consistently associated with advanced age (see the next section), but because of the intense focus on stimulus recognition memory in the neuropsychological studies of nonhuman primates, and the converging interest in examining the dysfunction of medial temporal lobe structures as a neural substrate for age-related cognitive decline, we begin the discussion of cognitive aging in nonhuman primates with this domain of cognitive function.

The most common test of stimulus recognition memory, visual delayed nonmatching-to-sample (DNMS), was developed in its standard form, using trial-unique objects and a performance test that increases memory demand by lengthening the delay between sample and choice, and presenting lists of sample items, by two different research groups in the mid-1970s. This test capitalizes on the monkey's natural propensity to explore novel objects. In the first phase of each DNMS trial, the monkey is allowed to view a sample object that covers the central well of a three-well test tray. The monkey displaces the object and obtains a food reward hidden underneath the object. An opaque screen is then lowered that occludes the monkey's view of the test tray. In the second phase of each trial, which takes place after a delay ranging from seconds to minutes, the opaque screen is raised and the monkey is allowed to choose between the sample object and a novel object, one covering each of the lateral wells of the test tray. The monkey can obtain another food reward by displacing the novel object. Ordinarily, a very large pool of objects (hundreds to thousands) is available for use as stimuli, and therefore, the objects the monkey encounters are functionally 'trial-unique' and repeated only rarely throughout the testing.

Because of the large number of studies exploring the effects of selective ablations of different brain areas on DNMS performance, this task provides a potentially rich source of information about the nature of functional impairment in aged monkeys; different brain areas may be involved in acquisition of the DNMS rule and in maintaining memory of the objects across delays. In detecting a memory deficit in such tasks, particular emphasis is placed on the presence of a statistical interaction between condition (age or lesion) and delay: if the effect of condition increases with extended delays, the impairment is interpreted as one of memory, because forgetting occurs more rapidly in the impaired group. If there is no such interaction, the impairment could be one of sensory processing, attention, motivation, or some other nonmnemonic factor. This analytic method has been criticized on a number of grounds, however, including the fact that the monkey often receives extensive training on the short delay and not at longer ones.

### Effects of Aging

Aged monkeys (22–29 years old across the different studies) as a group show impairments in acquisition of the DNMS rule as well as impaired performance when the delay interval between sample and choice is increased. Initial studies noted some individual differences in the performance of aged monkeys: certain aged monkeys acquired the DNMS rule as efficiently as young monkeys and performed as well across delays as young monkeys. Notably, however, there was no interaction between age and delay in the scores on the DNMS performance test, casting doubt on the interpretation of the DNMS impairment as one of memory. A subsequent analysis classified the mean performance of aged (22–27 years old) monkeys across all delays as 'impaired' or 'unimpaired,' depending on whether the score of each individual aged monkey fell within the range of young performance or not. The two subgroups of aged monkeys in this sample were of similar chronological age. The subpopulation of aged monkeys that were 'impaired' (2/3 of the sample) indeed demonstrated a differentially impaired performance across longer delays (that is, an age by delay interaction) that was not observed when the entire sample of aged monkeys was compared as a single group with the young monkeys. Interestingly, there was not a consistent relationship in this study population between impairment in delay performance and impairment in acquisition of the DNMS rule – both subgroups of aged monkeys demonstrated impairments in acquisition of the DNMS rule relative to the comparison group of young monkeys. More recent studies using larger samples of aged monkeys indicate that the incidence of impairment in DNMS acquisition seems to increase with advancing age, although there is no apparent relationship between chronological age and delay performance when aged monkeys are considered as a whole (**Figure 1**).

Altering the task demands of the DNMS task by testing DNMS performance using only a single pair of objects, rather than trial-unique objects, revealed a dramatic impairment in the aged monkeys. This suggests that while object recognition memory is relatively unimpaired in aged monkeys, memory for temporal order (which of two familiar objects was presented most recently) is profoundly affected by aging. These two cognitive processes have distinct neural substrates.

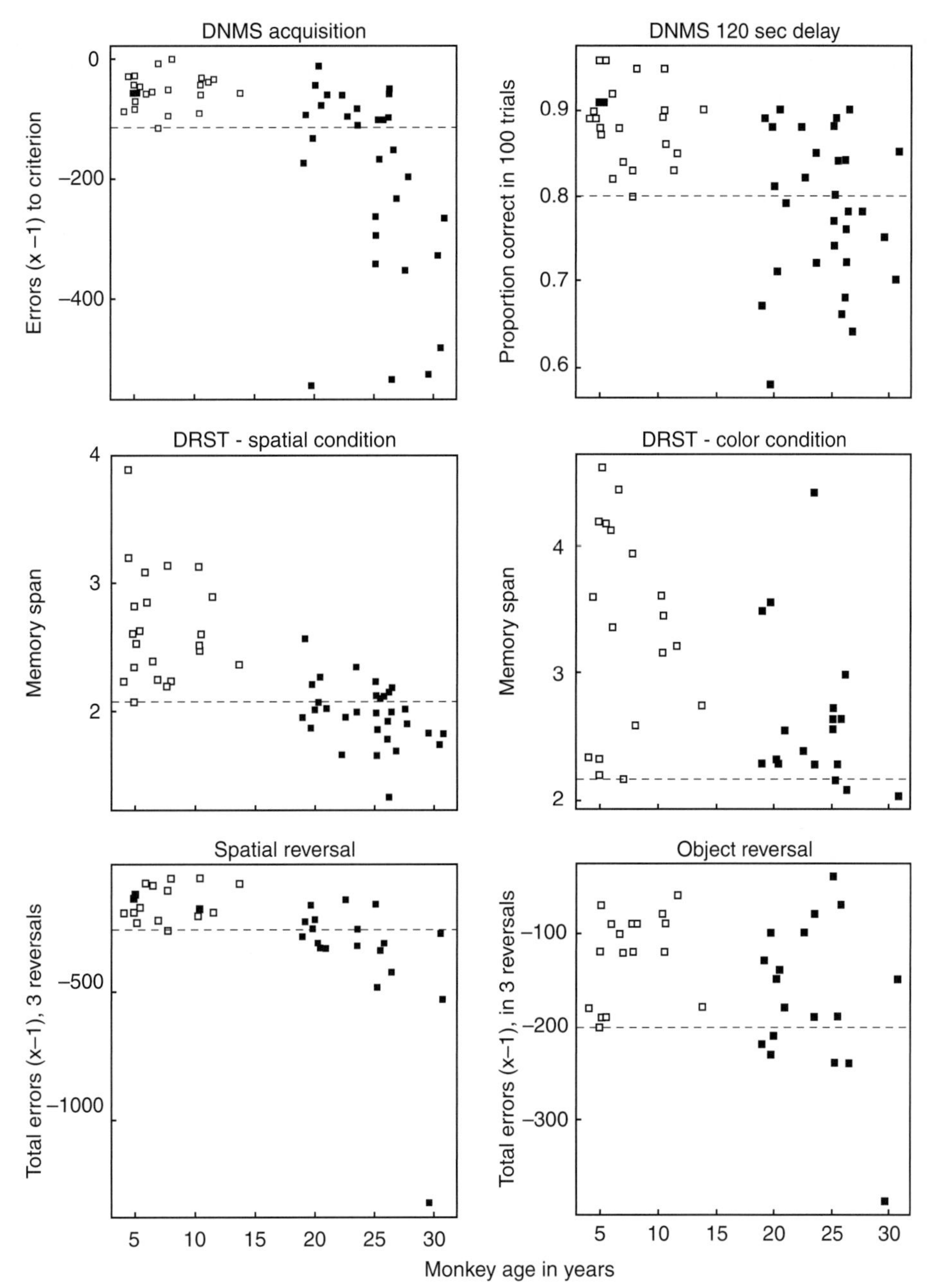


**Figure 1** Scatterplots of performance of rhesus monkeys on six different behavioral tasks as a function of age (Herndon et al., 1997): acquisition of DNMS, performance on a 120 s delay in DNMS, DRST spatial and color conditions, spatial and object reversals. The horizontal dotted line represents the score of the worse-performing young monkey (younger than 15 years), defining an operational criterion for impairment in the aged monkeys. Note the different patterns of impairment on different tasks, in terms of the prevalence of impaired aged monkeys and the relationship between task performance and chronological age. Reproduced from Herndon JG, Moss MB, Rosene DL, et al. (1997) Patterns of cognitive decline in aged rhesus monkeys. *Behavioural Brain Research* 87: 25–34, with permission from Elsevier.

Another procedure for assessing stimulus recognition memory, similar to DNMS, is the delayed recognition span task (DRST). In the object and color conditions of this task, either objects or colored discs are placed one-by-one on a test tray. Each object or colored disc covers a food well. When a new object or colored disc is placed on the test tray, it covers a food reward; the old stimuli are moved to new locations on the test tray but are not rebaited. Hence, the monkey must remember on each trial which stimuli it has seen previously, in order to choose the new object or colored disc and obtain a food reward. New objects or colored discs are added until the monkey makes a mistake. Monkeys older than 19 years of age appear to be impaired on the DRST-object condition, whereas impairment on the DRST-color condition does not appear to emerge until 25 years of age (**Figure 1**).

### Neural Basis

Since the initial demonstration of an impairment in DNMS produced by combined aspiration lesions of the amygdala and the hippocampus, approximating the severe anterograde amnesia produced by a similar neurosurgical operation in patient H.M., considerable effort has been devoted to fractionating the contributions of different neural structures to the performance of this task, particularly within the medial temporal lobe. Of the tasks used to investigate the neuropsychology of memory function in nonhuman primates, DNMS is probably the most widely investigated in terms of the particular neural systems responsible for good performance. Hence, a fairly rich database exists for the purpose of comparing the effects of aging on DNMS performance with the effects of selective damage to particular brain regions.

DNMS performance seems to involve the interaction of a network of three basic neural structures: the inferior temporal cortex, the mediodorsal thalamus, and the ventromedial prefrontal cortex, with a possible modulatory effect of the basal forebrain cholinergic system. Within the inferior temporal cortex, the perirhinal cortex seems to be the single most important component for supporting good recognition memory performance, with combined damage to the entorhinal and perirhinal cortex producing an even more severe deficit. Lesions of the entorhinal cortex alone produce only a mild deficit that appears to be transient. These deficits are characterized by an impairment in reacquiring the DNMS rule as well as by a more rapid forgetting across increased delays between sample and choice. Damage limited to the hippocampus produces little or no deficit in performance of this task; similarly, the amygdala appears to make no contribution to performance of this task in its standard form.

Less information is available about the contribution of specific thalamic and prefrontal structures to this task. Lesions of the mediodorsal thalamus also produce a severe impairment in DNMS performance, with an approximately equal contribution of anterior and posterior groups of thalamic nuclei. Lesions of the ventromedial prefrontal cortex produce a profound impairment in DNMS performance as well. Interestingly, lesions restricted to the inferior prefrontal convexity (ventrolateral prefrontal cortex) appear to produce a selective deficit in the reacquisition of the DNMS rule, without an impairment in memory across increasing delays, whereas damage to the orbital frontal cortex produces an impairment in memory across increasing delays with less of an impairment in the reacquisition of the DNMS rule. Direct interaction between the frontal cortex and perirhinal cortex, and between the mediodorsal thalamus and perirhinal cortex, is critical for visual recognition memory. Lesions of basal forebrain cholinergic neurons have little effect on the performance of this task, although periperhal administration of a muscarinic receptor antagonist produces DNMS impairments.

The overlap between deficits produced by different lesions makes it difficult to draw conclusions about defects in particular neural systems based solely on the pattern of DNMS performance. Decreased performance across delays, as well as impaired acquisition of the DNMS rule, is associated with damage to a number of different neural structures, although the relative magnitude of such deficits can sometimes be dissociated. There is a lack of consensus on the relationship between DNMS acqusition and delay performance deficits in aged monkeys; some studies with relatively small numbers of subjects see little association between acquisition of the task and delay performance, which would suggest that these two deficits must have their origins in different neural systems. However, a study with a larger sample size reported that acquisition and delay performance loaded onto the same factor in a principal components analysis, suggesting that age-related declines in these two processes are linked. A correlated impairment in DNMS rule acquisition and delay performance is consistent with a dysfunction in aging in the inferior temporal-prefrontal system that subserves visual recognition memory, affecting both DNMS learning and delay performance, an hypothesis that would also account for age-related deficits in stimulus-reward associative learning. However, impairments in learning the DNMS rule, without impairments in performance across increasing delays, might be suggestive of a selective impairment in prefrontal function, specifically ventrolateral prefrontal cortex.

Performance on DNMS with a single pair of objects appears to be more sensitive to aspects of prefrontal cortex function, likely tapping into similar cognitive processes that underlie performance on the delayed response (DR) task (see the next section). Monkeys with lesions of the hippocampus are not impaired on performance of DNMS with a small set of objects, and indeed, are facilitated in learning DNMS with a single pair of objects. However, monkeys with damage to ventrolateral prefrontal cortex are impaired at relearning delayed matching-to-sample (DMS) with a single pair of objects or color stimuli, although this impairment may be related to rule-learning and not stimulus memory per se.

## Spatial Memory

### Effects of Aging

In a commonly used procedure to assess spatial working memory, the spatial DR task, the monkey is allowed to watch while one of two food wells is baited with a reward. Then an opaque screen is lowered between the monkey and the food wells, and the wells are covered with identical gray plaques. The screen is then raised after some retention interval and the monkey is allowed to displace one of the plaques in an attempt to gain the food reward. The food is not moved in the retention interval, and therefore, the monkey must simply remember which well it saw baited, and displace the plaque covering that well to obtain the food reward.

Aged monkeys are impaired on the DR task. As with visual recognition memory, impaired DR performance is not an inevitable consequence of advanced age in macaque monkeys. Some aged monkeys have been reported to perform DR at levels comparable to young monkeys, whereas others fall to near-chance performance even at short retention intervals. Notably, in a more complex test of spatial working memory that involved more than two possible response locations, aged rhesus monkeys exhibited more substantial impairment, with no overlap between scores of young and aged monkeys. This suggests that spatial memory in aged monkeys may be particularly sensitive to the demand to maintain information about multiple different locations in memory, rather than simply remembering which of two locations is baited on the current trial.

Another related task is the spatial condition of the DRST. In this task the monkey is faced with a $3 \times 6$ array of wells on a test board. To begin a trial sequence, a brown disc covers a food reward placed in one well of the test board. The monkey displaces the disc and is allowed to obtain the food reward. An opaque screen is lowered between the monkey and the test board; the first disc is returned to its original position, now covering an unbaited well, and a new disc (visually identical to the first) is added, covering a food reward. The screen is then raised and the monkey is allowed to respond. The trial continues in this fashion, with one disc covering a new location and a food reward being added after each response, until the monkey makes a mistake. Thus, the monkey must maintain in memory the locations that have already been chosen, in order to correctly identify the new location on the test tray and obtain the food reward.

Rhesus monkeys demonstrate impairment on this task as early as 19 years of age. There appears to be a robust linear relationship between chronological age and DRST-spatial performance (see **Figure 1**). Thus, the pattern of performance in this task is quite similar to that from the classical DR task – an impairment in spatial working memory that increases with advanced age. Notably, within the relatively large sample studied by these authors, few monkeys of advanced age perform as well as young monkeys. This might suggest that, unlike DR, the DRST is more sensitive to an aspect of cognitive function that is more consistently impaired in advanced age (possibly, as mentioned before, the ability to monitor many spatial locations instead of just two), but investigation of this possibility awaits a direct test.

The foregoing tests measure memory for locations on a test board that is placed in a fixed position in front of the monkey. These tests are unable to distinguish between memory for allocentric (environment-centered) and egocentric (subject-centered) space – the monkey may remember the absolute position of the response locations within the testing environment, or may remember its own movements relative to the test board (reaching left or right). Aged monkeys have also been given a task that tests allocentric spatial memory. In this task, the monkey is allowed to move around on a large octagonal platform. At the center of each edge of the platform is a small well containing a food reward. The monkey is allowed to retrieve the rewards in any order, provided it returns to the center of the platform between choices (to prevent it from simply running around the edge to collect all the food rewards). This remarkable procedure is like the radial arm maze task used to test spatial memory in rats. Like rats, young monkeys employ the position of distal cues placed within the test room to locate the rewards. Aged monkeys (23–33 years old) were impaired in acquiring this task and performed very poorly when a delay was interposed between the first four choices and the last four choices. Interestingly, the same monkeys that showed impaired allocentric spatial memory were unimpaired on standard DNMS testing, suggesting a distinct neural substrate for the spatial memory impairment.

### Neural Basis

Impairments in spatial working memory are associated with damage to the dorsolateral prefrontal cortex. Monkeys with lesions of dorsolateral prefrontal cortex are unable to acquire the DR task. The integrity of catecholaminergic projections to prefrontal cortex appears critical for normal spatial working memory, and the dysfunction of these projections has been suggested as a neural substrate for the age-related impairment in DR performance. Bartus, in 1978, noted the similarity between the short-term memory impairment in aged monkeys and the impairment in short-term memory in young monkeys induced by the administration of scopolamine, an antagonist of muscarinic cholinergic receptors. A reduction in gray matter in a network of brain structures including dorsolateral and ventrolateral prefrontal cortex is associated with age-related impairment in DR in rhesus monkeys.

Hippocampal ablation does not impair the acquisition of the DR task but causes a deficit when longer retention intervals are interposed. Thus, a contribution of impaired hippocampal and/or medial temporal lobe function to impaired performance at longer delays in the DR task is possible. Indeed, an in vivo measure of hippocampal glucose metabolism correlated highly with aged monkeys' performance of the DR task. However, lesions restricted to the hippocampus, made with stereotaxically placed injections of the neurotoxin ibotenic acid, do not impair performance on a closely related task, spatial DNMS, which calls into question the specific involvement of the hippocampus in memory for spatial location. No lesion data are currently available for the allocentric spatial memory task of Rapp and colleagues, but the intact DNMS performance in these monkeys suggests that the locus of this deficit must lie outside of the inferior temporal-medial thalamic-ventromedial prefrontal system. Certainly, based on ample evidence in rats for hippocampal involvement in allocentric spatial memory, this task would be expected to depend on intact hippocampal function.

The involvement of the prefrontal cortex in spatial working memory in aging female monkeys may be particularly sensitive to ovarian hormones. Aged female monkeys that are peri- or postmenopausal show impairments in the DR task relative to monkeys of similar chronological age that are still premenopausal. Cyclic estrogen replacement in aged, ovariectomized female monkeys is effective in improving DR performance and increases the density of dendritic spines in dorsolateral prefrontal cortex. These effects may be dependent on the duration of estrogen deprivation: beneficial effects of estrogen on prefrontal function – including DR and a test of executive function – are not seen in aged monkeys that were ovariectomized earlier in life. They are also dependent on the age of the monkeys; DR is unimpaired by ovariectomy in young monkeys and estrogen replacement does not result in a net change in the number of dendritic spines per pyramidal neuron in young monkeys. The cholinergic innervation of prefrontal cortex may also be particularly affected by estrogen replacement, and loss of this input may be involved in age-related impairment in working memory.

## Stimulus-Reward Associative Learning

Stimulus-reward associative learning tasks include classical object and spatial discrimination tasks, in which one of two (or more) stimuli is consistently associated with a reinforcer. Tasks of this nature can be varied to examine a number of different aspects of cognitive function, by varying the rate of presentation of individual discrimination problems, or by reversing the response-reinforcement contingencies once they are learned. These reversal problems test behavioral flexibility, a component of 'executive' function, as well as the strength of stimulus-reinforcer bonds.

### Effects of Aging

Monkeys trained on single object discrimination problems appear to show no impairment in the acquisition of individual problems, or impairment in the retention of these problems, although they are impaired in pattern discrimination learning problems where the stimuli are less easy to discriminate. Aged monkeys are impaired in learning concurrent object discrimination problems, given as a list of 20 problems, each presented once per day, suggesting a selective age-related impairment in long-term memory for stimulus-reward associations. This may reflect a selective impairment in a slow-learning, 'habit'-like system for forming stimulus-reward associations. Aged monkeys are not impaired on spatial discrimination learning, an interesting contrast to their impairment in spatial working memory.

The reversal of stimulus-reward associations is markedly impaired in aged monkeys, although the precise nature of this impairment is somewhat unclear. Impairments in object reversal learning and spatial reversal learning have been reported, along with intact spatial reversal learning and object reversal learning. All of these studies used similar two-choice discrimination procedures, but those that administered both spatial and object reversal learning differed in the order in which the tasks were given. In these cases, it was the first reversal task administered (whether spatial or object) that was impaired. Thus, impairments in reversal learning may be overcome by experience with reversal problems, a phenomenon that is sometimes seen after prefrontal cortex lesions. As in other behavioral tests in aged monkeys, individual differences in reversal learning have been reported such that some aged monkeys are able to perform reversal tasks as well as young monkeys.

### Neural Basis

Discrimination learning problems may be acquired in different ways, typically either serially or concurrently. When discrimination problems are presented serially, a single pair of items is learned to criterion (or presented for a fixed number of trials), followed by a second pair, a third pair, and so on. When problems are learnt concurrently, a number of different pairs of problems are given intermixed with one another until the overall performance reaches a certain level; the most common form of this task is the '24-hour intertrial interval (ITI)' task, in which a set of object discrimination problems (often 20) is presented once a day until the overall performance reaches a 90% criterion level. The difference in neural substrates between these two versions of discrimination learning appears to relate to the formation of learning set: learning sets are acquired for serially presented single discriminations, but not for concurrently learnt discriminations. The acquisition, and even the reversal, of concurrently learned visual discriminations does not require interaction between the temporal cortex and the prefrontal cortex, although both of these areas are required for the application of a learning set to single discrimination or reversal problems. Bilateral damage to either the prefrontal or the inferotemporal cortex, of course, will impair visual discrimination learning regardless of whether problems are learnt individually or concurrently, but the requirement for intrahemispheric interaction between these cortical structures appears to be unique to situations in which a learning set is applied to speed the rate at which discrimination problems are acquired when they are learned one at a time. Because prefrontal–inferotemporal interaction is also necessary for visual recognition memory, it might be expected that if impairments in functional interaction occur in aging, tests of visual recognition memory and discrimination learning would be affected to similar degrees.

The ventrolateral and orbital areas of prefrontal cortex are particularly identified with reversal learning, to the extent that lesions of these areas tend to have little or no effect on the acquisition of discrimination problems but produce significant impairments in reversal learning. Perseverative behaviors – continuing to respond to the previously rewarded member of a stimulus pair, even though it is no longer correct – appear to be associated mainly with damage to the ventrolateral prefrontal cortex, whereas the orbital prefrontal cortex appears to be more associated with an impairment in acquiring the new stimulus-reward associations after the perseverative behavior has stopped.

## Relational Memory

### Effects of Aging

In an attempt to probe an aspect of memory function that might be more specifically related to the functional integrity of the hippocampus and its associated cortical regions than recognition memory, some recent investigations have looked specifically at relational memory processing in aged monkeys.

One task of interest tests transitive inference ability. The task takes the form of a set of ordered object discriminations such that the stimulus that is rewarded on a particular trial depends on what other stimulus is present on that trial. First, the monkey is taught a discrimination between two objects, A and B, where A is rewarded and B is not (A + B-). On the next discrimination problem, B is rewarded and a new object (C) is not (B + C-). Additional problems are learned (C + D- and D + E-) and then the problems are presented in a randomly intermixed fashion within a test session. The ordering of the discriminations sets up a hierarchy among the different objects: A > B > C > D > E so that the object highest in position in the hierarchy is chosen on any particular individual trial. It is worth noting that each of the objects in the middle of the list (B, C, D) is paired with reward 50% of the time, making it difficult to use the associative history of a particular object to guide choices.

Once these problems (referred to as 'premise problems') are learned, relational probe trials are introduced. On these trials, two objects that have not been paired during training are presented simultaneously. One such pair (AE) can be solved purely on the basis of reward histories of the individual objects; A is always reinforced during training and E is never reinforced, an hence, no inference is required. But if objects B and D are paired, each has been paired with reward 50% of the time, and therefore, this cannot guide the monkey's choice. The understanding of the relationship between the objects in the hierarchy – that B occupies a higher position than D – would produce a choice of B more often than D on such trials.

Aged monkeys trained on this task acquire the premise problems at a rate comparable to that of young monkeys, and demonstrate an equivalent level of inferential responding on BD trials – that is, they choose B over D ~80% of the time. Extension of the series of discriminations to include an additional two objects (so the objects are ordered such that A > B > C > D > E > F > G) permits the testing of 'symbolic distance' effects: items farther apart in the list should elicit greater degrees of inferential responding. For example, inferential responding on BF trials (separated by three objects in the list) should be greater than inferential responding on CE or DF trials (separated by one object in the list). Such effects were observed with both young and aged rhesus monkeys, and there was no

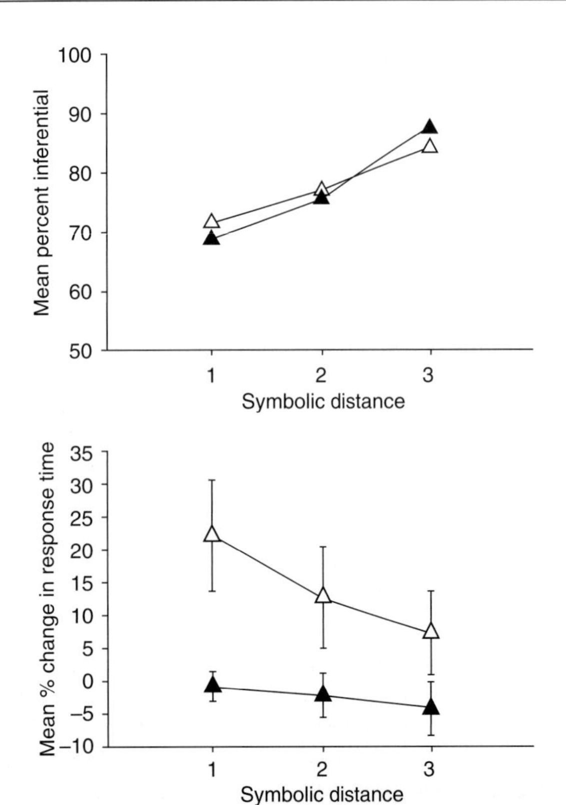


**Figure 2** Performance of young and aged monkeys on a relational memory task (Rapp et al., 1996). In the top panel, aged and young monkeys exhibit identical levels of inferential responding on probe trials with novel pairings of elements from the premise problems. Elements farther apart in the series (larger symbolic distance) result in higher levels of inferential responding. In the lower panel, young monkeys exhibit a reaction time signature of inferential processing (longer reaction times on trials where the symbolic distance is lower). Aged monkeys do not display this change in reaction time, suggesting that they are solving the inferential probe trials by a nonrelational strategy. Reproduced from Rapp PR, Kansky MT, and Eichenbaum H (1996) Learning and memory for hierarchical relationships in the monkey: Effects of aging. *Behavioral Neuroscience* 110: 887–897, with permission from the American Psychological Association.

age effect on the magnitude of the symbolic distance effect (**Figure 2**).

However, despite the preserved choice behavior in inferential probe trials, measures of response time on these trials suggested that the aged monkeys were using an alternative strategy to make these choices. Young monkeys took longer to make responses on inferential probe trials (by ~15% on average); indeed, this effect varied with symbolic distance, longer response times being elicited on probe trials with shorter symbolic distances. These response time phenomena are predicted by theories of performance in this task that emphasize cognitive representations of the ordering among the objects. Aged monkeys, despite levels of choice behavior on inferential probe trials equivalent to young monkeys, did not show these response time signatures of transitive inference (**Figure 2**). Such behavior would be consistent with the guiding of choice behavior by reinforcement histories of the individual objects, without recourse to a cognitive representation of the ordering of the objects. Hence, aged monkeys may have impairments in forming a cognitive representation of the ordered relationship between these objects, or in accessing this representation to guide choice behavior. Nevertheless, other behavioral mechanisms that do not involve formation of an ordered representation presumably remain intact to support accurate inferential choice behavior.

### Neural Basis

Only one study has examined the effect of cortical lesions on the performance of the relational memory task used to test aged monkeys, and this used a modified test procedure in which the discriminanda were cookies so that when the monkey chose an object, the object could be consumed as the reward for that trial. Incorrect items were spiked with a bitter-tasting flavor that made them unpalatable. This experiment found that bilateral damage to the entorhinal cortex impaired performance on relational (BD) probe trials, which did not differ from chance in monkeys with entorhinal lesions. This differs from the pattern of behavior shown by aged monkeys that performed normally on BD probe trials but did not show reaction time signatures of transitive inference, although it is possible that a milder impairment in temporal lobe function would express itself as a difference in reaction time rather than choice performance, or the difference in testing procedures between the two studies permits different strategies. Areas of the medial temporal lobe outside the entorhinal cortex are almost certainly involved in the performance of relational memory tasks. A related paradigm, transverse patterning, has been used to test relational memory in monkeys; in this task discriminations of the form A + B-/B + C-/C + A-are learned so that each object is reinforced 50% of the time and the identity of both objects on a particular trial must be taken into account in order to make a correct choice – strategies based on learning to approach or avoid particular individual objects will fail. Lesions of the perirhinal and parahippocampal cortex impair transverse patterning performance in monkeys; selective lesions of the hippocampus have a less consistent effect on transverse patterning. Therefore, although relational memory tasks do not appear to offer specific probes of the functional integrity of the entorhinal cortex, they are useful to test specific information-processing capacities of the medial temporal lobe. In this regard, it is important to note that monkeys with entorhinal cortex damage that were impaired on transitive inference were not impaired on other standard tests of memory including DNMS and object discrimination, and hence, relational memory tests may detect impairments in cognition that are missed by these other, more standard neuropsychological tasks.

## Attention/Executive Function

### Effects of Aging

Reversal learning is sometimes considered to be an indicator of executive function, primarily because of the extent to which it requires flexibility in behavioral performance and the ability to rapidly change responses based on reinforcement contingencies. Thus, impairments in reversal learning in aged monkeys, particularly those characterized by perseverative behavior, could be characterized as reflecting an impaired executive function in addition to an impaired stimulus-reward association learning. Other tasks have been devised to more specifically examine aspects of attention and executive function beyond the domain of flexibility of stimulus-reward associations.

The first explicit investigation of attentional processing in aged monkeys examined the performance of adult and aged monkeys on a test of spatial orienting of attention. In this task, the monkey must depress and hold the center panel in an array of three panels. A brief peripheral cue illuminates over one of the side panels; during this time, the monkey continues to depress the center panel. One of the side panels then illuminates, and the monkey releases the center panel and depresses the illuminated side panel in order to obtain a reward. On most trials, the illuminated side panel (the 'target') appears on the same side where the cue light appeared; these trials are referred to as 'valid' trials. On a small fraction of the trials, the target appears on the opposite side from the cue light; these trials are referred to as 'invalid' trials. The detection of the target is facilitated on valid trials, and slightly delayed on invalid trials, as reflected in the response time of the monkey to release the center panel in order to depress the side panel. The difference in reaction time between valid and invalid trials is an index of shifting or orienting of attention.

Aged monkeys performed identically to adult monkeys when tested in this task, showing no indication of impaired orienting of spatial attention. Furthermore, when 'neutral' trials were included that permitted a separate analysis of costs and benefits of cueing (on neutral trials, both cue lights are illuminated, and therefore, no information about the target location is provided), aged monkeys showed similar costs and benefits of cueing compared with adult monkeys. These observations are consistent with studies showing preserved spatial orienting of attention in normal aged humans, but not in aged humans with Alzheimer's disease. They also suggest that impairments in spatial attention are unlikely to contribute to age-related impairments in the performance of spatial memory tasks such as DR.

Another aspect of executive function that is impaired in aged humans is the ability to shift between behavioral strategies, or rules, as these change over the course of a task. This is characterized by an impaired performance on the Wisconsin Card Sorting Test, in which cards that are printed with different numbers of colored shapes are to be sorted into piles, based on either the color, shape, or number of symbols on each card. This task has been adapted in different ways for testing animals. One has been termed the 'conceptual set shifting task' or 'category set shifting task (CSST),' in which monkeys learn with a three-alternative discrimination problem between colored shapes. Each color and shape is used only once in each stimulus display – that is, there is never more than one red stimulus or more than one circle – but the pairings of color and shape can switch from trial to trial. The monkey must discover the category that governs correct responding on a particular block of the task: initially, it is always the red stimulus that is correct regardless of what shape it is. Once the monkey makes 10 consecutive correct responses, the category rule changes such that the triangle is always correct regardless of what color it is, and the monkey must discover the new rule. Aged monkeys are impaired on this task, making perseverative errors when the rule switches and showing more 'broken sets' (making 6–9 consecutive correct responses and then an error). These impairments are seen even in middle-aged rhesus monkeys, indicating that impairments in executive function are apparent relatively early in the aging process. Aged monkeys are able to learn normally three-choice discrimination problems in which one colored shape is consistently reinforced, and hence, poor performance in the CSST is not due to problems with stimulus-reward association learning or other nonspecific behavioral problems.

### Neural Basis

The cued reaction time task that tests the spatial orienting of attention is sensitive to damage to the basal forebrain in monkeys and to parietal cortex damage in humans. Basal forebrain lesions and damage to the parietal cortex increase the 'cost' of invalid trials, an effect interpreted as an inability to shift attention from the invalidly cued location. The lack of impairment of aged monkeys in this task suggests that the neural systems that underlie this form of attentional processing are functionally intact in aging, although the small sample size in that study precluded an analysis of individual differences in performance.

Little neuropsychological data are available on the CSST in which aged monkeys are impaired. Bilateral lesions of dorsolateral prefrontal cortex impair the performance in this task, apparently by disrupting the monkey's ability to discover the correct sorting rule, but it is not yet known how lesions to other cortical areas affect performance. Interestingly, dorsolateral prefrontal lesions do not cause an increase in perseverative errors, suggesting that the dysfunction outside the dorsolateral prefrontal cortex in aged monkeys must contribute to poor performance on the CSST.

## Integration/Conclusions about Neuropsychological Profile of Aged Nonhuman Primates

Aged monkeys generally appear to have a profile of impairments indicative of temporal lobe and frontal lobe function, evidenced by difficulties in recognition memory, spatial memory, relational memory, and executive function, with some attentional and stimulus-reward association learning capacities intact. Most tests of these abilities demonstrate individual differences in performance to some degree. Advanced chronological age does not necessarily imply that a particular aged monkey will be impaired on a particular task, although the incidence of age-related impairment on some tasks appears to be relatively high.

Techniques for assessing brain-wide neurobiological changes with aging have begun to shed light on the extent to which alterations in networks of structures may underlie age-related cognitive impairments. For example, voxel-based morphometry combined with multivariate statistical techniques reveals a network of frontal and temporal gray matter that is associated with age-related impairment in spatial memory. This goes beyond trying to match particular behavioral tasks to particular brain areas as probes of their functional integrity, although this approach has been useful in directing the search to particular brain areas for neurobiological changes in aging that cannot (yet) be easily assayed in a whole-brain way.

Caution also needs to be taken when changes in neurobiological variables are related to changes in behavior, when there are common main effects of age on both. There are many instances in which behavioral and neurobiological changes are both correlated with age, creating the possibility of a common age effect and no unique relationship of biological marker to behavior. In these instances, the neurobiological marker can serve as a proxy for age when the common relationship with age is not considered in the analysis.

The complex pattern of age-related deficits in nonhuman primates will likely preclude the identification of a single neurobiological substrate, or even a relatively limited number of substrates, for cognitive impairment in aging. In particular, the behavioral impairments of aged monkeys resemble no lesion of a single cortical region. This is consistent with neurobiological investigations in both rodents and monkeys that have found little support for the idea that frank neuronal loss is responsible for age-related cognitive impairment. Instead, these observations seem to be consistent with alterations in cortical connectivity, and more subtle alterations in neural systems with aging that may impair information processing capacity in less obvious ways.

Despite the complexity of the neuropsychological profile of aged nonhuman primates, these behavioral studies have provided a strong foundation for efforts to define the neurobiological substrates of particular aspects of age-related cognitive decline. As techniques develop for determining the status of neurobiological markers in vivo, the combination with behavioral assessments promise a model system for testing interventions that may ameliorate cognitive impairments associated with aging. Some caution may be warranted in terms of the repeated exposure to anesthesia that is required in order to collect in vivo imaging data in monkeys, but this is likely to be a very productive avenue of research in order to test specific hypotheses about how changes in brain function result in impairments in cognition and how these impairments may be moderated or reversed.

## Acknowledgments

My research is supported by a grant from the Wellcome Trust.

*See also:* Aging and Memory in Animals; Aging and Memory in Humans; Aging of the Brain; Cognition in Aging and Age-Related Disease; Declarative Memory System: Anatomy; Functional Neuroimaging Studies of Aging; Non-Primate Models of Normal Brain Aging; Rodent Aging; Short Term and Working Memory; Synaptic Plasticity: Learning and Memory in Normal Aging.

## Further Reading

Alexander GE, Chen K, Aschenbrenner M, et al. (2008) Age-related regional network of magnetic resonance imaging gray matter in the rhesus macaque. *Journal of Neuroscience* 28: 2710–2718.

Bachevalier J, Landis LS, Walker LC, et al. (1991) Aged monkeys exhibit behavioral deficits indicative of widespread cerebral dysfunction. *Neurobiology of Aging* 12: 99–111.

Bartus RT, Fleming D, and Johnson HR (1978) Aging in the rhesus monkey: Debilitating effects on short-term memory. *Journal of Gerontology* 33: 858–871.

Baxter MG and Voytko ML (1996) Spatial orienting of attention in adult and aged rhesus monkeys. *Behavioral Neuroscience* 110: 898–904.

Browning PG, Easton A, and Gaffan D (2007) Frontal-temporal disconnection abolishes object discrimination learning set in macaque monkeys. *Cerebral Cortex* 17: 859–864.

Buckmaster CA, Eichenbaum H, Amaral DG, et al. (2004) Entorhinal cortex lesions disrupt the relational organization of memory in monkeys. *Journal of Neuroscience* 24: 9811–9825.

Culley DJ, Baxter MG, Yukhananov R, et al. (2004) Long-term impairment of acquisition of a spatial memory task following isoflurane-nitrous oxide anesthesia in rats. *Anesthesiology* 100: 309–314.

Eberling JL, Roberts JA, Rapp PR, et al. (1997) Cerebral glucose metabolism and memory in aged rhesus macaques. *Neurobiology of Aging* 18: 437–443.

Hao J, Rapp PR, Janssen WG, et al. (2007) Interactive effects of age and estrogen on cognition and pyramidal neurons in monkey prefrontal cortex. *Proceedings of the National Academy of Sciences of the USA* 104: 11465–11470.

Hao J, Rapp PR, Leffler AE, et al. (2006) Estrogen alters spine number and morphology in prefrontal cortex of aged female rhesus monkeys. *Journal of Neuroscience* 26: 2571–2578.

Herndon JG, Moss MB, Rosene DL, et al. (1997) Patterns of cognitive decline in aged rhesus monkeys. *Behavioural Brain Research* 87: 25–34.

Killiany RJ, Moss MB, Rosene DL, et al. (2000) Recognition memory function in early senescent rhesus monkeys. *Psychobiology* 28: 45–56.

Kowalska DM, Bachevalier J, and Mishkin M (1991) The role of the inferior prefrontal convexity in performance of delayed nonmatching-to-sample. *Neuropsychologia* 29: 583–600.

Lacreuse A, Chhabra RK, Hall MJ, et al. (2004) Executive function is less sensitive to estradiol than spatial memory: Performance on an analog of the card sorting test in ovariectomized aged rhesus monkeys. *Behavioral Processes* 67: 313–319.

Lacreuse A and Herndon JG (2009) *Nonhuman Primate Models of Cognitive Aging*. In: Bizon JL and woods A (eds.) *Animal Models of Human Cognitive Aging*, pp. 29–58. Totowa, NJ: Humana Press.

Lacreuse A, Wilson ME, and Herndon JG (2002) Estradiol, but not raloxifene, improves aspects of spatial working memory in aged ovariectomized rhesus monkeys. *Neurobiology of Aging* 23: 589–600.

Mishkin M and Murray EA (1994) Stimulus recognition. *Current Opinion in Neurobiology* 4: 200–206.

Moore TL, Killiany RJ, Herndon JG, et al. (2006) Executive system dysfunction occurs as early as middle-age in the rhesus monkey. *Neurobiology of Aging* 27: 1484–1493.

Moore TL, Schettler SP, Killiany RJ, et al. (2009) Effects on executive function following damage to the prefrontal cortex in the rhesus monkey. *Behavioral Neuroscience* 123: 231–241.

Moss MB, Killiany RJ, Lai ZC, et al. (1997) Recognition memory span in rhesus monkeys of advanced age. *Neurobiology of Aging* 18: 13–19.

Murray EA and Gaffan D (2006) Prospective memory in the formation of learning sets by rhesus monkeys (*Macaca mulatta*). *Journal of Experimental Psychology: Animal Behavior Processes* 32: 87–90.

Murray EA and Mishkin M (1998) Object recognition and location memory in monkeys with excitotoxic lesions of the amygdala and hippocampus. *Journal of Neuroscience* 18: 6568–6582.

Rapp PR and Amaral DG (1991) Recognition memory deficits in a subpopulation of aged monkeys resemble the effects of medial temporal lobe damage. *Neurobiology of Aging* 12: 481–486.

Rapp PR, Kansky MT, and Eichenbaum H (1996) Learning and memory for hierarchical relationships in the monkey: Effects of aging. *Behavioral Neuroscience* 110: 887–897.

Rapp PR, Kansky MT, and Roberts JA (1997) Impaired spatial information processing in aged monkeys with preserved recognition memory. *NeuroReport* 8: 1923–1928.

Rapp PR, Morrison JH, and Roberts JA (2003) Cyclic estrogen replacement improves cognitive function in aged ovariectomized rhesus monkeys. *Journal of Neuroscience* 23: 5708–5714.

Roberts JA, Gilardi KV, Lasley B, et al. (1997) Reproductive senescence predicts cognitive decline in aged female monkeys. *NeuroReport* 8: 2047–2051.

Smith DE, Rapp PR, McKay HM, et al. (2004) Memory impairment in aged primates is associated with focal death of cortical neurons and atrophy of subcortical neurons. *Journal of Neuroscience* 24: 4373–4381.

Tinkler GP, Tobin JR, and Voytko ML (2004) Effects of two years of estrogen loss or replacement on nucleus basalis cholinergic neurons and cholinergic fibers to the dorsolateral prefrontal and inferior parietal cortex of monkeys. *Journal of Comparative Neurology* 469: 507–521.

Voytko ML (1997) Functional and neurobiological similarities of aging in monkeys and humans. *Age* 20: 29–44.

Voytko ML (2000) The effects of long-term ovariectomy and estrogen replacement therapy on learning and memory in monkeys (*Macaca fascicularis*). *Behavioral Neuroscience* 114: 1078–1087.

Walker ML and Herndon JG (2008) Menopause in nonhuman primates? *Biology of Reproduction* 79: 398–406.

Wilson CR and Gaffan D (2008) Prefrontal-inferotemporal interaction is not always necessary for reversal learning. *Journal of Neuroscience* 28: 5529–5538.

# Non-Primate Models of Normal Brain Aging

**C T Siwak-Tapp and P D Tapp**, University of California at Irvine, Irvine, CA, USA



## Introduction

An important goal of aging research is the preservation of cognitive function with age, which is becoming more critical as the proportion of aged people continues to rise. Despite successful modeling of numerous specific features of aging, no single animal model fully replicates all aspects of human aging. It is thus necessary to continue developing new models to explore the unique aspects of each known model and combine information from convergent studies. The most commonly used animal models of human brain aging have employed rats, transgenic mice, and nonhuman primates. Companion animals, such as the dog and cat, are now living longer due to advances in animal health care. Consequently, age-related cognitive and behavioral changes comparable to those in humans are now being observed in dogs and cats. The work described herein assesses normal brain aging in other mammalian species, with our primary focus on the dog and cat.

## Brain Aging in the Dog

Veterinarians have long been aware of geriatric behavior problems in pet dogs. The field of veterinary behavioral medicine has introduced the term 'canine cognitive dysfunction' to describe age-related behavioral changes not attributable to a medical condition in dogs. Canine cognitive dysfunction is a progressive neurodegenerative disorder of senior dogs characterized by a gradual decline in a variety of behaviors over a long period of time. Five general categories of behavioral changes have been defined: (1) signs of disorientation, that is, becoming trapped in corners or behind furniture; (2) signs of decreased social interactions with family members or changes in recognition; (3) disturbances in the sleep–wake cycle, such as increased daytime sleeping and less nighttime sleeping; (4) loss of house-training; and (5) decreased activity. These behaviors involve learning and memory processes, and changes in such behaviors are used as indicators of cognitive dysfunction.

Evaluation of these behaviors is usually achieved by owners completing a checklist of noticeable changes in their dogs' behavior. These behavioral signs are observed in dogs that were normal when younger; over 60% of owners with dogs aged 11 years and older report one or more behavioral changes consistent with cognitive dysfunction, after other medical conditions have been eliminated. The prevalence of canine cognitive dysfunction increases from 32% of 11-year-old dogs to 100% of 16-year-old dogs. No sex differences have been reported.

### Cognition and Behavior in the Aging Dog

The measurement of cognitive functions in the aging beagle dog using neuropsychological tests has been the focus of a series of laboratory studies for more than 15 years. Like humans, cognitive aging in the dog is not a uniform process and individual differences become more common with age. Some cognitive processes remain intact with age while others deteriorate, each at different rates. Additionally, learning and memory impairments can occur independently of one another. Cognitive decline is not, however, an inevitable consequence of aging, as some old dogs perform just as well as young dogs on some tasks. Others show a mild impairment or become severely impaired. This section will discuss the cognitive and behavioral changes that occur with age in dogs.

**Age-related learning impairments in dogs** In order to administer neuropsychological tests, the dogs must first learn how to perform the procedures. Learning is the process of acquiring knowledge or skills through study or experience and generally becomes more difficult with age. Procedural learning is the acquisition of new skills. The procedural learning task for the dog is designed to teach the animal the skills necessary to perform in the testing apparatus, a modified version of the Wisconsin General Test Apparatus (**Figure 1**) used for nonhuman primates, and consists of reward–approach and object–approach learning. Dogs learn that there is a food reward, that they can approach the food, and that they must move objects to obtain the food. The procedural learning performance of aged dogs varies with previous life experience. Old dogs raised as companion animals learn these procedures as quickly as do young dogs raised in the same way. Aged dogs raised in a kennel setting, however, are slower to acquire these tasks compared to young kennel-reared animals. This simple task reveals the importance of the rearing environment and previous experience in assessing cognitive ability in aged animals.

In a discrimination learning paradigm, used to assess associative learning, the animal must learn to associate a specific object or set of object characteristics with a food reward. Simple discrimination learning tasks, such as an object discrimination

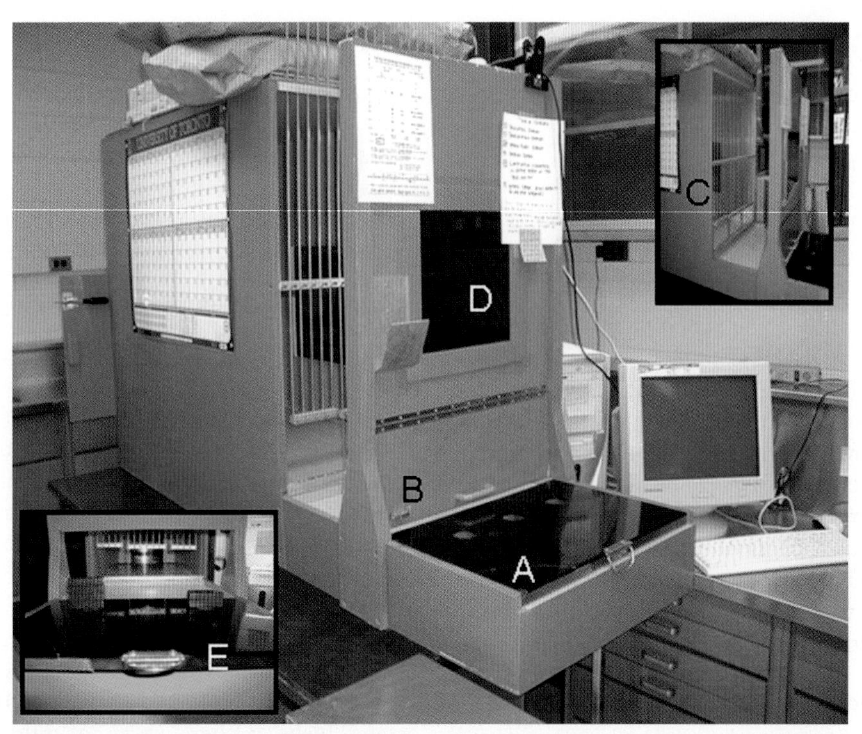


**Figure 1** Toronto general test apparatus used to collect cognitive data in dogs. Objects and food rewards are placed on a sliding tray (a) and presented to the animal by lifting a door (b). Adjustable gates (c) allow animals to make responses, which are observed through a one-way mirror (c). An example of the setup for a size discrimination task is shown (e).

whereby two different objects must be compared, are easily acquired by dogs and are not sensitive to age. Increasing the complexity of discrimination problems, however, makes these tasks more sensitive to age differences. For example, the size discrimination task uses two blocks that are identical in all attributes (e.g., color, shape, texture) except size – one is taller than the other. Old dogs are slower to learn this task and make more errors relative to young dogs. A second way to increase task complexity is to increase the number of choices available. The oddity discrimination task uses three objects, two that are identical and a third that has different physical attributes. On this task, the animal must learn to choose the odd object to receive a food reward. Aged dogs are slower to learn this task compared to young dogs. Further testing using objects that appear more and more similar exacerbates the age-related impairment.

Discrimination learning tasks are often followed by reversal learning tasks, whereby the reward contingencies of the objects are switched such that the previously correct object now becomes incorrect. Reversal learning tasks are age sensitive and are often used as measures of executive functions. This will be discussed in further detail in the section titled 'Executive functions in dogs.'

Object recognition learning is more difficult than discrimination learning because correct responses are based on rule learning and not just an association between an object and reward. Object recognition learning involves identifying a previously viewed object. In the delayed-non-matching-to-sample (DNMS) task, dogs are first shown a single object covering a food reward and are allowed to respond and obtain the food. After a delay period, two objects are presented – the original object and a novel object. Each trial of the task uses random sets of objects drawn from a large assortment. The dog must learn the rule of selecting the nonmatching object in the second presentation. The object recognition learning task is very difficult for dogs to acquire. Most young dogs are able to learn the task, but only some aged dogs are successful. The aged dogs that successfully learn the DNMS task perform more poorly relative to the young dogs.

The delayed-non-matching-to-position (DNMP) task is another rule-based learning task that assesses egocentric spatial learning – the ability to locate objects in space with reference to one's own body position. The simple version of the DNMP task is the two-choice (2c) DNMP. In this task, the dog is presented with a single object over one of two lateral food wells. After a delay period, a second identical object is presented over the remaining lateral food well along with the first object still in the same location. The dog must learn to choose the object in the novel location. All young dogs and some old dogs acquire this task, but a subset of aged dogs will show impairments in spatial learning and perform significantly more poorly. The complex version of the DNMP uses a three-choice (3c) paradigm. In the 3cDNMP, three locations are used (two lateral and one middle well). This version of the task is much more difficult for dogs to learn. Compared to the 2cDNMP, which all young dogs learn, 12% of young dogs are unable to solve the 3cDNMP problem. Similarly, only 18% of aged dogs fail the 2cDNMP but 83% are unable to learn the more complex 3cDNMP task.

The landmark task is another rule-based spatial learning task, but unlike the 2c- and 3cDNMP tasks, the landmark task measures allocentric spatial learning – the ability to locate objects in space with reference to another object or landmark. In the landmark task, dogs are rewarded for selecting the one object, of two objects, that is closest to an external landmark or referent. Thus, information about the correct response is only provided by the landmark. Like the egocentric spatial learning tasks, performance on the landmark discrimination task declines with age in dogs.

**Age-related memory impairments in dogs** An important observation in the canine model of cognitive aging is that learning and memory are differentially impaired with age. Some aged dogs that successfully learn a task may exhibit memory impairments. Two protocols used to examine memory function in the dog are the variable delay and maximal memory paradigms. Both protocols have been applied to the DNMP and DNMS tasks.

In the variable delay paradigm, the duration between the sample presentation and the matching phase of the task is varied between 20, 70, and 110 s. Each delay occurs an equal number of times per 12-trial session. Increasing the time between the sample and match presentation of a trial increases the demands placed on memory. On both the DNMP and the DNMS tasks, accuracy declines as the length of delay increases in both young and old dogs. The decline represents a working memory impairment since the animal successfully learned the task but has difficulty retaining the information over longer delay periods. Aged dogs perform more poorly than young dogs do at all delays.

The maximal memory paradigm uses a progressively increasing delay period until a preset number of sessions is reached. On the 2cDNMP, young dogs perform accurately at delays of over 210 s while old dogs only reach 30–50 s, indicating they have more difficulty retaining the spatial information as long as young dogs do. With the more difficult 3cDNMP, young dogs usually only reach 110 s and old dogs often fail to complete delay intervals longer than 5 s. Results from the maximal memory paradigm demonstrate that old dogs can acquire spatial information but are unable to retain information in memory for long periods of time. The age differences are more apparent on the DNMS task, in which young dogs respond accurately up to delays of 300 s.

**Executive functions in dogs** Executive functions describe a set of cognitive abilities that control and regulate other abilities and behaviors. These higher order cognitive processes include inhibitory control, concept abstraction, set shifting, set maintenance, and manipulation. Executive functions can be difficult to assess directly, and many of the tasks used to measure other abilities can be modified to evaluate executive functions.

Discrimination reversal tasks are often used to obtain measures of inhibitory control. Inhibitory control is the ability to inhibit interfering information or previously activated cognitive processes in order to focus attention on the relevant task requirements. On reversal tasks, animals must inhibit the previously correct response to learn a new stimulus–reward relationship using the same objects. In the dog, two important findings are reported on reversal tasks. First, aged dogs are much slower to acquire an object discrimination reversal and a size discrimination reversal compared to young dogs. Second, analysis of the types of errors made on the size discrimination reversal task indicates that the cause of reversal learning impairments changes with age. In old dogs (8–11 years old), reversal learning impairments are caused by an inability to learn the new stimulus–reward contingency. In senior dogs (over 11 years old), reversal deficits are caused by both an inability to learn new reward contingencies and an inability to inhibit a previous response. The perseverative responses observed in senior dogs suggest a deficit in inhibitory control, an executive function.

Another executive function impaired with age in the dog is the manipulation component of working memory. Maintenance involves actively holding information in working memory, while manipulation occurs when additional reorganization or updating of this information is required, an executive function. Tasks that require the manipulation of information held in working memory place greater demands on the cognitive process of working memory. The DNMP and DNMS tasks discussed earlier are tests of working memory performance. The previous protocols described varied the demands on working memory through the time required to hold the information in memory. A second way to change working memory demands is to increase the amount of information to be retained. The spatial list learning (SLL) task assesses the manipulation component of working memory in dogs.

In the SLL task, dogs are presented with a single red disk randomly located over one of three food wells during the first phase of a trial. After a response is made, a brief delay interval occurs, followed by the second phase of the trial in which two identical red disks are presented, one in the original location and a second in one of the two remaining spatial locations. Only responses to the novel location are rewarded on the second phase, and this component forces the dog to update or manipulate the information held in working memory. After a second delay period, the third phase of the task occurs in which the dog is presented with three identical red disks, two in the familiar locations and a third in the final novel location. Only responses to the novel location are rewarded. Locations of each object are randomized across trials.

Young and aged dogs successfully learn the task but old dogs perform significantly worse than young dogs do. Old dogs take longer to acquire the task and are unable to perform at delays longer than 10 s. Even old dogs that perform well on the 3cDNMP are unable to complete the SLL task. Modified versions of the SLL which place even greater demands on working memory manipulation exacerbate age differences between young and old dogs. Together, the results suggest that age-related impairments on the SLL reflect deficits in manipulation rather than maintenance, since many of the old dogs tested on the SLL task had performed the 3cDNMP with high accuracy. Thus executive processes associated with the manipulation of information in working memory can decline even if the ability to hold information in working memory does not.

Concept learning, the ability to categorize or sort objects based on a particular dimension, is also impaired with age in the dog. In the dog, the size concept task was developed to assess concept learning. On this task, dogs are trained on a series of 2c and 3c size discrimination tasks, using various combinations of the same set of four blocks to teach the animals that size is the feature of interest. Following this training, novel stimuli differing only in size are presented to determine if the dogs learn that size could be used conceptually to solve the new tasks. In tests of size concept learning, young dogs successfully transfer the concept of size to new tests. Older dogs are unable to solve the concept tests, suggesting that dogs are capable of learning conceptual dimensions but the ability declines significantly with age.

**Age-related changes in noncognitive behaviors in dogs** Changes in noncognitive behaviors with age are another manifestation of the underlying brain pathology that contributes to cognitive dysfunction. Many of the noncognitive changes observed in aging dogs correlate with cognitive status. Locomotor activity, for example, decreases with age in dogs, but this decrease depends upon several factors, including the testing situation, cognitive status, and life experience. Aged dogs raised in a kennel or laboratory-type setting do not show a decrease in locomotor activity in a novel testing environment, such as an open field arena, while pet dogs show a decrease in open field activity with age. In a home area test, however, kennel/laboratory-reared dogs show declines in locomotor activity with age. This suggests that the open field and home area tests are measuring different aspects of locomotor activity. The open field test occurs in a novel setting and locomotor responses may reflect behavioral reactivity rather than spontaneous motor activity. Moreover, aged dogs that are cognitively impaired are also hyperactive in both the novel and the familiar test situation, compared to aged unimpaired dogs. Laboratory/kennel-reared dogs are also more active compared to pet raised or companion animals. Thus, environmental enrichment over the life of a dog may interact with genetic factors and contribute to the extent of locomotor decline and cognitive dysfunction.

Locomotor activity in laboratory dogs follows a circadian activity–rest pattern whereby locomotor activity is high during the day and low at night. This circadian pattern of locomotor activity also varies as a function of age, housing environment, and cognitive status. In young dogs, activity levels are higher for a longer period of time during the day compared to older dogs. In an indoor housing area, aged dogs display shorter bouts of activity during the day, more frequent rest periods, and take longer to become active once the lights are turned on compared to young dogs. These age-related differences are not observed in dogs housed in an outdoor area, suggesting that bright sunlight is effective in synchronizing and consolidating the activity rhythm. Cognitively impaired aged dogs are hyperactive and show fewer periods of rest during the day compared to old unimpaired dogs.

In addition to locomotor activity, we adapted a series of behavioral tests to assess social and exploratory behaviors in the aging dog. These tests are modified versions of the open field test whereby specific social or exploratory stimuli are placed in the arena and the response of the dog is recorded. The expression of social and exploratory behaviors in dogs varies with age and cognitive status. In response to the presence of a person sitting in the open field arena, young dogs display active social responses, including initiating physical contact, climbing, sniffing, and licking. Age-unimpaired dogs are more passive and tend to sit or lay quietly beside the person.

The overall amount of time engaged in the social behaviors does not differ with age; only the types of behaviors show an age-related difference. Cognitively impaired dogs exhibit very little interest in the person in the room and do not engage in social behaviors.

We further measured reactions to conspecifics to assess how social responses within the species may change with age. Three different stimuli were used: a cardboard silhouette of a dog, a sandcast life-size model of a dog, and a mirror. In these tests, dogs treat the cardboard silhouette like a real dog by directing their sniffing to the face and anal regions, locations that are normally investigated when live dogs interact. Young and age-unimpaired dogs habituate rapidly to the silhouette, but cognitively impaired aged dogs spend significantly more time sniffing the cardboard figure. This could reflect a deficit in habituation or stimulus recognition. To make the conspecific more realistic, we used a life-size sandcast model of a golden retriever. Young dogs spend more time sniffing and attempting to interact with the model dog compared to old dogs. Old unimpaired and impaired dogs do not differ in the amount of time spent sniffing the model dog. To make the conspecific more interactive, we also assessed the reaction of the dogs to a mirror reflection of themselves. Reactions to the reflection in the mirror depend on the dogs' ability to recognize their own species. The reactions of the dogs toward the mirror reflection are initially other-directed, treating the reflection like another dog. Responses include jumping at, barking at, and trying to play with the dog in the mirror, as well as attempts to look behind the mirror to find the other dog. Unimpaired old and young dogs habituate to the reflected image rapidly. Old impaired dogs, however, spend significantly more time reacting to the reflection in the mirror, perhaps reflecting a deficit in habituation or stimulus recognition.

Exploratory behavior, or curiosity, is another non-cognitive behavior that changes with age. To assess changes in exploratory behavior in aging dogs, we placed several novel commercially available dog toys in a large test room and measured the amount of time the dogs spent exploring the unfamiliar objects. Results from this test show that young dogs spend more time playing with and sniffing toys than does either of two groups of old dogs. Old-unimpaired dogs show reactions similar to those of young dogs but spend less time exploring objects. Old-impaired dogs generally avoid the toys and express no interest in investigating the objects.

### Neuropathology

The aforementioned cognitive and behavioral measures are often combined with analyses of neuropathology to determine whether brain pathology is linked to the age-related changes observed. $\beta$-Amyloid deposition and neurofibrillary tangle formation are thought to be related to the cognitive and behavioral decline that typically occurs during aging and have received the most investigation in the dog.

**$\beta$-Amyloid deposition** The dog brain is very well suited for studying early stages of $\beta$-amyloid deposition in the aging brain. The earliest and most consistent distribution of $\beta$-amyloid (A$\beta$) plaques in the dog occurs in the prefrontal cortex starting around 8 years of age. By 14 years of age, the entorhinal and parietal cortices are also affected but the cerebellum remains relatively clear. Dogs accumulate the longer, more toxic, less soluble form of $A\beta_{1-42}$ prior to the shorter more soluble $A\beta_{1-40}$ type. Plaque morphology in the dog is generally diffuse and deposits are thioflavine-S negative, suggesting a lack of $\beta$-pleated sheet formation. We have found a strong correlation between $\beta$-amyloid deposition and cognitive decline in aged dogs.

**Tau phosphorylation** In contrast to aging humans, dogs do not develop neurofibrillary tangles in the brain but do show early stages of tangle formation characterized by tau phosphorylation and an intracellular punctuate distribution. Hyperphosphorylated tau has been observed in select neuronal populations in older dogs. The absence of neurofibrillary tangles in the dog brain could indicate that the tau protein in the dog is different from the human tau protein. The implication of this finding is that $\beta$-amyloid and not tau pathology may be sufficient to lead to cognitive and behavioral dysfunction in the aging dog.

**Neuron loss and function** Several studies have reported changes related to neuronal integrity with age in the dog. Age-related loss of neurons in the canine brain has been reported in the cingulate gyrus, superior colliculus, and claustrum. Selective neuron loss has also been observed in the prefrontal, temporal, and occipital cortex as well as in the hilus of the hippocampus of aged dogs.

Neurons in the prefrontal cortex of the aged dog brain show other morphological hallmarks of dysfunction, including distorted soma, loss of dendritic spines, shrinkage of dendritic branches, and tortuous apical dendrites. The presence of DNA damage and apoptosis is an additional indicator of neuron dysfunction. In dogs, the number of neurons with damaged DNA increases with age and is positively correlated with $\beta$-amyloid load in the frontal cortex.

**Oxidative damage** Oxidative damage to lipids and proteins in the brain is a common hallmark of aging and may compromise neuron function. In aging dogs, malondialdehyde levels, a marker of lipid peroxidation, increase with age in the prefrontal cortex and blood serum in the dog starting around 8 years of age. Glutamine synthetase activity, an endogenous antioxidant that protects against oxidative damage, decreases with age in the prefrontal cortex. These age-related changes in oxidative damage are concomitant with increased $\beta$-amyloid deposition in the prefrontal cortex.

A role of oxidative damage in age-related cognitive dysfunction was demonstrated in a longitudinal study in dogs. Cognitive function was significantly improved and maintained in aged dogs fed a food fortified with antioxidants and mitochondrial cofactors for almost 3 years, compared to aged dogs given a control food. The fortified food contained higher amounts of vitamin E, vitamin C, *l*-carnitine, and *dl*-lipoic acid as well as a 1% inclusion of spinach flakes, tomato pomace, grape pomace, carrot granules, and citrus pulp. In addition to improving cognition and behavior, the fortified food also led to a reduction in $\beta$-amyloid deposition in some brain regions in the aged dogs. This suggests that oxidative damage is an important contributor to cognitive dysfunction and neuropathology in normal aging, and dietary interventions may significantly reduce these events.

### Neuroanatomy and Hemodynamics

Noninvasive brain imaging technology permits analyses of metabolic, vascular, and structural changes associated with brain aging. Two common methods available to evaluate volumetric changes in the brain include region-of-interest (ROI) planimetry and voxel-based morphometry (VBM) using brain images acquired from magnetic resonance imaging (MRI). ROI planimetry involves manually tracing the particular brain region of interest on several consecutive images to obtain an estimate of total volume of that region. Studies using ROI planimetry reveal that global cortical atrophy and ventricular enlargement are characteristic features of the aged dog brain. Further analysis shows that the total volume of the brain, and in particular the volume of the frontal lobes and hippocampus, will decrease with age. The volume of the occipital cortex does not change with age. Moreover, these age-related decreases in frontal lobe volume correlate with increased $\beta$-amyloid deposition and measures of executive functions.

By contrast, VBM is an automated technique that performs a voxel-by-voxel comparison of local densities of gray and white matter on the brain images. In the dog brain, VBM analysis has revealed gray matter reductions bilaterally in the frontal lobes, parietal lobes, temporal lobes, thalamus, cerebellum, and brain stem. Many of the differences in structural brain aging are sexually dimorphic. In particular, aged male dogs showed a greater reduction in the frontal lobes while changes in the temporal lobes were more apparent in females. White-matter reductions with age were observed in the internal capsula and cranial nerve bundles of aged male dogs and the alveus of the hippocampus in aged females. The additional information provided by the VBM analysis suggests that brain aging in the dog varies regionally and is sex dependent.

MRI can also be used to monitor longitudinal changes in the aging dog brain using serial MRI techniques to coregister images from different time points. With this technique, we have found that dogs spontaneously develop lesions, resembling small lacunes, with age. The majority of lesions in the aged dog brain form in the frontal cortex and caudate nucleus.

A technique called dynamic susceptibility contrast (DSC)-enhanced MRI can be used to study vascular characteristics under the same conditions as anatomical MRI scans. Administration of contrast agents during DSC-MRI procedures permits kinetic modeling to obtain measures of vascular volume and permeability. In the dog, DSC-MRI revealed a decline in regional cerebral blood volume (rCBV) with age in the gray- and white-matter regions. These changes were greater in the white matter compared to the gray matter, even though gray-matter rCBV was consistently higher relative to white-matter rCBV. The decreased rCBV and increased blood–brain barrier permeability developed in tandem with cortical atrophy in this study.

## Brain Aging in the Cat

### Cognition and Behavior

Many of the behavioral changes typical of cognitive dysfunction in dogs also occur in elderly cats, but the age of onset may be later in cats. Information on behavioral changes with age in cats comes largely from observations in veterinary practice. Behavioral changes observed in aged cats that are consistent with feline cognitive dysfunction include signs of wandering, confusion, excessive vocalization, decreased grooming, lethargy, decreased awareness, greater time sleeping or other disruptions in the sleep/wake cycle, inappropriate elimination or loss of house-training, increased displays of affection or other alterations in social interactions, decreased recognition of owners, and increased anxiety. Estimates of prevalence suggest that 35% of aged cats, once those with possible underlying medical causes were removed, have clinical signs related to cognitive dysfunction. In cats 11–15 years old, the most common change was in social interactions, while changes in cats over 15 years of age primarily affected activity and vocalization. Laboratory studies of cats have reported altered patterns of habituation, increased reactivity to auditory stimuli, and superior performance on spatial reversal learning tests with increasing age. No decrements in fine motor coordination were observed and locomotor activity levels did not change with age.

### Neuropathology

Studies of brain pathology in cats indicate that cats develop $\beta$-amyloid and tau abnormalities. In aged cats, the $\beta$-amyloid plaques are of the diffuse variety and are localized in the prefrontal cortex, hippocampus, parahippocampal gyrus, occipital cortex, and parietal cortex. The diffuse plaques contain predominantly the $A\beta_{1-42}$ species, consistent with findings in dog and human brain tissue, but the morphology of the diffuse plaques in cat brain is different than what is usually observed in the human brain.

Hyperphosphorylated tau is also present in the aged cat brain. The pattern is consistent with that in the human brain, including area CA1 of the hippocampus, subiculum, and entorhinal/parahippocampal cortex. Aged cats with clinical signs of cognitive dysfunction are frequently positive for dystrophic neurites, but not all exhibit phosphorylated tau accumulation within neuronal cell bodies.

## Brain Aging in Other Mammals

The dog and cat are more widely studied than are other mammals, likely due to the prevalence of these animals as pets. Some isolated studies exist that have examined brain aging in other animals, and are described in the following sections.

### Rabbit

Eyeblink classical conditioning (EBCC) in older rabbits is used as a model system for studying the neurobiology of learning, memory, and the aging brain. Age-related impairments in EBCC occur relatively early in aging in both rabbits and humans. EBCC is also sensitive for identifying patients with Alzheimer's disease and is abnormal in Down's syndrome individuals over 35 years of age.

Investigation into the aged rabbit brain shows cell loss and gliosis in the cerebellum and hippocampus but no evidence of $\beta$-amyloid deposits or abnormal tau accumulations, even in rabbits over 7 years of age. Manipulating dietary cholesterol, however, induces neuronal accumulation of $\beta$-amyloid in New Zealand white rabbits, and this increases when copper is administered in the drinking water. Thioflavine-S-reactive features are also observed in animals supplemented with copper. Emerging evidence implicates cholesterol in the production and accumulation of $\beta$-amyloid by the 'mismetabolism' of the amyloid precursor protein. This mismetabolism causes the 42-amino-acid form of $\beta$-amyloid to form oligomers that, in turn, start a chain of events leading to the accumulation of amyloid plaques. Thus, the rabbit may be a valuable model to study initiating factors and perhaps the role of cholesterol in the deposition of $\beta$-amyloid.

### Horse

Learning studies with horses are designed to provide a better understanding of trainability in this animal. Discrimination learning ability has been examined in horses using visual cues in a 3c test with food as the reward. In this study, quarter horses learned faster than thoroughbreds, and both groups learned a second discrimination task faster than the first task. In both tasks, however, there was a significant negative correlation between age and rate of learning. Separate neuropathological analysis of the brains of horses up to 27 years of age showed that all regions were devoid of hyperphosphorylated tau. No reports have been published to our knowledge about the existence of $\beta$-amyloid in the brains of horses.

### $\beta$-Amyloid and Tau in Other Mammals

There are various reports of naturally occurring $\beta$-amyloid deposition or neurofibrillary tangle-like formations in several other mammals. $\beta$-Amyloid deposition has been reported in the brains of camels, coyotes, and polar bears. Tangle-like pathology has been observed in brains of sheep, goats, bears, wolverines, guanacos (similar to llamas), bison, and reindeer. To our knowledge there are no data available concerning behavioral changes with age in any of these animals.

## Conclusion

Additional animal models for human brain aging have new insights to offer. A complete replication of human aging does not exist in a single animal system, but a more comprehensive picture emerges when information from several animal models is consolidated and integrated.

*See also:* Aging of the Brain; Cognition in Aging and Age-Related Disease; Oxidative Damage in Neurodegeneration and Injury.

## Further Reading


Chan AD, Nippak PM, Murphey H, et al. (2002) Visuospatial impairments in aged canines (*Canis familiaris*): The role of cognitive-behavioral flexibility. *Behavioral Neuroscience* 116(3): 443–454.

Cotman CW, Head E, Muggenburg BA, et al. (2002) Brain aging in the canine: A diet enriched in antioxidants reduces cognitive dysfunction. *Neurobiology of Aging* 23(5): 809–818.

Head E, Callahan H, Cummings BJ, et al. (1997) Open field activity and human interaction as a function of age and breed in dogs. *Physiology & Behavior* 62(5): 963–971.

Head E, Milgram NW, and Cotman CW (2001) Neurobiological models of aging in the dog and other vertebrate species. In: Hof PR and Mobbs CV (eds.) *Functional Neurobiology of Aging*, pp. 457–465. San Diego: Academic Press.

Head E, Moffat K, Das P, et al. (2005) Beta-amyloid deposition and tau phosphorylation in clinically characterized aged cats. *Neurobiology of Aging* 26(5): 749–763.

Levine MS, Lloyd RL, Fisher RS, et al. (1987) Sensory, motor and cognitive alterations in aged cats. *Neurobiology of Aging* 8(3): 253–263.

Mader DR and Price EO (1980) Discrimination learning in horses: Effects of breed, age and social dominance. *Journal of Animal Science* 50(5): 962–965.

Milgram NW, Head E, Weiner E, et al. (1994) Cognitive functions and aging in the dog: Acquisition of nonspatial visual tasks. *Behavioral Neuroscience* 108(1): 57–68.

Milgram NW, Head E, Zicker SC, et al. (2005) Learning ability in aged beagle dogs is preserved by behavioral enrichment and dietary fortification: A two-year longitudinal study. *Neurobiology of Aging* 26(1): 77–90.

Ruehl WW and Hart BL (1998) Canine cognitive dysfunction. In: Dodman N and Shuster L (eds.) *Psychopharmacology of Animal Behavior Disorders*, pp. 283–304. Malden, MA: Blackwell Scientific Publications.

Siwak CT, Tapp PD, and Milgram NW (2001) Effect of age and level of cognitive function on spontaneous and exploratory behaviors in the beagle dog. *Learning & Memory* 8(6): 317–325.

Siwak CT, Tapp PD, Zicker SC, et al. (2003) Locomotor activity rhythms in dogs vary with age and cognitive status. *Behavioral Neuroscience* 117(4): 813–824.

Sparks LD (2004) Cholesterol, copper, and accumulation of thioflavine S-reactive Alzheimer's-like amyloid beta in rabbit brain. *Journal of Molecular Neuroscience* 24(1): 97–104.

Su MY, Tapp PD, Vu L, et al. (2005) A longitudinal study of brain morphometrics using serial magnetic resonance imaging analysis in a canine model of aging. *Progress in Neuro-Psychopharmacology & Biological Psychiatry* 29(3): 389–397.

Tapp PD, Siwak CT, Estrada J, et al. (2003) Size and reversal learning in the beagle dog as a measure of executive function and inhibitory control in aging. *Learning & Memory* 10(1): 64–73.

Tapp PD, Siwak CT, Gao FQ, et al. (2004) Frontal lobe volume, function, and beta-amyloid pathology in a canine model of aging. *Journal of Neuroscience* 24(38): 8205–8213.

Tapp PD, Chu Y, Araujo JA, et al. (2005) Effects of scopolamine challenge on regional cerebral blood volume. A pharmacological model to validate the use of contrast enhanced magnetic resonance imaging to assess cerebral blood volume in a canine model of aging. *Progress in Neuro-Psychopharmacology & Biological Psychiatry* 29(3): 399–406.

Tapp PD, Head K, Head E, et al. (2006) Application of an automated voxel-based morphometry technique to assess regional gray and white matter brain atrophy in a canine model of aging. *Neuroimage* 29(1): 234–244.

Tapp PD and Siwak CT (2006) The canine model of human brain aging: Cognition, behavior, and neuropathology. In: Conn PM (ed.) *Handbook of Models for Human Aging*, pp. 415–434. Amsterdam: Elsevier.

Woodruff-Pak DS and Trojanowski JQ (1996) The older rabbit as an animal model: Implications for Alzheimer's disease. *Neurobiology of Aging* 17(2): 283–290.

# Rodent Aging

**B Teter**, University of California, Los Angeles, CA, USA



## Introduction

The study of aging has incorporated such vast theoretical and experimental effort that a review of brain aging must necessarily cover a wide conceptual landscape, from evolution to biochemistry, teleology to stochasticism, and genetics to environment. Aging, determined by the entire life history and its evolutionary determinants, is influenced by aspects of fitness, disease, reproduction, mutation, genetics, and random events. Since the brain and aging are the two most complex biological phenomena, it is not surprising that our understanding of brain aging is limited. In regulating and coordinating most somatic processes, the brain is a nexus of control systems linking organ systems and the environment, and while it is rarely the organ of pathological death, its degeneration may be intimately involved in inducing, synergizing, or failing to protect the development of terminal diseases of other organs. The brain, therefore, plays an important role in aging and in life span determination.

Normal aging in the brain, defined in the absence of diseases associated with neuron loss, is not itself associated with extensive age-related neuron loss, but is recognized as sublethal, functional neurodegeneration caused by increased vulnerability to metabolic stress and trauma. Brain aging is influenced by processes and events that begin with genetic and epigenetic contributions from the parents, acting during development, and accelerate in middle age. Aging can be defined to begin when neuron birth and differentiation is surpassed by neurodegeneration that eventually affects cognitive functional ability (approximately after age 10–20 years of human life for most functions).

Animal models of brain aging address select aspects of aging to varying degrees, including animals with phenotypically accelerated aging or prolonged longevity, engineered transgenic, mutant, and knockout models that focus on a single gene's role (and *in vitro* cell and organotypic systems proximally extracted from them), and interventions and environmental manipulations that attempt to modify specific behavioral, anatomic, cellular, or biochemical targets. However, the complexities of aging are not well elucidated in the reductionist approach of animal models. The translational extrapolation to improve the human condition is where animal models receive their greatest challenge and achieve their greatest relevance. Congruencies between humans and rodents justify such extrapolation, particularly at the cellular and gene expression levels. However, evolutionarily, humans and rodents evolved with different life histories (high entropy (equilibrium) and low entropy (opportunistic), respectively) that ultimately determine their respective chronological longevities. This profoundly impacts the strength of inferences drawn from rodent models for the human condition.

## Rodents as Models of Aging

Rodent strains that vary in aging processes (qualitatively and quantitatively) and life span have revealed aspects of aging that control health and longevity. Inbred strains have similar or shorter life spans than hybrids (hybrid vigor), in part because they die from strain-specific disease patterns not active in hybrids. Therefore, the more diverse genetic background and greater interanimal variation of hybrids are more relevant to human populations.

It is important to distinguish the processes of aging, disease, and life span (death). 'Normal' aging is defined as time-related change in the absence of disease, yet both can be driven by the same processes and therefore difficult to completely experimentally (or conceptually) dissociate, for example, Alzheimer's disease and normal cognitive decline. Prepathological events play important roles in aging and longevity, including lifetime accumulation of 'events' involving inflammation, infection, and nutrition. Two aspects of brain development could establish the basis for later brain aging effects: genetic antagonistic pleiotropy and cognitive reserve. Antagonistic pleiotropy occurs when genes and events that are adaptively essential for brain development become deleterious with aging, for examples, the progressive decrease in brain plasticity at early ages which serves to focus the mind on the established (and, presumably, understood) environment that continues into adulthood resulting in cognitive decline with age, and age-related changes in inflammatory cytokines which modify responses to stress and reactive oxygen species (ROS) signaling and damage. In studying postdevelopmental influences on aging in mouse transgenic models, artifacts caused by transgene expression during development can be mitigated by the use of conditional-expression transgenics, where gene expression can be experimentally turned on in adult life. The use of cell type and tissue type-specific transgene expression and controlled genetic knockout (using the Cre-lox system) mitigates global expression issues, as well as developmental issues.

Lifetime random events including spontaneous mutation and epigenetic events can explain variation in longevity of lab rodents and may model human longevity variation. Long-term strain inbreeding within an institution can result in genetic drift that may alter strain-specific phenotypes and responses to manipulations. This random variation can be exploited to develop models of premature brain aging even within an inbred strain, using prematurely aging littermates or substrains that are 'biologically' but not chronologically older (see the section titled 'Accelerated aging models and the oxidative stress theory of aging'), illustrating that the phenotypes of aging and death do not necessarily develop chronologically.

Some lab rodent husbandry practices create selective pressure against disease-sensitive individuals by invoking the mandatory, humane euthanasia of 'suffering' animals in longevity studies. For example, age-related ulcerative dermatitis in C57Bl/6 mice, exacerbated by dominance hierarchy aggression, leads to submissive mice being culled from the population and dominant mice being 'selected.' Further, social housing strongly influences stress levels, both increasing them in male cages and decreasing them in female cages, with effects on learning and memory. The addition of an enriched environment to standard housing conditions can reverse the age-related decline in neurogenesis in the dentate gyrus. Uncontrolled variation in state of the animals at the time of (and, particularly, immediately before) sacrifice can blur effects on immediate-early responses, for example, phosphorylation cascades.

In humans, medicalization of aging (prevention or delay of lethal diseases, including that by hygiene) has contributed to increased longevity, which has also allowed expression of older age-related diseases like Alzheimer's. This raises the profound issue that 'normal' aging is clearly changing in modern humans. A link between lifelong inflammation and nutrition drove the adaptive influence of human diet changes in the distant past to increase longevity (see the section titled 'Apolipoprotein E'); however, this is drastically changing with the recent obesity epidemic in some Western cultures that will actually reduce the expected average human life span. A similar disconnect between evolved life history and current lifestyle of diet and medicalization occurs in laboratory rodents where they age in a stable caged environment at various pathogen-free levels (which limits immune system development and response) and are fed *ad libitum* low-fat, high-calorie, antioxidant diets containing soy estrogens and other compounds that are increasingly recognized for their ability to mask effects or create experimental artifacts.

## The Role of the Brain in Aging and Life Span

The brain is a nexus of control systems linking organ systems and the environment that regulates and coordinates most somatic processes. It evolved as a regulator of food acquisition, energy metabolism, stress responses, and reproduction. Lethal disease processes of other organs can be induced or synergized by neurodegeneration. The brain, therefore, plays an important role in aging and in life span determination, for example, brain-specific knockdown of insulin-like growth factor 1 receptor (IGF-1R) increased life span (see the section titled 'Retarded aging and longevity extension models').

### Rodent Strain Differences

Rodent strains differ in the pace of behavioral changes with age, and while ultimately phenotypically similar, might implicate fundamentally different processes. The types of strain-specific effects that would best reveal genes regulating behavioral aging are those that differ in both the rate of age-specific mortality and the rate of behavioral decline. For example, strains with shorter life spans like 129/SvJ and FVB mice seem more behaviorally vulnerable (temporal or specific vulnerability) to age-related changes than longer-lived C57Bl/6J. Further, neuronal aging at the levels of cell structure and neurochemistry, while largely studied in C57Bl/6J, show strain specific differences in vacuolar neuritic dystrophy, Ca-binding proteins, inclusion bodies, neurogenesis, neuron cell number in the hippocampal dentate gyrus, various neurotransmitters, and responses to the toxin kainic acid and ischemia.

Behavioral test performance is notoriously strain dependent. Even sublines of the C57Bl/6 strain, the J and Nnia substrains, may differ in age-related decline in Morris water maze learning, and 129/SvJ mice are poor performers in part due to their visual pathology. Further, mice and rats do not perform the Morris water maze in identical ways, particularly regarding the performance parameter that is used in probe trials. As a further testament to the value of rodent models in predicting human responses, the rodent-specific behavioral test, the Morris water maze of learning and memory, has been adapted to a virtual (computer-based) task for testing congruent age-related deficits in humans.

### Genetic Contributions

Genetic variation accounts for that about one-quarter of the variance in adult human life span. Several of the genes known to influence rodent life span have important effects in the brain, including insulin-regulated genes and *SIRT-1*, *daf2*, *age1* (PI-3-kinase), *pit1/prop1* (pituitary), growth hormone-related genes, *PARP-1*, and *apoE* (see sections titled 'Retarded Aging and longevity extension models' and 'Glial influences'), to name a few, and many other candidates identified by microarray analysis. Although some senescent phenotypes can be similar between humans and rodents, they are occurring over a 30-fold difference in time; is this simply a quantitative difference in rate or are mechanisms qualitatively different? Gene expression microarray analysis of genes whose expression changes with aging in rat, mouse, and human brain (in different regions) has shown a surprising degree of agreement across these species. Among the few hundred changing genes, as many as 33% were identical between rat and human. These and other genes were in concordant categories of aging-related inflammation, oxidation, and glial processes, as well as novel (not primarily aging-related) categories like developmental regulation, structural plasticity (including *apoE*), signaling, transcription factors, lipid metabolism, and protein processing. This conservation of genomic responses across short- and long-lived species strongly validates the relevance of rodents to study age-related genetic, and by inference, physiologic, changes.

### Hormonal Control

Two brain regions that interact through neuroendocrine modulation of learning and memory, the hypothalamus and hippocampus, are particularly central to aging and the control of longevity. Hypothalamic control of gonadal, adrenal, and endocrine hormones impacts functions of and aging in the hippocampus.

**The hypothalamus** The hypothalamus is the central regulator of endocrine system homeostasis, integrating somatic growth and development driven by pituitary and gonadal hormones, stress responses driven by glucocorticoid hormones, and nutritional and metabolic signals, all central players in the neuroendocrine regulation of life span. The two organ system axes, the hypothalamic–pituitary–ovary (HPO) axis (or more generally, hypothalamic–pituitary–gonad (HPG) and the hypothalamic–pituitary–adrenal (HPA) axis, are dramatically affected by and mediate age-related changes in the brain and fundamentally regulate aspects of somatic aging and longevity. The aging hypothalamus shows gene expression microarray changes in several gene ontogeny categories of neuronal structure, signaling, stress response, and metabolism (also see sections titled 'Accelerated aging models and the oxidative stress theory of aging,' 'Retarded aging and longevity extension models,' and 'Glial influences').

The HPO axis regulates reproductive aging. There is a well-established tradeoff between longevity and reproduction (simplistically recognized by Aristotle as each act of sex shortening life). Reproductive aging is central to life history evolution and the pleiotropic effects of gonadal hormones on brain functions of both neurons and glia. Reproductive senescence is driven by oocyte depletion, which depends on rodent strain and strain hybridization, and on random variation in embryonic oocyte number. Caloric restriction retards ovarian aging by reducing or delaying reproduction (see the section titled 'Caloric restriction'); this may be a selected response, adaptive in times of famine to preserve reproduction for later times of feast when reproduction success is greater.

Human postmenopausal loss of estrogen drives some cognitive changes. Postreproductive female rodents differ from humans most significantly in their elevated estradiol levels. Brain control of and responses to reproductive aging have been investigated using both hormone and genetic manipulation models. Modeling the human postmenopausal estrogen loss in rodents by ovariectomy (OVX) and comparing this to estrogen replacement has shown that estrogen loss accelerates reproductive senescence. Estrogen receptor knockout and transgenic mice have shown the short- and long-term effects of estrogen exposure on neurons and glia. Additional key elements identified in rodent reproductive decline are the age-related decline in follicle-stimulating hormone (FSH), which alters communication between the hypothalamus and pituitary, and the age-related change in luteinizing hormone (LH) secretion and loss of gonadotropin-releasing hormone (GnRH) neurons and systems that control them, which alter the HPG axis.

The HPA axis controls developmental growth and stress responses. This axis controls the lifelong basal and stress-induced glucocorticoid hormone levels, which are elevated with age in several rat strains, but not in C57Bl/6J mice, perhaps reflecting species-specific pathologies. Modeling adrenal hormone action by adrenalectomy (ADX) largely but incompletely eliminates glucocorticoid production. A well-studied target of glucocorticoid action is the hippocampus.

**The hippocampus** Aspects of age-related changes in hippocampal structure and function are partly understood in terms of their responses to glucocorticoid and gonadal hormones and their roles in longevity. Rodent models of age-related changes in learning and memory, as assessed by hippocampal long-term potentiation (LTP) and long-term depression (LTD), are dramatically different among rat strains, where F344 show increased LTD with age and outbred Long-Evans show decreased LTD. A canonical change across all mammals is the age-related decline in the *N*-methyl-D-aspartate receptor (NMDAR) subunit NR2B, which may contribute to hippocampal decline in LTP and learning.

Neuron number in the hippocampus is relatively preserved with aging across all mammals; small changes are relatively region specific and rodent strain dependent. Historically, conflicting data are attributed to nonstandardized cell counting methods. Therefore, hippocampal atrophy with aging is attributed to the neuron cell atrophy and neurite and synapse loss. Neurogenesis of immature hippocampal granule cells shows age- and exercise-dependent changes in rodent models.

The entorhinal cortex input to the hippocampal dentate granule cells degenerates with age in humans and rats, which causes compensatory neurite sprouting in the hippocampus. This is experimentally modeled in rodents, albeit acutely, by various afferent lesions, and this recapitulates aspects of normal aging, including modulation by estrogen and impairment in aging female rat. Age-related decline in sprouting is due to defects in the aged target to which the sprouting neurite synapses, not to defects in the aged sprouting neuron, as shown in the model of implanting superior cervical ganglion tissue near the hippocampus and reciprocally varying the age of the ganglionic tissue (neurite source) or the host (hippocampal target) animal.

Estrogen stimulates neuronal function at many cellular levels, including neurogenesis, sprouting of neurites and spines, and molecular pathways. For example, aged rats do not respond to estrogen with increased spine formation in hippocampal CA1 neurons, perhaps simply due to receptor downregulation. Glucocorticoids cause neuronal and glial changes in the hippocampus of humans and rodents that contribute to cognitive changes during aging. Modeling age-related glucocorticoid exposure in rodents includes both acute and chronic treatment (e.g., hormone pellet implantation in the brain), ADX, and genetic knockout of a cortisol-regenerating enzyme. Effects and correlates of these manipulations include hippocampal atrophy, neuron pyknosis (cell body atrophy) and cell loss (pyramidal neurons), decreased neurogenesis, glial activation (hypertrophy, altered gene expression including protective changes in glial fibrillary acidic protein (GFAP) and transforming growth factor $\beta$-1 (TGF$\beta$-1), and gliogenesis (see the section titled 'Glial influences')), and concomitant learning and memory impairments. Glucocorticoids are also the target of paradoxical regulation by caloric restriction (see the section titled 'Caloric restriction'). Rat strains differ in age-related losses of corticoid receptors in the hippocampus. The effects of developmental experience and stress on age-related hippocampal impairments are modeled in a paradigm of neonatal handling and maternal deprivation that results in reduced cortical brain-derived neurotropic factor (BDNF) in the adult rat. Therefore, uncontrolled differences in lab rodent rearing could also contribute to contradictory reports of neuronal loss with aging.

## Accelerated Aging Models and the Oxidative Stress Theory of Aging

Accelerating aging models, while showing phenotypes similar to normal aging, are not necessarily driven by the same mechanisms as aging. They display only a subset of normal aging phenotypes, underscoring the greater complexity of normal aging, and show a preponderance of only one or a few disease pathologies. For example, genetically autoimmune mice, which challenge the historical doctrine of brain immune privilege by the blood–brain barrier, show age-related cognitive disability in acquisition of avoidance response that is related to the age of onset of their autoimmunity (biological age), while their retention deficit is related to chronological age.

### Oxidative Stress Models and ROS

Normal brain aging is associated with decreased homeostatic reserve (allostatic load) in the metabolic triad of mitochondria-ROS-intracellular Ca. ROS is a molecular example of antagonistic pleiotropy, being controlled during normal development for its signaling effects, for example, those regulating neuroplasticity, but becoming deleterious in brain with age, due to declining resistance to its detrimental actions and its increased production (free radical theory of aging); these differ between humans and rodents. There is an optimal ROS level; therefore, drastic reductions in ROS metabolism, either genetically or chemically with high dose chronic antioxidants, can block its beneficial activities. Genetic models of oxidative stress in the aging brain include knockout of the excitatory amino acid carrier-1 gene which impairs glutathione metabolism, knockout of the p66-Shc gene (an adaptor/signaling protein regulating ROS metabolism) which extends life span 30% and prevents increased emotionality and pain sensitivity, and knockout of the hexosaminidase B gene which models the accelerated aging in human Sandhoff disease caused by lysosomal glycosphingolipid storage dysfunction, producing premature neurodegeneration involving microglia activation and oxidation, and which also partially responds to caloric restriction.

### The Senescence-Accelerated Mouse

The senescence-accelerated mouse (SAM) was established by phenotypic selection of spontaneous, prematurely senile, and short-lived littermates of the AKR/J strain; SAM-prone (SAMP) and SAM-resistant (SAMR) littermates were separately inbred. The SAMP8 and/or SAMP10 show age-related learning and memory deficits, mood disorders, other cognitive impairments with associated biochemical and synaptic changes, peripheral amyloidosis, altered blood–brain barrier, region-specific brain atrophy (mostly in cortex, but not hippocampus), and neurotransmitter-specific (cholinergic) neuron atrophy. Cognitive impairments in SAMP8 are correlated with oxidative stress and are corrected by antioxidants, implicating fundamental, age-related oxidative stress mechanisms involving defense enzymes (superoxide dismutase (SOD), glutathione, catalase), with downstream effects on neurotrophic factor expression (glial-derived neuronal factor (*GDNF*), neurotrophin-3 (*NT-3*), nerve growth factor (*NGF*)), lipid peroxidation, protein oxidation (as detected by proteomics), and defects in energy metabolism. Microarray and specific gene analyses show involvement of oxidative and stress response genes, including those regulating signaling, energy metabolism, immune response, cytoskeleton (increased tau and its hyperphosphorylation), and decreased apoE and glucocorticoid receptor subunit expression.

## Retarded Aging and Longevity Extension Models

Models that increase longevity may be more informative about aging mechanisms than accelerated aging models, because increased population life span is likely to be achieved by modulating multiple age-related mechanisms, particularly in hybrid or outbred strains that die of multiple causes (see sections tilted 'Rodents as models of aging' and 'Caloric restriction'). With the definition of normal aging that excludes disease, the distinction between slowing aging and inhibiting or delaying disease must be carefully considered, since preventing one relatively common lethal disease can apparently extend population life span without modifying fundamental aging processes.

### Genetic Models

Several genetic models with increased longevity involve the insulin/IGF-1/growth hormone (GH) pathways, suggesting that chronological longevity is extended by inhibiting growth pathways and by increased protection from stress. Life span extension has been achieved in mice by genetic mutations in the genes including p66, *FIRKO* (insulin receptor), and *IGF-1R* (in females only). IGF-1 control of brain cell growth, neuronal myelination and sprouting, and astrocyte activation (including GFAP induction) in development and injury-response regeneration has been shown to depend on the cellular site of its expression, by comparing neuron-specific and astrocyte-specific inducible-expression transgenics. In mice, the specific role of brain expression of IGF-1R on life span and mortality was elegantly shown by genetic knockdown of the IGF-1R gene (by brain-specific Cre-lox-mediated heterozygous knockout), producing a brain hyposensitive to IGF-1 that had increased life span and decreased mortality.

The pituitary is specifically implicated in growth-associated longevity effects. Various hypopituitary long-lived dwarf mice (Ames and Snell) support the model of antagonistic pleiotropy in normal brain aging and age-related decline in cognition by counteracting this effect through delayed puberty. *GH* knockout mice have preserved learning and memory even with aging, while *GH* transgenic mice age faster with earlier puberty and faster cognitive decline. Dwarf and GH-resistant mice (*GH* receptor knockout) implicate multiple complex mechanisms and abnormalities, including insulin supersensitivity (decreased insulin and hippocampal *IGF-1* expression), ROS, lipids, carbohydrates, stress, and age-related disease. Many effects are gender, diet, and strain dependent.

### Caloric Restriction

Caloric restriction (CR) is a well-established method of life span extension with significant, possibly contributing effects in the brain. CR may represent a more natural diet for rodents and *ad libitum* may cause overfeeding-induced stress (see the section titled 'Rodents as models of aging'). Its intense study has informed that energy metabolism (glucose and ketones), signaling pathways of insulin, IGF-1, and GH, and ROS metabolism impact fundamental aging processes as important subsets of normal aging changes. CR slows fundamental processes of aging that affect the rate of development of multiple diseases, although other mechanisms are implicated in brain. Life span extension by CR is influenced by strain-dependent pathologic diseases and lesions and by gender, including variation in reproductive aging and the hypothalamic and pituitary control of estrus cycling.

CR extends life span in all organisms up to rodents. However, in primates and humans, fundamental differences in life history evolution and the set points of metabolic stability and ROS metabolism have argued against predicted efficacy. A human clinical trial was initiated in 2002 by the National Institute on Aging (NIA) (called CALERIE) which has shown some neuroendocrine and pyschophysiologic benefits. CR mimetics or intermittent (hormetic) CR may be much more practical for humans, in particular for brain aging. Mimetics of CR can target glucose, insulin, oxidation, and inflammation and can be quickly identified by their induction of short-term (8-week) CR-like gene expression profile responses. The NIA tests candidate compounds for mean and maximal life extension and/or health improvement in hybrid mice with genetic heterogeneity (HET mice) and pathologic diversity, which will include cognitive testing.

Many of the beneficial effects of CR can be linked to effects on the nervous system with specific effects on age-related changes in neuroendocrine and sympathetic systems, glia, vasculature (constriction and microvessel loss), hippocampal volume (gliogenesis, but not neurogenesis), synaptic efficacy, and neurotrophic factors, particularly BDNF. CR attenuates or reverses several age-related glial responses, including the increase in hippocampal astrocyte activation and *GFAP* expression (see section titled 'Glial influences'), the region-specific modulations in *apoE* and *apoJ* mRNA in hybrid rats, and microglial activation, by several mechanisms including reducing proinflammatory inducers like advanced glycation end products and promoting anti-inflammatory effects of cortisol on cytokine gene expression. Glucocorticoids are paradoxically increased with CR in rats in response to reduced plasma cortisol binding globulin, which suggests that the neuroprotective effects of CR (including increased BDNF and attenuated microglial activation) outweigh deleterious effects of increased glucocorticoids; however, generalizations are limited because rodent species differ in the relationship between age-related cortisol and cognitive impairment. Further, while CR protects some age-related neuron loss in the periphery (enteric and retinal, and motor neurons), effects of CR on brain neuron number, particularly in relation to elevated glucocorticoids, have not been assessed. Limited studies show that CR has opposite effects on different neuronal population in the hypothalamus, protecting the age-related loss of IGF-1R neurons but inducing the loss of other populations, and in the hippocampus, CR does not impact age-related decline in neurogenesis of granule cells but does promote the survival of newborn glial cells. CR impacts aspects of oxidation and stress responses, including inducing the neuroprotective actions of heat shock protein-70 (HSP-70) and attenuating its age-related loss that involves Ca, ROS, and cell death. Proteomics of CR effects in brain have identified the striatum and hippocampus as particularly benefiting, with lowered oxidized proteins. Several neurodegenerative disease models show beneficial effects of CR, including region-specific neurotoxicity of 1-methyl-4-phenyl-1,2,3,6-tetrahydropyridire (MPTP) (substantia nigra) and kainate (hippocampus), both dependent on BDNF. The BDNF knockout (conditional) mouse shows alterations in several CR-relevant systems including stress, glucose, and insulin.

A central question is whether CR impacts the same mechanisms that drive normal brain aging. Comparison of age- and CR-related gene expression profile changes in brain showed that most of the changed neurotrophic-like genes were dissimilar, suggesting that CR achieves brain effects and perhaps longevity by mechanisms distinct from or only partially overlapping normal aging. A gene that could be central to mediating CR effects on longevity and in models of age-related neurodegeneration is sirtuin 1 (*Sirt1*), the homolog of the yeast longevity gene *sir2*, which regulates genes involved in nutrient sensing, cell cycle, and DNA repair.

## Glial Influences

Astrocyte and microglial phenotypes vary in a continuum from resting to reactive, playing both protective and deleterious roles in neurodegeneration and in aging. A major controversy is whether age-related changes in glia induce neurodegeneration or are simply a response to aging-induced neurodegeneration. The inability to co-localize activated microglia with neurodegeneration argues against the latter hypothesis, in some models. Microglial number and activation state increase with age across all mammals; in one strain (C57Bl/6NNIA), glial cells increased only in females. Astrocyte activity affects the permissiveness for neuroplasticity and repair in the brain, as suggested by the age-dependent activities of astrocyte genes *GFAP* and *apoE*. They act in membrane structure (GFAP) or in lipid trafficking and signaling (apoE), contributing to the multiple pathways for regeneration in the brain and its age-related manifestations, including the suppression of reactive gliosis by estrogen regulation of their genes.

### Glial Fibrillary Acidic Protein

GFAP is an intermediate filament protein integral to cytoskeletal dynamics regulating the morphology and function of astrocytes including astrocyte reactive responses during aging. GFAP is consistently upregulated with age across rodents and humans and is a surrogate marker of aging in the brain, as indicated by unbiased microarray analysis of various brain regions across rat, mouse, and human species. The conservation of transcription control elements in the *GFAP* promoter across rat, mouse, and human species implicates this gene in canonic features of brain cell aging across both short- and long-lived species. *GFAP* also responds in caloric restriction (see the section titled 'Caloric restriction'). Human and mouse GFAP protein have similar activities, as shown by the phenotypes of mice whose *GFAP* coding region is replaced by the human *GFAP* sequence.

In modeling the age-related increase in *GFAP* expression, *GFAP* overexpressing mice show an earlier onset of the gene expression profile changes that characterize normal brain aging in mice and age-related changes that suggest a temporal relation between astrocyte and microglial activation where the former precedes the latter. The profile of expression responses shows increased stress, immune activation, oxidation, and inflammation and decreased neuronal plasticity and development genes, indicating that increased *GFAP* expression can initiate a cascade of changes that affect all other cell types in the brain. Age-related phenotypic changes in astrocyte reactivity, including increased *GFAP* mRNA, persist in *in vitro* astrocyte culture and affect neurite sprouting of co-cultured neurons. This supports the direct role of age-related astrocyte dysfunction in neurodegeneration with age.

GFAP regulates neurite and synaptic morphology, since various means of reducing its expression increases neurite sprouting. *GFAP*-knockout (ko) mice show enhanced neurite sprouting and (in combination with vimentin-ko) show increased posttraumatic regeneration and neurogenesis, an effect reproducible in an *in vitro* lesion model. *GFAP*-ko mice show abnormal, late-onset myelination, which models aspects of the human disease (Alexander disease) caused by *GFAP* mutation. Estrogen downregulation of *GFAP* controls its activity in regulating morphologic changes of astrocyte end foot processes that displace hypothalamic neuron synapses or neurites that in turn regulate endocrine hormone release during the estrus cycle. Loss of gonadal hormones with age may result in increased reactive gliosis and reduced neuroplasticity, as shown in models of stimulated neurite sprouting in both *in vitro* and *in vivo* lesion recovery models using OVX and estrogen replacement.

### Apolipoprotein E

ApoE is a major lipoprotein that primarily regulates lipid metabolism but has pleiotropic activities. Of the three common human alleles, *apoE2* is rare and is associated with increased longevity, *apoE3* is the most common, and *apoE4* strongly increases risk for several diseases and disorders, including cardiovascular disease and neurodegenerative disorders, particularly Alzheimer's disease (AD). Extensive human studies show apoE4 effects on milder and mid-life aspects of cognitive impairment, which are modeled in apoE4 transgenic mice.

Human *apoE* alleles evolved as part of changing life history evolution, where the ancestral *apoE4* allele was adaptive for the lower calorie, lower fat diet of early humans, while the common allele *apoE3* became adaptive with increased dietary meat by decreasing cholesterol metabolism. This may underlie the current pandemic of obesity-induced cardiovascular disease caused by the mismatch between current diet and exercise trends and the adapted apoE3 activities and maladaptive apoE4 activities in lipid metabolism. In addition to causing a decline in human longevity, this could also alter 'normal' human brain aging.

**ApoE knockout and human isotype transgenic mouse models** ApoE function has been studied using the *apoE* knockout mouse (*APOE*-ko) and transgenics expressing human apoE isotypes, usually on the background of mouse *APOE*-ko. While *APOE*-ko and rare humans lacking apoE develop to adulthood with relatively minor cognitive impairments, in the aging brain, compensatory pathways may become ineffective or insufficient, increasing the reliance on apoE activity to maintain plasticity, revealing defects in apoE4. *APOE*-ko begin to show neurodegenerative changes at 5–6 months, with defective LTP and loss of dendritic cytoskeleton and synaptic markers; some effects are background strain dependent.

A major issue in understanding apoE biology in the brain is defining which cell types produce apoE, which is determined by the transcription promoter regulation. Humans express apoE in both glia and at low levels in neurons. Mouse apoE is expressed in basal conditions only in glia, but can be induced in neurons by cytotoxic challenge. There is a critical functional trafficking and cellular localization difference between receptor-mediated uptake of apoE and actual production of apoE within the neuron. The human-like neuron expression of apoE is replicated in mouse transgenics driven by the human apoE promoter (human genomic promoter, HGp transgenics). To evaluate which cellular production site is relevant to detrimental effects of apoE4, transgenics driven by cell type-specific promoters limit expression to neurons using the neuron-specific enolase (NSE) promoter (NSEp transgenics) or to astrocytes using the GFAP promoter (GFAPp transgenics). The caveat with these mice is that the cell-specific expression of apoE is either abnormally high in NSEp transgenics or unnaturally regulated in GFAPp transgenics. This caveat is partly addressed by the newest transgenics in which the endogenous mouse apoE promoter drives expression of the human apoE isotypes (the targeted replacement transgenics, TR). These mice are ideal in that they recapitulate the human brain region-specific levels of apoE, and equivalent levels among the three isotypes, but do not show the human-like neuronal basal expression of apoE.

All human apoE isotype transgenic mice (except TR) are maintained on the, *APOE*-ko background, which has severe peripheral hypercholesterolemia with a potential influence on brain and blood–brain barrier functioning. The brain cell-specific expression transgenics (GFAPp and NSEp) maintain the peripheral hypercholesterolemia, while the genomic transgenics (HGp) correct it; although TR mice are not on the, *APOE*-ko background, they show marked isotype-specific differences in peripheral apoE levels. *APOE*-ko show disturbed blood–brain barrier permeability and increased susceptibility to microvascular disease, which contributes to the more severe responses to acute ischemia and stroke models by, *APOE*-ko and apoE4 transgenics (HGp and TR), and which is relevant to similar responses in humans and to the regional glucose utilization defects seen in human positron emission tomography (PET) studies.

Considering the definition of 'normal' brain aging and the role of apoE isotypes, there is an important distinction in transgenic models between what is normal for a mouse (with mouse apoE) and what is normal for a human (defined by the phenotype of the common allele apoE3). Comparisons of wild-type C57Bl6/J mice with isogenic apoE3 transgenics (on, *APOE*-ko) have revealed some similarities in phenotypes, but also some very important differences indicating that mouse apoE is not consistently like human apoE3 or apoE4. However, some clear congruencies between apoE transgenics and humans have been identified, for example, apoE4 in elderly humans and middle-age transgenic mice both show reduced hippocampal neuron dendritic spine density.

Compared to, *APOE*-ko mice, apoE4 shows phenotypic gain-of-negative activities in inhibiting neurite sprouting, mortality from head injury, and amyloid deposition, and loss-of-function activities in inhibiting astrogliosis, age-related dendrite loss, age-related synapse marker loss, resistance to excitotoxicity, and defective LTP. For age-related memory loss, neuronal expression (NSEp) shows a loss-of-function while astrocyte expression (GFAPp and TR) show a gain-of-negative function. This phenotypic distinction has important implications for the design and administration of drugs that target apoE activity or expression in that a gain-of-negative function of apoE4 would be exacerbated by a drug that increased apoE expression, while a loss-of-function would be rescued. Indeed, pharmacologic attempts to correct age-related disease processes including cognitive disorders, for example, lipid normalization by statins, antioxidants, hormone replacement therapy, fish oil, insulin sensitizers, and other treatments for AD, have shown apoE genotype-dependent efficacy, where apoE4 are nonresponsive.

**Effects of human ApoE isotypes on aspects of brain aging** ApoE has pleiotropic effects on brain and neuronal functioning, including lipid metabolism, neurite sprouting, gender-specific effects, multiple signaling cascades, LTP, inflammation, oxidation, neurotoxicity, and neurotrauma. ApoE4 accelerates the age-related change in the differential distribution of lipids between cytofacial and exofacial leaflets of the bilayer, which affects membrane fluidity and lipid raft microdomains. Rodent models of brain aging should carefully consider dietary lipid composition for its effects both developmentally (maternal diet) and in aging, particularly the age-related reductions in the essential omega-3 fatty acid docosahexaenoic acid (DHA) (see the section titled 'Rodents as models of aging'). *APOE*-ko and apoE4 show defective neurite sprouting and other aspects of neuronal plasticity, including age-related loss of synaptic markers, shown at levels from *in vitro* neuron cultures to organotypic cultures of hippocampal slices (derived from apoE isotype transgenic mice), to *in vivo* reactive neurite sprouting responses in all the promoter forms of transgenic mice. This effect models aspects of human aging in cognitively normal individuals where dentate spine loss is greater, and compensatory sprouting (in AD) is reduced or nonexistent in apoE4.

Gonadal hormone regulation of the apoE gene and its interaction with the effects of the apoE4 isotype are a paradigm of life history evolution caused by adaptive changes in diet and reproduction (somatic theory of aging). Gonadal hormones, particularly estrogen, modify apoE4 effects, both in atherosclerosis and in cognitive disorders, resulting in a greater susceptibility of females in humans and in mice to the age-exacerbated impairments in learning and exploration seen in NSEp and GFAPp mice. This occurs in brain regions that undergo estrus cycle synaptic remodeling. Conversely, estrogen's neurotrophic effect of stimulating neurite sprouting is modified by the age-dependent increased apoE expression, as shown in a rat OVX/estrogen replacement model of sprouting where the absence of apoE (*APOE*-ko mice) prevents estrogen stimulation of neurite sprouting. In addition, effects of male androgens on learning, memory, and recognition are altered by apoE4 as shown by castration studies (NSEp) and in androgen receptor functions.

Inflammation, like lipid metabolism, is a common feature of apoE-related mechanisms in both vascular and cognitive disease with the involvement of the common cellular element, the macrophage or microglial cell, respectively; however, it is important to distinguish between adaptive versus pathological effects. ApoE is generally anti-inflammatory, its production being repressed during the cytotoxic phase of the innate immune response and promoted during the repair phase, and inflammatory challenges including age-dependent changes in the HPA axis have more severe behavioral effects (memory and anxiety) in *APOE*-ko and apoE4 mice. Inflammatory effects of apoE4 (TR) implicate activated microglial and astrocyte production of ROS, glutamate, proteases, and inflammatory immune factors like interferon gamma (INFg), NO, tumor necrosis factor-$\alpha$ (TNF-$\alpha$), and interleukin 1 beta (IL-1b), and their feedback regulation on apoE production, including regulation by anti-inflammatory factors TGF-$\beta$ and estrogen, and reversal of anti-inflammatory effects of testosterone.

Neurotoxicity of apoE4 is linked to several of its other pleiotropic activities, for examples, proteolytic fragments of apoE4 are formed and are neurotoxic only when apoE4 is expressed in neurons (NSEp) but not in astrocytes (GFAPp), hyperphosphorylation of tau (NSEp), poor antioxidant activity (where $E4 < E3 < E2$), greater microglial activation, and differential effects on cell signaling including extracellular signal-regulated kinase (ERK) and AKT. BDNF and its signaling effects are implicated in apoE4 effects on cognitive responses to environmental enrichment and head trauma in transgenic models.

## Conclusions and Relevance to Human Brain Aging

Rodent models of aspects of brain aging provide valid tools to understand the roles of select components and their interactions. However, studying the intersection of the two most complex biological phenomena is not well elucidated in the reductionist approach. Rodent models are strongly validated by the multiple levels of congruence with human brain aging, largely due to their common phylogeny. However, significant differences preclude the use of lab rodents to fully understand human aging, particularly fundamental differences in dietary energy sources and utilization, reproductive aging, and temporally accumulated stochastic events that feed evolutionarily established responses of antagonistic pleiotropy in middle and later ages. Differences even between and within rodent species and strains imply that human–rodent differences are much greater than is currently understood. Genetic manipulations intended to target adult-related processes can have developmental artifacts, particularly in gene knockout and quasi-controlled transgenics. Conditional knockouts and expression-controlled transgenics eliminate such developmental artifacts and control the temporal and dose-dependent influences of gene expression.

*See also:* Aging of the Brain; Brain Glucose Metabolism: Age, Alzheimer's Disease and ApoE Allele Effects; Gene Expression in Normal Aging Brain; Genomics of Brain Aging: Apolipoprotein E; Glial Cells: Microglia During Normal Brain Aging; Glial Cells: Astrocytes and Oligodendrocytes During Normal Brain Aging; Lipids and Membranes in Brain Aging; Synaptic Plasticity: Learning and Memory in Normal Aging; Transgenic Models of Neurodegenerative Disease.

## Further Reading

Christensen K, Johnson TE, and Vaupel JW (2006) The quest for genetic determinants of human longevity: Challenges and insights. *Nature Review Genetics* 7(6): 436–448.

Conde JR and Streit WJ (2006) Microglia in the aging brain. *Journal of Neuropathology and Experimental Neurology* 65(3): 199–203.

Crimmins EM and Finch CE (2006) Infection, inflammation, height, and longevity. *Proceedings of the National Academy of Sciences of the United States of America* 103(2): 498–503.

Demetrius L (2006) Aging in mouse and human systems: A comparative study. *Annals of the New York Academy of Science* 1067: 66–82.

Finch CE (2003) Neurons, glia, and plasticity in normal brain aging. *Neurobiology of Aging* 24(supplement 1): S123–127, S131.

Holliday R (2006) Aging is no longer an unsolved problem in biology. *Annals of the New York Academy of Science* 1067: 1–9.

Ingram DK and Jucker M (1999) Developing mouse models of aging: A consideration of strain differences in age-related behavioral and neural parameters. *Neurobiology of Aging* 20(2): 137–145.

Ji LL, Gomez-Cabrera MC, and Vina J (2006) Exercise and hormesis: Activation of cellular antioxidant signaling pathway. *Annals of the New York Academy of Science* 1067: 425–435.

Kirkwood TB and Shanley DP (2005) Food restriction, evolution and ageing. *Mechanisms of Ageing and Development* 126(9): 1011–1016.

Longo VD and Finch CE (2003) Evolutionary medicine: From dwarf model systems to healthy centenarians? *Science* 299 (5611): 1342–1346.

Martin B, Mattson MP, and Maudsley S (2006) Caloric restriction and intermittent fasting: Two potential diets for successful brain aging. *Ageing Research Reviews* 5(3): 332–353.

Martin GM (2005) Genetic engineering of mice to test the oxidative damage theory of aging. *Annals of the New York Academy of Science* 1055: 26–34.

McEwen BS (2000) Allostasis, allostatic load, and the aging nervous system: Role of excitatory amino acids and excitotoxicity. *Neurochemistry Research* 25(9–10): 1219–1231.

Nadon NL (2006) Exploiting the rodent model for studies on the pharmacology of lifespan extension. *Aging Cell* 5(1): 9–15.

NIA Interventions Testing, Program to test candidate compounds for mean, maximal life extension and/or health improvement in hybrid mice. http://www.nia.nih.gov/ResearchInformation/ScientificResources/ITPapp.htmand. http://grants1.nih.gov/grants/guide/notice-files/NOT-AG-06–001.html, (accessed March 2007).

Pan F, Chiu CH, Pulapura S, et al. (2007) Gene Aging Nexus: A web database and data mining platform for microarray data on aging. *Nucleic Acids Research* 35: D756–D759.

Teter B and Finch CE (2004) Caliban's heritance and the genetics of neuronal aging. *Trends in Neuroscience* 27(10): 627–632.

Warner HR (2005) Longevity genes: From primitive organisms to humans. *Mechanisms of Ageing and Development* 126(2): 235–242.

# Transgenic Models of Neurodegenerative Disease

**C Li**, Weill Medical College of Cornell University, New York, NY, USA



The rapid advances in human genetics have been accompanied by a major increase in transgenic animal research for the purpose of neurodegenerative disease study. It is impossible to cover all the transgenic models, their phenotypical analyses, and their usage in this research field. It is also unnecessary to do so because for any particular neurodegenerative disease of interest, a simple search in PUBMed or websites of pertinent disease research foundations will lead to many excellent review articles that are constantly being published with updated information. Therefore, this article introduces the transgenic approach rather than the particular animal models *per se*, explains the underlying principles of the transgenic approach, and compares different transgenic approaches and their advantages and disadvantages. For this purpose, only a small subset of the animal models are presented here.

The development of transgenic animal models has been closely linked to the revolution in human genetics. Most neurodegenerative diseases have either a genetic inheritance component or are purely genetic. Some examples of the former include Alzheimer's disease (AD), Parkinson's disease (PD), and amyotrophic lateral sclerosis, with approximately 10% of patients carrying genetic mutations. Examples of the latter include Huntington's disease (HD) and spinocerebellar ataxia type 1. For the purely genetic diseases, the reason for establishing and studying genetic animal models is self-explanatory. For diseases that are mostly sporadic and are only 10% genetic, why do we put such a major effort in studying the genetic forms of the diseases? First, although many changes or abnormalities have been found from studying the sporadic forms of diseases, in general it is difficult to determine whether these abnormalities are causal, innocuous, or compensatory. In other words, these abnormalities may be pathologic rather than pathogenic. In genetic forms of the diseases, the 'first pushes' of the disease pathogenesis have been unequivocally established – the mutations in disease genes. Thus, the subsequent research, albeit long and arduous, has a firm foundation in logic. Second, the pathogenic mechanisms of genetic forms of the diseases may shed light on the sporadic forms because the pathogenic pathways may converge at some points.

For the previously mentioned reasons, the study of purely genetic neurodegenerative diseases or the genetic forms of neurodegenerative diseases has been the focus of many researchers. In this line of work, although studies at molecular, cell culture levels are important, genetic animal models are clearly one of the most important tools for research.

The genetic diseases have two subtypes: recessive inheritance, which requires both chromosomes to be mutated, or dominant inheritance, which requires only one mutant allele. The former is usually caused by loss of function of gene products, and the latter is commonly caused by gain of function, hyperfunction, or haploinsufficiency. In neurodegenerative diseases, both types are common. For example, in PD, Park 1 and Park 8 have dominant inheritance due to the mutations in α-synuclein and LRRK2 genes, whereas Park 2, Park5, Park 6, and Park 7 have recessive inheritance due to the loss of function of Parkin, UCHL-1, DJ-1, and PINK 1, respectively.

To model these different types of genetic deficits, a variety of methods have been developed. For genetic loss-of-function mutations, researchers commonly use the gene targeting (i.e., knockout) approach in which the gene of interest is deleted from the genome of experimental animals. For genetic dominant diseases, scientists may choose the transgenic approach in which a mutated gene is introduced into the genome of experimental animals. Since the gene targeting approach is explained elsewhere in this encyclopedia, this article focuses on the transgenic approach. It is worth noting that many organisms have been used to generate transgenic models, including yeast, *Caenorhabditis elegans*, *Drosophila*, mice, rat, and sheep. The majority of work in neurodegenerative disease research, however, is mainly done in mouse because this experimental organism is a mammal with a brain structure similar to that of humans, and it is amenable to many highly sophisticated genetic engineering tools.

The generation of transgenic mice involves construction of DNA vectors carrying the gene of interest and subsequent microinjection of DNA vectors into fertilized mouse eggs. In a certain percentage of injected eggs, DNA vectors will integrate into the mouse genome as a transgene, and therefore transgenic mouse founder lines are obtained. It is important to note that the integration site of transgene is random, rather than targeted to the homologous chromosomal sites, and that the copy number of the transgene is random, ranging from a single copy to 100–200. A direct implication of the transgene random integration sites and variable copy numbers is that the expression of transgenes may vary widely among the founder lines, both in spatial and temporal patterns and in abundance. The random insertion of transgene into mouse genome may disrupt other genes in some cases. Therefore, studies with transgenic mice usually require multiple transgenic lines to ensure that the phenotypes are authentically caused by the transgenes.

The choice of transgenes deserves some discussion. There are two options: a human gene with human disease mutations or the mouse homologous gene with human mutations. Empirically, the former works better in recapitulating the human disease. For example, in HD mouse models, the one carrying mouse Huntingtin with 111 CAG repeats developed much milder phenotypes than another model with human Huntingtin of similar CAG repeat. In α-synuclein mouse models for PD, human α-synuclein gene in transgenic mouse was able to recapitulate disease pathology, whereas mouse α-synuclein gene failed. Conceptually, the most straightforward explanation is that evolution has accumulated enough divergence that a mutation in human protein may or may not cause the same pathogenic conformational change in a mouse homolog. In fact, the human A53T mutation in α-synuclein caused Park 1, whereas in mice the 53T is a wild-type amino acid sequence. Hence, it seems that human genes are more appropriate for modeling human diseases in transgenic mice.

The methods of generating transgenic mice have evolved and become more powerful and versatile. Among the newer ones are bacterial artificial chromosome (BAC)-mediated transgenesis and inducible transgenic system. The traditional method uses a small plasmid vector with a capacity of less than 20 kb of DNA and with promoters such as prion, CMV, and Thy1. The small capacity precludes the regulatory elements such as enhancers, suppressors, and insulators that are all important for gene regulation. When the transgenes integrate into the mouse genome, they are influenced strongly by the adjacent chromosomal areas due to the positional effects that often cause misexpression of the transgenes.Compared to the traditional method, BAC-mediated transgenesis has several advantages. First, the conventional method accommodates only 10 to 15 kb DNA inserts, whereas the BAC vector accommodates up to 400 kb DNA as transgenes, which are large enough for most of the mammalian genes. Second, the large BAC capacity allows the transgene to be insulated from positional effects of insertion sites on the chromosome and to be regulated by its native promoters, enhancers, and suppressors; intron–exon structures; and 5′ and 3′ untranslated regions. Finally, the BAC method does not require assumptions about the splicing variant to be used as a transgene. BAC transgenesis has been used increasingly more often in neurodegenerative disease research.

Another powerful method is the inducible tet-on/tet-off system that allows inducible expression or suppression of transgenes. An elegant example of this method is provided by Yamamoto et al., who demonstrated the possibility of reversing pathology in HD mice by suppressing the expression of mutant Huntingtin.

It is clear that the transgenic mice are increasingly being used in modeling neurodegenerative disease. A further development is that in addition to mimicking the genotypes of human diseases, researchers are generating and using transgenic mice at a new level of sophistication. In an HD study, for example, mice with specific mutations in Huntingtin were made to test the role of proteolysis in HD pathogenesis, and mice expressing mutant HD in particular brain regions and cell types were used to address important questions of cell-autonomous versus non-cell-autonomous degeneration. Also reported are studies using transgenic mice to test genetic modifiers of HD. As a result of the collective efforts from many researchers, there are many transgenic mouse models for each neurodegenerative disease. The vast number of existing mouse models are powerful tools for researchers in this field.

However, despite the overall successful use of transgenic mice to model neurodegenerative diseases, it is a common problem that mice are much more resilient and often do not exhibit the cardinal phenotype of neuronal cell death. For example, in PD and AD research, mouse models were unable to recapitulate this most important aspect of pathology. To combat this obstacle, the transgenic rat approach may be a worthy alternative to create new opportunities. Although closely related in evolution, rats are different from mice and are closer to humans in many physiological aspects. In many pharmacological studies, rats have much more similar pharmacokinetics to humans. It is well-known that cardiovascular diseases are modeled in rats and are unable to be modeled in mice. In Huntington's disease modeling, rat models recapitulated striatal neuronal death with 57 CAG repeat length similar to human mutation, whereas mouse models needed much stronger insults of more than 150 CAG repeats to elicit mutant phenotypes. In chemical models for PD, rats are much more sensitive to rotenone than mice. In an α-synuclein study, viral expression of α-synuclein in rat substantia nigra resulted in loss of dopaminergic neurons, which was not observed in mouse models. These data indicate that rats may be more sensitive to some pathological insults than mice.

In summary, the transgenic animal approach has provided researchers unprecedented powerful tools for mechanistic dissection of pathogenesis as well as development of therapeutics. The methodology, techniques, and usage of transgenics continue to improve. There is reason for optimism that deeper understanding of neurodegenerative disease pathogenesis will eventually help us fight these human diseases.

*See also:* Alzheimer's Disease: Neurodegeneration; Alzheimer's Disease: An Overview; Alzheimer's Disease: Molecular Genetics; Alzheimer's Disease: Transgenic Mouse Models; Animal Models of Motor and Sensory Neuron Disease; Animal Models of Alzheimer's Disease; Animal Models of Huntington's Disease; Animal Models of Parkinson's Disease; Huntington's Disease: Neurodegeneration; Parkinson's Disease: Alpha-Synuclein and Neurodegeneration.

## Further Reading

Alam M, Mayerhofer A, and Schmidt WJ (2004) The neurobehavioral changes induced by bilateral rotenone lesion in medial forebrain bundle of rats are reversed by L-DOPA. *Behavioural Brain Research* 151: 117–124.

Chandra S, Gallardo G, Fernandez-Chacon R, Schluter OM, and Sudhof TC (2005) Alpha-synuclein cooperates with CSPalpha in preventing neurodegeneration. *Cell* 123: 383–396.

Fleming SM, Fernagut PO, and Chesselet MF (2005) Genetic mouse models of parkinsonism: Strengths and limitations. *NeuroRx* 2: 495–503.

Hodgson JG, Agopyan N, Gutekunst CA, et al. (1999) A YAC mouse model for Huntington's disease with full-length mutant huntingtin, cytoplasmic toxicity, and selective striatal neurodegeneration. *Neuron* 23: 181–192.

Gong S, Yang XW, Li C, and Heintz N (2002) Highly efficient modification of bacterial artificial chromosomes (BACs) using novel shuttle vectors containing the R6Kgamma origin of replication. *Genome Research* 12: 1992–1998.

Graham RK, Deng Y, Slow EJ, et al. (2006) Cleavage at the caspase-6 site is required for neuronal dysfunction and degeneration due to mutant huntingtin. *Cell* 125: 1179–1191.

Gu X, Li C, Wei W, et al. (2005) Pathological cell–cell interactions elicited by a neuropathogenic form of mutant huntingtin contribute to cortical pathogenesis in HD mice. *Neuron* 46: 433–444.

Levine MS, Cepeda C, Hickey MA, Fleming SM, and Chesselet MF (2004) Genetic mouse models of Huntington's and Parkinson's diseases: illuminating but imperfect. *Trends in Neuroscience* 27: 691–697.

Lo Bianco C, Schneider BL, Bauer M, et al. (2004) Lentiviral vector delivery of parkin prevents dopaminergic degeneration in an alpha-synuclein rat model of Parkinson's disease. *Proceedings of the National Academy of Sciences of the United States of America* 101: 17510–17515.

Spires TL and Hyman BT (2005) Transgenic models of Alzheimer's disease: Learning from animals. *NeuroRx* 2: 423–437.

von Horsten S, Schmitt I, Nguyen HP, et al. (2003) Transgenic rat model of Huntington's disease. *Human Molecular Genetics* 12: 617–624.

Wheeler VC, Auerbach W, White JK, et al. (1999) Length-dependent gametic CAG repeat instability in the Huntington's disease knock-in mouse. *Human Molecular Genetics* 8: 115–122.

Yamada M, Iwatsubo T, Mizuno Y, and Mochizuki H (2004) Overexpression of α-synuclein in rat substantia nigra results in loss of dopaminergic neurons, phosphorylation of α-synuclein and activation of caspase-9: Resemblance to pathogenetic changes in Parkinson's disease. *Journal of Neurochemistry* 91: 451–461.

Yamamoto A, Lucas JJ, and Hen R (2000) Reversal of neuropathology and motor dysfunction in a conditional model of Huntington's disease. *Cell* 101: 57–66.

## Relevant Websites

http://www.alzforum.org – Alzheimer Research Forum.

http://www.hdfoundation.org – Hereditary Disease Foundation.

http://www.ncbi.nlm.nih.gov – National Center for Biotechnology Information.

http://www.nih.gov – National Institutes of Health.

# Animal Models of Alzheimer's Disease

**J Koenigsknecht-Talboo and D M Holtzman**, Washington University, St. Louis, MO, USA



## Introduction

Alzheimer's disease (AD) is the most common cause of dementia. The pathological hallmarks of this neurodegenerative disease include extracellular structures called diffuse and neuritic plaques as well as amyloid angiopathy. In addition, there are intracellular structures known as neurofibrillary tangles and neuropil threads. The plaques are comprised primarily of aggregated, nonfibrillar, and fibrillar forms of amyloid-$\beta$ (A$\beta$) peptide, a 38–43-amino-acid peptide that is a normal cleavage product of the amyloid precursor protein (APP). The neurofibrillary tangles and neuropil threads are composed of hyperphosphorylated, aggregated forms of the microtubule-associated protein tau. In addition to neuropil threads, there is also neuritic dystrophy consisting of large axonal and dendritic varicosities, surrounding fibrillar amyloid deposits in the brain parenchyma. Neuronal cell death and synaptic degeneration also occur in AD. Furthermore, gliosis invariably occurs whereby microglial cells and astrocytes accumulate and cluster around plaques of deposited A$\beta$. Genetic mutations in the APP and presenilin (PS) genes have been identified and are associated with rare forms of early onset, autosomal-dominant familial AD as well as cerebral amyloid angiopathy (CAA). The normal processing of APP or the sequence of A$\beta$ is altered in the presence of these mutations, thus leading to abnormal production or an increased aggregation propensity of A$\beta$ and its subsequent accumulation. While mutations in tau do not result in AD, they do cause certain forms of fronto-temporal dementia and tangle accumulation. These mutations appear to alter the likelihood that tau will aggregate leading to a tauopathy. Advances in research and further understanding of AD have increased in part due to the generation of animal models that mimic certain aspects of AD, CAA, and fronto-temporal dementia that take place in humans. Introduction of mutations in APP, PS, and tau using transgenic, knockout, and knockin techniques in mouse models has resulted in the production of animals that develop aspects of AD-like pathology and other changes (**Figure 1**). These transgenic animals directly test whether expression of mutant genes leads to AD-like pathology and behavioral phenotypes, provide insight into molecular mechanisms, and are useful for developing better diagnostic methods as well as testing the efficacy of new treatments.

The presence of a few important criteria is vital to model aspects of AD in animal models. An AD animal model should exhibit at least one if not more than one pathological hallmark of AD. Another criterion is observation of cognitive and behavioral deficits. In models that contain mutations that cause familial AD or that result in expression of genes that lead to AD-like changes, changes in phenotype should be more severe in genetically altered than in age-matched wild-type animals and occur in an age-dependent fashion. The findings seen in a good animal model should be reproducible and be confirmed by several different labs. The generation of a comprehensive animal model of AD has proved to be a challenge for various reasons. AD is a complex disease that thus far involves at least three proven causative genes and one gene that is a known risk factor. In addition, there are also other genes that are required for various aspects of AD pathology. Another obstacle for constructing an AD model is mimicking a human disease that normally occurs in the seventh decade of life or later in mice, which live only 2–3 years. It is also extremely difficult to reproduce all the complexities of human behaviors in rodents. Various model organisms have been used to try and replicate aspects of human AD. Each organism has its strengths and weaknesses in regard to this endeavor. Although the anatomy of a mouse brain is similar to other mammals including humans in many regions, the brain of a rodent is also very different than a primate brain. Mice can be induced to develop plaques and tangles in region-specific areas; however, mice do not normally develop plaques and tangles. Aged rhesus monkeys have been shown to contain similar brain pathology as AD including the presence of aggregated amyloid, reactive glia, and enlarged distal axons. However, monkeys are expensive to maintain and experiments can take a long time to complete. Both worms and flies provide the benefits of easy and fast breeding in addition to the ability to employ powerful genetic tools. Yet, the brain anatomy of worms and flies is vastly different than in humans and AD-like behavior is difficult to address in these organisms. Therefore, much of the advancement in AD research in animal models has come from studies utilizing genetically manipulated mice. These models are reviewed here.

## Genetically Modified Animal Models

### APP Mouse Models and A$\beta$-Related Pathology

While multiple different APP transgenic mice have been produced, several examples are described below

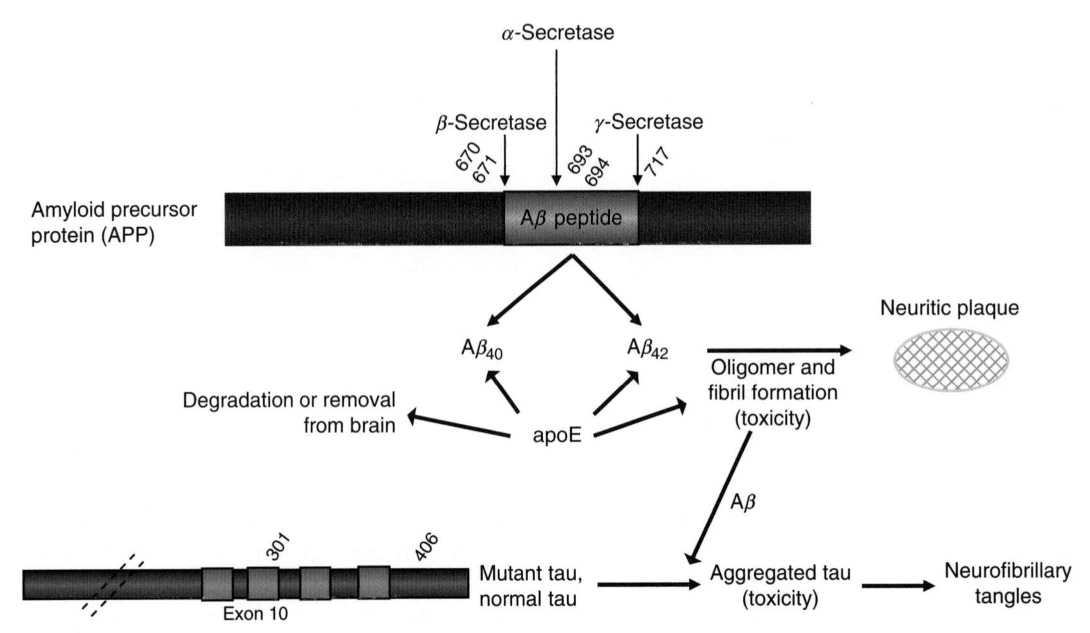


**Figure 1** Model for Aβ and tau leading to cell toxicity. Aβ is generated by cleavage of APP by β-secretase and γ-secretase. Oligomers and fibrils of Aβ form and likely lead to neuronal and synaptic damage. Aβ also is hypothesized to induce tau aggregation and toxicity. The numbered sites in APP and tau correspond to amino acids where mutations are present that cause familial autosomal dominant AD or fronto-temporal dementia, respectively, that have been introduced into the mouse models described herein.

to illustrate some key features of the different types of models. One of the first APP transgenic mouse models to develop AD-like pathology described was the PDAPP mouse model of AD. The PDAPP mouse was generated by employing a platelet-derived growth factor (PDGF)-β promoter to induce overexpression of a transgene construct of human APP that contains the V717F mutation predominantly in neurons in the central nervous system (CNS). This point mutation in APP results in a mutation first found in families in Indiana which causes a form of autosomal dominant familial AD. Human APP levels were reported to be 10 times higher than endogenous mouse APP levels in this animal model. PDAPP mice develop age-dependent Aβ accumulation and deposition in diffuse as well as fibrillar neuritic plaques in the hippocampus, cerebral cortex and corpus callosum, as well as to a lesser extent in cerebral arterioles. This pattern of deposition corresponds with the pattern of Aβ plaques revealed in the brains of AD patients. Aβ deposition generally develops between 6 and 9 months and continues to accumulate with age. Different forms of Aβ deposition occur in this animal model. Diffuse plaques with nonfibrillar as well as compacted plaques containing fibrillar Aβ are both observed. In this model, only about 10% of the plaques are fibrillar, as defined with dyes such as Thioflavin-S or Congo red. In addition to Aβ deposition, these animals also exhibit robust gliosis in which plaques are associated with reactive astrocytes and microglial cells. Neuritic dystrophy has been reported in the PDAPP mice; however, studies have not reported a loss of neurons in this model. A decrease in synaptic and dendritic density in the hippocampus has also been observed in this model. It is not clear whether this is a developmental or neurodegenerative phenotype in this model. While behavioral studies have revealed some cognitive deficits in aspects of learning and memory in these mice prior to AD pathology, there are also age-dependent and pathology-dependent behavior deficits in learning and memory, making this likely to be a useful model for certain aspects of cognitive deficits in AD as well as a model of the Aβ-related pathological changes.

The Tg2576 transgenic mouse model was also one of the first APP transgenic mouse models developed. Also known as APPsw, the Tg2576 mouse model was generated by overexpression of the human APP695 construct that contains a double Swedish mutation. This mutation was identified in a large Swedish family that presented with early onset AD. The Swedish mutations (K670N/M671L) are driven by the hamster prion protein in this model. This mouse model expresses 5–6 times the amount of human APP as endogenous mouse APP. Senile plaques with dense cores of fibrillar amyloid in addition to diffuse amyloid deposits have been reported in the Tg2576 animal model. Deposition of amyloid generally occurs between 9 and 12 months of age in Tg2576 mice. Clusters of astrocytes and microglia surround many of the amyloid deposits. These cells induce an inflammatory response surrounding the areas of amyloid plaques. Despite the presence of amyloid deposits, neurofibrillary tangles are not present in any of the APP transgenic models including the Tg2576 mouse model. Neuritic dystrophy is present in this and other APP transgenic mice but there is no marked neuronal loss. Neuronal loss in the immediate vicinity of fibrillar amyloid has been reported. CAA has been shown to be more prevalent in the Tg2576 model than in PDAPP mice. Notable CAA has also been observed in the APP23 transgenic mouse model along with micro- and sometimes macrohemorrhages. Cognitive decline has been demonstrated to increase with age in this mouse model. For example, spatial learning in a water maze was shown to be impaired in Tg2576 mice. Also, results of a spatial alternation task in this mouse model were consistent with hippocampal damage. These behavior deficits have been reported to occur just prior to the onset of plaque deposition, beginning at ~6 months of age. Learning and memory was normal at 3 months of age in the water maze. Deficits in synaptic plasticity such as long-term potentiation (LTP) have been reported in the hippocampus and the dentate gyrus. Since development of the PDAPP and Tg2576 mouse models, numerous other APP transgenic mouse models that develop A$\beta$ deposition and associated changes have been developed.

The PSAPP mouse model incorporates a PS mutation and was generated by crossing an APP mutated mouse (Tg2576) with a PS1 mutated mouse (M146L). This mouse model exhibits accelerated plaque pathology as well as robust gliosis most likely due to enhanced A$\beta$42 generation or an altered 42/40 ratio due to the overexpression of a PS1 mutation responsible for familial AD. Abundant diffuse and fibrillar A$\beta$ deposits form in the cortex and hippocampus of these double transgenic mice much earlier than their Tg2576 singly transgenic littermates. By 3 months of age, these mice contain fibrillar plaques within the cortex. Amyloid deposition accumulates and plaque size increases and extends to the hippocampus by 6–8 months of age. This double transgenic mouse model preferentially expresses A$\beta$42, and brains of these mice have been reported to have a 41% increase in A$\beta$42 as compared to singly transgenic Tg2576 mice. Neurofibrillary tangles have not been reported in the PSAPP mouse model. Despite the accelerated amyloid pathology in PSAPP mice, the clearest cognitive deficits appear at 15–17 months of age as reported by some. Other models combining expression of a PS mutation together with APP have been developed.

### Mouse Models with Mutations in the A$\beta$ Region

A few mouse models have been generated that contain mutations within the A$\beta$ sequence of APP. For example, the APP Dutch mouse model was created by inducing a point mutation, E693Q, in human APP. This mutation, in humans, results in an autosomal dominant form of CAA. Expression of APP Dutch results in a higher level of A$\beta$40 accumulation than A$\beta$42. Similar to the human cases, amyloid is deposited predominantly in cerebrovascular plaques in APP Dutch animals in contrast to the parenchymal A$\beta$ deposits that are characteristic of other APP transgenic animal models. The APP Dutch animal has been crossed with animals containing the G384A PS1 mutation, which is known to increase A$\beta$42 production. The expression of the G348A PS1 mutation in these mice results in a shift of amyloid deposition from the vessels into the brain parenchyma in the cortex and hippocampus. It is hypothesized that a high A$\beta$42/40 ratio favors parenchymal deposition, whereas a low A$\beta$42/40 ratio favors vascular A$\beta$ deposition. Similar to the Dutch mutation, the Iowa mutation (D694N) also leads to A$\beta$ accumulation in the brain vasculature. These two mutations have been reported to augment fibrillogenic properties of A$\beta$ *in vitro* and may also result in abnormal clearance of A$\beta$ from the brain leading to its accumulation.

The Arctic mutation, E693G, is another mutation that is within the A$\beta$ sequence and causes an AD phenotype in humans. Transgenic mice that contain the Arctic mutation in addition to the Swedish mutation have been reported to develop intraneuronal A$\beta$ aggregates prior to the deposition of extracellular A$\beta$. The intraneuronal accumulation of A$\beta$ is not grossly fibrillar. The Arctic lines promote accumulation of shorter species of A$\beta$. This was demonstrated by a higher A$\beta$1-38/A$\beta$1-40 ratio and a lower A$\beta$1-42/A$\beta$1-40 ratio. The Arctic mutation is extremely amyloidgenic; one of the Arctic lines generated was reported to contain extracellular neuritic plaque deposition of A$\beta$ as early as 2 months of age. A significant difference between the Arctic and the Dutch mutations is the location of A$\beta$ deposition. The Dutch mutation results in amyloid deposition primarily in vessels; however, the Arctic mutation generally results in parenchymal A$\beta$ plaques within the brain.

All of the APP models discussed result in production of different ratios and amounts of A$\beta$ species, making it difficult to determine the specific effects of A$\beta$40 versus A$\beta$42 on pathology and behavior. In addition, in some models the overexpression of APP may be responsible for some of the changes in the animals. An important advancement in AD animal models was achieved by creating mice that specifically produce A$\beta$40 or A$\beta$42, known as the BRI-A$\beta$ models. This model employs fusion constructs that contain 243 amino acids of BRI protein followed by a sequence that encodes either A$\beta$40 or A$\beta$42. A$\beta$ is produced when the BRI protein is cleaved by furin in a vesicular compartment in cells. Aβ40 is selectively produced, expressed, and secreted by BRI-Aβ40 mice, whereas Aβ42 is selectively generated, expressed, and secreted by BRI-Aβ42 mice. Only BRI-Aβ42 mice develop plaques demonstrating the critical importance of Aβ42 in the seeding and propagation of Aβ deposition. These models will be useful in sorting out specific effects of Aβ40 or Aβ42 without the confound of APP overexpression.

### ApoE Mouse Models

The only proven genetic risk factor for AD is apoE genotype. The ε4 allele of apoE has been demonstrated to be a risk factor for late onset AD and CAA, whereas the ε2 allele of apoE has been shown to be protective against AD. apoE avidly binds Aβ and co-localizes with neuritic plaques and CAA. Evidence suggests that one of the main reasons for the linkage between Aβ and apoE results from the ability of apoE to bind Aβ, acting as a chaperone to influence Aβ metabolism and structure.

In regard to a potential role for apoE in normal brain structure and function, apoE knockout mice have been shown in some studies but not others to contain less synaptophysin and staining with dendritic markers. Interestingly, when either PDAPP or Tg2576 mouse models are bred with apoE knockout mice, the resultant APP transgenic mice lacking apoE exhibit a striking decrease in Aβ deposition compared to animals that express apoE, thereby demonstrating the importance of apoE in amyloid deposition. In addition, the Aβ deposits in mice lacking apoE are not fibrillar until very old ages and contain little to no neuritic dystrophy, microglial inflammation, CAA, or CAA-associated hemorrhages. This suggests that apoE plays an important role in induction of fibrillar Aβ and its consequences.

To examine the effects of individual human apoE isoforms on the brain and on AD mouse models, mice have been generated that contain a specific apoE isoform knocked into the endogenous mouse *apoe* locus. These mice display apoE expression predominantly in glial cells, the normal pattern of endogenous apoE expression. The apoE2, apoE3, and apoE4 mice all express similar levels of apoE. Other mice have been generated that express specific apoE isoforms in astrocytes with a GFAP promoter (GFAP-apoE) and in neurons with a neuron-specific endolase promoter (NSE-apoE). Each of these mice were bred with mouse apoE knockout mice to eliminate the confound of endogenous mouse apoE. In the NSE-apoE4 mice, C-terminal truncated fragments of apoE accumulate in an age-dependent manner. Fragment accumulation occurs to a lesser extent in NSE-apoE3 mice and is not seen in the GFAP-apoE mice. Tau phosphorylation also accumulates in an age-dependent manner in NSE-apoE4 mice. As with apoE fragment accumulation, tau phosphorylation occurs to a lesser degree in NSE-apoE3 mice and is not observed in the GFAP-apoE mice. Co-cultures of hippocampal neurons and astrocytes derived from GFAP-apoE mice displayed greater neurite outgrowth in GFAP-apoE3 mice than in GFAP-apoE4 mice. Importantly, in regard to effects on Aβ, using human apoE isoform-specific transgenic or knockin mice bred to different APP transgenic mice, strong isoform-specific effects of apoE are seen. The isoform-specific effects on Aβ are such that earlier and more Aβ-related pathology occurs in the pattern E4 > E3 > E2. In addition to the isoform-specific pattern, all human apoE isoforms delay the development of Aβ deposition relative to the presence of murine apoE or no apoE. This suggests that human apoE plays a role in Aβ clearance or transport in addition to its role in fibrillogenesis. While the effects of apoE on Aβ are profound, experiments with animals that develop both Aβ deposition and tauopathy will be required to determine if apoE also influences tangle formation.

### Tau Mouse Models

Several AD animal models have focused on APP/Aβ; however, other models have been developed that examine other aspects of AD pathology. The first tau transgenic model was generated prior to the identification of pathogenic tau mutations. This model expresses the longest human brain tau isoform under the control of the human Thy-1 promoter. Although this model mimicked some aspects of AD including hyperphosphorylation of tau, neurofibrillary tangles did not develop in these animals. This AD animal model may represent an early AD-like phenotype that occurs prior to tangle formation.

Pathogenic mutations were later identified in the tau protein that cause fronto-temporal dementia and parkinsonism linked to chromosome 17 (FTDP-17). One tau mouse model employs the most common FTDP-17 mutation, P301L, driven by the prion protein promoter. The JNPL3 mouse develops neurofibrillary tangles in addition to astrogliosis, most prominently in the brain stem and spinal cord. Although tangles do not develop in the cortex or hippocampus of JNPL3 mice, there is pretangle pathology in these brain regions. Contrary to the APP mouse models of AD, this mouse model does not develop amyloid plaques. The majority of these mice developed motor and behavioral deficiencies by 10 months of age. The behavioral deficits described in JNPL3 mice include delayed righting reflex, decreased locomotion, and muscular weakness. Another tau mouse model contained a P301L mutation; however, this model employed the Thy1.2 promoter. These mice expressed P301L tau in neurons in the cortex and hippocampus. Hyperphosphorylation of tau occurs in these mice and neurofibrillary tangles are present. In addition, this mouse model also develops numerous TUNEL-positive neurons in the cortex suggesting neuronal damage.

Another tau mutation, R406W, was utilized to make a tau transgenic mouse model. This transgenic mouse expressed the R406W mutation driven by the α-calcium-calmodulin-dependent kinase-II promoter. Hyperphosphorylated tau inclusions are present in neurons in the forebrain of these animals by18 months of age. These tau inclusions were closely associated with disruptions in microtubules. Associative memory was reported to be impaired in this mouse model through contextual and cued fear conditioning tests. This tau mouse model was not reported to have obvious motor deficits, as has been the case with other tau mutations. A variety of mice with tau mutations causing fronto-temporal dementia have now been generated that have pathological features similar to those described above, although the location of the pathology varies depending on the promoter utilized as well as the mutation.

Htau mice were generated by crossing mice that express a human tau transgene and tau knockout mice. These mice express the normal sequence of human tau. The htau mice therefore express six isoforms of human tau, however, does not express mouse tau. These mice develop pathology despite the fact that they contain nonmutant tau. Hyperphosphorylated tau accumulates within the cell bodies and dendrites of neurons in the hippocampus. Tau aggregated and accumulated inside cells within the cortex and the hippocampus of this mouse model.

Recently, a novel tau model was described that expresses a mutant tau which can be suppressed with doxycycline. These mice overexpress the P301L tau mutation in the forebrain. The expression of mutant tau resulted in the accumulation of hyperphosphorylated tau and a loss of neurons that increased as the animals aged. Spatial memory was also impaired in association with age. Blocking tau expression of tau caused memory function to improve and neuron numbers to stabilize. Importantly, neurofibrillary tangles continued to accumulate, after the mutant tau was suppressed despite improved cognitive performance. Results from this mouse model therefore suggest that a soluble form of tau may be contributing strongly to toxicity.

### Other Mouse Models

An advance in AD mouse models was made with the introduction of a novel triple transgenic model (3x Tg-AD) which progressively developed both amyloid plaques and neurofibrillary tangles in brain regions that are affected in AD. The 3x Tg-AD model contained a PS1 mutation (M146V), an APP mutation (APPswe), and a tau mutation (P301L). These mice were generated by microinjecting two transgenes into single cell embryos derived from homozygous M146V knockin mice. One benefit of this method is that the APP and tau transgenes are co-integrated and transmit together to the offspring. Plaque formation was shown to develop in these animals prior to the production of tangles. This pathology is consistent with the amyloid cascade hypothesis. The neocortex of 3x Tg-AD mice has been reported to contain intracellular A$\beta$ immunoreactivity as early as 3–4 months of age which is followed by A$\beta$ aggregation in the hippocampus at 6 months of age. In a similar region-dependent manner, extracellular A$\beta$ deposits formed in the cortex prior to the hippocampus as early as 6 months of age and were easily detectable at the age of 12 months. In contrast to plaque formation, tau pathology of tangle formation appears first in the hippocampus and later is observed in the cortical area of the brains of 3x Tg-AD mice. Synaptic dysfunction was also observed in an age-dependent manner in this mouse model. Interestingly, removal of A$\beta$ by immunotherapy in these mice reverses early changes in tau such as hyperphosphorylation but not later changes such as tangle formation.

Another mouse model that results in both tau and amyloid pathology has been generated by crossing the Tg2576 mouse with the P301L tau mouse model. The resulting mutant tau and APP mice, TAPP mice, develop both amyloid plaques and neurofibrillary tangles. The plaques generated in the TAPP mice were similar in number and distribution as was reported in Tg2576 mice. A$\beta$ plaques were detected as early as 6 months of age and increased as the animals aged. Neurofibrillary tangles were seen in the TAPP mice as early as 3 months of age; however, they were more numerous as the animals aged. There was an acceleration of tau pathology in the TAPP mice as compared to singly transgenic mice only expressing tau with the P301L mutation suggesting that A$\beta$ may somehow exacerbate tangle pathology. Although plaques and tangles were shown to be located in the same brain regions, the neurofibrillary tangles did not form in the immediate environment of the amyloid plaques. Tangle pathology also occurs in brain areas not affected by A$\beta$ deposition in TAPP mice. Dystrophic neurites were observed around the amyloid deposits. Similar to the P301L mutated mice, the TAPP mice demonstrated some disturbances in motor skills. It was also observed in TAPP mice that neurofibrillary tangle pathology developed earlier and more robustly in female mice than in male mice.

## Conclusions

Various animal models have been developed which recapitulate one or more aspects of AD. While many of these models are useful, none of the models encompass all of the pathology and cognitive deficits observed in human AD. A key feature of AD that is not readily demonstrated in the mouse models is the presence of plaques and tangles combined with progressive and severe neuronal loss that is seen in certain brain regions in human AD. More models with AD-like pathology can also be bred with mice in which other genes are altered to assess their effects on aspects of AD-like pathology or behavior. Alternatively, expression of genes with viral vectors can be another way to both assess effects of potential gene therapies as well as better model aspects of AD. Important information has been gained from crossing AD models with apoE models and with mice lacking or overexpressing $\beta$-secretase, $\alpha$-secretase, ABCA1, neprilysin, and insulin degrading enzyme. There are many other proteins that potentially play a role in AD that can be assessed in a similar manner. Although there is not yet a perfect AD animal model, valuable information on AD pathogenesis and potential therapies such as secretase inhibitors and immunotherapy have been gleaned from the current available models. It is likely that future improvements in current models as well as a better understanding of AD pathogenesis will continue to be made as the full complement of genes involved in AD is unraveled. Rodent models will likely play a major role in this endeavor.

*See also:* Alzheimer's Disease: Neurodegeneration; Alzheimer's Disease: An Overview; Alzheimer's Disease: Transgenic Mouse Models; Axonal Transport and Alzheimer's Disease.

## Further Reading

Allen B, Ingram E, Takao M, et al. (2002) Abundant tau filaments and nonapoptotic neurodegeneration in transgenic mice expressing human P301S tau protein. *The Journal of Neuroscience* 22(21): 9340–9351.

Bales KR, Verina T, Dodel RC, et al. (1997) Lack of apolipoprotein E dramatically reduces amyloid beta-peptide deposition. *Nature Genetics* 17(3): 263–264.

Brendza RP, O'Brien C, Simmons K, et al. (2003) PDAPP; YFP double transgenic mice: A tool to study amyloid-beta associated changes in axonal, dendritic, and synaptic structures. *The Journal of Comparative Neurology* 456(4): 375–383.

Cheng IH, Palop JJ, Esposito LA, et al. (2004) Aggressive amyloidosis in mice expressing human amyloid peptides with the Arctic mutation. *Nature Medicine* 10(11): 1190–1192.

Davis J, Xu F, Deanes R, et al. (2004) Early-onset and robust cerebral microvascular accumulation of amyloid beta-protein in transgenic mice expressing low levels of a vasculotropic Dutch/Iowa mutant form of amyloid beta-protein precursor. *The Journal of Biological Chemistry* 279(19): 20296–20306.

Fagan AM, Murphy BA, Patel SN, et al. (1998) Evidence for normal aging of the septo-hippocampal cholinergic system in apoE (–/–) mice but impaired clearance of axonal degeneration products following injury. *Experimental Neurology* 151(2): 314–325.

Fryer JD, Taylor JW, DeMattos RB, et al. (2003) Apolipoprotein E markedly facilitates age-dependent cerebral amyloid angiopathy and spontaneous hemorrhage in amyloid precursor protein transgenic mice. *The Journal of Neuroscience* 23(2): 7889–7896.

Games D, Adams D, Alessandrini R, et al. (1995) Alzheimer-type neuropathology in transgenic mice overexpressing V717F beta-amyloid precursor protein. *Nature* 373(6514): 523–527.

Gotz J, Chen F, Barmettler R, and Nitsch RM (2001) Tau filament formation in transgenic mice expressing P301L tau. *The Journal of Biological Chemistry* 276(1): 529–534.

Herzig MC, Winkler DT, Burgermeister P, et al. (2004) Abeta is targeted to the vasculature in a mouse model of hereditary cerebral hemorrhage with amyloidosis. *Nature Neuroscience* 7(9): 954–960.

Holcomb L, Gordon MN, Mcgowan E, et al. (1998) Accelerated Alzheimer-type phenotype in transgenic mice carrying both mutant amyloid precursor protein and presenilin 1 transgenes. *Nature Medicine* 4(1): 97–100.

Holtzman DM, Bales KR, Tenkova T, et al. (2000) Apolipoprotein E isoform-dependent amyloid deposition and neuritic degeneration in a mouse model of Alzheimer's disease. *Proceedings of the National Academy of Sciences of the United States of America* 97(6): 2892–2897.

Hsiao K, Chapman P, Nilsen S, et al. (1996) Correlative memory deficits, Abeta elevation, and amyloid plaques in transgenic mice. *Science* 274(5284): 99–102.

Lewis J, Dickson DW, Lin W-L, et al. (2001) Enhanced neurofibrillary degeneration in transgenic mice expressing mutant tau and APP. *Science* 293(5534): 1487–1491.

Lewis J, McGowan E, Rockwood J, et al. (2000) Neurofibrillary tangles, amyotrophy and progressive motor disturbance in mice expressing mutant (P301L) tau protein. *Nature Genetics* 25(4): 402–405.

Lord A, Kalimo H, Eckman C, Zhang X-Q, Lannfelt L, and Ng Nilsson L (2006) The Arctic Alzheimer mutation facilitates early intraneuronal Abeta aggregation and senile plaque formation in transgenic mice. *Neurobiology of Aging* 27(1): 67–77.

Oddo S, Caccamo A, Shepherd JD, et al. (2003) Triple-transgenic model of Alzheimer's disease with plaques and tangles: Intracellular Abeta and synaptic dysfunction. *Neuron* 39(3): 409–421.

Santacruz K, Lewis J, Spires T, et al. (2005) Tau suppression in a neurodegenerative mouse model improves memory function. *Science* 309(5733): 476–481.

Sturchler-Pierrat C, Abramowski D, Duke M, et al. (1997) Two amyloid precursor protein transgenic mouse models with Alzheimer disease-like pathology. *Proceedings of the National Academy of Sciences of the United States of America* 94(24): 13287–13292.

Sun Y, Wu S, Bu G, et al. (1998) Glial fibrillary acidic protein-apolipoprotein E (apoE) transgenic mice: Astrocyte-specific expression and differing biological effects of astrocyte-secreted apoE3 and apoE4 lipoproteins. *The Journal of Neuroscience* 18(9): 3261–3272.

Tatebayashi Y, Miyasaka T, Chui D-H, et al. (2002) Tau filament formation and associative memory deficit in aged mice expressing mutant (R406W) human tau. *Proceedings of the National Academy of Sciences of the United States of America* 99(21): 13896–13901.

## Relevant Website

http://www.alzforum.org – Alzheimer Research Forum.

# Alzheimer's Disease: Transgenic Mouse Models

**K H Ashe**, University of Minnesota, Minneapolis, MN, USA



## Creation of Transgenic Mouse Models

### Biological Foundation

Three scientific breakthroughs made the creation of the earliest transgenic mouse models of Alzheimer's disease (AD) possible. The first was the isolation and sequencing of the amyloid-$\beta$ (A$\beta$) peptide in 1984. The second was the cloning of the amyloid precursor protein (APP) gene in 1987 and the elucidation of its role in generating the A$\beta$ peptide. Third was the discovery of the first mutation in autosomal dominant familial AD in APP in 1991 and the subsequent realization that all autosomal dominant mutations causing AD enhance the ability of the A$\beta$ protein to aggregate, either by increasing its overall production or by the generation of amino acid variants that potentiate its aggregation.

This information enabled investigators to create APP transgenic mice modeling AD. The earliest mouse models were developed in the first half of the previous decade.

Altogether about 20 such mice have been published, many of which show age-related amyloid plaque deposition and memory loss. However, APP transgenic mice are incomplete models of AD because they lack neurofibrillary tangles and develop few or no neurodegenerative changes, such as neuronal or synaptic loss. Transgenic mice that develop neurofibrillary tangles accompanied by significant neuronal loss emerged with the creation of tau transgenic mice.

Four important landmarks in tau biology made the generation of tau transgenic mice possible. First was the isolation and characterization in 1975 of tau, which is involved in promoting the aggregation and polymerization of tubulin to form microtubules. Second was the cloning in 1986 of the tau gene. Third was the recognition that tau is the principal protein forming the core of the paired helical filaments of neurofibrillary tangles. Fourth was the discovery in 1998 of mutations in tau linked to familial tauopathies, called frontotemporal dementia with parkinsonism (FDTP). These advances led to the development of tau transgenic mice in the first years of the current decade.

### Technical Methodology

The laboratory mouse remains the genetic model organism that is evolutionarily closest to the human. Mice pose advantages over other species, such as rats, dogs, pigs, or primates, in several aspects: (1) it is relatively easy to house large numbers of them; (2) the mouse genome is better characterized than that of any other mammal except humans; (3) many genetically altered transgenic knockout and knockin lines and genetically characterized mouse strains are available for cross-breeding and further genetic analysis; (4) cognitive studies can be done in appropriate strains of mice, such as C57B6/SJL and 129S6, making them especially relevant to the study of AD; and (5) studies in invertebrates, while useful, will lack some relevance in pharmacological studies and pathogenic investigations. To create transgenic mice with salient features of AD, investigators must take care in several aspects of design and development. The most important factors are the selection of the gene or gene variant to be expressed, the promoter driving transgene expression, the background strain of the mice, and the levels and distribution of transgenic protein expression achieved in the brain. The mouse lines with the most robust neuropathological phenotypes have been developed using the method of pronuclear microinjection of transgenes into fertilized oocytes (**Figure 1**).

Both wild-type and variant human APP and tau genes have been used to generate mice that develop neuropathology related to AD. Transgenes containing mutations linked to autosomal dominant AD or FDTP generate more-robust neuropathology than transgenes encoding wild-type genes. Although mutations in presenilin-1 and presenilin-2 genes lead to early-onset AD in humans, mice expressing presenilin gene variants do not develop neuropathology *per se*. In the presence of human APP transgenes, however, the presenilin gene variants accelerate plaque deposition in transgenic mice.

## Characteristics of Transgenic Mouse Models

### APP Transgenic Mice

APP transgenic mice often develop age-related cognitive deficits and recapitulate many of the neuropathological features of AD, including amyloid plaques, oxidative stress, astrogliosis, microgliosis, cytokine production, and dystrophic neurites (**Figure 2**). However, many important neuropathological features of AD are conspicuously absent in APP transgenic mice, including neurofibrillary tangles and gross atrophy.

There are variable degrees of neuronal and synaptic loss among the various transgenic lines, but no line exhibits the severity of neurodegeneration found in

Alzheimer's patients, despite amyloid plaque loads that often exceed those in human brain specimens. For example, Tg2576 mice are virtually devoid of neuronal or synaptic loss, while synaptic loss is present in J20 and PDAPP mice, and there is some neuronal loss in APP23 mice with massive quantities of amyloid plaques. All four of these lines of mice eventually develop amyloid plaque loads that exceed the amount found in typical Alzheimer patient brains at autopsy. Although we do not understand the factors which account for the variations in neurodegeneration between the mouse lines, it is clear that the differences are not related to the quantity of amyloid plaque deposition.

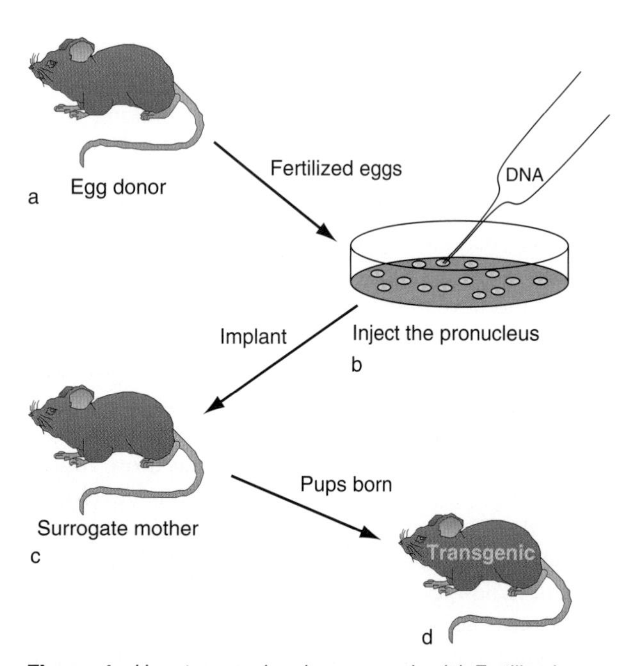


**Figure 1** How transgenic mice are made. (a) Fertilized eggs (oocytes) are harvested from a donor mouse. (b) DNA encoding genes linked to Alzheimer's disease is injected into the pronucleus of the fertilized oocytes. (c) The oocytes are implanted into surrogate mothers, where they mature to full-term pups. (d) Of the pups that are born, a certain fraction, usually 10–20%, contain the transgene.

## Tau Transgenic Mice

Tau transgenic mice generally develop age-related neurological abnormalities, neurofibrillary tangles, and neurodegeneration but invariably fail to form amyloid plaques. The earliest tau transgenic lines expressing tau ubiquitously in the neurons of the brain developed neurodegeneration mainly in the brain stem and spinal cord, leading to paralysis. Because the paralysis interfered with cognitive testing, tau transgenic mice in which variant tau expression was restricted to the forebrain were created and shown to develop age-related memory impairment.

Although the tau transgenic mice with the most robust neuropathological phenotypes express tau variants linked to FDTP (**Figure 3**), neurofibrillary tangles and neurodegeneration occur also in mice expressing wild-type human tau on a null tau background, suggesting a propensity for mouse tau to inhibit neurofibrillary tangle formation. Notably, these mice uniquely form the true paired-helical filaments characteristic of neurofibrillary tangles in AD. Straight filaments, rather than paired-helical filaments, are found in tau transgenic mice expressing FDTP tau variants. This distinction, once believed to be critical, has become less important with the discovery that neurofibrillary tangles contribute little or not at all

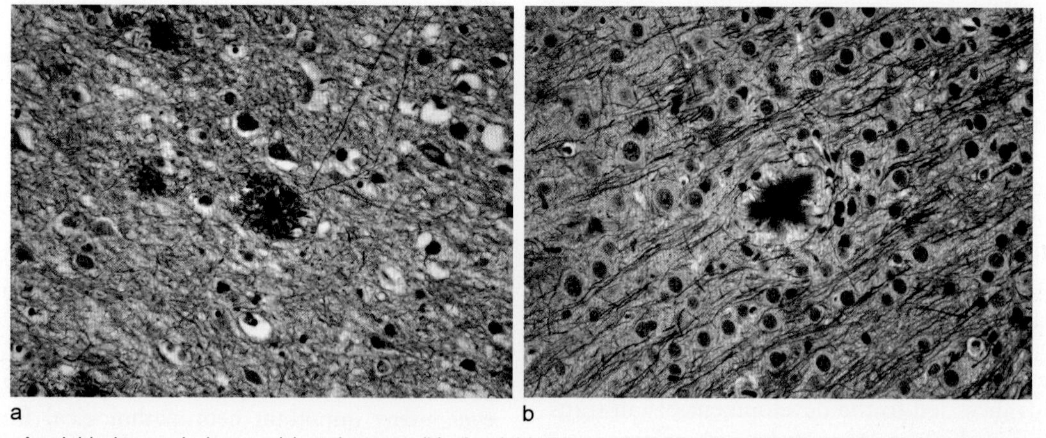


**Figure 2** Amyloid plaques in human (a) and mouse (b). Amyloid plaques that are generated in amyloid precursor protein (APP) transgenic mice closely resemble the shape, composition, and size of those found in humans with Alzheimer's disease. The plaque on the left is from a human specimen, while the one on the right is from an aged APP transgenic mouse. Despite the similarities in plaques in the two species, neurodegenerative changes that are clearly present in the human specimen are largely absent in the mouse specimen. Both specimens were stained with the Bielschowsky silver stain. The photomicrographs were taken at the same magnification (40×). Photomicrographs were kindly provided by Dr. Martin Ramsden.

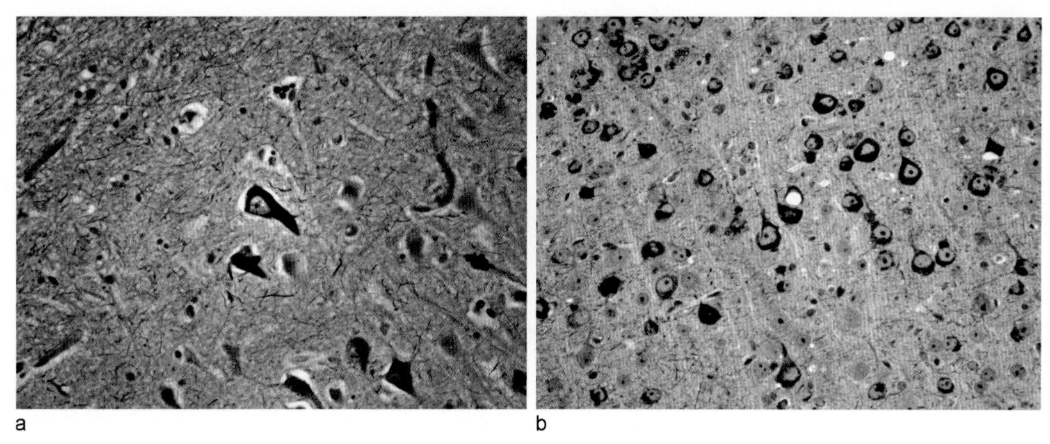


**Figure 3** Neurofibrillary tangles in (a) human and (b) mouse. Neurofibrillary tangles that are generated in tau transgenic mice resemble those found in humans with Alzheimer's disease. The tangles on the left are from a human specimen, while those on the right are from an aged tau transgenic mouse. Both specimens were stained with the Bielschowsky silver stain. The photomicrographs were taken at the same magnification (40×). Photomicrographs were kindly provided by Dr. Martin Ramsden.

to neuronal death and memory impairment, as discussed more fully in the section titled 'Mechanism of memory loss in transgenic mouse models.'

### Other Types of Transgenic Mice

Efforts to combine the cardinal neurological and neuropathological features of AD, namely amyloid plaques, neurofibrillary tangles, neurodegeneration, and cognitive impairment, within a single mouse resulted in the generation of transgenic mice expressing multiple tau and APP protein variants, sometimes in combination with presenilin 1 variants. Although some such mice develop amyloid plaques, neurofibrillary tangles, neuronal loss, and memory loss, they constitute essentially a hybrid of two separate disorders, AD and FDTP. A number of interesting studies have been carried out using these hybrid mice, but their relevance to AD is not entirely clear. A mouse exhibiting all the hallmarks of AD that does not depend on the use of genes not linked to familial AD remains an elusive creature.

## Utility of Transgenic Mouse Models

### Validity of Transgenic Mouse Models

The lack of several important features of AD in APP transgenic mice, such as neurofibrillary tangles and prominent neurodegeneration, along with the shortcomings of the tau transgenic mice and the APP-tau hybrid mice, have prompted some scientists, rightly, to challenge the validity of transgenic mouse models of AD. To address this challenge, criteria may be devised for validating Alzheimer's mouse models, three of which are discussed here. First, theoretical validity refers to whether the use of a given transgene is based on sound biological principles. Second, factual validity refers to how accurately various aspects of the human disease are represented. Third, predictive validity refers to whether studies using the model predict outcomes in human trials. Clearly, predictive validity is the main determinant of the value of a given mouse model. Thanks to rapid progress in bench-to-bedside research in AD, we know more about the predictive validity of APP transgenic mice than of transgenic mice modeling any other neurodegenerative illness.

A$\beta$ immunotherapy illustrates this point. Young PDAPP mice vaccinated with A$\beta$ fail to develop amyloid deposits and astrogliosis. This result, reported in 1999, quickly led to an international, multicenter, randomized trial of an experimental A$\beta$ vaccine in Alzheimer patients. The vaccine cleared amyloid plaques away from large regions of brain in Alzheimer patients. However, neurofibrillary tangles were unaffected, and neuronal loss persisted. Furthermore, when the effects of the vaccine on cognitive function were examined, it became clear that memory decline may have been slowed, but it was not restored. The failure to restore memory function in Alzheimer patients is in marked contrast to the effects of either active or passive A$\beta$ immunization on memory function in APP transgenic mice. Memory loss can be prevented, and preexisting memory loss can be reversed and memory function fully restored in several different lines of APP transgenic mice following A$\beta$ immunization treatments.

These studies show that when the effects of the A$\beta$ vaccine on amyloid plaques are studied, then the results in APP transgenic mice predict what occurs in humans. However, when memory function is examined, then there is a striking discrepancy between what occurs in the mice and what occurs in patients. Thus, the predictive validity of APP transgenic mice as regards amyloid plaques is good. In contrast, the predictive validity of APP transgenic mice in relation to memory function is poor.

### Relevance of Transgenic Mouse Models

The value of transgenic models of AD is as good as our understanding of how closely the models mimic various stages of the illness. To appreciate the context in which these mice have helped us study AD, it is useful delineate the natural history of the illness. AD has a very insidious onset; we do not know precisely when neural dysfunction begins. Recent work suggests that the disease process may begin long before symptoms are present or neurodegeneration has occurred. By the time AD is clinically diagnosed, neurodegeneration is already under way.

APP transgenic mice and tau transgenic mice mimic different stages of AD. Each one is useful for exploring the progression of cognitive decline and the pathology of AD. The picture of minimal loss of neurons in many lines of memory-impaired APP transgenic mice suggests a closer resemblance to preclinical stages of AD than to the disease itself. In contrast, the substantial neuronal loss found in tau transgenic mice implies that they may be better models for studying the neurodegenerative aspects of AD.

### Testing Experimental Therapies in Transgenic Mouse Models

A major effort in developing AD-modifying therapies has been directed at reducing the amyloid deposits and neurofibrillary tangles in the brain. Transgenic mice displaying plaques and tangles are useful models for testing the effects of therapies on these neuropathological abnormalities and have been used extensively for this purpose. However, as will be discussed in the next section, targeting molecules causing cognitive dysfunction and memory impairment may hold still better promise for treating the symptoms of AD.

### Mechanism of Memory Loss in Transgenic Mouse Models

Ever since Alois Alzheimer described the amyloid plaques and neurofibrillary tangles that were evident when he examined the brain of a demented patient in 1906, the scientific research on what has come to be known as AD has focused on how the accumulation of tau and A$\beta$ proteins form amyloid plaques and neurofibrillary tangles, which define the disease neuropathologically, rather than on the functional effects of these molecules on the brain. Why and how tau and A$\beta$ molecules cause memory loss and dementia are only now becoming understood. Our knowledge of the manner in which these molecules disrupt brain function lags a century behind the neuropathological studies of tau and A$\beta$ proteins because suitable tools for examining their effects on the workings of the living brain were lacking until recently. The creation of transgenic mouse models of AD helped fill this void and provided the vital reagents needed to begin to understand the adverse effects that the tau and A$\beta$ proteins exert on memory and cognitive function.

In the Tg2576 APP transgenic mouse model of AD, a major source of heterogeneity in A$\beta$ proteins arises from the aggregation of monomeric proteins to form higher-order structures. Monomeric and trimeric A$\beta$ protein assemblies are located within neurons. In the extracellular space, monomers and trimers coexist with larger A$\beta$ protein assemblies that have similar molecular weights as higher-order species, such as hexamers, nonamers, and dodecamers. Many of the monomeric and oligomeric forms of A$\beta$ protein in the brain are soluble in aqueous buffers. In contrast, amyloid plaques are composed of A$\beta$ fibrils that are insoluble in aqueous buffers and most detergents. In Tg2576 mice, amyloid plaques lag 3–6 months behind the initial loss of memory function, which occurs at 6 months of age, and they do not correlate with memory impairment, indicating that the plaques contribute little to brain dysfunction. Memory impairment is caused, instead, by a specific, soluble A$\beta$ protein assembly, called A$\beta$*56 (read as A-beta star 56). A$\beta$*56 correlates strongly with memory dysfunction, and when purified from the brains of impaired Tg2576 mice and administered to young, healthy animals, it transiently disrupts memory function in the healthy subjects. A$\beta$*56 disrupts memory by a mechanism that does not involve synaptic or neuronal loss. A$\beta$*56 is present in brain tissue from Alzheimer patients and may contribute to memory loss associated with AD.

Neurofibrillary tangles contribute little, if at all, to neuronal loss and brain dysfunction in a line of transgenic mice modeling FDTP, called rTg4510. By dint of identical hyperphosphorylated tau species in the neurofibrillary tangles of rTg4510 mice and Alzheimer's brain tissue, it may be supposed that the effects of neurofibrillary tangles on dementia and neuronal loss in AD are similarly inconsequential. However, the molecular mechanisms by which tau induces memory impairment and kills neurons in rTg4510 mice or AD are unknown; it is likely that specific, as yet unidentified, tau* (tau star) molecules are responsible for disrupting brain function and interfering with neuronal viability. Finally, the relative contributions of A$\beta$*56, tau, and neurodegeneration to memory loss, cognitive impairment, and dementia remain unresolved.

*See also:* Aging of the Brain and Alzheimer's Disease; Alzheimer's Disease: Neurodegeneration; Alzheimer's Disease: An Overview; Alzheimer's Disease: Molecular Genetics; Animal Models of Alzheimer's Disease; Axonal Transport and Alzheimer's Disease; Brain Glucose Metabolism: Age, Alzheimer's Disease and ApoE Allele Effects; Lipids and Membranes in Brain Aging.

## Further Reading


Andorfer C, Kress Y, Espinoza M, et al. (2003) Hyperphosphorylation and aggregation of tau in mice expressing normal human tau isoforms. *Journal of Neurochemistry* 86: 582–590.

Games D, Adams D, Alessandrini R, et al. (1995) Alzheimer-type neuropathology in transgenic mice overexpressing V717F beta-amyloid precursor protein. *Nature* 373: 523–527.

Hsiao K, Chapman P, Nilsen S, et al. (1996) Correlative memory deficits, A$\beta$ elevation, and amyloid plaques in transgenic mice. *Science* 274: 99–102.

Ishihara T, Hong M, Zhang B, et al. (1999) Age-dependent emergence and progression of a tauopathy in transgenic mice overexpressing the shortest human tau isoform. *Neuron* 24: 751–762.

Iyadurai SJP and Ashe KH (2005) Creating APP transgenic lines in mice. In: Xia W and Xu H (eds.) *Amyloid Precursor Protein: A Practical Approach*, pp. 185–200. Boca Raton, FL: CRC Press.

Lesné S, Koh MT, Kotilinek L, et al. (2006) A specific amyloid-$\beta$ assembly in the brain impairs memory. *Nature* 440: 352–357.

Lewis J, McGowan E, Rockwood J, et al. (2000) Neurofibrillary tangles, amyotrophy and progressive motor disturbance in mice expressing mutant (P301L) tau protein. *Nature Genetics* 25: 402–405.

Oddo S, Caccamo A, Shepherd JD, et al. (2003) Triple-transgenic model of Alzheimer's disease with plaques and tangles: Intracellular Abeta and synaptic dysfunction. *Neuron* 39: 409–421.

SantaCruz K, Lewis J, Spires T, et al. (2005) Tau suppression in a neurodegenerative mouse model improves memory function. *Science* 309: 476–481.

# Animal Models of Parkinson's Disease

**K E Soderstrom, G Baum, and J H Kordower**, Rush University Medical Center, Chicago, IL, USA



## Introduction

Parkinson's disease (PD) is the second-most-common neurodegenerative disorder (after Alzheimer's disease), affecting more than 1 000 000 people in the United States. While there is widespread degeneration in the central and peripheral nervous systems in PD, the hallmark pathology remains the dopaminergic striatal insufficiency secondary to degeneration of dopaminergic neurons in the substantia nigra (SN). While a small percentage of PD is of a genetic origin, the vast majority of cases are sporadic and of unknown origin. It is particularly challenging to model a disease whose etiology is, for the most part, unclear. Therefore, PD researchers have attempted to model specific aspects of PD pathology independent of etiology. These attempts have included pharmacological models aiming to replicate striatal dopamine (DA) depletion and neurotoxic models that replicate nigral cell loss. While excellent for evaluating the role of the nigrostriatal system in PD and for testing therapeutic compounds, most of these models are limited in their ability to replicate such important aspects of the human disease as Lewy body formation, nondopaminergic systems affected in PD, and the progressive nature of the disease. Novel models emerging in the field have attempted to model these important facets of the human disease more accurately. Genetic models, the lipopolysaccharide model, and the use of aged animals as a PD model are all emerging alternatives to traditional DA-centric models of PD. The validity, strengths, and weaknesses of these traditional and novel animal models of PD will be addressed in this article, as well as the importance of animal models in general in our understanding and treatment of PD.

## PD

PD is a progressive neurodegenerative disorder clinically characterized by the cardinal symptoms of cogwheel rigidity, resting tremor, bradykinesia, stooped posture, and shuffling gait. As stated above, there is a loss of dopaminergic cells in the SN pars compacta (SNpc) that results in insufficient DA innervation of the basal ganglia and subsequent increased inhibition of excitatory thalamo-cortical connections. Lewy bodies, intracellular inclusions principally containing α-synuclein, are also found in the remaining nigral neurons of PD patients.

While motor symptoms are often the most obvious deficits exhibited in PD, it is clear that numerous nonmotor and peripheral nervous system symptoms can significantly affect patient quality of life. Braak and co-workers recently used α-synuclein immunohistochemistry to hypothesize that PD pathology begins in the lower brain stem and olfactory bulb and then progresses in a caudal to rostral fashion. This progressive nature of PD and the involvement of nondopaminergic pathways in its development have traditionally been unrepresented in animal models of PD, despite the growing realization of their role in the human disease.

### DA Depletion

The ultimate result of cell loss and cell dysfunction in the SN is the depletion of the neurotransmitter DA in the basal ganglia. This insufficient DA innervation is principally localized to the postcommissural putamen and results in the overdrive of globus pallidus and subthalamic nuclear outputs. The resulting inhibition of thalamocortical function results in the characteristic bradykinesia experienced by PD patients. Researchers first attempted to model PD by replicating this DA depletion in animals. Early studies using pharmacological interventions aimed to selectively deplete monoamine neurotransmitters lost in PD patients. Of these pharmacological models, one of the earliest and most utilized has been the reserpine model, which prevents the storage of DA in presynaptic terminals.

### Reserpine

In the late 1950s, Arvid Carlsson demonstrated that the drug reserpine could cause reversible parkinsonism in animals, in a manner similar to that seen in humans. Reserpine blocks the storage of monoamines in catecholaminergic neurons, resulting in DA depletion in the striatum. Treating rabbits with reserpine causes a behavioral syndrome that modeled idiopathic PD and helped elucidate the critical role of DA in the pathogenesis of this disease. This new concept radically changed clinical neurology and lead to the discovery that the DA precursor, levodopa, as well as other prodopaminergic drugs, could profoundly benefit the lives of PD patients. Indeed, decades later, administration of levodopa remains the gold-standard treatment for PD patients.

While clearly groundbreaking, the reserpine model has significant limitations. Most obviously, reserpine produces only biochemical lesions, and therefore reserpine-treated animals do not model the neuronal dysfunction and degeneration seen in the human disease. In addition, the behavioral effects of reserpine are transient. While this can be advantageous in some circumstances in understanding nigrostriatal function, PD is a progressive chronic disorder, and if possible, progressive functional deficits are most often required for study. Last, reserpine depletes all monoamines, not exclusively DA. While norepinephrine depletion from the locus coeruleus is a facet of PD degeneration, depletion of all monoamines does not accurately model PD.

## Nigral Cell Loss

While pharmacological models helped elucidate the role of DA in PD, their inability to model the histopathology of the disease made them poor models for evaluating many novel therapies. More beneficial have been models that aim to replicate the loss of dopaminergic nigral neurons observed in Parkinson's patients. Recognized as a primary neuropathology responsible for PD, SNpc cellular loss has been the prime focus of neurotoxic models that aim to replicate this cellular pathology.

### 6-Hydroxydopamine

The DA analog 6-hydroxydopamine (6-OHDA), because of its similarity in molecular structure, can be taken up into dopaminergic terminals via the DA transporter. Once inside the cell, it is metabolized, resulting in the production of hydrogen peroxide and free radicals. Ultimately these toxic molecules induce neuronal death via mitochondrial dysfunction.

Like DA itself, 6-OHDA is not able to cross the blood–brain barrier and therefore must be delivered directly to the brain of experimental animals via stereotaxic surgery. In 1968, Ungerstedt and colleagues demonstrated the utility of 6-OHDA lesions as animal models of PD. In their study, 6-OHDA was unilaterally injected into the medial forebrain bundle, extensively depleting the nigrostriatal pathway on one side. Ungerstedt noted that lesioned animals rotated toward the side of their lesions spontaneously as well as after administration of the dopaminergic drug d-amphetamine. Conversely, apomorphine, a drug that acts upon upregulated DA receptors on the side ipsilateral to the lesion, induces rotations contralateral to the lesion.

The number of rotations performed by a lesioned animal can be quantified to serve as an index of the integrity of nigrostriatal function. Experimental therapeutic strategies, such as neural or stem cell transplantation and gene therapy, can use the number of rotations an animal performs as an index of the intervention's efficacy.

Since these initial observations, there have been modifications and improvements on the 6-OHDA model although it is still frequently used in its original form. By varying the position and extent of the lesion, different stages of human PD can be modeled. The late stages of PD have been modeled using 6-OHDA lesions to the medial forebrain bundle and SN (**Figure 1**). Using this model, neuronal loss is detected as soon as 12 h postinjection and peaks at 48 h. In addition, striatal fibers are found to degenerate between 1 and 7 days after 6-OHDA delivery, ultimately resulting in more than 90% striatal DA depletion. This provides an ideal environment to evaluate cellular replacement strategies. Alternatively, 6-OHDA delivery to the striatum can result in levels of DA depletion more representative of early-stage PD. Kirik and colleagues demonstrated the location of striatal injections, either 'terminal' (within the caudate–putamen) or 'preterminal' (at the caudate–putamen boundary), greatly affected the resulting lesion, with preterminal injections creating greater levels of DA depletion. In addition, they found variable reductions in tyrosine hydroxylase-positive (TH+) fiber densities and TH+ SN neurons after either single or multiple 6-OHDA intrastriatal injections. As injection site number

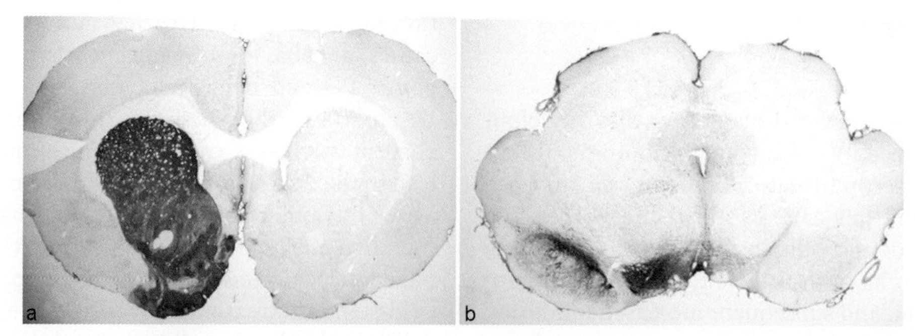


**Figure 1** Tyrosine hydroxylase (TH)-stained coronal sections of a 6-OHDA-treated rat striatum (a) and substantia nigra (b) in which the dopaminergic neurotoxin was injected into the right medial forebrain bundle. Note the comprehensive loss of TH-immunoreactive fibers staining in the striatum (a) and the loss of TH-immunoreactive neurons in the nigra ipsilateral to the lesion.

increased, so too did DA depletion. A single preterminal injection site elicited a 60–70% decrease in TH+ fiber density in the rostral striatum and a 25% decrease in TH+ fiber density in the caudal striatum. In contrast, four preterminal site injections resulted in decreases as high as 80% and 90% in TH+ fiber density in the rostral and caudal striatum, respectively. These intrastriatal 6-OHDA models that provide varying degrees of DA depletion have been particularly effective in the evaluation of neuroprotective strategies for human PD.

The 6-OHDA lesion produces a range of functional deficits that have been assessed by means of a number of behavioral tests. Rotational analyses in response to prodopaminergic drugs such as amphetamine or agonists such as apomorphine have been used extensively to assess the extent of an animal's lesion and of its subsequent recovery after therapeutic behavioral analyzes. Nonpharmacological interventions such as open field activity in which a rat is observed for a time within an activity-monitoring chamber are also altered after 6-OHDA. Analysis of the animal's movements can provide a researcher with information about the animal's overall activity, rearing behavior, and any stereotypic behavior. The cylinder test (or limb use asymmetry test) consists of rating the use (and disuse) of each forelimb while an animal navigates itself around a plexiglass cylinder. While an unlesioned rat will typically use each limb equally, a rat with a 6-OHDA lesion will preferentially use the limb ipsilateral to its lesion. The adjusting steps test and its variations involve the restriction of an animal's hindpaws and one forepaw. The unbound forepaw is then moved slowly sideways across a flat surface. In an unlesioned rat, both forepaws will take steps to adjust to the sideways motion. However, in the lesioned animal, the forepaw contralateral to the lesioned striatum will drag along without adjustment. These and many other behavioral tests have been instrumental in the assessment of 6-OHDA lesion-induced deficits and in determining the functional efficacy of therapeutic interventions.

To date, no model of PD has been used as often or, arguably, as effectively as the 6-OHDA model. However, its limitations must also be acknowledged and considered when one is interpreting data obtained from its use. As with any transmitter-specific model, the 6-OHDA lesion does not replicate the constellation of neurotransmitter-based pathologies seen in human PD. The depletion of norepinephrine, due to degeneration of the locus coeruleus seen prior to SN degeneration in human PD patients, is not replicated in the 6-OHDA model. In addition, dysfunction of serotonergic and peptidergic systems seen in PD patients is not mimicked in 6-OHDA-treated animals. Also problematic is the timing of the degenerative process. Whereas Parkinson's is now thought to be a progressive degenerative disorder beginning in the olfactory bulbs and caudal brain stem and progressing over many years rostrally to the midbrain, 6-OHDA lesions are regionally selective and exhibit cell loss in a matter of days to weeks. Finally, while lesioned animals show some of the cellular and behavioral deficits seen in the human disease, 6-OHDA-treated animals fail to develop Lewy bodies or α-synuclein aggregates, features that are pathoneumonic for PD. While 6-OHDA remains a convenient and effective model for the analysis of the nigrostriatal dysfunction and evaluation of many novel therapies, these important shortcomings must be taken into account when translating information obtained from the 6-OHDA-treated animal from the laboratory to the clinic.

### 1-Methyl-4-Phenyl-1,2,3,6-Tetrahydropyridine

One of the best models of PD results from the administration of 1-methyl-4-phenyl-1,2,3,6-tetrahydropyridine (MPTP), which selectively targets dopaminergic neurons in the brain. Its validity as a model is supported by the fact that MPTP is a definitive, albeit rare, environmental cause of PD in humans. Its neurotoxic effects were first reported in the early 1980s by Bill Langston and co-workers after drug users exposed to MPTP began to develop Parkinsonian symptoms. Taken up primarily by astrocytes, MPTP is converted to its metabolite MPP+. MPP+ can then be taken up by DA neurons, where it exerts its toxic effect through interactions with cytosolic proteins and through the interruption of the complex I component of the electron transport chain.

**The MPTP mouse model** Due to the relative economy and small size of mice, the cost and convenience of using them have made the MPTP mouse model a widely used model of PD. In addition, the ability to genetically modify the mouse genotype has allowed for the analysis of possible genetic factors that may contribute to MPTP sensitivity, a procedure presently not possible in primate models of PD. In this regard, transgenic α-synuclein knockout mice show an increased resistance to MPTP toxicity, while transgenic mice overexpressing α-synuclein show increased susceptibility.

As with 6-OHDA, varying regimens of MPTP delivery can produce varying degrees of nigrostriatal dysfunction. Acute dosing, consisting of a single injection or a short series of high dose injections ($\sim$10–40 mg kg$^{-1}$), can produce lesions that result in mild to moderate cell loss and adequately represent early stages in human PD. An early study of acute

MPTP toxicity found a 33% reduction of TH+ SN neurons and a 90% depletion in striatal DA in the C57/bl mouse after a single acute injection of MPTP (40 mg $kg^{-1}$). Alternatively, chronic dosing, consisting of long series of low-dose injections (~5–20 mg $kg^{-1}$), results in more-robust lesions representative of late-stage human PD. Jackson-Lewis and colleagues have demonstrated as much as 70% nigral cell loss and 90% striatal DA depletion in C57/BL6 mice after a chronic MPTP dosing regimen (80 mg $kg^{-1}$ total). Chronic administration is also associated with lower levels of spontaneous recovery. Research has suggested that neuronal loss and DA depletion incurred after MTPT administration occur through different mechanisms, with acute dosing resulting in necrotic cell loss and chronic dosing leading to apoptotic cell loss. It has also been shown that MPTP toxicity in the mouse can be enhanced through the delivery of MPTP-enhancing agents such as probenecid. In one study, the chronic delivery of MPTP (25 mg $kg^{-1}$; 10 doses over 5 weeks) with probenecid (250 mg $kg^{-1}$) to C57/bl mice resulted in the accumulation of α-synuclein immunoreactive aggregations, although these were not phenotypically similar to the Lewy bodies observed in human PD.

A number of behavioral tests have been developed to assess deficits and recovery of function in MPTP-treated mice. As with the 6-OHDA-treated rat, the open-field test has been used extensively in the mouse to assess overall behavioral activity. The rotarod test has also been used to assess the coordination and balance of MPTP-treated and control mice. In this test, mice are taught, prior to MPTP administration, to run on a rod rotating at varying speeds. The time that lesioned mice are able to remain on the rod then decreases with MPTP-induced nigrostriatal dysfunction. MPTP also impairs mouse performance on the pole test, a measure of bradykinesia that involves placing a mouse, head facing upwards, at the top of a textured pole. The amounts of time it takes the mouse to rotate its body toward the ground and to descend the pole have been taken as indices of the mouse's functional impairment.

While the MPTP mouse model has been invaluable in an understanding of the mechanisms of MPTP toxicity, there are some limitations to its uses. Due to the small size of mice, MPTP is typically administered to them systemically. While systemic delivery has the advantage of producing a bilateral lesion more representative of human PD, it is also associated with higher levels of animal morbidity and increases the animal maintenance required of caregivers. In addition, mice may show robust and spontaneous improvement after the initial lesions, making determinations of therapeutic efficacy difficult. These factors, combined with the mouse's insensitivity to MPTP relative to primates, result in the need for high doses to create stable lesions of clinical relevance in the mouse. Also problematic is the variability of MPTP sensitivity that exists between mouse strains. Muthane and colleagues have found that C57/bl mice innately possess fewer TH+ SN cells and furthermore show an increased sensitivity to the toxic effects of MPTP when compared with the CD-1 strain of mouse. Finally, while MPTP-treated mice show some occurrence of motor dysfunction, they have not been particularly useful in the analysis of parkinsonian behavior and recovery after therapeutic intervention.

**The MPTP primate model** Nonhuman primates are much more sensitive to MPTP toxicity and require smaller doses to create adequate stable nigrostriatal lesions. MPTP has been used experimentally in monkeys to create both bilateral and unilateral lesions. Systemic administration delivered via intraperitoneal, subcutaneous, intravenous, or intramuscular injections creates bilateral lesions that successfully model the cardinal symptoms associated with human PD. However, because these delivery methods are associated with increased animal morbidity and maintenance, researchers have also used primates that receive unilateral MPTP lesions delivered via intracarotid artery (ICA) injection (**Figure 2**). Despite the necessity of a more invasive surgical procedure, ICA lesions can be more attractive to researchers. Also, despite a small degree of damage to the contralateral hemisphere, it remains relatively intact in unilaterally injected animals and may serve as a useful control and allow for the occurrence of rotational behavior.

In an attempt to combine the benefits of ICA and systemic MPTP treatment and to avoid the weaknesses of each delivery method, recent studies have created an 'overlesioned' model in which an initial ICA unilateral injection is followed by subsequent intravenous injections. This approach elicits an asymmetrical bilateral lesion. In one study looking at the effects of overlesioning in rhesus monkeys, a comprehensive cell loss was seen in the SN ipsilateral to the ICA injection, with a subtotal cell loss in the contralateral SN. Researchers noted a positive correlation between the animals' behavioral deficits and the extent of their lesion contralateral to the ICA injection. The preferential degeneration of one hemisphere over another produced by the overlesioning method is much like the hemispheric biases seen in human PD patients. It is important to note that the deficits in these animals are bilateral and remain stable and that the animals are able to maintain themselves and do not require heroic veterinary intervention.

Again, the method of delivery and dosage of the MPTP toxin may be manipulated to create lesions representative of varying stages of human PD. Acute

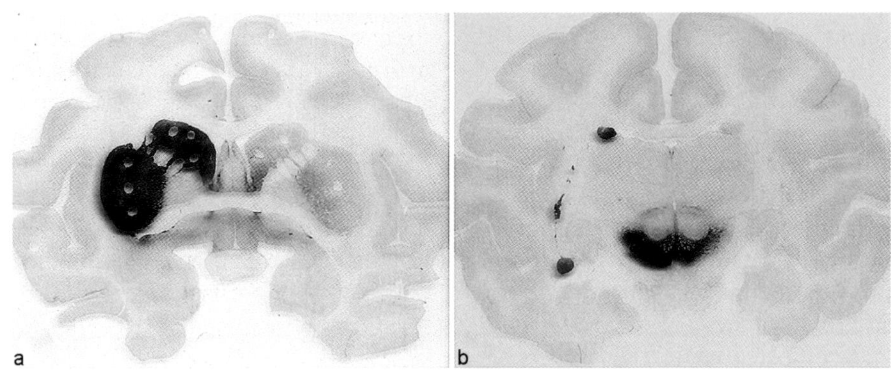


**Figure 2** Tyrosine hydroxylase (TH)-stained coronal sections through the striatum (a) and substantia nigra (b) from a rhesus monkey that received an injection of MPTP in the right internal carotid artery. Note the comprehensive loss of TH-immunoreactive fibers staining in the striatum (a) and the loss of TH-immunoreactive neurons in the nigra ipsilateral to the lesion.

MPTP administration results in a significant nigral cell loss, modeling late-stage human PD. In contrast, chronic MPTP administration results in more-moderate cell loss, resembling the middle stages of human PD. In this regard, squirrel monkeys display a 70% nigral cell loss and 95% striatal DA depletion after acute MPTP treatment compared with 40% nigral cell loss and 60–70% striatal DA depletion after chronic systemic administration. Chronic MPTP administration may also be used to model the more progressive nature of cell loss and behavioral deficit observed in the human disease.

As with idiopathic PD, MPTP-induced deficits can be evaluated using clinical rating scales. These assess the severity of symptoms such as tremor, posture, gait, bradykinesia, balance, defense reactions, freezing, and gross motor skills. Fine motor skills testing can also be analyzed through a number of behavioral tests. The forelimb reaching test involves training the monkey to reach toward a recurring target object on which a reward is given, usually food. The time it takes for the monkey to reach toward its reward is an index of its nigrostriatal function. A similar test is the bar-pressing test, in which monkeys are made to press a lever bar to receive a food reward. Tests have also been developed to assess the cognitive functioning of MPTP-treated monkeys. In the barrier–detour retrieval test, monkeys are made to retrieve a food reward from a transparent box with an opening on one side. The monkey must determine which side is the open side of the box before receiving the reward. Both the speed at which the monkey can attain the reward and the number of errors that occur for it to do so can be taken as indices of cognitive functioning.

The effectiveness of MPTP to create lesions of the nigrostriatal pathway can be remarkably variable. One cause for this variability is the age of the animal exposed to the toxin, with older animals showing a much greater sensitivity to MPTP toxicity than their younger counterparts do. This finding may be explained by older individuals' increased levels of MAO-B, the enzyme required for the conversion of MPTP to its toxic metabolite, MPP+. In addition, studies have shown a higher sensitivity to MPTP toxicity in males than in females, suggesting a possible protective effect of estrogen on dopaminergic neurons. Animal species is also a critical predictor of MPTP susceptibility, with some species, such as the nonhuman primate, showing a high sensitivity to MPTP toxicity and others, like the rat, showing little to no effect.

MPTP has done a remarkably better job than 6-OHDA in reproducing many of the nondopaminergic deficits observed in human PD. MPTP-treated monkeys show depletions in norepinephrine and serotonin metabolites and after chronic exposure, cell loss in the locus coeruleus similar to findings in human patients. While true Lewy body formations are not seen in MPTP models of PD, accumulations of $\alpha$-synuclein protein have been noted in the brains of baboons and squirrel monkeys after MPTP administration. MPTP treatment has been shown to elicit a resting tremor akin to the cardinal symptom manifested in human patients. Resting tremor has been observed in the MPTP-treated vervet monkey although it has not been reported in any other primate models. While the above cases demonstrate that protein aggregation and tremor are possible after MPTP exposure, it should be stressed that these important Parkinson's pathologies are not consistently found to occur within this model.

## Environmental Toxin Models

The success of MPTP in replicating the neuropathology of human PD has led researchers to examine other neurotoxins that may similarly affect the central nervous system. Two promising models to emerge have been the rotenone and combined paraquat and maneb models.

**Rotenone** Rotenone, an organic pesticide that interrupts complex I of the electron transport chain, also elicits mitochondrial dysfunction and ultimately cell loss in the nigrostriatal pathway. Rotenone may be delivered systemically to create a bilateral lesion or via intranigral infusion to create a unilateral lesion. Like other neurotoxic models, bilateral lesions, though similar to the human disease, frequently result in high animal morbidity, leading researchers to occasionally favor unilateral surgeries. After rotenone exposure, rats can show progressive nigral degeneration, ubiquitin and α-synuclein inclusions in remaining nigral cells, and behavioral deficits similar to those observed in human patients. Because rotenone exerts its effects in a DA transporter (DAT)-independent fashion, it elicits mitochondrial dysfunction in non-dopaminergic systems, although its neurotoxic effects are DA neuron selective.

While rotenone can accurately model human nigrostriatal degeneration in PD, the great variability in individual sensitivity to its effects has hindered its use in research. In one study, only 46% of animals treated with rotenone ($2\,mg\,kg^{-1}$ daily for 21 days) showed decreased TH reactivity, with only one animal showing any nigral cell loss. Recently another study has found that while C57/bl mice displayed some functional dysfunction after chronic rotenone treatment, they failed to develop the nigral degeneration associated with the rotenone rat model. This substantial variability among individuals and species is a major obstacle for the use of the rotenone model for the evaluation of therapies for PD.

**Paraquat and maneb** Paraquat (PQ; 1,1′-dimethyl-4-4′-bipyridinium) is another pesticide that has been examined as a possible neurotoxic animal model for PD. Paraquat is initially converted to a PQ cation, which is then reoxidized to from a parent compound as well as superoxide radicals. This allows for further redox cycling and further oxidative stress. Paraquat is often administered in combination with the fungicide manganese ethylene-bis-dithiocarbamate (maneb). Exposure to maneb has been linked to the development of parkinsonian symptoms in humans, and its combination with paraquat as an animal model has been useful in the investigation of the role of environmental toxins in the etiology of human PD.

Rodent studies have found significant nigral cell loss after combination paraquat and maneb exposure. Mice receiving biweekly PQ ($10\,mg\,kg^{-1}$) and maneb ($30\,mg\,kg^{-1}$) injections had 25–47% TH+ SN cell loss. PQ-exposed mice have also been shown to develop α-synuclein inclusions in remaining nigral neurons. Paraquat and maneb have also been used extensively in research looking at effects of gestational exposure to toxins. Researchers have found that mice exposed to maneb during gestation show increased sensitivity to PQ toxicity in adulthood, providing support for the 'multihit' hypothesis of PD and suggesting that an initial early insult can sensitize the nigrostriatal system to subsequent 'hits' in adulthood.

## Protein Aggregation

The presence of Lewy bodies, α-synuclein-rich intracellular inclusions, in the brains of parkinsonian patients has suggested a role for protein aggregation in the pathogenesis of PD. This has been further supported by the discovery that genetic mutations in the α-synuclein gene can cause genetic forms of PD. In addition, increased levels of α-synuclein in SN neurons have been observed in response to toxin exposure and oxidative stress, suggesting that protein aggregation may have a pivotal importance in the development of idiopathic PD as well. Despite evidence to suggest the central role of protein aggregation in PD development, animal models have traditionally been very poor at exhibiting this important pathology. However, with the development of transgenic animals and the advancement of viral vector delivery systems have come new models for this previously overlooked pathology.

### Transgenic Animals

Transgenic models of PD have been emerging that attempt to recreate the known genetic causes of PD. Particularly interesting have been models that knock out, overexpress, and mutate the α-synuclein gene to examine the role of protein aggregation in PD neurodegeneration. In addition, researchers have looked to mutate proteins involved in the ubiquitin–proteosome pathway known to be affected in familial cases to examine the role of proteosome dysfunction in PD.

**Alpha-synuclein transgenic models** Discovery of the A53T and A30P mutations of the α-synuclein gene in familial PD have led researchers to attempt to manipulate its expression in animals in order to evaluate its role in the human disease. Knockout mice are viable and show decreased striatal DA levels and reduced rearing in the open field, despite DA metabolism's being unaffected. These findings may be explained by the hypothesis that α-synuclein plays a role in synaptic vesicle function. Conversely, mice overexpressing α-synuclein develop intraneuronal inclusions and show decreases in striatal DA levels, despite showing no nigral cell loss. These effects are generally directly correlated to the amount of α-synuclein overexpressed. Mice expressing mutated forms of α-synuclein have shown varying degrees of pathology, dependent on the forms of construct utilized. Although transgenic animals expressing human α-synuclein with the platelet-derived growth factor β promoter show behavioral deficits and increased SN inclusions, other constructs have shown very little pathology or behavioral deficit.

A major strength of the transgenic α-synuclein models for PD is that expression alterations are not limited to dopaminergic neurons but are expressed throughout the nervous system. In addition, the slow accumulation of protein that occurs in transgenic animals provides a better model of the progression seen in human disease. However, despite these models; ability to replicate the Lewy body inclusions seen in human PD, a major failing is their inability to model SNpc cell loss, making them suboptimal for the evaluation of many novel therapies for PD. In addition, technical restrictions limit the use of transgenic models to the phylogenetically distant mouse, preventing these models' immediate application to the clinic.

**Parkin and ubiquitin C-terminal hydrolase L1 transgenic models** The discovery of mutations to parkin and ubiquitin C-terminal hydrolase L1 genes, both components of the ubiquitin–proteosome system, in familial cases of PD has led researchers to examine the role of proteosome dysfunction in PD and to develop models that target these genes.

### Viral Vector-Delivered α-Synuclein

While transgenic models have been limited to mice, adeno-associated (AAV) and lentiviral delivery systems have enabled the selective overexpression of α-synuclein within the nervous system of higher-ordered species. In rats, delivery of wild-type or mutant AAV-human α-synuclein to the SNpc resulted in cytoplasmic inclusions, behavioral dysfunction, and nigral cell loss. Similar cellular pathology was found in rats receiving lentiviral-delivered wild-type and mutant forms of human α-synuclein. The overexpression of α-synuclein via viral vectors has also been evaluated in the nonhuman primate. Marmosets receiving mutant AAV-human α-synuclein showed a 32–41% reduction of DA neurons in the SN. Behavioral deficits observed included head positional bias and rotation asymmetry, both of which are commonly seen in PD toxicity models.

The ability to overexpress α-synuclein has given researchers the opportunity to evaluate the role of protein aggregation in PD. In addition, it offers a relatively slowly progressing model with which to analyze the gradually progressive nature of the human disease, an advantage not offered by most toxicity models. While it is a seemingly powerful model that can encompass the majority of PD pathologies, the selective overexpression of α-synuclein to one specific brain region limits the observation of nonmotor effects such as olfactory impairments, gastrointestinal dysfunction, depression, sleep disturbances, and cognitive impairments that result from α-synuclein aggregation in other brain regions.

## Other PD Models

### Lipopolysaccharide

The immune response has been suggested as a major contributor to the progressive nature of nigral cell loss in PD. Researchers have tried to replicate this increased immune activation through exposure to the bacteriotoxin lipopolysaccharide (LPS). Stereotaxic injections of LPS have been used to induce nigral cell loss in the adult animals, and prenatal LPS exposure has been shown to elicit decreased DA neuron number in offspring. In addition, these animals show locus coeruleus cell loss, suggesting nondopaminergic effects, as well as intracellular inclusions. Researchers have also found that prenatal LPS treatment further prolongs the immune response to secondary LPS exposure in adulthood, suggesting a role for neuroinflammation in the 'multihit' hypothesis of PD.

### Aged Animals

Aging is the primary risk factor for the development of PD. In normal aging, striatal DA is lost, and there is a loss of TH immunoreactive and DA transporter immunoreactive nigral neurons. These losses are associated with age-related increases in α-synuclein. While controversial, emerging evidence suggests that PD represents the extreme end on a continuum of aging. Some experimenters therefore have utilized aged animals themselves as models for PD. Taken together, these findings suggest that aged animals may represent a useful model for evaluation of therapies aimed at the earliest stages of PD.

### Animal Models of Dyskinesia

In addition to mimicking the pathology of PD, models have been used to study treatment-related side effects of PD. In this regard, the laboratory of Angela Cenci has led the field in demonstrating the use of the 6-OHDA rat model of PD for the study of levodopa-induced dyskinesias. Dyskinesias are disabling, involuntary movements that commonly arise in PD patients after prolonged levodopa therapy. While dyskinesias can take several years to develop in patients, rats may be primed with levodopa to exhibit abnormal involuntary movements in a matter of weeks. The affordability and convenience of the rat model make it ideal for the evaluation of the molecular and cellular causes of this serious side effect.

While demanding much more cost and care, MPTP-treated monkeys have also been one of the best and most widely used animal models in which to study levodopa-induced dyskinesias because of their phylogenetic proximity to humans. Unlike the 6-OHDA-treated rat, the MPTP-treated primate model shows similarities to human anatomy and physiology.

## Conclusions

The modeling of PD in animals has been invaluable to our understanding of the human disease. By selectively targeting certain aspects of PD, researchers have been able to evaluate potential therapies aimed at specific pathologies. Pharmacological models, such as reserpine-treated animals, have provided a better understanding of the role of DA in PD. Through its experimental use, the reserpine model has served to evaluate DA replacement strategies such as levodopa treatment, still routinely used effectively to alleviate parkinsonian symptoms. Neurotoxin models, such as the 6-OHDA and MPTP models, replicate the nigral cell loss seen in PD patients and have been used successfully to evaluate neuroprotective strategies and novel therapies such as cell transplantation. A new wave of neurotoxic models, such as the rotenone and paraquat and maneb models, have been invaluable in the investigation of the role of environmental toxins in the etiology of PD.

While the experimental value of the models cannot be understated, they have so far failed to provide a comprehensive model of the human disease. Aspects of human PD that have been particularly poorly modeled by traditional models are Lewy body formation, pathologies of nondopaminergic systems, and the progressive nature of the disease. Novel genetic models mimic the previously unexamined pathology of protein aggregation in PD, allowing for the investigation of its role in PD development. LPS exposure provides an excellent model to observe the role of neuroinflammation in PD and the progressive loss of dopaminergic cells. Finally, aged animals may represent a truly global model for analyzing the very early stages of PD.

*See also:* Aging of the Brain; Cell Replacement Therapy: Parkinson's Disease; Deep Brain Stimulation and Parkinson's Disease; Parkinsonian Syndromes; Parkinson's Disease: Alpha-Synuclein and Neurodegeneration; Transgenic Models of Neurodegenerative Disease.

## Further Reading

Braak H, Del Tredici K, Rub U, de Vos RA, Jansen Steur EN, and Braak E (2003) Staging of brain pathology related to sporadic Parkinson's disease. *Neurobiology of Aging* 24(2): 197–211.

Carlsson A (1959) The occurrence, distribution and physiological role of catecholamines in the nervous system. *Pharmacology Review* 11: 490–493.

Chu Y and Kordower JH (2007) Age-associated increases of alpha-synuclein in monkeys and humans are associated with nigrostriatal dopamine depletion: Is this the target for Parkinson's disease? *Neurobiology of Disease* 25(1): 134–149.

Fleming SM, Zhu C, Fernagut PO, et al. (2004) Behavioral and immunohistochemical effects of chronic intravenous and subcutaneous infusions of varying doses of rotenone. *Experimental Neurology* 187(2): 418–429.

Herrera AJ, Castano A, Venero JL, Cano J, and Machado A (2000) A single intranigral injection of LPS as a new model for studying the selective effects of inflammatory reactions on the dopaminergic system. *Neurobiology of Disease* 7(4): 429–447.

Jackson-Lewis V, Jakowec M, Burke RE, and Przedborski S (1995) Time course and morphology of dopaminergic neuronal death caused by the neurotoxin 1-methyl-4-phenyl-1,2,3,6-tetrahydorpyridine. *Neurodegeneration* 4(3): 257–269.

Jellinger KA (1991) Pathology of Parkinson's disease: Changes other than the nigrostriatal pathway. *Molecular Chemical Neuropathology* 14: 153–197.

Jeon BS, Jackson-Lewis V, and Burke RE (1995) 6-Hydroxydopamine lesion of the rat substantia nigra: Time course and morphology of cell death. *Neurodegeneration* 4: 131–137.

Kirik D, Annett L, Burger C, Muzycka N, Mandel R, and Bjorklund A (2003) Nigrostriatal α-synucleinopathy induced by viral vector-mediated overexpression of human α-synuclein: A new primate model of Parkinson's disease. *Proceedings of the National Academy of Sciences of the United States of America* 100(5): 2884–2889.

Kirik D, Rosenblad C, and Bjorklund A (1998) Characterization of behavioral and neurodegenerative changes following partial lesions of the nigrostriatal dopamine system induced by intrastriatal 6-hydroxydopamine in the rat. *Experimental Neurology* 152: 259–277.

Langston JW, Ballard P, Tetrud JW, and Irwin I (1983) Chronic parkinsonism in humans due to a product of meperidine-analog synthesis. *Science* 219: 979–980.

Oiwa Y, Eberling JL, Nagy D, Pivirotto P, Emborg ME, and Bankiewicz KS (2003) Overlesioned hemiparkinsonian non human primate model: Correlation between clinical, neurochemical and histochemical changes. *Frontiers in Bioscience* 8: a155–a166.

Olanow CW and Tatton WG (1999) Etiology and pathogenesis of Parkinson's disease. *Annual Review of Neuroscience* 22: 123–144.

Przedborski S, Jackson-Lewis V, Naini AB, et al. (2001) The parkinsonian toxin 1-methyl-4-phenyl-1,2,3,6-tetrahydropyridine (MPTP): A technical review of its utility and safety. *Journal of Neurochemistry* 76: 1265–1274.

Przedborski S, Levivier M, Jiang H, et al. (1995) Dose-dependent lesions of the dopaminergic nigrostriatal pathway induced by intrastriatal injection of 6-hydroxydopamine. *Neuroscience* 67: 631–647.

Przedborski S and Vila M (2003) The 1-methyl-4-phenyl-1,2,3,6-tetrahydropyridine mouse model: A tool to explore the pathogenesis of Parkinson's disease. *Annals of New York Academy of Science* 991: 189–198.

Schultz W (1988) MPTP-induced parkinsonism in monkeys: Mechanism of action, selectivity and pathophysiology. *General Pharmacology* 19: 153–161.

Thiruchelvam M, Brockel BJ, Richfield EK, Baggs RB, and Cory-Slechta DA (2000) Potentiated and preferential effects of combined paraquat and maneb on nigrostriatal dopamine systems: Environmental risk factors for Parkinson's disease. *Brain Research* 873(2): 225–234.

Ungerstedt U (1968) 6-Hydroxydopamine induced degeneration of central monoamine neurons. *European Journal of Pharmacology* 5: 107–110.

van der Putten H, Wiederhold KH, Probst A, et al. (2000) Neuropathology in mice expressing human alpha-synuclein. *Journal of Neuroscience* 20: 6021–6029.

# Animal Models of Huntington's Disease

**J Alberch, E Pérez-Navarro, and J M Canals,**
University of Barcelona, Barcelona, Spain



The generation of animal models is necessary to understand the pathophysiology of neurodegenerative disorders, which will allow the development of new therapeutic strategies. Therefore, it is important to know when and what triggers the signals that induce the neurodegenerative process. The ideal animal model would have reproducible and well-defined behavioral abnormalities and neuropathological features, such as selective loss or dysfunction of specific neuronal populations, all of which must be the closest match to the human pathology. However, neurodegenerative disorders are chronic with different severity and onset, suggesting that there are different factors that modulate to activate the pathogenic mechanisms leading to selective neuronal dysfunction or death. Therefore, different animal models are generated to reproduce different stages of the disease. The generation of animals with a rapid onset and progression will help to determine the earliest molecular and cellular changes associated with the disease, although they can be too aggressive to test new treatments. However, animals with a late onset and milder disease can provide greater specificity and are potentially more useful to test neuroprotective approaches.

Huntington's disease (HD) is a devastating neurodegenerative disorder characterized by irrepressible abnormal movements, cognitive deterioration, and psychiatric disturbances. There is also striking specificity of neuronal loss localized in the striatum (caudate nucleus and putamen), which is the main coordinator of motor activity. The most sensitive cell populations are the striatal medium-sized spiny neurons that project to the globus pallidus and substantia nigra. Moreover, in more advanced stages of the disease, there is also a loss of cortical cells, affecting predominantly the large pyramidal neurons in layers III, V, and VI. Together with the anatomical and neurochemical substrate of HD, the etiology of the neurodegenerative process is well known. The disease is caused by an expansion of a CAG repeat region in the gene encoding huntingtin (IT15) and it is transmitted in an autosomal dominant manner. This mutation results in an increased stretch of glutamines in the N-terminal portion of the protein, which is widely expressed in the brain and peripheral tissues. Asymptomatic individuals have 35 or fewer CAG repeats, with longer expansions causing the illness. There is an inverse relationship between CAG repeat number and the age of onset of symptoms. Longer expansions (>55 repeats) usually give rise to a juvenile form of the disease. However, for medium-length expansions, the age of onset is unpredictable for any individual, which strongly suggests the existence of modifier genes or others factors that influence the age of onset of HD. Among these factors, excitotoxicity, mitochondrial dysfunction, neurotrophic factors, or transcriptional dysregulation have been proposed to modulate the neurodegenerative process.

The mutation has been known for over a decade, but there is still no effective treatment for HD because the pathogenic mechanism is not well understood. Although mutant huntingtin is expressed in all body tissues, the mutation preferentially induces the degeneration of specific neuronal populations in the striatum and cortex. Thus, the development of animal models for HD is required to study the pathogenic mechanisms and to test new treatments.

The first animal models for HD were achieved in the 1980s by either excitotoxic lesioning or metabolic impairment. The intrastriatal injection of glutamate receptor agonists, such as quinolinic acid, induces in rats a selective degeneration of striatal projection neurons, whereas interneurons are relatively spared. Similar results are observed after systemic administration of the mitochondrial toxin 3-nitropropionic acid. These models have been very useful because they produce a similar neuropathological profile as that observed in HD. However, they do not provide any data about the participation of mutant huntingtin in the degenerative process.

Other models have been developed in *Drosophila* and *Caenorhabditis elegans*. It has been shown that flies and worms expressing mutant huntingtin or polyQ peptides alone present a progressive degeneration. These models can be a powerful approach for genetic screens to identify genes that alleviate or modify the disease. However, the mutant protein is not expressed in neuronal types that degenerate in human HD. Hence, cell-specific effects are missing in these models.

However, breakthrough in HD research has been the development of transgenic mice expressing expanded polyglutamine repeats in the huntingtin gene. The first transgenic mice were developed in 1996. These models have been followed by many new HD transgenic and knockin mice that differ with regard to the type of mutation expressed, portion of the protein included in the transgene, promoter employed, and expression levels of mutant protein and background strain, making each of them unique and related to different degrees of the human pathology. These mouse models can fit in three broad categories: (1) mice that express exon-1 fragments of the human huntingtin gene containing polyglutamine expansions; (2) mice that express the full-length human HD gene; and (3) mice with pathogenic CAG repeats inserted into the endogenous CAG expansion (knockin mice) (**Figure 1**).

## N-Terminal Exon-1 Transgenic Mouse Models

R6/1 and R6/2 transgenic mice were the first mouse models developed to study HD. They both express exon 1 of the human HD gene with 115 and 150 CAG repeats, respectively. The transgene expression in these mice is driven by the human huntingtin promoter. The resulting levels of transgene expression are around 31% and 75% of the endogenous huntingtin

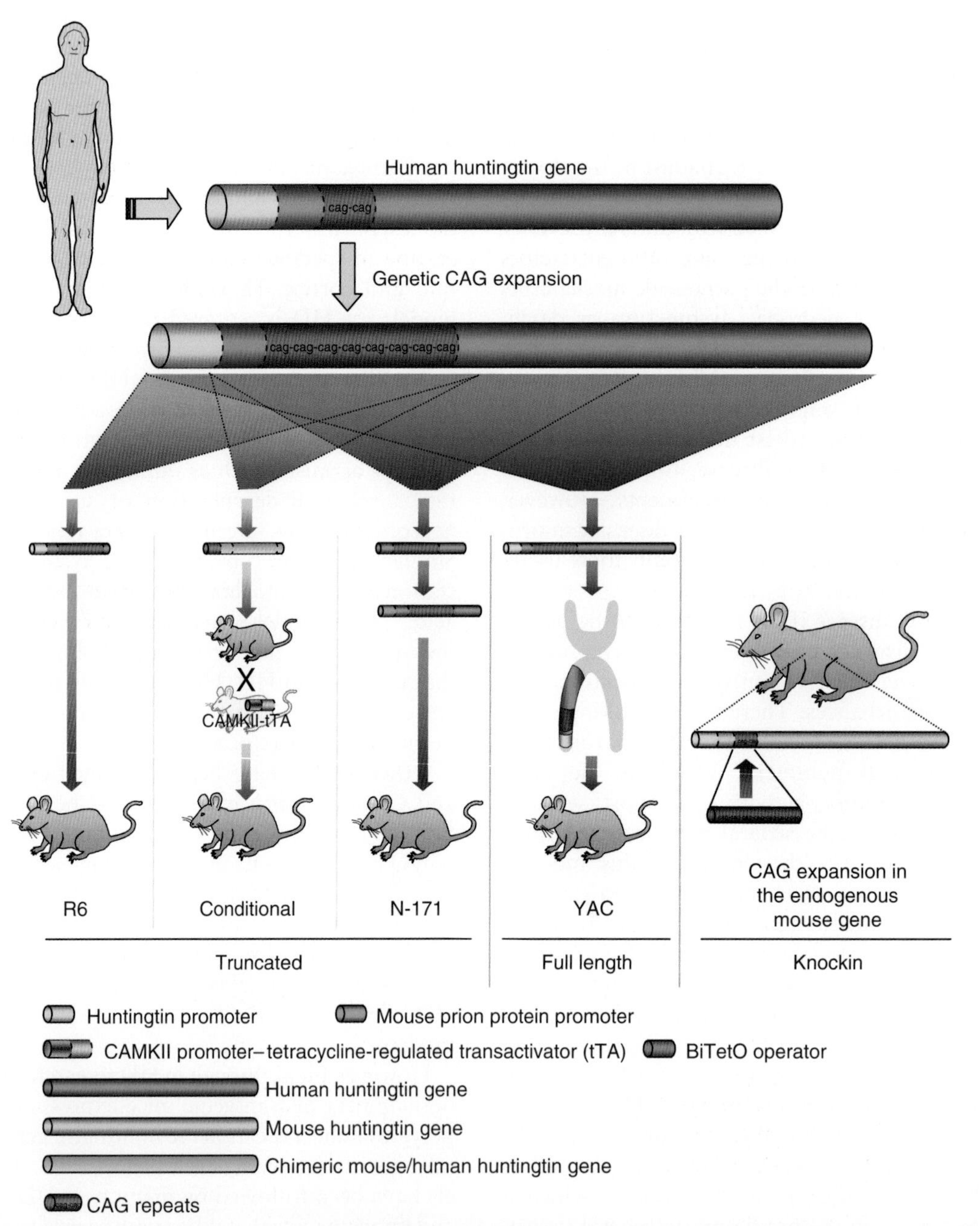


**Figure 1** Schematic representation of the human huntingtin gene and constructs used for the generation of mouse models of Huntington's disease. The CAG repeats are localized in the exon 1 of the huntingtin gene, and the normal range is between 19 and 35. In transgenic mouse models the huntingtin gene (A fragment: R6, Conditional, N-171 or full length: yeast artificial chromosome (YAC)) with an expanded CAG repeat is randomly inserted into the genome. Knockin mouse models are obtained by gene targeting of the endogenous mouse huntingtin gene.

in the R6/1 and R6/2, respectively. In accordance with human pathology, R6 mice with the longest CAG repeats (R6/2) have an earlier onset and more severe symptoms like a juvenile form of HD. R6/2 mice are severely impaired by 8–12 weeks of age, whereas R6/1 mice are around 15–20 weeks. A typical sign in these models is the abnormal paw-clasping response. When suspended by the tail, normal mice spread their four limbs, whereas R6 mice clasp their hind- and forelimbs tightly against their thorax and abdomen (**Figure 2**). Although the pathophysiology of this behavior is not understood, the paw-clasping test is often used in studies examining novel treatments. Other changes in motor function are stereotypical hindlimb grooming, changes in gait patterns, and the emergence of some involuntary movements. As a result, their motor coordination progressively deteriorates, which can be detected as a reduction in the time they can stay on a rotating rod (rotarod).

Neuropathological analysis shows that brain and striatal volume are markedly reduced in R6 mice. However, the significant reduction in brain volume is the result of atrophy of individual neurons with a massive decreased neuropil, because neuronal death is minimal in these mice. Comparison of the time courses of cell death and behavioral anomalies shows that changes in motor activity precede any evidence of neuronal death in these mice, by a matter of several weeks. However, cell death can be observed if the neurotrophin brain-derived neurotrophic factor (BDNF) is reduced in R6/1 mice. The generation and characterization of a double-mutant mouse obtained by crossing R6/1 mice and BDNF heterozygote mice demonstrate that the levels of endogenous BDNF together with mutant huntingtin modulate the onset and severity of motor dysfunction and the degeneration of striatopallidal neurons. Thus, the different levels of endogenous BDNF in various mouse models of HD may also regulate the severity of the phenotype.

The most prominent morphological feature of HD, in addition to the reduction in brain volume, is the development of neuronal intranuclear and intracytoplasmatic inclusions. These are aggregations formed by polyglutamine repeats in the N-terminal huntingtin fragment. They also contain other components such as ubiquitin, and are located within neuronal cell bodies, the nucleus, and their arbors. These inclusions were first observed in animal models with mutant huntingtin and later confirmed in postmortem brains of patients with HD. In contrast to the lack of cell death in early stages in R6 mice, the presence of huntingtin- and ubiquitin-positive aggregates in R6/2 mice is observed as early as postnatal day 1 in the striatum and other brain areas, with the greatest number at 90 days. Thus, there is a close correlation between the degree of motor impairment and the number of striatal neurons exhibiting intranuclear inclusions of mutant huntingtin in this model. However, the function of these aggregates is under discussion, because there is evidence suggesting that they can be either toxic or neuroprotective.

As observed in HD patients, R6 mice also show cognitive deficits in advance of classical motor symptoms. R6/2 mice between 3.5 and 8 weeks of age

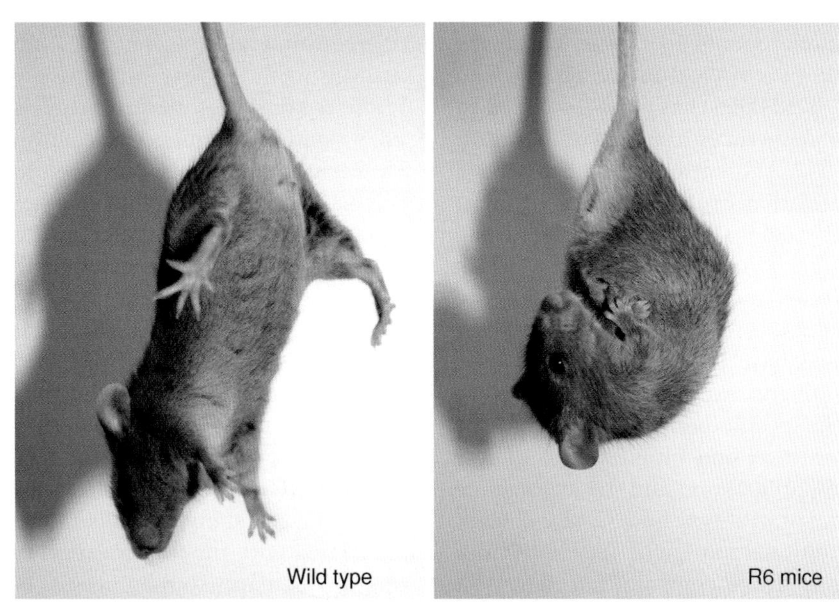


**Figure 2** Clasping phenotype of R6 mice. Photomicrographs showing 30-week-old wild-type and R6 mice during the tail suspension test. R6 mice display a characteristic full body clasp.

display progressive learning impairment on cognitive tasks sensitive to frontostriatal and hippocampal function, as shown by the Morris water maze, two-choice swim tank, and T-maze tests.

Although the R6 mice are the most extensively studied, other transgenic mouse lines expressing the truncated huntingtin gene with expanded CAG repeats have been generated. N-171 mice express a cDNA encoding an N-terminal fragment (171 amino acids) of human huntingtin with 82, 44, or 18 glutamines. The expression of the transgene is directed by the mouse prion protein promoter, which drives the expression of foreign genes in virtually every neuron of the central nervous system. Mice expressing relatively low steady-state levels of N-171 huntingtin with 82 glutamine repeats (N-171-82Q) develop behavioral abnormalities, including loss of coordination, tremors, hypokinesis, hind limb clasping, and abnormal gait by 3 or 4 months of age, before dying prematurely (4–6 months). In mice exhibiting these abnormalities, diffuse nuclear labeling to the N-terminus of huntingtin, intranuclear inclusions, and neuritic aggregates were found in cortex, striatum, hippocampus, cerebellum, and amygdala as early as 4 weeks of age.

Other lines have also been generated expressing the N-terminal one-third of huntingtin with expanded (HD46, HD100) glutamine repeats under the regulation of the rat neuron-specific enolase promoter. These HD mice exhibit motor deficits at 3 months of age with an exacerbation throughout their lifespan. Neuropathological features, such as accumulation of huntingtin, dysmorphic dendrites, and atrophy in the striatum and cortex, precede the onset of behavioral deficits.

An interesting model would be one in which the mutated gene could be switched off or on. Thus, a conditional model (Tet/HD94) has been developed by the expression of a chimeric mouse/human exon 1 of the HD gene with 94 CAG repeats, under the control of a BiTetO operator that can be switched off after adding tetracycline analogs to the drinking water. The tetracycline-regulated transactivator (tTA), which constitutively activates the BiTetO in the absence of tetracycline analogs, is under the control of the calcium/calmodulin kinase II$\alpha$ (CAMKII) promoter. Hence, mutant huntingtin expression is restricted to neurons localized in specific brain areas such as the striatum, septum, cortex, and hippocampus. These animals have a slow progression and the average life span is similar to that of wild-type mice (2 years). Until 4 weeks of age, conditional transgenic Tet/HD94 mice are indistinguishable from control littermates. At 4 weeks, some mice begin to clasp, and by 8 weeks all mice are clasping. In addition to this abnormal behavior, at 20 weeks Tet/HD94 mice begin to show a mild tremor, and at 36 weeks they are hypoactive. Neuropathological analysis shows that the brain and striatal volume at 18 weeks of age is smaller than in wild-type mice. This is accompanied by a gross enlargement of the ventricles, although cell death is not observed. The long survival of these animals has allowed the study of aged animals, revealing that at 17 months of age neuronal death is present in the striatum. Interestingly, shutting off expression of the mutant transgene produces significant changes in the neuropathological phenotype of these Tet/HD94 mice. When the gene is turned off, mutant huntingtin is rapidly cleared from the mouse brain and the motor symptoms are reversed. In aged animals, gene silencing can still be beneficial and full motor recovery is possible despite neuronal loss. This model provides evidence that treatments blocking mutant huntingtin expression in the early or even advanced stages of HD might enable the reversal of the clinical symptoms (**Table 1**).

## Full-Length Transgenic Mouse Models

Several mouse models have been generated using a full-length huntingtin gene as the transgene. Mice that express a full-length huntingtin cDNA clone with either 48 (HD48) or 89 (HD89) repeats driven by the cytomegalovirus (CMV) promoter show a progressive motor phenotype. Feet clasping following suspension by their tails is observed in all HD48 and HD89 mice at 8 weeks of age. At 24 weeks of age, decreased motor activity is observed which progresses to locomotor deterioration and akinesia. Importantly, in the hypokinetic stage, neuronal loss, gliosis, and degeneration also occur. Although the number of neurons with nuclear inclusions varies in different brain regions, less than 1% of striatal neurons of HD48 and HD89 mice show intranuclear inclusions. However, the presence of polyglutamine aggregation is seen as early as 12 weeks of age.

Similar features have been observed in other mouse models obtained by the expression of yeast artificial chromosome (YAC) that spans the entire human HD gene with 46 (YAC46), 72 (YAC72), or 128 (YAC128) CAG repeats. YAC46 and YAC72 mice show electrophysiological abnormalities prior to the presence of nuclear inclusions or neurodegeneration indicating a cytoplasmic dysfunction. By 12 months of age, YAC72 mice have a selective degeneration of medium-sized spiny neurons in the striatum. This is associated with the translocation of N-terminal huntingtin fragments to the nucleus. Neurodegeneration can be present in the absence of macro- or micro-aggregates, showing that aggregates are not essential

**Table 1** Mouse models of Huntington's disease

| | *Model design* | | | *Abnormal motor behavior (onset)* | *Neuropathology* | | *Life span* |
|---|---|---|---|---|---|---|---|
| | *Promoter* | *Gene size* | *CAG repeats* | | *Aggregates/inclusions* | *Cell loss* | |
| *Transgenic models* | | | | | | | |
| R6<br>Mangiarini et al. (1996) | IT15 | Exon 1 | 115 (R6/1) | Clasping (15–21 w)<br>Decline in rotarod performance (13–20 w) | Observed in cortex and striatum from 8 w | Dark cell degeneration in cortex and striatum<br>Reduction of medium-sized spiny neurons in the striatum | 32–40 w |
| | | | 144 (R6/2) | Clasping (8 w)<br>Locomotor hyperactivity (3 w)<br>Hypoactivity (8 w)<br>Decline in rotarod performance (8–12 w) | Present in cortex and striatum from 3–4 w | Dark cell degeneration in cortex and striatum<br>Reduction of medium-sized spiny neurons in the striatum | 13–16 w |
| HD<br>Laforet et al. (2001) | Rat neuron specific enolase | 3-kb N-terminal fragment | 48 | Clasping (3–6 m)<br>Impaired performance in rotarod (3–6 m) | Few intranuclear inclusions in cortex and striatum | Reduced number of neurons in the striatum in some animals | >12 m |
| | | | 100 | Clasping (3–6 m)<br>Impaired performance in rotarod (3–6 m) | Intranuclear inclusions in cortex and striatum | Reduced number of neurons in the striatum in some animals | |
| N-171<br>Schilling et al. (1999) | Mouse prion protein | First 171 amino acids from exon 1 | 44 | Normal up to 2 years of age | No | No | Normal |
| | | | 82 | Clasping (not reported)<br>Impaired performance in rotarod (3 m)<br>Abnormal gait (3–4 m)<br>Hypoactivity (5 m) | Cytoplasmatic aggregates at 6 m<br>Nuclear inclusions in the cortex, striatum, hippocampus, cerebellar granular cells, and amygdala at the end stage | Not severe loss of neurons in cortex, striatum, hippocampus, and cerebellum | 10–50 w |
| YAC46<br>YAC72<br>Hodgson et al. (1999) | IT15 | Full length | 46<br>72 | Normal up to 20 m<br>Clasping (nondescribed)<br>Hyperactivity (7 m)<br>Circling (9 m) | No<br>Cytoplasmic microaggregates in striatal neurons<br>Nuclear inclusions: high number of neurons in the striatum, olfactory tubercle, and nucleus accumbens; small number of neurons in the septum and granule cell layer of the cerebellum | No<br>Apoptotic cells in the striatum (12 m) | >12 m<br>>12 m |
| YAC128<br>Slow et al. (2003) | | | 128 | Hyperkinesia (3 m)<br>Motor deficit on the rotarod (6 m)<br>Hypokinesia (12 m) | Inclusions in striatal and cortical neurons | Reduced number of striatal neurons | >12 m |
| HD94-tet off<br>Yamamoto et al. (2000)<br>Diaz-Hernandez et al. (2005) | CAMKII$\alpha$-tTA | Exon 1 | 94 | Clasping (4 w)<br>Mild tremor (20 w)<br>Hypoactivity (36 w)<br>Progressive motor decline (10 m) | Nuclear inclusions in cortex and striatum<br>Extranuclear aggregates in cortex and with higher occurrence in the striatum | Dark cell degenerating neurons in the striatum (14 m)<br>Decreased number of striatal neurons (17 m) | Normal |

Continued

**Table 1** Continued

| | *Model design* | | | *Abnormal motor behavior (onset)* | *Neuropathology* | | *Life span* |
|---|---|---|---|---|---|---|---|
| | *Promoter* | *Gene size* | *CAG repeats* | | *Aggregates/inclusions* | *Cell loss* | |
| *Knockin models* | | | | | | | |
| HdhQ92 Wheeler et al. (2000) | Hdh | | 90 | No clasping (up to 17 m) | Nuclear inclusions in striatum, olfactory tubercle, and cortex<br>In older mice, nuclear aggregates are also observed in septum, olfactory bulb, nucleus accumbens, cerebellar granule cell layer, and hippocampus | No | Normal |
| Hdh94 Menalled et al. (2002a) | Hdh | | 94 | No clasping (up to 24 m)<br>Increased rearing (2 m)<br>Decreased locomotion (4 m) | Nuclear microaggregates in the striatum<br>Nuclear inclusions in the striatum (old mice)<br>Other brain areas nonexamined | No | Normal |
| HdhQ111 Wheeler et al. (2000) | Hdh | | 109 | Subtle gait deficits (24 m) | Nuclear inclusions in striatum, olfactory tubercle, and cortex<br>In older mice nuclear aggregates are also observed in septum, olfactory bulb, nucleus accumbens, cerebellar granule cell layer, and hippocampus | No | Normal |
| HdhQ140 Menalled et al. (2003) | Hdh | | 140 | Hyperactivity (1 m)<br>Decreased locomotor activity (4 m)<br>Gait anomalies (12 m) | Nuclear microaggregates in striatum, olfactory tubercle, cortex, and olfactory bulb<br>Nuclear inclusions in striatum, olfactory tubercle, cerebellum, and layer II of piriform cortex<br>Neuropil aggregates in striatum, globus pallidus, substantia nigra pars reticulata, cortex, cerebellum, olfactory tubercle, and olfactory bulb | Not reported | Normal |
| Hdh150 Lin et al. (2001) | Hdh | | 150 | Gait and rotarod deficits (15–40 w)<br>Clasping (15–40 w)<br>Hypokinesia (40 w) | Nuclear inclusions in the striatum and nucleus accumbens; less frequent in cortex, hippocampus, and cerebellum; sparse in the thalamus, hypothalamus, and brain stem | No loss of striatal neurons | Normal |

w, weeks; m, months.

to the initiation of neuronal death. YAC128 mice reveal a hyperkinetic phenotype first manifested at 3 months of age, followed by a progressive motor deficit on the rotarod at 6 months with eventual progression to hypokinesis by 12 months of age. These behavioral changes are followed by striatal atrophy clearly evident by 9 months, cortical atrophy at 12 months, and a progressive loss of striatal neurons accompanied by a decrease in their soma area. In contrast to N-terminal exon-1 transgenic models, the motor deficit in the YAC128 mice is highly correlated with neuronal loss. These mice demonstrate that initial neuronal cytoplasmic toxicity is followed by cleavage of huntingtin, nuclear translocation of huntingtin N-terminal fragments, and selective neurodegeneration.

Mild cognitive deficits have also been described in YAC128 mice that precede motor onset and progressively worsen with age. Rotarod testing reveals a motor learning deficit at 2 months, which progresses until 12 months of age, when YAC128 mice are unable to learn the rotarod task. These mice also show deficits in procedural learning, memory, sensorimotor gating, and strategy shifting.

## Knockin Models

Knockin mouse models have been generated by homologous recombination of CAG repeats in the endogenous mouse huntingtin gene. Therefore, knockin mice carry the mutation in its appropriate genomic and protein context, making them more faithful genetic models of the human condition than other transgenic lines, which have the transgene inserted randomly. Several knockin mice have been developed with different CAG expansions and they exhibit similar neuropathological phenotype, although at different ages. All these mice have a late onset and slow progression. However, there are differences in the magnitude of motor abnormalities in knockin and transgenic mice, but both lines show a shift from hyper- to hypoactivity. Thus, for example, knockin mice with 94 repeats display a biphasic motor behavior characterized by increased motor activity at 2 months, followed by hypoactivity at 4 months of age. Gait abnormalities are observed at 24 months. HdhQ111 mice do not differ from wild-type animals in paw clasping and rotarod at 15 months of age, but at 24 months motor function assessed by tunnel walks reveals a subtle gait deficit. In animals with longer CAG repeats, similar patterns of motor abnormalities are observed although at early ages. Gait abnormalities are already observed at 12 months of age. None of the knockin mice develop neuronal loss, although reactive gliosis is observed in the striatum. Nuclear staining and aggregates of the mutant huntingtin are only observed in various knockin models when the mice are old. In mice with 94 repeats, intranuclear inclusions are restricted to the striatum, whereas with longer expansions (140 and 150 CAG repeats) the distribution is more widespread.

## Conclusion

In the previous years, several models have been developed showing different phenotypes and timing depending on the genetic construct, length of CAG repeats, and background strain. They all produce aspects of the pathology observed in HD patients, such as movement abnormalities, specific atrophy of the striatum and cortex, and the presence of intranuclear and cytoplasmatic aggregates of mutant huntingtin. Hence, the availability of these mouse models has been an important step in research aimed at dissecting the pathogenesis of HD. Each model can be useful for different types of studies. Knockin models accurately replicate the underlying genetic defect of HD, but they have a slow progression with no robust motor deficits, and brain atrophy is only observed in aged animals. The transgenic mice expressing full-length mutant huntingtin have a less aggressive course of the disease than mice that express a truncated form. The varying degrees of striatal atrophy, brain weight loss, and motor deficit in these animal models allow the study of different stages and intensities (juvenile or adult) of the pathology. Therefore, the different animal models of HD are powerful tools to understand the pathogenic mechanisms and for the assessment of neuroprotective and other therapeutic interventions.

*See also:* Axonal Transport and Huntington's Disease; Huntington's Disease: Neurodegeneration; Transgenic Models of Neurodegenerative Disease; Triplicate Repeats: Huntington's disease.

## Further Reading

Beal MF and Ferrante RJ (2004) Experimental therapeutics in transgenic mouse models of Huntington's disease. *Nature Reviews Neuroscience* 5: 373–384.

Canals JM, Pineda JR, Torres-Peraza JF, et al. (2004) Brain-derived neurotrophic factor regulates the onset and severity of motor dysfunction associated with enkephalinergic neuronal degeneration in Huntington's disease. *Journal of Neuroscience* 24: 7727–7739.

Diaz-Hernandez M, Torres-Peraza J, Salvatori-Abarca A, et al. (2005) Full motor recovery despite striatal neuron loss and formation of irreversible amyloid-like inclusions in a conditional mouse model of Huntington's disease. *Journal of Neuroscience* 25: 9773–9781.

Hodgson JG, Agopyan N, Gutekunst CA, et al. (1999) A YAC mouse model for Huntington's disease with full-length mutant huntingtin, cytoplasmic toxicity, and selective striatal neurodegeneration. *Neuron* 23: 181–192.
Laforet GA, Sapp E, Chase K, et al. (2001) Changes in cortical and striatal neurons predict behavioral and electrophysiological abnormalities in a transgenic murine model of Huntington's disease. *Journal of Neuroscience* 21: 9112–9123.
Lin CH, Tallaksen-Greene S, Chien WM, et al. (2001) Neurological abnormalities in a knockin mouse model of Huntington's disease. *Human Molecular Genetics* 10: 137–144.
Mangiarini L, Sathasivam K, Seller M, et al. (1996) Exon 1 of the HD gene with an expanded CAG repeat is sufficient to cause a progressive neurological phenotype in transgenic mice. *Cell* 87: 493–506.
Menalled LB and Chesselet MF (2002) Mouse models of Huntington's disease. *Trends in Pharmacological Science* 23: 32–39.
Menalled LB, Sison JD, Dragatsis I, Zeitlin S, and Chesselet MF (2003) Time course of early motor and neuropathological anomalies in a knockin mouse model of Huntington's disease with 140 CAG repeats. *Journal of Comparative Neurology* 465: 11–26.
Menalled LB, Sison JD, Wu Y, et al. (2002) Early motor dysfunction and striosomal distribution of huntingtin microaggregates in Huntington's disease knockin mice. *Journal of Neuroscience* 22: 8266–8276.
Rubinsztein DC (2002) Lessons from animal models of Huntington's disease. *Trends in Genetics* 18: 202–209.
Schilling G, Becher MW, Sharp AH, et al. (1999) Intranuclear inclusions and neuritic aggregates in transgenic mice expressing a mutant N-terminal fragment of huntingtin. *Human Molecular Genetics* 8: 397–407.
Slow EJ, van Raamsdonk J, Rogers D, et al. (2003) Selective striatal neuronal loss in a YAC128 mouse model of Huntington disease. *Human Molecular Genetics* 12: 1555–1567.
Wheeler VC, White JK, Gutekunst CA, et al. (2000) Long glutamine tracts cause nuclear localization of a novel form of huntingtin in medium spiny striatal neurons in HdhQ92 and HdhQ111 knockin mice. *Human Molecular Genetics* 9: 503–513.
Yamamoto A, Lucas JJ, and Hen R (2000) Reversal of neuropathology and motor dysfunction in a conditional model of Huntington's disease. *Cell* 101: 57–66.

## Relevant Website

http://www.hdfoundation.org – The Hereditary Disease Foundation.

# Animal Models of Motor and Sensory Neuron Disease

**P Bomont and D W Cleveland**, University of California at San Diego, La Jolla, CA, USA



## Motor and Sensory Tracts

Movements of many parts of the body, from the lips and the eyelids to the hands and toes, have their origin in the brain in a specialized region called the motor cortex (**Figure 1**). Upper motor neurons receive information there that they transmit to spinal motor neurons, often through an intermediate interneuron. Firing of the lower motor neuron triggers muscle contraction. Upper motor neuron degeneration induces brisk tendon reflexes, spasticity, and hyperreflexia, whereas denervation of the muscle upon lower motor neuron degeneration or loss leads to muscle weakness and loss of tone.

The sensory system works in the opposite direction with information flowing to the brain from the periphery. Receptors, primarily located at the surface of the skin, provide initial sensory input (touch, pressure, vibration perception, pain, and temperature) to the axons of the sensory neurons which carry these signals into the spinal cord. The information is then transferred to the neurons located in the sensory cortex, where the information is decoded by the brain.

## Gene Deletion ('Knockout') and Transgenic Mouse Models

Two main principles are used to reproduce inherited human pathologies in the mouse (**Figure 2**). When the human disease is recessive, a situation usually caused by the absence of one particular protein, a genetic mimic in mice can be constructed by disrupting the corresponding gene in the mouse genome (**Figure 2(a)**). These models are frequently called 'knockout models,' referring to the disruption or removal of the targeted gene and its encoded protein. A variation of this method, called 'conditional knockout,' permits the removal of the gene of interest only in chosen tissues (**Figure 2**, box 1). This alternative is preferred in cases in which the deletion of the gene from the entire animal induces lethality in early development.

For dominant neurodegenerative disorders, disease is usually caused by the gain of a toxic property resulting from a modification in the protein (an example of this is amyotrophic lateral sclerosis (ALS) caused by mutation in the gene encoding superoxide dismutase). Two approaches are possible for generating a mouse model. The mouse gene can be replaced by the mutated one (**Figure 2(a)**), thereby creating a gene 'knock in' model. Alternatively, random insertion of the abnormal gene in the mouse genome can be achieved to produce a transgenic mouse (**Figure 2(b)**). The latter is largely preferred because it can be constructed more quickly. Importantly, transgene insertion can be readily achieved in mice or rats. An advantage to rat models in neuroscience is that the larger size permits more complex surgical approaches and tissue dissections. Additionally, for some classes of neurons, those derived from rats are more easily grown in culture. Finally, rats are amenable to some behavioral analyses that are more difficult in the mouse.

## Neurodegeneration in Human

### Motor Disease

**Amyotrophic lateral sclerosis** ALS, more familiarly known in the United States as Lou Gehrig's disease, is the most prominent adult motor neuron disease, with a typical age of disease onset of 50–60 years. Most (90%) incidences of ALS are referred to as sporadic because there is no family history of disease and hence no evidence for a major genetic component. Disease involves the selective alteration and death of both the upper and the lower motor neurons, resulting in spasticity, hyperreflexia, and progressive weakness of skeletal muscles, atrophy, and death due to respiratory muscle paralysis within 1–5 years after onset.

For the incidences (~10%) of ALS with a genetic origin, determination of the gene(s) responsible has been significantly challenged by the wide heterogeneity of the disease, genetically and clinically. Many genetic loci have been associated with different forms of motor neuron degeneration. Most prominently, mutations in the gene encoding for Cu/Zn superoxide dismutase (SOD1), an enzyme used to detoxify an aberrant oxygen species, have been shown to account for 20% of the familial cases of ALS. Dominantly inherited, analysis of enzymatic activity of mutated SOD1 in ALS patients together with genetic manipulations in mice have clearly revealed that SOD1-mediated toxicity is not due to a reduction in activity of this enzyme but, rather, due to a gain of a new toxic property or properties. Accordingly, transgenic rodent models have been generated by inserting into their genome a human SOD1 gene carrying different ALS-causing mutations. Many models have been created to dissect the disease-associated toxic mechanism(s), the most extensively used being transgenic for the

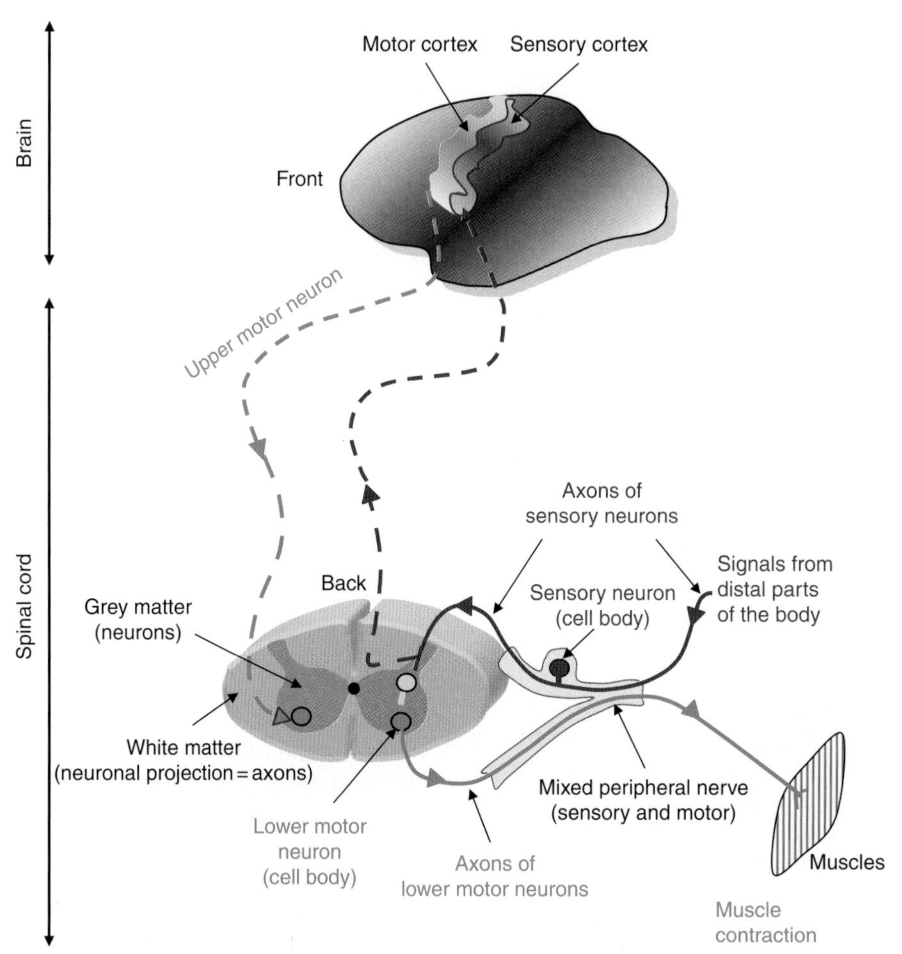


**Figure 1** Motor and sensory pathways. Sensory neurons receive information from different parts of the body and transmit it to the spinal cord and then to the brain through the ascending axons of sensory neurons (dark blue). The motor pathway originates in the motor cortex, where upper motor neurons (light blue) connect to transmit information to the lower motor neurons in the spinal cord, which in turn send and deliver signals to the muscles, inducing their contraction.

amino acid substitution mutations G93A (glycine substituted to alanine at position 93), G37R (glycine to arginine at position 37), and G85R (glycine to arginine at position 85). Although none of these completely recapitulate all features of the human disease, they all develop progressive motor neuron degeneration with slightly different symptoms and disease progressions that are characteristic for each mutation, with onset and survival determined by the level of mutant expression. All of these mice develop motor neuron degeneration, limb weakness associated with neurofilament misaccumulation, impaired axonal transport, and axonal swelling.

Mouse and rat models for ALS have been extensively used to test proposals for a wide range of pathological mechanisms. These include impairment of the chaperone/degradation machinery, dysfunction of mitochondria, oxidative stress, alteration of cytoskeletal architecture and axonal transport, glutamate-mediated excitotoxicity, aberrant growth factor signaling, microglial cell involvement, and inflammation. Another aspect particularly important to the identification of therapeutic targets is whether other cell types contribute to motor neuron degeneration. Motor neurons require support provided by several cell types in the central nervous system which also express the mutant SOD1. Use of a transgene that can be eliminated selectively from motor neurons has shown that mutant damage in motor neurons initiates disease onset and an early phase of disease progression. A similar approach has proven that mutant damage within microglial cells, the immune cells of the central nervous system, has no effect on disease onset but accelerates disease progression.

Additional forms of human ALS-like motor neuron disease, all associated with motor neuron degeneration and death, have been designated ALS2–ALS8. Significant divergence in the age of onset and clinical presentation has led to disagreement as to whether these should be classified as forms of ALS. This is

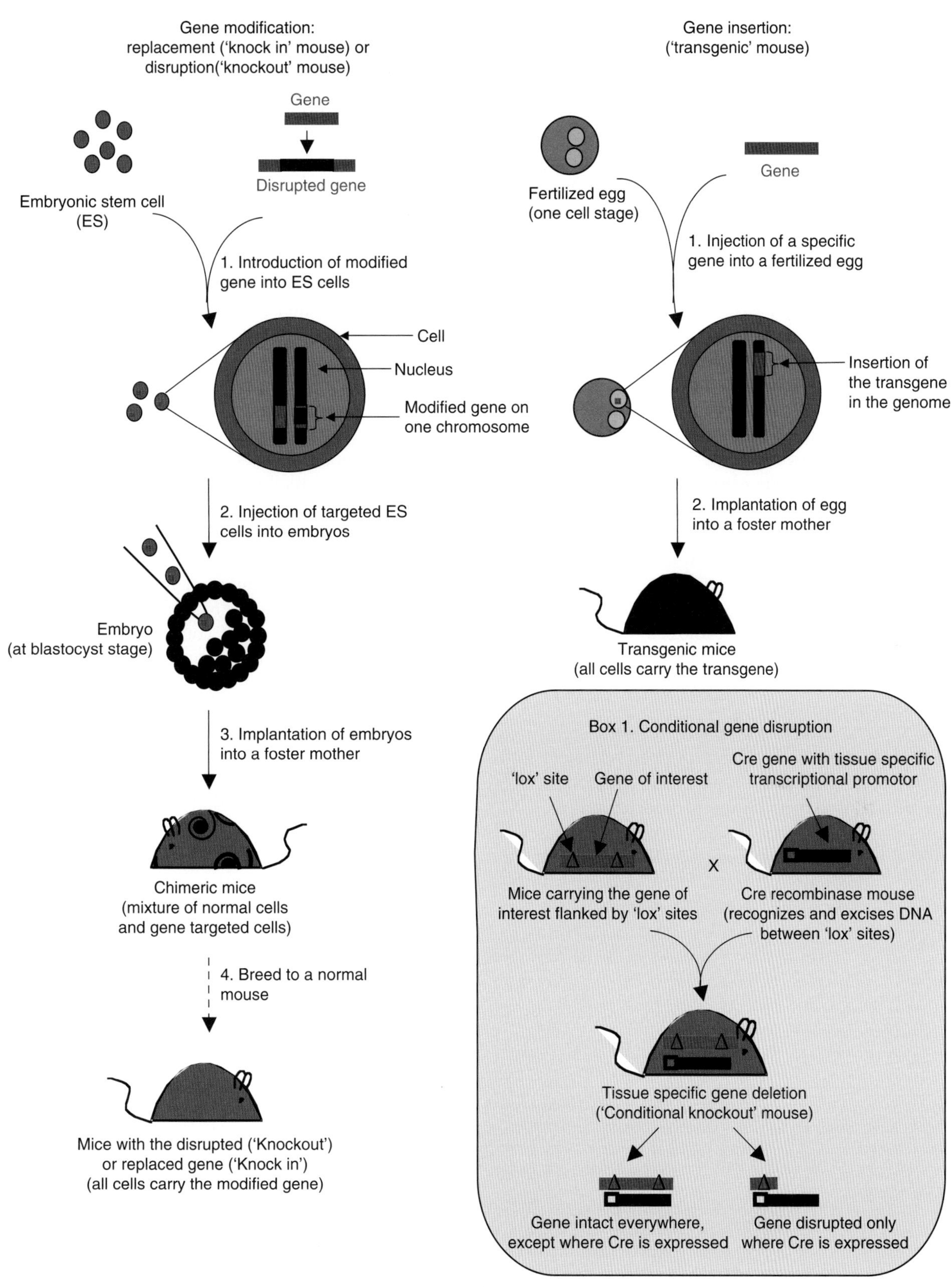


**Figure 2** Continued

especially so for ALS2, an infantile and juvenile-onset form of motor neuron disease which is caused by loss of an apparent guanine exchange factor enzyme, termed alsin, that can activate one or more enzymes from the family of small GTPases known as G-proteins. Proposed G-protein partners of alsin include Rab5 and Rac1. Alsin has been shown to modulate processes such as cytoskeleton organization and transport of membrane cargoes.

Recessive AL2 motor neuron disease has been reproduced in mice by deleting the corresponding gene (**Figure 2(a)**). This first revealed that the absence of alsin in the mouse does not produce a juvenile disease that has as severe a phenotype as found in human. The lack of perfect concordance between the human and mouse genetic defect has been proven to be true for many other diseases. Importantly, lower motor neurons (those that directly trigger contraction of skeletal muscles) are almost completely spared in mice deleted of the ALS2 gene, which do not develop muscle weakness. Analysis of voluntary running, however, revealed that these mice move slowly, a well-recognized clinical sign of an upper motor neuron defect. Additionally, the upper motor neurons degenerate. Thus, the animal model reinforces that the pathology and the phenotype of loss of ALS2 provoke an upper motor neuron disease quite distinct from ALS, resembling instead hereditary spastic paralysis.

ALS4 is a dominantly inherited juvenile form of ALS characterized by distal muscle weakness and atrophy before 25 years of age. Rare in humans, this disease is caused by mutation in the senataxin (SETX) gene. The function of the senataxin protein is not known, but its sequence suggests DNA/RNA helicase activity, potentially acting in DNA repair pathways and RNA processing. Senataxin mutations are also causative of a recessive disease termed AOA2 (ataxia–oculomotor apraxia 2); symptoms include balance difficulties and defects in coordination of movements (ataxia) associated with an alteration of eye movement (oculomotor apraxia).

Axonal transport has been shown to play an essential role in neuronal survival. Two families of motor proteins, the kinesins and dynein, which power transport of cargoes along the microtubules away from the nerve cell body (anterograde) or returning components from the nerve ending to the cell body (retrograde), respectively, have been shown to cause or contribute to motor neuron disease in humans. This has been replicated in mice. In humans, mutation in the subunit p150 of the dynactin complex (which interacts with the dynein complex) has been identified to be causative of an unusual form of lower motor neuron disease in early adulthood, characterized by vocal fold paralysis, progressive weakness, and muscle atrophy.

Although a mouse model of the p150 dynactin mutation has not been reported, it is widely established that alteration of the dynein–dynactin complex causes motor deficits. Indeed, overexpression of dynamitin (also called p50 of the dynactin complex) has been shown in mice to disrupt dynein–dynactin interaction leading to late-onset slowly progressive motor neuron degeneration, characterized by muscle weakness, trembling, abnormal gait, and deficits in strength and endurance. Progressive alterations of muscle function and motor coordination have been linked in mice to distinct mutations in the dynein gene. The corresponding mutants, *Loa* (Legs at odd angles) and *Cra1* (cramping 1), develop motor deficits, confirming that alteration in the dynein motor complex and its presumed effects on moving components through axons are of unusual importance for motor axons. A completely unexpected and unexplained finding is that both dynein mutations delay disease progression and increase the life span of mice that develop early onset motor neuron disease caused by expression of an ALS-causing SOD1 mutant. Understanding what is causing the motor deficits in the *Loa* and *Cra1* mice will certainly help us to understand what leads to this amelioration of disease in SOD1 mutant mice and might help to identify new therapeutic approaches for ALS.

**Figure 2** Mouse models for human neurodegenerative disease. Strategies for gene modification or replacement. (a) Gene modification. In this approach, the gene of interest is modified *in vitro* and then introduced into (brown) embryonic stem (ES) cells. Following recombination in the ES cells, the modified gene replaces the endogenous gene on one of the two chromosomes. The corresponding targeted ES cells are then injected into mouse embryos at the blastocyst stage and are implanted into a foster mother. Those females produce progeny mice that are mixtures of normal cells and modified cells. Mice carrying the modified gene in all cells (also called 'knockout mice' in the case of a gene disruption and 'knock in' when the gene is replaced) are obtained by standard breeding with normal mice. (Box 1) To selectively disrupt a gene in a specific tissue, the gene region to be deleted is flanked by two lox sites, which are small sequences recognized and cut by an enzyme called Cre recombinase. Breeding mice carrying the modified lox–gene–lox with mice expressing the Cre recombinase under a tissue-specific transcriptional promoter generates the gene disruption only in the tissues that express the Cre recombinase. (b) Gene insertion: 'transgenic' mouse. An exogenous gene is inserted randomly into the mouse genome after direct injection of the gene into the one cell stage fertilized egg. After implantation into a foster mother, the oocyte matures into mice whose cells all carry the transgene.

**Spinal muscular atrophy** After cystic fibrosis, spinal muscular atrophy (SMA) is the most common autosomal recessive disorder in humans and represents the most common genetic cause of infant mortality. In this disorder, specific neuronal loss of lower motor neurons in the spinal cord results in atrophy of proximal muscles of the trunk and the limbs. SMA cases are classified into three groups: Type I SMA is the most severe phenotype, with an age of onset before 6 months and death occurring before 2 years of age; type II SMA is intermediate, with an age of onset between 6 and 18 months; and type III SMA, which initiates after 18 months, is the mildest form, slowly progressing with a normal life span.

Identifying the genetic cause of SMA was challenging because of the complex and unstable nature of the region of human chromosome 5 where the defective gene resides. For all disease types, symptoms are caused by inactivating mutations in the *SMN1* (survival of motor neuron 1) gene, but the variability in the disease severity depends on the number of copies (ranging from one to five) of an almost identical copy gene, the *SMN2*, located just next to *SMN1*. Each copy of the *SMN2* gene produces only a small amount of functional product so that only a single copy leads to fatal, severe type I disease, and five copies lead to the more benign type III disease. Although the role of SMN protein in lower motor neuron survival is not fully understood, it has been implicated in processing within the nucleus of initial RNA copies of each expressed gene into their mature forms capable of translation into the corresponding proteins. A second likely function is as an RNA-binding protein for RNA transport into axons.

Because the mouse genome contains only the *SMN1* gene and no *SMN2* gene, mouse models for SMA have been attempted by deleting the *SMN1* locus. These animals are not viable, demonstrating an essential role for SMN1 in cell survival. Selective deletion of the *SMN1* gene solely from neurons (**Figure 2**, box 1) produced mice with motor abnormalities and skeletal muscle denervation secondary to motor axon loss, culminating in death at a mean age of 25 days. This is not a very satisfying model, but it does demonstrate an essential role for SMN1 in neurons. SMN1 expression has also been modified only in muscle cells. This has revealed that SMN expression in muscle prevents the SMA phenotype and increases the life span of the animals. This supports targeting muscle as a therapeutic strategy in SMA. Because in human the severity of disease is modulated by the *SMN2* gene, the *SMN2* gene has been introduced (by a transgene) to the mice with the 'neuronal' deletion of *SMN1*. This showed that *SMN2* gene number ameliorates SMN1 absence, closely mimicking what occurs in human disease.

## Motor and Sensory Disease

**Charcot–Marie–Tooth** Charcot–Marie–Tooth (CMT) diseases refer to a heterogeneous class of neuropathies that affect not only motor but also sensory nerves in the peripheral nervous system. In addition to the weakness and atrophy of the distal limb muscles, patients experience impaired sensation and absence of deep tendon reflexes. Representing the most common inherited disorder of the peripheral nervous system, this group of diseases, which are sometimes transmitted through dominant, recessive, or X-linked modes of inheritance, has been classified into two subgroups. The demyelinating form (also called CMT1) results in an impairment of the myelin sheath produced by Schwann cells. CMT2 forms are characterized by degeneration of the axon. The genetic heterogeneity of the disease has led to the identification of at least 17 genes involved in both demyelinating and axonal forms of CMT. The CMT1-causing genes (*PMP22*, *MPZ*, *LITAF/SIMPLE*, and *EGR2*) have revealed an essential role of myelin compaction, the regulation of myelin protein degradation, and the transcriptional control of myelination-specific genes in demyelination.

Axonal survival has been shown to depend on many different pathways. Among these, mutation in the gene encoding the mitochondrial GTPase mitosfusin 2 as a cause of CMT has proven an essential role of mitochondrial fusion/transport in axon survival. Maintenance of axonal architecture and transport (for motor neuron survival) is particularly important for sustaining neuron function. Forms of CMT2 are also caused by mutation in the neurofilament subunit NF-L or the motor kinesin KIF1b$\beta$. Impaired intracellular trafficking is also a common theme in axonal degeneration, as revealed by mutation in the regulator of vesicle trafficking Rab7 and the phosphatase MTMR2, which modulate membrane trafficking.

CMT1A is the most frequent form of CMT. Instead of a typical disease-causing mutation that alters or eliminates the encoded protein, CMT1A disease is caused by a duplication of the genomic region surrounding an otherwise normal *PMP22* gene so that affected individuals have an extra copy of the normal gene. Transgenic rat and mouse models containing multiple copies of the *PMP22* gene have proven that overexpression of only this gene is sufficient to cause peripheral demyelination and the symptoms associated with the human disease. This also demonstrated an increased disease severity with increased expression of PMP22. These animal models have been shown to be very useful to test the first rational experimental therapies of CMT1A. Indeed, a synthetic antagonist of the nuclear progesterone receptor was shown to reduce PMP22 expression and to ameliorate the clinical severity in those animals. Moreover, administration of ascorbic acid, an essential factor of *in vitro* myelination (already approved by the US Food and Drug Administration for other clinical indications), has prolonged survival and restoration of myelination in those models, suggesting its use in human clinical trials for CMT1A.

**Giant axonal neuropathy** Factors that establish the cytoarchitecture of the axon and its associated axonal transport components are critical for neuronal survival. Disorganized cytoskeletal intermediate filaments constitute the hallmark of another neurodegenerative disease termed giant axonal neuropathy (GAN). This is a rare, recessively inherited condition characterized by a structural deficit in both the central nervous system and the peripheral motor and sensory axons. With disease onset in infancy and marked by gait instability and frequent falls, patients develop diminution of deep tendon reflexes, muscle atrophy, and muscle weakness that progressively evolves to sensory loss and loss of ambulation. Other symptoms, including ataxia, language deficits, and mental retardation, reveal a later impairment of the central nervous system.

GAN is a progressive, fatal disease, with life expectancy of less than 30 years. Identification of the *GAN* gene led to recognition that its encoded protein, gigaxonin, plays a critical role in the ubiquitination machinery whose function, among others, is to tag proteins for degradation. Other errors in other components of this degradation apparatus are implicated in several neurodegenerative disorders, the most prominent of which is Parkinson's disease. Gigaxonin seems to link degradation and impairment of the axonal cytoskeleton. Loss of gigaxonin prevents normal degradation of proteins involved in the assembly and dynamics of microtubules, the protein polymers that serve as the tracks for transport of components up and down the axon. Whether altered microtubules properties are the disease causing damage and, if so, how this provokes the specific aggregation of intermediate filaments seen in patients are not known.

Developing animal models for rare disorders such as GAN is particularly crucial since insights from direct inspection of patient materials are very limited. The study of the mechanisms underlying neurodegeneration through construction and analysis of genetic models mimicking the loss of gigaxonin is at its earliest stages. Mice deleted of the *GAN* gene develop a progressive deterioration of motor function and ataxia, a sign of central nervous system impairment. As in human disease, the absence of gigaxonin generates axonal loss and neurofilament aggregation. Alteration of cytoskeleton architecture has been further confirmed by the decreased density of axonal microtubules and the increased abundance of proteins that affect microtubule assembly.

## Conclusion

Since essentially all forms of motor or sensory neuron degeneration are incurable, mouse and rat models represent an indispensable resource to address the origin of neuronal dysfunction, the mechanisms of degeneration, and the determinants of the selectivity of death of subpopulations of neurons. In turn, these key issues will power the design of strategies to ameliorate these life-long, frequently fatal conditions. Thus, for example, for ALS, this fundamental research has led to the development of many experimental therapies in these animals, including antiglutamate drugs, neurotrophic growth factor delivery, SOD1 silencing, and stem cell therapy, which will be tested in human clinical trials. Identifying the pathways of neuron survival will help in understanding the basis of motor/sensory neuron diseases in humans for which the majority of genetic causes remain unknown.

## Further Reading


Boillee S, Vande Velde C, and Cleveland DW (2006) ALS: A disease of motor neurons and their nonneuronal neighbors. *Neuron* 52: 39–59.

Boillee S, Yamanaka K, Copeland NG, et al. (2006) Motor neurons and microglia as determinants of disease onset and progression in inherited ALS. *Science* 312: 1389–1392.

Bomont P, Cavalier L, Blondeau F, et al. (2000) The gene encoding gigaxonin, a new member of the cytoskeletal BTB/kelch repeat family, is mutated in giant axonal neuropathy. *Nature Genetics* 26(3): 370–374.

Chen YZ, Bennett CL, Huynh HM, et al. (2004) DNA/RNA helicase gene mutations in a form of juvenile amyotrophic lateral sclerosis (ALS4). *American Journal of Human Genetics* 74(6): 1128–1135.

Ding J, Allen E, Wang W, et al. (2006) Gene targeting of GAN in mouse causes a toxic accumulation of microtubule-associated protein 8 and impaired retrograde axonal transport. *Human Molecular Genetics* 15(9): 1451–1463.

Monani UR (2005) Spinal muscular atrophy: A deficiency in a ubiquitous protein; A motor neuron-specific disease. *Neuron* 48(6): 885–896.

Rosen DR, Siddique T, Patterson D, et al. (1993) Mutations in Cu/Zn superoxide dismutase gene are associated with familial amyotrophic lateral sclerosis. *Nature* 362(6415): 59–62.

Shy ME (2004) Charcot–Marie–Tooth disease: An update. *Current Opinion in Neurology* 17(5): 579–585.

# Animal Models of Stroke

**L C Hoyte and A M Buchan**, University of Oxford, Oxford, UK



## Introduction

Among western countries, such as the United States, Canada, and the United Kingdom, stroke is the third leading cause of death and a major source of disability. The risk of having a stroke or a transient ischemic attack increases dramatically with age. These facts, combined with the current shift in age demographics toward a more elderly population in western countries, ensure that stroke will continue to be a health care burden. For example, in the United States, the estimated cost of stroke (from direct and indirect sources) in 2006 was predicted to be $57.9 billion (American Heart Association). It is vital to reduce brain damage and improve quality of life for stroke patients.

Human stroke occurs in two forms: ischemia and hemorrhage. The ischemic stroke results from an occlusive clot, this being from either a thrombus or an embolus that blocks blood flow within an artery of the brain. During a focal ischemic insult, there is a difference in cerebral blood flow (CBF) to the brain depending on the proximity to the clot and the level of collateral flow. The area that is within the main occluded territory is generally termed the core of the infarct, while the surrounding tissue is termed the penumbra. Within the core infarct there is extensive tissue damage, with pannecrosis, gliosis, inflammatory infiltration, breakdown of the blood–brain barrier, cytotoxic and vasogenic edema, and excitotoxic cell death. The penumbra is the area of the brain that has reduced CBF and metabolism, but damage is fundamentally reversible with reperfusion. Hemorrhagic stroke is due to a cerebral vessel bursting and the resultant bleeding into the brain. Hemorrhagic stroke can be further subdivided into two main categories: intracerebral hemorrhage (bleeding from cranial blood vessels into the brain tissue) and subarachnoid hemorrhage (bleeding from cranial blood vessels into the subarachnoid space).

A vast majority of human strokes are ischemic in nature; 85% of all strokes in western countries are ischemic, while 15% are hemorrhagic, however, there is a shift to increased hemorrhage rate among Asian populations. Due to the fact that most strokes are ischemic, experimental stroke research primarily attempts to discover ways to reduce morbidity and mortality among the aging population.

## Can Animal Models Mimic Human Stroke?

Most experimental stroke modeling is conducted using rodent models, with surgical procedures used to induce stroke. The rodent brain has a smooth cortex (lissencephalic), and different components of white and gray matter as compared to the human brain. As well, the rodent and human immune systems have marked differences. For example, a prevalence of infiltrating neutrophils is seen in rats following stroke, while in humans generally there is predominantly a macrophage infiltration. However, many rodent species, with the gerbil being a notable exception, possess a complete circle of Willis, similar to the human circulation.

This similarity in circulation allows for experimenters to examine stroke and CBF in the rodent brain and relate it to the human condition. Fundamental brain regions are similar from animal to human, allowing for insight into stroke recovery and plastic adaptation of the brain. However, there has been a plethora of failed human clinical trials; the animal work frequently did not translate to human stroke treatment in the past. Particularly, it has been seen that many experimental results were confounded by alterations in animal physiology; such that the improved outcome was due to an innate physiological response, and was not due to the specific pharmacological action of the drug on its original target. A classic example of such a phenomenon is the use of the *N*-methyl-D-aspartate (NMDA) noncompetitive receptor blocker MK-801. Treatment with this compound was found to be highly efficacious in reducing infarct volumes in the rat cerebral ischemia, but it was later proved that this drug caused a persistent hypothermia in the global model and that an improved collateral flow to the brain in the middle cerebral artery (MCA) occlusion models was responsible for improved outcome. This compound was moved to human clinical trial, without success, and there have been many more, similar examples since. This has led to some despair among researchers that animal modeling will not advance the treatment of human stroke.

However, there has been evidence of successful translation of animal work to the clinic. The only US Food and Drug Administration (FDA)-approved drug for the treatment of acute cerebral ischemia in the human, tissue plasminogen activator (tPA), was first shown to be efficacious in a rabbit model of small clot emboli. Currently there is clinical testing on another compound (NXY-059, a free radical spin trap agent) that shows positive experimental results and has also shown some promise in the clinic.

## Cerebral Ischemia Models: Global versus Focal Ischemia

Cerebral ischemia accounts for approximately 85% of the strokes seen in western society. As such, these are extensively studied experimentally. The two forms of cerebral ischemia are global (either complete or incomplete) and focal (either multifocal or focal) (**Figure 1**).

### Global Cerebral Ischemia

Global cerebral ischemia is used to mimic the clinical situation of cardiac arrest, and is also widely used to examine the selective vulnerability of neurons to cerebral ischemia. It is thought that any treatment that prevents the loss of these highly susceptible neurons might be well suited to act as a neuroprotective agent. Complete global ischemia, which can be produced by decapitation, cardiac arrest, neck cuff, or aortic occlusion, results in a loss of blood flow to the entire brain. Incomplete global ischemia; also called transient, severe forebrain ischemia, results in oligemia throughout the forebrain, including the cortex, striatum, and hippocampus. The blood flow to the brain stem is preserved in this model, allowing for the maintenance of heart rate and breathing by the animal, and possibly sparing the systemic effects of such complete global ischemia models as cardiac arrest.

Complete and incomplete global ischemia models are extremely severe, with very exceptional drops in cerebral blood flow. Consequently, the length of the ischemia in these models is generally very short in comparison with focal models. Global ischemia results in a characteristic cell death, without infarct, but with selective neuronal death. Susceptible neurons, such as the CA1 pyramidal cells of the hippocampus and select neurons of the striatum, and cortical layer III, V, and VI, die when exposed to the insult. The complexity arises from this model when one considers that adjacent to the CA1 pyramidal cells in the hippocampus are the CA3 cells – neurons that seem to have similar characteristics in location and cell type, but are resistant to the insult. It has been postulated that the CA3 are more robust than the CA1 due to a difference in the incoming afferents to these areas of the hippocampus, and to differences in the regulation of α-amino-3-hydroxy-5-methyl-4-isoxazole propionic acid (AMPA) receptor subunit GluR2 following global ischemia.

**Complete global ischemia** The most commonly studied form of complete global ischemia is the cardiac arrest model. This model is used in both gyroencephalic (convoluted cortex) and lissencephalic (smooth cortex) animals. Frequently, in canine models, it is achieved with exsanguinations, followed by mechanical

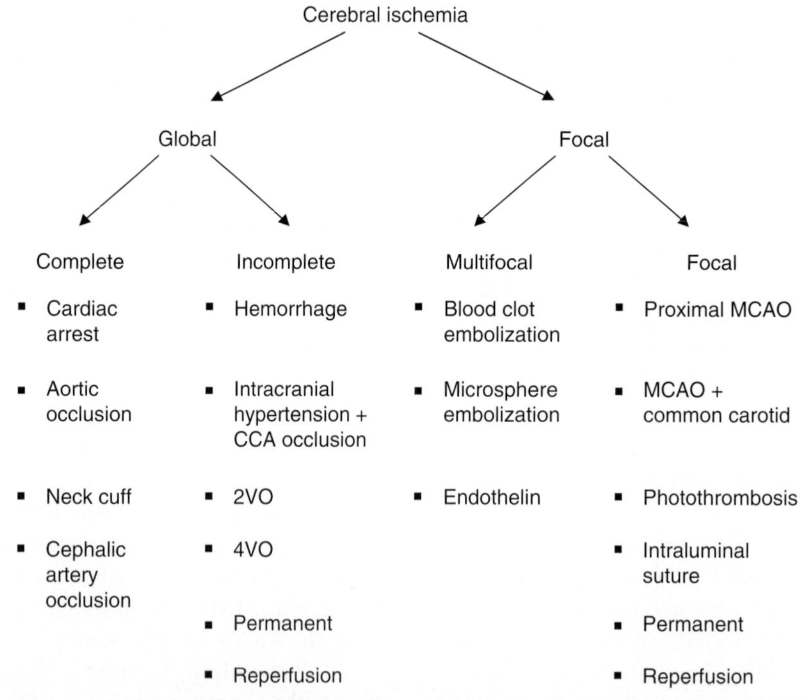


**Figure 1** Methods used to generate ischemia in animal models. These include both focal and global models and permanent and reperfusion models. MCAO, middle cerebral artery occlusion; 2VO, two-vessel occlusion; 4VO, four-vessel occlusion. Adapted from Traystman RJ (2003) Animal models of focal and global cerebral ischemia. *ILAR Journal* 44: 85–95 (p. 87). Reprinted with permission from the *ILAR Journal*, 44, Institute for Laboratory Animal Research, The Keck Center of the National Academies, 500 Fifth Street NW, Washington DC 20001 (www.national-academies.org/ilar).

or chemical resuscitation after the specified arrest time. Arrest of the heart can also be achieved by electrical fibrillation, in the cat, with consequent resuscitation. In rodents, the cardiac arrest model is frequently performed using KCl to arrest the heart, with resuscitation. Cardiac arrest models can be used to examine the efficacy of rapid induction of cerebral hypothermia, by aortic or jugular flushing with ice-cold saline. This can rapidly reduce brain temperature and improve outcome in dogs, even following very severe ischemia. The main drawback of the cardiac arrest models is that they generate systemic ischemia, which may, in turn, make interpretation of the cerebral ischemia more challenging due to damage occurring to other body organs.

**Incomplete global ischemia** Incomplete cerebral ischemia results in transient, but severe, forebrain ischemia, followed by reperfusion. This avoids the systemic effects of cardiac arrest, while still achieving very low cerebral blood flows of less than 10% of baseline. The use of four-vessel occlusion (4VO) is widespread in rat experiments, and can be achieved by occlusion of both vertebral arteries as well as both common carotid arteries (**Figure 2**). This ischemic insult results in a selective neuronal death that is dependent on ischemia duration. Pyramidal cells of the CA1 region of the hippocampus are especially sensitive to global ischemia, and duration of ischemia has a direct effect on progression of CA1 cell death; shorter durations of ischemia result in slower progression of CA1 damage. The model of global forebrain ischemia creates a lack of blood flow to the hippocampus, striatum, and cortex, while sparing flow to the brain stem. This oligemia can be reversed after 5, 10, 15, or 30 min and a recovery of blood flow and ATP levels is observed immediately. The degree of oligemia is not influenced by infusion of drug during the insult; however, hypothermia, even 6 or 24 h postischemia, is successful in increasing the long-term survival of the sensitive CA1 neurons. One major caveat to the 4VO model is the necessity of a 2-day procedure to generate the ischemia. This model relies on a preparatory day, which constitutes a surgery to isolate and prepare the common carotid arteries (CCAs) for the following day's occlusion along with a cauterization of the vertebral arteries and placement of sutures to limit collateral flow. The animals are fasted overnight, and the following day are reanesthetized to allow for occlusion of the CCAs. Some groups wish to avoid this preparatory day, and to achieve ischemia on the first day. This has led to the development of two-vessel occlusion (2VO) with hypotension.

The process of 2VO with hypotension involves a bilateral occlusion of the common carotid arteries, along with a systemic hypotension by exsanguinations. This model must achieve a level of hypotension sufficient to generate the global ischemia – reduction of blood pressure to 50 mmHg is sufficient in the rat. This generates a similar pattern of selective neuronal death as is seen in the 4VO and the cardiac arrest models, and is highly sensitive to hypothermia. 2VO is also used in the mouse, frequently without systemic hypotension to generate global ischemia, but the duration of ischemia tends to be longer to

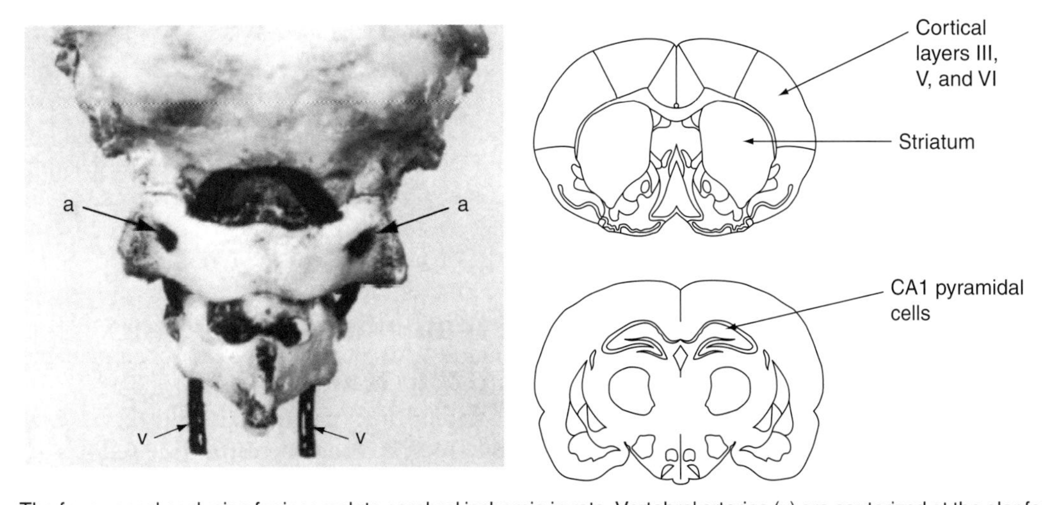


**Figure 2** The four-vessel occlusion for incomplete cerebral ischemia in rats. Vertebral arteries (v) are cauterized at the alar foramina (a) in the preparatory surgery, the animals are fasted overnight, and the following day bilateral common carotid artery occlusion generates the transient, but severe, forebrain ischemia. CA1 pyramidal neurons are very susceptible to the ischemia, along with select neurons of the striatum and cortical layers III, V, and VI; this model shows selective neuronal death. Adapted from Pulsinelli WA and Brierley JB (1979) A new model of bilateral hemispheric ischemia in the unanesthetized rat. *Stroke* 10: 267–272. Reproduced with permission from Lippincott Williams & Wilkins (http://lww.com).

compensate for the increased flow due to the posterior circulation.

Gerbils can be used to examine the effects of global ischemia, with the 2VO model, but without systemic hypotension. This is because, in general, gerbils lack a posterior communicating artery and do not have a complete circle of Willis. However, new evidence suggests the presence of a posterior communicating artery in some gerbils from specific suppliers, so experimenters must be careful in the assumption that bilateral CCA occlusion will generate global ischemia. The use of the gerbil model has verified the selective vulnerability of CA1 pyramidal cells, and has confirmed the protection of the CA1 by hypothermia. This protection has been shown to markedly improve survival of the CA1 for 18–20 months. One major advantage of this model is the relative ease of preparation and execution of the ischemia.

Currently, a new model for global cerebral ischemia has been developed in the mouse; this model mimics the effects of 4VO in the rat. The model involves the bilateral occlusion of the common carotid arteries as well as an occlusion of the basilar artery. It preserves blood flow to the brain stem while a transient oligemia is induced in the striatum, hippocampus, and cortex. The difficulties with induction of global ischemia in the mouse include an increased risk of respiratory failure due to manipulation of the trachea, which is necessary to access the basilar artery, and the possible breaking of the basilar artery following removal of the microaneurysm clip. The main advantage of generating this global model in mice is to study different transgenic strains following ischemia. This should allow for an in-depth examination of specific genes that might be implicated in selective neuronal death.

### Focal Cerebral Ischemia

Focal cerebral ischemia is used to study the tissue damage that results from stroke. Generally, focal cerebral ischemia models examine the blockage of one artery of the brain (e.g., the MCA). Focal cerebral ischemia can also be generated with multiple focal insults, such as occur with embolization. These models result in classic infarcts with core and penumbra areas. The core of the infarct is the area that is closest to the occlusion; this area has the lowest cerebral perfusion and the tissue contained within this area is presumed to be unsalvageable. Seminal work from the late 1970s demonstrated that there was a difference in cerebral perfusion depending on the distance from the occlusion, and the area surrounding the core infarct was termed the ischemic penumbra. The neurons within the penumbral area are structurally-intact, but metabolically paralyzed. There is a direct relationship to the cerebral blood flow and the electrical activity of the brain, such that during cerebral ischemia, the electrical activity of the brain will disappear; however, this loss of electrical activity occurs at a less severe ischemia than is necessary to generate failure of the cell's baseline functions. Thus, the core infarct contains cells that have become structurally and metabolically compromised, while neurons in the penumbra retain structural integrity. The infarct loses the ability for autoregulation, and over time, develops into a glial scar.

Focal ischemia models are considered to have either a single focus (focal) or to have multiple foci (multifocal). The focal insult is generally created by surgical occlusion of an artery of the brain; however, there are cases whereby an artery was occluded by direct placement of clot into the artery. Multifocal insults are produced by injection of emboli (either clot or microspheres) into the brain. This form of insult results in multiple infarcts and is far more variable than focal ischemia.

**Middle cerebral artery occlusion** By far the most common form of focal ischemia is achieved by occlusion of the MCA. This was first attempted by Harvey and Rasmussen in 1951, when they surgically occluded the MCA with a clip to produce a large infarct and marked hemiparesis in dogs. The same two experimenters showed that the duration of ischemia was important for the development of infarct in the monkey – a 50 min ischemia resulted in an infarct similar to that of permanent ischemia, while ischemia durations of less than 30 min did not produce gross cortical damage. Currently, MCA occlusion is used to induce stroke in many animal species, including rats, mice, pigs, cats, dogs, and nonhuman primates. Rodents are by far the most commonly used experimental animal in MCA occlusion studies due to ethical and financial constraints of using higher order mammals. The occlusion of the MCA can be induced either proximally, such as in the case of intraluminal suture techniques, or more distally by a clipping of the distal MCA.

One of the more commonly used methods for generating MCA ischemia is the intraluminal suture technique. This method achieves ischemia by introduction of an occlusive suture into the internal carotid artery, whereby it is advanced to block the origin of the MCA. The carotid artery is accessed from the ventral cervical surface, and craniotomy is avoided in this model. The occlusion is generated by accessing and isolating the common carotid artery, external carotid artery (ECA), and internal carotid artery (ICA). Originally, the suture was passed directly into the ICA through a small puncture, while the arteries (CCA,

ICA, and ECA) were tied off to prevent bleeding. Further refinements were established to allow for introduction of the suture into the ECA and temporary placement of the suture in the CCA, followed by a movement of the suture into the ICA. This refinement necessitated the permanent ligation and cutting of the ECA, so that the artery could be moved, allowing for advancement of the suture into the ICA. Upon introduction of the suture into the ICA, it is advanced until it blocks the proximal MCA (**Figure 3**). Various small alterations in this technique exist. They include using different sizes of sutures to block the MCA with a heat-blunted tip, silicone coating of the suture tip, or the use of polylysine-coated sutures. The intraluminal suture model induces severe ischemia in the striatum with milder cortical ischemia. Commonly, the intraluminal suture model results in ischemia within the hypothalamus, and, in the rat, this results in hyperthermia. This hyperthermia can markedly enhance infarct progression and increases the variability of this model. Proximal MCA occlusion in the mouse is conducted in a manner similar to that in the rat, which again results in a severe ischemia of the striatum with milder ischemia in the cortex. The intraluminal mouse model can be variable due to the differences in patency of the posterior communicating artery among strains, and even among animals of the same strain. The possible use of transgenic animals is a great advantage of using the mouse model. However, differences in patency of posterior communicating arteries make the use of transgenic mice more difficult if the background strain used in not the same as the control strain.

Mice are more delicate than rats and frequently succumb to the stress of surgical procedures, and the survival rate of experimental mice is less than that of rats. Physiology of rodents must be carefully controlled, and following focal ischemia by intraluminal suture, mice tend to become hypothermic. This results in smaller infarct volumes and better histological and behavioral scores as compared to temperature-regulated mice.

In rats, direct placement of a clot into the proximal MCA has been shown to generate consistent ischemia, while introducing an occlusive thrombus. In this model, the clot is placed into the MCA by introducing tubing containing the clot into the ECA and advancing it through the ICA, in a manner similar to the placement of the suture in the intraluminal technique. However, the clot is ejected from the tubing, effectively obstructing the MCA. The main advantage of this model is the placement of the clot within the MCA. This should allow for testing of the efficacy of thrombolytic drugs on clot lysis in this model, which is not possible in the intraluminal suture models.

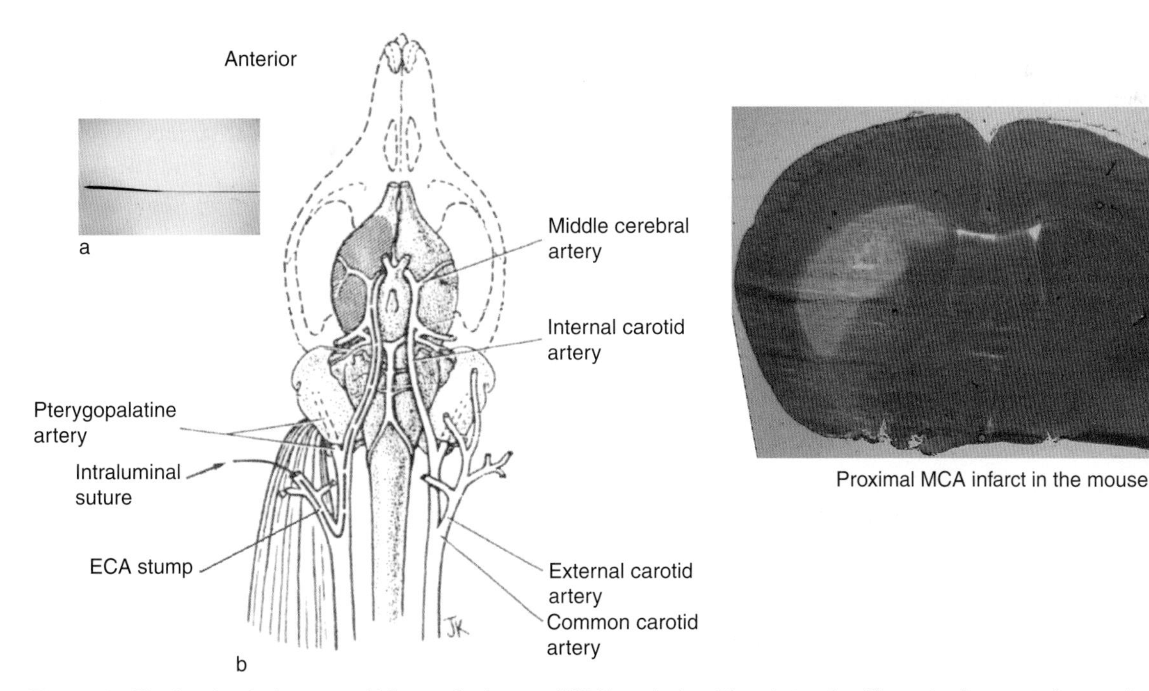


**Figure 3** The intraluminal suture middle cerebral artery (MCA) occlusion. The obstructive filament, often coated to reach a specified diameter (a), is introduced in the external carotid artery (ECA) and advanced briefly into the common carotid artery. The ECA is cut to allow for manipulation of the suture. The suture is withdrawn slightly, is maneuvered into the internal carotid artery, and then advanced until it blocks the proximal middle cerebral artery (b). This model generally creates a large striatal infarct, with some cortical involvement. Adapted from Longa EZ, Weinstein PR, Carlson S, et al. (1989) Reversible middle cerebral artery occlusion without craniectomy in rats. *Stroke* 20: 84–91.

Distal MCA occlusion is achieved by accessing the distal MCA, either by craniotomy over the parietal/ frontal cortex, in rodents; or by orbital rim approach following enucleation, in gyroencephalic mammals. In the rat model, the skin between the eye and the ear is cut, and the underlying muscles are dissected to expose the skull. Two small burr holes are drilled into the skull, one to access the artery for monitoring of CBF and the other for occlusion of the artery. The distal MCA is then directly visualized through the operating microscope and the more distal burr hole is used for occlusion of the MCA. The dura over the MCA is opened and the MCA is occluded along with bilateral or ipsilateral occlusion of the CCA, depending on the strain. This model was first described in 1986 by Chen and colleagues, who occluded the MCA by ligation with consequent occlusion of one or both CCAs, generating large cortical insults. A newer model of distal MCA occlusion was created that also generated a large cortical insult, but which relied on the placement of a microsurgical clip on the distal MCA, increasing the ease of producing ischemia with reperfusion (**Figure** 4). One major advantage of this clip model is that it generates the ischemia with only brief periods of anesthesia during the clip placement and removal. Occlusion of one or both CCAs is dependent on rat strain: Wistars require both CCAs to be occluded for successful ischemia, while spontaneously hypertensive rats require only ipsilateral occlusion. The cortical infarction induced by these models correlates with a breakdown of the blood–brain barrier and edema.

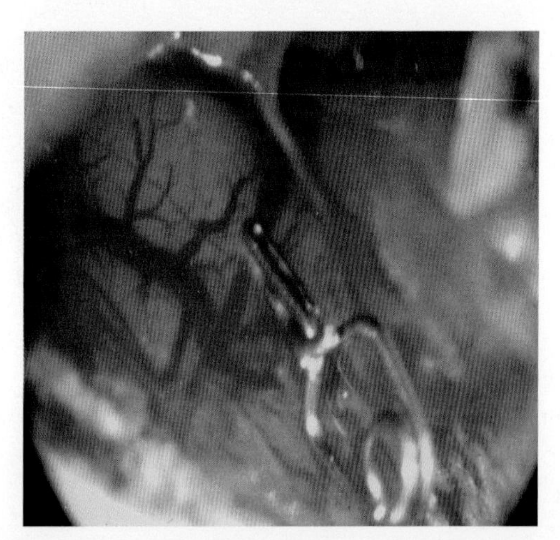

**Figure 4** Distal middle cerebral artery occlusion; clip placement on the middle cerebral artery. This model requires a craniotomy and incision through the dura to access the distal middle cerebral artery. The rat is anesthetized during the clip placement and occlusion of one or both common carotid arteries, depending on the strain. The animal is awake during the ischemia, but is reanesthetized for clip removal.

Gyroencephalic models generally attempt to address questions regarding ischemia in an animal with brain structure more similar to that of humans. A new model to assess MCA occlusion in higher mammals is achieved in the miniature pig; the approach to the MCA is over the frontotemporal cortex, accessing the MCA via an orbital rim osteotomy. The cat has also been used to study focal ischemia and the effects of hypothermia on the brain. Generally, the nonhuman primate model, such as in the baboon, generates occlusion of the MCA by transorbital approach. The MCA is located following enucleation and an inflatable silastic balloon cuff is placed around the MCA. This allows for inflation of the balloon and induction of ischemia in an awake animal. Nonhuman primates are important to assess complex behaviors and recovery following stroke and allow for an assessment of injury in a brain that is very similar to a human brain. Nonhuman primate models should only be used to evaluate neuroprotective agents when positive results in rodent models generate enthusiasm before transferal into clinical trial.

**Common carotid artery occlusion** Occlusion of the common carotid artery can be used to generate focal ischemia in the gerbil, the adult rat with systemic hypotension, or the young rat with hypoxia. By far the simplest method for generating a focal insult is by clipping one of the CCAs in the gerbil, due to their lack of the posterior communicating artery, but investigations must be carefully conducted due to the possibility of gerbils having a PCA, which might confound results. The ventral cervical surface is opened surgically, under general anesthetic, and the CCA is isolated. The artery is then occluded for a specified period of time, with reperfusion. The gerbil model has been used to evaluate motor deficits and examine magnetic resonance image changes following focal ischemia. The main advantage of the gerbil model is the ease of inducing focal ischemia; however, to limit variability in this model, one must be careful to ensure that the posterior communicating arteries are not present in the supplies of gerbils.

Occlusion of one CCA in the rat with systemic hypotension can be used to generate focal insults with minimal surgical invasiveness. The ventral cervical surface of the rat is opened, and the CCA located and isolated. This artery is then occluded and systemic hypotension is achieved by exsanguinations to a blood pressure of 40 mmHg. This model approaches the ease of the gerbil model, but introduces more difficulty due to the removal of blood to generate hypotension. In neonatal rats, it is possible to generate a focal insult by a combination of hypoxia and ischemia. Again, the ventral cervical surface is opened and the CCA located and dissected. The artery is occluded and hypoxia is created by placing the rat in a low-oxygen environment. This model is very important for modeling neonatal or fetal ischemia in humans; among children, ischemia *in utero* or during parturition is a common mechanism for brain damage.

**Pial strip** The pial strip method for creating a localized, specific cortical lesion is achieved by removal of the meninges over a particular area of the cortex. This method is generally used in the rat, and is able to generate lesions consistent in size and location. A midline incision in the scalp in created and a small craniotomy is used to access the meninges. The dura and pia mater are removed, thus removing the collateral vessels from that particular portion of cortex, which will go on to die. The main advantage of the pial strip is the reproducibility and the ability to selectively place the lesion on a specific brain region (i.e., motor cortex), creating a specific deficit. This allows for greater power in behavioral studies, and assessment of different treatments in very specific and precise motor behavioral tests. A disadvantage of this model is that it is not a classical focal ischemia/reperfusion model, with a core and penumbral areas of CBF reduction and infarct, so it may be difficult to assess some aspects of ischemic cell death.

**Photothrombosis** Photothrombosis is a method to induce stroke in rodents; it relies on the injection of light-sensitive dye followed by a laser pulse to create clot within the artery. To generate ischemia in this model, a photosensitizing dye is first injected intravenously into the animal. Following this, the skin and muscle over the skull are retracted, and the area of the brain to undergo ischemia is irradiated through the intact skull, using green light at 560 nm. This model has been shown to create thrombotic plugs and stasis of the red blood cells in vessels, with platelet aggregates. The associated infarction is due to the formation of a thrombus within the vessels and generates a reproducible ischemia, making this model very attractive for investigations requiring a thrombus. As well, this process is much less invasive than the other surgical models and the application of the laser can be directed to ensure ischemia in one particular brain area, such as the motor cortex, making this an attractive model for consequent behavioral testing. A major disadvantage of this model is that it is a permanent ischemia. There is no reperfusion aspect to the injury, limiting the evaluation of reperfusion injury that is possible with other models. This model has also been adapted for use in the mouse, using rose bengal and a krypton laser beam (568 nm). The region of the MCA is irradiated along with a ligation of the CCA to generate a consistent, reproducible ischemia.

**Embolic** The embolic model is created by introducing a cannula containing occluding substances, such as a clot or microspheres, into the CCA, and injecting these into the artery; the occluding substances generate ischemia in different locations dependent on their size. These embolic models are frequently used in rabbit ischemia. Generally, the rabbit model of focal ischemia induces an occlusion through the injection of a clot into the common carotid artery. This includes both small microclot and larger clot models. The rabbit small clot embolus model (RSCEM) introduces obstructive emboli into random small arteries of the brain, resulting in a heterogeneous stroke, while the rabbit large clot embolus model (RLCEM) causes an obstructive embolus in the MCA. Following embolus injection into the rabbit, behavioral changes can be used to assess damage. The RSCEM can also be used to determine clot load needed to affect 50% of rabbits, which can be used to assess neuroprotection. Because this model involves thrombus formation, it is very useful for determining the effects of proposed thrombolytics. In fact, tissue plasminogen activator was first evaluated as useful therapy for ischemic stroke in the rabbit model.

Rabbit physiology is very stable following the preparatory surgery to induce RSCEM or RLCEM. This is in contrast to the rat and mouse models, which require constant monitoring to maintain physiology within normal levels.

**Endothelin** The endothelin (ET-1) model of cerebral ischemia is achieved by stereotaxic injection of endothelin, a potent vasoconstrictor that occludes arteries in the area of injection. The application of ET-1 onto the cerebral cortex has been shown to create an ischemic injury that is reproducible, with consistent lesions following injection, and which shows a dose-dependent reduction of flow. ET-1 can be used to induce ischemia in several different brain locations, depending on the delivery. The drug can be applied topically to the cerebral cortex, directly over the MCA, or injected into various brain regions, including the white matter, by advancing a needle into the brain to the correct region using stereotaxic coordinates. Upon application of the ET-1 there is a rapid, but not immediate, reduction in blood flow, which shows gradual reperfusion over time. This mimics the aspect of human stroke without thrombolysis. Increased variability can be introduced into the MCA occlusion model of ET-1 ischemia due to the difficulty in accurate application over the MCA.

Another difficulty with the MCA occlusion model is the necessity for craniotomy, which might cause some trauma to the cortex. The intracerebral injections limit this trauma because craniotomy is not required for the injections. A final consideration must be made on the appropriate application site for experimental study, because certain application sites (i.e., MCA) do not allow for a sufficient motor cortex infarction, necessary for assessment of behavior.

## Hemorrhagic Stroke Models

Hemorrhagic stroke models usually attempt to address the effects of cerebral bleeding into either the intracerebral tissue or in the surrounding meninges. These models are usually achieved by either an injection of autologous blood into the brain or injection of collagenase, a chemical that breaks down the vessel walls.

The injection of collagenase into the cerebral tissue will break down type 4 collagen, a major component of the vasculature, and create intracerebral hemorrhage, with a core infarct and a surrounding penumbral area. The core of the hemorrhage has high levels of fluid, red blood cells, and fibrin.

Injection of autologous blood can be performed in various brain areas, including directly into the cerebral tissue or into spaces between the meninges. Following creation of subarachnoid hemorrhage in the monkey, there is a consequent chronic vasospasm. Immediately following hemorrhaging into the brain, there is a consequent reduction of perfusion pressure to the brain and also a reduced CBF. The newest model to generate hemorrhage is by ultrashort laser pulses, which will allow for creation of small, discrete hemorrhage.

## Conclusion

Stroke modeling, both ischemic and hemorrhagic, is constantly evolving in an effort to create animal models that most closely mimic human stroke. While these models approximate the human condition, they are not absolute. Thus, it is important to show the efficacy of treatment in more than one model, and for extended durations of animal survival, with extensive behavioral testing. All animal physiology must be carefully monitored and kept within the physiologic levels to ensure that the outcome seen is due to the treatment and not to physiologic perturbations. A clear understanding of the models themselves and what they are capable of testing is also required, and the suitability of the chosen animal model for testing a hypothesis must be carefully considered.

*See also:* Stroke: Injury Mechanisms; Stroke: Neonate vs. Adult.

## Further Reading

Astrup J, Siesjo BK, and Symon L (1981) Thresholds in cerebral ischemia – the ischemic penumbra. *Stroke* 12: 723–725.

Barber PA, Demchuk AM, Hirt L, et al. (2003) Biochemistry of ischemic stroke. *Advances in Neurology* 92: 151–164.

Carmichael ST (2005) Rodent models of focal stroke: Size, mechanism, and purpose. *NeuroRx* 2: 396–409.

del Zoppo GJ (1998) Clinical trials in acute stroke: Why have they not been successful? *Neurology* 51: S59–S61.

Dirnagl U, Iadecola C, and Moskowitz MA (1999) Pathobiology of ischaemic stroke: An integrated view. *Trends in Neurosciences* 22: 391–397.

Fisher M (2004) The ischemic penumbra: Identification, evolution and treatment concepts. *Cerebrovascular Disease* 17(supplement 1): 1–6.

Fukuda S and del Zoppo GJ (2003) Models of focal cerebral ischemia in the nonhuman primate. *ILAR Journal* 44: 96–104.

Hoyte L, Kaur J, and Buchan AM (2004) Lost in translation: Taking neuroprotection from animal models to clinical trials. *Experimental Neurology* 188: 200–204.

Longa EZ, Weinstein PR, Carlson S, et al. (1989) Reversible middle cerebral artery occlusion without craniectomy in rats. *Stroke* 20: 84–91.

Pulsinelli WA and Brierley JB (1979) A new model of bilateral hemispheric ischemia in the unanesthetized rat. *Stroke* 10: 267–272.

Richard GA, Odergren T, and Ashwood T (2003) Animal models of stroke: Do they have value for discovering neuroprotective agents? *Trends in Pharmacological Sciences* 24: 402–408.

Traystman RJ (2003) Animal models of focal and global cerebral ischemia. *ILAR Journal* 44: 85–95.

Xi G, Keep RF, and Hoff JT (2006) Mechanisms of brain injury after intracerebral haemorrhage. *Lancet Neurology* 5: 53–63.

Zivin JA, Fisher M, DeGirolami U, et al. (1985) Tissue plasminogen activator reduces neurological damage after cerebral embolism. *Science* 230: 1289–1292.

# Cell Replacement Therapy: Parkinson's Disease

**P Brundin and A Björklund**, Wallenberg Neuroscience Center, Lund, Sweden



## Introduction

Cell replacement therapy for Parkinson's disease (PD) is based on the idea that implanted immature dopamine neurons may be able to substitute for the lost nigrostriatal neurons. The principal limitations of this approach are the practical and ethical problems associated with the use of tissue derived from aborted human embryos or fetuses and also the large numbers of donors needed to obtain good therapeutic effects. To date, differentiated neuroblasts and young postmitotic dopamine neurons have been used for transplantation. However, progenitors taken at earlier stages of development might prove more effective in the future. Efforts are under way to expand multipotent neural stem or progenitor cells *in vitro* and control their phenotypic differentiation into a dopaminergic neuronal fate. Initial results suggest that *in vitro* expanded cells can survive and function after transplantation to the striatum in the rat PD model, but the number of surviving dopamine neurons has in most cases been very low. With further development, expanded progenitors or dopamine neuron precursors, possibly modified by cell engineering techniques, may offer new sources of cells for replacement therapy in PD.

## History

The first studies on the transplantability of brain catecholaminergic (including dopaminergic) neurons were conducted by Seiger and Olson in the early 1970s. These studies in rats showed that immature, developing dopamine neurons from the ventral midbrain can survive transplantation to the anterior eye chamber and extend an extensive axonal terminal network on the underlying iris. The experiments showed that dopamine neurons have to be obtained at embryonic or fetal stages of development, ideally at the time of development when they are generated. Inspired by these early observations, in 1979 Björklund and Stenevi and Perlow et al. showed that fetal dopamine neurons can survive, grow, and function after transplantation to the brain in rats subjected to a 6-hydroxydopamine (6-OHDA) lesion of the intrinsic nigrostriatal dopamine system. Extensive animal experiments in rodent and primate models of PD during the subsequent decades have shown that immature dopamine neurons and neuroblasts can survive and reinnervate the denervated striatum, as well as restore dopamine synthesis and release in the area. Based on these animal studies, the first clinical trials using cells obtained from embryonic human ventral mesencephalon (VM) were carried out in 1987–89. The observations from these initial trials were quite encouraging. To date, several hundred patients have received dopamine cell transplants in open-label and double-blind, placebo-controlled trials with mixed results.

## Studies in Rodents and Primates

The dopamine neurons used for transplantation are harvested from the VM (i.e., the region containing the developing substantia nigra and the ventral tegmental area) at the time when the dopamine neurons are born – that is, from 13- to 15-day-old fetuses in the rat and 6- to 8-week-old human aborted embryos. In most cases, the cells are transplanted into the striatum (i.e., close to the denervated target cells). Thereby they can extend a new terminal network in the host striatum and establish functional synaptic contacts with the striatal target neurons. Only approximately 10% of the dopamine neurons that are implanted survive the graft surgery, and various neuroprotective agents and growth factors have been employed to boost this survival rate to some extent. Coadministration of growth factors such as glial cell line-derived neurotrophic factor can promote axonal outgrowth from the transplants.

Experiments performed in rodents and primates in which the nigrostriatal dopamine system is destroyed either by the 6-OHDA toxin (in rats) or by MPTP (in monkeys) have shown that grafted dopaminergic neuroblasts reverse some, but not all, motor symptoms associated with the loss of the intrinsic nigrostriatal system. The functional effect develops gradually: the first transplant-induced effects on motor behavior are seen after 3–6 weeks with rodent cells and after a few months with human dopamine neurons. Functional recovery is correlated with growth of axonal processes from the grafted cells and establishment of a new dopamine-containing terminal network in the host striatum. Rodent studies have shown that the grafted dopamine neurons are spontaneously active and can restore a tonic release of dopamine to near-normal levels within the reinnervated area. Several observations clearly indicate that the functional recovery is dependent on a specific interaction between dopamine neurons and the host striatum: (1) the recovered function is maintained only with continued survival of the grafted neurons, (2) dopamine neurons transplanted outside the basal ganglia induce no improvement, and (3) non-dopaminergic neurons grafted to otherwise effective striatal sites exert no effect.

In rats with 6-OHDA lesions, intrastriatal VM grafts containing between 1000 and 2000 surviving dopamine neurons (approximately 10–15% of the number of dopamine neurons normally present in the adult substantia nigra) are known to be sufficient to reverse some behavioral deficits. They can restore striatal dopamine levels to approximately 10–30% of normal, *in vivo* striatal dopamine synthesis to approximately 20–30% of normal levels, and striatal dopamine release (as measured by microdialysis in the area densely innervated by the grafted neurons) up to 40–100% of the levels seen in the intact striatum. Using multiple graft placements, it has been possible to obtain a striatal dopamine innervation density in the range of 40–75% of normal (as assessed by tyrosine hydroxylase immunoreactivity). Even with such large grafts, the recovery of lesion-induced motor deficits is incomplete and varies depending on the type of test. Thus, motor asymmetry assessed by drug-induced turning behavior is completely reversed, whereas several aspects of spontaneous motor behavior, such as stepping, forelimb use, and postural balance, are only partly restored.

There are several possible reasons why graft-induced functional recovery is limited to certain basic aspects of motor function, whereas impairments in more complex motor behavior are influenced to a lesser extent, if at all. The transplants are placed in an ectopic location (i.e., in striatum rather than substantia nigra) in order to obtain extensive reinnervation of the striatum from the grafted neurons. This may impose limitations on their functionality because the grafted dopamine neurons are likely to lack crucial afferent regulatory inputs. In the intact brain, baseline striatal dopamine neurotransmission is normally maintained by tonic synaptic and nonsynaptic dopamine release, which is largely independent of changes in neuronal impulse flow in the nigrostriatal pathway. In grafted animals, the establishment of a new dopamine-storing terminal network in the host striatum is closely correlated with the onset of graft-derived dopamine release and reversal of dopamine receptor supersensitivity. This suggests that the grafted neurons are spontaneously active and release dopamine at both synaptic and nonsynaptic sites at a near-normal rate. The grafted neurons are probably not entirely devoid of host afferent control, but these contacts might not be fully appropriate for normal function. In 1991, Fisher et al. showed that more than 50% of the grafted neurons respond to stimulation of either frontal cortex or striatum, and burst firing, which normally depends on a functional cortical input, is present in a portion of them. Neuroanatomical data also indicate that some afferent inputs (e.g., from premotor cortex and/or striatum) may contribute to the functional efficacy of intrastriatal dopamine neuron transplants. An alternative explanation for incomplete functional recovery seen in animals with intrastriatal dopamine neuron transplants is that the graft-derived innervation is confined to the striatum, whereas the denervating lesion also involves other nonstriatal areas. The inability of intrastriatal VM grafts to provide complete functional recovery may be due to the fact that limbic and cortical areas, as well as other parts of the basal ganglia, remain denervated and thus unaffected by the transplanted cells. In support of this view, in 2001 Kirik et al. showed that graft-reduced functional recovery is more pronounced in animals with partial lesions of the nigrostriatal pathway, in which the denervation is confined to the striatum, than in animals with lesions which involve both striatal and nonstriatal areas. This indicates that it may be functionally important that part of the nonstriatal dopamine projections remains intact. These spared portions of the host dopamine system, particularly those innervating nonstriatal forebrain areas, may be necessary for the VM grafts to exert optimal functional effects.

## Clinical Trials

In 1990, for the first time, neural transplants were reported to survive and reduce neurological symptoms in a PD patient. The donor tissue was obtained from embryos at routine abortions. The trials were preceded by an ethical debate. In many countries, it was considered justifiable to use donor tissue from abortions if clearly separated from the transplantation surgery and the women undergoing abortions had given their informed consent. In the first trial, tissue was dissected from the VM of multiple 6- to 8-week-old human embryos, mechanically dissociated, and then injected into the striatum using stereotactic surgery on one side of the brain in a patient receiving immunosuppression. After 4 or 5 months, the patient started to exhibit a progressive reduction in bradykinesia and rigidity, whereas the tremor remained largely unaffected. Using positron emission tomography, it was possible to monitor an increase in fluorodopa uptake in the graft site, indicating that dopaminergic neurons in the implant had survived. Several subsequent reports confirmed that grafted embryonic neurons can alleviate symptoms and replace the dopaminergic neurotransmission that is lost in PD. In extreme cases, the fluorodopa uptake approached that of a normal striatum, comparable to the 10–30% of normal levels seen in the PD patients prior to surgery or in unoperated parts of their brains. Ten years after surgery, which is the longest reported follow-up, transplanted neurons have been shown to still be alive and capable of releasing dopamine. Functional positron emission tomography studies have indicated that the grafted cells integrate with basal ganglia – cortical circuits that participate in fine motor control. This functional integration seems to take several months longer than for the grafted neurons to start producing dopamine and appears to coincide with further improvements of symptoms.

The brains of a few patients who died several years postoperatively (unrelated to the graft surgery) have been examined microscopically. These have revealed that approximately 5–10% of the grafted dopaminergic neurons survive and extend a dense network of axons into the host brain. Interestingly, some of the grafts are infiltrated with immunocompetent cells. Although the brain is an immunologically privileged transplantation site (i.e., histoincompatible grafts survive longer in the brain than in peripheral sites), it is clear that transplant rejection can take place in the brain. The level of immunosuppressive treatment that is required to avoid rejection with certainty is not known. Consequently, different clinical trials have employed widely different strategies to deal with the risk of immune rejection.

Approximately 300 patients have undergone cell transplantation surgery in several small clinical trials. However, not all patients benefited to the same extent. Approximately one-third of patients demonstrated remarkable recovery and were able to stop taking anti-parkinsonian medication, one-third were moderately improved, and one-third exhibited no clear lasting benefit from the surgery. Variation in outcome was evident even among patients operated at the same surgical center. The small trials were conducted in independent centers and therefore not coordinated with each other. As a consequence, there were several potentially important variations in the experimental protocols regarding, for example, disease stage and age of patients, donor tissue age, numbers of donors, graft storage and preparation technique, surgical methodology, and the use of immunosuppression. Therefore, it is difficult to compare the results from the different centers and devise a protocol that would have a good chance of reproducible success. Another shortcoming of the initial neural transplantation trials was that they were open-label (i.e., both the patient and the investigator were aware that the patient was receiving a graft). This meant that there was a risk that the reported beneficial effects of the transplants were due to a placebo effect or observer bias.

In the mid-1990s, the US National Institutes of Health (NIH) sponsored two large randomized, double-blind, placebo-controlled clinical trials with neural transplants in PD. The outcomes of these trials, published in 2001 and 2003, were largely disappointing to patients and researchers. In brief, there was no significant symptomatic relief in the grafted patients compared to the control subjects, according to primary outcome parameters. The positron emission tomography scans and postmortem morphological examinations showed that grafted dopamine neurons had survived, despite the lack of clear beneficial effects of the surgery. Although individual patients in these two trials may have experienced graft-related improvement, these findings clearly question whether cell transplantation is a viable strategy for PD. It is important to note, however, that there are several outstanding issues regarding how the trials were performed. First, the donor tissue was stored prior to surgery for between 2 days and 4 weeks, which may have added variability in the quality of the implants. Second, in one of the trials an unconventional frontal (as opposed to dorsal) stereotactic approach was used to target the putamen. The impact of this surgical technique on graft function is not known. Third, in one of the trials no immunusuppression was given, and in the other it was limited to a 6-month administration of cyclosporine. Fourth, there is an indication that younger PD patients who are still responsive to anti-parkinsonian medication respond better to the surgery than aged patients, and selection of relatively old patients may have affected the overall outcome. Taken together, the two NIH-sponsored trials provided very valuable information but clearly showed that neural grafts do not reliably provide clinically valuable effects with the particular protocols they employed.

Aside from the lack of graft-induced benefit, the first of the two NIH-sponsored trials also reported that some patients (approximately 15%) experienced involuntary movements (dyskinesias) that were caused by the implants. This finding was unexpected and it was particularly disturbing that the dyskinesias did not improve when the anti-parkinsonian medication was terminated. Some of the patients from open-label trials and the patients in the second NIH-sponsored trial were also found to have developed graft-induced dyskinesias of varying severity. The reports of graft-induced dyskinesias and problems in obtaining suitable donor tissue in a reproducible manner brought clinical transplantation trials in PD to a standstill in the early 2000s. The pathogenetic mechanisms underlying the involuntary movements are not understood. Several possible explanations have been proposed, including that there is too much or unregulated dopamine release from the implants in certain striatal regions, that non-dopaminergic neurons in the implants influence the striatum, or that inflammation due to an immune response triggers the movements. Concerted efforts in experimental animals are currently under way to elucidate why neural transplants can sometimes elicit this undesirable side effect. It is of vital importance to understand when, how, and why graft dyskinesias occur because they could conceivably also develop following transplantation of stem cell-derived neurons.

## Future Cell Replacement Strategies

It is important to emphasize that so far no systematic stem cell transplantation trials have been performed for PD. The cells used in clinical trials have all been postmitotic immature neurons that have already been committed to the dopaminergic neuron fate. Stem cells are characterized by their ability to self-renew and to differentiate into many different forms of cells. There are various types of stem cells related to the age of the organism in which they are found and their anatomical location. Research is ongoing in experimental models of PD aimed at developing a therapy based on neurons derived from stem cells.

An existing proposal is to utilize endogenous neurogenesis in the adult brain as a source of new dopamine neurons. Under normal conditions, neurogenesis takes place in the adult mammalian brain in at least two regions, namely the dentate gyrus of the hippocampus and the subventricular zone, which supplies the olfactory bulb with new neurons. Claims that new dopaminergic neurons can be generated in the substantia nigra of adult rodents and that lesion-induced motor deficits can be reversed are highly controversial, with several studies disputing these findings. Extensive basic research is needed before adult neurogenesis can be proposed as a strategy for brain repair in PD.

Currently, it seems more feasible that it will be possible to transplant neurons derived from stem cells in PD. These stem cells are essentially one of two transplantable types – adult-derived or embryonic stem (ES) cells. Concerning stem cells obtained from adults, several investigators have suggested that it may be possible to transplant dopamine neurons derived from stem cells located outside the central nervous system. These neurons would be the result of a phenomenon called transdifferentiation; that is, the stem cells differentiate into a cell type that is not associated with the anatomical structure in which they are located. For example, it has been suggested that neurons suitable for transplantation can be generated from stem cells residing in bone marrow or skin. These ideas have been questioned, and it is currently unclear if neurons can be generated by transdifferentiation. In some experimental paradigms in which transdifferentiation was initially suggested to have taken place, the host neurons appear to have undergone fusion with injected stem cells. Another option is to harvest stem cells from the developing or adult nervous system and coax them by specific culture media or genetic modification into becoming dopamine neurons. Although there exist several cell culture protocols that allow propagation of central nervous system stem cells *in vitro*, it is difficult to manipulate these to differentiate into dopamine-producing neurons that can survive grafting. Therefore, new experimental strategies need to be developed before neural stem cells can provide a reliable source of dopamine neurons for transplantation in PD.

The most promising option for a stem cell therapy in PD is to employ ES cells. The ES cells are derived from the inner cell mass of a blastocyst and can be propagated in large numbers *in vitro*. Mouse ES cells can be induced to differentiate into dopamine-producing neurons. These neurons have successfully been grafted to rodent models of PD, in which they innervate the brain and reverse lesion-induced motor deficits. Dopamine neurons have also been derived from nonhuman primate ES cells and successfully grafted to monkeys. In addition, research has demonstrated that it is possible to generate dopaminergic neurons from human ES cells in culture. Unfortunately, the success rate has been low when transplanting dopamine neurons derived from human ES cells into rat models of PD. This problem needs to be resolved before translation into a clinical protocol can continue. Possible explanations are that the dopaminergic neurons die because they are particularly sensitive to the grafting procedure and/or that they downregulate some key phenotypic or regulatory proteins after implantation into the adult brain.

At least three other important issues need to be resolved before dopaminergic neurons derived from human ES cells can be tested in PD patients. First, ES cells are pluripotent and therefore they can form teratomas (tumors containing tissues from all three germ layers) upon grafting. A clinical protocol would require that all pluripotent stem cells are purged from the transplant preparation. Even a single remaining pluripotent or otherwise tumorigenic stem cell could have devastating consequences for a PD patient receiving an intracerebral implant. Second, most current cell culture protocols for human ES cells include animal-derived products (culture media, feeder cells, etc.). To eliminate the risk of transfer of potentially dangerous zoonooses (e.g., species-specific retroviruses) and to reduce the risk of graft rejection due to transfer of animal-derived molecules to the graft cells, the ES cells should be grown without any animal products present. Third, further research is required to analyze the type of dopaminergic neurons that are generated from ES cells. It is important that the dopamine neurons used for cell replacement therapy have a midbrain-specific phenotype. To be fully functional in the striatum, it is likely that they need to possess the properties of substantia nigra dopaminergic neurons because the specific physiological features of nigral neurons differ from those of dopaminergic neurons from other brain regions. Advances in transcriptional control have provided information that can help devise protocols that generate large numbers of midbrain dopamine neurons from human ES cells. Overexpression of key regulatory transcription factors, for example, can be used to achieve a high yield of dopamine neurons from mouse ES cells. Future research will reveal if that is also the case for human ES cells.

*See also:* Animal Models of Parkinson's Disease; Neurogenesis in the Intact Adult Brain; Neurotrophic Factor Therapy: GDNF and CNTF; Parkinson's Disease: Alpha-Synuclein and Neurodegeneration.

## Further Reading

Alvarez-Dolado M, Pardal R, Garcia-Verdugo JM, et al. (2003) Fusion of bone-marrow-derived cells with Purkinje neurons, cardiomyocytes and hepatocytes. *Nature* 425: 968–973.

Annett L (1994) *Functional Studies of Neural Grafts in Parkinsonian Primates.* New York: Raven Press.

Bjorklund A, Dunnett SB, Brundin P, et al. (2003) Neural transplantation for the treatment of Parkinson's disease. *Lancet Neurology* 2: 437–445.

Bjorklund A and Stenevi U (1979) Reconstruction of the nigrostriatal dopamine pathway by intracerebral nigral transplants. *Brain Research* 177: 555–560.

Brundin P, Duan WM, and Sauer H (1994) *Functional Effects of Mesencephalic Dopamine Neurons and Adrenal Chrafffin Cells Grafted to the Rodent Striatum.* New York: Raven Press.

Brundin P, Karlsson J, Emgard M, et al. (2000) Improving the survival of grafted dopaminergic neurons: A review over current approaches. *Cell Transplantation* 9: 179–195.

Cenci MA and Hagell P (2006) Dyskinesias and neural grafting in Parkinson's disease. In: Brundin P and Olanow CW (eds.) *Restorative Therapies in Parkinson's Disease*, pp. 184–224. New York: Springer.

Doucet G, Murata Y, Brundin P, et al. (1989) Host afferents into intrastriatal transplants of fetal ventral mesencephalon. *Experimental Neurology* 106: 1–19.

Fisher LJ, Young SJ, Tepper JM, Groves PM, and Gage FH (1991) Electrophysiological characteristics of cells within mesencephalon suspension grafts. *Neuroscience* 40: 109–122.

Freed CR, Breeze RE, Rosenberg NL, et al. (1992) Survival of implanted fetal dopamine cells and neurologic improvement 12 to 46 months after transplantation for Parkinson's disease. *New England Journal of Medicine* 327: 1549–1555.

Freed CR, Greene PE, Breeze RE, et al. (2001) Transplantation of embryonic dopamine neurons for severe Parkinson's disease. *New England Journal of Medicine* 344: 710–719.

Frielingsdorf H, Schwarz K, Brundin P, and Mohapel P (2004) No evidence for new dopaminergic neurons in the adult mammalian substantia nigra. *Proceedings of the National Academy of Sciences of the United States of America* 101: 10177–10182.

Kim JH, Auerbach JM, Rodriguez-Gomez JA, et al. (2002) Dopamine neurons derived from embryonic stem cells function in an animal model of Parkinson's disease. *Nature* 418: 50–56.

Kirik D, Winkler C, and Bjorklund A (2001) Growth and functional efficacy of intrastriatal nigral transplants depend on the extent of nigrostriatal degeneration. *Journal of Neuroscience* 21: 2889–2896.

Kordower JH, Freeman TB, Snow BJ, et al. (1995) Neuropathological evidence of graft survival and striatal reinnervation after the transplantation of fetal mesencephalic tissue in a patient with Parkinson's disease. *New England Journal of Medicine* 332: 1118–1124.

Kordower JH, Styren S, Clarke M, DeKosky ST, Olanow CW, and Freeman TB (1997) Fetal grafting for Parkinson's disease: Expression of immune markers in two patients with functional fetal nigral implants. *Cell Transplantation* 6: 213–219.

Lindvall O, Brundin P, Widner H, et al. (1990) Grafts of fetal dopamine neurons survive and improve motor function in Parkinson's disease. *Science* 247: 574–577.

Olanow CW, Goetz CG, Kordower JH, et al. (2003) A double-blind controlled trial of bilateral fetal nigral transplantation in Parkinson's disease. *Annals of Neurology* 54: 403–414.

Olson L and Seiger A (1972) Brain tissue transplanted to the anterior chamber of the eye: 1. Fluorescence histochemistry of immature catecholamine and 5-hydroxytryptamine neurons reinnervating the rat iris. *Zeitschrift fur Zellforschung und Mikroskopische Anatomie* 135: 175–194.

Perlow MJ, Freed WJ, Hoffer BJ, Seiger A, Olson L, and Wyatt RJ (1979) Brain grafts reduce motor abnormalities produced by destruction of nigrostriatal dopamine system. *Science* 204: 643–647.

Perrier AL, Tabar V, Barberi T, et al. (2004) Derivation of midbrain dopamine neurons from human embryonic stem cells. *Proceedings of the National Academy of Sciences of the United States of America* 101: 12543–12548.

Piccini P, Brooks DJ, Bjorklund A, et al. (1999) Dopamine release from nigral transplants visualized *in vivo* in a Parkinson's patient. *Nature Neuroscience* 2: 1137–1140.

Piccini P, Lindvall O, Bjorklund A, et al. (2000) Delayed recovery of movement-related cortical function in Parkinson's disease after striatal dopaminergic grafts. *Annals of Neurology* 48: 689–695.

Rosenblad C, Martinez-Serrano A, and Bjorklund A (1996) Glial cell line-derived neurotrophic factor increases survival, growth and function of intrastriatal fetal nigral dopaminergic grafts. *Neuroscience* 75: 979–985.

Takagi Y, Takahashi J, Saiki H, et al. (2005) Dopaminergic neurons generated from monkey embryonic stem cells function in a Parkinson primate model. *Journal of Clinical Investigation* 115: 102–109.

# Neurogenesis in the Intact Adult Brain

**G Kempermann**, Max Delbrück Center for Molecular Medicine (MDC) Berlin-Buch, Berlin, Germany



Adult neurogenesis is the development of new neurons from resident neural precursor cells in the adult brain. The term 'adult neurogenesis' encompasses the entire process of neuronal development from the division of the precursor cell to the existence of a functionally integrated new neuron. Adult neurogenesis is rare and locally restricted in mammals but frequent and widespread in lower vertebrates; it seems that over the course of evolution with increasing brain complexity adult neurogenesis became more and more limited. Quantitatively, most adult neurogenesis occurs relatively early in life, but adult neurogenesis persists lifelong on a very low level. There is no evidence that adult neurogenesis would primarily contribute to a neuronal turnover in that it would replace damage or lost neurons. Rather, the new neurons are added to the persisting networks.

In mammals, physiological adult neurogenesis appears to be restricted to two canonical neurogenic regions, the dentate gyrus of the hippocampus and the olfactory bulb (**Figure 1**). Adult hippocampal neurogenesis generates new excitatory granule cells, the principal cells of the dentate gyrus, whose axons form the mossy fiber tract to hippocampal subregion CA3. Adult olfactory neurogenesis produces two types of interneurons in the granule cell layer and the periglomerular regions of the olfactory bulb.

Adult neurogenesis in other brain regions, most notably the neocortex, remains controversial, and if it exists will range on a minute scale. Initial reports about the production of large numbers of cortical neurons were not confirmed by others. In contrast, the question of whether small numbers of cortical interneurons, primarily in the deep cortical layers, might be generated in adulthood is more difficult to prove or disprove. In addition, under particular circumstances (e.g., in the face of locally limited pathology), regenerative or targeted neurogenesis in physiologically nonneurogenic regions seems to be possible. Such exceptions might question a rigid conceptual distinction between neurogenic and nonneurogenic regions. For most practical purposes, the distinction, however, is valid. In contrast to the nonneurogenic areas, the canonical neurogenic regions contain a morphologically distinct neurogenic or stem cell niche that provides a microenvironment permissive for precursor cell function and neuronal development. The niche consists of the precursor cells and their progeny, supporting cells, immune cells, vasculature, and a specialized extracellular matrix, whose exact composition is still unknown. The structure of the precursor cell niches of the adult brain resemble similar niches, for example, in bone marrow, testes, and olfactory epithelium.

Precursor cells can also be found outside the neurogenic regions, albeit at very low density. No structural evidence of a niche is found in these regions and physiologically no (or extremely limited) neurogenesis appears to take place. The function of precursor cells outside the neurogenic regions is unknown. It is also not clear how homogenous the population of neural precursor cells in the adult brain is. Concepts of neuroanatomists like Wilhelm His, Alfred Schaper, and Wilder Penfield, who proposed the existence of a 'spongioblast' in the brain parenchyma, are thereby revived. Currently, the best candidates for the parenchymal precursor cells outside the neurogenic regions are slowly proliferative cells expressing the proteoglycan NG2. The neurogenic precursor cells in the hippocampus and olfactory system, in contrast, are NG2-negative but express markers including nestin, doublecortin, brain lipid binding protein, glial fibriallary acidic protein, Sox2, and others, but neither sensitivity nor specificity are sufficiently clear.

Meaning and relevance of markers that are associated with neurogenesis in the neurogenic regions (nestin, doublecortin, calretinin, etc.) are not known for the nonneurogenic regions. The presence of these markers alone should not be taken as evidence of a neurogenic potential or even ongoing neurogenesis.

Adult neurogenesis also takes place in the olfactory epithelium (as part of the peripheral nervous system), where olfactory receptor neurons are continuously replaced from local precursor cells.

## History

Adult hippocampal neurogenesis was first described by Joseph Altman, then at the Massachusetts Institute of Technology, in the early 1960s. Altman also described postnatal neurogenesis in the olfactory bulb in 1969. The response by the scientific community was skeptic, largely for methodological reasons but also because of the fact that no stem cell population was known in the brain and the origin of new neurons thus remained speculative. In the 1980s, Fernando Nottebohm and colleagues described adult neurogenesis in those brain nuclei of canary birds that are responsible for song learning. These studies provided the first link between adult neurogenesis

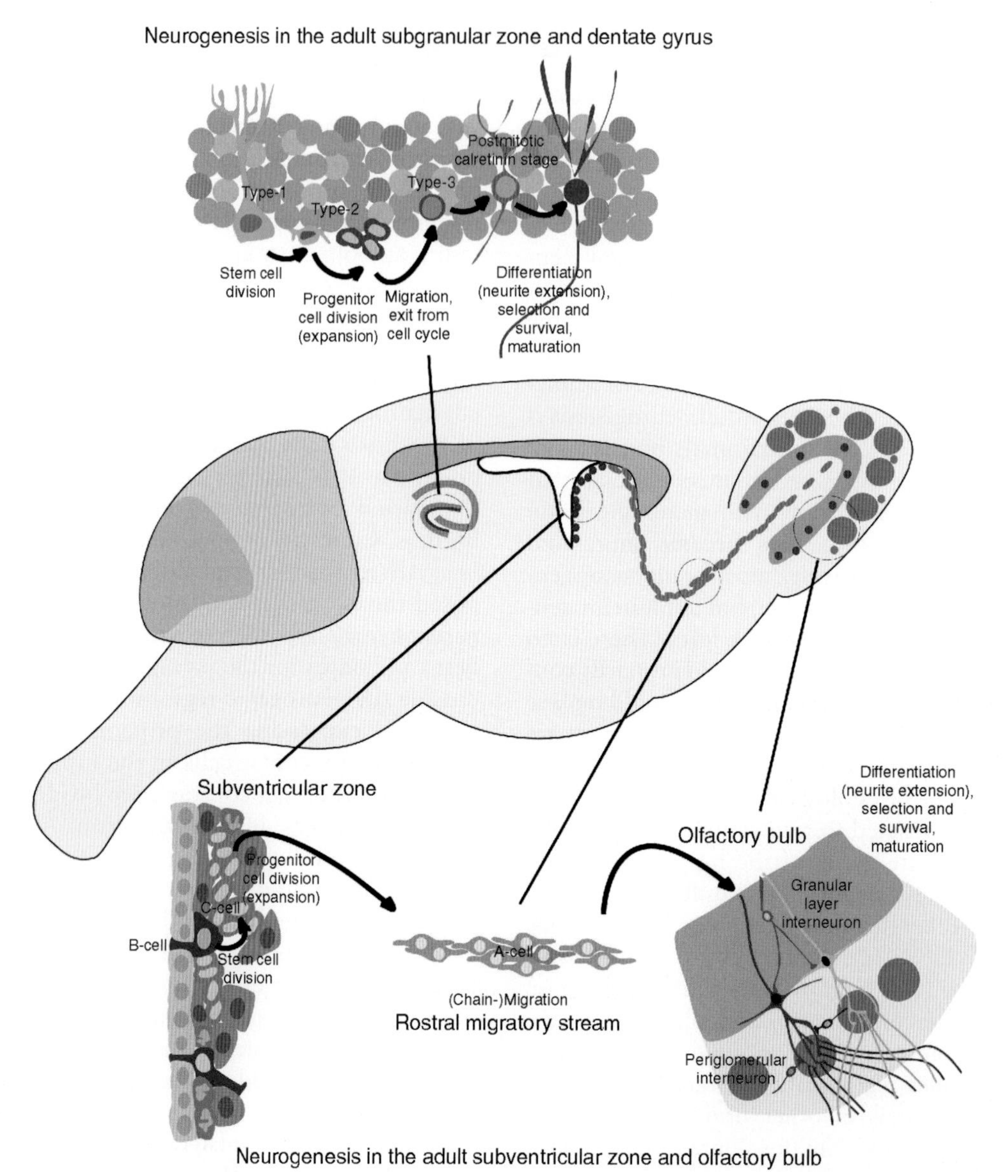


**Figure 1** Adult neurogenesis occurs in two canonical neurogenic regions: the olfactory system and the hippocampus.

and behavior. After the discovery of neural stem cells in the adult mammalian brain in 1992, research on adult neurogenesis gained momentum. The isolation of stem and progenitor cells in the olfactory system was first reported by Brent Reynolds and Sam Weiss. Stem cells in the adult hippocampus followed in 1995 with the work of Theo D Palmer, Jasodarah Ray, and Fred H Gage. In the early 1990s, Elizabeth Could and co-workers reported that corticosteroids negatively affect adult neurogenesis. In 1997 and 1998, Fred H Gage's group showed that environmental enrichment and physical exercise positively influence adult hippocampal neurogenesis. The occurrence of adult neurogenesis in the human dentate gyrus was confirmed in 1998 by Eriksson and colleagues.

## Adult Hippocampal Neurogenesis

The precursor cells driving adult hippocampal neurogenesis reside in the subgranular zone (SGZ) of the adult dentate gyrus. An astrocyte-like stem cell with radial morphology (type 1 cell) gives rise to highly proliferative intermediate progenitor cells (type 2 cells) and lineage-determined neuroblast-like cells (type 3 cells) that exit from the cell cycle and go through an immature postmitotic stage. During this stage, which coincides with the transient expression of calretinin, the new neurons fully extend their neurites and go through a phase of increased synaptic plasticity. The distinction of the cell types by numbers is preliminary and should one day be exchanged to a functional nomenclature.

Regulation of adult neurogenesis mainly occurs in two stages, an expansion phase on the level of the precursor cells and a phase of selective survival on a postmitotic stage. A transient expression of doublecortin lasts from the type 2 cell level to the calretinin phase. Doublecortin has become a surrogate marker of adult neurogenesis that is even used for quantification. The equalization of doublecortin expression with hippocampal neurogenesis, however, is not without problems because doublecortin expression ends before the cells are fully integrated. The newly generated cells receive first input through GABAergic synapses, which are excitatory. This input is involved in the induction of full maturation. The new cells receive glutamatergic input within days to weeks and go through a transient phase of increased synaptic plasticity (i.e., a lowered threshold for the induction of long-term potentiation). After approximately 7 weeks, the cells have become largely indistinguishable from the older granule cells and remain stably integrated into the network.

## Adult Olfactory Neurogenesis

The precursor cells for adult olfactory neurogenesis are found in the subventricular zone (SVZ) of the lateral ventricle. The stem cell of this region has a cilia-bearing process touching the ventricular surface between the ependymal cells. The exact relationship between these stem cells (B cells) and the ependymal cells (E cells) is not clear; in particular, it remains controversial whether under certain conditions ependymal cells can act as stem cells. Undoubtedly, however, the astrocyte-like B cells do serve as stem cells and give rise to highly proliferative progenitor cells (C cells) that generate migratory neuroblasts (A cells). The A cells migrate in a specialized form of migration, homonymic chain migration, through the rostral migratory stream (RMS) to the core of the olfactory bulb, where they change for a radial migratory pattern and reach the granule cell layer and the glomerular and periglomerular regions. Precursor cells with a neurogenic potential can be isolated not only from the SVZ but also from the RMS and the olfactory bulb. All new neurons in the olfactory bulb are inhibitory interneurons and thus GABAergic. A subset of new periglomerular neurons co-expresses dopamine as neurotransmitter. Maturation of the new cells is delayed until they have reached their final position. The new cells in the granular layer express calretinin. The exact lineage relationship between the different types of new olfactory bulb neurons is not clear.

## Regulation

Regulation of adult neurogenesis mainly falls into two categories. Numerous relatively nonspecific stimuli can affect the proliferation of precursor cells. Acutely this increase can lead to an increase in net neurogenesis, but this does not necessarily have to be the case. For many factors, the question of whether a particular stimulus has long-term effects on net neurogenesis has not yet been specifically addressed. Many of the nonspecific stimuli are associated with 'activity,' and many examples of pathology (ischemia, trauma, etc.) can upregulate neurogenesis. Other stimuli such as stress and inflammation downregulate precursor cell proliferation and neurogenesis. For many such regulators, complex dose–response relationships might exist so that generalization is difficult. Age can be considered a strong negative regulator early in life but apparently loses this strong impact, because with increasing age adult neurogenesis settles at a very low level.

Stimuli that appear to be of more specific relevance to the hippocampus or the olfactory system act on the selective survival of the new cells. Under physiological conditions, these stimuli tend to be associated with the function of the hippocampus (memory, learning, etc.) or the olfactory system (olfaction). Genetical studies indicate that regulation on the survival and differentiation level has a larger quantitative (and presumably qualitative) impact on the control of adult neurogenesis than the regulation of cell proliferation.

On a mechanistic level, regulation involves several neurotransmitter systems that provide input to the SGZ and SVZ. For the SGZ, the research focus has been on the GABAergic, the serotonergic, and the glutamatergic system. High levels of glucocorticoids acutely suppress neurogenesis. Presumably because the expression of corticosteroid receptor expression changes in the course of neuronal development in the adult, the net effect under prolonged exposure to glucocorticoids is complex. Many situations with elevated levels of glucocorticoids are thus associated with increased levels of neurogenesis. The details of this regulation are unknown. The role of other hormone systems is even more ambiguous.

Growth factors and neurotrophic factors, most notably fibroblast growth factor 2 (FGF2) and epidermal growth factor (EGF), strongly affect adult neurogenesis in a complex way. Insulin-like growth factor 1 (IGF1) and vascular endothelial growth factor (VEGF) have been discussed as centrally involved in the activity-dependent regulation of adult neurogenesis. Brain-derived neurotrophic factor (BDNF) is widely proposed to play a major role in controlling adult neurogenesis but experimental evidence supporting this assumption so far remains limited.

Among the short-range-acting signaling molecules with effects on adult neurogenesis contributions of the ephrins, bone morphogenic proteins (BMPs) and the Wnt system have been described. More are likely to follow. It is assumed that the general and intrinsic developmental principles regulating neuronal development in the adult more or less recapitulate the processes prevalent in the fetus. Extrinsic and activity-dependent regulation, in contrast, are likely to set adult neurogenesis apart from fetal neurogenesis. In addition, differences between the two neurogenic regions might exist with regard to the contribution of individual regulatory factors.

## Function

The fact that adult neurogenesis is the exception and not the rule has raised the question of why dentate gyrus and olfactory bulb in contrast to all other brain regions not only accommodate the integration of new neurons but actually call for it in a function-dependent manner. In both systems, adult neurogenesis does not add bulk numbers of new neurons. Rather, the contribution of adult neurogenesis might be qualitative and allow a possibly subtle modification of the existing neuronal network. In the hippocampus, adult neurogenesis allows the activity-dependent optimization of the mossy fiber connection between the dentate gyrus and CA3. Integration is triggered during a phase of increased synaptic plasticity and leads to a cumulative adaptation of the network to a level of complexity and novelty. Alternative hypotheses that mostly suggest a transient functional contribution of the immature neurons in the sense of a direct contribution to memory formation are questioned by the lacking evidence for substantial neuronal turnover and the scarcity of new neurons. Mechanistically, it has been proposed that the new neurons help the hippocampal network to avoid catastrophic interference in the dentate gyrus, that is, the destruction of codings for older information by incoming new information.

Similarly, adult olfactory neurogenesis has been brought into connection with novelty detection, here the coding of novel odors. It seems that new neurons are more sensitive to novel odors and might help to adjust the network to the processing of a new range of olfactory stimuli.

A general principle might be that adult neurogenesis is used in bottleneck situations, where the insertion of new neurons is strategic and allows the incremental optimization of a network that in general tends to remain as lean as possible. Networks in nonneurogenic areas might not exist under this priority of size limitation.

## Medical Relevance

Adult neurogenesis might be of medical relevance either because a failure of adult neurogenesis might contribute to brain disease or because a reactive increase in adult neurogenesis might be part of endogenous regenerative attempts. A role for adult neurogenesis in brain pathology, among others, is discussed for temporal lobe epilepsy, major depression, and dementias. In temporal lobe epilepsy, cell proliferation is massively increased, many new neurons are found in ectopic locations, and aberrant projections from the new neurons might support the chronification of the disease. In the context of major depression, a reduction in hippocampal volume has been noted (that is, however, much larger to be explained by a lacking contribution of adult neurogenesis) and all antidepressants have been found to increase adult hippocampal neurogenesis. As the most extreme position in the context, it has been suggested that antidepressants might even require neurogenesis for their action. Dementias might at least partly be explained by a reduction in the contribution that new neurons make to hippocampal function – a concept that is complicated by the fact that the functional relevance of adult neurogenesis itself is not yet fully understood. As yet, no similar concrete links to pathology have been proposed for neurogenesis in the olfactory bulb.

Numerous pathological stimuli, most notably trauma and ischemia, induce the proliferation of precursor cells and in many cases net adult neurogenesis. These effects appear to be indirect because they are found even in the absence of direct damage to the dentate gyrus or SVZ. In addition, after middle cerebral artery occlusion in rats, precursor cells migrated from the SVZ into the striatum and differentiated into local neurons. The massive infusion of growth factors supported the replacement of ischemic CA1 neurons by SVZ precursor cells. Phototoxically induced circumscript cell death in the cortex also led to the replacement of neurons.

Envisioned therapeutic strategies that are based on the transplantation of neurons that have been generated from neural precursor cells benefit from the knowledge accumulating from research on adult neurogenesis, because here nature exemplifies how neuronal development is possible under the conditions of the otherwise nonneurogenic adult brain.

Medical relevance also depends on the occurrence of adult neurogenesis in humans. Adult hippocampal neurogenesis has been demonstrated for humans and nonhuman primates. There is evidence for neurogenesis in the adult olfactory bulb of nonhuman primates but not yet unambiguously of humans. Precursor cells, however, have been isolated from ventricle biopsies and resection specimens from the adult human hippocampus, as well as postmortem brain tissue of infants.

*See also:* Cell Replacement Therapy: Parkinson's Disease; Exercise: Optimizing Function and Survival at the Cellular Level; Lipids and Membranes in Brain Aging; Synaptic Plasticity: Neuronogenesis and Stem Cells in Normal Brain Aging.

## Further Reading


Altman J and Das GD (1965) Autoradiographic and histological evidence of postnatal hippocampal neurogenesis in rats. *Journal of Comparative Neurology* 124: 319–335.

Alvarez-Buylla A, Garcia-Verdugo JM, and Tramontin AD (2001) A unified hypothesis on the lineage of neural stem cells. *Nature Reviews Neuroscience* 2: 287–293.

Alvarez-Buylla A and Lim DA (2004) For the long run: Maintaining germinal niches in the adult brain. *Neuron* 4: 683–686.

Cameron HA, Woolley CS, McEwen BS, and Gould E (1993) Differentiation of newly born neurons and glia in the dentate gyrus of the adult rat. *Neuroscience* 56: 337–344.

Emsley JG, Mitchell BD, Kempermann G, and Macklis JD (2005) Adult neurogenesis and repair of the adult CNS with neural progenitors, precursors, and stem cells. *Progress in Neurobiology* 75: 321–341.

Eriksson PS, Perfilieva E, Bjork-Eriksson T, et al. (1998) Neurogenesis in the adult human hippocampus. *Nature Medicine* 4: 1313–1317.

Gould E and Tanapat P (1999) Stress and hippocampal neurogenesis. *Biological Psychiatry* 46: 1472–1479.

Kempermann G (2006) *Adult Neurogenesis – Stem Cells and Neuronal Development in the Adult Brain.* New York: Oxford University Press.

Kempermann G, Jessberger S, Steiner B, and Kronenberg G (2004) Milestones of neuronal development in the adult hippocampus. *Trends in Neurosciences* 27: 447–452.

Kempermann G, Kuhn HG, and Gage FH (1997) More hippocampal neurons in adult mice living in an enriched environment. *Nature* 386: 493–495.

Lie DC, Song H, Colamarino SA, Ming GL, and Gage FH (2004) Neurogenesis in the adult brain: New strategies for central nervous system diseases. *Annual Review of Pharmacology and Toxicology* 44: 399–421.

Lledo PM, Alonso M, and Grubb MS (2006) Adult neurogenesis and functional plasticity in neuronal circuits. *Nature Reviews Neuroscience* 7: 179–193.

Lledo PM and Saghatelyan A (2005) Integrating new neurons into the adult olfactory bulb: Joining the network, life–death decisions, and the effects of sensory experience. *Trends in Neurosciences* 28: 248–254.

Magavi SS, Leavitt BR, and Macklis JD (2000) Induction of neurogenesis in the neocortex of adult mice. *Nature* 405: 951–955.

Nottebohm F (2004) The road we travelled: Discovery, choreography, and significance of brain replaceable neurons. *Annals of the New York Academy of Sciences* 1016: 628–658.

Palmer TD, Ray J, and Gage FH (1995) FGF-2-responsive neuronal progenitors reside in proliferative and quiescent regions of the adult rodent brain. *Molecular and Cellular Neurosciences* 6: 474–486.

# Neurotrophic Factor Therapy: GDNF and CNTF

**E-Y Chen, D Fox, and J H Kordower**, Rush University Medical Center, Chicago, IL, USA



## Introduction

During the past decade, there has been a major increase in research associated with neurotrophic factors (NTFs), molecules that provide survival-promoting effects to developing and adult central nervous system (CNS) neurons. From a clinical standpoint, it has been hoped that research would ultimately lead to the development of novel therapeutic approaches for the treatment of various neurodegenerative disorders, such as Alzheimer's disease, Parkinson's disease (PD), Huntington's disease (HD), and amyotrophic lateral sclerosis (ALS). To date, such clinical promise has not been realized, but optimism remains high due to the potency of the NTFs to protect and augment the function of vulnerable neuronal systems, combined with novel methods for their delivery.

## Glial Cell Line-Derived Neurotrophic Factor

### Identification and Role

The glial cell-derived neurotrophic factor (GDNF) family is distantly related to the transforming growth factor superfamily and includes GDNF and three structurally related members named neurturin, persephin, and artemin. GDNF, the first-identified member of this family, was originally isolated from the rat B49 glial cell line. GDNF binds to a multisubunit receptor system composed of the GFR$\alpha$1 subunit and the RET subunit, which function as the ligand binding and signaling components, respectively. Ret is a transmembrane receptor tyrosine kinase (encoded by the proto-oncogene c-ret) that triggers many intracellular signaling pathways, including the Ras-MAPK, the phosphoinositol 3-kinase, and Jun N-terminal kinase- and PLC$\gamma$-dependent pathways. After the initial discovery that GDNF protects embryonic nigral neurons in culture, a plethora of *in vitro* and *in vivo* studies demonstrated that this factor prevents the degeneration and augments the phenotypic expression of dopaminergic midbrain neurons. With regard to GDNF and dopaminergic systems, there have been a series of pivotal experiments. A single intraventricular injection of GDNF to a normal rat induces a long-term increase in striatal dopamine. Most critically, GDNF has been able to prevent the structural and/or functional consequences that are engendered in virtually every rodent model of PD, including those with lesions induced by 6-hydroxydopamine, methamphetamine, or axotomy, as well as the degenerative changes associated with aging. Furthermore, CNS infusions of GDNF augment motor function and reverse nigrostriatal degeneration in parkinsonian and aged rhesus monkeys, and lentiviral delivery of GDNF prevents nigrostriatal degeneration in 1-methyl-4-phenyl-1,2,3,6-tetrahydropyridire (MPTP)-treated monkeys and augments dopaminergic function in aged monkeys. An exception to this rule is provided by one study in which GDNF failed to prevent the structural and functional consequences that result from overexpression of $\alpha$-synuclein in a rat model of PD. GDNF also has well-documented effects on the survival and function of cranial and spinal cord motor neurons, brain stem noradrenergic neurons, basal forebrain cholinergic neurons, Purkinje cells, and specific groups of dorsal ganglia and sympathetic neurons.

### Distribution of GDNF in the CNS

Using polymerase chain reaction (PCR) and *in situ* hybridization techniques, the regional distribution and cellular localization of GDNF have been examined in the rat and human CNS. GDNF transcripts have been identified in all major regions of the rat CNS, such as the striatum, hippocampus, cortex, cerebellum, and spinal cord. Reverse-transcriptase polymerase chain reaction (RT-PCR) analysis revealed that GDNF mRNA is also observed in the human CNS in regions such as the striatum, hippocampus, cortex, and spinal cord. Using immunohistochemical techniques, it has been shown that GDNF immunoreactive neurons are widely distributed in the normal human brain and spinal cord, including cerebral cortex, hippocampus, basal forebrain, hypothalamus, thalamus, cerebellum, and brain stem. GDNF and its signaling receptor RET are also found in human peripheral nerve axons and DRG neurons, and GDNF is found in Schwann cells. GDNF immunostaining has also been found in melanin-containing neuronal perikarya of the substantia nigra pars compacta (SNpc) in normal human brain. This region undergoes extensive degeneration in PD. Experimental studies performed using RT-PCR techniques indicate that GDNF and its receptor components are broadly expressed in the rat and human hippocampus from early embryonic age to adult life. Available quantitative data show that in the rat hippocampus, levels of GDNF mRNA and protein steadily increase from the early stages to birth, reaching a maximum during the first week after birth and gradually decreasing to prenatal levels in adult. The occurrence and pattern of overall distribution of GDNF-positive neurons in human hippocampus appear consistently similar throughout life stages from 29 weeks of gestation to adulthood. However, the relative numerical density of labeled neurons changes with age.

### GDNF and PD

PD is a neurodegenerative disorder that affects more than 1 million people in North America. PD is characterized by the selective degeneration of nigral dopaminergic neurons in the SNpc and subsequent depletion of striatal dopamine which leads to motor dysfunction. PD patients can present with a variety of symptoms (bradykinesia, muscle rigidity, resting tremor, and loss of postural reflexes), two of which are required for diagnosis of the disease. The diagnosis of PD is made on the basis of clinical criteria. Confirmation of the clinical diagnosis occurs upon neuropathologic examination. Available pharmacological treatments do not stop the progressive loss of dopamine neurons observed during the course of PD. The progressive nature of the disease provides the possibility of therapeutic interventions that slow, or completely halt, the ongoing degenerative process. The use of trophic factors has emerged as a potential therapeutic approach because of their ability to regulate the survival of specific neuronal populations in the CNS. GDNF, a member of the transforming growth factor-$\beta$ superfamily, has been shown to be one of the most potent neuroprotective factors for dopaminergic neurons. Many studies have demonstrated that GDNF controls the survival and physiologic properties of mesencephalic dopaminergic neurons both *in vitro* and *in vivo*. Therefore, reduction of GDNF expression in the SNpc might lead to neuronal loss in PD, although this concept is controversial. In one early *in situ* hybridization study, GDNF mRNA expression failed to show any labeling in the mesencephalon and striatum of control subjects or patients with PD. RT-PCR analysis revealed the highest expression of GDNF mRNA in the human caudate, with low levels in the putamen and no detectable mRNA in the SN, suggesting that GDNF is synthesized and secreted by striatal neurons and then retrogradely transported to the SN in a target-derived retrograde manner. Receptor studies found that tissue sections from the SNpc of control subjects and PD patients localized RET immunoreactivity to the dopaminergic neurons. RET mRNA expression was also observed in many surviving neurons in all PD patients. GDNF is present in the adult striatum, and the dopaminergic neurons of the SNpc express mRNAs encoding the GDNF receptors, RET and GFR$\alpha$1. GDNF levels were significantly reduced (approximately 20%) in all sampled surviving melanin-containing neurons in the SNpc of PD patients compared with age-matched controls, supporting a role for decreased availability of GDNF as part of the process of degeneration of SNc neurons in PD. Together, these findings indicate that GDNF may play a role in the pathology of PD and may be the most potent neuroprotective factor for PD.

These and other studies prompted researchers to investigate putative protective effects in different PD models as well as PD patients. As discussed previously, direct injection of the recombinant GDNF protein has potent neuroprotective effects in rodent and primate models of PD. In polymer-encapsulated GDNF-secreting cells, continuous release of low levels of GDNF protects the nigral dopaminergic neurons from an axotomy-induced lesion and improves dopamine graft survival in 6-OHDA-treated rats. Gene therapy is another method that has great potential for delivering trophic factors continually to focal regions of the CNS. In this regard, viral vectors are attractive candidates for delivering GDNF to the PD brain given the nature of the disease. There are numerous studies using gene therapy in animal models of PD. Overexpression of GDNF to the striatum and/or SN has been achieved primarily using adeno-associated viruses or lentiviruses. One of the most promising studies demonstrated that delivery of GDNF using a primate-derived lentiviral vector provided almost complete neuroprotection against MPTP-mediated toxicity in monkeys and augmented dopaminergic function in aged monkeys (**Figure 1**). These encouraging preclinical observations led to a series of pilot human trials in which GDNF was used in PD patients. A double-blind, placebo-controlled, sequential cohort study performed for 8 months compared the effects of monthly intracerebroventricular administration of placebo and GDNF in 50 moderately advanced PD patients. These trials were suspended because of lack of observed efficacy and frequent occurrence of nausea, anorexia, tingling (Lhermitte sign), hallucinations, and depression. GDNF did not improve parkinsonism, possibly because GDNF did not reach the target tissues – putamen and SN. A phase I safety trial reported that 1 year after long-term infusion of GDNF directly via a pump and catheter into the posterodorsal putamen of 5 PD patients, there were no serious clinical side effects, and there was a 39% improvement in the off-medication motor subscore of the Unified Parkinson's Disease Rating Scale (UPDRS) and a 61% improvement in the activities of daily living subscore. Furthermore, levodopa-induced dyskinesias were reduced by 64%. This clinical improvement was accompanied by a 28% increase in putamen PET

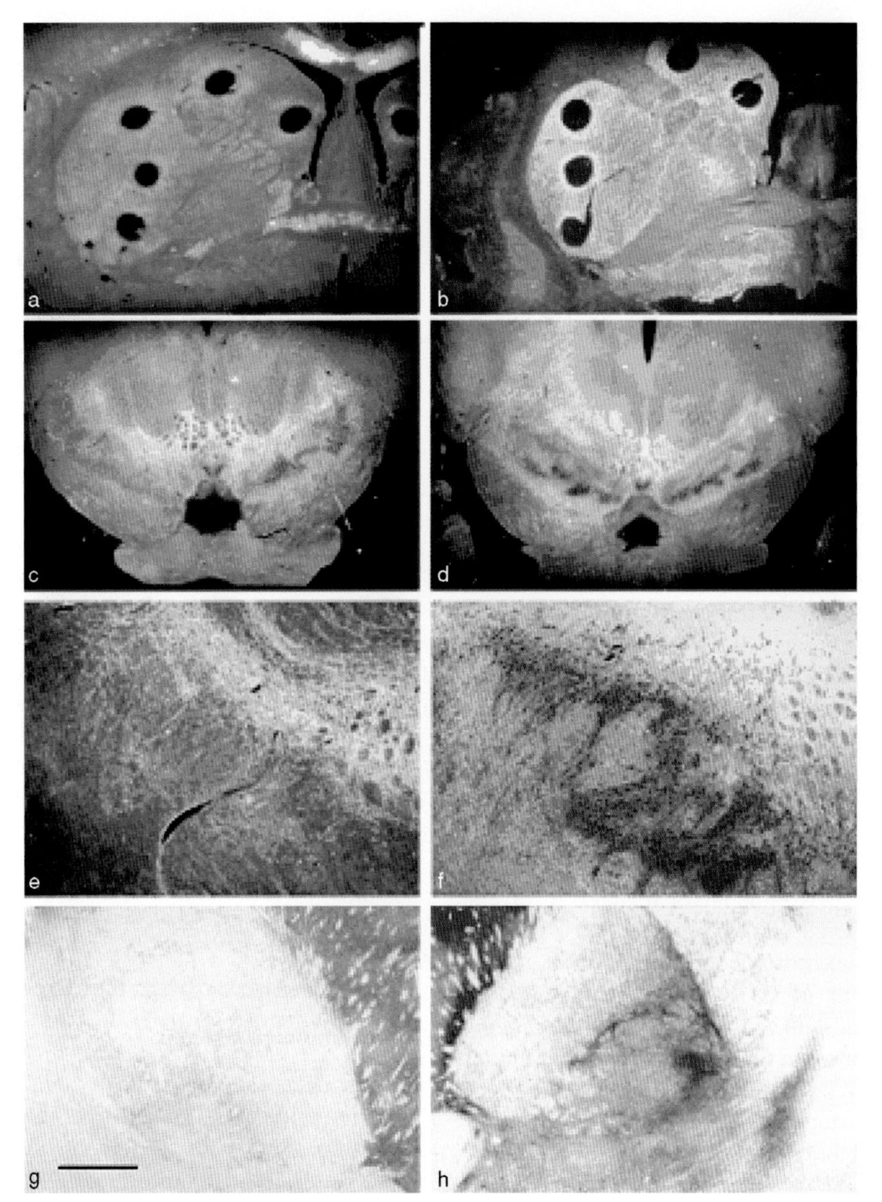


**Figure 1** (a and b) Low-power dark field photomicrographs through the right striatum of TH immunostained sections of MPTP-treated monkeys treated with (a) lenti-$\beta$Gal or (b) lenti-GDNF. (a) There was a comprehensive loss of TH immunoreactivity in the caudate and putamen of lenti-βGal-treated animal. In contrast, a near normal level of TH immunoreactivity is seen in lenti-GDNF-treated animals. Low-power (c, d) and medium-power (e, f) photomicrographs of TH immunostained section through the substantia nigra of animals treated with lenti-$\beta$Gal (c, e) and lenti-GDNF (d, f). Note the loss of TH immunoreactive neurons in the lenti-$\beta$Gal-treated animals on the side of the MPTP injection. TH immunoreactive sprouting fibers, as well as a supranormal number of TH immunoreactive nigral perikarya, are seen in lenti-GDNF-treated animals on the side of the MPTP injection. (g, h) Bright field low-power photomicrographs of a TH immunostained section from a lenti-GDNF-treated monkey. (g) Note the normal TH immunoreactive fiber density through the globus pallidus on the intact side that was not treated with lenti-GDNF. (h) In contrast, an enhanced network of TH immunoreactive fibers is seen on the side treated with both MPTP and lenti-GDNF. Scale bar in (g) represents the following: 3500 μm (a–d); 1150 μm (e–h).

F-DOPA uptake and evidence of neuronal sprouting in an autopsied brain. After 2 years of continuous GDNF infusion, there were still no serious clinical side effects and no significant detrimental effects on cognition. Patients showed a 57% and 63% improvement in their off-medication motor and activities of daily living subscores of the UPDRS, respectively, and health-related quality-of-life measures showed general improvement over time. A multicenter trial, however, showed no benefit from GDNF, and further studies have been suspended because of evidence of the development of GDNF antibodies in some patients. In addition, cerebellar neurodegeneration involving both Purkinjie cells and the granular cell layer in monkeys exposed to high doses of GDNF has been found. Based on these data, clinical trials delivering GDNF protein to the putamen have been stopped.

### GDNF and HD

Another potential therapeutic role of GDNF is in HD. HD is an autosomal dominant neurodegenerative disorder resulting from an expanded trinucleotide (CAG) repeat at the IT15 locus on chromosome 4 within the huntingtin gene. The abnormal DNA is translated into mutant huntingtin with an expanded glutamine stretch at the N-terminus of the protein. The excessive number of glutamine repeats is responsible for the misfolding of huntingtin and the subsequent formation of neuronal inclusions, degeneration of striatal and cortical neurons, and a triad of symptoms including severe motor, cognitive, and psychological disturbances that are ultimately fatal. To date, HD remains incurable. Toward this end, GDNF has shown promise in animal models of several different neurodegenerative disorders. In addition to its potent trophic effects on dopaminergic midbrain neurons, GDNF has also been shown to protect striatal, medium-sized spiny GABA projection neurons, the neuronal population most vulnerable in HD. Moreover, the expression of GDNF's receptors (GFR$\alpha$1 and Ret) is upregulated in striatal neurons and astrocytes after injury, supporting its role as a potential trophic factor for HD. Intraventricularly administered GDNF protected striatal neurons from both excitotoxic and metabolic insult. Intrastriatal grafting of a GDNF-producing cell line or GDNF-secreting stem cells prevented the degeneration of striatal neurons in the excitotoxic rodent model of HD. Injection of adeno-associated virus (AAV) vector-mediated gene transfer of GDNF into the striatum improves behavior and provides neuronal protection in both rat and transgenic HD mice, the gold-standard animal model for HD.

## GDNF and Cerebral Ischemia

Ischemic brain injury is a major cause of death and permanent disability worldwide. Cell injury and death after cerebral vascular occlusion have been postulated to result from a number of interacting pathophysiologic factors, including depolarization-induced calcium entry via *N*-methyl-D-aspartate receptors, intracellular free radical generation, damage to mitochondrial respiratory enzymes, and induction of programmed cell death. GDNF was initially considered to be a specific trophic factor and neuroprotective agent for dopaminergic neurons. It was later found that GDNF has effects on other systems. GDNF diminishes damage produced by kainate-induced tonic–clonic convulsions and axotomy of brain stem motor neurons. Since GDNF has an antiapoptotic effect in dopamine and nondopaminergic neurons after injury, it was hypothesized that GDNF could attenuate ischemic brain injury through the same mechanism. Exogenous administration of GDNF reduces the toxic effects of excitatory amino acids, attenuates nitric oxide production, and lowers apoptosis/cell death in stroke animals. Intracerebroventricular and intraparenchymal administration of GDNF potently protected the cerebral hemispheres from damage induced by middle cerebral artery (MCA) occlusion in rats. A reduction of ischemic brain injury by topical application of GDNF after permanent MCA occlusion in rats has also been reported. Intravenous application of a blood–brain barrier-penetrating form of GDNF, TAT-GDNF, reduces infarct size after ischemia in mice. Topically applied to the ischemic brain after MCA ligation, GDNF-fibrin glue reduces ischemic brain injury in chronic focal cerebral ischemia in conscious rats. Both recombinant AAV vector and herpes simplex virus amplicon-based vector expressing GDNF reduce ischemia-induced damage. These data support the concept that GDNF may play a therapeutic role in the treatment of ischemia. However, a remaining challenge is to determine how to deliver this trophic molecule in a manner in which it can be clinically useful.

## Ciliary Neurotrophic Factor

### Identification and Role

Ciliary neurotrophic factor (CNTF) is structurally and functionally related to members of a family of cytokines that includes leukemia inhibitory factor (LIF), interleukin-6, interleukin-11, and oncostatin M. CNTF was initially identified due to its ability to support the survival of parasympathetic neurons of the chick ciliary ganglion *in vitro* and subsequently purified to homogeneity from sciatic nerves. CNTF is synthesized by astrocytes and is generally thought to provide autocrine and paracrine signals from astrocytes which induce cellular protection and hypertrophy in response to injury. CNTF exerts its biological effects by binding to a ligand-specific $\alpha$ subunit, CNTF receptor $\alpha$ (CNTFR $\alpha$ ), which leads to the subsequent recruitment of two signal-transducing $\beta$ components, LIFR$\beta$ and gp130. Following dimerization of the $\beta$ components, preassociated cytoplasmic Jak and Tyk kinases are activated, resulting in tyrosine phosphorylation of the $\beta$ components and other cellular proteins. CNTF plays an important role in the maintenance and regeneration of the adult nervous system.

CNTF prolongs neuronal survival of various culture systems and enhances the survival of sensory neurons, motor neurons, cerebral neurons, and hippocampal neurons. CNTF also promotes cholinergic differentiation of noradrenergic sympathetic neurons, the differentiation of glial progenitors into astrocytes, and the maturation and survival of oligodendrocytes. Some peripheral tissues, notably skeletal muscle, are also known to respond to CNTF. Many studies have demonstrated that CNTF administration prevents injury-induced cell death and degeneration of cholinergic, dopaminergic, GABAergic, and thalamocortical neurons in various lesion models of CNS diseases. However, the majority of studies have linked CNTF with motor function in animal models of ALS, HD, and sensory function such as retinopathies.

### Distribution of CNTF in the CNS

CNTF was originally identified as a survival-promoting activity for ciliary ganglion neurons in chick eye tissues. Subsequently, it was cloned from mammalian nervous tissue. The protein is mainly expressed in glial cells of both the peripheral nerve system and the CNS and has been implicated in many developmental processes in the nervous system, including cell fate determination of neural stem cells, the proliferation and transmitter choice of neuronal precursor cells, and the survival and differentiation of developing neurons as well as glial cells. In addition, it acts on a variety of mature peripheral and central neurons preventing injury-induced cell death and degeneration. Demonstrations of retrograde axonal transport of radiolabeled CNTF in the sensory neurons of the sciatic nerve suggest that CNTF may function by target synthesis and retrograde transport in central neurons. CNTF was observed only after birth and found in both the neurons and the neuroglia of the SN, mesencephalon, cerebellum, and the spinal cord from postnatal day 1 to postnatal week 3 in developing Mongolian gerbils. CNTF has been localized primarily in glial cells, with highest concentrations in Schwann cells, and in white matter tract astrocytes of the mouse CNS. The expression of CNTFR $\alpha$ is widely distributed in both rat and primate CNS motor, sensory, and autonomic neurons. There are no published data on the distribution of CNTF in human brain.

### CNTF and Amyotrophic Lateral Sclerosis

ALS is the most common adult form of motor neuron disease. The clinical hallmarks of ALS are progressive muscle weakness, atrophy, and spasticity, which reflect the degeneration and death of upper and lower motor neurons in the cortex, brain stem, and spinal cord. Generally fatal within 1–5 years of onset, ALS has a prevalence of 1 or 2 per 100 000 people. In 90% of cases, there is no apparent genetic linkage (sporadic ALS), but in the remaining 10% the disease is inherited (familial ALS), often in an autosomal-dominant manner. Among familial cases of ALS, approximately 20% are caused by mutations in the gene encoding Cu, Zn-superoxide dismutase (SOD1). The role of CNTF in the etiology and pathogenesis of ALS is not clear. Using a sensitive enzyme-linked immunoassay, CNTF protein in the lateral corticospinal tract was markedly lower in ALS patients than in controls. In the spinal cord of ALS patients, there is a marked increase in the hybridization signal for CNTFR $\alpha$ subunit mRNA overlying motor neurons. CNTF improves survival in the progressive motor neuronopathy (pmn) and wobbler mutant mouse models of motor neuron disease, even though these mice do not have an obvious defect in CNTF or its receptor. Conversely, mice bred for the null mutation of CNTF or its receptor do develop motor neuron degeneration. CNTF also appears to be effective for retarding cellular and functional losses in the mouse models of ALS. CNTF significantly reduced motor neuron loss, improved motor function, and increased survival time. These observations led to the unsuccessful use of CNTF in a therapeutic clinical trial in ALS. This trial was designed to evaluate the safety, tolerability, and efficacy of subcutaneous administration of recombinant human CNTF. Unfortunately, it failed to ameliorate symptoms of ALS while producing adverse effects, such as injection site reactions, cough, asthenia, nausea, anorexia, weight loss, and increased salivation.

### CNTF and HD

CNTF has also been tested in multiple animal models of HD. CNTF has been shown to protect striatal output neurons in a rat model of HD. In one study, encapsulated cells containing baby hamster kidney cells genetically modified to deliver hCNTF were implanted into the lateral ventricle and provided dose-dependent neurochemical and functional protection in a rat model of HD. Human neural stem cell transplants improved motor function in a rat model of HD, and recovery was superior in rats receiving CNTF-treated human cells compared to rats receiving transplants of untreated cells. Intrastriatal injections of adenovirus-mediated CNTF provided neuroprotection for corticostriatopallidal neurons in a rat model of progressive striatal degeneration. In nonhuman primate models of HD, intrastriatal implantation encapsulated CNTF-producing cells prevented the degeneration of vulnerable striatal populations and cortical–striatal basal ganglia circuits, and in a

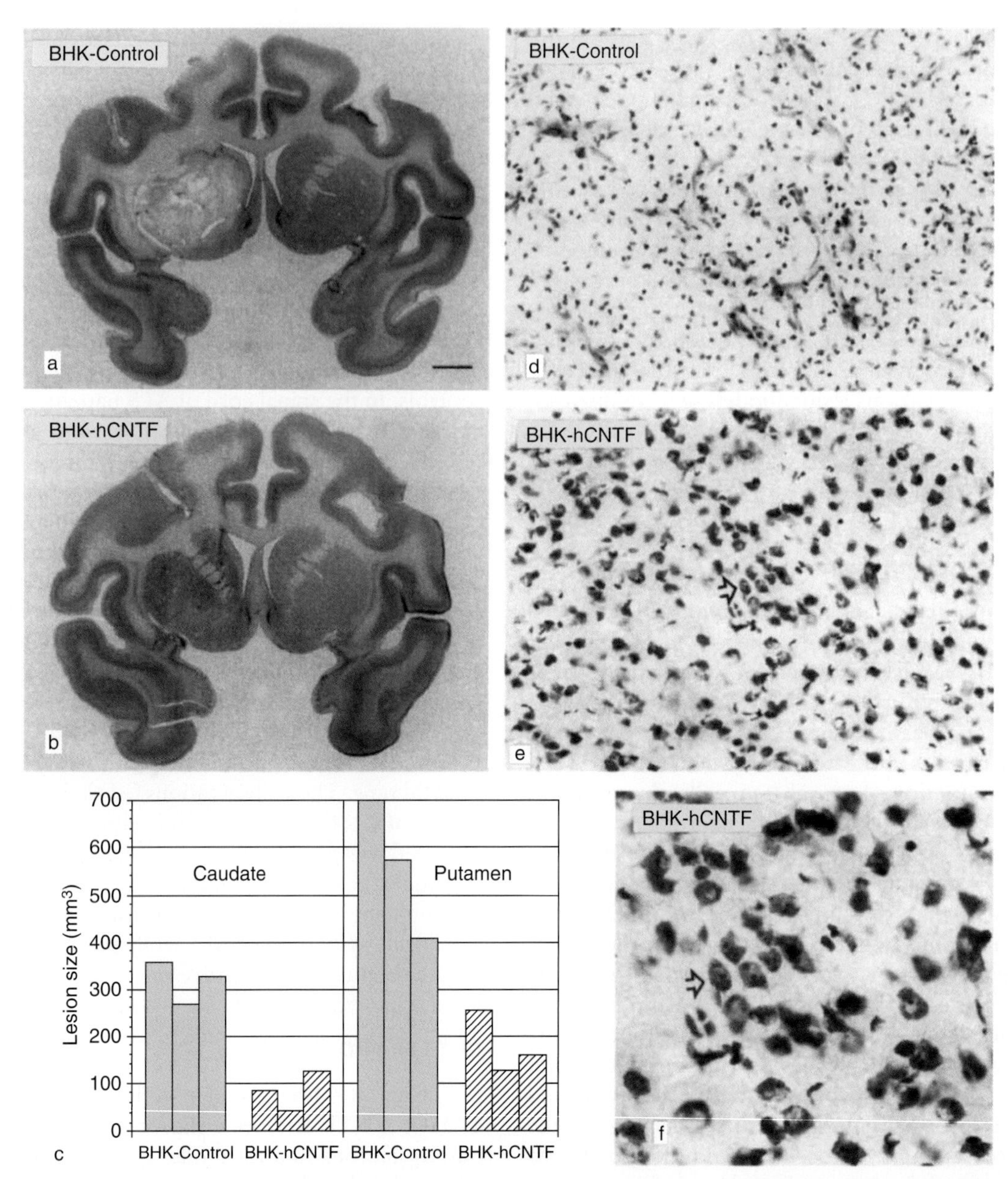


**Figure 2** Low-power photomicrographs of Nissl stained sections through the forebrain of monkeys receiving unilateral intracaudate and intraputamenal lesions with QA and encapsulated (a) BHK-control or (b) BHK-hCNTF implants. Note the extensive lesion area on the left side which encompasses much of the striatum in a. In contrast, the damage in b is negligible for a monkey treated with hCNTF. (c) The size of the lesion in both the caudate nucleus and putamen as quantified on Nissl stained sections was significantly decreased in BHK-hCNTF implanted monkeys (striped bars) relative to BHK-control implanted animals (gray bars). Each bar represents data from an individual animal. (d, e) Medium- and high-power (f) photomicrographs of Nissl stained sections through the striatum of BHK-control (d) and BHK-hCNTF implanted (e, f) monkeys. (d) A paucity of neurons is seen in the striatum of control implanted monkeys. (e,f) In contrast, numerous healthy neurons can be seen near a QA injection site in BHK-hCNTF implanted animals. For reference, the arrows in (e) and (f) point to the same location. Scale bar in (a) represents the following: 4000 μm (a, b); 50 μm (d, e); 25 μm (f).

primate model of HD they restored cognitive and motor functions (**Figure 2**). In 2004, a human clinical trial of CNTF for HD was reported. This phase I trial evaluated the safety of intracerebral administration of CNTF protein in stage 1 or 2 HD patients using a device formed by a semipermeable membrane encapsulating a BHK cell line engineered to synthesize CNTF. No sign of CNTF-induced toxicity was reported, and improvements in electrophysiological results were observed.

### CNTF and Retinal Degenerative Diseases

Retinal degeneration in its various forms is responsible for vision loss in a large number of young adults. Retinitis pigmentosa (RP), for example, affects 50 000–100 000 people in the United States and an estimated 1.5 million people worldwide. RP characterized by a progressive loss of photoreceptors through-mechanisms is not fully understood. Subsequent to photoreceptor loss, there is both vascular and neural remodeling. The pathogenesis of retinal degeneration is complicated. With the exception of vitamin A nutritional supplementation, no treatments have been shown to be effective across the range of these disorders. Regardless of the initial causative genetic defect, the end result is photoreceptor cell death. The multiplicity of mechanisms stimulated a search for therapeutic agents that are effective in slowing photoreceptor death regardless of the causative genetic mutation. Intervention studies have indicated the possibility of using neurotrophic factors as therapeutic agents for RP. CNTF is a cytokine with neurotrophic activity across a broad spectrum of peripheral and CNS cells. CNTF was first identified as a survival factor in studies involving ciliary ganglion neurons in the chick eye. CNTF mRNA expression in the retina during postnatal development increased with postnatal age. CNTF upregulation plays a role in retinal protection against light-induced damage after optic nerve section. CNTF has been shown to modulate the survival of retinal neuronal cells *in vivo* and *in vitro*. Repeated injections of a CNTF analog lead to long-term photoreceptor survival in hereditary retinal degeneration. Encapsulated cell-based delivery of CNTF reduces photoreceptor degeneration in canine models of RP, protects photoreceptors in a canine model of retinal degeneration, and does not adversely affect either rod or cone ERG function of normal rabbit retina. AAV-mediated delivery of CNTF prolongs photoreceptor survival in the rhodopsin knockout mouse and rat models of RP. Intraocular gene transfer of CNTF prevents photoreceptor degeneration in Royal College of Surgeons rats and mice. However, intraocular gene delivery of CNTF results in significant loss of retinal function in both normal mice and Prph2Rd2/Rd2 retinal degeneration mice. A phase I human clinical trial of CNTF for retinal neurodegeneration indicates that CNTF is safe for the human retina even with severely compromised photoreceptors. The approach to delivering therapeutic CNTF proteins to degenerating retinas using encapsulated cell implants may have application beyond disease caused by genetic mutations.

## Conclusion

Trophic factors such as GDNF and CTNF hold great promise for the treatment of degenerative disease. However, significant challenges remain with regard to establishing biomarkers for these diseases that will allow for interventions much earlier in the disease process than are currently available. Furthermore, additional studies aimed at developing optimal delivery methods for these potent compounds would ultimately benefit their clinical use.

*See also:* Animal Models of Parkinson's Disease; Cell Replacement Therapy: Parkinson's Disease; Parkinson's Disease: Alpha-Synuclein and Neurodegeneration.

## Further Reading

Aebischer P, Schluep M, Deglon N, et al. (1996) Intrathecal delivery of CNTF using encapsulated genetically modified xenogeneic cells in amyotrophic lateral sclerosis patients. *Nature Medicine* 2: 696–699.

Cayouette M, Behn D, Sendtner M, Lachapelle P, and Gravel C (1998) Intraocular gene transfer of ciliary neurotrophic factor prevents death and increases responsiveness of rod photoreceptors in the retinal degeneration slow mouse. *Journal of Neuroscience* 18: 9282–9293.

Clatterbuck RE, Price DL, and Koliatsos VE (1993) Ciliary neurotrophic factor prevents retrograde neuronal death in the adult central nervous system. *Proceedings of the National Academy of Sciences of the United States of America* 90: 2222–2226.

Emerich DF, Winn SR, Hantraye PM, et al. (1997) Protective effect of encapsulated cells producing neurotrophic factor CNTF in a monkey model of Huntington's disease. *Nature* 386: 395–399.

Gash DM, Zhang Z, Ovadia A, et al. (1996) Functional recovery in parkinsonian monkeys treated with GDNF. *Nature* 380: 252–255.

Gill SS, Patel NK, Hotton GR, et al. (2003) Direct brain infusion of glial cell line-derived neurotrophic factor in Parkinson disease. *Nature Medicine* 9: 589–595.

Henderson CE, Phillips HS, Pollock RA, et al. (1994) GDNF: A potent survival factor for motoneurons present in peripheral nerve and muscle. *Science* 266: 1062–1064.

Kordower JH, Emborg ME, Bloch J, et al. (2000) Neurodegeneration prevented by lentiviral vector delivery of GDNF in primate models of Parkinson's disease. *Science* 290: 767–773.

Lang AE, Gill S, Patel NK, et al. (2006) Randomized controlled trial of intraputamenal glial cell line-derived neurotrophic factor infusion in Parkinson disease. *Annals of Neurology* 59: 459–466.

Lin LF, Doherty DH, Lile JD, Bektesh S, and Collins F (1993) GDNF: A glial cell line-derived neurotrophic factor for midbrain dopaminergic neurons. *Science* 260: 1130–1132.

McBride JL, Ramaswamy S, Gasmi M, et al. (2006) Viral delivery of glial cell line-derived neurotrophic factor improves behavior and protects striatal neurons in a mouse model of Huntington's disease. *Proceedings of the National Academy of Sciences of the United States of America* 103: 9345–9350.

Mittoux V, Joseph JM, Conde F, et al. (2000) Restoration of cognitive and motor functions by ciliary neurotrophic factor in a primate model of Huntington's disease. *Human Gene Therapy* 11: 1177–1187.

Patel NK, Bunnage M, Plaha P, et al. (2005) Intraputamenal infusion of glial cell line-derived neurotrophic factor in PD: A two-year outcome study. *Annals of Neurology* 57: 298–302.

Wang Y, Lin SZ, Chiou AL, Williams LR, and Hoffer BJ (1997) Glial cell line-derived neurotrophic factor protects against ischemia-induced injury in the cerebral cortex. *Journal of Neuroscience* 17: 4341–4348.

Yan Q, Matheson C, and Lopez OT (1995) *In vivo* neurotrophic effects of GDNF on neonatal and adult facial motor neurons. *Nature* 373: 341–344.

# Exercise: Optimizing Function and Survival at the Cellular Level

**A Russo-Neustadt**, California State University, Los Angeles, CA, USA



## Introduction

For many years, consistent physical activity has been associated with improved mental health as well as physical health and longevity. Through recorded history, several nonscientific assertions have been made that 'an active body makes for a healthy mind.' In the past 30 years, several reports from the sports medicine field have documented acute alterations in mood and cognitive functioning immediately following a bout of vigorous physical activity. Both aerobic and weight-bearing forms of activity have been associated with decreased anxiety, improved mood, and a subjective or actual experience of sharper mental functioning. In more recent years, formal, long-term exercise studies have been conducted, reporting improved cognitive functioning, heightened coping abilities, and decreased incidence of mental disorders such as dementia and depression. Importantly, many of these studies have focused on the aging population.

Intervention studies of the effects of exercise on mental functioning in older adults have revealed prevention of memory decline, greater attentional abilities, and substantial improvements in executive functions such as the ability to plan activities, focus on task-relevant information, and perform multiple skills concurrently. Recent prospective cohort studies indicate that regular exercise is associated with reduced risk and delayed onset of age-associated cognitive impairments and dementia. A direct imaging study of older adults has revealed decreased brain tissue volume in several brain areas with age, and a striking reduction in brain tissue loss with regular activity.

Exercise has not only shown promise as an intervention to preserve cognitive functioning, but also has been demonstrated to improve emotional functioning and mood. For example, long-term exercise has been linked with improved morale in normal elderly populations. Exercise has also been introduced as an adjunct for antidepressant medication treatment or as a single intervention in controlled clinical trials for depression, and significant improvements in symptoms have been reported. Studies have shown that a long-term (16 weeks) exercise treatment program was equally as effective as medication treatment for major depression. Regular activity has also been associated with a reduced risk for developing depression in elderly populations.

It is thought that natural selection of human ancestors, after many generations of survival as hunter-gatherers, led to the evolution of humans for whom regular vigorous activity improves and prolongs survival and function. It is likely that normal human cellular function has evolved in such a way that a threshold of physical activity is required for optimal physiologic gene expression in many tissues throughout the body, including the brain. As discussed in the following sections, exercise maintains the activity of a number of important central neurotransmitter systems, and this activation state sets into motion key signal transduction pathways promoting survival and functioning of neurons at the cellular level. It appears that the activation state of neurons (as well as the entire organism) enhances processes associated with growth and plasticity.

## Exercise and Central Nervous System Activation: Neurotransmitter Function

A wealth of evidence from both animal and human studies indicates that general aerobic activity leads to enhanced secretion of a number of important neurotransmitters and neuromodulators in the central nervous system (CNS). Such findings serve as neurochemical evidence for increased activation of the CNS (both global excitation of the brain and stimulation within specific brain regions). Enhanced concentrations of neurotransmitters and neurotransmitter metabolites have been measured in blood and cerebrospinal fluid (CSF) of humans. In animal models, microdialysis studies have allowed investigators to observe dynamic neurotransmitter release within specific brain areas during (and following) exercise. These specific forms of neurotransmitter activation have important implications for vulnerability to stress and depression, and impact survival and plasticity at the cellular level.

The monoamines, most notably serotonin, norepinephrine, and dopamine, are essential for multiple CNS functions, including arousal, cognition, and the regulation of mood. The activity of these systems is compromised with chronic stress and depression, and most antidepressant treatments in current practice are designed to enhance monoaminergic activity. In animal studies, exercise has been observed to enhance norepinephrine (NE) release in the brain stem, hypothalamus, and limbic brain regions. In humans, increased NE release into plasma has been observed following acute exercise. These changes prevent NE depletion during stress, and thus have been demonstrated to modulate physiological vulnerability to stressors. Some evidence also exists for changes in serotonin (5-hydroxytryptamine; 5-HT) release and metabolism during exercise. Clinical studies have indicated enhanced 5-HT availability and/or release in physically trained men, or after acute exercise, and another study has demonstrated enhanced release of dopamine and tryptophan (a metabolic precursor of serotonin). It has been proposed that enhanced dopamine neurotransmission occurring during exercise may contribute to enhanced central as well as peripheral function. Adequate monoaminergic (particularly serotonergic) function optimizes behavioral neuromodulation during various stressors.

The excitatory neurotransmitter, glutamate, plays a very important role in functional neuronal plasticity, signaling changes in synaptic function and morphology occurring during learning and other forms of neuronal activation. Studies have revealed that long-term exercise increases the levels or activated forms of several postsynaptic glutamatergic receptors in the cortex and hippocampus of rodents. Exercise-enhanced neurogenesis (discussed later) has been demonstrated to be dependent upon glutamatergic (*N*-methyl-D-aspartate; NMDA) receptor function. Neuropeptides such as vasopressin and galanin have also been studied in exercise research. Investigators have demonstrated changes in affinity of vasopressin, along with noradrenergic, receptors following exercise training. Both voluntary wheel-running and treadmill exercise training increase galanin gene expression in the rat locus coeruleus. Increased serum galanin secretion has also been detected after acute exercise in humans. Galanin may have an important neuromodulatory role in the NE system adaptations occurring during exercise.

In addition to an acute impact on intracellular activity and metabolism, cellular neurotransmission, as stimulated by exercise, also directs gene expression via the activation of signal transduction pathways. A key example is the activation of protein kinase A (PKA) and cyclic adenosine monophosphate (AMP) response element-binding protein (CREB) by monoamines via G-protein-coupled receptors, subsequently directing the expression of neurotrophins (discussed later) and other molecules promoting neuronal growth and survival.

## Exercise and Growth Factor Expression

The neurotrophins are a family of molecules, including nerve growth factor (NGF), brain-derived neurotrophic factor (BDNF), and neurotrophins (NT-3 and NT-4/5). These compounds promote survival, growth, and differentiation of neurons during development and throughout the life span. Recent evidence has indicated that BDNF may play an important role in supporting normal cognitive and emotional function into old age and senescence, and that its expression is influenced by neuronal activity.

It has become evident that physical activity increases the expression of BDNF and other growth factors in several areas of the CNS. In animal models, both voluntary wheel-running and forced treadmill training (under conditions controlling stress) rapidly and robustly enhance BDNF mRNA and protein expression in the hippocampus. It is known that environmental enrichment, or the addition of visual, motor, tactile, and social stimulation to the rodents' cage environment, enhances BDNF expression in the hippocampus and neocortex, and that activity has been suggested to be the single most important component of this model for the observed increases in neurotrophin expression and CNS function. Increased hippocampal BDNF mRNA occurs in as little as 6 h and is sustained with long-term (several weeks) activity. Increases in BDNF protein are evident within 1–2 weeks and are sustained for at least 3 months. Increases in both BDNF mRNA and BDNF protein are highly correlated with running distance. Increased peripheral motor activity has been demonstrated to be required for these observed changes, which are thought to mediate many of the prosurvival effects of exercise. Confocal microscopic analysis revealed that short-term (7 days) activity caused a redistribution of BDNF from the cell body to the dendrites in neurons of the lumbar spinal cord, whereas longer term (28 days) exercise also led to increased perikaryonal BDNF.

The phenomenon of increased hippocampal BDNF with exercise is very much intact throughout the life span of rodents. Even with relatively small amounts of activity, aged rats demonstrate an ability to significantly enhance BDNF mRNA and protein to an extent no less than that achieved with antidepressant medications. Voluntary wheel running has also greatly enhanced neurogenesis and hippocampal (learning) function in aged mice without prior exercise experience. Recent evidence also indicates that swimming exercise of a mother rat during pregnancy enhances hippocampal BDNF mRNA expression and neurogenesis of the offspring. Exercise can therefore represent a powerful, nonpharmacological means for maintaining neurotrophin levels and neuronal health throughout development, and throughout the life span.

It is becoming increasingly evident that exercise, and the resulting enhancement in neurotrophin expression, can improve neuronal function. Exercise has been demonstrated to increase the levels of proteins involved in vesicle formation and neurotransmitter release, such as synaptophysin and synapsin I. Enhanced synaptic function, such as that occurring during development or learning, can thus result from exercise. One cellular phenomenon that is highly dependent upon BDNF, and that in turn stimulates additional BDNF expression, is long-term potentiation (LTP). LTP is activated in the hippocampus during the process of spatial learning. An increase in hippocampal LTP has been observed in rats undergoing voluntary wheel-running activity, which appears to reflect an alteration in the induction threshold for synaptic plasticity occurring during exercise. Voluntary wheel-running activity for 3 weeks has been shown to enhance performance of rats in the Morris water maze, a widely studied spatial learning paradigm. The acquisition of another challenging spatial learning task, the radial arm maze, has been facilitated by long-term voluntary exercise. In aged (23 months) rats, long-term (7 weeks) mild, forced treadmill exercise improved performance in the Morris water maze, demonstrating both a shorter path length and significantly decreased latency to find the hidden platform.

Evidence exists that the BDNF-increasing effects of exercise in the hippocampus follow, and are dependent upon, noradrenergic activation. In addition, some cognitive benefits of exercise, such as the augmentation of contextual fear conditioning, are attenuated by blockade of $\beta$-adrenoreceptors. Stimulation of the PKA pathway following adrenergic G-protein-coupled receptor activation mediates increased BDNF expression via the transcription factor CREB. Abundant evidence exists for increased CREB phosphorylation (activation) and expression following exercise. The enhanced neurotrophin expression brought about by exercise subsequently activates tyrosine kinase (Trk) receptors and key intracellular signaling pathways mediating survival and growth. One important transcriptional target of these pathways is, once again, CREB. CREB, along with other transcription factors (including nuclear factor-kappa B, which has been shown to be increased with exercise), enhances the expression of BDNF along with several other prosurvival genes. In this way, BDNF expression can be increased via a positive feedback mechanism. This type of mechanism may explain the very rapid and robust increase in BDNF expression observed with exercise. Researchers in several laboratories have found evidence for neurotrophin expression being enhanced via a type of positive feedback loop. One of their reports cited direct evidence for changes in neurotrophin expression being driven by neurotrophin pulses in neuronal culture. This mechanism is discussed further in the following section.

In addition to the striking changes in hippocampal BDNF, evident with exercise, increases in BDNF and/or NGF and NT-3 have been demonstrated in the hippocampus, neocortex, basal forebrain, striatum, and spinal cord. The spinal cord has been a focus area for study of the effects of activity on neurotrophin expression, survival, and recovery from injury. Motor activity (treadmill walking or stepping) has been shown to enhance the expression of several neurotrophins in the spinal cord, and to enhance neuronal outgrowth and recovery in injured areas. In intact rats, locomotor exercise increases levels of BDNF and its receptor, TrkB, in the lumbar spinal cord, and also increases levels of NT-3 and its receptor, TrkC. In models of spinal cord injury, exercise also compensated for reductions in BDNF occurring in the lumbar spinal cord after hemisection to the midthoracic region, and increased BDNF was associated with reduction in sensory symptoms such as allodynia. In the latter study, increasing BDNF protein levels were highly correlated with running distance. Exercise was also shown to enhance regrowth of axons after a nerve-crush injury; and sensory (dorsal root ganglion) neurons from animals undergoing exercise also showed increased neurite outgrowth. Exercise interventions, therefore, show much promise to promote the recovery of motor function following spinal cord injury.

Another highly important growth factor for the mediation of exercise-induced changes within the brain is insulin-like growth factor (IGF). Investigators have shown that the uptake of peripheral IGF into the CNS occurs during voluntary exercise, and that this action is essential for both increased CNS expression of neurotrophins and functional recovery from stress. Recent evidence indicates that IGF from central as well as peripheral sources participates in exercise-induced increases in hippocampal BDNF expression as well as augmentation of recall in the Morris water maze, and also increases progenitor cell proliferation in the dentate gyrus. Much evidence is accumulating that IGF plays a very important role in antiaging and anticellular stress phenomena occurring with exercise, environmental enrichment, and dietary restriction.

It is becoming increasingly evident that interaction between multiple growth factors may be playing a role in exercise-associated neurochemical changes. Glial maturation factor (GMF), production of which is increased with exercise, appears to also be essential for exercise induction of BDNF expression. Vascular endothelial growth factor (VEGF) has been demonstrated to be particularly important for exercise-induced adult hippocampal neurogenesis.

## Exercise and Cellular Survival Signaling

The survival of neurons is dependent upon the activity of a number of finely tuned cellular signaling cascades. Two primary signaling pathways known to promote neuronal growth and survival are the phosphatidylinositol 3-kinase (PI3K)/Akt and mitogen-activated protein kinase (MAPK) pathways, which influence gene expression through activation of transcription factors, such as CREB. The MAPK and PI3K pathways are regulated by neurotrophin/Trk receptor signaling, and therefore represent two of the major signaling pathways mediating BDNF-induced TrkB activation. In cultured cortical neurons subjected to apoptotic conditions, BDNF has been shown to promote survival and neuroprotection by activating the MAPK and PI3K pathways. Consistently, BDNF stimulates the phosphorylation of Akt, which plays a critical role in controlling the balance between survival and apoptosis. Akt is activated via the PI3K pathway, following stimulation by insulin and several other growth/survival factors, and promotes cell survival by inhibiting apoptosis through its ability to phosphorylate and inactivate several substrates, including BAD, the forkhead transcription factors, caspase-9, and GSK-3$\beta$. In addition to increasing cellular survival, the PI3K and MAPK pathways play principal roles in promoting neurite growth, synaptic strength, and plasticity. Via these pathways, excitatory neural stimulation via exercise may therefore be an important means to enhance neuronal growth or recovery in an activity-dependent manner.

Microarray analyses of brain tissue from rodents undergoing physical activity have provided an excellent preliminary view of the genes and intracellular molecules activated by exercise. The 'exercise profile' includes a number of molecules along the PI3K survival pathway, and several other genes promoting cellular growth and survival. Some recent studies have also directly demonstrated increased concentrations of proteins along the PI3K and MAPK pathways in the brains of rodents undergoing chronic or subchronic voluntary activity.

Exercise enhances by two- to threefold the generation and survival of new neurons derived from progenitor cells within the adult hippocampus. Several of the key humoral factors outlined in the previous section, such as IGF-1 and BDNF, are thought to mediate exercise-stimulated neurogenesis. Evidence indicates that the MAPK signaling pathway is required for IGF-1-stimulated proliferation, whereas the PI3K/Akt pathway plays a major role in maintaining survival and inhibiting apoptosis. Recent evidence also indicates that neurogranin, a specific substrate of protein kinase C, is required for exercise-induced neurogenesis and resulting cognitive enhancements. Regarding the birth of new neurons, exercise appears to have a more profound effect upon proliferation than on differentiation or survival. Anatomical examination has revealed that the proportion of cells exhibiting an immature morphology is increased within the hippocampal dentate gyrus following exercise. Subsequent differentiation of these progenitors into mature neurons, and integration into the neuronal circuitry, may be stimulated by environmental enrichment and learning. Exercise may therefore benefit hippocampal function by increasing the potential for neurogenesis. Indeed, human evidence exists that cognitive enrichment complements physical fitness training in promoting life span and mental health.

The effectiveness of antidepressant medications for improving mood and emotional function has been attributed to increased neurotrophin expression, stable signaling changes, and increased neurogenesis within the hippocampus, all of which occur after 2–3 weeks of treatment (the time frame often required for symptom relief in humans). The relatively rapid effectiveness of exercise for improving neurotrophin expression (within hours to days) may be attributed to the simultaneous activation of several survival- and plasticity-promoting signaling cascades, including the PI3K and MAPK pathways, pathway cross-talk, and positive feedback. Several clinical studies have indicated that exercise is an effective antidepressant or adjunctive agent. Electroconvulsive therapy (ECT), a treatment of choice for medication-resistant patients with major depression, has been demonstrated to activate the expression of multiple growth-promoting genes, including the 'exercise profile.'

Investigations into the intracellular mechanisms of survival-promoting interventions such as exercise have also been pursued in cell culture models. It has been revealed that primary neurons harvested from exercising animals retain an activated or stimulated state. In one study, dorsal root ganglion neurons showed an increase in neurite outgrowth when cultured from animals that had undergone 3 or 7 days of exercise. Since much evidence has pointed to the possibility that NE activation is an essential component of the cascade of events leading to exercise-induced increases in neurotrophin expression, we studied cellular signaling within a model system of NE applied to primary hippocampal neurons. NE application to hippocampal neurons in culture significantly increased BDNF levels in a time- and dose-dependent manner, and the phenomenon appeared to be dependent upon Trk receptor activation as well as on activity of the PI3K and MAPK pathways. The NE-induced activation of the MAPK cascade via Trk activation has also been demonstrated in primary cortical neurons.

## Neurodegeneration: How Exercise Can Decrease Risk and Contribute to CNS Repair Mechanisms

The aging process is associated with metabolic dysfunction and an impaired ability of the organism to cope with adversity. With normal aging, the efficiency of metabolic processes decreases along with the efficacy of systems designed to remove molecules associated with oxidative stress. Oxidative stress can lead to neurodegeneration in many areas of the brain. The hippocampus is also particularly vulnerable to the effects of psychosocial stress and the secretion of stress hormones (glucocorticoids), since it is an area with very high glucocorticoid receptor numbers. Severe, prolonged stress is thought to play an important role in the pathophysiology of several mental disorders such as major depression. Age-associated neurological disorders such as Alzheimer's disease and Parkinson's disease involve a great pronouncement of neurodegeneration beyond the normal aging process. Exercise has shown increasing promise as a preventive or reparative intervention for these various forms of neurodegeneration.

Although exercise imposes increased metabolic demands on the organism, evidence exists that it improves metabolic efficiency and decreases cellular vulnerability. As noted earlier, exercise is capable of increasing the expression of hippocampal BDNF throughout the life span of rodents. Via neurotransmitter activation and enhanced neurotrophin production, exercise can protect neurons by stimulating the production of proteins that suppress oxyradical production, stabilize cellular calcium homeostasis, and inhibit apoptotic biochemical cascades. Decreased oxidative damage and improvement in cognitive function have been demonstrated in the rat brain following exercise. Exercise also modifies the age-related decrease of hippocampal growth-associated protein, GAP-43, and synaptophysin, and the age-related decrease in antioxidant enzyme activities. It has also been shown that exercise potentiates the expression of the heat shock protein, Hsp72, and that this could contribute to neuroprotection from stress. In addition, although exercise is associated with increased glucocorticoid secretion, voluntary activity can still decrease glucocorticoid levels and hippocampal damage associated with ongoing physical/psychological stress. Prior exercise was also demonstrated to be protective against reductions in hippocampal BDNF protein caused by acute immobilization stress, even though increased corticosterone levels were sustained. Exercise has also been demonstrated to reverse several detrimental effects of a high-fat diet, including increased levels of reactive oxygen species, decreased neurotrophin (BDNF) expression, decreased expression of downstream effectors of synaptic function, and behavioral (spatial learning) deficits.

Specific evidence is accumulating that exercise can provide benefits in the face of neurodegenerative disease. Investigators have shown that exercise almost completely protects the brain against experimentally induced parkinsonism. This finding may be associated with increased glial-derived neurotrophic factor levels. Moderate exercise has been demonstrated to induce neurotrophin production in humans (evident in serum levels), and this may underlie the benefits of exercise for multiple sclerosis. Exercise has also shown much promise as a therapeutic strategy against amyotrophic lateral sclerosis. Even brief exposures to voluntary exercise improved motor function and survival in a mouse model of this disorder. Evidence also exists that increased physical activity reduces stroke damage. Preischemic treadmill activity reduced neurologic deficits and brain infarct volume in rats; increased neurotrophin levels and microvessel density were associated findings. Microvessel integrity within the brain may be a key factor for the prevention of many types of neurodegenerative disorders. Exercise stimulates brain vessel density in experimental animals in a manner that is dependent upon IGF-1.

Aging is associated with reductions in both hippocampal neurogenesis and IGF-1 levels, and administration of IGF-1 to old rats reverses age-related changes in central energy metabolism/function, increases neurogenesis, and reverses cognitive impairments. A reduction in the activity of the growth hormone/IGF-1 axis with age is well documented in humans. Analogs of IGF-1, along with the molecules of the PI3K pathway, are present in a wide variety of organisms from nematodes to humans, and evidence is strong that their function is closely tied to stress resistance and survival. Insulin-like signaling, along with the actions of neurotransmitters and neurotrophins, therefore link energy metabolism to cellular survival in an evolutionarily conserved manner.

## Conclusion

In mammals, it is becoming increasingly evident that optimal cellular function within the nervous system and throughout the body is favored by heightened activity. The evidence outlined in this article has shown us that various forms of physical activity, such as voluntary aerobic exercise, activate highly specific intracellular systems that optimize function and survival. These systems are put into motion by molecules such as monoaminergic neurotransmitters, IGF-1, and the neurotrophin BDNF. This brings us, at the cellular level, to a full-circle connection to our earlier discussion of evolution of humans as active hunter-gatherers, and how cellular machinery likely similarly evolved in such a way that optimal function would be favored by increased activity. The aforementioned studies have suggested that, at the molecular level, monoamines, BDNF, and IGFs can stimulate the production of proteins involved in cellular stress adaptation, survival, growth/repair, and neurogenesis. All of these actions appear to be dependent upon synaptic activity, which is enhanced during exercise. These mechanisms may also compensate for changes taking place with aging and situations of severe stress. There is much yet to be learned about the specific mechanisms involved in exercise-induced cellular adaptations, but we have much evidence that exercise is an essential intervention to maintain brain health throughout life.

*See also:* Exercise in Neurodegenerative Disease and Stroke; Neurotrophic Factor Therapy: GDNF and CNTF.

## Further Reading

Aberg ND, Brywe KG, and Isgaard J (2006) Aspects of growth hormone and insulin-like growth factor-I related to neuroprotection, regeneration, and functional plasticity in the adult brain. *Scientific World Journal* 6: 53–80.

Barbour KA and Blumenthal JA (2005) Exercise training and depression in older adults. *Neurobiology of Aging* 26(supplement 1): 119–123.

Colcombe SJ, Kramer AF, McAuley E, et al. (2004) Neurocognitive aging and cardiovascular fitness: Recent findings and future directions. *Journal of Molecular Neuroscience* 24(1): 9–14.

Mattson MP (2002) Brain evolution and lifespan regulation: Conservation of signal transduction pathways that regulate energy metabolism. *Mechanisms of Ageing and Development* 123: 947–953.

Mattson MP, Maudsley S, and Martin B (2004) A neural signaling triumvirate that influences ageing and age-related disease: Insulin/IGF-1, BDNF and serotonin. *Ageing Research Reviews* 3(4): 445–464.

McEwen BS (2002) The neurobiology and neuroendocrinology of stress. Implications for post-traumatic stress disorder from a basic science perspective. *Psychiatric Clinics of North America* 25(2): 469–494, ix.

Meeusen R and De Meirleir K (1995) Exercise and brain neurotransmission. *Sports Medicine* 20(3): 160–188.

Pereira AC, Huddleston DE, Brickman AM, et al. (2007) An *in vivo* correlate of exercise-induced neurogenesis in the adult dentate gyrus. *Proceedings of the National Academy of Sciences of the United States of America* 104(13): 5638–5643.

Popoli M, Gennarelli M, and Racagni G (2002) Modulation of synaptic plasticity by stress and antidepressants. *Bipolar Disorder* 4(3): 166–182.

Russo-Neustadt A (2003) Brain-derived neurotrophic factor, behavior, and new directions for the treatment of mental disorders. *Seminars Clinical Neuropsychiatry* 8(2): 109–118.

Sapolsky RM (2000) The possibility of neurotoxicity in the hippocampus in major depression: A primer on neuron death. *Biological Psychiatry* 48(8): 755–765.

Vaidya VA and Duman RS (2001) Depression – Emerging insights from neurobiology. *British Medical Bulletin* 57: 61–79.

# Exercise in Neurodegenerative Disease and Stroke

**C W Cotman**, University of California, Irvine, CA, USA
**A D Smith**, University of Pittsburgh, Pittsburgh, PA, USA
**T Schallert**, University of Texas at Austin, Austin, TX, USA
**M J Zigmond**, University of Pittsburgh, Pittsburgh, PA, USA



## Introduction

It has long been known that exercise is essential to good health. Individuals who lead a sedentary lifestyle greatly increase their risk of many diseases, including osteoporosis, cardiovascular disease, and diabetes. During the past decade, it has become apparent that physical activity also is essential to a healthy brain, particularly in aging populations. In addition to its generally beneficial effects, there is reason to believe that exercise and other forms of motor enrichment can be beneficial with regard to several neurological disorders, including Parkinson's disease (PD) and Alzheimer's disease (AD), and can even be helpful in reducing the deficits induced by stroke.

PD and AD have much in common. They involve a progressive degeneration of neurons in the brain that often goes undetected for many years. Because of this long preclinical gestation and perhaps because of a reduction in the capacity of the brain to defend itself against insults, the emergence of these conditions as clinical entities increases sharply as individuals age, with the great majority of cases emerging in people aged 60 and older. In addition, whereas their pathological and clinical manifestations are quite different at the outset, there often is an increasing overlap as PD and AD progress.

This article briefly reviews the evidence for the neuroprotective effects of exercise, discusses supportive data from animal and cellular models, and then comments on possible underlying mechanisms and implications for the causes and treatments of these neurodegenerative diseases, as well as for the treatment of stroke.

## PD

### The Condition

Originally identified as a clinical syndrome by James Parkinson in 1817, PD is typically characterized by its motor signs – slowness of movement (bradykinesia), stooped posture, rigidity, and tremor. In reality, however, these appear to reflect an intermediate sign of the underlying neurodegenerative process that may also affect cardiovascular function, olfactory sensitivity, speech, fatigue, and gastrointestinal function and may eventually be accompanied by cognitive dysfunction, including dementia. The prevalence of PD is about 1%. Diagnosis typically occurs at about age 60, though the disease probably has a 10- to 20-year 'preclinical' phase, during which symptoms are either not detected or not associated with PD. On the other hand, 5–10% of patients with PD are diagnosed before the age of 40. Men, particularly younger individuals, are at a slightly greater risk for the condition.

In the early 1960s, the majority of the motor deficits in PD were attributed by Oleh Hornykiewicz and his colleagues to the reduction in dopamine (DA) in the caudate and putamen (collectively referred to as the striatum). Subsequently it was shown that the DA in these regions was located in the terminals of neurons that project from the substantia nigra pars compacta (SNpc), a group of about half a million pigmented neurons per side located in the ventral mesencephalon. By the time PD is diagnosed, about half of the SNpc neurons have been lost, along with about 80% of the DA in the striatum. However, as with the symptoms of PD, the neurodegeneration underlying the condition is quite broad and eventually may include the loss of many other specific groups of neurons both caudal and rostral to the SNpc, as has been suggested most convincingly by Heiko Braak, beginning in the late 1980s. It seems likely that many of the nonmotor consequences of PD result at least in part from pathology in these extradopaminergic regions.

### Clinical Evidence for the Beneficial Effects of Exercise in PD

Many drugs are used in the treatment of PD; however, virtually all of them strive directly or indirectly to replace the impact of the DA that is lost as part of the disease. Thus, current treatments target motor symptoms rather than the full complex of the disease. These medications can be highly effective for many years and usually permit patients with PD to have a normal life span. However, there is little evidence that the actual neurodegenerative process is affected by such treatments. This is also the case for surgical treatments, which have included pallidotomy and the now more common approach of deep brain stimulation of the subthalamic nucleus.

In contrast to more conventional therapies, physical exercise appears to reduce the incidence of PD and to improve the motor, cognitive, and affective deficits in patients with mild to moderate PD. A lifetime of exercise seems to reduce the incidence of PD, at least in men. In addition, a program of aerobic activity has been reported to improve movement initiation, whereas resistance and balance training have been found to improve balance.

Despite the encouragement such studies provide, virtually all of them suffer from one of two failings. Those studies examining long-term effects of exercise derive their results from surveys. For example, a negative correlation between PD and the level of physical activity throughout life has been reported. However, whether these data are collected retrospectively or over a long period through regular questionnaires, such studies always suffer from a lack of precision. Moreover, in the case of PD, they leave open the possibility that the results reflect a tendency toward inactivity of individuals who are in the preclinical stages of the disorder.

A few studies have tried to address this problem with prospective studies, separating the patients into an exercise and a control group. However, here, too, there are serious problems. Such studies are typically very brief (weeks or at best a few months), involve relatively few participants, may not provide adequate controls for the effect of the increased attention that the exercise group receives, and fail to distinguish between symptomatic changes and an actual influence on underlying disease progression. Of course, these problems hold not only for studies of PD but also for studies of AD, described in the section titled 'AD.' Thus, although such data are encouraging, they should be viewed with caution, and for the present, studies of experimental animals form the basis for assessing the neurobiological benefits of this therapy.

## Animal Studies

Although animal models of neurodegenerative diseases are of necessity imperfect, several models of PD are in use. Most rely on the fact that a major component of PD pathology is the loss of DA neurons and that these neurons can be selectively destroyed in many animals through the use of DA-specific neurotoxins. The first such toxin to be introduced was 6-hydroxydopamine (6-OHDA), which enters via DA uptake sites and, when combined with the noradrenaline uptake blocker desipramine, is selectively concentrated in DA neurons, causing toxic oxidative stress. Later, 1-methyl-4-phenyl-1,2,3,6-tetrahydropyridine (MPTP) was introduced as a tool in the study of PD. MPTP is converted to the DA neurotoxin MPP+, which, like 6-OHDA, is concentrated in DA neurons, where it inhibits complex I of the mitochondria.

When the toxin-induced loss of DA is severe, rodents and nonhuman primates lesioned in this way show several motor deficits that are reminiscent of PD. For example, rats treated with 6-OHDA become bradykinetic or even akinetic, whereas monkeys treated with MPTP provide an even closer model of the motor symptoms of PD, exhibiting tremor, stooped posture, and rigidity.

The protective effect of constraint therapy was first used with some success in patients with stroke (see the section titled 'Stroke') and has recently been applied to 6-OHDA-treated rats as well. Adult rats receiving unilateral injections of 6-OHDA had their 'normal,' ipsilateral limb constrained by a plaster-of-paris cast for 7 days. When assessed for behavioral and neurodegenerative abnormalities several weeks later, these deficits were greatly attenuated. This was true whether the forced exercise occurred immediately before or immediately after the administration of 6-OHDA. However, if the placement of the cast on the forelimb was delayed for a week after 6-OHDA, the therapy was not effective (**Figure 1**). Moreover, if animals previously protected against 6-OHDA by limb immobilization had their 'protected' limb constrained in a cast for a 1 week period 2–3 weeks later, behavioral and neurochemical evidence of DA loss reemerged. These results indicate that there is a brief window of opportunity during which forced exercise is protective. They also suggest that continued limb use is essential to maintaining the DA neurons in a viable state.

Housing in a more complex environment or caloric restriction can also protect DA neurons. For example, the loss of DA neurons incurred from MPTP as assessed by cells immunoreactive for the rate-limiting enzyme tyrosine hydroxylase (TH) was significantly reduced by providing mice with daily access to tunnels, opportunities for climbing, running wheels, and toys for 2 months (**Figure 1**). These results have since been replicated by Dr. Richard Smeyne and colleagues by giving mice continuous access to a running wheel, suggesting that a major impact of environmental complexity is to increase exercise. Likewise, reducing caloric intake by 30–40% in young mice for 3 months prior to MPTP administration improved motor performance and caused a near-complete protection of DA cells and terminals. Similar results of caloric restriction have been obtained in a hemiparkinsonian primate MPTP model and in other brain-injury models.

The protective effects of complex environments and caloric restriction may be due at least in part to the increase in motor activity that is associated with these treatments. At least one report has indicated that the protective effects of environmental enrichment in PD models could be duplicated by housing animals with a running wheel. Moreover, running has been clearly linked to neuroprotection in studies of mouse models of AD and of stroke.

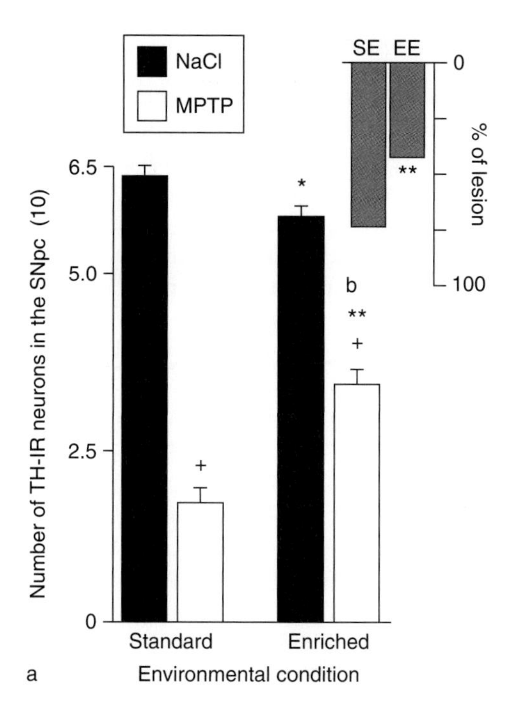


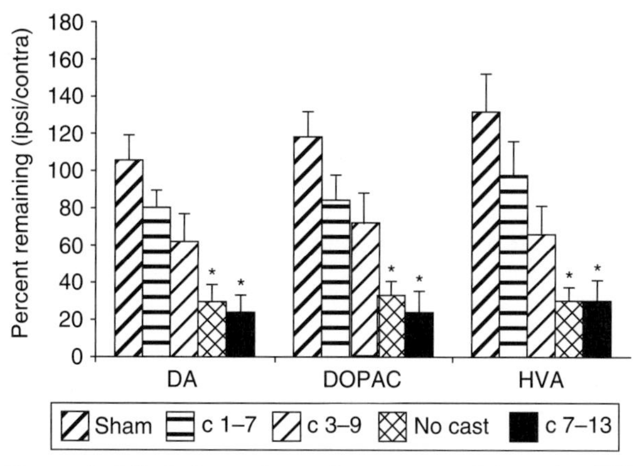


**Figure 1** Effect of exercise on the response of dopamine (DA) neurons to neurotoxins in models of Parkinson's disease (PD). Upper panel: At weaning, mice were placed in either an enriched or a standard environment and examined 2 months later. Shown are the number of tyrosine hydroxylase immunoreactive (TH-IR) neurons in the substantia nigra pars compacta (SNpc) of mice from the two groups 7 days after treatment with either saline (NaCl; dark bars) or 1-methyl-4-phenyl-1,2,3,6-tetrahydropyridine (MPTP; 20 mg $kg^{-1}$ intraperitoneally; open bars). From Bezard E, Dovero S, Belin D, et al. (2003) Enriched environment confers resistance to 1-methyl-4-phenyl-1,2,3,6-tetrahydropyridine and cocaine: Involvement of dopamine transporter and trophic factors. *Journal of Neuroscience* 23: 10999–11007. Lower panel: Adult male rats were given a unilateral injection of 6-hydroxydopamine (6-OHDA; 10 μg) along the medial forebrain bundle. In some groups of rats, a unilateral cast was used to restrict the use of the ipsilateral forelimb for 7 days, thereby forcing the animals to rely on their contralateral limb for movement during this time. The cast was placed immediately after the 6-OHDA (c 1–7), 3 days later (c 3–9), or 7 days later (c 7–13). The animals were sacrificed 65–80 days later for analysis of DA, DOPAC, or HVA levels by high-performance liquid chromatography. Animals receiving early casts and sham animals did not show significant differences in DA, DOPAC, or HVA levels; animals receiving casts on days 3–9 showed intermediate DA levels, although still not significantly different from sham. Animals with late casts (days 7–13) and animals not receiving casts showed significantly lower DA levels when compared with sham-treated animals. These data parallel performance on a variety of behavioral tasks (data not shown). From Tillerson JL, Cohen AD, Philhower J, Miller GW, Zigmond MJ, and Schallert T (2001) Forced limb-use effects on the behavioral and neurochemical effects of 6-hydroxydopamine. *Journal of Neuroscience* 21: 4427–4435.

## AD

### The Condition

AD is the most common form of dementia. Like PD, it is a progressive, degenerative disorder that has an extended preclinical phase and becomes prominent with age. Whereas the prevalence rate of AD is only slightly higher than that of PD (~1.5%), about half of those who reach the age of 85 will be affected. Moreover, although patients with PD have a near-normal life span if properly treated, patients with AD seldom live for more than a decade beyond diagnosis. The deficits typically begin with memory loss and proceed to affect thinking and language. The condition has been associated with neuronal loss, particularly among subcortical neurons that project to the telencephalon. By the time that AD is identified, this loss is usually sufficient to have caused a gross reduction in brain size. At a cellular level, two abnormalities can be detected and are diagnostic for the condition: neuritic plaques containing amyloid-$\beta$ (A$\beta$) and neurofibrillary tangles composed largely of tau protein. The condition and its neuropathological hallmarks were first described in 1906 by Alois Alzheimer.

Areas of the brain that are affected by AD are widespread. However, one group of neurons that often is targeted early is the cholinergic neurons of the basal nucleus of Meynert. This, in turn, has led to the use of inhibitors of acetylcholine esterase in the treatment of the condition, an approach that has limited, symptomatic benefit at best, with no evidence for any effect on the neurodegenerative process. As in the case of PD, many drugs are under development that are designed to influence the underlying pathophysiology of the disease, but they are still in early experimental stages. However, again as with PD, there is reason to believe that exercise can help hold AD at bay or slow its progression.

### Clinical Findings on Efficacy of Exercise in AD

Evidence suggests that the incidence and progression of cognitive decline in later life are reduced by exercise. For example, individuals who engaged in strenuous walking three times a week showed improved executive function and increased brain activation, as monitored by functional magnetic resonance imaging, compared with those who undertook stretching and nonaerobic activity. However, as in the case of PD, the field still awaits long-term clinical trials with enough power to permit definitive conclusions about exercise and cognition, as well as clear statements regarding the amounts and types of exercise that are most effective. Thus, we must again turn to animal studies.

### Animal Studies

The first requirement for an animal study is the establishment of a model. As noted above, the classic hallmarks of AD are accumulation of A$\beta$ and the development of neurofibrillary tangles. Transgenic mice with mutations in the amyloid precursor protein accumulate A$\beta$. This accumulation is correlated with cognitive impairment and a reduction in indications of neuroplasticity, including learning. Conversely, reductions in A$\beta$ by pharmacological intervention improve cognitive function and synaptic plasticity. These data suggest that A$\beta$ is causally linked to the cognitive difficulties present in these mutant, transgenic mice and that reducing A$\beta$ is important for healthy brain aging.

Housing in a complex environment can decrease A$\beta$ accumulation (**Figure 2**) and improve cognitive performance in mouse models of AD. However, at least one study indicates that these effects cannot always be duplicated by voluntary running alone and thus may be associated with more cognitive aspects of such enrichment. This would not be surprising given the clinical literature linking cognitive activity to a reduction in the incidence and progression of dementia.

## Stroke

A stroke commonly results from either blockage (ischemic) or bursting (hemorrhagic) of an artery supplying blood to a region of the brain, with the former accounting for about 85% of all incidents. In major industrialized countries, stroke is the third leading cause of death, after heart disease and cancer, and the major cause of adult disability. Unlike the slow neurodegenerative process associated with PD and AD, hemorrhagic or ischemic injury is usually immediately severe, leading to total or near-total cell death within affected brain tissue. However, this primary injury is typically followed by slow secondary degeneration around the damage (termed the penumbra) as well as in remote regions, and this secondary degeneration can continue to develop for months. It is this component of the brain damage that is the main target for treatment.

Pharmacological treatment of stroke is limited. In the case of ischemic stroke, tissue plasminogen activator, a clot-dissolving drug, may be useful if given within the first 2–3 h of the incident. Glutamate antagonists have been examined extensively in stroke treatment, the rationale being that excitotoxicity appears to play a major role in the associated cell death. Unfortunately, although such agents appeared very promising in studies with animal models, they have not proven effective in clinical trials, in part because of their behavioral side effects.

### Exercise

Physical therapy has been used in the treatment of stroke for decades. The concept of constraint therapy

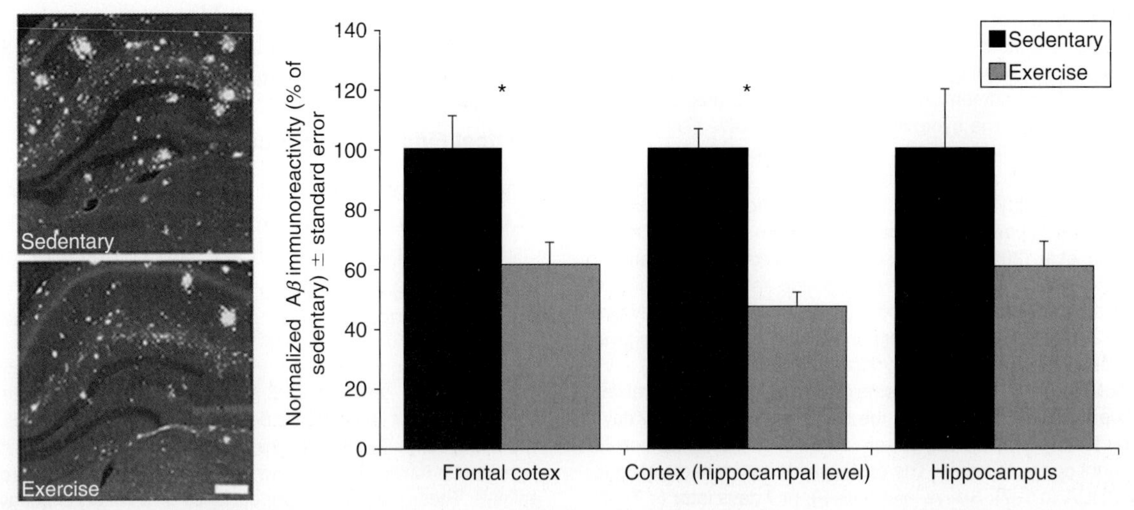


**Figure 2** Effect of exercise on amyloid-$\beta$ (A$\beta$) load in three different brain regions of transgenic mice that normally overexpress this protein (TgCRND8). Exercising animals showed a reduction in A$\beta$ immunoreactivity in cortical regions and in hippocampus. The photomicrograph shows a representative section from both groups. Scale bar = 200 μm. From Adlard PA, Perreau VM, Pop V, and Cotman CW (2005) Voluntary exercise decreases amyloid load in a transgenic model of Alzheimer's disease. *Journal of Neuroscience* 25: 4217–4221.

in the treatment of motor deficits in stroke victims was introduced in the early 1980s, largely because of the work of Edward Taub. The concept was to increase the use of the affected limb by constraining the use of the normal limb for several hours each day for up to 2 weeks. Some restoration of function was observed, and the approach is now often used in patients with at least some capacity to move the stroke-impaired wrist or digits of the upper extremity, presumably in part because the injury to the corresponding brain region is partial. Other types of exercise have included weight-supported and unsupported treadmill and cycling. There is no evidence that the primary injury can be reduced with poststroke exercise. However, whereas the initial effects of severe ischemic injury may not be responsive to exercise, compensatory behavioral adaptations can be promoted with constraint and other forms of motor-enrichment therapies. These treatments presumably lead to attenuation of slow secondary degeneration (though see discussion below) and promote adaptive rewiring.

Exercise of the lower extremities in human stroke patients, including facilitated treadmill walking, can improve walking ability, in contrast to stretching and other control procedures. The improvement has been linked to behavior-related induction of more normal patterns of activation in cortical and subcortical motor regions. There is some evidence that poststroke training can ameliorate speech deficits, particularly when damage to speech areas in the cortex is subtotal. However, it has been shown that intense speech therapy starting immediately after a stroke can degrade function compared with a control rehabilitative strategy involving simple social contact with limited conversation.

### Animal Studies

There are three major animal models of stroke: global ischemia (blockade of all major blood vessels to forebrain), focal ischemia (blockade of specific blood vessels), and hemorrhage (disruption of blood vessels). Focal ischemia, caused by occlusion of the middle cerebral artery is the most common type of stroke, can readily be modeled in animals. Two approaches are in use, brief occlusion followed by vascular reperfusion and permanent vessel occlusion. The former approach leads to more extensive and slower secondary degeneration of peri-injury and remote tissue.

Although research suggests that stroke outcome can sometimes be improved modestly by rehabilitative training, including exercise, neurodegeneration in animals can actually be worsened by either very early forced motor inactivity or overactivity. It is currently unclear what the optimal strategies might be for different degrees, locations, or types of brain injury. Slow secondary degeneration may present a more feasible target for neuroprotection than does rapid acute cell loss.

Constraint-induced movement therapy has been examined after minor focal ischemia or hemorrhage in nonhuman primates and rodents. Where examined, such forced forelimb rehabilitative experience has been found to improve function, reduce loss of cortical motor maps of the upper extremity, and expand such maps into novel regions of cortex. Housing animals in an enriched environment (**Figure 3**) or providing skill training after focal ischemic injury can enhance dendritic arborization and synaptogenesis in remaining tissue in both the peri-injury and homotopic brain regions.

The timing of the exercise is important. In some studies, exercise (usually treadmill running) has been provided before the ischemic insult in order to mimic a more active lifestyle. These studies have generally shown a 'preconditioning' effect, that is, a reduction in the size of the ischemic damage because of prior exercise. Intense rehabilitation during the first few days after injury can worsen the injury even as it improves certain functions. The exaggeration of damage, which may take many weeks to be readily detectable, does not appear to be due to a general effect on brain because forced forelimb overuse exaggerates only injury to the forelimb region of the brain but not to injury of the visual cortex. Intense forelimb use may increase the extent of delayed secondary degeneration of vulnerable peri-injury tissue via local hyperthermia, excitotoxicity, or metabolic demands. Sudden social isolation from cagemates at the time of injury can combine with the effects of limb overuse to cause damage that is even more extensive. This may be important clinically because stroke patients often have the additional stress of being confined to a relatively isolated hospital environment rather than quickly returning home to be surrounded by family and friends.

Despite exaggerating neural injury, intense exercise and motor training can improve function or delay dysfunction through learned compensatory motor behaviors in animals. Indeed, motor enrichment can overcome deficits so effectively that the adverse effects of secondary degeneration are difficult to observe with behavioral outcome measures alone. Ideally, the rehabilitation training would be applied so that the benefits accrue without tissue loss. Based on animal models, motor enrichment might best be applied sparingly during the acute poststroke period and gradually increased in intensity during subacute periods and chronically.

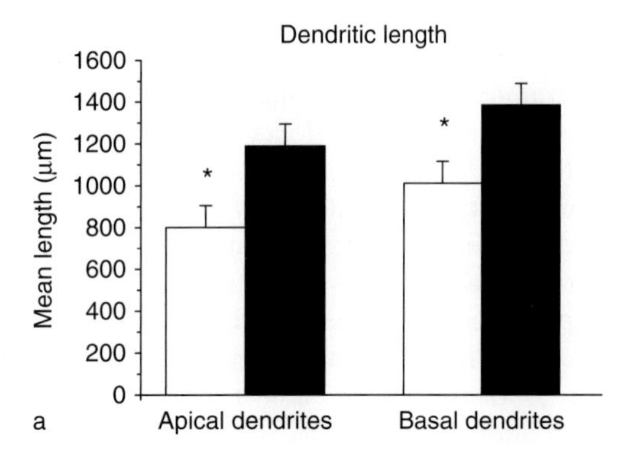


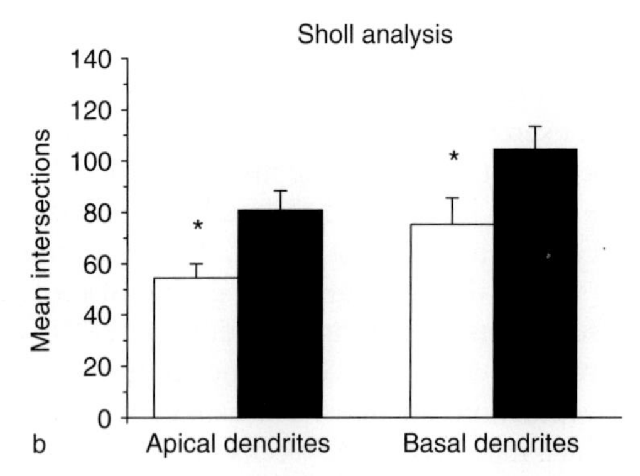


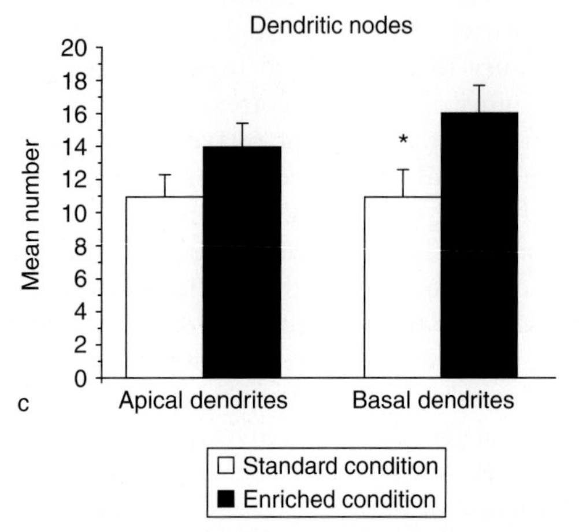


**Figure 3** The effects of enriched rearing on (a) maximal dendritic length, (b) the number of intersections, and (c) branching of layer-III parietal pyramidal neurons. Rat pups were placed in an enriched environment at weaning (21 days) and examined at 3 months. As shown, environmental enrichment provoked both increased dendritic arborization and the density of dendritic spines in layer-III parietal pyramidal neurons. In addition, the rats raised in the enriched environment exhibited improved performance in a radial arm maze and a Morris water maze (data not shown). From Leggio MG, Mandolesi L, Federico F, et al. (2005) Environmental enrichment promotes improved spatial abilities and enhanced dendritic growth in the rat. *Behavioural Brain Research* 163: 78–90.

## How Might Exercise Work?

It is now clear from rodent research that exercise improves learning, increases resistance to depression, and enhances synaptic plasticity such as long-term potentiation. This is accompanied in the brain by a series of biochemical events associated with enhanced plasticity (**Table 1**). Much of these data have come from studies of the hippocampus and cortex, but other brain regions can show plastic changes, too. In short, there are several effects of exercise on brain function, each of which could contribute to its apparent ability to reduce the incidence of neurodegeneration as well as the progression of neurodegeneration if it does occur. Thus, the question is not how exercise can be neuroprotective but which of the many changes are most critical to the neuroprotective effects of exercise and how the changes interact with each other. One of the variables in these relationships is deemed to be particularly important – trophic factors.

**Table 1** Selected effects of exercise on brain[a]

| *Type of change* | *Examples* |
|---|---|
| Transmitter release | Increased serotonin, dopamine, $\gamma$-aminobutyric acid, glutamate; decreased acetylcholinesterase |
| Increased levels of messenger RNA and/or protein for trophic factors or their receptors | Brain-derived neurotrophic factor, basic fibroblast growth factor, fibroblast growth factor-2, glial cell-derived neurotrophic factor, insulin-like growth factor 1, nerve growth factor, vascular endothelial growth factor, tyrosine receptor kinase B |
| Reduced inflammatory markers | Tumor necrosis factor-$\alpha$ (TNF-$\alpha$), TNF-$\alpha$ receptors, intercellular adhesion molecule-1 |
| Increased amounts of synaptic proteins | Synapsin, synaptophysin, NR2B, and glutamate receptor 1 |
| Stimulation of morphological changes | Gliogenesis, neurogenesis, synaptogenesis, and vascularization (angiogenesis) |
| Activation of survival cascades and associated transcription factors | Ras/extracellular signal-regulated kinase and phosphoinositide-3 kinase/protein kinase B; cyclic adenosine monophosphate responsive element binding protein |
| Increases in antioxidant defenses | Superoxide dismutase, catalase (decreased reactive oxygen species) |
| Modification of brain circuitry | Increased area devoted to task; changes in anatomical circuitry, changes in activity of existing neurons and circuits |

[a]Some entries are based on only one or two publications. Excluded from this list are variables that as yet have been examined only in peripheral tissue (often muscle).

### Exercise Regulates Trophic Factors

Exercise is a powerful regulator of gene expression and protein levels for several trophic factors. The largest amount of data on this issue comes from studies of hippocampal brain-derived neurotrophic factor (BDNF). The neurotrophin BDNF promotes neuronal survival and resistance to insults and is essential for brain plasticity mechanisms, including certain forms of long-term potentiation. Exercise increases BDNF messenger RNA expression in hippocampal neurons. The gene upregulation is rapid (after several hours of exercise; **Figure 4**), occurs in both male and female rats and mice, is sustained even after several months of exercise, and is paralleled by increased amounts of BDNF protein. Once induced, protein levels remain elevated for several days after exercise has stopped, then decay back to baseline. It is interesting that an exercise regime of alternating days of exercise is as effective as daily activity for increasing hippocampal BDNF protein levels. The practical significance of these findings is that even intermittent exercise is a powerful enough stimulus to activate the molecular machinery that drives synaptic plasticity.

Can increases in trophic factors explain neuroprotection? The evidence for neuroprotective effects of trophic factors is considerable. Trophic factors such as glial cell-derived neurotrophic factor and BDNF can protect against some of the spontaneous neuronal loss that occurs *in vitro* and that is associated with normal aging. These and other trophic factors can also reduce the damage normally caused by several neurotoxins and stroke. Both of these types of neuroprotective effects may, in turn, be due in part to the ability of trophic factors to stimulate survival cascades. Further evidence for a role of trophic factors in preventing neurodegeneration comes

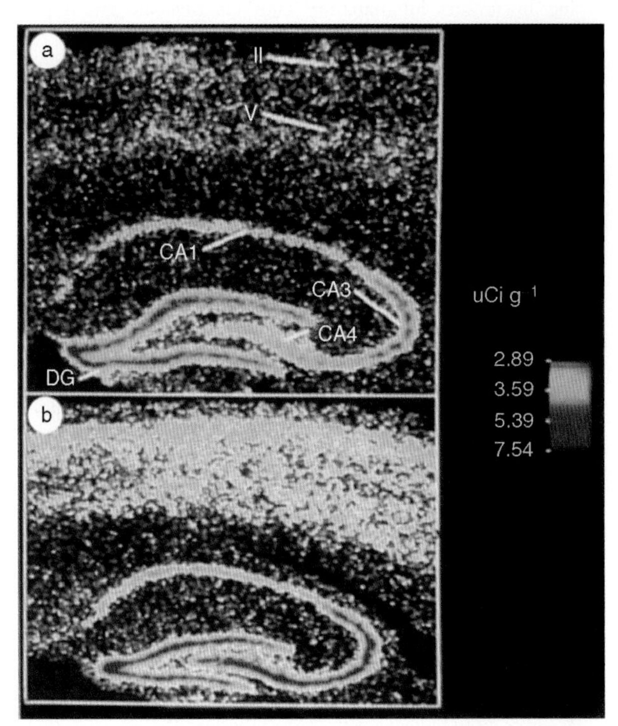


**Figure 4** Brain-derived neurotrophic factor (BDNF) messenger RNA in rat brains obtained under (a) basal, sedentary conditions and (b) after 7 days of exercise. Neocortical layers II and V, as well as hippocampal areas CA1, CA3, and CA4 and dentate gyrus (DG) are shown. This *in situ* analysis of exercise-induced increases in BDNF message may have been the first observation of an effect of exercise on a neurotrophic factor in brain. Reprinted by permission from Macmillan Publishers Ltd: *Nature* (Neeper SA, Gómez-Pinilla F, Choi J, and Cotman C (1995) Exercise and brain neurotrophins. *Nature* 373: 109), © 1995.

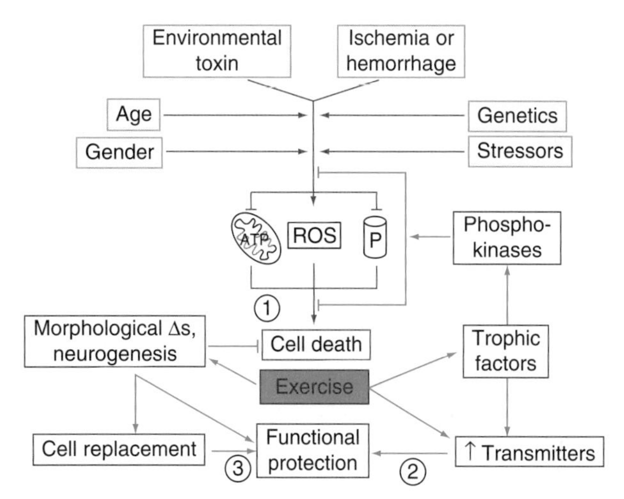


**Figure 5** Schematic indicating multiple ways by which exercise may alter brain. Shown (orange) are several of the factors that can influence the vulnerability of neurons, which might include age, gender, environmental toxins, or other insults (e.g., head injury), hemorrhage, ischemia, genetic predisposition, or stressors. These factors can combine to cause cell death through one or more mechanisms, including inhibition of mitochondrial function (leading, in part, to a reduction in adenosine triphosphate (ATP) production), an increase in oxidative stress, and disruption of the processes involved in handling misfolded proteins (e.g., via inhibition of the proteasome $\mathcal{P}$). Exercise may counteract these events, blocking neuronal death and/or reducing its rate of progression (①). This could result from an increase in the expression of trophic factors (e.g., glial cell-derived neurotrophic factor and brain-derived neurotrophic factor) leading to an increase in prosurvival cascades, such as those involving the kinases, extracellular signal-regulated kinase and protein kinase B. Other factors include morphological changes, such as synaptogenesis, neurogenesis (and associated cell replacement), and angiogenesis (and associated increases in oxygen and nutrients) (②). Among the other changes that could ameliorate cell death are neurochemical changes that could lead to an exercise-induced increase in the activity of residual neurons within a damaged system, and changes in neuronal circuitry leading to a transfer of function from damaged neurons to healthy neurons (③). ROS, reactive oxygen species.

from observations of reductions in trophic factors in brains of patients with PD and AD.

## Summary and Conclusions

Stroke and neurodegenerative diseases such as PD and AD are a major health problem throughout the world. Moreover, since the principal risk factor for these conditions is age, the burden of these conditions will only increase in years to come unless neuroprotective treatments can be found. Although a great deal of research is ongoing to develop pharmacotherapies, there is not yet convincing evidence that such drugs are now in hand. Moreover, in the condition in which drug development and testing have been most intense, the results have been extremely disappointing. This work is continuing, as it should. In the meantime, however, the evidence for neuroprotective effects of exercise, at least in animal models, is considerable and mounting. Indeed, it is now fair to say that in addition to its many other beneficial effects, exercise also is a key to brain health.

Although this discussion enumerated many possible factors (**Table 1**) and then focused on a role for trophic factors, this focus should suggest that changes in trophic factors exist apart from the other changes enumerated. On the contrary, changes in the availability of trophic factors may be a consequence of some of the effects listed (e.g., gliogenesis) and in turn may be the stimulus for other effects (e.g., synaptogenesis and increased antioxidant defenses; **Figure 5**).

In the years ahead, it will be essential to learn more about the mechanism of these effects, to move carefully with research-based treatment strategies from animal models to humans, and to determine how best to optimize an exercise intervention to promote healthy aging as well as to reduce the progression of disease.

*See also:* Aging of the Brain and Alzheimer's Disease; Alzheimer's Disease: Neurodegeneration; Animal Models of Alzheimer's Disease; Animal Models of Parkinson's Disease; Animal Models of Stroke; Exercise: Optimizing Function and Survival at the Cellular Level; Neurotrophic Factor Therapy: GDNF and CNTF; Parkinson's Disease: Alpha-Synuclein and Neurodegeneration.

## Further Reading

Adlard PA, Perreau VM, Pop V, and Cotman CW (2005) Voluntary exercise decreases amyloid load in a transgenic model of Alzheimer's disease. *Journal of Neuroscience* 25: 4217–4221.

Berchtold NC, Chinn G, Chou M, Kesslak JP, and Cotman CW (2005) Exercise primes a molecular memory for brain-derived neurotrophic factor protein induction in the rat hippocampus. *Neuroscience* 133: 853–861.

Bergen JL, Toole T, Elliott RG III, Wallace B, Robinson K, and Maitland CG (2002) Aerobic exercise intervention improves aerobic capacity and movement initiation in Parkinson's disease patients. *NeuroRehabilitation* 17: 161–168.

Bezard E, Dovero S, Belin D, et al. (2003) Enriched environment confers resistance to 1-methyl-4-phenyl-1,2,3,6-tetrahydropyridine and cocaine: Involvement of dopamine transporter and trophic factors. *Journal of Neuroscience* 23: 10999–11007.

Chen H, Zhang SM, Schwarzschild MA, Hernan MA, and Ascherio A (2005) Physical activity and the risk of Parkinson disease. *Neurology* 64: 664–669.

Cotman CW and Berchtold NC (2002) Exercise: A behavioral intervention to enhance brain health and plasticity. *Trends in Neurosciences* 25: 295–301.

Crizzle AM and Newhouse IJ (2006) Is physical exercise beneficial for persons with Parkinson's disease? *Clinical Journal of Sport Medicine* 16: 422–425.

Debow SB, Davies ML, Clarke HL, and Colbourne F (2003) Constraint-induced movement therapy and rehabilitation exercises lessen motor deficits and volume of brain injury after striatal hemorrhagic stroke in rats. *Stroke* 34: 1021–1026.

Ding Y-H, Young CN, Luan X, et al. (2005) Exercise preconditioning ameliorates inflammatory injury in ischemic rats during reperfusion. *Acta Neuropathologica* 109: 237–246.

Dishman RK, Berthoud HR, Booth FW, et al. (2006) Neurobiology of exercise. *Obesity* 14: 345–356.

Faherty CJ, Raviie Shepherd K, Herasimtschuk A, and Smeyne RJ (2005) Environmental enrichment in adulthood eliminates neuronal death in experimental Parkinsonism. *Brain Research Molecular Brain Research* 134: 170–179.

Hirsch MA, Toole T, Maitland CG, and Rider RA (2003) The effects of balance training and high-intensity resistance training on persons with idiopathic Parkinson's disease. *Archives of Physical Medicine and Rehabilitation* 84: 1109–1117.

Ivanco TL and Greenough WT (2000) Physiological consequences of morphologically detectable synaptic plasticity: Potential uses for examining recovery following damage. *Neuropharmacology* 39: 765–776.

Jankowsky JL, Slunt HH, Gonzales V, et al. (2005) Persistent amyloidosis following suppression of Abeta production in a transgenic model of Alzheimer disease. *PLoS Medicine* 2: e355.

Kleim JA, Jones TA, and Schallert T (2003) Motor enrichment and the induction of plasticity before and after brain injury. *Neurochemical Research* 28: 1757–1769.

Kramer AF, Bherer L, Colcombe SJ, Dong W, and Greenough WT (2004) Environmental influences on cognitive and brain plasticity during aging. *Journals of Gerontology, Series A: Biological Sciences and Medical Sciences* 59: M940–M957.

Lazarov O, Robinson J, Tang YP, et al. (2005) Environmental enrichment reduces A-beta levels and amyloid deposition in transgenic mice. *Cell* 120: 701–713.

Mattson MP (2000) Neuroprotective signaling and the aging brain: Take away my food and let me run. *Brain Research* 886: 47–53.

Schallert T and Whishaw IQ (1978) Two types of aphagia and two types of sensorimotor impairment after lateral hypothalamic lesions: Observations in normal weight, dieted and fattened

rats. *Journal of Comparative and Physiological Psychology* 92: 720–741.
Smith AD and Zigmond MJ (2003) Can the brain be protected through exercise? Lessons from an animal model of parkinsonism. *Experimental Neurology* 184: 31–39.
Tillerson JL, Cohen AD, Philhower J, Miller GW, Zigmond MJ, and Schallert T (2001) Forced limb-use effects on the behavioral and neurochemical effects of 6-hydroxydopamine. *Journal of Neuroscience* 21: 4427–4435.
Woodlee MT and Schallert T (2006) The impact of motor activity and inactivity on the brain: Implications for the prevention and treatment of nervous system disorders. *Current Directions in Psychological Science* 15: 203–206.

# Hormones and Memory

**J S Janowsky and D R Roalf**, Oregon Health and Science University, Portland, OR, USA



## Background

Compelling neurobiological evidence from animal models suggests that estrogen and testosterone influence the structure and function of brain regions important for memory. These regions include the hippocampus, amygdala, basal ganglia, and prefrontal cortex, all of which have traditional androgen or estrogen receptors or have what are considered to be rapid 'membrane mechanisms' of hormone action. A variety of physiological changes occur in these regions when sex hormone levels are manipulated. These findings have shaped studies of sex hormone actions on memory in humans; however, findings from human studies are as yet ambiguous. The literature can best be characterized as the 'the good, the bad, and the ugly.' There are several situations through which hormone action in humans has been studied. These include effects on memory of hormone changes across the menstrual cycle, age and menopause-related hormone loss, hormone therapy in healthy men and women, hormone supplementation in neurodegenerative disease, and hormone loss for prostate cancer treatment in men. Recent studies of memory effects from these models will be briefly reviewed here with attention to the specific forms of memory affected. That is the good news. Most studies of animal models use 17-$\beta$-estradiol, testosterone, or a metabolite of these sex steroids. Human studies must utilize medications and application methods approved for human use. Thus, the studies reviewed examined effects from a variety of medications that modify the hormonal milieu of the brain including estradiol, conjugated equine estrogen, testosterone, gonadotropin-releasing hormone agonists, administered by pill, injection, patch, gel, or after surgical oopherectomy. Differences in memory effects among these different methods of hormone intervention are not well established in the literature. That is the bad news. Finally, data come from naturalistic, cross-sectional, and placebo-controlled trials. Thus, differences in effects or lack of effects among studies may well be due to the difference in medications, sample, and design. That is ugly, but part and parcel of the constraints of human studies.

## Estrogen and Declarative Memory

### Effects of Estrogen on Memory during Normal Menstrual Cycling

Definitive studies of memory effects during menstrual cycling and their relationship to sex hormone levels in young women are sparse. The phase of the menstrual cycle (e.g., high estradiol and progesterone in the luteal phase versus low sex hormones in the menstrual phase) does not modify memory in women with mild or severe premenstrual symptoms nor does it modify cortisol-induced memory impairments. Similarly, explicit verbal memory is not affected by menstrual phase in healthy young women despite effects in other cognitive domains including visual priming and mental rotation. Visual memory is poorer at the low estrogen menstrual phase as opposed to the luteal phase, but other measures of memory are unaffected. Factors that qualify studies in the literature are the lack of confirmation of menstrual phase with estradiol levels, very small sample sizes, and oral contraceptive use by women in some studies. The lack of definitive effects of estradiol modulation, particularly of verbal memory in younger women, is in contrast to effects in aging.

### Effects of Estrogen in the Perimenopausal Period

During the transition to menopause, estrogen production gradually declines and women report memory difficulties. This transition causes physiological symptoms, such as hot flushes and sleep disturbances. Estradiol therapy relieves these physiological symptoms. Studies suggest that women on estrogen therapy to relieve symptoms of the menopause and particularly after surgical menopause, when hormone levels drop precipitously to negligible levels, have superior recall of verbal information. What is still unclear is whether this is a critical period for long-term estrogen effects in aging, or whether the critical factor is the severity or rate of hormone loss. A placebo-controlled trial showed that women (mean age 51.2 years) on conjugated equine estrogen for 21 days had numerically higher verbal memory performance than controls, but this did not reach significance when scores were corrected for general intelligence differences among treatment groups. Furthermore, five other randomized clinical trials show that women under the age of 65 receiving estrogen are more likely to show better verbal memory performance later in life than those not on estrogen. In addition, a novel intranasal estradiol preparation administered for 2 years did improve women's perception of their memory in a dose-dependent manner. These studies are in contrast to no effects or poorer performance in studies where hormone therapy is initiated late in life, after the age of 65. Verbal memory does not differ in women at different stages of the menopausal transition, nor between current versus past hormone users in a population-based observational study. Similarly, a placebo-controlled study finds that the severity of menopausal symptoms, their relief, or estradiol treatment during the transitional phase of natural menopause affects memory. However, a significant association occurs such that those who began hormone use during the menopausal transition have better memory subsequently than those who begin hormone use only after the final menstrual period. Neither verbal nor nonverbal memory differed among women who were never, current, or past users of hormone therapy when the data were adjusted for age, education, and other cognitive performance differences among hormone-use groups in another cross-sectional study. Together these studies suggest that verbal memory is affected by estrogen when it is begun during but not after the menopausal transition and this may be a narrow window. Thus, there may be a critical period during which the influence of estrogen on memory, particularly verbal memory, is most significant. Similar temporally limited effects of estrogen replacement after gonadectomy are found in rodent models. This would suggest that memory areas change in their responsiveness to estradiol depending on age or the prior hormonal milieu of the brain.

### Effects of Estrogen on Older (Postmenopausal) Women

Hormone therapy in older postmenopausal women may have a protective effect against age-related memory decline, and/or the promotion of memory performance over and above age-related changes in memory. For instance, an early randomized controlled study found that a year of estradiol therapy improved performance on the Wechsler memory scale as compared to a placebo control group or pretreatment performance in older women confined to a nursing home. In contrast, recent data from women in their 70s find that there is no effect of estrogen in combination with progesterone on verbal or nonverbal (figural) memory approximately 3 years after initiation of hormone therapy. However, estrogen and progesterone treatment are negatively associated with verbal memory and positively associated with nonverbal memory rates of change over the approximately 3-year follow-up interval. Estradiol alone for a shorter duration (20 weeks) in a similar sample of older women does not affect cognition. Blinded controlled data on the long-term effects of estradiol alone are as yet not available. The female advantage for verbal memory tasks remains when older men and women with similar estradiol levels are examined. This suggests that sex differences are not dependent on circulating estradiol levels in adulthood.

Case-control studies compare women who are users versus nonusers of hormone therapy and who are matched for a variety of other demographic variables. This is the only method that is available to understand whether estradiol has effects in older women when used over much of their postmenopausal life. Healthy, older postmenopausal women on estrogen therapy perform significantly better than nonusers of estrogen on word-list recall and on both immediate and delayed paragraph recall. Users of hormone therapy had better visual memory on the Benton visual retention task as compared to women not using hormone therapy. Visual memory, as measured by the visual reproduction test of the Weschler memory scale, is also significantly better in women on hormone therapy. However, not all studies find such effects. For instance, users of hormone therapy perform similarly to never users of hormone therapy on a test of story memory in a large sample of healthy older women. Hormone use is not related to rate of memory decline in older women over a 2-year study period, and in fact hormone users have a numerically faster rate of memory decline compared to never users.

Estradiol levels are positively related to verbal memory in elderly women (age 61–91 years) even if they are not on hormone therapy but are not associated or are negatively associated with visual memory. The negative association of estradiol levels and visual memory in hormone users as opposed to the positive effects observed in hormone users suggests that there may be a threshold over which estradiol has positive effects on nonverbal memory in elderly women.

Brain activity during memory tasks is also associated with hormone therapy in older women. For example, measurements of blood flow (using positron emission tomography (PET)) in brain regions that subserve verbal and nonverbal memory processing, including the parahippocampus and right frontal regions, are higher in women using hormone therapy as compared to women without hormone therapy. A 2-year follow-up PET study finds increases in regional cerebral blood flow (rCBF) in estrogen users as compared to nonusers in areas associated with a verbal recognition memory test, including the hippocampus, parahippocampus, and middle temporal lobe. This parallels findings of improved memory performance over an 18-month period in hormone users as compared to nonusers.

## Estrogen and Working Memory

Estrogen does not affect working memory performance in middle-aged women (mean age 50.8 years) but does result in increased brain activity in the inferior parietal lobe during storage of verbal information and decreased brain activity in the same region during storage of nonverbal material. Another placebo-controlled study finds no influence of estrogen therapy on working memory in older women using the subject-ordered pointing test. Cross-sectional studies find middle-aged women on estrogen perform better than nonusers on working memory tasks comprised of both verbal and spatial components. Young women have better verbal working memory (span task) during the high estrogen phase as opposed to the low estrogen phase of the menstrual cycle as inferred by body temperature, while the performance of women on birth control, and estradiol levels, do not vary but remain constant.

## Estrogen and Implicit Learning

Higher estradiol levels or performance at the high estrogen phase of the menstrual cycle in young, premenopausal women is related to faster motor dexterity on the grooved pegboard task, and motor sequence learning, but not simple reaction time using a task associated with the striatum. Visual priming may be lessened at the high estrogen phase, although this was confounded with the order of testing across menstrual phase.

## Sex Hormone Effects in Men

Testosterone supplementation is most commonly used to treat hypogonadism in older men. In addition, testosterone can be converted to estrogen in the brain. Thus, studies using inhibitors to block the metabolism to estrogen, or studies of complete blockade of the androgen pathway, as is a common treatment to prevent the progression of prostate cancer, will result in very low estradiol in men. Very high-dose estradiol has also been studied as a novel treatment for prostate cancer. Studies using these models show that significant changes of testosterone or estradiol levels affect memory in older men. Testosterone deprivation results in poorer verbal memory in younger men, as well as older men on hormonal therapy for prostate cancer. Replacement with very high-dose estradiol, that also maintains an androgen blockade, results in improved verbal memory in men with prostate cancer. Similarly, testosterone supplementation in hypogonadal men results in improved verbal memory, while dihydrotestosterone, an androgen that does not metabolize to estradiol, results in improved spatial memory. Blocking the conversion of testosterone to estradiol blocks the enhancement of verbal memory in older men. Together these studies suggest that estradiol is important for verbal memory enhancement, at least in older men. Memory is not affected by testosterone supplementation in men with congenital disorders that cause hypogonadism. Very brief testosterone supplementation by a single injection or dose does not affect memory in younger men, although object location memory is improved in younger women.

Working memory as assessed with the subject-ordered pointing test is improved with testosterone supplementation in older men, and cross-sectional studies show that higher free (unbound) testosterone levels are related to lower rates of decline in verbal and nonverbal memory as well as a working memory measure, Trails B, performance as men age. The relationship between testosterone levels and memory performance gets stronger as men age. Like estrogen, cross-sectional studies suggest that low free testosterone is a premorbid risk factor for Alzheimer's disease in men and is associated with poorer performance on screening measures for dementia in women. Together these studies suggest that testosterone or its metabolites may play a role in maintaining brain health in aging.

## Concluding Comments on Sex Hormones and Memory

Estradiol does not affect declarative memory in younger cycling women, and the effects on memory in older women are unclear. The timing, dose, and form of sex hormones influence the outcome of studies. Still unclear is whether these hormones activate memory performance or have a neuroprotective role when cognition declines. If it were the latter, then effects would only be seen with comparisons among people with age-related or disease-related cognitive decline. These lifetime factors complicate the ability to study effects in humans versus effects found in animal models. Regardless, the effects in humans are subtle as compared to the effects of gonadectomy and hormone replacement in animal models. This suggests that the day-to-day role of sex hormones in humans is to modulate memory formation, while severe and sudden hormone deprivation, particularly in critical periods for age-related decline, may amplify memory impairments and risk for neurodegenerative disease.

*See also:* Aging and Memory in Animals; Aging and Memory in Humans.

## Further Reading

Cherrier MM, Matsumoto AM, Amory JK, et al. (2005) The role of aromatization in testosterone supplementation: Effects on cognition in older men. *Neurology* 64(2): 290–296.

Henderson VW, Guthrie JR, Dudley EC, Burger HG, and Dennerstein L (2003) Estrogen exposures and memory at midlife: A population-based study of women. *Neurology* 60(8): 1369–1371.

Janowsky JS (2006) Thinking with your gonads: Testosterone and cognition. *Trends in Cognitive Science* 10(2): 77–82.

Janowsky JS, Chavez B, and Orwoll E (2000) Sex steroids modify working memory. *Journal of Cognitive Neuroscience* 12(3): 407–414.

LeBlanc ES, Janowsky J, Chan BK, and Nelson HD (2001) Hormone replacement therapy and cognition: Systematic review and meta-analysis. *JAMA* 285(11): 1489–1499.

Maki PM (2006) Hormone therapy and cognitive function: Is there a critical period for benefit? *Neuroscience* 138(3): 1027–1030.

McEwen B (1999) The molecular and neuroanatomical basis for estrogen effects in the central nervous system. *Journal of Clinical Endocrinology and Metabolism* 84(6): 1790–1797.

Moffat SD, Zonderman AB, Metter EJ, Blackman MR, Harman SM, and Resnick SM (2002) Longitudinal assessment of serum free testosterone concentration predicts memory performance and cognitive status in elderly men. *Journal of Clinical Endocrinology and Metabolism* 87(11): 5001–5007.

Moffat SD, Zonderman AB, Metter EJ, et al. (2004) Free testosterone and risk for Alzheimer disease in older men. *Neurology* 62(2): 188–193.

Muller M, Aleman A, Grobbee DE, de Haan EH, and van der Schouw YT (2005) Endogenous sex hormone levels and cognitive function in aging men: Is there an optimal level? *Neurology* 64(5): 866–871.

Parducz A, Hajszan T, Maclusky NJ, et al. (2006) Synaptic remodeling induced by gonadal hormones: Neuronal plasticity as a mediator of neuroendocrine and behavioral responses to steroids. *Neuroscience* 138(3): 977–985.

Resnick SM, Maki PM, Rapp SR, et al. (2006) Effects of combination estrogen plus progestin hormone treatment on cognition and affect. *Journal of Clinical Endocrinology and Metabolism* 91: 1802–1810.

Sherwin BB (2003) Estrogen and cognitive functioning in women. *Endocrine Reviews* 24(2): 133–151.

Sherwin BB (2006) Estrogen and cognitive aging in women. *Neuroscience* 138: 1021–1026.

Yonker JE, Eriksson E, Nilsson LG, and Herlitz A (2003) Sex differences in episodic memory: Minimal influence of estradiol. *Brain Cognition* 52(2): 231–238.

# Deep Brain Stimulation and Movement Disorder Treatment

**A W Laxton, C Hamani, E Moro, and A M Lozano,**
University of Toronto, Toronto, ON, Canada



## History and Development

The first attempts at therapeutic deep brain stimulation (DBS) in humans began in the 1950s and 1960s. At this time, reliable surgical stereotactic frames were being developed to enable the precise localization of subcortical targets for lesioning procedures used to treat movement disorders. Electrical stimulation was used in these early procedures to identify desirable target locations for lesion placement. Over time, electrical stimulation itself was used to treat movement disorders, psychiatric disorders, epilepsy, cerebral palsy, and pain. Long-term application of these therapies was limited by the unavailability of suitable implantable electrodes and batteries, however. Furthermore, the success of levodopa to treat Parkinson's disease (PD) in the early 1970s led to a general decline in functional neurosurgery until the late 1980s. In the 1990s, DBS became an accepted therapeutic option for a variety of movement disorders, including PD, levodopa-induced dyskinesias, essential tremor (ET), and dystonia. More than 30 000 DBS operations for movement disorders have been performed worldwide.

## Description

The goal of DBS is to improve neurological function through the targeted application of electrical current to specific brain regions or structures. It involves the placement of electrode leads into desired brain regions (see **Figure 1**). These electrode leads are then connected to a programmable internal pulse generator (IPG). The IPG is placed subcutaneously in the subclavicular region. Each currently available DBS lead has four electrode contacts at its tip, typically 1.5 mm wide and 0.5 or 1.5 mm apart. The DBS leads are made out of polyurethane tubing, and the electrodes are made out of platinum and iridium. The IPG consists of a silver vanadium oxide or lithium ion cell and integrated circuits encased in titanium with an insulating coating.

The typical wave shape of the delivered current of stimulation is square. Stimulation parameters are commonly set to a frequency in the range of 130–185 Hz, a pulse width in the range of 60–120 μs, and an amplitude in the range of 2–4 V, according to the type of movement disorder treated and the type of IPG implanted. Although stimulation can be delivered continuously or in cycles, the continuous mode is used most commonly. The longevity of the IPG battery depends on the stimulation parameters set for each patient. With typical stimulation parameters, the IPG needs to be replaced every 3–5 years.

## Indications and Targets

DBS is appropriate for PD patients who have shown a positive response to dopamine agonist medications, such as levodopa, in the past, whose response to dopaminergic agents is of limited or decreasing duration, and who experience significant motor fluctuations or dyskinesias in response to levodopa. The primary goals of DBS are to reestablish the maximal benefit patients have previously achieved with levodopa or to allow a levodopa dose reduction which will diminish drug-induced dyskinesias while maintaining the patient in a good 'on' state throughout the day. DBS may also be used for patients with severe, disabling tremor such as ET or primary and to a lesser extent secondary dystonia that is unresponsive to medical treatments.

The subthalamic nucleus (STN) is the most commonly used target in patients with PD. For patients with ET or tremor secondary to multiple sclerosis, head injury, cerebellar disease or stroke, the most commonly used DBS target is the ventral intermediate nucleus of the thalamus (Vim). For segmental or generalized dystonia, the internal segment of the globus pallidus (GPi) is currently used most commonly although the pallidal-receiving thalamus and the STN may also be useful (see **Figure 2**).

More recently, pedunculopontine nucleus DBS has been investigated in PD patients, with promising preliminary results. There have been a few reports of the efficacious application of GPi DBS in patients with Huntington's disease and other types of chorea. Patients with Tourette's syndrome have also recently been treated with thalamic and GPi DBS, with variable results.

## Anatomy and Pathophysiology

Structural lesions, neural degeneration, problems in connectivity, neurotransmitter deficiencies, and intrinsic disruptions in neural activity are likely contributors to the disturbances in normal motor physiology that underlie movement disorders. These disturbances are funneled to the output structures of the basal ganglia and cerebellum and conveyed to the thalamus and then to the corresponding cortical terminal fields to cause

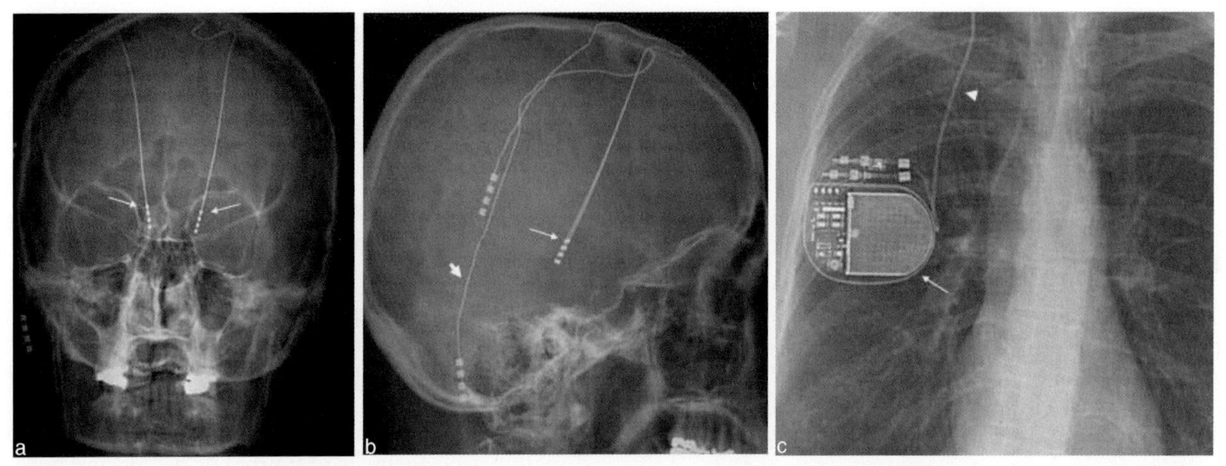


**Figure 1** X-ray images of a patient who had subthalamic nucleus deep brain stimulation electrodes implanted. Anteroposterior (a) and lateral (b) skull X-rays showing the bilaterally implanted electrodes with four contacts each (small arrows are pointing at the most dorsal contacts). The large arrow in (b) is pointing to one of the extension cables that connect the electrodes to the internal pulse generator (IPG). (c) Chest X-ray showing the extension cables (arrowhead) that connect the IPG (small arrow) to the electrodes.

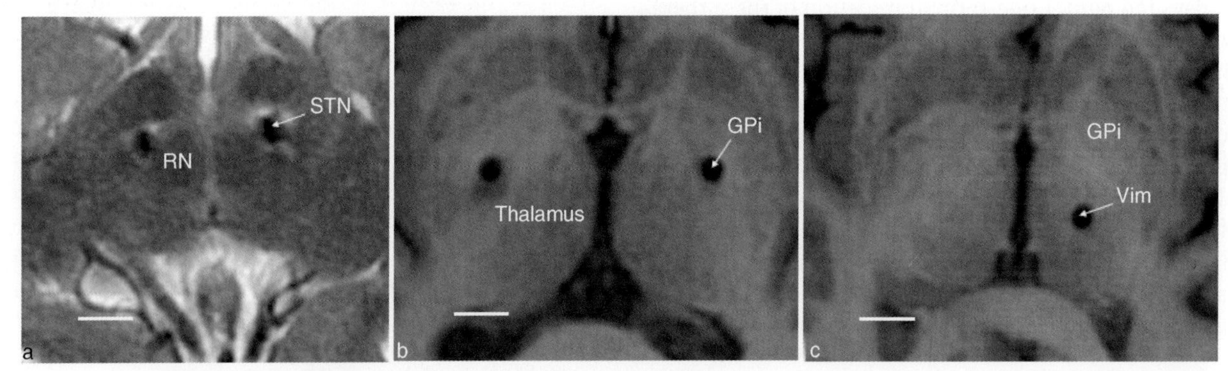


**Figure 2** Magnetic resonance images of electrodes (arrows) (a) implanted in the subthalamic nucleus (STN), (b) globus pallidus internus (GPi), and (c) ventral intermediate nucleus of the thalamus (Vim). RN, red nucleus.

the widespread disturbance in neural activity and function that is characteristic of movement disorders. Several theories have been proposed for how the activity within this circuit is regulated. There is evidence for alterations in neuronal firing rates, changes in patterns of discharges and neuronal bursting behavior, the emergence of oscillatory behavior among specific groups of neurons, enhanced correlated activity across neurons, and the expansion of sensorimotor receptive fields as potential contributors to the pathophysiology of movement disorders.

Movement disorders arise when the normal activity within the motor circuit is disturbed. In PD, the activity is disrupted to a large extent because dopaminergic output from the degenerating substantia nigra pars compacta (SNc) is decreased. In primary generalized dystonia, the pattern of GPi neuronal discharge is altered for unknown reasons, leading, it is thought, to dysregulated thalamocortical activity. In ET, groups of thalamic neurons fire synchronously with the peripheral limb tremor. It is believed that DBS interferes with this dysfunctional activity, leading to more-normal motor function.

## Mechanism of Action

Electrical brain stimulation affects a broad range of neural elements, including neuronal cell bodies, afferent and efferent axons, and glia. The electrophysiological properties associated with each of these elements vary depending on anatomical location. The effects of electrical stimulation on these elements are incompletely understood but are thought to vary with the intrinsic properties of the stimulated elements and with the stimulation parameters used (i.e., frequency, amplitude, and pulse width).

Because high-frequency stimulation (HFS) of certain targets mimics the effects of lesions to those targets, it had been believed that stimulation was inhibitory. However, this simple view ignores the possibility of stimulation's activating inhibitory circuits, the potential complex polysynaptic downstream effects of local activation, the possible differential effect of stimulation on cell bodies versus axons, and the possibility that HFS may lead to neurotransmitter depletion or synaptic failure. To better account for the mechanisms of DBS, several explanations have been proposed.

It has been observed that single pulses in the vicinity of GPi neurons produce a cessation of spontaneous activity for 15–25 ms. Because of the latency and duration of this effect, it has been suggested that $\gamma$-aminobutyric acidergic (GABAergic) synaptic transmission may be involved. That is, stimulation within GPi causes GABA release from either pallidal (pars externa) or striatal axons projecting onto GPi neurons, or local dendritic release of GABA, resulting in GPi neuronal inhibition. Although the GPi also receives glutamatergic afferents from the STN, the larger GABAergic input is believed to trump these excitatory signals. Further supporting the synaptic transmission proposal, studies have demonstrated increased firing rates in the GPi following STN HFS, and decreased firing rates in the thalamus following GPi HFS. Similarly, glutamate release in the downstream GPi has also been found to be increased following STN HFS. Computer modeling studies have suggested that DBS may inhibit neuronal cell bodies and excite axons. Functional imaging also corroborates the role of HFS in activating axonal elements as increased upstream (antidromic) and downstream (orthodromic) activity has been demonstrated. A variation on the synaptic transmission proposal states that although HFS is excitatory, the stores of neurotransmitter are soon depleted, resulting in functional inhibition.

The ability of HFS to alter intrinsic membrane properties has also been posited as a potential mechanism of the DBS effect. Some *in vitro* studies suggest that HFS transiently depresses voltage-gated $Ca^{2+}$ channels, producing a 'depolarization block.' Alternatively, HFS may also directly activate $Na^{+}$ channels, leading to depolarization. The relevance of these *in vitro* findings awaits further investigation.

The seemingly contradictory explanations for DBS effects may actually be complementary. That is, the effect of DBS may depend on several mechanisms, and the effects may vary depending on what structures are being stimulated and how far a structure is from the stimulating electrode. In addition, since the pathophysiology of most of the movement disorders is still not entirely known, the stimulation of the same target in different diseases may require different parameters of stimulation (e.g., low-frequency GPi stimulation in Huntington's disease and high-frequency in PD). The net effect of DBS, however, appears to be to disrupt the dysfunctional neuronal activity associated with movement disorders.

## Operative Procedure

DBS surgery involves brain imaging, target identification, and the insertion of DBS electrodes and an IPG. The STN and GPi can be visualized with magnetic resonance imaging. The Vim target is estimated indirectly, relative to the position of the anterior and posterior commissures. Intraoperative electrophysiological mapping is used to characterize the location of the recording electrode in relation to the desired target, to assess the therapeutic effect of an electrode in a particular location, and to identify the potential unwanted side effects associated with stimulation in a given location. Approaches may involve one or more microelectrodes (single unit recording; local field potentials; microstimulation) or macroelectrodes (usually, the DBS electrode itself; local field potentials; macrostimulation). The majority of DBS centers use a combination of these techniques.

Standard mapping of the STN begins with microelectrode recordings starting in the range of 10–15 mm above the chosen target. The single unit activity recorded depends on the specific trajectory chosen but may include the following structures, in order (see **Figure 3**): the thalamus (with its characteristic bursting cells), the zona incerta (exhibiting a low firing rate), the STN (with its characteristic dense cellular activity in the range of 30–50 Hz, and occasional 4–6 Hz bursting activity synchronized to the patient's tremor), and then the substantia nigra pars reticulata (characterized by a more regular firing rate in the range of 50–70 Hz, without bursting or tremor cells). When the STN has been reached, testing begins to determine whether the cellular activity changes in response to the patient's active or passive limb movements. The presence of movement-related activity suggests that the electrode may be in a therapeutically desirable location. At this point, microstimulation may also be performed within the STN to check for tremor arrest or benefit to the other clinical signs or to detect the presence of paresthesias, muscle contractions, double vision, or other unwanted side effects (related to current diffusion into the areas surrounding the STN, which aids in the identification of the correct electrode locations). The microelectrodes may be repositioned to sample other STN areas and to identify the portion of the STN populated by movement-responsive neurons – the 'motor territory.' The microelectrodes are then removed and replaced by the DBS electrode. This insertion is done using X-ray guidance to ensure that the electrode is placed in the desired location. Once in place, each contact of the DBS electrode

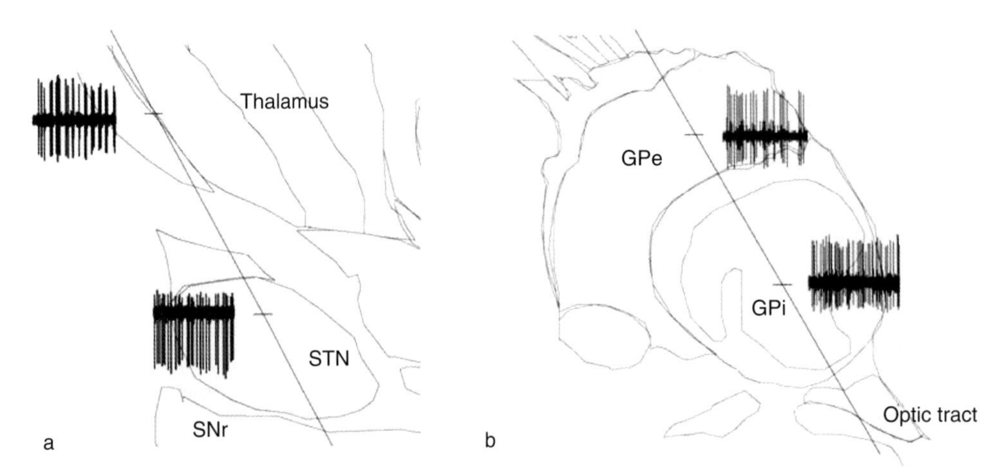


**Figure 3** Schematic representation of sagittal sections of the brain that are 12 mm (a) and 20 mm (b) from the midline. The oblique line in the center of the figures represents a typical microelectrode trajectory during the physiological exploration of the subthalamic nucleus (STN) (a) and globus pallidus (b) in patients with Parkinson's disease. (a) During the exploration of the subthalamic region, neuronal activity may be recorded from the thalamus, STN, and substantia nigra pars reticulata (SNr). Examples of a thalamic bursting cell and an irregularly firing STN cell are shown on the left side of the trajectory line. (b) Neuronal activity recorded from the globus pallidus externus (GPe) and globus pallidus internus (GPi). Stimulation of the optic tract often induces phosphenes and maps the bottom of the trajectory. For details on the neuronal firing characteristics in each of these regions, reader is referred to the text.

is stimulated with incrementally increasing voltage to check the threshold for stimulation-induced clinical benefit and unwanted side effects. The DBS electrode position can be adjusted to minimize adverse effects.

The general procedure and rationale for intra-operative electrophysiological mapping of the GPi and Vim are similar to those described for the STN, except for differences in regional anatomy and associated cellular characteristics. The microelectrode trajectory to the GPi traverses the striatum, external segment of the globus pallidus (firing activity around 50 Hz), the GPi (high-frequency activity around 80 Hz with occasional 4–6 Hz tremor-associated bursts), and then either the optic tract (inferiorly) or the internal capsule (medially and posteriorly). As with the STN, movement-related cells are sought within the GPi, and microstimulation is used to identify side effects such as muscle contractions (posterior limb of the internal capsule) or visual phosphenes (optic tract). To reach the Vim, the microelectrode may traverse the following thalamic nuclei, depending on trajectory: ventral oralis anterior/posterior (Voa/Vop; tonic firing rate around 20 Hz; active movement-responsive activity), Vim (around 25 Hz; passive movement-responsive activity), ventral caudalis (Vc; cellular activity similar to Vim but displays tactile/sensory-responsive activity), and finally beyond the thalamus into the medial lemniscus. Stimulation of the Voa/Vop and Vim should produce tremor suppression; stimulation of the Vc and medial lemniscus elicits paresthesias.

## Advantages, Disadvantages, and Alternatives

Prior to the development of DBS, lesioning procedures were the standard surgical approach to treat movement disorders. Lesions are made with high-radiofrequency stimulation to the same targets used in DBS therapy. While lesion surgery is still practiced, bilateral lesions can be associated with a high incidence of adverse effects (e.g., cognitive impairment, dysarthria, and dysphagia) and is only rarely used. The adjustability and reversibility of DBS are two of its main advantages over lesioning procedures. With DBS therapy, stimulation parameters can be adjusted to enhance or maximize a positive effect or to decrease unwanted side effects (e.g., paresthesias, muscle contractions, and dysarthria). Postoperative programming of the DBS system involves adjustments to the following parameters: choice of contact, type of stimulation (monopolar or bipolar), mode of stimulation (continuous or cycled), amplitude (voltage), frequency, and pulse width. Another interesting advantage of DBS over lesioning is the possibility of performing a bilateral procedure with a lower incidence of the severe complications that have been found with bilateral ablative surgery.

The major disadvantages of DBS include its expense; the need to replace IPGs (batteries) every 3–5 years, which involves minor surgery; and because it entails implanted foreign bodies, an increased infection risk relative to lesions. In addition, DBS surgery requires the involvement and coordination of a large multidisciplinary team (neurologist, psychiatrist, neuropsychologist, neurophysiologist, and neurosurgeon) in order to select the appropriate candidates before surgery and to manage medication reductions, DBS programming, and possible DBS complications after surgery.

## Outcomes

For patients with PD, STN DBS has been shown to produce improvements in the range of 60–70% in overall motor functioning (based on the motor component of the Unified Parkinson's Disease Rating Scale). Stimulation-associated improvements in the off-medication state vary across the various motor domains of PD and are approximately as follows: tremor, 90%; rigidity, 75%; gait, 60%; bradykinesia, 50%; postural instability, 40%; and dystonia, 30–50%. Some of these improvements, particularly in tremor and rigidity, can persist for 5 or more years. The benefits in gait and postural disability, however, wane by 2 years. STN DBS improves dyskinesias by allowing dopaminergic medication dosages to be reduced (on average by approximately 50%) and through a direct antidyskinetic effect. STN DBS has also been found to improve patients' activities of daily living and quality of life. The benefits of GPi DBS in PD may approach those seen with STN DBS, but the outcomes are less well characterized.

Vim stimulation is associated with up to 80–90% tremor reduction in PD and ET patients. In comparison with STN and GPI stimulation, however, Vim DBS does not improve the bradykinesia and rigidity associated with PD. For patients with ET, Vim stimulation may be more effective at treating hand tremor than voice and head tremor.

For primary generalized dystonia, GPi DBS is associated with an approximately 50% improvement in motor functioning and reduction in disability, on the basis of the Burke–Fahn–Marsden Dystonia Scale, and an improved quality of life related to general health and physical functioning. These beneficial effects can be expected to last for 1 or more years. GPi DBS may be more effective in improving motor function in the neck, trunk, and limbs than in improving facial movements and speech. The emergence of DBS over the last decade is a major advance in the surgical treatment of movement disorders.

*See also:* Animal Models of Parkinson's Disease; Deep Brain Stimulation and Parkinson's Disease; Parkinsonian Syndromes; Parkinson's Disease: Alpha-Synuclein and Neurodegeneration.

## Further Reading

Abosch A and Lozano A (2003) Stereotactic neurosurgery for movement disorders. *Canadian Journal of Neurological Sciences* 30: S72–S82.

Benabid AL, Pollak P, Gervason C, et al. (1991) Long-term suppression of tremor by chronic stimulation of the ventral intermediate thalamic nucleus. *Lancet* 337: 403–406.

Deuschl G, Schade-Brittinger C, Krack P, et al. (2006) A randomized trial of deep-brain stimulation for Parkinson's disease. *New England Journal of Medicine* 355: 896–908.

Dostrovsky JO and Lozano AM (2002) Mechanisms of deep brain stimulation. *Movement Disorders* 17: S63–S68.

Gildenberg PL (2005) Evolution of neuromodulation. *Stereotactic and Functional Neurosurgery* 83: 71–79.

Haber SN and Gdowski MJ (2004) The basal ganglia. In: Paxinos G and Mai JK (eds.) *The Human Nervous System,* 2nd edn., pp. 676–738. London: Elsevier.

Hamani C, Richter E, Schwalb JM, and Lozano AM (2005) Bilateral subthalamic nucleus stimulation for Parkinson's disease: A systematic review of the clinical literature. *Neurosurgery* 56: 1313–1321; discussion 1321–1324.

Hassler R, Riechert T, and Mundinger F (1960) Physiological observations in stereotaxic operations in extrapyramidal motor disturbances. *Brain* 83: 337–350.

Krack P, Batir A, Van Blercom N, et al. (2003) Five-year follow-up of bilateral stimulation of the subthalamic nucleus in advanced Parkinson's disease. *New England Journal of Medicine* 349: 1925–1934.

Kupsdch A, Beneke R, Muller J, et al. (2006) Pallidal deep-brain stimulation in primary generalized or segmental dystonia. *New England Journal of Medicine* 355: 1978–1990.

Lang AE, Deuschl G, and Rezai AR (eds.) (2006) Deep brain stimulation for Parkinson's disease. *Movement Disorders* 21 (supplement 14): S167–S327.

Lozano AM, Lang AE, Galvez-Jiminez N, et al. (1995) Effect of GPi pallidotomy on motor function in Parkinson's disease. *Lancet* 346: 1383–1387.

Perlmutter JS and Mink JW (2006) Deep brain stimulation. *Annual Review of Neuroscience* 29: 229–257.

Rodriguez-Oroz MC, Obeso JA, Lang AE, et al. (2005) Bilateral deep brain stimulation in Parkinson's disease: A multicentre study with 4 years follow-up. *Brain* 128: 2240–2249.

Vidailhet M, Vercueil L, Houeto J-L, et al. (2005) Bilateral deep-brain stimulation of the globus pallidus in primary generalized dystonia. *New England Journal of Medicine* 352: 459–467.

# Deep Brain Stimulation and Parkinson's Disease

**M C Rodriguez-Oroz, J M Matsubara, P Clavero, J Guridi, and J A Obeso**, University Clinic and Medical School, University of Navarra, Pamplona, Spain



## Introduction

Deep brain stimulation (DBS) of the ventral intermediate nucleus of the thalamus (Vim) in Parkinson's disease (PD) has been performed since the late 1980s, mimicking the effect of the lesion against tremor. Bilateral approaches with DBS and bilateral thalamotomy were similarly effective, but the incidence of adverse events was much lower with DBS. Thus, DBS of the Vim replaced thalamotomy in patients with tremor. The realization that DBS delivered similar benefits with fewer adverse events than did lesions, the need for bilateral surgery in the majority of severe PD patients, the reversibility of this approach in contrast to the lesion, and a study in 1-methyl, 4-phenyl-1,2,3,6 tetrahydropyridine monkeys showing high antiparkinsonian efficacy using high-frequency stimulation of the subthalamic nucleus (STN) led to the introduction of DBS of the STN in the therapeutic armamentarium for advanced PD patients. Concomitantly, this technique was also applied to the globus pallidus pars interna (GPi) in an attempt to achieve the benefit of a bilateral approach without the undesirable adverse effects induced by the lesions. Initial results were positive, leading to a multicenter trial begun in 1996 to assess the safety, viability, and short-term efficacy of DBS of both STN and GPi. The positive findings of this trial allowed a prompt spreading of this therapy, which continues to the present, with stimulation of the STN predominating. This article reviews the most important aspects of bilateral DBS of the STN and GPi.

## Methodological Aspects

The success of surgical treatment with DBS in PD depends on several factors. The selection of the appropriate candidate for DBS is the first critical point. Thus, the clinician has to evaluate in every patient the known negative and positive predictive factors for the outcome of this therapy. The second key factor is the accuracy in targeting the STN or GPi during the surgical procedure and prior to the implant of the therapeutic electrode. The precise location of the electrode in the nucleus determines a better antiparkinsonian benefit without stimulation-induced adverse events. Last, the careful exploration of the benefit obtained with the different programming of the electrodes, the adequate selection of the parameters for continuous stimulation, and the concomitant adjustment of medication guarantee an optimization of the therapy.

### Selection of PD Patients for Surgical Therapy with DBS

The experience gained in the last decade with PD patients submitted to DBS has allowed us to define a profile of patients with a higher probability of obtaining better results with this therapy. Numerous factors need to be taken into account in the evaluation of whether a patient with PD is a good candidate to be treated with DBS.

Advanced age, concomitant severe medical illness, and neuropsychological and psychiatric disturbances can be a drawback in the outcome of this therapy. Although age-related comorbidities seem to be more important than age *per se*, and so no specific age cutoff has been defined, in practice, the vast majority of centers exclude patients older than 75. Serious systemic or neurologic disease other than PD should be regarded as a contraindication for DBS, as they would significantly compromise DBS benefits or amplify surgical risks. Previous functional surgery should not exclude additional DBS procedures. Neurobehavioral contraindications may include primary psychotic disorders, primary and compensated bipolar disorders, severe treatment refractory depression, severe substance abuse, and severe personality disorders.

It is important to note that factors related to the PD itself and levodopa-induced side effects have to be carefully analyzed. Dementia, frequently seen in advanced PD patients, should be considered a contraindication for surgery despite the lack of outcome studies. Levodopa-resistant symptoms (i.e., freezing of gate, instability, speech disturbances) are not significantly improved by DBS and, although not a formal contraindication, should be considered in every particular case. As a rule, DBS improves PD signs that are alleviated with levodopa therapy. Preoperative levodopa response is a predictive factor for the positive response to surgery, with the possible exception of drug-refractory but otherwise classical rest tremor. Currently, all studies that have evaluated predictors of benefit from DBS have found that levodopa response is the best one. Thus, a test evaluating the motor improvement in response to a high dose of levodopa is crucial to estimate the achievable antiparkinsonian benefit with DBS. A minimum 25–50% reduction in the Unified Parkinson's Disease Rating Scale (UPDRS) motor part (III) score following the administration of an effective dose of levodopa in the practically defined off-medication state (a minimum of 12 h without antiparkinsonian medication) is considered a significant response. There is some controversy about the dose of the levodopa challenge used for the test: the first dose of levodopa in the morning or a suprathreshold dose (50% more of this dose) is the most frequently employed.

Another important issue is the outcome of DBS surgery in patients with dopaminergic-induced psychiatric disturbances. Although some publications indicate patients with gambling and dopamine dysregulation syndrome can improve both the parkinsonism and the dopaminergic side effects with DBS of the STN, there is not enough data in the literature at present to draw conclusions. Thus, no recommendation can be made on this particular and relevant aspect of PD until more experience is accumulated.

In addition to the previous criteria, it is recommended to operate on patients with a score higher than 30 in the motor UPDRS scale in the off medication state and with a disease duration of at least 5 years (to allow for atypical forms of parkinsonism to become evident).

A question recently raised is the benefit of this therapy in patients with genetic forms of PD. The available data indicate these patients can be successfully operated on with bilateral DBS of the STN. The inclusion and exclusion criteria are the same as for idiopathic PD.

### Surgical Issues

Accuracy in targeting the STN or GPi is critical for the outcome of DBS in PD, as previously explained. Stereotactic functional surgery is the procedure universally employed. Thus, the first step of the intervention is the stereotactic targeting of the nucleus by using computed tomography, magnetic resonance imaging, venticulography, or neuronavigational systems. They are all safe and reliable techniques that have been used effectively by experienced groups. Image fusion may improve the accuracy but is not mandatory. Imaging techniques should be coupled with physiological and/or clinical intraoperative testing for target refinement. The physiological definition of the nuclei is usually undertaken with microelectrode or semimicro electrode recording of the neuronal activity of the nucleus (STN or GPi) (**Figure 1**). Macroelectrodes can also be used for recording but provide very poor spatial resolution. When physiological methods are employed, once the target is precisely identified, microstimulation with the electrode used for micro or semimicrorecording and/or macrostimulation must beused for refining the final target location. In those centers that do not use physiological

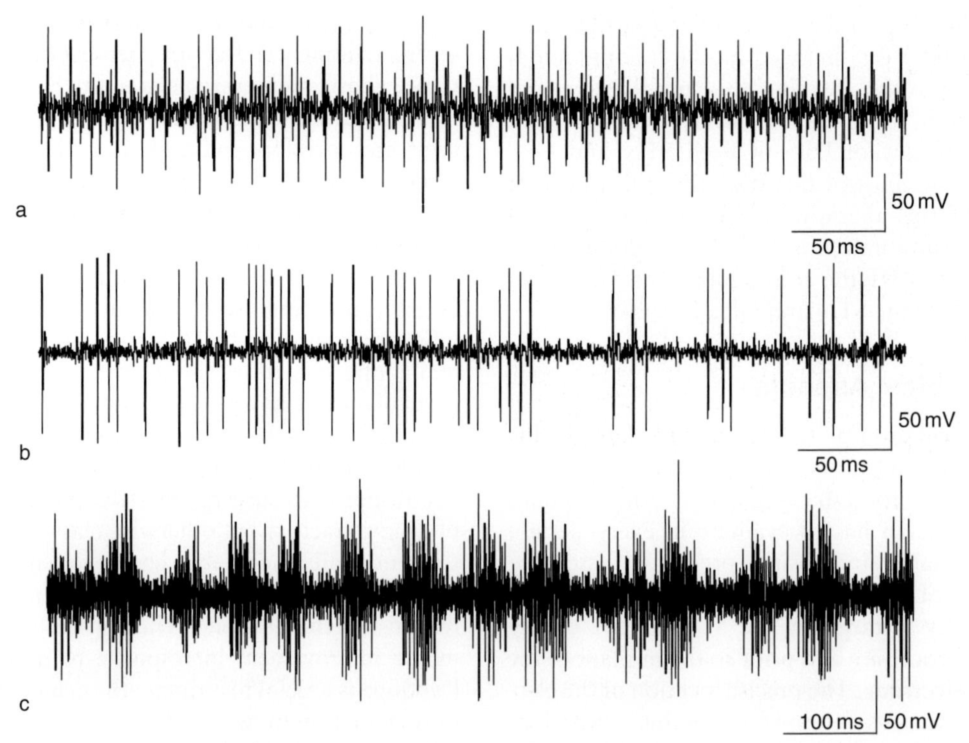


**Figure 1** Microrecording of the neuronal activity of the STN in a patient with Parkinson's disease. Example of the three types of neuronal discharges: (a) tonic, high-frequency pattern; (b) irregular activity with long pauses; (c) oscillatory, slow activity.

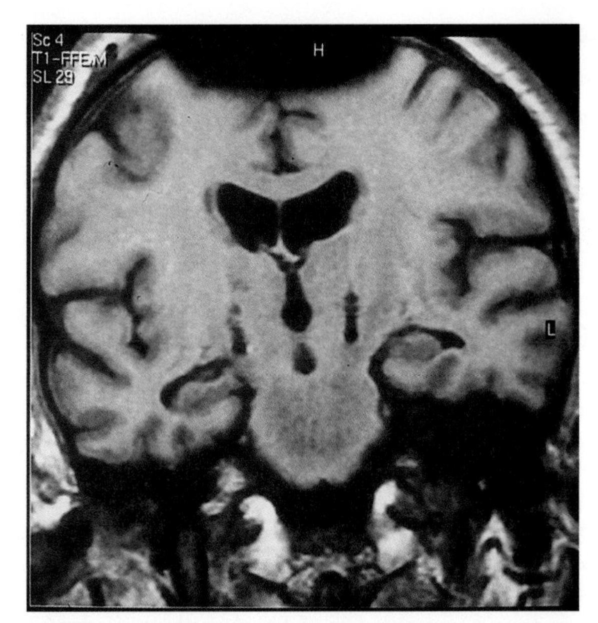


**Figure 2** Coronal magnetic resonance image of a patient which leads for DBS bilaterally implanted in the STN.

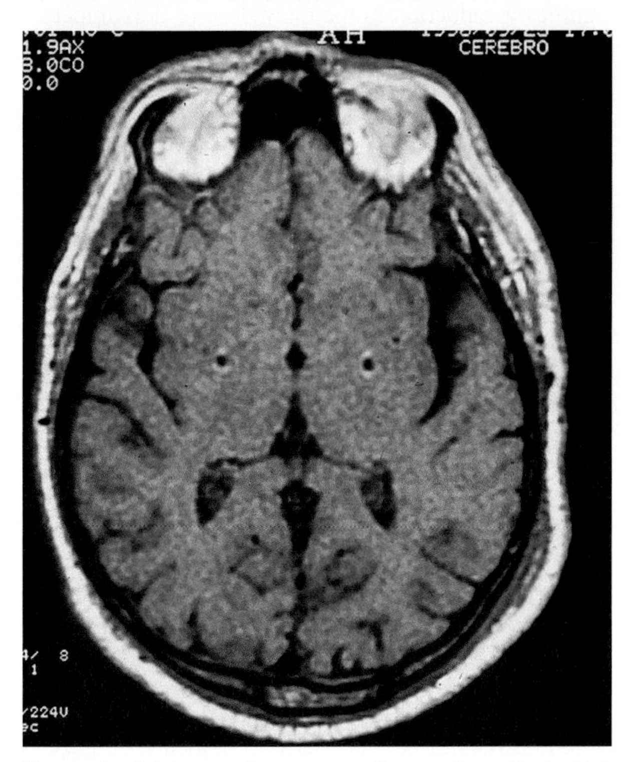


**Figure 3** Axial magnetic resonance image of a patient which leads for DBS bilaterally implanted in the Gpi.

parameters to better define the intended target, clinical intraoperative testing with a macroelectrode or with the DBS electrode can also be performed directly. In any case, and independent of the method used, once the therapeutic electrode has been positioned (**Figures 2** and **3**), intraoperative DBS stimulation is conducted to assess the benefit of the therapy against parkinsonian signs and to test the possible induction of stimulation-related adverse events that could thereafter preclude the utilization of this treatment. It has to be known that in some instances, there is an 'impact' or 'lesionlike' effect that consists of an amelioration of the parkinsonian signs after the mechanical introduction of the therapeutic electrode into the nucleus. In these cases, the intraoperative stimulation with the DBS electrode does not provide any relevant information.

Patients remain awake during the whole intervention in off-medical condition. Local anesthesia is used in the area of the burn hole. In some particular cases (i.e., severe tremor or dystonia), antiparkinsonian medication, sedation, or general anesthesia is administered. This situation is far from ideal because it interferes with the neuronal activity of the nucleus and its recognition by the physician while recording and because it does not allow testing of the antiparkinsonian efficacy and side effects of the stimulation.

Once the therapeutic electrode has been placed in the nucleus and the effect of the stimulation has been evaluated, the next step is to confirm its location by intraoperative fluoroscopy before securing it to the burn hole with a specific device.

Apart from the therapeutic electrode(s), the rest of the hardware (external cable and battery) has to be placed under the subcutaneous cellular tissue and connected to the lead already implanted (**Figure 4**). The battery or internal pulse generator (IPG) is placed in a tissue pocket located in the upper chest or abdomen and connected to the therapeutic electrode through the external cable. This procedure is undertaken with general anesthesia and can be done in the same surgery or later.

Another important decision when bilateral procedures are planned is whether they should be done in the same surgery. Available data show no significant differences between surgical time and patient clinical outcome, and every case should be individually evaluated.

## Postoperative Management

Postoperative antiparkinsonian treatment should be restarted as soon as possible to relieve discomfort and limit the risk of acute dopaminergic withdrawal and malignant hyperthermia. The postoperative levodopa equivalent dose (LED) should be the same as the presurgical dose and composed primarily of levodopa to facilitate simple titration during the process of programming the stimulation. A confusional state may be present in some patients, but it generally lasts a few days and can be managed easily with atypical antipsychotics such as clozapine or quetiapine.

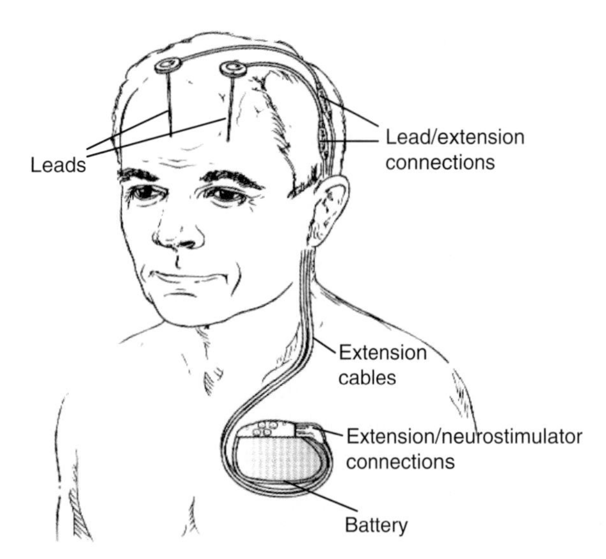


**Figure 4** Schematic representation of the implanted device for DBS.

Neuroimaging to confirm the electrode location and to rule out intracranial adverse events is recommended.

Programming can be initiated at any time after the whole device has been implanted depending on the center and the situation of the patient. When a lesion-like effect is present, the programming can be delayed until the parkinsonian symptoms reappear. Programming should be started in the off-medication state to select and set up the parameters with the highest antiparkinsonian efficacy. Parameters inducing permanent adverse events should be ruled out. Stimulation with the greatest efficacy and fewest side effects is selected for therapy. Once programming has been selected, gradual reduction of drug doses has to be initiated to adjust the medication to the new motor state. Levodopa and STN stimulation have similar and additive effects in the antiparkinsonian and dyskinetic states. Thus, in PD patients treated with DBS of the STN, medication titration is needed for fine adjustment of the motor situation in the majority of patients. In cases with DBS of the GPi, this additive effect of stimulation and medication is not as strong, and LED cannot be reduced.

Treatment should be maintained with levodopa to allow the programming to be adjusted adequately to both off- and on-medication motor conditions. Once programming is stable, dopaminergic agonists can help to achieve a more stable motor benefit than levodopa.

After this first adjustment of stimulation and medication, the frequency of visits to the neurologist depends on the patient's needs and on the available resources of each center. In any case, follow-up visits are recommended.

## Results of Bilateral DBS in PD

The multicenter study conducted in 1996 to assess the safety, viability, and efficacy of DBS of the STN and GPi was not randomized, and no specific guidelines were issued for allocating patients to STN or GPi surgery. Both groups showed significant improvement induced by stimulation. Over the years, many groups all over the world have applied DBS of the STN and GPi to treat PD, but studies properly comparing the therapeutic efficacy of the two targets are limited.

Nevertheless, STN DBS has become the preferred approach in most centers. There appear to be three main reasons. First, the impression that the clinical relief of the cardinal features of PD obtained by STN DBS was greater than that obtained by GPi DBS spread quickly after the initial trial despite absence of validated data. Second, from a surgical point of view, it may be easier to physiologically define the sensorimotor region and surrounding structures of the STN than those of the GPi. Finally, and as an established fact, STN DBS allows a significant reduction in the daily dose of dopaminergic drugs that is not achieved with GPi DBS. This is a very attractive clinical possibility considering the profusion of motor problems (i.e., dyskinesia) and nonmotor problems (e.g., psychiatric complications, sleep disturbance) associated with chronic levodopa treatment. Overall, the issue of advantages and disadvantages of either target remains unresolved.

### Bilateral DBS of the STN

Bilateral DBS of the STN has become the most frequent surgical therapy for PD patients.

Since the first patients operated on by the Grenoble group early in the 1990s, a huge number of publications have described the effect of this therapy in PD patients at different points in the postoperative follow-up. Some of them, because of the number of patients included or long-term follow-up, are more indicative of the benefit–risk ratio of this treatment. A meta-analysis of the outcome of DBS of the STN based on a review of the literature from 1993 until June 2004, while acknowledging between-study heterogeneity, reached the conclusion that the treatment is highly effective in PD patients. It improves UPDRS-II or activities of daily living (ADL) scores by 50%, UPDRS-III scores by 52%, dyskinesia scores by 69.1%, daily off-medication time by 68.2%, and quality of life by 34.5%. LED was reduced in 55.9%. After June 2004, and therefore not included in this meta-analysis, many more articles about DBS of the STN in PD have been published with similar findings. Because of the profusion of data, only those studies with long-term follow-up (4–5 years) published after the meta-analysis are reviewed in this article (**Table 1**).

**Table 1** Summary of the most relevant long-term follow-up studies of DBS of the STN

| *Study* | *Follow-up (months)* | *Age at surgery (years)* | *Mean H&Y (off)* | *Disease duration (years)* | *Preoperative levodopa response (UPDRS III off/on: % change)* | *UPDRS II score (postop stim on and med off vs. preop med off; % change)* | *UPDRS III score (postop stim on and med off vs. preop med off; % change)* | *LED (% change)* | *Dyskinesia (% change)* | *Duration of daily 'off' (% change)* |
|---|---|---|---|---|---|---|---|---|---|---|
| Rodriguez-Oroz (*N*=49) | 36–48 | 59.8 ± 9.8 | 4.3 ± 0.8 | 15.4 ± 6.3 | 59.78 | 43 | 50 | 35 | 59 | 56 |
| Visser-Vandewalle (*N*=20) | 53.6 ± 2.6 | 60.9 ± 8.1 | NR | 15 ± 4.4 | NR | 59 | 42.8 | 47.2 | 79 | 78 |
| Østergaard (*N*=18) | 48 ± 2 | 63 ± 8 | 3.7 ± 0.8 | 19 ± 5 | 54.19 | 42 | 55 | 29 | 90 | 67 |
| Shupbach (*N*=30) | 60 | 54.9 ± 9.1 | 5 | 15.2 ± 5.3 | 69 | NR | 49.51 | 54.56 | 79 | 67 |

[a]Data reported as mean ± SD/range.

H & Y, Hoehn and Yahr scale; LED, levodopa equivalent dose; *N*, effective sample size; NR, not reported; UPDRS, Unified Parkinson's Disease Rating Scale; II, UPDRS activities of daily living scale; III, UPDRS motor scale.

Rodriguez-Oroz et al. reported the results of a multicenter study that is the follow-up of the first multicenter trial aimed at assessing the safety, viability, and efficacy of DBS of either the STN or the GPi. Overall, 49 patients treated with DBS of the STN were available for study after 4 years. Stimulation of the STN in off-medication condition induced a reduction of 50% in the motor UPDRS score and of 43% in the ADL score compared with the preoperative medication off state. All cardinal features were also improved except speech. In the on-medication situation and in comparison with the preoperative on-medication state, stimulation did not improve either the motor or the ADL-UPDRS scores. Motor fluctuations were reduced with an increment of the time spent in the on-medication condition without dyskinesias. Dyskinesia severity was reduced by 59% with respect to baseline and LED in 35%. Comparison of the improvement induced by STN stimulation at 1 year and at 4 years showed a worsening in axial signs (speech, postural stability, and gait) in both off- and on-medication conditions. This result is similar to the one obtained in a cohort of 49 patients after 5 years of follow-up by Krack et al. that was included in the meta-analysis. Adverse events were frequent (53% of patients) and will be detailed below.

In addition to the multicenter study, there are other studies with very similar results (**Table 1**). Visser-Vandewalle reported that the benefit of stimulation of the STN in the off-medication condition was maintained after 4 years in the UPDRS motor score and in all cardinal signs of PD (tremor, rigidity, bradykinesia and gait) when compared with the off-medication state at baseline. Antiparkinsonian benefit was lower at 4 years than at 3 months (i.e., total motor UPDRS score reduction of 54% at 3 months and 43% at 4 years). In the on-medication state, stimulation-linked improvement in total UPDRS motor score, rigidity, and akinesia at 3 months vanished at 4 years, and gait worsened significantly. Dyskinesias, motor fluctuations, and ADL were similarly improved by 79%, 78%, and 59%, during the 4 years of follow-up, and LED was reduced by 42.8%.

Another cohort of 18 patients has shown similar results after 4 years of follow-up. Stimulation induced significant benefit in the off-medication situation compared with the off-medication preoperative state in ADL and motor UPDRS scores except for the axial signs (postural stability, gait, speech), which also were significantly worsened compared with the 1-year and basal condition. Stimulation in the on-medication state did not improve the total motor score, and there was even a worsening of speech and postural stability compared with the on-medication condition at baseline. The reduction in LED was 29%. It is interesting that the benefit of levodopa in axial signs and total motor UPDRS was significantly lower at 4 years.

Shüpbach et al. showed a 49.5% UPDRS motor score benefit of stimulation in the off-medication state, a less-striking improvement in the axial signs, and a 40% benefit in the ADL. Motor complications and dyskinesias were reduced by 67% and 79%, respectively, although motor complications were significantly higher at the 5-year evaluation than at the 6-month follow-up. The reduction in LED was 54%.

Recently the effect of DBS of the STN has been compared with the best medical management in two trials. Deuschl et al. studied 156 patients in a multicenter, randomized-pairs trial with 6 months of follow-up. Stimulation produced greater improvement in the quality of life and the various motor evaluations (UPDRS, motor fluctuations, etc.) than the best medical treatment. In the other trial, only young patients (younger than age 55) with no severe PD were included in a study aimed at assessing the benefit of this surgery in comparison with the best medical treatment at an early stage of the disease. Similar to the previous trial, the patients who underwent surgery early after the onset of PD had an outcome superior to that of patients treated with the best medical therapy.

In summary, bilateral stimulation of the STN produces a significant benefit on the off-medication motor and ADL UPDRS scores and in cardinal features of PD patients after long-term follow-up, a benefit that is less striking in axial signs. In the on-medication condition, stimulation tends to lose efficacy in the motor and ADL scores. Remarkably, a significant decline in the effect of stimulation in the off-medication states between the short- and long-term evaluations was observed in all the studies, mainly in the axial signs. However, this worsening in axial features was also seen in the on-medication condition, indicating that it may reflect the progression of the disease and extension of the pathological process beyond the nigrostriatal dopaminergic system. Moreover, in terms of quality of life, parkinsonian signs, and motor complications, this therapy seems to be more beneficial than the best medical therapy.

#### Bilateral DBS of the GPi

Bilateral stimulation of the STN became the preferred approach for DBS in PD. Thus, the data about the efficacy of DBS of the GPi are less abundant. With the exception of the multicenter study initially undertaken in 1996 and the follow-up of this trial at 4 years after surgery, the majority of studies conducted to assess the outcome of this intervention are open trials, the numbers of patients are small, and the follow-up

**Table 2** Summary of the most relevant studies evaluating the antiparkinsonian benefit of deep brain stimulation of the GPi

| *Study* | *Follow-up (months)* | *Age at surgery (years)* | *Mean H&Y (off)* | *Disease duration (years)* | *Pre-operative levodopa response (UPDRS III off/on; % change)* | *UPDRS II score (postop stim on/med off vs. preop med off; % change)* | *UPDRS III score (postop stim on/med off vs. preop med off; % change)* | *LED (% change)* | *Dyskinesia (% change)* | *Duration of daily 'off' (% change)* |
|---|---|---|---|---|---|---|---|---|---|---|
| Rodrigues ($N=11$) | 24.8 | 61.45 | 2.7 | 14.54 | 50 | NR | 46 | UCH | 76 | NR |
| Volkmann ($N=11$)[b] | 12 ($n=10$) | NR | NR | 11.2 ± 2.7 | 44 | 50 | 56 | UCH | 58 | NR |
| | 36 ($n=9$) | | | | | 30 | 43 | | 63 | |
| | 60 ($n=5$) | | | | | 10 | 24 | | 64 | |
| Durif ($N=5$) | 36 | 64 | 4.3 ± 0.2 | 15 ± 2 | 72 | 9 | 32 | UCH | 50 | 10 |
| Ghika ($N=6$)[c] | 24 | 55 | 4.2 | 15.5 | 33 minimum | >50 | >50 | UCH | 75 | >50 |
| Obeso ($N=36$) | 6 | 55.7 ± 9.8 | NR | 14.5 | NR | 36 | 33.3 | UCH | 67 | 35 |
| Rodriguez-Oroz ($N=20$) | 46.8 | 55.8 ± 9.4 | NR | 15.4 | NR | 28 | 39 | UCH | 76 | 45 |
| Loher ($N=10$) | 12 | 65 | 4.2 ± 0.8 | | NR | 34 | 41 | UCH | 71 | NR |
| Kumar ($N=17$) | 2 | 52.7 ± 8.5 | NR | 14.8 ± 6.1 | NR | 39 | 31 | UCH | 68 | NR |
| Anderson ($N=10$) | 10 | 54 ± 12 | 4 | 10.3 ± 2 | 57 | 23 | 39 | 3 | 89 | NR |

[a]Data reported as mean ± SD/range.
[b]Four patients who experienced no efficacy were operated on in the STN after 2 years ($n=2$) and 3 years ($n=2$).
[c]Three patients experienced worsening after 1 year with stimulation-resistant gait ignition failure.
H&Y, Hoehn and Yahr scale; LED, levodopa equivalent dose; NR, not reported; *N*, effective sample size; UCH, unchanged; UPDRS, Unified Parkinson's Disease Rating Scale; II, UPDRS activities of daily living scale; III, UPDRS motor scale.

after surgery is variable. The most important studies of DBS of the GPi are summarized in **Table 2**.

In the first multicenter study, the final evaluations were undertaken after 6 months of surgery in 36 patients bilaterally treated with DBS of the GPi. Stimulation in the off-medication condition compared with the same state at baseline induced a significant improvement of 33% and 35% in motor UPDRS and ADL scores, respectively. All cardinal features were also improved. In the on-medication state compared with baseline on-medication, stimulation provided 26.6% and 30% reduction in the motor UPDRS and the ADL scores, respectively. Not all cardinal features were improved in this medication condition. Dyskinesias were reduced by 66.6%, and the time spent in the on-medication condition without dyskinesias was increased by 56%. The total dose of antiparkinsonian medication was unchanged.

An extended follow-up of this multicenter study that evaluated the effect of this treatment in 20 patients after 4 years showed the benefit of stimulation was maintained during this period of time in the off-medication state. There was a reduction of 39% and 28% in motor UPDRS and ADL scores, respectively, with respect to the preoperative off-medication state. The improvement on postural stability and speech initially observed was lost at 4 years in the off-medication state. Furthermore, the on-medication assessment at 4 years showed a progressive decline in motor UPDRS, gait, and ADL scores with respect to the 1-year situation in the on-medication state, with no significant changes in the off-medication scores. Dyskinesias were reduced by 76% with respect to the baseline situation. These data indicate DBS of the GPi maintains its efficacy in the long-term follow-up in the off-medication state but not in the on-medication situation, with a worsening in the motor state, predominantly in the axial features. The antidiskinetic effect was maintained, and the dose of medication was unchanged with respect to baseline (a slight increment was not significant).

In contrast to this study, Volkmann et al. reported that in 11 patients treated with bilateral DBS of the GPi, there was a significant improvement in the parkinsonian condition in the off-medication situation in the first year after the surgery, but that it started to decline after 3 years and was not maintained at 5 years. Cardinal features were not equally improved, and the axial signs (speech, swallowing, gait, and posture) were the least improved. Actually, there were four patients with lack of efficacy mainly due to gait deterioration after 1 year who required a new-intervention with bilateral stimulation of the STN, with significant improvement. In on-medication condition, stimulation induced a significant benefit in the motor and ADL UPDRS scores but no benefit in individual cardinal features with respect to the preoperative on-medication status in the short term. After 5 years, the benefit had vanished. On the other hand, the reduction in on-condition dyskinesias was maintained during the 5 years. The dose of antiparkinsonian medication was unchanged.

Some other studies with a smaller number of patients and variable follow-up report similar results. Ghika et al. reported six patients in whom bilateral stimulation of the GPi improved ADL and motor UPDRS scores in the off-medication state, the time spent in off condition during the day, and the on dyskinesias after 2 years. During this follow-up period, four patients developed motor fluctuations and needed additional medication, and three patients developed gait initiation failure resistant to stimulation of the GPi. Stimulation did not improve the on-medication state at any interval after surgery. The improvement in the severity of on-condition dyskinesias was significant during the whole period. Another study, also conducted with a small number of patients, found similar results in the off-medication state in the short term but found that after 3 years, the motor scores showed a progressive decline. There was no improvement in the on-medication condition at any point. Benefit was maintained in the on-medication dyskinesias during the entire period.

Recently another study of 11 patients treated with bilateral DBS of the GPi for 2 years reported similar results. There was a benefit in the off-medication state in the motor UPDRS score, with all cardinal signs improved. There was a slight benefit of 18% in the on-medication status in comparison with the on-medication state preoperatively. Dyskinesias were improved by 76%, and no change in the medication dosage was found.

In summary, all studies with short-term follow-up have shown that bilateral GPi stimulation induces a dramatic decrease of on-medication dyskinesia, with or without change in the on-medication motor and ADL UPDRS scores, and a significant improvement in the off-medication motor and ADL UPDRS scores, with different degrees of benefit in the cardinal features. It is important to note that in spite of the antiparkinsonian benefit, the dopaminergic treatment cannot be reduced. In the long-term follow-up, the antidyskinetic effect is maintained, but there is a tendency to lose the antiparkinsonian benefit in the on-medication state and, in many reports, also in the off-medication state, mainly in the axial signs. Actually, for a significant number of patients, the antiparkinsonian benefit is no longer sustained. It is interesting that some of them have had successful surgery on the STN. Although nowadays the STN is the preferred target for most of the surgical teams, until the results of a randomized-to-either-target, prospective, double-blind, multicenter study are available, the GPi could be considered in some patients in whom dyskinesias are the most relevant problem.

## Adverse Events

Complications or adverse events (AEs) associated with DBS can be categorized as related to the surgery itself, to the implanted device, or to the stimulation and/or disease progression.

### Surgical and Device-Related AEs

The most severe complication associated with the surgical procedure is the intraoperative bleeding which occurs in 1–5% of operated patients. Hemorrhages should be handled according to accepted neurosurgical practice. Misplacement of leads is an infrequent complication of the surgery. It is associated with lack of efficacy and significant stimulation-induced side effects. Another intervention for re-placement of the electrode may be needed.

Superficial infections of the surgical wounds can occur immediately after surgery. Erosions or infections in the skin covering the device (extension cable, IPG, cap of the burn hole) are not infrequent (from 1% to 15%), even months or years following surgery. They must be treated with intravenous or oral antibiotics with good tissue penetration. Deep infections involving purulent collections around hardware must be surgically treated, often requiring device removal. They may be related to systemic infections (sepsis), cellulitis, or skin erosion exposing the implant.

Device-related or hardware complications such as lead or extension wire fracture and consumption or malfunction of the IPG are frequent long-term complications. In these circumstances, loss of effect, intermittent stimulation, or tolerance can be the signs indicating the abnormal functioning of the device. Once the broken or malfunctioning component is identified, it must be replaced.

### Stimulation and/or Disease-Related AEs

Stimulation AEs are problems that clearly arise because of the stimulation of the structures surrounding the nucleus or the nucleus itself such as dyskinesias in the case of STN and sometimes GPi. It has been reported that GPi DBS may induce dyskinetic movements when stimulating with more dorsal contacts. On the contrary, stimulation-induced akinesia was observed in the same patients when activating the most ventral contacts of the electrode. Persistent AEs, such as muscle contraction due to stimulation of the corticospinal fibers, can also be seen in both STN and GPi DBS. This effect is unbearable, requiring this particular programming to be ruled out.

The majority of stimulation-induced AEs has been observed in STN DBS. Among the most frequent ones are paresthesias, speech dysfunction (dysarthria and hypophonia), eyelid apraxia, and ocular and visual disturbances. In these cases, modulation of the parameters of stimulation can ameliorate the AEs, but if the result is not satisfactory, the parameters should be changed.

Weight gain is a common AE in STN-treated patients. Candidates for a surgical procedure may be given preoperative nutritional counseling to prevent excessive weight gain. Uncontrolled long-term observations suggest that the weight of most patients normalizes after 1–2 years.

Some other AEs are present with and without stimulation and are also frequent in PD patients not treated with DBS. Thus, the relation to stimulation or to the disease is not clear at this moment. Psychiatric, cognitive, and behavioral disturbances have been associated with STN DBS. Depression, apathy, abulia, desinhibition, mania and hypomania, hallucination, and other disturbances have been described. One important point to consider is the possibility that a large number of patients might already have psychiatric complications or cognitive deficits which could be worsened after surgery. Some studies indicate that patients with depression after STN DBS had depressive episodes before surgery. Nevertheless, cognitive and mood changes have been reported more frequently in STN DBS than in GPi DBS and may be related to the spreading of the current to limbic or associative portions of the nucleus or areas surrounding the STN.

Memory decline, psychiatric disturbances, and axial problems (gait and speech difficulties, disequilibrium, and falls) are also frequently described as AEs in patients treated with DBS but seem to be more frequent among those patients treated with DBS STN, as reported in the multicenter study at 4-year follow-up. It is interesting that all these complications are also frequently seen in advanced PD patients treated chronically only with pharmacological means, so they may be related to the disease. In this context, the multicenter study shows that for the STN DBS-treated patients, depression, thought disorders, rising from a chair, and gait were more affected at baseline in those patients who were to exhibit AEs at 4 years' follow-up. These data support the idea that the AEs exhibited by STN-treated patients could be mainly due to the disease progression rather than to stimulation. At present no conclusion can be reached, and well-designed studies will be needed to clarify these crucial aspects in relation to the disease and to DBS therapy.

In summary, DBS for PD is a successful therapy with a good long-term evolution result. Adverse events associated with this therapy are frequent, but in the majority of cases, they do not preclude the continuation of the therapy. The greatest motor benefits associated with DBS of the STN have to be prospectively evaluated in studies with patients randomly operated in the STN or in the GPi. The recent impression that DBS of the STN produces more AEs than DBS of the GPi also needs to be confirmed. It is necessary to rule out the possibility that the disease evolution is the major contributor to AE.

*See also:* Animal Models of Parkinson's Disease; Deep Brain Stimulation and Movement Disorder Treatment; Parkinsonian Syndromes; Parkinson's Disease: Alpha-Synuclein and Neurodegeneration.

## Further Reading

Anderson VC, Burchiel KJ, Hogarth P, Favre J, and Hammerstad JP (2005) Pallidal vs subthalamic nucleus deep brain stimulation in Parkinson disease. *Archives of Neurology* 62: 554–560.

Deuschl G, Herzog J, Kleiner-Fisman G, et al. (2006) Deep brain stimulation: Postoperative issues. *Movement Disorders* 21(supplement 14): S219–S237.

Deuschl G, Schade-Brittinger C, Krack P, et al. (2006) A randomized trial of deep-brain stimulation for Parkinson's disease. *New England Journal of Medicine* 355: 896–908.

Durif F, Lemaire JJ, Debilly B, and Dordain G (2002) Long-term follow-up of globus pallidus chronic stimulation in advanced Parkinson's disease. *Movement Disorders* 17: 803–807.

Ghika J, Villemure JG, Fankhauser H, Favre J, Assal G, and Ghika-Schmid F (1998) Efficiency and safety of bilateral contemporaneous pallidal stimulation (deep brain stimulation) in levodopa-responsive patients with Parkinson's disease with severe motor fluctuations: A 2-year follow-up review. *Journal of Neurosurgery* 89: 713–718.

Gross RE, Krack P, Rodriguez-Oroz MC, Rezai AR, and Benabid AL (2006) Electrophysiological mapping for the implantation of deep brain stimulators for Parkinson's disease and tremor. *Movement Disorders* 21(supplement 14): S259–S283.

Hariz MI (2002) Complications of deep brain stimulation surgery. *Movement Disorders* 17(supplement 3): S162–S166.

Joint C, Nandi D, Parkin S, Gregory R, and Aziz T (2002) Hardware-related problems of deep brain stimulation. *Movement Disorders* 17(supplement 3): S175–S180.

Kleiner-Fisman G, Herzog J, Fisman DN, et al. (2006) Subthalamic nucleus deep brain stimulation: Summary and meta-analysis of outcomes. *Movement Disorders* 21(supplement 14): S290–S304.

Kumar R, Lang AE, Rodriguez-Oroz MC, et al. (2000) Deep brain stimulation of the globus pallidus pars interna in advanced Parkinson's disease. *Neurology* 55(12 supplement 6): S34–S39.

Lang AE, Houeto JL, Krack P, et al. (2006) Deep brain stimulation: Preoperative issues. *Movement Disorders* 21(supplement 14): S171–S196.

Loher TJ, Burgunder JM, Pohle T, Weber S, Sommerhalder R, and Krauss JK (2002) Long-term pallidal deep brain stimulation in patients with advanced Parkinson disease: 1-year follow-up study. *Journal of Neurosurgery* 96: 844–853.

Østergaard K and AaSunde N (2006) Evolution of Parkinson's disease during 4 years of bilateral deep brain stimulation of the subthalamic nucleus. *Movement Disorders* 21: 624–631.

Rezai AR, Kopell BH, Gross RE, et al. (2006) Deep brain stimulation for Parkinson's disease: Surgical issues. *Movement Disorders* 21(supplement 14): S197–S218.

Rodriguez-Oroz MC, Obeso JA, Lang AE, et al. (2005) Bilateral deep brain stimulation in Parkinson's disease: A multicentre study with 4 years follow-up. *Brain* 128: 2240–2249.

Schupbach WM, Maltete D, Houeto JL, et al. (2007) Neurosurgery at an earlier stage of Parkinson disease: A randomized, controlled trial. *Neurology* 68: 267–271.

Visser-Vandewalle V, van der Linden C, Temel Y, et al. (2005) Long-term effects of bilateral subthalamic nucleus stimulation in advanced Parkinson disease: A 4 year follow-up study. *Parkinsonism & Related Disorders* 11: 157–165.

Volkmann J, Allert N, Voges J, Sturm V, Schnitzler A, and Freund HJ (2004) Long-term results of bilateral pallidal stimulation in Parkinson's disease. *Annals of Neurology* 55: 871–875.

Voon V, Kubu C, Krack P, Houeto JL, and Troster AI (2006) Deep brain stimulation: Neuropsychological and neuropsychiatric issues. *Movement Disorders* 21(supplement 14): S305–S327.

Welter ML, Houeto JL, Tezenas du Montcel S, et al. (2002) Clinical predictive factors of subthalamic stimulation in Parkinson's disease. *Brain* 125: 575–583.

# Subject Index

## Notes

Cross-reference terms in italics are general cross-references, or refer to subentry terms within the main entry (the main entry is not repeated to save space). Readers are also advised to refer to the end of each article for additional cross-references – not all of these cross-references have been included in the index cross-references.

The index is arranged in set-out style with a maximum of three levels of heading. Major discussion of a subject is indicated by bold page numbers. Page numbers suffixed by T and F refer to Tables and Figures respectively. vs. indicates a comparison.

This index is in letter-by-letter order, whereby hyphens and spaces within index headings are ignored in the alphabetization. Prefixes and terms in parentheses are excluded from the initial alphabetization.

## A

## B

# C

## D

## E

## F

## G

## H

## I

## J

## K



## L

## M

## N

## O

## P

## Q



## R

## S

# T

## U



## V